## TABLE IV
### Values of $t_\alpha$

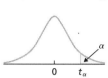

**NOTE:** *See the version of Table IV in Appendix A for additional values of $t_\alpha$.*

| df | $t_{0.10}$ | $t_{0.05}$ | $t_{0.025}$ | $t_{0.01}$ | $t_{0.005}$ | df |
|---|---|---|---|---|---|---|
| 1 | 3.078 | 6.314 | 12.706 | 31.821 | 63.657 | 1 |
| 2 | 1.886 | 2.920 | 4.303 | 6.965 | 9.925 | 2 |
| 3 | 1.638 | 2.353 | 3.182 | 4.541 | 5.841 | 3 |
| 4 | 1.533 | 2.132 | 2.776 | 3.747 | 4.604 | 4 |
| 5 | 1.476 | 2.015 | 2.571 | 3.365 | 4.032 | 5 |
| 6 | 1.440 | 1.943 | 2.447 | 3.143 | 3.707 | 6 |
| 7 | 1.415 | 1.895 | 2.365 | 2.998 | 3.499 | 7 |
| 8 | 1.397 | 1.860 | 2.306 | 2.896 | 3.355 | 8 |
| 9 | 1.383 | 1.833 | 2.262 | 2.821 | 3.250 | 9 |
| 10 | 1.372 | 1.812 | 2.228 | 2.764 | 3.169 | 10 |
| 11 | 1.363 | 1.796 | 2.201 | 2.718 | 3.106 | 11 |
| 12 | 1.356 | 1.782 | 2.179 | 2.681 | 3.055 | 12 |
| 13 | 1.350 | 1.771 | 2.160 | 2.650 | 3.012 | 13 |
| 14 | 1.345 | 1.761 | 2.145 | 2.624 | 2.977 | 14 |
| 15 | 1.341 | 1.753 | 2.131 | 2.602 | 2.947 | 15 |
| 16 | 1.337 | 1.746 | 2.120 | 2.583 | 2.921 | 16 |
| 17 | 1.333 | 1.740 | 2.110 | 2.567 | 2.898 | 17 |
| 18 | 1.330 | 1.734 | 2.101 | 2.552 | 2.878 | 18 |
| 19 | 1.328 | 1.729 | 2.093 | 2.539 | 2.861 | 19 |
| 20 | 1.325 | 1.725 | 2.086 | 2.528 | 2.845 | 20 |
| 21 | 1.323 | 1.721 | 2.080 | 2.518 | 2.831 | 21 |
| 22 | 1.321 | 1.717 | 2.074 | 2.508 | 2.819 | 22 |
| 23 | 1.319 | 1.714 | 2.069 | 2.500 | 2.807 | 23 |
| 24 | 1.318 | 1.711 | 2.064 | 2.492 | 2.797 | 24 |
| 25 | 1.316 | 1.708 | 2.060 | 2.485 | 2.787 | 25 |
| 26 | 1.315 | 1.706 | 2.056 | 2.479 | 2.779 | 26 |
| 27 | 1.314 | 1.703 | 2.052 | 2.473 | 2.771 | 27 |
| 28 | 1.313 | 1.701 | 2.048 | 2.467 | 2.763 | 28 |
| 29 | 1.311 | 1.699 | 2.045 | 2.462 | 2.756 | 29 |
| 30 | 1.310 | 1.697 | 2.042 | 2.457 | 2.750 | 30 |
| 35 | 1.306 | 1.690 | 2.030 | 2.438 | 2.724 | 35 |
| 40 | 1.303 | 1.684 | 2.021 | 2.423 | 2.704 | 40 |
| 50 | 1.299 | 1.676 | 2.009 | 2.403 | 2.678 | 50 |
| 60 | 1.296 | 1.671 | 2.000 | 2.390 | 2.660 | 60 |
| 70 | 1.294 | 1.667 | 1.994 | 2.381 | 2.648 | 70 |
| 80 | 1.292 | 1.664 | 1.990 | 2.374 | 2.639 | 80 |
| 90 | 1.291 | 1.662 | 1.987 | 2.369 | 2.632 | 90 |
| 100 | 1.290 | 1.660 | 1.984 | 2.364 | 2.626 | 100 |
| 1000 | 1.282 | 1.646 | 1.962 | 2.330 | 2.581 | 1000 |
| 2000 | 1.282 | 1.646 | 1.961 | 2.328 | 2.578 | 2000 |

| $z_{0.10}$ | $z_{0.05}$ | $z_{0.025}$ | $z_{0.01}$ | $z_{0.005}$ |
|---|---|---|---|---|
| 1.282 | 1.645 | 1.960 | 2.326 | 2.576 |

# Procedure Index

Following is an index that provides page-number references for the various statistical procedures discussed in the book.

# Elementary Statistics

Seventh Edition

# Elementary Statistics
## Seventh Edition

**Neil A. Weiss**

*Department of Mathematics and Statistics*
*Arizona State University*

Biographies by Carol A. Weiss

San Francisco  Boston  New York
Cape Town  Hong Kong  London  Madrid  Mexico City
Montreal  Munich  Paris  Singapore  Sydney  Tokyo  Toronto

Publisher: Greg Tobin
Editor-in-Chief: Deirdre Lynch
Project Editor: Joanne Ha
Developmental Editor: Lenore Parens
Associate Editors: Elizabeth Bernardi, Sara Oliver Gordus
Senior Managing Editor: Karen Wernholm
Production Supervisor: Sheila Spinney
Senior Designer: Barbara T. Atkinson
Photo Researcher: Beth Anderson
Digital Assets Manager: Marianne Groth
Media Producer: Christine Stavrou
Software Development: Bob Carroll, Mary Durnwald
Marketing Manager: Phyllis Hubbard
Marketing Assistant: Caroline Celano
Senior Author Support/Technology Specialist: Joe Vetere
Senior Prepress Supervisor: Caroline Fell
Manufacturing Manager: Evelyn Beaton
Media Buyer: Ginny Michaud
Cover Design: Night & Day Design
Text Design: Jeanne Calabrese Design, Inc.
Production Coordination: WestWords, Inc.
Composition: Techsetters, Inc.
Illustration: Techsetters, Inc.

Cover Photograph: Field of wildflowers © Getty Images

*Library of Congress Cataloging-in-Publication Data*

Weiss, N. A. (Neil A.)
    Elementary statistics / Neil A. Weiss; biographies by Carol A. Weiss. – 7th ed.
        p. cm.
    Includes index.
        ISBN 0-321-42209-0
    1. Statistics. I. Title.

QA276.12.W445 2007
519.5–dc22

2006043506

2 3 4 5 6 7 8 9 10-DOW-11 10 09 08

*To my father
and the memory
of my mother*

## About the Author

Neil A. Weiss received his Ph.D. from UCLA in 1970 and subsequently accepted an assistant-professor position at Arizona State University (ASU), where he was ultimately promoted to the rank of full professor. Dr. Weiss has taught statistics, probability, and mathematics—from the freshman level to the advanced graduate level—for more than 30 years. In recognition of his excellence in teaching, he received the *Dean's Quality Teaching Award* from the ASU College of Liberal Arts and Sciences. Dr. Weiss' comprehensive knowledge and experience ensures that his texts are mathematically and statistically accurate, as well as pedagogically sound.

In addition to his numerous research publications, Dr. Weiss is the author of *A Course in Probability* (Addison-Wesley, 2006). He has also authored or coauthored books in finite mathematics, statistics, and real analysis, and is currently working on a new book on applied regression analysis and the analysis of variance. His texts—well known for their precision, readability, and pedagogical excellence—are used worldwide.

Dr. Weiss is a pioneer of the integration of statistical software into textbooks and the classroom, first providing such integration over 20 years ago in the book *Introductory Statistics* (Addison-Wesley, 1982). Weiss and Addison-Wesley continue that pioneering spirit to this day with the inclusion of some of the most comprehensive Web sites in the field.

In his spare time, Dr. Weiss enjoys walking, studying and practicing meditation, and playing hold 'em poker. He is married and has two sons.

# Contents

*indicates optional material

*indicates optional material

## Part IV    Inferential Statistics    335

---

*indicates optional material

## Appendixes

*indicates optional material

# Preface

Using and understanding statistics and statistical procedures have become required skills in virtually every profession and academic discipline. The purpose of this book is to help students grasp basic statistical concepts and techniques, and to present real-life opportunities for applying them.

## About This Book

The text is intended for a one-quarter or one-semester course. Instructors can easily fit the text to the pace and depth they prefer. Introductory high school algebra is a sufficient prerequisite.

Although mathematically and statistically sound, the approach doesn't require students to examine complex concepts such as probability theory and random variables. Students need only understand basic ideas such as percentages and histograms.

This edition continues the book's tradition of being on the cutting edge of statistical pedagogy, technology, and data analysis. It includes hundreds of new exercises with real data from journals, magazines, newspapers, and Web sites.

We support and adhere to the following Guidelines for Assessment and Instruction in Statistics Education (GAISE), funded and endorsed by the American Statistical Association:

- Emphasize statistical literacy and develop statistical thinking.
- Use real data.
- Stress conceptual understanding rather than mere knowledge of procedures.
- Foster active learning in the classroom.
- Use technology for developing conceptual understanding and analyzing data.
- Use assessments to improve and evaluate student learning.

## Highlights of the Approach

**ASA/MAA-Guidelines Compliant.** We follow ASA/MAA guidelines to stress the interpretation of statistical results, the contemporary applications of statistics, and the importance of critical thinking.

**Unique Variable-Centered Approach.** By consistent and proper use of the terms *variable* and *population*, we unified and clarified the various statistical concepts.

**Data Analysis and Exploration.**   We emphasize the use of data analysis, both for exploratory purposes and to check assumptions required for inference. Recognizing that not all readers have access to technology, we provide ample opportunity to analyze and explore data without the use of a computer or statistical calculator.

**Detailed and Careful Explanations.**   We include every step of explanation that a typical reader might need. Our guiding principle is to avoid cognitive jumps, making the learning process smooth and enjoyable. We believe that detailed and careful explanations result in better understanding.

**Emphasis on Application.**   We concentrate on the application of statistical techniques to the analysis of data. Although statistical theory has been kept to a minimum, we provide a thorough explanation of the rationale for the use of each statistical procedure.

**Critical Thinking.**   Throughout the text, we emphasize critical thinking. Many of the exercises involve critical thinking, particularly those in the Extending the Concepts and Skills sections of the exercises.

**Parallel Critical-Value/$P$-Value Approaches.**   Through a parallel presentation, the book offers complete flexibility in the coverage of the critical-value and $P$-value approaches to hypothesis testing. Instructors can concentrate on either approach, or they can cover and compare both approaches.

**Parallel Presentations of Technology.**   The book's technology coverage is completely flexible, and includes options for the use of Minitab®, Excel®, and the TI-83/84 Plus. Instructors can concentrate on one technology or cover and compare two or more technologies.

## New and Hallmark Features

**UPDATED!**   **Chapter-Opening Features.**   Each chapter begins with a general description of the chapter, an explanation of how the chapter relates to the text as a whole, and a chapter outline. A classic or contemporary case study highlights the real-world relevance of the material. Each case study is reviewed and discussed at the end of the chapter.

**UPDATED!**   **Real-World Examples.**   Every concept discussed in the text is illustrated by at least one detailed example. Based on real-life situations, these examples are interesting as well as illustrative.

**Interpretations.**   This feature presents the meaning and significance of statistical results in everyday language and highlights the importance of interpreting answers and results.

**What Does It Mean?**   This margin feature states in "plain English" the meanings of definitions, formulas, key facts, and some discussions.

**NEW!** **Data Sets.** In most examples and many exercises, we present both raw data and summary statistics. This practice gives a more realistic view of statistics and lets students solve problems by computer or statistical calculator. Hundreds of data sets are included, many of which are new or updated. The data sets are available in multiple formats on the WeissStats CD, which accompanies new copies of the book.

**Procedure Boxes: Why, When, and How.** To help students learn statistical procedures, we developed easy-to-follow, step-by-step methods for carrying them out. Each step is highlighted and presented again within the illustrating example. This approach shows how the procedure is applied and helps students master its steps.

The procedure boxes include the "why, when, and how" of the methods. Usually, a procedure has a brief identifying title followed by a statement of its purpose (why it's used), the assumptions for its use (when it's used), and the steps for applying the procedure (how it's used). The dual procedures, which provide both the critical-value and $P$-value approaches to a hypothesis-testing method, are combined in a side-by-side, easy-to-use format.

**EXPANDED!** **The Technology Center.** The in-text, statistical-technology presentation discusses three of the most popular applications: Minitab, Excel, and the TI-83/84 Plus graphing calculators. We have expanded the coverage to include instructions and output for all appropriate built-in routines in these three applications. The Technology Centers are integrated as optional material.

**Computer Simulations.** Computer simulations, appearing in both the text and the exercises, serve as pedagogical aids for understanding complex concepts such as sampling distributions.

**NEW!** **Exercises.** We now present more than 1500 end-of-section exercises (an increase of almost 45% from the previous edition) and more than 325 review problems. We updated existing exercises wherever appropriate.

Most exercises provide current, real-world applications, constructed from an extensive variety of articles in newspapers, magazines, statistical abstracts, journals, and Web sites; sources are explicitly cited. Section exercise sets are divided into the following three categories:

- *Understanding the Concepts and Skills* exercises help students master the concepts and skills explicitly discussed in the section. These exercises can be done with or without the use of a statistical technology, at the instructor's discretion. At the request of users, we have added routine exercises on statistical inferences that allow students to practice fundamentals.
- *Working with Large Data Sets* exercises are intended to be done with a statistical technology and let students apply and interpret the computing and statistical capabilities of Minitab, Excel, the TI-83/84 Plus, SPSS®, or any other statistical technology.
- *Extending the Concepts and Skills* exercises invite students to extend their skills by examining material not necessarily covered in the text. These exercises include many critical thinking problems.

**Note:** Exercises related to optional materials are marked with asterisks, unless the entire section is optional.

**NEW!**    **You Try It!**   This new feature, which follows most examples, allows students to immediately check their understanding by asking them to work a similar exercise.

**NEW!**    **Technology Appendixes.**   Appendixes for Minitab, Excel, and the TI-83/84 Plus introduce these three statistical technologies, explain how to input data, and discuss how to perform other basic tasks. These appendixes, which are entitled *Getting Started with ...*, are located in the Technology Basics folder of the WeissStats CD.

**Formula/Table Card.**   The book's detachable formula/table card (FTC) contains all the formulas and many of the tables that appear in the text. The FTC is helpful for quick-reference purposes; many instructors also find it convenient for use with examinations.

**Procedure Index.**   A *Procedure Index* (located near the front of the book) provides a quick and easy way to find the right procedure for performing any statistical analysis.

**EXPANDED!**    **WeissStats CD.**   This PC- and Macintosh-compatible CD, included with every new copy of the book, contains a wealth of resources. Its ReadMe file contains a complete contents list, including the following items:

- *Data Sets* files for all appropriate data sets in the book in multiple formats, including Excel (xls), Minitab (mtw), TI (8xl), text (txt), JMP® (jmp), and SPSS (sav). Nearly 800 electronic data files are supplied in most formats.
- *Data Desk/XL (DDXL™)* software, an Excel add-in from Data Description, Inc. that enhances Excel's standard statistics and graphics capabilities.
- *Probability and random variable* sections that provide optional coverage of contingency tables, joint and marginal probabilities, conditional probability, the multiplication rule, independence, Bayes's rule, counting rules, and the Poisson distribution.
- *Regression* and *ANOVA* chapters that provide optional coverage of multiple regression analysis, model building in regression, and design of experiments and two-way analysis of variance.

## End-of-Chapter Features

**Chapter Reviews.**   Each chapter review includes *chapter objectives*, a list of *key terms* with page references, and *review problems* to help students review and study the chapter. Items related to optional materials are marked with asterisks, unless the entire chapter is optional.

**EXPANDED!**    **Focusing on Data Analysis.**   This feature lets students work with large data sets, practice using technology, and discover the many methods of exploring and analyzing data. The *Focus database*, which is in a file named Focus in the Focus Database folder on the WeissStats CD, contains information on 13 variables for the undergraduate students attending the University of Wisconsin - Eau Claire (UWEC). Those students constitute the population of interest in the *Focusing on Data Analysis* sections at the end of each chapter.

New to this edition, the Focus Database folder also includes a file named FocusSample that contains data on the same 13 variables for a simple random sample of 200 of the undergraduate students at UWEC. Those 200 students constitute a sample that can be used for making statistical inferences. We call these sample data the *Focus sample*.

Instructors may use the Focus database and Focus sample in a variety of ways:

- The entire Focus database (Focus) can be used when performing statistical analyses that require population data.
- The Focus database (Focus) can be sampled when performing statistical analyses that require sample data.
- Alternatively, the Focus sample (FocusSample) can be used as a required sample.

**Case Study Discussion.** At the end of each chapter, the chapter-opening case study is reviewed and discussed in light of the chapter's major points, and then problems are presented for students to solve.

**Biographical Sketches.** Each chapter ends with a brief biography of a famous statistician. Besides being of general interest, these biographies help students obtain a perspective on the development of the science of statistics.

**NEW!** **StatCrunch in MyStatLab.** StatCrunch is on-line statistical software available through the Weiss MyStatLab Course. The StatCrunch Activities folder of the WeissStats CD contains chapter-by-chapter materials that explain and illustrate the use of StatCrunch to perform statistical analyses discussed in the book. Exercises encourage students to further apply StatCrunch to other statistical analyses examined in the book.

**Award-Winning Internet Projects.** Each chapter has an Internet Project to engage students in active and collaborative learning through simulations, demonstrations, and other activities, and to guide them through applications by using Internet links to access data and other information. The Internet Projects are featured on the Weiss Web site at www.aw-bc.com/weiss.

## Flexible Syllabus

The text offers a great deal of flexibility in choosing material to cover. The flowchart on page xviii indicates chapter-coverage flexibility. Here are some additional noteworthy items.

**Option for Brief Sampling Coverage.** In this book, the only sampling design required for study is simple random sampling, which is presented in Section 1.2. Further sampling designs (systematic random sampling, stratified sampling, cluster sampling, and multistage sampling), found in Section 1.3, are optional.

**Option for Brief Experimental Design Coverage.** Coverage of experimental designs (in Section 1.4) is optional. It contains an introduction to the principles of experimental design, the terminology of experimental design, and basic statistical designs.

**Option for Placement of Regression Coverage.**    Chapter 4, on descriptive methods in regression and correlation, is written so that it can be covered either early or late in the course, specifically, anywhere after Chapter 3 or before Chapter 14.

**Option for Brief Probability Coverage.**    Only a rudimentary coverage of probability is required, mostly for the frequentist interpretation of probability and for standard statistical terminology such as Type I and Type II error probabilities and $P$-values.   Coverage of more probability, including probability theory and random variables, is optional.   The probability concepts required for statistical inference can now be covered in two or three class periods.   To cover probability briefly, omit sections marked as optional (with an asterisk) in Chapter 5.

**NEW!**  **Option for Further Probability Coverage.**   New to this edition, we provide additional (optional) coverage of probability and random variables in Module P on the WeissStats CD. This coverage includes contingency tables, joint and marginal probabilities, conditional probability, the multiplication rule, independence, Bayes's rule, counting rules, and the Poisson distribution.

## Organization

Following is a brief chapter-by-chapter summary that also includes some important revisions and additions.

- Chapter 1 presents the nature of statistics, sampling designs, and an introduction to experimental designs. The material in Section 1.3 (Other Sampling Designs) is optional, as is the material on experimental designs in Section 1.4. The optional chapter *Design of Experiments and Analysis of Variance* (Module C), located in the Regression-ANOVA Modules folder on the WeissStats CD, provides a more comprehensive treatment of experimental designs.
- Chapter 2 discusses various ways to organize and display data. This edition includes stem-and-leaf diagrams in the section on graphs and charts; furthermore, it now covers only one type of stem-and-leaf diagram (referred to as an ordered stem-and-leaf diagram in the previous edition). Also new to this edition is a presentation of using technology to generate stem-and-leaf diagrams, pie charts, and bar graphs.
- Chapter 3 examines descriptive measures. We have reorganized the presentation by combining the first two sections of the previous edition, and we now cover only one type of boxplot (referred to as a modified boxplot in the previous edition). We have also provided an in-text example that shows how boxplots can be used to compare two or more data sets.
- Chapter 4 gives an informal, but precise, treatment of descriptive methods in regression and correlation, relying on intuitive and graphical presentations of important concepts. The placement is flexible—this chapter can be covered anywhere between Chapters 3 and 14. This edition gives a more detailed discussion of scatterplots (scatter diagrams). Also new is optional

material on the Spearman rank correlation coefficient, which appears in the Extending the Concepts and Skills exercises.

- Chapter 5 examines probability and optional material on discrete random variables. Only the first three sections of Chapter 5 are prerequisite to coverage of inferential statistics. New to Chapter 5 is a discussion of cumulative probabilities.

- Chapter 6 provides a concise discussion of the normal distribution. We have standardized the construction of normal probability plots by always using the following rule: If two or more observations in a sample are equal, we treat them as slightly different from one another for purposes of obtaining their normal scores.

- Chapter 7 introduces the concept of sampling distributions and presents an improved and simplified introduction to the sampling distribution of the sample mean. We include an in-text example of using the sampling distribution of the sample mean to explicitly evaluate sampling error.

- Chapters 8 and 9 give easily accessible presentations of confidence intervals and hypothesis tests for one population mean by using the terminology of variables and avoiding formal probability. Both chapters use the $\sigma$-known versus $\sigma$-unknown criterion for deciding which parametric procedure to use, which makes confidence intervals and hypothesis tests easier to understand and apply and provides a method consistent with most statistical software. New to this edition, in the Extending the Concepts and Skills exercises, we discuss one-sided confidence intervals and the general relationship between hypothesis tests and confidence intervals.

- Chapter 10 examines inferences for two population means. It contains a detailed discussion of the meaning of independent samples, including graphics for easy understanding. We cover pooled $t$-procedures (as well as nonpooled $t$-procedures) because they provide valuable motivation for one-way ANOVA. New to this edition, in the Extending the Concepts and Skills exercises, we discuss the general relationship between hypothesis tests and confidence intervals for two population means.

- Chapter 11 presents inferences for one and two population proportions, and contains discussions of margin of error and sample-size determination.

- Chapter 12 introduces the chi-square goodness-of-fit test and the chi-square independence test. Included are a section on grouping bivariate data into contingency tables and an easy-to-understand presentation of the concept of association. New to this edition, we show how to use technology to perform a chi-square goodness-of-fit test, to group bivariate data into a contingency table, and to obtain conditional and marginal distributions for bivariate data.

- Chapter 13 presents one-way analysis-of-variance (ANOVA), using the language of variables to simplify and unify assumptions.

- Chapter 14 examines inferential methods in regression and correlation. New to this edition, we show how to use technology to conduct a correlation $t$-test.

The following flowchart summarizes the preceding discussion and shows the interdependence among chapters. In the flowchart, the prerequisites for a given chapter consist of all chapters that have a path that leads to that chapter. Optional sections and chapters can be identified by consulting the table of contents.

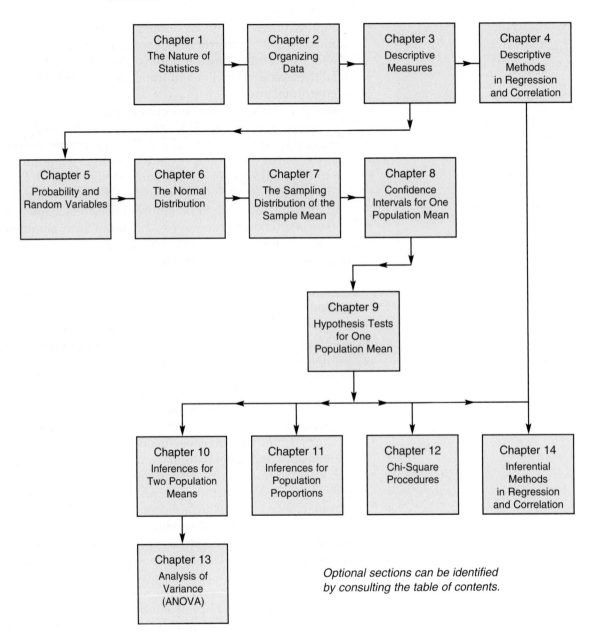

# Acknowledgments

For this and the previous few editions of the book, it is our pleasure to thank the following reviewers, whose comments and suggestions resulted in significant improvements:

James Albert
*Bowling Green State University*

John F. Beyers, II
*University of Maryland,
  University College*

Yvonne Brown
*Pima Community College*

Beth Chance
*California Polytechnic State
  University*

Brant Deppa
*Winona State University*

Carol DeVille
*Louisiana Tech University*

Jacqueline Fesq
*Raritan Valley Community College*

Richard Gilman
*Holy Cross College*

Donna Gorton
*Butler Community College*

Joel Haack
*University of Northern Iowa*

Bernard Hall
*Newbury College*

Susan Herring
*Sonoma State University*

David Holmes
*The College of New Jersey*

Michael Hughes
*Miami University*

Satish Iyengar
*University of Pittsburgh*

Jann-Huei Jinn
*Grand Valley State University*

Christopher Lacke
*Rowan University*

Tze-San Lee
*Western Illinois University*

Ennis Donice McCune
*Stephen F. Austin State University*

Jackie Miller
*The Ohio State University*

Bernard J. Morzuch
*University of Massachusetts,
  Amherst*

Dennis M. O'Brien
*University of Wisconsin, La Crosse*

Dwight M. Olson
*John Carroll University*

JoAnn Paderi
*Lourdes College*

Melissa Pedone
*Valencia Community College*

Alan Polansky
*Northern Illinois University*

Cathy D. Poliak
*Northern Illinois University*

Kimberley A. Polly
*Indiana University Bloomington*

Geetha Ramachandran
*California State University*

B. Madhu Rao
*Bowling Green State University*

Gina F. Reed
*Gainesville College*

Steven E. Rigdon
*Southern Illinois University,
  Edwardsville*

Kevin M. Riordan
*South Suburban College*

Sharon Ross
*Georgia Perimeter College*

Edward Rothman
*University of Michigan*

George W. Schultz
*St. Petersburg Jr. College*

Arvind Shah
*University of South Alabama*

Cid Srinivasan
*University of Kentucky, Lexington*

W. Ed Stephens
*McNeese State University*

Kathy Taylor
*Clackamas Community College*

Bill Vaughters
*Valencia Community College*

Roumen Vesselinov
*University of South Carolina*

Brani Vidakovic
*Georgia Institute of Technology*

Jackie Vogel
*Austin Peay State University*

Daniel Weiner
*Boston University*

Dawn White
*California State University,
    Bakersfield*

Marlene Will
*Spalding University*

Matthew Wood
*University of Missouri, Columbia*

Nicholas A. Zaino Jr.
*University of Rochester*

Our thanks as well are extended to Professor Michael Driscoll for his help in selecting the statisticians for the biographical sketches; and Professors Fuchun Huang, Charles Kaufman, Sharon Lohr, Richard Marchand, Kathy Prewitt, Walter Reid, and Bill Steed with whom we have had several illuminating discussions. Thanks also go to Professors Matthew Hassett and Ronald Jacobowitz for their many helpful comments and suggestions.

Several other people provided useful input and resources. They include Professor Thomas A. Ryan, Jr., Professor Webster West, Dr. William Feldman and Mr. Frank Crosswhite, Dr. Lawrence W. Harding, Jr., Dr. George McManus and Mr. Gregory Weiss, Professor Jeanne Sholl, Professor R. B. Campbell, Mr. Howard Blaut, Mr. Rick Hanna, Ms. Alison Stern-Dunyak, Mr. Dale Phibrick, Ms. Christine Sarris, and Ms. Maureen Quinn. Our sincere thanks to all of them for their help in making a better book.

To Professor Larry Griffey, we give our appreciation for his formula/table card. We are grateful to the following people for preparing the technology manuals to accompany the book: Professors Jacqueline Fesq (*Minitab Manual*), Ellen Fischer (*TI-83/84 Plus Manual*), James Ball (*SPSS Manual*), and Ian Walters (*Excel Manual*). Our gratitude goes as well to Professor David Lund for writing the *Instructor's Solutions Manual* and the *Student's Solutions Manual*.

We express our appreciation to Professor Dennis Young for his linear models modules and for his collaboration on numerous statistical and pedagogical issues. Thanks also go to Paul Lorczak for updating the Internet Projects and case-study extensions. For checking the accuracy of the entire text, we extend our gratitude to Professor Jackie Miller, and for checking the answers to the exercises, to Professors Jane L. Harvill, Lorraine Hughes, Jackie Miller, and Kimberley Polly.

We are also grateful to Professor David Lund and Ms. Patricia Lee for obtaining the database for the Focusing on Data Analysis sections. Our thanks are extended to the following people for their research in finding myriad new and interesting statistical studies and data for the examples, exercises, and case studies: Ms. Toni Coombs, Ms. Traci Gust, Professor David Lund, Ms. Jelena Milovanovic, and Mr. Gregory Weiss.

Many thanks to Ms. Christine Stavrou for directing the development and construction of the WeissStats CD and the Weiss Web site, and to Professor James Ball, Ms. Cindy Bowles, and Ms. Carol Weiss for constructing the data files. Our appreciation also goes to our software editors Ms. Mary Durnwald and Mr. Bob Carroll.

We are grateful to Pat McCutcheon of WestWords, Inc., who, along with Deirdre Lynch, Sheila Spinney, and Joanne Ha of Addison-Wesley, coordinated the development and production of the book. Thanks as well to our copyeditor Gordon LaTourette of Write With, Inc., and to our proofreaders Cindy Bowles, Carol Sawyer of The Perfect Proof, and Carol Weiss.

A special thanks is due to our developmental editor, Lenore Parens, for her advice, suggestions, and encouragement. She did a marvelous job on an extremely difficult task.

To Barbara Atkinson of Addison-Wesley, Jeanne Calabrese Design, Inc. (interior design), and Night & Day Design (cover design), we express our thanks for an awesome design and cover. And our sincere thanks go as well to all the people at Techsetters, Inc., for a terrific job of composition and illustration, in particular, John Rogosich (LaTeX design creation), Rena Lam (project manager, quality control, typesetting, and page makeup), Carl DiStefano (illustrations), and Lois Gaine (proofreading). We also thank Beth Anderson for her photo research.

Without the help of many people at Addison-Wesley, this book and its numerous ancillaries would not have been possible; to all of them go our heartfelt thanks. We would, however, like to give special thanks to Greg Tobin, Deirdre Lynch, and to the following other people at Addison-Wesley: Sara Oliver Gordus, Elizabeth Bernardi, Joanne Ha, Ron Hampton, Joe Vetere, Sheila Spinney, Yolanda Cossio, Phyllis Hubbard, Caroline Celano, Caroline Fell, and Evelyn Beaton.

Finally, we convey our appreciation to Carol A. Weiss. Apart from writing the text, she was involved in every aspect of development and production. Moreover, Carol did a superb job of researching and writing the biographies.

Prescott, Arizona                                                                                     *N.A.W.*

# Supplements

## Student Supplements

### Technology Manuals

The following technology manuals include instructions, examples from the main text, and interpretations to complement those given in the text.

- *Minitab Manual*, written by Jacqueline Fesq.
  ISBN: 0-321-43811-6
- *Excel Manual*, written by Ian C. Walters.
  ISBN: 0-321-43622-9
- *TI-83/84 Plus Manual*, written by Ellen Fischer.
  ISBN: 0-321-43171-5
- *SPSS Manual*, written by James J. Ball.
  ISBN: 0-321-43172-3

### Student's Solutions Manual

- Written by David Lund, this supplement contains detailed, worked-out solutions to the odd-numbered section exercises (Understanding the Concepts and Skills, Working with Large Data Sets, and Extending the Concepts and Skills) and all Review Problems.
- ISBN: 0-321-43555-9

### Weiss Web Site

- Includes data sets, the Formula/Table card, and access to the StatCrunch pages, Internet Projects, and Case Study Extensions.
- The URL is www.aw-bc.com/weiss.

### Addison-Wesley Math and Statistics Tutor Center

- Provides tutoring through a registration number that can be packaged with a new textbook or purchased separately.
- Staffed by qualified college mathematics instructors.
- Accessible via toll-free telephone, toll-free fax, e-mail, and the Internet.
- The URL is www.aw-bc.com/tutorcenter.

## Instructor Supplements

### Instructor's Edition

- This version of the text includes the answers to all the Understanding the Concepts and Skills exercises. (The Student's Edition contains the answers to only the odd-numbered ones.)
- ISBN: 0-321-43333-5

### Instructor's Solutions Manual

- Written by David Lund, this supplement contains detailed, worked-out solutions to all section exercises (Understanding the Concepts and Skills, Working with Large Data Sets, and Extending the Concepts and Skills), the Review Problems, the Focusing on Data Analysis exercises, and the Case Study Discussion exercises.
- ISBN: 0-321-43556-7

### Printed Test Bank

- Prepared by Michael Butros, this supplement provides three printed examinations for each chapter of the text.
- Answer keys are included.
- ISBN: 0-321-43810-8

### TestGen®

- Enables instructors to build, edit, print, and administer tests.
- Features a computerized bank of questions developed to cover all text objectives.
- Available on a dual-platform Windows®/Macintosh® CD-ROM.
- ISBN: 0-321-42974-5

### PowerPoint Lecture Presentation

- Classroom presentation slides are geared specifically to the sequence of this textbook.
- Available within MyStatLab or on the Internet at www.aw-bc.com/irc.

### New! Adjunct Support Center

- Offers consultation on suggested syllabi, helpful tips on using the textbook support package, assistance with content, and advice on classroom strategies.
- Available Sunday–Thursday evenings from 5 P.M. to midnight EST; telephone: 1-800-435-4084; e-mail: AdjunctSupport@aw.com; fax: 1-877-262-9774.

# Technology Resources

## The Student Edition of MINITAB

This condensed edition of the Professional release of Minitab statistical software still offers the full range of its statistical methods and graphical capabilities, along with worksheets that can include up to 10,000 data points.

- A user's manual includes case studies and hands-on-tutorials, and is perfect for use in any introductory statistics course.
- The currently available Student Edition included with the user's manual is *The Student Guide to Minitab Release 14* by John D. McKenzier, Jr. and Robert Goldman. ISBN: 0-201-77469-0
- Individual copies of the software may also be bundled with the Weiss book. ISBN: 0-321-11313-6.

## ActivStats®

Developed by Paul Velleman and Data Description, Inc., ActivStats presents a complete introductory statistics course on CD-ROM.

- Integrating video, simulation, animation, narration, text, interactive experiments, www access, and Data Desk®, this product gives students a rich learning environment.
- (PC and Mac) ISBN: 0-321-30364-4
- Also available are:
  *ActivStats for Minitab* (PC) ISBN: 0-321-30373-3
  *ActivStats for Excel* (Mac and PC) ISBN: 0-321-30375-X
  *ActivStats for SPSS* (PC) ISBN: 0-321-30372-5
  *ActivStats for JMP* (Mac and PC) ISBN: 0-321-30374-1

## MathXL® for Statistics

MathXL for Statistics is a powerful online homework, tutorial, and assessment system that accompanies Addison-Wesley textbooks in statistics. With MathXL for Statistics, instructors can create, edit, and assign online homework and tests using algorithmically generated exercises correlated at the objective level to the textbook. They can also create and assign their own online exercises and import TestGen tests for added flexibility. All student work is tracked in MathXL's online gradebook. Students can take chapter tests in MathXL and receive personalized study plans based on their test results. The study plan diagnoses weaknesses and links students directly to tutorial exercises for the objectives they need to study and retest. Students can also access supplemental animations directly from selected exercises. MathXL for Statistics is available to qualified adopters. For more information, instructors can visit www.mathxl.com or contact their Addison-Wesley sales representative.

## MyStatLab

MyStatLab (part of the MyMathLab and MathXL product family) is a text-specific, easily customizable online course that integrates interactive multimedia instruction with the textbook content. Powered by CourseCompass™ (Pearson Education's online teaching and learning environment) and MathXL (our online homework, tutorial, and assessment system), MyStatLab gives instructors the tools they need to deliver all or a portion of their course online, whether students are in a lab setting or working from home. MyStatLab provides a rich and flexible set of course materials, featuring free-response tutorial exercises for unlimited practice and mastery. Students can also use online tools, such as animations and a multimedia textbook, to independently improve their understanding and performance. Instructors can use MyStatLab's homework and test managers to select and assign online exercises correlated directly to the textbook, and they can also create and assign their own online exercises and import TestGen tests for added flexibility. MyStatLab's online gradebook—designed specifically for mathematics and statistics—automatically tracks students' homework and test results and gives the instructor control over how to calculate final grades. Instructors can also add offline (paper-and-pencil) grades to the gradebook. MyStatLab is available to qualified adopters. For more information, instructors can visit www.mystatlab.com or contact their Addison-Wesley sales representative.

## StatCrunch (available within MyStatLab)

StatCrunch is a powerful online tool that provides an interactive environment for doing statistics. You can use StatCrunch for both numerical and graphical data analysis, taking advantage of interactive graphics to help you see the connection between objects selected in a graph and the underlying data. In MyStatLab, all of the data sets from your textbook are pre-loaded into StatCrunch, and StatCrunch is also available as a tool from all of the online homework and practice exercises.

# Data Sources

AAA Daily Fuel Gauge Report
AAA Foundation for Traffic Safety
AAUP Annual Report on the Economic Status of the Profession
ABCNEWS Poll
ABCNews.com
A.C. Nielsen Company
Academic Libraries
Accident Facts
Acta Opthalmologica
Advances in Cancer Research
Alcohol Consumption and Related Problems: Alcohol and Health Monograph 1
All About Diabetes
Alzheimer's Care Quarterly
AMATYC Review
America's Network Telecom Investor Supplement
American Association of University Professors
American Automobile Manufacturers Association
American Bar Foundation
American Demographics
American Express Retail Index
American Film Institute
American Freshman
American Hospital Association
American Housing Survey for the United States
American Industrial Hygiene Association
American Industrial Hygiene Association Journal
American Journal of Clinical Nutrition
American Journal of Obstetrics and Gynecology
American Journal of Political Science
American Laboratory
American Scientist
American Statistician
Amstat News
Amusement Business
Analytical Chemistry
Animal Behaviour
Annals of Epidemiology
Annals of the Association of American Geographers
Appetite
Aquaculture
Archives of Physical Medicine and Rehabilitation
Arizona Chapter of the American Lung Association
Arizona Republic
Arizona Residential Property Valuation System
Arizona State University Enrollment Summary
Arthritis Today
Associated Press

Australian Journal of Rural Health
Auto Trader
Avis Rent-A-Car
BARRON'S
Beer Institute
Beer Institute Online
Behavior Research Center
Behavioral Risk Factor Surveillance System Summary Prevalence Report
Billboard Online
Biological Conservation
Biomaterials
Biometrika
BioScience
Board of Governors of the Federal Reserve System
Bottom Line/Personal
Bowker Annual Library and Book Trade Almanac
Boyce Thompson Southwestern Arboretum
Bride's Magazine
British Journal of Educational Psychology
British Journal of Haematology
British Medical Journal
Brokerage Report
Bureau of Crime Statistics and Research of Australia
Bureau of Justice Statistics Special Report
Bureau of Labor Statistics
Bureau of Transportation Statistics
Business Journal
Business Times
California Agriculture
California Wild: Natural Sciences for Thinking Animals
Census of Agriculture
Centers for Disease Control and Prevention
Chance
Characteristics of New Housing
Cheetah Conservation of Southern Africa
Chesapeake and Ohio Railroad Company
Climates of the World
Climatography of the United States
CNN/USA TODAY
CNN/USA TODAY/Gallup Poll
College Bound Seniors
College Entrance Examination Board
Comerica Auto Affordability Index
Communications Industry Forecast & Report
Comparative Climatic Data
Congressional Directory
Conservation Biology
Consumer Expenditure Survey
Consumer Expenditures
Consumer Profile

Consumer Reports
Contributions to Boyce Thompson Institute
Controlling Road Rage: A Literature Review and Pilot Study
Crime in the United States
Current Housing Reports
Current Population Reports
Daily Courier
Daily Racing Form
Dallas Mavericks Roster
Dave Leip's Atlas of U.S. Presidential Elections
Deep Sea Research Part I: Oceanographic Research Papers
Demographic Profiles
Demography
Desert Samaritan Hospital
Design and Analysis of Factorial Experiments
Detection of Psychiatric Illness by Questionnaire
Diet for a New America
Dietary Guidelines for Americans
Dietary Reference Intakes
Digest of Education Statistics
Directions Research, Inc.
Discover
Early Medieval Europe
Eau Claire Leader Telegram
Ecology
Economic Development Corporation Report
Economic Journal
Employment and Earnings
Environmental Geology Journal
Environmental Pollution (Series A)
ESPN
Europe-Asia Studies
Experimental Agriculture
Fédération Internationale de Football Association
Family Planning Perspectives
Fatality Analysis Reporting System (FARS)
Federal Reserve Bulletin
Financial Planning
FlightStats
Florida Department of Environmental Protection
Food Consumption, Prices, and Expenditures
Footwear News
Forbes Magazine
Forest Mensuration
Forrester Research
Fortune Magazine
Fuel Economy Guide
Gallup Organization
Geography

Global Financial Data
Golf Digest
Golf Laboratories, Inc.
Governors' Political Affiliations & Terms of Office
Graduating Student and Alumni Survey
Handbook of Small Data Sets
Harris Poll
Health, United States
Health Letter
High School Profile Report
High Speed Services for Internet Access
Higher Education Research Institute
Highway Statistics
Hilton Hotels Corporation
Historical Income Tables
History of Statistics
HIV/AIDS Surveillance Report
Hospital Statistics
Human Biology
Hydrobiologia
Industry Research
Information and Communications, University
    of California
Information Please Almanac
Injury Prevention
Inside MS
International Classification of Diseases
International Communications Research
International Data Base
International Shark Attack File
International Waterpower & Dam Construction
    Handbook
Japan Automobile Manufacturer's
    Association
Japan's Motor Vehicle Statistics, Total Exports
    by Year
Journal of Abnormal Psychology
Journal of Advertising Research
Journal of the American College of Cardiology
Journal of American College Health
Journal of the American Geriatrics Society
Journal of the American Medical Association
Journal of the American Public Health
    Association
Journal of Anatomy
Journal of Applied Ecology
Journal of Arachnology
Journal of Bone and Joint Surgery
Journal of Chemical Ecology
Journal of Chronic Diseases
Journal of Clinical Endocrinology & Metabolism
Journal of Clinical Oncology
Journal of College Science Teaching
Journal of Counseling Psychology
Journal of Early Adolescence
Journal of Environmental Psychology
Journal of Environmental Science and Health
Journal of Geography
Journal of Herpetology
Journal of Human Evolution
Journal of Nutrition
Journal of Organizational Behavior
Journal of Paleontology
Journal of Pediatrics
Journal of Prosthetic Dentistry
Journal of Real Estate and Economics

Journal of Statistics Education
Journal of Sustainable Tourism
Journal of Tropical Ecology
Journal of Zoology, London
Kansas City Star
Kelley Blue Book
Lancet
Land Economics
Lawlink
Lawyer Statistical Report
Leonard Martin Movie Guide
Life Insurers Fact Book
Limnology and Oceanography
Literary Digest
Los Angeles Times
Manufactured Housing Statistics
Marine Ecology Progress Series
Mediamark Research Inc.
Medical Biology and Etruscan Origins
Medical Principles and Practice
Merck Manual
MLB Roster Analysis
Monitoring the Future
Monthly Labor Review
Monthly Tornado Statistics
Morbidity and Mortality Weekly Report
Motor Gasoline Price Survey
Motor Vehicle Facts and Figures
Motor Vehicle Manufacturers Association of
    the United States
National Academy of Sciences
National Aeronautics and Space
    Administration
National Association of Colleges and
    Employers
National Association of REALTORS
National Association of State Racing
    Commissioners
National Basketball Association
National Biennial RCRA Hazardous Waste
    Report
National Center for Health Statistics
National Corrections Reporting Program
National Geographic
National Geographic Traveler
National Governors Association
National Health and Nutrition Examination
    Survey
National Household Survey on Drug Abuse
National Household Travel Survey, Summary of
    Travel Trends
National Institute on Alcohol Abuse and
    Alcoholism
National Institute on Drug Abuse
National Institute of Mental Health
National Low Income Housing Coalition
National Mortgage News
National Oceanic and Atmospheric
    Administration
National Safety Council
National Science Foundation
National Survey of Salaries and Wages in Public
    Schools
National Transportation Statistics
National Vital Statistics Report
Nature

NCAAsports.com
New Car Ratings and Review
New England Journal of Medicine
New Scientist
Newsweek
Nielsen Media Research
Nielsen/NetRatings
Nielsen Ratings
Nielsen Report on Television
NOAA Technical Memorandum
Nutrition
Obstetrics & Gynecology
OECD in Figures
Office of Educational Research and
    Improvement
O'Neil Associates
Opinion Research Corporation
Organization for Economic Cooperation and
    Development
Origin of Species
Osteoporosis International
Out of Reach
Parade magazine
Payless ShoeSource
Pediatrics
Pediatrics Journal
Perspectives
Perspectives on Sexual and Reproductive Health
Phoenix Gazette
Plant Disease, An International Journal of
    Applied Plant Pathology
PLOS Biology
Pollstar
Popular Mechanics
Popular Science
Population-at-Risk Rates and Selected Crime
    Indicators
Preventative Medicine
pricewatch.com
Proceedings of the 6th Berkeley Symposium on
    Mathematics and Statistics, VI
Proceedings of the National Academy of
    Science USA
Psychology of Addictive Behaviors
Public Citizen's Health Research Group
    Newsletter
Radio Advertising Bureau of New York
Radio Facts
Reader's Digest/Gallup Survey
Recording Industry Association of America
Regional Markets, Vol. 2/Households
Research Quarterly for Exercise and Sport
Research Resources, Inc.
Residential Energy Consumption Survey:
    Consumption and Expenditures
Residential Energy Consumption Survey:
    Household Energy Consumption and
    Expenditures
R. Jacobowitz, Ph.D. & G. Vishteh, M.D.
Roper Starch Worldwide
R. R. Bowker Company of New York
Rubber Age
Runner's World
Salary Survey
San Francisco Giants Roster
Scarborough Research

*Science*
*Science and Engineering Indicators*
*Science News*
*Scientific American*
*Scottish Executive*
*Semi-annual Wireless Survey*
*Sexually Transmitted Disease Surveillance*
*Signs of Progress*
*Social Forces*
South Carolina Budget and Control Board
*South Carolina Statistical Abstract*
*Sports Illustrated*
*SportsCenturyRetrospective*
*Statistical Abstract of the United States*
*Statistical Report*
*Statistical Yearbook*
*Statistics of Income, Individual Income Tax Returns*
*Statistics Norway*
Storm Prediction Center
*Survey of Current Business*
*Survey of Graduate Science Engineering Students and Postdoctorates*
*TalkBack Live*
*Teaching Issues and Experiments in Ecology*
*Technometrics*
Television Bureau of Advertising, Inc.
*Tempe Daily News*
*The Earth: Structure, Composition and Evolution*

*The Marathon: Physiological, Medical, Epidemiological, and Psychological Studies*
*The Methods of Statistics*
Thomas Stanley, Georgia State University
*Thoroughbred Times*
*Time* magazine
*Time Spent Viewing*
*Time Style and Design*
*Trade Environment Database*
*Travel + Leisure Golf*
*Trends in Television*
*Tropical Biodiversity*
*TV Basics*
*Urban Studies*
*U.S. Agricultural Trade Update*
U.S. Air Force, National Oceanic and Atmospheric Administration
U.S. Bureau of Labor Statistics
U.S. Bureau of Prisons
U.S. Census Bureau
U.S. Coast Guard
U.S. Congress, Joint Committee on Printing
U.S. Department of Agriculture
U.S. Department of Commerce
U.S. Department of Health and Human Services
U.S. Department of Housing and Urban Development
U.S. Department of Justice
U.S. Energy Information Administration

U.S. Environmental Protection Agency
U.S. Federal Bureau of Investigation
U.S. Federal Highway Administration
U.S. Geological Survey
U.S. Internal Revenue Service
U.S. Justice Department
U.S. National Center for Education Statistics
*U.S. News and World Report*
U.S. Substance Abuse and Mental Health Services Administration
*U.S. Women's Open*
*USA TODAY*
*USA TODAY/CNN Gallup Poll*
*USA TODAY Online*
*Vegetarian Journal*
Vegetarian Resource Group
VentureOne Corporation
*Vital and Health Statistics*
*Vital Statistics of the United States*
*Wall Street Journal*
*Washington Post*
*Webster's New World Dictionary*
*Western Journal of Medicine*
*Wichita Eagle*
*Women and Cardiovascular Disease Hospitalizations*
*WONDER database*
*World Almanac*
*World Series Overview*
Zogby International Poll

# PART I

## Introduction

# 1

# The Nature of Statistics

**Chapter Objectives**

What does the word *statistics* bring to mind? To most people, it suggests numerical facts or data, such as unemployment figures, farm prices, or the number of marriages and divorces. *Webster's New World Dictionary* gives two definitions of the word *statistics:*

1. facts or data of a numerical kind, assembled, classified, and tabulated so as to present significant information about a given subject.
2. [construed as sing.], the science of assembling, classifying, and tabulating such facts or data.

But statisticians also analyze data for the purpose of making generalizations and decisions. For example, a political analyst can use data from a portion of the voting population to predict the political preferences of the entire voting population. And a city council can decide where to build a new airport runway based on environmental impact statements and demographic reports that include a variety of statistical data.

In this chapter, we introduce some basic terminology so that the various meanings of the word *statistics* will become clear to you. We also examine two primary ways of producing data, namely, through sampling and experimentation. We discuss sampling designs in Sections 1.2 and 1.3, and experimental designs in Section 1.4.

# Greatest American Screen Legends

As part of its ongoing effort to lead the nation to discover and rediscover the classics, the American Film Institute (AFI) conducted a survey on the greatest American screen legends. AFI defines an *American screen legend* as "...an actor or a team of actors with a significant screen presence in American feature-length films whose screen debut occurred in or before 1950, or whose screen debut occurred after 1950 but whose death has marked a completed body of work."

AFI polled 1800 leaders from the American film community, including artists, historians, critics, and other cultural dignitaries. Each of these leaders was asked to choose the greatest American screen legends from a list of 250 nominees in each gender category, as compiled by AFI historians.

After tallying the responses, AFI compiled a list of the 50 greatest American screen legends—the top 25 women and the top 25 men—naming Katharine Hepburn and Humphrey Bogart the number one legends. The following table provides the complete list. At the end of this chapter, you will be asked to analyze further this AFI poll.

| Men | | Women | |
|---|---|---|---|
| 1. Humphrey Bogart | 14. Laurence Olivier | 1. Katharine Hepburn | 14. Ginger Rogers |
| 2. Cary Grant | 15. Gene Kelly | 2. Bette Davis | 15. Mae West |
| 3. James Stewart | 16. Orson Welles | 3. Audrey Hepburn | 16. Vivien Leigh |
| 4. Marlon Brando | 17. Kirk Douglas | 4. Ingrid Bergman | 17. Lillian Gish |
| 5. Fred Astaire | 18. James Dean | 5. Greta Garbo | 18. Shirley Temple |
| 6. Henry Fonda | 19. Burt Lancaster | 6. Marilyn Monroe | 19. Rita Hayworth |
| 7. Clark Gable | 20. The Marx Brothers | 7. Elizabeth Taylor | 20. Lauren Bacall |
| 8. James Cagney | 21. Buster Keaton | 8. Judy Garland | 21. Sophia Loren |
| 9. Spencer Tracy | 22. Sidney Poitier | 9. Marlene Dietrich | 22. Jean Harlow |
| 10. Charlie Chaplin | 23. Robert Mitchum | 10. Joan Crawford | 23. Carole Lombard |
| 11. Gary Cooper | 24. Edward G. Robinson | 11. Barbara Stanwyck | 24. Mary Pickford |
| 12. Gregory Peck | 25. William Holden | 12. Claudette Colbert | 25. Ava Gardner |
| 13. John Wayne | | 13. Grace Kelly | |

## 1.1 Statistics Basics

You probably already know something about statistics. If you read newspapers, surf the Web, watch the news on television, or follow sports, you see and hear the word *statistics* frequently. In this section, we use familiar examples such as baseball statistics and voter polls to introduce the two major types of statistics: **descriptive statistics** and **inferential statistics.** We also introduce terminology that helps differentiate among various types of statistical studies.

### Descriptive Statistics

Each spring in the late 1940s, President Harry Truman officially opened the major league baseball season by throwing out the "first ball" at the opening game of the Washington Senators. We use the 1948 baseball season to illustrate the first major type of statistics, descriptive statistics.

| | |
|---|---|
| **Example 1.1** | **Descriptive Statistics** |

*The 1948 Baseball Season* In 1948, the Washington Senators played 153 games, winning 56 and losing 97. They finished seventh in the American League and were led in hitting by Bud Stewart, whose batting average was .279. Baseball statisticians compiled these and many other statistics by organizing the complete records for each game of the season.

Although fans take baseball statistics for granted, much time and effort is required to gather and organize them. Moreover, without such statistics, baseball would be much harder to follow. For instance, imagine trying to select the best hitter in the American League given only the official score sheets for each game. (More than 600 games were played in 1948; the best hitter was Ted Williams, who led the league with a batting average of .369.)

• • •

The work of baseball statisticians is an illustration of *descriptive statistics*.

| | |
|---|---|
| **Definition 1.1** | **Descriptive Statistics** |

**Descriptive statistics** consists of methods for organizing and summarizing information.

Descriptive statistics includes the construction of graphs, charts, and tables and the calculation of various descriptive measures such as averages, measures of variation, and percentiles. We discuss descriptive statistics in detail in Chapters 2 and 3.

### Inferential Statistics

We use the 1948 presidential election to introduce the other major type of statistics, inferential statistics.

**Example 1.2** | **Inferential Statistics**

*The 1948 Presidential Election* In the fall of 1948, President Truman was concerned about statistics. The *Gallup Poll* taken just prior to the election predicted that he would win only 44.5% of the vote and be defeated by the Republican nominee, Thomas E. Dewey. But the statisticians had predicted incorrectly. Truman won more than 49% of the vote and, with it, the presidency. The Gallup Organization modified some of its procedures and has correctly predicted the winner ever since.

• • •

Political polling provides an example of inferential statistics. Interviewing everyone of voting age in the United States on their voting preferences would be expensive and unrealistic. Statisticians who want to gauge the sentiment of the entire **population** of U.S. voters can afford to interview only a carefully chosen group of a few thousand voters. This group is called a **sample** of the population. Statisticians analyze the information obtained from a sample of the voting population to make inferences (draw conclusions) about the preferences of the entire voting population. Inferential statistics provides methods for drawing such conclusions.

The terminology just introduced in the context of political polling is used in general in statistics.

**Definition 1.2** | **Population and Sample**

**Population:** The collection of all individuals or items under consideration in a statistical study.

**Sample:** That part of the population from which information is obtained.

Figure 1.1 depicts the relationship between a population and a sample from the population.

**FIGURE 1.1**
Relationship between population and sample

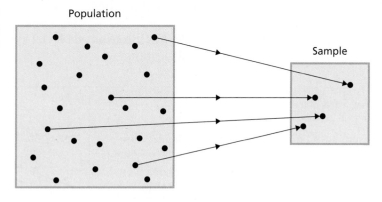

Now that we have discussed the terms *population* and *sample,* we can define *inferential statistics.*

| Definition 1.3 | **Inferential Statistics** |
|---|---|
| | **Inferential statistics** consists of methods for drawing and measuring the reliability of conclusions about a population based on information obtained from a sample of the population. |

Descriptive statistics and inferential statistics are interrelated. You must almost always use techniques of descriptive statistics to organize and summarize the information obtained from a sample before carrying out an inferential analysis. Furthermore, as you will see, the preliminary descriptive analysis of a sample often reveals features that lead you to the choice of (or to a reconsideration of the choice of) the appropriate inferential method.

## Classifying Statistical Studies

As you proceed through this book, you will obtain a thorough understanding of the principles of descriptive and inferential statistics. In this section, you will classify statistical studies as either descriptive or inferential. In doing so, you should consider the purpose of the statistical study.

If the purpose of the study is to examine and explore information for its own intrinsic interest only, the study is descriptive. However, if the information is obtained from a sample of a population and the purpose of the study is to use that information to draw conclusions about the population, the study is inferential.

Thus, a descriptive study may be performed either on a sample or on a population. Only when an inference is made about the population, based on information obtained from the sample, does the study become inferential.

Examples 1.3 and 1.4 further illustrate the distinction between descriptive and inferential studies. In each example, we present the result of a statistical study and classify the study as either descriptive or inferential. Classify each study yourself before reading our explanation.

| Example 1.3 | **Classifying Statistical Studies** |
|---|---|

*The 1948 Presidential Election* Table 1.1 displays the voting results for the 1948 presidential election.

**TABLE 1.1**
Final results of the 1948 presidential election

| Ticket | Votes | Percentage |
|---|---|---|
| Truman–Barkley (Democratic) | 24,179,345 | 49.7 |
| Dewey–Warren (Republican) | 21,991,291 | 45.2 |
| Thurmond–Wright (States Rights) | 1,176,125 | 2.4 |
| Wallace–Taylor (Progressive) | 1,157,326 | 2.4 |
| Thomas–Smith (Socialist) | 139,572 | 0.3 |

You try it!

Exercise 1.7
on page 9

*Classification* This study is descriptive. It is a summary of the votes cast by U.S. voters in the 1948 presidential election. No inferences are made.

• • •

| Example 1.4 | **Classifying Statistical Studies** |

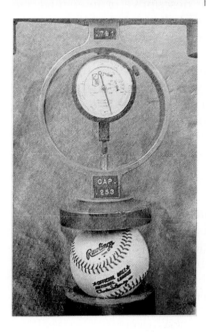

*Testing Baseballs* For the 101 years preceding 1977, the major leagues purchased baseballs from the Spalding Company. In 1977, that company stopped manufacturing major league baseballs, and the major leagues then bought their baseballs from the Rawlings Company.

Early in the 1977 season, pitchers began to complain that the Rawlings ball was "livelier" than the Spalding ball. They claimed it was harder, bounced farther and faster, and gave hitters an unfair advantage. Indeed, in the first 616 games of 1977, 1033 home runs were hit, compared to only 762 home runs hit in the first 616 games of 1976.

*Sports Illustrated* magazine sponsored a study of the liveliness question and published the results in the article "They're Knocking the Stuffing Out of It" (*Sports Illustrated*, June 13, 1977, pp. 23–27) by Larry Keith. In this study, an independent testing company randomly selected 85 baseballs from the current (1977) supplies of various major league teams. It measured the bounce, weight, and hardness of the chosen baseballs, and compared these measurements with measurements obtained from similar tests on baseballs used in 1952, 1953, 1961, 1963, 1970, and 1973.

The conclusion was that "...the 1977 Rawlings ball is livelier than the 1976 Spalding, but not as lively as it could be under big league rules, or as the ball has been in the past."

*Classification*  This study is inferential. The independent testing company used a sample of 85 baseballs from the 1977 supplies of major league teams to make an inference about the population of all such baseballs. (An estimated 360,000 baseballs were used by the major leagues in 1977.)

**You try it!**

Exercise 1.9
on page 9

• • •

The *Sports Illustrated* study also shows that it is often not feasible to obtain information for the entire population. Indeed, after the bounce and hardness tests, all of the baseballs sampled were taken to a butcher in Plainfield, New Jersey, to be sliced in half so that researchers could look inside them. Clearly, testing every baseball in this way would not have been practical.

## The Development of Statistics

Historically, descriptive statistics appeared before inferential statistics. Censuses were taken as long ago as Roman times. Over the centuries, records of such things as births, deaths, marriages, and taxes led naturally to the development of descriptive statistics.

Inferential statistics is a newer arrival. Major developments began to occur with the research of Karl Pearson (1857–1936) and Ronald Fisher (1890–1962), who published their findings in the early years of the twentieth century. Since the work of Pearson and Fisher, inferential statistics has evolved rapidly and is now applied in a myriad of fields.

Familiarity with statistics will help you make sense of many things you read in newspapers and magazines and on the Internet. For instance, could the *Sports Illustrated* baseball test (Example 1.4), which used a sample of only 85 baseballs, legitimately draw a conclusion about 360,000 baseballs? After working through Chapter 9, you will understand why such inferences are reasonable.

**What Does It Mean?**

An understanding of statistical reasoning and of the basic concepts of descriptive and inferential statistics has become mandatory for virtually everyone, in both their private and professional lives.

### Observational Studies and Designed Experiments

Besides classifying statistical studies as either descriptive or inferential, we often need to classify them as either *observational studies* or *designed experiments*. In an **observational study,** researchers simply observe characteristics and take measurements, as in a sample survey. In a **designed experiment,** researchers impose treatments and controls (discussed in Section 1.4) and then observe characteristics and take measurements. Observational studies can reveal only *association*, whereas designed experiments can help establish *causation*.

Examples 1.5 and 1.6 illustrate some major differences between observational studies and designed experiments.

| Example 1.5 | An Observational Study |

*Vasectomies and Prostate Cancer*  Approximately 450,000 vasectomies are performed each year in the United States. In this surgical procedure for contraception, the tube carrying sperm from the testicles is cut and tied.

Several studies have been conducted to analyze the relationship between vasectomies and prostate cancer. The results of one such study by E. Giovannucci et al. appeared in the paper "A Retrospective Cohort Study of Vasectomy and Prostate Cancer in U.S. Men" (*The Journal of the American Medical Association*, Vol. 269(7), pp. 878–882).

Dr. Giovannucci, study leader and epidemiologist at Harvard-affiliated Brigham and Women's Hospital, said that "...we found 113 cases of prostate cancer among 22,000 men who had a vasectomy. This compares to a rate of 70 cases per 22,000 among men who didn't have a vasectomy."

The study shows about a 60% elevated risk of prostate cancer for men who have had a vasectomy, thereby revealing an association between vasectomy and prostate cancer. But does it establish causation: that having a vasectomy causes an increased risk of prostate cancer?

The answer is no, because the study is observational. The researchers simply observed two groups of men, one with vasectomies and the other without. Thus, although an association was established between vasectomy and prostate cancer, the association might be due to other factors (e.g., temperament) that make some men more likely to have vasectomies and also put them at greater risk of prostate cancer.

**You try it!**

Exercise 1.19
on page 11

• • •

| Example 1.6 | A Designed Experiment |

*Folic Acid and Birth Defects*  For several years, evidence had been mounting that folic acid reduces major birth defects. Drs. Andrew E. Czeizel and Istvan Dudas of the National Institute of Hygiene in Budapest directed a study that provided the strongest evidence to date. Their results were published in the paper "Prevention of the First Occurrence of Neural-Tube Defects by Periconceptional Vitamin Supplementation" (*The New England Journal of Medicine*, Vol. 327(26), p. 1832).

For the study, the doctors enrolled 4753 women prior to conception, and divided them randomly into two groups. One group took daily multivitamins containing 0.8 mg of folic acid, whereas the other group received only trace elements. A drastic reduction in the rate of major birth defects occurred among the women who took folic acid: 13 per 1000 as compared to 23 per 1000 for those women who did not take folic acid.

In contrast to the observational study considered in Example 1.5, this is a designed experiment and does help establish causation. The researchers did not simply observe two groups of women but, instead, randomly assigned one group to take daily doses of folic acid and the other group to take only trace elements.

Exercise 1.21
on page 11

• • •

## Exercises 1.1

### Understanding the Concepts and Skills

**1.1** Define the following terms:
**a.** Population          **b.** Sample

**1.2** What are the two major types of statistics? Describe them in detail.

**1.3** Identify some methods used in descriptive statistics.

**1.4** Explain two ways in which descriptive statistics and inferential statistics are interrelated.

**1.5** Define the following terms:
**a.** Observational study      **b.** Designed experiment

**1.6** Fill in the following blank: Observational studies can reveal only association, whereas designed experiments can help establish _____.

*In Exercises 1.7–1.12, classify each of the studies as either descriptive or inferential. Explain your answers.*

**1.7 TV Viewing Times.** The A. C. Nielsen Company collects and publishes information on the television viewing habits of Americans. Data from a sample of Americans yielded the following estimates of average TV viewing time per week for all Americans. The times are in hours and minutes. [SOURCE: Nielsen Media Research, *Nielsen Report on Television*.]

| Group (by age) | | Time |
|---|---|---|
| *Average all persons* | | *30:14* |
| Women | Total 18+ | 34:47 |
| | 18–24 | 28:54 |
| | 25–54 | 31:05 |
| | 55+ | 44:11 |
| Men | Total 18+ | 30.41 |
| | 18–24 | 23:31 |
| | 25–54 | 28:44 |
| | 55+ | 38:47 |
| Teens | 12–17 | 21:50 |
| Children | 2–11 | 23:01 |

**1.8 Professional Athlete Salaries.** In the *Statistical Abstract of the United States*, average professional athletes' salaries in baseball, basketball, and football were compiled and compared for the years 1993 and 2003.

| | Average salary ($1000) | |
|---|---|---|
| **Sport** | **1993** | **2003** |
| Baseball (MLB) | 1,076 | 2,372 |
| Basketball (NBA) | 1,300 | 3,950 |
| Football (NFL) | 619 | 1,259 |

**1.9 Geography Performance Assessment.** In an article titled "Teaching and Assessing Information Literacy in a Geography Program" (*Journal of Geography*, Vol. 104, No. 1, pp. 17–23), Dr. Mary Kimsey and S. Lynn Cameron reported results from an on-line assessment instrument given to senior geography students at one institution of higher learning. The results for level of performance of 22 senior geography majors in 2003 and 29 senior geography majors in 2004 are presented in the following table.

| Level of performance | Percent in 2003 | Percent in 2004 |
|---|---|---|
| Met the standard: 36–48 items correct | 82% | 93% |
| Passed at the advanced level: 41–48 items correct | 50% | 59% |
| Failed: 0–35 items correct | 18% | 7% |

**1.10 Drug Use.** The U.S. Substance Abuse and Mental Health Services Administration collects and publishes data on drug use, by type of drug and age group, in *National Household Survey on Drug Abuse*. The following table provides data for the years 1990 and 2000. The percentages shown are estimates for the entire nation based on information obtained from a sample.

| Type of drug | Percentage, 12 years old and over | | | |
| --- | --- | --- | --- | --- |
| | Ever used | | Current user | |
| | 1990 | 2000 | 1990 | 2000 |
| Marijuana | 30.5 | 34.2 | 5.4 | 4.8 |
| Cocaine | 11.2 | 11.2 | 0.9 | 0.5 |
| Inhalants | 5.7 | 7.5 | 0.4 | 0.3 |
| Hallucinogens | 7.9 | 11.7 | 0.4 | 0.4 |
| Heroin | 0.8 | 1.2 | — | 0.1 |
| Stimulants[1] | 5.5 | 6.6 | 0.6 | 0.4 |
| Sedatives[1] | 2.8 | 3.2 | 0.2 | 0.1 |
| Tranquilizers[1] | 4.0 | 5.8 | 0.6 | 0.4 |
| Pain relievers[1] | 6.3 | 8.6 | 0.9 | 1.2 |
| Alcohol | 82.2 | 81.0 | 52.6 | 46.6 |

1 = Nonmedical use

**1.11 Dow Jones Industrial Averages.** The following table provides the closing values of the Dow Jones Industrial Averages as of the end of December for the years 1997–2004. [SOURCE: Global Financial Data.]

| Year | Closing value |
| --- | --- |
| 1997 | 7,908.25 |
| 1998 | 9,181.43 |
| 1999 | 11,497.12 |
| 2000 | 10,786.85 |
| 2001 | 10,021.50 |
| 2002 | 8,341.63 |
| 2003 | 10,453.92 |
| 2004 | 10,783.01 |

**1.12 The Music People Buy.** Results of monthly telephone surveys yielded the percentage estimates of all music expenditures shown in the following table. These statistics were published in *2004 Consumer Profile*. [SOURCE: Recording Industry Association of America, Inc.]

| Music type | Expenditure (%) |
| --- | --- |
| Rock | 23.9 |
| Country | 13.0 |
| Rap/Hip-hop | 12.1 |
| R&B/Urban | 11.3 |
| Pop | 10.0 |
| Religious | 6.0 |
| Children's | 2.8 |
| Jazz | 2.7 |
| Classical | 2.0 |
| Oldies | 1.4 |
| Soundtracks | 1.1 |
| New Age | 1.0 |
| Other | 8.9 |
| Unknown | 3.8 |

**1.13 Thoughts on Evolution.** In an article titled "Who has designs on your student's minds?" (*Nature*, Vol. 434, pp. 1062–1065), author Geoff Brumfiel postulates that support for Darwinism increases with level of education. The following table provides percentages of U.S. adults, by educational level, who believe that evolution is a scientific theory well supported by evidence.

| Education | Percentage |
| --- | --- |
| Postgraduate education | 65% |
| College graduate | 52% |
| Some college education | 32% |
| High school or less | 20% |

a. Do you think that this study is descriptive or inferential? Explain your answer.
b. If, in fact, the study is inferential, identify the sample and population.

**1.14 Pricey Gasoline.** An Associated Press/AOL poll of 1000 U.S. adults, taken April 18–20, 2005 and appearing in the *Eau Claire Leader Telegram* on April 22, asked about some of the consequences of the rising cost of gasoline. Of those sampled, 58% had reduced their driving, 57% had cut back on other expenses, 41% had planned vacations closer to home, and 41% said that they may buy a more fuel-efficient vehicle.
a. Identify the population and sample for this study.
b. Are the percentages provided descriptive statistics or inferential statistics? Explain your answer.

**1.15 Organically Grown Produce.** A *Newsweek* poll of a sample of Americans revealed that "84% of those surveyed would choose organically grown produce over produce grown using chemical fertilizers, pesticides, and herbicides."
a. Is the statement in quotes an inferential or a descriptive statement? Explain your answer.
b. Based on the same information, what if the statement had been "84% of Americans would choose organically grown produce over produce grown using chemical fertilizers, pesticides, and herbicides"?

**1.16 Vasectomies and Prostate Cancer.** Refer to the vasectomy/prostate cancer study discussed in Example 1.5 on page 8.
a. How could the study be modified to make it a designed experiment?
b. Comment on the feasibility of the designed experiment that you described in part (a).

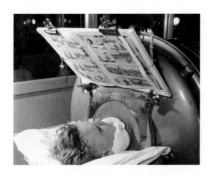

**1.17 The Salk Vaccine.** In the 1940s and early 1950s, the public was greatly concerned about polio. In an attempt to prevent this disease, Jonas Salk of the University of Pittsburgh developed a polio vaccine. In a test of the vaccine's efficacy, involving nearly 2 million grade-school children, half of the children received the Salk vaccine; the other half received a placebo, in this case an injection of salt dissolved in water. Neither the children nor the doctors performing the diagnoses knew which children belonged to which group, but an evaluation center did. The center found that the incidence of polio was far less among the children inoculated with the Salk vaccine. From that information, the researchers concluded that the vaccine would be effective in preventing polio for all U.S. school children; consequently, it was made available for general use. Is this investigation an observational study or a designed experiment? Justify your answer.

**1.18 Do Left-Handers Die Earlier?** According to a study published in the *Journal of the American Public Health Association*, left-handed people do not die at an earlier age than right-handed people, contrary to the conclusion of a highly publicized report done 2 years earlier. The investigation involved a 6-year study of 3800 people in East Boston older than age 65. Researchers at Harvard University and the National Institute of Aging found that the "lefties" and "righties" died at exactly the same rate. "There was no difference, period," said Dr. Jack Guralnik, an epidemiologist at the institute and one of the coauthors of the report. Is this investigation an observational study or a designed experiment? Justify your answer.

**1.19 Skinfold Thickness.** A study titled "Body Composition of Elite Class Distance Runners" was conducted by M. L. Pollock et al. to determine whether elite distance runners actually are thinner than other people. Their results were published in *The Marathon: Physiological, Medical, Epidemiological, and Psychological Studies*, P. Milvey (ed.), New York: New York Academy of Sciences, p. 366. The researchers measured skinfold thickness, an indirect indicator of body fat, of runners and nonrunners in the same age group. Is this investigation a designed experiment or an observational study? Explain your answer.

**1.20 Aspirin and Cardiovascular Disease.** In an article by P. Ridker et al. titled "A Randomized Trial of Low-dose Aspirin in the Primary Prevention of Cardiovascular Disease in Women" (*New England Journal of Medicine*, Vol. 352, pp. 1293–1304), the researchers noted that "We randomly assigned 39,876 initially healthy women 45 years of age or older to receive 100 mg of aspirin or placebo on alternate days and then monitored them for 10 years for a first major cardiovascular event (i.e., nonfatal myocardial infarction, nonfatal stroke, or death from cardiovascular causes)." Is this investigation a designed experiment or an observational study? Explain your answer.

**1.21 Treating Heart Failure.** In the paper "Cardiac-Resynchronization Therapy with or without an Implantable Defibrillator in Advanced Chronic Heart Failure" (*New England Journal of Medicine*, Vol. 350, pp. 2140–2150), M. Bristow et al. reported the results of a study of methods for treating patients who had advanced heart failure due to ischemic or nonischemic cardiomyopathies. A total of 1520 patients were randomly assigned in a 1:2:2 ratio to receive optimal pharmacologic therapy alone or in combination with either a pacemaker or a pacemaker–defibrillator combination. The patients were then observed until they died or were hospitalized for any cause. Is this investigation a designed experiment or an observational study? Explain your answer.

**1.22 Starting Salaries.** The National Association of Colleges and Employers (NACE) compiles information on salary offers to new college graduates and publishes the results in *Salary Survey*. Are these statistical studies designed experiments or observational studies? Explain your answer.

## Extending the Concepts and Skills

**1.23 Ballistic Fingerprinting.** In an on-line press release, *ABCNews.com* reported that "…73 percent of Americans…favor a law that would require every gun sold in the United States to be test-fired first, so law enforcement would have its fingerprint in case it were ever used in a crime."
**a.** Do you think that the statement in the press release is inferential or descriptive? Can you be sure?
**b.** Actually, ABCNews.com conducted a telephone survey of a random national sample of 1032 adults and determined that 73% of them favored a law that would require every gun sold in the United States to be test-fired first, so law enforcement would have its fingerprint in case it were ever used in a crime. How would you rephrase the statement in the press release to make clear that it is a descriptive statement? an inferential statement?

**1.24 Causes of Death.** The U.S. National Center for Health Statistics published the following data on the leading causes of death in 2001 in *Vital Statistics of the United States*. Deaths are classified according to the tenth revision of the *International Classification of Diseases*. Rates are per 100,000 population. Do you think that these rates are descriptive statistics or inferential statistics? Explain your answer.

| Cause | Rate |
|---|---|
| Diseases of heart | 245.8 |
| Malignant neoplasms | 194.4 |
| Cerebrovascular diseases | 57.4 |
| Chronic lower respiratory diseases | 43.2 |

**1.25 Highway Fatalities.** An Associated Press news article, appearing in the *Kansas City Star* on April 22, 2005, stated that "The highway fatality rate sank to a record low last year, the government estimated Thursday. But the overall number of traffic deaths increased slightly, leading the Bush administration to urge a national focus on seat belt use…. Overall, 42,800 people died on the nation's highways in 2004, up from 42,643 in 2003, according to projections from the National Highway Traffic Safety Administration (NHTSA)." Answer the following questions and explain your answers.
a. Is the figure 42,800 a descriptive statistic or an inferential statistic?

b. Is the figure 42,643 a descriptive statistic or an inferential statistic?

**1.26 Motor Vehicle Facts.** Refer to Exercise 1.25. In 2004, the number of vehicles registered grew to 235.4 million from 230.9 million in 2003. Vehicle miles traveled increased from 2.89 trillion in 2003 to 2.92 trillion in 2004. Answer the following questions and explain your answers.
a. Are the numbers of registered vehicles descriptive statistics or inferential statistics?
b. Are the vehicle miles traveled descriptive statistics or inferential statistics?
c. How do you think the NHTSA determined the number of vehicle miles traveled?
d. The highway fatality rate dropped from 1.48 deaths per 100 million vehicle miles traveled in 2003 to 1.46 deaths per 100 million vehicle miles traveled in 2004. It was the lowest rate since records were first kept in 1966. Are the highway fatality rates descriptive statistics or inferential statistics?

# 1.2 Simple Random Sampling

Throughout this book, we present examples of organizations or people conducting studies: A consumer group wants information about the gas mileage of a particular make of car, so it performs mileage tests on a sample of such cars; a teacher wants to know about the comparative merits of two teaching methods, so she tests those methods on two groups of students. This approach reflects a healthy attitude: To obtain information about a subject of interest, plan and conduct a study.

Suppose, however, that a study you are considering has already been done. Repeating it would be a waste of time, energy, and money. Therefore, before planning and conducting a study, do a literature search. You do not necessarily need to go through the entire library or make an extensive Internet search. Instead, you might use an information collection agency that specializes in finding studies on specific topics.

**What Does It Mean?**

You can often avoid the effort and expense of a study if someone else has already done that study and published the results.

### Census, Sampling, and Experimentation

If the information you need is not already available from a previous study, you might acquire it by conducting a **census**—that is, by obtaining information for the entire population of interest. However, conducting a census may be time consuming, costly, impractical, or even impossible.

Two methods other than a census for obtaining information are **sampling** and **experimentation**. In much of this book, we concentrate on sampling. However, we introduce experimentation in Section 1.4, discuss it sporadically throughout the text, and examine it in detail in the chapter *Design of Experiments and Analysis of Variance* (Module C) on the WeissStats CD accompanying this book.

If sampling is appropriate, you must decide how to select the sample; that is, you must choose the method for obtaining a sample from the population.

Because the sample will be used to draw conclusions about the entire population, it should be a **representative sample**—that is, it should reflect as closely as possible the relevant characteristics of the population under consideration.

For instance, using the average weight of a sample of professional football players to make an inference about the average weight of all adult males would be unreasonable. Nor would it be reasonable to estimate the median income of California residents by sampling the incomes of Beverly Hills residents.

To see what can happen when a sample is not representative, consider the presidential election of 1936. Before the election, the *Literary Digest* magazine conducted an opinion poll of the voting population. Its survey team asked a sample of the voting population whether they would vote for Franklin D. Roosevelt, the Democratic candidate, or for Alfred Landon, the Republican candidate.

Based on the results of the survey, the magazine predicted an easy win for Landon. But when the actual election results were in, Roosevelt won by the greatest landslide in the history of presidential elections! What happened?

- The sample was obtained from among people who owned a car or had a telephone. In 1936, that group included only the more well-to-do people, and historically such people tend to vote Republican.
- The response rate was low (less than 25% of those polled responded), and there was a nonresponse bias (a disproportionate number of those who responded to the poll were Landon supporters).

The sample obtained by the *Literary Digest* was not representative.

Most modern sampling procedures involve the use of **probability sampling.** In probability sampling, a random device—such as tossing a coin, consulting a table of random numbers, or employing a random number generator—is used to decide which members of the population will constitute the sample instead of leaving such decisions to human judgment.

The use of probability sampling may still yield a nonrepresentative sample. However, probability sampling eliminates unintentional selection bias and permits the researcher to control the chance of obtaining a nonrepresentative sample. Furthermore, the use of probability sampling guarantees that the techniques of inferential statistics can be applied. In this section and the next, we examine the most important probability-sampling methods.

## Simple Random Sampling

The inferential techniques considered in this book are intended for use with only one particular sampling procedure: **simple random sampling.**

---

**Definition 1.4**

### Simple Random Sampling; Simple Random Sample

**Simple random sampling:**   A sampling procedure for which each possible sample of a given size is equally likely to be the one obtained.

**Simple random sample:**   A sample obtained by simple random sampling.

**What Does It Mean?**

Simple random sampling corresponds to our intuitive notion of random selection by lot.

---

There are two types of simple random sampling. One is **simple random sampling with replacement,** whereby a member of the population can be selected more than once; the other is **simple random sampling without**

**replacement,** whereby a member of the population can be selected at most once. *Unless we specify otherwise, assume that simple random sampling is done without replacement.*

In Example 1.7, we chose a very small population—the five top Oklahoma state officials—to illustrate simple random sampling. In practice, we would not sample from such a small population but would instead take a census. Using a small population here makes understanding the concept of simple random sampling easier.

| | |
|---|---|
| **Example 1.7** | **Simple Random Samples** |

*Sampling Oklahoma State Officials* As reported by *The World Almanac*, the top five state officials of Oklahoma are as shown in Table 1.2. Consider these five officials a population of interest.

**TABLE 1.2**
Five top Oklahoma state officials

| |
|---|
| Governor (G) |
| Lieutenant Governor (L) |
| Secretary of State (S) |
| Attorney General (A) |
| Treasurer (T) |

a.   List the possible samples (without replacement) of two officials from this population of five officials.

b.   Describe a method for obtaining a simple random sample of two officials from this population of five officials.

c.   For the sampling method described in part (b), what are the chances that any particular sample of two officials will be the one selected?

d.   Repeat parts (a)–(c) for samples of size 4.

**Solution**   For convenience, we represent the officials in Table 1.2 by using the letters in parentheses.

**TABLE 1.3**
The 10 possible samples of two officials

| | | | | |
|---|---|---|---|---|
| G, L | G, S | G, A | G, T | L, S |
| L, A | L, T | S, A | S, T | A, T |

a.   Table 1.3 lists the 10 possible samples of two officials from this population of five officials.

b.   To obtain a simple random sample of size 2, we could write the letters that correspond to the five officials (G, L, S, A, and T) on separate pieces of paper. After placing these five slips of paper in a box and shaking it, we could, while blindfolded, pick two slips of paper.

**TABLE 1.4**
The five possible samples of four officials

| | |
|---|---|
| G, L, S, A | G, L, S, T |
| G, L, A, T | G, S, A, T |
| L, S, A, T | |

c.   The procedure described in part (b) will provide a simple random sample. Consequently, each of the possible samples of two officials is equally likely to be the one selected. There are 10 possible samples, so the chances are $\frac{1}{10}$ (1 in 10) that any particular sample of two officials will be the one selected.

d.   Table 1.4 lists the five possible samples of four officials from this population of five officials. A simple random sampling procedure, such as picking four slips of paper out of a box, gives each of these samples a 1 in 5 chance of being the one selected.

• • •

*You try it!*

Exercise 1.37
on page 17

### Random-Number Tables

Obtaining a simple random sample by picking slips of paper out of a box is usually impractical, especially when the population is large. Fortunately, we can use several practical procedures to get simple random samples. One common method involves a **table of random numbers**—a table of randomly chosen digits, as illustrated in Example 1.8.

## Example 1.8 | Random-Number Tables

*Sampling Student Opinions* Student questionnaires, known as "teacher evaluations," gained widespread use in the late 1960s and early 1970s. Generally, professors hand out evaluation forms a week or so before the final.

That practice, however, poses several problems. On some days, less than 60% of students registered for a class may attend. Moreover, many of those who are present complete their evaluation forms in a hurry in order to prepare for other classes. A better method, therefore, might be to select a simple random sample of students from the class and interview them individually.

During one semester, Professor Hassett wanted to sample the attitudes of the students taking college algebra at his school. He decided to interview 15 of the 728 students enrolled in the course. Using a registration list on which the 728 students were numbered 1–728, he obtained a simple random sample of 15 students by randomly selecting 15 numbers between 1 and 728. To do so, he used the random-number table that appears in Appendix A as Table I and here as Table 1.5.

**TABLE 1.5**
Random numbers

| Line number | 00–09 | | 10–19 | | 20–29 | | 30–39 | | 40–49 | |
|---|---|---|---|---|---|---|---|---|---|---|
| 00 | 15544 | 80712 | 97742 | 21500 | 97081 | 42451 | 50623 | 56071 | 28882 | 28739 |
| 01 | 01011 | 21285 | 04729 | 39986 | 73150 | 31548 | 30168 | 76189 | 56996 | 19210 |
| 02 | 47435 | 53308 | 40718 | 29050 | 74858 | 64517 | 93573 | 51058 | 68501 | 42723 |
| 03 | 91312 | 75137 | 86274 | 59834 | 69844 | 19853 | 06917 | 17413 | 44474 | 86530 |
| 04 | 12775 | 08768 | 80791 | 16298 | 22934 | 09630 | 98862 | 39746 | 64623 | 32768 |
| 05 | 31466 | 43761 | 94872 | 92230 | 52367 | 13205 | 38634 | 55882 | 77518 | 36252 |
| 06 | 09300 | 43847 | 40881 | 51243 | 97810 | 18903 | 53914 | 31688 | 06220 | 40422 |
| 07 | 73582 | 13810 | 57784 | 72454 | 68997 | 72229 | 30340 | 08844 | 53924 | 89630 |
| 08 | 11092 | 81392 | 58189 | 22697 | 41063 | 09451 | 09789 | 00637 | 06450 | 85990 |
| 09 | 93322 | 98567 | 00116 | 35605 | 66790 | 52965 | 62877 | 21740 | 56476 | 49296 |
| 10 | 80134 | 12484 | 67089 | 08674 | 70753 | 90959 | 45842 | 59844 | 45214 | 36505 |
| 11 | 97888 | 31797 | 95037 | 84400 | 76041 | 96668 | 75920 | 68482 | 56855 | 97417 |
| 12 | 92612 | 27082 | 59459 | 69380 | 98654 | 20407 | 88151 | 56263 | 27126 | 63797 |
| 13 | 72744 | 45586 | 43279 | 44218 | 83638 | 05422 | 00995 | 70217 | 78925 | 39097 |
| 14 | 96256 | 70653 | 45285 | 26293 | 78305 | 80252 | 03625 | 40159 | 68760 | 84716 |
| 15 | 07851 | 47452 | 66742 | 83331 | 54701 | 06573 | 98169 | 37499 | 67756 | 68301 |
| 16 | 25594 | 41552 | 96475 | 56151 | 02089 | 33748 | 65289 | 89956 | 89559 | 33687 |
| 17 | 65358 | 15155 | 59374 | 80940 | 03411 | 94656 | 69440 | 47156 | 77115 | 99463 |
| 18 | 09402 | 31008 | 53424 | 21928 | 02198 | 61201 | 02457 | 87214 | 59750 | 51330 |
| 19 | 97424 | 90765 | 01634 | 37328 | 41243 | 33564 | 17884 | 94747 | 93650 | 77668 |

**TABLE 1.6**
Registration numbers
of students interviewed

| | | | | |
|---|---|---|---|---|
| 69 | 303 | 458 | 652 | 178 |
| 386 | 97 | 9 | 694 | 578 |
| 539 | 628 | 36 | 24 | 404 |

You
try it!

Exercise 1.43(a)
on page 18

To select 15 random numbers between 1 and 728, we first pick a random starting point, say, by closing our eyes and placing a finger on Table 1.5. Then, beginning with the three digits under the finger, we go down the table and record the numbers as we go. Because we want numbers between 1 and 728 only, we discard the number 000 and numbers between 729 and 999. To avoid repetition, we also eliminate duplicate numbers. If we have not found enough numbers by the time we reach the bottom of the table, we move over to the next column of three-digit numbers and go up.

Using this procedure, Professor Hassett began with 069, circled in Table 1.5. Reading down from 069 to the bottom of Table 1.5 and then up the next column of three-digit numbers, he found the 15 random numbers displayed in Fig. 1.2 and in Table 1.6. Professor Hassett then interviewed the 15 students whose registration numbers are shown in Table 1.6.

• • •

**FIGURE 1.2**
Procedure used by Professor Hassett
to obtain 15 random numbers
between 1 and 728 from Table 1.5

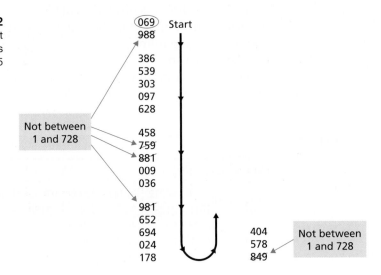

Simple random sampling, the basic type of probability sampling, is also the foundation for the more complex types of probability sampling, which we explore in Section 1.3.

## Random-Number Generators

Nowadays, statisticians prefer statistical software packages or graphing calculators, rather than random-number tables, to obtain simple random samples. The built-in programs for doing so are called **random-number generators.** When using random-number generators, be aware of whether they provide samples with replacement or samples without replacement.

The technology manuals that accompany this book discuss the use of random-number generators for obtaining simple random samples.

## Exercises 1.2

### Understanding the Concepts and Skills

**1.27** Explain why a census is often not the best way to obtain information about a population.

**1.28** Identify two methods other than a census for obtaining information.

**1.29** In sampling, why is obtaining a representative sample important?

**1.30 Memorial Day Poll.** An on-line poll conducted over one Memorial Day Weekend asked people what they were doing to observe Memorial Day. The choices were: (1) stay home and relax, (2) vacation outdoors over the weekend, or (3) visit a military cemetery. More than 22,000 people participated in the poll, with 86% selecting option 1. Discuss this poll with regard to its suitability.

**1.31 Estimating Median Income.** Explain why a sample of 30 dentists from Seattle taken to estimate the median income of all Seattle residents is not representative.

**1.32** Provide a scenario of your own in which a sample is not representative.

**1.33** Regarding probability sampling:
**a.** What is it?
**b.** Does probability sampling always yield a representative sample? Explain your answer.
**c.** Identify some advantages of probability sampling.

**1.34** Regarding simple random sampling:
**a.** What is simple random sampling?
**b.** What is a simple random sample?
**c.** Identify two forms of simple random sampling and explain the difference between the two.

**1.35** The inferential procedures discussed in this book are intended for use with only one particular sampling procedure. What sampling procedure is that?

**1.36** Identify two methods for obtaining a simple random sample.

**1.37 Oklahoma State Officials.** The five top Oklahoma state officials are displayed in Table 1.2 on page 14. Use that table to solve the following problems.
**a.** List the 10 possible samples (without replacement) of size 3 that can be obtained from the population of five officials.
**b.** If a simple random sampling procedure is used to obtain a sample of three officials, what are the chances that it is the first sample on your list in part (a)? the second sample? the tenth sample?

**1.38 Best-Selling Albums.** *Billboard Online* provides data on the best-selling albums of all time. As of June, 2005, the top six best-selling albums of all time, by artist, are the Eagles (E), Michael Jackson (M), Pink Floyd (P), Led Zeppelin (L), AC/DC (A), and Billy Joel (B). [SOURCE: Recording Industry Association of America, Inc.]
**a.** List the 15 possible samples (without replacement) of two artists that can be selected from the six. For brevity, use the initial provided.
**b.** Describe a procedure for taking a simple random sample of two artists from the six.
**c.** If a simple random sampling procedure is used to obtain two artists, what are the chances of selecting P and A? M and E?

**1.39 Best-Selling Albums.** Refer to Exercise 1.38.
**a.** List the 15 possible samples (without replacement) of four artists that can be selected from the six.
**b.** Describe a procedure for taking a simple random sample of four artists from the six.
**c.** If a simple random sampling procedure is used to obtain four artists, what are the chances of selecting E, A, L, and B? P, B, M, and A?

**1.40 Best-Selling Albums.** Refer to Exercise 1.38.
**a.** List the 20 possible samples (without replacement) of three artists that can be selected from the six.
**b.** Describe a procedure for taking a simple random sample of three artists from the six.
**c.** If a simple random sampling procedure is used to obtain three artists, what are the chances of selecting M, A, and L? P, L, and E?

**1.41 Unique National Parks.** In a recent issue of *National Geographic Traveler* (Vol. 22, No. 1, pp. 53, 100–105), Paul Martin gives a list of five unique National Parks that he recommends visiting. They are Crater Lake in Oregon (C), Wolf Trap in Virginia (W), Hot Springs in Arkansas (H), Cuyahoga Valley in Ohio (V), and American Samoa in the Samoan Islands of the South Pacific (A).
**a.** Suppose you want to sample three of these national parks to visit. List the 10 possible samples (without replacement) of size 3 that can be selected from the five. For brevity, use the parenthetical abbreviations provided.
**b.** If a simple random sampling procedure is used to obtain three parks, what are the chances of selecting C, H, and A? V, H, and W?

**1.42 Megacities Risk.** In an issue of *Discover* (Vol. 26, No. 5, p. 14), Anne Casselman looks at the natural hazards risk index of megacities to evaluate potential loss from catastrophes such as earthquakes, storms, and volcanic eruptions. Urban areas have more to lose from natural perils, technological risks, and environmental hazards than rural areas. The top 10 megacities in the world are Tokyo, San Francisco, Los Angeles, Osaka, Miami, New York, Hong Kong, Manila, London, and Paris.

a. There are 45 possible samples (without replacement) of size 2 that can be obtained from these 10 megacities. If a simple random sampling procedure is used, what is the chance of selecting Manila and Miami?

b. There are 252 possible samples (without replacement) of size 5 that can be obtained from these 10 megacities. If a simple random sampling procedure is used, what is the chance of selecting Tokyo, Los Angeles, Osaka, Miami, and London?

c. Suppose that you decide to take a simple random sample of five of these 10 megacities. Use Table I in Appendix A to obtain five random numbers that you can use to specify your sample.

d. If you have access to a random-number generator, use it to solve part (c).

**1.43 The International 500.** Each year, *Fortune Magazine* publishes an article titled "The International 500" that provides a ranking by sales of the top 500 firms outside the United States. Suppose that you want to examine various characteristics of successful firms. Further suppose that, for your study, you decide to take a simple random sample of 10 firms from *Fortune Magazine*'s list of "The International 500."

a. Use Table I in Appendix A to obtain 10 random numbers that you can use to specify your sample.

b. If you have access to a random-number generator, use it to solve part (a).

**1.44 Keno.** In the game of keno, 20 balls are selected at random from 80 balls, numbered 1–80.

a. Use Table I in Appendix A to simulate one game of keno by obtaining 20 random numbers between 1 and 80.

b. If you have access to a random-number generator, use it to solve part (a).

## Extending the Concepts and Skills

**1.45 Oklahoma State Officials.** Refer to Exercise 1.37.

a. List the possible samples of size 1 that can be obtained from the population of five officials.

b. What is the difference between obtaining a simple random sample of size 1 and selecting one official at random?

**1.46 Oklahoma State Officials.** Refer to Exercise 1.37.

a. List the possible samples (without replacement) of size 5 that can be obtained from the population of five officials.

b. What is the difference between obtaining a simple random sample of size 5 and taking a census?

**1.47 Flu Vaccine.** Leading up to the winter of 2004–2005, there was a shortage of flu vaccine in the United States due to impurities found in the supplies of one major vaccine supplier. *The Harris Poll* took a survey to determine the effects of that shortage and posted the results on the Harris Poll Web site. Following the posted results were two paragraphs concerning the methodology, of which the first one is shown here. Did this poll use simple random sampling? Explain your answer.

> The *Harris Poll*® was conducted online within the United States between March 8 and 14, 2005 among a nationwide cross section of 2630 adults aged 18 and over, of whom 698 got a flu shot before the winter of 2004/2005. Figures for age, sex, race, education, region and household income were weighted where necessary to bring the sample of adults into line with their actual proportions in the population. Propensity score weighting was also used to adjust for respondents' propensity to be online.

**1.48 Random-Number Generators.** A random-number generator makes it possible to automatically obtain a list of random numbers within any specified range. Often a random-number generator returns a real number, $r$, between 0 and 1. To obtain random integers (whole numbers) in an arbitrary range, $m$ to $n$, inclusive, apply the conversion formula $m + (n - m + 1)r$ and round down to the nearest integer. Explain how to use this type of random-number generator to solve

a. Exercise 1.43(b).          b. Exercise 1.44(b).

---

## 1.3    Other Sampling Designs*

Simple random sampling is the most natural and easily understood method of probability sampling—it corresponds to our intuitive notion of random selection by lot. However, simple random sampling does have drawbacks. For instance, it may fail to provide sufficient coverage when information about subpopulations is required and may be impractical when the members of the population are widely scattered geographically.

In this section, we examine some commonly used sampling procedures that are often more appropriate than simple random sampling. Remember, though, the inferential procedures discussed in this book must be modified before they can be applied to data that are obtained by sampling procedures other than simple random sampling.

## Systematic Random Sampling

One method that takes less effort to implement than simple random sampling is **systematic random sampling.**

**Example 1.9** | ## Systematic Random Sampling

*Sampling Student Opinions* Recall Example 1.8, in which Professor Hassett wanted a sample of 15 of the 728 students enrolled in college algebra at his school. Let's use systematic random sampling to obtain the sample.

**TABLE 1.7**
Numbers obtained by systematic random sampling

| 22 | 166 | 310 | 454 | 598 |
|----|-----|-----|-----|-----|
| 70 | 214 | 358 | 502 | 646 |
| 118 | 262 | 406 | 550 | 694 |

**Solution**   To begin, we divide the population size by the sample size and round the answer down to the nearest whole number: $\frac{728}{15} = 48$ (rounded down). Next, we select a number at random between 1 and 48 by using, say, a table of random numbers. Suppose that we do so and obtain the number 22. Then, we list every 48th number, starting at 22, until we have 15 numbers. This method yields the 15 numbers displayed in Table 1.7.

If the professor had used systematic random sampling and had begun with the number 22, he would have interviewed the 15 students whose registration numbers are shown in Table 1.7.

• • •

Procedure 1.1 formalizes the steps used in Example 1.9.

**Procedure 1.1**    ## Systematic Random Sampling

*You try it!*

Exercise 1.49
on page 23

**STEP 1  Divide the population size by the sample size and round the result down to the nearest whole number, $m$.**

**STEP 2  Use a random-number table (or a similar device) to obtain a number, $k$, between 1 and $m$.**

**STEP 3  Select for the sample those members of the population that are numbered $k, k + m, k + 2m, \ldots.$**

Systematic random sampling is easier to execute than simple random sampling and usually provides comparable results. The exception is the presence of some kind of cyclical pattern in the listing of the members of the population (e.g., male, female, male, female, …), a phenomenon that is relatively rare.

### Cluster Sampling

Another sampling method is **cluster sampling,** which is particularly useful when the members of the population are widely scattered geographically.

| | |
|---|---|
| **Example 1.10** | ### Cluster Sampling |

*Bike Paths Survey* Many years ago, citizens' groups pressured the city council of Tempe, Arizona, to install bike paths in the city. The council members wanted to be sure that they were supported by a majority of the taxpayers, so they decided to poll the city's homeowners.

Their first survey of public opinion was a questionnaire mailed out with the city's 18,000 homeowner water bills. Unfortunately, this method did not work very well. Only 19.4% of the questionnaires were returned, and a large number of those had written comments that indicated they came from avid bicyclists or from people who strongly resented bicyclists. The city council realized that the questionnaire generally had not been returned by the average voter.

An employee in the city's planning department had sample survey experience, so the council asked her to do a survey. She was given two assistants to help her interview voters and 10 days to complete the project.

The planner first considered taking a simple random sample of 300 voters: 100 interviews for herself and for each of her two assistants. However, the city was so spread out that an interviewer of 100 randomly scattered voters would have to drive an average of 18 minutes from one interview to the next. Doing so would require approximately 30 hours of driving time for each interviewer and could delay completion of the report. She needed a different sampling design.

To save time, the planner decided to use cluster sampling. The residential portion of the city was divided into 947 blocks, each containing approximately 20 houses, as shown in Fig. 1.3.

**FIGURE 1.3**
A typical block of homes

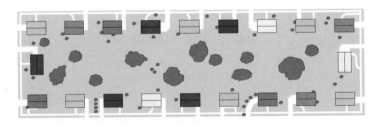

The planner numbered the blocks (clusters) on the city map from 1 to 947 and then used a table of random numbers to obtain a simple random sample of 15 of the 947 blocks. Each of the three interviewers was then assigned five of these 15 blocks. This method gave each interviewer roughly 100 homes to visit but saved much travel time; an interviewer could complete the interviews on one block before driving to another neighborhood. The report was finished on time.

· · ·

Procedure 1.2 formalizes the steps used in Example 1.10.

---

**Procedure 1.2**  **Cluster Sampling**

You try it!

Exercise 1.51(a) on page 23

**STEP 1**  **Divide the population into groups (clusters).**

**STEP 2**  **Obtain a simple random sample of the clusters.**

**STEP 3**  **Use all the members of the clusters obtained in Step 2 as the sample.**

---

Although cluster sampling can save time and money, it does have disadvantages. Ideally, each cluster should mirror the entire population. In practice, however, members of a cluster may be more homogeneous than the members of the entire population, which can cause problems.

For instance, consider a simplified small town, as depicted in Fig. 1.4. The town council wants to build a town swimming pool. A town planner needs to sample voter opinion about using public funds to build the pool. Many upper-income and middle-income homeowners may say "No" if they own or can access pools. Many low-income voters may say "Yes" if they do not have access to pools.

**FIGURE 1.4**
Clusters for a small town

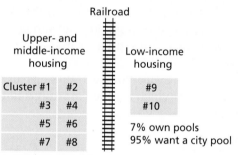

If the planner uses cluster sampling and interviews the voters of, say, three randomly selected clusters, there is a good chance that no low-income voters will be interviewed.[†] And if no low-income voters are interviewed, the results of the survey will be misleading. If, for instance, the planner surveyed clusters #3, #5, and #8, then his survey would show that only about 30% of the voters want a pool. But that is not true because more than 40% of the voters actually want a pool. The clusters most strongly in favor of the pool would not have been included in the survey.

In this hypothetical example, the town is so small that common sense indicates that a cluster sample may not be representative. However, in situations with hundreds of clusters, such problems may be difficult to detect.

---

[†]There are 120 possible three-cluster samples, and 56 of those contain neither of the low-income clusters, #9 and #10. In other words, 46.7% of the possible three-cluster samples contain neither of the low-income clusters.

### Stratified Sampling

Another sampling method, known as **stratified sampling,** is often more reliable than cluster sampling. In stratified sampling the population is first divided into subpopulations, called **strata,** and then sampling is done from each stratum. Ideally, the members of each stratum should be homogeneous relative to the characteristic under consideration, as illustrated in Example 1.11.

---

**Example 1.11** | ### Stratified Sampling

*Town Swimming Pool* Consider again the town swimming pool situation. In stratified sampling, the town planner could divide voters into three strata: upper income, middle income, and low income. The planner could then take a simple random sample from each of the three strata.

This stratified sampling procedure ensures that no income group is missed. It also improves the precision of the statistical estimates (because the voters within each income group tend to be homogeneous) and makes it possible to estimate the separate opinions of each of the three strata.

• • •

In stratified sampling, the strata are often sampled in proportion to their size, which is called **proportional allocation.** For instance, suppose that the strata consisting of the three income groups (upper, middle, and low) in Example 1.11 comprise, respectively, 10%, 70%, and 20% of the town. Then, for a sample size of, say, 50, the number of upper-income, middle-income, and low-income individuals sampled would be, respectively, 5 (10% of 50), 35 (70% of 50), and 10 (20% of 50).

Procedure 1.3 formalizes the steps used in this simplest type of stratified sampling, which is called **stratified random sampling with proportional allocation.**

---

**Procedure 1.3** **Stratified Random Sampling with Proportional Allocation**

**You try it!**
Exercise 1.51(c) on page 23

**STEP 1 Divide the population into subpopulations (strata).**

**STEP 2 From each stratum, obtain a simple random sample of size proportional to the size of the stratum; that is, the sample size for a stratum equals the total sample size times the stratum size divided by the population size.**

**STEP 3 Use all the members obtained in Step 2 as the sample.**

---

### Multistage Sampling

Most large-scale surveys combine one or more of simple random sampling, systematic random sampling, cluster sampling, and stratified sampling. Such **multistage sampling** is used frequently by pollsters and government agencies.

For instance, the U.S. National Center for Health Statistics conducts surveys of the civilian noninstitutional U.S. population to obtain information on illnesses, injuries, and other health issues. Data collection is by a multistage probability sample of approximately 42,000 households. Information obtained from the surveys is published in the *National Health Interview Survey*.

## Exercises 1.3

### Understanding the Concepts and Skills

**1.49 The International 500.** In Exercise 1.43 on page 18, you used simple random sampling to obtain a sample of 10 firms from *Fortune Magazine*'s list of "The International 500."
a. Use systematic random sampling to accomplish that same task.
b. Which method is easier: simple random sampling or systematic random sampling?
c. Does it seem reasonable to use systematic random sampling to obtain a representative sample? Explain your answer.

**1.50 Keno.** In the game of keno, 20 balls are selected at random from 80 balls, numbered 1–80. In Exercise 1.44 on page 18, you used simple random sampling to simulate one game of keno.
a. Use systematic random sampling to obtain a sample of 20 of the 80 balls.
b. Which method is easier: simple random sampling or systematic random sampling?
c. Does it seem reasonable to use systematic random sampling to simulate one game of keno? Explain your answer.

**1.51 Sampling Dorm Residents.** Students in the dormitories of a university in the state of New York live in clusters of four double rooms, called *suites*. There are 48 suites, with eight students per suite.
a. Describe a cluster sampling procedure for obtaining a sample of 24 dormitory residents.
b. Students typically choose friends from their classes as suitemates. With that in mind, do you think cluster sampling is a good procedure for obtaining a representative sample of dormitory residents? Explain your answer.
c. The university housing office has separate lists of dormitory residents by class level. The number of dormitory residents in each class level is as follows.

| Class level | Number of dorm residents |
|---|---|
| Freshman | 128 |
| Sophomore | 112 |
| Junior | 96 |
| Senior | 48 |

Use the table to design a procedure for obtaining a stratified sample of 24 dormitory residents. Use stratified random sampling with proportional allocation.

**1.52 Best High Schools.** In an issue of *Newsweek* (Vol. CXLV, No. 20, pp. 48–57), Barbara Kantrowitz lists "The 100 best high schools in America" according to a ranking devised by Jay Mathews. Another characteristic measured from the high school is the percent free lunch, which is the percentage of student body that is eligible for free and reduced-price lunches, an indicator of socioeconomic status. A percentage of 40% or more generally signifies a high concentration of children in poverty. The top 100 schools, grouped according to their percent free lunch, is as follows.

| Percent free lunch (x) | Number of top 100 ranked high schools |
|---|---|
| $0 \le x < 10$ | 50 |
| $10 \le x < 20$ | 18 |
| $20 \le x < 30$ | 11 |
| $30 \le x < 40$ | 8 |
| $40 \le x$ | 13 |

a. Use the table to design a procedure for obtaining a stratified sample of 25 high schools from the list of the top 100 ranked high schools.
b. If stratified random sampling with proportional allocation is used to select the sample of 25 high schools, how many would be selected from the stratum with a percent-free-lunch value of $30 \le x < 40$?

**1.53 Ghost of Speciation Past.** In the article, "Ghost of Speciation Past" (*Nature*, Vol. 435, pp. 29–31), Thomas D. Kocher looks at the origins of a diverse flock of cichlid fishes in the lakes of southeast Africa. Suppose that you wanted to select a sample from the hundreds of species of cichlid fishes that live in the lakes of southeast Africa. If a simple random sample were taken from the species of each lake, which type of sampling design would have been used? Explain your answer.

### Extending the Concepts and Skills

**1.54 Flu Vaccine.** Leading up to the winter of 2004–2005, there was a shortage of flu vaccine in the United States due to impurities found in the supplies of one major vaccine supplier. *The Harris Poll* took a survey to determine the effects of that shortage and posted the results on the Harris

Poll Web site. Following the posted results were two paragraphs concerning the methodology, of which the second one is shown here.

> In theory, with probability samples of this size, one could say with 95 percent certainty that the results have a sampling error of plus or minus 2 percentage points. Sampling error for the various subsample results is higher and varies. Unfortunately, there are several other possible sources of error in all polls or surveys that are probably more serious than theoretical calculations of sampling error. They include refusals to be interviewed (non-response), question wording and question order, and weighting. It is impossible to quantify the errors that may result from these factors. This online sample is not a probability sample.

**a.** Note the last sentence. Why do you think that this sample is not a probability sample?
**b.** Is the sampling process any one of the other sampling designs discussed in this section: systematic random sampling, cluster sampling, stratified sampling, or multistage sampling? For each sampling design, explain your answer.

**1.55 The Terri Schiavo Case.** In the early part of 2005, the Terri Schiavo case received national attention as her husband sought to have life support removed and her parents sought to maintain that life support. The courts allowed the life support to be removed and her death ensued. A *Harris Poll* of 1010 U.S. adults was taken by telephone on April 21, 2005 to determine how common it is for life support systems to be removed. Those questioned in the sample were asked: (1) Has one of your parents, a close friend, or a family member died in the last ten years? (2) Before (this death/these deaths) happened, was this person/were any of these people, kept alive by any support system? (3) Did this person die while on a life support system, or had it been withdrawn? Respondents were also asked questions about age, sex, race, education, region, and household income to ensure that results represented a cross section of U.S. adults.
**a.** What kind of sampling design was used in this survey? Explain your answer.

**b.** If 78% of the respondents answered the first question in the affirmative, what was the approximate sample size for the second question?
**c.** If 28% of those responding to the second question answered "yes," what was the approximate sample size for the third question?

**1.56** In simple random sampling, all samples of a given size are equally likely. Is that true in systematic random sampling? Explain your answer.

**1.57 White House Ethics.** On June 27, 1996, an article appeared in *The Wall Street Journal* presenting the results of a nationwide poll regarding the White House procurement of FBI files on prominent Republicans and related ethical controversies. The article was headlined "White House Assertions on FBI Files Are Widely Rejected, Survey Shows." At the end of the article, the following explanation of the sampling procedure was given. Discuss the different aspects of sampling that appear in this explanation.

> The Wall Street Journal/NBC News poll was based on nationwide telephone interviews of 2,010 adults, including 1,637 registered voters, conducted Thursday to Tuesday by the polling organizations of Peter Hart and Robert Teeter. Questions related to politics were asked only of registered voters; questions related to economics and health were asked of all adults.
>
> The sample was drawn from 520 randomly selected geographic points in the continental U.S. Each region was represented in proportion to its population. Households were selected by a method that gave all telephone numbers, listed and unlisted, an equal chance of being included.
>
> One adult, 18 years or older, was selected from each household by a procedure to provide the correct number of male and female respondents.
>
> Chances are 19 of 20 that if all adults with telephones in the U.S. had been surveyed, the finding would differ from these poll results by no more than 2.2 percentage points in either direction among all adults and 2.5 among registered voters. Sample tolerances for subgroups are larger.

## 1.4 Experimental Designs*

As we mentioned earlier, two methods for obtaining information, other than a census, are sampling and experimentation. In Sections 1.2 and 1.3, we discussed some of the basic principles and techniques of sampling. Now, we do the same for experimentation.

### Principles of Experimental Design

The study presented in Example 1.6 on page 8 illustrates three basic principles of experimental design: **control, randomization,** and **replication.**

- *Control:* The doctors compared the rate of major birth defects for the women who took folic acid to that for the women who took only trace elements.
- *Randomization:* The women were divided randomly into two groups to avoid unintentional selection bias.
- *Replication:* A large number of women were recruited for the study to make it likely that the two groups created by randomization would be similar and also to increase the chances of detecting any effect due to the folic acid.

In the language of experimental design, each woman in the folic acid study is an **experimental unit,** or a **subject.** More generally, we have the following definition.

---

**Definition 1.5** **Experimental Units; Subjects**

In a designed experiment, the individuals or items on which the experiment is performed are called **experimental units.** When the experimental units are humans, the term **subject** is often used in place of experimental unit.

---

In the folic acid study, both dosages of folic acid (0.8 mg and essentially none) are called *treatments* in the context of experimental design. Generally, each experimental condition is called a **treatment,** of which there may be several.

Now that we have introduced the terms *experimental unit* and *treatment,* we can present the three basic principles of experimental design in a general setting.

---

**Key Fact 1.1** **Principles of Experimental Design**

The following principles of experimental design enable a researcher to conclude that differences in the results of an experiment not reasonably attributable to chance are likely caused by the treatments.

- **Control:** Two or more treatments should be compared.
- **Randomization:** The experimental units should be randomly divided into groups to avoid unintentional selection bias in constituting the groups.
- **Replication:** A sufficient number of experimental units should be used to ensure that randomization creates groups that resemble each other closely and to increase the chances of detecting any differences among the treatments.

---

One of the most common experimental situations involves a specified treatment and *placebo,* an inert or innocuous medical substance. Technically, both the specified treatment and placebo are treatments. The group receiving the specified treatment is called the **treatment group,** and the group receiving placebo is called the **control group.** In the folic acid study, the women who took folic acid constituted the treatment group and those who took only trace elements constituted the control group.

## Terminology of Experimental Design

In the folic acid study, the researchers were interested in the effect of folic acid on major birth defects. Birth-defect classification (whether major or not) is the **response variable** for this study. The daily dosage of folic acid is called the **factor.** In this case, the factor has two **levels,** namely, 0.8 mg and essentially none.

When there is only one factor, as in the folic acid study, the treatments are the same as the levels of the factor. But, if a study has more than one factor, each treatment is a combination of levels of the various factors. Example 1.12 presents an experiment with two factors.

| Example 1.12 | Experimental Design |

*Weight Gain of Golden Torch Cacti* The Golden Torch Cactus (*Trichocereus spachianus*), a cactus native to Argentina, has excellent landscape potential. William Feldman and Frank Crosswhite, two researchers at the Boyce Thompson Southwestern Arboretum, investigated the optimal method for producing these cacti.

The researchers examined, among other things, the effects of a hydrophilic polymer and irrigation regime on weight gain. Hydrophilic polymers are used as soil additives to keep moisture in the root zone. For this study, the researchers chose Broadleaf P-4 polyacrylamide, abbreviated P4. The hydrophilic polymer was either used or not used, and five irrigation regimes were employed: none, light, medium, heavy, and very heavy. Identify the

a. experimental units.    b. response variable.    c. factors.

d. levels of each factor.    e. treatments.

### Solution

a. The experimental units are the cacti used in the study.

b. The response variable is weight gain.

c. The factors are hydrophilic polymer and irrigation regime.

d. Hydrophilic polymer has two levels: with and without. Irrigation regime has five levels: none, light, medium, heavy, and very heavy.

e. Each treatment is a combination of a level of hydrophilic polymer and a level of irrigation regime. Table 1.8 depicts the 10 treatments for this experiment. In the table, we abbreviated "very heavy" as "Xheavy."

**TABLE 1.8**
Schematic for the 10 treatments in the cactus study

| | | Irrigation regime | | | | |
|---|---|---|---|---|---|---|
| | | None | Light | Medium | Heavy | Xheavy |
| Polymer | No P4 | No water No P4 (Treatment 1) | Light water No P4 (Treatment 2) | Medium water No P4 (Treatment 3) | Heavy water No P4 (Treatment 4) | Xheavy water No P4 (Treatment 5) |
| | With P4 | No water With P4 (Treatment 6) | Light water With P4 (Treatment 7) | Medium water With P4 (Treatment 8) | Heavy water With P4 (Treatment 9) | Xheavy water With P4 (Treatment 10) |

• • •

We now formally define some terms in experimental design.

| Definition 1.6 | **Response Variable, Factors, Levels, and Treatments** |
|---|---|

**Response variable:**   The characteristic of the experimental outcome that is to be measured or observed.

**Factor:**   A variable whose effect on the response variable is of interest in the experiment.

**Levels:**   The possible values of a factor.

**Treatment:**   Each experimental condition. For one-factor experiments, the treatments are the levels of the single factor. For multifactor experiments, each treatment is a combination of levels of the factors.

*You try it!*

Exercise 1.63
on page 29

### Statistical Designs

Once we have chosen the treatments, we must decide how the experimental units are to be assigned to the treatments (or vice versa). The women in the folic acid study were randomly divided into two groups; one group received folic acid and the other only trace elements. In the cactus study, 40 cacti were divided randomly into 10 groups of four cacti each and then each group was assigned a different treatment from among the 10 depicted in Table 1.8. Both of these experiments used a **completely randomized design.**

| Definition 1.7 | **Completely Randomized Design** |
|---|---|

In a **completely randomized design,** all the experimental units are assigned randomly among all the treatments.

Although the completely randomized design is commonly used and simple, it is not always the best design. Several alternatives to that design exist.

For instance, in a **randomized block design,** experimental units that are similar in ways that are expected to affect the response variable are grouped in **blocks.** Then the random assignment of experimental units to the treatments is made block by block.

| Definition 1.8 | **Randomized Block Design** |
|---|---|

In a **randomized block design,** the experimental units are assigned randomly among all the treatments separately within each block.

Example 1.13 contrasts completely randomized designs and randomized block designs.

| Example 1.13 | **Statistical Designs** |
|---|---|

*Golf Ball Driving Distances* Suppose we want to compare the driving distances for five different brands of golf ball. For 40 golfers, discuss a method of comparison based on

**a.** a completely randomized design.       **b.** a randomized block design.

**Solution** Here the experimental units are the golfers, the response variable is driving distance, the factor is brand of golf ball, and the levels (and treatments) are the five brands.

a. For a completely randomized design, we would randomly divide the 40 golfers into five groups of 8 golfers each and then randomly assign each group to drive a different brand of ball, as illustrated in Fig. 1.5.

**FIGURE 1.5**
Completely randomized design for golf ball experiment

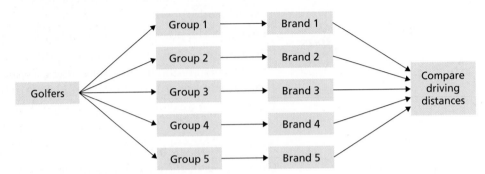

b. Because driving distance is affected by gender, using a randomized block design that blocks by gender is probably a better approach. We could do so by using 20 men golfers and 20 women golfers. We would randomly divide the 20 men into five groups of 4 men each and then randomly assign each group to drive a different brand of ball, as shown in Fig. 1.6. Likewise, we would randomly divide the 20 women into five groups of 4 women each and then randomly assign each group to drive a different brand of ball, as also shown in Fig. 1.6.

**FIGURE 1.6**  Randomized block design for golf ball experiment

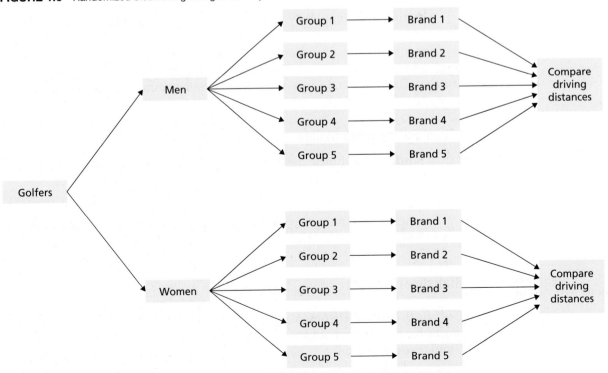

**You try it!**

Exercise 1.66
on page 30

By blocking, we can isolate and remove the variation in driving distances between men and women and thereby make it easier to detect any differences in driving distances among the five brands of golf ball. Additionally, blocking permits us to analyze separately the differences in driving distances among the five brands for men and women.

• • •

As illustrated in Example 1.13, blocking can isolate and remove systematic differences among blocks, thereby making any differences among treatments easier to detect. Blocking also makes possible the separate analysis of treatment effects on each block.

In this section, we introduced some of the basic terminology and principles of experimental design. However, we have just scratched the surface of this vast and important topic to which entire courses and books are devoted. Further discussion of experimental design is provided in the chapter *Design of Experiments and Analysis of Variance* (Module C) on the WeissStats CD accompanying this book.

## Exercises 1.4

### Understanding the Concepts and Skills

**1.58** State and explain the significance of the three basic principles of experimental design.

**1.59** In a designed experiment,
a. what are the experimental units?
b. if the experimental units are humans, what term is often used in place of experimental unit?

**1.60 Adverse Effects of Prozac.** Prozac (fluoxetine hydrochloride), a product of Eli Lilly and Company, is used for the treatment of depression, obsessive–compulsive disorder (OCD), and bulimia nervosa. An issue of the magazine *Arthritis Today* contained an advertisement reporting on the "…treatment-emergent adverse events that occurred in 2% or more patients treated with Prozac and with incidence greater than placebo in the treatment of depression, OCD, or bulimia." In the study, 2444 patients took Prozac and 1331 patients were given placebo. Identify the
a. treatment group.          b. control group.
c. treatments.

**1.61 Treating Heart Failure.** In the paper "Cardiac-Resynchronization Therapy with or without an Implantable Defibrillator in Advanced Chronic Heart Failure" (*New England Journal of Medicine*, Vol. 350, pp. 2140–2150), M. Bristow et al. reported the results of a study of methods for treating patients who had advanced heart failure due to ischemic or nonischemic cardiomyopathies. A total of 1520 patients were randomly assigned in a 1:2:2 ratio to receive optimal pharmacologic therapy alone or in combination with either a pacemaker or a pacemaker–defibrillator combination. The patients were then observed until they died or were hospitalized for any cause.

a. How many treatments were there?
b. Which group would be considered the control group?
c. How many treatment groups were there? Which treatments did they receive?
d. How many patients were in each of the three groups studied?
e. Explain how a table of random numbers or a random-number generator could be used to divide the patients into the three groups.

*In Exercises **1.62–1.65**, we present descriptions of designed experiments. In each case, identify the*
*a. experimental units.*
*b. response variable.*
*c. factor(s).*
*d. levels of each factor.*
*e. treatments.*

**1.62 Storage of Perishable Items.** Storage of perishable items is an important concern for many companies. One study examined the effects of storage time and storage temperature on the deterioration of a particular item. Three different storage temperatures and five different storage times were used.

**1.63 Increasing Unit Sales.** Supermarkets are interested in strategies to increase temporarily the unit sales of a product. In one study, researchers compared the effect of display type and price on unit sales for a particular product. The following display types and pricing schemes were employed.

• Display types: normal display space interior to an aisle, normal display space at the end of an aisle, and enlarged display space.
• Pricing schemes: regular price, reduced price, and cost.

**1.64 Oat Yield and Manure.** In a classic study, described by F. Yates in *The Design and Analysis of Factorial Experiments*, the effect on oat yield was compared for three different varieties of oats and four different concentrations of manure (0, 0.2, 0.4, and 0.6 cwt per acre).

**1.65 The Lion's Mane.** In a study by Peyton M. West, titled "The Lion's Mane" (*American Scientist*, Vol. 93, No. 3, pp. 226–236), the effects of the mane of a male lion as a signal of quality to mates and rivals was explored. Four life-sized dummies of male lions provided a tool for testing female response to the unfamiliar lions whose manes varied by length (long or short) and color (blonde or dark). The female lions were observed to see whether they approached each of the four life-sized dummies.

**1.66 Lifetimes of Flashlight Batteries.** Two different options are under consideration for comparing the lifetimes of four brands of flashlight battery, using 20 flashlights.
a. One option is to randomly divide 20 flashlights into four groups of 5 flashlights each and then randomly assign each group to use a different brand of battery. Would this statistical design be a completely randomized design or a randomized block design? Explain your answer.

b. Another option is to use 20 flashlights—five different brands of 4 flashlights each—and randomly assign the 4 flashlights of each brand to use a different brand of battery. Would this statistical design be a completely randomized design or a randomized block design? Explain your answer.

## Extending the Concepts and Skills

**1.67 The Salk Vaccine.** Exercise 1.17 on page 11 discussed the Salk vaccine experiment. The experiment utilized a technique called *double-blinding* because neither the children nor the doctors involved knew which children had been given the vaccine and which had been given placebo. Explain the advantages of using double-blinding in the Salk vaccine experiment.

**1.68** In sampling from a population, state which type of sampling design corresponds to each of the following experimental designs:
a. Completely randomized design
b. Randomized block design

## Chapter in Review

### You Should be Able to

1. classify a statistical study as either descriptive or inferential.

2. identify the population and the sample in an inferential study.

3. explain the difference between an observational study and a designed experiment.

4. classify a statistical study as either an observational study or a designed experiment.

5. explain what is meant by a representative sample.

6. describe simple random sampling.

7. use a table of random numbers to obtain a simple random sample.

*8. describe systematic random sampling, cluster sampling, and stratified sampling.

*9. state the three basic principles of experimental design.

*10. identify the treatment group and control group in a study.

*11. identify the experimental units, response variable, factor(s), levels of each factor, and treatments in a designed experiment.

*12. distinguish between a completely randomized design and a randomized block design.

### Key Terms

blocks,* 27
census, 12
cluster sampling,* 21
completely randomized design,* 27
control,* 25
control group,* 25
descriptive statistics, 4
designed experiment, 8
experimental unit,* 25

experimentation, 12
factor,* 27
inferential statistics, 6
levels,* 27
multistage sampling,* 22
observational study, 8
population, 5
probability sampling, 13
proportional allocation,* 22

randomization,* 25
randomized block design,* 27
random-number generator, 16
replication,* 25
representative sample, 13
response variable,* 27
sample, 5
sampling, 12
simple random sample, 13

## Review Problems

### Understanding the Concepts and Skills

**1.** In a newspaper or magazine, or on the Internet, find an example of
**a.** a descriptive study.    **b.** an inferential study.

**2.** Almost any inferential study involves aspects of descriptive statistics. Explain why.

**3. Baseball Scores.** On September 3, 2005, the following baseball scores were printed in *The Daily Courier*. Is this study descriptive or inferential? Explain your answer.

| Major League Baseball | |
|---|---|
| Giants 6, D'backs 3 | Blue Jays 4, Devil Rays 3 |
| Cubs 7, Pirates 3 | Orioles 7, Red Sox 3 |
| Marlins 4, Mets 2 | White Sox 9, Tigers 1 |
| Phillies 7, Nationals 1 | Indians 6, Twins 1 |
| Braves 7, Reds 4 | Rangers 8, Royals 7 |
| Brewers 12, Padres 2 | A's 12, Yankees 0 |
| Astros 6, Cardinals 5 | Angels 4, Mariners 1 |
| Rockies 11, Dodgers 3 | |

**4. Serious Energy Situation.** In a *USA TODAY/CNN Gallup Poll*, 94% of those surveyed said that the United States faced a serious energy situation, but, by 47% to 35%, they preferred an emphasis on conservation rather than on more production. Is this study descriptive or inferential? Explain your answer.

**5. British Backpacker Tourists.** Research by Gustav Visser and Charles Barker in "A Geography of British Backpacker Tourists in South Africa" (*Geography*, Vol. 89, No. 3, pp. 226–239) reflects on the impact of British backpacker tourists visiting South Africa. A sample of British backpackers was interviewed. The information obtained from the sample was used to construct the following table for the age distribution of all British backpackers. Classify this study as descriptive or inferential, and explain your answer.

| Age (yrs) | Percentage |
|---|---|
| Less than 21 | 9% |
| 21–25 | 46% |
| 26–30 | 27% |
| 31–35 | 10% |
| 36–40 | 4% |
| Over 40 | 4% |

**6. Teen Drug Abuse.** In an article dated April 24, 2005, *USA TODAY* reported on the 17th annual study on teen drug abuse, conducted by the Partnership for a Drug-Free America. According to the survey of 7300 teens, the most popular prescription drug abused by teens was Vicodin, with 18%—or about 4.3 million youths—reporting that they had used it to get high. OxyContin and drugs for attention deficit disorder, such as Ritalin/Adderall, followed with one in 10 teens reporting that they had tried them. Answer the following questions and explain your answers.
**a.** Is the statement about 18% of youths abusing Vicodin inferential or descriptive?
**b.** Is the statement about 4.3 million youths abusing Vicodin inferential or descriptive?

**7.** Regarding observational studies and designed experiments:
**a.** Describe each type of statistical study.
**b.** With respect to possible conclusions, what important difference exists between these two types of statistical studies?

**8. Persistent Poverty and IQ.** An article appearing in an issue of *The Arizona Republic* reported on a study conducted by Greg Duncan of the University of Michigan. According to the report, "Persistent poverty during the first 5 years of life leaves children with IQs 9.1 points lower at age 5 than children who suffer no poverty during that period...." Is this statistical study an observational study or is it a designed experiment? Explain your answer.

**9. Wasp Hierarchical Status.** In the February 2005 issue of *Discover* (Vol. 26, No. 2, pp. 10–11), Jesse Netting describes the research of Elizabeth Tibbetts of the University of Arizona in the article, "The Kind of Face Only a Wasp Could Trust." Tibbetts found that wasps signal their strength and status with the number of black splotches on their yellow faces, with more splotches denoting higher status. Tibbetts decided to see if she could cheat the system. She painted some of the insects' faces to make their status appear higher or lower than it really was. She then placed the painted wasps with a group of female wasps to see if painting the faces altered their hierarchical status. Was this investigation an observational study or a designed experiment? Justify your answer.

**10.** Before planning and conducting a study to obtain information, what should be done?

**11.** Explain the meaning of
a. a representative sample.
b. probability sampling.
c. simple random sampling.

**12. Incomes of College Students' Parents.** A researcher wants to estimate the average income of parents of college students. To accomplish that, he surveys a sample of 250 students at Yale. Is this a representative sample? Explain your answer.

**13.** Which of the following sampling procedures involve the use of probability sampling?
a. A college student is hired to interview a sample of voters in her town. She stays on campus and interviews 100 students in the cafeteria.
b. A pollster wants to interview 20 gas station managers in Baltimore. He posts a list of all such managers on his wall, closes his eyes, and tosses a dart at the list 20 times. He interviews the people whose names the dart hits.

**14. On-Time Airlines.** Conducive Technology Corporation compiles information on the on-time performance of all scheduled passenger flights arriving in the United States for a period of time, typically 1 month. Results are published in *FlightStats*. For the month of August, 2005, the five airlines with the highest percentage of on-time arrivals were Hawaiian Airlines (HA), Corporate Express (3C), Allegiant Air (G4), Air Midwest (ZV), and Frontier Airlines (F9).
a. List the 10 possible samples (without replacement) of size 3 that can be obtained from the population of five airlines. Use the parenthetical abbreviations in your list.
b. If a simple random sampling procedure is used to obtain a sample of three of these five airlines, what are the chances that it is the first sample on your list in part (a)? the second sample? the tenth sample?
c. Describe three methods for obtaining a simple random sample of three of these five airlines.
d. Use one of the methods that you described in part (c) to obtain a simple random sample of three of these five airlines.

**15. Top North American Athletes.** As part of ESPN's *SportsCenturyRetrospective*, a panel chosen by ESPN ranked the top 100 North American athletes of the twentieth century. For a class project, you are to obtain a simple random sample of 15 of these 100 athletes and briefly describe their athletic feats.
a. Explain how you can use Table I in Appendix A to obtain the simple random sample.
b. Starting at the three-digit number in line number 10 and column numbers 7–9 of Table I, read down the column, up the next, and so on, to find 15 numbers that you can use to identify the athletes to be considered.
c. If you have access to a random-number generator, use it to obtain the required simple random sample.

**\*16.** Describe each of the following sampling methods and indicate conditions under which each is appropriate.
a. Systematic random sampling

b. Cluster sampling
c. Stratified random sampling with proportional allocation

**\*17. Top North American Athletes.** Refer to Problem 15.
a. Use systematic random sampling to obtain a sample of 15 athletes.
b. In this case, is systematic random sampling an appropriate alternative to simple random sampling? Explain your answer.

**\*18. Surveying the Faculty.** The faculty of a college consists of 820 members. A new president has just been appointed. The president wants to get an idea of what the faculty considers the most important issues currently facing the school. She does not have time to interview all the faculty members and so decides to stratify the faculty by rank and use stratified random sampling with proportional allocation to obtain a sample of 40 faculty members. There are 205 full professors, 328 associate professors, 246 assistant professors, and 41 instructors.
a. How many faculty members of each rank should be selected for interviewing?
b. Use Table I in Appendix A to obtain the required sample. Explain your procedure in detail.

**19. QuickVote.** *TalkBack Live* conducts on-line surveys on various issues. The following photo shows the result of a *quickvote* taken on July 5, 2000, that asked whether a person would vote for a third-party candidate. Beneath the vote tally is a statement regarding the sampling procedure. Discuss this statement in light of what you have learned in this chapter.

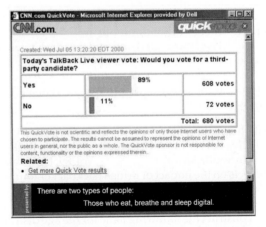

**\*20. AVONEX and MS.** An issue of *Inside MS* contained an article describing AVONEX (Interferon beta-1a), a drug used in the treatment of relapsing forms of multiple sclerosis. Included in the article was a report on "…adverse events and selected laboratory abnormalities that occurred at an incidence of 2% or more among the 158 multiple sclerosis patients treated with 30 mcg of AVONEX once weekly by IM injection." In the study, 158 patients took AVONEX and 143 patients were given placebo.
a. Is this study observational or is it a designed experiment?
b. Identify the treatment group, control group, and treatments.

*21. Identify and explain the significance of the three basic principles of experimental design.

*22. **Plant Density and Tomato Yield.** In the paper "Effects of Plant Density on Tomato Yields in Western Nigeria" (*Experimental Agriculture*, Vol. 12(1), pp. 43–47), B. Adelana reported on the effect of tomato variety and planting density on yield. Identify the

a. experimental units.
b. response variable.
c. factor(s).
d. levels of each factor.
e. treatments.

*23. **Child-Proof Bottles.** Designing medication packaging that resists opening by children, but yields readily to adults, presents numerous challenges. In the article "Painful Design" (*American Scientist*, Vol. 93, No. 2, pp. 113–118), Henry Petroski examined the packaging used for Aleve, a brand of pain reliever. Three new container designs were given to a panel of children aged 42 months to 51 months. For each design, the children were handed the bottle, shown how to open it, and then left alone with it. If more than 20% of the children succeed in opening the bottle on their own within 10 minutes, even if by using their teeth, the bottle failed to qualify as child resistant. Identify the

a. experimental units
b. response variable
c. factor(s)
d. levels of each factor
e. treatments

*24. **Doughnuts and Fat.** A classic study, conducted in 1935 by B. Lowe at the Iowa Agriculture Experiment Station, analyzed differences in the amount of fat absorbed by doughnuts in cooking with four different fats. For the experiment, 24 batches of doughnuts were randomly divided into four groups of 6 batches each. The four groups were then randomly assigned to the four fats. What type of statistical design was used for this study? Explain your answer.

*25. **Comparing Gas Mileages.** An experiment is to be conducted to compare four different brands of gasoline for gas mileage.

a. Suppose that you randomly divide 24 cars into four groups of 6 cars each and then randomly assign the four groups to the four brands of gasoline, one group per brand. Is this experimental design a completely randomized design or a randomized block design? If it is the latter, what are the blocks?

b. Suppose, instead, that you use six different models of car whose varying characteristics (e.g., weight and horsepower) affect gas mileage. Four cars of each model are randomly assigned to the four different brands of gasoline. Is this experimental design a completely randomized design or a randomized block design? If it is the latter, what are the blocks?

c. Which design is better, the one in part (a) or the one in part (b)? Explain your answer.

26. **USA TODAY Polls.** The following explanation of *USA TODAY* polls and surveys was obtained from the *USA TODAY* Web site. Discuss the explanation in detail.

> USATODAY.com frequently publishes the results of both scientific opinion polls and online reader surveys. Sometimes the topics of these two very different types of public opinion sampling are similar but the results appear very different. It is important that readers understand the difference between the two.
>
> USA TODAY/CNN/Gallup polling is a scientific phone survey taken from a random sample of U.S. residents and weighted to reflect the population at large. This is a process that has been used and refined for more than 50 years. Scientific polling of this type has been used to predict the outcome of elections with considerable accuracy.
>
> Online surveys, such as USATODAY.com's "Quick Question," are not scientific and reflect the views of a self-selected slice of the population. People using the Internet and answering online surveys tend to have different demographics than the nation as a whole and as such, results will differ---sometimes dramatically---from scientific polling.
>
> USATODAY.com will clearly label results from the various types of surveys for the convenience of our readers.

27. **Crosswords and Dementia.** An article appearing in the *Los Angeles Times* on June 21, 2003, discussed a report by researchers from a recent issue of the *New England Journal of Medicine*. The article, titled "Crosswords Reduce Risk of Dementia," stated that "Elderly people who frequently read, do crossword puzzles, practice a musical instrument or play board games cut their risk of Alzheimer's and other forms of dementia by nearly two-thirds compared with people who seldom do such activities..." Comment on the statement in quotes, keeping in mind the type of study for which causation can be reasonably inferred.

# Focusing on Data Analysis  UWEC Undergraduates

The file Focus.txt in the Focus Database folder of the WeissStats CD contains information on the undergraduate students at the University of Wisconsin - Eau Claire (UWEC). Those students constitute the population of interest in the *Focusing on Data Analysis* sections that appear at the end of each chapter of the book.[†]

Thirteen variables are considered. Table 1.9 lists the variables and the names used for those variables in the data files. We call the database of information for those variables the **Focus database.**

**TABLE 1.9**

Variables and variable names for the Focus database

| Variable | Variable name |
|---|---|
| Sex | SEX |
| High school percentile | HSP |
| Cumulative GPA | GPA |
| Age | AGE |
| Total earned credits | CREDITS |
| Classification | CLASS |
| School/college | COLLEGE |
| Primary major | MAJOR |
| Residency | RESIDENCY |
| Admission type | TYPE |
| ACT English score | ENGLISH |
| ACT math score | MATH |
| ACT composite score | COMP |

Also provided in the Focus Database folder is a file called FocusSample.txt that contains data on the same 13 variables for a simple random sample of 200 of the undergraduate students at UWEC. Those 200 students constitute a sample that can be used for making statistical inferences in the *Focusing on Data Analysis* sections. We call this sample data the **Focus sample.**

Large data sets are almost always analyzed by computer, and that is how you should handle both the Focus database and the Focus sample. We have supplied the Focus database and Focus sample in several file formats in the Focus Database folder of the WeissStats CD.

If you use a statistical software package for which we have not supplied a Focus database file, you should (1) input the file Focus.txt into that software, (2) name the variables as indicated in Table 1.9, and (3) save the worksheet to a file named Focus in the format suitable to your software, that is, with the appropriate file extension. Then, any time that you want to analyze the Focus database, you can simply retrieve your Focus worksheet. These same remarks apply to the Focus sample as well as to the Focus database.

# Case Study Discussion  Greatest American Screen Legends

At the beginning of this chapter, we discussed the results of a survey by the American Film Institute (AFI). Now that you have learned some of the basic terminology of statistics, we want you to examine that survey in greater detail.

Answer each of the following questions pertaining to the survey. In doing so, you may want to reread the description of the survey given on page 3.

**a.** Identify the population.
**b.** Identify the sample.
**c.** Is the sample representative of the population of all U.S. moviegoers? Explain your answer.

**d.** Consider the following statement: "Among the 1800 artists, historians, critics, and other cultural dignitaries polled by AFI, the top-ranking male and female American screen legends were Humphrey Bogart and Katharine Hepburn." Is this statement descriptive or inferential? Explain your answer.
**e.** Suppose that the statement in part (d) is changed to: "Based on the AFI poll, Humphrey Bogart and Katharine Hepburn are the top-ranking male and female American screen legends among all artists, historians, critics, and other cultural dignitaries." Is this statement descriptive or inferential? Explain your answer.

[†]We have restricted attention to those undergraduate students at UWEC with complete records for all the variables under consideration.

## Biography    FLORENCE NIGHTINGALE: Lady of the Lamp

**Florence Nightingale** (1820–1910), the founder of modern nursing, was born in Florence, Italy, into a wealthy English family. In 1849, over the objections of her parents, she entered the Institution of Protestant Deaconesses at Kaiserswerth, Germany, which "…trained country girls of good character to nurse the sick."

The Crimean War began in March, 1854, when England and France declared war on Russia. After serving as superintendent of the Institution for the Care of Sick Gentlewomen in London, Nightingale was appointed by the English Secretary of State at War, Sidney Herbert, to be in charge of 38 nurses who were to be stationed at military hospitals in Turkey.

Nightingale found the conditions in the hospitals appalling—overcrowded, filthy, and without sufficient facilities. In addition to the administrative duties she undertook to alleviate those conditions, she spent many hours tending patients. After 8:00 P.M. she allowed none of her nurses in the wards, but made rounds herself every night, a deed that earned her the epithet Lady of the Lamp.

Nightingale was an ardent believer in the power of statistics and used statistics extensively to gain an understanding of social and health issues. She lobbied to introduce statistics into the curriculum at Oxford and invented the coxcomb chart, a type of pie chart. Nightingale felt that charts and diagrams were a means of making statistical in-

formation understandable to people who would otherwise be unwilling to digest the dry numbers.

In May 1857, as a result of Nightingale's interviews with officials ranging from the Secretary of State to Queen Victoria herself, the Royal Commission on the Health of the Army was established. Under the auspices of the commission, the Army Medical School was founded. In 1860, Nightingale used a fund set up by the public to honor her work in the Crimean War to create the Nightingale School for Nurses at St. Thomas's Hospital. During that same year, at the International Statistical Congress in London, she authored one of the three papers discussed in the Sanitary Section and also met Adolphe Quetelet (see Chapter 2 biography) who had greatly influenced her work.

After 1857, Nightingale lived as an invalid, although it has never been determined that she had any specific illness. In fact, many speculated that her invalidism was a stratagem she employed to devote herself to her work.

Nightingale was elected an Honorary Member of the American Statistical Association in 1874. In 1907, she was presented the Order of Merit for meritorious service by King Edward VII; she was the first woman to receive that award.

Florence Nightingale died in 1910. An offer of a national funeral and burial at Westminster Abbey was declined, and, according to her wishes, Nightingale was buried in the family plot in East Mellow, Hampshire, England.

## StatCrunch in MyStatLab
**Analyzing Data Online**

StatCrunch online statistical software offers an easy-to-use interface customized for this book. The StatCrunch feature for each chapter illustrates the use of the software to perform a statistical analysis discussed in the chapter. Exercises are provided to further apply StatCrunch to other statistical analyses examined in the chapter. Go to the WeissStats CD or to the Weiss Web site at www.aw-bc.com/weiss to access StatCrunch instructions and data sets. To access StatCrunch statistical software, go to the student content area of your Weiss MyStatLab course.

## Internet Projects
**Exploring Data Online**

The Internet project for each chapter provides simulations, demonstrations, or activities that enhance the topics covered in the chapter. The project materials come from universities, individuals, governments, and companies from all over the world. To access the Internet projects on the Web, go to www.aw-bc.com/weiss. From this Web page, you can reach the Internet Projects Page, which we suggest that you bookmark for easy access in the future.

# PART II

## Descriptive Statistics

# 2

# Organizing Data

## Chapter Objectives

In Chapter 1, we introduced two major interrelated branches of statistics: descriptive statistics and inferential statistics. In this chapter, you will begin your study of descriptive statistics, which consists of methods for organizing and summarizing information.

In Section 2.1, we show you how to classify data by type. Data type can help you choose the correct statistical method. In Section 2.2, we explain how to group data so that they are easier to work with and understand. In Section 2.3, we demonstrate various ways to "picture" data graphically. That section also introduces stem-and-leaf diagrams—one of an arsenal of statistical tools known collectively as **exploratory data analysis.**

In Section 2.4, we discuss the identification of the shape of a data set. And, in Section 2.5, we present tips for avoiding confusion when you read and interpret graphical displays.

# Preventing Infant Mortality

The infant mortality rate (IMR) is the number of deaths of children under 1 year old per 1000 live births during a calendar year. In 1987, the U.S. Congress established the National Commission to Prevent Infant Mortality, whose charge is to create a national strategy for reducing the IMR of the United States.

The *Statistical Abstract of the United States* gives the 2003 infant mortality rates for nations with 12 million or more population. The 30 countries with the lowest IMRs are as shown in the following table.

At the end of this chapter, you will apply some of your newly learned statistical skills to analyze these infant mortality rates.

| Rank | Country | IMR | Rank | Country | IMR |
|------|---------|-----|------|---------|-----|
| 1 | Japan | 3.3 | 16 | Sri Lanka | 15.2 |
| 2 | Germany | 4.2 | 17 | Argentina | 16.2 |
| 3 | France | 4.4 | 18 | Russia | 17.4 |
| 4 | Spain | 4.5 | 19 | Malaysia | 19.0 |
| 5 | Australia | 4.8 | 20 | Ukraine | 20.9 |
| 6 | Canada | 4.9 | 21 | Thailand | 21.8 |
| 7 | Netherlands | 5.2 | 22 | Colombia | 22.5 |
| 8 | United Kingdom | 5.3 | 23 | Mexico | 22.5 |
| 9 | Italy | 6.2 | 24 | Venezuela | 23.8 |
| 10 | Taiwan | 6.7 | 25 | Philippines | 25.0 |
| **11** | **United States** | **6.8** | 26 | Ecuador | 25.4 |
| 12 | South Korea | 7.3 | 27 | North Korea | 25.7 |
| 13 | Poland | 9.0 | 28 | China | 26.4 |
| 14 | Chile | 9.3 | 29 | Romania | 28.1 |
| 15 | Saudi Arabia | 14.2 | 30 | Vietnam | 30.8 |

## 2.1 Variables and Data

A characteristic that varies from one person or thing to another is called a **variable.** Examples of variables for humans are height, weight, number of siblings, sex, marital status, and eye color. The first three of these variables yield numerical information and are examples of **quantitative variables;** the last three yield nonnumerical information and are examples of **qualitative variables,** also called **categorical variables.**[†]

Quantitative variables can be classified as either *discrete* or *continuous*. A **discrete variable** is a variable whose possible values can be listed, even though the list may continue indefinitely. This property holds, for instance, if either the variable has only a finite number of possible values or its possible values are some collection of whole numbers.[‡] A discrete variable usually involves a count of something, such as the number of siblings a person has, the number of cars owned by a family, or the number of students in an introductory statistics class.

A **continuous variable** is a variable whose possible values form some interval of numbers. Typically, a continuous variable involves a measurement of something, such as the height of a person, the weight of a newborn baby, or the length of time a car battery lasts.

The preceding discussion is summarized graphically in Fig. 2.1 and verbally in the following definition.

**Definition 2.1**

**What Does It Mean?**

A discrete variable usually involves a count of something, whereas a continuous variable usually involves a measurement of something.

### Variables

**Variable:** A characteristic that varies from one person or thing to another.

**Qualitative variable:** A nonnumerically valued variable.

**Quantitative variable:** A numerically valued variable.

**Discrete variable:** A quantitative variable whose possible values can be listed.

**Continuous variable:** A quantitative variable whose possible values form some interval of numbers.

**FIGURE 2.1**
Types of variables

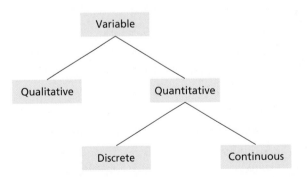

The values of a variable for one or more people or things yield **data.** Thus the information collected, organized, and analyzed by statisticians is data.

---

[†]Values of a qualitative variable are sometimes coded with numbers—for example, zip codes, which represent geographical locations. We cannot do arithmetic with such numbers, in contrast to those of a quantitative variable.

[‡]Mathematically speaking, a discrete variable is any variable whose possible values form a *countable set*, a set that is either finite or countably infinite.

Data, like variables, can be classified as **qualitative data, quantitative data, discrete data,** and **continuous data.**

| Definition 2.2 | **Data** |

**Data:** Values of a variable.

**Qualitative data:** Values of a qualitative variable.

**Quantitative data:** Values of a quantitative variable.

**Discrete data:** Values of a discrete variable.

**Continuous data:** Values of a continuous variable.

**What Does It Mean?**

Data are classified according to the type of variable from which they were obtained.

Each individual piece of data is called an **observation,** and the collection of all observations for a particular variable is called a **data set.**[†] We illustrate various types of variables and data in Examples 2.1–2.4.

### Example 2.1 | Variables and Data

*The 109th Boston Marathon* At noon on April 18, 2005, roughly 20,000 men and women set out to run 26 miles and 385 yards from rural Hopkinton to Boston. Thousands of people lining the streets leading into Boston and millions more on television watched this 109th running of the Boston Marathon.

The Boston Marathon provides examples of different types of variables and data. The classification of each entrant as either male or female illustrates the simplest type of variable. "Gender" is a qualitative variable because its possible values (male or female) are nonnumerical. Thus, for instance, the information that Hailu Negussie is a male and Catherine Ndereba is a female is qualitative data.

"Place of finish" is a quantitative variable, which is also a discrete variable because it makes sense to talk only about first place, second place, and so on—there are only a finite number of possible finishing places. Thus, the information that, among the women, Catherine Ndereba and Elfenesh Alemu finished first and second, respectively, is discrete, quantitative data.

"Finishing time" is a quantitative variable, which is also a continuous variable because the finishing time of a runner can conceptually be any positive number. The information that Hailu Negussie won the men's competition in 2:11:45 and Catherine Ndereba won the women's in 2:25:13 is continuous, quantitative data.

*You try it!*

Exercise 2.7 on page 43

• • •

### Example 2.2 | Variables and Data

*Human Blood Types* Human beings have one of four blood types: A, B, AB, or O. What kind of data do you receive when you are told your blood type?

**Solution** Blood type is a qualitative variable because its possible values are nonnumerical. Therefore your blood type is qualitative data.

• • •

---

[†]Sometimes *data set* is used to refer to all the data for all the variables under consideration.

## Example 2.3 | Variables and Data

*Household Size* The U.S. Census Bureau collects data on household size and publishes the information in *Current Population Reports*. What kind of data is the number of people in your household?

**Solution** Household size is a quantitative variable, which is also a discrete variable because its possible values are 1, 2, .... Therefore the number of people in your household is discrete, quantitative data.

• • •

## Example 2.4 | Variables and Data

*The World's Highest Waterfalls* The *Information Please Almanac* lists the world's highest waterfalls. The list shows that Angel Falls in Venezuela is 3281 feet high, or more than twice as high as Ribbon Falls in Yosemite, California, which is 1612 feet high. What kind of data are these heights?

**Solution** Height is a quantitative variable, which is also a continuous variable because height can conceptually be any positive number. Therefore the waterfall heights are continuous, quantitative data.

• • •

### Classification and the Choice of a Statistical Method

Some of the statistical procedures that you will study are valid for only certain types of data. This limitation is one reason why you must be able to classify data. The classifications we have discussed are sufficient for most applications, even though statisticians sometimes use additional classifications.

Data classification can be difficult; even statisticians occasionally disagree over data type. For example, some classify amounts of money as discrete data; others say it is continuous data. In most cases, however, data classification is fairly clear and will help you choose the correct statistical method for analyzing the data.

## Exercises 2.1

### Understanding the Concepts and Skills

**2.1** Give an example, other than those presented in this section, of a
a. qualitative variable.
b. discrete, quantitative variable.
c. continuous, quantitative variable.

**2.2** Explain the meaning of
a. qualitative variable.
b. discrete, quantitative variable.
c. continuous, quantitative variable.

**2.3** Explain the meaning of
a. qualitative data.　　b. discrete, quantitative data.
c. continuous, quantitative data.

**2.4** Provide a reason why the classification of data is important.

**2.5** Of the variables you have studied so far, which type yields nonnumerical data?

*For each part of Exercises 2.6–2.10, classify the data as either qualitative or quantitative; if quantitative, further classify it as discrete or continuous. Also, identify the variable under consideration in each case.*

**2.6 Doctor Disciplinary Actions.** In the June 2003 issue of *Health Letter*, the Public Citizen Health Research Group ranked the states by serious doctor disciplinary actions per

1000 doctors for the year 2002. Here are data for the 10 states with the lowest rates.

| State | No. of actions | No. of doctors | Actions per 1000 doctors |
|-------|---------------|----------------|--------------------------|
| Hawaii | 4 | 3,746 | 1.07 |
| Delaware | 3 | 2,219 | 1.35 |
| Wisconsin | 20 | 14,241 | 1.40 |
| Tennessee | 22 | 14,954 | 1.47 |
| South Carolina | 17 | 9,607 | 1.77 |
| Maryland | 39 | 21,883 | 1.78 |
| North Carolina | 43 | 20,851 | 2.06 |
| Florida | 93 | 44,747 | 2.08 |
| Pennsylvania | 82 | 39,052 | 2.10 |
| Minnesota | 30 | 14,218 | 2.11 |

Identify the type of data provided by the information in each of the following columns of the table.
**a.** first          **b.** second          **c.** third

**2.7 How Hot Does It Get?** The highest temperatures on record for selected cities are collected by the U.S. National Oceanic and Atmospheric Administration and published in *Comparative Climatic Data*. The following table displays data for years through 2002.

| City | Rank | Highest temp. (°F) |
|------|------|--------------------|
| Phoenix, AZ | 1 | 122 |
| Sacramento, CA | 2 | 115 |
| El Paso, TX | 3 | 114 |
| Omaha, NE | 3 | 114 |
| Dallas–Fort Worth | 5 | 113 |
| Wichita, KS | 5 | 113 |

**a.** What type of data is presented in the second column of the table?
**b.** What type of data is provided in the third column of the table?
**c.** What type of data is provided by the information that Phoenix is in Arizona?

**2.8 Earnings from the Crypt.** Drawing on *Forbes'* 18 years of wealth-estimating experience, four ABCNews.com reporters calculated pretax earnings to the estates of deceased celebrities from licensing agreements and book and record sales for the 12-month period from June 2001 to June 2002. The following table shows those deceased celebrities with the top 13 earnings during that time period.

| Rank | Name | Earnings ($ millions) |
|------|------|-----------------------|
| 1 | Elvis Presley | 37 |
| 2 | Charles Schulz | 28 |
| 3 | John Lennon | 20 |
| 4 | Dale Earnhardt | 20 |
| 5 | Theodor Geisel | 19 |
| 6 | George Harrison | 17 |
| 7 | J.R.R. Tolkien | 12 |
| 8 | Bob Marley | 10 |
| 9 | Jimi Hendrix | 8 |
| 10 | Tupac Shakur | 7 |
| 11 | Marilyn Monroe | 7 |
| 12 | Jerry Garcia | 5 |
| 13 | Robert Ludlum | 5 |

**a.** What type of data is presented in the first column of the table?
**b.** What type of data is provided by the information in the third column of the table?

**2.9 Top Broadband Cities.** According to research by *Nielsen/NetRatings*, as of August, 2004, the top 10 wired local markets in the United States connected via broadband access are as follows.

| Rank | Local market | Percent broadband |
|------|--------------|-------------------|
| 1 | San Diego | 69.6 |
| 2 | Phoenix | 68.4 |
| 3 | Detroit | 67.0 |
| 4 | New York | 66.8 |
| 5 | Sacramento | 64.9 |
| 6 | Orlando | 64.7 |
| 7 | Seattle | 63.0 |
| 8 | San Francisco | 63.0 |
| 9 | Los Angeles | 61.6 |
| 10 | Boston | 61.4 |

What type of data is provided by the information in the
**a.** first column of the table?
**b.** second column of the table?
**c.** third column of the table? (*Hint:* The possible ratios of positive whole numbers can be listed.)

**2.10 Shipment Values.** For the year 2004, the *Recording Industry Association of America* reported the following manufacturers' shipment values, in millions of dollars, for various products.

| Product | Shipment value ($ millions) |
|---|---|
| CD | 11,446.5 |
| CD single | 14.9 |
| Cassette | 23.6 |
| LP/EP | 19.2 |
| Vinyl single | 19.8 |
| Music video | 607.2 |
| DVD audio | 6.4 |
| SA CD | 16.6 |
| DVD video | 561.1 |

What kind of data is provided by the information in the
a. first column of the table?
b. second column of the table?

**2.11 Top Broadcast Shows.** The following table gives the top five television shows, as determined by the *Nielsen Ratings* for the week of April 18–24, 2005. Identify the type of data provided by the information in each column of the table.

| Rank | Season to date | Show title | Network | Viewers (millions) |
|---|---|---|---|---|
| 1 | 1 | CSI | CBS | 27.0 |
| 2 | 2 | American Idol, Tues | Fox | 24.1 |
| 3 | 4 | Desperate Housewives | ABC | 23.9 |
| 4 | 3 | American Idol, Wed | Fox | 22.7 |
| 5 | 5 | CSI: Miami | CBS | 20.1 |

**2.12 Medicinal Plants Workshop.** The Medicinal Plants of the Southwest summer workshop is an inquiry-based learning approach to increase interest and skills in biomedical research, as described by Mary O'Connell and Antonio Lara in the *Journal of College Science Teaching* (January/February 2005, pp. 26–30). Following is some information obtained from the 20 students who participated in the 2003 workshop. Discuss the types of data provided by this information.

- **Duration:** 6 weeks
- **Number of students:** 20
- **Gender:** 3 males, 17 females
- **Ethnicity:** 14 Hispanic, 1 African American, 2 Native American, 3 other
- **Number of Web reports:** 6

**2.13 Smartphones.** In the May 2005 issue of *Popular Science* (Vol. 266, No. 5, pp. 77–86), Suzanne Kirschner provides a buyer's guide to smartphones. Cell phones make calls, but smartphones do whatever you want them to—with PDA functions, Internet access, and the ability to run hundreds of applications. Following are the specifications for the top 10 smartphones. Identify the type of data provided by the information in each column of the table.

| Smartphone | Size (inches) | Battery (hours) | Voice recog. | Weight (ounces) |
|---|---|---|---|---|
| HP iPaq h6315 | 2.90 × 2.25 | 5 | No | 6.7 |
| Audiovox XV6600 | 2.90 × 2.25 | 2.5 | Yes | 7.4 |
| Samsung i730 | 2.30 × 1.75 | N/A | Yes | 5.5 |
| Nokia 6620 | 1.75 × 1.50 | 4 | Yes | 4.4 |
| Audiovox SMT5600 | 1.75 × 1.40 | 4 | No | 3.6 |
| PalmOne Treo 650 | 1.80 × 1.80 | 6 | No | 6.3 |
| Nokia 9300 | 1.25 × 3.90 | 5 | No | 5.9 |
| Motorola MPx220 | 1.25 × 1.50 | 6 | Yes | 3.9 |
| Sony Ericsson P910 | 1.60 × 2.40 | 13 | Yes | 5.5 |
| Sierra Wireless Voq | 1.75 × 1.40 | 6 | Yes | 4.9 |

### Extending the Concepts and Skills

**2.14 Ordinal data.** Another important type of data is *ordinal data*, which is data about order or rank given on a scale such as 1, 2, 3, … or A, B, C, …. Following are several variables. Which, if any, yield ordinal data? Explain your answer.
a. Height    b. Weight    c. Age    d. Sex
e. Number of siblings    f. Religion
g. Place of birth    h. High school class rank

## 2.2 Grouping Data

Some situations generate an overwhelming amount of data. For example, *The World Almanac* lists U.S. colleges and universities with information on enrollment, number of teachers, highest degree offered, and governing official. These data occupy 28 pages of small type!

We can often make a large and complicated set of data more compact and easier to understand by organizing it. In this section, we discuss **grouping,** or putting data into groups rather than treating each observation individually. First we consider the grouping of quantitative data, both continuous and discrete. Later in this section, we examine the grouping of qualitative data.

## Grouping Quantitative Data

To begin, consider the grouping of quantitative data presented in Example 2.5.

---

**Example 2.5**

## Grouping Quantitative Data

*Days to Maturity for Short-Term Investments*  Table 2.1 displays the number of days to maturity for 40 short-term investments. The data are from *BARRON'S* magazine.

Getting a clear picture of these data is difficult, but is much easier if we group them into categories, or **classes.** The first step is to decide on the classes. One convenient way to group these data is by 10s.

Because the shortest maturity period is 36 days, our first class is for maturity periods from 30 days up to, but not including, 40 days. We use the symbol $\prec$ as a shorthand for "up to, but not including," so our first class is denoted $30 \prec 40$.[†] The longest maturity period is 99 days, so grouping by 10s results in the seven classes given in the first column of Table 2.2.

The final step for grouping the data is to determine the number of investments in each class. We do so by placing a tally mark for each investment from Table 2.1 on the appropriate line of Table 2.2. For instance, the first investment in Table 2.1 has a 70-day maturity period, calling for a tally mark on the line for the class $70 \prec 80$. The results of the tallying procedure are shown in the second column of Table 2.2. Now we count the tallies for each class and record each total in the third column of Table 2.2.

By simply glancing at Table 2.2, we can easily obtain various pieces of useful information. For instance, more investments are in the 60- to 69-days range than in any other.

• • •

**TABLE 2.1**
Days to maturity for 40 short-term investments

| | | | | | | | |
|---|---|---|---|---|---|---|---|
| 70 | 64 | 99 | 55 | 64 | 89 | 87 | 65 |
| 62 | 38 | 67 | 70 | 60 | 69 | 78 | 39 |
| 75 | 56 | 71 | 51 | 99 | 68 | 95 | 86 |
| 57 | 53 | 47 | 50 | 55 | 81 | 80 | 98 |
| 51 | 36 | 63 | 66 | 85 | 79 | 83 | 70 |

**TABLE 2.2**
Classes and counts for the days-to-maturity data in Table 2.1

| Days to maturity | Tally | Number of investments |
|---|---|---|
| $30 \prec 40$ | III | 3 |
| $40 \prec 50$ | I | 1 |
| $50 \prec 60$ | LHI III | 8 |
| $60 \prec 70$ | LHI LHI | 10 |
| $70 \prec 80$ | LHI II | 7 |
| $80 \prec 90$ | LHI II | 7 |
| $90 \prec 100$ | IIII | 4 |
| | | 40 |

**What Does It Mean?**

Comparing Tables 2.1 and 2.2 clearly shows that grouping the data makes the data much easier to read and understand.

Example 2.5 exemplifies three commonsense and important guidelines for grouping:

**1.** *The number of classes should be small enough to provide an effective summary but large enough to display the relevant characteristics of the data.*

In Example 2.5, we used seven classes. A rule of thumb is that the number of classes should be between 5 and 20.

**2.** *Each observation must belong to one, and only one, class.*

Careless planning in Example 2.5 could have led to classes such as 30–40, 40–50, 50–60, and so on. Then, for instance, it would be unclear to which class the investment with a 50-day maturity period would belong. The classes in

---

[†]We formed the $\prec$ symbol by superimposing a less-than symbol over a dash. The notation $30 \prec 40$ is less awkward than the notation 30–< 40 or 30–under 40.

Table 2.2 do not cause such confusion; they cover all maturity periods and do not overlap.

**3.** *Whenever feasible, all classes should have the same width.*

All the classes in Table 2.2 have a width of 10 days. Among other things, choosing classes of equal width facilitates the graphical display of the data.

The list could go on, but for our purposes these three guidelines provide a solid basis for grouping data.

## Frequency and Relative-Frequency Distributions

The number of observations that fall into a particular class is called the **frequency** (or **count**) of that class. For instance, in Table 2.2, the frequency of the class $50 \leqslant 60$ is 8 because eight investments are in the 50- to 59-days range. A table that provides all classes and their frequencies is called a **frequency distribution**. The first and third columns of Table 2.2 constitute a frequency distribution for the days-to-maturity data.

In addition to the frequency of a class, we are often interested in the **percentage** of a class. We find the percentage by first dividing the frequency of the class by the total number of observations and then multiplying the result by 100. From Table 2.2, the percentage of investments in the class $50 \leqslant 60$ is

$$\frac{8}{40} = 0.20 \quad \text{or} \quad 20\%.$$

Thus 20% of the investments have a number of days to maturity in the 50s.

The percentage of a class, expressed as a decimal, is called the **relative frequency** of the class. For the class $50 \leqslant 60$, the relative frequency is 0.20. A table that provides all classes and their relative frequencies is called a **relative-frequency distribution**. Table 2.3 displays a relative-frequency distribution for the days-to-maturity data. Note that the relative frequencies sum to 1 (100%).

Relative-frequency distributions are better than frequency distributions for comparing two data sets. The reason is that relative frequencies always fall between 0 and 1 and hence provide a standard for comparison.

**TABLE 2.3**

Relative-frequency distribution for the days-to-maturity data in Table 2.1

| Days to maturity | Relative frequency | | |
|---|---|---|---|
| $30 \leqslant 40$ | 0.075 | ← | 3/40 |
| $40 \leqslant 50$ | 0.025 | ← | 1/40 |
| $50 \leqslant 60$ | 0.200 | ← | 8/40 |
| $60 \leqslant 70$ | 0.250 | ← | 10/40 |
| $70 \leqslant 80$ | 0.175 | ← | 7/40 |
| $80 \leqslant 90$ | 0.175 | ← | 7/40 |
| $90 \leqslant 100$ | 0.100 | ← | 4/40 |
| | 1.000 | | |

## Grouping Terminology

Let's return to the days-to-maturity data to learn more grouping terms. Consider, for example, the class $50 \leqslant 60$. The smallest maturity period that could go in this class is 50; this value is called the **lower cutpoint** of the class. The smallest maturity period that could go in the next higher class is 60; this value is called the **upper cutpoint** of the $50 \leqslant 60$ class and, of course, is also the lower cutpoint of the next higher class, $60 \leqslant 70$.

The number in the middle of the class $50 \leqslant 60$ is $(50 + 60)/2 = 55$ and is called the **midpoint** of the class. Midpoints provide single numbers for representing classes and are often used in graphical displays and for computing descriptive measures. The **width** of the class $50 \leqslant 60$, obtained by subtracting its lower cutpoint from its upper cutpoint, is $60 - 50 = 10$.

**Definition 2.3**

## Terms Used in Grouping

**Classes:** Categories for grouping data.

**Frequency:** The number of observations that fall in a class.

**Frequency distribution:** A listing of all classes and their frequencies.

**Relative frequency:** The ratio of the frequency of a class to the total number of observations.

**Relative-frequency distribution:** A listing of all classes and their relative frequencies.

**Lower cutpoint:** The smallest value that could go in a class.

**Upper cutpoint:** The smallest value that could go in the next higher class (equivalent to the lower cutpoint of the next higher class).

**Midpoint:** The middle of a class, found by averaging its cutpoints.

**Width:** The difference between the cutpoints of a class.

A table that provides the classes, frequencies, relative frequencies, and midpoints of a data set is called a **grouped-data table.** Table 2.4 is a grouped-data table for the days-to-maturity data and Example 2.6 illustrates another.

**TABLE 2.4**
Grouped-data table for the days-to-maturity data

| Days to maturity | Frequency | Relative frequency | Midpoint |
|---|---|---|---|
| $30 \leqslant 40$ | 3 | 0.075 | 35 |
| $40 \leqslant 50$ | 1 | 0.025 | 45 |
| $50 \leqslant 60$ | 8 | 0.200 | 55 |
| $60 \leqslant 70$ | 10 | 0.250 | 65 |
| $70 \leqslant 80$ | 7 | 0.175 | 75 |
| $80 \leqslant 90$ | 7 | 0.175 | 85 |
| $90 \leqslant 100$ | 4 | 0.100 | 95 |
| | 40 | 1.000 | |

**Example 2.6** | ## Grouped-Data Tables

**TABLE 2.5**
Weights of 37 males, aged 18–24 years

| | | | | |
|---|---|---|---|---|
| 129.2 | 185.3 | 218.1 | 182.5 | 142.8 |
| 155.2 | 170.0 | 151.3 | 187.5 | 145.6 |
| 167.3 | 161.0 | 178.7 | 165.0 | 172.5 |
| 191.1 | 150.7 | 187.0 | 173.7 | 178.2 |
| 161.7 | 170.1 | 165.8 | 214.6 | 136.7 |
| 278.8 | 175.6 | 188.7 | 132.1 | 158.5 |
| 146.4 | 209.1 | 175.4 | 182.0 | 173.6 |
| 149.9 | 158.6 | | | |

*Weights of 18–24-Year-Old Males* The U.S. National Center for Health Statistics publishes data on weights and heights by age and sex in *Vital and Health Statistics*. The weights shown in Table 2.5, given to the nearest tenth of a pound, were obtained from a sample of 18–24-year-old males. Construct a grouped-data table for these weights. Use a class width of 20 and a first cutpoint of 120.

**Solution** A class width of 20 and a first cutpoint of 120 gives us a first class of $120 \leqslant 140$, as shown in the first column of Table 2.6 (next page). The largest weight in Table 2.5 is 278.8 lb, so the last class in Table 2.6 is $260 \leqslant 280$.

Tallying the data in Table 2.5 gives us the frequencies in the second column of Table 2.6. To illustrate the methods for finding the entries in the third and fourth columns of Table 2.6, we use the class $160 \leqslant 180$:

$$\text{relative frequency} = \frac{14}{37} = 0.378 \quad \text{and} \quad \text{midpoint} = \frac{160 + 180}{2} = 170.$$

**TABLE 2.6**
Grouped-data table for the weights of 37 males, aged 18–24 years

| Weight (lb) | Frequency | Relative frequency | Midpoint |
|---|---|---|---|
| 120 ≺ 140 | 3 | 0.081 | 130 |
| 140 ≺ 160 | 9 | 0.243 | 150 |
| 160 ≺ 180 | 14 | 0.378 | 170 |
| 180 ≺ 200 | 7 | 0.189 | 190 |
| 200 ≺ 220 | 3 | 0.081 | 210 |
| 220 ≺ 240 | 0 | 0.000 | 230 |
| 240 ≺ 260 | 0 | 0.000 | 250 |
| 260 ≺ 280 | 1 | 0.027 | 270 |
| | 37 | 0.999 | |

You try it!

Exercise 2.25
on page 53

Although relative frequencies must always sum to 1, their sum in Table 2.6 is given as 0.999. This discrepancy occurs because each relative frequency is rounded to three decimal places and, in this case, the resulting sum differs from 1 by a little. Such a discrepancy is called *rounding error* or *roundoff error*.

• • •

## An Alternate Method for Depicting Classes

We often use an alternate method to depict classes, especially when the data are expressed as whole numbers. Consider that the third class for the days-to-maturity data, 50 ≺ 60, is for periods from 50 days up to, but not including, 60 days. Because the periods are to the nearest whole day, we could also characterize the third class as being from 50 days up to and including 59 days.

Thus, we can depict 50 ≺ 60 as 50–59. The smallest maturity period that could go in this class is 50 and, in this context, is called the **lower limit** of the class. The largest maturity period that can go in this class is 59 and, in this context, is called the **upper limit** of the class.

You try it!

Exercise 2.27
on page 53

When using class limits to depict classes, we generally use *marks* instead of midpoints to represent the classes. The **mark** of a class is the average of its lower and upper limits. The mark of the class 50–59 is $(50 + 59)/2 = 54.5$.

Table 2.7 provides a grouped-data table for the days-to-maturity data that uses class limits and class marks. Compare Tables 2.4 and 2.7.

**TABLE 2.7**
Grouped-data table for the days-to-maturity data, using class limits and class marks

| Days to maturity | Frequency | Relative frequency | Mark |
|---|---|---|---|
| 30–39 | 3 | 0.075 | 34.5 |
| 40–49 | 1 | 0.025 | 44.5 |
| 50–59 | 8 | 0.200 | 54.5 |
| 60–69 | 10 | 0.250 | 64.5 |
| 70–79 | 7 | 0.175 | 74.5 |
| 80–89 | 7 | 0.175 | 84.5 |
| 90–99 | 4 | 0.100 | 94.5 |
| | 40 | 1.000 | |

The weight data in Table 2.5 are rounded to one decimal place. Consequently, the alternate method for depicting the classes shown in the first column of Table 2.6 would be 120–139.9, 140–159.9, 160–179.9, and so on. The marks for these classes are, respectively, 129.95, 149.95, 169.95, and so on.

Both methods of depicting classes and their representatives are commonly used, and each has its advantages and disadvantages. We use both methods throughout this book, so be sure you understand them.

### Single-Value Grouping

Up to this point, each class that we have used represents a range of possible values. For instance, in Table 2.6, the first class represents weights from 120 lb up to, but not including, 140 lb. In some cases, however, classes that represent a single possible value are more appropriate. Such classes are particularly suitable for discrete data in which there are only a few distinct observations.

**Example 2.7** | **Single-Value Grouping**

*TVs per Household* *Trends in Television*, published by the Television Bureau of Advertising, provides information on television ownership. Table 2.8 gives the number of TV sets per household for 50 randomly selected households. Use classes based on a single value to construct a grouped-data table for these data.

**TABLE 2.8**
Number of TV sets in each of 50 randomly selected households

| | | | | | | | | | |
|---|---|---|---|---|---|---|---|---|---|
| 1 | 1 | 1 | 2 | 6 | 3 | 3 | 4 | 2 | 4 |
| 3 | 2 | 1 | 5 | 2 | 1 | 3 | 6 | 2 | 2 |
| 3 | 1 | 1 | 4 | 3 | 2 | 2 | 2 | 2 | 3 |
| 0 | 3 | 1 | 2 | 1 | 2 | 3 | 1 | 1 | 3 |
| 3 | 2 | 1 | 2 | 1 | 1 | 3 | 1 | 5 | 1 |

**Solution** Because each class is to represent a single numerical value, the classes are 0, 1, 2, 3, 4, 5, and 6. Table 2.9 displays these classes in its first column.

Tallying the data in Table 2.8, we obtain the frequencies in the second column of Table 2.9. Dividing each frequency by the total number of observations, 50, we get the relative frequencies shown in the third column of Table 2.9. That table indicates, for example, that 14 of the 50 households, or 0.280 (28.0%), have exactly two television sets.

**TABLE 2.9**
Grouped-data table for number of TV sets

| Number of TVs | Frequency | Relative frequency |
|---|---|---|
| 0 | 1 | 0.020 |
| 1 | 16 | 0.320 |
| 2 | 14 | 0.280 |
| 3 | 12 | 0.240 |
| 4 | 3 | 0.060 |
| 5 | 2 | 0.040 |
| 6 | 2 | 0.040 |
| | 50 | 1.000 |

**You try it!**

Exercise 2.31 on page 53

Because each class is based on a single value, the midpoint (and mark) of each class is the same as the class. For instance, the midpoint (and mark) of the class 3 is 3. Therefore a midpoint (or mark) column in Table 2.9 is unnecessary, as such a column would be identical to the first column. In other words, Table 2.9 can serve as a grouped-data table.[†]

• • •

---

[†]For single-value grouped data, the upper and lower limits of each class are also identical to the class. And, although it is generally unnecessary, we can also obtain the cutpoints. For the grouped data in Table 2.9, the cutpoints are −0.5, 0.5, 1.5, 2.5, 3.5, 4.5, 5.5, and 6.5.

## Grouping Qualitative Data

The concepts of cutpoints and midpoints are inappropriate for qualitative data, such as data that categorize people as male or female. We can, of course, still group qualitative data and compute frequencies and relative frequencies for the classes. For qualitative data, the classes coincide with the observed values of the corresponding variable.

**Example 2.8** | **Grouping Qualitative Data**

**TABLE 2.10**
Political party affiliations of the students in introductory statistics

| D | R | O | R | R | R | R | R |
| D | O | R | D | O | O | R | D |
| D | R | O | D | R | R | O | R |
| D | O | D | D | D | R | O | D |
| O | R | D | R | R | R | R | D |

*Political Party Affiliations* Professor Weiss asked his introductory statistics students to state their political party affiliations as Democratic (D), Republican (R), or Other (O). The responses are given in Table 2.10. Determine the frequency and relative-frequency distributions for these data.

**Solution** The classes for grouping the data are "Democratic," "Republican," and "Other." Tallying the data in Table 2.10, we obtain the frequency distribution displayed in the first two columns of Table 2.11.

**TABLE 2.11**
Frequency and relative-frequency distributions for political party affiliations

| Party | Frequency | Relative frequency |
|-------|-----------|--------------------|
| Democratic | 13 | 0.325 |
| Republican | 18 | 0.450 |
| Other | 9 | 0.225 |
| | 40 | 1.000 |

Dividing each frequency in the second column of Table 2.11 by the total number of students, 40, we get the relative frequencies in the third column. The first and third columns of Table 2.11 provide the relative-frequency distribution for the data.

**Interpretation** From Table 2.11, we see that, of the 40 students in the class, 13 are Democrats, 18 are Republicans, and 9 are Other. Equivalently, 32.5% are Democrats, 45.0% are Republicans, and 22.5% are Other.

Exercise 2.33
on page 53

• • •

# The Technology Center

**Introduction to the Technology Centers** Today, programs for conducting statistical and data analyses are available in dedicated statistical software packages, general-use spreadsheet software, and graphing calculators. In this book, we discuss three of the most popular technologies for doing statistics: **Minitab, Excel,** and the **TI-83/84 Plus.**[†]

---

[†]For brevity, we write TI-83/84 Plus for TI-83 Plus and/or TI-84 Plus. Keystrokes and output remain the same from the TI-83 Plus to the TI-84 Plus. Thus, instructions and output given in the book apply to both calculators.

For Excel, we mostly use **Data Desk/XL (DDXL)** from Data Description, Inc. This statistics add-in complements Excel's standard statistics capabilities; it is included on the **WeissStats CD,** which comes with your book.

At the end of most sections of the book, in subsections titled "The Technology Center," we present and interpret output from the three technologies that provides technology solutions to problems solved by hand earlier in the section. For this aspect of The Technology Center, you need neither a computer nor a graphing calculator, nor do you need working knowledge of any of the technologies.

Another aspect of The Technology Center provides step-by-step instructions for using any of the three technologies to obtain the output presented. When studying this material, you will get the best results by actually performing the steps described.

Successful technology use requires knowing how to input data. We discuss doing that and several other basic tasks for Minitab, Excel, and the TI-83/84 Plus in the documents contained in the Technology Basics folder on the WeissStats CD.

**Using Technology to Group Data**  Grouping data can be tedious if done by hand. You can avoid the tedium by using a computer. Here, we present output and step-by-step instructions for grouping data into single-value classes. Refer to the technology manuals for other grouping methods.

Some statistical technologies have programs for single-value grouping of quantitative data or grouping of qualitative data, but others do not. At the time of this writing, the TI-83/84 Plus does not have such a program.

**Example 2.9**  **Using Technology to Obtain a Grouped-Data Table**

*TVs per Household*    Table 2.8 on page 49 displays the number of TV sets per household for 50 randomly selected households. Use Minitab or Excel to obtain a grouped-data table for those data, based on single-value classes.

**Solution**  We applied the grouping programs to the data, resulting in Output 2.1. Steps for generating that output are presented in Instructions 2.1.

**OUTPUT 2.1**

Grouped-data tables for the data on the number of TVs per household

**MINITAB**

**Tally for Discrete Variables: TVs**

| TVs | Count | Percent |
|---|---|---|
| 0 | 1 | 2.00 |
| 1 | 16 | 32.00 |
| 2 | 14 | 28.00 |
| 3 | 12 | 24.00 |
| 4 | 3 | 6.00 |
| 5 | 2 | 4.00 |
| 6 | 2 | 4.00 |
| N= | 50 | |

**EXCEL**

Summary of TVs

| | | |
|---|---|---|
| Total Cases | 50 | |
| Number of Categories | 7 | |

TVs Frequency Table

| Group | Count | % |
|---|---|---|
| 0 | 1 | 2 |
| 1 | 16 | 32 |
| 2 | 14 | 28 |
| 3 | 12 | 24 |
| 4 | 3 | 6 |
| 5 | 2 | 4 |
| 6 | 2 | 4 |

Compare Output 2.1 to the grouped-data table obtained by hand in Table 2.9 on page 49. Note that both Minitab and Excel use percents instead of relative frequencies.

• • •

**INSTRUCTIONS 2.1**
Steps for generating Output 2.1

**MINITAB**

1 Store the data from Table 2.8 in a column named TVs
2 Choose **Stat ➤ Tables ➤ Tally Individual Variables...**
3 Specify TVs in the **Variables** text box
4 Select the **Counts** and **Percents** check boxes from the **Display** list
5 Click **OK**

**EXCEL**

1 Store the data from Table 2.8 in a range named TVs
2 Choose **DDXL ➤ Tables**
3 Select **Frequency Table** from the **Function type** drop-down list box
4 Specify TVs in the **Categorical Variable** text box
5 Click **OK**

**Note:** The steps in Instructions 2.1 are specifically for the data set in Table 2.8 on the number of TVs per household. To apply those steps for a different data set, simply make the necessary changes in the instructions to reflect the different data set—in this case, to steps 1 and 3 in Minitab and to steps 1 and 4 in Excel. Similar comments hold for all technology instructions throughout the book.

# Exercises 2.2

## Understanding the Concepts and Skills

**2.15** Identify an important reason for grouping data.

**2.16** Do the concepts of cutpoints and midpoints make sense for qualitative data? Explain your answer.

**2.17** State three of the most important guidelines in choosing the classes for grouping a data set.

**2.18** Explain the difference between
a. frequency and relative frequency.
b. percentage and relative frequency.

**2.19** Answer true or false to each of the statements in parts (a) and (b), and explain your reasoning.
a. Two data sets that have identical frequency distributions have identical relative-frequency distributions.
b. Two data sets that have identical relative-frequency distributions have identical frequency distributions.
c. Use your answers to parts (a) and (b) to explain why relative-frequency distributions are better than frequency distributions for comparing two data sets.

**2.20** What are the four elements of a grouped-data table? Explain the meaning of each element.

**2.21** With regard to grouping quantitative data into classes that each represent a range of possible values, we discussed two methods for depicting the classes. Identify the two methods and explain the relative advantages and disadvantages of each.

**2.22** For grouping quantitative data, we examined three types of classes: (1) $a < b$, (2) $a$–$b$, and (3) single-value grouping. For each type of data given, decide which of these three types is usually best. Explain your answers.
a. Continuous data displayed to one or more decimal places
b. Discrete data in which there are relatively few distinct observations

**2.23** When you group quantitative data into classes that each represent a single possible numerical value, why is it unnecessary to include a midpoint column in the grouped-data table?

**2.24 Residential Energy Consumption.** The U.S. Energy Information Administration collects data on residential energy consumption and expenditures. Results are published in the document *Residential Energy Consumption Survey: Consumption and Expenditures*. The following table gives one year's energy consumptions for a sample of 50 households in the South. Data are in millions of BTUs.

| | | | | | | | | | |
|---|---|---|---|---|---|---|---|---|---|
| 130 | 55 | 45 | 64 | 155 | 66 | 60 | 80 | 102 | 62 |
| 58 | 101 | 75 | 111 | 151 | 139 | 81 | 55 | 66 | 90 |
| 97 | 77 | 51 | 67 | 125 | 50 | 136 | 55 | 83 | 91 |
| 54 | 86 | 100 | 78 | 93 | 113 | 111 | 104 | 96 | 113 |
| 96 | 87 | 129 | 109 | 69 | 94 | 99 | 97 | 83 | 97 |

Use classes of equal width beginning with $40 < 50$ to construct a grouped-data table for these data.

**2.25 Clocking the Cheetah.** The cheetah (*Acinonyx jubatus*) is the fastest land mammal on Earth and is highly specialized to run down prey. According to the Cheetah Conservation of Southern Africa *Trade Environment Database*, the cheetah often exceeds speeds of 60 mph and has been clocked at speeds of more than 70 mph. The following table, based on information in the database, gives the speeds, in miles per hour, over a 1/4 mile for 35 cheetahs.

| | | | | | | |
|---|---|---|---|---|---|---|
| 57.3 | 57.5 | 59.0 | 56.5 | 61.3 | 57.6 | 59.2 |
| 65.0 | 60.1 | 59.7 | 62.6 | 52.6 | 60.7 | 62.3 |
| 65.2 | 54.8 | 55.4 | 55.5 | 57.8 | 58.7 | 57.8 |
| 60.9 | 75.3 | 60.6 | 58.1 | 55.9 | 61.6 | 59.6 |
| 59.8 | 63.4 | 54.7 | 60.2 | 52.4 | 58.3 | 66.0 |

Use 52 as the first cutpoint and classes of equal width 2 to construct a grouped-data table for these speeds.

**2.26 Residential Energy Consumption.** Redo Exercise 2.24 by using the alternative method for grouping data based on class limits and class marks.

**2.27 Clocking the Cheetah.** Redo Exercise 2.25 by using the alternative method for grouping data based on class limits and class marks.

**2.28 Super Bowl Ratings.** Super Bowl television broadcasts rate among the highest in all television broadcasts. The Television Bureau of Advertising published a report titled *TV Basics*. Part of the report listed the television ratings for 32 Super Bowls, as shown in the following table.

| | | | | | | | |
|---|---|---|---|---|---|---|---|
| 49.1 | 44.4 | 42.7 | 42.3 | 46.4 | 45.8 | 44.5 | 41.9 |
| 40.4 | 41.3 | 45.5 | 47.2 | 46.3 | 43.3 | 41.6 | 40.2 |
| 43.3 | 45.1 | 48.6 | 44.4 | 42.4 | 40.7 | 40.3 | 44.2 |
| 46.0 | 48.3 | 43.5 | 47.1 | 41.9 | 40.4 | 41.4 | 46.4 |

Construct a grouped-data table for these ratings. Use classes of equal width 2, starting with the cutpoint 40.0.

**2.29 Top Broadcast Shows.** The viewing audiences, in millions, for the top 20 television shows, as determined by the *Nielsen Ratings* for the week of April 18–24, 2005, are shown in the following table.

| | | | | |
|---|---|---|---|---|
| 27.0 | 24.1 | 23.9 | 22.7 | 20.1 |
| 19.3 | 17.9 | 17.5 | 17.1 | 17.1 |
| 17.0 | 16.7 | 16.1 | 16.0 | 14.3 |
| 13.9 | 12.7 | 12.6 | 12.3 | 11.8 |

Construct a grouped-data table for these viewing audiences. Use classes of equal width 3, with a first class midpoint of 12.5.

**2.30 Household Size.** The U.S. Census Bureau conducts nationwide surveys on characteristics of U.S. households and publishes the results in *Current Population Reports*. Following are data on the number of people per household for a sample of 40 households.

| | | | | | | | |
|---|---|---|---|---|---|---|---|
| 2 | 5 | 2 | 1 | 1 | 2 | 3 | 4 |
| 1 | 4 | 4 | 2 | 1 | 4 | 3 | 3 |
| 7 | 1 | 2 | 2 | 3 | 4 | 2 | 2 |
| 6 | 5 | 2 | 5 | 1 | 3 | 2 | 5 |
| 2 | 1 | 3 | 3 | 2 | 2 | 3 | 3 |

Construct a grouped-data table for these household sizes. Use classes based on a single value.

**2.31 Number of Siblings.** Professor Weiss asked his introductory statistics students to state how many siblings they have. The responses are shown in the following table.

| | | | | | | | |
|---|---|---|---|---|---|---|---|
| 1 | 3 | 2 | 1 | 1 | 0 | 1 | 1 |
| 3 | 0 | 2 | 2 | 1 | 2 | 0 | 2 |
| 1 | 2 | 2 | 1 | 0 | 1 | 1 | 1 |
| 1 | 1 | 0 | 2 | 0 | 3 | 4 | 2 |
| 0 | 2 | 1 | 1 | 2 | 1 | 1 | 0 |

Construct a grouped-data table for these numbers of siblings. Use classes based on a single value.

**2.32 Top Broadcast Shows.** The networks for the top 20 television shows, as determined by the *Nielsen Ratings* for the week of April 18–24, 2005, are shown in the following table. Construct a frequency distribution and a relative-frequency distribution for these network data.

| | | | | |
|---|---|---|---|---|
| CBS | Fox | ABC | Fox | CBS |
| CBS | ABC | Fox | CBS | ABC |
| CBS | CBS | NBC | CBS | NBC |
| NBC | CBS | CBS | NBC | NBC |

**2.33 NCAA Wrestling Champs.** From *NCAAsports.com*—the official Web site for NCAA sports—we obtained the National Collegiate Athletic Association wrestling champions for the years 1981–2005. They are shown in the following table. Construct a frequency distribution and a relative-frequency distribution for these wrestling-champion data.

| Iowa | Arizona St. | Oklahoma St. | Iowa |
|------|-------------|--------------|------|
| Iowa | Oklahoma St. | Iowa | Minnesota |
| Iowa | Oklahoma St. | Iowa | Minnesota |
| Iowa | Iowa | Iowa | Oklahoma St. |
| Iowa | Iowa | Iowa | Oklahoma St. |
| Iowa | Iowa | Iowa | Oklahoma St. |
| Iowa St. | | | |

## Working With Large Data Sets

**2.34 U.S. Regions.** The U.S. Census Bureau divides the states in the United States into four regions: Northeast, Midwest, South, and West. The data providing the region of each state is presented on the WeissStats CD.

**a.** Use the technology of your choice to obtain a frequency distribution and a relative-frequency distribution of the regions.

**b.** Interpret your results from part (a).

**2.35 The Great White Shark.** In an article titled "Great White, Deep Trouble" (*National Geographic*, Vol. 197(4), pp. 2–29), Peter Benchley—the author of *JAWS*—discussed various aspects of the Great White Shark (*Carcharodon carcharias*). Data on the number of pups borne in a lifetime by each of 80 Great White Shark females are provided on the WeissStats CD.

**a.** Use the technology of your choice to obtain an appropriate grouped-data table for these data.

**b.** Interpret your results from part (a).

**2.36 Road Rage.** The report *Controlling Road Rage: A Literature Review and Pilot Study* was prepared for the AAA Foundation for Traffic Safety by D. Rathbone and J. Huckabee. The authors discuss the results of a literature review and pilot study on how to prevent aggressive driving and road rage. As described in the study, *road rage* is criminal behavior by motorists characterized by uncontrolled anger that results in violence or threatened violence on the road. One of the goals of the study was to determine when road rage occurs most often. The days on which 69 road rage incidents occurred are provided on the WeissStats CD.

**a.** Use the technology of your choice to construct a frequency distribution and a relative-frequency distribution for these data.

**b.** Interpret your results from part (a).

**2.37 Presidential Election.** In a study summarized in *Chance* (Vol. 17, No. 4, pp. 43–46), H. Wainer provides yet another characterization of the states that supported George Bush versus Al Gore in the 2000 U.S. presidential election. Data on the WeissStats CD include National Assessment of Educational Progress (NAEP) score and the result of the 2000 presidential election for each of the 50 states. The NAEP scores reflect the academic performance for a random sample of children from each state. Use the technology of your choice to do the following.

**a.** Construct an appropriate grouped-data table for the NAEP scores.

**b.** Construct a frequency distribution and a relative-frequency distribution for the 2000 election results.

## Extending the Concepts and Skills

*The following table gives the closing Dow Jones Industrial Averages (Dow) on September 16, 2005. Use these data in Exercises 2.38–2.40.*

| Symbol | Last ($) | Change ($) | Volume (100s) |
|--------|----------|------------|---------------|
| AA | 26.68 | 0.30 | 61,045 |
| AIG | 61.30 | 0.35 | 131,773 |
| AXP | 59.46 | 1.90 | 162,653 |
| BA | 64.80 | −0.28 | 51,987 |
| C | 45.45 | 0.42 | 241,982 |
| CAT | 57.72 | −0.02 | 48,726 |
| DD | 40.51 | 0.41 | 79,254 |
| DIS | 24.11 | 0.11 | 117,519 |
| GE | 34.47 | 0.09 | 401,186 |
| GM | 32.48 | −0.07 | 76,402 |
| HD | 39.90 | −0.44 | 143,060 |
| HON | 39.29 | 0.69 | 55,331 |
| HPQ | 28.34 | 0.47 | 173,368 |
| IBM | 80.33 | 0.32 | 76,249 |
| INTC | 24.81 | 0.26 | 725,725 |
| JNJ | 65.18 | 0.81 | 129,953 |
| JPM | 34.99 | 0.86 | 195,574 |
| KO | 43.40 | −0.23 | 198,752 |
| MCD | 34.24 | 0.79 | 251,246 |
| MMM | 73.35 | 0.04 | 39,595 |
| MO | 73.14 | 0.89 | 118,553 |
| MRK | 28.90 | 0.14 | 117,614 |
| MSFT | 26.07 | −0.20 | 1,914,908 |
| PFE | 25.90 | 0.20 | 309,366 |
| PG | 55.95 | 0.45 | 136,346 |
| SBC | 24.23 | 0.32 | 191,932 |
| UTX | 51.38 | 0.70 | 64,848 |
| VZ | 32.81 | 0.42 | 150,953 |
| WMT | 43.87 | −0.45 | 966,782 |
| XOM | 63.70 | 1.24 | 303,785 |

**2.38 Dow Closing Prices.** The column headed "Last" in the Dow table gives the closing price per share, in dollars, for each of the Dow stocks. Construct a grouped-data table for these closing prices. Use equal-width classes and begin with the class $15 \leqslant 25$.

**2.39 Dow Volume.** The column headed "Volume" in the Dow table shows the number of shares sold, in hundreds, for each of the Dow stocks.

a. Construct a grouped-data table for these sales volumes. Use the classes $0 \leqslant 5$, $5 \leqslant 10$, ..., $40 \leqslant 45$, and "45 & over," where the values are in millions of sales.

b. Why is there no midpoint for the last class?

**2.40 Dow Gains and Losses.** The column headed "Change" in the Dow table provides the difference, in dollars, between the closing price per share given in the second column of the table and the closing price per share on the previous trading day. Construct a grouped-data table for the changes, using classes of your choice. Explain your choice of classes.

**2.41 Exam Scores.** The exam scores for the students in an introductory statistics class are as follows.

| | | | | |
|---|---|---|---|---|
| 88 | 82 | 89 | 70 | 85 |
| 63 | 100 | 86 | 67 | 39 |
| 90 | 96 | 76 | 34 | 81 |
| 64 | 75 | 84 | 89 | 96 |

a. Group these exam scores, using the classes 30–39, 40–49, 50–59, 60–69, 70–79, 80–89, and 90–100.

b. What are the widths of the classes?

c. If you wanted all the classes to have the same width, what classes would you use?

**Contingency Tables.** The methods presented in this section apply to grouping data obtained from observing values of one variable of a population. Such data are called *univariate data*. For instance, in Example 2.6 on page 47, we examined data obtained from observing the values of the variable "weight" for a sample of 18–24-year-old males; those data are univariate. We could have considered not only the weights of the males but also their heights. Then, we would have data on two variables, height and weight. Data obtained from observing values of two variables of a population are called *bivariate data*. Tables called *contingency tables* can be used to group bivariate data, as explained in Exercise 2.42.

**2.42 Age and Sex.** The following bivariate data on age (in years) and sex were obtained from the students in a freshman calculus course. The data show, for example, that the first student on the list is 21 years old and is a male.

| Age | Sex | Age | Sex | Age | Sex | Age | Sex | Age | Sex |
|---|---|---|---|---|---|---|---|---|---|
| 21 | M | 29 | F | 22 | M | 23 | F | 21 | F |
| 20 | M | 20 | M | 23 | M | 44 | M | 28 | F |
| 42 | F | 18 | F | 19 | F | 19 | M | 21 | F |
| 21 | M | 21 | M | 21 | M | 21 | F | 21 | F |
| 19 | F | 26 | M | 21 | F | 19 | M | 24 | F |
| 21 | F | 24 | F | 21 | F | 25 | M | 24 | F |
| 19 | F | 19 | M | 20 | F | 21 | M | 24 | F |
| 19 | M | 25 | M | 20 | F | 19 | M | 23 | M |
| 23 | M | 19 | F | 20 | F | 18 | F | 20 | F |
| 20 | F | 23 | M | 22 | F | 18 | F | 19 | M |

a. Group these data in the following contingency table. For the first student, place a tally mark in the box labeled by the "21–25" column and the "Male" row, as indicated. Tally the data for the other 49 students.

| | | Age (yr) | | | |
|---|---|---|---|---|---|
| | | Under 21 | 21–25 | Over 25 | Total |
| **Sex** | Male | | &#124; | | |
| | Female | | | | |
| | Total | | | | |

b. Construct a table like the one in part (a) but with frequencies replacing the tally marks. Add the frequencies in each row and column of your table and record the sums in the proper "Total" boxes.

c. What do the row and column totals in your table in part (b) represent?

d. Add the row totals and add the column totals. Why are those two sums equal, and what does their common value represent?

e. Construct a table that shows the relative frequencies for the data. (*Hint:* Divide each frequency obtained in part (b) by the total of 50 students.)

f. Interpret the entries in your table in part (e) as percentages.

## 2.3    Graphs and Charts

Besides grouping, another method for organizing and summarizing data is to draw a picture of some kind. The old saying "a picture is worth a thousand words" has particular relevance in statistics—a graph or chart of a data set often provides the simplest and most efficient display. In this section, we examine various techniques for organizing and summarizing data with graphs and charts, beginning with histograms.

**Example 2.10**

### Histograms

**TABLE 2.12**

Frequency and relative-frequency distributions for the days-to-maturity data

| Days to maturity | Freq. | Relative freq. |
|---|---|---|
| 30 ⊰ 40 | 3 | 0.075 |
| 40 ⊰ 50 | 1 | 0.025 |
| 50 ⊰ 60 | 8 | 0.200 |
| 60 ⊰ 70 | 10 | 0.250 |
| 70 ⊰ 80 | 7 | 0.175 |
| 80 ⊰ 90 | 7 | 0.175 |
| 90 ⊰ 100 | 4 | 0.100 |
| | 40 | 1.000 |

*Days to Maturity for Short-Term Investments* Table 2.12 repeats the first three columns of Table 2.4 in Section 2.2. Obtain graphical displays for these grouped data.

**Solution** One way to display these grouped data pictorially is to construct a graph, called a **frequency histogram,** that depicts the classes on the horizontal axis and the frequencies on the vertical axis, shown in Fig. 2.2(a). Here are some important observations about Fig. 2.2(a).

- The height of each bar is equal to the frequency of the class it represents.
- The bar for each class extends horizontally from that class's lower cutpoint to its upper cutpoint.[†]
- Each axis of the frequency histogram has a label, and the frequency histogram as a whole has a title.

A frequency histogram displays the frequencies of the classes. To display the relative frequencies (or percentages), we use a **relative-frequency histogram,** which is similar to a frequency histogram. The only difference

**FIGURE 2.2**

Days-to-maturity:
(a) frequency histogram;
(b) relative-frequency histogram

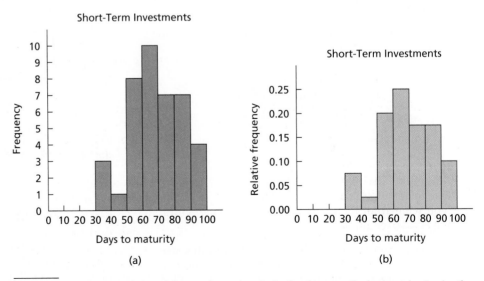

(a)

(b)

---

[†]This method is only one of several that can be used to depict the classes on the horizontal axis. Another common method is to use midpoints instead of cutpoints to label the horizontal axis. In that case, each bar is centered over the midpoint of the class it represents.

is that the height of each bar in a relative-frequency histogram is equal to the relative frequency of the class instead of the frequency of the class. Figure 2.2(b) shows a relative-frequency histogram for the days-to-maturity data.

Note that the shapes of the relative-frequency histogram in Fig. 2.2(b) and the frequency histogram in Fig. 2.2(a) are identical. The reason is that the frequencies and relative frequencies are proportional.

**You try it!**

Exercise 2.53 on page 66

• • •

---

**Definition 2.4**

**?**

**What Does It Mean?**

A histogram provides a graph of the values of the observations and how often they occur.

### Frequency and Relative-Frequency Histograms

**Frequency histogram:** A graph that displays the classes on the horizontal axis and the frequencies of the classes on the vertical axis. The frequency of each class is represented by a vertical bar whose height is equal to the frequency of the class.

**Relative-frequency histogram:** A graph that displays the classes on the horizontal axis and the relative frequencies of the classes on the vertical axis. The relative frequency of each class is represented by a vertical bar whose height is equal to the relative frequency of the class.

---

Relative-frequency histograms are better than frequency histograms for comparing two data sets. The same vertical scale is used for all relative-frequency histograms—a minimum of 0 and a maximum of 1—making direct comparison easy. In contrast, the vertical scale of a frequency histogram depends on the number of observations, making comparison more difficult.

### Histograms for Single-Value Grouping

In Figs. 2.2(a) and 2.2(b), the histogram bar for each class extends over a range. When data are grouped into classes based on a single value, however, we center each bar over the only possible value in the class.

---

**Example 2.11** | **Histograms for Single-Value Grouped Data**

**TABLE 2.13**

Frequency and relative-frequency distributions for number of TV sets

| Number of TVs | Freq. | Relative freq. |
|---|---|---|
| 0 | 1 | 0.020 |
| 1 | 16 | 0.320 |
| 2 | 14 | 0.280 |
| 3 | 12 | 0.240 |
| 4 | 3 | 0.060 |
| 5 | 2 | 0.040 |
| 6 | 2 | 0.040 |
|  | 50 | 1.000 |

*TVs per Household* Table 2.13 repeats Table 2.9 in Section 2.2. Construct a frequency histogram and a relative-frequency histogram for these grouped data.

**Solution** For single-value grouping, we place the middle of each histogram bar directly over the single value represented by the class. Figures 2.3(a) and (b) at the top of the next page show the frequency and relative-frequency histograms for the grouped data in Table 2.13.

Note the symbol // on the horizontal axes in Figs. 2.3(a) and (b). This symbol indicates that the zero point on that axis is not in its usual position at the intersection of the horizontal and vertical axes. Whenever any such modification is made, whether on the horizontal axis or vertical axis, the symbol // or some similar symbol should be used to indicate that fact.

• • •

**FIGURE 2.3**

Number of TVs per household:
(a) frequency histogram;
(b) relative-frequency histogram

Exercise 2.57
on page 67

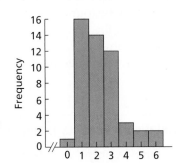

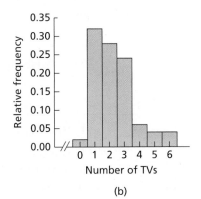

## Dotplots

Another type of graphical display for quantitative data is the **dotplot.** Dotplots are particularly useful for showing the relative positions of the data in a data set or for comparing two or more data sets.

**Example 2.12** | **Dotplots**

**TABLE 2.14**

Prices, in dollars, of 16 DVD players

| | | | |
|---|---|---|---|
| 210 | 219 | 214 | 197 |
| 224 | 219 | 199 | 199 |
| 208 | 209 | 215 | 199 |
| 212 | 212 | 219 | 210 |

*Prices of DVD Players* One of Professor Weiss's sons wanted to add a new DVD player to his home theater system. He used the Internet to shop and went to *pricewatch.com*. There he found 16 quotes on different brands and styles of DVD players. Table 2.14 lists the prices, in dollars. Construct a dotplot for these data.

**Solution** To construct a dotplot for the data in Table 2.14, we begin by drawing a horizontal axis that displays the possible prices. Then we record each price by placing a dot over the appropriate value on the horizontal axis. For instance, the first price is $210, which calls for a dot over the "210" on the horizontal axis. Figure 2.4 shows the dotplot for the data.

**FIGURE 2.4**

Dotplot for prices of DVD players

Exercise 2.59
on page 67

Note that dotplots are similar to histograms. In fact, when data are grouped in classes based on a single value, a dotplot and a frequency histogram are essentially identical. However, for single-value grouped data that involve decimals, dotplots are generally preferable to histograms because they are easier to construct and use.

### Stem-and-Leaf Diagrams

Statisticians continue to invent ways to display data. One method, developed in the 1960s by the late Professor John Tukey of Princeton University, is called a *stem-and-leaf diagram*. This ingenious diagram is often easier to construct than either a frequency distribution or a histogram and generally displays more information.

**Example 2.13** | ### Stem-and-Leaf Diagrams

**TABLE 2.15**

Days to maturity for 40 short-term investments

| | | | | | | | |
|---|---|---|---|---|---|---|---|
| 70 | 64 | 99 | 55 | 64 | 89 | 87 | 65 |
| 62 | 38 | 67 | 70 | 60 | 69 | 78 | 39 |
| 75 | 56 | 71 | 51 | 99 | 68 | 95 | 86 |
| 57 | 53 | 47 | 50 | 55 | 81 | 80 | 98 |
| 51 | 36 | 63 | 66 | 85 | 79 | 83 | 70 |

*Days to Maturity for Short-Term Investments* Table 2.15 repeats the data on the number of days to maturity for 40 short-term investments. Previously, we grouped these data with a frequency distribution (Table 2.2 on page 45) and graphed them with a frequency histogram (Fig. 2.2(a) on page 56). Now let's construct a stem-and-leaf diagram, which simultaneously groups the data and provides a graphical display similar to a histogram.

**Solution**  First, we list the leading digits of the numbers in Table 2.15 $(3, 4, \ldots, 9)$ in a column, as shown to the left of the vertical rule in Fig. 2.5(a).

**FIGURE 2.5**

Obtaining a stem-and-leaf diagram for the days-to-maturity data

```
                          Stems  Leaves
3 | 8 6 9                  3 | 6 8 9

4 | 7                      4 | 7

5 | 7 1 6 3 5 1 0 5        5 | 0 1 1 3 5 5 6 7

6 | 2 4 7 3 6 4 0 9 8 5    6 | 0 2 3 4 4 5 6 7 8 9

7 | 0 5 1 0 9 8 0          7 | 0 0 0 1 5 8 9

8 | 5 9 1 7 0 3 6          8 | 0 1 3 5 6 7 9

9 | 9 9 5 8                9 | 5 8 9 9

        (a)                        (b)
```

Next, we write the final digit of each number from Table 2.15 to the right of the vertical rule in Fig. 2.5(a) in the row containing the appropriate leading digit: The first number is 70, which calls for a 0 to the right of the leading digit 7. Reading down the first column of Table 2.15, we find that the second number is 62, which calls for a 2 to the right of the leading digit 6.

We continue in this manner until all observations in Table 2.15 have been accounted for. The result is the diagram displayed in Fig. 2.5(a).

Then, in each row, we arrange the digits to the right of the vertical rule in ascending order (i.e., from smallest to largest). The result is the diagram displayed in Fig. 2.5(b). As indicated in this figure, the leading digits are called **stems** and the final digits **leaves;** the entire diagram in Fig. 2.5(b) is called a **stem-and-leaf diagram,** or **stemplot.**

The stem-and-leaf diagram for the days-to-maturity data is similar to a frequency histogram for those data because the length of the row of leaves for a class equals the frequency of the class. [Turn the stem-and-leaf diagram 90° counterclockwise, and compare it to the frequency histogram shown in Fig. 2.2(a) on page 56.]

• • •

| Procedure 2.1 | To Construct a Stem-and-Leaf Diagram |
|---|---|

Exercise 2.63
on page 68

**STEP 1** Think of each observation as a stem—consisting of all but the rightmost digit—and a leaf, the rightmost digit.

**STEP 2** Write the stems from smallest to largest in a vertical column to the left of a vertical rule.

**STEP 3** Write each leaf to the right of the vertical rule in the row that contains the appropriate stem.

**STEP 4** Arrange the leaves in each row in ascending order.

Note that, in general, stems may use as many digits as required, but each leaf must contain only one digit. In Example 2.14, we describe the use of the stem-and-leaf diagram for three-digit numbers and also introduce the technique of using more than one line per stem.

**Example 2.14** | **Stem-and-Leaf Diagrams**

*Cholesterol Levels* A pediatrician tested the cholesterol levels of several young patients and was alarmed to find that many had levels higher than 200 mg per 100 mL. Table 2.16 presents the readings of 20 patients with high levels. Construct a stem-and-leaf diagram for these data by using

**a.** one line per stem.      **b.** two lines per stem.

**Solution** Because these observations are three-digit numbers, we use the first two digits of each number as the stem and the third digit as the leaf.

**a.** Using one line per stem and applying Procedure 2.1, we obtain the stem-and-leaf diagram displayed in Fig. 2.6(a).

**TABLE 2.16**
Cholesterol levels for 20 high-level patients

| 210 | 209 | 212 | 208 |
|---|---|---|---|
| 217 | 207 | 210 | 203 |
| 208 | 210 | 210 | 199 |
| 215 | 221 | 213 | 218 |
| 202 | 218 | 200 | 214 |

**FIGURE 2.6**
Stem-and-leaf diagram for cholesterol levels: (a) one line per stem; (b) two lines per stem

```
19 | 9
20 | 0 2 3 7 8 8 9
21 | 0 0 0 0 2 3 4 5 7 8 8
22 | 1

        (a)
```

```
19 |
19 | 9
20 | 0 2 3
20 | 7 8 8 9
21 | 0 0 0 0 2 3 4
21 | 5 7 8 8
22 | 1
22 |

        (b)
```

**b.** The stem-and-leaf diagram in Fig. 2.6(a) is only moderately helpful because there are so few stems. Figure 2.6(b) is a better stem-and-leaf diagram for these data. It uses two lines for each stem, with the first line for the leaf digits 0–4 and the second line for the leaf digits 5–9.

Exercise 2.65
on page 68

• • •

Although stem-and-leaf diagrams have several advantages over the more classical techniques for grouping and graphing, they do have some drawbacks. For instance, they are generally not useful with large data sets and can be awkward with data containing many digits; histograms are usually preferable to stem-and-leaf diagrams in such cases.

## Graphical Displays for Qualitative Data

Histograms, dotplots, and stem-and-leaf diagrams are designed for use with quantitative data. We use different techniques for qualitative data: two common methods are the *pie chart* and the *bar graph*.

**Example 2.15**  **Pie Charts and Bar Graphs**

*Political Party Affiliations* Table 2.17 repeats the frequency and relative-frequency distributions for the political party affiliations of Professor Weiss's introductory statistics students. Display the relative-frequency distribution of these qualitative data with a

**a.** pie chart.

**b.** bar graph.

**TABLE 2.17**
Frequency and relative-frequency distributions for political party affiliations in one statistics class

| Party | Freq. | Relative freq. |
|-------|-------|------|
| Democratic | 13 | 0.325 |
| Republican | 18 | 0.450 |
| Other | 9 | 0.225 |
|  | 40 | 1.000 |

### Solution

**a.** A **pie chart** is a disk divided into wedge-shaped pieces that are proportional to the frequencies or relative frequencies. In this case, we need to divide a disk into three wedge-shaped pieces that comprise 32.5%, 45.0%, and 22.5% of the disk. We do so by using a protractor and the fact that there are 360° in a circle. Thus, for instance, the first piece of the disk is obtained by marking off 117° (32.5% of 360°). Figure 2.7(a) shows the pie chart for this relative-frequency distribution.

**FIGURE 2.7**
Political party affiliations in one statistics class: (a) pie chart; (b) bar graph

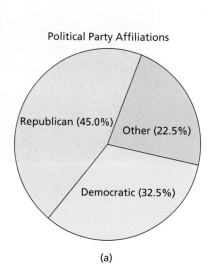

(a)

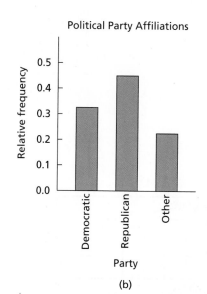

(b)

You try it!

Exercise 2.67 on page 68

b. A **bar graph** is like a histogram. However, to avoid confusing bar graphs and histograms, we position the bars in a bar graph so that they do not touch each other. Figure 2.7(b) shows the bar graph for this relative-frequency distribution.

• • •

Histograms, dotplots, stem-and-leaf diagrams, pie charts, and bar graphs are only a few of the ways that data can be portrayed pictorially. We will consider some additional graphical displays in the exercises for this section.

# The Technology Center

Some statistical technologies have programs that automatically generate a particular type of graph or chart, but others do not. In this subsection, we present output and step-by-step instructions for such programs, whenever available in Minitab, Excel, or the TI-83/84 Plus.

**Example 2.16    Using Technology to Obtain a Histogram**

*Days to Maturity for Short-Term Investments*    Table 2.1 on page 45 displays the numbers of days to maturity for 40 short-term investments. Use Minitab, Excel, or the TI-83/84 Plus to obtain a frequency histogram of those data.

**Solution**    We applied the histogram programs to the data, resulting in Output 2.2. Steps for generating that output are presented in Instructions 2.2.

**OUTPUT 2.2**    Histograms for the days-to-maturity data

**MINITAB**

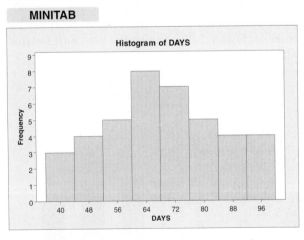

**EXCEL**

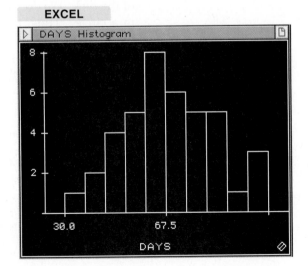

**TI-83/84 PLUS**

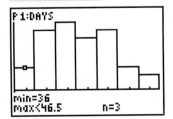

Some technologies require the user to specify a histogram's classes; others automatically choose the classes; others allow the user to specify the classes or to let the program choose them.

We generated all three histograms in Output 2.2 by letting the programs automatically choose the classes, which explains why the three histograms differ from each other and from the histogram we constructed by hand in Fig. 2.2(a) on page 56. To generate histograms based on user-specified classes, refer to the technology manuals.

• • •

**INSTRUCTIONS 2.2**  Steps for generating Output 2.2

| MINITAB | EXCEL | TI-83/84 PLUS |
|---|---|---|
| 1 Store the data from Table 2.1 in a column named DAYS<br>2 Choose **Graph ➤ Histogram…**<br>3 Select the **Simple** histogram and click **OK**<br>4 Specify DAYS in the **Graph variables** text box<br>5 Click **OK** | 1 Store the data from Table 2.1 in a range named DAYS<br>2 Choose **DDXL ➤ Charts and Plots**<br>3 Select **Histogram** from the **Function type** drop-down list box<br>4 Specify DAYS in the **Quantitative Variable** text box<br>5 Click **OK** | 1 Store the data from Table 2.1 in a list named DAYS<br>2 Press **2ND ➤ STAT PLOT** and then press **ENTER** twice<br>3 Arrow to the third graph icon and press **ENTER**<br>4 Press the down-arrow key<br>5 Press **2ND ➤ LIST**<br>6 Arrow down to DAYS and press **ENTER**<br>7 Press **ZOOM** and then **9** (and then **TRACE**, if desired) |

**Example 2.17**  **Using Technology to Obtain a Stem-and-Leaf Diagram**

*Cholesterol Levels*  Table 2.16 on page 60 provides the cholesterol levels of 20 patients with high levels. Apply Minitab to obtain a stem-and-leaf diagram for those data by using (a) one line per stem and (b) two lines per stem.

**Solution**  We applied Minitab's stem-and-leaf program to the data by using one line per stem, resulting in Output 2.3(a), and by using two lines per stem, resulting in Output 2.3(b). Steps for generating those stem-and-leaf diagrams are presented in Instructions 2.3.

**OUTPUT 2.3**  Stem-and-leaf diagrams for cholesterol levels: (a) one line per stem; (b) two lines per stem

MINITAB

```
Stem-and-Leaf Display: LEVEL

Stem-and-leaf of LEVEL   N  = 20
Leaf Unit = 1.0


  1     19   9
  8     20   0237889
 (11)   21   00002345788
  1     22   1
```

```
Stem-and-Leaf Display: LEVEL

Stem-and-leaf of LEVEL   N  = 20
Leaf Unit = 1.0


  1     19   9
  4     20   023
  8     20   7889
 (7)    21   0000234
  5     21   5788
  1     22   1
```

(a)                                           (b)

For each stem-and-leaf diagram in Output 2.3, the second and third columns give the stems and leaves, respectively. See Minitab's Help or the *Minitab Manual* for other aspects of these stem-and-leaf diagrams.

**INSTRUCTIONS 2.3**
Steps for generating Output 2.3

| MINITAB |
| --- |
| 1 Store the data from Table 2.16 in a column named LEVEL |
| 2 Choose **Graph ➤ Stem-and-Leaf...** |
| 3 Specify LEVEL in the **Graph variables** text box |
| 4 Type <u>10</u> in the **Increment** text box |
| 5 Click **OK** |

(a)

| MINITAB |
| --- |
| 1 Store the data from Table 2.16 in a column named LEVEL |
| 2 Choose **Graph ➤ Stem-and-Leaf...** |
| 3 Specify LEVEL in the **Graph variables** text box |
| 4 Type <u>5</u> in the **Increment** text box |
| 5 Click **OK** |

(b)

The *increment* specifies the difference between the smallest possible number on one line and the smallest possible number on the preceding line, and thereby controls the number of lines per stem. You can let Minitab choose the number of lines per stem automatically by leaving the **Increment** text box blank.

• • •

**Example 2.18**    **Using Technology to Obtain a Pie Chart**

*Political Party Affiliations*    Table 2.10 on page 50 displays the political party affiliations of the students in Professor Weiss's introductory statistics class. Use Minitab or Excel to obtain a pie chart for those data.

**Solution**    We applied the pie-chart programs to the data, resulting in Output 2.4. Steps for generating that output are presented in Instructions 2.4.

**OUTPUT 2.4**    Pie charts for political party affiliations

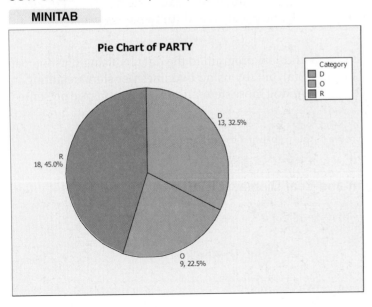

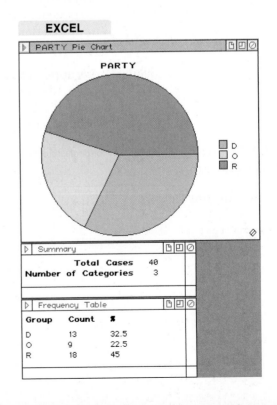

**INSTRUCTIONS 2.4**

Steps for generating Output 2.4

**MINITAB**

1 Store the data from Table 2.10 in a column named PARTY
2 Choose **Graph ➤ Pie Chart…**
3 Select the **Chart counts of unique values** option button
4 Specify PARTY in the **Categorical variables** text box
5 Click the **Labels…** button
6 Click the **Slice Labels** tab
7 Select all three check boxes from the **Label pie slices with** list
8 Click **OK** twice

**EXCEL**

1 Store the data from Table 2.10 in a range named PARTY
2 Choose **DDXL ➤ Charts and Plots**
3 Select **Pie Chart** from the **Function type** drop-down list box
4 Specify PARTY in the **Categorical Variable** text box
5 Click **OK**

• • •

**Example 2.19**   **Using Technology to Obtain a Bar Graph**

*Political Party Affiliations*   Table 2.10 on page 50 displays the political party affiliations of the students in Professor Weiss's introductory statistics class. Use Minitab or Excel to obtain a bar graph for those data.

Solution   We applied the bar-graph programs to the data, resulting in Output 2.5. Steps for generating the output are presented in Instructions 2.5.

**OUTPUT 2.5**   Bar graphs for political party affiliations

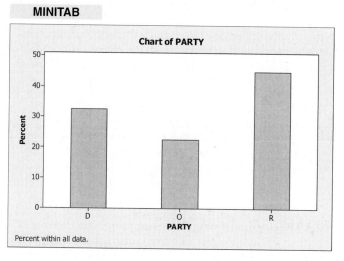

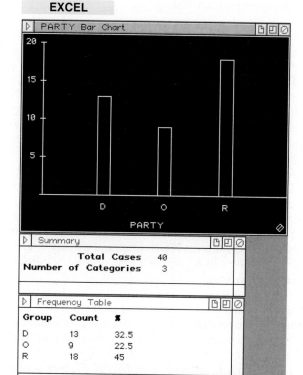

Compare Output 2.5 to the bar graph obtained by hand in Fig. 2.7(b) on page 61. Note that, by default, both Minitab and Excel arrange the categories in alphabetical order (D, O, and R).

• • •

**INSTRUCTIONS 2.5**
Steps for generating Output 2.5

**MINITAB**

1 Store the data from Table 2.10 in a column named PARTY
2 Choose **Graph ➤ Bar Chart...**
3 Select **Counts of unique values** from the **Bars represent** drop-down list box
4 Select the **Simple** bar chart and click **OK**
5 Specify PARTY in the **Categorical Variables** text box
6 Click the **Chart Options...** button
7 Select the **Show Y as Percent** check box
8 Click **OK** twice

**EXCEL**

1 Store the data from Table 2.10 in a range named PARTY
2 Choose **DDXL ➤ Charts and Plots**
3 Select **Bar Chart** from the **Function type** drop-down list box
4 Specify PARTY in the **Categorical Variable** text box
5 Click **OK**

▌ ▌ ▌

# Exercises 2.3

## Understanding the Concepts and Skills

**2.43** Explain the difference between a frequency histogram and a relative-frequency histogram.

**2.44** Explain the advantages and disadvantages of histograms relative to grouped-data tables.

**2.45** For data that are grouped in classes based on more than a single value, cutpoints are used on the horizontal axis of a histogram for depicting the classes. But, midpoints can also be used, in which case each bar is centered over the midpoint of the class it represents. Explain the advantages and disadvantages of each method.

**2.46** Discuss the relative advantages and disadvantages of stem-and-leaf diagrams versus frequency histograms.

**2.47** Suppose that you have a data set that contains a large number of observations. Which graphical display is generally preferable: a histogram or a stem-and-leaf diagram? Explain your answer.

**2.48** Suppose that you have constructed a stem-and-leaf diagram and discover that it is only moderately useful because there are too few stems. How can you remedy the problem?

**2.49** In a bar graph, unlike in a histogram, the bars do not abut. Give a reason for that.

**2.50** Some users of statistics prefer pie charts to bar graphs because people are accustomed to having the horizontal axis of a graph show order. For example, someone might infer from Fig. 2.7(b) on page 61 that "Republican" is less than "Other" because "Republican" is shown to the left of "Other" on the horizontal axis. Pie charts do not lead to such inferences. Give other advantages and disadvantages of each method.

**2.51 DVD Players.** Refer to Example 2.12 on page 58.
a. Explain why a frequency histogram of the DVD prices with classes based on a single value would be essentially identical to the dotplot shown in Fig. 2.4.
b. Would the dotplot and a frequency histogram be essentially identical if the classes for the histogram are not each based on a single value? Explain your answer.

**2.52 Residential Energy Consumption.** In Exercise 2.24 on page 52, you were asked to construct a grouped-data table for observations on one year's energy consumptions for a sample of 50 households in the South. Based on the grouped-data table obtained there, do the following:
a. Construct a frequency histogram.
b. Construct a relative-frequency histogram.

**2.53 Clocking the Cheetah.** In Exercise 2.25 on page 53, you were asked to construct a grouped-data table for the speeds, in miles per hour, of a sample of 35 cheetahs. Based on the grouped-data table obtained there, do the following:

**a.** Construct a frequency histogram.
**b.** Construct a relative-frequency histogram.

**2.54 Super Bowl Ratings.** In Exercise 2.28 on page 53, you were asked to construct a grouped-data table for the television ratings of 32 Super Bowls. Based on the grouped-data table obtained there, do the following:
**a.** Construct a frequency histogram.
**b.** Construct a relative-frequency histogram.

**2.55 Top Broadcast Shows.** In Exercise 2.29 on page 53, you were asked to construct a grouped-data table for the viewing audiences, in millions, of the top 20 television shows for the week of April 18–24, 2005. Based on the grouped-data table obtained there, do the following:
**a.** Construct a frequency histogram.
**b.** Construct a relative-frequency histogram.

**2.56 Household Size.** In Exercise 2.30 on page 53, you were asked to construct a grouped-data table, with classes based on a single value, for the number of people per household for a sample of 40 U.S. households. Based on the grouped-data table obtained there, do the following:
**a.** Construct a frequency histogram.
**b.** Construct a relative-frequency histogram.

**2.57 Number of Siblings.** In Exercise 2.31 on page 53, you were asked to construct a grouped-data table, with classes based on a single value, for the number of siblings for each of the 40 students in Professor Weiss's introductory statistics class. Based on the grouped-data table obtained there, do the following:
**a.** Construct a frequency histogram.
**b.** Construct a relative-frequency histogram.

**2.58 Exam Scores.** Construct a dotplot for the following exam scores of the students in an introductory statistics class.

| | | | | |
|---|---|---|---|---|
| 88 | 82 | 89 | 70 | 85 |
| 63 | 100 | 86 | 67 | 39 |
| 90 | 96 | 76 | 34 | 81 |
| 64 | 75 | 84 | 89 | 96 |

**2.59 Ages of Trucks.** The Motor Vehicle Manufacturers Association of the United States publishes information in *Motor Vehicle Facts and Figures* on the ages of cars and trucks currently in use. A sample of 37 trucks provided the ages, in years, displayed in the following table. Construct a dotplot for the ages.

| | | | | | | | |
|---|---|---|---|---|---|---|---|
| 8 | 12 | 14 | 16 | 15 | 5 | 11 | 13 |
| 4 | 12 | 12 | 15 | 12 | 3 | 10 | 9 |
| 11 | 3 | 18 | 4 | 9 | 11 | 17 | |
| 7 | 4 | 12 | 12 | 8 | 9 | 10 | |
| 9 | 9 | 1 | 7 | 6 | 9 | 7 | |

**2.60 Stressed-Out Bus Drivers.** Frustrated passengers, congested streets, time schedules, and air and noise pollution are just some of the physical and social pressures that lead many urban bus drivers to retire prematurely with disabilities such as coronary heart disease and stomach disorders. An intervention program designed by the Stockholm Transit District was implemented to improve the work conditions of the city's bus drivers. Improvements were evaluated by Evans et al. who collected physiological and psychological data for bus drivers who drove on the improved routes (intervention) and for drivers who were assigned the normal routes (control). Their findings were published in the article "Hassles on the Job: A Study of a Job Intervention With Urban Bus Drivers" (*Journal of Organizational Behavior*, Vol. 20, pp. 199–208). Following are data, based on the results of the study, for the heart rates, in beats per minute, of the intervention and control drivers.

| Intervention | | Control | | | | | | |
|---|---|---|---|---|---|---|---|---|
| 68 | 66 | 74 | 52 | 67 | 63 | 77 | 57 | 80 |
| 74 | 58 | 77 | 53 | 76 | 54 | 73 | 54 | |
| 69 | 63 | 60 | 77 | 63 | 60 | 68 | 64 | |
| 68 | 73 | 66 | 71 | 66 | 55 | 71 | 84 | |
| 64 | 76 | 63 | 73 | 59 | 68 | 64 | 82 | |

**a.** Obtain dotplots for each of the two data sets, using the same scales.
**b.** Use your result from part (a) to compare the two data sets.

**2.61 Acute Postoperative Days.** Several neurosurgeons wanted to determine whether a dynamic system (Z-plate) reduced the number of acute postoperative days in the hospital relative to a static system (ALPS plate). R. Jacobowitz, Ph.D., an Arizona State University professor, along with G. Vishteh, M.D., and other neurosurgeons, obtained the following data on the number of acute postoperative days in the hospital using the dynamic and static systems.

| Dynamic | | | | | | | Static | | |
|---|---|---|---|---|---|---|---|---|---|
| 7 | 5 | 8 | 8 | 6 | 7 | 7 | 6 | 18 | 9 |
| 9 | 10 | 7 | 7 | 7 | 7 | 8 | 7 | 14 | 9 |

**a.** Obtain dotplots for each of the two data sets, using the same scales.
**b.** Use your result from part (a) to compare the two data sets.

**2.62 Contents of Soft Drinks.** A soft-drink bottler fills bottles with soda. For quality assurance purposes, filled bottles are sampled to ensure that they contain close to the content indicated on the label. A sample of 30 "one-liter" bottles of soda contain the amounts, in milliliters, shown in the following table. Construct a stem-and-leaf diagram for these data.

| | | | | |
|---|---|---|---|---|
| 1025 | 977 | 1018 | 975 | 977 |
| 990 | 959 | 957 | 1031 | 964 |
| 986 | 914 | 1010 | 988 | 1028 |
| 989 | 1001 | 984 | 974 | 1017 |
| 1060 | 1030 | 991 | 999 | 997 |
| 996 | 1014 | 946 | 995 | 987 |

**2.63 Women in the Workforce.** In the April 22, 2005 issue of *Science* (Vol. 308, No. 5721, p. 483), D. Normile reports on a study from Japan's Statistics Bureau of the 30 industrialized countries in the Organization for Economic Cooperation and Development (OECD) titled, "Japan Mulls Workforce Goals for Women." Following are the percentages of women in their scientific workforces for a sample of 17 countries. Construct a stem-and-leaf diagram for these percentages.

| | | | | | | | | |
|---|---|---|---|---|---|---|---|---|
| 26 | 12 | 44 | 13 | 40 | 18 | 39 | 21 | 35 |
| 27 | 34 | 28 | 34 | 28 | 33 | 29 | 29 | |

**2.64 University Patents.** The number of patents a university receives is an indicator of the research level of the university. From a study titled, *Science and Engineering Indicators*, issued by the National Science Foundation, we found the number of U.S. patents awarded to a sample of 36 private and public universities to be as follows.

| | | | | | | | | |
|---|---|---|---|---|---|---|---|---|
| 93 | 27 | 11 | 30 | 9 | 30 | 35 | 20 | 9 |
| 35 | 24 | 19 | 14 | 29 | 11 | 2 | 55 | 15 |
| 35 | 2 | 15 | 4 | 16 | 79 | 16 | 22 | 49 |
| 3 | 69 | 23 | 18 | 41 | 11 | 7 | 34 | 16 |

Construct a stem-and-leaf diagram for these data using
**a.** one line per stem.  **b.** two lines per stem.
**c.** Which stem-and-leaf diagram do you find more useful? Why?

**2.65 Ages of Baseball Players.** From *MLB Roster Analysis* on the ESPN Web site, we found the average age of the players on each of the 30 major league baseball teams, as of May 2, 2005, to be as follows.

| | | | | | |
|---|---|---|---|---|---|
| 26.6 | 27.9 | 27.9 | 29.9 | 29.3 | 28.1 |
| 28.4 | 28.9 | 27.7 | 28.7 | 30.5 | 29.8 |
| 28.5 | 27.9 | 30.9 | 29.3 | 28.8 | 28.6 |
| 29.1 | 31.0 | 30.7 | 30.3 | 29.7 | 31.0 |
| 29.4 | 29.8 | 29.4 | 32.7 | 34.0 | 31.8 |

Construct a stem-and-leaf diagram for these data using
**a.** one line per stem.  **b.** two lines per stem.
**c.** Which stem-and-leaf diagram do you find more useful? Why?

**2.66 Top Broadcast Shows.** In Exercise 2.32 on page 53, you were asked to construct a frequency distribution and a relative-frequency distribution for the networks of the top 20 television shows for the week of April 18–24, 2005.
**a.** Draw a pie chart for the relative frequencies.
**b.** Construct a bar graph for the relative frequencies.

**2.67 NCAA Wrestling Champs.** In Exercise 2.33 on page 53, you were asked to construct a frequency distribution and a relative-frequency distribution for the National Collegiate Athletic Association wrestling champions for the years 1981–2005.
**a.** Draw a pie chart for the relative frequencies.
**b.** Construct a bar graph for the relative frequencies.

**2.68 Robbery Locations.** The Department of Justice and the Federal Bureau of Investigation publish a compilation on crime statistics for the United States in *Crime in the United States*. The following table provides a frequency distribution for robbery type during a one-year period.

| Robbery type | Frequency |
|---|---|
| Street/highway | 179,296 |
| Commercial house | 60,493 |
| Gas or service station | 11,362 |
| Convenience store | 25,774 |
| Residence | 56,641 |
| Bank | 9,504 |
| Miscellaneous | 70,333 |

**a.** Draw a pie chart for the relative frequencies.
**b.** Construct a bar graph for the relative frequencies.

**2.69 M&M Colors.** Observing that the proportion of blue M&Ms in his bowl of candy appeared to be less than that of the other colors, Ronald D. Fricker, Jr., decided to compare the color distribution in randomly chosen bags of M&Ms to the theoretical distribution reported by M&M/MARS consumer affairs. Fricker published his findings in the article "The Mysterious Case of the Blue M&Ms" (*Chance*, Vol. 9(4), pp. 19–22). For his study, Fricker bought three bags of M&Ms from local stores and counted the number of each color. The average number of each color in the three bags was distributed as follows.

| Color | Frequency |
|-------|-----------|
| Brown | 152 |
| Yellow | 114 |
| Red | 106 |
| Orange | 51 |
| Green | 43 |
| Blue | 43 |

**a.** Draw a pie chart for the relative frequencies.
**b.** Construct a bar graph for the relative frequencies.

**2.70 Adjusted Gross Incomes.** The Internal Revenue Service (IRS) publishes data on adjusted gross incomes in *Statistics of Income, Individual Income Tax Returns*. The following relative-frequency histogram shows one year's individual income tax returns for adjusted gross incomes of less than $50,000.

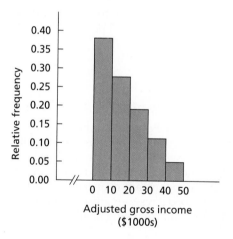

Use the histogram and the fact that adjusted gross incomes are expressed to the nearest whole dollar to answer each of the following questions.

**a.** Approximately what percentage of the individual income tax returns had an adjusted gross income between $10,000 and $19,999, inclusive?
**b.** Approximately what percentage had an adjusted gross income of less than $30,000?
**c.** The IRS reported that 89,928,000 individual income tax returns had an adjusted gross income of less than $50,000. Approximately how many had an adjusted gross income between $30,000 and $49,999, inclusive?

**2.71 Cholesterol Levels.** A pediatrician who tested the cholesterol levels of several young patients was alarmed to find that many had levels higher than 200 mg per 100 mL. The following relative-frequency histogram shows the readings for some patients who had high cholesterol levels.

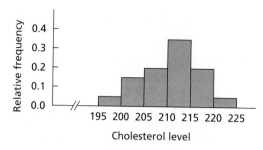

Use the graph to answer the following questions. Note that cholesterol levels are always expressed as whole numbers.

**a.** What percentage of the patients have cholesterol levels between 205 and 209, inclusive?
**b.** What percentage of the patients have levels of 215 or higher?
**c.** If the number of patients is 20, how many have levels between 210 and 214, inclusive?

## Working With Large Data Sets

**2.72 U.S. Regions.** The U.S. Census Bureau divides the states in the United States into four regions: Northeast, Midwest, South, and West. The data providing the region of each state is presented on the WeissStats CD. Use the technology of your choice to obtain a
**a.** pie chart for the regions.
**b.** bar graph for the regions.
**c.** Interpret your results from parts (a) and (b).

**2.73 The Great White Shark.** In an article titled "Great White, Deep Trouble" (*National Geographic*, Vol. 197(4), pp. 2–29), Peter Benchley—the author of *JAWS*—discussed various aspects of the Great White Shark (*Carcharodon carcharias*). Data on the number of pups borne in a lifetime by each of 80 Great White Shark females are provided on the WeissStats CD.
**a.** Use the technology of your choice to obtain either a frequency histogram or a relative-frequency histogram for these data. If possible, use classes based on a single value.
**b.** Interpret your results from part (a).

**2.74 Road Rage.** The report *Controlling Road Rage: A Literature Review and Pilot Study* was prepared for the AAA Foundation for Traffic Safety by D. Rathbone and J. Huckabee. The authors discuss the results of a literature review and pilot study on how to prevent aggressive driving and road rage. As described in the study, *road rage* is criminal behavior by motorists characterized by uncontrolled anger that results in violence or threatened violence on the road. One of the goals of the study was to determine when road rage occurs most often. The days on which 69 road rage incidents occurred are provided on the WeissStats CD. Use the technology of your choice to obtain a
**a.** pie chart for the days.     **b.** bar graph for the days.
**c.** Interpret your results from parts (a) and (b).

**2.75 Presidential Election.** In a study summarized in *Chance* (Vol. 17, No. 4, pp. 43–46), H. Wainer provides yet another characterization of the states that supported George Bush versus Al Gore in the 2000 U.S. presidential election. Data on the WeissStats CD include National Assessment of Educational Progress (NAEP) score and the result of the 2000 presidential election for each of the 50 states. The NAEP scores reflect the academic performance for a random sample of children from each state. Use the technology of your choice to obtain and interpret

a. a frequency histogram or a relative-frequency histogram for the NAEP scores.

b. a pie chart for the election results.

c. a bar graph for the election results.

**2.76 Top Recording Artists.** From the Recording Industry Association of America Web site, we obtained data on the number of albums sold, in millions, for the top recording artists (U.S. sales only) as of June 20, 2005. Those data are provided on the WeissStats CD. Use the technology of your choice to obtain and interpret

a. a frequency histogram or a relative-frequency histogram of the data.

b. a dotplot of the data.

c. Compare your graphs from parts (a) and (b).

**2.77 High School Completion Rates.** As reported by the U.S. Census Bureau in *Current Population Reports*, the percentage of adults in each state and the District of Columbia that have completed high school is as provided on the WeissStats CD. Use the technology of your choice to construct a stem-and-leaf diagram of the percentages using

a. one line per stem.     b. two lines per stem.

c. five lines per stem.

d. Which stem-and-leaf diagram do you consider most useful? Explain your answer.

**2.78 Crime Rates.** The U.S. Federal Bureau of Investigation publishes the annual crime rates for each state and the District of Columbia in the document *Crime in the United States*. Those rates, given per 1000 population, are provided on the WeissStats CD. Use the technology of your choice to construct a stem-and-leaf diagram of the rates using

a. one line per stem.     b. two lines per stem.

c. five lines per stem.

d. Which stem-and-leaf diagram do you consider most useful? Explain your answer.

**2.79 Body Temperature.** A study by researchers at the University of Maryland addressed the question of whether the mean body temperature of humans is 98.6°F. The results of the study by P. Mackowiak et al. appeared in the article "A Critical Appraisal of 98.6°F, the Upper Limit of the Normal Body Temperature, and Other Legacies of Carl

Reinhold August Wunderlich" (*Journal of the American Medical Association*, Vol. 268, pp. 1578–1580). Among other data, the researchers obtained the body temperatures of 93 healthy humans, as provided on the WeissStats CD. Use the technology of your choice to obtain and interpret

a. a frequency histogram or a relative-frequency histogram of the temperatures.

b. a dotplot of the temperatures.

c. a stem-and-leaf diagram of the temperatures.

d. Compare your graphs from parts (a)–(c). Which do you find most useful?

## Extending the Concepts and Skills

**Relative-Frequency Polygons.** Another graphical display commonly used is the relative-frequency polygon. In a *relative-frequency polygon,* a point is plotted above each class midpoint at a height equal to the relative frequency of the class. Then the points are connected with lines. For instance, the grouped days-to-maturity data given in Table 2.4 on page 47 yields the following relative-frequency polygon.

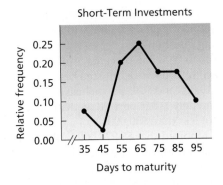

Short-Term Investments

**2.80 Residential Energy Consumption.** Construct a relative-frequency polygon for the energy-consumption data given in Exercise 2.24 on page 52. Use the classes specified in that exercise.

**Ogives.** Cumulative information can be portrayed using a graph called an *ogive* (ō'jīv). To construct an ogive, first make a table that displays cumulative frequencies and cumulative relative frequencies, as shown in Table 2.18 for the days-to-maturity data.

**TABLE 2.18**
Cumulative information for days-to-maturity data

| Less than | Cumulative frequency | Cumulative relative frequency |
|---|---|---|
| 30 | 0 | 0.000 |
| 40 | 3 | 0.075 |
| 50 | 4 | 0.100 |
| 60 | 12 | 0.300 |
| 70 | 22 | 0.550 |
| 80 | 29 | 0.725 |
| 90 | 36 | 0.900 |
| 100 | 40 | 1.000 |

The first column of Table 2.18 gives the cutpoints of the classes and the second column gives the cumulative frequencies. A *cumulative frequency* is obtained by summing the frequencies of all classes representing values less than the specified cutpoint. For instance, by referring to Table 2.12 on page 56, we can find the cumulative frequency of investments with a maturity period of less than 50 days:

$$\text{cumulative frequency} = 3 + 1 = 4.$$

That is, four of the investments have a maturity period of less than 50 days.

The third column of Table 2.18 gives the cumulative relative frequencies. A *cumulative relative frequency* is found by dividing the corresponding cumulative frequency by the total number of observations which, in this case, is 40. For instance, the cumulative relative frequency of investments with a maturity period of less than 50 days is

$$\text{cumulative relative frequency} = \frac{4}{40} = 0.100.$$

That is, 10% of the investments have a maturity period of less than 50 days.

Using Table 2.18, we can now construct an ogive for the days-to-maturity data. In an ogive, a point is plotted above each cutpoint at a height equal to the cumulative relative frequency. Then the points are connected with lines. The ogive for the days-to-maturity data is as follows.

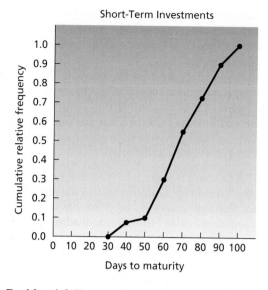

Short-Term Investments

**2.81 Residential Energy Consumption.** Refer to Exercise 2.24 on page 52.

a. Construct a table similar to Table 2.18 for the energy-consumption data, based on the classes specified in Exercise 2.24. Interpret your results.

b. Draw an ogive for the data.

**Further Stem-and-Leaf Techniques.** We mentioned earlier that the use of stem-and-leaf diagrams can be awkward with data that contain many digits. In such cases, we can either round or truncate each observation to a suitable number of digits. Exercises 2.82–2.84 involve rounding and truncating numbers for use in stem-and-leaf diagrams.

**2.82 Weights of Males.** The U.S. National Center for Health Statistics publishes data on weights and heights by age and sex in *Vital and Health Statistics*. The following weights, to the nearest tenth of a pound, were obtained from a sample of 18–24-year-old males.

| | | | | |
|---|---|---|---|---|
| 129.2 | 185.3 | 218.1 | 182.5 | 142.8 |
| 155.2 | 170.0 | 151.3 | 187.5 | 145.6 |
| 167.3 | 161.0 | 178.7 | 165.0 | 172.5 |
| 191.1 | 150.7 | 187.0 | 173.7 | 178.2 |
| 161.7 | 170.1 | 165.8 | 214.6 | 136.7 |
| 278.8 | 175.6 | 188.7 | 132.1 | 158.5 |
| 146.4 | 209.1 | 175.4 | 182.0 | 173.6 |
| 149.9 | 158.6 | | | |

a. Round each observation to the nearest pound and then construct a stem-and-leaf diagram of the rounded data.

b. Truncate each observation by dropping the decimal part and then construct a stem-and-leaf diagram of the truncated data.

c. Compare the stem-and-leaf diagrams that you obtained in parts (a) and (b).

**2.83 Contents of Soft Drinks.** Refer to Exercise 2.62.

a. Round each observation to the nearest 10 mL, drop the terminal 0s, and then obtain a stem-and-leaf diagram of the resulting data.

b. Truncate each observation by dropping the units digit and then construct a stem-and-leaf diagram of the truncated data.

c. Compare the stem-and-leaf diagrams that you obtained in parts (a) and (b) with each other and with the one obtained in Exercise 2.62.

**2.84 NBA Leading Scorers.** The National Basketball Association (NBA) provides information on average points per game for leading scorers on the Web site www.nba.com. A random sample of 24 NBA scoring leaders for the 2004–2005 regular season gave the following data on average points per game (PPG).

| Player | PPG | Player | PPG |
|--------|-----|--------|-----|
| Drew Gooden | 14.4 | Tracy McGrady | 25.7 |
| Kevin Garnett | 22.2 | Michael Redd | 23.0 |
| Paul Pierce | 21.6 | Dirk Nowitzki | 26.1 |
| Chris Bosh | 16.8 | Jalen Rose | 18.5 |
| Joe Johnson | 17.1 | Dwyane Wade | 24.1 |
| Damon Stoudamire | 15.8 | Bobby Simmons | 16.4 |
| Quentin Richardson | 14.9 | Troy Murphy | 15.4 |
| Tayshaun Prince | 14.7 | Jason Richardson | 21.7 |
| Elton Brand | 20.0 | Ray Allen | 23.9 |
| Mike Bibby | 19.6 | Steve Francis | 21.3 |
| Caron Butler | 15.5 | Shaquille O'Neal | 22.9 |
| Kobe Bryant | 27.6 | Desmond Mason | 17.2 |

The following Minitab printout displays a stem-and-leaf diagram for the data on average number of points per game. The second column gives the stems and the third column gives the leaves.

```
Stem-and-Leaf Display: PPG

Stem-and-leaf of PPG  N = 24
Leaf Unit = 1.0

  6    1    444555
 10    1    6677
 12    1    89
 12    2    0111
  8    2    2233
  4    2    45
  2    2    67
```

Did Minitab use rounding or truncation to obtain this stem-and-leaf diagram? Explain your answer.

## 2.4 Distribution Shapes

In this section, we discuss *distributions* and their associated properties.

**Definition 2.5**

### Distribution of a Data Set

The **distribution of a data set** is a table, graph, or formula that provides the values of the observations and how often they occur.

Up to now, we have portrayed distributions of data sets by frequency distributions, relative-frequency distributions, frequency histograms, relative-frequency histograms, dotplots, stem-and-leaf diagrams, pie charts, and bar graphs.

An important aspect of the distribution of a quantitative data set is its shape. Indeed, as we demonstrate in later chapters, the shape of a distribution frequently plays a role in determining the appropriate method of statistical analysis. To identify the shape of a distribution, the best approach usually is to use a smooth curve that approximates the overall shape.

For instance, Fig. 2.8 displays a relative-frequency histogram for the heights of the 3264 female students who attend a midwestern college. Also included in Fig. 2.8 is a smooth curve that approximates the overall shape of the distribution. Both the histogram and the smooth curve show that this distribution of heights is bell shaped (or mound shaped), but the smooth curve makes seeing the shape a little easier.

**FIGURE 2.8**

Relative-frequency histogram and approximating smooth curve for the distribution of heights

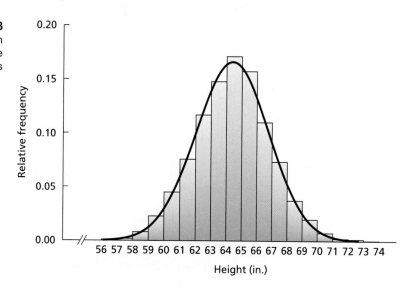

Another advantage of using smooth curves to identify distribution shapes is that we need not worry about minor differences in shape. Instead we can concentrate on overall patterns, which, in turn, allows us to classify most distributions by designating relatively few shapes.

## Distribution Shapes

Figure 2.9 displays some common distribution shapes: **bell shaped, triangular, uniform, reverse J shaped, J shaped, right skewed, left skewed, bimodal,** and

**FIGURE 2.9**

Common distribution shapes

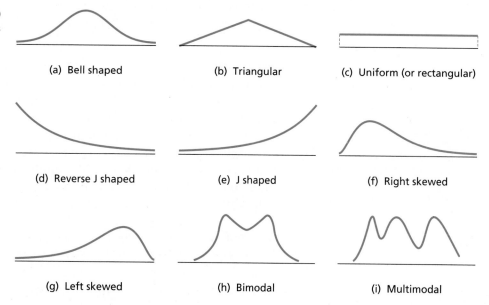

**multimodal.** A distribution doesn't have to have one of these exact shapes in order to take the name: it need only approximate the shape, especially if the data set is small. So, for instance, we describe the distribution of heights in Fig. 2.8 as bell shaped, even though the histogram does not form a perfect bell.

| Example 2.20 | **Identifying Distribution Shapes** |

*Household Size* The relative-frequency histogram for household size in the United States shown in Fig. 2.10(a) is based on data contained in *Current Population Reports*, a publication of the U.S. Census Bureau.[†] Identify the distribution shape for sizes of U.S. households.

**FIGURE 2.10**
Relative-frequency histogram for household size

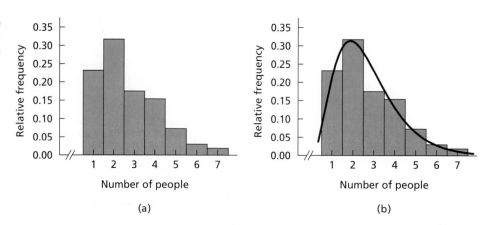

(a)                    (b)

**You try it!**

Exercise 2.93(a) on page 78

**Solution** First, we draw a smooth curve through the histogram shown in Fig. 2.10(a) to get Fig. 2.10(b). Then, by referring to Fig. 2.9, we find that the distribution of household sizes is right skewed.

• • •

Distribution shapes other than those shown in Fig. 2.9 exist, but the types shown in Fig. 2.9 are the most common and are all we need for this book.

### Modality

When considering the shape of a distribution, you should observe its number of peaks (highest points). A distribution is **unimodal** if it has one peak; **bimodal** if it has two peaks; and **multimodal** if it has three or more peaks.

The distribution of heights in Fig. 2.8 is unimodal. More generally, we see from Fig. 2.9 that bell-shaped, triangular, reverse J-shaped, J-shaped, right-skewed, and left-skewed distributions are unimodal. Representations of bimodal and multimodal distributions are displayed in Figs. 2.9(h) and (i), respectively.[‡]

**What Does It Mean?**

Technically, a distribution is bimodal or multimodal only if the peaks are the same height. However, in practice, distributions with pronounced but not necessarily equal-height peaks are often called bimodal or multimodal.

---

[†]Actually, the class 7 portrayed in Fig. 2.10 is for seven or more people.

[‡]A uniform distribution has either no peaks or infinitely many peaks, depending on how you look at it. In any case, we do not classify a uniform distribution according to modality.

## Symmetry and Skewness

Each of the three distributions in Figs. 2.9(a)–(c) can be divided into two pieces that are mirror images of one another. A distribution with that property is called **symmetric.** Therefore bell-shaped, triangular, and uniform distributions are symmetric. The bimodal distribution pictured in Fig. 2.9(h) also happens to be symmetric, but it is not always true that bimodal or multimodal distributions are symmetric. Figure 2.9(i) shows an asymmetric multimodal distribution.

Again, when classifying distributions, we must be flexible. Thus exact symmetry is not required to classify a distribution as symmetric. For example, the distribution of heights in Fig. 2.8 is considered symmetric.

A unimodal distribution that is not symmetric is either right skewed, as in Fig. 2.9(f), or left skewed, as in Fig. 2.9(g). A right-skewed distribution rises to its peak rapidly and comes back toward the horizontal axis more slowly—its "right tail" is longer than its "left tail." A left-skewed distribution rises to its peak slowly and comes back toward the horizontal axis more rapidly—its "left tail" is longer than its "right tail." Note that reverse J-shaped distributions (Fig. 2.9(d)) and J-shaped distributions (Fig. 2.9(e)) are special types of right-skewed and left-skewed distributions, respectively.

## Population and Sample Distributions

Recall that a variable is a characteristic that varies from one person or thing to another and that values of a variable yield data. Distinguishing between data for an entire population and data for a sample of a population is an essential aspect of statistics.

---

**Definition 2.6**    **Population and Sample Data**

**Population data:** The values of a variable for the entire population.
**Sample data:** The values of a variable for a sample of the population.

---

**Note:** Population data is also called **census data.**

To distinguish between the distribution of population data and the distribution of sample data, we use the terminology presented in Definition 2.7.

---

**Definition 2.7**    **Population and Sample Distributions; Distribution of a Variable**

The distribution of population data is called the **population distribution,** or the **distribution of the variable.**
The distribution of sample data is called a **sample distribution.**

---

For a particular population and variable, sample distributions vary from sample to sample. However, there is only one population distribution, namely, the distribution of the variable under consideration on the population under consideration. The following example illustrates this point and some others as well.

### Example 2.21 | Population and Sample Distributions

*Household Size* In Example 2.20, we considered the distribution of household size for U.S. households. Here the variable is household size, and the population consists of all U.S. households. We repeat the graph for that example in Fig. 2.11(a). This graph is a relative-frequency histogram of household size for the population of all U.S. households; it gives the population distribution or, equivalently, the distribution of the variable "household size."

**FIGURE 2.11** Population distribution and six sample distributions for household size

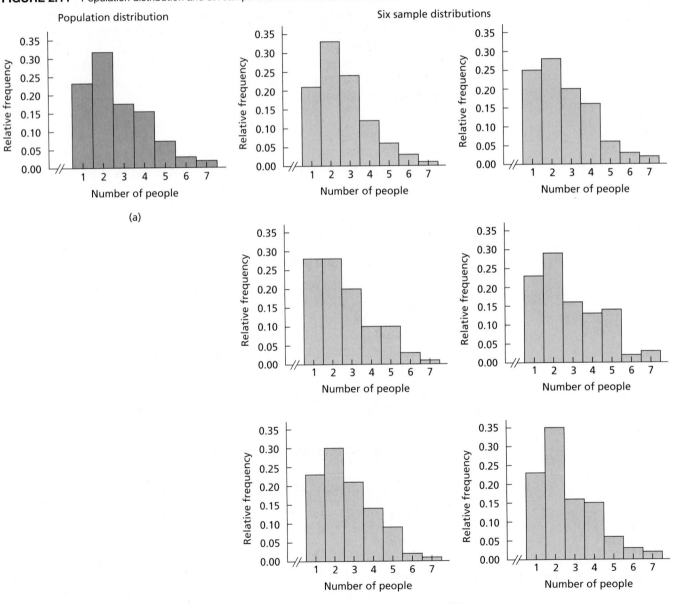

We simulated six simple random samples of 100 households each from the population of all U.S. households. Figure 2.11(b) shows relative-frequency histograms of household size for all six samples. Compare the six sample distributions in Fig. 2.11(b) to each other and to the population distribution in Fig. 2.11(a).

**Solution**  The distributions of the six samples are similar, but have definite differences. This result is not surprising because we would expect variation from one sample to another. Nonetheless, the overall shapes of the six sample distributions are roughly the same and also are similar in shape to the population distribution—all of these distributions are right skewed.

• • •

In practice, we usually do not know the population distribution. As Example 2.21 suggests, however, we can use the distribution of a simple random sample from the population to get a rough idea of the population distribution.

---

**Key Fact 2.1**    **Population and Sample Distributions**

For a simple random sample, the sample distribution approximates the population distribution (i.e., the distribution of the variable under consideration). The larger the sample size, the better the approximation tends to be.

---

## Exercises 2.4

### Understanding the Concepts and Skills

**2.85** Explain the meaning of
a. distribution of a data set.      b. sample data.
c. population data.                 d. census data.
e. sample distribution.             f. population distribution.
g. distribution of a variable.

**2.86** Give two reasons why the use of smooth curves to describe shapes of distributions is helpful.

**2.87** Suppose that a variable of a population has a bell-shaped distribution. If you take a large simple random sample from the population, roughly what shape would you expect the distribution of the sample to be? Explain your answer.

**2.88** Suppose that a variable of a population has a reverse J-shaped distribution and that two simple random samples are taken from the population.
a. Would you expect the distributions of the two samples to have roughly the same shape? If so, what shape?
b. Would you expect some variation in shape for the distributions of the two samples? Explain your answer.

**2.89** Identify and sketch three distribution shapes that are symmetric.

*In each of Exercises 2.90–2.99, we have provided a graphical display of a data set. For each exercise,*
a. *identify the overall shape of the distribution by referring to Fig. 2.9 on page 73.*
b. *state whether the distribution is (roughly) symmetric, right skewed, or left skewed.*

**2.90 Children of U.S. Presidents.** The *Information Please Almanac* provides the number of children of each of the U.S. presidents. A frequency histogram for number of children by president, through President George W. Bush, is as follows.

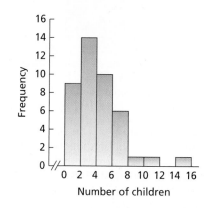

**2.91 Clocking the Cheetah.** The cheetah (*Acinonyx jubatus*) is the fastest land mammal on Earth and is highly specialized to run down prey. According to the Cheetah Conservation of Southern Africa *Trade Environment Database*, the cheetah often exceeds speeds of 60 mph and has been clocked at speeds of more than 70 mph. Following is a frequency histogram for the speeds, in miles per hour, of a sample of 35 cheetahs.

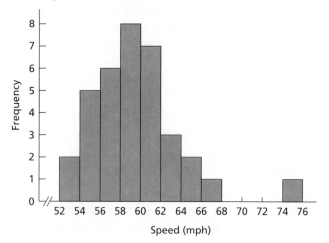

**2.92 Malnutrition and Poverty.** R. Reifen et al. studied various nutritional measures of Ethiopian school children and published their findings in the paper "Ethiopian-Born and Native Israeli School Children Have Different Growth Patterns" (*Nutrition*, Vol. 19, pp. 427–431). The study, conducted in Azezo, North West Ethiopia, found that malnutrition is prevalent in primary and secondary school children because of economic poverty. A frequency histogram for the weights, in kilograms (kg), of 60 randomly selected male Ethiopian-born school children, ages 12–15 years old, is as follows.

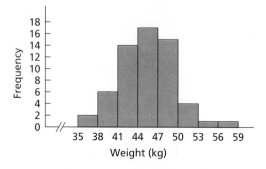

**2.93 The Coruro's Burrow.** The subterranean coruro (*Spalacopus cyanus*) is a social rodent that lives in large colonies in underground burrows that can reach lengths of up to 600 meters. Zoologists S. Begall and M. Gallardo studied the characteristics of the burrow systems of the subterranean coruro in central Chile and published their findings in the *Journal of Zoology, London*, (Vol. 251, pp. 53–60).

A sample of 51 burrows, whose depths were measured in centimeters, yielded the following frequency histogram.

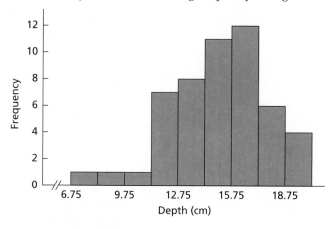

**2.94 San Francisco Giants.** From the *San Francisco Giants Roster* on the ESPN Web site, we obtained the heights of the players on that baseball team as of October 4, 2005. A dotplot of those heights, in inches, is as follows.

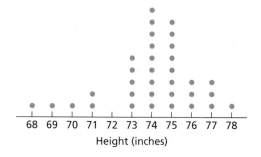

**2.95 PCBs and Pelicans.** Polychlorinated biphenyls (PCBs), industrial pollutants, are known to be a great danger to natural ecosystems. In a study by R. W. Risebrough titled "Effects of Environmental Pollutants Upon Animals Other Than Man" (*Proceedings of the 6th Berkeley Symposium on Mathematics and Statistics, VI*, University of California Press, pp. 443–463), 60 Anacapa pelican eggs were collected and measured for their shell thickness, in millimeters (mm), and concentration of PCBs, in parts per million (ppm). Following is a relative-frequency histogram of the PCB concentration data.

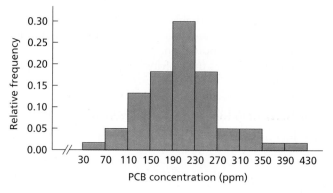

**2.96 Adjusted Gross Incomes.** The Internal Revenue Service (IRS) publishes data on adjusted gross incomes in *Statistics of Income, Individual Income Tax Returns*. The following relative-frequency histogram shows one year's individual income tax returns for adjusted gross incomes of less than $50,000.

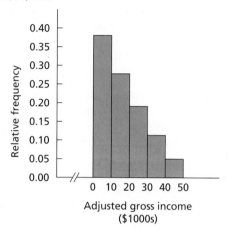

**2.97 Cholesterol Levels.** A pediatrician who tested the cholesterol levels of several young patients was alarmed to find that many had levels higher than 200 mg per 100 mL. The following relative-frequency histogram shows the readings for some patients who had high cholesterol levels.

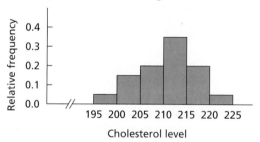

**2.98 Sickle Cell Disease.** A study published by E. Anionwu et al. in the *British Medical Journal* (Vol. 282, pp. 283–286) measured the steady state hemoglobin levels of patients with three different types of sickle cell disease. Following is a stem-and-leaf diagram of the data.

```
 7 | 2 7
 8 | 0 1 1 3 4 4 5 6 7
 9 | 1 1 1 2 8
10 | 0 1 3 4 6 7 9
11 | 1 3 5 6 7 8 9
12 | 0 0 1 1 3 6 6
13 | 3 3 8 9
```

**2.99 Stays in Europe and the Mediterranean.** The Bureau of Economic Analysis gathers information on the length of stay in Europe and the Mediterranean by U.S. travelers. Data are published in *Survey of Current Business*. The following stem-and-leaf diagram portrays the length of stay, in days, of a sample of 36 U.S. residents who traveled to Europe and the Mediterranean last year.

```
0 | 1 1 1 2 3 3 3 5 5 6 8
1 | 0 0 0 1 2 2 2 3 4 5 6 7 8
2 | 0 1 1 1 7
3 | 1 2
4 | 1 4 8
5 | 6
6 | 4
```

**2.100 Airport Passengers.** A report titled *National Transportation Statistics*, sponsored by the Bureau of Transportation Statistics, provides statistics on travel in the United States. During one year, the total number of passengers, in millions, for a sample of 40 airports was as follows.

| | | | | | | | |
|---|---|---|---|---|---|---|---|
| 38.3 | 7.6 | 13.5 | 4.8 | 3.5 | 7.0 | 11.5 | 6.0 |
| 17.2 | 8.7 | 5.1 | 13.3 | 10.0 | 25.1 | 4.3 | 3.6 |
| 15.8 | 16.7 | 7.1 | 4.8 | 15.1 | 16.7 | 9.0 | 6.6 |
| 3.6 | 10.6 | 11.0 | 16.9 | 5.7 | 3.8 | 7.4 | 4.4 |
| 6.2 | 27.8 | 15.2 | 8.7 | 3.9 | 3.5 | 11.4 | 7.0 |

**a.** Construct a frequency histogram for these data. Use classes of equal width 4 and a first midpoint of 2.
**b.** Identify the overall shape of the distribution.
**c.** State whether the distribution is symmetric, right skewed, or left skewed.

**2.101 Snow Goose Nests.** In the article, "Trophic Interaction Cycles in Tundra Ecosystems and the Impact of Climate Change" (*BioScience*, Vol. 55, No. 4, pp. 311–321), R. Ims and E. Fuglei provide an overview of animal species in the northern tundra. One threat to the snow goose in arctic Canada is the lemming. Snowy owls act as protection to the snow goose breeding grounds. For the years 1993 and 1996, the following graphs give relative frequency histograms of the distances, in meters, of snow goose nests to the nearest snowy owl nest.

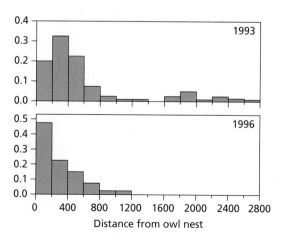

For each histogram,

a. identify the overall shape of the distribution.

b. state whether the distribution is symmetric, right skewed, or left skewed.

c. Compare the two distributions.

## Working With Large Data Sets

*In Exercises 2.102–2.107, you are asked to use the technology of your choice to identify distribution shapes. Note that answers may vary depending on the type of graph that you obtain for the data and on the technology that you use.*

**2.102 The Great White Shark.** In an article titled "Great White, Deep Trouble" (*National Geographic*, Vol. 197(4), pp. 2–29), Peter Benchley—the author of *JAWS*—discussed various aspects of the Great White Shark (*Carcharodon carcharias*). Data on the number of pups borne in a lifetime by each of 80 Great White Shark females are provided on the WeissStats CD.

a. Use the technology of your choice to obtain either a frequency histogram or a relative-frequency histogram for these data. If possible, use classes based on a single value.

b. Identify the overall shape of the distribution.

c. Classify the distribution as symmetric, right skewed, or left skewed.

**2.103 Presidential Election.** In a study summarized in *Chance* (Vol. 17, No. 4, pp. 43–46), H. Wainer provides yet another characterization of the states that supported George Bush versus Al Gore in the 2000 U.S. presidential election. Data on the WeissStats CD include National Assessment of Educational Progress (NAEP) score and the result of the 2000 presidential election for each of the 50 states. The NAEP scores reflect the academic performance for a random sample of children from each state.

a. Use the technology of your choice to obtain either a frequency histogram or a relative-frequency histogram for the NAEP scores.

b. Identify the overall shape of the distribution.

c. Classify the distribution as symmetric, right skewed, or left skewed.

**2.104 Top Recording Artists.** From the Recording Industry Association of America Web site, we obtained data on the number of albums sold, in millions, for the top recording artists (U.S. sales only) as of June 20, 2005. Those data are provided on the WeissStats CD. Use the technology of your choice to do the following.

a. Identify and interpret the overall shape of the distribution of number of albums sold.

b. Classify the distribution as symmetric, right skewed, or left skewed.

**2.105 High School Completion Rates.** As reported by the U.S. Census Bureau in *Current Population Reports*, the percentage of adults in each state and the District of Columbia who have completed high school is as provided on the WeissStats CD. Use the technology of your choice to do the following.

a. Identify and interpret the overall shape of the distribution of high school completion rates.

b. Classify the distribution as symmetric, right skewed, or left skewed.

**2.106 Crime Rates.** The U.S. Federal Bureau of Investigation publishes the annual crime rates for each state and the District of Columbia in the document *Crime in the United States*. Those rates, given per 1000 population, are provided on the WeissStats CD. Use the technology of your choice to do the following.

a. Identify and interpret the overall shape of the distribution of crime rates.

b. Classify the distribution as symmetric, right skewed, or left skewed.

**2.107 Body Temperature.** A study by researchers at the University of Maryland addressed the question of whether the mean body temperature of humans is 98.6°F. The results of the study by P. Mackowiak et al. appeared in the article "A Critical Appraisal of 98.6°F, the Upper Limit of the Normal Body Temperature, and Other Legacies of Carl Reinhold August Wunderlich" (*Journal of the American Medical Association*, Vol. 268, pp. 1578–1580). Among other data, the researchers obtained the body temperatures of 93 healthy humans, as provided on the WeissStats CD. Use the technology of your choice to do the following.

a. Identify and interpret the overall shape of the distribution of body temperatures.

b. Classify the distribution as symmetric, right skewed, or left skewed.

## Extending the Concepts and Skills

**2.108 Class Project: Number of Siblings.** This exercise is a class project and works best in relatively large classes.

a. Determine the number of siblings for each student in the class.

b. Obtain a relative-frequency histogram for the number of siblings. Use single-value grouping.

c. Obtain a simple random sample of about one-third of the students in the class.

d. Determine the number of siblings for each student in the sample.

e. Obtain a relative-frequency histogram for the number of siblings for the sample. Use single-value grouping.

f. Repeat parts (c)–(e) three more times.

g. Compare the histograms for the samples to each other and to that for the entire population. Relate your observations to Key Fact 2.1.

**2.109 Class Project: Random Digits.** This exercise can be done individually or, better yet, as a class project.

a. Use a table of random numbers or a random-number generator to obtain 50 random integers between 0 and 9.

b. Without graphing the distribution of the 50 numbers you obtained, guess its shape. Explain your reasoning.

c. Construct a relative-frequency histogram based on single-value grouping for the 50 numbers that you obtained in part (a). Is its shape about what you expected?

d. If your answer to part (c) was "no," provide an explanation.

e. What would you do to make getting a "yes" answer to part (c) more plausible?

f. If you are doing this exercise as a class project, repeat parts (a)–(c) for 1000 random integers.

**Simulation.** For purposes of both understanding and research, simulating variables is often useful. Simulating a variable involves the use of a computer or statistical calculator to generate observations of the variable. In Exercises 2.110 and 2.111, the use of simulation will enhance your understanding of distribution shapes and the relation between population and sample distributions.

**2.110 Random Digits.** In this exercise, use technology to work Exercise 2.109, as follows:

a. Use the technology of your choice to obtain 50 random integers between 0 and 9.

b. Use the technology of your choice to get a relative-frequency histogram based on single-value grouping for the numbers that you obtained in part (a).

c. Repeat parts (a) and (b) five more times.

d. Are the shapes of the distributions that you obtained in parts (a)–(c) about what you expected?

e. Repeat parts (a)–(d), but generate 1000 random integers each time instead of 50.

**2.111 Standard Normal Distribution.** One of the most important distributions in statistics is the *standard normal distribution.* We discuss this distribution in detail in Chapter 6.

a. Use the technology of your choice to generate a sample of 3000 observations from a variable that has the standard normal distribution, that is, a normal distribution with mean 0 and standard deviation 1.

b. Use the technology of your choice to get a relative-frequency histogram for the 3000 observations that you obtained in part (a).

c. Based on the histogram you obtained in part (b), what shape does the standard normal distribution have? Explain your reasoning.

## 2.5 Misleading Graphs

Graphs and charts are frequently misleading, sometimes intentionally and sometimes inadvertently. Regardless of intent, we need to read and interpret graphs and charts with a great deal of care. In this section, we examine some misleading graphs and charts.

**Example 2.22** | **Truncated Graphs**

*Unemployment Rates* Figure 2.12(a) shows a bar graph from an article in a major metropolitan newspaper. The graph displays the unemployment rates in the United States from September of one year through March of the next year.

**FIGURE 2.12**
Unemployment rates:
(a) truncated graph;
(b) nontruncated graph

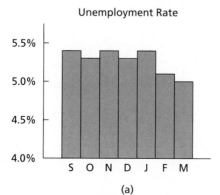

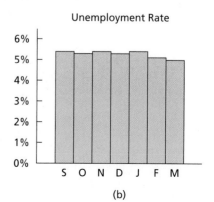

Because the bar for March is about three-fourths as large as the bar for January, a quick look at Fig. 2.12(a) might lead you to conclude that the unemployment rate dropped by roughly one-fourth between January and March. In reality, however, the unemployment rate dropped by less than one-thirteenth, from 5.4% to 5.0%. Let's analyze the graph more carefully to discover what it truly represents.

Figure 2.12(a) is an example of a **truncated graph** because the vertical axis, which should start at 0%, starts at 4% instead. Thus the part of the graph from 0% to 4% has been cut off, or truncated. This truncation causes the bars to be out of proportion and hence creates a misleading impression.

Figure 2.12(b) is a nontruncated version of Fig. 2.12(a). Although the nontruncated version provides a correct graphical display, the "ups" and "downs" in the unemployment rates are not as easy to spot as they are in the truncated graph.

• • •

Truncated graphs have long been a target of statisticians, and many statistics books warn against their use. Nonetheless, as illustrated by Example 2.22, truncated graphs are still used today, even in reputable publications.

However, Example 2.22 also suggests that cutting off part of the vertical axis of a graph may allow relevant information to be conveyed more easily. In such cases, though, the illustrator should include a special symbol, such as //, to signify that the vertical axis has been modified.

The two graphs shown in Fig. 2.13 provide an excellent illustration. Both portray the number of new single-family homes sold per month over several months. The graph in Fig. 2.13(a) is truncated—most likely in an attempt to present a clear visual display of the variation in sales. The graph in Fig. 2.13(b)

**FIGURE 2.13**
New single-family home sales

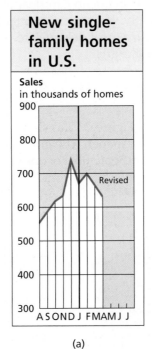

(a)

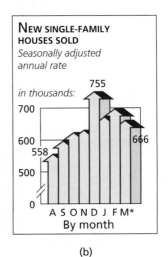

(b)

---

SOURCES: Figure 2.13(a) reprinted by permission of Tribune Media Services. Figure 2.13(b) data from U.S. Department of Commerce and U.S. Department of Housing and Urban Development.

accomplishes the same result but is less subject to misinterpretation; you are aptly warned by the slashes that part of the vertical axis between 0 and 500 has been removed.

### Improper Scaling

Misleading graphs and charts can also result from *improper scaling*.

**Example 2.23 | Improper Scaling**

*Home Building* A developer is preparing a brochure to attract investors for a new shopping center to be built in an area of Denver, Colorado. The area is growing rapidly; this year twice as many homes will be built there as last year. To illustrate that fact, the developer draws a **pictogram** (a symbol representing an object or concept by illustration), as shown in Fig. 2.14.

The house on the left represents the number of homes built last year. Because the number of homes that will be built this year is double the number built last year, the developer makes the house on the right twice as tall and twice as wide as the house on the left. However, this **improper scaling** gives the visual impression that four times as many homes will be built this year as last. Thus the developer's brochure may mislead the unwary investor.

**FIGURE 2.14**
Pictogram for home building

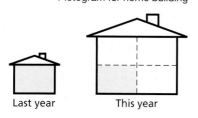

Last year        This year

• • •

Graphs and charts can be misleading in countless ways besides the two that we discussed. Many more examples of misleading graphs can be found in the entertaining and classic book *How to Lie with Statistics* by Darrell Huff (New York: Norton, 1993). The main purpose of this section has been to show you to construct and read graphs and charts carefully.

## Exercises 2.5

### Understanding the Concepts and Skills

**2.112** Give one reason why constructing and reading graphs and charts carefully is important.

**2.113** This exercise deals with truncated graphs.
a. What is a truncated graph?
b. Give a legitimate motive for truncating the axis of a graph.
c. If you have a legitimate motive for truncating the axis of a graph, how can you correctly obtain that objective without creating the possibility of misinterpretation?

**2.114** In a current newspaper or magazine, find two examples of graphs that might be misleading. Explain why you think the graphs are potentially misleading.

**2.115 Reading Skills.** Each year the director of the reading program in a school district administers a standard test of reading skills. Then the director compares the average score for his district with the national average. Figure 2.15 was presented to the school board in the year 2005.

**FIGURE 2.15**
Average reading scores

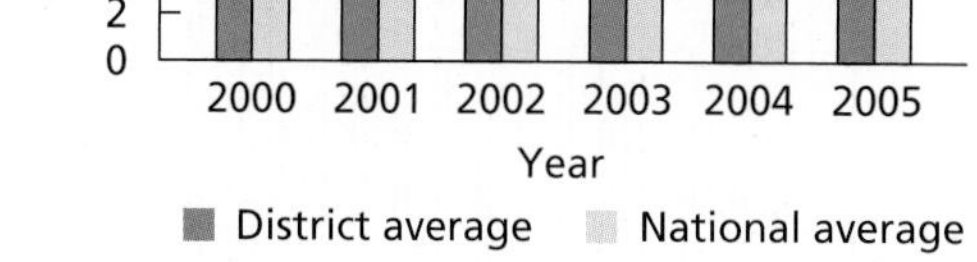

a. Obtain a truncated version of Fig. 2.15 by sliding a piece of paper over the bottom of the graph so that the bars start at 16.
b. Repeat part (a) but have the bars start at 18.

c. What misleading impression about the year 2005 scores is given by the truncated graphs obtained in parts (a) and (b)?

**2.116 America's Melting Pot.** The following bar graph is based on a newspaper article entitled "Immigrants add seasoning to America's melting pot." [Used with permission from American Demographics, Ithaca, NY.]

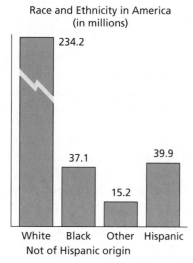

**Race and Ethnicity in America**
(in millions)

234.2 · 37.1 · 15.2 · 39.9

White   Black   Other   Hispanic
Not of Hispanic origin

Data from Census Bureau June 2004 Estimates

a. Explain why a break is shown in the first bar.
b. Why was the graph constructed with a broken bar?
c. Is this graph potentially misleading? Explain your answer.

**2.117 M2 Money Supply.** The following bar graph, taken from *The Arizona Republic*, provides data on the M2 money supply over several months. M2 consists of cash in circulation, deposits in checking accounts, nonbank traveler's checks, accounts such as savings deposits, and money-market mutual funds.

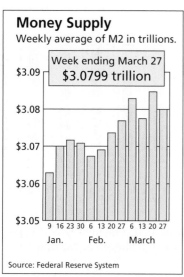

**Money Supply**
Weekly average of M2 in trillions.

Week ending March 27
$3.0799 trillion

$3.09 · $3.08 · $3.07 · $3.06 · $3.05

9  16 23 30  6  13 20 27  6  13 20 27
Jan.       Feb.        March

Source: Federal Reserve System

a. What is wrong with the bar graph?
b. Construct a version of the bar graph with a nontruncated and unmodified vertical axis.
c. Construct a version of the bar graph in which the vertical axis is modified in an acceptable manner.

**2.118 Drunk-Driving Fatalities.** Drunk-driving fatalities represent the total number of people (occupants and non-occupants) killed in motor vehicle traffic crashes in which at least one driver had a blood alcohol content (BAC) of 0.08 or higher. The following graph, titled "Drunk Driving Fatalities Down 38% Despite a 31% Increase in Licensed Drivers," was taken from page 13 of the document *Signs of Progress* on the Web site of The Beer Institute.

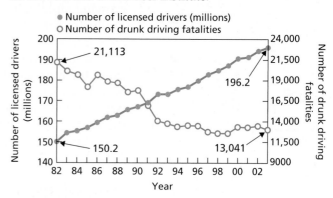

● Number of licensed drivers (millions)
○ Number of drunk driving fatalities

21,113 · 196.2 · 150.2 · 13,041

Number of licensed drivers (millions)
Number of drunk driving fatalities

82  84  86  88  90  92  94  96  98  00  02
Year

a. What features of the graph are potentially misleading?
b. Do you think that it was necessary to incorporate those features in order to display the data?
c. What could be done to more correctly display the data?

**2.119 Oil Prices.** The following graph was obtained from the *Los Angeles Times* on September 26, 2005. It shows the cost of a barrel of crude oil over a period of several weeks.

**Oil prices drop**

Near-term crude oil futures in New York, per barrel, daily closes

$70 · 65 · 60

Sunday: $63.05
Aug. 3: $60.86

August    September

Source: Bloomberg News

*Los Angeles Times*

a. Cover the numbers on the vertical axis of the graph with a piece of paper.

b. What impression does the graph convey regarding the percentage drop in oil prices from the peak to the last day shown on the graph?

c. Now remove the piece of paper from the graph. Use the vertical scale to find the actual percentage drop in oil prices from the peak to the last day shown on the graph.

d. Why is the graph potentially misleading?

e. What can be done to make the graph less potentially misleading?

## Extending the Concepts and Skills

**2.120 Home Building.** Refer to Example 2.23 on page 83. Suggest a way in which the developer can accurately illustrate that twice as many homes will be built in the area this year as last.

**2.121 Marketing Golf Balls.** A golf ball manufacturer has determined that a newly developed process results in a ball that lasts roughly twice as long as a ball produced by the current process. To illustrate this advance graphically, she designs a brochure showing a "new" ball having twice the radius of the "old" ball.

Old ball          New ball

a. What is wrong with this depiction?

b. How can the manufacturer accurately illustrate the fact that the "new" ball lasts twice as long as the "old" ball?

---

## Chapter in Review

### You Should be Able to

1. classify variables and data as either qualitative or quantitative.

2. distinguish between discrete and continuous variables and data.

3. identify terms associated with the grouping of data.

4. group data into a frequency distribution and a relative-frequency distribution.

5. construct a grouped-data table.

6. draw a frequency histogram and a relative-frequency histogram.

7. construct a dotplot.

8. construct a stem-and-leaf diagram.

9. draw a pie chart and a bar graph.

10. identify the shape and modality of the distribution of a data set.

11. specify whether a unimodal distribution is symmetric, right skewed, or left skewed.

12. understand the relationship between sample distributions and the population distribution (distribution of the variable under consideration).

13. identify and correct misleading graphs.

### Key Terms

bar graph, *62*
bell shaped, *73*
bimodal, *73, 74*
categorical variables, *40*
census data, *75*
classes, *47*
continuous data, *41*
continuous variable, *40*
count, *46*
data, *41*
data set, *41*
discrete data, *41*
discrete variable, *40*
distribution of a data set, *72*
distribution of a variable, *75*
dotplot, *58*

exploratory data analysis, *38*
frequency, *47*
frequency distribution, *47*
frequency histogram, *57*
grouped-data table, *47*
grouping, *45*
improper scaling, *83*
J shaped, *73*
leaves, *59*
left skewed, *73*
lower cutpoint, *47*
lower limit, *48*
mark, *48*
midpoint, *47*
multimodal, *74*
observation, *41*

percentage, *46*
pictogram, *83*
pie chart, *61*
population data, *75*
population distribution, *75*
qualitative data, *41*
qualitative variables, *40*
quantitative data, *41*
quantitative variables, *40*
relative frequency, *47*
relative-frequency distribution, *47*
relative-frequency histogram, *57*
reverse J shaped, *73*
right skewed, *73*
sample data, *75*
sample distribution, *75*

## Review Problems

### Understanding the Concepts and Skills

**1.** This problem is about variables and data.
**a.** What is a variable?
**b.** Identify two main types of variables.
**c.** Identify the two types of quantitative variables.
**d.** What are data?
**e.** How is data type determined?

**2.** Explain why grouping data is important.

**3.** To which type of data do the concepts of cutpoints and midpoints not apply? Explain your answer.

**4.** A quantitative data set has been grouped with equal-width classes of width 8.
**a.** If the midpoint of the first class is 10, what are its lower and upper cutpoints?
**b.** What is the midpoint of the second class?
**c.** What are the lower and upper cutpoints of the third class?
**d.** Into which class would an observation of 22 go?

**5.** A quantitative data set has been grouped with equal-width classes.
**a.** If the lower and upper cutpoints of the first class are 5 and 15, respectively, what is the common class width?
**b.** What is the midpoint of the second class?
**c.** What are the lower and upper cutpoints of the third class?

**6.** When is the use of single-value grouping particularly appropriate?

**7.** In each of the following cases, explain the relative positioning of the bars in a histogram to the numbers that label the horizontal axis.
**a.** Cutpoints are used to label the horizontal axis.
**b.** Midpoints are used to label the horizontal axis.

**8.** Identify two main types of graphical displays that are used for qualitative data.

**9.** Which is preferable as a graphical display for a large, quantitative data set: a histogram or a stem-and-leaf diagram? Explain your answer.

**10.** Sketch the curve corresponding to each of the following distribution shapes.
**a.** Bell shaped
**b.** Right skewed
**c.** Reverse J shaped
**d.** Uniform

**11.** Make an educated guess as to the distribution shape of each of the following variables. Explain your answers.
**a.** Height of American adult males
**b.** Annual income of U.S. households
**c.** Age of full-time college students
**d.** Cumulative GPA of college seniors

**12.** A variable of a population has a left-skewed distribution.
**a.** If a large simple random sample is taken from the population, roughly what shape will the distribution of the sample have? Explain your answer.
**b.** If two simple random samples are taken from the population, would you expect the two sample distributions to have identical shapes? Explain your answer.
**c.** If two simple random samples are taken from the population, would you expect the two sample distributions to have similar shapes? If so, what shape would that be? Explain your answers.

**13. Largest Hydroelectric Plants.** The world's five largest hydroelectric plants, based on ultimate capacity, are as shown in the following table. Capacities are in megawatts. [SOURCE: T. W. Mermel, *International Waterpower & Dam Construction Handbook*.]

| Rank | Name | Country | Capacity |
|------|------|---------|----------|
| 1 | Turukhansk | Russia | 20,000 |
| 2 | Three Gorges | China | 18,200 |
| 3 | Itaipu | Brazil/Para. | 13,320 |
| 4 | Grand Coulee | U.S.A. | 10,830 |
| 5 | Guri | Venezuela | 10,300 |

**a.** What type of data is given in the first column of the table?
**b.** What type of data is given in the fourth column?
**c.** What type of data is given in the third column?

**14. Inauguration Ages.** From the *Information Please Almanac*, we obtained the ages at inauguration for the first 43 presidents of the United States (from George Washington to George W. Bush).

| President | Age at inaug. | President | Age at inaug. |
|-----------|------|-----------|------|
| G. Washington | 57 | B. Harrison | 55 |
| J. Adams | 61 | G. Cleveland | 55 |
| T. Jefferson | 57 | W. McKinley | 54 |
| J. Madison | 57 | T. Roosevelt | 42 |
| J. Monroe | 58 | W. Taft | 51 |
| J. Q. Adams | 57 | W. Wilson | 56 |
| A. Jackson | 61 | W. Harding | 55 |
| M. Van Buren | 54 | C. Coolidge | 51 |
| W. Harrison | 68 | H. Hoover | 54 |
| J. Tyler | 51 | F. Roosevelt | 51 |
| J. Polk | 49 | H. Truman | 60 |
| Z. Taylor | 64 | D. Eisenhower | 62 |
| M. Fillmore | 50 | J. Kennedy | 43 |
| F. Pierce | 48 | L. Johnson | 55 |
| J. Buchanan | 65 | R. Nixon | 56 |
| A. Lincoln | 52 | G. Ford | 61 |
| A. Johnson | 56 | J. Carter | 52 |
| U. Grant | 46 | R. Reagan | 69 |
| R. Hayes | 54 | G. Bush | 64 |
| J. Garfield | 49 | W. Clinton | 46 |
| C. Arthur | 50 | G. W. Bush | 54 |
| G. Cleveland | 47 | | |

a. Construct a grouped-data table for the inauguration ages in the preceding table. Use equal-width classes and begin with the class 40–44.
b. Identify the lower and upper cutpoints of the first class. (*Hint:* Be careful!)
c. Identify the common class width.
d. Draw a frequency histogram for the inauguration ages based on your grouping in part (a).
e. Identify the overall shape of the distribution of inauguration ages for the first 43 presidents of the United States.
f. State whether the distribution is (roughly) symmetric, right skewed, or left skewed.

**15. Inauguration Ages.** Refer to Problem 14. Construct a dotplot for the ages at inauguration of the first 43 presidents of the United States.

**16. Inauguration Ages.** Refer to Problem 14. Construct a stem-and-leaf diagram for the inauguration ages of the first 43 presidents of the United States.
a. Use one line per stem.
b. Use two lines per stem.
c. Which of the two stem-and-leaf diagrams that you just constructed corresponds to the frequency distribution of Problem 14(a)?

**17. Busy Bank Tellers.** The Prescott National Bank has six tellers available to serve customers. The data in the following table provide the number of busy tellers observed during 25 spot checks.

| | | | | |
|---|---|---|---|---|
| 6 | 5 | 4 | 1 | 5 |
| 6 | 1 | 5 | 5 | 5 |
| 3 | 5 | 2 | 4 | 3 |
| 4 | 5 | 0 | 6 | 4 |
| 3 | 4 | 2 | 3 | 6 |

a. Construct a grouped-data table for these data. Use single-value grouping.
b. Draw a relative-frequency histogram for the data based on the grouping in part (a).
c. Identify the overall shape of the distribution of these numbers of busy tellers.
d. State whether the distribution is (roughly) symmetric, right skewed, or left skewed.
e. Construct a dotplot for the data on the number of busy tellers.
f. Compare the dotplot that you obtained in part (e) to the relative-frequency histogram that you drew in part (b).

**18. Old Ballplayers.** From the ESPN Web site, we obtained the age of the oldest player on each of the major league baseball teams as of May 2, 2005. Here are the data.

| | | | | | |
|---|---|---|---|---|---|
| 33 | 37 | 36 | 40 | 36 | 36 |
| 40 | 36 | 37 | 36 | 40 | 42 |
| 37 | 42 | 38 | 39 | 35 | 37 |
| 40 | 44 | 39 | 40 | 46 | 38 |
| 37 | 40 | 37 | 42 | 41 | 41 |

a. Construct a dotplot for these data.
b. Use your dotplot from part (a) to identify the overall shape of the distribution of these ages.
c. State whether the distribution is (roughly) symmetric, right skewed, or left skewed.

**19. Handguns Buyback.** In the article "Missing the Target: A Comparison of Buyback and Fatality Related Guns" (*Injury Prevention*, Vol. 8, pp. 143–146), Kuhn et al. examined the relationship between the types of guns that were bought back by the police and the types of guns that were used in homicides in Milwaukee during the year 2002. The following table provides the details.

| Caliber | Buybacks | Homicides |
|---------|----------|-----------|
| Small | 719 | 75 |
| Medium | 182 | 202 |
| Large | 20 | 40 |
| Other | 20 | 52 |

a. Construct a pie chart for the relative frequencies of the types of guns that were bought back by the police in Milwaukee during 2002.
b. Construct a pie chart for the relative frequencies of the types of guns that were used in homicides in Milwaukee during 2002.

**c.** Discuss and compare your pie charts from parts (a) and (b).

**20. Student Class Levels.** The class levels of the students in Professor Weiss's introductory statistics course are shown in the following table. The abbreviations Fr, So, Ju, and Se represent Freshman, Sophomore, Junior, and Senior, respectively.

| Fr | So | Ju | So | Ju | Ju | Se | Ju |
|----|----|----|----|----|----|----|----|
| Se | So | Fr | Ju | So | Ju | So | Se |
| So | So | Se | So | So | Se | So | Fr |
| Ju | So | Ju | Fr | Fr | Ju | Ju | Fr |
| So | Se | Ju | Ju | So | So | So | Se |

**a.** Obtain frequency and relative-frequency distributions for these data.
**b.** Draw a pie chart of the data that displays the percentage of students at each class level.
**c.** Draw a bar graph of the data that displays the relative frequency of students at each class level.

**21. Dow Jones Annual Highs.** According to *The World Almanac*, the highs for the Dow Jones Industrial Averages for 1969–2004 are as follows.

| Year | High | Year | High |
|------|------|------|------|
| 1969 | 968.85 | 1987 | 2722.42 |
| 1970 | 842.00 | 1988 | 2183.50 |
| 1971 | 950.82 | 1989 | 2791.41 |
| 1972 | 1036.27 | 1990 | 2999.75 |
| 1973 | 1051.70 | 1991 | 3168.83 |
| 1974 | 891.66 | 1992 | 3413.21 |
| 1975 | 881.81 | 1993 | 3794.33 |
| 1976 | 1014.79 | 1994 | 3978.36 |
| 1977 | 999.75 | 1995 | 5216.47 |
| 1978 | 907.74 | 1996 | 6560.91 |
| 1979 | 897.61 | 1997 | 8259.31 |
| 1980 | 1000.17 | 1998 | 9547.94 |
| 1981 | 1024.05 | 1999 | 11568.80 |
| 1982 | 1071.55 | 2000 | 11722.98 |
| 1983 | 1287.20 | 2001 | 11337.92 |
| 1984 | 1286.64 | 2002 | 10635.25 |
| 1985 | 1553.10 | 2003 | 10494.44 |
| 1986 | 1955.57 | 2004 | 10895.10 |

**a.** Construct a grouped-data table for the highs. Use classes of equal width and start with the class $0 \leqslant 1000$.
**b.** Draw a relative-frequency histogram for the highs based on your result in part (a).

**22.** Draw a smooth curve that represents a symmetric trimodal (three-peak) distribution.

**23. Clean Fossil Fuels.** In the article, "Squeaky Clean Fossil Fuels" (*New Scientist*, Vol. 186, No. 2497, p. 26), F. Pearce reported on the benefits of using clean fossil fuels that release no carbon dioxide ($CO_2$), helping to reduce the threat of global warming. One technique of slowing down global warming caused by $CO_2$, is to bury the $CO_2$ underground in old oil or gas wells, coal mines, or porous rocks filled with salt water. Global estimates are that 11,000 billion tonnes of $CO_2$ could be disposed of underground, several times more than the likely emissions of $CO_2$ from burning fossil fuels in the coming century. This could give the world extra time to give up its reliance on fossil fuels. The following bar graph shows the distribution of space available to bury $CO_2$ gas underground.

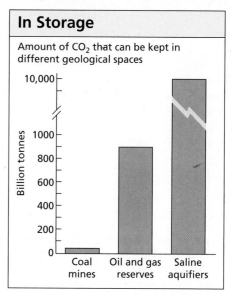

**a.** Explain why the break is found in the third bar.
**b.** Why was the graph constructed with a broken bar?

**24. Reshaping the Labor Force.** The following graph is based on one that appeared in a newspaper article entitled "Hand that rocked cradle turns to work as women reshape U.S. labor force." The graph depicts the labor force participation rates for the years 1960, 1980, and 2000.

Working Men and Women by Age, 1960–2000

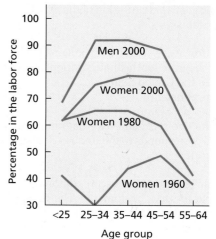

a. Cover the numbers on the vertical axis of the graph with a piece of paper.
b. Look at the 1960 and 2000 graphs for women, focusing on the 35–44-year-old age group. What impression does the graph convey regarding the ratio of the percentages of women in the labor force for 1960 and 2000?
c. Now remove the piece of paper from the graph. Use the vertical scale to find the actual ratio of the percentages of 35–44-year-old women in the labor force for 1960 and 2000.
d. Why is the graph potentially misleading?
e. What can be done to make the graph less potentially misleading?

## Working With Large Data Sets

**25. U.S. Divisions.** The U.S. Census Bureau divides the states in the United States into nine divisions. The data giving the division of each state are presented on the WeissStats CD.
a. Identify the population and variable under consideration.
b. Use the technology of your choice to obtain both a frequency distribution and a relative-frequency distribution of the divisions.
c. Use the technology of your choice to obtain a pie chart for the divisions.
d. Use the technology of your choice to obtain a bar graph for the divisions.
e. Interpret your results.

*In Problems 26–28, do the following:*
*a. Identify the population and variable under consideration.*
*b. Use the technology of your choice to obtain and interpret a frequency histogram or a relative-frequency histogram of the data.*
*c. Use the technology of your choice to obtain a dotplot of the data.*
*d. Use the technology of your choice to obtain a stem-and-leaf diagram of the data.*
*e. Identify the overall shape of the distribution.*
*f. State whether the distribution is (roughly) symmetric, right skewed, or left skewed.*

**26. Agricultural Exports.** The U.S. Department of Agriculture collects data pertaining to the value of agricultural exports and publishes its findings in *U.S. Agricultural Trade Update*. For one year, the values of these exports, by state, are provided on the WeissStats CD. Data are in millions of dollars.

**27. Life Expectancy.** From the U.S. Census Bureau, in the document *International Data Base*, we obtained data on the expectation of life (in years) for people in various countries and areas. Those data are presented on the WeissStats CD.

**28. High and Low Temperatures.** The U.S. National Oceanic and Atmospheric Administration publishes temperature data in *Climatography of the United States*. According to that document, the annual average maximum and minimum temperatures for selected cities in the United States are as provided on the WeissStats CD. [*Note:* Do parts (a)–(f) for both the maximum and minimum temperatures.]

# Focusing on Data Analysis   UWEC Undergraduates

Recall from Chapter 1 (see page 34) that the Focus database and Focus sample contain information on the undergraduate students at the University of Wisconsin - Eau Claire (UWEC). Now would be a good time for you to review the discussion about these data sets.

a. For each of the following variables, make an educated guess at its distribution shape: high school percentile, cumulative GPA, age, ACT English score, ACT math score, and ACT composite score.

b. Open the Focus sample (FocusSample) in the statistical software package of your choice and then obtain and interpret histograms for each of the samples corresponding to the variables in part (a). Compare your results with the educated guesses that you made in part (a).

c. If your statistical software package will accommodate the entire Focus database (Focus), open that worksheet and then obtain and interpret histograms for each of the variables in part (a). Compare your results with the educated guesses that you made in part (a). Also, discuss and explain the relationship between the histograms that you obtained in this part and those that you obtained in part (b).

d. Open the Focus sample and then determine and interpret pie charts and bar graphs of the samples for the variables sex, classification, residency, and admission type.

e. If your statistical software package will accommodate the entire Focus database, open that worksheet and then obtain and interpret pie charts and bar graphs for each of the variables in part (d). Also, discuss and explain the relationship between the pie charts and bar graphs that you obtained in this part and those that you obtained in part (d).

# Case Study Discussion   Preventing Infant Mortality

Recall that the infant mortality rate (IMR) of a nation is the number of deaths of children under 1 year of age per 1000 live births in a calendar year. At the beginning of this chapter, we presented data on 2003 IMRs for the 30 nations with the lowest rates among those with 12 million or more population. Refer to that data table on page 39 and solve each of the following problems.

a. What type of data is displayed in the second column of the table?

b. What type of data is given by the statement that the United States ranks 11th in infant mortality rate?

c. Construct a grouped-data table for the IMRs. Use classes of equal width and start with the class $0 \le 4$.

d. Construct a frequency histogram for the IMRs based on your grouping in part (c).

e. Construct a stem-and-leaf diagram for the IMRs.

f. Obtain a dotplot for the IMRs.

## Biography    ADOLPHE QUETELET: On "The Average Man"

**Lambert Adolphe Jacques Quetelet** was born in Ghent, Belgium, on February 22, 1796. He attended school locally and, in 1819, received the first doctorate of science degree granted at the newly established University of Ghent. In that same year, he obtained a position as a professor of mathematics at the Brussels Athenaeum.

Quetelet was elected to the Belgian Royal Academy in 1820 and served as its secretary from 1834 until his death in 1874. He was founder and director of the Royal Observatory in Brussels, founder and a major contributor to the journal *Correspondance Mathématique et Physique*, and, according to Stephen M. Stigler in *The History of Statistics*, was "...active in the founding of more statistical organizations than any other individual in the nineteenth century." Among the organizations he established was the International Statistical Congress, initiated in 1853.

In 1835, Quetelet wrote a two-volume set titled *A Treatise on Man and the Development of His Faculties*, the publication in which he introduced his concept of the "average man" and that firmly established his international reputation as a statistician and sociologist. A review in the *Athenaeum* stated, "We consider the appearance of these volumes as forming an epoch in the literary history of civilization."

In 1855, Quetelet suffered a stroke that limited his work but not his popularity. He died on February 17, 1874. His funeral was attended by royalty and famous scientists from around the world. A monument to his memory was erected in Brussels in 1880.

## StatCrunch in MyStatLab
### Analyzing Data Online

StatCrunch online statistical software offers an easy-to-use interface customized for this book. The StatCrunch feature for each chapter illustrates the use of the software to perform a statistical analysis discussed in the chapter. Exercises are provided to further apply StatCrunch to other statistical analyses examined in the chapter. Go to the WeissStats CD or to the Weiss Web site at www.aw-bc.com/weiss to access StatCrunch instructions and data sets. To access StatCrunch statistical software, go to the student content area of your Weiss MyStatLab course.

## Internet Projects
### Exploring Data Online

The Internet project for each chapter provides simulations, demonstrations, or activities that enhance the topics covered in the chapter. The project materials come from universities, individuals, governments, and companies from all over the world. To access the Internet projects on the Web, go to www.aw-bc.com/weiss. From this Web page, you can reach the Internet Projects Page, which we suggest that you bookmark for easy access in the future.

# 3

# Descriptive Measures

**Chapter Objectives**

In Chapter 2, you began your study of descriptive statistics. There you learned how to organize data into tables and summarize data with graphs.

Another method of summarizing data is to compute numbers, such as averages and percentiles, that describe the data set. Numbers that are used to describe data sets are called **descriptive measures.** In this chapter, we continue our discussion of descriptive statistics by examining some of the most commonly used descriptive measures.

In Section 3.1, we present *measures of center*—descriptive measures that indicate the center, or most typical value, in a data set. Next, in Section 3.2, we examine *measures of variation*—descriptive measures that indicate the amount of variation or spread in a data set.

The five-number summary, which we discuss in Section 3.3, includes descriptive measures that can be used to obtain both measures of center and measures of variation. That summary also provides the basis for a widely used graphical display, the boxplot.

In Section 3.4, we examine descriptive measures of populations. We also illustrate how sample data can be used to provide estimates of descriptive measures of populations when census data are unavailable.

# The Triple Crown

The Triple Crown is the most prestigious title that a 3-year-old thoroughbred racehorse can win. It is garnered by winning the Kentucky Derby, Preakness Stakes, and Belmont Stakes. These three races are open only to 3-year-olds; thus a horse has only one chance to win the Triple Crown.

*Triple Crown* was coined by sportswriter Charles Hatton in 1930. Sir Barton, in 1919, was the first horse to win all three races. Since then, only 10 horses have won the Triple Crown. Two trainers, James Fitzsimmons and Ben A. Jones, have trained two Triple Crown champions. Eddie Arcaro is the only jockey to ride two Triple Crown winners, on Whirlaway in 1941 and on Citation in 1948.

The Kentucky Derby, run on the first Saturday in May, has always been a $1\frac{1}{4}$ mile race. The Preakness Stakes, run 2 weeks later, is a distance of $1\frac{3}{16}$ miles. Three weeks then elapse until the grueling $1\frac{1}{2}$ miles of the Belmont Stakes.

The following table provides the year, name, and times of the 11 Triple Crown winners. Note that, in 1919, the distances for the Preakness and Belmont were different from those for 1930 and thereafter.

In this chapter, we demonstrate several additional techniques to help you analyze data. At the end of the chapter, you will apply those techniques to analyze the times of the Triple Crown winners.

| Year | Horse | Time (minutes:seconds) | | |
|------|-------|------------------------|---|---|
| | | **Kentucky Derby** | **Preakness Stakes** | **Belmont Stakes** |
| 1919 | Sir Barton | 2:09$\frac{4}{5}$ | 1:53 ($1\frac{1}{8}$ mi.) | 2:17$\frac{2}{5}$ ($1\frac{3}{8}$ mi.) |
| 1930 | Gallant Fox | 2:07$\frac{3}{5}$ | 2:00$\frac{3}{5}$ | 2:30$\frac{3}{5}$ |
| 1935 | Omaha | 2:05 | 1:58$\frac{2}{5}$ | 2:30$\frac{3}{5}$ |
| 1937 | War Admiral | 2:03$\frac{1}{5}$ | 1:58$\frac{2}{5}$ | 2:28$\frac{3}{5}$ |
| 1941 | Whirlaway | 2:01$\frac{2}{5}$ | 1:58$\frac{4}{5}$ | 2:31 |
| 1943 | Count Fleet | 2:04 | 1:57$\frac{2}{5}$ | 2:28$\frac{1}{5}$ |
| 1946 | Assault | 2:06$\frac{3}{5}$ | 2:01$\frac{2}{5}$ | 2:30$\frac{4}{5}$ |
| 1948 | Citation | 2:05$\frac{2}{5}$ | 2:02$\frac{2}{5}$ | 2:28$\frac{1}{5}$ |
| 1973 | Secretariat | 1:59$\frac{2}{5}$ | 1:54$\frac{2}{5}$ | 2:24 |
| 1977 | Seattle Slew | 2:02$\frac{1}{5}$ | 1:54$\frac{2}{5}$ | 2:29$\frac{3}{5}$ |
| 1978 | Affirmed | 2:01$\frac{2}{5}$ | 1:54$\frac{2}{5}$ | 2:26$\frac{4}{5}$ |

## 3.1    Measures of Center

Descriptive measures that indicate where the center or most typical value of a data set lies are called **measures of central tendency** or, more simply, **measures of center.** Measures of center are often called *averages.*

In this section, we discuss the three most important measures of center: the *mean, median,* and *mode.* The mean and median apply only to quantitative data, whereas the mode can be used with either quantitative or qualitative (categorical) data.

### The Mean

The most commonly used measure of center is the *mean.* When people speak of taking an average, they are most often referring to the mean.

Definition 3.1

**Mean of a Data Set**

The **mean** of a data set is the sum of the observations divided by the number of observations.

**What Does It Mean?**

The mean of a data set is its arithmetic average.

Example 3.1

### The Mean

*Weekly Salaries* Professor Hassett spent one summer working for a small mathematical consulting firm. The firm employed a few senior consultants, who made between $800 and $1050 per week; a few junior consultants, who made between $400 and $450 per week; and several clerical workers, who made $300 per week.

**TABLE 3.1**
Data Set I

| | | | | |
|---|---|---|---|---|
| $300 | 300 | 300 | 940 | 300 |
| 300 | 400 | 300 | 400 | |
| 450 | 800 | 450 | 1050 | |

The firm required more employees during the first half of the summer than the second half. Tables 3.1 and 3.2 list typical weekly earnings for the two halves of the summer. Find the mean of each of the two data sets.

**Solution** Data Set I has 13 observations. The sum of those observations is $6290, so

$$\text{Mean of Data Set I} = \frac{\$6290}{13} = \$483.85 \text{ (rounded to the nearest cent).}$$

**TABLE 3.2**
Data Set II

| | | | | |
|---|---|---|---|---|
| $300 | 300 | 940 | 450 | 400 |
| 400 | 300 | 300 | 1050 | 300 |

Similarly,

$$\text{Mean of Data Set II} = \frac{\$4740}{10} = \$474.00.$$

**You try it!**

Exercise 3.9(a) on page 102

**Interpretation** The employees who worked in the first half of the summer earned more, on average (a mean salary of $483.85), than those who worked in the second half (a mean salary of $474.00).

• • •

## The Median

Another frequently used measure of center is the median. Essentially, the *median* of a data set is the number that divides the bottom 50% of the data from the top 50%. A more precise definition of the median follows.

**Definition 3.2**

### Median of a Data Set

Arrange the data in increasing order.

- If the number of observations is odd, then the **median** is the observation exactly in the middle of the ordered list.
- If the number of observations is even, then the **median** is the mean of the two middle observations in the ordered list.

In both cases, if we let $n$ denote the number of observations, then the median is at position $(n + 1)/2$ in the ordered list.

**What Does It Mean?**

The median of a data set is the middle value in its ordered list.

**Example 3.2** | **The Median**

*Weekly Salaries* Consider again the two sets of salary data shown in Tables 3.1 and 3.2. Determine the median of each of the two data sets.

**Solution** To find the median of Data Set I, we first arrange the data in increasing order:

300    300    300    300    300    300    **400**    400    450    450    800    940    1050

The number of observations is 13, so $(n + 1)/2 = (13 + 1)/2 = 7$. Consequently, the median is the seventh observation in the ordered list, which is 400 (shown in boldface).

To find the median of Data Set II, we first arrange the data in increasing order:

300    300    300    300    **300**    **400**    400    450    940    1050

The number of observations is 10, so $(n + 1)/2 = (10 + 1)/2 = 5.5$. Consequently, the median is halfway between the fifth and sixth observations (shown in boldface) in the ordered list, which is 350.

**Interpretation** Again, the analysis shows that the employees who worked in the first half of the summer tended to earn more (a median salary of $400) than those who worked in the second half (a median salary of $350).

**You try it!**

Exercise 3.9(b)
on page 102

• • •

To determine the median of a data set, you must first arrange the data in increasing order. Constructing a stem-and-leaf diagram as a preliminary step to ordering the data is often helpful.

### The Mode

The final measure of center that we discuss here is the *mode.*

---

**Definition 3.3**

**What Does It Mean?**

The mode of a data set is its most frequently occurring value.

**Mode of a Data Set**

Find the frequency of each value in the data set.

• If no value occurs more than once, then the data set has *no mode.*

• Otherwise, any value that occurs with the greatest frequency is a **mode** of the data set.

---

**Example 3.3**

**TABLE 3.3**
Frequency distribution for Data Set I

| Salary | Frequency |
|--------|-----------|
| 300 | 6 |
| 400 | 2 |
| 450 | 2 |
| 800 | 1 |
| 940 | 1 |
| 1050 | 1 |

Exercise 3.9(c)
on page 102

### The Mode

*Weekly Salaries*  Determine the mode(s) of each of the two sets of salary data given in Tables 3.1 and 3.2 on page 94.

**Solution**  Referring to Table 3.1, we obtain the frequency of each value in Data Set I, as shown in Table 3.3. From Table 3.3, we see that the greatest frequency is 6, and that 300 is the only value that occurs with that frequency. So the mode is $300.

Proceeding in the same way, we find that, for Data Set II, the greatest frequency is 5 and that 300 is the only value that occurs with that frequency. So the mode is $300.

**Interpretation**  The most frequent salary was $300 both for the employees who worked in the first half of the summer and those who worked in the second half.

• • •

A data set will have more than one mode if more than one of its values occurs with the greatest frequency. For instance, suppose the first two $300-per-week employees who worked in the first half of the summer were promoted to $400-per-week jobs. Then the weekly earnings for the 13 employees would be as follows.

| | | | | |
|------|-----|-----|------|-----|
| $400 | 400 | 300 | 940 | 300 |
| 300 | 400 | 300 | 400 | |
| 450 | 800 | 450 | 1050 | |

Now, both the value 300 and the value 400 would occur with greatest frequency, 4. This new data set would thus have two modes, 300 and 400.

### Comparison of the Mean, Median, and Mode

The mean, median, and mode of a data set are often different. Table 3.4 summarizes the definitions of these three measures of center and gives their values for Data Set I and Data Set II, which we computed in Examples 3.1–3.3.

In both Data Sets I and II, the mean is larger than the median. The reason is that the mean is strongly affected by the few large salaries in each data set. In general, the mean is sensitive to extreme (very large or very small) observations,

**TABLE 3.4**
Means, medians, and modes of salaries
in Data Set I and Data Set II

| Measure of center | Definition | Data Set I | Data Set II |
|---|---|---|---|
| Mean | $\dfrac{\text{Sum of observations}}{\text{Number of observations}}$ | $483.85 | $474.00 |
| Median | Middle value in ordered list | $400.00 | $350.00 |
| Mode | Most frequent value | $300.00 | $300.00 |

whereas the median is not. Consequently, when the choice for the measure of center is between the mean and the median, the median is usually preferred for data sets that have extreme observations.

Figure 3.1 shows the relative positions of the mean and median for right-skewed, symmetric, and left-skewed distributions. Note that the mean is pulled in the direction of skewness, that is, in the direction of the extreme observations. For a right-skewed distribution, the mean is greater than the median; for a symmetric distribution, the mean and the median are equal; and, for a left-skewed distribution, the mean is less than the median.

**FIGURE 3.1**
Relative positions of the mean
and median for (a) right-skewed,
(b) symmetric, and (c) left-skewed
distributions

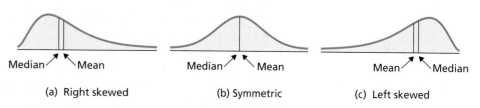

A **resistant measure** is not sensitive to the influence of a few extreme observations. The median is a resistant measure of center, but the mean is not. A *trimmed mean* can improve the resistance of the mean: removing a percentage of the smallest and largest observations before computing the mean gives a **trimmed mean.** In Exercise 3.42, we discuss trimmed means in more detail.

The mode for each of Data Sets I and II differs from both the mean and the median. Whereas the mean and the median are aimed at finding the center of a data set, the mode is really not—the value that occurs most frequently may not be near the center.

It should now be clear that the mean, median, and mode generally provide different information. There is no simple rule for deciding which measure of center to use in a given situation. Even experts may disagree about the most suitable measure of center for a particular data set.

**Example 3.4** | **Selecting an Appropriate Measure of Center**

a. A student takes four exams in a biology class. His grades are 88, 75, 95, and 100. Which measure of center is the student likely to report?

b. The National Association of REALTORS publishes data on resale prices of U.S. homes. Which measure of center is most appropriate for such resale prices?

c. The 2005 Boston Marathon had two categories of official finishers: male and female, of which there were 10,894 and 6,655, respectively. Which measure of center should be used here?

### Solution

a. Chances are that the student would report the mean of his scores, which is 89.5. The mean is probably the most suitable measure of center for the student to use because it takes into account the numerical value of each score and therefore indicates his overall performance.

b. The most appropriate measure of center for resale home prices is the median because it is aimed at finding the center of the data on resale home prices and because it is not strongly affected by the relatively few homes with extremely high resale prices. Thus the median provides a better indication of the "typical" resale price than either the mean or the mode.

c. The only suitable measure of center for these data is the mode, which is "male." Each observation in this data set is either "male" or "female." There is no way to compute a mean or median for such data. *Of the mean, median, and mode, the mode is the only measure of center that can be used for qualitative data.*

Exercise 3.17
on page 103

•  •  •

Many measures of center that appear in newspapers or that are reported by government agencies are medians, as is the case for household income and number of years of school completed. In an attempt to provide a clearer picture, some reports include both the mean and the median. For instance, the National Center for Health Statistics does so for daily intake of nutrients in the publication *Vital and Health Statistics*.

### Population Mean and Sample Mean

Recall that a variable is a characteristic that varies from one person or thing to another and that values of a variable yield data. The values of a variable for an entire population are called *population data;* the values of a variable for a sample of the population are called *sample data.*

The mean of population data is called the **population mean** or the **mean of the variable;** the mean of sample data is called a **sample mean.** The same terminology is used for the median and mode and, for that matter, any descriptive measure. Figure 3.2 shows the two ways in which the mean of a data set can be interpreted.

**FIGURE 3.2**
Possible interpretations
for the mean of a data set

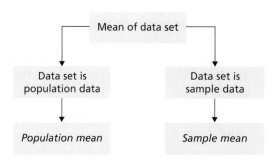

Note the following for a particular variable on a particular population:

- There is only one population mean—namely, the mean of all possible observations of the variable for the entire population.
- There are many sample means—one for each possible sample of the population.

In the remainder of this section and in Sections 3.2 and 3.3, we concentrate on descriptive measures of samples. In Section 3.4, we discuss descriptive measures of populations and their relationship to descriptive measures of samples.

## Summation Notation

In statistics, as in algebra, letters such as $x$, $y$, and $z$ are used to denote variables. So, for instance, in a study of heights and weights of college students, we might let $x$ denote the variable "height" and $y$ denote the variable "weight."

We can often use notation for variables, along with other mathematical notations, to express statistics definitions and formulas concisely. Of particular importance, in this regard, is *summation notation*.

| Example 3.5 | **Introducing Summation Notation** |
|---|---|

*Exam Scores* The exam scores for the student in Example 3.4(a) are 88, 75, 95, and 100.

**a.** Use mathematical notation to represent the individual exam scores.

**b.** Use summation notation to express the sum of the four exam scores.

**Solution** Let $x$ denote the variable "exam score."

**a.** We use the symbol $x_i$ (read as "$x$ sub $i$") to represent the $i$th observation of the variable $x$. Thus, for the exam scores,

$$x_1 = \text{score on Exam } 1 = 88;$$

$$x_2 = \text{score on Exam } 2 = 75;$$

$$x_3 = \text{score on Exam } 3 = 95;$$

$$x_4 = \text{score on Exam } 4 = 100.$$

More simply, we can just write $x_1 = 88$, $x_2 = 75$, $x_3 = 95$, and $x_4 = 100$. The numbers 1, 2, 3, and 4 written below the $x$s are called **subscripts.** Subscripts do not necessarily indicate order but, rather, provide a way of keeping the observations distinct.

**b.** We can use the notation in part (a) to write the sum of the exam scores as

$$x_1 + x_2 + x_3 + x_4.$$

Summation notation, which uses the uppercase Greek letter $\Sigma$ (sigma), provides a shorthand description for that sum. The letter $\Sigma$ corresponds to the uppercase English letter S and is used here as an abbreviation for the phrase "the sum of." So, in place of $x_1 + x_2 + x_3 + x_4$, we can use **summation notation,** $\Sigma x_i$, read as "summation $x$ sub $i$" or "the sum of the observations of the variable $x$." For the exam-score data,

$$\Sigma x_i = x_1 + x_2 + x_3 + x_4 = 88 + 75 + 95 + 100 = 358.$$

**Interpretation**  The sum of the student's four exam scores is 358 points.

• • •

Note the following about summation notation:

- When no confusion can arise, we sometimes write $\Sigma x_i$ even more simply as $\Sigma x$.
- For clarity, we sometimes use **indices** to write $\Sigma x_i$ as $\sum\limits_{i=1}^{n} x_i$, which is read as "summation $x$ sub $i$ from $i$ equals 1 to $n$," where $n$ stands for the number of observations.

### Notation for a Sample Mean

The symbol used for a sample mean is a bar over the letter representing the variable. So, for a variable $x$, we denote a sample mean as $\bar{x}$, read as "$x$ bar." If we also use the letter $n$ to denote the **sample size** or, equivalently, the number of observations, we can express the definition of a sample mean concisely.

---

**Definition 3.4**

| **Sample Mean** |

> **?**
> **What Does It Mean?**
>
> A sample mean is the arithmetic average (mean) of sample data.

For a variable $x$, the mean of the observations for a sample is called a **sample mean** and is denoted $\bar{x}$.  Symbolically,

$$\bar{x} = \frac{\Sigma x_i}{n}$$

where $n$ is the sample size.

---

**Example 3.6** | **The Sample Mean**

*Children of Diabetic Mothers*  The paper "Correlations Between the Intrauterine Metabolic Environment and Blood Pressure in Adolescent Offspring of Diabetic Mothers" (*The Journal of Pediatrics*, Vol. 136, Issue 5, pp. 587–592) by Cho et al. presents findings of research on children of diabetic mothers. Past studies have shown that maternal diabetes results in obesity, blood pressure, and glucose tolerance complications in the offspring.

**TABLE 3.5**
Arterial blood pressures
of 16 children of diabetic mothers

| | | | |
|---|---|---|---|
| 81.6 | 84.1 | 87.6 | 82.8 |
| 82.0 | 88.9 | 86.7 | 96.4 |
| 84.6 | 104.9 | 90.8 | 94.0 |
| 69.4 | 78.9 | 75.2 | 91.0 |

Table 3.5 presents the arterial blood pressures, in millimeters of mercury (mm Hg), for a sample of 16 children of diabetic mothers. Determine the sample mean of these arterial blood pressures.

**Solution**  Let $x$ denote the variable "arterial blood pressure." We want to find the mean, $\bar{x}$, of the 16 observations of $x$ shown in Table 3.5. The sum of those observations is $\Sigma x_i = 1378.9$.  The sample size (or number of observations) is 16, so $n = 16$. Thus,

$$\bar{x} = \frac{\Sigma x_i}{n} = \frac{1378.9}{16} = 86.18.$$

You try it!

Exercise 3.25
on page 104

**Interpretation** The mean arterial blood pressure of the sample of 16 children of diabetic mothers is 86.18 mm Hg.

• • •

## The Technology Center

All statistical technologies have programs that automatically compute descriptive measures. In this subsection, we present output and step-by-step instructions for such programs.

**Example 3.7**   **Using Technology to Obtain Descriptive Measures**

*Weekly Salaries*   Use Minitab, Excel, or the TI-83/84 Plus to find the mean and median of the salary data for Data Set I, displayed in Table 3.1 on page 94.

**Solution**   We applied the descriptive-measures programs to the data, resulting in Output 3.1. Steps for generating that output are presented in Instructions 3.1.

**OUTPUT 3.1**   Descriptive measures for Data Set I

MINITAB

### Descriptive Statistics: SALARY

| Variable | N | N* | Mean | SE Mean | StDev | Minimum | Q1 | Median | Q3 | Maximum |
|----------|---|----|------|---------|-------|---------|-----|--------|-----|---------|
| SALARY | 13 | 0 | 483.8 | 73.7 | 265.8 | 300.0 | 300.0 | 400.0 | 625.0 | 1050.0 |

EXCEL

TI-83/84 PLUS

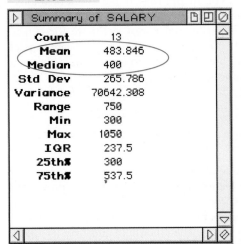

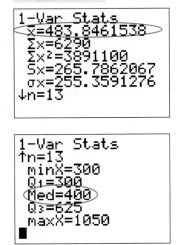

As shown in Output 3.1, the mean and median of the salary data for Data Set I are 483.8 (to one decimal place) and 400, respectively.

• • •

**INSTRUCTIONS 3.1** Steps for generating Output 3.1

| **MINITAB** | **EXCEL** | **TI-83/84 PLUS** |
|---|---|---|
| 1 Store the data from Table 3.1 in a column named SALARY<br>2 Choose **Stat ➤ Basic Statistics ➤ Display Descriptive Statistics...**<br>3 Specify SALARY in the **Variables** text box<br>4 Click **OK** | 1 Store the data from Table 3.1 in a range named SALARY<br>2 Choose **DDXL ➤ Summaries**<br>3 Select **Summary of One Variable** from the **Function type** drop-down box<br>4 Specify SALARY in the **Quantitative Variable** text box<br>5 Click **OK** | 1 Store the data from Table 3.1 in a list named SAL<br>2 Press **STAT**<br>3 Arrow over to **CALC**<br>4 Press **1**<br>5 Press **2nd ➤ LIST**<br>6 Arrow down to SAL and press **ENTER** twice |

▪ ▪ ▪

# Exercises 3.1

## Understanding the Concepts and Skills

**3.1** Explain in detail the purpose of a measure of center.

**3.2** Name and describe the three most important measures of center.

**3.3** Of the mean, median, and mode, which is the only one appropriate for use with qualitative data?

**3.4** True or false: The mean, median, and mode can all be used with quantitative data.

**3.5** Consider the data set 1, 2, 3, 4, 5, 6, 7, 8, 9.
**a.** Obtain the mean and median of the data.
**b.** Replace the 9 in the data set by 99 and again compute the mean and median. Decide which measure of center works better here and explain your answer.
**c.** For the data set in part (b), the mean is neither central nor typical for the data. The lack of what property of the mean accounts for this result?

**3.6** Complete the following statement: A descriptive measure is *resistant* if ....

**3.7** **Floor Space.** The U.S. Department of Housing and Urban Development and the U.S. Census Bureau compile information on new, privately owned single-family houses. According to the document *Characteristics of New Housing*, in 2003 the mean floor space of such homes was 2330 sq ft and the median was 2137 sq ft. Which measure of center do you think is more appropriate? Justify your answer.

**3.8** **Net Worth.** The Board of Governors of the Federal Reserve System publishes information on family net worth in the *Federal Reserve Bulletin*. In 2001, the mean net worth of families in the United States was $395.5 thousand and the median net worth was $86.1 thousand. Which measure of center do you think is more appropriate? Explain your answer.

*In Exercises 3.9–3.16, find the*
*a. mean.*   *b. median.*   *c. mode(s).*
*For the mean and the median, round each answer to one more decimal place than that used for the observations.*

**3.9** **Amphibian Embryos.** In a study of the effects of radiation on amphibian embryos titled "Shedding Light on Ultraviolet Radiation and Amphibian Embryos" (*BioScience*, Vol. 53, No. 6, pp. 551–561), L. Licht recorded the time it took for a sample of seven different species of frogs' and toads' eggs to hatch. The following table shows the times to hatch, in days.

| 6 | 7 | 11 | 6 | 5 | 5 | 11 |
|---|---|---|---|---|---|---|

**3.10** **Hurricanes.** An article by D. Schaefer et al. (*Journal of Tropical Ecology*, Vol. 16, pp. 189–207) reported on a long-term study of the effects of hurricanes on tropical streams of the Luquillo Experimental Forest in Puerto Rico. The study shows that Hurricane Hugo had a significant impact on stream water chemistry. The following table shows a sample of 10 ammonia fluxes in the first year after Hugo. Data are in kilograms per hectare per year.

| 96 | 66 | 147 | 147 | 175 |
|---|---|---|---|---|
| 116 | 57 | 154 | 88 | 154 |

**3.11** **Tornado Touchdowns.** Each year, tornadoes that touch down are recorded by the Storm Prediction Center and published in *Monthly Tornado Statistics*. The following table gives the number of tornadoes that touched down in the United States during each month of one year. [SOURCE: National Oceanic and Atmospheric Administration.]

| 3 | 2 | 47 | 118 | 204 | 97 |
|---|---|---|---|---|---|
| 68 | 86 | 62 | 57 | 98 | 99 |

**3.12** **Technical Merit.** In the 2002 Winter Olympics, Michelle Kwan competed in the Short Program ladies singles event. From nine judges, she received scores ranging from 1 (poor) to 6 (perfect). The following table provides the scores that the judges gave her on technical merit, found in an article by S. Berry (*Chance*, Vol. 15, No. 2, pp. 14–18).

| | | | | | | | | |
|---|---|---|---|---|---|---|---|---|
| 5.8 | 5.7 | 5.9 | 5.7 | 5.5 | 5.7 | 5.7 | 5.7 | 5.6 |

**3.13 Billionaires' Club.** Each year, *Forbes* magazine compiles a list of the 400 richest Americans. For 2005, the top five on the list are as shown in the following table.

| Person | Wealth ($ billions) |
|---|---|
| Bill Gates | 51 |
| Warren Buffett | 40 |
| Paul Allen | 22.5 |
| Michael Dell | 18 |
| Larry Ellison | 17 |

**3.14 AML and the Cost of Labor.** Active Management of Labor (AML) was introduced in the 1960s to reduce the amount of time a woman spends in labor during the birth process. R. Rogers et al. conducted a study to determine whether AML also translates into a reduction in delivery cost to the patient. They reported their findings in the paper "Active Management of Labor: A Cost Analysis of a Randomized Controlled Trial" (*Western Journal of Medicine*, Vol. 172, pp. 240–243). The following table displays the costs, in dollars, of eight randomly sampled AML deliveries.

| | | | |
|---|---|---|---|
| 3141 | 2873 | 2116 | 1684 |
| 3470 | 1799 | 2539 | 3093 |

**3.15 Fuel Economy.** Every year, *Consumer Reports* publishes a magazine titled *New Car Ratings and Review* that looks at vehicle profiles for the year's models. It lets you see in one place how, within each category, the vehicles compare. One category of interest, especially when fuel prices are rising, is fuel economy, measured in miles per gallon (mpg). Following is a list of overall mpg for 14 different full-sized and compact pickups.

| | | | | | | |
|---|---|---|---|---|---|---|
| 14 | 13 | 14 | 13 | 14 | 14 | 11 |
| 12 | 15 | 15 | 17 | 14 | 15 | 16 |

**3.16 Router Horsepower.** In the article "Router Roundup" (*Popular Mechanics*, Vol. 180, No. 12, pp. 104–109), T. Klenck reports on tests of seven fixed-base routers for performance, features, and handling. The following table gives the horsepower (hp) for each of the seven routers tested.

| | | | | | | |
|---|---|---|---|---|---|---|
| 1.75 | 2.25 | 2.25 | 2.25 | 1.75 | 2.00 | 1.50 |

**3.17 Medieval Cremation Burials.** In the article "Material Culture as Memory: Combs and Cremations in Early Medieval Britain" (*Early Medieval Europe*, Vol. 12, Issue 2, pp. 89–128), H. Williams discussed the frequency of cremation burials found in 17 archaeological sites in eastern England. Here are the data.

| | | | | | | | | |
|---|---|---|---|---|---|---|---|---|
| 83 | 64 | 46 | 48 | 523 | 35 | 34 | 265 | 2484 |
| 46 | 385 | 21 | 86 | 429 | 51 | 258 | 119 | |

a. Obtain the mean, median, and mode of these data.
b. Which measure of center do you think works best here? Explain your answer.

**3.18 Monthly Motorcycle Casualties.** The *Scottish Executive*, Analytical Services Division Transport Statistics, compiles data on motorcycle casualties. During one year, monthly casualties from motorcycle accidents in Scotland for built-up roads and non-built-up roads were as follows.

| Month | Built up | Non built up |
|---|---|---|
| January | 25 | 16 |
| February | 38 | 9 |
| March | 38 | 26 |
| April | 56 | 48 |
| May | 61 | 73 |
| June | 52 | 72 |
| July | 50 | 91 |
| August | 90 | 69 |
| September | 67 | 71 |
| October | 51 | 28 |
| November | 64 | 19 |
| December | 40 | 12 |

a. Find the mean, median, and mode of the number of motorcycle casualties for built-up roads.
b. Find the mean, median, and mode of the number of motorcycle casualties for non-built-up roads.
c. If you had a list of only the month of each casualty, what month would be the modal month for each type of road?

**3.19 Daily Motorcycle Accidents.** The *Scottish Executive*, Analytical Services Division Transport Statistics, compiles data on motorcycle accidents. During one year, the numbers of motorcycle accidents in Scotland were tabulated by day of the week for built-up roads and non-built-up roads and resulted in the following data.

| Day | Built up | Non built up |
|---|---|---|
| Monday | 88 | 70 |
| Tuesday | 100 | 58 |
| Wednesday | 76 | 59 |
| Thursday | 98 | 53 |
| Friday | 103 | 56 |
| Saturday | 85 | 94 |
| Sunday | 69 | 102 |

**a.** Find the mean and median of the number of accidents for built-up roads.

**b.** Find the mean and median of the number of accidents for non-built-up roads.

**c.** If you had a list of only the day of the week for each accident, what day would be the modal day for each type of road?

**d.** What might explain the difference in the modal days for the two types of roads?

**3.20** Explain what each symbol represents.
**a.** $\Sigma$          **b.** $n$          **c.** $\bar{x}$

**3.21** For a particular population, is the population mean a variable? What about a sample mean?

**3.22** Consider these sample data: $x_1 = 1$, $x_2 = 7$, $x_3 = 4$, $x_4 = 5$, $x_5 = 10$.
**a.** Find $n$.          **b.** Compute $\Sigma x_i$.          **c.** Determine $\bar{x}$.

**3.23** Consider these sample data: $x_1 = 12$, $x_2 = 8$, $x_3 = 9$, $x_4 = 17$.
**a.** Find $n$.          **b.** Compute $\Sigma x_i$.          **c.** Determine $\bar{x}$.

*In each of Exercises 3.24–3.27, do the following:*
*a. Find n.*
*b. Compute $\Sigma x_i$.*
*c. Determine the sample mean. Round your answer to one more decimal place than that used for the observations.*

**3.24 Honeymoons.** Popular destinations for the newly-weds of today are the Caribbean and Hawaii. According to *Bride's Magazine*, a honeymoon, on average, lasts 9 days and costs $3657. A sample of 12 newlyweds reported the following lengths of stay of their honeymoons.

| | | | | | |
|---|---|---|---|---|---|
| 5 | 14 | 7 | 10 | 6 | 8 |
| 12 | 9 | 10 | 9 | 7 | 11 |

**3.25 Sleep.** In 1908, W. S. Gosset published the article "The Probable Error of a Mean" (*Biometrika*, Vol. 6, pp. 1–25). It is in this pioneering paper, written under the pseudonym "Student," that Gosset introduced what later became known as Student's *t*-distribution, which we discuss in a later chapter. Gosset used the following data set, which shows the additional sleep in hours obtained by a sample of 10 patients given laevohysocyamine hydrobromide.

| | | | | |
|---|---|---|---|---|
| 1.9 | 0.8 | 1.1 | 0.1 | −0.1 |
| 4.4 | 5.5 | 1.6 | 4.6 | 3.4 |

**3.26 Pesticides in Pakistan.** Pesticides are chemicals often used in agriculture to control pests. In Pakistan, 70% of the population depends on agriculture, and pesticide use there has increased rapidly. In the article, "Monitoring Pesticide Residues in Fresh Fruits Marketed in Peshawar, Pakistan"

(*American Laboratory*, Vol. 37, No. 7, pp. 22–24), J. Shah, et al. sampled the most commonly used fruit in Pakistan and analyzed the pesticide residues in the fruit. The amounts, in mg/kg, of the pesticide Dichlorovos for a sample of apples, guavas, and mangos were as follows.

| | | | | | |
|---|---|---|---|---|---|
| 0.2 | 1.6 | 4.0 | 5.4 | 5.7 | 11.4 |
| 0.2 | 3.4 | 2.4 | 6.6 | 4.2 | 2.7 |

**3.27 Hazardous Wastes.** The Environmental Protection Agency (EPA) collects information on generation, management, and final disposal of hazardous wastes. In a study issued by the EPA, titled *National Biennial RCRA Hazardous Waste Report*, data is filed with states by waste generating, treatment, storage, and disposal facilities. A sample of hazardous waste shipping companies in Tennessee yielded the following amounts, in tons, of waste shipped during 2001.

| | | | | | | |
|---|---|---|---|---|---|---|
| 3202 | 226 | 207 | 203 | 3718 | 362 | 772 |
| 1252 | 2093 | 3388 | 358 | 187 | 281 | 209 |

*In each of Exercises 3.28–3.31, do the following:*
*a. Determine the mode of the data.*
*b. Decide whether it would be appropriate to use either the mean or the median as a measure of center. Explain your answer.*

**3.28 Top Broadcast Shows.** The networks for the top 20 television shows, as determined by the *Nielsen Ratings* for the week of April 18–24, 2005, are shown in the following table.

| | | | | |
|---|---|---|---|---|
| CBS | Fox | ABC | Fox | CBS |
| CBS | ABC | Fox | CBS | ABC |
| CBS | CBS | NBC | CBS | NBC |
| NBC | CBS | CBS | NBC | NBC |

**3.29 NCAA Wrestling Champs.** From *NCAAsports.com*—the official Web site for NCAA sports—we obtained the National Collegiate Athletic Association wrestling champions for the years 1981–2005. They are shown in the following table.

| | | | |
|---|---|---|---|
| Iowa | Arizona St. | Oklahoma St. | Iowa |
| Iowa | Oklahoma St. | Iowa | Minnesota |
| Iowa | Oklahoma St. | Iowa | Minnesota |
| Iowa | Iowa | Iowa | Oklahoma St. |
| Iowa | Iowa | Iowa | Oklahoma St. |
| Iowa | Iowa | Iowa | Oklahoma St. |
| Iowa St. | | | |

**3.30 Robbery Locations.** The Department of Justice and the Federal Bureau of Investigation publish a compilation on crime statistics for the United States in *Crime in the United*

*States.* The following table provides a frequency distribution for robbery type during a one-year period.

| Robbery type | Frequency |
|---|---|
| Street/highway | 179,296 |
| Commercial house | 60,493 |
| Gas or service station | 11,362 |
| Convenience store | 25,774 |
| Residence | 56,641 |
| Bank | 9,504 |
| Miscellaneous | 70,333 |

**3.31 M&M Colors.** Observing that the proportion of blue M&Ms in his bowl of candy appeared to be less than that of the other colors, Ronald D. Fricker, Jr., decided to compare the color distribution in randomly chosen bags of M&Ms to the theoretical distribution reported by M&M/MARS consumer affairs. Fricker published his findings in the article "The Mysterious Case of the Blue M&Ms" (*Chance*, Vol. 9(4), pp. 19–22). For his study, Fricker bought three bags of M&Ms from local stores and counted the number of each color. The average number of each color in the three bags was distributed as follows.

| Color | Frequency |
|---|---|
| Brown | 152 |
| Yellow | 114 |
| Red | 106 |
| Orange | 51 |
| Green | 43 |
| Blue | 43 |

## Working With Large Data Sets

*In each of Exercises 3.32–3.39, use the technology of your choice to obtain the measures of center that are appropriate from among the mean, median, and mode. Discuss your results and decide which measure of center is most appropriate. Provide a reason for your answer.*

**3.32 U.S. Regions.** The U.S. Census Bureau divides the states in the United States into four regions: Northeast, Midwest, South, and West. The data providing the region of each state is presented on the WeissStats CD.

**3.33 The Great White Shark.** In an article titled "Great White, Deep Trouble" (*National Geographic*, Vol. 197(4), pp. 2–29), Peter Benchley—the author of *JAWS*—discussed various aspects of the Great White Shark (*Carcharodon carcharias*). Data on the number of pups borne in a lifetime by each of 80 Great White Shark females are provided on the WeissStats CD.

**3.34 Road Rage.** The report *Controlling Road Rage: A Literature Review and Pilot Study* was prepared for the AAA Foundation for Traffic Safety by D. Rathbone and J. Huckabee. The authors discuss the results of a literature review and pilot study on how to prevent aggressive driving and road rage. As described in the study, *road rage* is criminal behavior by motorists characterized by uncontrolled anger that results in violence or threatened violence on the road. One of the goals of the study was to determine when road rage occurs most often. The days on which 69 road-rage incidents occurred are provided on the WeissStats CD.

**3.35 Presidential Election.** In a study summarized in *Chance* (Vol. 17, No. 4, pp. 43–46), H. Wainer provides yet another characterization of the states that supported George Bush versus Al Gore in the 2000 U.S. presidential election. Data on the WeissStats CD include National Assessment of Educational Progress (NAEP) score and the result of the 2000 presidential election for each of the 50 states. The NAEP scores reflect the academic performance for a random sample of children from each state.

**3.36 Top Recording Artists.** From the Recording Industry Association of America Web site, we obtained data on the number of albums sold, in millions, for the top recording artists (U.S. sales only) as of June 20, 2005. Those data are provided on the WeissStats CD.

**3.37 High School Completion Rates.** As reported by the U.S. Census Bureau in *Current Population Reports*, the percentage of adults in each state and the District of Columbia who have completed high school is as provided on the WeissStats CD.

**3.38 Crime Rates.** The U.S. Federal Bureau of Investigation publishes the annual crime rates for each state and the District of Columbia in the document *Crime in the United States*. Those rates, given per 1000 population, are provided on the WeissStats CD.

**3.39 Body Temperature.** A study by researchers at the University of Maryland addressed the question of whether the mean body temperature of humans is 98.6°F. The results of the study by P. Mackowiak et al. appeared in the article "A Critical Appraisal of 98.6°F, the Upper Limit of the Normal Body Temperature, and Other Legacies of Carl Reinhold

August Wunderlich" (*Journal of the American Medical Association*, Vol. 268, pp. 1578–1580). Among other data, the researchers obtained the body temperatures of 93 healthy humans, as provided on the WeissStats CD.

**3.40 Sex and Direction.** In the paper "The Relation of Sex and Sense of Direction to Spatial Orientation in an Unfamiliar Environment" (*Journal of Environmental Psychology*, Vol. 20, pp. 17–28), Sholl et al. published the results of examining the sense of direction of 30 male and 30 female students. After being taken to an unfamiliar wooded park, the students were given a number of spatial orientation tests, including pointing to south, which tested their absolute frame of reference. Pointing to south was done by moving a pointer attached to a 360° protractor. We have supplied on the WeissStats CD the absolute pointing errors, in degrees, for both the male and female participants.

a. Use the technology of your choice to obtain the mean and median of each of the two data sets.

b. Use the results that you obtained in part (a) to compare the two data sets.

## Extending the Concepts and Skills

**3.41 Food Choice.** As you discovered earlier, *ordinal data* are data about order or rank given on a scale such as $1, 2, 3, \ldots$ or A, B, C, .... Most statisticians recommend using the median to indicate the center of an ordinal data set, but some researchers also use the mean. In the paper "Measurement of Ethical Food Choice Motives" (*Appetite*, Vol. 34, pp. 55–59), research psychologists M. Lindeman and M. Väänänen of the University of Helsinki published a study on the factors that most influence people's choice of food. One of the questions asked of the participants was how important, on a scale of 1 to 4 (1 = not at all important, 4 = very important), is ecological welfare in food choice motive, where ecological welfare includes animal welfare and environmental protection. Here are the ratings given by 14 of the participants.

| | | | | | | |
|---|---|---|---|---|---|---|
| 2 | 4 | 1 | 2 | 4 | 3 | 3 |
| 2 | 2 | 1 | 2 | 4 | 2 | 3 |

a. Compute the mean of the data.

b. Compute the median of the data.

c. Decide which of the two measures of center is best.

**3.42 Outliers and Trimmed Means.** Some data sets contain *outliers*, observations that fall well outside the overall pattern of the data. (We discuss outliers in more detail in Section 3.3.) Suppose, for instance, that you are interested in the ability of high school algebra students to compute square roots. You decide to give a square-root exam to 10 of these students. Unfortunately, one of the students had a fight with his girlfriend and cannot concentrate—he gets a 0. The 10 scores are displayed in increasing order in the following table. The score of 0 is an outlier.

| | | | | | | | | | |
|---|---|---|---|---|---|---|---|---|---|
| 0 | 58 | 61 | 63 | 67 | 69 | 70 | 71 | 78 | 80 |

Statisticians have a systematic method for avoiding extreme observations and outliers when they calculate means. They compute *trimmed means*, in which high and low observations are deleted or "trimmed off" before the mean is calculated. For instance, to compute the 10% trimmed mean of the test-score data, we first delete both the bottom 10% and the top 10% of the ordered data, that is, 0 and 80. Then we calculate the mean of the remaining data. Thus the 10% trimmed mean of the test-score data is

$$\frac{58 + 61 + 63 + 67 + 69 + 70 + 71 + 78}{8} = 67.1.$$

The following table displays a set of scores for a 40-question algebra final exam.

| | | | | | | | | | |
|---|---|---|---|---|---|---|---|---|---|
| 2 | 15 | 16 | 16 | 19 | 21 | 21 | 25 | 26 | 27 |
| 4 | 15 | 16 | 17 | 20 | 21 | 24 | 25 | 27 | 28 |

a. Do any of the scores look like outliers?

b. Compute the usual mean of the data.

c. Compute the 5% trimmed mean of the data.

d. Compute the 10% trimmed mean of the data.

e. Compare the means you obtained in parts (b)–(d). Which of the three means provides the best measure of center for the data?

**3.43** Explain the difference between the quantities $(\Sigma x_i)^2$ and $\Sigma x_i^2$. Construct an example to show that, in general, those two quantities are unequal.

**3.44** Explain the difference between the quantities $\Sigma x_i y_i$ and $(\Sigma x_i)(\Sigma y_i)$. Provide an example to show that, in general, those two quantities are unequal.

## 3.2 Measures of Variation

Up to this point, we have discussed only descriptive measures of center, specifically, the mean, median, and mode. However, two data sets can have the same mean, median, or mode and still differ in other respects. For example, consider the heights of the five starting players on each of two men's college basketball teams, as shown in Fig. 3.3.

**FIGURE 3.3**
Five starting players on two basketball teams

| Team I | | | | | Team II | | | | |
|---|---|---|---|---|---|---|---|---|---|
| Feet and inches | 6' | 6'1" | 6'4" | 6'4" | 6'6" | 5'7" | 6' | 6'4" | 6'4" | 7' |
| Inches | 72 | 73 | 76 | 76 | 78 | 67 | 72 | 76 | 76 | 84 |

The two teams have the same mean heights, 75 inches (6' 3"); the same median heights, 76 inches (6' 4"); and the same modes, 76 inches (6' 4"). Nonetheless, the two data sets clearly differ. In particular, the heights of the players on Team II vary much more than those on Team I. To describe that difference quantitatively, we use a descriptive measure that indicates the amount of variation, or spread, in a data set. Such descriptive measures are referred to as **measures of variation** or **measures of spread.**

Just as there are several different measures of center, there are also several different measures of variation. In this section, we examine two of the most frequently used measures of variation: the *range* and *sample standard deviation.*

## The Range

The contrast between the height difference of the two teams is clear if we place the shortest player on each team next to the tallest, as in Fig. 3.4.

**FIGURE 3.4**
Shortest and tallest starting players on the teams

| Team I | | Team II | |
|---|---|---|---|
| Feet and inches | 6' | 6'6" | 5'7" | 7' |
| Inches | 72 | 78 | 67 | 84 |

**You try it!**

Exercise 3.51(a) on page 116

The **range** of a data set is the difference between the maximum (largest) and minimum (smallest) observations. From Fig. 3.4,

Team I: Range = 78 − 72 = 6 inches,
Team II: Range = 84 − 67 = 17 inches.

**Interpretation** The difference between the heights of the tallest and shortest players on Team I is 6 inches, whereas that difference for Team II is 17 inches.

**Definition 3.5**

## Range of a Data Set

The **range** of a data set is given by the formula

$$\text{Range} = \text{Max} - \text{Min},$$

where Max and Min denote the maximum and minimum observations, respectively.

The range of a data set is easy to compute, but takes into account only the largest and smallest observations. For that reason, two other measures of variation, the *standard deviation* and the *interquartile range*, are generally favored over the range. We discuss the standard deviation in this section and consider the interquartile range in Section 3.3.

### The Sample Standard Deviation

In contrast to the range, the standard deviation takes into account all the observations. It is the preferred measure of variation when the mean is used as the measure of center.

Roughly speaking, the **standard deviation** measures variation by indicating how far, on average, the observations are from the mean. For a data set with a large amount of variation, the observations will, on average, be far from the mean; so the standard deviation will be large. For a data set with a small amount of variation, the observations will, on average, be close to the mean; so the standard deviation will be small.

The formulas for the standard deviations of sample data and population data differ slightly. In this section, we concentrate on the sample standard deviation. We discuss the population standard deviation in Section 3.4.

The first step in computing a sample standard deviation is to find the **deviations from the mean,** that is, how far each observation is from the mean.

**Example 3.8** | **The Deviations From the Mean**

*Heights of Starting Players* The heights, in inches, of the five starting players on Team I are 72, 73, 76, 76, and 78, as we saw in Fig. 3.3. Find the deviations from the mean.

**Solution** The mean height of the starting players on Team I is

$$\bar{x} = \frac{\Sigma x_i}{n} = \frac{72 + 73 + 76 + 76 + 78}{5} = \frac{375}{5} = 75 \text{ inches.}$$

To find the deviation from the mean for an observation, $x_i$, we subtract the mean from it; that is, we compute $x_i - \bar{x}$. For instance, the deviation from the mean for the height of 72 inches is $x_i - \bar{x} = 72 - 75 = -3$. The deviations from the mean for all five observations are given in the second column of Table 3.6 and are represented by arrows in Fig. 3.5.

• • •

**TABLE 3.6**
Deviations from the mean

| Height $x$ | Deviation from mean $x - \bar{x}$ |
|---|---|
| 72 | −3 |
| 73 | −2 |
| 76 | 1 |
| 76 | 1 |
| 78 | 3 |

**FIGURE 3.5**
Observations (shown by dots) and deviations from the mean (shown by arrows)

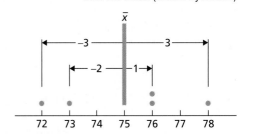

The second step in computing a sample standard deviation is to obtain a measure of the total deviation from the mean for all the observations. Although the quantities $x_i - \bar{x}$ represent deviations from the mean, adding them to get a total deviation from the mean is of no value because their sum, $\Sigma(x_i - \bar{x})$, always equals zero. Summing the second column of Table 3.6 illustrates this fact for the height data of Team I.

To obtain quantities that do not sum to zero, we square the deviations from the mean. The sum of the squared deviations from the mean, $\Sigma(x_i - \bar{x})^2$, is called the **sum of squared deviations** and gives a measure of total deviation from the mean for all the observations. We show how to calculate it next.

**Example 3.9** | **The Sum of Squared Deviations**

*Heights of Starting Players* Compute the sum of squared deviations for the heights of the starting players on Team I.

**Solution**   To get Table 3.7, we added a column for $(x - \bar{x})^2$ to Table 3.6.

**TABLE 3.7**
Table for computing the sum of squared deviations for the heights of Team I

| Height $x$ | Deviation from mean $x - \bar{x}$ | Squared deviation $(x - \bar{x})^2$ |
|---|---|---|
| 72 | −3 | 9 |
| 73 | −2 | 4 |
| 76 | 1 | 1 |
| 76 | 1 | 1 |
| 78 | 3 | 9 |
| | | 24 |

From the third column of Table 3.7, $\Sigma(x_i - \bar{x})^2 = 24$. The sum of squared deviations is 24 inches$^2$.

• • •

The third step in computing a sample standard deviation is to take an average of the squared deviations. We do so by dividing the sum of squared deviations by $n - 1$, or 1 less than the sample size. The resulting quantity is called a **sample variance** and is denoted $s_x^2$ or, when no confusion can arise, $s^2$. In symbols,

$$s^2 = \frac{\Sigma(x_i - \bar{x})^2}{n - 1}.$$

**Note:** If we divided by $n$ instead of by $n - 1$, the sample variance would be the mean of the squared deviations. Although dividing by $n$ seems more natural, we divide by $n - 1$ for the following reason. One of the main uses of the sample variance is to estimate the population variance (defined in Section 3.4). Division by $n$ tends to underestimate the population variance, whereas division by $n - 1$ gives, on average, the correct value.

| Example 3.10 | The Sample Variance |

*Heights of Starting Players*  Determine the sample variance of the heights of the starting players on Team I.

**Solution**  From Example 3.9, the sum of squared deviations is 24 inches². Because $n = 5$,

$$s^2 = \frac{\Sigma(x_i - \bar{x})^2}{n - 1} = \frac{24}{5 - 1} = 6.$$

The sample variance is 6 inches².

• • •

As we have just seen, a sample variance is in units that are the square of the original units, the result of squaring the deviations from the mean. Because descriptive measures should be expressed in the original units, the final step in computing a sample standard deviation is to take the square root of the sample variance, which gives us the following definition.

| Definition 3.6 | Sample Standard Deviation |

**What Does It Mean?**

Roughly speaking, the sample standard deviation indicates how far, on average, the observations in the sample are from the mean of the sample.

For a variable $x$, the standard deviation of the observations for a sample is called a **sample standard deviation.** It is denoted $s_x$ or, when no confusion will arise, simply $s$. We have

$$s = \sqrt{\frac{\Sigma(x_i - \bar{x})^2}{n - 1}},$$

where $n$ is the sample size.

| Example 3.11 | The Sample Standard Deviation |

*Heights of Starting Players*  Determine the sample standard deviation of the heights of the starting players on Team I.

**Solution**  From Example 3.10, the sample variance is 6 inches². Thus the sample standard deviation is

$$s = \sqrt{\frac{\Sigma(x_i - \bar{x})^2}{n - 1}} = \sqrt{6} = 2.4 \text{ inches.}$$

**Interpretation**  Roughly speaking, on average, the heights of the players on Team I vary from the mean height of 75 inches by 2.4 inches.

• • •

For teaching purposes, we spread our calculations of a sample standard deviation over four separate examples. Now we summarize the procedure with three steps.

**STEP 1** Calculate the sample mean, $\bar{x}$.

**STEP 2** Construct a table to obtain the sum of squared deviations, $\Sigma(x_i - \bar{x})^2$.

**STEP 3** Apply Definition 3.6 to determine the sample standard deviation, $s$.

| Example 3.12 | **The Sample Standard Deviation** |

*Heights of Starting Players* The heights, in inches, of the five starting players on Team II are 67, 72, 76, 76, and 84. Determine the sample standard deviation of these heights.

**Solution** We apply the three-step procedure just described.

**STEP 1** Calculate the sample mean, $\bar{x}$.

**TABLE 3.8**

Table for computing the sum of squared deviations for the heights of Team II

| $x$ | $x - \bar{x}$ | $(x - \bar{x})^2$ |
|-----|-----|-----|
| 67 | −8 | 64 |
| 72 | −3 | 9 |
| 76 | 1 | 1 |
| 76 | 1 | 1 |
| 84 | 9 | 81 |
| | | 156 |

We have

$$\bar{x} = \frac{\Sigma x_i}{n} = \frac{67 + 72 + 76 + 76 + 84}{5} = \frac{375}{5} = 75 \text{ inches.}$$

**STEP 2** Construct a table to obtain the sum of squared deviations, $\Sigma(x_i - \bar{x})^2$.

Table 3.8 provides columns for $x$, $x - \bar{x}$, and $(x - \bar{x})^2$. The third column shows that $\Sigma(x_i - \bar{x})^2 = 156$ inches$^2$.

**STEP 3** Apply Definition 3.6 to determine the sample standard deviation, $s$.

Because $n = 5$ and $\Sigma(x_i - \bar{x})^2 = 156$, the sample standard deviation is

$$s = \sqrt{\frac{\Sigma(x_i - \bar{x})^2}{n - 1}} = \sqrt{\frac{156}{5 - 1}} = \sqrt{39} = 6.2 \text{ inches.}$$

**You try it!**

Exercise 3.51(b) on page 116

**Interpretation** Roughly speaking, on average, the heights of the players on Team II vary from the mean height of 75 inches by 6.2 inches.

• • •

In Examples 3.11 and 3.12, we found that the sample standard deviations of the heights of the starting players on Teams I and II are 2.4 inches and 6.2 inches, respectively. Hence Team II, which has more variation in height than Team I, also has a larger standard deviation.

| Key Fact 3.1 | **Variation and the Standard Deviation** |

The more variation that there is in a data set, the larger is its standard deviation.

Key Fact 3.1 shows that the standard deviation satisfies the basic criterion for a measure of variation; in fact, the standard deviation is the most commonly used measure of variation. However, the standard deviation does have its drawbacks. For instance, it is not resistant: its value can be strongly affected by a few extreme observations.

### A Computing Formula for *s*

Next, we present an alternative formula for obtaining a sample standard deviation, which we call the *computing formula* for *s*. We call the original formula given in Definition 3.6 the *defining formula* for *s*.

---

**Formula 3.1**

### Computing Formula for a Sample Standard Deviation

A sample standard deviation can be computed using the formula

$$s = \sqrt{\frac{\Sigma x_i^2 - (\Sigma x_i)^2/n}{n-1}},$$

where $n$ is the sample size.

---

**Note:** In the numerator of the computing formula, the division of $(\Sigma x_i)^2$ by $n$ should be performed before the subtraction from $\Sigma x_i^2$. In other words, first compute $(\Sigma x_i)^2/n$ and then subtract the result from $\Sigma x_i^2$.

The computing formula for *s* is equivalent to the defining formula—both formulas give the same answer, although differences owing to roundoff error are possible. However, the computing formula is usually faster and easier for doing calculations by hand and also reduces the chance for roundoff error.

Before illustrating the computing formula for *s*, let's investigate its expressions, $\Sigma x_i^2$ and $(\Sigma x_i)^2$. The expression $\Sigma x_i^2$ represents the sum of the squares of the data; to find it, first square each observation and then sum those squared values. The expression $(\Sigma x_i)^2$ represents the square of the sum of the data; to find it, first sum the observations and then square that sum.

---

**Example 3.13** | **Computing Formula for a Sample Standard Deviation**

**TABLE 3.9**

Table for computation of *s*, using the computing formula

| $x$ | $x^2$ |
|-----|-------|
| 67  | 4,489 |
| 72  | 5,184 |
| 76  | 5,776 |
| 76  | 5,776 |
| 84  | 7,056 |
| 375 | 28,281 |

*Heights of Starting Players* Find the sample standard deviation of the heights for the five starting players on Team II by using the computing formula.

**Solution** We need the sums $\Sigma x_i^2$ and $(\Sigma x_i)^2$, which Table 3.9 shows to be 375 and 28,281, respectively. Now applying Formula 3.1, we get

$$s = \sqrt{\frac{\Sigma x_i^2 - (\Sigma x_i)^2/n}{n-1}} = \sqrt{\frac{28{,}281 - (375)^2/5}{5-1}}$$

$$= \sqrt{\frac{28{,}281 - 28{,}125}{4}} = \sqrt{\frac{156}{4}} = \sqrt{39} = 6.2 \text{ inches,}$$

which is the same value that we got by using the defining formula.

• • •

Exercise 3.51(c)
on page 116

## Rounding Basics

Here is an important rule to remember when you use only basic calculator functions to obtain a sample standard deviation or any other descriptive measure.

> **Rounding Rule:** Do not perform any rounding until the computation is complete; otherwise, substantial roundoff error can result.

Another common rounding rule is to round final answers that contain units to one more decimal place than the raw data. Although we usually abide by this convention, occasionally we vary from it for pedagogical reasons. In general, you should stick to this rounding rule as well.

## Further Interpretation of the Standard Deviation

Again, the standard deviation is a measure of variation—the more variation there is in a data set, the larger is its standard deviation. Table 3.10 contains two data sets, each with 10 observations. Notice that Data Set II has more variation than Data Set I.

**TABLE 3.10**
Data sets that have different variation

| Data Set I | 41 | 44 | 45 | 47 | 47 | 48 | 51 | 53 | 58 | 66 |
|---|---|---|---|---|---|---|---|---|---|---|
| Data Set II | 20 | 37 | 48 | 48 | 49 | 50 | 53 | 61 | 64 | 70 |

**TABLE 3.11**
Means and standard deviations of the data sets in Table 3.10

| Data Set I | Data Set II |
|---|---|
| $\bar{x} = 50.0$ | $\bar{x} = 50.0$ |
| $s = 7.4$ | $s = 14.2$ |

We computed the sample mean and sample standard deviation of each data set and summarized the results in Table 3.11. As expected, the standard deviation of Data Set II is larger than that of Data Set I.

To enable you to compare visually the variations in the two data sets, we produced the graphs shown in Figs. 3.6 and 3.7. On each graph, we marked the observations with dots. In addition, we located the sample mean, $\bar{x} = 50$, and measured intervals equal in length to the standard deviation: 7.4 for Data Set I and 14.2 for Data Set II.

**FIGURE 3.6**   Data Set I; $\bar{x} = 50$, $s = 7.4$

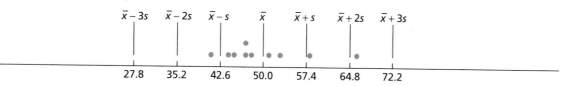

**FIGURE 3.7**   Data Set II; $\bar{x} = 50$, $s = 14.2$

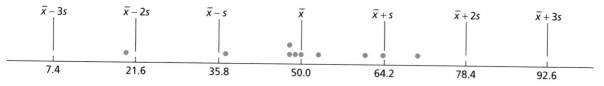

In Fig. 3.6, note that the horizontal position labeled $\bar{x} + 2s$ represents the number that is two standard deviations to the right of the mean, which in this

case is

$$\bar{x} + 2s = 50.0 + 2 \cdot 7.4 = 50.0 + 14.8 = 64.8.^{\dagger}$$

Likewise, the horizontal position labeled $\bar{x} - 3s$ represents the number that is three standard deviations to the left of the mean, which in this case is

$$\bar{x} - 3s = 50.0 - 3 \cdot 7.4 = 50.0 - 22.2 = 27.8.$$

Figure 3.7 is interpreted in a similar manner.

The graphs shown in Figs. 3.6 and 3.7 vividly illustrate that Data Set II has more variation than Data Set I. They also show that for each data set all observations lie within a few standard deviations to either side of the mean. This result is no accident.

---

| Key Fact 3.2 | **Three-Standard-Deviations Rule** |
|---|---|
| | Almost all the observations in any data set lie within three standard deviations to either side of the mean. |

---

A data set with a great deal of variation has a large standard deviation, so three standard deviations to either side of its mean will be extensive, as shown in Fig. 3.7. A data set with little variation has a small standard deviation, and hence three standard deviations to either side of its mean will be narrow, as shown in Fig. 3.6.

The three-standard-deviations rule is vague—what does "almost all" mean? It can be made more precise in several ways, two of which we now briefly describe.

We can apply **Chebychev's rule,** which is valid for all data sets and implies, in particular, that at least 89% of the observations lie within three standard deviations to either side of the mean. If the distribution of the data set is approximately bell shaped, we can apply the **empirical rule** which implies, in particular, that roughly 99.7% of the observations lie within three standard deviations to either side of the mean. Both Chebychev's rule and the empirical rule are discussed in detail in the exercises of this section.

---

## The Technology Center

In Section 3.1, we showed how to use Minitab, Excel, and the TI-83/84 Plus to obtain several descriptive measures. We can apply those same programs to obtain the range and sample standard deviation.

### Example 3.14   Using Technology to Obtain Descriptive Measures

*Heights of Starting Players*   The first column of Table 3.9 on page 112 gives the heights of the players on Team II. Use Minitab, Excel, or the TI-83/84 Plus to find the range and sample standard deviation of those heights.

---

$^{\dagger}$Recall that the rules for the order of arithmetic operations say to multiply and divide before adding and subtracting. So, to evaluate $a + b \cdot c$, find $b \cdot c$ first and then add the result to $a$. Similarly, to evaluate $a - b \cdot c$, find $b \cdot c$ first and then subtract the result from $a$.

**Solution**   We applied the descriptive-measures programs to the data, resulting in Output 3.2. Steps for generating that output are presented in Instructions 3.2.

**OUTPUT 3.2**   Descriptive measures for the heights of the players on Team II

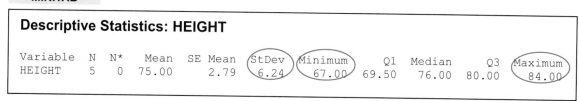

MINITAB

**Descriptive Statistics: HEIGHT**

| Variable | N | N* | Mean | SE Mean | StDev | Minimum | Q1 | Median | Q3 | Maximum |
|----------|---|----|------|---------|-------|---------|----|--------|----|---------|
| HEIGHT | 5 | 0 | 75.00 | 2.79 | 6.24 | 67.00 | 69.50 | 76.00 | 80.00 | 84.00 |

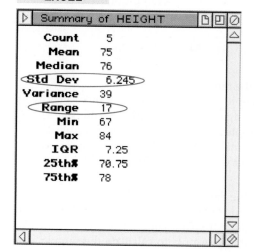

EXCEL

| Summary of HEIGHT | |
|-------------------|------|
| Count | 5 |
| Mean | 75 |
| Median | 76 |
| Std Dev | 6.245 |
| Variance | 39 |
| Range | 17 |
| Min | 67 |
| Max | 84 |
| IQR | 7.25 |
| 25th⌧ | 70.75 |
| 75th⌧ | 78 |

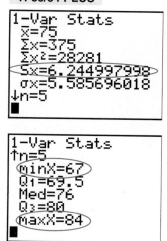

TI-83/84 PLUS

```
1-Var Stats
 x̄=75
 Σx=375
 Σx²=28281
 Sx=6.244997998
 σx=5.585696018
↓n=5
■
```

```
1-Var Stats
↑n=5
 minX=67
 Q₁=69.5
 Med=76
 Q₃=80
 maxX=84
■
```

As shown in Output 3.2, the sample standard deviation of the heights for the players on Team II is 6.24 inches (to two decimal places). The Excel output also shows that the range of the heights is 17 inches. We can get the range from the Minitab or TI-83/84 Plus output by subtracting the minimum from the maximum.

**INSTRUCTIONS 3.2**   Steps for generating Output 3.2

MINITAB

1 Store the height data from Table 3.9 in a column named HEIGHT
2 Choose **Stat ➤ Basic Statistics ➤ Display Descriptive Statistics...**
3 Specify HEIGHT in the **Variables** text box
4 Click **OK**

EXCEL

1 Store the height data from Table 3.9 in a range named HEIGHT
2 Choose **DDXL ➤ Summaries**
3 Select **Summary of One Variable** from the **Function type** drop-down box
4 Specify HEIGHT in the **Quantitative Variable** text box
5 Click **OK**

TI-83/84 PLUS

1 Store the height data from Table 3.9 in a list named HT
2 Press **STAT**
3 Arrow over to **CALC**
4 Press **1**
5 Press **2nd ➤ LIST**
6 Arrow down to HT and press **ENTER** twice

*Note to Minitab users:* The range is optionally available from the **Display Descriptive Statistics** dialog box by first clicking the **Statistics...** button and then checking the **Range** check box.

■ ■ ■

# Exercises 3.2

## Understanding the Concepts and Skills

**3.45** Explain the purpose of a measure of variation.

**3.46** Why is the standard deviation preferable to the range as a measure of variation?

**3.47** When you use the standard deviation as a measure of variation, what is the reference point?

**3.48** The following dartboards represent darts thrown by two players, Tracey and Joan.

Tracey

Joan

For the variable "distance from the center," which player's board represents data with a smaller sample standard deviation? Explain your answer.

**3.49** Consider the data set 1, 2, 3, 4, 5, 6, 7, 8, 9.
a. Use the defining formula to obtain the sample standard deviation.
b. Replace the 9 in the data set by 99 and again use the defining formula to compute the sample standard deviation.
c. Compare your answers in parts (a) and (b). The lack of what property of the standard deviation accounts for its extreme sensitivity to the change of 9 to 99?

**3.50** Consider the following four data sets.

| Data Set I | | Data Set II | | Data Set III | | Data Set IV | |
|---|---|---|---|---|---|---|---|
| 1 | 5 | 1 | 9 | 5 | 5 | 2 | 4 |
| 1 | 8 | 1 | 9 | 5 | 5 | 4 | 4 |
| 2 | 8 | 1 | 9 | 5 | 5 | 4 | 4 |
| 2 | 9 | 1 | 9 | 5 | 5 | 4 | 10 |
| 5 | 9 | 1 | 9 | 5 | 5 | 4 | 10 |

a. Compute the mean of each data set.
b. Although the four data sets have the same means, in what respect are they quite different?
c. Which data set appears to have the least variation? the greatest variation?
d. Compute the range of each data set.

e. Use the defining formula to compute the sample standard deviation of each data set.
f. From your answers to parts (d) and (e), which measure of variation better distinguishes the spread in the four data sets: the range or the standard deviation? Explain your answer.
g. Are your answers from parts (c) and (e) consistent?

**3.51 Age of U.S. Residents.** The U.S. Census Bureau publishes information about ages of people in the United States in *Current Population Reports*. A sample of five U.S. residents have the following ages, in years.

| 21 | 54 | 9 | 45 | 51 |
|---|---|---|---|---|

a. Determine the range of these ages.
b. Find the sample standard deviation of these ages by using the defining formula, Definition 3.6 on page 110.
c. Find the sample standard deviation of these ages by using the computing formula, Formula 3.1 on page 112.
d. Compare your work in parts (b) and (c).

**3.52** Consider the data set 3, 3, 3, 3, 3, 3.
a. Guess the value of the sample standard deviation without calculating it. Explain your reasoning.
b. Use the defining formula to calculate the sample standard deviation.
c. Complete the following statement and explain your reasoning: If all observations in a data set are equal, the sample standard deviation is _____.
d. Complete the following statement and explain your reasoning: If the sample standard deviation of a data set is 0, then….

*In Exercises 3.53–3.60, determine the range and sample standard deviation for each of the data sets. For the sample standard deviation, round each answer to one more decimal place than that used for the observations.*

**3.53 Amphibian Embryos.** In a study of the effects of radiation on amphibian embryos titled "Shedding Light on Ultraviolet Radiation and Amphibian Embryos" (*BioScience*, Vol. 53, No. 6, pp. 551–561), L. Licht recorded the time it took for a sample of seven different species of frogs' and toads' eggs to hatch. The following table shows the times to hatch, in days.

| 6 | 7 | 11 | 6 | 5 | 5 | 11 |
|---|---|---|---|---|---|---|

| Person | Wealth ($ billions) |
|---|---|
| Bill Gates | 51 |
| Warren Buffett | 40 |
| Paul Allen | 22.5 |
| Michael Dell | 18 |
| Larry Ellison | 17 |

**3.54 Hurricanes.** An article by D. Schaefer et al. (*Journal of Tropical Ecology*, Vol. 16, pp. 189–207) reported on a long-term study of the effects of hurricanes on tropical streams of the Luquillo Experimental Forest in Puerto Rico. The study shows that Hurricane Hugo had a significant impact on stream water chemistry. The following table shows a sample of 10 ammonia fluxes in the first year after Hugo. Data are in kilograms per hectare per year.

| | | | | |
|---|---|---|---|---|
| 96 | 66 | 147 | 147 | 175 |
| 116 | 57 | 154 | 88 | 154 |

**3.55 Tornado Touchdowns.** Each year, tornadoes that touch down are recorded by the Storm Prediction Center and published in *Monthly Tornado Statistics*. The following table gives the number of tornadoes that touched down in the United States during each month of one year. [SOURCE: National Oceanic and Atmospheric Administration.]

| | | | | | |
|---|---|---|---|---|---|
| 3 | 2 | 47 | 118 | 204 | 97 |
| 68 | 86 | 62 | 57 | 98 | 99 |

**3.56 Technical Merit.** In the 2002 Winter Olympics, Michelle Kwan competed in the Short Program ladies singles event. From nine judges, she received scores ranging from 1 (poor) to 6 (perfect). The following table provides the scores that the judges gave her on technical merit, found in an article by S. Berry (*Chance*, Vol. 15, No. 2, pp. 14–18).

| | | | | | | | | |
|---|---|---|---|---|---|---|---|---|
| 5.8 | 5.7 | 5.9 | 5.7 | 5.5 | 5.7 | 5.7 | 5.7 | 5.6 |

**3.57 Billionaires' Club.** Each year, *Forbes* magazine compiles a list of the 400 richest Americans. For 2005, the top five on the list are as shown in the following table.

**3.58 AML and the Cost of Labor.** Active Management of Labor (AML) was introduced in the 1960s to reduce the amount of time a woman spends in labor during the birth process. R. Rogers et al. conducted a study to determine whether AML also translates into a reduction in delivery cost to the patient. They reported their findings in the paper "Active Management of Labor: A Cost Analysis of a Randomized Controlled Trial" (*Western Journal of Medicine*, Vol. 172, pp. 240–243). The following table displays the costs, in dollars, of eight randomly sampled AML deliveries.

| | | | |
|---|---|---|---|
| 3141 | 2873 | 2116 | 1684 |
| 3470 | 1799 | 2539 | 3093 |

**3.59 Fuel Economy.** Every year, *Consumer Reports* publishes a magazine titled *New Car Ratings and Review* that looks at vehicle profiles for the year's models. It lets you see in one place how, within each category, the vehicles compare. One category of interest, especially when fuel prices are rising, is fuel economy, measured in miles per gallon (mpg). Following is a list of overall mpg for 14 different full-sized and compact pickups.

| | | | | | | |
|---|---|---|---|---|---|---|
| 14 | 13 | 14 | 13 | 14 | 14 | 11 |
| 12 | 15 | 15 | 17 | 14 | 15 | 16 |

**3.60 Router Horsepower.** In the article "Router Roundup" (*Popular Mechanics*, Vol. 180, No. 12, pp. 104–109), T. Klenck reports on tests of seven fixed-base routers for performance, features, and handling. The following table gives the horsepower for each of the seven routers tested.

| | | | | | | |
|---|---|---|---|---|---|---|
| 1.75 | 2.25 | 2.25 | 2.25 | 1.75 | 2.00 | 1.50 |

**3.61 Medieval Cremation Burials.** In the article "Material Culture as Memory: Combs and Cremations in Early Medieval Britain" (*Early Medieval Europe*, Vol. 12, Issue 2, pp. 89–128), H. Williams discussed the frequency of cremation burials found in 17 archaeological sites in eastern England. Here are the data.

| | | | | | | | | |
|---|---|---|---|---|---|---|---|---|
| 83 | 64 | 46 | 48 | 523 | 35 | 34 | 265 | 2484 |
| 46 | 385 | 21 | 86 | 429 | 51 | 258 | 119 | |

a. Obtain the sample standard deviation of these data.
b. Do you think that, in this case, the sample standard deviation provides a good measure of variation? Explain your answer.

**3.62 Monthly Motorcycle Casualties.** The *Scottish Executive*, Analytical Services Division Transport Statistics, compiles data on motorcycle casualties. During one year, monthly casualties resulting from motorcycle accidents in Scotland for built-up roads and non-built-up roads were as follows.

| Month | Built up | Non built up |
|-------|----------|--------------|
| January | 25 | 16 |
| February | 38 | 9 |
| March | 38 | 26 |
| April | 56 | 48 |
| May | 61 | 73 |
| June | 52 | 72 |
| July | 50 | 91 |
| August | 90 | 69 |
| September | 67 | 71 |
| October | 51 | 28 |
| November | 64 | 19 |
| December | 40 | 12 |

a. Without doing any calculations, make an educated guess at which of the two data sets, built up or non built up, has the greater variation.
b. Find the range and sample standard deviation of each of the two data sets. Compare your results here to the educated guess that you made in part (a).

**3.63 Daily Motorcycle Accidents.** The *Scottish Executive*, Analytical Services Division Transport Statistics, compiles data on motorcycle accidents. During one year, the numbers of motorcycle accidents in Scotland were tabulated by day of the week for built-up roads and non-built-up roads and resulted in the following data.

| Day | Built up | Non built up |
|-----|----------|--------------|
| Monday | 88 | 70 |
| Tuesday | 100 | 58 |
| Wednesday | 76 | 59 |
| Thursday | 98 | 53 |
| Friday | 103 | 56 |
| Saturday | 85 | 94 |
| Sunday | 69 | 102 |

a. Without doing any calculations, make an educated guess at which of the two data sets, built up or non built up, has the greater variation.
b. Find the range and sample standard deviation of each of the two data sets. Compare your results here to the educated guess that you made in part (a).

## Working With Large Data Sets

*In each of Exercises 3.64–3.71, use the technology of your choice to obtain and interpret the range and sample standard deviation for those data sets to which those concepts apply. If those concepts don't apply, explain why.*

**3.64 U.S. Regions.** The U.S. Census Bureau divides the states in the United States into four regions: Northeast, Midwest, South, and West. The data providing the region of each state is presented on the WeissStats CD.

**3.65 The Great White Shark.** In an article titled "Great White, Deep Trouble" (*National Geographic*, Vol. 197(4), pp. 2–29), Peter Benchley—the author of *JAWS*—discussed various aspects of the Great White Shark (*Carcharodon carcharias*). Data on the number of pups borne in a lifetime by each of 80 Great White Shark females are provided on the WeissStats CD.

**3.66 Road Rage.** The report *Controlling Road Rage: A Literature Review and Pilot Study* was prepared for the AAA Foundation for Traffic Safety by D. Rathbone and J. Huckabee. The authors discuss the results of a literature review and pilot study on how to prevent aggressive driving and road rage. As described in the study, *road rage* is criminal behavior by motorists characterized by uncontrolled anger that results in violence or threatened violence on the road. One of the goals of the study was to determine when road rage occurs most often. The days on which 69 road-rage incidents occurred are provided on the WeissStats CD.

**3.67 Presidential Election.** In a study summarized in *Chance* (Vol. 17, No. 4, pp. 43–46), H. Wainer provides yet another characterization of the states that supported George Bush versus Al Gore in the 2000 U.S. presidential election. Data on the WeissStats CD include National Assessment of Educational Progress (NAEP) score and the result of the 2000 presidential election for each of the 50 states. The NAEP scores reflect the academic performance for a random sample of children from each state.

**3.68 Top Recording Artists.** From the Recording Industry Association of America Web site, we obtained data on the number of albums sold, in millions, for the top recording artists (U.S. sales only) as of June 20, 2005. Those data are provided on the WeissStats CD.

**3.69 High School Completion Rates.** As reported by the U.S. Census Bureau in *Current Population Reports*, the percentage of adults in each state and the District of Columbia that have completed high school is as provided on the WeissStats CD.

**3.70 Crime Rates.** The U.S. Federal Bureau of Investigation publishes the annual crime rates for each state and the District of Columbia in the document *Crime in the United States*. Those rates, given per 1000 population, are provided on the WeissStats CD.

**3.71 Body Temperature.** A study by researchers at the University of Maryland addressed the question of whether the mean body temperature of humans is 98.6°F. The results of the study by P. Mackowiak et al. appeared in the article "A Critical Appraisal of 98.6°F, the Upper Limit of the Normal Body Temperature, and Other Legacies of Carl Reinhold August Wunderlich" (*Journal of the American Medical Association*, Vol. 268, pp. 1578–1580). Among other data, the researchers obtained the body temperatures of 93 healthy humans, as provided on the WeissStats CD.

**3.72 Sex and Direction.** In the paper "The Relation of Sex and Sense of Direction to Spatial Orientation in an Unfamiliar Environment" (*Journal of Environmental Psychology*, Vol. 20, pp. 17–28), Sholl et al. published the results of examining the sense of direction of 30 male and 30 female students. After being taken to an unfamiliar wooded park, the students were given a number of spatial orientation tests, including pointing to south, which tested their absolute frame of reference. Pointing to south was done by moving a pointer attached to a 360° protractor. We have supplied on the WeissStats CD the absolute pointing errors, in degrees, for both the male and female participants.

**a.** Use the technology of your choice to obtain the range and sample standard deviation of each of the two data sets.

**b.** Use the results that you obtained in part (a) to compare the two data sets.

## Extending the Concepts and Skills

**3.73 Outliers.** In Exercise 3.42 on page 106, we discussed *outliers,* or observations that fall well outside the overall pattern of the data. The following table contains two data sets. Data Set II was obtained by removing the outliers from Data Set I.

| Data Set I | | | | | Data Set II | | | |
|---|---|---|---|---|---|---|---|---|
| 0 | 12 | 14 | 15 | 23 | 10 | 14 | 15 | 17 |
| 0 | 14 | 15 | 16 | 24 | 12 | 14 | 15 | |
| 10 | 14 | 15 | 17 | | 14 | 15 | 16 | |

**a.** Compute the sample standard deviation of each of the two data sets.

**b.** Compute the range of each of the two data sets.

**c.** What effect do outliers have on variation? Explain your answer.

**Grouped-Data Formulas.** When data are grouped in a frequency distribution, we use the following formulas to obtain the sample mean and sample standard deviation.

---

**Grouped-Data Formulas**

$$\bar{x} = \frac{\Sigma x_i f_i}{n} \quad \text{and} \quad s = \sqrt{\frac{\Sigma (x_i - \bar{x})^2 f_i}{n-1}},$$

where $x_i$ denotes class midpoint, $f_i$ denotes class frequency, and $n\ (= \Sigma f_i)$ denotes sample size. The sample standard deviation can also be obtained by using the computing formula

$$s = \sqrt{\frac{\Sigma x_i^2 f_i - (\Sigma x_i f_i)^2/n}{n-1}}.$$

---

In general, these formulas yield only approximations to the actual sample mean and sample standard deviation. We ask you to apply the grouped-data formulas in Exercises 3.74 and 3.75.

**3.74 Weekly Salaries.** In the following table, we repeat the salary data in Data Set II from Example 3.1.

| | | | | |
|---|---|---|---|---|
| 300 | 300 | 940 | 450 | 400 |
| 400 | 300 | 300 | 1050 | 300 |

a. Use Definitions 3.4 and 3.6 on pages 100 and 110, respectively, to obtain the sample mean and sample standard deviation of this (ungrouped) data set.

b. A frequency distribution for Data Set II, based on single-value grouping, is presented in the first two columns of the following table. The third column of the table is for the $xf$-values, that is, class midpoint (which here is the same as the class) times class frequency. Complete the missing entries in the table and then use the grouped-data formula to obtain the sample mean.

| Salary $x$ | Frequency $f$ | Salary · Frequency $xf$ |
|---|---|---|
| 300 | 5 | 1500 |
| 400 | 2 | |
| 450 | 1 | |
| 940 | 1 | |
| 1050 | 1 | |

c. Compare the answers that you obtained for the sample mean in parts (a) and (b). Explain why the grouped-data formula always yields the actual sample mean when the data are grouped in classes that are each based on a single value. (*Hint:* What does $xf$ represent for each class?)

d. Construct a table similar to the one in part (b) but with columns for $x, f, x - \bar{x}, (x - \bar{x})^2$, and $(x - \bar{x})^2 f$. Use the table and the grouped-data formula to obtain the sample standard deviation.

e. Compare your answers for the sample standard deviation in parts (a) and (d). Explain why the grouped-data formula always yields the actual sample standard deviation when the data are grouped in classes that are each based on a single value.

**3.75 Days to Maturity.** Following is a grouped-data table for the days to maturity for 40 short-term investments, as found in *BARRON'S*.

| Days to maturity | Frequency $f$ | Relative frequency | Midpoint $x$ |
|---|---|---|---|
| 30 ⋖ 40 | 3 | 0.075 | 35 |
| 40 ⋖ 50 | 1 | 0.025 | 45 |
| 50 ⋖ 60 | 8 | 0.200 | 55 |
| 60 ⋖ 70 | 10 | 0.250 | 65 |
| 70 ⋖ 80 | 7 | 0.175 | 75 |
| 80 ⋖ 90 | 7 | 0.175 | 85 |
| 90 ⋖ 100 | 4 | 0.100 | 95 |

a. Use the grouped-data formulas to estimate the sample mean and sample standard deviation of the days-to-maturity data. Round your final answers to one decimal place.

b. The following table gives the raw days-to-maturity data.

| | | | | | | | |
|---|---|---|---|---|---|---|---|
| 70 | 64 | 99 | 55 | 64 | 89 | 87 | 65 |
| 62 | 38 | 67 | 70 | 60 | 69 | 78 | 39 |
| 75 | 56 | 71 | 51 | 99 | 68 | 95 | 86 |
| 57 | 53 | 47 | 50 | 55 | 81 | 80 | 98 |
| 51 | 36 | 63 | 66 | 85 | 79 | 83 | 70 |

Using Definitions 3.4 and 3.6 on pages 100 and 110, respectively, gives the true sample mean and sample standard deviation of the days-to-maturity data as 68.3 and 16.7, rounded to one decimal place. Compare these actual values of $\bar{x}$ and $s$ to the estimates from part (a). Explain why the grouped-data formulas generally yield only approximations to the sample mean and sample standard deviation for non–single-value grouping.

**Chebychev's Rule.** A more precise version of the three-standard-deviations rule (Key Fact 3.2 on page 114) can be obtained from *Chebychev's rule*, which we state as follows.

---

**Chebychev's Rule**

For any data set and any real number $k > 1$, at least $100(1 - 1/k^2)\%$ of the observations lie within $k$ standard deviations to either side of the mean.

---

Two special cases of Chebychev's rule are applied frequently, namely, when $k = 2$ and $k = 3$. These state, respectively, that:

- At least 75% of the observations in any data set lie within two standard deviations to either side of the mean.
- At least 89% of the observations in any data set lie within three standard deviations to either side of the mean.

You are asked to examine Chebychev's rule in Exercises 3.76–3.79.

**3.76** Verify that the two statements in the preceding bulleted list are indeed special cases of Chebychev's rule with $k = 2$ and $k = 3$, respectively.

**3.77** Consider the data sets portrayed in Figs. 3.6 and 3.7 on page 113.

a. Chebychev's rule says that at least 75% of the observations lie within two standard deviations to either side of the mean. What percentage of the observations portrayed in Fig. 3.6 actually lie within two standard deviations to either side of the mean?

b. Chebychev's rule says that at least 89% of the observations lie within three standard deviations to either side of the mean. What percentage of the observations portrayed in Fig. 3.6 actually lie within three standard deviations to either side of the mean?

c. Repeat parts (a) and (b) for the data portrayed in Fig. 3.7.

d. From parts (a)–(c), we see that Chebychev's rule provides a lower bound on, rather than a precise estimate for, the percentage of observations that lie within a specified number of standard deviations to either side of the mean. Nonetheless, Chebychev's rule is quite important for several reasons. Can you think of some?

**3.78 Exam Scores.** Consider the following sample of exam scores, arranged in increasing order.

| | | | | | |
|---|---|---|---|---|---|
| 28 | 57 | 58 | 64 | 69 | 74 |
| 79 | 80 | 83 | 85 | 85 | 87 |
| 87 | 89 | 89 | 90 | 92 | 93 |
| 94 | 94 | 95 | 96 | 96 | 97 |
| 97 | 97 | 97 | 98 | 100 | 100 |

**Note:** The sample mean and sample standard deviation of these exam scores are, respectively, 85 and 16.1.

a. Use Chebychev's rule to obtain a lower bound on the percentage of observations that lie within two standard deviations to either side of the mean.

b. Use the data to obtain the exact percentage of observations that lie within two standard deviations to either side of the mean. Compare your answer here to that in part (a).

c. Use Chebychev's rule to obtain a lower bound on the percentage of observations that lie within three standard deviations to either side of the mean.

d. Use the data to obtain the exact percentage of observations that lie within three standard deviations to either side of the mean. Compare your answer here to that in part (c).

**3.79 Book Costs.** Chebychev's rule also permits you to make pertinent statements about a data set when only its mean and standard deviation are known. Here is an example of that use of Chebychev's rule. Information Today, Inc. publishes information on costs of new books in *The Bowker Annual Library and Book Trade Almanac*. A sample of 40 sociology books has a mean cost of $76.75 and a standard deviation of $10.42. Use this information and the two aforementioned special cases of Chebychev's rule to complete the following statements.

a. At least 30 of the 40 sociology books cost between _____ and _____.

b. At least _____ of the 40 sociology books cost between $45.49 and $108.01.

**The Empirical Rule.** For data sets with approximately bell-shaped distributions, we can improve on the estimates given by Chebychev's rule by using the *empirical rule*, which is as follows.

---

**Empirical Rule**

For any data set having roughly a bell-shaped distribution:

- Approximately 68% of the observations lie within one standard deviation to either side of the mean.
- Approximately 95% of the observations lie within two standard deviations to either side of the mean.
- Approximately 99.7% of the observations lie within three standard deviations to either side of the mean.

---

You are asked to examine the empirical rule in Exercises 3.80–3.83.

**3.80** In this exercise, you will compare Chebychev's rule and the empirical rule.

a. Compare the estimates given by the two rules for the percentage of observations that lie within two standard deviations to either side of the mean. Comment on the differences.

b. Compare the estimates given by the two rules for the percentage of observations that lie within three standard deviations to either side of the mean. Comment on the differences.

**3.81 Malnutrition and Poverty.** R. Reifen et al. studied various nutritional measures of Ethiopian school children and published their findings in the paper "Ethiopian-Born and Native Israeli School Children Have Different Growth Patterns" (*Nutrition*, Vol. 19, pp. 427–431). The study, conducted in Azezo, North West Ethiopia, found that malnutrition is prevalent in primary and secondary school children because of economic poverty. The weights, in kilograms (kg), of 60 randomly selected male Ethiopian-born school children, ages 12–15 years old, are presented in increasing order in the following table.

| | | | | | | | | |
|---|---|---|---|---|---|---|---|---|
| 36.3 | 37.7 | 38.0 | 38.8 | 38.9 | 39.0 | 39.3 | 40.9 | 41.1 |
| 41.3 | 41.5 | 41.8 | 42.0 | 42.0 | 42.1 | 42.5 | 42.5 | 42.8 |
| 42.9 | 43.3 | 43.4 | 43.5 | 44.0 | 44.4 | 44.7 | 44.8 | 45.2 |
| 45.2 | 45.2 | 45.4 | 45.5 | 45.7 | 45.9 | 45.9 | 46.2 | 46.3 |
| 46.5 | 46.6 | 46.8 | 47.2 | 47.4 | 47.5 | 47.8 | 47.9 | 48.1 |
| 48.2 | 48.3 | 48.4 | 48.5 | 48.6 | 48.9 | 49.1 | 49.2 | 49.5 |
| 50.9 | 51.4 | 51.8 | 52.8 | 53.8 | 56.6 | | | |

*Note:* The sample mean and sample standard deviation of these weights are, respectively, 45.30 kg and 4.16 kg.

a. Use the empirical rule to estimate the percentages of the observations that lie within one, two, and three standard deviations to either side of the mean.

**b.** Use the data to obtain the exact percentages of the observations that lie within one, two, and three standard deviations to either side of the mean.

**c.** Compare your answers in parts (a) and (b).

**d.** A histogram for these weights is shown in Exercise 2.92 on page 78. Based on that histogram, comment on your comparisons in part (c).

**e.** Is it appropriate to use the empirical rule for these data? Explain your answer.

**3.82 Exam Scores.** Refer to the exam scores displayed in Exercise 3.78.

**a.** Use the empirical rule to estimate the percentages of the observations that lie within one, two, and three standard deviations to either side of the mean.

**b.** Use the data to obtain the exact percentages of the observations that lie within one, two, and three standard deviations to either side of the mean.

**c.** Compare your answers in parts (a) and (b).

**d.** Construct a histogram or a stem-and-leaf diagram for the exam scores. Based on your graph, comment on your comparisons in part (c).

**e.** Is it appropriate to use the empirical rule for these data? Explain your answer.

**3.83 Book Costs.** Refer to Exercise 3.79. Assuming that the distribution of costs for the 40 sociology books is approximately bell shaped, apply the empirical rule to complete the following statements, and compare your answers to those obtained in Exercise 3.79, where Chebychev's rule was used.

**a.** Approximately 38 of the 40 sociology books cost between _____ and _____.

**b.** Approximately _____ of the 40 sociology books cost between $45.49 and $108.01.

## 3.3 The Five-Number Summary; Boxplots

So far, we have focused on the mean and standard deviation to measure center and variation. We now examine several descriptive measures based on percentiles.

Unlike the mean and standard deviation, descriptive measures based on percentiles are *resistant*—they are not sensitive to the influence of a few extreme observations. For this reason, descriptive measures based on percentiles are often preferred over those based on the mean and standard deviation.

### Quartiles

As you learned in Section 3.1, the median of a data set divides the data into two equal parts: the bottom 50% and the top 50%. The **percentiles** of a data set divide it into hundredths, or 100 equal parts. A data set has 99 percentiles, denoted $P_1, P_2, \ldots, P_{99}$. Roughly speaking, the first percentile, $P_1$, is the number that divides the bottom 1% of the data from the top 99%; the second percentile, $P_2$, is the number that divides the bottom 2% of the data from the top 98%; and so on. Note that the median is also the 50th percentile.

Certain percentiles are particularly important: the **deciles** divide a data set into tenths (10 equal parts), the **quintiles** divide a data set into fifths (five equal parts), and the **quartiles** divide a data set into quarters (four equal parts).

Quartiles are the most commonly used percentiles. A data set has three quartiles, which we denote $Q_1$, $Q_2$, and $Q_3$. Roughly speaking, the **first quartile, $Q_1$,** is the number that divides the bottom 25% of the data from the top 75%; the **second quartile, $Q_2$,** is the median, which, as you know, is the number that divides the bottom 50% of the data from the top 50%; and the **third quartile, $Q_3$,** is the number that divides the bottom 75% of the data from the top 25%. Note that the first and third quartiles are the 25th and 75th percentiles, respectively.

Figure 3.8 depicts the quartiles for uniform, bell-shaped, right-skewed, and left-skewed distributions.

**FIGURE 3.8**
Quartiles for (a) uniform, (b) bell-shaped, (c) right-skewed, and (d) left-skewed distributions

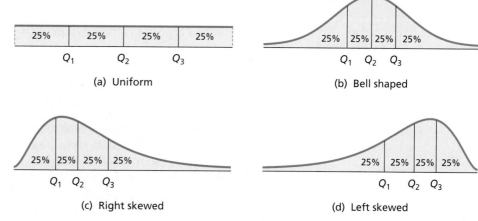

(a) Uniform

(b) Bell shaped

(c) Right skewed

(d) Left skewed

---

**Definition 3.7**

### Quartiles

Arrange the data in increasing order and determine the median.

- The **first quartile** is the median of the part of the entire data set that lies at or below the median of the entire data set.
- The **second quartile** is the median of the entire data set.
- The **third quartile** is the median of the part of the entire data set that lies at or above the median of the entire data set.

> **What Does It Mean?**
>
> The quartiles divide a data set into quarters (four equal parts).

**Note:** Not all statisticians define quartiles in exactly the same way. Other definitions may lead to different values but, in practice, the differences tend to be small with large data sets.

---

**Example 3.15** | **Quartiles**

**TABLE 3.12**
Weekly TV-viewing times

| 25 | 41 | 27 | 32 | 43 |
|----|----|----|----|----|
| 66 | 35 | 31 | 15 | 5 |
| 34 | 26 | 32 | 38 | 16 |
| 30 | 38 | 30 | 20 | 21 |

*Weekly TV-Viewing Times* The A. C. Nielsen Company publishes information on the TV-viewing habits of Americans in *Nielsen Report on Television*. A sample of 20 people yielded the weekly viewing times, in hours, displayed in Table 3.12. Determine and interpret the quartiles for these data.

**Solution** First, we arrange the data in Table 3.12 in increasing order:

5 15 16 20 21 25 26 27 30 **30 31** 32 32 34 35 38 38 41 43 66

Next, we determine the median of the entire data set. The number of observations is 20, so the median is at position $(20 + 1)/2 = 10.5$, halfway between the tenth and eleventh observations (shown in boldface) in the ordered list. Thus, the median of the entire data set is $(30 + 31)/2 = 30.5$.

Because the median of the entire data set is 30.5, the part of the entire data set that lies at or below the median of the entire data set is

5 15 16 20 **21 25** 26 27 30 30

This data set has 10 observations, so its median is at position $(10 + 1)/2 = 5.5$, halfway between the fifth and sixth observations (shown in boldface) in the ordered list. Thus the median of this data set—and hence the first quartile—is $(21 + 25)/2 = 23$; that is, $Q_1 = 23$.

The second quartile is the median of the entire data set, or 30.5. Therefore, we have $Q_2 = 30.5$.

Because the median of the entire data set is 30.5, the part of the entire data set that lies at or above the median of the entire data set is

31 32 32 34 35 38 38 41 43 66

This data set has 10 observations, so its median is at position $(10 + 1)/2 = 5.5$, halfway between the fifth and sixth observations (shown in boldface) in the ordered list. Thus the median of this data set—and hence the third quartile—is $(35 + 38)/2 = 36.5$; that is, $Q_3 = 36.5$.

In summary, the three quartiles for the TV-viewing times in Table 3.12 are $Q_1 = 23$ hours, $Q_2 = 30.5$ hours, and $Q_3 = 36.5$ hours.

**You try it!**

Exercise 3.93(a) on page 132

**Interpretation** We see that 25% of the TV-viewing times are less than 23 hours, 25% are between 23 hours and 30.5 hours, 25% are between 30.5 hours and 36.5 hours, and 25% are greater than 36.5 hours.

• • •

In Example 3.15, the number of observations is 20, which is even. To illustrate how to find quartiles when the number of observations is odd, we consider the TV-viewing-time data again, but this time without the largest observation, 66. In this case, the ordered list of the entire data set is

5 15 16 20 21 25 26 27 30 30 31 32 32 34 35 38 38 41 43

The median of the entire data set (also the second quartile) is 30, shown in boldface. The first quartile is the median of the 10 observations from 5 through the boldfaced 30, which is $(21 + 25)/2 = 23$. The third quartile is the median of the 10 observations from the boldfaced 30 through 43, which is $(34 + 35)/2 = 34.5$. Thus, for this data set, we have $Q_1 = 23$ hours, $Q_2 = 30$ hours, and $Q_3 = 34.5$ hours.

### The Interquartile Range

Next, we discuss the *interquartile range*. Because quartiles are used to define the interquartile range, it is the preferred measure of variation when the median is used as the measure of center. Like the median, the interquartile range is a resistant measure.

---

**Definition 3.8**

**What Does It Mean?**

Roughly speaking, the IQR gives the range of the middle 50% of the observations.

### Interquartile Range

The **interquartile range,** or **IQR,** is the difference between the first and third quartiles; that is, $\text{IQR} = Q_3 - Q_1$.

---

In Example 3.16, we show how to obtain the interquartile range for the data on TV-viewing times.

**Example 3.16** | **The Interquartile Range**

*Weekly TV-Viewing Times* Find the IQR for the TV-viewing-time data given in Table 3.12 on page 123.

**Solution** As we discovered in Example 3.15, the first and third quartiles are 23 and 36.5, respectively. Therefore,

$$\text{IQR} = Q_3 - Q_1 = 36.5 - 23 = 13.5 \text{ hours.}$$

**Interpretation** The middle 50% of the TV-viewing times are spread out over a 13.5-hour interval, roughly.

*You try it!*

Exercise 3.93(b)
on page 132

•  •  •

### The Five-Number Summary

From the three quartiles, we can obtain a measure of center (the median, $Q_2$) and measures of variation of the two middle quarters of the data, $Q_2 - Q_1$ for the second quarter and $Q_3 - Q_2$ for the third quarter. But the three quartiles don't tell us anything about the variation of the first and fourth quarters.

To gain that information, we need only include the minimum and maximum observations as well. Then the variation of the first quarter can be measured as the difference between the minimum and the first quartile, $Q_1 - \text{Min}$, and the variation of the fourth quarter can be measured as the difference between the third quartile and the maximum, $\text{Max} - Q_3$.

Thus the minimum, maximum, and quartiles together provide, among other things, information on center and variation.

**Definition 3.9**

**Five-Number Summary**

The **five-number summary** of a data set is Min, $Q_1$, $Q_2$, $Q_3$, Max.

---

**What Does It Mean?**

The five-number summary of a data set consists of the minimum, maximum, and quartiles, written in increasing order.

---

In Example 3.17, we show how to obtain and interpret the five-number summary of a set of data.

**Example 3.17** | **The Five-Number Summary**

*Weekly TV-Viewing Times* Find and interpret the five-number summary for the TV-viewing-time data given in Table 3.12 on page 123.

**Solution** From the ordered list of the entire data set (see page 123), Min = 5 and Max = 66. Furthermore, as we showed earlier, $Q_1 = 23$, $Q_2 = 30.5$, and $Q_3 = 36.5$. Consequently, the five-number summary of the data on TV-viewing times is 5, 23, 30.5, 36.5, and 66 hours. The variations of the four quarters of the TV-viewing-time data are therefore 18, 7.5, 6, and 29.5 hours, respectively.

*You try it!*

Exercise 3.93(c)
on page 132

**Interpretation** There is less variation in the middle two quarters of the TV-viewing times than in the first and fourth quarters, and the fourth quarter has the greatest variation of all.

•  •  •

### Outliers

In data analysis, the identification of **outliers**—observations that fall well outside the overall pattern of the data—is important. An outlier requires special attention. It may be the result of a measurement or recording error, an observation from a different population, or an unusual extreme observation. Note that an extreme observation need not be an outlier; it may instead be an indication of skewness.

As an example of an outlier, consider the data set consisting of the individual wealths (in dollars) of all U.S. residents. For this data set, the wealth of Bill Gates is an outlier—in this case, an unusual extreme observation.

Whenever you observe an outlier, try to determine its cause. If an outlier is caused by a measurement or recording error, or if for some other reason it clearly does not belong in the data set, the outlier can simply be removed. However, if no explanation for the outlier is apparent, the decision whether to retain it in the data set can be a difficult judgment call.

We can use quartiles and the IQR to identify potential outliers, that is, as a diagnostic tool for spotting observations that may be outliers. To do so, we first define the *lower limit* and the *upper limit* of a data set.

**Definition 3.10**

**Lower and Upper Limits**

The **lower limit** and **upper limit** of a data set are

$$\text{Lower limit} = Q_1 - 1.5 \cdot \text{IQR};$$
$$\text{Upper limit} = Q_3 + 1.5 \cdot \text{IQR}.$$

**What Does It Mean?**

The lower limit is the number that lies 1.5 IQRs below the first quartile; the upper limit is the number that lies 1.5 IQRs above the third quartile.

Observations that lie below the lower limit or above the upper limit are **potential outliers**. To determine whether a potential outlier is truly an outlier, you should perform further data analyses by constructing a histogram, stem-and-leaf diagram, and other appropriate graphics that we present later.

**Example 3.18** | **Outliers**

*Weekly TV-Viewing Times* For the TV-viewing-time data in Table 3.12 on page 123,

**a.** obtain the lower and upper limits.

**b.** determine potential outliers, if any.

**Solution**

**a.** As before, $Q_1 = 23$, $Q_3 = 36.5$, and IQR $= 13.5$. Therefore

$$\text{Lower limit} = Q_1 - 1.5 \cdot \text{IQR} = 23 - 1.5 \cdot 13.5 = 2.75 \text{ hours};$$
$$\text{Upper limit} = Q_3 + 1.5 \cdot \text{IQR} = 36.5 + 1.5 \cdot 13.5 = 56.75 \text{ hours}.$$

These limits are shown in Fig. 3.9.

**FIGURE 3.9**
Lower and upper limits for
TV-viewing times

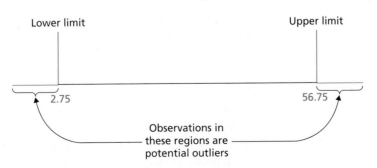

**b.** The ordered list of the entire data set on page 123 reveals one observation, 66, that lies outside the lower and upper limits—specifically, above the upper limit. Consequently, 66 is a potential outlier. A histogram and a stem-and-leaf diagram both indicate that the observation of 66 hours is truly an outlier.

Exercise 3.93(d)
on page 132

**Interpretation** The weekly viewing time of 66 hours lies outside the overall pattern of the other 19 viewing times in the data set.

• • •

### Boxplots

A **boxplot,** also called a **box-and-whisker diagram,** is based on the five-number summary and can be used to provide a graphical display of the center and variation of a data set. These diagrams, like stem-and-leaf diagrams, were invented by Professor John Tukey.[†]

To construct a boxplot, we also need the concept of *adjacent values.* The **adjacent values** of a data set are the most extreme observations that still lie within the lower and upper limits; they are the most extreme observations that are not potential outliers. Note that, if a data set has no potential outliers, the adjacent values are just the minimum and maximum observations.

---

**Procedure 3.1**    **To Construct a Boxplot**

**STEP 1  Determine the quartiles.**

**STEP 2  Determine potential outliers and the adjacent values.**

**STEP 3  Draw a horizontal axis on which the numbers obtained in Steps 1 and 2 can be located. Above this axis, mark the quartiles and the adjacent values with vertical lines.**

**STEP 4  Connect the quartiles to make a box and then connect the box to the adjacent values with lines.**

**STEP 5  Plot each potential outlier with an asterisk.**

---

**Note:**

- In a boxplot, the two lines emanating from the box are called **whiskers.**
- Boxplots are frequently drawn vertically instead of horizontally.
- Symbols other than an asterisk are often used to plot potential outliers.

**Example 3.19** | **Boxplots**

*Weekly TV-Viewing Times* The weekly TV-viewing times for a sample of 20 people are given in Table 3.12 on page 123. Construct a boxplot for these data.

---

[†]Several types of boxplots are in common use. Here we discuss a type that displays any potential outliers, sometimes called a **modified boxplot.**

**Solution** We apply Procedure 3.1. For easy reference, we repeat here the ordered list of the TV-viewing times.

5  15  16  20  21  25  26  27  30  30  31  32  32  34  35  38  38  41  43  66

**STEP 1 Determine the quartiles.**

In Example 3.15, we found the quartiles for the TV-viewing times to be $Q_1 = 23$, $Q_2 = 30.5$, and $Q_3 = 36.5$.

**STEP 2 Determine potential outliers and the adjacent values.**

As we found in Example 3.18(b), the TV-viewing times contain one potential outlier, 66. Therefore, from the ordered list of the data, we see that the adjacent values are 5 and 43.

**STEP 3 Draw a horizontal axis on which the numbers obtained in Steps 1 and 2 can be located. Above this axis, mark the quartiles and the adjacent values with vertical lines.**

See Fig. 3.10(a).

**FIGURE 3.10** Constructing a boxplot for the TV-viewing times

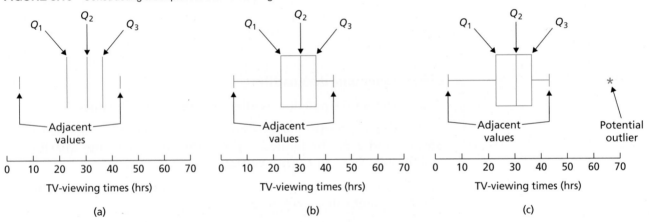

**STEP 4 Connect the quartiles to make a box and then connect the box to the adjacent values with lines.**

See Fig. 3.10(b).

**STEP 5 Plot each potential outlier with an asterisk.**

As we noted in Step 2, this data set contains one potential outlier—namely, 66. It is plotted with an asterisk in Fig. 3.10(c).

Figure 3.10(c) is a boxplot for the TV-viewing times data. Because the ends of the combined box are at the quartiles, the width of that box equals the interquartile range, IQR. Notice also that the left whisker represents the spread of the first quarter of the data, the two individual boxes represent the spreads of the second and third quarters, and the right whisker and asterisk represent the spread of the fourth quarter.

You try it!

Exercise 3.93(e) on page 132

**Interpretation** There is less variation in the middle two quarters of the TV-viewing times than in the first and fourth quarters, and the fourth quarter has the greatest variation of all.

• • •

## Other Uses of Boxplots

Boxplots are especially suited for comparing two or more data sets. In doing so, the same scale should be used for all the boxplots.

**Example 3.20** | ## Comparing Data Sets by Using Boxplots

*Skinfold Thickness* A study titled "Body Composition of Elite Class Distance Runners" was conducted by M. Pollock et al. to determine whether elite distance runners are actually thinner than other people. Their results were published in *The Marathon: Physiological, Medical, Epidemiological, and Psychological Studies* (P. Milvey (ed.), New York: New York Academy of Sciences, p. 366). The researchers measured skinfold thickness, an indirect indicator of body fat, of samples of runners and nonrunners in the same age group. The sample data, in millimeters (mm), presented in Table 3.13, are based on their results. Use boxplots to compare these two data sets, paying special attention to center and variation.

**TABLE 3.13**
Skinfold thickness (mm) for samples of elite runners and others

| Runners | | | Others | | | |
|---|---|---|---|---|---|---|
| 7.3 | 6.7 | 8.7 | 24.0 | 19.9 | 7.5 | 18.4 |
| 3.0 | 5.1 | 8.8 | 28.0 | 29.4 | 20.3 | 19.0 |
| 7.8 | 3.8 | 6.2 | 9.3 | 18.1 | 22.8 | 24.2 |
| 5.4 | 6.4 | 6.3 | 9.6 | 19.4 | 16.3 | 16.3 |
| 3.7 | 7.5 | 4.6 | 12.4 | 5.2 | 12.2 | 15.6 |

**Solution** Figure 3.11 displays boxplots for the two data sets, using the same scale.

**FIGURE 3.11**
Boxplots of the data sets in Table 3.13

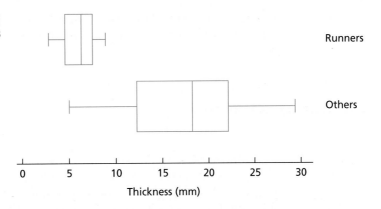

From Fig. 3.11, it is apparent that, on average, the elite runners sampled have smaller skinfold thickness than the other people sampled. Furthermore,

**You try it!**

Exercise 3.103
on page 133

there is much less variation in skinfold thickness among the elite runners sampled than among the other people sampled. By the way, when you study inferential statistics, you will be able to decide whether these descriptive properties of the samples can be extended to the populations from which the samples were drawn.

• • •

You can also use a boxplot to identify the approximate shape of the distribution of a data set. Figure 3.12 displays some common distribution shapes and their corresponding boxplots. Pay particular attention to how box width and whisker length relate to skewness and symmetry.

**FIGURE 3.12**
Distribution shapes and boxplots
for (a) uniform, (b) bell-shaped,
(c) right-skewed, and (d) left-skewed
distributions

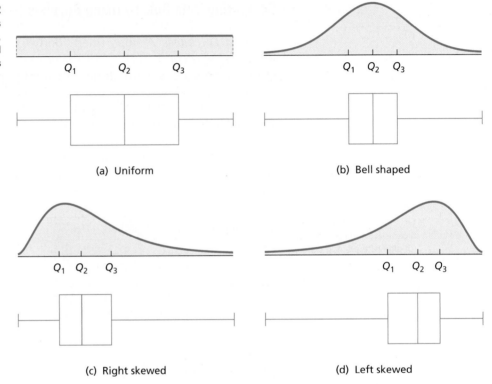

(a) Uniform

(b) Bell shaped

(c) Right skewed

(d) Left skewed

Employing boxplots to identify the shape of a distribution is most useful with large data sets. For small data sets, boxplots can be unreliable in identifying distribution shape; using a stem-and-leaf diagram or a dotplot is generally better.

## The Technology Center

In Sections 3.1 and 3.2, we showed how to use Minitab, Excel, and the TI-83/84 Plus to obtain several descriptive measures. You can apply those same programs to obtain the five-number summary of a data set, as you can see by referring to Outputs 3.1 and 3.2 on pages 101 and 115, respectively.

But remember that not all statisticians, statistical software packages, or statistical calculators define quartiles in exactly the same way. The results you obtain for quartiles by using the definitions in this book may therefore differ from those you obtain by using technology.

Most statistical technologies have programs that automatically produce boxplots. In this subsection, we present output and step-by-step instructions for such programs.

**Example 3.21**

## Using Technology to Obtain a Boxplot

*Weekly TV-Viewing Times*    Use Minitab, Excel, or the TI-83/84 Plus to obtain a boxplot for the TV-viewing times given in Table 3.12 on page 123.

**Solution**    We applied the boxplot programs to the data, resulting in Output 3.3. Steps for generating that output are presented in Instructions 3.3.

**OUTPUT 3.3**
Boxplot for the TV-viewing times

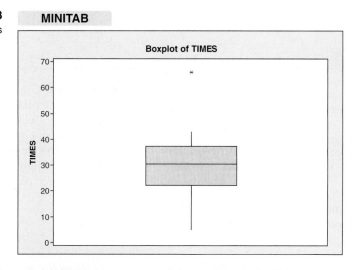

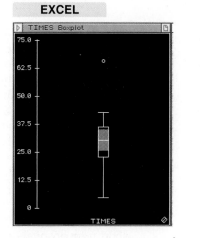

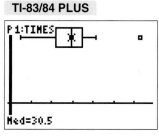

Notice that, as we mentioned earlier, some of these boxplots are drawn vertically instead of horizontally. Compare the boxplots in Output 3.3 to the one in Fig. 3.10(c) on page 128.

• • •

**INSTRUCTIONS 3.3** Steps for generating Output 3.3

| MINITAB | EXCEL | TI-83/84 PLUS |
|---|---|---|
| 1 Store the TV-viewing times from Table 3.12 in a column named TIMES<br>2 Choose **Graph ➤ Boxplot...**<br>3 Select the **Simple** boxplot and click **OK**<br>4 Specify TIMES in the **Graph variables** text box<br>5 Click **OK** | 1 Store the TV-viewing times from Table 3.12 in a range named TIMES<br>2 Choose **DDXL ➤ Charts and Plots**<br>3 Select **Boxplot** from the **Function type** drop-down box<br>4 Specify TIMES in the **Quantitative Variables** text box<br>5 Click **OK** | 1 Store the TV-viewing times from Table 3.12 in a list named TIMES<br>2 Press **2nd ➤ STAT PLOT** and then press **ENTER** twice<br>3 Arrow to the fourth graph icon and press **ENTER**<br>4 Press the down-arrow key<br>5 Press **2nd ➤ LIST**<br>6 Arrow down to TIMES and press **ENTER**<br>7 Press **ZOOM** and then **9** (and then **TRACE**, if desired) |

▮ ▮ ▮

# Exercises 3.3

## Understanding the Concepts and Skills

**3.84** Identify by name three important groups of percentiles.

**3.85** Identify an advantage that the median and interquartile range have over the mean and standard deviation, respectively.

**3.86** Explain why the minimum and maximum observations are added to the three quartiles to describe better the variation in a data set.

**3.87** Is an extreme observation necessarily an outlier? Explain your answer.

**3.88** Under what conditions are boxplots useful for identifying the shape of a distribution?

**3.89** Regarding the interquartile range,
a. what type of descriptive measure is it?
b. what does it measure?

**3.90** Identify a use of the lower and upper limits.

**3.91** When are the adjacent values just the minimum and maximum observations?

**3.92** Which measure of variation is preferred when
a. the mean is used as a measure of center?
b. the median is used as a measure of center?

*In Exercises 3.93–3.100, do the following for each of the data sets:*
*a. Obtain and interpret the quartiles.*
*b. Determine and interpret the interquartile range.*
*c. Find and interpret the five-number summary.*
*d. Identify potential outliers, if any.*
*e. Construct and interpret a boxplot.*

**3.93 The Great Gretzky.** Wayne Gretzky, a retired professional hockey player, played 20 seasons in the National Hockey League (NHL), from 1980 through 1999. S. Berry explored some of Gretzky's accomplishments in "A Statistician Reads the Sports Pages" (*Chance*, Vol. 16, No. 1, pp. 49–54). The following table shows the number of games in which Gretzky played during each of his 20 seasons in the NHL.

| | | | | |
|---|---|---|---|---|
| 79 | 80 | 80 | 80 | 74 |
| 80 | 80 | 79 | 64 | 78 |
| 73 | 78 | 74 | 45 | 81 |
| 48 | 80 | 82 | 82 | 70 |

**3.94 Parenting Grandparents.** In the article "Grandchildren Raised by Grandparents, a Troubling Trend" (*California Agriculture*, Vol. 55, No. 2, pp. 10–17), M. Blackburn considered the rates of children (under 18 years of age) living in California with grandparents as their primary caretakers. A sample of 14 California counties yielded the following percentages of children under 18 living with grandparents.

| | | | | | | |
|---|---|---|---|---|---|---|
| 5.9 | 4.0 | 5.7 | 5.1 | 4.1 | 4.4 | 6.5 |
| 4.4 | 5.8 | 5.1 | 6.1 | 4.5 | 4.9 | 4.9 |

**3.95 Hospital Stays.** The U.S. National Center for Health Statistics compiles data on the length of stay by patients in short-term hospitals and publishes its findings in *Vital and Health Statistics*. A random sample of 21 patients yielded the following data on length of stay, in days.

| | | | | | | |
|---|---|---|---|---|---|---|
| 4 | 4 | 12 | 18 | 9 | 6 | 12 |
| 3 | 6 | 15 | 7 | 3 | 55 | 1 |
| 10 | 13 | 5 | 7 | 1 | 23 | 9 |

**3.96 Miles Driven.** The U.S. Federal Highway Administration conducts studies on motor vehicle travel by type of vehicle. Results are published annually in *Highway Statistics*. A sample of 15 cars yields the following data on number of miles driven, in thousands, for last year.

| | | | | |
|---|---|---|---|---|
| 13.2 | 13.3 | 11.9 | 15.7 | 11.3 |
| 12.2 | 16.7 | 10.7 | 3.3 | 13.6 |
| 14.8 | 9.6 | 11.6 | 8.7 | 15.0 |

**3.97 Hurricanes.** An article by D. Schaefer et al. (*Journal of Tropical Ecology*, Vol. 16, pp. 189–207) reported on a long-term study of the effects of hurricanes on tropical streams of the Luquillo Experimental Forest in Puerto Rico. The study shows that Hurricane Hugo had a significant impact on stream water chemistry. The following table shows a sample of 10 ammonia fluxes in the first year after Hugo. Data are in kilograms per hectare per year.

| | | | | |
|---|---|---|---|---|
| 96 | 66 | 147 | 147 | 175 |
| 116 | 57 | 154 | 88 | 154 |

**3.98 Sky Guide.** The publication *California Wild: Natural Sciences for Thinking Animals* has a monthly feature called the "Sky Guide" that keeps track of the sunrise and sunset, for the first day of each month, in San Francisco, CA. Over several issues, B. Quock, from the Morrison Planetarium, recorded the following sunrise times from the first of July of one year through the first of June of the next year. The times are given in minutes past midnight.

| | | | | | |
|---|---|---|---|---|---|
| 352 | 374 | 400 | 426 | 396 | 427 |
| 445 | 434 | 400 | 354 | 374 | 349 |

**3.99 Capital Spending.** An issue of *Brokerage Report* discussed the capital spending of telecommunications companies in the United States and Canada. The capital spending, in thousands of dollars, for each of 27 telecommunications companies is shown in the following table.

| | | | | | | |
|---|---|---|---|---|---|---|
| 9,310 | 2,515 | 3,027 | 1,300 | 1,800 | 70 | 3,634 |
| 656 | 664 | 5,947 | 649 | 682 | 1,433 | 389 |
| 17,341 | 5,299 | 195 | 8,543 | 4,200 | 7,886 | 11,189 |
| 1,006 | 1,403 | 1,982 | 21 | 125 | 2,205 | |

**3.100 Medieval Cremation Burials.** In the article "Material Culture as Memory: Combs and Cremations in Early Medieval Britain" (*Early Medieval Europe*, Vol. 12, Issue 2, pp. 89–128), H. Williams discussed the frequency of cremation burials found in 17 archaeological sites in eastern England. Here are the data.

| | | | | | | | | |
|---|---|---|---|---|---|---|---|---|
| 83 | 64 | 46 | 48 | 523 | 35 | 34 | 265 | 2484 |
| 46 | 385 | 21 | 86 | 429 | 51 | 258 | 119 | |

**3.101 Nicotine Patches.** In the paper "The Smoking Cessation Efficacy of Varying Doses of Nicotine Patch Delivery Systems 4 to 5 Years Post-Quit Day" (*Preventative Medicine*, 28, pp. 113–118), D. Daughton et al. discussed the long-term effectiveness of transdermal nicotine patches on participants who had previously smoked at least 20 cigarettes per day. A sample of 15 participants in the Transdermal Nicotine Study Group (TNSG) reported that they now smoke the following number of cigarettes per day.

| | | | | |
|---|---|---|---|---|
| 10 | 9 | 10 | 8 | 7 |
| 6 | 10 | 9 | 10 | 8 |
| 9 | 10 | 8 | 8 | 10 |

**a.** Determine the quartiles for these data.
**b.** Remark on the usefulness of quartiles with respect to this data set.

**3.102 Starting Salaries.** The National Association of Colleges and Employers (NACE) conducts surveys of salary offers to new college graduates and publishes the results in *Salary Survey*. The following diagram provides boxplots for the starting annual salaries, in thousands of dollars, obtained from samples of 35 business administration graduates (top boxplot) and 32 liberal arts graduates (bottom boxplot). Use the boxplots to compare the starting salaries of the sampled business administration graduates and liberal arts graduates, paying special attention to center and variation.

**3.103 Obesity.** Researchers in obesity wanted to compare the effectiveness of dieting with exercise against dieting without exercise. Seventy-three patients were randomly divided into two groups. Group 1, composed of 37 patients, was put on a program of dieting with exercise. Group 2, composed of 36 patients, dieted only. The results for weight loss, in pounds, after 2 months are summarized in the following boxplots. The top boxplot is for Group 1 and the bottom boxplot is for Group 2. Use the boxplots to compare the weight losses for the two groups, paying special attention to center and variation.

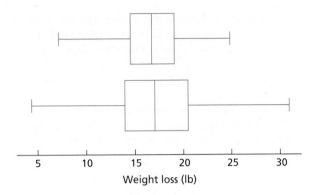

**3.104 Cuckoo Care.** Many species of cuckoos are brood parasites. The females lay their eggs in the nests of smaller bird species who then raise the young cuckoos at the expense of their own young. Data on the lengths, in millimeters (mm), of cuckoo eggs found in the nests of three bird species—the Tree Pipit, Hedge Sparrow, and Pied Wagtail—were collected by the late O. M. Latter in 1902 and used by L. H. C. Tippett in his text *The Methods of Statistics* (New York: Wiley, 1952, p. 176). Use the following boxplots to compare the lengths of cuckoo eggs found in the nests of the three bird species, paying special attention to center and variation.

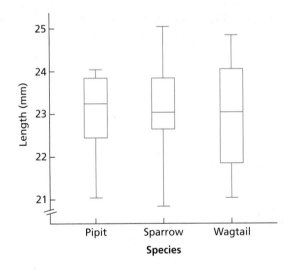

**3.105 Sickle Cell Disease.** A study published by E. Anionwu et al. in the *British Medical Journal* (Vol. 282, pp. 283–286) examined the steady state hemoglobin levels of patients with three different types of sickle cell disease: HB SC, HB SS, and HB ST. Use the following boxplots to compare the hemoglobin levels for the three groups of patients, paying special attention to center and variation.

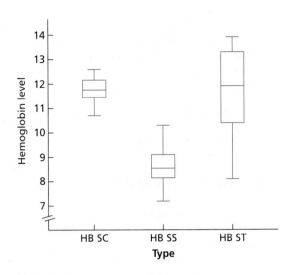

**3.106** Each of the following boxplots was obtained from a very large data set. Use the boxplots to identify the approximate shape of the distribution of each data set.

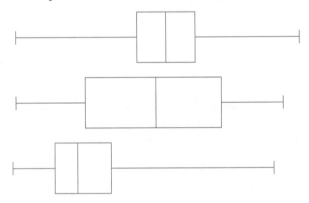

**3.107** What can you say about the boxplot of a symmetric distribution?

## Working With Large Data Sets

*In Exercises **3.108–3.113**, use the technology of your choice to do the following:*
*a. Obtain and interpret the quartiles.*
*b. Determine and interpret the interquartile range.*
*c. Find and interpret the five-number summary.*
*d. Identify potential outliers, if any.*
*e. Obtain and interpret a boxplot.*

**3.108 Women Students.** The U.S. Department of Education sponsors a report on educational institutions, including colleges and universities, titled *Digest of Education Statistics*. Among many of the statistics provided are the numbers of men and women enrolled in 2-year and 4-year degree-granting institutions. During one year, the percentage of full-time enrolled students that were women, for each of the 50 states and the District of Columbia, is as presented on the WeissStats CD.

**3.109 The Great White Shark.** In an article titled "Great White, Deep Trouble" (*National Geographic*, Vol. 197(4), pp. 2–29), Peter Benchley—the author of *JAWS*—discussed various aspects of the Great White Shark (*Carcharodon carcharias*). Data on the number of pups borne in a lifetime by each of 80 Great White Shark females are provided on the WeissStats CD.

**3.110 Top Recording Artists.** From the Recording Industry Association of America Web site, we obtained data on the number of albums sold, in millions, for the top recording artists (U.S. sales only) as of June 20, 2005. Those data are provided on the WeissStats CD.

**3.111 High School Completion Rates.** As reported by the U.S. Census Bureau in *Current Population Reports*, the percentage of adults in each state and the District of Columbia that have completed high school is as provided on the WeissStats CD.

**3.112 Crime Rates.** The U.S. Federal Bureau of Investigation publishes the annual crime rates for each state and the District of Columbia in the document *Crime in the United States*. Those rates, given per 1000 population, are provided on the WeissStats CD.

**3.113 Body Temperature.** A study by researchers at the University of Maryland addressed the question of whether the mean body temperature of humans is 98.6°F. The results of the study by P. Mackowiak et al. appeared in the article "A Critical Appraisal of 98.6°F, the Upper Limit of the Normal Body Temperature, and Other Legacies of Carl Reinhold August Wunderlich" (*Journal of the American Medical Association*, Vol. 268, pp. 1578–1580). Among other data, the researchers obtained the body temperatures of 93 healthy humans, as provided on the WeissStats CD.

*In each of Exercises 3.114–3.117, use the technology of your choice to do the following:*

*a. Obtain boxplots for the data sets, using the same scale.*

*b. Use the boxplots obtained in part (a) to compare the data sets, paying special attention to center and variation.*

**3.114 Sex and Direction.** In the paper "The Relation of Sex and Sense of Direction to Spatial Orientation in an Unfamiliar Environment" (*Journal of Environmental Psychology*,

Vol. 20, pp. 17–28), Sholl et al. published the results of examining the sense of direction of 30 male and 30 female students. After being taken to an unfamiliar wooded park, the students were given a number of spatial orientation tests, including pointing to south, which tested their absolute frame of reference. Pointing to south was done by moving a pointer attached to a 360° protractor. We have supplied on the WeissStats CD the absolute pointing errors, in degrees, for both the male and female participants.

**3.115 Vegetarians and Omnivores.** Philosophical and health issues are prompting an increasing number of Taiwanese to switch to a vegetarian lifestyle. A study by Lu et al., published in the *Journal of Nutrition* (Vol. 130, pp. 1591–1596), compared the daily intake of nutrients by vegetarians and omnivores living in Taiwan. Among the nutrients considered was protein. Too little protein stunts growth and interferes with all bodily functions; too much protein puts a strain on the kidneys, can cause diarrhea and dehydration, and can leach calcium from bones and teeth. The daily protein intakes, in grams, for 51 female vegetarians and 53 female omnivores are provided on the WeissStats CD.

**3.116 Magazine Ads.** Advertising researchers Shuptrine and McVicker wanted to determine whether there were significant differences in the readability of magazine advertisements. Thirty magazines were classified based on their educational level—high, mid, or low—and then three magazines were randomly selected from each level. From each magazine, six advertisements were randomly chosen and examined for readability. In this particular case, readability was characterized by the numbers of words, sentences, and words of three syllables or more in each ad. The researchers published their findings in the *Journal of Advertising Research* (Vol. 21, p. 47). The number of words of three syllables or more in each ad are provided on the WeissStats CD.

**3.117 Prolonging Life.** Vitamin C (ascorbate) boosts the human immune system and is effective in preventing a variety of illnesses. In a study by E. Cameron and L. Pauling, published in the *Proceedings of the National Academy of Science USA* (Vol. 75, pp. 4538–4542), patients in advanced stages of cancer were given a Vitamin C supplement. Patients were grouped according to the organ affected by cancer: stomach, bronchus, colon, ovary, or breast. The study yielded the survival times, in days, given on the WeissStats CD.

## 3.4 Descriptive Measures for Populations; Use of Samples

In this section, we discuss several descriptive measures for *population data*—the data obtained by observing the values of a variable for an entire population. Although, in reality, we often don't have access to population data, it is nonetheless helpful to become familiar with the notation and formulas used for descriptive measures of such data.

### The Population Mean

Recall that, for a variable $x$ and a sample of size $n$ from a population, the sample mean is

$$\bar{x} = \frac{\Sigma x_i}{n}.$$

**TABLE 3.14**
Notation used for a sample and for the population

| | Size | Mean |
|---|---|---|
| **Sample** | $n$ | $\bar{x}$ |
| **Population** | $N$ | $\mu$ |

First, we sum the observations of the variable for the sample, and then we divide by the size of the sample.

We can find the mean of a finite population similarly: first, we sum all possible observations of the variable for the entire population, and then we divide by the size of the population. However, to distinguish the *population mean* from a sample mean, we use the Greek letter $\mu$ (pronounced "mew") to denote the population mean. We also use the uppercase English letter $N$ to represent the size of the population. Table 3.14 summarizes the notation that is used for both a sample and the population.

**Definition 3.11**

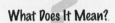

**What Does It Mean?**

A population mean (mean of a variable) is the arithmetic average (mean) of population data.

### Population Mean (Mean of a Variable)

For a variable $x$, the mean of all possible observations for the entire population is called the **population mean** or **mean of the variable** $x$. It is denoted $\mu_x$ or, when no confusion will arise, simply $\mu$. For a finite population,

$$\mu = \frac{\Sigma x_i}{N}$$

where $N$ is the population size.

**Example 3.22**

### The Population Mean

*U.S. Women's World Cup Soccer Team* From FIFA.com, the official Web site of the *Fédération Internationale de Football Association*, we obtained data for the players on the 2003 U.S. Women's World Cup soccer team, as shown in Table 3.15. Heights are given in centimeters (cm) and weights in kilograms (kg). Find the population mean weight of these soccer players.

**Solution** Here the variable is weight, and the population consists of the players on the 2003 U.S. Women's World Cup soccer team. The sum of the weights in the fourth column of Table 3.15 is 1251 kg. Because there are 20 players, $N = 20$. Consequently,

$$\mu = \frac{\Sigma x_i}{N} = \frac{1251}{20} = 62.55 \text{ kg.}$$

**TABLE 3.15**
U.S. Women's World Cup
soccer team, 2003

| Name | Position | Height (cm) | Weight (kg) | College |
|------|----------|-------------|-------------|---------|
| Scurry, Briana | GK | 175 | 68 | UMass |
| Bivens, Kylie | DF | 165 | 59 | Santa Clara |
| Pearce, Christie | DF | 165 | 61 | Monmouth |
| Reddick, Cat | DF | 170 | 68 | UNC |
| Roberts, Tiffany | MF | 161 | 51 | UNC |
| Chastain, Brandi | DF | 171 | 58 | Santa Clara |
| Boxx, Shannon | MF | 173 | 67 | Notre Dame |
| MacMillan, Shannon | FW | 164 | 61 | Portland |
| Hamm, Mia | FW | 163 | 61 | UNC |
| Wagner, Aly | MF | 165 | 61 | Santa Clara |
| Foudy, Julie | MF | 168 | 59 | Stanford |
| Parlow, Cindy | FW | 180 | 70 | UNC |
| Lilly, Kristine | MF | 163 | 57 | UNC |
| Fawcett, Joy | DF | 170 | 61 | California |
| Sobrero, Kate | DF | 170 | 61 | Notre Dame |
| Milbrett, Tiffeny | FW | 157 | 61 | Portland |
| Slaton, Danielle | DF | 168 | 66 | Santa Clara |
| Mullinix, Siri | GK | 173 | 64 | UNC |
| Hucles, Angela | MF | 168 | 64 | Virginia |
| Wambach, Abby | FW | 180 | 73 | Florida |

**You try it!**

Exercise 3.127(a)
on page 145

**Interpretation** The population mean weight of the players on the 2003 U.S. Women's World Cup soccer team is 62.55 kg.

• • •

## Using a Sample Mean to Estimate a Population Mean

In inferential studies, we analyze sample data. Nonetheless, the objective is to describe the entire population. We use samples because they are usually more practical, as illustrated in the next example.

**Example 3.23** | **A Use of a Sample Mean**

*Estimating Mean Household Income* The U.S. Census Bureau reports the mean (annual) income of U.S. households in the publication *Current Population Reports*. To obtain the population data—the incomes of all U.S. households—would be extremely expensive and time consuming. It is also unnecessary because accurate estimates of the mean income of all U.S. households can be obtained from the mean income of a sample of such households. The Census Bureau samples only 60,000 households from a total of more than 100 million.

Here are the basic elements for this problem, also summarized in Fig. 3.13:

- *Variable:* income
- *Population:* all U.S. households
- *Population data:* incomes of all U.S. households
- *Population mean:* mean income, $\mu$, of all U.S. households
- *Sample:* 60,000 U.S. households sampled by the Census Bureau
- *Sample data:* incomes of the 60,000 U.S. households sampled
- *Sample mean:* mean income, $\bar{x}$, of the 60,000 U.S. households sampled

**FIGURE 3.13**
Population and sample for incomes
of U.S. households

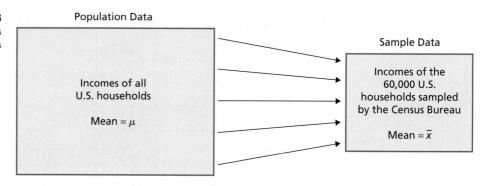

The Census Bureau uses the sample mean income, $\bar{x}$, of the 60,000 U.S. households sampled to estimate the population mean income, $\mu$, of all U.S. households.

• • •

## The Population Standard Deviation

Recall that, for a variable $x$ and a sample of size $n$ from a population, the sample standard deviation is

$$s = \sqrt{\frac{\sum(x_i - \bar{x})^2}{n - 1}}.$$

The standard deviation of a finite population is obtained in a similar, but slightly different, way. To distinguish the *population standard deviation* from a sample standard deviation, we use the Greek letter $\sigma$ (pronounced "sigma") to denote the population standard deviation.

**Definition 3.12**

**What Does It Mean?**

Roughly speaking, the population standard deviation indicates how far, on average, the observations in the population are from the mean of the population.

## Population Standard Deviation (Standard Deviation of a Variable)

For a variable $x$, the standard deviation of all possible observations for the entire population is called the **population standard deviation** or **standard deviation of the variable** $x$. It is denoted $\sigma_x$ or, when no confusion will arise, simply $\sigma$. For a finite population, the defining formula is

$$\sigma = \sqrt{\frac{\sum(x_i - \mu)^2}{N}}$$

where $N$ is the population size.

The population standard deviation can also be found from the computing formula

$$\sigma = \sqrt{\frac{\sum x_i^2}{N} - \mu^2}.$$

**Note:** Just as $s^2$ is called a sample variance, $\sigma^2$ is called the **population variance** (or **variance of the variable**).

## Example 3.24 | The Population Standard Deviation

***U.S. Women's World Cup Soccer Team*** Calculate the population standard deviation of the weights of the players on the 2003 U.S. Women's World Cup soccer team.

**Solution** We apply the computing formula given in Definition 3.12. To do so, we need the sum of the squares of the weights and the population mean weight, $\mu$. From Example 3.22, $\mu = 62.55$ kg. Squaring each weight in Table 3.15 and adding the results yields $\Sigma x_i^2 = 78{,}737$. Recalling that there are 20 players, we have

$$\sigma = \sqrt{\frac{\Sigma x_i^2}{N} - \mu^2} = \sqrt{\frac{78{,}737}{20} - (62.55)^2} = 4.9 \text{ kg.}$$

**You try it!**

Exercise 3.127(b) on page 145

**Interpretation** The population standard deviation of the weights of the players on the 2003 U.S. Women's World Cup soccer team is 4.9 kg. Roughly speaking, the weights of the individual players fall, on average, 4.9 kg from their mean weight of 62.55 kg.

• • •

### Using a Sample Standard Deviation to Estimate a Population Standard Deviation

We have shown that a sample mean can be used to estimate a population mean. Likewise, a sample standard deviation can be used to estimate a population standard deviation, as illustrated in the next example.

## Example 3.25 | A Use of a Sample Standard Deviation

***Estimating Variation in Bolt Diameters*** A hardware manufacturer produces "10-millimeter (mm)" bolts. The manufacturer knows that the diameters of the bolts produced vary somewhat from 10 mm and also from each other. But even if he is willing to accept some variation in bolt diameters, he cannot tolerate too much variation—if the variation is too large, too many of the bolts will be unusable (too narrow or too wide).

To evaluate the variation in bolt diameters, the manufacturer needs to know the population standard deviation, $\sigma$, of bolt diameters. Because, in this case, $\sigma$ cannot be determined exactly (do you know why?), the manufacturer must use the standard deviation of the diameters of a sample of bolts to estimate $\sigma$. He decides to take a sample of 20 bolts.

Here are the basic elements for this problem, also summarized in Fig. 3.14:

- *Variable:* diameter
- *Population:* all "10-mm" bolts produced by the manufacturer
- *Population data:* diameters of all bolts produced
- *Population standard deviation:* standard deviation, $\sigma$, of the diameters of all bolts produced
- *Sample:* 20 bolts sampled by the manufacturer
- *Sample data:* diameters of the 20 bolts sampled by the manufacturer
- *Sample standard deviation:* standard deviation, $s$, of the diameters of the 20 bolts sampled

**FIGURE 3.14**
Population and sample
for bolt diameters

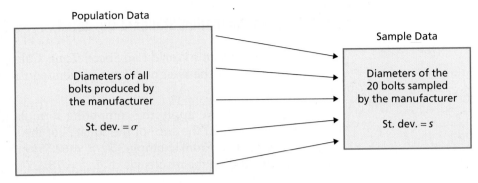

The manufacturer can use the sample standard deviation, $s$, of the diameters of the 20 bolts sampled to estimate the population standard deviation, $\sigma$, of the diameters of all bolts produced.

• • •

## Parameter and Statistic

The following terminology helps us distinguish between descriptive measures for populations and samples.

---

**Definition 3.13**  **Parameter and Statistic**

**Parameter:**  A descriptive measure for a population.

**Statistic:**  A descriptive measure for a sample.

---

Thus, for example, $\mu$ and $\sigma$ are parameters, whereas $\bar{x}$ and $s$ are statistics.

## Standardized Variables

From any variable $x$, we can form a new variable $z$, defined as follows.

---

**Definition 3.14**  **Standardized Variable**

For a variable $x$, the variable

$$z = \frac{x - \mu}{\sigma}$$

is called the **standardized version** of $x$ or the **standardized variable** corresponding to the variable $x$.

**What Does It Mean?**

The standardized version of a variable $x$ is obtained by first subtracting from $x$ its mean and then dividing by its standard deviation.

---

A standardized variable always has mean 0 and standard deviation 1. For this and other reasons, standardized variables play an important role in many aspects of statistical theory and practice. We present a few applications of standardized variables in this section; several others appear throughout the rest of the book.

| Example 3.26 | **Standardized Variables** |

*Understanding the Basics* Let's consider a simple variable $x$—namely, one with possible observations shown in the first row of Table 3.16.

**TABLE 3.16**
Possible observations of $x$ and $z$

| $x$ | −1 | 3 | 3 | 3 | 5 | 5 |
|---|---|---|---|---|---|---|
| $z$ | −2 | 0 | 0 | 0 | 1 | 1 |

a.  Determine the standardized version of $x$.

b.  Find the observed value of $z$ corresponding to an observed value of $x$ of 5.

c.  Calculate all possible observations of $z$.

d.  Find the mean and standard deviation of $z$ using Definitions 3.11 and 3.12. Was it necessary to do these calculations to obtain the mean and standard deviation?

e.  Show dotplots of the distributions of both $x$ and $z$. Interpret the results.

**Solution**

a.  Using Definitions 3.11 and 3.12, we find that the mean and standard deviation of $x$ are $\mu = 3$ and $\sigma = 2$. Consequently, the standardized version of $x$ is

$$z = \frac{x - 3}{2}.$$

b.  The observed value of $z$ corresponding to an observed value of $x$ of 5 is

$$z = \frac{x - 3}{2} = \frac{5 - 3}{2} = 1.$$

c.  Applying the formula $z = (x - 3)/2$ to each of the possible observations of the variable $x$ shown in the first row of Table 3.16, we obtain the possible observations of the standardized variable $z$ shown in the second row of Table 3.16.

d.  From the second row of Table 3.16,

$$\mu_z = \frac{\Sigma z_i}{N} = \frac{0}{6} = 0$$

and

$$\sigma_z = \sqrt{\frac{\Sigma (z_i - \mu_z)^2}{N}} = \sqrt{\frac{6}{6}} = 1.$$

The results of these two computations illustrate that the mean of a standardized variable is always 0 and its standard deviation is always 1. We didn't need to perform these calculations.

e.  Figures 3.15(a) and 3.15(b) at the top of the next page show dotplots of the distributions of $x$ and $z$, respectively.

**Interpretation** The two dotplots in Fig. 3.15 show how standardizing shifts a distribution so the new mean is 0 and changes the scale so the new standard deviation is 1.

• • •

**FIGURE 3.15**

Dotplots of the distributions of $x$ and its standardized version $z$

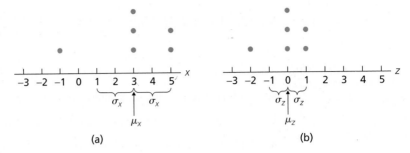

(a)                                        (b)

### z-Scores

An important concept associated with standardized variables is that of the *z-score*, or *standard score*.

---

**Definition 3.15**

**What Does It Mean?**

The z-score of an observation tells us the number of standard deviations that the observation is from the mean, that is, how far the observation is from the mean in units of standard deviation.

**z-Score**

For an observed value of a variable $x$, the corresponding value of the standardized variable $z$ is called the **z-score** of the observation. The term **standard score** is often used instead of *z-score*.

---

A negative z-score indicates that the observation is below (less than) the mean, whereas a positive z-score indicates that the observation is above (greater than) the mean. Example 3.27 illustrates calculation and interpretation of z-scores.

**Example 3.27** | **z-Scores**

*U.S. Women's World Cup Soccer Team* The weight data for the U.S. Women's World Cup soccer team is given in the fourth column of Table 3.15 on page 137. We determined earlier that the mean and standard deviation of the weights are 62.55 kg and 4.9 kg, respectively. So, in this case, the standardized variable is

$$z = \frac{x - 62.55}{4.9}.$$

a. Find and interpret the z-score of Tiffany Roberts's weight of 51 kg.

b. Find and interpret the z-score of Cindy Parlow's weight of 70 kg.

c. Construct a graph showing the results obtained in parts (a) and (b).

**Solution**

a. The z-score for Tiffany Roberts's weight of 51 kg is

$$z = \frac{x - 62.55}{4.9} = \frac{51 - 62.55}{4.9} = -2.36.$$

**Interpretation** Tiffany Roberts's weight is 2.36 standard deviations below the mean.

b. The z-score for Cindy Parlow's weight of 70 kg is

$$z = \frac{x - 62.55}{4.9} = \frac{70 - 62.55}{4.9} = 1.52.$$

**Interpretation**  Cindy Parlow's weight is 1.52 standard deviations above the mean.

    **c.**  In Fig. 3.16, we marked Tiffany Roberts's weight of 51 kg with a green dot and Cindy Parlow's weight of 70 kg with a red dot. Additionally, we located the mean, $\mu = 62.55$ kg, and measured intervals equal in length to the standard deviation, $\sigma = 4.9$ kg.

**FIGURE 3.16**   Graph showing Tiffany Roberts's weight (green dot) and Cindy Parlow's weight (red dot)

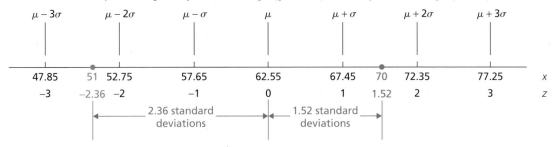

In Fig. 3.16, the numbers in the row labeled $x$ represent weights in kilograms and the numbers in the row labeled $z$ represent $z$-scores (i.e., number of standard deviations from the mean).

**You try it!**

Exercise 3.131
on page 146

$\bullet \ \bullet \ \bullet$

## The z-Score as a Measure of Relative Standing

The three-standard-deviations rule (Key Fact 3.2 on page 114) states that almost all the observations in any data set lie within three standard deviations to either side of the mean. Thus, for any variable, almost all possible observations have $z$-scores between $-3$ and $3$.

    The $z$-score of an observation, therefore, can be used as a rough measure of its relative standing among all the observations comprising a data set. For instance, a $z$-score of 3 or more indicates that the observation is larger than most of the other observations; a $z$-score of $-3$ or less indicates that the observation is smaller than most of the other observations; and a $z$-score near 0 indicates that the observation is located near the mean.

    The use of $z$-scores as a measure of relative standing can be refined and made more precise by applying Chebychev's rule, as you are asked to explore in Exercises 3.140 and 3.141. Moreover, if the distribution of the variable under consideration is roughly bell shaped, then, as you will see in Chapter 6, the use of $z$-scores as a measure of relative standing can be improved even further.

    Percentiles usually give a more exact method of measuring relative standing than do $z$-scores. However, if only the mean and standard deviation of a variable are known, $z$-scores provide a feasible alternative to percentiles for measuring relative standing.

## Other Descriptive Measures for Populations

Up to this point, we have concentrated on the mean and standard deviation in our discussion of descriptive measures for populations. The reason is that many of the classical inference procedures for center and variation concern those two parameters.

However, modern statistical analyses also rely heavily on descriptive measures based on percentiles. Quartiles, the IQR, and other descriptive measures based on percentiles are defined in the same way for (finite) populations as they are for samples. For simplicity and with one exception, we use the same notation for descriptive measures based on percentiles whether we are considering a sample or a population. The exception is that we use $M$ to denote a sample median and $\eta$ (eta) to denote a population median.

## Exercises 3.4

### Understanding the Concepts and Skills

**3.118** Identify each quantity as a parameter or a statistic.
**a.** $\mu$      **b.** $s$      **c.** $\bar{x}$      **d.** $\sigma$

**3.119** Although, in practice, sample data are generally analyzed in inferential studies, what is the ultimate objective of such studies?

**3.120 Microwave Popcorn.** For a specific brand of microwave popcorn, what property is desirable for the population standard deviation of the cooking time? Explain your answer.

**3.121** Complete the following sentences.
**a.** A standardized variable always has mean _____ and standard deviation _____.
**b.** The $z$-score corresponding to an observed value of a variable tells you _____.
**c.** A positive $z$-score indicates that the observation is _____ the mean, whereas a negative $z$-score indicates that the observation is _____ the mean.

**3.122** Identify the statistic that is used to estimate
**a.** a population mean.
**b.** a population standard deviation.

**3.123 Women's Soccer.** Earlier in this section, we found that the population mean weight of the players on the 2003 U.S. Women's World Cup soccer team is 62.55 kg. In this context, is the number 62.55 a parameter or a statistic? Explain your answer.

**3.124 Heights of Basketball Players.** In Section 3.2, we analyzed the heights of the starting five players on each of two men's college basketball teams. The heights, in inches, of the players on Team II are 67, 72, 76, 76, and 84. Regarding the five players as a sample of all male starting college basketball players,
**a.** compute the sample mean height, $\bar{x}$.
**b.** compute the sample standard deviation, $s$.
Regarding the players now as a population,
**c.** compute the population mean height, $\mu$.

**d.** compute the population standard deviation, $\sigma$.
Comparing your answers from parts (a) and (c) and from parts (b) and (d),
**e.** why are the values for $\bar{x}$ and $\mu$ equal?
**f.** why are the values for $s$ and $\sigma$ different?

**3.125 Age of U.S. Residents.** The U.S. Census Bureau collects information about the ages of people in the United States. Results are published in *Current Population Reports*.
**a.** Identify the variable and population under consideration.
**b.** A sample of six U.S. residents yielded the following data on ages (in years). Determine the mean and median of these age data. Decide whether those descriptive measures are parameters or statistics, and use statistical notation to express the results.

| 29 | 54 | 9 | 45 | 51 | 7 |
|----|----|----|----|----|----|

**c.** By consulting the most recent census data, we found that the mean age and median age of all U.S. residents are 35.8 years and 35.3 years, respectively. Decide whether those descriptive measures are parameters or statistics, and use statistical notation to express the results.

**3.126 Back to Pinehurst.** In the June 2005 issue of *Golf Digest* is a preview of the 2005 U.S. Open, titled "Back to Pinehurst." Included is information on the course, Pinehurst in North Carolina. The following table lists the lengths, in yards, of the 18 holes at Pinehurst.

| 401 | 469 | 336 | 565 | 472 | 220 | 404 | 467 | 175 |
|-----|-----|-----|-----|-----|-----|-----|-----|-----|
| 607 | 476 | 449 | 378 | 468 | 203 | 492 | 190 | 442 |

**a.** Obtain and interpret the population mean of the hole lengths at Pinehurst.
**b.** Obtain and interpret the population standard deviation of the hole lengths at Pinehurst.

**3.127 Hurricane Hunters.** The Air Force Reserve's 53rd Weather Reconnaissance Squadron, better known as the *Hurricane Hunters,* fly into the eye of tropical cyclones in their WC-130 Hercules aircraft to collect and report vital meteorological data for advance storm warnings. The data are relayed to the National Hurricane Center in Miami, Florida, for broadcasting emergency storm warnings on land. Over the Atlantic, 2004 was a busy year for the Hurricane Hunters with 15 tropical storms, 8 of which were hurricanes. The maximum winds were recorded for each cyclone, and are shown in the following table, where TS = tropical storm and Hu = hurricane. [SOURCE: *U.S. Air Force, National Oceanic and Atmospheric Administration.*]

| Cyclone | Date | Max wind (mph) |
|---|---|---|
| Hu Alex | 07/31–08/06 | 120 |
| TS Bonnie | 08/03–08/12 | 65 |
| Hu Charley | 08/09–08/15 | 145 |
| Hu Danielle | 08/13–08/21 | 105 |
| TS Earl | 08/13–08/16 | 45 |
| Hu Frances | 08/25–09/10 | 145 |
| TS Gaston | 08/27–09/01 | 70 |
| TS Hermine | 08/29–08/31 | 50 |
| Hu Ivan | 09/02–09/18 | 165 |
| Hu Jeanne | 09/13–09/27 | 115 |
| Hu Karl | 09/16–09/24 | 140 |
| Hu Lisa | 09/19–10/03 | 75 |
| TS Matthew | 10/08–10/10 | 45 |
| TS Nicole | 10/10–10/11 | 50 |
| TS Otto | 11/30–12/02 | 50 |

Consider these cyclones a population of interest. Obtain the following parameters for the maximum wind speeds. Use the appropriate mathematical notation for the parameters to express your answers.
a. Mean
b. Standard deviation
c. Median
d. Mode
e. IQR

**3.128 Dallas Mavericks.** From the ESPN Web site, in the *Dallas Mavericks Roster,* we obtained the following weights, in pounds, for the players on that basketball team as of May 2, 2005.

| | | | | |
|---|---|---|---|---|
| 235 | 180 | 280 | 265 | 200 |
| 225 | 185 | 240 | 210 | 245 |
| 245 | 260 | 218 | 180 | 240 |

Obtain the following parameters for these weights. Use the appropriate mathematical notation for the parameters to express your answers.
a. Mean
b. Standard deviation
c. Median
d. Mode
e. IQR

**3.129 STD Surveillance.** The Centers for Disease Control and Prevention compiles reported cases and rates of diseases in United States cities and outlying areas. In a document titled, *Sexually Transmitted Disease Surveillance,* the number of reported cases of all stages of syphilis is provided for cities, including Phoenix and Los Angeles. Following is the number of reported cases of syphilis for those two cities for the years 1999 through 2003.

| Phoenix | 722 | 957 | 737 | 855 | 829 |
|---|---|---|---|---|---|
| **Los Angeles** | 1189 | 1857 | 1339 | 1626 | 1752 |

a. Obtain the individual population means of the number of cases for both cities.
b. Without doing any calculations, decide for which city the population standard deviation of the number of cases is smaller. Explain your answer.
c. Obtain the individual population standard deviations of the number of cases for both cities.
d. Are your answers to parts (b) and (c) consistent? Why or why not?

**3.130 Dart Doubles.** The top two players in the 2001–2002 Professional Darts Corporation World Championship were Phil Taylor and Peter Manley. Taylor and Manley dominated the competition with a record number of doubles. A *double* is a throw that lands in either the outer ring of the dartboard or the outer ring of the bull's-eye. The following table provides the number of doubles thrown by each of the two players during the five rounds of competition, as found in *Chance* (Vol. 15, No. 3, pp. 48–55).

| Taylor | 21 | 18 | 18 | 19 | 13 |
|---|---|---|---|---|---|
| **Manley** | 5 | 24 | 20 | 26 | 14 |

a. Obtain the individual population means of the number of doubles.
b. Without doing any calculations, decide for which player the standard deviation of the number of doubles is smaller. Explain your answer.

c. Obtain the individual population standard deviations of the number of doubles.

d. Are your answers to parts (b) and (c) consistent? Why or why not?

**3.131 Doing Time.** According to *Statistical Report*, published by the U.S. Bureau of Prisons, the mean time served by prisoners released from federal institutions for the first time is 16.3 months. Assume the standard deviation of the times served is 17.9 months. Let $x$ denote time served by a prisoner released for the first time from a federal institution.

a. Find the standardized version of $x$.

b. Find the mean and standard deviation of the standardized variable.

c. Determine the $z$-scores for prison times served of 64.7 months and 4.2 months. Round your answers to two decimal places.

d. Interpret your answers in part (c).

e. Construct a graph similar to Fig. 3.16 on page 143 that depicts your results from parts (b) and (c).

**3.132 Gestation Periods of Humans.** Gestation periods of humans have a mean of 266 days and a standard deviation of 16 days. Let $y$ denote the variable "gestation period" for humans.

a. Find the standardized variable corresponding to $y$.

b. What are the mean and standard deviation of the standardized variable?

c. Obtain the $z$-scores for gestation periods of 227 days and 315 days. Round your answers to two decimal places.

d. Interpret your answers in part (c).

e. Construct a graph similar to Fig. 3.16 on page 143 that shows your results from parts (b) and (c).

**3.133 Frog Thumb Length.** W. Duellman and J. Kohler explore a new species of frog in the article "New Species of Marsupial Frog (Hylidae: Hemiphractinae: *Gastrotheca*) from the Yungas of Bolivia" (*Journal of Herpetology*, Vol. 39, No. 1, pp. 91–100). These two museum researchers collected information on the lengths and widths of different body parts for the male and female *Gastrotheca piperata*. Thumb length for the female *Gastrotheca piperata* has a mean of 6.71 mm and a standard deviation of 0.67 mm. Let $x$ denote thumb length for a female specimen.

a. Find the standardized version of $x$.

b. Determine and interpret the $z$-scores for thumb lengths of 5.2 mm and 8.1 mm. Round your answers to two decimal places.

**3.134 Low Birth Weight Hospital Stays.** Data on low birth weight babies were collected over a 2-year period by 14 participating centers of the National Institute of Child Health and Human Development Neonatal Research Network. Results were reported by J. Lemons et al. in the on-line

paper "Very Low Birth Weight Outcomes of the National Institute of Child Health and Human Development Neonatal Research Network" (*Pediatrics*, Vol. 107, No. 1, p. e1). For the 1084 surviving babies whose birth weights were 751–1000 grams, the average length of stay in the hospital was 86 days, although one center had an average of 66 days and another had an average of 108 days.

a. Are the mean lengths of stay sample means or population means? Explain your answer.

b. Assuming that the population standard deviation is 12 days, determine the $z$-score for a baby's length of stay of 86 days at the center where the mean was 66 days.

c. Assuming that the population standard deviation is 12 days, determine the $z$-score for a baby's length of stay of 86 days at the center where the mean was 108 days.

d. What can you conclude from parts (b) and (c) about an infant with a length of stay equal to the mean at all centers if that infant was born at a center with a mean of 66 days? mean of 108 days?

**3.135 Low Gas Mileage.** Suppose you buy a new car whose advertised mileage is 25 miles per gallon (mpg). After driving your car for several months, you find that its mileage is 21.4 mpg. You telephone the manufacturer and learn that the standard deviation of gas mileages for all cars of the model you bought is 1.15 mpg.

a. Find the $z$-score for the gas mileage of your car, assuming the advertised claim is correct.

b. Does it appear that your car is getting unusually low gas mileage? Explain your answer.

**3.136 Exam Scores.** Suppose that you take an exam with 400 possible points and are told that the mean score is 280 and that the standard deviation is 20. You are also told that you got 350. Did you do well on the exam? Explain your answer.

## Extending the Concepts and Skills

**Population and Sample Standard Deviations.** In Exercises 3.137–3.139, you examine the numerical relationship between the population standard deviation and the sample standard deviation computed from the same data. This relationship is helpful when the computer or statistical calculator being used has a built-in program for sample standard deviation but not for population standard deviation.

**3.137** Consider the following three data sets.

| Data Set 1 | | Data Set 2 | | | | Data Set 3 | | | | |
|---|---|---|---|---|---|---|---|---|---|---|
| 2 | 4 | 7 | 5 | 5 | 3 | 4 | 7 | 8 | 9 | 7 |
| 7 | 3 | 9 | 8 | 6 | | 4 | 5 | 3 | 4 | 5 |

**a.** Assuming that each of these data sets is sample data, compute the standard deviations. (Round your final answers to two decimal places.)

**b.** Assuming that each of these data sets is population data, compute the standard deviations. (Round your final answers to two decimal places.)

**c.** Using your results from parts (a) and (b), make an educated guess about the answer to the question: If both $s$ and $\sigma$ are computed for the same data set, will they tend to be closer together if the data set is large or if it is small?

**3.138** Consider a data set with $m$ observations. If the data are sample data, you compute the sample standard deviation, $s$, whereas if the data are population data, you compute the population standard deviation, $\sigma$.

**a.** Derive a mathematical formula that gives $\sigma$ in terms of $s$ when both are computed for the same data set. (*Hint:* First note that, numerically, the values of $\bar{x}$ and $\mu$ are identical. Consider the ratio of the defining formula for $\sigma$ to the defining formula for $s$.)

**b.** Refer to the three data sets in Exercise 3.137. Verify that your formula in part (a) works for each of the three data sets.

**c.** Suppose that a data set consists of 15 observations. You compute the sample standard deviation of the data and obtain $s = 38.6$. Then you realize that the data are actually population data and that you should have obtained the population standard deviation instead. Use your formula from part (a) to obtain $\sigma$.

**3.139 Women's Soccer.** Refer to the heights of the 2003 U.S. Women's World Cup soccer team in the third column of Table 3.15 on page 137. Use the technology of your choice to obtain

**a.** the population mean height.

**b.** the population standard deviation of the heights. *Note:* Depending on the technology that you're using, you may need to refer to the formula derived in Exercise 3.138(a).

**Estimating Relative Standing.** On page 120, we stated Chebychev's rule: For any data set and any real number $k > 1$, at least $100(1 - 1/k^2)\%$ of the observations lie within $k$ standard deviations to either side of the mean. You can use $z$-scores and Chebychev's rule to estimate the relative standing of an observation.

To see how, let us consider again the weights of the players on the 2003 U.S. Women's World Cup soccer team given in the fourth column of Table 3.15 on page 137. Earlier, we found that the population mean and standard deviation of these weights are 62.55 kg and 4.9 kg, respectively. We also found that the $z$-score for Cindy Parlow's weight of 70 kg is 1.52.

Applying Chebychev's rule to that $z$-score (that is, with $k = 1.52$), we conclude that at least $100(1 - 1/1.52^2)\%$,

or 56.7%, of the weights lie within 1.52 standard deviations to either side of the mean. Therefore, Cindy Parlow's weight, which is 1.52 standard deviations above the mean, is greater than at least 56.7% of the other players' weights.

**3.140 Stewed Tomatoes.** A company produces cans of stewed tomatoes with an advertised weight of 14 oz. The standard deviation of the weights is known to be 0.4 oz. A quality-control engineer selects a can of stewed tomatoes at random and finds its net weight to be 17.28 oz.

**a.** Estimate the relative standing of that can of stewed tomatoes, assuming the true mean weight is 14 oz. Use the $z$-score and Chebychev's rule.

**b.** Does the quality-control engineer have reason to suspect that the true mean weight of all cans of stewed tomatoes being produced is not 14 oz? Explain your answer.

**3.141 Buying a Home.** Suppose that you are thinking of buying a resale home in a large tract. The owner is asking $205,500. Your realtor obtains the sale prices of comparable homes in the area that have sold recently. The mean of the prices is $220,258 and the standard deviation is $5,237. Does it appear that the home you are contemplating buying is a bargain? Explain your answer using the $z$-score and Chebychev's rule.

**Comparing Relative Standing.** If two distributions have the same shape or, more generally, if they differ only by center and variation, then $z$-scores can be used to compare the relative standings of two observations from those distributions. The two observations can be of the same variable from different populations or they can be of different variables from the same population. Consider Exercise 3.142.

**3.142 SAT Scores.** Each year, thousands of high school students bound for college take the Scholastic Assessment Test (SAT). This test measures the verbal and mathematical abilities of prospective college students. Student scores are reported on a scale that ranges from a low of 200 to a high of 800. Summary results for the scores are published by the College Entrance Examination Board in *College Bound Seniors*. In one high school graduating class, the mean SAT math score is 528 with a standard deviation of 105; the mean SAT verbal score is 475 with a standard deviation of 98. A student in the graduating class scored 740 on the SAT math and 715 on the SAT verbal.

**a.** Under what conditions would it be reasonable to use $z$-scores to compare the standings of the student on the two tests relative to the other students in the graduating class?

**b.** Assuming that a comparison using $z$-scores is legitimate, relative to the other students in the graduating class, on which test did the student do better?

## Chapter in Review

### You Should be Able to

1. use and understand the formulas in this chapter.

2. explain the purpose of a measure of center.

3. obtain and interpret the mean, the median, and the mode(s) of a data set.

4. choose an appropriate measure of center for a data set.

5. use and understand summation notation.

6. define, compute, and interpret a sample mean.

7. explain the purpose of a measure of variation.

8. define, compute, and interpret the range of a data set.

9. define, compute, and interpret a sample standard deviation.

10. define percentiles, deciles, and quartiles.

11. obtain and interpret the quartiles, IQR, and five-number summary of a data set.

12. obtain the lower and upper limits of a data set and identify potential outliers.

13. construct and interpret a boxplot.

14. use boxplots to compare two or more data sets.

15. use a boxplot to identify distribution shape for large data sets.

16. define the population mean (mean of a variable).

17. define the population standard deviation (standard deviation of a variable).

18. compute the population mean and population standard deviation of a finite population.

19. distinguish between a parameter and a statistic.

20. understand how and why statistics are used to estimate parameters.

21. define and obtain standardized variables.

22. obtain and interpret z-scores.

### Key Terms

adjacent values, *127*
box-and-whisker diagram, *127*
boxplot, *127*
Chebychev's rule, *114, 120*
deciles, *122*
descriptive measures, *92*
deviations from the mean, *108*
empirical rule, *114, 121*
first quartile ($Q_1$), *123*
five-number summary, *125*
indices, *100*
interquartile range (IQR), *124*
lower limit, *126*
mean, *94*
mean of a variable ($\mu$), *136*
measures of center, *94*
measures of central tendency, *94*
measures of spread, *107*

measures of variation, *107*
median, *95*
mode, *96*
outliers, *125*
parameter, *140*
percentiles, *122*
population mean ($\mu$), *136*
population standard deviation ($\sigma$), *138*
population variance ($\sigma^2$), *138*
potential outlier, *126*
quartiles, *123*
quintiles, *122*
range, *108*
resistant measure, *97*
sample mean ($\bar{x}$), *100*
sample size ($n$), *100*
sample standard deviation ($s$), *110*

sample variance ($s^2$), *109*
second quartile ($Q_2$), *123*
standard deviation, *108*
standard deviation of a variable ($\sigma$), *138*
standard score, *142*
standardized variable, *140*
standardized version, *140*
statistic, *140*
subscripts, *99*
sum of squared deviations, *109*
summation notation, *99*
third quartile ($Q_3$), *123*
trimmed means, *97, 106*
upper limit, *126*
variance of a variable ($\sigma^2$), *138*
whiskers, *127*
z-score, *142*

## Review Problems

### Understanding the Concepts and Skills

**1.** Define
**a.** descriptive measures.
**b.** measures of center.
**c.** measures of variation.

**2.** Identify the two most commonly used measures of center for quantitative data. Explain the relative advantages and disadvantages of each.

**3.** Among the measures of center discussed, which is the only one appropriate for qualitative data?

**4.** Identify the most appropriate measure of variation corresponding to each of the following measures of center.
**a.** Mean                    **b.** Median

**5.** Specify the mathematical symbol used for each of the following descriptive measures.
**a.** Sample mean
**b.** Sample standard deviation
**c.** Population mean
**d.** Population standard deviation

**6.** Data Set A has more variation than Data Set B. Decide which of the following statements are necessarily true.
**a.** Data Set A has a larger mean than Data Set B.
**b.** Data Set A has a larger standard deviation than Data Set B.

**7.** Complete the statement: Almost all the observations in any data set lie within _____ standard deviations to either side of the mean.

**8.** Regarding the five-number summary:
**a.** Identify its components.
**b.** How can it be employed to describe center and variation?
**c.** What graphical display is based on it?

**9.** Regarding outliers:
**a.** What is an outlier?
**b.** Explain how you can identify potential outliers, using only the first and third quartiles.

**10.** Regarding $z$-scores:
**a.** How is a $z$-score obtained?
**b.** What is the interpretation of a $z$-score?
**c.** An observation has a $z$-score of 2.9. Roughly speaking, what is the relative standing of the observation?

**11. Party Time.** An integral part of doing business in the dot-com culture of the late 1990s was frequenting the party circuit centered in San Francisco. Here high-tech companies threw as many as five parties a night to recruit or retain talented workers in a highly competitive job market. With as many as 700 guests at a single party, the food and booze

flowed with an average alcohol cost per guest of $15–$18 and an average food bill of $75–$150. A sample of guests at a dot-com party yielded the following data on number of alcoholic drinks consumed per person. [SOURCE: *USA TODAY* Online.]

| | | | | |
|---|---|---|---|---|
| 4 | 4 | 1 | 0 | 5 |
| 1 | 1 | 2 | 4 | 3 |
| 1 | 5 | 3 | 0 | 2 |
| 2 | 2 | 1 | 2 | 4 |

**a.** Find the mean, median, and mode of these data.
**b.** Which measure of center do you think is best here? Explain your answer.

**12. Duration of Marriages.** The National Center for Health Statistics publishes information on the duration of marriages in *Vital Statistics of the United States*. Which measure of center is more appropriate for data on the duration of marriages, the mean or the median? Explain your answer.

**13. Causes of Death.** Death certificates provide data on the causes of death. Which of the three main measures of center is appropriate here? Explain your answer.

**14. Fossil Argonauts.** In the article "Fossil Argonauts (Mollusca: Cephalopoda: Octopodida) from Late Miocene Silt-stones of the Los Angeles Basin, California" (*Journal of Paleontology*, Vol. 79, No. 3, pp. 520–531), paleontologists L. Saul and C. Stadum discussed fossilized Argonaut egg cases from the late Miocene period found in California. A sample of 10 fossilized egg cases yielded the following data on height, in millimeters. Obtain the mean, median, and mode(s) of these data.

| | | | | |
|---|---|---|---|---|
| 37.5 | 31.5 | 27.4 | 21.0 | 32.0 |
| 33.0 | 33.0 | 38.0 | 17.4 | 34.5 |

**15. Road Patrol.** In the paper "Injuries and Risk Factors in a 100-Mile (161-km) Infantry Road March" (*Preventative Medicine*, Vol. 28, pp. 167–173), Reynolds et al. reported on a study commissioned by the U.S. Army. The purpose of the study was to improve medical planning and identify risk factors during multiple-day road patrols by examining the acute effects of long-distance marches by light-infantry soldiers. Each soldier carried a standard U.S. Army rucksack, Meal-Ready-to-Eat packages, and other field equipment. A sample of 10 participating soldiers revealed the following data on total load mass, in kilograms.

| | | | | |
|---|---|---|---|---|
| 48 | 50 | 45 | 49 | 44 |
| 47 | 37 | 54 | 40 | 43 |

**a.** Obtain the sample mean of these 10 load masses.
**b.** Obtain the range of the load masses.
**c.** Obtain the sample standard deviation of the load masses.

**16. Millionaires.** Dr. Thomas Stanley of Georgia State University has collected information on millionaires, including their ages, since 1973. A sample of 36 millionaires has a mean age of 58.5 years and a standard deviation of 13.4 years.
**a.** Complete the following graph.

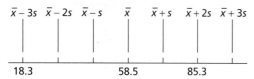

$\bar{x} - 3s \quad \bar{x} - 2s \quad \bar{x} - s \quad \bar{x} \quad \bar{x} + s \quad \bar{x} + 2s \quad \bar{x} + 3s$

18.3          58.5          85.3

**b.** Fill in the blanks: Almost all the 36 millionaires are between _____ and _____ years old.

**17. Millionaires.** Refer to Problem 16. The ages of the 36 millionaires sampled are arranged in increasing order in the following table.

| | | | | | | | | |
|---|---|---|---|---|---|---|---|---|
| 31 | 38 | 39 | 39 | 42 | 42 | 45 | 47 | 48 |
| 48 | 48 | 52 | 52 | 53 | 54 | 55 | 57 | 59 |
| 60 | 61 | 64 | 64 | 66 | 66 | 67 | 68 | 68 |
| 69 | 71 | 71 | 74 | 75 | 77 | 79 | 79 | 79 |

**a.** Determine the quartiles for the data.
**b.** Obtain and interpret the interquartile range.
**c.** Find and interpret the five-number summary.
**d.** Calculate the lower and upper limits.
**e.** Identify potential outliers, if any.
**f.** Construct and interpret a boxplot.

**18. Oxygen Distribution.** In the article "Distribution of Oxygen in Surface Sediments from Central Sagami Bay, Japan: In Situ Measurements by Microelectrodes and Planar Optodes" (*Deep Sea Research Part I: Oceanographic Research Papers*, Vol. 52, Issue 10, pp. 1974–1987), R. Glud et al. explore the distributions of oxygen in surface sediments from central Sagami Bay. The oxygen distribution gives important information on the general biogeochemistry of marine sediments. Measurements were performed at 16 sites. A sample of 22 depths yielded the following data,

in millimoles per square meter per day (mmol m$^{-2}$ d$^{-1}$), on diffusive oxygen uptake (DOU).

| | | | | | | | |
|---|---|---|---|---|---|---|---|
| 1.8 | 2.0 | 1.8 | 2.3 | 3.8 | 3.4 | 2.7 | 1.1 |
| 3.3 | 1.2 | 3.6 | 1.9 | 7.6 | 2.0 | 1.5 | 2.0 |
| 1.1 | 0.7 | 1.0 | 1.8 | 1.8 | 6.7 | | |

**a.** Obtain the five-number summary for these data.
**b.** Identify potential outliers, if any.
**c.** Construct a boxplot.

**19. Traffic Fatalities.** From the *Fatality Analysis Reporting System (FARS)* of the National Highway Traffic Safety Administration, we obtained data on the numbers of traffic fatalities in Wisconsin and New Mexico for the years 1982–2003. Use the following boxplots for those data to compare the traffic fatalities in the two states, paying special attention to center and variation.

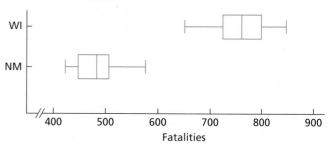

**20. UC Enrollment.** According to *Information and Communications, University of California*, the Fall 2004 enrollment figures for undergraduates at the University of California campuses were as follows.

| Campus | Enrollment (1000s) |
|---|---|
| Berkeley | 22.9 |
| Davis | 23.2 |
| Irvine | 20.0 |
| Los Angeles | 24.9 |
| Riverside | 15.2 |
| San Diego | 20.3 |
| Santa Barbara | 18.1 |
| Santa Cruz | 13.7 |

**a.** Compute the population mean enrollment, $\mu$, of the UC campuses. (Round your answer to two decimal places.)
**b.** Compute $\sigma$. (Round your answer to two decimal places.)
**c.** Letting $x$ denote enrollment, specify the standardized variable, $z$, corresponding to $x$.
**d.** Without performing any calculations, give the mean and standard deviation of $z$. Explain your answers.
**e.** Construct dotplots for the distributions of both $x$ and $z$. Interpret your graphs.

**f.** Obtain and interpret the $z$-scores for the enrollments at the Los Angeles and Riverside campuses.

**21. Gasoline Prices.** The U.S. Energy Information Administration reports weekly figures on retail gasoline prices in *Motor Gasoline Price Survey*. Every Monday, retail prices for all three grades of gasoline are collected by telephone from a sample of approximately 800 retail gasoline outlets out of a total of approximately 115,000 retail gasoline outlets. For the 800 stations sampled on October 24, 2005, the mean price per gallon for unleaded regular gasoline was $2.603.

**a.** Is the mean price given here a sample mean or a population mean? Explain your answer.

**b.** What letter or symbol would you use to designate the mean of $2.603?

**c.** Is the mean price given here a statistic or a parameter? Explain your answer.

## Working With Large Data Sets

**22. U.S. Divisions and Regions.** The U.S. Census Bureau classifies the states in the United States by region and division. The data giving the region and division of each state are presented on the WeissStats CD. Use the technology of your choice to do the following:

**a.** Determine the mode(s) of the regions.

**b.** Determine the mode(s) of the divisions.

*In Problems 23–25, use the technology of your choice to do the following:*

*a. Obtain the mean, median, and mode(s) of the data. Determine which of these measures of center is best and explain your answer.*

*b. Determine the range and sample standard deviation of the data.*

*c. Find the five-number summary and interquartile range of the data.*

*d. Identify potential outliers, if any.*

*e. Obtain and interpret a boxplot.*

**23. Agricultural Exports.** The U.S. Department of Agriculture collects data pertaining to the value of agricultural exports and publishes its findings in *U.S. Agricultural Trade Update*. For one year, the values of these exports, by state, are provided on the WeissStats CD. Data are in millions of dollars.

**24. Life Expectancy.** From the U.S. Census Bureau, in the document *International Data Base*, we obtained data on the expectation of life (in years) for people in various countries and areas. Those data are presented on the WeissStats CD.

**25. High and Low Temperatures.** The U.S. National Oceanic and Atmospheric Administration publishes temperature data in *Climatography of the United States*. According to that document, the annual average maximum and minimum temperatures for selected cities in the United States are as provided on the WeissStats CD. [*Note:* Do parts (a)–(e) for both the maximum and minimum temperatures.]

**26. World Series.** The two teams that played in the 2005 World Series were the Chicago White Sox and the Houston Astros. From the Major League Baseball Web site, MLB.com, we obtained the heights and weights of the players (both active and non-active) on these two teams and stored those data on the WeissStats CD. Heights are in inches and weights are in pounds.

**a.** Obtain boxplots for the two sets of height data, using the same scale.

**b.** Use the boxplots obtained in part (a) to compare the heights of the players on the two teams, paying special attention to center and variation.

**c.** Obtain boxplots for the two sets of weight data, using the same scale.

**d.** Use the boxplots obtained in part (c) to compare the weights of the players on the two teams, paying special attention to center and variation.

# Focusing on Data Analysis UWEC Undergraduates

Recall from Chapter 1 (see page 34) that the Focus database and Focus sample contain information on the undergraduate students at the University of Wisconsin - Eau Claire (UWEC). Now would be a good time for you to review the discussion about these data sets.

a. Open the Focus sample (FocusSample) in the statistical software package of your choice and then obtain the mean and standard deviation of the ages of the sample of 200 UWEC undergraduate students. Are these descriptive measures parameters or statistics? Explain your answer.

b. If your statistical software package will accommodate the entire Focus database (Focus), open that worksheet and then obtain the mean and standard deviation of the ages of all UWEC undergraduate students. (*Answers:* 20.75 years and 1.87 years) Are these descriptive measures parameters or statistics? Explain your answer.

c. Compare your means and standard deviations from parts (a) and (b). What do these results illustrate?

d. If you used a different simple random sample of 200 UWEC undergraduate students than the one in the Focus sample, would you expect the mean and standard deviation of the ages to be the same as that in part (a)? Explain your answer.

e. Open the Focus sample and then obtain the mode of the classifications (class levels) of the sample of 200 UWEC undergraduate students.

f. If your statistical software package will accommodate the entire Focus database, open that worksheet and then obtain the mode of the classifications of all UWEC undergraduate students. (*Answer:* Senior)

g. From parts (e) and (f), you found that the mode of the classifications is the same for both the population and sample of UWEC undergraduate students. Would this necessarily always be the case? Explain your answer.

h. Open the Focus sample and then obtain the five-number summary of the ACT math scores, individually for males and females. Use those statistics to compare the two samples of scores, paying particular attention to center and variation.

i. Open the Focus sample and then obtain the five-number summary of the ACT English scores, individually for males and females. Use those statistics to compare the two samples of scores, paying particular attention to center and variation.

j. Open the Focus sample and then obtain boxplots of the cumulative GPAs, individually for males and females. Use those statistics to compare the two samples of cumulative GPAs, paying particular attention to center and variation.

k. Open the Focus sample and then obtain boxplots of the cumulative GPAs, individually for each classification (class level). Use those statistics to compare the four samples of cumulative GPAs, paying particular attention to center and variation.

# Case Study Discussion The Triple Crown

The table on page 93 displays the year, name, and times for each of the 11 Triple Crown winners. Do the following separately for the 11 times in each of the three races—the Kentucky Derby, Preakness Stakes, and Belmont Stakes.

a. Convert all times to speeds, in miles per hour, and use those data in the parts that follow. *Hint:* Divide each distance run by the running time in seconds and then multiply the result by 3600 (the number of seconds in an hour).

b. Determine the mean and median of the speeds. Explain any difference between these two measures of center.

c. Obtain the range and population standard deviation of the speeds.

d. Find and interpret the $z$-scores for the speeds of Citation and Secretariat.

e. Determine and interpret the quartiles of the speeds.

f. Find the lower and upper limits. Use them to identify potential outliers.

g. Construct a boxplot for the speeds and interpret your result in terms of the variation in the speeds.

h. Use the technology of your choice to solve parts (a)–(g).

# Biography    JOHN TUKEY: A Pioneer of EDA

**John Wilder Tukey** was born on June 16, 1915, in New Bedford, Massachusetts. After earning bachelor's and master's degrees in chemistry from Brown University in 1936 and 1937, respectively, he enrolled in the mathematics program at Princeton University, where he received a master's degree in 1938 and a doctorate in 1939.

After graduating, Tukey was appointed Henry B. Fine Instructor in Mathematics at Princeton; 10 years later he was advanced to a full professorship. In 1965, Princeton established a department of statistics, and Tukey was named its first chairperson. In addition to his position at Princeton, he was a member of the Technical Staff at AT&T Bell Laboratories where he served as Associate Executive Director, Research in the Information Sciences Division, from 1945 until his retirement in 1985.

Tukey was among the leaders in the field of exploratory data analysis (EDA), which provides techniques such as stem-and-leaf diagrams for effectively investigating data. He also made fundamental contributions to the areas of robust estimation and time series analysis. Tukey wrote numerous books and more than 350 technical papers on mathematics, statistics, and other scientific subjects. Addi-

tionally, he coined the word *bit,* a contraction of *binary digit* (a unit of information, often as processed by a computer).

Tukey's participation in educational, public, and government service was most impressive. He was appointed to serve on the President's Science Advisory Committee by President Eisenhower; was chairperson of the committee that prepared "Restoring the Quality of our Environment" in 1965; helped develop the National Assessment of Educational Progress; and was a member of the Special Advisory Panel on the 1990 Census of the U.S. Department of Commerce, Bureau of the Census—to name only a few of his involvements.

Among many honors, Tukey received the National Medal of Science, the IEEE Medal of Honor, Princeton University's James Madison Medal, and Foreign Member, The Royal Society (London). He was the first recipient of the Samuel S. Wilks Award of the American Statistical Association. Until his death, Tukey remained on the faculty at Princeton as Donner Professor of Science, Emeritus; Professor of Statistics, Emeritus; and Senior Research Statistician. Tukey died on July 26, 2000, after a short illness. He was 85 years old.

## StatCrunch in MyStatLab
### Analyzing Data Online

StatCrunch online statistical software offers an easy-to-use interface customized for this book. The StatCrunch feature for each chapter illustrates the use of the software to perform a statistical analysis discussed in the chapter. Exercises are provided to further apply StatCrunch to other statistical analyses examined in the chapter. Go to the WeissStats CD or to the Weiss Web site at www.aw-bc.com/weiss to access StatCrunch instructions and data sets. To access StatCrunch statistical software, go to the student content area of your Weiss MyStatLab course.

## Internet Projects
### Exploring Data Online

The Internet project for each chapter provides simulations, demonstrations, or activities that enhance the topics covered in the chapter. The project materials come from universities, individuals, governments, and companies from all over the world. To access the Internet projects on the Web, go to www.aw-bc.com/weiss. From this Web page, you can reach the Internet Projects Page, which we suggest that you bookmark for easy access in the future.

# 4 Descriptive Methods in Regression and Correlation

## Chapter Objectives

We often want to know whether two or more variables are related and, if they are, how they are related. In this chapter, we discuss relationships between two quantitative variables. In Chapter 12, we examine relationships between two qualitative (categorical) variables.

*Linear regression* and *correlation* are two commonly used methods for examining the relationship between quantitative variables and for making predictions. We discuss descriptive methods in linear regression and correlation in this chapter and consider inferential methods in Chapter 14.

To prepare for our discussion of linear regression, we review linear equations with one independent variable in Section 4.1. In Section 4.2, we explain how to determine the *regression equation*, the equation of the line that best fits a set of data points.

In Section 4.3, we examine the coefficient of determination, a descriptive measure of the utility of the regression equation for making predictions. In Section 4.4, we discuss the linear correlation coefficient, which provides a descriptive measure of the strength of the linear relationship between two quantitative variables.

# Shoe Size and Height

Most of us have heard that tall people generally have larger feet than short people. Is that really true and, if so, what is the precise relationship between height and foot length? To examine the relationship, Professor Dennis Young obtained data on shoe size and height for a sample of students at Arizona State University. We have displayed the results obtained by Professor Young in the table below, where height is measured in inches.

At the end of this chapter, after you have studied the fundamentals of descriptive methods in regression and correlation, you will be asked to analyze these data to determine the relationship between shoe size and height and to ascertain the strength of that relationship. In particular, you will discover how shoe size can be used to predict height.

| Shoe size | Height | Sex | Shoe size | Height | Sex |
|-----------|--------|-----|-----------|--------|-----|
| 6.5 | 66.0 | F | 13.0 | 77.0 | M |
| 9.0 | 68.0 | F | 11.5 | 72.0 | M |
| 8.5 | 64.5 | F | 8.5 | 59.0 | F |
| 8.5 | 65.0 | F | 5.0 | 62.0 | F |
| 10.5 | 70.0 | M | 10.0 | 72.0 | M |
| 7.0 | 64.0 | F | 6.5 | 66.0 | F |
| 9.5 | 70.0 | F | 7.5 | 64.0 | F |
| 9.0 | 71.0 | F | 8.5 | 67.0 | M |
| 13.0 | 72.0 | M | 10.5 | 73.0 | M |
| 7.5 | 64.0 | F | 8.5 | 69.0 | F |
| 10.5 | 74.5 | M | 10.5 | 72.0 | M |
| 8.5 | 67.0 | F | 11.0 | 70.0 | M |
| 12.0 | 71.0 | M | 9.0 | 69.0 | M |
| 10.5 | 71.0 | M | 13.0 | 70.0 | M |

## 4.1 Linear Equations with One Independent Variable

To understand linear regression, let's first review linear equations with one independent variable. The general form of a **linear equation** with one independent variable can be written as

$$y = b_0 + b_1 x,$$

where $b_0$ and $b_1$ are constants (fixed numbers), $x$ is the independent variable, and $y$ is the dependent variable.[†]

The graph of a linear equation with one independent variable is a **straight line,** or simply a **line;** furthermore, any nonvertical line can be represented by such an equation. Examples of linear equations with one independent variable are $y = 4 + 0.2x$, $y = -1.5 - 2x$, and $y = -3.4 + 1.8x$. The graphs of these three linear equations are shown in Fig. 4.1.

**FIGURE 4.1**
Graphs of three linear equations

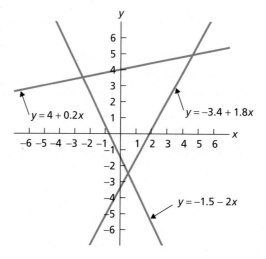

Linear equations with one independent variable occur frequently in applications of mathematics to many different fields, including the management, life, and social sciences, as well as the physical and mathematical sciences.

| Example 4.1 | **Linear Equations** |

*Word-Processing Costs* $CJ^2$ Business Services offers its clients word processing at a rate of $20 per hour plus a $25 disk charge. The total cost to a customer depends, of course, on the number of hours needed to complete the job. Find the equation that expresses the total cost in terms of the number of hours needed to complete the job.

---

[†]You may be familiar with the form $y = mx + b$ instead of the form $y = b_0 + b_1 x$. Statisticians prefer the latter form because it allows a smoother transition to multiple regression, in which there is more than one independent variable. Material on multiple regression is provided in the chapters *Multiple Regression Analysis* and *Model Building in Regression* on the WeissStats CD accompanying this book.

**Solution**  Because the rate for word processing is $20 per hour, a job that takes $x$ hours will cost $20x plus the $25 disk charge. Hence the total cost, $y$, of a job that takes $x$ hours is $y = 25 + 20x$.

• • •

The equation $y = 25 + 20x$ is linear; here $b_0 = 25$ and $b_1 = 20$. This equation gives us the exact cost for a job if we know the number of hours required. For instance, a job that takes 5 hours will cost $y = 25 + 20 \cdot 5 = \$125$; a job that takes 7.5 hours will cost $y = 25 + 20 \cdot 7.5 = \$175$. Table 4.1 displays these costs and a few others.

As we have already mentioned, the graph of a linear equation, such as $y = 25 + 20x$, is a line. To obtain the graph of $y = 25 + 20x$, we first plot the points displayed in Table 4.1 and then connect them with a line, as shown in Fig. 4.2.

**FIGURE 4.2**
Graph of $y = 25 + 20x$, obtained from the points displayed in Table 4.1

**TABLE 4.1**
Times and costs for five word-processing jobs

| Time (hr) | Cost ($) |
| $x$ | $y$ |
| --- | --- |
| 5.0 | 125 |
| 7.5 | 175 |
| 15.0 | 325 |
| 20.0 | 425 |
| 22.5 | 475 |

*You try it!*

Exercise 4.5 on page 160

The graph in Fig. 4.2 is useful for quickly estimating cost. For example, a glance at the graph shows that a 10-hour job will cost somewhere between $200 and $300. The exact cost is $y = 25 + 20 \cdot 10 = \$225$.

## Intercept and Slope

For a linear equation $y = b_0 + b_1x$, the number $b_0$ is the $y$-value of the point of intersection of the line and the $y$-axis. The number $b_1$ measures the steepness of the line; more precisely, $b_1$ indicates how much the $y$-value changes when the $x$-value increases by 1 unit. Figure 4.3 illustrates these relationships.

**FIGURE 4.3**
Graph of $y = b_0 + b_1x$

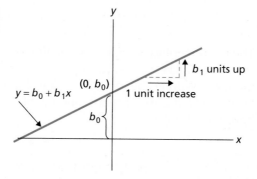

The numbers $b_0$ and $b_1$ have special names that reflect these geometric interpretations.

---

**Definition 4.1**

### *y*-Intercept and Slope

For a linear equation $y = b_0 + b_1 x$, the number $b_0$ is called the **$y$-intercept** and the number $b_1$ is called the **slope**.

---

In the next example, we apply the concepts of $y$-intercept and slope to the illustration of word-processing costs.

**Example 4.2** | *y*-Intercept and Slope

*Word-Processing Costs* In Example 4.1, we found the linear equation that expresses the total cost, $y$, of a word-processing job in terms of the number of hours, $x$, required to complete the job. The equation is $y = 25 + 20x$.

**a.** Determine the $y$-intercept and slope of that linear equation.

**b.** Interpret the $y$-intercept and slope in terms of the graph of the equation.

**c.** Interpret the $y$-intercept and slope in terms of word-processing costs.

**Solution**

**a.** The $y$-intercept for the equation is $b_0 = 25$, and the slope is $b_1 = 20$.

**b.** The $y$-intercept $b_0 = 25$ is the $y$-value where the line intersects the $y$-axis, as shown in Fig. 4.4. The slope $b_1 = 20$ indicates that the $y$-value increases by 20 units for every increase in $x$ of 1 unit.

**FIGURE 4.4**
Graph of $y = 25 + 20x$

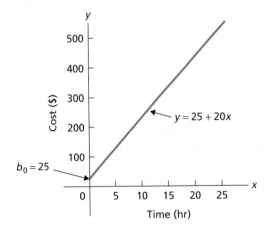

**c.** The $y$-intercept $b_0 = 25$ represents the total cost of a job that takes 0 hours. In other words, the $y$-intercept of $25 is a fixed cost that is charged no matter how long the job takes. The slope $b_1 = 20$ represents the cost per hour of $20; it is the amount that the total cost goes up for every additional hour the job takes.

**You try it!**

Exercise 4.9 on page 160

• • •

A line is determined by any two distinct points that lie on it. Thus, to draw the graph of a linear equation, first substitute two different $x$-values into the equation to get two distinct points; then connect those two points with a line.

For example, to graph the linear equation $y = 5 - 3x$, we can use the $x$-values 1 and 3 (or any other two $x$-values). The $y$-values corresponding to those two $x$-values are $y = 5 - 3 \cdot 1 = 2$ and $y = 5 - 3 \cdot 3 = -4$, respectively. Therefore the graph of $y = 5 - 3x$ is the line that passes through the two points $(1, 2)$ and $(3, -4)$, as shown in Fig. 4.5.

**FIGURE 4.5**

Graph of $y = 5 - 3x$

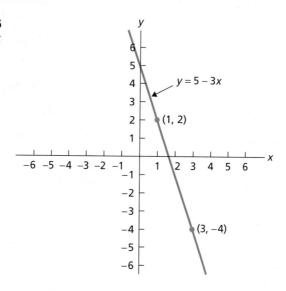

Note that the line in Fig. 4.5 slopes downward—the $y$-values decrease as $x$ increases—because the slope of the line is negative: $b_1 = -3 < 0$. Now look at the line in Fig. 4.4, the graph of the linear equation $y = 25 + 20x$. That line slopes upward—the $y$-values increase as $x$ increases—because the slope of the line is positive: $b_1 = 20 > 0$.

Key Fact 4.1    **Graphical Interpretation of Slope**

The graph of the linear equation $y = b_0 + b_1 x$ slopes upward if $b_1 > 0$, slopes downward if $b_1 < 0$, and is horizontal if $b_1 = 0$, as shown in Fig. 4.6.

**FIGURE 4.6**

Graphical interpretation of slope

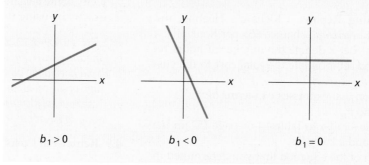

# Exercises 4.1

## Understanding the Concepts and Skills

**4.1** Regarding linear equations with one independent variable, answer the following questions:
a. What is the general form of such an equation?
b. In your expression in part (a), which letters represent constants and which represent variables?
c. In your expression in part (a), which letter represents the independent variable and which represents the dependent variable?

**4.2** Fill in the blank. The graph of a linear equation with one independent variable is a _____.

**4.3** Consider the linear equation $y = b_0 + b_1 x$.
a. Identify and give the geometric interpretation of $b_0$.
b. Identify and give the geometric interpretation of $b_1$.

**4.4** Answer true or false to each statement and explain your answers.
a. The graph of a linear equation slopes upward unless the slope is 0.
b. The value of the $y$-intercept has no effect on the direction that the graph of a linear equation slopes.

**4.5 Rental-Car Costs.** During one month, the Avis Rent-A-Car rate for renting a Buick LeSabre in Mobile, Alabama, was $68.22 per day plus 25¢ per mile. For a 1-day rental, let $x$ denote the number of miles driven and let $y$ denote the total cost, in dollars.
a. Find the equation that expresses $y$ in terms of $x$.
b. Determine $b_0$ and $b_1$.
c. Construct a table similar to Table 4.1 on page 157 for the $x$-values 50, 100, and 250 miles.
d. Draw the graph of the equation that you determined in part (a) by plotting the points from part (c) and connecting them with a line.
e. Apply the graph from part (d) to estimate visually the cost of driving the car 150 miles. Then calculate that cost exactly by using the equation from part (a).

**4.6 Air-Conditioning Repairs.** Richard's Heating and Cooling in Prescott, Arizona, charges $55 per hour plus a $30 service charge. Let $x$ denote the number of hours required for a job and let $y$ denote the total cost to the customer.
a. Find the equation that expresses $y$ in terms of $x$.
b. Determine $b_0$ and $b_1$.
c. Construct a table similar to Table 4.1 on page 157 for the $x$-values 0.5, 1, and 2.25 hours.
d. Draw the graph of the equation that you determined in part (a) by plotting the points from part (c) and connecting them with a line.

e. Apply the graph from part (d) to estimate visually the cost of a job that takes 1.75 hours. Then calculate that cost exactly by using the equation from part (a).

**4.7 Measuring Temperature.** The two most commonly used scales for measuring temperature are the Fahrenheit and Celsius scales. If you let $y$ denote Fahrenheit temperature and $x$ denote Celsius temperature, you can express the relationship between those two scales with the linear equation $y = 32 + 1.8x$.
a. Determine $b_0$ and $b_1$.
b. Find the Fahrenheit temperatures corresponding to the Celsius temperatures $-40°$, $0°$, $20°$, and $100°$.
c. Graph the linear equation $y = 32 + 1.8x$, using the four points found in part (b).
d. Apply the graph obtained in part (c) to estimate visually the Fahrenheit temperature corresponding to a Celsius temperature of $28°$. Then calculate that temperature exactly by using the linear equation $y = 32 + 1.8x$.

**4.8 A Law of Physics.** A ball is thrown straight up in the air with an initial velocity of 64 feet per second (ft/sec). According to the laws of physics, if you let $y$ denote the velocity of the ball after $x$ seconds, $y = 64 - 32x$.
a. Determine $b_0$ and $b_1$ for this linear equation.
b. Determine the velocity of the ball after 1, 2, 3, and 4 sec.
c. Graph the linear equation $y = 64 - 32x$, using the four points obtained in part (b).
d. Use the graph from part (c) to estimate visually the velocity of the ball after 1.5 sec. Then calculate that velocity exactly by using the linear equation $y = 64 - 32x$.

*In Exercises 4.9–4.12, do the following.*
*a. Find the y-intercept and slope of the specified linear equation.*
*b. Explain what the y-intercept and slope represent in terms of the graph of the equation.*
*c. Explain what the y-intercept and slope represent in terms relating to the application.*

**4.9 Rental-Car Costs.** $y = 68.22 + 0.25x$ (from Exercise 4.5)

**4.10 Air-Conditioning Repairs.** $y = 30 + 55x$ (from Exercise 4.6)

**4.11 Measuring Temperature.** $y = 32 + 1.8x$ (from Exercise 4.7)

**4.12 A Law of Physics.** $y = 64 - 32x$ (from Exercise 4.8)

*In Exercises 4.13–4.22, we give linear equations. For each equation, do the following.*
*a. Find the y-intercept and slope.*
*b. Determine whether the line slopes upward, slopes downward, or is horizontal, without graphing the equation.*
*c. Use two points to graph the equation.*

**4.13** $y = 3 + 4x$      **4.14** $y = -1 + 2x$

**4.15** $y = 6 - 7x$      **4.16** $y = -8 - 4x$

**4.17** $y = 0.5x - 2$      **4.18** $y = -0.75x - 5$

**4.19** $y = 2$      **4.20** $y = -3x$

**4.21** $y = 1.5x$      **4.22** $y = -3$

*In Exercises 4.23–4.30, we identify the y-intercepts and slopes, respectively, of lines. For each line, do the following.*
*a. Determine whether it slopes upward, slopes downward, or is horizontal, without graphing the equation.*
*b. Find its equation.*
*c. Use two points to graph the equation.*

**4.23** 5 and 2      **4.24** −3 and 4

**4.25** −2 and −3      **4.26** 0.4 and 1

**4.27** 0 and −0.5      **4.28** −1.5 and 0

**4.29** 3 and 0      **4.30** 0 and 3

## Extending the Concepts and Skills

**4.31 Hooke's Law.** According to *Hooke's law* for springs, developed by Robert Hooke (1635–1703), the force exerted by a spring that has been compressed to a length $x$ is given by the formula $F = -k(x - x_0)$, where $x_0$ is the natural length of the spring and $k$ is a constant, called the *spring constant*. A certain spring exerts a force of 32 lb when compressed to a length of 2 ft, and a force of 16 lb when compressed to a length of 3 ft. For this spring, find the following.
a. The linear equation that relates the force exerted to the length compressed
b. The spring constant
c. The natural length of the spring

**4.32 Road Grade.** The *grade* of a road is defined as the distance it rises (or falls) to the distance it runs horizontally, usually expressed as a percentage. Consider a road with positive grade, $g$. Suppose that you begin driving on that road at an altitude $a_0$.
a. Find the linear equation that expresses the altitude, $a$, when you have driven a distance, $d$, along the road. (*Hint:* Draw a graph and apply the Pythagorean Theorem.)
b. Identify and interpret the $y$-intercept and slope of the linear equation in part (a).
c. Apply your results in parts (a) and (b) to a road with a 5% grade and an initial altitude of 1 mile. Express your answer for the slope to four decimal places.
d. For the road in part (c), what altitude will you reach after driving 10 miles along the road?
e. For the road in part (c), how far along the road must you drive to reach an altitude of 3 miles?

**4.33** In this section, we stated that any nonvertical line can be described by an equation of the form $y = b_0 + b_1 x$.
a. Explain in detail why a vertical line can't be expressed in this form.
b. What is the form of the equation of a vertical line?
c. Does a vertical line have a slope? Explain your answer.

## 4.2 The Regression Equation

In Examples 4.1 and 4.2, we discussed the linear equation $y = 25 + 20x$, which expresses the total cost, $y$, of a word-processing job in terms of the time in hours, $x$, required to complete it. Given the amount of time required, $x$, we can use the equation to determine the *exact* cost of the job, $y$.

Real-life applications are seldom as simple as the word-processing example, in which one variable (cost) can be predicted exactly in terms of another variable (time required). Rather, we must often rely on rough predictions. For instance, we cannot predict the exact price, $y$, of a particular make and model of car just by knowing its age, $x$. Indeed, even for a fixed age, say, 3 years old, price varies from car to car. We must be content with making a rough prediction for the price of a 3-year-old car of the particular make and model or with an estimate of the mean price of all such 3-year-old cars.

**TABLE 4.2**

Age and price data
for a sample of 11 Orions

| Car | Age (yr) x | Price ($100) y |
|-----|-----|-----|
| 1 | 5 | 85 |
| 2 | 4 | 103 |
| 3 | 6 | 70 |
| 4 | 5 | 82 |
| 5 | 5 | 89 |
| 6 | 5 | 98 |
| 7 | 6 | 66 |
| 8 | 6 | 95 |
| 9 | 2 | 169 |
| 10 | 7 | 70 |
| 11 | 7 | 48 |

Table 4.2 displays data on age and price for a sample of cars of a particular make and model. We refer to the car as the Orion, but the data, obtained from the *Asian Import* edition of the *Auto Trader* magazine, is for a real car. Ages are in years; prices are in hundreds of dollars, rounded to the nearest hundred dollars.

Plotting the data in a *scatterplot* helps us visualize any apparent relationship between age and price. Generally speaking, a **scatterplot** (or **scatter diagram**) is a graph of data from two quantitative variables of a population.[†] To construct a scatterplot, we use a horizontal axis for the observations of one variable and a vertical axis for the observations of the other. Each pair of observations is then plotted as a point.

Figure 4.7 shows a scatterplot for the age–price data in Table 4.2. Note that we use a horizontal axis for ages and a vertical axis for prices. Each age–price observation is plotted as a point. For instance, the second car in Table 4.2 is 4 years old and has a price of 103 ($10,300). We plot this age–price observation as the point (4, 103), shown in magenta in Fig. 4.7.

**FIGURE 4.7**

Scatterplot for the age and price
data of Orions from Table 4.2

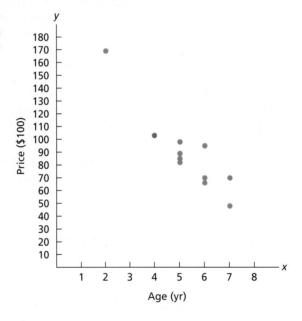

Although the age–price data points do not fall exactly on a line, they appear to cluster about a line. We want to fit a line to the data points and use that line to predict the price of an Orion based on its age.

Because we could draw many different lines through the cluster of data points, we need a method to choose the "best" line. The method, called the *least-squares criterion*, is based on an analysis of the errors made in using a line to fit the data points. To introduce the least-squares criterion, we use a very simple data set in Example 4.3. We return to the Orion data shortly.

---

[†]Data from two quantitative variables of a population are called **bivariate quantitative data.**

## Example 4.3 | Introducing the Least-Squares Criterion

Consider the problem of fitting a line to the four data points in Table 4.3, whose scatterplot is shown in Fig. 4.8. Many (in fact, infinitely many) lines can "fit" those four data points. Two possibilities are shown in Figs. 4.9(a) and 4.9(b).

**FIGURE 4.8**
Scatterplot for the data points in Table 4.3

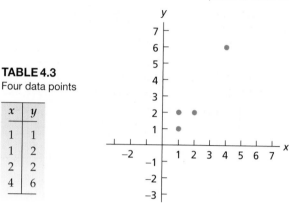

**TABLE 4.3**
Four data points

| x | y |
|---|---|
| 1 | 1 |
| 1 | 2 |
| 2 | 2 |
| 4 | 6 |

**FIGURE 4.9**
Two possible lines to fit the data points in Table 4.3

Line A: $y = 0.50 + 1.25x$

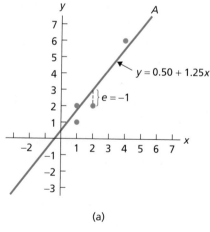

(a)

Line B: $y = -0.25 + 1.50x$

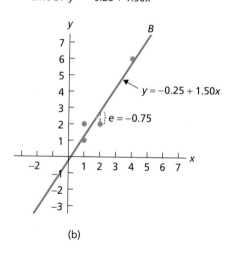

(b)

To avoid confusion, we use $\hat{y}$ to denote the $y$-value predicted by a line for a value of $x$. For instance, the $y$-value predicted by Line A for $x = 2$ is

$$\hat{y} = 0.50 + 1.25 \cdot 2 = 3,$$

and the $y$-value predicted by Line B for $x = 2$ is

$$\hat{y} = -0.25 + 1.50 \cdot 2 = 2.75.$$

To measure quantitatively how well a line fits the data, we first consider the errors, $e$, made in using the line to predict the $y$-values of the data points. For instance, as we have just demonstrated, Line A predicts a $y$-value of $\hat{y} = 3$

when $x = 2$. The actual $y$-value for $x = 2$ is $y = 2$ (see Table 4.3). So, the error made in using Line $A$ to predict the $y$-value of the data point $(2, 2)$ is

$$e = y - \hat{y} = 2 - 3 = -1,$$

as seen in Fig. 4.9(a). In general, an **error, $e$,** is the signed vertical distance from the line to a data point. The fourth column of Table 4.4(a) shows the errors made by Line $A$ for all four data points; the fourth column of Table 4.4(b) shows the same for Line $B$.

**TABLE 4.4**
Determining how well the data points in Table 4.3 are fit by (a) Line $A$ and (b) Line $B$

Line A: $y = 0.50 + 1.25x$

| $x$ | $y$ | $\hat{y}$ | $e$ | $e^2$ |
|---|---|---|---|---|
| 1 | 1 | 1.75 | −0.75 | 0.5625 |
| 1 | 2 | 1.75 | 0.25 | 0.0625 |
| 2 | 2 | 3.00 | −1.00 | 1.0000 |
| 4 | 6 | 5.50 | 0.50 | 0.2500 |
| | | | | 1.8750 |

(a)

Line B: $y = -0.25 + 1.50x$

| $x$ | $y$ | $\hat{y}$ | $e$ | $e^2$ |
|---|---|---|---|---|
| 1 | 1 | 1.25 | −0.25 | 0.0625 |
| 1 | 2 | 1.25 | 0.75 | 0.5625 |
| 2 | 2 | 2.75 | −0.75 | 0.5625 |
| 4 | 6 | 5.75 | 0.25 | 0.0625 |
| | | | | 1.2500 |

(b)

You try it!

Exercise 4.41
on page 173

To decide which line, Line $A$ or Line $B$, fits the data better, we first compute the sum of the squared errors, $\Sigma e_i^2$, in the final column of Table 4.4(a) and Table 4.4(b). The line having the smaller sum of squared errors, in this case Line $B$, is the one that fits the data better. And, among all lines, the **least-squares criterion** is that the line having the smallest sum of squared errors is the one that fits the data best.

• • •

**Key Fact 4.2**  **Least-Squares Criterion**

The **least-squares criterion** is that the line that best fits a set of data points is the one having the smallest possible sum of squared errors.

Next we present the terminology used for the line (and corresponding equation) that best fits a set of data points according to the least-squares criterion.

**Definition 4.2**  **Regression Line and Regression Equation**

**Regression line:** The line that best fits a set of data points according to the least-squares criterion.

**Regression equation:** The equation of the regression line.

Although the least-squares criterion states the property that the regression line for a set of data points must satisfy, it does not tell us how to find that line. This task is accomplished by Formula 4.1. In preparation, we introduce some notation that will be used throughout our study of regression and correlation.

**Definition 4.3**    **Notation Used in Regression and Correlation**

For a set of $n$ data points, the defining and computing formulas for $S_{xx}$, $S_{xy}$, and $S_{yy}$ are as follows.

| Quantity | Defining formula | Computing formula |
|----------|-----------------|-------------------|
| $S_{xx}$ | $\Sigma(x_i - \bar{x})^2$ | $\Sigma x_i^2 - (\Sigma x_i)^2/n$ |
| $S_{xy}$ | $\Sigma(x_i - \bar{x})(y_i - \bar{y})$ | $\Sigma x_i y_i - (\Sigma x_i)(\Sigma y_i)/n$ |
| $S_{yy}$ | $\Sigma(y_i - \bar{y})^2$ | $\Sigma y_i^2 - (\Sigma y_i)^2/n$ |

**Formula 4.1**    **Regression Equation**

The regression equation for a set of $n$ data points is $\hat{y} = b_0 + b_1 x$, where

$$b_1 = \frac{S_{xy}}{S_{xx}} \quad \text{and} \quad b_0 = \frac{1}{n}(\Sigma y_i - b_1 \Sigma x_i) = \bar{y} - b_1\bar{x}.$$

**Note:** Although we have not used $S_{yy}$ in Formula 4.1, we will use it later in this chapter.

**Example 4.4**    **The Regression Equation**

**TABLE 4.5**
Table for computing the regression equation for the Orion data

| Age (yr) $x$ | Price ($100) $y$ | $xy$ | $x^2$ |
|---|---|---|---|
| 5 | 85 | 425 | 25 |
| 4 | 103 | 412 | 16 |
| 6 | 70 | 420 | 36 |
| 5 | 82 | 410 | 25 |
| 5 | 89 | 445 | 25 |
| 5 | 98 | 490 | 25 |
| 6 | 66 | 396 | 36 |
| 6 | 95 | 570 | 36 |
| 2 | 169 | 338 | 4 |
| 7 | 70 | 490 | 49 |
| 7 | 48 | 336 | 49 |
| 58 | 975 | 4732 | 326 |

*Age and Price of Orions* In the first two columns of Table 4.5, we repeat our data on age and price for a sample of 11 Orions.

**a.** Determine the regression equation for the data.

**b.** Graph the regression equation and the data points.

**c.** Describe the apparent relationship between age and price of Orions.

**d.** Interpret the slope of the regression line in terms of prices for Orions.

**e.** Use the regression equation to predict the price of a 3-year-old Orion and a 4-year-old Orion.

**Solution**

**a.** We first need to compute $b_1$ and $b_0$ by using Formula 4.1. We did so by constructing a table of values for $x$ (age), $y$ (price), $xy$, $x^2$, and their sums in Table 4.5.

The slope of the regression line therefore is

$$b_1 = \frac{S_{xy}}{S_{xx}} = \frac{\Sigma x_i y_i - (\Sigma x_i)(\Sigma y_i)/n}{\Sigma x_i^2 - (\Sigma x_i)^2/n} = \frac{4732 - (58)(975)/11}{326 - (58)^2/11} = -20.26.$$

The $y$-intercept is

$$b_0 = \frac{1}{n}(\Sigma y_i - b_1 \Sigma x_i) = \frac{1}{11}[975 - (-20.26) \cdot 58] = 195.47.$$

So the regression equation is $\hat{y} = 195.47 - 20.26x$.

**Note:** The usual warnings about rounding apply. When computing the slope, $b_1$, of the regression line, do not round until the computation is finished. And when computing the $y$-intercept, $b_0$, do not use the rounded value of $b_1$; instead, keep full calculator accuracy.

**b.** To graph the regression equation, we need to substitute two different $x$-values in the regression equation to obtain two distinct points. Let's use the $x$-values 2 and 8. The corresponding $y$-values are

$$\hat{y} = 195.47 - 20.26 \cdot 2 = 154.95 \quad \text{and} \quad \hat{y} = 195.47 - 20.26 \cdot 8 = 33.39.$$

Therefore, the regression line goes through the two points $(2, 154.95)$ and $(8, 33.39)$. In Fig. 4.10, we plotted these two points with hollow dots. Drawing a line through the two hollow dots yields the regression line, the graph of the regression equation. Figure 4.10 also shows the data points from the first two columns of Table 4.5.

**FIGURE 4.10**

Regression line and data points for Orion data

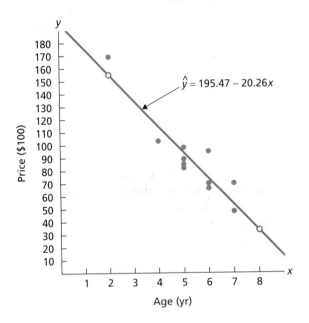

**c.** Because the slope of the regression line is negative, price tends to decrease as age increases, which is no particular surprise.

**d.** Because $x$ represents age, in years, and $y$ represents price, in hundreds of dollars, the slope of $-20.26$ indicates that Orions depreciate an estimated $2026 per year, at least in the 2- to 7-year-old range.

**e.** For a 3-year-old Orion, $x = 3$, and the regression equation yields the predicted price of

$$\hat{y} = 195.47 - 20.26 \cdot 3 = 134.69.$$

Similarly, the predicted price for a 4-year-old Orion is

$$\hat{y} = 195.47 - 20.26 \cdot 4 = 114.43.$$

**Interpretation** The estimated price of a 3-year-old Orion is $13,469, and the estimated price of a 4-year-old Orion is $11,443.

We discuss questions concerning the accuracy and reliability of such predictions later in this chapter and also in Chapter 14.

**You try it!**

Exercise 4.51
on page 174

• • •

## Predictor Variable and Response Variable

For a linear equation $y = b_0 + b_1 x$, $y$ is the dependent variable and $x$ is the independent variable. However, in the context of regression analysis, we usually call $y$ the **response variable** and $x$ the **predictor variable** or **explanatory variable** (because it is used to predict or explain the values of the response variable). For the Orion example, then, age is the predictor variable and price is the response variable.

## Extrapolation

Suppose that a scatterplot indicates a linear relationship between two variables. Then, within the range of the observed values of the predictor variable, we can reasonably use the regression equation to make predictions for the response variable. However, to do so outside that range, which is called **extrapolation,** may not be reasonable because the linear relationship between the predictor and response variables may not hold there.

Grossly incorrect predictions can result from extrapolation. The Orion example is a case in point. Its observed ages (values of the predictor variable) range from 2 to 7 years old. But suppose that we extrapolate to predict the price of an 11-year-old Orion. Using the regression equation, the predicted price is

$$\hat{y} = 195.47 - 20.26 \cdot 11 = -27.39,$$

or $-\$2739$. Clearly, this result is ridiculous: no one is going to pay us \$2739 to take away their 11-year-old Orion.

Consequently, although the relationship between age and price of Orions appears to be linear in the range from 2 to 7 years old, it is definitely not so in the range from 2 to 11 years old. Figure 4.11 summarizes the discussion on extrapolation as it applies to age and price of Orions.

**FIGURE 4.11**
Extrapolation in the Orion example

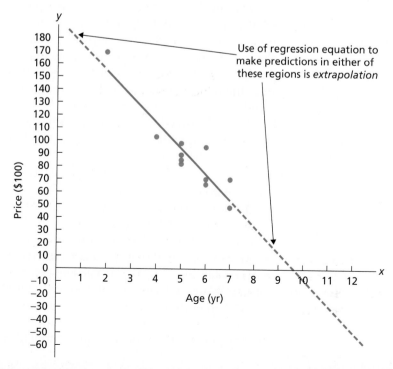

To help avoid extrapolation, some researchers include the range of the observed values of the predictor variable with the regression equation. For the Orion example, we would write

$$\hat{y} = 195.47 - 20.26x, \qquad 2 \le x \le 7.$$

Writing the regression equation in this way makes clear that using it to predict price for ages outside the range from 2 to 7 years old is extrapolation.

## Outliers and Influential Observations

Recall that an outlier is an observation that lies outside the overall pattern of the data. In the context of regression, an **outlier** is a data point that lies far from the regression line, relative to the other data points. Figure 4.10 on page 166 shows that the Orion data have no outliers.

An outlier can sometimes have a significant effect on a regression analysis. Thus, as usual, we need to identify outliers and remove them from the analysis when appropriate—for example, if we find that an outlier is a measurement or recording error.

We must also watch for *influential observations*. In regression analysis, an **influential observation** is a data point whose removal causes the regression equation (and line) to change considerably. A data point separated in the $x$-direction from the other data points is often an influential observation because the regression line is "pulled" toward such a data point without counteraction by other data points.

If an influential observation is due to a measurement or recording error or for some other reason it clearly does not belong in the data set, it can be removed without further consideration. However, if no explanation for the influential observation is apparent, the decision whether to retain it is often difficult and calls for a judgment by the researcher.

For the Orion data, Fig. 4.10 on page 166 (or Table 4.5 on page 165) shows that the data point $(2, 169)$ might be an influential observation because the age of 2 years appears separated from the other observed ages. Removing that data point and recalculating the regression equation yields $\hat{y} = 160.33 - 14.24x$. Figure 4.12 reveals that this equation differs markedly from the regression equation based on the full data set. The data point $(2, 169)$ is indeed an influential observation.

The influential observation $(2, 169)$ is not a recording error; it is a legitimate data point. Nonetheless, we may need either to remove it—thus limiting the analysis to Orions between 4 and 7 years old—or to obtain additional data on 2- and 3-year-old Orions so that the regression analysis is not so dependent on one data point.

We added data for one 2-year-old and three 3-year-old Orions and obtained the regression equation $\hat{y} = 193.63 - 19.93x$. This regression equation differs little from our original regression equation, $\hat{y} = 195.47 - 20.26x$. Therefore we could justify using the original regression equation to analyze the relationship between age and price of Orions between 2 and 7 years of age, even though the corresponding data set contains an influential observation.

An outlier may or may not be an influential observation; and an influential observation may or may not be an outlier. Many statistical software packages identify potential outliers and influential observations.

**FIGURE 4.12**
Regression lines with and without the influential observation removed

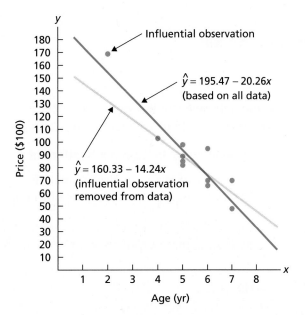

A Warning on the Use of Linear Regression

The idea behind finding a regression line is based on the assumption that the data points are scattered about a line.[†] Frequently, however, the data points are scattered about a curve instead of a line, as depicted in Fig. 4.13(a).

**FIGURE 4.13**
(a) Data points scattered about a curve; (b) inappropriate line fit to the data points

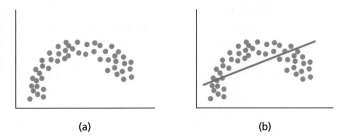

(a)                          (b)

One can still compute the values of $b_0$ and $b_1$ to obtain a regression line for these data points. The result, however, will yield an inappropriate fit by a line, as shown in Fig. 4.13(b), when in fact a curve should be used. For instance, the regression line suggests that $y$-values in Fig. 4.13(a) will keep increasing when they have actually begun to decrease.

---

Key Fact 4.3 | **Criterion for Finding a Regression Line**

Before finding a regression line for a set of data points, draw a scatterplot. If the data points do not appear to be scattered about a line, do not determine a regression line.

---

Techniques are available for fitting curves to data points that show a curved pattern, such as the data points plotted in Fig. 4.13(a). We discuss those techniques, referred to as **curvilinear regression,** in the chapter *Model Building in Regression* on the WeissStats CD accompanying this book.

---

[†]We discuss this assumption in detail and make it more precise in Section 14.1.

# The Technology Center

Most statistical technologies have programs that automatically generate a scatterplot and determine a regression line. In this subsection, we present output and step-by-step instructions for such programs.

**Example 4.5** **Using Technology to Obtain a Scatterplot**

*Age and Price of Orions* Use Minitab, Excel, or the TI-83/84 Plus to obtain a scatterplot for the age and price data in Table 4.2 on page 162.

**Solution** We applied the scatterplot programs to the data, resulting in Output 4.1. Steps for generating that output are presented in Instructions 4.1.

**OUTPUT 4.1**
Scatterplots for the age and price data of 11 Orions

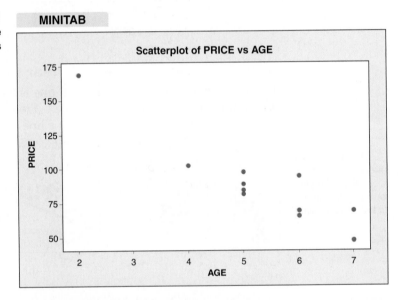

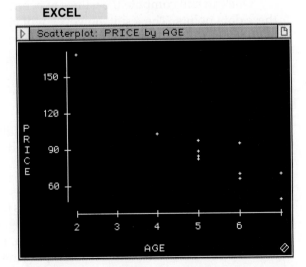

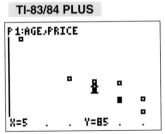

As shown in Output 4.1, the data points are scattered about a line. So, we can reasonably find a regression line for these data.

$\bullet$ $\bullet$ $\bullet$

**INSTRUCTIONS 4.1**   Steps for generating Output 4.1

| MINITAB | EXCEL | TI-83/84 PLUS |
|---|---|---|
| 1 Store the age and price data from Table 4.2 in columns named AGE and PRICE, respectively | 1 Store the age and price data from Table 4.2 in ranges named AGE and PRICE, respectively | 1 Store the age and price data from Table 4.2 in lists named AGE and PRICE, respectively |
| 2 Choose **Graph ➤ Scatterplot...** | 2 Choose **DDXL ➤ Charts and Plots** | 2 Press **2nd ➤ STAT PLOT** and then press **ENTER** twice |
| 3 Select the **Simple** scatterplot and click **OK** | 3 Select **Scatterplot** from the **Function type** drop-down list box | 3 Arrow to the first graph icon and press **ENTER** |
| 4 Specify PRICE in the **Y variables** text box | 4 Specify AGE in the **x-Axis Variable** text box | 4 Press the down-arrow key |
| 5 Specify AGE in the **X variables** text box | 5 Specify PRICE in the **y-Axis Variable** text box | 5 Press **2nd ➤ LIST**, arrow down to AGE, and press **ENTER** twice |
| 6 Click **OK** | 6 Click **OK** | 6 Press **2nd ➤ LIST**, arrow down to PRICE, and press **ENTER** twice |
| | | 7 Press **ZOOM** and then **9** (and then **TRACE**, if desired) |

### Example 4.6   Using Technology to Obtain a Regression Line

*Age and Price of Orions*   Use Minitab, Excel, or the TI-83/84 Plus to determine the regression equation for the age and price data in Table 4.2 on page 162.

**Solution**   We applied the regression programs to the data, resulting in Output 4.2 on the following page. Steps for generating that output are presented in Instructions 4.2 on page 173.

As shown in Output 4.2 (see the items circled in red), the $y$-intercept and slope of the regression line are 195.47 and $-20.261$, respectively. Thus the regression equation is $\hat{y} = 195.47 - 20.261x$.

$\bullet$ $\bullet$ $\bullet$

We can also use Minitab, Excel, or the TI-83/84 Plus to generate a scatterplot of the age and price data with a superimposed regression line, similar to the graph in Fig. 4.10 on page 166. To do so, proceed as follows.

- *Minitab:* In the third step of Instructions 4.1, select the **With Regression** scatterplot instead of the **Simple** scatterplot.
- *Excel:* See the complete DDXL output that results from applying the steps in Instructions 4.2.
- *TI-83/84 Plus:* After executing the steps in Instructions 4.2, press **GRAPH** and then **TRACE**.

**OUTPUT 4.2**

Regression analysis for the age
and price data of 11 Orions

**MINITAB**

## Regression Analysis: PRICE versus AGE

```
The regression equation is
PRICE = 195 - 20.3 AGE


Predictor       Coef    SE Coef        T       P
Constant      195.47      15.24    12.83   0.000
AGE          -20.261       2.800    -7.24   0.000


S = 12.5766    R-Sq = 85.3%    R-Sq(adj) = 83.7%


Analysis of Variance

Source          DF        SS        MS       F       P
Regression       1    8285.0    8285.0   52.38   0.000
Residual Error   9    1423.5     158.2
Total           10    9708.5
```

**EXCEL**

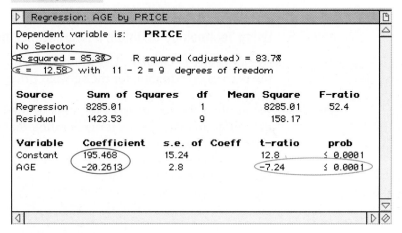

| ▷ | Regression: AGE by PRICE | | | | | 🖹 |

Dependent variable is:   **PRICE**
No Selector
R squared = 85.3%      R squared (adjusted) = 83.7%
s =   12.58   with  11 - 2 = 9  degrees of freedom

| **Source** | **Sum of Squares** | **df** | **Mean Square** | **F-ratio** |
|---|---|---|---|---|
| Regression | 8285.01 | 1 | 8285.01 | 52.4 |
| Residual | 1423.53 | 9 | 158.17 | |

| **Variable** | **Coefficient** | **s.e. of Coeff** | **t-ratio** | **prob** |
|---|---|---|---|---|
| Constant | 195.468 | 15.24 | 12.8 | ≤ 0.0001 |
| AGE | -20.2613 | 2.8 | -7.24 | ≤ 0.0001 |

**TI-83/84 PLUS**

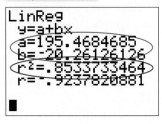

```
LinReg
 y=a+bx
 a=195.4684685
 b=-20.26126126
 r²=.8533733464
 r=-.9237820881
■
```

**INSTRUCTIONS 4.2**   Steps for generating Output 4.2

| MINITAB |
|---|
| 1  Store the age and price data from Table 4.2 in columns named AGE and PRICE, respectively |
| 2  Choose **Stat ➤ Regression ➤ Regression...** |
| 3  Specify PRICE in the **Response** text box |
| 4  Specify AGE in the **Predictors** text box |
| 5  Click the **Results...** button |
| 6  Select the **Regression equation, table of coefficients, s, R-squared, and basic analysis of variance** option button |
| 7  Click **OK** twice |

| EXCEL |
|---|
| 1  Store the age and price data from Table 4.2 in ranges named AGE and PRICE, respectively |
| 2  Choose **DDXL ➤ Regression** |
| 3  Select **Simple regression** from the **Function type** drop-down list box |
| 4  Specify PRICE in the **Response Variable** text box |
| 5  Specify AGE in the **Explanatory Variable** text box |
| 6  Click **OK** |

| TI-83/84 PLUS |
|---|
| 1  Store the age and price data from Table 4.2 in lists named AGE and PRICE, respectively |
| 2  Press **2nd ➤ CATALOG** and then press **D** |
| 3  Arrow down to **DiagnosticOn** and press **Enter** twice |
| 4  Press **STAT**, arrow over to **CALC**, and press **8** |
| 5  Press **2nd ➤ LIST**, arrow down to AGE, and press **ENTER** |
| 6  Press , ➤ **2nd ➤ LIST**, arrow down to PRICE, and press **ENTER** |
| 7  Press , ➤ **VARS**, arrow over to **Y-VARS**, and press **ENTER** three times |

▌ ▌ ▌

# Exercises 4.2

## Understanding the Concepts and Skills

**4.34**  Regarding a scatterplot,
a. identify one of its uses.
b. what property should it have to obtain a regression line for the data?

**4.35**  Regarding the criterion used to decide on the line that best fits a set of data points,
a. what is that criterion called?
b. specifically, what is the criterion?

**4.36**  Regarding the line that best fits a set of data points,
a. what is that line called?
b. what is the equation of that line called?

**4.37**  Regarding the two variables under consideration in a regression analysis,
a. what is the dependent variable called?
b. what is the independent variable called?

**4.38**  Using the regression equation to make predictions for values of the predictor variable outside the range of the observed values of the predictor variable is called _____.

**4.39**  Fill in the blanks.
a. In the context of regression, an _____ is a data point that lies far from the regression line, relative to the other data points.

b. In regression analysis, an _____ is a data point whose removal causes the regression equation to change considerably.

*In Exercises **4.40** and **4.41**,*
*a. graph the linear equations and data points.*
*b. construct tables for x, y, ŷ, e, and e² similar to Table 4.4 on page 164.*
*c. determine which line fits the set of data points better, according to the least-squares criterion.*

**4.40**  Line $A$: $y = 1.5 + 0.5x$
Line $B$: $y = 1.125 + 0.375x$

| $x$ | 1 | 1 | 5 | 5 |
|---|---|---|---|---|
| $y$ | 1 | 3 | 2 | 4 |

**4.41**  Line $A$: $y = 3 - 0.6x$
Line $B$: $y = 4 - x$

| $x$ | 0 | 2 | 2 | 5 | 6 |
|---|---|---|---|---|---|
| $y$ | 4 | 2 | 0 | -2 | 1 |

**4.42**  For a data set consisting of two data points:
a. Identify the regression line.
b. What is the sum of squared errors for the regression line? Explain your answer.

**4.43** Refer to Exercise 4.42. For each of the following sets of data points, determine the regression equation both without and with the use of Formula 4.1 on page 165.

**a.**

| x | 2 | 4 |
|---|---|---|
| y | 1 | 3 |

**b.**

| x | 1 | 5 |
|---|---|---|
| y | 3 | -3 |

*In each of Exercises 4.44–4.49, do the following.*
*a. Find the regression equation for the data points.*
*b. Graph the regression equation and the data points.*

**4.44**

| x | 2 | 4 | 3 |
|---|---|---|---|
| y | 3 | 5 | 7 |

**4.45**

| x | 3 | 1 | 2 |
|---|---|---|---|
| y | -4 | 0 | -5 |

**4.46**

| x | 0 | 4 | 3 | 1 | 2 |
|---|---|---|---|---|---|
| y | 1 | 9 | 8 | 4 | 3 |

**4.47**

| x | 3 | 4 | 1 | 2 |
|---|---|---|---|---|
| y | 4 | 5 | 0 | -1 |

**4.48** The data points in Exercise 4.40

**4.49** The data points in Exercise 4.41

*In each of Exercises 4.50–4.55, do the following.*
*a. Find the regression equation for the data points.*
*b. Graph the regression equation and the data points.*
*c. Describe the apparent relationship between the two variables under consideration.*
*d. Interpret the slope of the regression line.*
*e. Identify the predictor and response variables.*
*f. Identify outliers and potential influential observations.*
*g. Predict the values of the response variable for the specified values of the predictor variable, and interpret your results.*

**4.50 Tax Efficiency.** *Tax efficiency* is a measure, ranging from 0 to 100, of how much tax due to capital gains stock or mutual funds investors pay on their investments each year; the higher the tax efficiency, the lower the tax. In the article "At the Mercy of the Manager" (*Financial Planning*, Vol. 30(5), pp. 54–56), C. Israelsen examined the relationship between investments in mutual fund portfolios and their associated tax efficiencies. The following table shows percentage of investments in energy securities (x) and tax efficiency (y) for 10 mutual fund portfolios. For part (g), predict the tax efficiency of a mutual fund portfolio with

5.0% of its investments in energy securities and one with 7.4% of its investments in energy securities.

| x | 3.1 | 3.2 | 3.7 | 4.3 | 4.0 | 5.5 | 6.7 | 7.4 | 7.4 | 10.6 |
|---|---|---|---|---|---|---|---|---|---|---|
| y | 98.1 | 94.7 | 92.0 | 89.8 | 87.5 | 85.0 | 82.0 | 77.8 | 72.1 | 53.5 |

**4.51 Corvette Prices.** The *Kelley Blue Book* provides information on wholesale and retail prices of cars. Following are age and price data for 10 randomly selected Corvettes between 1 and 6 years old. Here, x denotes age, in years, and y denotes price, in hundreds of dollars. For part (g), predict the prices of a 2-year-old Corvette and a 3-year-old Corvette.

| x | 6 | 6 | 6 | 2 | 2 | 5 | 4 | 5 | 1 | 4 |
|---|---|---|---|---|---|---|---|---|---|---|
| y | 270 | 260 | 275 | 405 | 364 | 295 | 335 | 308 | 405 | 305 |

**4.52 Custom Homes.** Hanna Properties specializes in custom-home resales in the Equestrian Estates, an exclusive subdivision in Phoenix, Arizona. A random sample of nine custom homes currently listed for sale provided the following information on size and price. Here, x denotes size, in hundreds of square feet, rounded to the nearest hundred, and y denotes price, in thousands of dollars, rounded to the nearest thousand. For part (g), predict the price of a 2600 sq ft home in the Equestrian Estates.

| x | 26 | 27 | 33 | 29 | 29 | 34 | 30 | 40 | 22 |
|---|---|---|---|---|---|---|---|---|---|
| y | 540 | 555 | 575 | 577 | 606 | 661 | 738 | 804 | 496 |

**4.53 Plant Emissions.** Plants emit gases that trigger the ripening of fruit, attract pollinators, and cue other physiological responses. N. Agelopolous et al. examined factors that affect the emission of volatile compounds by the potato plant *Solanum tuberosom* and published their findings in the *Journal of Chemical Ecology* (Vol. 26, Issue 2, pp. 497–511). The volatile compounds analyzed were hydrocarbons used by other plants and animals. Following are data on plant weight (x), in grams, and quantity of volatile compounds emitted (y), in hundreds of nanograms, for 11 potato plants. For part (g), predict the quantity of volatile compounds emitted by a potato plant that weighs 75 grams.

| x | 57 | 85 | 57 | 65 | 52 | 67 | 62 | 80 | 77 | 53 | 68 |
|---|---|---|---|---|---|---|---|---|---|---|---|
| y | 8.0 | 22.0 | 10.5 | 22.5 | 12.0 | 11.5 | 7.5 | 13.0 | 16.5 | 21.0 | 12.0 |

**4.54 Crown-Rump Length.** In the article "The Human Vomeronasal Organ. Part II: Prenatal Development" (*Journal of Anatomy*, Vol. 197, Issue 3, pp. 421–436), T. Smith and K. Bhatnagar examined the controversial issue of the human vomeronasal organ, regarding its structure, function, and identity. The following table shows the age of fetuses (x), in weeks, and length of crown-rump ($y$), in millimeters. For part (g), predict the crown-rump length of a 19-week-old fetus.

| $x$ | 10 | 10 | 13 | 13 | 18 | 19 | 19 | 23 | 25 | 28 |
|-----|----|----|----|----|----|----|----|----|----|----|
| $y$ | 66 | 66 | 108 | 106 | 161 | 166 | 177 | 228 | 235 | 280 |

**4.55 Study Time and Score.** An instructor at Arizona State University asked a random sample of eight students to record their study times in a beginning calculus course. She then made a table for total hours studied ($x$) over 2 weeks and test score ($y$) at the end of the 2 weeks. Here are the results. For part (g), predict the score of a student who studies for 15 hours.

| $x$ | 10 | 15 | 12 | 20 | 8 | 16 | 14 | 22 |
|-----|----|----|----|----|----|----|----|----|
| $y$ | 92 | 81 | 84 | 74 | 85 | 80 | 84 | 80 |

**4.56** For which of the following sets of data points can you reasonably determine a regression line? Explain your answer.

**4.57** For which of the following sets of data points can you reasonably determine a regression line? Explain your answer.

**4.58 Tax Efficiency.** In Exercise 4.50, you determined a regression equation that relates the variables percentage of investments in energy securities and tax efficiency for mutual fund portfolios.
a. Should that regression equation be used to predict the tax efficiency of a mutual fund portfolio with 6.4% of its investments in energy securities? with 15% of its investments in energy securities? Explain your answers.
b. For which percentages of investments in energy securities is use of the regression equation to predict tax efficiency reasonable?

**4.59 Corvette Prices.** In Exercise 4.51, you determined a regression equation that can be used to predict the price of a Corvette, given its age.
a. Should that regression equation be used to predict the price of a 4-year-old Corvette? a 10-year-old Corvette? Explain your answers.
b. For which ages is use of the regression equation to predict price reasonable?

**4.60 Palm Beach Fiasco.** The 2000 U.S. presidential election brought great controversy to the election process. Many voters in Palm Beach, Florida, claimed that they were confused by the ballot format and may have accidentally voted for Pat Buchanan when they intended to vote for Al Gore. Professors Greg D. Adams of Carnegie Mellon University and Chris Fastnow of Chatham College compiled and analyzed data on election votes in Florida, by county, for both 1996 and 2000. What conclusions would you draw from the following scatterplots constructed by the researchers? Explain your answers.

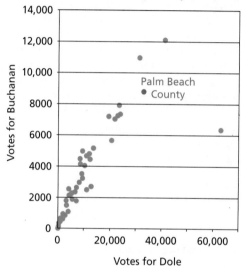

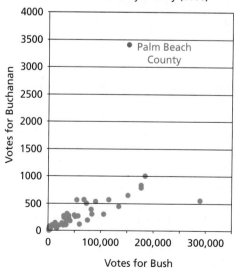

Source: *Prof. Greg D. Adams, Dept. of Social & Decision Sciences, Carnegie Mellon University, and Prof. Chris Fastnow, Director, Center for Women in Politics in Pennsylvania, Chatham College*

**4.61 Study Time and Score.** The negative relation between study time and test score, found in Exercise 4.55, has been discovered by many investigators. Provide a possible explanation for it.

**4.62 Age and Price of Orions.** In Table 4.2, we provided data on age and price for a sample of 11 Orions between 2 and 7 years old. On the WeissStats CD, we have given the ages and prices for a sample of 31 Orions between 1 and 11 years old.
a. Obtain a scatterplot for the data.
b. Is it reasonable to find a regression line for the data? Explain your answer.

**4.63 Wasp Mating Systems.** In the paper "Mating System and Sex Allocation in the Gregarious Parasitoid *Cotesia glomerata*" (*Animal Behaviour*, Vol. 66, pp. 259–264), H. Gu and S. Dorn reported on various aspects of the mating system and sex allocation strategy of the wasp *C. glomerata.* One part of the study involved the investigation of the percentage of male wasps dispersing before mating in relation to the brood sex ratio (proportion of males). The data obtained by the researchers are on the WeissStats CD.
a. Obtain a scatterplot for the data.
b. Is it reasonable to find a regression line for the data? Explain your answer.

## Working With Large Data Sets

*In Exercises 4.64–4.74, use the technology of your choice to do the following.*
*a. Obtain a scatterplot for the data.*
*b. Decide whether finding a regression line for the data is reasonable. If so, then also do parts (c)–(f).*
*c. Determine and interpret the regression equation for the data.*
*d. Identify potential outliers and influential observations.*
*e. In case a potential outlier is present, remove it and discuss the effect.*
*f. In case a potential influential observation is present, remove it and discuss the effect.*

**4.64 Birdies and Score.** How important are birdies (one under par) in determining the final score of a woman golfer? From the *U.S. Women's Open* Web site, we obtained data on number of birdies during a tournament and final score for 63 women golfers. The data are presented on the WeissStats CD.

**4.65 U.S. Presidents.** The *Information Please Almanac* provides data on the ages at inauguration and of death for the presidents of the United States. We have given those data on the WeissStats CD for those presidents who are not still living at the time of this writing.

**4.66 Health Care.** From the *Statistical Abstract of the United States*, we obtained data on percentage of gross domestic product (GDP) spent on health care and life expectancy, in years, for selected countries. Those data are provided on the WeissStats CD.

**4.67 Acreage and Value.** The document *Arizona Residential Property Valuation System*, published by the Arizona Department of Revenue, describes how county assessors use computerized systems to value single family residential properties for property tax purposes. On the WeissStats CD are data on lot size (in acres) and assessed value (in thousands of dollars) for a sample of homes in a particular area.

**4.68 Home Size and Value.** On the WeissStats CD are data on home size (in square feet) and assessed value (in thousands of dollars) for the same homes as in Exercise 4.67.

**4.69 High and Low Temperature.** The U.S. National Oceanic and Atmospheric Administration publishes temperature information of cities around the world in *Climates of the World*. A random sample of 50 cities gave the data on average high and low temperatures in January shown on the WeissStats CD.

**4.70 PCBs and Pelicans.** Polychlorinated biphenyls (PCBs), industrial pollutants, are known to be a great danger to natural ecosystems. In a study by R. W. Risebrough titled "Effects of Environmental Pollutants Upon Animals Other Than Man" (*Proceedings of the 6th Berkeley Symposium on Mathematics and Statistics, VI,* University of California Press, pp. 443–463), 60 Anacapa pelican eggs were collected and measured for their shell thickness, in millimeters (mm), and concentration of PCBs, in parts per million (ppm). The data are on the WeissStats CD.

**4.71 More Money More Beer?** Does a higher state per capita income equate to a higher per capita beer consumption? Data downloaded from *The Beer Institute Online* on per capita income, in dollars, and per capita beer consumption, in gallons, for the 50 states and Washington, D.C., are provided on the WeissStats CD.

**4.72 Gas Guzzlers.** The magazine *Consumer Reports* publishes information on automobile gas mileage and variables that affect gas mileage. In one issue, data on gas mileage (in miles per gallon) and engine displacement (in liters) were published for 121 vehicles. Those data are available on the WeissStats CD.

**4.73 Top Wealth Managers.** The September 15, 2003 issue of *BARRON'S* presented information on top wealth managers in the United States, based on individual clients with accounts of $1 million or more. Data were given for various variables, two of which were number of private client managers and private client assets. Those data are provided on the WeissStats CD, where private client assets are in billions of dollars.

**4.74 Shortleaf Pines.** The ability to estimate the volume of a tree based on a simple measurement, such as the tree's diameter, is important to the lumber industry, ecologists, and conservationists. Data on volume, in cubic feet, and diameter at breast height, in inches, for 70 shortleaf pines were reported in C. Bruce and F. X. Schumacher's *Forest Mensuration* (New York: McGraw-Hill, 1935) and analyzed by A. C. Akinson in the article "Transforming Both Sides of a Tree" (*The American Statistician*, Vol. 48, pp. 307–312). The data are presented on the WeissStats CD.

## Extending the Concepts and Skills

**Sample Covariance.** For a set of $n$ data points, the **sample covariance**, $s_{xy}$, is given by

$$s_{xy} = \underbrace{\frac{\Sigma(x_i - \bar{x})(y_i - \bar{y})}{n - 1}}_{\text{defining formula}} = \underbrace{\frac{\Sigma x_i y_i - (\Sigma x_i)(\Sigma y_i)/n}{n - 1}}_{\text{computing formula}} \quad (4.1)$$

The sample covariance can be used as an alternative method for finding the slope and $y$-intercept of a regression line. The formulas are

$$b_1 = s_{xy}/s_x^2 \quad \text{and} \quad b_0 = \bar{y} - b_1\bar{x} \quad (4.2)$$

where $s_x$ denotes the sample standard deviation of the $x$-values.

*In each of Exercises* **4.75** *and* **4.76**, *do the following for the data points in the specified exercise.*
*a. Use Equation (4.1) to determine the sample covariance.*
*b. Use Equations (4.2) and your answer from part (a) to find the regression equation. Compare your result to that found in the specified exercise.*

**4.75** Exercise 4.47.

**4.76** Exercise 4.46.

**Time Series.** A collection of observations of a variable $y$ taken at regular intervals over time is called a **time series.** Economic data and electrical signals are examples of time series. We can think of a time series as providing data points $(x_i, y_i)$, where $x_i$ is the $i$th observation time and $y_i$ is the observed value of $y$ at time $x_i$. If a time series exhibits a linear trend, we can find that trend by determining the regression equation for the data points. We can then use the regression equation for forecasting purposes.

*Exercises* **4.77** *and* **4.78** *concern time series. In each exercise, do the following.*
*a. Obtain a scatterplot for the data.*
*b. Find and interpret the regression equation.*
*c. Make the specified forecasts.*

**4.77 U.S. Population.** The U.S. Census Bureau publishes information on the population of the United States in *Current Population Reports*. The following table gives the resident U.S. population, in millions of persons, for the years 1990–2004. Forecast the U.S. population in the years 2005 and 2006.

| Year | Population (millions) | Year | Population (millions) |
|------|-----------------------|------|-----------------------|
| 1990 | 250 | 1998 | 276 |
| 1991 | 253 | 1999 | 279 |
| 1992 | 257 | 2000 | 282 |
| 1993 | 260 | 2001 | 285 |
| 1994 | 263 | 2002 | 288 |
| 1995 | 266 | 2003 | 291 |
| 1996 | 269 | 2004 | 294 |
| 1997 | 273 |  |  |

**4.78 Global Warming.** Is there evidence of global warming in the records of ice cover on lakes? If the earth is getting warmer, lakes that freeze over in the winter should be covered with ice for shorter periods of time as the earth gradually warms. R. Bohanan examined records of ice duration for Lake Mendota at Madison, WI, in the paper "Changes in Lake Ice: Ecosystem Response to Global Change" (*Teaching Issues and Experiments in Ecology*, Vol. 3). The data are presented on the WeissStats CD and should be analyzed with the technology of your choice. Forecast the ice duration in the years 2006 and 2007.

## 4.3 The Coefficient of Determination

In Example 4.4, we determined the regression equation, $\hat{y} = 195.47 - 20.26x$, for data on age and price of a sample of 11 Orions, where $x$ represents age, in years, and $\hat{y}$ represents predicted price, in hundreds of dollars. We also applied the regression equation to predict the price of a 4-year-old Orion

$$\hat{y} = 195.47 - 20.26 \cdot 4 = 114.43,$$

or $11,443. But how valuable are such predictions? Is the regression equation useful for predicting price, or could we do just as well by ignoring age?

In general, several methods exist for evaluating the utility of a regression equation for making predictions. One method is to determine the percentage of variation in the observed values of the response variable that is explained by the regression (or predictor variable), as discussed below. To find this percentage, we need to define two measures of variation: (1) the total variation in the observed values of the response variable and (2) the amount of variation in the observed values of the response variable that is explained by the regression.

### Sums of Squares and Coefficient of Determination

To measure the total variation in the observed values of the response variable, we use the sum of squared deviations of the observed values of the response variable from the mean of those values. This measure of variation is called the **total sum of squares, SST.** Thus, $SST = \Sigma(y_i - \bar{y})^2$. If we divide $SST$ by $n - 1$, we get the sample variance of the observed values of the response variable. So, $SST$ really is a measure of total variation.

To measure the amount of variation in the observed values of the response variable that is explained by the regression, we first look at a particular observed value of the response variable, say, corresponding to the data point $(x_i, y_i)$, as shown in Fig. 4.14.

**FIGURE 4.14** Decomposing the deviation of an observed $y$-value from the mean into the deviations explained and not explained by the regression

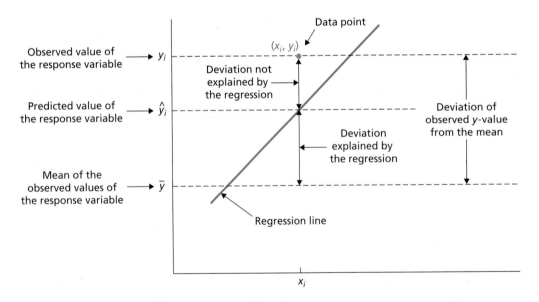

The total variation in the observed values of the response variable is based on the deviation of each observed value from the mean value, $y_i - \bar{y}$. As shown in Fig. 4.14, each such deviation can be decomposed into two parts: the deviation explained by the regression line, $\hat{y}_i - \bar{y}$, and the remaining unexplained deviation, $y_i - \hat{y}_i$. Hence the amount of variation (squared deviation) in the observed values of the response variable that is explained by the regression is $\Sigma(\hat{y}_i - \bar{y})^2$. This measure of variation is called the **regression sum of squares, SSR.** Thus, $SSR = \Sigma(\hat{y}_i - \bar{y})^2$.

Using the total sum of squares and the regression sum of squares, we can determine the percentage of variation in the observed values of the response variable that is explained by the regression, namely, $SSR/SST$. This quantity is called the **coefficient of determination** and is denoted $r^2$. Thus, $r^2 = SSR/SST$.

Before applying the coefficient of determination, let's consider the remaining deviation portrayed in Fig. 4.14: the deviation not explained by the regression, $y_i - \hat{y}_i$. The amount of variation (squared deviation) in the observed values of the response variable that is not explained by the regression is $\Sigma(y_i - \hat{y}_i)^2$. This measure of variation is called the **error sum of squares, SSE.** Thus, $SSE = \Sigma(y_i - \hat{y}_i)^2$.

---

**Definition 4.4**

## Sums of Squares in Regression

**Total sum of squares, SST:** The total variation in the observed values of the response variable: $SST = \Sigma(y_i - \bar{y})^2$.

**Regression sum of squares, SSR:** The variation in the observed values of the response variable explained by the regression: $SSR = \Sigma(\hat{y}_i - \bar{y})^2$.

**Error sum of squares, SSE:** The variation in the observed values of the response variable not explained by the regression: $SSE = \Sigma(y_i - \hat{y}_i)^2$.

---

**Definition 4.5**

## Coefficient of Determination

### What Does It Mean?

The coefficient of determination is a descriptive measure of the utility of the regression equation for making predictions.

The **coefficient of determination, $r^2$,** is the proportion of variation in the observed values of the response variable explained by the regression:

$$r^2 = \frac{SSR}{SST}.$$

The coefficient of determination, $r^2$, always lies between 0 and 1. A value of $r^2$ near 0 suggests that the regression equation is not very useful for making predictions, whereas a value of $r^2$ near 1 suggests that the regression equation is quite useful for making predictions.

---

**Example 4.7** | **The Coefficient of Determination**

*Age and Price of Orions* The scatterplot and regression line for the age and price data of 11 Orions are repeated in Fig. 4.15 at the top of the next page.

The scatterplot reveals that the prices of the 11 Orions vary widely, ranging from a low of 48 ($4800) to a high of 169 ($16,900). But Fig. 4.15 also shows that much of the price variation is "explained" by the regression (or age); that is, the regression line, with age as the predictor variable, predicts a sizeable portion of the type of variation found in the prices. Make this qualitative statement precise by finding and interpreting the coefficient of determination for the Orion data.

**Solution** We need the total sum of squares and the regression sum of squares, as given in Definition 4.4.

**FIGURE 4.15**
Scatterplot and regression
line for Orion data

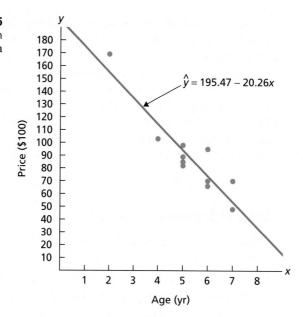

$\hat{y} = 195.47 - 20.26x$

To compute the total sum of squares, $SST$, we must first find the mean of the observed prices. Referring to the second column of Table 4.6, we get

$$\bar{y} = \frac{\Sigma y_i}{n} = \frac{975}{11} = 88.64.$$

After constructing the third column of Table 4.6, we calculate the entries for the fourth column and then find the total sum of squares:

$$SST = \Sigma(y_i - \bar{y})^2 = 9708.5,[†]$$

which is the total variation in the observed prices.

**TABLE 4.6**
Table for computing $SST$
for the Orion price data

| Age (yr) $x$ | Price ($100) $y$ | $y - \bar{y}$ | $(y - \bar{y})^2$ |
|---|---|---|---|
| 5 | 85 | −3.64 | 13.2 |
| 4 | 103 | 14.36 | 206.3 |
| 6 | 70 | −18.64 | 347.3 |
| 5 | 82 | −6.64 | 44.0 |
| 5 | 89 | 0.36 | 0.1 |
| 5 | 98 | 9.36 | 87.7 |
| 6 | 66 | −22.64 | 512.4 |
| 6 | 95 | 6.36 | 40.5 |
| 2 | 169 | 80.36 | 6458.3 |
| 7 | 70 | −18.64 | 347.3 |
| 7 | 48 | −40.64 | 1651.3 |
| | 975 | | 9708.5 |

[†]Values in Table 4.6 and all other tables in this section are displayed to various numbers of decimal places, but computations were done with full calculator accuracy.

To compute the regression sum of squares, $SSR$, we need the predicted prices and the mean of the observed prices. We have already computed the mean of the observed prices. Each predicted price is obtained by substituting the age of the Orion in question for $x$ in the regression equation $\hat{y} = 195.47 - 20.26x$. The third column of Table 4.7 shows the predicted prices for all 11 Orions.

**TABLE 4.7**

Table for computing $SSR$ for the Orion data

| Age (yr) $x$ | Price ($100) $y$ | $\hat{y}$ | $\hat{y} - \bar{y}$ | $(\hat{y} - \bar{y})^2$ |
|---|---|---|---|---|
| 5 | 85 | 94.16 | 5.53 | 30.5 |
| 4 | 103 | 114.42 | 25.79 | 665.0 |
| 6 | 70 | 73.90 | −14.74 | 217.1 |
| 5 | 82 | 94.16 | 5.53 | 30.5 |
| 5 | 89 | 94.16 | 5.53 | 30.5 |
| 5 | 98 | 94.16 | 5.53 | 30.5 |
| 6 | 66 | 73.90 | −14.74 | 217.1 |
| 6 | 95 | 73.90 | −14.74 | 217.1 |
| 2 | 169 | 154.95 | 66.31 | 4397.0 |
| 7 | 70 | 53.64 | −35.00 | 1224.8 |
| 7 | 48 | 53.64 | −35.00 | 1224.8 |
| | | | | 8285.0 |

Recalling that $\bar{y} = 88.64$, we construct the fourth column of Table 4.7. We then calculate the entries for the fifth column and obtain the regression sum of squares:

$$SSR = \Sigma(\hat{y}_i - \bar{y})^2 = 8285.0,$$

which is the variation in the observed prices explained by the regression.

From $SST$ and $SSR$, we compute the coefficient of determination, the percentage of variation in the observed prices explained by the regression (i.e., by the linear relationship between age and price for the sampled Orions):

$$r^2 = \frac{SSR}{SST} = \frac{8285.0}{9708.5} = 0.853 \quad (85.3\%).$$

**Interpretation** Evidently, age is quite useful for predicting price because 85.3% of the variation in the observed prices is explained by the regression of price on age.

• • •

Shortly, we will also want the error sum of squares for the Orion data. To compute $SSE$, we need the observed prices and the predicted prices. Both quantities are displayed in Table 4.7 and are repeated in the second and third columns of Table 4.8.

**TABLE 4.8**

Table for computing SSE
for the Orion data

| Age (yr)  $x$ | Price ($100)  $y$ | $\hat{y}$ | $y - \hat{y}$ | $(y - \hat{y})^2$ |
|---|---|---|---|---|
| 5 | 85 | 94.16 | −9.16 | 83.9 |
| 4 | 103 | 114.42 | −11.42 | 130.5 |
| 6 | 70 | 73.90 | −3.90 | 15.2 |
| 5 | 82 | 94.16 | −12.16 | 147.9 |
| 5 | 89 | 94.16 | −5.16 | 26.6 |
| 5 | 98 | 94.16 | 3.84 | 14.7 |
| 6 | 66 | 73.90 | −7.90 | 62.4 |
| 6 | 95 | 73.90 | 21.10 | 445.2 |
| 2 | 169 | 154.95 | 14.05 | 197.5 |
| 7 | 70 | 53.64 | 16.36 | 267.7 |
| 7 | 48 | 53.64 | −5.64 | 31.8 |
| | | | | 1423.5 |

**You try it!**

Exercise 4.85(a)
on page 185

From the final column of Table 4.8, we get the error sum of squares:

$$SSE = \Sigma(y_i - \hat{y}_i)^2 = 1423.5,$$

which is the variation in the observed prices not explained by the regression. Because the regression line is the line that best fits the data according to the least squares criterion, SSE is also the smallest possible sum of squared errors among all lines.

### The Regression Identity

For the Orion data, $SST = 9708.5$, $SSR = 8285.0$, and $SSE = 1423.5$. Because $9708.5 = 8285.0 + 1423.5$, we see that $SST = SSR + SSE$. This equation is always true and is called the **regression identity.**

**Key Fact 4.4**

### Regression Identity

The total sum of squares equals the regression sum of squares plus the error sum of squares: $SST = SSR + SSE$.

**What Does It Mean?**

The total variation in the observed values of the response variable can be partitioned into two components, one representing the variation explained by the regression and the other representing the variation not explained by the regression.

Because of the regression identity, we can also express the coefficient of determination in terms of the total sum of squares and the error sum of squares:

$$r^2 = \frac{SSR}{SST} = \frac{SST - SSE}{SST} = 1 - \frac{SSE}{SST}.$$

This formula shows that we can also interpret the coefficient of determination as the percentage reduction obtained in the total squared error by using the regression equation instead of the mean, $\bar{y}$, to predict the observed values of the response variable. See Exercise 4.107 (page 186).

### Computing Formulas for the Sums of Squares

Calculating the three sums of squares—SST, SSR, and SSE—with the defining formulas is time consuming and can lead to significant roundoff error unless full accuracy is retained. For those reasons, we usually use computing formulas or a computer to obtain the sums of squares. Here are the computing formulas.

**Formula 4.2**

## Computing Formulas for the Sums of Squares

The three sums of squares can be expressed as follows in terms of the quantities $S_{xx}$, $S_{xy}$, and $S_{yy}$ given in Definition 4.3 on page 165:

$$SST = S_{yy}, \qquad SSR = \frac{S_{xy}^2}{S_{xx}}, \qquad \text{and} \qquad SSE = S_{yy} - \frac{S_{xy}^2}{S_{xx}}.$$

In particular, we can use the following formulas to compute the three sums of squares:

$$SST = \Sigma y_i^2 - (\Sigma y_i)^2/n, \qquad SSR = \frac{[\Sigma x_i y_i - (\Sigma x_i)(\Sigma y_i)/n]^2}{\Sigma x_i^2 - (\Sigma x_i)^2/n},$$

and $SSE = SST - SSR$.

**Example 4.8**

## Computing Formulas for the Sums of Squares

*Age and Price of Orions*  The age and price data for a sample of 11 Orions are repeated in the first two columns of Table 4.9. Use the computing formulas in Formula 4.2 to determine the three sums of squares.

**Solution**  To apply the computing formulas, we need a table of values for $x$ (age), $y$ (price), $xy$, $x^2$, $y^2$, and their sums, as shown in Table 4.9.

**TABLE 4.9**
Table for obtaining the three sums of squares for the Orion data by using the computing formulas

| Age (yr) $x$ | Price ($100) $y$ | $xy$ | $x^2$ | $y^2$ |
|---|---|---|---|---|
| 5 | 85 | 425 | 25 | 7,225 |
| 4 | 103 | 412 | 16 | 10,609 |
| 6 | 70 | 420 | 36 | 4,900 |
| 5 | 82 | 410 | 25 | 6,724 |
| 5 | 89 | 445 | 25 | 7,921 |
| 5 | 98 | 490 | 25 | 9,604 |
| 6 | 66 | 396 | 36 | 4,356 |
| 6 | 95 | 570 | 36 | 9,025 |
| 2 | 169 | 338 | 4 | 28,561 |
| 7 | 70 | 490 | 49 | 4,900 |
| 7 | 48 | 336 | 49 | 2,304 |
| 58 | 975 | 4732 | 326 | 96,129 |

Using the last row of Table 4.9 and Formula 4.2, we can now find the three sums of squares for the Orion data. The total sum of squares is

$$SST = \Sigma y_i^2 - (\Sigma y_i)^2/n = 96{,}129 - (975)^2/11 = 9708.5;$$

the regression sum of squares is

$$SSR = \frac{[\Sigma x_i y_i - (\Sigma x_i)(\Sigma y_i)/n]^2}{\Sigma x_i^2 - (\Sigma x_i)^2/n} = \frac{[4732 - (58)(975)/11]^2}{326 - (58)^2/11} = 8285.0;$$

**You try it!**

and, from the two preceding results, the error sum of squares is

$$SSE = SST - SSR = 9708.5 - 8285.0 = 1423.5.$$

Exercise 4.89 on page 185

• • •

## The Technology Center

Most statistical technologies have programs to compute the coefficient of determination, $r^2$, and the three sums of squares, $SST$, $SSR$, and $SSE$. In fact, many statistical technologies present those four statistics as part of the output for a regression equation. In the next example, we concentrate on the coefficient of determination. Refer to the technology manuals for a discussion of the three sums of squares.

**Example 4.9**    **Using Technology to Obtain a Coefficient of Determination**

*Age and Price of Orions*   The age and price data for a sample of 11 Orions are given in Table 4.2 on page 162. Use Minitab, Excel, or the TI-83/84 Plus to obtain the coefficient of determination, $r^2$, for those data.

**Solution**   In Section 4.2, we used the three statistical technologies to find the regression equation for the age and price data. The results, displayed in Output 4.2 on page 172, also give the coefficient of determination. See the items circled in blue. Thus, to three decimal places, $r^2 = 0.853$.

■ ■ ■

## Exercises 4.3

### Understanding the Concepts and Skills

**4.79** In this section, we introduced a descriptive measure of the utility of the regression equation for making predictions. Do the following for that descriptive measure.
a. Identify the term and symbol.
b. Provide an interpretation.

**4.80** Fill in the blanks.
a. A measure of total variation in the observed values of the response variable is the _____. The mathematical abbreviation for it is _____.
b. A measure of the amount of variation in the observed values of the response variable explained by the regression is the _____. The mathematical abbreviation for it is _____.
c. A measure of the amount of variation in the observed values of the response variable not explained by the regression is the _____. The mathematical abbreviation for it is _____.

**4.81** For a particular regression analysis, $SST = 8291.0$ and $SSR = 7626.6$.
a. Obtain and interpret the coefficient of determination.
b. Determine $SSE$.

*In Exercises 4.82–4.87, we repeat the data and provide the regression equations for Exercises 4.44–4.49. Do the following.*
a. *Compute the three sums of squares, SST, SSR, and SSE, using the defining formulas (page 179).*
b. *Verify the regression identity, SST = SSR + SSE.*
c. *Compute the coefficient of determination.*
d. *Determine the percentage of variation in the observed values of the response variable that is explained by the regression.*
e. *State how useful the regression equation appears to be for making predictions. (Answers for this part may vary, owing to differing interpretations.)*

**4.82**

| x | 2 | 4 | 3 |
|---|---|---|---|
| y | 3 | 5 | 7 |

$\hat{y} = 2 + x$

**4.83**

| x | 3 | 1 | 2 |
|---|---|---|---|
| y | −4 | 0 | −5 |

$\hat{y} = 1 - 2x$

**4.84**

| x | 0 | 4 | 3 | 1 | 2 |
|---|---|---|---|---|---|
| y | 1 | 9 | 8 | 4 | 3 |

$\hat{y} = 1 + 2x$

**4.85**

| $x$ | 3 | 4 | 1 | 2 |
|---|---|---|---|---|
| $y$ | 4 | 5 | 0 | -1 |

$\hat{y} = -3 + 2x$

**4.86**

| $x$ | 1 | 1 | 5 | 5 |
|---|---|---|---|---|
| $y$ | 1 | 3 | 2 | 4 |

$\hat{y} = 1.75 + 0.25x$

**4.87**

| $x$ | 0 | 2 | 2 | 5 | 6 |
|---|---|---|---|---|---|
| $y$ | 4 | 2 | 0 | -2 | 1 |

$\hat{y} = 2.875 - 0.625x$

*For Exercises 4.88–4.93, do the following.*
*a. Compute SST, SSR, and SSE, using Formula 4.2 on page 183.*
*b. Compute the coefficient of determination, $r^2$.*
*c. Determine the percentage of variation in the observed values of the response variable explained by the regression, and interpret your answer.*
*d. State how useful the regression equation appears to be for making predictions.*

**4.88 Tax Efficiency.** Following are the data on percentage of investments in energy securities and tax efficiency from Exercise 4.50.

| $x$ | 3.1 | 3.2 | 3.7 | 4.3 | 4.0 | 5.5 | 6.7 | 7.4 | 7.4 | 10.6 |
|---|---|---|---|---|---|---|---|---|---|---|
| $y$ | 98.1 | 94.7 | 92.0 | 89.8 | 87.5 | 85.0 | 82.0 | 77.8 | 72.1 | 53.5 |

**4.89 Corvette Prices.** Following are the age and price data for Corvettes from Exercise 4.51:

| $x$ | 6 | 6 | 6 | 2 | 2 | 5 | 4 | 5 | 1 | 4 |
|---|---|---|---|---|---|---|---|---|---|---|
| $y$ | 270 | 260 | 275 | 405 | 364 | 295 | 335 | 308 | 405 | 305 |

**4.90 Custom Homes.** Following are the size and price data for custom homes from Exercise 4.52.

| $x$ | 26 | 27 | 33 | 29 | 29 | 34 | 30 | 40 | 22 |
|---|---|---|---|---|---|---|---|---|---|
| $y$ | 540 | 555 | 575 | 577 | 606 | 661 | 738 | 804 | 496 |

**4.91 Plant Emissions.** Following are the data on plant weight and quantity of volatile emissions from Exercise 4.53.

| $x$ | 57 | 85 | 57 | 65 | 52 | 67 | 62 | 80 | 77 | 53 | 68 |
|---|---|---|---|---|---|---|---|---|---|---|---|
| $y$ | 8.0 | 22.0 | 10.5 | 22.5 | 12.0 | 11.5 | 7.5 | 13.0 | 16.5 | 21.0 | 12.0 |

**4.92 Crown-Rump Length.** Following are the data on age and crown-rump length for fetuses from Exercise 4.54.

| $x$ | 10 | 10 | 13 | 13 | 18 | 19 | 19 | 23 | 25 | 28 |
|---|---|---|---|---|---|---|---|---|---|---|
| $y$ | 66 | 66 | 108 | 106 | 161 | 166 | 177 | 228 | 235 | 280 |

**4.93 Study Time and Score.** Following are the data on study time and score for calculus students from Exercise 4.55.

| $x$ | 10 | 15 | 12 | 20 | 8 | 16 | 14 | 22 |
|---|---|---|---|---|---|---|---|---|
| $y$ | 92 | 81 | 84 | 74 | 85 | 80 | 84 | 80 |

## Working With Large Data Sets

*In Exercises 4.94–4.105, use the technology of your choice to do the following.*
*a. Decide whether finding a regression line for the data is reasonable. If so, then also do parts (b)–(d).*
*b. Obtain the coefficient of determination.*
*c. Determine the percentage of variation in the observed values of the response variable explained by the regression, and interpret your answer.*
*d. State how useful the regression equation appears to be for making predictions.*

**4.94 Birdies and Score.** The data from Exercise 4.64 for number of birdies during a tournament and final score for 63 women golfers are on the WeissStats CD.

**4.95 U.S. Presidents.** The data from Exercise 4.65 for the ages at inauguration and of death for the presidents of the United States are on the WeissStats CD.

**4.96 Health Care.** The data from Exercise 4.66 for percentage of gross domestic product (GDP) spent on health care and life expectancy, in years, for selected countries are on the WeissStats CD.

**4.97 Acreage and Value.** The data from Exercise 4.67 for lot size (in acres) and assessed value (in thousands of dollars) for a sample of homes in a particular area are on the WeissStats CD.

**4.98 Home Size and Value.** The data from Exercise 4.68 for home size (in square feet) and assessed value (in thousands of dollars) for the same homes as in Exercise 4.97 are on the WeissStats CD.

**4.99 High and Low Temperature.** The data from Exercise 4.69 for average high and low temperatures in January for a random sample of 50 cities are on the WeissStats CD.

**4.100 PCBs and Pelicans.** The data for shell thickness and concentration of PCBs for 60 Anacapa pelican eggs from Exercise 4.70 are on the WeissStats CD.

**4.101 More Money More Beer?** The data for per capita income and per capita beer consumption for the 50 states and Washington, D.C., from Exercise 4.71 are on the WeissStats CD.

**4.102 Gas Guzzlers.** The data for gas mileage and engine displacement for 121 vehicles from Exercise 4.72 are on the WeissStats CD.

**4.103 Shortleaf Pines.** The data from Exercise 4.74 for volume, in cubic feet, and diameter at breast height, in inches, for 70 shortleaf pines are on the WeissStats CD.

**4.104 Body Fat.** In the paper "Total Body Composition by Dual-Photon ($^{153}$Gd) Absorptiometry" (*American Journal of Clinical Nutrition*, Vol. 40, pp. 834–839), R. Mazess et al. studied methods for quantifying body composition. Eighteen randomly selected adults were measured for percentage of body fat, using dual-photon absorptiometry. Each adult's age and percentage of body fat are shown on the WeissStats CD.

**4.105 Estriol Level and Birth Weight.** J. Greene and J. Touchstone conducted a study on the relationship between the estriol levels of pregnant women and the birth weights of their children. Their findings, "Urinary Tract Estriol: An Index of Placental Function," were published in the *American Journal of Obstetrics and Gynecology* (Vol. 85(1), pp. 1–9). The data from the study are provided on the WeissStats CD, where estriol levels are in mg/24 hr and birth weights are in hectograms.

### Extending the Concepts and Skills

**4.106** What can you say about *SSE*, *SSR*, and the utility of the regression equation for making predictions if
**a.** $r^2 = 1$?      **b.** $r^2 = 0$?

**4.107** As we noted, because of the regression identity, we can express the coefficient of determination in terms of the total sum of squares and the error sum of squares as $r^2 = 1 - SSE/SST$.
**a.** Explain why this formula shows that the coefficient of determination can also be interpreted as the percentage reduction obtained in the total squared error by using the regression equation instead of the mean, $\bar{y}$, to predict the observed values of the response variable.
**b.** Refer to Exercise 4.89. What percentage reduction is obtained in the total squared error by using the regression equation instead of the mean of the observed prices to predict the observed prices?

## 4.4 Linear Correlation

We often hear statements pertaining to the correlation or lack of correlation between two variables: "There is a positive correlation between advertising expenditures and sales" or "IQ and alcohol consumption are uncorrelated." In this section, we explain the meaning of such statements.

Several statistics can be used to measure the correlation between two variables. The statistic most commonly used is the **linear correlation coefficient, *r*,** which is also called the **Pearson product moment correlation coefficient,** in honor of its developer, Karl Pearson.

**Definition 4.6**

## Linear Correlation Coefficient

For a set of $n$ data points, the **linear correlation coefficient, $r$,** is defined by

$$r = \frac{\frac{1}{n-1}\Sigma(x_i - \bar{x})(y_i - \bar{y})}{s_x s_y}$$

where $s_x$ and $s_y$ denote the sample standard deviations of the $x$-values and $y$-values, respectively. It can be expressed as $r = S_{xy}/\sqrt{S_{xx}S_{yy}}$. Thus, we can also use the following computing formula to obtain the linear correlation coefficient:

$$r = \frac{\Sigma x_i y_i - (\Sigma x_i)(\Sigma y_i)/n}{\sqrt{[\Sigma x_i^2 - (\Sigma x_i)^2/n][\Sigma y_i^2 - (\Sigma y_i)^2/n]}}.$$

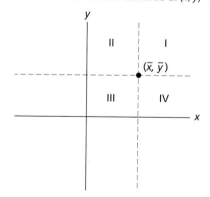

**What Does It Mean?**

The linear correlation coefficient is a descriptive measure of the strength of the linear (straight-line) relationship between two variables.

The computing formula presented in Definition 4.6 is almost always preferred for hand calculations, but the defining formula reveals the meaning and basic properties of the linear correlation coefficient. For instance, because of the division by the sample standard deviations, $s_x$ and $s_y$, in the defining formula for $r$, we can conclude that $r$ is independent of the choice of units and always lies between $-1$ and $1$.

### Understanding the Linear Correlation Coefficient

We now discuss some other important properties of the linear correlation coefficient, $r$. Keep in mind that $r$ measures the strength of the *linear* relationship between two variables and that the following properties of $r$ are meaningful only when the data points are scattered about a line.

- *$r$ reflects the slope of the scatterplot.* The linear correlation coefficient is positive when the scatterplot shows a positive slope and is negative when the scatterplot shows a negative slope. To demonstrate why this property is true, we refer to the defining formula in Definition 4.6 and to Fig. 4.16, where we have drawn a coordinate system with a second set of axes centered at point $(\bar{x}, \bar{y})$.

  If the scatterplot shows a positive slope, the data points, on average, will lie either in Region I or Region III. For such a data point, the deviations from the means, $x_i - \bar{x}$ and $y_i - \bar{y}$, will either both be positive or both be negative. This condition implies that, on average, the product $(x_i - \bar{x})(y_i - \bar{y})$ will be positive and consequently that the correlation coefficient will be positive.

  If the scatterplot shows a negative slope, the data points, on average, will lie either in Region II or Region IV. For such a data point, one of the deviations from the mean will be positive and the other negative. This condition implies that, on average, the product $(x_i - \bar{x})(y_i - \bar{y})$ will be negative and consequently that the correlation coefficient will be negative.

- *The magnitude of $r$ indicates the strength of the linear relationship.* A value of $r$ close to $-1$ or to $1$ indicates a strong linear relationship between the variables and that the variable $x$ is a good linear predictor of the variable $y$ (i.e., the regression equation is extremely useful for making predictions). A value of $r$ near $0$ indicates at most a weak linear relationship between the variables and that the variable $x$ is a poor linear predictor of the variable $y$

**FIGURE 4.16**

Coordinate system with a second set of axes centered at $(\bar{x}, \bar{y})$

(i.e., the regression equation is either useless or not very useful for making predictions).

- *The sign of r suggests the type of linear relationship.* A positive value of $r$ suggests that the variables are **positively linearly correlated,** meaning that $y$ tends to increase linearly as $x$ increases, with the tendency being greater the closer that $r$ is to 1. A negative value of $r$ suggests that the variables are **negatively linearly correlated,** meaning that $y$ tends to decrease linearly as $x$ increases, with the tendency being greater the closer that $r$ is to $-1$.
- *The sign of r and the sign of the slope of the regression line are identical.* If $r$ is positive, so is the slope of the regression line (i.e., the regression line slopes upward); if $r$ is negative, so is the slope of the regression line (i.e., the regression line slopes downward).

To graphically portray the meaning of the linear correlation coefficient, we present various degrees of linear correlation in Fig. 4.17.

**FIGURE 4.17**
Various degrees of linear correlation

(a) Perfect positive linear correlation $r = 1$

(b) Strong positive linear correlation $r = 0.9$

(c) Weak positive linear correlation $r = 0.4$

(d) Perfect negative linear correlation $r = -1$

(e) Strong negative linear correlation $r = -0.9$

(f) Weak negative linear correlation $r = -0.4$

(g) No linear correlation (linearly uncorrelated) $r = 0$

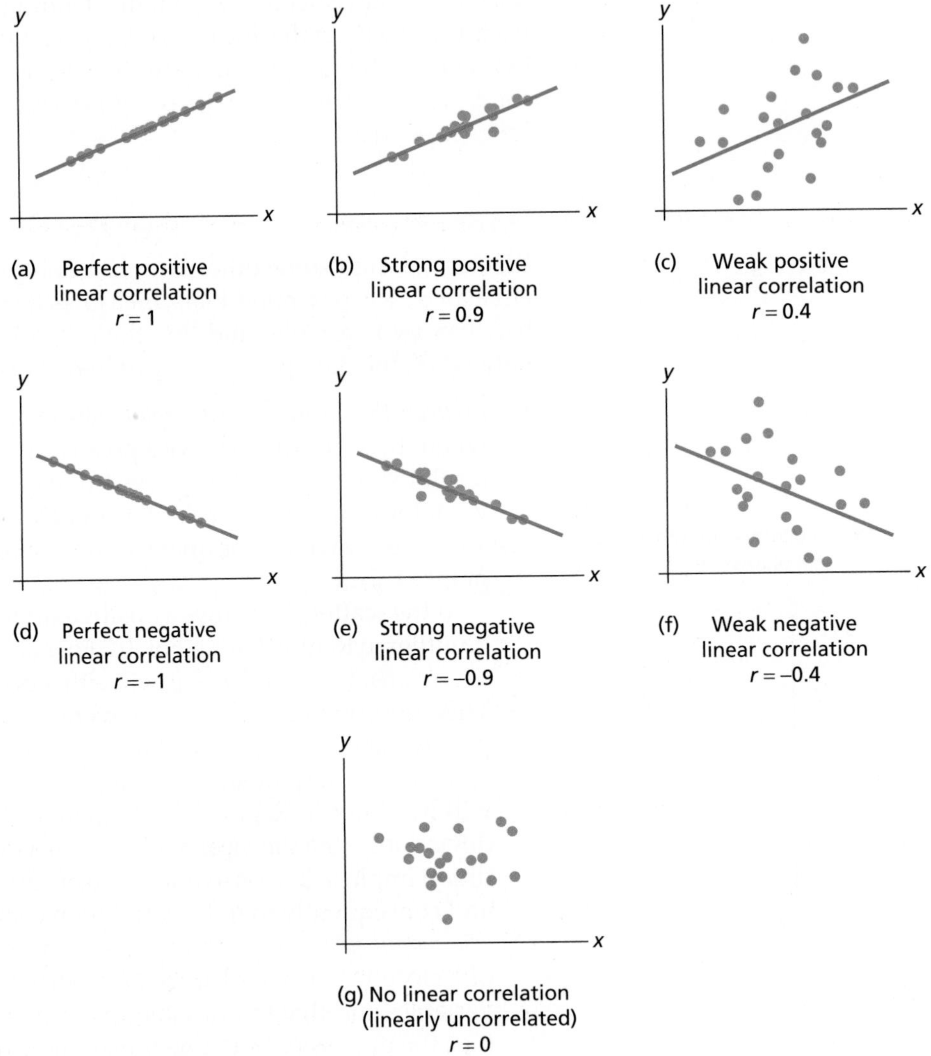

If $r$ is close to $\pm 1$, the data points are clustered closely about the regression line, as shown in Fig. 4.17(b) and (e). If $r$ is farther from $\pm 1$, the data points are more widely scattered about the regression line, as shown in Fig. 4.17(c) and (f). If $r$ is near 0, the data points are essentially scattered about a horizontal line, as shown in Fig. 4.17(g), indicating at most a weak linear relationship between the variables.

## Computing and Interpreting the Linear Correlation Coefficient

We demonstrate how to compute and interpret the linear correlation coefficient by returning to the data on age and price for a sample of Orions.

**Example 4.10** | **The Linear Correlation Coefficient**

*Age and Price of Orions*  The age and price data for a sample of 11 Orions are repeated in the first two columns of Table 4.10.

**TABLE 4.10**
Table for obtaining the linear correlation coefficient for the Orion data by using the computing formula

| Age (yr) $x$ | Price (\$100) $y$ | $xy$ | $x^2$ | $y^2$ |
|---|---|---|---|---|
| 5 | 85 | 425 | 25 | 7,225 |
| 4 | 103 | 412 | 16 | 10,609 |
| 6 | 70 | 420 | 36 | 4,900 |
| 5 | 82 | 410 | 25 | 6,724 |
| 5 | 89 | 445 | 25 | 7,921 |
| 5 | 98 | 490 | 25 | 9,604 |
| 6 | 66 | 396 | 36 | 4,356 |
| 6 | 95 | 570 | 36 | 9,025 |
| 2 | 169 | 338 | 4 | 28,561 |
| 7 | 70 | 490 | 49 | 4,900 |
| 7 | 48 | 336 | 49 | 2,304 |
| 58 | 975 | 4732 | 326 | 96,129 |

a. Compute the linear correlation coefficient, $r$, of the data.

b. Interpret the value of $r$ obtained in part (a) in terms of the linear relationship between the variables age and price of Orions.

c. Discuss the graphical implications of the value of $r$.

**Solution**  First recall that the scatterplot shown in Fig. 4.7 on page 162 indicates that the data points are scattered about a line. Hence it is meaningful to obtain the linear correlation coefficient of these data.

a. We apply the computing formula in Definition 4.6 on page 187 to find the linear correlation coefficient. To do so, we need a table of values for $x$, $y$,

$xy$, $x^2$, $y^2$, and their sums, as shown in Table 4.10. Referring to the last row of Table 4.10, we get

$$r = \frac{\Sigma x_i y_i - (\Sigma x_i)(\Sigma y_i)/n}{\sqrt{\left[\Sigma x_i^2 - (\Sigma x_i)^2/n\right]\left[\Sigma y_i^2 - (\Sigma y_i)^2/n\right]}}$$

$$= \frac{4732 - (58)(975)/11}{\sqrt{\left[326 - (58)^2/11\right]\left[96,129 - (975)^2/11\right]}} = -0.924.$$

b.  **Interpretation**  The linear correlation coefficient, $r = -0.924$, suggests a strong negative linear correlation between age and price of Orions. In particular, it indicates that as age increases there is a strong tendency for price to decrease, which is not surprising. It also implies that the regression equation, $\hat{y} = 195.47 - 20.26x$, is extremely useful for making predictions.

You try it!

Exercise 4.123
on page 194

c.  Because the correlation coefficient, $r = -0.924$, is quite close to $-1$, the data points should be clustered closely about the regression line. Figure 4.15 on page 180 shows that to be the case.

• • •

### Relationship Between the Correlation Coefficient and the Coefficient of Determination

In Section 4.3, we discussed the coefficient of determination, $r^2$, a descriptive measure of the utility of the regression equation for making predictions. In this section, we introduced the linear correlation coefficient, $r$, as a descriptive measure of the strength of the linear relationship between two variables.

We expect the strength of the linear relationship also to indicate the usefulness of the regression equation for making predictions. In other words, there should be a relationship between the linear correlation coefficient and the coefficient of determination—and there is. The relationship is precisely the one suggested by the notation used.

---

**Key Fact 4.5**

### Relationship Between the Correlation Coefficient and the Coefficient of Determination

The coefficient of determination equals the square of the linear correlation coefficient.

---

In Example 4.10, we found that the linear correlation coefficient for the data on age and price of a sample of 11 Orions is $r = -0.924$. From this result and Key Fact 4.5, we can easily obtain the coefficient of determination: $r^2 = (-0.924)^2 = 0.854$. As expected, this value is the same (except for roundoff error) as the value we found for $r^2$ on page 181 by using the defining formula $r^2 = SSR/SST$. In general, we can find the coefficient of determination either by using the defining formula or by first finding the linear correlation coefficient and then squaring the result.

Likewise, we can find the linear correlation coefficient, $r$, either by using Definition 4.6 or from the coefficient of determination, $r^2$, provided we also know the direction of the regression line. Specifically, the square root of $r^2$ gives the magnitude of $r$; the sign of $r$ is the same as that of the slope of the regression line.

## Warnings on the Use of the Linear Correlation Coefficient

Because the linear correlation coefficient describes the strength of the *linear* relationship between two variables, it should be used as a descriptive measure only when a scatterplot indicates that the data points are scattered about a line.

For instance, in general, we cannot say that a value of $r$ near 0 implies that there is no relationship between the two variables under consideration, nor can we say that a value of $r$ near $\pm 1$ implies that a linear relationship exists between the two variables. Such statements are meaningful only when a scatterplot indicates that the data points are scattered about a line. See Exercises 4.129 and 4.130 for more on these issues.

When using the linear correlation coefficient, you must also watch for outliers and influential observations. Such data points can sometimes unduly affect $r$ because sample means and sample standard deviations are not resistant to outliers and other extreme values.

## Correlation and Causation

Two variables may have a high correlation without being causally related. For example, Table 4.11 displays data on total pari-mutuel turnover (money wagered) at U.S. racetracks and college enrollment for five randomly selected years. [SOURCE: *National Association of State Racing Commissioners* and *U.S. National Center for Education Statistics.*]

**TABLE 4.11**
Pari-mutuel turnover and college enrollment for five randomly selected years

| Pari-mutuel turnover ($ millions) $x$ | College enrollment (thousands) $y$ |
|---|---|
| 5,977 | 8,581 |
| 7,862 | 11,185 |
| 10,029 | 11,260 |
| 11,677 | 12,372 |
| 11,888 | 12,426 |

The linear correlation coefficient of the data points in Table 4.11 is $r = 0.931$, suggesting a strong positive linear correlation between pari-mutuel wagering and college enrollment. But this result doesn't mean that a causal relationship exists between the two variables, such as that when people go to racetracks they are somehow inspired to go to college. On the contrary, we can only infer that the two variables have a strong tendency to increase (or decrease) simultaneously and that total pari-mutuel turnover is a good predictor of college enrollment.

**What Does It Mean?**

Correlation does not imply causation!

Two variables may be strongly correlated because they are both associated with other variables, called **lurking variables,** that cause changes in the two variables under consideration.  For example, a study showed that teachers' salaries and the dollar amount of liquor sales are positively linearly correlated. A possible explanation for this curious fact might be that both variables are tied to other variables, such as the rate of inflation, that pull them along together.

## The Technology Center

Most statistical technologies have programs that automatically determine a linear correlation coefficient.  In this subsection, we present output and step-by-step instructions for such programs.

**Example 4.11**   **Using Technology to Find a Linear Correlation Coefficient**

*Age and Price of Orions*   Use Minitab, Excel, or the TI-83/84 Plus to determine the linear correlation coefficient of the age and price data in the first two columns of Table 4.10 on page 189.

**Solution**   We applied the linear correlation coefficient programs to the data, resulting in Output 4.3. Steps for generating that output are presented in Instructions 4.3.

**OUTPUT 4.3**

Linear correlation coefficient for the age and price data of 11 Orions

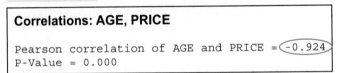

**MINITAB**

**Correlations: AGE, PRICE**

Pearson correlation of AGE and PRICE = -0.924
P-Value = 0.000

**EXCEL**

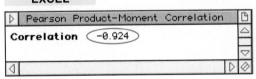

| ▷ | Pearson Product-Moment Correlation | 🗅 |
|---|---|---|
| **Correlation** | -0.924 | |

**TI-83/84 PLUS**

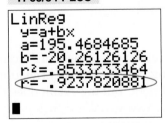

```
LinReg
  y=a+bx
  a=195.4684685
  b=-20.26126126
  r²=.8533733464
  r=-.9237820881
■
```

As shown in Output 4.3, the linear correlation coefficient for the age and price data is −0.924.

• • •

**INSTRUCTIONS 4.3** Steps for generating Output 4.3

| MINITAB |
|---|
| 1 Store the age and price data from Table 4.10 in columns named AGE and PRICE, respectively |
| 2 Choose **Stat ➤ Basic Statistics ➤ Correlation…** |
| 3 Specify AGE and PRICE in the **Variables** text box |
| 4 Click **OK** |

| EXCEL |
|---|
| 1 Store the age and price data from Table 4.10 in ranges named AGE and PRICE, respectively |
| 2 Choose **DDXL ➤ Regression** |
| 3 Select **Correlation** from the **Function type** drop-down list box |
| 4 Specify AGE in the **x-Axis Quantitative Variable** text box |
| 5 Specify PRICE in the **y-Axis Quantitative Variable** text box |
| 6 Click **OK** |

| TI-83/84 PLUS |
|---|
| 1 Store the age and price data from Table 4.10 in lists named AGE and PRICE, respectively |
| 2 Press **2nd ➤ CATALOG** and then press **D** |
| 3 Arrow down to **DiagnosticOn** and press **Enter** twice |
| 4 Press **STAT**, arrow over to **CALC**, and press **8** |
| 5 Press **2nd ➤ LIST**, arrow down to AGE, and press **ENTER** |
| 6 Press **,** ➤ **2nd ➤ LIST**, arrow down to PRICE, and press **ENTER** twice |

# Exercises 4.4

## Understanding the Concepts and Skills

**4.108** What is one purpose of the linear correlation coefficient?

**4.109** The linear correlation coefficient is also known by another name. What is it?

**4.110** Fill in the blanks.
**a.** The symbol used for the linear correlation coefficient is ____.
**b.** A value of $r$ close to $\pm 1$ indicates that there is a ____ linear relationship between the variables.
**c.** A value of $r$ close to ____ indicates that there is either no linear relationship between the variables or a weak one.

**4.111** Fill in the blanks.
**a.** A value of $r$ close to ____ indicates that the regression equation is extremely useful for making predictions.
**b.** A value of $r$ close to 0 indicates that the regression equation is either useless or ____ for making predictions.

**4.112** Fill in the blanks.
**a.** If $y$ tends to increase linearly as $x$ increases, the variables are ____ linearly correlated.
**b.** If $y$ tends to decrease linearly as $x$ increases, the variables are ____ linearly correlated.
**c.** If there is no linear relationship between $x$ and $y$, the variables are linearly ____.

**4.113** Answer true or false to the following statement and provide a reason for your answer: If there is a very strong positive correlation between two variables, a causal relationship exists between the two variables.

**4.114** The linear correlation coefficient of a set of data points is 0.846.
**a.** Is the slope of the regression line positive or negative? Explain your answer.
**b.** Determine the coefficient of determination.

**4.115** The coefficient of determination of a set of data points is 0.709 and the slope of the regression line is $-3.58$. Determine the linear correlation coefficient of the data.

*In Exercises 4.116–4.121, we repeat data from exercises in Section 4.2. For each exercise, determine the linear correlation coefficient by using the defining formula in Definition 4.6 on page 187.*

**4.116**

| $x$ | 2 | 4 | 3 |
|---|---|---|---|
| $y$ | 3 | 5 | 7 |

**4.117**

| $x$ | 3 | 1 | 2 |
|---|---|---|---|
| $y$ | $-4$ | 0 | $-5$ |

**4.118**

| $x$ | 0 | 4 | 3 | 1 | 2 |
|---|---|---|---|---|---|
| $y$ | 1 | 9 | 8 | 4 | 3 |

**4.119**

| x | 3 | 4 | 1 | 2 |
|---|---|---|---|---|
| y | 4 | 5 | 0 | −1 |

**4.120**

| x | 1 | 1 | 5 | 5 |
|---|---|---|---|---|
| y | 1 | 3 | 2 | 4 |

**4.121**

| x | 0 | 2 | 2 | 5 | 6 |
|---|---|---|---|---|---|
| y | 4 | 2 | 0 | −2 | 1 |

*In Exercises 4.122–4.127, we repeat data from exercises in Section 4.2. For each exercise here, do the following.*

a. *Obtain the linear correlation coefficient by using the computing formula in Definition 4.6 on page 187.*

b. *Interpret the value of r in terms of the linear relationship between the two variables in question.*

c. *Discuss the graphical interpretation of the value of r and verify that it is consistent with the graph you obtained in the corresponding exercise in Section 4.2.*

d. *Square r and compare the result with the value of the coefficient of determination you obtained in the corresponding exercise in Section 4.3.*

**4.122  Tax Efficiency.** Following are the data on percentage of investments in energy securities and tax efficiency from Exercises 4.50 and 4.88.

| x | 3.1 | 3.2 | 3.7 | 4.3 | 4.0 | 5.5 | 6.7 | 7.4 | 7.4 | 10.6 |
|---|---|---|---|---|---|---|---|---|---|---|
| y | 98.1 | 94.7 | 92.0 | 89.8 | 87.5 | 85.0 | 82.0 | 77.8 | 72.1 | 53.5 |

**4.123  Corvette Prices.** Following are the age and price data for Corvettes from Exercises 4.51 and 4.89.

| x | 6 | 6 | 6 | 2 | 2 | 5 | 4 | 5 | 1 | 4 |
|---|---|---|---|---|---|---|---|---|---|---|
| y | 270 | 260 | 275 | 405 | 364 | 295 | 335 | 308 | 405 | 305 |

**4.124  Custom Homes.** Following are the size and price data for custom homes from Exercises 4.52 and 4.90.

| x | 26 | 27 | 33 | 29 | 29 | 34 | 30 | 40 | 22 |
|---|---|---|---|---|---|---|---|---|---|
| y | 540 | 555 | 575 | 577 | 606 | 661 | 738 | 804 | 496 |

**4.125  Plant Emissions.** Following are the data on plant weight and quantity of volatile emissions from Exercises 4.53 and 4.91.

| x | 57 | 85 | 57 | 65 | 52 | 67 | 62 | 80 | 77 | 53 | 68 |
|---|---|---|---|---|---|---|---|---|---|---|---|
| y | 8.0 | 22.0 | 10.5 | 22.5 | 12.0 | 11.5 | 7.5 | 13.0 | 16.5 | 21.0 | 12.0 |

**4.126  Crown-Rump Length.** Following are the data on age and crown-rump length for fetuses from Exercises 4.54 and 4.92.

| x | 10 | 10 | 13 | 13 | 18 | 19 | 19 | 23 | 25 | 28 |
|---|---|---|---|---|---|---|---|---|---|---|
| y | 66 | 66 | 108 | 106 | 161 | 166 | 177 | 228 | 235 | 280 |

**4.127  Study Time and Score.** Following are the data on study time and score for calculus students from Exercises 4.55 and 4.93.

| x | 10 | 15 | 12 | 20 | 8 | 16 | 14 | 22 |
|---|---|---|---|---|---|---|---|---|
| y | 92 | 81 | 84 | 74 | 85 | 80 | 84 | 80 |

**4.128  Height and Score.** A random sample of 10 students was taken from an introductory statistics class. The following data were obtained, where $x$ denotes height, in inches, and $y$ denotes score on the final exam.

| x | 71 | 68 | 71 | 65 | 66 | 68 | 68 | 64 | 62 | 65 |
|---|---|---|---|---|---|---|---|---|---|---|
| y | 87 | 96 | 66 | 71 | 71 | 55 | 83 | 67 | 86 | 60 |

a. What sort of value of $r$ would you expect to find for these data? Explain your answer.

b. Compute $r$.

**4.129** Consider the following set of data points.

| x | −3 | −2 | −1 | 0 | 1 | 2 | 3 |
|---|---|---|---|---|---|---|---|
| y | 9 | 4 | 1 | 0 | 1 | 4 | 9 |

a. Compute the linear correlation coefficient, $r$.

b. Can you conclude from your answer in part (a) that the variables $x$ and $y$ are unrelated? Explain your answer.

c. Draw a scatterplot for the data.

d. Is use of the linear correlation coefficient as a descriptive measure for the data appropriate? Explain your answer.

e. Show that the data are related by the quadratic equation $y = x^2$. Graph that equation and the data points.

**4.130** Consider the following set of data points.

| x | −3 | −2 | −1 | 0 | 1 | 2 | 3 |
|---|---|---|---|---|---|---|---|
| y | −27 | −8 | −1 | 0 | 1 | 8 | 27 |

a. Compute the linear correlation coefficient, $r$.

b. Can you conclude from your answer in part (a) that the variables $x$ and $y$ are linearly related? Explain your answer.

c. Draw a scatterplot for the data.

d. Is use of the linear correlation coefficient as a descriptive measure for the data appropriate? Explain your answer.

e. Show that the data are related by the cubic equation $y = x^3$. Graph that equation and the data points.

**4.131** Determine whether $r$ is positive, negative, or zero for each of the following data sets.

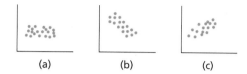

(a)          (b)          (c)

## Working With Large Data Sets

*In Exercises 4.132–4.144, use the technology of your choice to do the following.*

*a. Decide whether use of the linear correlation coefficient as a descriptive measure for the data is appropriate. If so, then also do parts (b) and (c).*

*b. Obtain the linear correlation coefficient.*

*c. Interpret the value of $r$ in terms of the linear relationship between the two variables in question.*

**4.132 Birdies and Score.** The data from Exercise 4.64 for number of birdies during a tournament and final score for 63 women golfers are on the WeissStats CD.

**4.133 U.S. Presidents.** The data from Exercise 4.65 for the ages at inauguration and of death for the presidents of the United States are on the WeissStats CD.

**4.134 Health Care.** The data from Exercise 4.66 for percentage of gross domestic product (GDP) spent on health care and life expectancy, in years, for selected countries are on the WeissStats CD.

**4.135 Acreage and Value.** The data from Exercise 4.67 for lot size (in acres) and assessed value (in thousands of dollars) for a sample of homes in a particular area are on the WeissStats CD.

**4.136 Home Size and Value.** The data from Exercise 4.68 for home size (in square feet) and assessed value (in thousands of dollars) for the same homes as in Exercise 4.135 are on the WeissStats CD.

**4.137 High and Low Temperature.** The data from Exercise 4.69 for average high and low temperatures in January for a random sample of 50 cities are on the WeissStats CD.

**4.138 PCBs and Pelicans.** The data on shell thickness and concentration of PCBs for 60 Anacapa pelican eggs from Exercise 4.70 are on the WeissStats CD.

**4.139 More Money More Beer?** The data for per capita income and per capita beer consumption for the 50 states and Washington, D.C., from Exercise 4.71 are on the WeissStats CD.

**4.140 Gas Guzzlers.** The data for gas mileage and engine displacement for 121 vehicles from Exercise 4.72 are on the WeissStats CD.

**4.141 Shortleaf Pines.** The data from Exercise 4.74 for volume, in cubic feet, and diameter at breast height, in inches, for 70 shortleaf pines are on the WeissStats CD.

**4.142 Body Fat.** The data from Exercise 4.104 for age and percentage of body fat for 18 randomly selected adults are on the WeissStats CD.

**4.143 Estriol Level and Birth Weight.** The data for estriol levels of pregnant women and birth weights of their children from Exercise 4.105 are on the WeissStats CD.

**4.144 Fiber Density.** In the article "Comparison of Fiber Counting by TV Screen and Eyepieces of Phase Contrast Microscopy" (*American Industrial Hygiene Association Journal*, Vol. 63, pp. 756–761), I. Moa et al. reported on determining fiber density by two different methods. Twenty samples of varying fiber density were each counted by 10 viewers by means of an eyepiece method and a television-screen method to determine the relationship between the counts done by each method. The results, in fibers per square millimeter, are presented on the WeissStats CD.

## Extending the Concepts and Skills

**4.145** The coefficient of determination of a set of data points is 0.716.

a. Can you determine the linear correlation coefficient? If yes, obtain it. If no, why not?

b. Can you determine whether the slope of the regression line is positive or negative? Why or why not?

c. If we tell you that the slope of the regression line is negative, can you determine the linear correlation coefficient? If yes, obtain it. If no, why not?

d. If we tell you that the slope of the regression line is positive, can you determine the linear correlation coefficient? If yes, obtain it. If no, why not?

**4.146 Country Music Blues.** A Knight-Ridder News Service article in an issue of the *Wichita Eagle* discussed a study on the relationship between country music and suicide. The results of the study, coauthored by Steven Stack and James Gundlach, appeared as the paper "The Effect of Country Music on Suicide" (*Social Forces*, Vol. 71 Issue 1, pp. 211–218). According to the article, "…analysis of 49 metropolitan areas shows that the greater the airtime devoted to country music, the greater the white suicide rate." (Suicide rates in the black population were found to be uncorrelated with the amount of country music airtime.)

a. Use the terminology introduced in this section to describe the statement quoted above.

b. One of the conclusions stated in the journal article was that country music "nurtures a suicidal mood" by dwelling on marital status and alienation from work. Is this conclusion warranted solely on the basis of the positive correlation found between airtime devoted to country music and white suicide rate? Explain your answer.

**Rank Correlation.** The **rank correlation coefficient**, $r_s$, is a nonparametric alternative to the linear correlation coefficient. It was developed by Charles Spearman (1863–1945) and therefore is also known as the **Spearman rank correlation coefficient.** To determine the rank correlation coefficient, we first rank the $x$-values among themselves and the $y$-values among themselves, and then compute the linear correlation coefficient of the rank pairs. An advantage of the rank correlation coefficient over the linear correlation coefficient is that the former can be used to describe the strength of a positive or negative nonlinear (as well as

linear) relationship between two variables. Ties are handled as usual: if two or more $x$-values (or $y$-values) are tied, each is assigned the mean of the ranks they would have had if there were no ties.

*In each of Exercises **4.147** and **4.148**, do the following for the specified exercise.*
*a. Construct a scatterplot for the data.*
*b. Decide whether using the rank correlation coefficient is reasonable.*
*c. Decide whether using the linear correlation coefficient is reasonable.*
*d. Find and interpret the rank correlation coefficient.*

**4.147 Study Time and Score.** Exercise 4.127.

**4.148 Shortleaf Pines.** Exercise 4.141. (*Note:* Use technology here.)

## Chapter in Review

### You Should be Able to

1. use and understand the formulas in this chapter.

2. define and apply the concepts related to linear equations with one independent variable.

3. explain the least-squares criterion.

4. obtain and graph the regression equation for a set of data points, interpret the slope of the regression line, and use the regression equation to make predictions.

5. define and use the terminology *predictor variable* and *response variable*.

6. understand the concept of extrapolation.

7. identify outliers and influential observations.

8. know when obtaining a regression line for a set of data points is appropriate.

9. calculate and interpret the three sums of squares, SST, SSE, and SSR, and the coefficient of determination, $r^2$.

10. find and interpret the linear correlation coefficient, $r$.

11. identify the relationship between the linear correlation coefficient and the coefficient of determination.

### Key Terms

bivariate quantitative data, *162*
coefficient of determination ($r^2$), *179*
curvilinear regression, *169*
error, *164*
error sum of squares (SSE), *179*
explanatory variable, *167*
extrapolation, *167*
influential observation, *168*
least-squares criterion, *164*
line, *156*
linear correlation coefficient ($r$),
    *186, 187*

linear equation, *156*
lurking variables, *192*
negatively linearly correlated
    variables, *188*
outlier, *168*
Pearson product moment correlation
    coefficient, *186*
positively linearly correlated
    variables, *188*
predictor variable, *167*
regression equation, *164, 165*
regression identity, *182*

regression line, *164*
regression sum of squares (SSR),
    *178, 179*
response variable, *167*
scatter diagram, *162*
scatterplot, *162*
slope, *158*
straight line, *156*
total sum of squares (SST), *178, 179*
$y$-intercept, *158*

## Review Problems

### Understanding the Concepts and Skills

**1.** For a linear equation $y = b_0 + b_1 x$, identify the
**a.** independent variable.    **b.** dependent variable.
**c.** slope.                     **d.** $y$-intercept.

**2.** Consider the linear equation $y = 4 - 3x$.
**a.** At what $y$-value does its graph intersect the $y$-axis?
**b.** At what $x$-value does its graph intersect the $y$-axis?
**c.** What is its slope?
**d.** By how much does the $y$-value on the line change when the $x$-value increases by 1 unit?
**e.** By how much does the $y$-value on the line change when the $x$-value decreases by 2 units?

**3.** Answer true or false to each statement and explain your answers.
**a.** The $y$-intercept of a line has no effect on the steepness of the line.
**b.** A horizontal line has no slope.
**c.** If a line has a positive slope, $y$-values on the line decrease as the $x$-values decrease.

**4.** What kind of plot is useful for deciding whether finding a regression line for a set of data points is reasonable?

**5.** Identify one use of a regression equation.

**6.** Regarding the variables in a regression analysis,
**a.** what is the independent variable called?
**b.** what is the dependent variable called?

**7.** Fill in the blanks.
**a.** Based on the least-squares criterion, the line that best fits a set of data points is the one having the _____ possible sum of squared errors.
**b.** The line that best fits a set of data points according to the least-squares criterion is called the _____ line.
**c.** Using a regression equation to make predictions for values of the predictor variable outside the range of the observed values of the predictor variable is called _____.

**8.** In the context of regression analysis, what is an
**a.** outlier?    **b.** influential observation?

**9.** Identify a use of the coefficient of determination as a descriptive measure.

**10.** For each of the sums of squares in regression, state its name and what it measures.
**a.** *SST*    **b.** *SSR*    **c.** *SSE*

**11.** Fill in the blanks.
**a.** One use of the linear correlation coefficient is as a descriptive measure of the strength of the _____ relationship between two variables.

**b.** A positive linear relationship between two variables means that one variable tends to increase linearly as the other _____.
**c.** A value of $r$ close to $-1$ suggests a strong _____ linear relationship between the variables.
**d.** A value of $r$ close to _____ suggests at most a weak linear relationship between the variables.

**12.** Answer true or false to the following statement and explain your answer: A strong correlation between two variables doesn't necessarily mean that they're causally related.

**13. Equipment Depreciation.** A small company has purchased a microcomputer system for $7200 and plans to depreciate the value of the equipment by $1200 per year for 6 years. Let $x$ denote the age of the equipment, in years, and $y$ denote the value of the equipment, in hundreds of dollars.
**a.** Find the equation that expresses $y$ in terms of $x$.
**b.** Find the $y$-intercept, $b_0$, and slope, $b_1$, of the linear equation in part (a).
**c.** Without graphing the equation in part (a), decide whether the line slopes upward, slopes downward, or is horizontal.
**d.** Find the value of the computer equipment after 2 years; after 5 years.
**e.** Obtain the graph of the equation in part (a) by plotting the points from part (d) and connecting them with a line.
**f.** Use the graph from part (e) to visually estimate the value of the equipment after 4 years. Then calculate that value exactly, using the equation from part (a).

**14. Graduation Rates.** Graduation rate—the percentage of entering freshmen, attending full time and graduating within 5 years—and what influences it have become a concern in U.S. colleges and universities. *U.S. News and World Report*'s "College Guide" provides data on graduation rates for colleges and universities as a function of the percentage of freshmen in the top 10% of their high school class, total spending per student, and student-to-faculty ratio. A random sample of 10 universities gave the following data on student-to-faculty ratio (S/F ratio) and graduation rate (grad rate).

| S/F ratio $x$ | Grad rate $y$ | S/F ratio $x$ | Grad rate $y$ |
|---|---|---|---|
| 16 | 45 | 17 | 46 |
| 20 | 55 | 17 | 50 |
| 17 | 70 | 17 | 66 |
| 19 | 50 | 10 | 26 |
| 22 | 47 | 18 | 60 |

a. Draw a scatterplot of the data.
b. Is finding a regression line for the data reasonable? Explain your answer.
c. Determine the regression equation for the data and draw its graph on the scatterplot you drew in part (a).
d. Describe the apparent relationship between student-to-faculty ratio and graduation rate.
e. What does the slope of the regression line represent in terms of student-to-faculty ratio and graduation rate?
f. Use the regression equation to predict the graduation rate of a university having a student-to-faculty ratio of 17.
g. Identify outliers and potential influential observations.

**15. Graduation Rates.** Refer to Problem 14.
a. Determine *SST*, *SSR*, and *SSE* by using the computing formulas.
b. Obtain the coefficient of determination.
c. Obtain the percentage of the total variation in the observed graduation rates that is explained by student-to-faculty ratio (i.e., by the regression line).
d. State how useful the regression equation appears to be for making predictions.

**16. Graduation Rates.** Refer to Problem 14.
a. Compute the linear correlation coefficient, *r*.
b. Interpret your answer from part (a) in terms of the linear relationship between student-to-faculty ratio and graduation rate.
c. Discuss the graphical implications of the value of the linear correlation coefficient, *r*.
d. Use your answer from part (a) to obtain the coefficient of determination.

## Working With Large Data Sets

**17. Exotic Plants.** In the article "Effects of Human Population, Area, and Time on Non-native Plant and Fish Diversity in the United States" (*Biological Conservation*, Vol. 100, No. 2, pp. 243–252), M. McKinney investigated the relationship of various factors on the number of exotic plants in each state. On the WeissStats CD, you will find the data on population (in millions), area (in thousands of square miles), and number of exotic plants for each state. Use the technology of your choice to determine the linear correlation coefficient between each of the following:
a. population and area
b. population and number of exotic plants
c. area and number of exotic plants
d. Interpret and explain the results you got in parts (a)–(c).

*In Problems 18–21, use the technology of your choice to do the following.*
*a. Construct and interpret a scatterplot for the data.*
*b. Decide whether finding a regression line for the data is reasonable. If so, then also do parts (c)–(f).*
*c. Determine and interpret the regression equation.*
*d. Make the indicated predictions.*
*e. Compute and interpret the correlation coefficient.*
*f. Identify potential outliers and influential observations.*

**18. IMR and Life Expectancy.** From the *Statistical Abstract of the United States*, we obtained data on infant mortality rate (IMR) and life expectancy, in years, for countries with 12 million or more population in 2003. The data are presented on the WeissStats CD. For part (d), predict the life expectancy of a country with an IMR of 30.

**19. High Temperature and Precipitation.** The U.S. National Oceanic and Atmospheric Administration publishes temperature and precipitation information for cities around the world in *Climates of the World*. Data on average high temperature (in degrees Fahrenheit) in July and average precipitation (in inches) in July for 48 cities are on the WeissStats CD. For part (d), predict the average July precipitation of a city with an average July temperature of 83°F.

**20. Fat Consumption and Prostate Cancer.** Researchers have asked whether there is a relationship between nutrition and cancer, and many studies have shown that there is. In fact, one of the conclusions of a study by B. Reddy et al., "Nutrition and Its Relationship to Cancer" (*Advances in Cancer Research*, Vol. 32, pp. 237–345), was that "…none of the risk factors for cancer is probably more significant than diet and nutrition." One dietary factor that has been studied for its relationship with prostate cancer is fat consumption. On the WeissStats CD, you will find data on per capita fat consumption (in grams per day) and prostate cancer death rate (per 100,000 males) for nations of the world. The data were obtained from a graph—adapted from information in the article mentioned—in John Robbins' classic book *Diet for a New America* (Walpole, NH: Stillpoint, 1987, p. 271). For part (d), predict the prostate cancer death rate for a nation with a per capita fat consumption of 92 grams per day.

**21. Payroll and Success.** Steve Galbraith, Morgan Stanley's Chief Investment Officer for U.S. Equity Research, examined several financial aspects of baseball in the article "Finding the Financial Equivalent of a Walk" (*Perspectives*, October 2003, pp. 1–3). The article includes a table containing the 2003 payrolls (in millions of dollars) and winning percentages (through August 3) of major league baseball teams. We have reproduced the data on the WeissStats CD. For part (d), predict the winning percentage of a team with a payroll of $50 million; $100 million.

## Focusing on Data Analysis    UWEC Undergraduates

Recall from Chapter 1 (see page 34) that the Focus database and Focus sample contain information on the undergraduate students at the University of Wisconsin - Eau Claire (UWEC). Now would be a good time for you to review the discussion about these data sets.

Open the Focus sample worksheet (FocusSample) in the technology of your choice and do the following.

a. Find the linear correlation coefficient between cumulative GPA and high school percentile for the 200 UWEC undergraduate students in the Focus sample.
b. Repeat part (a) for cumulative GPA and each of ACT English score, ACT math score, and ACT composite score.

c. Among the variables high school percentile, ACT English score, ACT math score, and ACT composite score, identify the one that appears to be the best predictor of cumulative GPA. Explain your reasoning.

Now perform a regression analysis on cumulative GPA, using the predictor variable identified in part (c), as follows.

d. Obtain and interpret a scatterplot.
e. Find and interpret the regression equation.
f. Find and interpret the coefficient of determination.
g. Determine and interpret the three sums of squares SSR, SSE, and SST.

## Case Study Discussion    Shoe Size and Height

At the beginning of this chapter, we presented data on shoe size and height for a sample of students at Arizona State University. Now that you have studied regression and correlation, you can analyze the relationship between those two variables. We recommend that you use statistical software or a graphing calculator to solve the following problems, but they can also be done by hand.

a. Separate the data in the table on page 155 into two tables, one for males and the other for females. Parts (b)–(k) are for the male data.
b. Draw a scatterplot for the data on shoe size and height for males.
c. Does obtaining a regression equation for the data appear reasonable? Explain your answer.
d. Find the regression equation for the data, using shoe size as the predictor variable.

e. Interpret the slope of the regression line.
f. Use the regression equation to predict the height of a male student who wears a size $10\frac{1}{2}$ shoe.
g. Obtain and interpret the coefficient of determination.
h. Compute the correlation coefficient of the data and interpret your result.
i. Identify outliers and potential influential observations, if any.
j. If there are outliers, first remove them and then repeat parts (b)–(h).
k. Decide whether any potential influential observation that you detected is in fact an influential observation. Explain your reasoning.
l. Repeat parts (b)–(k) for the data on shoe size and height for females. For part (f), do the prediction for the height of a female student who wears a size 8 shoe.

## Biography ADRIEN LEGENDRE: Introducing the Method of Least-Squares

**Adrien-Marie Legendre** was born in Paris, France, on September 18, 1752, the son of a moderately wealthy family. He studied at the Collège Mazarin and received degrees in mathematics and physics in 1770 at the age of 18.

Although Legendre's financial assets were sufficient to allow him to devote himself to research, he took a position teaching mathematics at the École Militaire in Paris from 1775 to 1780. In March 1783, he was elected to the Academie des Sciences in Paris and, in 1787, he was assigned to a project undertaken jointly by the observatories at Paris and at Greenwich, England. At that time, he became a fellow of the Royal Society.

As a result of the French Revolution, which began in 1789, Legendre lost his "small fortune" and was forced to find work. He held various positions during the early 1790s as, for example, commissioner of astronomical operations for the Academie des Sciences, professor of pure mathematics at the Institut de Marat, and head of the National Executive Commission of Public Instruction. During this same period, Legendre wrote a geometry book that became the major text used in elementary geometry courses for nearly a century.

Legendre's major contribution to statistics was the publication, in 1805, of the first statement and the first application of the most widely used, nontrivial technique of statistics: the method of least squares. In his book, *The History of Statistics: The Measurement of Uncertainty Before 1900* (Cambridge, Mass.: Belknap Press of Harvard University Press, 1986), Stephen M. Stigler writes "[Legendre's] presentation … must be counted as one of the clearest and most elegant introductions of a new statistical method in the history of statistics."

Because Gauss also claimed the method of least squares, there was strife between the two men. Although evidence shows that Gauss was not successful in any communication of the method prior to 1805, his development of the method was crucial to its usefulness.

In 1813, Legendre was appointed Chief of the Bureau des Longitudes. He remained in that position until his death, following a long illness, in Paris on January 10, 1833.

## StatCrunch in MyStatLab
### Analyzing Data Online

StatCrunch online statistical software offers an easy-to-use interface customized for this book. The StatCrunch feature for each chapter illustrates the use of the software to perform a statistical analysis discussed in the chapter. Exercises are provided to further apply StatCrunch to other statistical analyses examined in the chapter. Go to the WeissStats CD or to the Weiss Web site at www.aw-bc.com/weiss to access StatCrunch instructions and data sets. To access StatCrunch statistical software, go to the student content area of your Weiss MyStatLab course.

## Internet Projects
### Exploring Data Online

The Internet project for each chapter provides simulations, demonstrations, or activities that enhance the topics covered in the chapter. The project materials come from universities, individuals, governments, and companies from all over the world. To access the Internet projects on the Web, go to www.aw-bc.com/weiss. From this Web page, you can reach the Internet Projects Page, which we suggest that you bookmark for easy access in the future.

# PART III

## Probability, Random Variables, and Sampling Distributions

# 5 Probability and Random Variables

## Chapter Outline

## Chapter Objectives

Until now, we have concentrated on descriptive statistics—methods for organizing and summarizing data. Another important aspect of this text is to present the fundamentals of inferential statistics—methods of drawing conclusions about a population based on information from a sample of the population.

Because inferential statistics involves using information from part of a population (a sample) to draw conclusions about the entire population, we can never be certain that our conclusions are correct; that is, uncertainty is inherent in inferential statistics. Consequently, you need to become familiar with uncertainty before you can understand, develop, and apply the methods of inferential statistics.

The science of uncertainty is called **probability theory.** It enables you to evaluate and control the likelihood that a statistical inference is correct. More generally, probability theory provides the mathematical basis for inferential statistics.

The theory of probability also allows us to extend concepts that apply to variables of finite populations—such as relative-frequency distribution, mean, and standard deviation—to other types of variables. In doing so, we are led to the notion of a *random variable* and its *probability distribution*.

In Sections 5.1–5.3, we present the fundamentals of probability. Next, in Sections 5.4 and 5.5, we discuss the basics of discrete random variables and their probability distributions. Then, as an application of those two sections, we examine, in Section 5.6, the most important discrete probability distribution, namely, the *binomial distribution*.

# The Powerball

The Powerball is a lottery that was introduced in 1992. It is now played in 27 states, Washington D.C., and the U.S. Virgin Islands. Profits from Powerball tickets stay in the state where the ticket is sold and each state uses its own computer system to issue and validate Powerball tickets.

Although the Powerball is a multistate lottery game, it isn't the first. That distinction goes to Lotto*America, which was created in 1987 when Iowa and six other states joined forces to offer a game with a large jackpot. The more people who play, the bigger the jackpots tend to be, so multistate lotteries offer larger prizes than those of individual states.

A Powerball jackpot grows if no one wins it. Because the chance of winning a jackpot is small, the jackpot often grows to huge amounts, sometimes more than $300 million. When there are multiple winners, the jackpot is divided equally among them. Drawings take place on Wednesday and Saturday evenings at 10:59 P.M. Eastern Time.

To play the basic Powerball, a player first selects five numbers from the numbers 1–55 and then chooses a PowerBall number, which can be any number between 1 and 42. A ticket costs $1. In the drawing, five white balls are drawn randomly from 55 white balls numbered 1–55; then one red PowerBall is drawn randomly from 42 red balls numbered 1–42.

To win the jackpot, a ticket must match all the balls drawn; smaller prizes are awarded for matching some but not all the balls drawn. What are the chances of winning the jackpot? What are the chances of winning any prize at all? After studying probability, you will be able to answer these and similar questions. You will be asked to do so when you revisit the Powerball at the end of this chapter.

## 5.1    Probability Basics

Although most applications of probability theory to statistical inference involve large populations, we will explain the fundamental concepts of probability in this chapter with examples that involve relatively small populations or games of chance.

### The Equal-Likelihood Model

We discussed an important aspect of probability when we examined probability sampling in Chapter 1. The following example returns to the illustration of simple random sampling from Example 1.7 on page 14.

---

**Example 5.1** | **Introducing Probability**

*Oklahoma State Officials*  As reported by *The World Almanac*, the top five state officials of Oklahoma are as shown in Table 5.1. Suppose that we take a simple random sample without replacement of two officials from the five officials.

**a.** Find the probability that we obtain the governor and treasurer.

**b.** Find the probability that the attorney general is included in the sample.

**Solution**  For convenience, we use the letters in parentheses after the titles in Table 5.1 to represent the officials. As we discovered in Example 1.7, there are 10 possible samples of two officials from the population of five officials. They are listed in Table 5.2. If we take a simple random sample of size 2, each of the possible samples of two officials is equally likely to be the one selected.

**a.** Because there are 10 possible samples, the probability is $\frac{1}{10}$, or 0.1, of selecting the governor and treasurer (G, T). Another way of looking at this result is that 1 out of 10, or 10%, of the samples include both the governor and treasurer; hence the probability of obtaining such a sample is 10%, or 0.1. The same goes for any other two particular officials.

**b.** Table 5.2 shows that the attorney general (A) is included in 4 of the 10 possible samples of size 2. As each of the 10 possible samples is equally likely to be the one selected, the probability is $\frac{4}{10}$, or 0.4, that the attorney general is included in the sample. Another way of looking at this result is that 4 out of 10, or 40%, of the samples include the attorney general; hence the probability of obtaining such a sample is 40%, or 0.4.

• • •

**TABLE 5.1**

Five top Oklahoma state officials

| |
| --- |
| Governor (G) |
| Lieutenant Governor (L) |
| Secretary of State (S) |
| Attorney General (A) |
| Treasurer (T) |

**TABLE 5.2**

The 10 possible samples of two officials

| | | | | |
| --- | --- | --- | --- | --- |
| G, L | G, S | G, A | G, T | L, S |
| L, A | L, T | S, A | S, T | A, T |

**You try it!**

Exercise 5.9
on page 209

The essential idea in Example 5.1 is that when outcomes are equally likely, probabilities are nothing more than percentages (relative frequencies).

**Definition 5.1**

## Probability for Equally Likely Outcomes (*f/N* Rule)

Suppose an experiment has $N$ possible outcomes, all equally likely. An event that can occur in $f$ ways has probability $f/N$ of occurring:

Number of ways event can occur

$$\text{Probability of an event} = \frac{f}{N}.$$

Total number of possible outcomes

In stating Definition 5.1, we used the terms *experiment* and *event* in their intuitive sense. Basically, by an **experiment,** we mean an action whose outcome cannot be predicted with certainty. By an **event,** we mean some specified result that may or may not occur when an experiment is performed.

For instance, in Example 5.1 the experiment consists of taking a random sample of size 2 from the five officials. It has 10 possible outcomes ($N = 10$), all equally likely. In part (b), the event is that the sample obtained includes the attorney general, which can occur in four ways ($f = 4$); hence its probability equals

$$\frac{f}{N} = \frac{4}{10} = 0.4,$$

as we noted in Example 5.1(b).

**Example 5.2**

## Probability for Equally Likely Outcomes

*Family Income* The U.S. Census Bureau compiles data on family income and publishes its findings in *Current Population Reports*. Table 5.3 gives a frequency distribution of annual income for U.S. families.

A U.S. family is selected **at random,** meaning that each family is equally likely to be the one obtained (simple random sample of size 1). Determine the probability that the family selected has an annual income of

**a.** between $80,000 and $99,999, inclusive.

**b.** between $20,000 and $79,999, inclusive.

**c.** under $40,000.

**Solution** The second column of Table 5.3 shows that there are 75,617 thousand U.S. families; so $N = 75,617$ thousand.

**a.** The event in question is that the family selected makes between $80,000 and $99,999. Table 5.3 shows that the number of such families is 7,534 thousand, so $f = 7,534$ thousand. Applying the $f/N$ rule, we find that the probability the family selected makes between $80,000 and $99,999 is

$$\frac{f}{N} = \frac{7,534}{75,617} = 0.100.$$

**Interpretation** 10.0% of families in the United States make between $80,000 and $99,999, inclusive.

**TABLE 5.3**
Frequency distribution of annual income for U.S. families

| Income | Frequency (1000s) |
|---|---|
| Under $20,000 | 11,370 |
| $20,000–$39,999 | 17,370 |
| $40,000–$59,999 | 14,436 |
| $60,000–$79,999 | 11,410 |
| $80,000–$99,999 | 7,534 |
| $100,000–$199,999 | 11,219 |
| $200,000 & above | 2,278 |
| | 75,617 |

**b.** The event in question is that the family selected makes between $20,000 and $79,999. Table 5.3 reveals that the number of such families is $17,370 + 14,436 + 11,410$, or 43,216 thousand. Consequently, $f = 43,216$ thousand, and the required probability is

$$\frac{f}{N} = \frac{43,216}{75,617} = 0.572.$$

**Interpretation** 57.2% of families in the United States make between $20,000 and $79,999, inclusive.

**c.** Proceeding as in parts (a) and (b), we find that the probability that the family selected makes under $40,000 is

$$\frac{f}{N} = \frac{11,370 + 17,370}{75,617} = \frac{28,740}{75,617} = 0.380.$$

**Interpretation** 38.0% of families in the United States make less than $40,000.

You
try it!

Exercise 5.13
on page 210

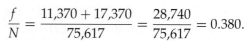

**Example 5.3** | **Probability for Equally Likely Outcomes**

*Dice* When two balanced dice are rolled, 36 equally likely outcomes are possible, as depicted in Fig. 5.1. Find the probability that

**a.** the sum of the dice is 11.

**b.** doubles are rolled; that is, both dice come up the same number.

**FIGURE 5.1**
Possible outcomes for rolling
a pair of dice

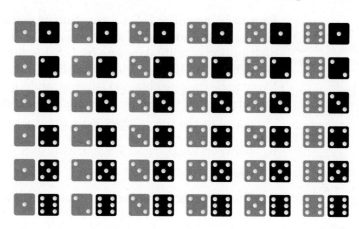

**Solution** For this experiment, $N = 36$.

**a.** The sum of the dice can be 11 in two ways, as is apparent from Fig. 5.1. Hence the probability that the sum of the dice is 11 equals $f/N = 2/36 = 0.056$.

**Interpretation** There is a 5.6% chance of a sum of 11 when two balanced dice are rolled.

You
try it!

Exercise 5.19
on page 211

**b.** Figure 5.1 also shows that doubles can be rolled in six ways. Consequently, the probability of rolling doubles equals $f/N = 6/36 = 0.167$.

**Interpretation**   There is a 16.7% chance of doubles when two balanced dice are rolled.

• • •

### The Meaning of Probability

Essentially, probability is a generalization of the concept of percentage. When we select a member at random from a finite population, as we did in Example 5.2, probability is nothing more than percentage. But, in general, how do we interpret probability? For instance, what do we mean when we say that

- the probability is 0.314 that the gestation period of a woman will exceed 9 months or
- the probability is 0.667 that the favorite in a horse race finishes in the money (first, second, or third place) or
- the probability is 0.40 that a traffic fatality involves an intoxicated or alcohol-impaired driver or nonoccupant?

Some probabilities are easy to interpret: A probability near 0 indicates that the event in question is very unlikely to occur when the experiment is performed, whereas a probability near 1 (100%) suggests that the event is quite likely to occur. More generally, the **frequentist interpretation of probability** construes the probability of an event to be the proportion of times it occurs in a large number of repetitions of the experiment.

Consider, for instance, the simple experiment of tossing a balanced coin once. Because the coin is balanced, we reason that there is a 50–50 chance the coin will land with heads facing up. Consequently, we attribute a probability of 0.5 to that event. The frequentist interpretation is that in a large number of tosses, the coin will land with heads facing up about half the time.

We used a computer to perform two simulations of tossing a balanced coin 100 times. The results are displayed in Fig. 5.2. Each graph shows the number of tosses of the coin versus the proportion of heads. Both graphs seem to corroborate the frequentist interpretation.

**FIGURE 5.2**
Two computer simulations of tossing a balanced coin 100 times

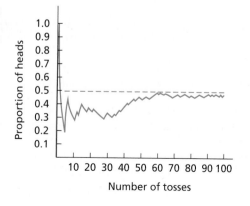
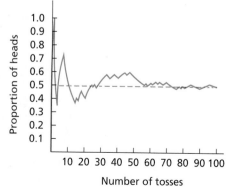

Although the frequentist interpretation is helpful for understanding the meaning of probability, it cannot be used as a definition of probability. One common way to define probabilities is to specify a **probability model**—a mathematical description of the experiment based on certain primary aspects and assumptions.

The **equal-likelihood model** discussed earlier in this section is an example of a probability model. Its primary aspect and assumption are that all possible outcomes are equally likely to occur. We discuss other probability models later in this and subsequent chapters.

### Basic Properties of Probabilities

Some basic properties of probabilities are as follows.

---

**Key Fact 5.1**    **Basic Properties of Probabilities**

**Property 1:** The probability of an event is always between 0 and 1, inclusive.

**Property 2:** The probability of an event that cannot occur is 0. (An event that cannot occur is called an **impossible event.**)

**Property 3:** The probability of an event that must occur is 1. (An event that must occur is called a **certain event.**)

---

Property 1 indicates that numbers such as 5 or $-0.23$ could not possibly be probabilities. Example 5.4 illustrates Properties 2 and 3.

**Example 5.4**  |  **Basic Properties of Probabilities**

*Dice*  Let's return to Example 5.3, wherein two balanced dice are rolled. Determine the probability that

a.  the sum of the dice is 1.

b.  the sum of the dice is 12 or less.

**Solution**

a.  Figure 5.1 on page 206 shows that the sum of the dice must be 2 or more. Thus the probability that the sum of the dice is 1 equals $f/N = 0/36 = 0$.

   **Interpretation**  Getting a sum of 1 when two balanced dice are rolled is impossible and hence has probability 0.

b.  From Fig. 5.1, the sum of the dice must be 12 or less. Thus the probability of that event equals $f/N = 36/36 = 1$.

   **Interpretation**  Getting a sum of 12 or less when two balanced dice are rolled is certain and hence has probability 1.

*You try it!*

Exercise 5.21 on page 211

• • •

# Exercises 5.1

## Understanding the Concepts and Skills

**5.1** Roughly speaking, what is an experiment? an event?

**5.2** Concerning the equal-likelihood model of probability,
**a.** what is it?
**b.** how is the probability of an event found?

**5.3** What is the difference between selecting a member at random from a finite population and taking a simple random sample of size 1?

**5.4** If a member is selected at random from a finite population, probabilities are identical to _____.

**5.5** State the frequentist interpretation of probability.

**5.6** Interpret each of the following probability statements, using the frequentist interpretation of probability.
**a.** The probability is 0.487 that a newborn baby will be a girl.
**b.** The probability of a single ticket winning a prize in the Powerball lottery is 0.028.
**c.** If a balanced dime is tossed three times, the probability that it will come up heads all three times is 0.125.

**5.7** Which of the following numbers could not possibly be probabilities? Justify your answer.
**a.** 0.462       **b.** −0.201       **c.** 1
**d.** $\frac{5}{6}$       **e.** 3.5       **f.** 0

**5.8 Oklahoma State Officials.** Refer to Table 5.1, presented on page 204.
**a.** List the possible samples without replacement of size 3 that can be obtained from the population of five officials. (*Hint:* There are 10 possible samples.)
If a simple random sample without replacement of three officials is taken from the five officials, determine the probability that
**b.** the governor, attorney general, and treasurer are obtained.
**c.** the governor and treasurer are included in the sample.
**d.** the governor is included in the sample.

**5.9 Oklahoma State Officials.** Refer to Table 5.1, presented on page 204.
**a.** List the possible samples without replacement of size 4 that can be obtained from the population of five officials. (*Hint:* There are five possible samples.)
If a simple random sample without replacement of four officials is taken from the five officials, determine the probability that
**b.** the governor, attorney general, and treasurer are obtained.
**c.** the governor and treasurer are included in the sample.
**d.** the governor is included in the sample.

*In the following exercises, express your probability answers as a decimal rounded to three places.*

**5.10 Educated CEOs.** Reporter D. McGinn looked at the changing demographics for successful chief executive officers (CEOs) of America's top companies in the article, "Fresh Ideas" (*Newsweek*, June 13, 2005, pp. 42–46). The following frequency distribution reports the highest education level achieved by Standard and Poor's top 500 CEOs.

| Level      | Frequency |
|------------|-----------|
| No college | 14        |
| B.S./B.A.  | 164       |
| M.B.A.     | 191       |
| J.D.       | 50        |
| Other      | 81        |

Find the probability that a randomly selected CEO from Standard and Poor's top 500 achieved the educational level of
**a.** B.S./B.A.
**b.** either M.B.A. or J.D.
**c.** at least some college.

**5.11 Prospects for Democracy.** In the journal article "The 2003–2004 Russian Elections and Prospects for Democracy" (*Europe-Asia Studies*, Vol. 57, No. 3, pp. 369–398), R. Sakwa examined the fourth electoral cycle that independent Russia entered in 2003. The following frequency table lists the candidates and numbers of votes from the presidential election on March 14, 2004.

| Candidate           | Votes      |
|---------------------|------------|
| Putin, Vladimir     | 49,565,238 |
| Kharitonov, Nikolai | 9,513,313  |
| Glaz'ev, Sergei     | 2,850,063  |
| Khakamada, Irina    | 2,671,313  |
| Malyshkin, Oleg     | 1,405,315  |
| Mironov, Sergei     | 524,324    |

Find the probability that a randomly selected voter voted for
**a.** Vladimir Putin.
**b.** either Malyshkin or Mironov.
**c.** someone other than Vladimir Putin.

**5.12 Cardiovascular Hospitalizations.** From the Florida State Health Department's Health Outcome Series, *Women and Cardiovascular Disease Hospitalization*, we obtained the following table showing the number of female hospitalizations for cardiovascular disease, by age group, during one year.

| Age group | Number |
|-----------|--------|
| 0–19 | 810 |
| 20–39 | 5,029 |
| 40–49 | 10,977 |
| 50–59 | 20,983 |
| 60–69 | 36,884 |
| 70–79 | 65,017 |
| 80 and over | 69,167 |

One of these case records is selected at random. Find the probability that the woman was

**a.** in her 50s.
**b.** less than 50 years old.
**c.** between 40 and 69 years old, inclusive.
**d.** 70 years old or older.

**5.13 Housing Units.** The U.S. Census Bureau publishes data on housing units in *American Housing Survey for the United States*. The following table provides a frequency distribution for the number of rooms in U.S. housing units. The frequencies are in thousands.

| Rooms | No. of units |
|-------|--------------|
| 1 | 520 |
| 2 | 1,425 |
| 3 | 10,943 |
| 4 | 23,363 |
| 5 | 27,976 |
| 6 | 24,646 |
| 7 | 14,670 |
| 8+ | 17,234 |

A U.S. housing unit is selected at random. Find the probability that the housing unit obtained has

**a.** four rooms.
**b.** more than four rooms.
**c.** one or two rooms.
**d.** fewer than one room.
**e.** one or more rooms.

**5.14 Murder Victims.** As reported by the Federal Bureau of Investigation in *Crime in the United States*, the age distribution of murder victims between 20 and 59 years old is as shown in the following table.

| Age | Frequency |
|-------|-----------|
| 20–24 | 2810 |
| 25–29 | 2077 |
| 30–34 | 1617 |
| 35–39 | 1356 |
| 40–44 | 1152 |
| 45–49 | 864 |
| 50–54 | 574 |
| 55–59 | 360 |

A murder case in which the person murdered was between 20 and 59 years old is selected at random. Find the probability that the murder victim was

**a.** between 40 and 44 years old, inclusive.
**b.** at least 25 years old, that is, 25 years old or older.
**c.** between 45 and 59 years old, inclusive.
**d.** under 30 or over 54.

**5.15 Occupations in Seoul.** The population of Seoul was studied in an article by B. Lee and J. McDonald, "Determinants of Commuting Time and Distance for Seoul Residents: The Impact of Family Status on the Commuting of Women" (*Urban Studies*, Vol. 40, No. 7, pp. 1283–1302). The authors examined the different occupations for males and females in Seoul. Following is a frequency distribution of occupation type for males taking part in a survey. (*Note:* M = manufacturing, N = nonmanufacturing)

| Occupation | Frequency |
|------------|-----------|
| Administrative/M | 2,197 |
| Administrative/N | 6,450 |
| Technical/M | 2,166 |
| Technical/N | 6,677 |
| Clerk/M | 1,640 |
| Clerk/N | 4,538 |
| Production workers/M | 5,721 |
| Production workers/N | 10,266 |
| Service | 9,274 |
| Agriculture | 159 |

If one of these males is selected at random, find the probability that his occupation is

**a.** service.
**b.** administrative.
**c.** manufacturing.
**d.** not manufacturing.

**5.16 Nobel Prize Winners.** The National Science Foundation collects data on Nobel Prize Laureates in the field of science and the date and location of their award-winning research. A frequency distribution for the number of winners, by country, for the years 1901–2003 is as follows.

| Country | Winners |
|---------|---------|
| United States | 219 |
| United Kingdom | 76 |
| Germany | 63 |
| France | 25 |
| Soviet Union | 12 |
| Japan | 8 |
| Other Countries | 91 |

Suppose that a recipient of a Nobel Prize in science between 1901–2003 is selected at random. Find the probability that the Nobel Laureate is from

a. Japan.
b. either France or Germany.
c. any country other than the United States.

**5.17 Graduate Science Students.** According to *Survey of Graduate Science Engineering Students and Postdoctorates*, published by the U.S. National Science Foundation, the distribution of graduate science students in doctorate-granting institutions is as follows. Frequencies are in thousands.

| Field | Frequency |
|---|---|
| Physical sciences | 31.3 |
| Environmental | 13.1 |
| Mathematical sciences | 16.4 |
| Computer sciences | 47.6 |
| Agricultural sciences | 11.8 |
| Biological sciences | 57.2 |
| Psychology | 40.8 |
| Social sciences | 78.8 |

A graduate science student who is attending a doctorate-granting institution is selected at random. Determine the probability that the field of the student obtained is
a. psychology.
b. physical or social science.
c. not computer science.

**5.18 Family Size.** A *family* is defined to be a group of two or more persons related by birth, marriage, or adoption and residing together in a household. According to *Current Population Reports*, published by the U.S. Census Bureau, the size distribution of U.S. families is as follows. Frequencies are in thousands.

| Size | Frequency |
|---|---|
| 2 | 33,706 |
| 3 | 16,652 |
| 4 | 15,149 |
| 5 | 6,619 |
| 6 | 2,298 |
| 7+ | 1,173 |

A U.S. family is selected at random. Find the probability that the family obtained has
a. two persons.
b. more than three persons.
c. between one and three persons, inclusive.
d. one person.
e. one or more persons.

**5.19 Dice.** Two balanced dice are rolled. Refer to Fig. 5.1 on page 206 and determine the probability that the sum of the dice is
a. 6.
b. even.
c. 7 or 11.
d. 2, 3, or 12.

**5.20 Coin Tossing.** A balanced dime is tossed three times. The possible outcomes can be represented as follows.

| | | | |
|---|---|---|---|
| HHH | HTH | THH | TTH |
| HHT | HTT | THT | TTT |

Here, for example, HHT means that the first two tosses come up heads and the third tails. Find the probability that
a. exactly two of the three tosses come up heads.
b. the last two tosses come up tails.
c. all three tosses come up the same.
d. the second toss comes up heads.

**5.21 Housing Units.** Refer to Exercise 5.13. Which, if any, of the events in parts (a)–(e) are certain? impossible?

**5.22 Family Size.** Refer to Exercise 5.18. Which, if any, of the events in parts (a)–(e) are certain? impossible?

## Extending the Concepts and Skills

**5.23** Explain what is wrong with the argument: When two balanced dice are rolled, the sum of the dice can be 2, 3, 4, 5, 6, 7, 8, 9, 10, 11, or 12, giving 11 possibilities. Therefore the probability is $\frac{1}{11}$ that the sum is 12.

**5.24** Why can't the frequentist interpretation be used as a definition of probability?

**Odds.** Closely related to probabilities are *odds*. Newspapers, magazines, and other popular publications often express likelihood in terms of odds instead of probabilities, and odds are used much more than probabilities in gambling contexts. If the probability that an event occurs is $p$, the odds that the event occurs are $p$ to $1 - p$. This fact is also expressed by saying that the odds are $p$ to $1 - p$ *in favor of the event* or that the odds are $1 - p$ to $p$ *against the event*. Conversely, if the odds in favor of an event are $a$ to $b$ (or, equivalently, the odds against it are $b$ to $a$), the probability the event occurs is $a/(a + b)$. For example, if an event has probability 0.75 of occurring, the odds that the event occurs are 0.75 to 0.25, or 3 to 1; if the odds against an event are 3 to 2, the probability that the event occurs is $2/(2 + 3)$, or 0.4. We examine odds in Exercises 5.25–5.28.

**5.25 Roulette.** An American roulette wheel contains 38 numbers, of which 18 are red, 18 are black, and 2 are green. When the roulette wheel is spun, the ball is equally likely to land on any of the 38 numbers. For a bet on red, the house pays even odds (i.e., 1 to 1). What should the odds actually be to make the bet fair?

**5.26 Cyber Affair.** As found in *USA TODAY*, results of a survey by International Communications Research revealed that roughly 75% of adult women believe that a romantic relationship over the Internet while in an exclusive relationship in the real world is cheating. What are the odds against randomly selecting an adult female Internet user who believes that having a "cyber affair" is cheating?

**5.27 The Triple Crown.** Funny Cide, winner of both the 2003 Kentucky Derby and the 2003 Preakness Stakes, was the even-money (1-to-1 odds) favorite to win the 2003 Belmont Stakes and thereby capture the coveted Triple Crown of thoroughbred horseracing. The second favorite and actual winner of the 2003 Belmont Stakes, Empire Maker, posted odds at 2 to 1 (against) to win the race. Based on the posted odds, determine the probability that the winner of the race would be

a. Funny Cide.  b. Empire Maker.

**5.28 Lightning Casualties.** An issue of *Travel + Leisure Golf* magazine (May/June, 2005, p. 36) reported several facts about lightning. Here are three of them.

- The odds of an individual being struck by lightning in a year in the United States are about 280,000 to 1 (against).
- The odds of an individual being struck by lightning in a year in Florida—the state with the most golf courses—are about 80,000 to 1 (against).
- About 5% of all lightning fatalities occur on golf courses.

Based on these data, answer the following questions.

a. What is the probability of a person being struck by lightning in a year in the United States? Express your answer as a decimal rounded to eight places.
b. What is the probability of a person being struck by lightning in a year in Florida? Express your answer as a decimal rounded to seven decimal places.
c. If a person dies from being hit by lightning, what are the odds that the fatality did not occur on a golf course?

## 5.2 Events

Before continuing, we need to discuss events in greater detail. In Section 5.1, we used the word *event* intuitively. More precisely, an event is a collection of outcomes, as illustrated in Example 5.5.

**Example 5.5** | **Introducing Events**

*Playing Cards* A deck of playing cards contains 52 cards, as displayed in Fig. 5.3. When we perform the experiment of randomly selecting one card from the deck, we will get one of these 52 cards. The collection of all 52 cards—the possible outcomes—is called the **sample space** for this experiment.

**FIGURE 5.3**
A deck of playing cards

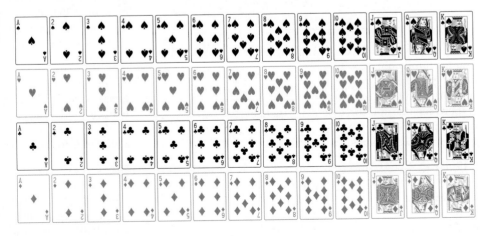

Many different events can be associated with this card-selection experiment. Let's consider four:

a. The event that the card selected is the king of hearts.

b. The event that the card selected is a king.

c. The event that the card selected is a heart.

d. The event that the card selected is a face card.

List the outcomes constituting each of these four events.

**FIGURE 5.4**
The event the king of hearts is selected

**Solution**

a. The event that the card selected is the king of hearts consists of the single outcome "king of hearts," as pictured in Fig. 5.4.

**FIGURE 5.5**
The event a king is selected

b. The event that the card selected is a king consists of the four outcomes "king of spades," "king of hearts," "king of clubs," and "king of diamonds," as depicted in Fig. 5.5.

c. The event that the card selected is a heart consists of the 13 outcomes "ace of hearts," "two of hearts,"..., "king of hearts," as shown in Fig. 5.6.

**FIGURE 5.6**
The event a heart is selected

d. The event that the card selected is a face card consists of 12 outcomes, namely, the 12 face cards shown in Fig. 5.7.

**FIGURE 5.7**
The event a face card is selected

**You try it!**

Exercise 5.33
on page 218

When the experiment of randomly selecting a card from the deck is performed, a specified event **occurs** if that event contains the card selected. For instance, if the card selected turns out to be the king of spades, the second and fourth events (Figs. 5.5 and 5.7) occur, whereas the first and third events (Figs. 5.4 and 5.6) do not.

• • •

**Definition 5.2**   **Sample Space and Event**

**Sample space:** The collection of all possible outcomes for an experiment.

**Event:** A collection of outcomes for the experiment, that is, any subset of the sample space.

**Note:** The term *sample space* reflects the fact that, in statistics, the collection of possible outcomes often consists of the possible samples of a given size, as illustrated in Table 5.2 on page 204.

## Notation and Graphical Displays for Events

For convenience, we use letters such as $A, B, C, D, \ldots$ to represent events. In the card-selection experiment of Example 5.5, for instance, we might let

$A$ = event the card selected is the king of hearts,
$B$ = event the card selected is a king,
$C$ = event the card selected is a heart, and
$D$ = event the card selected is a face card.

**FIGURE 5.8**
Venn diagram for event $E$

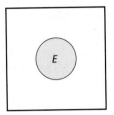

**Venn diagrams,** named after English logician John Venn (1834–1923), are one of the best ways to portray events and relationships among events visually. The sample space is depicted as a rectangle, and the various events are drawn as disks (or other geometric shapes) inside the rectangle. In the simplest case, only one event is displayed, as shown in Fig. 5.8, with the colored portion representing the event.

## Relationships Among Events

Each event $E$ has a corresponding event defined by the condition that "$E$ does not occur." That event is called the **complement** of $E$ and is denoted **(not $E$).** Event (not $E$) consists of all outcomes not in $E$, as shown in the Venn diagram in Fig. 5.9(a).

**FIGURE 5.9**
Venn diagrams for (a) event (not $E$),
(b) event ($A$ & $B$), and (c) event ($A$ or $B$)

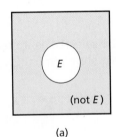

(a)

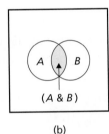

(b)

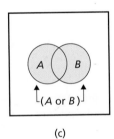

(c)

With any two events, say, $A$ and $B$, we can associate two new events. One new event is defined by the condition that "both event $A$ and event $B$ occur" and is denoted **($A$ & $B$).** Event ($A$ & $B$) consists of all outcomes common to both event $A$ and event $B$, as illustrated in Fig. 5.9(b).

The other new event associated with $A$ and $B$ is defined by the condition that "either event $A$ or event $B$ or both occur" or, equivalently, that "at least one of events $A$ and $B$ occurs." That event is denoted **($A$ or $B$)** and consists of all outcomes in either event $A$ or event $B$ or both, as Fig. 5.9(c) shows.

**Definition 5.3**

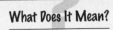

**What Does It Mean?**

Event (not $E$) consists of all outcomes not in event $E$; event ($A$ & $B$) consists of all outcomes common to event $A$ and event $B$; event ($A$ or $B$) consists of all outcomes either in event $A$ or in event $B$ or both.

## Relationships Among Events

**(not $E$):** The event "$E$ does not occur"

**($A$ & $B$):** The event "both $A$ and $B$ occur"

**($A$ or $B$):** The event "either $A$ or $B$ or both occur"

**Note:** Because the event "both $A$ and $B$ occur" is the same as the event "both $B$ and $A$ occur," event ($A$ & $B$) is the same as event ($B$ & $A$). Similarly, event ($A$ or $B$) is the same as event ($B$ or $A$).

## Example 5.6 | Relationships Among Events

*Playing Cards*  For the experiment of randomly selecting one card from a deck of 52, let

$A$ = event the card selected is the king of hearts,

$B$ = event the card selected is a king,

$C$ = event the card selected is a heart, and

$D$ = event the card selected is a face card.

We showed the outcomes for each of those four events in Figs. 5.4–5.7, respectively, in Example 5.5. Determine the following events.

**a.** (not $D$)  **b.** ($B$ & $C$)  **c.** ($B$ or $C$)  **d.** ($C$ & $D$)

### Solution

**a.** (not $D$) is the event $D$ does not occur—the event that a face card is not selected. Event (not $D$) consists of the 40 cards in the deck that are not face cards, as depicted in Fig. 5.10.

**FIGURE 5.10**
Event (not $D$)

**FIGURE 5.11**
Event ($B$ & $C$)

**b.** ($B$ & $C$) is the event both $B$ and $C$ occur—the event that the card selected is both a king and a heart. Consequently, ($B$ & $C$) is the event that the card selected is the king of hearts and consists of the single outcome shown in Fig. 5.11.

**Note:** Event ($B$ & $C$) is the same as event $A$, so we can write $A = (B$ & $C)$.

**c.** ($B$ or $C$) is the event either $B$ or $C$ or both occur—the event that the card selected is either a king or a heart or both. Event ($B$ or $C$) consists of 16 outcomes—namely, the 4 kings and the 12 non-king hearts—as illustrated in Fig. 5.12.

**FIGURE 5.12**
Event ($B$ or $C$)

**Note:** Event ($B$ or $C$) can occur in 16, not 17, ways because the outcome "king of hearts" is common to both event $B$ and event $C$.

**FIGURE 5.13**
Event (C & D)

**d.** (C & D) is the event both C and D occur—the event that the card selected is both a heart and a face card. For that event to occur, the card selected must be the jack, queen, or king of hearts. Thus event (C & D) consists of the three outcomes displayed in Fig. 5.13. These three outcomes are those common to events C and D.

• • •

You try it!

Exercise 5.37
on page 219

In the previous example, we described events by listing their outcomes. Sometimes, describing events verbally is more appropriate, as in the next example.

**Example 5.7** | **Relationships Among Events**

**TABLE 5.4**
Frequency distribution
for students' ages

*Student Ages* A frequency distribution for the ages of the 40 students in Professor Weiss's introductory statistics class is presented in Table 5.4. One student is selected at random. Let

$A$ = event the student selected is under 21,
$B$ = event the student selected is over 30,
$C$ = event the student selected is in his or her 20s, and
$D$ = event the student selected is over 18.

| Age (yrs) | Frequency |
|:---:|:---:|
| 17 | 1 |
| 18 | 1 |
| 19 | 9 |
| 20 | 7 |
| 21 | 7 |
| 22 | 5 |
| 23 | 3 |
| 24 | 4 |
| 26 | 1 |
| 35 | 1 |
| 36 | 1 |

Determine the following events.

**a.** (not D)     **b.** (A & D)     **c.** (A or D)     **d.** (B or C)

**Solution**

**a.** (not D) is the event D does not occur—the event that the student selected is not over 18, that is, is 18 or under. From Table 5.4, (not D) comprises the two students in the class who are 18 or under.

**b.** (A & D) is the event both A and D occur—the event that the student selected is both under 21 and over 18, that is, is either 19 or 20. Event (A & D) comprises the 16 students in the class who are 19 or 20.

**c.** (A or D) is the event either A or D or both occur—the event that the student selected is either under 21 or over 18 or both. But every student in the class is either under 21 or over 18. Consequently, event (A or D) comprises all 40 students in the class and is certain to occur.

You try it!

Exercise 5.41
on page 219

**d.** (B or C) is the event either B or C or both occur—the event that the student selected is either over 30 or in his or her 20s. Table 5.4 shows that (B or C) comprises the 29 students in the class who are 20 or over.

• • •

## Mutually Exclusive Events

Next, we introduce the concept of *mutually exclusive events.*

**Definition 5.4**

## Mutually Exclusive Events

Two or more events are **mutually exclusive events** if no two of them have outcomes in common.

---

**What Does It Mean?**

Events are mutually exclusive if no two of them can occur simultaneously or, equivalently, if at most one of the events can occur when the experiment is performed.

---

The Venn diagrams shown in Fig. 5.14 portray the difference between two events that are mutually exclusive and two events that are not mutually exclusive. In Fig. 5.15, we show three mutually exclusive events and two cases of three events that are not mutually exclusive.

**FIGURE 5.14**
(a) Two mutually exclusive events;
(b) two non–mutually exclusive events

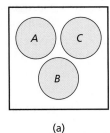
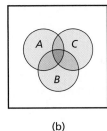

(a)                    (b)

**FIGURE 5.15**
(a) Three mutually exclusive events;
(b) three non–mutually exclusive events;
(c) three non–mutually exclusive events

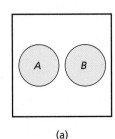
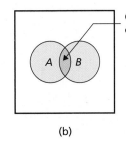
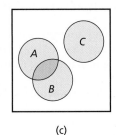

(a)                    (b)                    (c)

**Example 5.8** | **Mutually Exclusive Events**

*Playing Cards*  For the experiment of randomly selecting one card from a deck of 52, let

$$C = \text{event the card selected is a heart,}$$
$$D = \text{event the card selected is a face card,}$$
$$E = \text{event the card selected is an ace,}$$
$$F = \text{event the card selected is an 8, and}$$
$$G = \text{event the card selected is a 10 or a jack.}$$

Which of the following collections of events are mutually exclusive?

**a.**  $C$ and $D$          **b.**  $C$ and $E$          **c.**  $D$ and $E$

**d.**  $D$, $E$, and $F$          **e.**  $D$, $E$, $F$, and $G$

**Solution**

**a.**  Event $C$ and event $D$ are not mutually exclusive because they have the common outcomes "king of hearts," "queen of hearts," and "jack of hearts." Both events occur if the card selected is the king, queen, or jack of hearts.

b. Event $C$ and event $E$ are not mutually exclusive because they have the common outcome "ace of hearts." Both events occur if the card selected is the ace of hearts.

c. Event $D$ and event $E$ are mutually exclusive because they have no common outcomes. They cannot both occur when the experiment is performed because selecting a card that is both a face card and an ace is impossible.

d. Events $D, E$, and $F$ are mutually exclusive because no two of them can occur simultaneously.

e. Events $D$, $E$, $F$, and $G$ are not mutually exclusive because event $D$ and event $G$ both occur if the card selected is a jack.

**You try it!**

Exercise 5.47
on page 220

• • •

## Exercises 5.2

### Understanding the Concepts and Skills

**5.29** What type of graphical displays are useful for portraying events and relationships among events?

**5.30** Construct a Venn diagram representing each event.
a. (not $E$)     b. ($A$ or $B$)     c. ($A$ & $B$)
d. ($A$ & $B$ & $C$)     e. ($A$ or $B$ or $C$)     f. ((not $A$) & $B$)

**5.31** What does it mean for two events to be mutually exclusive? for three events?

**5.32** Answer true or false to each statement, and give reasons for your answers.
a. If event $A$ and event $B$ are mutually exclusive, so are events $A$, $B$, and $C$ for every event $C$.
b. If event $A$ and event $B$ are not mutually exclusive, neither are events $A$, $B$, and $C$ for every event $C$.

**5.33 Dice.** When one die is rolled, the following six outcomes are possible:

List the outcomes constituting

$A$ = event the die comes up even,

$B$ = event the die comes up 4 or more,

$C$ = event the die comes up at most 2, and

$D$ = event the die comes up 3.

**5.34 Horse Racing.** In a horse race, the odds against winning are as shown in the following table. For example, the odds against winning are 8 to 1 for horse #1.

| Horse | #1 | #2 | #3 | #4 | #5 | #6 | #7 | #8 |
|-------|----|----|----|----|----|----|----|----|
| Odds  | 8  | 15 | 2  | 3  | 30 | 5  | 10 | 5  |

List the outcomes constituting

$A$ = event one of the top two favorites wins (the top two favorites are the two horses with the lowest odds against winning),

$B$ = event the winning horse's number is above 5,

$C$ = event the winning horse's number is at most 3, that is, 3 or less, and

$D$ = event one of the two long shots wins (the two long shots are the two horses with the highest odds against winning).

**5.35 Committee Selection.** A committee consists of five executives, three women and two men. Their names are Maria (M), John (J), Susan (S), Will (W), and Holly (H). The committee needs to select a chairperson and a secretary. It decides to make the selection randomly by drawing straws. The person getting the longest straw will be appointed chairperson, and the one getting the shortest straw will be appointed secretary. The possible outcomes can be represented in the following manner.

| MS | SM | HM | JM | WM |
|----|----|----|----|----|
| MH | SH | HS | JS | WS |
| MJ | SJ | HJ | JH | WH |
| MW | SW | HW | JW | WJ |

Here, for example, MS represents the outcome that Maria is appointed chairperson and Susan is appointed secretary. List the outcomes constituting each of the following four events.

$A$ = event a male is appointed chairperson,

$B$ = event Holly is appointed chairperson,

$C$ = event Will is appointed secretary,

$D$ = event only females are appointed.

**5.36 Coin Tossing.** When a dime is tossed four times, there are the following 16 possible outcomes.

| | | | |
|---|---|---|---|
| HHHH | HTHH | THHH | TTHH |
| HHHT | HTHT | THHT | TTHT |
| HHTH | HTTH | THTH | TTTH |
| HHTT | HTTT | THTT | TTTT |

Here, for example, HTTH represents the outcome that the first toss is heads, the next two tosses are tails, and the fourth toss is heads. List the outcomes constituting each of the following four events.

> $A$ = event exactly two heads are tossed,
> $B$ = event the first two tosses are tails,
> $C$ = event the first toss is heads,
> $D$ = event all four tosses come up the same.

**5.37 Dice.** Refer to Exercise 5.33. For each of the following events, list the outcomes that constitute the event and describe the event in words.
**a.** (not $A$)    **b.** ($A$ & $B$)    **c.** ($B$ or $C$)

**5.38 Horse Racing.** Refer to Exercise 5.34. For each of the following events, list the outcomes that constitute the event and describe the event in words.
**a.** (not $C$)    **b.** ($C$ & $D$)    **c.** ($A$ or $C$)

**5.39 Committee Selection.** Refer to Exercise 5.35. For each of the following events, list the outcomes that constitute the event, and describe the event in words.
**a.** (not $A$)    **b.** ($B$ & $D$)    **c.** ($B$ or $C$)

**5.40 Coin Tossing.** Refer to Exercise 5.36. For each of the following events, list the outcomes that constitute the event, and describe the event in words.
**a.** (not $B$)    **b.** ($A$ & $B$)    **c.** ($C$ or $D$)

**5.41 Diabetes Prevalence.** In a report titled *Behavioral Risk Factor Surveillance System Summary Prevalence Report*, the Centers for Disease Control and Prevention discusses the prevalence of diabetes in the United States. The following frequency distribution provides a diabetes prevalence frequency distribution for the 50 U.S. states.

| Diabetes (%) | Frequency |
|---|---|
| 4 ≤ 5 | 8 |
| 5 ≤ 6 | 10 |
| 6 ≤ 7 | 15 |
| 7 ≤ 8 | 10 |
| 8 ≤ 9 | 5 |
| 9 ≤ 10 | 1 |
| 10 ≤ 11 | 1 |

For a randomly selected state, let

> $A$ = event that the state has a diabetes prevalence percentage of at least 8%,
> $B$ = event that the state has a diabetes prevalence percentage of less than 7%,
> $C$ = event that the state has a diabetes prevalence percentage of at least 5% but less than 10%, and
> $D$ = event that the state has a diabetes prevalence percentage of less than 9%.

Describe each of the following events in words and determine the number of outcomes (states) that constitute each event.
**a.** (not $C$)    **b.** ($A$ & $B$)    **c.** ($C$ or $D$)    **d.** ($C$ & $B$)

**5.42 Family Planning.** The following table provides a frequency distribution for the ages of adult women seeking pregnancy tests at public health facilities in Missouri during a 3-month period. It appeared in the article "Factors Affecting Contraceptive Use in Women Seeking Pregnancy Tests" (*Family Planning Perspectives*, Vol. 32, No. 3, pp. 124–131) by M. Sable et al.

| Age | Frequency |
|---|---|
| 18–19 | 89 |
| 20–24 | 130 |
| 25–29 | 66 |
| 30–39 | 26 |

For one of these woman selected at random, let

> $A$ = event the woman is at least 25 years old,
> $B$ = event the woman is at most 29 years old,
> $C$ = event that the woman is between 18 and 29 years old, and
> $D$ = event that the woman is at least 20 years old.

Describe the following events in words and determine the number of outcomes (women) that constitute each event.
**a.** (not $D$)    **b.** ($B$ & $D$)    **c.** ($C$ or $A$)    **d.** ($A$ & $B$)

**5.43 Hospitalization Payments.** From the Florida State Health Department's Health Outcome Series, *Women and Cardiovascular Disease Hospitalization*, we obtained the following frequency distribution showing who paid for the hospitalization of female cardiovascular patients between the ages of 0 and 64 in Florida during one year.

| Payer | Frequency |
|---|---|
| Medicare | 9,983 |
| Medicaid | 8,142 |
| Private insurance | 26,825 |
| Other government | 1,777 |
| Self pay/charity | 5,512 |
| Other | 150 |

For one of these cases selected at random, let

$A$ = event that Medicare paid the bill,

$B$ = event that some government agency paid the bill,

$C$ = event that private insurance did not pay the bill, and

$D$ = event that the bill was paid by the patient or by a charity.

Describe each of the following events in words and determine the number of outcomes that constitute each event.

**a.** $(A$ or $D)$        **b.** (not $C$)

**c.** $(B$ & (not $A$))        **d.** (not $(C$ or $D$))

**5.44 Naturalization.** The U.S. Bureau of Citizenship and Immigration Services collects and reports information about naturalized persons in *Statistical Yearbook*. Suppose that a naturalized person is selected at random. Define events as follows:

$A$ = the person is under 20 (years old),

$B$ = the person is between 30 and 64, inclusive,

$C$ = the person is 50 or older, and

$D$ = the person is older than 64.

Determine the following events:

**a.** (not $A$)      **b.** $(B$ or $D)$      **c.** $(A$ & $C)$

Which of the following collections of events are mutually exclusive?

**d.** $B$ and $C$             **e.** $A$, $B$, and $D$

**f.** (not $A$) and (not $D$)

**5.45 Housing Units.** The U.S. Census Bureau publishes data on housing units in *American Housing Survey for the United States*. The following table provides a frequency distribution for the number of rooms in U.S. housing units. The frequencies are in thousands.

| Rooms | No. of units |
|-------|--------------|
| 1 | 520 |
| 2 | 1,425 |
| 3 | 10,943 |
| 4 | 23,363 |
| 5 | 27,976 |
| 6 | 24,646 |
| 7 | 14,670 |
| 8+ | 17,234 |

For a U.S. housing unit selected at random, let

$A$ = event the unit has at most four rooms,

$B$ = event the unit has at least two rooms,

$C$ = event the unit has between five and seven rooms, inclusive, and

$D$ = event the unit has more than seven rooms.

Describe each of the following events in words and determine the number of outcomes (housing units) that constitute each event.

**a.** (not $A$)      **b.** $(A$ & $B)$      **c.** $(C$ or $D)$

**5.46 Protecting the Environment.** A survey was conducted in Canada to ascertain public opinion about a major national park region in the Banff-Bow Valley. One question asked the amount that respondents would be willing to contribute per year to protect the environment in the Banff-Bow Valley region. The following frequency distribution was found in an article by J. Ritchie, et al. titled "Public Reactions to Policy Recommendations from the Banff-Bow Valley Study" (*Journal of Sustainable Tourism*, Vol. 10, No. 4, pp. 295–308).

| Contribution ($) | Frequency |
|------------------|-----------|
| 0 | 85 |
| 1–50 | 116 |
| 51–100 | 59 |
| 101–200 | 29 |
| 201–300 | 5 |
| 301–500 | 7 |
| 501–1000 | 3 |

For a respondent selected at random, let

$A$ = event that the respondent would be willing to contribute at least $101,

$B$ = event that the respondent would not be willing to contribute more than $50,

$C$ = event that the respondent would be willing to contribute between $1 and $200, and

$D$ = event that the respondent would be willing to contribute at least $1.

Describe the following events in words and determine the number of outcomes (respondents) that make up each event.

**a.** (not $D$)    **b.** $(A$ & $B)$    **c.** $(C$ or $A)$    **d.** $(B$ & $D)$

**5.47 Dice.** Refer to Exercise 5.33.

**a.** Are events $A$ and $B$ mutually exclusive?

**b.** Are events $B$ and $C$ mutually exclusive?

**c.** Are events $A$, $C$, and $D$ mutually exclusive?

**d.** Are there three mutually exclusive events among $A$, $B$, $C$, and $D$? four?

**5.48 Horse Racing.** Each part of this exercise contains events from Exercise 5.34. In each case, decide whether the events are mutually exclusive.

**a.** $A$ and $B$       **b.** $B$ and $C$       **c.** $A$, $B$, and $C$

**d.** $A$, $B$, and $D$       **e.** $A$, $B$, $C$, and $D$

**5.49 Housing Units.** Refer to Exercise 5.45. Among the events $A$, $B$, $C$, and $D$, identify the collections of events that are mutually exclusive.

**5.50 Protecting the Environment.** Refer to Exercise 5.46. Among the events $A$, $B$, $C$, and $D$, identify the collections of events that are mutually exclusive.

**5.51** Draw a Venn diagram portraying four mutually exclusive events.

## Extending the Concepts and Skills

**5.52** Construct a Venn diagram that portrays four events, $A$, $B$, $C$, and $D$ that have the following properties: Events $A$, $B$, and $C$ are mutually exclusive; events $A$, $B$, and $D$ are mutually exclusive; no other three of the four events are mutually exclusive.

**5.53** Suppose that $A$, $B$, and $C$ are three events that cannot all occur simultaneously. Does this condition necessarily imply that $A$, $B$, and $C$ are mutually exclusive? Justify your answer and illustrate it with a Venn diagram.

## 5.3 Some Rules of Probability

In this section, we discuss several rules of probability after introducing an additional notation used in probability.

### Example 5.9 | Probability Notation

*Dice* When a balanced die is rolled once, six equally likely outcomes are possible, as shown in Fig. 5.16. Use probability notation to express the probability that the die comes up an even number.

**FIGURE 5.16**
Sample space for rolling a die once

**Solution** The event that the die comes up an even number can occur in three ways—namely, if 2, 4, or 6 is rolled. Because $f/N = 3/6 = 0.5$, *the probability that the die comes up even is 0.5.* We want to express the italicized phrase using probability notation.

Let $A$ denote the event that the die comes up even. We use the notation $P(A)$ to represent the probability that event $A$ occurs. Hence we can rewrite the italicized statement simply as $P(A) = 0.5$, which is read "the probability of $A$ is 0.5."

• • •

**What Does It Mean?**

Keep in mind that $A$ refers to the event that the die comes up even, whereas $P(A)$ refers to the probability of that event occurring.

### Definition 5.5 | Probability Notation

If $E$ is an event, then $P(E)$ represents the probability that event $E$ occurs. It is read "the probability of $E$."

**FIGURE 5.17**
Two mutually exclusive events

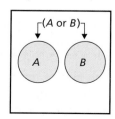

### The Special Addition Rule

The first rule of probability that we present is the **special addition rule**, which states that, for mutually exclusive events, the probability that one or another of the events occurs equals the sum of the individual probabilities.

We use the Venn diagram in Fig. 5.17, which shows two mutually exclusive events $A$ and $B$, to illustrate the special addition rule. If you think of the colored regions as probabilities, the colored disk on the left is $P(A)$, the colored disk on the right is $P(B)$, and the total colored region is $P(A \text{ or } B)$. Because events $A$ and $B$ are mutually exclusive, the total colored region equals the sum of the two colored disks; that is, $P(A \text{ or } B) = P(A) + P(B)$.

---

**Formula 5.1**

**What Does It Mean?**

For mutually exclusive events, the probability that at least one occurs equals the sum of their individual probabilities.

### The Special Addition Rule

If event $A$ and event $B$ are mutually exclusive, then

$$P(A \text{ or } B) = P(A) + P(B).$$

More generally, if events $A, B, C, \ldots$ are mutually exclusive, then

$$P(A \text{ or } B \text{ or } C \text{ or } \cdots) = P(A) + P(B) + P(C) + \cdots.$$

---

Example 5.10 illustrates use of the special addition rule.

**Example 5.10** | **The Special Addition Rule**

**TABLE 5.5**
Size of farms in the United States

| Size (acres) | Relative freq. | Event |
|---|---|---|
| Under 10 | 0.084 | A |
| 10 ≺ 50 | 0.265 | B |
| 50 ≺ 100 | 0.161 | C |
| 100 ≺ 180 | 0.149 | D |
| 180 ≺ 260 | 0.077 | E |
| 260 ≺ 500 | 0.106 | F |
| 500 ≺ 1000 | 0.076 | G |
| 1000 ≺ 2000 | 0.047 | H |
| 2000 & over | 0.035 | I |

*Size of Farms* The first two columns of Table 5.5 show a relative-frequency distribution for the size of farms in the United States. The U.S. Department of Agriculture, National Agricultural Statistics Service, compiled this information and published it in *Census of Agriculture*.

In the third column of Table 5.5, we introduce events that correspond to the size classes. For example, if a farm is selected at random, $D$ denotes the event that the farm has between 100 and 180 acres, that is, at least 100 acres but less than 180 acres. The probabilities of the events in the third column of Table 5.5 equal the relative frequencies in the second column. For instance, the probability is 0.149 that a randomly selected farm has between 100 and 180 acres: $P(D) = 0.149$.

Use Table 5.5 and the special addition rule to determine the probability that a randomly selected farm has between 100 and 500 acres.

**Solution** The event that the farm selected has between 100 and 500 acres can be expressed as $(D \text{ or } E \text{ or } F)$. Because events $D$, $E$, and $F$ are mutually exclusive, the special addition rule gives

$$P(D \text{ or } E \text{ or } F) = P(D) + P(E) + P(F)$$
$$= 0.149 + 0.077 + 0.106 = 0.332.$$

**You try it!**

Exercise 5.55 on page 226

The probability that a randomly selected U.S. farm has between 100 and 500 acres is 0.332.

**Interpretation** 33.2% of U.S. farms have between 100 and 500 acres.

• • •

**FIGURE 5.18**
An event and its complement

$E$

(not $E$)

### The Complementation Rule

The second rule of probability that we discuss is the **complementation rule**. It states that the probability an event occurs equals 1 minus the probability the event does not occur.

We use the Venn diagram in Fig. 5.18, which shows an event $E$ and its complement (not $E$), to illustrate the complementation rule. If you think of the regions as probabilities, the entire region enclosed by the rectangle is the probability of the sample space, or 1. Furthermore, the colored region is $P(E)$ and the uncolored region is $P(\text{not } E)$. Thus, $P(E) + P(\text{not } E) = 1$ or, equivalently, $P(E) = 1 - P(\text{not } E)$.

---

**Formula 5.2**

### The Complementation Rule

For any event $E$,

$$P(E) = 1 - P(\text{not } E).$$

**What Does It Mean?**

The probability that an event occurs equals 1 minus the probability that it does not occur.

The complementation rule is useful because sometimes computing the probability that an event does not occur is easier than computing the probability that it does occur. In such cases, we can subtract the former from 1 to find the latter.

---

**Example 5.11** | ### The Complementation Rule

*Size of Farms*  We saw that the first two columns of Table 5.5 provide a relative-frequency distribution for the size of U.S. farms. Find the probability that a randomly selected farm has

**a.**  less than 2000 acres.        **b.**  50 acres or more.

**Solution**

**a.**  Let

$$J = \text{event the farm selected has less than 2000 acres.}$$

To determine $P(J)$, we apply the complementation rule because $P(\text{not } J)$ is easier to compute than $P(J)$. Note that (not $J$) is the event the farm obtained has 2000 or more acres, which is event $I$ in Table 5.5. Thus $P(\text{not } J) = P(I) = 0.037$. Applying the complementation rule yields

$$P(J) = 1 - P(\text{not } J) = 1 - 0.035 = 0.965.$$

The probability that a randomly selected U.S. farm has less than 2000 acres is 0.965.

**Interpretation**  96.5% of U.S. farms have less than 2000 acres.

**b.**  Let

$$K = \text{event the farm selected has 50 acres or more.}$$

We apply the complementation rule to find $P(K)$. Now, (not $K$) is the event the farm obtained has less than 50 acres. From Table 5.5, event (not $K$) is the same as event ($A$ or $B$). Because events $A$ and $B$ are mutually exclusive, the special addition rule implies that

$$P(\text{not } K) = P(A \text{ or } B) = P(A) + P(B) = 0.084 + 0.265 = 0.349.$$

Using this result and the complementation rule, we conclude that

$$P(K) = 1 - P(\text{not } K) = 1 - 0.349 = 0.651.$$

**You try it!**

The probability that a randomly selected U.S. farm has 50 acres or more is 0.651.

**Interpretation**  65.1% of U.S. farms have at least 50 acres.

Exercise 5.61 on page 227

• • •

### The General Addition Rule

**FIGURE 5.19**

Non–mutually exclusive events

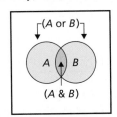

The special addition rule concerns mutually exclusive events. For events that are not mutually exclusive, we must use the **general addition rule.** To introduce it, we use the Venn diagram shown in Fig. 5.19.

If you think of the colored regions as probabilities, the colored disk on the left is $P(A)$, the colored disk on the right is $P(B)$, and the total colored region is $P(A \text{ or } B)$. To obtain the total colored region, $P(A \text{ or } B)$, we first sum the two colored disks, $P(A)$ and $P(B)$. When we do so, however, we count the common colored region, $P(A \And B)$, twice. Thus, we must subtract $P(A \And B)$ from the sum. So, we see that $P(A \text{ or } B) = P(A) + P(B) - P(A \And B)$.

**Formula 5.3**

### The General Addition Rule

If $A$ and $B$ are any two events, then

$$P(A \text{ or } B) = P(A) + P(B) - P(A \And B).$$

#### ? What Does It Mean?

For any two events, the probability that at least one occurs equals the sum of their individual probabilities less the probability that both occur.

In the next example, we consider a situation where a required probability can be computed both with and without use of the general addition rule.

**Example 5.12** | ### The General Addition Rule

*Playing Cards* Consider again the experiment of selecting one card at random from a deck of 52 playing cards. Find the probability that the card selected is either a spade or a face card

**a.** without using the general addition rule.

**b.** by using the general addition rule.

**Solution**

**a.** Let

$$E = \text{event the card selected is either a spade or a face card.}$$

Event $E$ consists of 22 cards—namely, the 13 spades plus the other nine face cards that are not spades—as shown in Fig. 5.20. So, by the $f/N$ rule,

$$P(E) = \frac{f}{N} = \frac{22}{52} = 0.423.$$

The probability that a randomly selected card is either a spade or a face card is 0.423.

**FIGURE 5.20**

Event $E$

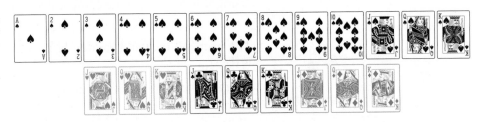

5.3 Some Rules of Probability

**b.** To determine $P(E)$ by using the general addition rule, we first note that we can write $E = (C \text{ or } D)$, where

$$C = \text{event the card selected is a spade, and}$$
$$D = \text{event the card selected is a face card.}$$

Event $C$ consists of the 13 spades, and event $D$ consists of the 12 face cards. Also, event $(C \& D)$ consists of the three spades that are face cards—the jack, queen, and king of spades. Applying the general addition rule gives

$$P(E) = P(C \text{ or } D) = P(C) + P(D) - P(C \& D)$$
$$= \frac{13}{52} + \frac{12}{52} - \frac{3}{52} = 0.250 + 0.231 - 0.058 = 0.423,$$

which agrees with the answer found in part (a).

**You try it!**

Exercise 5.63
on page 227

• • •

Computing the probability in the previous example was simpler without using the general addition rule. Frequently, however, the general addition rule is the easier or the only way to compute a probability, as illustrated in the next example.

**Example 5.13** | **The General Addition Rule**

*Characteristics of People Arrested* Data on people who have been arrested are published by the U.S. Federal Bureau of Investigation in *Crime in the United States*. Records for one year show that 77.0% of the people arrested were male, 16.5% were under 18 years of age, and 11.8% were males under 18 years of age. If a person arrested that year is selected at random, what is the probability that that person is either male or under 18?

**Solution** Let

$$M = \text{event the person obtained is male, and}$$
$$E = \text{event the person obtained is under 18.}$$

We can represent the event that the selected person is either male or under 18 as $(M \text{ or } E)$. To find the probability of that event, we apply the general addition rule to the data provided:

$$P(M \text{ or } E) = P(M) + P(E) - P(M \& E)$$
$$= 0.770 + 0.165 - 0.118 = 0.817.$$

**You try it!**

The probability that the person obtained is either male or under 18 is 0.817.

**Interpretation** 81.7% of those arrested during the year in question were either male or under 18 years of age (or both).

Exercise 5.67
on page 228

• • •

Note the following:

- The general addition rule is consistent with the special addition rule—if two events are mutually exclusive, both rules yield the same result.
- There are also general addition rules for more than two events. For instance, the general addition rule for three events is

$$P(A \text{ or } B \text{ or } C) = P(A) + P(B) + P(C) - P(A \& B) - P(A \& C) - P(B \& C)$$
$$+ P(A \& B \& C).$$

## Exercises 5.3

### Understanding the Concepts and Skills

**5.54 A Lottery.** Suppose that you hold 20 out of a total of 500 tickets sold for a lottery. The grand-prize winner is determined by the random selection of one of the 500 tickets. Let G be the event that you win the grand prize. Find the probability that you win the grand prize. Express your answer in probability notation.

**5.55 Ages of Senators.** According to the *Congressional Directory*, the age distribution for senators in the 109th U.S. Congress is as follows.

| Age (yrs) | No. of senators |
|-----------|-----------------|
| Under 50 | 12 |
| 50–59 | 33 |
| 60–69 | 32 |
| 70–79 | 18 |
| 80 and over | 5 |

Suppose that a senator from the 109th U.S. Congress is selected at random. Let

$A$ = event the senator is under 50,

$B$ = event the senator is in his or her 50s,

$C$ = event the senator is in his or her 60s, and

$S$ = event the senator is under 70.

**a.** Use the table and the $f/N$ rule to find $P(S)$.
**b.** Express event S in terms of events A, B, and C.
**c.** Determine $P(A)$, $P(B)$, and $P(C)$.
**d.** Compute $P(S)$, using the special addition rule and your answers from parts (b) and (c). Compare your answer with that in part (a).

**5.56 Sales Tax Receipts.** The State of Texas maintains records pertaining to the economic development of corporations in the state. From *The Economic Development Corporation Report*, published by the Texas Comptroller of Public Accounts, we obtained the following frequency distribution summarizing the sales tax receipts from the state's Type 4A development corporations during one fiscal year.

| Receipts | Frequency |
|----------|-----------|
| $0–24,999 | 25 |
| $25,000–49,999 | 23 |
| $50,000–74,999 | 21 |
| $75,000–99,999 | 11 |
| $100,000–199,999 | 34 |
| $200,000–499,999 | 44 |
| $500,000–999,999 | 17 |
| $1,000,000 & over | 32 |

Suppose that one of these Type 4A development corporations is selected at random. Let

$A$ = event the receipts are less than $25,000,

$B$ = event the receipts are between $25,000 and $49,999,

$C$ = event the receipts are between $500,000 and $999,999,

$D$ = event the receipts are at least $1,000,000, and

$R$ = event the receipts are either less than $50,000 or at least $500,000.

**a.** Use the table and the $f/N$ rule to find $P(R)$.
**b.** Express event R in terms of events A, B, C, and D.
**c.** Determine $P(A)$, $P(B)$, $P(C)$, and $P(D)$.
**d.** Compute $P(R)$ by using the special addition rule and your answers from parts (b) and (c). Compare your answer with that in part (a).

**5.57 Twelfth-Grade Smokers.** The National Institute on Drug Abuse issued the report *Monitoring the Future*, which addresses the issue of drinking, cigarette, and smokeless tobacco use for eighth, tenth, and twelfth graders. During one year, 12,900 twelfth graders were asked the question "How frequently have you smoked cigarettes during the past 30 days?" Based on their responses, we constructed the following percentage distribution for all twelfth graders.

| Cigarettes per day | Percentage | Event |
|--------------------|------------|-------|
| None | 73.3 | A |
| Some, but less than 1 | 9.8 | B |
| 1–5 | 7.8 | C |
| 6–14 | 5.3 | D |
| 15–25 | 2.8 | E |
| 26–34 | 0.7 | F |
| 35 or more | 0.3 | G |

Find the probability that, within the last 30 days, a randomly selected twelfth grader
**a.** smoked.
**b.** smoked at least one cigarette per day.
**c.** smoked between 6 and 34 cigarettes per day, inclusive.

**5.58 Home Internet Access.** The on-line publication *CyberStats*, by Mediamark Research, Inc., reports Internet access and usage. The following is a percentage distribution of household income for households with home Internet access only.

| Household income | Percentage | Event |
|------------------|------------|-------|
| Under $50,000 | 31.8 | A |
| $50,000 < $75,000 | 24.4 | B |
| $75,000 < $150,000 | 34.0 | C |
| $150,000 or above | 9.8 | D |

Suppose that a household with home Internet access only is selected at random. Let $A$ denote the event the household has an income under $50,000, $B$ denote the event the household has an income between $50,000 and $75,000, and so on (see the third column of the table). Apply the special addition rule to find the probability that the household obtained has an income

a. under $75,000.
b. $50,000 or above.
c. between $50,000 and $150,000.
d. Interpret each of your answers in parts (a)–(c) in terms of percentages.

**5.59 Oil Spills.** The U.S. Coast Guard maintains a database of the number, source, and location of oil spills in U.S. navigable and territorial waters. The following is a probability distribution for location of oil spill events. [SOURCE: *Statistical Abstract of the United States*.]

| Location | Probability |
|---|---|
| Atlantic Ocean | 0.011 |
| Pacific Ocean | 0.064 |
| Gulf of Mexico | 0.229 |
| Great Lakes | 0.014 |
| Other lakes | 0.005 |
| Rivers and canals | 0.223 |
| Bays and sounds | 0.151 |
| Harbors | 0.118 |
| Other | 0.185 |

Apply the special addition rule to find the percentage of oil spills in U.S. navigable and territorial waters that
a. occur in an ocean.
b. occur in a lake or harbor.
c. do not occur in a lake, ocean, river, or canal.

**5.60 Religion in America.** R. Doyle diagrammed the religious preferences of Americans for the years 1940–2000 in an article titled "Religion in America," published in the February 2003 issue of *Scientific American*. Following is a relative-frequency distribution for the religious preference of Americans for the year 2000.

| Preference | Relative frequency |
|---|---|
| Protestant | 0.560 |
| Catholic | 0.250 |
| Jewish | 0.025 |
| Other | 0.015 |
| None | 0.150 |

Find the probability that the religious preference of a randomly selected American is
a. Catholic or Protestant.
b. not Jewish.
c. not Catholic, Protestant, or Jewish.

**5.61 Ages of Senators.** Refer to Exercise 5.55. Use the complementation rule to find the probability that a randomly selected senator in the 109th Congress is
a. 50 years old or older.    b. under 70 years old.

**5.62 Home Internet Access.** Solve part (b) of Exercise 5.58 by using the complementation rule. Compare your work here to that in Exercise 5.58(b) where you used the special addition rule.

**5.63 Day Laborers.** Mary Sheridan, a reporter for *The Washington Post*, wrote about a study describing the characteristics of day laborers in the Washington area (June 23, 2005, pp. A1, A12). The study, funded by the Ford and Rockefeller foundations, interviewed 476 day laborers—who are becoming common in the Washington area due to increase in construction and immigration—in 2004. The following table provides a percentage distribution for the number of years the day laborers lived in the United States at the time of the interview.

| Years in U.S. | Percentage |
|---|---|
| Less than 1 | 17 |
| 1–2 | 30 |
| 3–5 | 21 |
| 6–10 | 12 |
| 11–20 | 13 |
| 21 or more | 7 |

Suppose that one of these day laborers is randomly selected.
a. Without using the general addition rule, determine the probability that the day laborer obtained has lived in the United States either between 1 and 20 years, inclusive, or less than 11 years.
b. Obtain the probability in part (a) by using the general addition rule.
c. Which method did you find easier?

**5.64 Naturalization.** The U.S. Bureau of Citizenship and Immigration Services collects and reports information about naturalized persons in *Statistical Yearbook*. Following is an age distribution for persons naturalized during one year.

| Age (yrs) | Frequency | Age (yrs) | Frequency |
|-----------|-----------|-----------|-----------|
| 18–19 | 5,958 | 45–49 | 42,820 |
| 20–24 | 50,905 | 50–54 | 32,574 |
| 25–29 | 58,829 | 55–59 | 25,534 |
| 30–34 | 64,735 | 60–64 | 18,767 |
| 35–39 | 69,844 | 65–74 | 25,528 |
| 40–44 | 57,834 | 75 & over | 9,872 |

Suppose that one of these naturalized persons is selected at random.

a. Without using the general addition rule, determine the probability that the age of the person obtained is either between 30 and 64, inclusive, or at least 50.
b. Obtain the probability in part (a) by using the general addition rule.
c. Which method did you find easier?

**5.65 Craps.** In the game of *craps*, a player rolls two balanced dice. Thirty-six equally likely outcomes are possible, as shown in Fig. 5.1 on page 206. Let

$A$ = event the sum of the dice is 7,

$B$ = event the sum of the dice is 11,

$C$ = event the sum of the dice is 2,

$D$ = event the sum of the dice is 3,

$E$ = event the sum of the dice is 12,

$F$ = event the sum of the dice is 8, and

$G$ = event doubles are rolled.

a. Compute the probability of each of the seven events.
b. The player wins on the first roll if the sum of the dice is 7 or 11. Find the probability of that event by using the special addition rule and your answers from part (a).
c. The player loses on the first roll if the sum of the dice is 2, 3, or 12. Determine the probability of that event by using the special addition rule and your answers from part (a).
d. Compute the probability that either the sum of the dice is 8 or doubles are rolled, without using the general addition rule.
e. Compute the probability that either the sum of the dice is 8 or doubles are rolled by using the general addition rule and compare your answer to the one you obtained in part (d).

**5.66 Gender and Divorce.** According to *Current Population Reports*, published by the U.S. Census Bureau, 51.8% of U.S. adults are female, 10.2% are divorced, and 6.0% are divorced females. For a U.S. adult selected at random, let

$F$ = event the person is female, and

$D$ = event the person is divorced.

a. Obtain $P(F)$, $P(D)$, and $P(F \& D)$.
b. Determine $P(F \text{ or } D)$ and interpret your answer in terms of percentages.
c. Find the probability that a randomly selected adult is male.

**5.67 School Enrollment.** The U.S. National Center for Education Statistics publishes information about school enrollment in *Digest of Education Statistics*. According to that document, 85.8% of students attend public schools, 22.3% attend college, and 17.1% attend public colleges. What percentage of students attend either public school or college?

**5.68** Suppose $A$ and $B$ be events such that $P(A) = 1/3$, $P(A \text{ or } B) = 1/2$, and $P(A \& B) = 1/10$. Determine $P(B)$.

**5.69** Let $A$ and $B$ be events such that $P(A) = \frac{1}{4}$, $P(B) = \frac{1}{3}$, and $P(A \text{ or } B) = \frac{1}{2}$.
a. Are events $A$ and $B$ mutually exclusive? Explain your answer.
b. Determine $P(A \& B)$.

## Extending the Concepts and Skills

**5.70** Suppose that $A$ and $B$ are mutually exclusive events.
a. Use the special addition rule to express $P(A \text{ or } B)$ in terms of $P(A)$ and $P(B)$.
b. Show that the general addition rule gives the same answer as that in part (a).

**5.71 Secrets of Success.** Gerald Kushel, Ed.D., was interviewed by *Bottom Line/Personal* on the secrets of successful people. To study success, Kushel questioned 1200 people, among whom were lawyers, artists, teachers, and students. He found that 15% enjoy neither their jobs nor their personal lives, 80% enjoy their jobs but not their personal lives, and 4% enjoy both their jobs and their personal lives. Determine the percentage of the 1200 people interviewed who
a. enjoy either their jobs or their personal lives.
b. enjoy their personal lives but not their jobs.

**5.72 General Addition Rule Extended.** The general addition rule for two events is presented in Formula 5.3 on page 224 and that for three events is displayed on page 225.
a. Verify the general addition rule for three events.
b. Write the general addition rule for four events and explain your reasoning.

## 5.4  Discrete Random Variables and Probability Distributions*

In this section, we introduce discrete random variables and probability distributions. As you will discover, these concepts are natural extensions of the ideas of variables and relative-frequency distributions.

---

**Example 5.14** | **Introducing Random Variables**

**TABLE 5.6**
Grouped-data table for number of siblings for students in introductory statistics

| Siblings $x$ | Freq. $f$ | Relative freq. |
|:---:|:---:|:---:|
| 0 | 8 | 0.200 |
| 1 | 17 | 0.425 |
| 2 | 11 | 0.275 |
| 3 | 3 | 0.075 |
| 4 | 1 | 0.025 |
| | 40 | 1.000 |

*Number of Siblings*  Professor Weiss asked his introductory statistics students to state how many siblings they have. Table 5.6 presents a grouped-data table for that information. The table shows, for instance, that 11 of the 40 students, or 27.5%, have two siblings. Discuss the "number of siblings" in the context of randomness.

**Solution**  Because the "number of siblings" varies from student to student, it is a variable. Suppose now that a student is selected at random. Then the "number of siblings" of the student obtained is called a *random variable* because its value depends on chance—namely, on which student is selected.

• • •

Keeping the previous example in mind, we now present the definition of a random variable.

---

**Definition 5.6** | **Random Variable**

A **random variable** is a quantitative variable whose value depends on chance.

---

Example 5.14 shows how random variables arise naturally as (quantitative) variables of finite populations in the context of randomness. Specifically, as you learned in Chapter 2, a variable is a characteristic that varies from one member of a population to another. When one or more members are selected at random from the population, the variable, in that context, is called a random variable.

But there are random variables that are not quantitative variables of finite populations in the context of randomness. Four examples of such random variables are

• the sum of the dice when a pair of fair dice are rolled,
• the number of puppies in a litter,
• the return on an investment, and
• the lifetime of a flashlight battery.

As you also learned in Chapter 2, a *discrete variable* is a variable whose possible values can be listed, even though the list may continue indefinitely. This property holds, for instance, if either the variable has only a finite number of possible values or its possible values are some collection of whole numbers. The variable "number of siblings" in Example 5.14 is therefore a discrete variable. We use the adjective *discrete* for random variables in the same way that we do for variables—hence the term *discrete random variable*.

**Definition 5.7**

### Discrete Random Variable

A **discrete random variable** is a random variable whose possible values can be listed.

## Random-Variable Notation

Recall that we use lowercase letters such as $x$, $y$, and $z$ to denote variables. To represent random variables, however, we usually use uppercase letters. For instance, we could use $x$ to denote the variable "number of siblings" but we would generally use $X$ to denote the random variable "number of siblings."

Random-variable notation is a useful shorthand for discussing and analyzing random variables. For example, let $X$ denote the number of siblings of a randomly selected student. Then we can represent the event that the selected student has two siblings by $\{X = 2\}$, read "$X$ equals two," and the probability of that event as $P(X = 2)$, read "the probability that $X$ equals two."

## Probability Distributions and Histograms

Recall that the relative-frequency distribution or relative-frequency histogram of a discrete variable gives the possible values of the variable and the proportion of times each value occurs. Using the language of probability, we can extend the notions of relative-frequency distribution and relative-frequency histogram—concepts applying to variables of finite populations—to any discrete random variable. In doing so, we use the terms *probability distribution* and *probability histogram*.

---

**Definition 5.8**

### Probability Distribution and Probability Histogram

**Probability distribution:** A listing of the possible values and corresponding probabilities of a discrete random variable, or a formula for the probabilities.

**Probability histogram:** A graph of the probability distribution that displays the possible values of a discrete random variable on the horizontal axis and the probabilities of those values on the vertical axis. The probability of each value is represented by a vertical bar whose height equals the probability.

---

**Example 5.15** | **Probability Distributions and Histograms**

*Number of Siblings* Refer to Example 5.14 and let $X$ denote the number of siblings of a randomly selected student.

**a.** Determine the probability distribution of the random variable $X$.

**b.** Construct a probability histogram for the random variable $X$.

**Solution**

**a.** We want to determine the probability of each of the possible values of the random variable $X$. To obtain, for instance, $P(X = 2)$, the probability that

**TABLE 5.7**
Probability distribution of the random
variable *X*, the number of siblings
of a randomly selected student

| Siblings $x$ | Probability $P(X = x)$ |
|:---:|:---:|
| 0 | 0.200 |
| 1 | 0.425 |
| 2 | 0.275 |
| 3 | 0.075 |
| 4 | 0.025 |
| | 1.000 |

the student selected has two siblings, we apply the $f/N$ rule. From Table 5.6 on page 229, we find that

$$P(X = 2) = \frac{f}{N} = \frac{11}{40} = 0.275.$$

The other probabilities are found in the same way. Table 5.7 displays these probabilities and provides the probability distribution of the random variable *X*.

**b.** To construct a probability histogram for *X*, we plot its possible values on the horizontal axis and display the corresponding probabilities as vertical bars. Referring to Table 5.7, we get the probability histogram of the random variable *X*, as shown in Fig. 5.21.

**FIGURE 5.21**
Probability histogram for the random
variable *X*, the number of siblings
of a randomly selected student

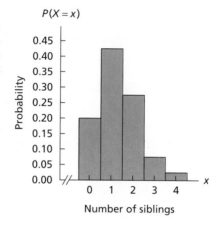

You
try it!

Exercise 5.79
on page 235

The probability histogram provides a quick and easy way to visualize how the probabilities are distributed.

• • •

The variable "number of siblings" is a variable of a finite population, so its probabilities are identical to its relative frequencies. As a consequence, its probability distribution, given in the first and second columns of Table 5.7, is the same as its relative-frequency distribution, shown in the first and third columns of Table 5.6. Apart from labeling, the variable's probability histogram is identical to its relative-frequency histogram. These statements hold for any variable of a finite population.

Note also that the probabilities in the second column of Table 5.7 sum to 1, which is always the case for discrete random variables.

**Key Fact 5.2**

### Sum of the Probabilities of a Discrete Random Variable

For any discrete random variable *X*, we have $\Sigma P(X = x) = 1.$[†]

**What Does It Mean?**

The sum of the probabilities of the
possible values of a discrete random
variable equals 1.

Examples 5.16 and 5.17 provide additional illustrations of random-variable notation and probability distributions.

_____

[†]The sum $\Sigma P(X = x)$ represents adding the individual probabilities, $P(X = x)$, for all possible values, $x$, of the random variable *X*.

**Example 5.16** | **Random Variables and Probability Distributions**

*Elementary-School Enrollment*  The U.S. National Center for Education Statistics compiles enrollment data on U.S. public schools and publishes the results in the *Digest of Education Statistics*.  Table 5.8 displays a frequency distribution for the enrollment by grade level in public elementary schools, where $0 = $ kindergarten, $1 = $ first grade, and so on.  Frequencies are in thousands of students.

Let $Y$ denote the grade level of a randomly selected elementary-school student.  Then $Y$ is a discrete random variable whose possible values are 0, 1, 2, ..., 8.

**a.** Use random-variable notation to represent the event that the selected student is in the fifth grade.

**b.** Determine $P(Y = 5)$ and express the result in terms of percentages.

**c.** Determine the probability distribution of $Y$.

**TABLE 5.8**

Frequency distribution for enrollment by grade level in U.S. public elementary schools

| Grade level $y$ | Frequency $f$ |
|:---:|:---:|
| 0 | 4,248 |
| 1 | 3,615 |
| 2 | 3,595 |
| 3 | 3,654 |
| 4 | 3,696 |
| 5 | 3,728 |
| 6 | 3,770 |
| 7 | 3,722 |
| 8 | 3,619 |
| | 33,647 |

**Solution**

**a.** The event that the selected student is in the fifth grade can be represented as $\{Y = 5\}$.

**b.** $P(Y = 5)$ is the probability that the selected student is in the fifth grade. Using Table 5.8 and the $f/N$ rule, we get

$$P(Y = 5) = \frac{f}{N} = \frac{3{,}728}{33{,}647} = 0.111.$$

**Interpretation**  11.1% of elementary-school students in the United States are in the fifth grade.

**c.** The probability distribution of $Y$ is obtained by computing $P(Y = y)$ for $y = 0, 1, 2, ..., 8$. We have already done that for $y = 5$. The other probabilities are computed similarly and are displayed in Table 5.9.

**Note:** In Table 5.9, the sum of the probabilities is given as 1.001. Key Fact 5.2, however, states that the sum of the probabilities must be exactly 1. Our computation is off slightly because we rounded the probabilities for $Y$.

● ● ●

**TABLE 5.9**

Probability distribution of the random variable $Y$, the grade level of a randomly selected elementary-school student

| Grade level $y$ | Probability $P(Y = y)$ |
|:---:|:---:|
| 0 | 0.126 |
| 1 | 0.107 |
| 2 | 0.107 |
| 3 | 0.109 |
| 4 | 0.110 |
| 5 | 0.111 |
| 6 | 0.112 |
| 7 | 0.111 |
| 8 | 0.108 |
| | 1.001 |

Once we have the probability distribution of a discrete random variable, we can easily determine any probability involving that random variable. The basic tool for accomplishing this is the special addition rule, Formula 5.1 on page 222.[†] We illustrate this technique in part (e) of the next example.

---

[†]Specifically, to find the probability that a discrete random variable takes a value in some set of real numbers, we simply sum the individual probabilities of that random variable over the values in the set. In symbols, if $X$ is a discrete random variable and $A$ is a set of real numbers, then

$$P(X \in A) = \sum_{x \in A} P(X = x),$$

where the sum on the right represents adding the individual probabilities, $P(X = x)$, for all possible values, $x$, of the random variable $X$ that belong to the set $A$.

**Example 5.17** | **Random Variables and Probability Distributions**

**TABLE 5.10**
Possible outcomes

| | | | |
|---|---|---|---|
| HHH | HTH | THH | TTH |
| HHT | HTT | THT | TTT |

*Coin Tossing*  When a balanced dime is tossed three times, eight equally likely outcomes are possible, as shown in Table 5.10. Here, for instance, HHT means that the first two tosses are heads and the third is tails. Let $X$ denote the total number of heads obtained in the three tosses. Then $X$ is a discrete random variable whose possible values are 0, 1, 2, and 3.

a.   Use random-variable notation to represent the event that exactly two heads are tossed.

b.   Determine $P(X = 2)$.

c.   Find the probability distribution of $X$.

d.   Use random-variable notation to represent the event that at most two heads are tossed.

e.   Find $P(X \leq 2)$.

**Solution**

a.   The event that exactly two heads are tossed can be represented as $\{X = 2\}$.

b.   $P(X = 2)$ is the probability that exactly two heads are tossed. Table 5.10 shows that there are three ways to get exactly two heads and that there are eight possible (equally likely) outcomes altogether. So, by the $f/N$ rule,

$$P(X = 2) = \frac{f}{N} = \frac{3}{8} = 0.375.$$

The probability that exactly two heads are tossed is 0.375.

**Interpretation**  There is a 37.5% chance of obtaining exactly two heads in three tosses of a balanced dime.

**TABLE 5.11**
Probability distribution of the random variable $X$, the number of heads obtained in three tosses of a balanced dime

| No. of heads $x$ | Probability $P(X = x)$ |
|---|---|
| 0 | 0.125 |
| 1 | 0.375 |
| 2 | 0.375 |
| 3 | 0.125 |
| | 1.000 |

c.   The remaining probabilities for $X$ are computed as in part (b) and are shown in Table 5.11.

d.   The event that at most two heads are tossed can be represented as $\{X \leq 2\}$, read as "$X$ is less than or equal to two."

e.   $P(X \leq 2)$ is the probability that at most two heads are tossed. The event that at most two heads are tossed can be expressed as

$$\{X \leq 2\} = (\{X = 0\} \text{ or } \{X = 1\} \text{ or } \{X = 2\}).$$

Because the three events on the right are mutually exclusive, we use the special addition rule and Table 5.11 to conclude that

$$P(X \leq 2) = P(X = 0) + P(X = 1) + P(X = 2)$$
$$= 0.125 + 0.375 + 0.375 = 0.875.$$

The probability that at most two heads are tossed is 0.875.

**Interpretation**  There is an 87.5% chance of obtaining two or fewer heads in three tosses of a balanced dime.

• • •

**You try it!**

Exercise 5.83
on page 236

## Interpretation of Probability Distributions

Recall that the frequentist interpretation of probability construes the probability of an event to be the proportion of times it occurs in a large number of independent repetitions of the experiment. Using that interpretation, we clarify the meaning of a probability distribution.

**Example 5.18** | ## Interpreting a Probability Distribution

*Coin Tossing* Suppose we repeat the experiment of observing the number of heads, $X$, obtained in three tosses of a balanced dime a large number of times. Then the proportion of those times in which, say, no heads are obtained ($X = 0$) should approximately equal the probability of that event [$P(X = 0)$]. The same statement holds for the other three possible values of the random variable $X$. Use simulation to verify these facts.

**TABLE 5.12**

Frequencies and proportions for the numbers of heads obtained in three tosses of a balanced dime for 1000 observations

| No. of heads $x$ | Freq. $f$ | Proportion $f/1000$ |
|---|---|---|
| 0 | 136 | 0.136 |
| 1 | 377 | 0.377 |
| 2 | 368 | 0.368 |
| 3 | 119 | 0.119 |
| | 1000 | 1.000 |

**Solution** Simulating a random variable means that we use a computer or statistical calculator to generate observations of the random variable. In this instance, we used a computer to simulate 1000 observations of the random variable $X$, the number of heads obtained in three tosses of a balanced dime.

Table 5.12 shows the frequencies and proportions for the numbers of heads obtained in the 1000 observations. For example, 136 of the 1000 observations resulted in no heads out of three tosses, which gives a proportion of 0.136.

As expected, the proportions in the third column of Table 5.12 are fairly close to the true probabilities in the second column of Table 5.11. This result is more easily seen if we compare the proportion histogram to the probability histogram of the random variable $X$, as shown in Fig. 5.22.

**FIGURE 5.22**

(a) Histogram of proportions for the numbers of heads obtained in three tosses of a balanced dime for 1000 observations; (b) probability histogram for the number of heads obtained in three tosses of a balanced dime

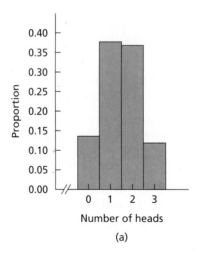
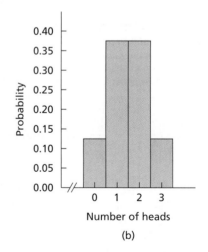

If we simulated, say, 10,000 observations instead of 1000, the proportions that would appear in the third column of Table 5.12 would most likely be even closer to the true probabilities listed in the second column of Table 5.11.

• • •

Key Fact 5.3     **Interpretation of a Probability Distribution**

In a large number of independent observations of a random variable $X$, the proportion of times each possible value occurs will approximate the probability distribution of $X$; or, equivalently, the proportion histogram will approximate the probability histogram for $X$.

# Exercises 5.4

## Understanding the Concepts and Skills

**5.73** Fill in the blanks.
**a.** A relative-frequency distribution is to a variable as a _____ distribution is to a random variable.
**b.** A relative-frequency histogram is to a variable as a _____ histogram is to a random variable.

**5.74** Provide an example (other than one discussed in the text) of a random variable that does not arise from a quantitative variable of a finite population in the context of randomness.

**5.75** Let $X$ denote the number of siblings of a randomly selected student. Explain the difference between $\{X = 3\}$ and $P(X = 3)$.

**5.76** Fill in the blank. For a discrete random variable, the sum of the probabilities of its possible values equals _____.

**5.77** Suppose that you make a large number of independent observations of a random variable and then construct a table giving the possible values of the random variable and the proportion of times each value occurs. What will this table resemble?

**5.78** What rule of probability permits you to obtain any probability for a discrete random variable by simply knowing its probability distribution?

**5.79 Space Shuttles.** The National Aeronautics and Space Administration (NASA) compiles data on space-shuttle launches and publishes them on its Web site. The following table displays a frequency distribution for the number of crew members on each shuttle mission from April 1981 to July 2000.

| Crew size | 2 | 3 | 4 | 5 | 6 | 7 | 8 |
|---|---|---|---|---|---|---|---|
| Frequency | 4 | 1 | 2 | 36 | 18 | 33 | 2 |

Let $X$ denote the crew size of a randomly selected shuttle mission between April 1981 and July 2000.
**a.** What are the possible values of the random variable $X$?
**b.** Use random-variable notation to represent the event that the shuttle mission obtained has a crew size of 7.
**c.** Find $P(X = 4)$; interpret in terms of percentages.
**d.** Obtain the probability distribution of $X$.
**e.** Construct a probability histogram for $X$.

**5.80 Persons per Housing Unit.** From the document *American Housing Survey for the United States*, published by the U.S. Census Bureau, we obtained the following frequency distribution for the number of persons per occupied housing unit, where we have used "7" in place of "7 or more." Frequencies are in millions of housing units.

| Persons | 1 | 2 | 3 | 4 | 5 | 6 | 7 |
|---|---|---|---|---|---|---|---|
| Frequency | 27.9 | 34.4 | 17.0 | 15.5 | 6.8 | 2.3 | 1.4 |

For a randomly selected housing unit, let $Y$ denote the number of persons living in that unit.
**a.** Identify the possible values of the random variable $Y$.
**b.** Use random-variable notation to represent the event that a housing unit has exactly three persons living in it.
**c.** Determine $P(Y = 3)$; interpret in terms of percentages.
**d.** Determine the probability distribution of $Y$.
**e.** Construct a probability histogram for $Y$.

**5.81 Color TVs.** The Television Bureau of Advertising, Inc., publishes information on color television ownership in *Trends in Television*. Following is a probability distribution for the number of color TVs, $Y$, owned by a randomly selected household with annual income between $15,000 and $29,999.

| $y$ | 0 | 1 | 2 | 3 | 4 | 5 |
|---|---|---|---|---|---|---|
| $P(Y = y)$ | 0.009 | 0.376 | 0.371 | 0.167 | 0.061 | 0.016 |

Use random-variable notation to represent each of the following events. The households owns
a. at least one color TV.
b. exactly two color TVs.
c. between one and three, inclusive, color TVs.
d. an odd number of color TVs.

Use the special addition rule and the probability distribution to determine
e. $P(Y \geq 1)$.
f. $P(Y = 2)$.
g. $P(1 \leq Y \leq 3)$.
h. $P(Y = 1 \text{ or } 3 \text{ or } 5)$.

**5.82 Children Gender.** A certain couple is equally likely to have either a boy child or a girl child. If the family has four children, let $X$ denote the number of girls.
a. Identify the possible values of the random variable $X$.
b. Determine the probability distribution of $X$. (*Hint:* There are 16 possible equally-likely outcomes. One is GBBB, meaning the first born is a girl and the next three born are boys.)

Use random-variable notation to represent each of the following events. Also use the special addition rule and the probability distribution obtained in part (b) to determine each event's probability. The couple has
c. exactly two girls.
d. at least two girls.
e. at most two girls.
f. between one and three girls, inclusive.
g. children all of the same gender.

**5.83 Dice.** When two balanced dice are rolled, 36 equally likely outcomes are possible, as depicted in Fig. 5.1 on page 206. Let $Y$ denote the sum of the dice.
a. What are the possible values of the random variable $Y$?
b. Use random-variable notation to represent the event that the sum of the dice is 7.
c. Find $P(Y = 7)$.
d. Find the probability distribution of $Y$. Leave your probabilities in fraction form.
e. Construct a probability histogram for $Y$.

In the game of craps, a first roll of a sum of 7 or 11 wins, whereas a first roll of a sum of 2, 3, or 12 loses. To win with any other first sum, that sum must be repeated before a sum of 7 is rolled. Determine the probability of
f. a win on the first roll.
g. a loss on the first roll.

**5.84 World Series.** The World Series in baseball is won by the first team to win four games (ignoring the 1903 and 1919–1921 World Series, when it was a best of nine). Thus it takes at least four games and no more than seven games to establish a winner. As found on the Major League Baseball Web site in *World Series Overview*, historically, the lengths of the World Series are as given in the following table.

| Number of games | Frequency | Relative frequency |
|---|---|---|
| 4 | 17 | 0.175 |
| 5 | 23 | 0.237 |
| 6 | 22 | 0.227 |
| 7 | 35 | 0.361 |

a. If $X$ denotes the number of games that it takes to complete a World Series, identify the possible values of the random variable $X$.
b. Do the first and third columns of the table provide a probability distribution for $X$? Explain your answer.
c. Historically, what is the most likely number of games it takes to complete a series?
d. Historically, for a randomly chosen series, what is the probability that it ends in five games?
e. Historically, for a randomly chosen series, what is the probability that it ends in five or more games?
f. The data in the table exhibit a statistical oddity. If the two teams in a series are evenly matched and one team is ahead three games to two, either team has the same chance of winning game number six. Thus there should be about an equal number of six and seven game series. If the teams are not evenly matched, the series should tend to be shorter, ending in six or fewer games, not seven games. Can you explain why the series tend to last longer than expected?

**5.85 Archery.** An archer shoots an arrow into a square target 6 feet on a side whose center we call the origin. The outcome of this random experiment is the point in the target hit by the arrow. The archer scores 10 points if she hits the bull's eye—a disk of radius 1 foot centered at the origin; she scores 5 points if she hits the ring with inner radius 1 foot and outer radius 2 feet centered at the origin; and she scores 0 points otherwise. Assume that the archer will actually hit the target and is equally likely to hit any portion of the target. For one arrow shot, let $S$ be the score.
a. Obtain and interpret the probability distribution of the random variable $S$. (*Hint:* The area of a square is the square of its side length; the area of a disk is the square of its radius times $\pi$.)
b. Use the special addition rule and the probability distribution obtained in part (a) to determine and interpret the probability of each of the following events: $\{S = 5\}$; $\{S > 0\}$; $\{S \leq 7\}$; $\{5 < S \leq 15\}$; $\{S < 15\}$; and $\{S < 0\}$.

**5.86 Solar Eclipses.** The *World Almanac* provides information on past and projected total solar eclipses from 1955–2015. Unlike total lunar eclipses, observing a total solar eclipse from Earth is rare because it can be seen along only a very narrow path and for only a short period of time.

a. Let $X$ denote the duration, in minutes, of a total solar eclipse. Is $X$ a discrete random variable? Explain your answer.
b. Let $Y$ denote the duration, to the nearest minute, of a total solar eclipse. Is $Y$ a discrete random variable? Explain your answer.

### Extending the Concepts and Skills

**5.87** Suppose that $P(Z > 1.96) = 0.025$. Find $P(Z \leq 1.96)$. (*Hint:* Use the complementation rule.)

**5.88** Suppose that $T$ and $Z$ are random variables.

a. If $P(T > 2.02) = 0.05$ and $P(T < -2.02) = 0.05$, obtain $P(-2.02 \leq T \leq 2.02)$.
b. Suppose that $P(-1.64 \leq Z \leq 1.64) = 0.90$ and also that $P(Z > 1.64) = P(Z < -1.64)$. Find $P(Z > 1.64)$.

**5.89** Let $c > 0$ and $0 \leq \alpha \leq 1$. Also let $X$, $Y$, and $T$ be random variables.
a. If $P(X > c) = \alpha$, determine $P(X \leq c)$ in terms of $\alpha$.
b. If $P(Y > c) = \alpha/2$ and $P(Y < -c) = P(Y > c)$, obtain $P(-c \leq Y \leq c)$ in terms of $\alpha$.
c. Suppose that $P(-c \leq T \leq c) = 1 - \alpha$ and, moreover, that $P(T < -c) = P(T > c)$. Find $P(T > c)$ in terms of $\alpha$.

**5.90 Simulation.** Refer to the probability distribution displayed in Table 5.11 on page 233.
a. Use the technology of your choice to repeat the simulation done in Example 5.18 on page 234.
b. Obtain the proportions for the number of heads in three tosses and compare it to the probability distribution in Table 5.11.
c. Obtain a histogram of the proportions and compare it to the probability histogram in Fig. 5.22(b) on page 234.
d. What do parts (b) and (c) illustrate?

## 5.5 The Mean and Standard Deviation of a Discrete Random Variable*

In this section, we introduce the mean and standard deviation of a discrete random variable. As you will see, the mean and standard deviation of a discrete random variable are analogous to the population mean and population standard deviation.

### Mean of a Discrete Random Variable

Recall that, for a variable $x$, the mean of all possible observations for the entire population is called the *population mean* or *mean of the variable $x$*. In Section 3.4, we gave a formula for the mean of a variable $x$:

$$\mu = \frac{\Sigma x_i}{N}.$$

Although this formula applies only to variables of finite populations, we can use it and the language of probability to extend the concept of the mean to any discrete variable. We show how to do so in Example 5.19.

| Example 5.19 | Introducing the Mean of a Discrete Random Variable |

**TABLE 5.13**
Ages of eight students

| | | | |
|---|---|---|---|
| 19 | 20 | 20 | 19 |
| 21 | 27 | 20 | 21 |

*Student Ages* Consider a population of eight students whose ages are those given in Table 5.13. Let $X$ denote the age of a randomly selected student. From a relative-frequency distribution of the age data in Table 5.13, we get the probability distribution of the random variable $X$ shown in Table 5.14. Express the mean age of the students in terms of the probability distribution of $X$.

**TABLE 5.14**
Probability distribution of $X$, the age of a randomly selected student

| Age $x$ | Probability $P(X = x)$ | |
|---------|------------------------|------------|
| 19 | 0.250 | ← 2/8 |
| 20 | 0.375 | ← 3/8 |
| 21 | 0.250 | ← 2/8 |
| 27 | 0.125 | ← 1/8 |

**Solution**  Referring first to Table 5.13 and then to Table 5.14, we get

$$\mu = \frac{\Sigma x_i}{N} = \frac{19 + 20 + 20 + 19 + 21 + 27 + 20 + 21}{8}$$

$$= \frac{\overbrace{19 + 19}^{2} + \overbrace{20 + 20 + 20}^{3} + \overbrace{21 + 21}^{2} + \overbrace{27}^{1}}{8}$$

$$= \frac{19 \cdot 2 + 20 \cdot 3 + 21 \cdot 2 + 27 \cdot 1}{8}$$

$$= 19 \cdot \frac{2}{8} + 20 \cdot \frac{3}{8} + 21 \cdot \frac{2}{8} + 27 \cdot \frac{1}{8}$$

$$= 19 \cdot P(X = 19) + 20 \cdot P(X = 20) + 21 \cdot P(X = 21) + 27 \cdot P(X = 27)$$

$$= \Sigma x P(X = x).$$

• • •

The previous example shows that we can express the mean of a variable of a finite population in terms of the probability distribution of the corresponding random variable: $\mu = \Sigma x P(X = x)$. Because the expression on the right of this equation is meaningful for any discrete random variable, we can define the *mean of a discrete random variable* as follows.

---

**Definition 5.9**

**?**

**What Does It Mean?**

To obtain the mean of a discrete random variable, multiply each possible value by its probability and then add those products.

### Mean of a Discrete Random Variable

The **mean of a discrete random variable $X$** is denoted $\mu_X$ or, when no confusion will arise, simply $\mu$. It is defined by

$$\mu = \Sigma x P(X = x).$$

The terms **expected value** and **expectation** are commonly used in place of the term *mean*.[†]

---

**Example 5.20** | **The Mean of a Discrete Random Variable**

**TABLE 5.15**
Table for computing the mean of the random variable $X$, the number of tellers busy with customers

| $x$ | $P(X = x)$ | $xP(X = x)$ |
|-----|------------|-------------|
| 0 | 0.029 | 0.000 |
| 1 | 0.049 | 0.049 |
| 2 | 0.078 | 0.156 |
| 3 | 0.155 | 0.465 |
| 4 | 0.212 | 0.848 |
| 5 | 0.262 | 1.310 |
| 6 | 0.215 | 1.290 |
| | | 4.118 |

*Busy Tellers*  Prescott National Bank has six tellers available to serve customers. The number of tellers busy with customers at, say, 1:00 P.M. varies from day to day and depends on chance; hence it is a random variable, say, $X$. Past records indicate that the probability distribution of $X$ is as shown in the first two columns of Table 5.15. Find the mean of the random variable $X$.

**Solution**  The third column of Table 5.15 provides the products of $x$ with $P(X = x)$, which, in view of Definition 5.9, are required to determine the mean of $X$. Summing that column gives

$$\mu = \Sigma x P(X = x) = 4.118.$$

**Interpretation**  The mean number of tellers busy with customers is 4.118.

• • •

---

[†]The formula in Definition 5.9 extends the concept of population mean to any discrete variable. We could also extend that concept to any continuous variable and, using integral calculus, develop an analogous formula.

You
try it!

Exercise 5.97(a)
on page 241

## Interpretation of the Mean of a Random Variable

Recall that the mean of a variable of a finite population is the arithmetic average of all possible observations. A similar interpretation holds for the mean of a random variable.

For instance, in the previous example, the random variable $X$ is the number of tellers busy with customers at 1:00 P.M., and the mean is 4.118. Of course, there never will be a day when 4.118 tellers are busy with customers at 1:00 P.M. Over many days, however, the average number of busy tellers at 1:00 P.M. will be about 4.118.

This interpretation holds in all cases. It is commonly known as the **law of averages** and in mathematical circles as the **law of large numbers.**

---

**Key Fact 5.4**

**What Does It Mean?**

The mean of a random variable can be considered the long-run-average value of the random variable in repeated independent observations.

### Interpretation of the Mean of a Random Variable

In a large number of independent observations of a random variable $X$, the average value of those observations will approximately equal the mean, $\mu$, of $X$. The larger the number of observations, the closer the average tends to be to $\mu$.

We used a computer to simulate the number of busy tellers at 1:00 P.M. on 100 randomly selected days; that is, we obtained 100 independent observations of the random variable $X$. The data are displayed in Table 5.16.

**TABLE 5.16**
One hundred observations of the random variable $X$, the number of tellers busy with customers

| 5 | 3 | 5 | 3 | 4 | 3 | 4 | 3 | 6 | 5 | 6 | 4 | 5 | 4 | 3 | 5 | 4 | 5 | 6 | 3 |
|---|---|---|---|---|---|---|---|---|---|---|---|---|---|---|---|---|---|---|---|
| 4 | 1 | 6 | 5 | 3 | 6 | 3 | 5 | 5 | 4 | 6 | 4 | 1 | 6 | 5 | 3 | 3 | 6 | 4 | 5 |
| 3 | 4 | 2 | 5 | 5 | 6 | 5 | 4 | 6 | 2 | 4 | 5 | 4 | 6 | 4 | 5 | 5 | 3 | 4 | 6 |
| 1 | 5 | 4 | 6 | 4 | 4 | 4 | 5 | 6 | 2 | 5 | 4 | 5 | 1 | 3 | 3 | 6 | 4 | 6 | 4 |
| 5 | 6 | 5 | 5 | 3 | 2 | 4 | 6 | 6 | 1 | 5 | 1 | 3 | 6 | 5 | 3 | 5 | 4 | 3 | 6 |

The average value of the 100 observations in Table 5.16 is 4.25. This value is quite close to the mean, $\mu = 4.118$, of the random variable $X$. If we made, say, 1000 observations instead of 100, the average value of those 1000 observations would most likely be even closer to 4.118.

Figure 5.23(a) shows a plot of the average number of busy tellers versus the number of observations for the data in Table 5.16. The dashed line is at

**FIGURE 5.23**
Graphs showing the average number of busy tellers versus the number of observations for two simulations of 100 observations each

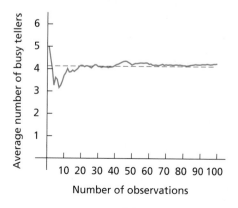

(a)

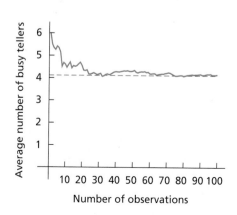

(b)

$\mu = 4.118$. Figure 5.23(b) depicts a plot for a different simulation of the number of busy tellers at 1:00 P.M. on 100 randomly selected days. Both plots suggest that, as the number of observations increases, the average number of busy tellers approaches the mean, $\mu = 4.118$, of the random variable $X$.

### Standard Deviation of a Discrete Random Variable

Similar reasoning also lets us extend the concept of population standard deviation (standard deviation of a variable) to any discrete variable.

**Definition 5.10**

**What Does It Mean?**

Roughly speaking, the standard deviation of a random variable $X$ indicates how far, on average, an observed value of $X$ is from its mean. In particular, the smaller the standard deviation of $X$, the more likely that an observed value of $X$ will be close to its mean.

### Standard Deviation of a Discrete Random Variable

The **standard deviation of a discrete random variable** $X$ is denoted $\sigma_X$ or, when no confusion will arise, simply $\sigma$. It is defined as

$$\sigma = \sqrt{\Sigma(x - \mu)^2 P(X = x)}.$$

The standard deviation of a discrete random variable can also be obtained from the computing formula

$$\sigma = \sqrt{\Sigma x^2 P(X = x) - \mu^2}.$$

**Note:** The square of the standard deviation, $\sigma^2$, is called the **variance** of $X$.

**Example 5.21** | **The Standard Deviation of a Discrete Random Variable**

*Busy Tellers* Recall Example 5.20, where $X$ denotes the number of tellers busy with customers at 1:00 P.M. Find the standard deviation of $X$.

**Solution** We apply the computing formula given in Definition 5.10. To use that formula, we need the mean of $X$, which we found in Example 5.20 to be 4.118, and columns for $x^2$ and $x^2 P(X = x)$, which are presented in the last two columns of Table 5.17.

**TABLE 5.17**
Table for computing the standard deviation of the random variable $X$, the number of tellers busy with customers

| $x$ | $P(X = x)$ | $x^2$ | $x^2 P(X = x)$ |
|---|---|---|---|
| 0 | 0.029 | 0 | 0.000 |
| 1 | 0.049 | 1 | 0.049 |
| 2 | 0.078 | 4 | 0.312 |
| 3 | 0.155 | 9 | 1.395 |
| 4 | 0.212 | 16 | 3.392 |
| 5 | 0.262 | 25 | 6.550 |
| 6 | 0.215 | 36 | 7.740 |
| | | | 19.438 |

From the final column of Table 5.17, $\Sigma x^2 P(X = x) = 19.438$. Thus

$$\sigma = \sqrt{\Sigma x^2 P(X = x) - \mu^2} = \sqrt{19.438 - (4.118)^2} = 1.6.$$

**You try it!**

Exercise 5.97(b)
on page 241

**Interpretation** Roughly speaking, on average, the number of busy tellers is 1.6 from the mean of 4.118 busy tellers.

• • •

# Exercises 5.5

## Understanding the Concepts and Skills

**5.91** What concept does the mean of a discrete random variable generalize?

**5.92 Comparing Investments.** Suppose that the random variables $X$ and $Y$ represent the amount of return on two different investments. Further suppose that the mean of $X$ equals the mean of $Y$ but that the standard deviation of $X$ is greater than the standard deviation of $Y$.
a. On average, is there a difference between the returns of the two investments? Explain your answer.
b. Which investment is more conservative? Why?

*In Exercises 5.93–5.97, we have provided the probability distributions of the random variables considered in Exercises 5.79–5.83 of Section 5.4. For each exercise, do the following.*
*a. Find and interpret the mean of the random variable.*
*b. Obtain the standard deviation of the random variable by using one of the formulas given in Definition 5.10 on page 240.*
*c. Draw a probability histogram for the random variable; locate the mean; and show one, two, and three standard-deviation intervals.*

**5.93 Space Shuttles.** The random variable $X$ is the crew size of a randomly selected shuttle mission between April 1981 and July 2000. Its probability distribution is as follows.

| $x$ | 2 | 3 | 4 | 5 | 6 | 7 | 8 |
|---|---|---|---|---|---|---|---|
| $P(X = x)$ | 0.042 | 0.010 | 0.021 | 0.375 | 0.188 | 0.344 | 0.021 |

**5.94 Persons per Housing Unit.** The random variable $Y$ is the number of persons living in a randomly selected occupied housing unit. Its probability distribution is as follows.

| $y$ | 1 | 2 | 3 | 4 | 5 | 6 | 7 |
|---|---|---|---|---|---|---|---|
| $P(Y = y)$ | 0.265 | 0.327 | 0.161 | 0.147 | 0.065 | 0.022 | 0.013 |

**5.95 Color TVs.** The random variable $Y$ is the number of color television sets owned by a randomly selected household with annual income between $15,000 and $29,999. Its probability distribution is as follows.

| $y$ | 0 | 1 | 2 | 3 | 4 | 5 |
|---|---|---|---|---|---|---|
| $P(Y = y)$ | 0.009 | 0.376 | 0.371 | 0.167 | 0.061 | 0.016 |

**5.96 Children Gender.** The random variable $X$ is the number of girls of four children born to a couple that is equally likely to have either a boy child or a girl child. Its probability distribution is as follows.

| $x$ | 0 | 1 | 2 | 3 | 4 |
|---|---|---|---|---|---|
| $P(X = x)$ | 0.0625 | 0.2500 | 0.3750 | 0.2500 | 0.0625 |

**5.97 Dice.** The random variable $Y$ is the sum of the dice when two balanced dice are rolled. Its probability distribution is as follows.

| $y$ | 2 | 3 | 4 | 5 | 6 | 7 | 8 | 9 | 10 | 11 | 12 |
|---|---|---|---|---|---|---|---|---|---|---|---|
| $P(Y = y)$ | $\frac{1}{36}$ | $\frac{1}{18}$ | $\frac{1}{12}$ | $\frac{1}{9}$ | $\frac{5}{36}$ | $\frac{1}{6}$ | $\frac{5}{36}$ | $\frac{1}{9}$ | $\frac{1}{12}$ | $\frac{1}{18}$ | $\frac{1}{36}$ |

**5.98 World Series.** The World Series in baseball is won by the first team to win four games (ignoring the 1903 and 1919–1921 World Series, when it was a best of nine). As found on the Major League Baseball Web site in *World Series Overview*, historically, the lengths of the World Series are as given in the following table.

| Number of games | Frequency | Relative frequency |
|---|---|---|
| 4 | 17 | 0.175 |
| 5 | 23 | 0.237 |
| 6 | 22 | 0.227 |
| 7 | 35 | 0.361 |

Let $X$ denote the number of games that it takes to complete a World Series, and let $Y$ denote the number of games that it took to complete a randomly selected World Series from among those considered in the table.
a. Determine the mean and standard deviation of the random variable $Y$. Interpret your results.
b. Provide an estimate for the mean and standard deviation of the random variable $X$. Explain your reasoning.

**5.99 Archery.** An archer shoots an arrow into a square target 6 feet on a side whose center we call the origin. The outcome of this random experiment is the point in the target hit by the arrow. The archer scores 10 points if she hits the bull's eye—a disk of radius 1 foot centered at the origin; she scores 5 points if she hits the ring with inner radius 1 foot and outer radius 2 feet centered at the origin; and she scores 0 points otherwise. Assume that the archer will actually hit the target and is equally likely to hit any portion of the target. For one arrow shot, let $S$ be the score. A probability distribution for the random variable $S$ is as follows.

| $s$ | 0 | 5 | 10 |
|---|---|---|---|
| $P(S = s)$ | 0.651 | 0.262 | 0.087 |

**a.** On average, how many points will the archer score per arrow shot?

**b.** Obtain and interpret the standard deviation of the score per arrow shot.

**5.100 High-Speed Internet Lines.** The Federal Communications Commission publishes a semiannual report on providers and services for Internet access titled *High Speed Services for Internet Access*. The report published in February 2002 included the following information on the percentage of zip codes with a specified number of high-speed Internet lines in service.

| Number of lines | Percentage of zip codes | Number of lines | Percentage of zip codes |
|---|---|---|---|
| 0 | 33.0 | 6 | 2.5 |
| 1 | 25.9 | 7 | 1.7 |
| 2 | 17.8 | 8 | 0.8 |
| 3 | 9.2 | 9 | 0.4 |
| 4 | 4.9 | 10 | 0.4 |
| 5 | 3.4 | | |

Let $X$ denote the number of high-speed lines in service for a randomly selected zip code.

**a.** Find the mean of $X$.

**b.** How many high-speed Internet lines would you expect to find in service for a randomly selected zip code?

**c.** Obtain and interpret the standard deviation of $X$.

**Expected Value.** *As noted in Definition 5.9 on page 238, the mean of a random variable is also called its expected value. This terminology is especially useful in gambling and decision theory, as illustrated in Exercises 5.101 and 5.102.*

**5.101 Roulette.** An American roulette wheel contains 38 numbers: 18 are red, 18 are black, and 2 are green. When the roulette wheel is spun, the ball is equally likely to land on any of the 38 numbers. Suppose that you bet $1 on red. If the ball lands on a red number, you win $1; otherwise you lose your $1. Let $X$ be the amount you win on your $1 bet. Then $X$ is a random variable whose probability distribution is as follows.

| $x$ | 1 | −1 |
|---|---|---|
| $P(X = x)$ | 0.474 | 0.526 |

**a.** Verify that the probability distribution shown in the table is correct.

**b.** Find the expected value of the random variable $X$.

**c.** On average, how much will you lose per play?

**d.** Approximately how much would you expect to lose if you bet $1 on red 100 times? 1000 times?

**e.** Is roulette a profitable game to play? Explain.

**5.102 Evaluating Investments.** An investor plans to put $50,000 in one of four investments. The return on each investment depends on whether next year's economy is strong or weak. The following table summarizes the possible payoffs, in dollars, for the four investments.

| | | Next year's economy | |
|---|---|---|---|
| | | Strong | Weak |
| Investment | Certificate of deposit | 6,000 | 6,000 |
| | Office complex | 15,000 | 5,000 |
| | Land speculation | 33,000 | −17,000 |
| | Technical school | 5,500 | 10,000 |

Let $V$, $W$, $X$, and $Y$ denote the payoffs for the certificate of deposit, office complex, land speculation, and technical school, respectively. Then $V$, $W$, $X$, and $Y$ are random variables. Assume that next year's economy has a 40% chance of being strong and a 60% chance of being weak.

**a.** Find the probability distribution of each random variable $V$, $W$, $X$, and $Y$.

**b.** Determine the expected value of each random variable.

**c.** Which investment has the best expected payoff? Which has the worst?

**d.** Which investment would you select? Explain.

**5.103 Equipment Breakdowns.** A factory manager collected data on the number of equipment breakdowns per day. From those data, she derived the probability distribution shown in the following table, where $W$ denotes the number of breakdowns on a given day.

| $w$ | 0 | 1 | 2 |
|---|---|---|---|
| $P(W = w)$ | 0.80 | 0.15 | 0.05 |

**a.** Determine $\mu_W$ and $\sigma_W$.

**b.** On average, how many breakdowns occur per day?

**c.** About how many breakdowns are expected during a 1-year period, assuming 250 work days per year?

## Extending the Concepts and Skills

**5.104 Simulation.** Let $X$ be the value of a randomly selected decimal digit, that is, a whole number between 0 and 9, inclusive.

a. Use simulation to estimate the mean of $X$. Explain your reasoning.

b. Obtain the exact mean of $X$ by applying Definition 5.9 on page 238. Compare your result with that in part (a).

**5.105 Queuing Simulation.** Benny's Barber Shop in Cleveland has five chairs for waiting customers. The number of customers waiting is a random variable $Y$ with the following probability distribution.

| $y$ | 0 | 1 | 2 | 3 | 4 | 5 |
|---|---|---|---|---|---|---|
| $P(Y = y)$ | 0.424 | 0.161 | 0.134 | 0.111 | 0.093 | 0.077 |

a. Compute and interpret the mean of the random variable $Y$.

b. In a large number of independent observations, how many customers will be waiting, on average?

c. Use the technology of your choice to simulate 500 observations of the number of customers waiting.

d. Obtain the mean of the observations in part (c) and compare it to $\mu_Y$.

e. What does part (d) illustrate?

**5.106 Mean as Center of Gravity.** Let $X$ be a discrete random variable with a finite number of possible values, say, $x_1, x_2, \ldots, x_m$. For convenience, set $p_k = P(X = x_k)$, for $k = 1, 2, \ldots, m$. Think of a horizontal axis as a seesaw and each $p_k$ as a mass placed at point $x_k$ on the seesaw. The *center of gravity* of these masses is defined to be the point $c$ on the horizontal axis at which a fulcrum could be placed to balance the seesaw.

Relative to the center of gravity, the torque acting on the seesaw by the mass $p_k$ is proportional to the product of that mass with the signed distance of the point $x_k$ from $c$, that is, to $(x_k - c) \cdot p_k$. Show that the center of gravity equals the mean of the random variable $X$. (*Hint:* To balance, the total torque acting on the seesaw must be 0.)

---

## 5.6 The Binomial Distribution*

Many applications of probability and statistics concern the repetition of an experiment. We call each repetition a **trial,** and we are particularly interested in cases where the experiment (each trial) has only two possible outcomes. Here are three examples.

- Testing the effectiveness of a drug: Several patients take the drug (the trials), and for each patient the drug is either effective or not effective (the two possible outcomes).
- Weekly sales of a car salesperson: The salesperson has several customers during the week (the trials), and for each customer the salesperson either makes a sale or does not make a sale (the two possible outcomes).
- Taste tests for colas: A number of people taste two different colas (the trials), and for each person the preference is either for the first cola or for the second cola (the two possible outcomes).

To analyze repeated trials of an experiment that has two possible outcomes requires knowledge of factorials, binomial coefficients, Bernoulli trials, and the binomial distribution. We begin with factorials.

### Factorials

*Factorials* are defined as follows.

**Definition 5.11**

### Factorials

The product of the first $k$ positive integers (counting numbers) is called $k$ **factorial** and is denoted $k!$. In symbols,

$$k! = k(k-1)\cdots 2 \cdot 1.$$

We also define $0! = 1$.

We illustrate the calculation of factorials in the next example.

**Example 5.22** | ### Factorials

*Doing the Calculations* Determine $3!$, $4!$, and $5!$.

**You try it!**

Exercise 5.109 on page 255

**Solution** Applying Definition 5.11 gives $3! = 3 \cdot 2 \cdot 1 = 6$, $4! = 4 \cdot 3 \cdot 2 \cdot 1 = 24$, and $5! = 5 \cdot 4 \cdot 3 \cdot 2 \cdot 1 = 120$.

• • •

Notice that $6! = 6 \cdot 5!$, $6! = 6 \cdot 5 \cdot 4!$, $6! = 6 \cdot 5 \cdot 4 \cdot 3!$, and so on. In general, if $j \leq k$, then $k! = k(k-1)\cdots(k-j+1)(k-j)!$.

### Binomial Coefficients

You may have already encountered *binomial coefficients* in algebra when you studied the binomial expansion, the expansion of $(a+b)^n$.

**Definition 5.12**

### Binomial Coefficients

If $n$ is a positive integer and $x$ is a nonnegative integer less than or equal to $n$, then the **binomial coefficient** $\binom{n}{x}$ is defined as

$$\binom{n}{x} = \frac{n!}{x!\,(n-x)!}.$$

**Example 5.23** | ### Binomial Coefficients

*Doing the Calculations* Determine the value of each binomial coefficient.

a. $\binom{6}{1}$  b. $\binom{5}{3}$  c. $\binom{7}{3}$  d. $\binom{4}{4}$

**Solution** We apply Definition 5.12.

a. $\binom{6}{1} = \frac{6!}{1!\,(6-1)!} = \frac{6!}{1!\,5!} = \frac{6 \cdot 5!}{1!\,5!} = \frac{6}{1} = 6$

b. $\binom{5}{3} = \frac{5!}{3!\,(5-3)!} = \frac{5!}{3!\,2!} = \frac{5 \cdot 4 \cdot 3!}{3!\,2!} = \frac{5 \cdot 4}{2} = 10$

You try it!

Exercise 5.111
on page 255

c. $\dbinom{7}{3} = \dfrac{7!}{3!\,(7-3)!} = \dfrac{7!}{3!\,4!} = \dfrac{7 \cdot 6 \cdot 5 \cdot \cancel{4!}}{3!\,\cancel{4!}} = \dfrac{7 \cdot 6 \cdot 5}{6} = 35$

d. $\dbinom{4}{4} = \dfrac{4!}{4!\,(4-4)!} = \dfrac{4!}{4!\,0!} = \dfrac{\cancel{4!}}{\cancel{4!}\,0!} = \dfrac{1}{1} = 1$

•  •  •

## Bernoulli Trials

Next we define *Bernoulli trials* and some related concepts.

**Definition 5.13**

### Bernoulli Trials

**What Does It Mean?**

Bernoulli trials are identical and independent repetitions of an experiment with two possible outcomes.

Repeated trials of an experiment are called **Bernoulli trials** if the following three conditions are satisfied:

1.  The experiment (each trial) has two possible outcomes, denoted generically *s*, for **success**, and *f*, for **failure**.
2.  The trials are independent.
3.  The probability of a success, called the **success probability** and denoted *p*, remains the same from trial to trial.

## Introducing the Binomial Distribution

The **binomial distribution** is the probability distribution for the number of successes in a sequence of Bernoulli trials.

**Example 5.24** | ### Introducing the Binomial Distribution

*Mortality* Mortality tables enable actuaries to obtain the probability that a person at any particular age will live a specified number of years. Insurance companies and others use such probabilities to determine life-insurance premiums, retirement pensions, and annuity payments.

According to tables provided by the U.S. National Center for Health Statistics in *Vital Statistics of the United States*, a person aged 20 has about an 80% chance of being alive at age 65. Suppose that three people aged 20 are selected at random.

a.  Formulate the process of observing which people are alive at age 65 as a sequence of three Bernoulli trials.

b.  Obtain the possible outcomes of the three Bernoulli trials.

c.  Determine the probability of each outcome in part (b).

d.  Find the probability that exactly two of the three people will be alive at age 65.

e.  Obtain the probability distribution of the number of people of the three who are alive at age 65.

**Solution**

a.  Each trial consists of observing whether a person currently aged 20 is alive at age 65 and has two possible outcomes: alive or dead. The trials are

**TABLE 5.18**
Possible outcomes

| | | | |
|---|---|---|---|
| sss | ssf | sfs | sff |
| fss | fsf | ffs | fff |

independent. If we let a success, $s$, correspond to being alive at age 65, the success probability is 0.8 (80%); that is, $p = 0.8$.

**b.** The possible outcomes of the three Bernoulli trials are shown in Table 5.18 ($s$ = success = alive, $f$ = failure = dead). For instance, $ssf$ represents the outcome that at age 65 the first two people are alive and the third is not.

**c.** As Table 5.18 indicates, eight outcomes are possible. However, because these eight outcomes are not equally likely, we cannot use the $f/N$ rule to determine their probabilities; instead, we must proceed as follows. First of all, by part (a), the success probability equals 0.8, or

$$P(s) = p = 0.8.$$

Therefore the failure probability is

$$P(f) = 1 - p = 1 - 0.8 = 0.2.$$

**TABLE 5.19**

Outcomes and probabilities for observing whether each of three people is alive at age 65

| Outcome | Probability |
|---|---|
| sss | $(0.8)(0.8)(0.8) = 0.512$ |
| ssf | $(0.8)(0.8)(0.2) = 0.128$ |
| sfs | $(0.8)(0.2)(0.8) = 0.128$ |
| sff | $(0.8)(0.2)(0.2) = 0.032$ |
| fss | $(0.2)(0.8)(0.8) = 0.128$ |
| fsf | $(0.2)(0.8)(0.2) = 0.032$ |
| ffs | $(0.2)(0.2)(0.8) = 0.032$ |
| fff | $(0.2)(0.2)(0.2) = 0.008$ |

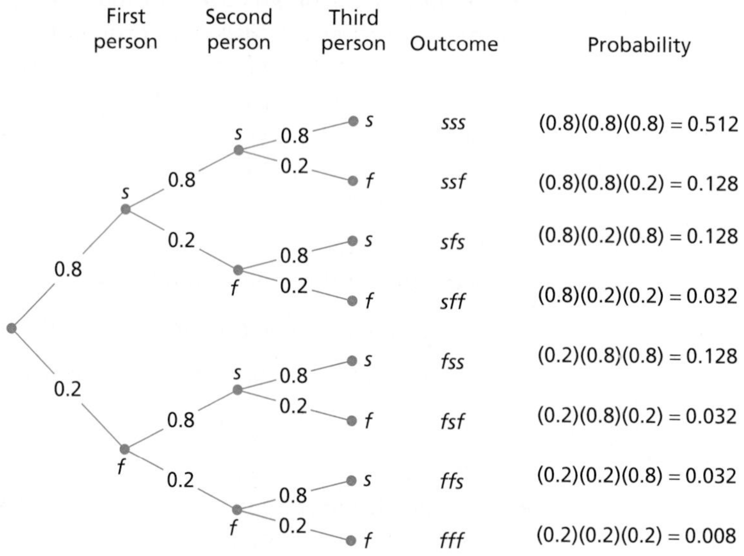

Because the trials are independent, we can obtain the probability of each three-trial outcome by multiplying the probabilities for each trial.[†] For instance, the probability of the outcome $ssf$ is

$$P(ssf) = P(s) \cdot P(s) \cdot P(f) = 0.8 \cdot 0.8 \cdot 0.2 = 0.128.$$

All eight possible outcomes and their probabilities are shown in Table 5.19. Note that outcomes containing the same number of successes have the same probability. For instance, the three outcomes containing exactly two successes—$ssf$, $sfs$, and $fss$—have the same probability: 0.128. Each probability is the product of two success probabilities of 0.8 and one failure probability of 0.2.

A tree diagram is useful for organizing and summarizing the possible outcomes of this experiment and their probabilities. See Fig. 5.24.

**FIGURE 5.24**

Tree diagram corresponding to Table 5.19

---

[†]Mathematically, this procedure is justified by the *special multiplication rule* of probability. See, for instance, Formula 4.7 on page 201 of *Introductory Statistics, 8/e*, by Neil A. Weiss (Boston: Addison-Wesley, 2008).

**d.** Table 5.19 shows that the event that exactly two of the three people are alive at age 65 consists of the outcomes *ssf*, *sfs*, and *fss*. So, by the special addition rule (Formula 5.1 on page 222),

$$P(\text{Exactly two will be alive}) = P(ssf) + P(sfs) + P(fss)$$
$$= \underbrace{0.128 + 0.128 + 0.128}_{3 \text{ times}} = 3 \cdot 0.128 = 0.384.$$

The probability that exactly two of the three people will be alive at age 65 is 0.384.

**e.** Let $X$ denote the number of people of the three who are alive at age 65. In part (d), we found $P(X = 2)$. We can proceed in the same way to find the remaining three probabilities: $P(X = 0)$, $P(X = 1)$, and $P(X = 3)$. The results are given in Table 5.20 and also in the probability histogram in Fig. 5.25. Note for future reference that this probability distribution is left skewed.

**FIGURE 5.25**
Probability histogram for the random variable *X*, the number of people of three who are alive at age 65

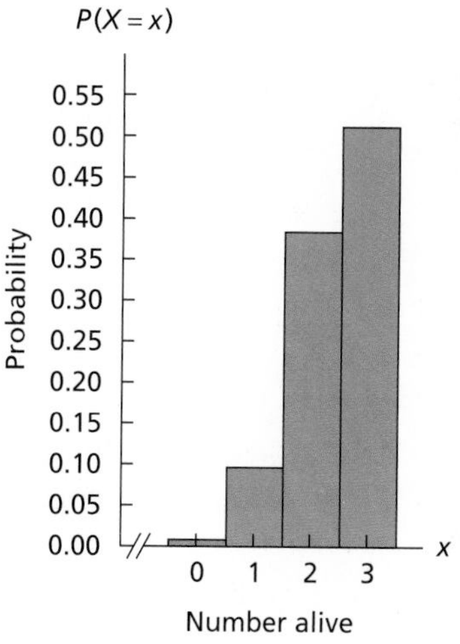

**TABLE 5.20**
Probability distribution of the random variable *X*, the number of people of three who are alive at age 65

| Number alive $x$ | Probability $P(X = x)$ |
|:---:|:---:|
| 0 | 0.008 |
| 1 | 0.096 |
| 2 | 0.384 |
| 3 | 0.512 |

You try it!

Exercise 5.115 on page 256

$P(X = x)$

• • •

## The Binomial Probability Formula

We obtained the probability distribution in Table 5.20 by using a tabulation method (Table 5.19), which required much work. In most practical applications, the amount of work required would be even more and often would be prohibitive because the number of trials is generally much larger than three. For instance, given twenty, rather than three, 20-year-olds, there would be over 1 million possible outcomes. The tabulation method certainly would not be feasible in that case.

The good news is that a relatively simple formula will give us binomial probabilities. Before we develop that formula, we need the following fact.

**Key Fact 5.5**

**Number of Outcomes Containing a Specified Number of Successes**

In $n$ Bernoulli trials, the number of outcomes that contain exactly $x$ successes equals the binomial coefficient $\binom{n}{x}$.

**? What Does It Mean?**

There are $\binom{n}{x}$ ways of getting exactly $x$ successes in $n$ Bernoulli trials.

We won't stop to prove Key Fact 5.5, but let's check it against the results in Example 5.24. For instance, in Table 5.18, we saw that there are three outcomes in which exactly two of the three people are alive at age 65. Key Fact 5.5 gives us that information more easily:

$$\begin{bmatrix} \text{Number of outcomes} \\ \text{comprising the event} \\ \text{exactly two alive} \end{bmatrix} = \binom{3}{2} = \frac{3!}{2!\,(3-2)!} = \frac{3!}{2!\,1!} = 3.$$

We can now develop a probability formula for the number of successes in Bernoulli trials. We illustrate how that formula is derived by referring to Example 5.24. For instance, to determine the probability that exactly two of the three people will be alive at age 65, $P(X = 2)$, we reason as follows:

1. Any particular outcome in which exactly two of the three people are alive at age 65 (e.g., *sfs*) has probability:

$$\underset{\underset{\substack{\uparrow \\ \text{Probability} \\ \text{alive}}}{\text{Two alive}}}{(0.8)^2} \cdot \underset{\underset{\substack{\uparrow \\ \text{Probability} \\ \text{dead}}}{\text{One dead}}}{(0.2)^1} = 0.64 \cdot 0.2 = 0.128.$$

2. By Key Fact 5.5, the number of outcomes in which exactly two of the three people are alive at age 65 is

$$\underset{\underset{\substack{\uparrow \\ \text{Number alive}}}{}}{\binom{\overset{\overset{\text{Number of trials}}{\downarrow}}{3}}{2}} = \frac{3!}{2!\,(3-2)!} = 3.$$

3. By the special addition rule, the probability that exactly two of the three people will be alive at age 65 is

$$P(X = 2) = \binom{3}{2} \cdot (0.8)^2 (0.2)^1 = 3 \cdot 0.128 = 0.384.$$

Of course, this result is the same as that obtained in Example 5.24(d). However, this time we found the probability without tabulating and listing. More important, the reasoning we used applies to any sequence of Bernoulli trials and leads to the *binomial probability formula*.

Formula 5.4    **Binomial Probability Formula**

Let $X$ denote the total number of successes in $n$ Bernoulli trials with success probability $p$. Then the probability distribution of the random variable $X$ is given by

$$P(X = x) = \binom{n}{x} p^x (1 - p)^{n-x}, \qquad x = 0, 1, 2, \dots, n.$$

The random variable $X$ is called a **binomial random variable** and is said to have the **binomial distribution** with parameters $n$ and $p$.

To determine a binomial probability formula in specific problems, having a well-organized strategy, such as the one presented in Procedure 5.1, is useful.

Procedure 5.1    **To Find a Binomial Probability Formula**

*Assumptions*

1. $n$ trials are to be performed.
2. Two outcomes, success or failure, are possible for each trial.
3. The trials are independent.
4. The success probability, $p$, remains the same from trial to trial.

**STEP 1  Identify a success.**

**STEP 2  Determine $p$, the success probability.**

**STEP 3  Determine $n$, the number of trials.**

**STEP 4  The binomial probability formula for the number of successes, $X$, is given by**

$$P(X = x) = \binom{n}{x} p^x (1 - p)^{n-x}.$$

In the following example, we illustrate this procedure by applying it to the random variable considered in Example 5.24.

Example 5.25 | **Obtaining Binomial Probabilities**

*Mortality* According to tables provided by the U.S. National Center for Health Statistics in *Vital Statistics of the United States*, there is roughly an 80% chance that a person aged 20 will be alive at age 65. Suppose that three people aged 20 are selected at random. Find the probability that the number alive at age 65 will be

**a.** exactly two.        **b.** at most one.        **c.** at least one.

**d.** Determine the probability distribution of the number alive at age 65.

**Solution** Let $X$ denote the number of people of the three who are alive at age 65. To solve parts (a)–(d), we first apply Procedure 5.1.

**STEP 1 Identify a success.**

A success is that a person currently aged 20 will be alive at age 65.

**STEP 2 Determine $p$, the success probability.**

The probability that a person currently aged 20 will be alive at age 65 is 80%, so $p = 0.8$.

**STEP 3 Determine $n$, the number of trials.**

The number of trials is the number of people in the study, which is three, so $n = 3$.

**STEP 4 The binomial probability formula for the number of successes, $X$, is**

$$P(X = x) = \binom{n}{x} p^x (1 - p)^{n-x}.$$

Because $n = 3$ and $p = 0.8$, the formula becomes

$$P(X = x) = \binom{3}{x} (0.8)^x (0.2)^{3-x}.$$

We see that $X$ is a binomial random variable and has the binomial distribution with parameters $n = 3$ and $p = 0.8$. Now we can solve parts (a)–(d) relatively easily.

a. Applying the binomial probability formula with $x = 2$ yields

$$P(X = 2) = \binom{3}{2}(0.8)^2(0.2)^{3-2} = \frac{3!}{2!\,(3-2)!}(0.8)^2(0.2)^1 = 0.384.$$

**Interpretation** Chances are 38.4% that exactly two of the three people will be alive at age 65.

b. The probability that at most one person will be alive at age 65 is

$$P(X \le 1) = P(X = 0) + P(X = 1)$$
$$= \binom{3}{0}(0.8)^0(0.2)^{3-0} + \binom{3}{1}(0.8)^1(0.2)^{3-1}$$
$$= 0.008 + 0.096 = 0.104.$$

**Interpretation** Chances are 10.4% that one or fewer of the three people will be alive at age 65.

c. The probability that at least one person will be alive at age 65 is $P(X \ge 1)$, which we can obtain by first using the fact that

$$P(X \ge 1) = P(X = 1) + P(X = 2) + P(X = 3)$$

and then applying the binomial probability formula to calculate each of the three individual probabilities. However, using the complementation rule is easier:

$$P(X \geq 1) = 1 - P(X < 1) = 1 - P(X = 0)$$

$$= 1 - \binom{3}{0}(0.8)^0(0.2)^{3-0} = 1 - 0.008 = 0.992.$$

**Interpretation**    Chances are 99.2% that one or more of the three people will be alive at age 65.

**d.** To obtain the probability distribution of the random variable $X$, we need to use the binomial probability formula to compute $P(X = x)$, for $x = 0, 1, 2,$ and 3. We have already done so for $x = 0, 1,$ and 2 in parts (a) and (b). For $x = 3$, we have

$$P(X = 3) = \binom{3}{3}(0.8)^3(0.2)^{3-3} = (0.8)^3 = 0.512.$$

Thus the probability distribution of $X$ is as shown in Table 5.20 on page 247. But this time we computed the probabilities quickly and easily by using the binomial probability formula.

You try it!

Exercise 5.121(a)–(e) on page 256

•  •  •

**Note:** The probability, $P(X \leq 1)$, required in part (b) of the previous example is a *cumulative probability*. In general, a **cumulative probability** is the probability that a random variable is less than or equal to a specified number, that is, a probability of the form $P(X \leq x)$. The concept of cumulative probability applies to any random variable, not just binomial random variables.

We can express the probability that a random variable lies between two specified numbers—say, $a$ and $b$—in terms of cumulative probabilities:

$$P(a < X \leq b) = P(X \leq b) - P(X \leq a).$$

## Shape of a Binomial Distribution

Figure 5.25 on page 247 shows that, for three people currently 20 years old, the probability distribution of the number who will be alive at age 65 is left skewed. The reason is that the success probability, $p = 0.8$, exceeds 0.5.

More generally, *a binomial distribution is right skewed if $p < 0.5$, is symmetric if $p = 0.5$, and is left skewed if $p > 0.5$.* Figure 5.26 illustrates these facts for three different binomial distributions with $n = 6$.

**FIGURE 5.26**
Probability histograms for three different binomial distributions with parameter $n = 6$

You try it!

Exercise 5.121(f)–(g) on pages 256–257

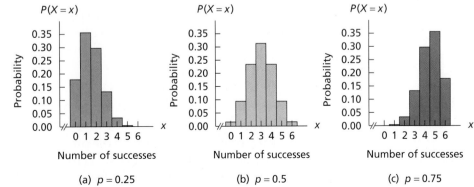

(a)  $p = 0.25$
Right skewed

(b)  $p = 0.5$
Symmetric

(c)  $p = 0.75$
Left skewed

## Mean and Standard Deviation of a Binomial Random Variable

In Section 5.5, we discussed the mean and standard deviation of a discrete random variable. We presented formulas to compute these parameters in Definition 5.9 on page 238 and Definition 5.10 on page 240.

Because these formulas apply to any discrete random variable, they work for a binomial random variable. Hence we can determine the mean and standard deviation of a binomial random variable by first using the binomial probability formula to obtain its probability distribution and then applying Definitions 5.9 and 5.10.

But there is an easier way. If we substitute the binomial probability formula into the formulas for the mean and standard deviation of a discrete random variable and then simplify mathematically, we obtain the following.

**Formula 5.5**

### Mean and Standard Deviation of a Binomial Random Variable

The mean and standard deviation of a binomial random variable with parameters $n$ and $p$ are

$$\mu = np \quad \text{and} \quad \sigma = \sqrt{np(1-p)},$$

respectively.

**? What Does It Mean?**

To obtain the mean of a binomial random variable, multiply the number of trials by the success probability. To obtain the standard deviation, multiply the number of trials by the success probability by the failure probability, and take the square root of the result.

In the next example, we apply the two formulas in Formula 5.5 to determine the mean and standard deviation of the binomial random variable considered in the mortality illustration.

**Example 5.26**

### Mean and Standard Deviation of a Binomial Random Variable

*Mortality* For three randomly selected 20-year-olds, let $X$ denote the number who are still alive at age 65. Find the mean and standard deviation of $X$.

**Solution** As we stated in the previous example, $X$ is a binomial random variable with parameters $n = 3$ and $p = 0.8$. Applying Formula 5.5 gives

$$\mu = np = 3 \cdot 0.8 = 2.4$$

and

$$\sigma = \sqrt{np(1-p)} = \sqrt{3 \cdot 0.8 \cdot 0.2} = 0.69.$$

You try it!

Exercise 5.121(h)–(i)
on page 257

**Interpretation** On average, 2.4 of every three 20-year-olds will still be alive at age 65. And, roughly speaking, on average, the number of three 20-year-olds still alive at age 65 will differ from the mean number of 2.4 by 0.69.

• • •

## Binomial Approximation to the Hypergeometric Distribution

We often want to determine the proportion (percentage) of members of a finite population that have a specified attribute. For instance, we might be interested in the proportion of U.S. adults that have Internet access. Here the population consists of all U.S. adults, and the attribute is "has Internet access." Or we might want to know the proportion of U.S. businesses that are minority owned. In this case, the population consists of all U.S. businesses, and the attribute is "minority owned."

Generally, the population under consideration is too large for the population proportion to be found by taking a census. Imagine, for instance, trying to interview every U.S. adult to determine the proportion that have Internet access. So, in practice, we rely mostly on sampling and use the sample data to estimate the population proportion.

Suppose that a simple random sample of size $n$ is taken from a population in which the proportion of members that have a specified attribute is $p$. Then a random variable of primary importance in the estimation of $p$ is the number of members sampled that have the specified attribute, which we denote $X$. The exact probability distribution of $X$ depends on whether the sampling is done with or without replacement.

If sampling is done with replacement, the sampling process constitutes Bernoulli trials: Each selection of a member from the population corresponds to a trial. A success occurs on a trial if the member selected in that trial has the specified attribute; otherwise, a failure occurs. The trials are independent because the sampling is done with replacement. The success probability remains the same from trial to trial—it always equals the proportion of the population that has the specified attribute. Therefore the random variable $X$ has the binomial distribution with parameters $n$ (the sample size) and $p$ (the population proportion).

In reality, however, sampling is ordinarily done without replacement. Under these circumstances, the sampling process does not constitute Bernoulli trials because the trials are not independent and the success probability varies from trial to trial. In other words, the random variable $X$ does not have a binomial distribution. Its distribution is important, however, and is referred to as a **hypergeometric distribution.**

We won't present the hypergeometric probability formula here because, in practice, a hypergeometric distribution can usually be approximated by a binomial distribution. The reason is that, if the sample size does not exceed 5% of the population size, there is little difference between sampling with and without replacement. We summarize the previous discussion as follows.

---

**Key Fact 5.6**

## What Does It Mean?

When a simple random sample is taken from a finite population, you can use a binomial distribution for the number of members obtained having a specified attribute, regardless of whether the sampling is with or without replacement, provided that, in the latter case, the sample size is small relative to the population size.

### Sampling and the Binomial Distribution

Suppose that a simple random sample of size $n$ is taken from a finite population in which the proportion of members that have a specified attribute is $p$. Then the number of members sampled that have the specified attribute

- has exactly a binomial distribution with parameters $n$ and $p$ if the sampling is done with replacement and
- has approximately a binomial distribution with parameters $n$ and $p$ if the sampling is done without replacement and the sample size does not exceed 5% of the population size.

For example, according to the U.S. Census Bureau publication *Current Population Reports*, 84.6% of U.S. adults have completed high school. Suppose that eight U.S. adults are to be randomly selected without replacement. Let $X$ denote the number of those sampled that have completed high school. Then, because the sample size does not exceed 5% of the population size, the random variable $X$ has approximately a binomial distribution with parameters $n = 8$ and $p = 0.846$.

### Other Discrete Probability Distributions

The binomial distribution is the most important and most widely used discrete probability distribution. Other common discrete probability distributions are the Poisson, hypergeometric, and geometric distributions, which you are asked to consider in the exercises.

## The Technology Center

Almost all statistical technologies include programs that determine binomial probabilities. In this subsection, we present output and step-by-step instructions for such programs.

**Example 5.27**  **Using Technology to Obtain Binomial Probabilities**

*Mortality*   Consider once again the mortality illustration discussed in Example 5.25 on page 249. Use Minitab, Excel, or the TI-83/84 Plus to determine the probability that exactly two of the three people will be alive at age 65.

**Solution**   Recall that, of three randomly selected people aged 20, the number, $X$, who are alive at age 65 has a binomial distribution with parameters $n = 3$ and $p = 0.8$. We want the probability that exactly two of the three people will be alive at age 65—that is, $P(X = 2)$.

We applied the binomial probability programs, resulting in Output 5.1. Steps for generating that output are presented in Instructions 5.1. As shown in Output 5.1, the required probability is 0.384.

• • •

You can also obtain cumulative probabilities for a binomial distribution by using Minitab, Excel, or the TI-83/84 Plus. To do so, modify Instructions 5.1 as follows:

- For Minitab, in step 2, select the **Cumulative probability** option button instead of the **Probability** option button.
- For Excel, in step 8, type <u>TRUE</u> instead of <u>FALSE</u>.
- For the TI-83/84 Plus, in step 2, arrow down to **binomcdf(** instead of **binompdf(**.

∎ ∎ ∎

**OUTPUT 5.1**  Probability that exactly two of the three people will be alive at age 65

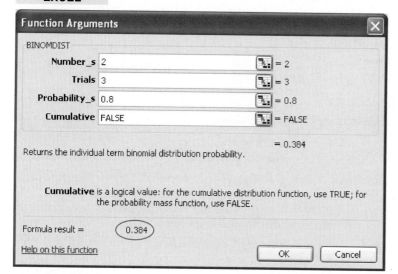

MINITAB

**Probability Density Function**

```
Binomial with n = 3 and p = 0.8

x   P( X = x )
2        0.384
```

EXCEL

TI-83/84 PLUS

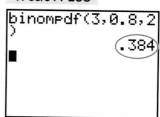

```
binompdf(3,0.8,2
)
                .384
■
```

**INSTRUCTIONS 5.1**  Steps for generating Output 5.1

**MINITAB**

1 Choose **Calc ➤ Probability Distributions ➤ Binomial…**
2 Select the **Probability** option button
3 Click in the **Number of trials** text box and type 3
4 Click in the **Event probability** text box and type 0.8
5 Select the **Input constant** option button
6 Click in the **Input constant** text box and type 2
7 Click **OK**

**EXCEL**

1 Click *f<sub>x</sub>* on the button bar
2 Select **Statistical** from the **Or select a category** drop down list box
3 Select **BINOMDIST** from the **Select a function** list
4 Click **OK**
5 Type 2 in the **Number_s** text box
6 Click in the **Trials** text box and type 3
7 Click in the **Probability_s** text box and type 0.8
8 Click in the **Cumulative** text box and type FALSE

**TI-83/84 PLUS**

1 Press **2nd ➤ DISTR**
2 Arrow down to **binompdf(** and press **ENTER**
3 Type 3,0.8,2) and press **ENTER**

# Exercises 5.6

## Understanding the Concepts and Skills

**5.107**  Give two examples of Bernoulli trials other than those presented in the text.

**5.108**  What does the "bi" in "binomial" signify?

**5.109**  Compute 3!, 7!, 8!, and 9!.

**5.110**  Find 1!, 2!, 4!, and 6!.

**5.111**  Determine the value of each of the following binomial coefficients.

a. $\binom{5}{2}$   b. $\binom{7}{4}$   c. $\binom{10}{3}$   d. $\binom{12}{5}$

**5.112** Evaluate the following binomial coefficients.

**a.** $\binom{3}{2}$ **b.** $\binom{6}{0}$ **c.** $\binom{6}{6}$ **d.** $\binom{7}{3}$

**5.113** Evaluate the following binomial coefficients.

**a.** $\binom{4}{1}$ **b.** $\binom{6}{2}$ **c.** $\binom{8}{3}$ **d.** $\binom{9}{6}$

**5.114** Determine the value of each binomial coefficient.

**a.** $\binom{5}{3}$ **b.** $\binom{10}{0}$ **c.** $\binom{10}{10}$ **d.** $\binom{9}{5}$

**5.115 Pinworm Infestation.** Pinworm infestation, commonly found in children, can be treated with the drug pyrantel pamoate. According to the *Merck Manual*, the treatment is effective in 90% of cases. Suppose that three children with pinworm infestation are given pyrantel pamoate.

**a.** Considering a success in a given case to be "a cure," formulate the process of observing which children are cured and which children are not cured as a sequence of three Bernoulli trials.

**b.** Construct a table similar to Table 5.19 on page 246 for the three cases. Display the probabilities to three decimal places.

**c.** Draw a tree diagram for this problem similar to the one shown in Fig. 5.24 on page 246.

**d.** List the outcomes in which exactly two of the three children are cured.

**e.** Find the probability of each outcome in part (d). Why are those probabilities all the same?

**f.** Use parts (d) and (e) to determine the probability that exactly two of the three children will be cured.

**g.** Without using the binomial probability formula, obtain the probability distribution of the random variable $X$, the number of children out of three who are cured.

**5.116 Psychiatric Disorders.** The National Institute of Mental Health reports that there is a 20% chance of an adult American suffering from a psychiatric disorder. Four randomly selected adult Americans are examined for psychiatric disorders.

**a.** If you let a success correspond to an adult American having a psychiatric disorder, what is the success probability, $p$? (*Note:* The use of the word *success* in Bernoulli trials need not reflect its usually positive connotation.)

**b.** Construct a table similar to Table 5.19 on page 246 for the four people examined. Display the probabilities to four decimal places.

**c.** Draw a tree diagram for this problem similar to the one shown in Fig. 5.24 on page 246.

**d.** List the outcomes in which exactly three of the four people examined have a psychiatric disorder.

**e.** Find the probability of each outcome in part (d). Why are those probabilities all the same?

**f.** Use parts (d) and (e) to determine the probability that exactly three of the four people examined have a psychiatric disorder.

**g.** Without using the binomial probability formula, obtain the probability distribution of the random variable $Y$,

the number of adults out of four who have a psychiatric disorder.

**5.117 Pinworm Infestation.** Use Procedure 5.1 on page 249 to solve part (g) of Exercise 5.115.

**5.118 Psychiatric Disorders.** Use Procedure 5.1 on page 249 to solve part (g) of Exercise 5.116.

**5.119** For each of the following probability histograms of binomial distributions, specify whether the success probability is less than, equal to, or greater than 0.5. Explain your answers.

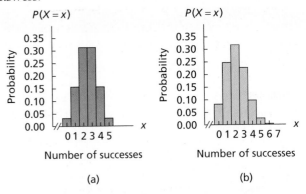

(a)                                    (b)

**5.120** For each of the following probability histograms of binomial distributions, specify whether the success probability is less than, equal to, or greater than 0.5. Explain your answers.

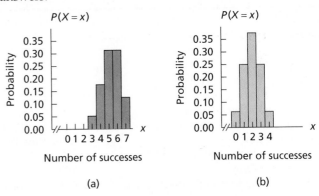

(a)                                    (b)

**5.121 Horse Racing.** According to the *Daily Racing Form*, the probability is about 0.67 that the favorite in a horse race will finish in the money (first, second, or third place). In the next five races, what is the probability that the favorite finishes in the money

**a.** exactly twice?     **b.** exactly four times?

**c.** at least four times?

**d.** between two and four times, inclusive?

**e.** Determine the probability distribution of the random variable $X$, the number of times the favorite finishes in the money in the next five races.

**f.** Identify the probability distribution of $X$ as right skewed, symmetric, or left skewed without consulting its probability distribution or drawing its probability histogram.

**g.** Draw a probability histogram for $X$.

**h.** Use your answer from part (e) and Definitions 5.9 and 5.10 on pages 238 and 240, respectively, to obtain the mean and standard deviation of the random variable $X$.

**i.** Use Formula 5.5 on page 252 to obtain the mean and standard deviation of the random variable $X$.

**j.** Interpret your answer for the mean in words.

**5.122 Gestation Periods.** The probability is 0.314 that the gestation period of a woman will exceed 9 months. In six human births, what is the probability that the number in which the gestation period exceeds 9 months is

**a.** exactly three.    **b.** exactly five.    **c.** at least five.

**d.** between three and five, inclusive.

**e.** Determine the probability distribution of the random variable $X$, the number of six human births in which the gestation period exceeds 9 months.

**f.** Identify the probability distribution of $X$ as right skewed, symmetric, or left skewed without consulting its probability distribution or drawing its probability histogram.

**g.** Draw a probability histogram for $X$.

**h.** Use your answer from part (e) and Definitions 5.9 and 5.10 on pages 238 and 240, respectively, to obtain the mean and standard deviation of the random variable $X$.

**i.** Use Formula 5.5 on page 252 to obtain the mean and standard deviation of the random variable $X$.

**j.** Interpret your answer for the mean in words.

**5.123 Traffic Fatalities and Intoxication.** The National Safety Council publishes information about automobile accidents in *Accident Facts*. According to that document, the probability is 0.40 that a traffic fatality involves an intoxicated or alcohol-impaired driver or nonoccupant. In eight traffic fatalities, find the probability that the number, $Y$, which involve an intoxicated or alcohol-impaired driver or nonoccupant is

**a.** exactly three; at least three; at most three.

**b.** between two and four, inclusive.

**c.** Find and interpret the mean of the random variable $Y$.

**d.** Obtain the standard deviation of $Y$.

**5.124 Multiple-Choice Exams.** A student takes a multiple-choice exam with 10 questions, each with four possible selections for the answer. A passing grade is 60% or better. Suppose that the student was unable to find time to study for the exam and just guesses at each question. Find the probability that the student

**a.** gets at least one question correct.

**b.** passes the exam.

**c.** receives an "A" on the exam (90% or better).

**d.** How many questions would you expect the student to get correct?

**e.** Obtain the standard deviation of the number of questions that the student gets correct.

**5.125 Love Stinks?** J. Fetto, in the article "Love Stinks" (*American Demographics*, Vol. 25, No. 1, pp. 10–11), reports

that Americans split with their significant other for many reasons—including indiscretion, infidelity, and simply "growing apart." According to the article, 35% of American adults have experienced a breakup at least once during the last 10 years. Of nine randomly selected American adults, find the probability that the number, $X$, who have experienced a breakup at least once during the last 10 years is

**a.** exactly five; at most five; at least five.

**b.** at least one; at most one.

**c.** between six and eight, inclusive.

**d.** Determine the probability distribution of the random variable $X$.

**e.** Strictly speaking, why is the probability distribution that you obtained in part (d) only approximately correct? What is the exact distribution called?

**5.126 Women Engineering Doctorates.** The Office of Educational Research and Improvement profiles persons receiving doctor's degrees and publishes its findings in *Digest of Education Statistics*. According to that document, 14.9% of those who have received a doctor's degree in engineering are women. Suppose that six people who have received their doctor's degree in engineering are randomly sampled. What is the probability that

**a.** exactly two are women?    **b.** exactly four are women?

**c.** at least two are women?

**d.** Determine the probability distribution of the number of women in a sample of six persons who have received their doctor's degree in engineering.

**e.** Strictly speaking, why is the probability distribution that you obtained in part (d) only approximately correct? What is the exact distribution called?

**5.127 Health Insurance.** According to the Centers for Disease Control and Prevention publication *Health, United States, 2004*, in 2002, 16.5% of persons under the age of 65 had no health insurance coverage. Suppose that, today, four persons under the age of 65 are randomly selected.

**a.** Assuming that the uninsured rate is the same today as it was in 2002, determine the probability distribution for the number, $X$, who have no health insurance coverage.

**b.** Determine and interpret the mean of $X$.

**c.** If, in fact, exactly three of the four people selected have no health insurance coverage, would you be inclined to conclude that the uninsured rate today has increased from the 16.5% rate in 2002? Explain your reasoning. *Hint:* First consider the probability $P(X \geq 3)$.

**d.** If, in fact, exactly two of the four people selected have no health insurance coverage, would you be inclined to conclude that the uninsured rate today has increased from the 16.5% rate in 2002? Explain your reasoning.

**5.128 Recidivism.** In the *Scientific American* article "Reducing Crime: Rehabilitation is Making a Comeback," R. Doyle examined rehabilitation of felons. One aspect of the article discussed recidivism of juvenile prisoners between 14 and 17 years old, indicating that 82% of those released in 1994 were rearrested within 3 years. Suppose that, today, six newly released juvenile prisoners between 14 and 17 years old are selected at random.

**a.** Assuming that the recidivism rate is the same today as it was in 1994, determine the probability distribution for the number, $Y$, who are rearrested within 3 years.

**b.** Determine and interpret the mean of $Y$.

**c.** If, in fact, exactly two of the six newly released juvenile prisoners are rearrested within 3 years, would you be inclined to conclude that the recidivism rate today has decreased from the 82% rate in 1994? Explain your reasoning. *Hint:* First consider the probability $P(Y \le 2)$.

**d.** If, in fact, exactly four of the six newly released juvenile prisoners are rearrested within 3 years, would you be inclined to conclude that the recidivism rate today has decreased from the 82% rate in 1994? Explain your reasoning.

## Extending the Concepts and Skills

**5.129 Roulette.** A success, $s$, in Bernoulli trials is often derived from a collection of outcomes. For example, an American roulette wheel consists of 38 numbers, of which 18 are red, 18 are black, and 2 are green. When the roulette wheel is spun, the ball is equally likely to land on any one of the 38 numbers. If you are interested in which number the ball lands on, each play at the roulette wheel has 38 possible outcomes. Suppose, however, that you are betting on red. Then you are interested only in whether the ball lands on a red number. From this point of view, each play at the wheel has only two possible outcomes—either the ball lands on a red number or it doesn't. Hence successive bets on red constitute a sequence of Bernoulli trials with success probability $\frac{18}{38}$. In four plays at a roulette wheel, what is the probability that the ball lands on red

**a.** exactly twice?     **b.** at least once?

**5.130 Lotto.** A previous Arizona state lottery, called *Lotto*, is played as follows: The player selects six numbers from the numbers 1–42 and buys a ticket for $1. There are six winning numbers, which are selected at random from the numbers 1–42. To win a prize, a *Lotto* ticket must contain three or more of the winning numbers. A probability distribution for the number of winning numbers for a single ticket is shown in the following table.

| Number of winning numbers | Probability |
|:---:|:---:|
| 0 | 0.3713060 |
| 1 | 0.4311941 |
| 2 | 0.1684352 |
| 3 | 0.0272219 |
| 4 | 0.0018014 |
| 5 | 0.0000412 |
| 6 | 0.0000002 |

**a.** If you buy one *Lotto* ticket, determine the probability that you win a prize. Round your answer to three decimal places.

**b.** If you buy one *Lotto* ticket per week for a year, determine the probability that you win a prize at least once in the 52 tries.

**5.131 Sickle Cell Anemia.** Sickle cell anemia is an inherited blood disease that occurs primarily in blacks. In the United States, about 15 of every 10,000 black children have sickle cell anemia. The red blood cells of an affected person are abnormal; the result is severe chronic anemia (inability to carry the required amount of oxygen), which causes headaches, shortness of breath, jaundice, increased risk of pneumococcal pneumonia and gallstones, and other severe problems. Sickle cell anemia occurs in children who inherit an abnormal type of hemoglobin, called hemoglobin S, from both parents. If hemoglobin S is inherited from only one parent, the person is said to have sickle cell trait and is generally free from symptoms. There is a 50% chance that a person who has sickle cell trait will pass hemoglobin S to an offspring.

**a.** Obtain the probability that a child of two people who have sickle cell trait will have sickle cell anemia.

**b.** If two people who have sickle cell trait have five children, determine the probability that at least one of the children will have sickle cell anemia.

**c.** If two people who have sickle cell trait have five children, find the probability distribution of the number of those children who will have sickle cell anemia.

**d.** Construct a probability histogram for the probability distribution in part (c).

**e.** If two people who have sickle cell trait have five children, how many can they expect will have sickle cell anemia?

**5.132 Tire Mileage.** A sales representative for a tire manufacturer claims that the company's steel-belted radials last at least 35,000 miles. A tire dealer decides to check that claim by testing eight of the tires. If 75% or more of the eight tires he tests last at least 35,000 miles, he will purchase tires from the sales representative. If, in fact, 90% of the steel-belted radials produced by the manufacturer last at least 35,000 miles, what is the probability that the tire dealer will purchase tires from the sales representative?

**5.133 Restaurant Reservations.** From past experience, the owner of a restaurant knows that, on average, 4% of the parties that make reservations never show. How many reservations can the owner accept and still be at least 80% sure that all parties that make a reservation will show?

**5.134 Sampling and the Binomial Distribution.** Refer to the discussion on the binomial approximation to the hypergeometric distribution that begins on page 252.

**a.** If sampling is with replacement, explain why the trials are independent and the success probability remains the same from trial to trial—always the proportion of the population that has the specified attribute.

**b.** If sampling is without replacement, explain why the trials are not independent and the success probability varies from trial to trial.

**5.135 Sampling and the Binomial Distribution.** Following is a gender frequency distribution for students in Professor Weiss's introductory statistics class.

| Sex | Frequency |
|--------|-----------|
| Male | 17 |
| Female | 23 |

Two students are selected at random. Find the probability that both students are male if the selection is done

**a.** with replacement.

**b.** without replacement.

**c.** Compare the answers obtained in parts (a) and (b).

Suppose that Professor Weiss's class had 10 times the students, but in the same proportions, that is, 170 males and 230 females.

**d.** Repeat parts (a)–(c), using this hypothetical distribution of students.

**e.** In which case is there less difference between sampling without and with replacement? Explain why this is so.

**5.136 The Hypergeometric Distribution.** In this exercise, we discuss the *hypergeometric distribution* in more detail. When sampling is done without replacement from a finite population, the hypergeometric distribution is the exact probability distribution for the number of members sampled that have a specified attribute. The hypergeometric probability formula is

$$P(X = x) = \frac{\binom{Np}{x}\binom{N(1-p)}{n-x}}{\binom{N}{n}},$$

where $X$ denotes the number of members sampled that have the specified attribute, $N$ is the population size, $n$ is the sample size, and $p$ is the population proportion.

To illustrate, suppose that a customer purchases 4 fuses from a shipment of 250, of which 94% are not defective. Let a success correspond to a fuse that is not defective.

**a.** Determine $N$, $n$, and $p$.

**b.** Use the hypergeometric probability formula to find the probability distribution of the number of nondefective fuses the customer gets.

Key Fact 5.6 shows that a hypergeometric distribution can be approximated by a binomial distribution provided the sample size does not exceed 5% of the population size. In particular, you can use the binomial probability formula

$$P(X = x) = \binom{n}{x}p^x(1-p)^{n-x},$$

with $n = 4$ and $p = 0.94$, to approximate the probability distribution of the number of nondefective fuses that the customer gets.

**c.** Obtain the binomial distribution with parameters $n = 4$ and $p = 0.94$.

**d.** Compare the hypergeometric distribution that you obtained in part (b) with the binomial distribution that you obtained in part (c).

**5.137 The Geometric Distribution.** In this exercise, we discuss the *geometric distribution,* the probability distribution for the number of trials until the first success in Bernoulli trials. The geometric probability formula is

$$P(X = x) = p(1-p)^{x-1},$$

where $X$ denotes the number of trials until the first success and $p$ the success probability. Using the geometric probability formula and Definition 5.9 on page 238, we can show that the mean of the random variable $X$ is $1/p$.

To illustrate, again consider the Arizona state lottery, *Lotto,* as described in Exercise 5.130. Suppose that you buy one *Lotto* ticket per week. Let $X$ denote the number of weeks until you win a prize.

**a.** Find and interpret the probability formula for the random variable $X$. (*Note:* The appropriate success probability was obtained in Exercise 5.130(a).)

**b.** Compute the probability that the number of weeks until you win a prize is exactly 3; at most 3; at least 3.

**c.** On average, how long will it be until you win a prize?

**5.138 The Poisson Distribution.** Another important discrete probability distribution is the *Poisson distribution,* named in honor of the French mathematician and physicist Simeon Poisson (1781–1840). This probability distribution is often used to model the frequency with which a specified event occurs during a particular period of time. The Poisson probability formula is

$$P(X = x) = e^{-\lambda}\frac{\lambda^x}{x!},$$

where $X$ is the number of times the event occurs and $\lambda$ is a parameter equal to the mean of $X$. The number $e$ is the base of natural logarithms and is approximately equal to 2.7183.

To illustrate, consider the following problem: Desert Samaritan Hospital, located in Mesa, Arizona, keeps records of emergency room traffic. Those records reveal that the number of patients who arrive between 6:00 P.M. and 7:00 P.M. has a Poisson distribution with parameter $\lambda = 6.9$. Determine the probability that, on a given day, the number of patients who arrive at the emergency room between 6:00 P.M. and 7:00 P.M. will be

**a.** exactly four.

**b.** at most two.

**c.** between four and 10, inclusive.

## Chapter in Review

### You Should be Able to

1. use and understand the formulas in this chapter.

2. compute probabilities for experiments having equally likely outcomes.

3. interpret probabilities, using the frequentist interpretation of probability.

4. state and understand the basic properties of probability.

5. construct and interpret Venn diagrams.

6. find and describe (not E), (A & B), and (A or B).

7. determine whether two or more events are mutually exclusive.

8. understand and use probability notation.

9. state and apply the special addition rule.

10. state and apply the complementation rule.

11. state and apply the general addition rule.

*12. determine the probability distribution of a discrete random variable.

*13. construct a probability histogram.

*14. describe events using random-variable notation, when appropriate.

*15. use the frequentist interpretation of probability to understand the meaning of the probability distribution of a random variable.

*16. find and interpret the mean and standard deviation of a discrete random variable.

*17. compute factorials and binomial coefficients.

*18. define and apply the concept of Bernoulli trials.

*19. assign probabilities to the outcomes in a sequence of Bernoulli trials.

*20. obtain binomial probabilities.

*21. compute the mean and standard deviation of a binomial random variable.

### Key Terms

(A & B), 214
(A or B), 214
at random, 205
Bernoulli trials,* 245
binomial coefficients,* 244
binomial distribution,* 245, 249
binomial probability formula,* 249
binomial random variable,* 249
certain event, 208
complement, 214
complementation rule, 223
cumulative probability,* 251
discrete random variable,* 230
equal-likelihood model, 208
event, 205, 213
expectation,* 238

expected value,* 238
experiment, 205
factorials,* 244
failure,* 245
$f/N$ rule, 205
frequentist interpretation of
   probability, 207
general addition rule, 224
hypergeometric distribution,* 253
impossible event, 208
law of averages,* 239
law of large numbers,* 239
mean of a discrete random
   variable,* 238
mutually exclusive events, 217
(not E), 214

P(E), 221
probability distribution,* 230
probability histogram,* 230
probability model, 208
probability theory, 202
random variable,* 229
sample space, 213
special addition rule, 222
standard deviation of a discrete
   random variable,* 240
success,* 245
success probability,* 245
trial,* 243
variance of a discrete random
   variable,* 240
Venn diagrams, 214

## Review Problems

### Understanding the Concepts and Skills

1. Why is probability theory important to statistics?

2. Regarding the equal-likelihood model,
a. what is it?
b. how are probabilities computed?

3. What meaning is given to the probability of an event by the frequentist interpretation of probability?

4. Decide which of these numbers could not possibly be probabilities. Explain your answers.
a. 0.047　　　b. −0.047　　　c. 3.5　　　d. 1/3.5

**5.** Identify a commonly used graphical technique for portraying events and relationships among events.

**6.** What does it mean for two or more events to be mutually exclusive?

**7.** Suppose that $E$ is an event. Use probability notation to represent
**a.** the probability that event $E$ occurs.
**b.** the probability that event $E$ occurs is 0.436.

**8.** Answer true or false to each statement and explain your answers.
**a.** For any two events, the probability that one or the other of the events occurs equals the sum of the two individual probabilities.
**b.** For any event, the probability that it occurs equals 1 minus the probability that it does not occur.

**9.** Identify one reason why the complementation rule is useful.

**10. Adjusted Gross Incomes.** The U.S. Internal Revenue Service compiles data on income tax returns and summarizes its findings in *Statistics of Income*. The first two columns of Table 5.21 show a frequency distribution (number of returns) for adjusted gross income (AGI) from federal individual income tax returns, where K = thousand.

**TABLE 5.21**
Adjusted gross incomes

| Adjusted gross income | Frequency (1000s) | Event | Probability |
|---|---|---|---|
| Under $10K | 26,015 | A | |
| $10K < 20K | 23,296 | B | |
| $20K < 30K | 18,373 | C | |
| $30K < 40K | 13,957 | D | |
| $40K < 50K | 10,452 | E | |
| $50K < 100K | 26,915 | F | |
| $100K & over | 11,415 | G | |
| | 130,423 | | |

A federal individual income tax return is selected at random.
**a.** Determine $P(A)$, the probability that the return selected shows an AGI under $10K.
**b.** Find the probability that the return selected shows an AGI between $30K and $100K (i.e., at least $30K but less than $100K).
**c.** Compute the probability of each of the seven events in the third column of Table 5.21, and record those probabilities in the fourth column.

**11. Adjusted Gross Incomes.** Refer to Problem 10. A federal individual income tax return is selected at random. Let

$H$ = event the return shows an AGI between $20K and $100K,

$I$ = event the return shows an AGI of less than $50K,

$J$ = event the return shows an AGI of less than $100K, and

$K$ = event the return shows an AGI of at least $50K.

Describe each of the following events in words and determine the number of outcomes (returns) that constitute each event.
**a.** (not $J$)  **b.** ($H$ & $I$)
**c.** ($H$ or $K$)  **d.** ($H$ & $K$)

**12. Adjusted Gross Incomes.** For the following groups of events from Problem 11, determine which are mutually exclusive.
**a.** $H$ and $I$  **b.** $I$ and $K$
**c.** $H$ and (not $J$)  **d.** $H$, (not $J$), and $K$

**13. Adjusted Gross Incomes.** Refer to Problems 10 and 11.
**a.** Use the second column of Table 5.21 and the $f/N$ rule to compute the probability of each of the events $H$, $I$, $J$, and $K$.
**b.** Express each of the events $H$, $I$, $J$, and $K$ in terms of the mutually exclusive events displayed in the third column of Table 5.21.
**c.** Compute the probability of each of the events $H$, $I$, $J$, and $K$, using your answers from part (b), the special addition rule, and the fourth column of Table 5.21, which you completed in Problem 10(c).

**14. Adjusted Gross Incomes.** Consider the events (not $J$), ($H$ & $I$), ($H$ or $K$), and ($H$ & $K$) discussed in Problem 11.
**a.** Find the probability of each of those four events, using the $f/N$ rule and your answers from Problem 11.
**b.** Compute $P(J)$, using the complementation rule and your answer for $P$(not $J$) from part (a).
**c.** In Problem 13(a), you found that $P(H) = 0.534$ and $P(K) = 0.294$; and, in part (a) of this problem, you found that $P(H \& K) = 0.206$. Using those probabilities and the general addition rule, find $P(H$ or $K)$.
**d.** Compare the answers that you obtained for $P(H$ or $K)$ in parts (a) and (c).

*15. Fill in the blanks.
  a. A _____ is a quantitative variable whose value depends on chance.
  b. A discrete random variable is a random variable whose possible values _____.

*16. What does the probability distribution of a discrete random variable tell you?

*17. How do you graphically portray the probability distribution of a discrete random variable?

*18. If you sum the probabilities of the possible values of a discrete random variable, the result always equals _____.

*19. A random variable X equals 2 with probability 0.386.
  a. Use probability notation to express that fact.
  b. If you make repeated independent observations of the random variable X, in approximately what percentage of those observations will you observe the value 2?
  c. Roughly how many times would you expect to observe the value 2 in 50 observations? 500 observations?

*20. A random variable X has mean 3.6. If you make a large number of repeated independent observations of the random variable X, the average value of those observations will be approximately _____.

*21. Two random variables, X and Y, have standard deviations 2.4 and 3.6, respectively. Which one is more likely to take a value close to its mean? Explain your answer.

*22. List the three requirements for repeated trials of an experiment to constitute Bernoulli trials.

*23. What is the relationship between Bernoulli trials and the binomial distribution?

*24. In 10 Bernoulli trials, how many outcomes contain exactly three successes?

*25. Explain how the special formulas for the mean and standard deviation of a binomial random variable are derived.

*26. Suppose that a simple random sample of size $n$ is taken from a finite population in which the proportion of members having a specified attribute is $p$. Let X be the number of members sampled that have the specified attribute.
  a. If the sampling is done with replacement, identify the probability distribution of X.
  b. If the sampling is done without replacement, identify the probability distribution of X.
  c. Under what conditions is it acceptable to approximate the probability distribution in part (b) by the probability distribution in part (a)? Why is it acceptable?

*27. **ASU-Main Enrollment.** According to the *Arizona State University Enrollment Summary*, a frequency distribution for the number of undergraduate students attending the main campus one year, by class level, is as shown in the following table. Here, 1 = freshman, 2 = sophomore, 3 = junior, and 4 = senior.

| Class level | 1 | 2 | 3 | 4 |
|---|---|---|---|---|
| No. of students | 8,722 | 8,538 | 10,028 | 12,361 |

Let X denote the class level of a randomly selected ASU undergraduate.
  a. What are the possible values of the random variable X?
  b. Use random-variable notation to represent the event that the student selected is a junior (class-level 3).
  c. Determine $P(X = 3)$ and interpret your answer in terms of percentages.
  d. Determine the probability distribution of the random variable X.
  e. Construct a probability histogram for the random variable X.

*28. **Busy Phone Lines.** An accounting office has six incoming telephone lines. The probability distribution of the number of busy lines, Y, is as follows. Use random-variable notation to express each of the following events. The number of busy lines is
  a. exactly four.          b. at least four.
  c. between two and four, inclusive.
  d. at least one.

| $y$ | $P(Y = y)$ |
|---|---|
| 0 | 0.052 |
| 1 | 0.154 |
| 2 | 0.232 |
| 3 | 0.240 |
| 4 | 0.174 |
| 5 | 0.105 |
| 6 | 0.043 |

Apply the special addition rule and the probability distribution to determine
  e. $P(Y = 4)$.          f. $P(Y \geq 4)$.
  g. $P(2 \leq Y \leq 4)$.          h. $P(Y \geq 1)$.

*29. **Busy Phone Lines.** Refer to the probability distribution displayed in the table in Problem 28.
  a. Find the mean of the random variable Y.
  b. On average, how many lines are busy?
  c. Compute the standard deviation of Y.
  d. Construct a probability histogram for Y; locate the mean; and show one, two, and three standard deviation intervals.

*30. Determine 0!, 3!, 4!, and 7!.

*31. Determine the value of each binomial coefficient.
  a. $\binom{8}{3}$   b. $\binom{8}{5}$   c. $\binom{6}{6}$   d. $\binom{10}{2}$   e. $\binom{40}{4}$   f. $\binom{100}{0}$

**\*32. Craps.** The game of craps is played by rolling two balanced dice. A first roll of a sum of 7 or 11 wins; and a first roll of a sum of 2, 3, or 12 loses. To win with any other first sum, that sum must be repeated before a sum of 7 is thrown. It can be shown that the probability is 0.493 that a player wins a game of craps. Suppose we consider a win by a player to be a success, $s$.

a. Identify the success probability, $p$.
b. Construct a table showing the possible win–lose results and their probabilities for three games of craps. Round each probability to three decimal places.
c. Draw a tree diagram for part (b).
d. List the outcomes in which the player wins exactly two out of three times.
e. Determine the probability of each of the outcomes in part (d). Explain why those probabilities are equal.
f. Find the probability that the player wins exactly two out of three times.
g. Without using the binomial probability formula, obtain the probability distribution of the random variable $Y$, the number of times out of three that the player wins.
h. Identify the probability distribution in part (g).

**\*33. Booming Pet Business.** The pet industry has undergone a surge in recent years, surpassing even the $20 billion a year toy industry. According to *U.S. News & World Report*, 60% of U.S. households live with one or more pets. If four U.S. households are selected at random without replacement, determine the (approximate) probability that the number living with one or more pets will be

a. exactly three.    b. at least three.    c. at most three.
d. Find the probability distribution of the random variable $X$, the number of U.S. households in a random sample of four that live with one or more pets.
e. Without referring to the probability distribution obtained in part (d) or constructing a probability histogram, decide whether the probability distribution is right skewed, symmetric, or left skewed. Explain your answer.
f. Draw a probability histogram for $X$.
g. Strictly speaking, why is the probability distribution that you obtained in part (d) only approximately correct? What is the exact distribution called?
h. Determine and interpret the mean of the random variable $X$.
i. Determine the standard deviation of $X$.

**\*34.** Following are two probability histograms of binomial distributions. For each, specify whether the success probability is less than, equal to, or greater than 0.5.

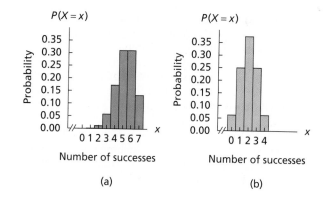

(a)                                    (b)

---

# Focusing on Data Analysis    UWEC Undergraduates

Recall from Chapter 1 (see page 34) that the Focus database and Focus sample contain information on the undergraduate students at the University of Wisconsin - Eau Claire (UWEC). Now would be a good time for you to review the discussion about these data sets.

The following problems are designed for use with the entire Focus database (Focus). If your statistical software package won't accommodate the entire Focus database, use the Focus sample (FocusSample) instead. Of course, in that case, your results will apply to the 200 UWEC undergrad-

uate students in the Focus sample rather than to all UWEC undergraduate students.

a. Obtain a relative-frequency distribution for the classification (class-level) data.
b. Using your answer from part (a), determine the probability that a randomly selected UWEC undergraduate student is a freshman.
c. Consider the experiment of selecting a UWEC undergraduate student at random and observing the classi-

fication of the student obtained. Simulate that experiment 1000 times. (*Hint:* The simulation is equivalent to taking a random sample of size 1000 with replacement.)

d. Referring to the simulation performed in part (c), in approximately what percentage of the 1000 experiments would you expect a freshman to be selected? Compare that percentage with the actual percentage of the 1000 experiments in which a freshman was selected.

e. Repeat parts (b)–(d) for sophomores; juniors; seniors.

*f. Let X denote the age of a randomly selected undergraduate student at UWEC. Obtain the probability distribution of the random variable X. Display the probabilities to six decimal places.

*g. Obtain a probability histogram or similar graphic for the random variable X.

*h. Determine the mean and standard deviation of the random variable X.

*i. Simulate 100 observations of the random variable X.

*j. Roughly, what would you expect the average value of the 100 observations obtained in part (i) to be? Explain your reasoning.

*k. In actuality, what is the average value of the 100 observations obtained in part (i)? Compare this value to the value you expected, as answered in part (j).

*l. Consider the experiment of randomly selecting 10 UWEC undergraduates with replacement and observing the number of those selected who are 21 years old. Simulate that experiment 1000 times. (*Hint:* Simulate an appropriate binomial distribution.)

*m. Referring to the simulation in part (l), in approximately what percentage of the 1000 experiments would you expect exactly 3 of the 10 students selected to be 21 years old? Compare that percentage to the actual percentage of the 1000 experiments in which exactly 3 of the 10 students selected are 21 years old.

---

## Case Study Discussion    The Powerball

At the beginning of this chapter on page 203, we discussed the Powerball lottery and described some of its rules. Recall that, for a single ticket, a player first selects five numbers from the numbers 1–55 and then chooses a PowerBall number, which can be any number between 1 and 42. A ticket costs $1. In the drawing, five white balls are drawn randomly from 55 white balls numbered 1–55, and one red PowerBall is drawn randomly from 42 red balls numbered 1–42.

To win the jackpot, a ticket must match all the balls drawn. Prizes are also given for matching some but not all the balls drawn. Table 5.22 displays the number of matches, the prizes given, and the probabilities of winning.

Here are some things to note about Table 5.22.

- The prize amount for the jackpot depends on how recently it has been won, how many people win it, and choice of jackpot payment.
- Each probability in the fifth column is given to nine decimal places and can be obtained by using the $f/N$ rule and special counting techniques, as discussed in Section 4.8 of *Introductory Statistics, 8/e*, by Neil A. Weiss (Boston: Addison-Wesley, 2008).

Let E be an event having probability p. In independent repetitions of the experiment, it takes, on average, $1/p$ times until event E occurs. We apply this fact to the Powerball.

Suppose that you were to purchase one Powerball ticket per week. How long should you expect to wait before winning the jackpot? The fifth column of Table 5.22 shows that, for the jackpot, $p = 0.000000007$, and therefore we have $1/0.000000007 = 146{,}107{,}962$. So, if you purchased one Powerball ticket per week, you should expect to wait approximately 146,107,962 weeks, or roughly 2.8 million years, before winning the jackpot.

a. If you purchase one ticket, what is the probability that you win a prize?

b. If you purchase one ticket, what is the probability that you don't win a prize?

c. If you were to buy one ticket per week, approximately how long should you expect to wait before getting a ticket with exactly three winning numbers and no PowerBall?

d. If you were to buy one ticket per week, approximately how long should you expect to wait before winning a prize?

**TABLE 5.22**
Powerball winning combinations, prizes, and probabilities

| White matches | PowerBall match | Prize | Ways | Probability | 1 in |
|---|---|---|---|---|---|
| 5 | yes | Jackpot | 1 | 0.000000007 | 146,107,962 |
| 5 | no | $200,000 | 41 | 0.000000281 | 3,563,609 |
| 4 | yes | $10,000 | 250 | 0.000001711 | 584,432 |
| 4 | no | $100 | 10250 | 0.000070154 | 14,254 |
| 3 | yes | $100 | 12250 | 0.000083842 | 11,927 |
| 3 | no | $7 | 502250 | 0.003437527 | 291 |
| 2 | yes | $7 | 196000 | 0.001341474 | 745 |
| 1 | yes | $4 | 1151500 | 0.007881158 | 127 |
| 0 | yes | $3 | 2118760 | 0.014501332 | 69 |

## Biography   ANDREI KOLMOGOROV: Father of Modern Probability Theory

**Andrei Nikolaevich Kolmogorov** was born on April 25, 1903, in Tambov, Russia. At the age of 17, Kolmogorov entered Moscow State University, from which he graduated in 1925. His contributions to the world of mathematics, many of which appear in his numerous articles and books, encompass a formidable range of subjects.

Kolmogorov revolutionized probability theory with the introduction of the modern axiomatic approach to probability and by proving many of the fundamental theorems that are a consequence of that approach. He also developed two systems of partial differential equations, which bear his name. Those systems extended the development of probability theory and allowed its broader application to the fields of physics, chemistry, biology, and civil engineering.

In 1938, Kolmogorov published an extensive article entitled "Mathematics," which appeared in the first edition of the *Bolshaya Sovyetskaya Entsiklopediya* (Great Soviet Encyclopedia). In this article he discussed the development of mathematics from ancient to modern times and interpreted it in terms of dialectical materialism, the philosophy originated by Karl Marx and Friedrich Engels.

Kolmogorov became a member of the faculty at Moscow State University in 1925, at the age of 22. In 1931, he was promoted to professor; in 1933, he was appointed a director of the Institute of Mathematics of the university; and in 1937, he became Head of the University.

In addition to his work in higher mathematics, Kolmogorov was interested in the mathematical education of schoolchildren. He was chairman of the Commission for Mathematical Education under the Presidium of the Academy of Sciences of the U.S.S.R. During his tenure as chairman, he was instrumental in the development of a new mathematics training program that was introduced into Soviet schools.

Kolmogorov remained on the faculty at Moscow State University until his death in Moscow on October 20, 1987.

## StatCrunch in MyStatLab
### Analyzing Data Online

StatCrunch online statistical software offers an easy-to-use interface customized for this book. The StatCrunch feature for each chapter illustrates the use of the software to perform a statistical analysis discussed in the chapter. Exercises are provided to further apply StatCrunch to other statistical analyses examined in the chapter. Go to the WeissStats CD or to the Weiss Web site at www.aw-bc.com/weiss to access StatCrunch instructions and data sets. To access StatCrunch statistical software, go to the student content area of your Weiss MyStatLab course.

## Internet Projects
### Exploring Data Online

The Internet project for each chapter provides simulations, demonstrations, or activities that enhance the topics covered in the chapter. The project materials come from universities, individuals, governments, and companies from all over the world. To access the Internet projects on the Web, go to www.aw-bc.com/weiss. From this Web page, you can reach the Internet Projects Page, which we suggest that you bookmark for easy access in the future.

# 6 The Normal Distribution

## Chapter Objectives

In this chapter, we discuss the most important distribution in statistics—the *normal distribution.* As you will see, its importance lies in the fact that it appears again and again in both theory and practice.

A variable is said to be *normally distributed* or to have a *normal distribution* if its distribution has the shape of a normal curve, a special type of bell-shaped curve. In Section 6.1, we introduce normally distributed variables, show that percentages (or probabilities) for such a variable are equal to areas under its associated normal curve, and explain how all normal distributions can be converted to a single normal distribution—the *standard normal distribution.*

In Section 6.2, we demonstrate how to determine areas under the *standard normal curve,* the normal curve corresponding to a variable that has the standard normal distribution. Then, in Section 6.3, we describe an efficient procedure for finding percentages (or probabilities) for any normally distributed variable from areas under the standard normal curve.

We present a method for graphically assessing whether a variable is normally distributed—*the normal probability plot*—in Section 6.4.

# Foot Length and Shoe Size for Women

According to research done on foot length of women, the mean length of women's feet is 9.58 inches with a standard deviation of 0.51 inch. Moreover, the distribution of foot length for women is bell shaped and, in fact, has the *normal distribution* shown in the graph below. (You will study normal distributions in this chapter.)

This distribution is useful to shoe manufacturers, shoe stores, and related merchants because it permits them to make informed decisions about shoe production, inventory, and so forth. Along these lines, the following table provides a foot-length–to–shoe-size conversion, obtained from *Payless ShoeSource*.

At the end of this chapter, you will be asked to answer several questions about foot length and shoe size for women, based on the information provided here.

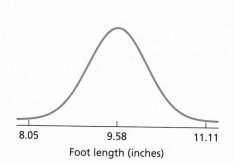

Foot length (inches)

| Length (inches) | Size (USA) | Length (inches) | Size (USA) |
|---|---|---|---|
| 8 | 3 | 10 | 9 |
| $8\frac{3}{16}$ | $3\frac{1}{2}$ | $10\frac{3}{16}$ | $9\frac{1}{2}$ |
| $8\frac{5}{16}$ | 4 | $10\frac{5}{16}$ | 10 |
| $8\frac{8}{16}$ | $4\frac{1}{2}$ | $10\frac{8}{16}$ | $10\frac{1}{2}$ |
| $8\frac{11}{16}$ | 5 | $10\frac{11}{16}$ | 11 |
| $8\frac{13}{16}$ | $5\frac{1}{2}$ | $10\frac{13}{16}$ | $11\frac{1}{2}$ |
| 9 | 6 | 11 | 12 |
| $9\frac{3}{16}$ | $6\frac{1}{2}$ | $11\frac{3}{16}$ | $12\frac{1}{2}$ |
| $9\frac{5}{16}$ | 7 | $11\frac{5}{16}$ | 13 |
| $9\frac{8}{16}$ | $7\frac{1}{2}$ | $11\frac{8}{16}$ | $13\frac{1}{2}$ |
| $9\frac{11}{16}$ | 8 | $11\frac{11}{16}$ | 14 |
| $9\frac{13}{16}$ | $8\frac{1}{2}$ | | |

## 6.1 Introducing Normally Distributed Variables

**FIGURE 6.1**
A normal curve

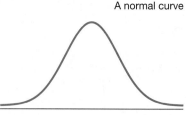

In everyday life, people deal with and use a wide variety of variables. Some of these variables—such as aptitude-test scores, heights of women, and wheat yield—share an important characteristic: Their distributions have roughly the shape of a **normal curve,** that is, a special type of bell-shaped curve like the one shown in Fig. 6.1.

Why the word "normal"? Because, in the last half of the nineteenth century, researchers discovered that it is quite usual, or "normal," for a variable to have a distribution shaped like that in Fig. 6.1. So, following the lead of noted British statistician Karl Pearson, such a distribution began to be referred to as a *normal distribution.*

**Definition 6.1**

### Normally Distributed Variable

A variable is said to be a **normally distributed variable** or to have a **normal distribution** if its distribution has the shape of a normal curve.

Here is some important terminology associated with normal distributions.

- If a variable of a population is normally distributed and is the only variable under consideration, common practice is to say that the **population is normally distributed** or that it is a **normally distributed population.**
- In practice, a distribution is unlikely to have exactly the shape of a normal curve. If a variable's distribution is shaped roughly like a normal curve, we say that the variable is an **approximately normally distributed variable** or that it has **approximately a normal distribution.**

A normal distribution (and hence a normal curve) is completely determined by the mean and standard deviation; that is, two normally distributed variables having the same mean and standard deviation must have the same distribution. We often identify a normal curve by stating the corresponding mean and standard deviation and calling those the **parameters** of the normal curve.[†]

A normal distribution is symmetric about and centered at the mean of the variable, and its spread depends on the standard deviation of the variable—the larger the standard deviation, the flatter and more spread out is the distribution. Figure 6.2 displays three normal distributions.

**FIGURE 6.2**
Three normal distributions

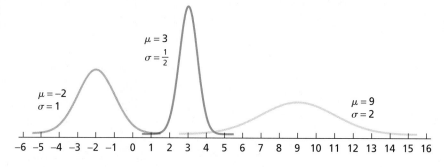

$\mu = 3$
$\sigma = \frac{1}{2}$

$\mu = -2$
$\sigma = 1$

$\mu = 9$
$\sigma = 2$

-6 -5 -4 -3 -2 -1 0 1 2 3 4 5 6 7 8 9 10 11 12 13 14 15 16

[†]The equation of the normal curve with parameters $\mu$ and $\sigma$ is $y = e^{-(x-\mu)^2/2\sigma^2}/(\sqrt{2\pi}\sigma)$, where $e \approx 2.718$ and $\pi \approx 3.142$.

When applied to a variable, the three-standard-deviations rule (Key Fact 3.2 on page 114) states that almost all the possible observations of the variable lie within three standard deviations to either side of the mean. This rule is illustrated by the three normal distributions in Fig. 6.2: Each normal curve is close to the horizontal axis outside the range of three standard deviations to either side of the mean.

For instance, the third normal distribution in Fig. 6.2 has mean $\mu = 9$ and standard deviation $\sigma = 2$. Three standard deviations to the left of the mean is

$$\mu - 3\sigma = 9 - 3 \cdot 2 = 3,$$

and three standard deviations to the right of the mean is

$$\mu + 3\sigma = 9 + 3 \cdot 2 = 15.$$

As shown in Fig. 6.2, the corresponding normal curve is close to the horizontal axis outside the range from 3 to 15.

In summary, the normal curve associated with a normal distribution is

- bell shaped,
- centered at $\mu$, and
- close to the horizontal axis outside the range from $\mu - 3\sigma$ to $\mu + 3\sigma$,

as depicted in Figs. 6.2 and 6.3. This information helps us sketch a normal distribution.

**FIGURE 6.3**
Graph of generic normal distribution

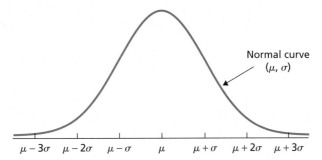

Example 6.1 illustrates a normally distributed variable and discusses some additional properties of such variables.

## Example 6.1 | A Normally Distributed Variable

*Heights of Female College Students* A midwestern college has an enrollment of 3264 female students. Records show that the mean height of these students is 64.4 inches and that the standard deviation is 2.4 inches. Here the variable is height, and the population consists of the 3264 female students attending the college. Frequency and relative-frequency distributions for these heights appear in Table 6.1 at the top of the following page. The table shows, for instance, that 7.35% (0.0735) of the students are between 67 and 68 inches tall.

**a.** Show that the variable "height" is approximately normally distributed for this population.

**TABLE 6.1**
Frequency and relative-frequency distributions for heights

| Height (inches) | Freq. $f$ | Relative freq. |
|---|---|---|
| 56 < 57 | 3 | 0.0009 |
| 57 < 58 | 6 | 0.0018 |
| 58 < 59 | 26 | 0.0080 |
| 59 < 60 | 74 | 0.0227 |
| 60 < 61 | 147 | 0.0450 |
| 61 < 62 | 247 | 0.0757 |
| 62 < 63 | 382 | 0.1170 |
| 63 < 64 | 483 | 0.1480 |
| 64 < 65 | 559 | 0.1713 |
| 65 < 66 | 514 | 0.1575 |
| 66 < 67 | 359 | 0.1100 |
| 67 < 68 | 240 | 0.0735 |
| 68 < 69 | 122 | 0.0374 |
| 69 < 70 | 65 | 0.0199 |
| 70 < 71 | 24 | 0.0074 |
| 71 < 72 | 7 | 0.0021 |
| 72 < 73 | 5 | 0.0015 |
| 73 < 74 | 1 | 0.0003 |
|  | 3264 | 1.0000 |

b. Identify the normal curve associated with the variable "height" for this population.

c. Discuss the relationship between the percentage of female students whose heights lie within a specified range and the corresponding area under the associated normal curve.

## Solution

a. Figure 6.4 shows a relative-frequency histogram for the heights of the female students. It shows that the distribution of heights has roughly the shape of a normal curve and, consequently, that the variable "height" is approximately normally distributed for this population.

b. The associated normal curve is the one whose parameters are the same as the mean and standard deviation of the variable, which are 64.4 and 2.4, respectively. Thus the required normal curve has parameters $\mu = 64.4$ and $\sigma = 2.4$. It is superimposed on the histogram in Fig. 6.4.

c. Consider, for instance, the students who are between 67 and 68 inches tall. According to Table 6.1, their exact percentage is 7.35%, or 0.0735. Note that 0.0735 also equals the area of the cross-hatched bar in Fig. 6.4 because the bar has height 0.0735 and width 1. Now look at the area under the curve between 67 and 68, shaded in Fig. 6.4. This area approximates the area of the cross-hatched bar. Thus we can approximate the percentage of students between 67 and 68 inches tall by the area under the normal curve between 67 and 68. This result holds in general.

**FIGURE 6.4**
Relative-frequency histogram for heights with superimposed normal curve

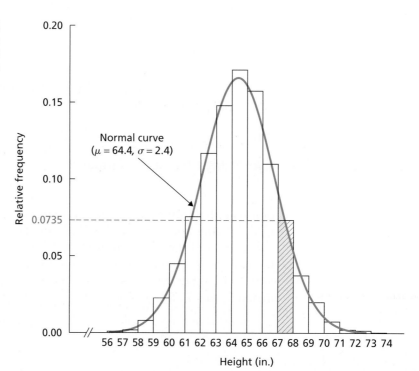

Exercise 6.13
on page 274

**Interpretation**   The percentage of female students whose heights lie within any specified range can be approximated by the corresponding area under the normal curve associated with the variable "height" for this population of female students.

• • •

**Key Fact 6.1**

## Normally Distributed Variables and Normal-Curve Areas

For a normally distributed variable, the percentage of all possible observations that lie within any specified range equals the corresponding area under its associated normal curve, expressed as a percentage. This result holds approximately for a variable that is approximately normally distributed.

**Note:** For brevity, we often paraphrase the content of Key Fact 6.1 with the statement "percentages for a normally distributed variable are equal to areas under its associated normal curve."

### Standardizing a Normally Distributed Variable

Now the question is: How do we find areas under a normal curve? Conceptually, we need a table of areas for each normal curve. This, of course, is impossible because there are infinitely many different normal curves—one for each choice of $\mu$ and $\sigma$. The way out of this difficulty is standardizing, which transforms every normal distribution into one particular normal distribution, the *standard normal distribution.*

**Definition 6.2**

## Standard Normal Distribution; Standard Normal Curve

**FIGURE 6.5**
Standard normal distribution

A normally distributed variable having mean 0 and standard deviation 1 is said to have the **standard normal distribution.** Its associated normal curve is called the **standard normal curve,** which is shown in Fig. 6.5.

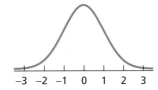

$$-3 \quad -2 \quad -1 \quad 0 \quad 1 \quad 2 \quad 3$$

Recall from Chapter 3 (page 140) that we standardize a variable $x$ by subtracting its mean and then dividing by its standard deviation. The resulting variable, $z = (x - \mu)/\sigma$, is called the standardized version of $x$ or the standardized variable corresponding to $x$.

The standardized version of any variable has mean 0 and standard deviation 1. A normally distributed variable furthermore has a normally distributed standardized version.

**Key Fact 6.2**

## Standardized Normally Distributed Variable

**What Does It Mean?**

Subtracting from a normally distributed variable its mean and then dividing by its standard deviation results in a variable with the standard normal distribution.

The standardized version of a normally distributed variable $x$,

$$z = \frac{x - \mu}{\sigma},$$

has the standard normal distribution.

We can interpret Key Fact 6.2 in several ways. Theoretically, it says that standardizing converts all normal distributions to the standard normal distribution, as depicted in Fig. 6.6.

**FIGURE 6.6**

Standardizing normal distributions

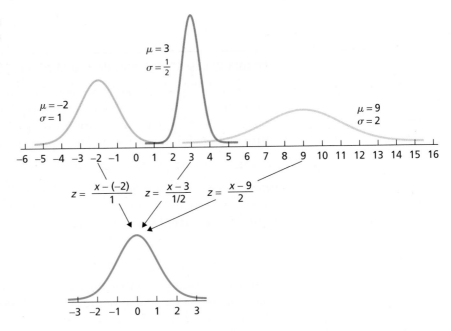

But we need a more practical interpretation of Key Fact 6.2. Let $x$ be a normally distributed variable with mean $\mu$ and standard deviation $\sigma$, and let $a$ and $b$ be real numbers with $a < b$. The percentage of all possible observations of $x$ that lie between $a$ and $b$ is the same as the percentage of all possible observations of $z$ that lie between $(a - \mu)/\sigma$ and $(b - \mu)/\sigma$. And, in light of Key Fact 6.2, this latter percentage equals the area under the standard normal curve between $(a - \mu)/\sigma$ and $(b - \mu)/\sigma$. We summarize these ideas graphically in Fig. 6.7.

**FIGURE 6.7**

Finding percentages for a normally distributed variable from areas under the standard normal curve

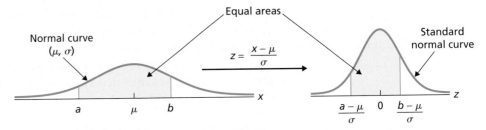

Consequently, for a normally distributed variable, we can find the percentage of all possible observations that lie within any specified range by

1. expressing the range in terms of $z$-scores, and
2. determining the corresponding area under the standard normal curve.

You already know how to convert to $z$-scores. Therefore you need only learn how to find areas under the standard normal curve, which we demonstrate in Section 6.2.

### Simulating a Normal Distribution

For understanding and for research, simulating a variable is often useful. Doing so involves use of a computer or statistical calculator to generate observations of the variable.

When we simulate a normally distributed variable, a histogram of the observations will have roughly the same shape as that of the normal curve associated with the variable. The shape of the histogram will tend to look more like that of the normal curve when the number of observations is large. We illustrate the simulation of a normally distributed variable in the next example.

**Example 6.2** | **Simulating a Normally Distributed Variable**

**OUTPUT 6.1**
Histogram of 1000 simulated human gestation periods with superimposed normal curve

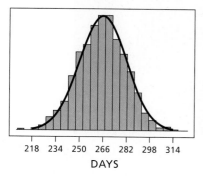

218   234   250   266   282   298   314
DAYS

*Gestation Periods of Humans*  Gestation periods of humans are normally distributed with a mean of 266 days and a standard deviation of 16 days. Simulate 1000 human gestation periods, obtain a histogram of the simulated data, and interpret the results.

**Solution**  Here the variable $x$ is gestation period and, for humans, it is normally distributed with mean $\mu = 266$ days and standard deviation $\sigma = 16$ days. We used a computer to simulate 1000 observations of the variable $x$ for humans. Output 6.1 shows a histogram for those observations. Note that we have superimposed the normal curve associated with the variable—namely, the one with parameters $\mu = 266$ and $\sigma = 16$.

The shape of the histogram in Output 6.1 is quite close to that of the normal curve, as we would expect because of the large number of simulated observations. If you do the simulation, your histogram should be similar to the one shown in Output 6.1.

• • •

## The Technology Center

Most statistical software packages and some graphing calculators have built-in procedures to simulate observations of normally distributed variables. In Example 6.2, we used Minitab, but Excel and the TI-83/84 Plus can also be used to conduct that simulation and obtain the histogram. Refer to the technology manuals for details.

▮ ▮ ▮

## Exercises 6.1

### Understanding the Concepts and Skills

**6.1** A variable is approximately normally distributed. If you draw a histogram of the distribution of the variable, roughly what shape will it have?

**6.2** Precisely what is meant by the statement that a population is normally distributed?

**6.3** Two normally distributed variables have the same means and the same standard deviations. What can you say about their distributions? Explain your answer.

**6.4** Which normal distribution has a wider spread: the one with mean 1 and standard deviation 2 or the one with mean 2 and standard deviation 1? Explain your answer.

**6.5** Consider two normal distributions, one with mean −4 and standard deviation 3 and the other with mean 6 and standard deviation 3. Answer true or false to each statement and explain your answers.
**a.** The two normal distributions have the same shape.
**b.** The two normal distributions are centered at the same place.

**6.6** Consider two normal distributions, one with mean −4 and standard deviation 3 and the other with mean −4 and standard deviation 6. Answer true or false to each statement and explain your answers.
**a.** The two normal distributions have the same shape.
**b.** The two normal distributions are centered at the same place.

**6.7** True or false: The mean of a normal distribution has no effect on its shape. Explain your answer.

**6.8** What are the parameters for a normal curve?

**6.9** Sketch the normal distribution with
**a.** $\mu = 3$ and $\sigma = 3$.
**b.** $\mu = 1$ and $\sigma = 3$.
**c.** $\mu = 3$ and $\sigma = 1$.

**6.10** Sketch the normal distribution with
**a.** $\mu = -2$ and $\sigma = 2$.
**b.** $\mu = -2$ and $\sigma = 1/2$.
**c.** $\mu = 0$ and $\sigma = 2$.

**6.11** For a normally distributed variable, what is the relationship between the percentage of all possible observations that lie between 2 and 3 and the area under the associated normal curve between 2 and 3? What if the variable is only approximately normally distributed?

**6.12** The area under a particular normal curve between 10 and 15 is 0.6874. A normally distributed variable has the same mean and standard deviation as the parameters for this normal curve. What percentage of all possible observations of the variable lie between 10 and 15? Explain your answer.

**6.13 Female College Students.** Refer to Example 6.1 on page 269.
**a.** Use the relative-frequency distribution in Table 6.1 to obtain the percentage of female students who are between 60 and 65 inches tall.
**b.** Use your answer from part (a) to estimate the area under the normal curve having parameters $\mu = 64.4$ and $\sigma = 2.4$ that lies between 60 and 65. Why do you get only an estimate of the true area?

**6.14 Female College Students.** Refer to Example 6.1 on page 269.

**a.** The area under the standard normal curve with parameters $\mu = 64.4$ and $\sigma = 2.4$ that lies to the left of 61 is 0.0783. Use this information to estimate the percentage of female students who are shorter than 61 inches.
**b.** Use the relative-frequency distribution in Table 6.1 to obtain the exact percentage of female students who are shorter than 61 inches.
**c.** Compare your answers from parts (a) and (b).

**6.15 Giant Tarantulas.** One of the larger species of tarantulas is the *Grammostola mollicoma*, whose common name is the Brazilian Giant Tawny Red. A tarantula has two body parts. The anterior part of the body is covered above by a shell, or carapace. From a recent article by F. Costa and F. Perez–Miles titled, "Reproductive Biology of Uruguayan Theraphosids" (*The Journal of Arachnology*, Vol. 30, No. 3, pp. 571–587), we find that the carapace length of the adult male *G. mollicoma* is normally distributed with a mean of 18.14 mm and a standard deviation of 1.76 mm. Let $x$ denote carapace length for the adult male *G. mollicoma*.
**a.** Sketch the distribution of the variable $x$.
**b.** Obtain the standardized version, $z$, of $x$.
**c.** Identify and sketch the distribution of $z$.
**d.** The percentage of adult male *G. mollicoma* that have carapace lengths between 16 mm and 17 mm is equal to the area under the standard normal curve between _____ and _____.
**e.** The percentage of adult male *G. mollicoma* that have carapace lengths exceeding 19 mm is equal to the area under the standard normal curve that lies to the _____ of _____.

**6.16 Serum Cholesterol Levels.** According to the *National Health and Nutrition Examination Survey*, the serum (noncellular portion of blood) total cholesterol level of U.S. females 20 years old or older is normally distributed with a mean

of 206 mg/dL (milligrams per deciliter) and a standard deviation of 44.7 mg/dL. Let $x$ denote serum total cholesterol level for U.S. females 20 years old or older.
a. Sketch the distribution of the variable $x$.
b. Obtain the standardized version, $z$, of $x$.
c. Identify and sketch the distribution of $z$.
d. The percentage of U.S. females 20 years old or older who have a serum total cholesterol level between 150 mg/dL and 250 mg/dL is equal to the area under the standard normal curve between _____ and _____.
e. The percentage of U.S. females 20 years old or older who have a serum total cholesterol level below 220 mg/dL is equal to the area under the standard normal curve that lies to the _____ of _____.

**6.17 New York City 10 km Run.** As reported in *Runner's World* magazine, the times of the finishers in the New York City 10 km run are normally distributed with mean 61 minutes and standard deviation 9 minutes. Let $x$ denote finishing time for finishers in this race.
a. Sketch the distribution of the variable $x$.
b. Obtain the standardized version, $z$, of $x$.
c. Identify and sketch the distribution of $z$.
d. The percentage of finishers with times between 50 and 70 minutes is equal to the area under the standard normal curve between _____ and _____.
e. The percentage of finishers with times less than 75 minutes is equal to the area under the standard normal curve that lies to the _____ of _____.

**6.18 Green Sea Urchins.** From the paper "Effects of Chronic Nitrate Exposure on Gonad Growth in Green Sea Urchin *Strongylocentrotus Droebachiensis*" (*Aquaculture*, Vol. 242, No. 1–4, pp. 357–363) by S. Siikavuopio et al., we found that weights of adult green sea urchins are normally distributed with mean 52.0 g and standard deviation 17.2 g. Let $x$ denote weight of adult green sea urchins.
a. Sketch the distribution of the variable $x$.
b. Obtain the standardized version, $z$, of $x$.
c. Identify and sketch the distribution of $z$.
d. The percentage of adult green sea urchins with weights between 50 g and 60 g is equal to the area under the standard normal curve between _____ and _____.
e. The percentage of adult green sea urchins with weights above 40 g is equal to the area under the standard normal curve that lies to the _____ of _____.

**6.19 Ages of Mothers.** From the document *National Vital Statistics Reports*, a publication of the U.S. National Center for Health Statistics, we obtained the following frequency distribution for the ages of women who became mothers during one year.

| Age (yrs) | Frequency |
|---|---|
| 10 ◄ 15 | 7,315 |
| 15 ◄ 20 | 425,493 |
| 20 ◄ 25 | 1,022,106 |
| 25 ◄ 30 | 1,060,391 |
| 30 ◄ 35 | 951,219 |
| 35 ◄ 40 | 453,927 |
| 40 ◄ 45 | 95,788 |
| 45 ◄ 50 | 54,872 |

a. Obtain a relative-frequency histogram of these age data.
b. Based on your histogram, do you think that the ages of women who became mothers that year are approximately normally distributed? Explain your answer.

**6.20 Birth Rates.** The U.S. National Center for Health Statistics publishes information about birth rates (per 1000 population) in the document *National Vital Statistics Report*. The following table provides a frequency distribution for birth rates during one year for the 50 states and the District of Columbia.

| Rate | Frequency | Rate | Frequency |
|---|---|---|---|
| 10 ◄ 11 | 2 | 16 ◄ 17 | 1 |
| 11 ◄ 12 | 3 | 17 ◄ 18 | 1 |
| 12 ◄ 13 | 10 | 18 ◄ 19 | 0 |
| 13 ◄ 14 | 17 | 19 ◄ 20 | 0 |
| 14 ◄ 15 | 9 | 20 ◄ 21 | 0 |
| 15 ◄ 16 | 7 | 21 ◄ 22 | 1 |

a. Obtain a frequency histogram of these birth-rate data.
b. Based on your histogram, do you think that birth rates for the 50 states and the District of Columbia are approximately normally distributed? Explain your answer.

## Working With Large Data Sets

**6.21 SAT Scores.** Each year, thousands of high school students bound for college take the Scholastic Assessment Test (SAT). This test measures the verbal and mathematical abilities of prospective college students. Student scores are reported on a scale that ranges from a low of 200 to a high of 800. Summary results for the scores are published by the College Entrance Examination Board in *College Bound Seniors*. In one high school graduating class, the SAT scores are as provided on the WeissStats CD. Use the technology of your choice to answer the following questions.
a. Do the SAT verbal scores for this class appear to be approximately normally distributed? Explain your answer.
b. Do the SAT math scores for this class appear to be approximately normally distributed? Explain your answer.

**6.22 Fertility Rates.** From the U.S. Census Bureau, in the document *International Data Base*, we obtained data on the total fertility rates for women in various countries. Those data are presented on the WeissStats CD. The total fertility rate gives the average number of children that would be born if all women in a given country lived to the end of their childbearing years and, at each year of age, they experienced the birth rates occurring in the specified year. Use the technology of your choice to decide whether total fertility rates for countries appear to be approximately normally distributed. Explain your answer.

## Extending the Concepts and Skills

**6.23 Chips Ahoy! 1,000 Chips Challenge.** Students in an introductory statistics course at the U.S. Air Force Academy participated in Nabisco's "Chips Ahoy! 1,000 Chips Challenge" by confirming that there were at least 1000 chips in every 18-ounce bag of cookies that they examined. As part of their assignment, they concluded that the number of chips per bag is approximately normally distributed. Could the number of chips per bag be exactly normally distributed? Explain your answer. [SOURCE: B. Warner and J. Rutledge, "Checking the Chips Ahoy! Guarantee," *Chance*, Vol. 12(1), pp. 10–14]

**6.24** Consider a normal distribution with mean 5 and standard deviation 2.
**a.** Sketch the associated normal curve.
**b.** Use the footnote on page 268 to write the equation of the associated normal curve.
**c.** Use the technology of your choice to graph the equation obtained in part (b).
**d.** Compare the curves that you obtained in parts (a) and (c).

**6.25 Gestation Periods of Humans.** Refer to the simulation of human gestation periods discussed in Example 6.2 on page 273.
**a.** Sketch the normal curve for human gestation periods.

**b.** Simulate 1000 human gestation periods. (*Note:* Users of the TI-83/84 Plus should simulate 500 human gestation periods.)
**c.** Approximately what values would you expect for the sample mean and sample standard deviation of the 1000 observations? Explain your answers.
**d.** Obtain the sample mean and sample standard deviation of the 1000 observations and compare your answers to your estimates in part (c).
**e.** Roughly what would you expect a histogram of the 1000 observations to look like? Explain your answer.
**f.** Obtain a histogram of the 1000 observations and compare your result to your expectation in part (e).

**6.26 Delaying Adulthood.** In the paper, "Delayed Metamorphosis of a Tropical Reef Fish (*Acanthurus triostegus*): A Field Experiment" (*Marine Ecology Progress Series*, Vol. 176, pp. 25–38), M. McCormick studied larval duration of the convict surgeonfish, a common tropical reef fish. This fish has been found to delay metamorphosis into adulthood by extending its larval phase, a delay that often leads to enhanced survivorship in the species by increasing the chances of finding suitable habitat. Duration of the larval phase for convict surgeonfish is normally distributed with mean 53 days and standard deviation 3.4 days. Let $x$ denote larval-phase duration for convict surgeonfish.
**a.** Sketch the normal curve for the variable $x$.
**b.** Simulate 1500 observations of $x$. (*Note:* Users of the TI-83/84 Plus should simulate 750 observations.)
**c.** Approximately what values would you expect for the sample mean and sample standard deviation of the 1500 observations? Explain your answers.
**d.** Obtain the sample mean and sample standard deviation of the 1500 observations and compare your answers to your estimates in part (c).
**e.** Roughly what would you expect a histogram of the 1500 observations to look like? Explain your answer.
**f.** Obtain a histogram of the 1500 observations and compare your result to your expectation in part (e).

## 6.2 Areas Under the Standard Normal Curve

In Section 6.1, we demonstrated, among other things, that we can obtain the percentage of all possible observations of a normally distributed variable that lie within any specified range by (1) expressing the range in terms of z-scores and (2) determining the corresponding area under the standard normal curve.

You already know how to convert to z-scores. In this section, you will discover how to implement the second step—determining areas under the standard normal curve.

## Basic Properties of the Standard Normal Curve

We first need to discuss some of the basic properties of the standard normal curve. Recall that this curve is the one associated with the standard normal distribution, which has mean 0 and standard deviation 1. Figure 6.8 again shows the standard normal distribution and the standard normal curve.

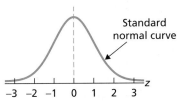

**FIGURE 6.8**
Standard normal distribution and standard normal curve

In Section 6.1, we showed that a normal curve is bell shaped, is centered at $\mu$, and is close to the horizontal axis outside the range from $\mu - 3\sigma$ to $\mu + 3\sigma$. Applied to the standard normal curve, these characteristics mean that it is bell shaped, is centered at 0, and is close to the horizontal axis outside the range from $-3$ to 3. Thus the standard normal curve is symmetric about 0. All of these properties are reflected in Fig. 6.8.

Another property of the standard normal curve is that the total area under it is 1. In fact, the total area under any curve that represents the distribution of a variable is equal to 1.

**Key Fact 6.3**

### Basic Properties of the Standard Normal Curve

**Property 1:** The total area under the standard normal curve is 1.

**Property 2:** The standard normal curve extends indefinitely in both directions, approaching, but never touching, the horizontal axis as it does so.

**Property 3:** The standard normal curve is symmetric about 0; that is, the part of the curve to the left of the dashed line in Fig. 6.8 is the mirror image of the part of the curve to the right of it.

**Property 4:** Almost all the area under the standard normal curve lies between $-3$ and 3.

Because the standard normal curve is the associated normal curve for a standardized normally distributed variable, we labeled the horizontal axis in Fig. 6.8 with the letter $z$ and refer to numbers on that axis as $z$-scores. For these reasons, the standard normal curve is sometimes called the **z-curve.**

## Using the Standard Normal Table (Table II)

Areas under the standard normal curve are so important that we have tables of those areas. Table II, located inside the back cover of this book and in Appendix A, is such a table.

A typical four-decimal-place number in the body of Table II gives the area under the standard normal curve that lies to the left of a specified $z$-score. The left page of Table II is for negative $z$-scores and the right page is for positive $z$-scores.

**Example 6.3** | ### Finding the Area to the Left of a Specified *z*-Score

Determine the area under the standard normal curve that lies to the left of 1.23, as shown in Fig. 6.9(a).

**FIGURE 6.9**

Finding the area under the standard normal curve to the left of 1.23

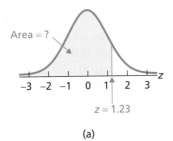

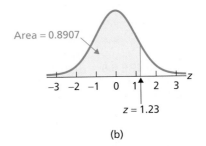

(a)                                                            (b)

**You try it!**

Exercise 6.37 on page 283

**Solution**   We use the right page of Table II because 1.23 is positive. First, we go down the left-hand column, labeled $z$, to "1.2." Then, going across that row to the column labeled "0.03," we reach 0.8907. This number is the area under the standard normal curve that lies to the left of 1.23, as shown in Fig. 6.9(b).

• • •

We can also use Table II to find the area to the right of a specified $z$-score and to find the area between two specified $z$-scores.

## Example 6.4 | Finding the Area to the Right of a Specified $z$-Score

Determine the area under the standard normal curve that lies to the right of 0.76, as shown in Fig. 6.10(a).

**FIGURE 6.10**

Finding the area under the standard normal curve to the right of 0.76

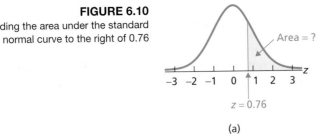

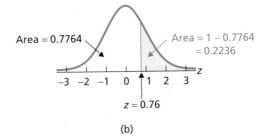

(a)                                                            (b)

**You try it!**

Exercise 6.39 on page 283

**Solution**   Because the total area under the standard normal curve is 1 (Property 1 of Key Fact 6.3), the area to the right of 0.76 equals 1 minus the area to the left of 0.76. We find this latter area as in the previous example, by first going down the $z$ column to "0.7." Then, going across that row to the column labeled "0.06," we reach 0.7764, which is the area under the standard normal curve that lies to the left of 0.76. Thus, the area under the standard normal curve that lies to the right of 0.76 is $1 - 0.7764 = 0.2236$, as shown in Fig. 6.10(b).

• • •

## Example 6.5 | Finding the Area Between Two Specified $z$-Scores

Determine the area under the standard normal curve that lies between $-0.68$ and 1.82, as shown in Fig. 6.11(a).

**FIGURE 6.11**
Finding the area under the standard
normal curve that lies between −0.68
and 1.82

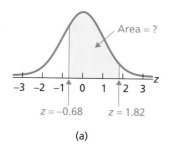

Area = ?

$z = -0.68$    $z = 1.82$

(a)

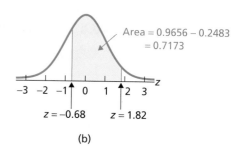

Area = 0.9656 − 0.2483
= 0.7173

$z = -0.68$    $z = 1.82$

(b)

**You try it!**

Exercise 6.41
on page 283

**Solution** The area under the standard normal curve that lies between −0.68 and 1.82 equals the area to the left of 1.82 minus the area to the left of −0.68. Table II shows that these latter two areas are 0.9656 and 0.2483, respectively. So the area we seek is $0.9656 - 0.2483 = 0.7173$, as shown in Fig 6.11(b).

• • •

The discussion presented in Examples 6.3–6.5 is summarized by the three graphs in Fig. 6.12.

**FIGURE 6.12**
Using Table II to find the area
under the standard normal curve that
lies (a) to the left of a specified *z*-score,
(b) to the right of a specified *z*-score,
and (c) between two specified *z*-scores

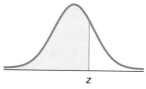

$z$

(a) Shaded area:
Area to left of $z$

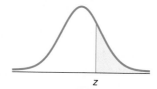

$z$

(b) Shaded area:
1 − (Area to left of $z$)

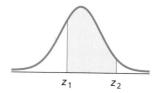

$z_1$    $z_2$

(c) Shaded area:
(Area to left of $z_2$)
− (Area to left of $z_1$)

## A Note Concerning Table II

The first area given in Table II, 0.0000, is for $z = -3.90$. This entry does not mean that the area under the standard normal curve that lies to the left of −3.90 is exactly 0, but only that it is 0 to four decimal places (the area is 0.0000481 to seven decimal places). Indeed, because the standard normal curve extends indefinitely to the left without ever touching the axis, the area to the left of any *z*-score is greater than 0.

Similarly, the last area given in Table II, 1.0000, is for $z = 3.90$. This entry does not mean that the area under the standard normal curve that lies to the left of 3.90 is exactly 1, but only that it is 1 to four decimal places (the area is 0.9999519 to seven decimal places). Indeed, the area to the left of any *z*-score is less than 1.

## Finding the *z*-score for a Specified Area

So far, we have used Table II to find areas. Now we show how to use Table II to find the *z*-score(s) corresponding to a specified area under the standard normal curve.

**Example 6.6** | **Finding the *z*-Score Having a Specified Area to Its Left**

Determine the *z*-score having an area of 0.04 to its left under the standard normal curve, as shown in Fig. 6.13(a).

**FIGURE 6.13**
Finding the *z*-score having
an area of 0.04 to its left

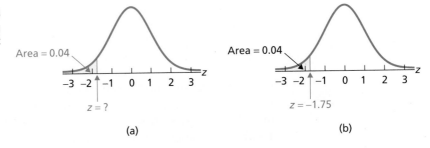

(a)        (b)

**Solution** Use Table II, a portion of which is given in Table 6.2.

**TABLE 6.2**
Areas under the standard normal curve

| | | | | Second decimal place in *z* | | | | | |
|---|---|---|---|---|---|---|---|---|---|
| *0.09* | *0.08* | *0.07* | *0.06* | *0.05* | *0.04* | *0.03* | *0.02* | *0.01* | *0.00* | *z* |
| · | · | · | · | · | · | · | · | · | · | · |
| · | · | · | · | · | · | · | · | · | · | · |
| 0.0233 | 0.0239 | 0.0244 | 0.0250 | 0.0256 | 0.0262 | 0.0268 | 0.0274 | 0.0281 | 0.0287 | −1.9 |
| 0.0294 | 0.0301 | 0.0307 | 0.0314 | 0.0322 | 0.0329 | 0.0336 | 0.0344 | 0.0351 | 0.0359 | −1.8 |
| 0.0367 | 0.0375 | 0.0384 | 0.0392 | 0.0401 | 0.0409 | 0.0418 | 0.0427 | 0.0436 | 0.0446 | −1.7 |
| 0.0455 | 0.0465 | 0.0475 | 0.0485 | 0.0495 | 0.0505 | 0.0516 | 0.0526 | 0.0537 | 0.0548 | −1.6 |
| 0.0559 | 0.0571 | 0.0582 | 0.0594 | 0.0606 | 0.0618 | 0.0630 | 0.0643 | 0.0655 | 0.0668 | −1.5 |
| · | · | · | · | · | · | · | · | · | · | · |
| · | · | · | · | · | · | · | · | · | · | · |

**You try it!**

Exercise 6.51
on page 284

Search the body of the table for the area 0.04. There is no such area, so use the area closest to 0.04, which is 0.0401. The *z*-score corresponding to that area is −1.75. Thus the *z*-score having area 0.04 to its left under the standard normal curve is roughly −1.75, as shown in Fig. 6.13(b).

• • •

The previous example shows that, when no area entry in Table II equals the one desired, we take the *z*-score corresponding to the closest area entry as an approximation of the required *z*-score. Two other cases are possible.

If an area entry in Table II equals the one desired, we of course use its corresponding *z*-score. If two area entries are equally closest to the one desired, we take the mean of the two corresponding *z*-scores as an approximation of the required *z*-score. Both of these cases are illustrated in the next example.

Finding the *z*-score that has a specified area to its right is often necessary. We have to make this determination so frequently that we use a special notation, $z_\alpha$.

**Definition 6.3**

**FIGURE 6.14**
The $z_\alpha$ notation

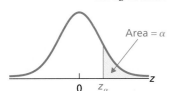

### The $z_\alpha$ Notation

The symbol $z_\alpha$ is used to denote the z-score that has an area of $\alpha$ (alpha) to its right under the standard normal curve, as illustrated in Fig. 6.14. Read "$z_\alpha$" as "z sub $\alpha$" or more simply as "z $\alpha$."

In the following two examples, we illustrate the $z_\alpha$ notation in a couple of different ways.

**Example 6.7** | **Finding $z_\alpha$**

Use Table II to find

**a.** $z_{0.025}$.

**b.** $z_{0.05}$.

**Solution**

**a.** $z_{0.025}$ is the z-score that has an area of 0.025 to its right under the standard normal curve, as shown in Fig. 6.15(a). Because the area to its right is 0.025, the area to its left is $1 - 0.025 = 0.975$, as shown in Fig. 6.15(b). Table II contains an entry for the area 0.975; its corresponding z-score is 1.96. Thus, $z_{0.025} = 1.96$, as shown in Fig. 6.15(b).

**FIGURE 6.15**
Finding $z_{0.025}$

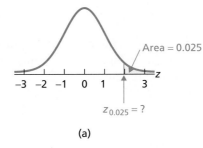

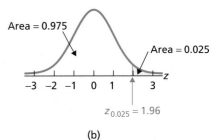

(a)

(b)

**b.** $z_{0.05}$ is the z-score that has an area of 0.05 to its right under the standard normal curve, as shown in Fig. 6.16(a). Because the area to its right is 0.05, the area to its left is $1 - 0.05 = 0.95$, as shown in Fig. 6.16(b). Table II does not contain an entry for the area 0.95 and has two area entries equally closest to 0.95—namely, 0.9495 and 0.9505. The z-scores corresponding to those two areas are 1.64 and 1.65, respectively. So our approximation of $z_{0.05}$ is the mean of 1.64 and 1.65; that is, $z_{0.05} = 1.645$, as shown in Fig. 6.16(b).

**FIGURE 6.16**
Finding $z_{0.05}$

**You try it!**

Exercise 6.57
on page 284

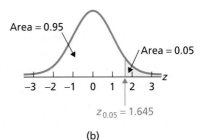

(a)

(b)

• • •

The next example shows how to find the two z-scores that divide the area under the standard normal curve into three specified areas.

## Example 6.8 | Finding the z-Scores for a Specified Area

Find the two z-scores that divide the area under the standard normal curve into a middle 0.95 area and two outside 0.025 areas, as shown in Fig. 6.17(a).

**FIGURE 6.17**
Finding the two z-scores that divide the area under the standard normal curve into a middle 0.95 area and two outside 0.025 areas

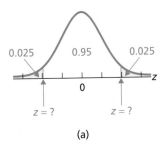

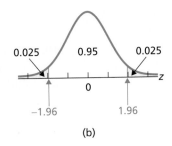

(a)                                    (b)

You try it!

Exercise 6.59
on page 284

**Solution**   The area of the shaded region on the right in Fig. 6.17(a) is 0.025. In Example 6.7(a), we found that the corresponding z-score, $z_{0.025}$, is 1.96. Because the standard normal curve is symmetric about 0, the z-score on the left is $-1.96$. Therefore the two required z-scores are $\pm 1.96$, as shown in Fig. 6.17(b).

• • •

**Note:**  We could also solve the previous example by first using Table II to find the z-score on the left in Fig. 6.17(a), which is $-1.96$, and then applying the symmetry property to obtain the z-score on the right, which is 1.96. Can you think of a third way to solve the problem?

## Exercises 6.2

### Understanding the Concepts and Skills

**6.27**  Explain why being able to obtain areas under the standard normal curve is important.

**6.28**  With which normal distribution is the standard normal curve associated?

**6.29**  Without consulting Table II, explain why the area under the standard normal curve that lies to the right of 0 is 0.5.

**6.30**  According to Table II, the area under the standard normal curve that lies to the left of $-2.08$ is 0.0188. Without further consulting Table II, determine the area under the standard normal curve that lies to the right of 2.08. Explain your reasoning.

**6.31**  According to Table II, the area under the standard normal curve that lies to the left of 0.43 is 0.6664. Without

further consulting Table II, determine the area under the standard normal curve that lies to the right of 0.43. Explain your reasoning.

**6.32**  According to Table II, the area under the standard normal curve that lies to the left of 1.96 is 0.975. Without further consulting Table II, determine the area under the standard normal curve that lies to the left of $-1.96$. Explain your reasoning.

**6.33**  Property 4 of Key Fact 6.3 states that most of the area under the standard normal curve lies between $-3$ and 3. Use Table II to determine precisely the percentage of the area under the standard normal curve that lies between $-3$ and 3.

**6.34**  Why is the standard normal curve sometimes referred to as the z-curve?

**6.35** Explain how Table II is used to determine the area under the standard normal curve that lies
**a.** to the left of a specified z-score.
**b.** to the right of a specified z-score.
**c.** between two specified z-scores.

**6.36** The area under the standard normal curve that lies to the left of a z-score is always strictly between _____ and _____.

*Use Table II to obtain the areas under the standard normal curve required in Exercises **6.37–6.44**. Sketch a standard normal curve and shade the area of interest in each problem.*

**6.37** Determine the area under the standard normal curve that lies to the left of
**a.** 2.24.
**b.** −1.56.
**c.** 0.
**d.** −4.

**6.38** Determine the area under the standard normal curve that lies to the left of
**a.** −0.87.
**b.** 3.56.
**c.** 5.12.

**6.39** Find the area under the standard normal curve that lies to the right of
**a.** −1.07.
**b.** 0.6.
**c.** 0.
**d.** 4.2.

**6.40** Find the area under the standard normal curve that lies to the right of
**a.** 2.02.
**b.** −0.56.
**c.** −4.

**6.41** Determine the area under the standard normal curve that lies between
**a.** −2.18 and 1.44.
**b.** −2 and −1.5.
**c.** 0.59 and 1.51.
**d.** 1.1 and 4.2.

**6.42** Determine the area under the standard normal curve that lies between
**a.** −0.88 and 2.24.
**b.** −2.5 and −2.
**c.** 1.48 and 2.72.
**d.** −5.1 and 1.

**6.43** Determine the area under the standard normal curve that lies
**a.** either to the left of −2.12 or to the right of 1.67.
**b.** either to the left of 0.63 or to the right of 1.54.

**6.44** Find the area under the standard normal curve that lies
**a.** either to the left of −1 or to the right of 2.
**b.** either to the left of −2.51 or to the right of −1.

**6.45** Use Table II to obtain each shaded area under the standard normal curve.

**a.**
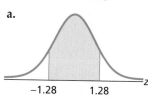
−1.28    1.28

**b.**
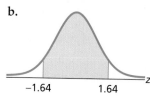
−1.64    1.64

**c.**
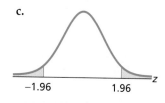
−1.96    1.96

**d.**
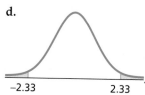
−2.33    2.33

**6.46** Use Table II to obtain each shaded area under the standard normal curve.

**a.**
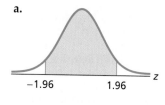
−1.96    1.96

**b.**
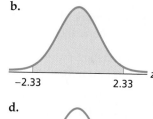
−2.33    2.33

**c.**
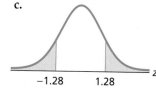
−1.28    1.28

**d.**
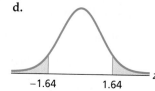
−1.64    1.64

**6.47** In each part, find the area under the standard normal curve that lies between the specified z-scores, sketch a standard normal curve, and shade the area of interest.
**a.** −1 and 1
**b.** −2 and 2
**c.** −3 and 3

**6.48** The total area under the following standard normal curve is divided into eight regions.

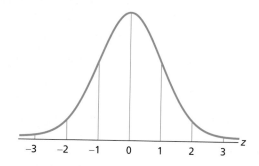
−3   −2   −1   0   1   2   3

**a.** Determine the area of each region.
**b.** Complete the following table.

| Region | Area | Percentage of total area |
|--------|------|--------------------------|
| $-\infty$ to $-3$ | 0.0013 | 0.13 |
| $-3$ to $-2$ | | |
| $-2$ to $-1$ | | |
| $-1$ to $0$ | | |
| $0$ to $1$ | 0.3413 | 34.13 |
| $1$ to $2$ | | |
| $2$ to $3$ | | |
| $3$ to $\infty$ | | |
| | 1.0000 | 100.00 |

*In Exercises **6.49–6.60**, use Table II to obtain the required z-scores. Illustrate your work with graphs.*

**6.49** Obtain the z-score for which the area under the standard normal curve to its left is 0.025.

**6.50** Determine the z-score for which the area under the standard normal curve to its left is 0.01.

**6.51** Find the z-score that has an area of 0.75 to its left under the standard normal curve.

**6.52** Obtain the z-score that has area 0.80 to its left under the standard normal curve.

**6.53** Obtain the z-score that has an area of 0.95 to its right.

**6.54** Obtain the z-score that has area 0.70 to its right.

**6.55** Determine $z_{0.33}$.

**6.56** Determine $z_{0.015}$.

**6.57** Find the following z-scores.
**a.** $z_{0.03}$          **b.** $z_{0.005}$

**6.58** Obtain the following z-scores.
**a.** $z_{0.20}$          **b.** $z_{0.06}$

**6.59** Determine the two z-scores that divide the area under the standard normal curve into a middle 0.90 area and two outside 0.05 areas.

**6.60** Determine the two z-scores that divide the area under the standard normal curve into a middle 0.99 area and two outside 0.005 areas.

**6.61** Complete the following table.

| $z_{0.10}$ | $z_{0.05}$ | $z_{0.025}$ | $z_{0.01}$ | $z_{0.005}$ |
|-----------|-----------|------------|-----------|------------|
| 1.28 | | | | |

## Extending the Concepts and Skills

**6.62** In this section, we mentioned that the total area under any curve representing the distribution of a variable equals 1. Explain why.

**6.63** Let $0 < \alpha < 1$. Determine the
**a.** z-score having an area of $\alpha$ to its right in terms of $z_\alpha$.
**b.** z-score having an area of $\alpha$ to its left in terms of $z_\alpha$.
**c.** two z-scores that divide the area under the curve into a middle $1 - \alpha$ area and two outside areas of $\alpha/2$.
**d.** Draw graphs to illustrate your results in parts (a)–(c).

# 6.3 Working With Normally Distributed Variables

You now know how to find the percentage of all possible observations of a normally distributed variable that lie within any specified range: First express the range in terms of z-scores and then determine the corresponding area under the standard normal curve. More formally, use Procedure 6.1.

| Procedure 6.1 | **To Determine a Percentage or Probability for a Normally Distributed Variable** |
|---------------|-------------------------------------------------------|

**STEP 1** Sketch the normal curve associated with the variable.

**STEP 2** Shade the region of interest and mark its delimiting x-value(s).

**STEP 3** Compute the z-score(s) for the delimiting x-value(s) found in Step 2.

**STEP 4** Use Table II to find the area under the standard normal curve delimited by the z-score(s) found in Step 3.

**FIGURE 6.18**
Graphical portrayal of Procedure 6.1

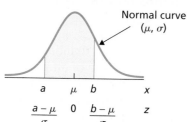

The steps in Procedure 6.1 are illustrated in Fig. 6.18, with the specified range lying between two numbers, $a$ and $b$. If the specified range is to the left (or right) of a number, it is represented similarly. However, there will be only one $x$-value, and the shaded region will be the area under the normal curve that lies to the left (or right) of that $x$-value.

**Note:** When computing $z$-scores in Step 3 of Procedure 6.1, round to two decimal places, the precision provided in Table II.

**Example 6.9** | **Percentages for a Normally Distributed Variable**

*Intelligence Quotients* Intelligence quotients (IQs) measured on the Stanford Revision of the Binet–Simon Intelligence Scale are normally distributed with a mean of 100 and a standard deviation of 16. Determine the percentage of people who have IQs between 115 and 140.

**Solution** Here the variable is IQ, and the population consists of all people. Because IQs are normally distributed, we can determine the required percentage by applying Procedure 6.1.

**STEP 1 Sketch the normal curve associated with the variable.**

Here $\mu = 100$ and $\sigma = 16$. The normal curve associated with the variable is shown in Fig. 6.19. Note that the tick marks are 16 units apart; that is, the distance between successive tick marks is equal to the standard deviation.

**FIGURE 6.19**
Determination of the percentage of people having IQs between 115 and 140

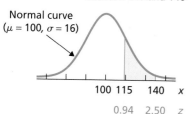

**STEP 2 Shade the region of interest and mark its delimiting $x$-values.**

Figure 6.19 shows the required shaded region and its delimiting $x$-values, which are 115 and 140.

**STEP 3 Compute the $z$-scores for the delimiting $x$-values found in Step 2.**

We need to compute the $z$-scores for the $x$-values 115 and 140:
$$x = 115 \quad \longrightarrow \quad z = \frac{115 - \mu}{\sigma} = \frac{115 - 100}{16} = 0.94,$$
and
$$x = 140 \quad \longrightarrow \quad z = \frac{140 - \mu}{\sigma} = \frac{140 - 100}{16} = 2.50.$$
These $z$-scores are marked beneath the $x$-values in Fig. 6.19.

**STEP 4 Use Table II to find the area under the standard normal curve delimited by the $z$-scores found in Step 3.**

We need to find the area under the standard normal curve that lies between 0.94 and 2.50. The area to the left of 0.94 is 0.8264, and the area to the left of 2.50 is 0.9938. The required area, shaded in Fig. 6.19, is therefore $0.9938 - 0.8264 = 0.1674$.

*You try it!*

Exercise 6.69(a)–(b) on page 292

**Interpretation** 16.74% of all people have IQs between 115 and 140. Equivalently, the probability is 0.1674 that a randomly selected person will have an IQ between 115 and 140.

• • •

### Visualizing a Normal Distribution

We now present a rule that helps us "visualize" a normally distributed variable. This rule gives the percentages of all possible observations that lie within one, two, and three standard deviations to either side of the mean.

Recall that the z-score of an observation tells us how many standard deviations the observation is from the mean. Thus the percentage of all possible observations that lie within one standard deviation to either side of the mean equals the percentage of all observations whose z-scores lie between $-1$ and $1$. For a normally distributed variable, that percentage is the same as the area under the standard normal curve between $-1$ and $1$, which is $0.6826$ or $68.26\%$. Proceeding similarly, we get the following rule.

**Key Fact 6.4**

## The 68.26-95.44-99.74 Rule

Any normally distributed variable has the following properties.

**Property 1:** 68.26% of all possible observations lie within one standard deviation to either side of the mean, that is, between $\mu - \sigma$ and $\mu + \sigma$.

**Property 2:** 95.44% of all possible observations lie within two standard deviations to either side of the mean, that is, between $\mu - 2\sigma$ and $\mu + 2\sigma$.

**Property 3:** 99.74% of all possible observations lie within three standard deviations to either side of the mean, that is, between $\mu - 3\sigma$ and $\mu + 3\sigma$.

These properties are illustrated in Fig. 6.20.

**FIGURE 6.20**
The 68.26-95.44-99.74 rule

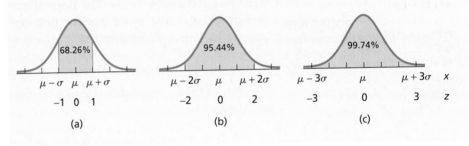

**Example 6.10**

## The 68.26-95.44-99.74 Rule

*Intelligence Quotients* Apply the 68.26-95.44-99.74 rule to IQs.

**Solution** Recall that IQs (measured on the Stanford Revision of the Binet–Simon Intelligence Scale) are normally distributed with a mean of 100 and a standard deviation of 16. In particular, we have $\mu = 100$ and $\sigma = 16$.

Property 1 of the 68.26-95.44-99.74 rule says 68.26% of all people have IQs within one standard deviation to either side of the mean. One standard deviation below the mean is $\mu - \sigma = 100 - 16 = 84$; one standard deviation above the mean is $\mu + \sigma = 100 + 16 = 116$.

**Interpretation** 68.26% of all people have IQs between 84 and 116, as illustrated in Fig. 6.21(a).

Property 2 of the rule says 95.44% of all people have IQs within two standard deviations to either side of the mean; that is, from $\mu - 2\sigma = 100 - 2 \cdot 16 = 68$ to $\mu + 2\sigma = 100 + 2 \cdot 16 = 132$.

**Interpretation**   95.44% of all people have IQs between 68 and 132, as illustrated in Fig. 6.21(b).

Property 3 of the rule says 99.74% of all people have IQs within three standard deviations to either side of the mean; that is, from $\mu - 3\sigma = 100 - 3 \cdot 16 = 52$ to $\mu + 3\sigma = 100 + 3 \cdot 16 = 148$.

**Interpretation**   99.74% of all people have IQs between 52 and 148, as illustrated in Fig. 6.21(c).

**FIGURE 6.21**
Graphical display of the
68.26-95.44-99.74 rule for IQs

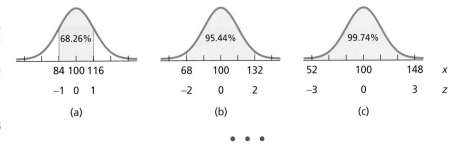

**You try it!**

Exercise 6.75
on page 293

As illustrated in the previous example, the 68.26-95.44-99.74 rule allows us to obtain useful information about a normally distributed variable quickly and easily. Note, however, that similar facts are obtainable for any number of standard deviations. For instance, Table II reveals that, for any normally distributed variable, 86.64% of all possible observations lie within 1.5 standard deviations to either side of the mean.

Experience has shown that the 68.26-95.44-99.74 rule works reasonably well for any variable having approximately a bell-shaped distribution, regardless of whether the variable is normally distributed. This fact is referred to as the **empirical rule,** which we alluded to earlier in Chapter 3 (see page 114) in our discussion of the three-standard-deviations rule.

### Finding the Observations for a Specified Percentage

Procedure 6.1 shows how to determine the percentage of all possible observations of a normally distributed variable that lie within any specified range. Frequently, however, we want to carry out the reverse procedure, that is, to find the observations corresponding to a specified percentage. Procedure 6.2 allows us to do that.

| Procedure 6.2 | **To Determine the Observations Corresponding to a Specified Percentage or Probability for a Normally Distributed Variable** |
|---|---|
| | **STEP 1**  Sketch the normal curve associated with the variable. |
| | **STEP 2**  Shade the region of interest. |
| | **STEP 3**  Use Table II to obtain the $z$-score(s) delimiting the region found in Step 2. |
| | **STEP 4**  Find the $x$-value(s) having the $z$-score(s) found in Step 3. |

**Note:** To find each $x$-value in Step 4 from its $z$-score in Step 3, use the formula

$$x = \mu + z \cdot \sigma,$$

where $\mu$ and $\sigma$ are the mean and standard deviation, respectively, of the variable under consideration.

Among other things, we can use Procedure 6.2 to obtain quartiles, deciles, or any other percentile for a normally distributed variable. Example 6.11 shows how to find percentiles by this method.

## Example 6.11 | Obtaining Percentiles for a Normally Distributed Variable

*Intelligence Quotients* Obtain and interpret the 90th percentile for IQs.

**Solution** The 90th percentile, $P_{90}$, is the IQ that is higher than those of 90% of all people. As IQs are normally distributed, we can determine the 90th percentile by applying Procedure 6.2.

**STEP 1 Sketch the normal curve associated with the variable.**

Here $\mu = 100$ and $\sigma = 16$. The normal curve associated with IQs is shown in Fig. 6.22.

**STEP 2 Shade the region of interest.**

**FIGURE 6.22**
Finding the 90th percentile for IQs

See the shaded region in Fig. 6.22.

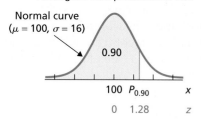

Normal curve
($\mu = 100$, $\sigma = 16$)

0.90

100  $P_{0.90}$    $x$

0   1.28    $z$

**STEP 3 Use Table II to obtain the $z$-score delimiting the region found in Step 2.**

The $z$-score corresponding to $P_{90}$ is the one having an area of 0.90 to its left under the standard normal curve. From Table II, that $z$-score is 1.28, approximately, as shown in Fig. 6.22.

**STEP 4 Find the $x$-value having the $z$-score found in Step 3.**

*You try it!*

Exercise 6.69(c)–(d)
on page 292

We must find the $x$-value having the $z$-score 1.28—the IQ that is 1.28 standard deviations above the mean. It is $100 + 1.28 \cdot 16 = 100 + 20.48 = 120.48$.

**Interpretation** The 90th percentile for IQs is 120.48. Thus, 90% of people have IQs below 120.48 and 10% have IQs above 120.48.

• • •

## The Technology Center

Most statistical technologies have programs that carry out the procedures discussed in this section—namely, to obtain for a normally distributed variable

- the percentage of all possible observations that lie within any specified range and
- the observations corresponding to a specified percentage.

In this subsection, we present output and step-by-step instructions for such programs.

Each of Minitab, Excel, and the TI-83/84 Plus has a program for determining the area under the associated normal curve of a normally distributed variable that lies to the left of a specified value. Such an area corresponds to a **cumulative probability,** the probability that the variable will be less than or equal to the specified value.

### Example 6.12   Using Technology to Obtain Normal Percentages

*Intelligence Quotients*    Recall that IQs are normally distributed with mean 100 and standard deviation 16. Use Minitab, Excel, or the TI-83/84 Plus to find the percentage of people who have IQs between 115 and 140.

**Solution**    We applied the cumulative-normal programs, resulting in Output 6.2. Steps for generating that output are presented in Instructions 6.1 (next page).

**OUTPUT 6.2**    The percentage of people with IQs between 115 and 140

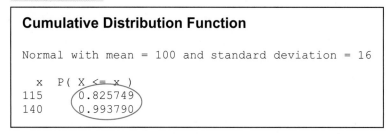

```
MINITAB

Cumulative Distribution Function

Normal with mean = 100 and standard deviation = 16

    x    P( X <= x )
  115     0.825749
  140     0.993790
```

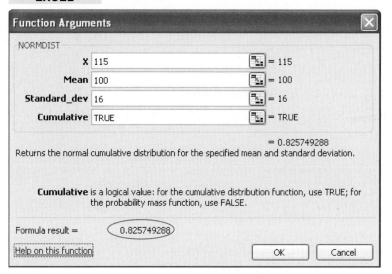

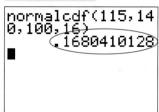

```
TI-83/84 PLUS

normalcdf(115,14
0,100,16)
      .1680410128
■
```

Note: Replacing 115 by 140 in the **X** text box yields 0.993790335.

We get the required percentage from Output 6.2 as follows.

- *Minitab:* Subtract the two cumulative probabilities: $0.993790 - 0.825749 = 0.168041$, or 16.80%.
- *Excel:* Subtract the two cumulative probabilities, the one circled in red from the one in the note: $0.993790355 - 0.825749288 = 0.168041047$, or 16.80%.
- *TI-83/84 Plus:* Direct from the output: 0.1680410128, or 16.80%.

Note that the percentages obtained by the three technologies differ slightly from the percentage of 16.74% that we found in Example 6.9. The differences reflect the fact that the technologies retain more accuracy than we can get from Table II.

• • •

**INSTRUCTIONS 6.1**  Steps for generating Output 6.2

| MINITAB | EXCEL | TI-83/84 PLUS |
|---|---|---|
| 1 Store the delimiting IQs, 115 and 140, in a column named IQ<br>2 Choose **Calc ➤ Probability Distributions ➤ Normal...**<br>3 Select the **Cumulative probability** option button<br>4 Click in the **Mean** text box and type 100<br>5 Click in the **Standard deviation** text box and type 16<br>6 Select the **Input column** option button<br>7 Click in the **Input column** text box and specify IQ<br>8 Click **OK** | 1 Click $f_x$ on the button bar<br>2 Select **Statistical** from the **Or select a category** drop down list box<br>3 Select **NORMDIST** from the **Select a function** list<br>4 Click **OK**<br>5 Type 115 in the **X** text box<br>6 Click in the **Mean** text box and type 100<br>7 Click in the **Standard_dev** text box and type 16<br>8 Click in the **Cumulative** text box and type TRUE<br>9 To obtain the cumulative probability for 140, replace 115 by 140 in the **X** text box | 1 Press **2nd ➤ DISTR**<br>2 Arrow down to **normalcdf(** and press **ENTER**<br>3 Type 115,140,100,16) and press **ENTER** |

Each of Minitab, Excel, and the TI-83/84 Plus also has a program for determining the observation that has a specified area to its left under the associated normal curve of a normally distributed variable. Such an observation corresponds to an **inverse cumulative probability,** the observation whose cumulative probability is the specified area.

**Example 6.13**  **Using Technology to Obtain Normal Percentiles**

*Intelligence Quotients*  Use Minitab, Excel, or the TI-83/84 Plus to find the 90th percentile for IQs.

**Solution**  We applied the inverse-cumulative-normal programs, resulting in Output 6.3. Steps for generating that output are presented in Instructions 6.2.

As shown in Output 6.3, the 90th percentile for IQs is 120.505. Note that this value differs slightly from the value of 120.48 that we obtained in Example 6.11. The difference reflects the fact that the three technologies retain more accuracy than we can get from Table II.

• • •

**OUTPUT 6.3**  The 90th percentile for IQs

### MINITAB

**Inverse Cumulative Distribution Function**

Normal with mean = 100 and standard deviation = 16

P( X <= x )           x
        0.9  120.505

### EXCEL

**Function Arguments**                                      ☒

NORMINV
    **Probability** | 0.90 |           = 0.9
        **Mean** | 100 |           = 100
**Standard_dev** | 16 |           = 16

                              = 120.504825
Returns the inverse of the normal cumulative distribution for the specified mean and standard deviation.

    **Standard_dev** is the standard deviation of the distribution, a positive number.

Formula result =          120.504825
Help on this function              [ OK ]  [ Cancel ]

### TI-83/84 PLUS

```
invNorm(0.90,100
,16)
      120.5048251
■
```

**INSTRUCTIONS 6.2**  Steps for generating Output 6.3

### MINITAB

1 Choose **Calc ➤ Probability Distributions ➤ Normal...**
2 Select the **Inverse cumulative probability** option button
3 Click in the **Mean** text box and type 100
4 Click in the **Standard deviation** text box and type 16
5 Select the **Input constant** option button
6 Click in the **Input constant** text box and type 0.90
7 Click **OK**

### EXCEL

1 Click $f_x$ on the button bar
2 Select **Statistical** from the **Or select a category** drop down list box
3 Select **NORMINV** from the **Select a function** list
4 Click **OK**
5 Type 0.90 in the **Probability** text box
6 Click in the **Mean** text box and type 100
7 Click in the **Standard_dev** text box and type 16

### TI-83/84 PLUS

1 Press **2nd ➤ DISTR**
2 Arrow down to **invNorm(** and press **ENTER**
3 Type 0.90,100,16) and press **ENTER**

■ ■ ■

# Exercises 6.3

## Understanding the Concepts and Skills

**6.64** Briefly, for a normally distributed variable, how do you obtain the percentage of all possible observations that lie within a specified range?

**6.65** Explain why the percentage of all possible observations of a normally distributed variable that lie within two standard deviations to either side of the mean equals the area under the standard normal curve between $-2$ and $2$.

**6.66** What does the empirical rule say?

**6.67 Giant Tarantulas.** One of the larger species of tarantulas is the *Grammostola mollicoma*, whose common name is the Brazilian Giant Tawny Red. A tarantula has two body parts. The anterior part of the body is covered above by a shell, or carapace. From a recent article by F. Costa and F. Perez–Miles titled "Reproductive Biology of Uruguayan Theraphosids" (*The Journal of Arachnology*, Vol. 30, No. 3, pp. 571–587), we find that the carapace length of the adult male *G. mollicoma* is normally distributed with mean 18.14 mm and standard deviation 1.76 mm.
a. Find the percentage of adult male *G. mollicoma* that have carapace lengths between 16 mm and 17 mm.
b. Find the percentage of adult male *G. mollicoma* that have carapace lengths exceeding 19 mm.
c. Determine and interpret the quartiles for carapace length of the adult male *G. mollicoma*.
d. Obtain and interpret the 95th percentile for carapace length of the adult male *G. mollicoma*.

**6.68 Serum Cholesterol Levels.** According to the *National Health and Nutrition Examination Survey*, the serum (noncellular portion of blood) total cholesterol level of U.S. females 20 years old or older is normally distributed with a mean of 206 mg/dL (milligrams per deciliter) and a standard deviation of 44.7 mg/dL.
a. Determine the percentage of U.S. females 20 years old or older who have a serum total cholesterol level between 150 mg/dL and 250 mg/dL.
b. Determine the percentage of U.S. females 20 years old or older who have a serum total cholesterol level below 220 mg/dL.
c. Obtain and interpret the quartiles for serum total cholesterol level of U.S. females 20 years old or older.
d. Find and interpret the fourth decile for serum total cholesterol level of U.S. females 20 years old or older.

**6.69 New York City 10 km Run.** As reported in *Runner's World* magazine, the times of the finishers in the New York City 10 km run are normally distributed with mean 61 minutes and standard deviation 9 minutes.
a. Determine the percentage of finishers with times between 50 and 70 minutes.
b. Find the percentage of finishers with times less than 75 minutes.
c. Obtain and interpret the 40th percentile for the finishing times.
d. Find and interpret the eighth decile for the finishing times.

**6.70 Green Sea Urchins.** From the paper "Effects of Chronic Nitrate Exposure on Gonad Growth in Green Sea Urchin *Strongylocentrotus Droebachiensis*" (*Aquaculture*, Vol. 242, No. 1–4, pp. 357–363) by S. Siikavuopio et al., we found that weights of adult green sea urchins are normally distributed with mean 52.0 g and standard deviation 17.2 g.
a. Find the percentage of adult green sea urchins with weights between 50 g and 60 g.
b. Obtain the percentage of adult green sea urchins with weights above 40 g.
c. Determine and interpret the 90th percentile for the weights.
d. Find and interpret the sixth decile for the weights.

**6.71 Drive for Show, Putt for Dough.** An article by Scott M. Berry titled "Drive for Show and Putt for Dough" (*Chance*, Vol. 12(4), pp. 50–54) discussed driving distances of PGA players. The mean distance for tee shots on the 1999 men's PGA tour is 272.2 yards with a standard deviation of 8.12 yards. Assuming that the 1999 tee-shot distances are normally distributed, find the percentage of such tee shots that went
a. between 260 and 280 yards.
b. more than 300 yards.

**6.72 Metastatic Carcinoid Tumors.** A study of sizes of metastatic carcinoid tumors in the heart was conducted by U. Pandya et al. and published as the article, "Metastatic Carcinoid Tumor to the Heart: Echocardiographic-Pathologic Study of 11 Patients" (*Journal of the American College of Cardiology*, Vol. 40, pp. 1328–1332). Based on that study, we assume that lengths of metastatic carcinoid tumors in the heart are normally distributed with mean 1.8 cm and standard deviation 0.5 cm. Find the percentage of metastatic carcinoid tumors in the heart that
a. are between 1 cm and 2 cm long.
b. exceed 3 cm in length.

**6.73 Gibbon Song Duration.** A preliminary behavioral study of the Jingdong black gibbon, a primate endemic to the Wuliang Mountains in China, found that the mean song bout duration in the wet season is 12.59 minutes with a standard deviation of 5.31 minutes. [SOURCE: L. Sheeran et al., "Preliminary Report on the Behavior of the Jingdong Black Gibbon (*Hylobates concolor jingdongensis*)," *Tropical Biodiversity*, Vol. 5(2), pp. 113–125] Assuming that song bout is normally distributed, determine the percentage of song bouts that have durations within

a. one standard deviation to either side of the mean.
b. two standard deviations to either side of the mean.
c. three standard deviations to either side of the mean.

**6.74 Friendship Motivation.** In the article "Assessing Friendship Motivation During Preadolescence and Early Adolescence" (*Journal of Early Adolescence*, Vol. 25, No. 3, pp. 367–385), J. Richard and B. Schneider describe the properties of the Friendship Motivation Scale for Children (FMSC), a scale designed to assess children's desire for friendships. Some interesting conclusions are that friends generally report similar levels of the FMSC and girls tend to score higher on the FMSC than boys. Boys in the seventh grade scored a mean of 9.32 with a standard deviation of 1.71, and girls in the seventh grade scored a mean of 10.04 with a standard deviation of 1.83. Assuming that FMSC scores are normally distributed, determine the percentage of seventh-grade boys who have FMSC scores within

a. one standard deviation to either side of the mean.
b. two standard deviations to either side of the mean.
c. three standard deviations to either side of the mean.
d. Repeat parts (a)–(c) for seventh-grade girls.

**6.75 Brain Weights.** In 1905, R. Pearl published the article "Biometrical Studies on Man. I. Variation and Correlation in Brain Weight" (*Biometrika*, Vol. 4, pp. 13–104). According to the study, brain weights of Swedish men are normally distributed with mean 1.40 kg and standard deviation 0.11 kg. Apply the 68.26-95.44-99.74 rule to fill in the blanks.

a. 68.26% of Swedish men have brain weights between _____ and _____.
b. 95.44% of Swedish men have brain weights between _____ and _____.

c. 99.74% of Swedish men have brain weights between _____ and _____.
d. Draw graphs similar to those in Fig. 6.21 on page 287 to portray your results.

**6.76 Children Watching TV.** The A. C. Nielsen Company reports in the *Nielsen Report on Television* that the mean weekly television viewing time for children aged 2–11 years is 24.50 hours. Assume that the weekly television viewing times of such children are normally distributed with a standard deviation of 6.23 hours and apply the 68.26-95.44-99.74 rule to fill in the blanks.

a. 68.26% of all such children watch between _____ and _____ hours of TV per week.
b. 95.44% of all such children watch between _____ and _____ hours of TV per week.
c. 99.74% of all such children watch between _____ and _____ hours of TV per week.
d. Draw graphs similar to those in Fig. 6.21 on page 287 to portray your results.

**6.77 Heights of Female Students.** Refer to Example 6.1 on page 269. The heights of the 3264 female students attending a midwestern college are approximately normally distributed with mean 64.4 inches and standard deviation 2.4 inches. Thus we can use the normal distribution with $\mu = 64.4$ and $\sigma = 2.4$ to approximate the percentage of these students having heights within any specified range. In each part, (i) obtain the exact percentage from Table 6.1, (ii) use the normal distribution to approximate the percentage, and (iii) compare your answers.

a. The percentage of female students with heights between 62 and 63 inches.
b. The percentage of female students with heights between 65 and 70 inches.

## Extending the Concepts and Skills

**6.78 Polychaete Worms.** *Opisthotrochopodus* n. sp. is a polychaete worm that inhabits deep sea hydrothermal vents along the Mid-Atlantic Ridge. According to the article "Reproductive Biology of Free-Living and Commensal Polynoid Polychaetes at the Lucky Strike Hydrothermal Vent Field (Mid-Atlantic Ridge)" (*Marine Ecology Progress Series*, Vol. 181, pp. 201–214) by C. Van Dover et al., the lengths of female polychaete worms are normally distributed with mean 6.1 mm and standard deviation 1.3 mm. Let $X$ denote the length of a randomly selected female polychaete worm. Determine and interpret

a. $P(X \leq 3)$.          b. $P(5 < X < 7)$.

**6.79 Booted Eagles.** The rare booted eagle of western Europe was the focus of a study by S. Suarez et al. to identify optimal nesting habitat for this raptor. According to their paper "Nesting Habitat Selection by Booted Eagles (*Hieraaetus pennatus*) and Implications for Management" (*Journal of Applied Ecology*, Vol. 37, pp. 215–223), the distances of such nests to the nearest marshland are normally distributed with mean 4.66 km and standard deviation 0.75 km. Let $Y$ be the distance of a randomly selected nest to the nearest marshland. Determine and interpret

**a.** $P(Y > 5)$.    **b.** $P(3 \leq Y \leq 6)$.

**6.80** For a normally distributed variable, fill in the blanks.
**a.** _____% of all possible observations lie within 1.96 standard deviations to either side of the mean.
**b.** _____% of all possible observations lie within 1.64 standard deviations to either side of the mean.

**6.81** For a normally distributed variable, fill in the blanks.
**a.** 99% of all possible observations lie within _____ standard deviations to either side of the mean.
**b.** 80% of all possible observations lie within _____ standard deviations to either side of the mean.

**6.82 Emergency Room Traffic.** Desert Samaritan Hospital, in Mesa, Arizona, keeps records of emergency room traffic. Those records reveal that the times between arriving patients have a mean of 8.7 minutes with a standard deviation of 8.7 minutes. Based solely on the values of these two parameters, explain why it is unreasonable to assume that the times between arriving patients is normally distributed or even approximately so.

**6.83** Let $0 < \alpha < 1$. For a normally distributed variable, show that $100(1 - \alpha)\%$ of all possible observations lie within $z_{\alpha/2}$ standard deviations to either side of the mean, that is, between $\mu - z_{\alpha/2} \cdot \sigma$ and $\mu + z_{\alpha/2} \cdot \sigma$.

**6.84** Let $x$ be a normally distributed variable with mean $\mu$ and standard deviation $\sigma$.
**a.** Express the quartiles, $Q_1$, $Q_2$, and $Q_3$, in terms of $\mu$ and $\sigma$.
**b.** Express the $k$th percentile, $P_k$, in terms of $\mu$, $\sigma$, and $k$.

## 6.4    Assessing Normality; Normal Probability Plots

You have now seen how to work with normally distributed variables. For instance, you know how to determine the percentage of all possible observations that lie within any specified range and how to obtain the observations corresponding to a specified percentage.

Another problem involves deciding whether a variable is normally distributed, or at least approximately so, based on a sample of observations. Such decisions often play a major role in subsequent analyses—from percentage or percentile calculations to statistical inferences.

From Key Fact 2.1 on page 77, if a simple random sample is taken from a population, the distribution of the observed values of a variable will approximate the distribution of the variable—and the larger the sample, the better the approximation tends to be. We can use this fact to help decide whether a variable is normally distributed.

If a variable is normally distributed, then, for a large sample, a histogram of the observations should be roughly bell shaped; for a very large sample, even moderate departures from a bell shape cast doubt on the normality of the variable. However, for a relatively small sample, ascertaining a clear shape in a histogram and, in particular, whether it is bell shaped is often difficult. These comments also hold for stem-and-leaf diagrams and dotplots.

Thus, for relatively small samples, a more sensitive graphical technique than the ones we have presented so far is required for assessing normality. *Normal probability plots* provide such a technique.

The idea behind a normal probability plot is simple: Compare the observed values of the variable to the observations expected for a normally distributed variable. More precisely, a **normal probability plot** is a plot of the observed values of the variable versus the **normal scores**—the observations expected for a variable having the standard normal distribution. If the variable is normally distributed, the normal probability plot should be roughly linear (i.e., fall roughly in a straight line) and vice versa.

When you use a normal probability plot to assess the normality of a variable, you must remember two things: (1) that the decision of whether a normal probability plot is roughly linear is a subjective one, and (2) that you are using a sample of observations of the variable to make a judgment about all possible observations of the variable. Keep these considerations in mind when using the following guidelines.

**Key Fact 6.5**

---

**What Does It Mean?**

Roughly speaking, a normal probability plot that falls nearly in a straight line indicates a normal variable, and one that does not indicates a nonnormal variable.

---

### Guidelines for Assessing Normality Using a Normal Probability Plot

To assess the normality of a variable using sample data, construct a normal probability plot.

- If the plot is roughly linear, you can assume that the variable is approximately normally distributed.
- If the plot is not roughly linear, you can assume that the variable is not approximately normally distributed.

These guidelines should be interpreted loosely for small samples, but usually strictly for large samples.

In practice, normal probability plots are generated by computer. However, to better understand these plots, constructing a few by hand is helpful. Table III in Appendix A gives the normal scores for sample sizes from 5 to 30. In the next example, we explain how to use Table III to obtain a normal probability plot.

**Example 6.14**    **Normal Probability Plots**

**TABLE 6.3**
Adjusted gross incomes ($1000s)

| | | | |
|---|---|---|---|
| 9.7 | 93.1 | 33.0 | 21.2 |
| 81.4 | 51.1 | 43.5 | 10.6 |
| 12.8 | 7.8 | 18.1 | 12.7 |

*Adjusted Gross Incomes* The Internal Revenue Service publishes data on federal individual income tax returns in *Statistics of Income, Individual Income Tax Returns*. A simple random sample of 12 returns from last year revealed the adjusted gross incomes, in thousands of dollars, shown in Table 6.3. Construct a normal probability plot for these data, and use the plot to assess the normality of adjusted gross incomes.

**Solution**    Here the variable is adjusted gross income, and the population consists of all last year's federal individual income tax returns. To construct a normal probability plot, we first arrange the data in increasing order and obtain the normal scores from Table III. The ordered data are shown in the first column of Table 6.4; the normal scores, from the $n = 12$ column of Table III, are shown in the second column of Table 6.4.

Next, we plot the points in Table 6.4, using the horizontal axis for the adjusted gross incomes and the vertical axis for the normal scores. For instance, the first point plotted has a horizontal coordinate of 7.8 and a vertical coordinate of $-1.64$. Figure 6.23, at the top of the following page, shows all 12 points from Table 6.4. This graph is the normal probability plot for the sample of adjusted gross incomes. Note that the normal probability plot in Fig. 6.23 is curved, not linear.

**TABLE 6.4**
Ordered data and normal scores

| Adjusted gross income | Normal score |
|---|---|
| 7.8 | −1.64 |
| 9.7 | −1.11 |
| 10.6 | −0.79 |
| 12.7 | −0.53 |
| 12.8 | −0.31 |
| 18.1 | −0.10 |
| 21.2 | 0.10 |
| 33.0 | 0.31 |
| 43.5 | 0.53 |
| 51.1 | 0.79 |
| 81.4 | 1.11 |
| 93.1 | 1.64 |

**FIGURE 6.23**
Normal probability plot for the sample of adjusted gross incomes

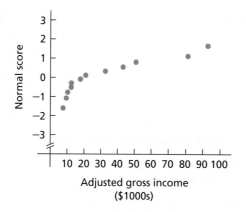

You try it!

Exercise 6.89(a) & (c) on page 299

**Interpretation** In light of Key Fact 6.5, last year's adjusted gross incomes apparently are not (approximately) normally distributed.

• • •

**Note:** If two or more observations in a sample are equal, you can think of them as slightly different from one another for purposes of obtaining their normal scores.

In some books and statistical technologies, you may encounter one or more of the following differences in normal probability plots:

- The vertical axis is used for the data and the horizontal axis for the normal scores.
- A probability or percent scale is used instead of normal scores.
- An averaging process is used to assign equal normal scores to equal observations.

### Detecting Outliers with Normal Probability Plots

Recall that outliers are observations that fall well outside the overall pattern of the data. We can also use normal probability plots to detect outliers.

**Example 6.15** | **Using Normal Probability Plots to Detect Outliers**

*Chicken Consumption* The U.S. Department of Agriculture publishes data on U.S. chicken consumption in *Food Consumption, Prices, and Expenditures*. The annual chicken consumption, in pounds, for 17 randomly selected people are displayed in Table 6.5. A normal probability plot for these observations is presented in Fig. 6.24(a). Use the plot to discuss the distribution of chicken consumption and to detect any outliers.

**TABLE 6.5**
Sample of last year's chicken consumption (lb)

| | | | | | |
|----|----|----|----|----|----|
| 47 | 39 | 62 | 49 | 50 | 70 |
| 59 | 53 | 55 | 0  | 65 | 63 |
| 53 | 51 | 50 | 72 | 45 |    |

**Solution** Figure 6.24(a) reveals that the normal probability plot falls roughly in a straight line except for the point corresponding to 0 lb, which falls well outside the overall pattern of the plot.

**FIGURE 6.24**

Normal probability plots for chicken consumption: (a) original data; (b) data with outlier removed

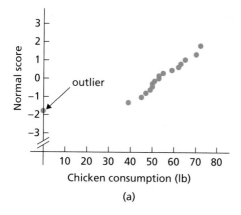

(a)

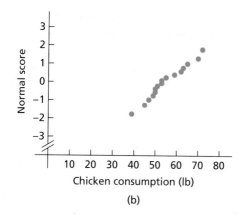

(b)

**Interpretation** The observation of 0 lb is an outlier, which might be a recording error or due to a person in the sample who does not eat chicken, such as a vegetarian.

If we remove the outlier 0 lb from the sample data and draw a new normal probability plot, Fig. 6.24(b) shows that this plot is quite linear.

**Interpretation** It appears plausible that, among people who eat chicken, the amounts they consume annually are (approximately) normally distributed.

Exercise 6.89(b) on page 299

• • •

Although the visual assessment of normality that we studied in this section is subjective, it is sufficient for most statistical analyses.

# The Technology Center

Most statistical technologies have programs that automatically construct normal probability plots. In this subsection, we present output and step-by-step instructions for such programs.

**Example 6.16**  **Using Technology to Obtain Normal Probability Plots**

*Adjusted Gross Incomes*  Use Minitab, Excel, or the TI-83/84 Plus to obtain a normal probability plot for the adjusted gross incomes in Table 6.3 on page 295.

**Solution**  We applied the normal-probability-plot programs to the data, resulting in Output 6.4, shown at the top of the next page. Steps for generating that output are presented in Instructions 6.3.

As we mentioned earlier and as you can see from the Excel output, normal probability plots sometimes use the vertical axis for the data and the horizontal axis for the normal scores.

• • •

**OUTPUT 6.4**

Normal probability plots for the sample of adjusted gross incomes

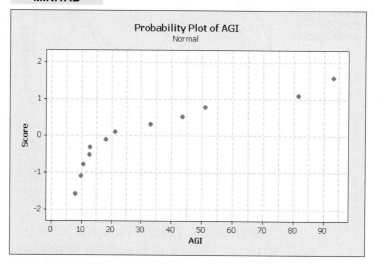

**MINITAB**

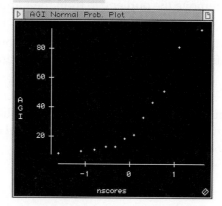

**EXCEL**

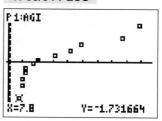

**TI-83/84 PLUS**

**INSTRUCTIONS 6.3**    Steps for generating Output 6.4

**MINITAB**

1  Store the data from Table 6.3 in a column named AGI
2  Choose **Graph ➤ Probability Plot...**
3  Select the **Single** plot and click **OK**
4  Specify AGI in the **Graph variables** text box
5  Click the **Distribution...** button
6  Click the **Data Display** tab, select the **Symbols only** option button from the **Data Display** list, and click **OK**
7  Click the **Scale...** button
8  Click the **Y-Scale Type** tab, select the **Score** option button from the **Y-Scale Type** list, and click **OK**
9  Click **OK**

**EXCEL**

1  Store the data from Table 6.3 in a range named AGI
2  Choose **DDXL ➤ Charts and Plots**
3  Select **Normal Probability Plot** from the **Function type** drop-down box
4  Specify AGI in the **Quantitative Variable** text box
5  Click **OK**

**TI-83/84 PLUS**

1  Store the data from Table 6.3 in a list named AGI
2  Press **2nd ➤ STAT PLOT** and then press **ENTER** twice
3  Arrow to the sixth graph icon and press **ENTER**
4  Press the down-arrow key
5  Press **2nd ➤ LIST**
6  Arrow down to AGI and press **ENTER**
7  Press **ZOOM** and then **9** (and then **TRACE**, if desired)

■ ■ ■

# Exercises 6.4

## Understanding the Concepts and Skills

**6.85** Explain why assessing the normality of a variable is often important.

**6.86** Under what circumstances is using a normal probability plot to assess the normality of a variable usually better than using a histogram, stem-and-leaf diagram, or dotplot?

**6.87** Regarding normal probability plots:
**a.** Explain in detail what a normal probability plot is and how it is used to assess the normality of a variable.
**b.** How is a normal probability plot used to detect outliers?

**6.88** Explain how to obtain normal scores from Table III in Appendix A when a sample contains equal observations.

*In Exercises 6.89–6.92, do the following:*
*a. Use Table III in Appendix A to construct a normal probability plot of the given data.*
*b. Use part (a) to identify any outliers.*
*c. Use part (a) to assess the normality of the variable under consideration.*

**6.89 Exam Scores.** A sample of the final exam scores in a large introductory statistics course is as follows.

| | | | | |
|---|---|---|---|---|
| 88 | 67 | 64 | 76 | 86 |
| 85 | 82 | 39 | 75 | 34 |
| 90 | 63 | 89 | 90 | 84 |
| 81 | 96 | 100 | 70 | 96 |

**6.90 Cell Phone Rates.** In the February 2003 issue of *Consumer Reports*, different cell phone providers and plans were compared. The monthly fees, in dollars, for a sample of the providers and plans are shown in the following table.

| | | | | |
|---|---|---|---|---|
| 40 | 110 | 90 | 30 | 70 |
| 70 | 30 | 60 | 60 | 50 |
| 60 | 70 | 35 | 80 | 75 |

**6.91 Thoroughbred Racing.** The following table displays finishing times, in seconds, for the winners of fourteen 1-mile thoroughbred horse races, as found in two recent issues of *Thoroughbred Times*.

| | | | | | | |
|---|---|---|---|---|---|---|
| 94.15 | 93.37 | 103.02 | 95.57 | 97.73 | 101.09 | 99.38 |
| 97.19 | 96.63 | 101.05 | 97.91 | 98.44 | 97.47 | 95.10 |

**6.92 Beverage Expenditures.** The Bureau of Labor Statistics publishes information on average annual expenditures by consumers in the *Consumer Expenditure Survey*. In 2000, the mean amount spent by consumers on nonalcoholic beverages was $250. A random sample of 12 consumers yielded the following data, in dollars, on last year's expenditures on nonalcoholic beverages.

| | | | |
|---|---|---|---|
| 361 | 176 | 184 | 265 |
| 259 | 281 | 240 | 273 |
| 259 | 249 | 194 | 258 |

*In Exercises 6.93–6.96, do the following:*
*a. Obtain a normal probability plot of the given data.*
*b. Use part (a) to identify any outliers.*
*c. Use part (a) to assess the normality of the variable under consideration.*

**6.93 Shoe and Apparel E-Tailers.** In the special report "Mousetrap: The Most-Visited Shoe and Apparel E-tailers" (*Footwear News*, Vol. 58, No. 3, p. 18), we found the following data on the average time, in minutes, spent per user per month from January to June of one year for a sample of 15 shoe and apparel retail Web sites.

| | | | | |
|---|---|---|---|---|
| 13.3 | 9.0 | 11.1 | 9.1 | 8.4 |
| 15.6 | 8.1 | 8.3 | 13.0 | 17.1 |
| 16.3 | 13.5 | 8.0 | 15.1 | 5.8 |

**6.94 Hotels and Motels.** The following table provides the daily charges, in dollars, for a sample of 15 hotels and motels operating in South Carolina. The data were found in the report *South Carolina Statistical Abstract*, sponsored by the South Carolina Budget and Control Board.

| | | | | |
|---|---|---|---|---|
| 81.05 | 69.63 | 74.25 | 53.39 | 57.48 |
| 47.87 | 61.07 | 51.40 | 50.37 | 106.43 |
| 47.72 | 58.07 | 56.21 | 130.17 | 95.23 |

**6.95 Oxygen Distribution.** In the article "Distribution of Oxygen in Surface Sediments from Central Sagami Bay, Japan: In Situ Measurements by Microelectrodes and Planar Optodes" (*Deep Sea Research Part I: Oceanographic Research Papers*, Vol. 52, Issue 10, pp. 1974–1987), R. Glud et al. explore the distributions of oxygen in surface sediments from central Sagami Bay. The oxygen distribution gives important information on the general biogeochemistry of marine sediments. Measurements were performed at 16 sites. A sample of 22 depths yielded the following data, in millimoles per square meter per day (mmol $m^{-2}$ $d^{-1}$), on diffusive oxygen uptake (DOU).

| | | | | | | | |
|---|---|---|---|---|---|---|---|
| 1.8 | 2.0 | 1.8 | 2.3 | 3.8 | 3.4 | 2.7 | 1.1 |
| 3.3 | 1.2 | 3.6 | 1.9 | 7.6 | 2.0 | 1.5 | 2.0 |
| 1.1 | 0.7 | 1.0 | 1.8 | 1.8 | 6.7 | | |

**6.96 Medieval Cremation Burials.** In the article "Material Culture as Memory: Combs and Cremations in Early Medieval Britain" (*Early Medieval Europe*, Vol. 12, Issue 2, pp. 89–128), H. Williams discussed the frequency of cremation burials found in 17 archaeological sites in eastern England. Here are the data.

| | | | | | | | | |
|---|---|---|---|---|---|---|---|---|
| 83 | 64 | 46 | 48 | 523 | 35 | 34 | 265 | 2484 |
| 46 | 385 | 21 | 86 | 429 | 51 | 258 | 119 | |

## Working With Large Data Sets

**6.97 Body Temperature.** A study by researchers at the University of Maryland addressed the question of whether the mean body temperature of humans is 98.6°F. The results of the study by P. Mackowiak et al. appeared in the article "A Critical Appraisal of 98.6°F, the Upper Limit of the Normal Body Temperature, and Other Legacies of Carl Reinhold August Wunderlich" (*Journal of the American Medical Association*, Vol. 268, pp. 1578–1580). Among other data, the researchers obtained the body temperatures of 93 healthy humans, as provided on the WeissStats CD. Use the technology of your choice to do the following.

a. Obtain a histogram of the data and use it to assess the (approximate) normality of the variable under consideration.
b. Obtain a normal probability plot of the data and use it to assess the (approximate) normality of the variable under consideration.
c. Compare your results in parts (a) and (b).

**6.98 Vegetarians and Omnivores.** Philosophical and health issues are prompting an increasing number of Taiwanese to switch to a vegetarian lifestyle. A study by Lu et al., published in the *Journal of Nutrition* (Vol. 130, pp. 1591–1596), compared the daily intake of nutrients by vegetarians and omnivores living in Taiwan. Among the nutrients considered was protein. Too little protein stunts growth and interferes with all bodily functions; too much protein puts a strain on the kidneys, can cause diarrhea and dehydration, and can leach calcium from bones and teeth. The daily protein intakes, in grams, for 51 female vegetarians and 53 female omnivores are provided on the WeissStats CD. Use the technology of your choice to do the following for each of the two sets of sample data.

a. Obtain a histogram of the data and use it to assess the (approximate) normality of the variable under consideration.
b. Obtain a normal probability plot of the data and use it to assess the (approximate) normality of the variable under consideration.
c. Compare your results in parts (a) and (b).

**6.99 Chips Ahoy! 1,000 Chips Challenge.** Students in an introductory statistics course at the U.S. Air Force Academy participated in Nabisco's "Chips Ahoy! 1,000 Chips Challenge" by confirming that there were at least 1000 chips in every 18-ounce bag of cookies that they examined. As part of their assignment, they concluded that the number of chips per bag is approximately normally distributed. Their conclusion was based on the data provided on the WeissStats CD, which gives the number of chips per bag for 42 bags. Do you agree with the conclusion of the students? Explain your answer. [SOURCE: B. Warner and J. Rutledge, "Checking the Chips Ahoy! Guarantee," *Chance*, Vol. 12(1), pp. 10–14]

## Extending the Concepts and Skills

**6.100 Finger Length of Criminals.** In 1902, W. R. Macdonell published the article "On Criminal Anthropometry and the Identification of Criminals" (*Biometrika*, 1, pp. 177–227). Among other things, the author presented data on the left middle finger length, in centimeters. The following table provides the midpoints and frequencies of the finger length classes used.

| Midpoint (cm) | Frequency | Midpoint (cm) | Frequency |
|---|---|---|---|
| 9.5 | 1 | 11.6 | 691 |
| 9.8 | 4 | 11.9 | 509 |
| 10.1 | 24 | 12.2 | 306 |
| 10.4 | 67 | 12.5 | 131 |
| 10.7 | 193 | 12.8 | 63 |
| 11.0 | 417 | 13.1 | 16 |
| 11.3 | 575 | 13.4 | 3 |

Use these data and the technology of your choice to assess the normality of middle finger length of criminals by using

a. a histogram.
b. a normal probability plot. Explain your procedure and reasoning in detail.

**6.101 Gestation Periods of Humans.** For humans, gestation periods are normally distributed with a mean of 266 days and a standard deviation of 16 days.

a. Use the technology of your choice to simulate four random samples of 50 human gestation periods each.
b. Obtain a normal probability plot of each sample in part (a).
c. Are the normal probability plots in part (b) what you expected? Explain your answer.

**6.102 Emergency Room Traffic.** Desert Samaritan Hospital in Mesa, Arizona, keeps records of emergency room traffic. Those records reveal that the times between arriving patients have a special type of reverse J-shaped distribution called an *exponential distribution.* The records also show that the mean time between arriving patients is 8.7 minutes.

a. Use the technology of your choice to simulate four random samples of 75 interarrival times each.

b. Obtain a normal probability plot of each sample in part (a).

c. Are the normal probability plots in part (b) what you expected? Explain your answer.

## Chapter in Review

### You Should be Able to

1. use and understand the formulas in this chapter.

2. explain what it means for a variable to be normally distributed or approximately normally distributed.

3. explain the meaning of the parameters for a normal curve.

4. identify the basic properties of and sketch a normal curve.

5. identify the standard normal distribution and the standard normal curve.

6. use Table II to determine areas under the standard normal curve.

7. use Table II to determine the z-score(s) corresponding to a specified area under the standard normal curve.

8. use and understand the $z_\alpha$ notation.

9. determine a percentage or probability for a normally distributed variable.

10. state and apply the 68.26-95.44-99.74 rule.

11. determine the observations corresponding to a specified percentage or probability for a normally distributed variable.

12. explain how to assess the normality of a variable with a normal probability plot.

13. construct a normal probability plot with the aid of Table III.

14. use a normal probability plot to detect outliers.

### Key Terms

68.26-95.44-99.74 rule, *286*
approximately normally distributed variable, *268*
cumulative probability, *289*
empirical rule, *287*
inverse cumulative probability, *290*
normal curve, *268*

normal distribution, *268*
normal probability plot, *294*
normal scores, *294*
normally distributed population, *268*
normally distributed variable, *268*
parameters, *268*
standard normal curve, *271*

standard normal distribution, *271*
standardized normally distributed variable, *271*
$z_\alpha$, *281*
z-curve, *277*

## Review Problems

### Understanding the Concepts and Skills

1. State two of the main reasons for studying the normal distribution.

2. Define
a. normally distributed variable.
b. normally distributed population.
c. parameters for a normal curve.

3. Answer true or false to each statement. Give reasons for your answers.
a. Two variables that have the same mean and standard deviation have the same distribution.

b. Two normally distributed variables that have the same mean and standard deviation have the same distribution.

4. Explain the relationship between percentages for a normally distributed variable and areas under the corresponding normal curve.

5. Identify the distribution of the standardized version of a normally distributed variable.

6. Answer true or false to each statement. Explain your answers.

**a.** Two normal distributions that have the same mean are centered at the same place, regardless of the relationship between their standard deviations.
**b.** Two normal distributions that have the same standard deviation have the same shape, regardless of the relationship between their means.

**7.** Consider the normal curves that have the parameters $\mu = 1.5$ and $\sigma = 3$; $\mu = 1.5$ and $\sigma = 6.2$; $\mu = -2.7$ and $\sigma = 3$; $\mu = 0$ and $\sigma = 1$.
**a.** Which curve has the largest spread?
**b.** Which curves are centered at the same place?
**c.** Which curves have the same shape?
**d.** Which curve is centered farthest to the left?
**e.** Which curve is the standard normal curve?

**8.** What key fact permits you to determine percentages for a normally distributed variable by first converting to $z$-scores and then determining the corresponding area under the standard normal curve?

**9.** Explain how to use Table II to determine the area under the standard normal curve that lies
**a.** to the left of a specified $z$-score.
**b.** to the right of a specified $z$-score.
**c.** between two specified $z$-scores.

**10.** Explain how to use Table II to determine the $z$-score that has a specified area to its
**a.** left under the standard normal curve.
**b.** right under the standard normal curve.

**11.** What does the symbol $z_\alpha$ signify?

**12.** State the 68.26-95.44-99.74 rule.

**13.** Roughly speaking, what are the normal scores corresponding to a sample of observations?

**14.** If you observe the values of a normally distributed variable for a sample, a normal probability plot should be roughly _____.

**15.** Sketch the normal curve having the parameters
**a.** $\mu = -1$ and $\sigma = 2$.  **b.** $\mu = 3$ and $\sigma = 2$.
**c.** $\mu = -1$ and $\sigma = 0.5$.

**16. Forearm Length.** In 1903, K. Pearson and A. Lee published a paper entitled "On the Laws of Inheritance in Man. I. Inheritance of Physical Characters" (*Biometrika*, Vol. 2, pp. 357–462). From information presented in that paper, forearm lengths of men, measured from the elbow to the middle fingertip, are (roughly) normally distributed with a mean of 18.8 inches and a standard deviation of 1.1 inches. Let $x$ denote forearm length, in inches, for men.
**a.** Sketch the distribution of the variable $x$.
**b.** Obtain the standardized version, $z$, of $x$.
**c.** Identify and sketch the distribution of $z$.

**d.** The area under the normal curve with parameters 18.8 and 1.1 that lies between 17 and 20 is 0.8115. Determine the probability that a randomly selected man will have a forearm length between 17 inches and 20 inches.
**e.** The percentage of men who have forearm lengths less than 16 inches equals the area under the standard normal curve that lies to the _____ of _____.

**17.** According to Table II, the area under the standard normal curve that lies to the left of 1.05 is 0.8531. Without further reference to Table II, determine the area under the standard normal curve that lies
**a.** to the right of 1.05.
**b.** to the left of $-1.05$.
**c.** between $-1.05$ and 1.05.

**18.** Determine and sketch the area under the standard normal curve that lies
**a.** to the left of $-3.02$.  **b.** to the right of 0.61.
**c.** between 1.11 and 2.75.  **d.** between $-2.06$ and 5.02.
**e.** between $-4.11$ and $-1.5$.
**f.** either to the left of 1 or to the right of 3.

**19.** For the standard normal curve, find the $z$-score(s)
**a.** that has area 0.30 to its left.
**b.** that has area 0.10 to its right.
**c.** $z_{0.025}$, $z_{0.05}$, $z_{0.01}$, and $z_{0.005}$.
**d.** that divide the area under the curve into a middle 0.99 area and two outside 0.005 areas.

**20. Birth Weights.** The *WONDER database*, maintained by the Centers for Disease Control and Prevention, provides a single point of access to a wide variety of reports and numeric public health data. From that database, we obtained the following data for one year's birth weights of male babies who weighed under 5000 grams (about 11 pounds).

| Weight (g) | Frequency |
|---|---|
| $0 \leftarrow 500$ | 2,025 |
| $500 \leftarrow 1000$ | 8,400 |
| $1000 \leftarrow 1500$ | 10,215 |
| $1500 \leftarrow 2000$ | 19,919 |
| $2000 \leftarrow 2500$ | 67,068 |
| $2500 \leftarrow 3000$ | 274,913 |
| $3000 \leftarrow 3500$ | 709,110 |
| $3500 \leftarrow 4000$ | 609,719 |
| $4000 \leftarrow 4500$ | 191,826 |
| $4500 \leftarrow 5000$ | 31,942 |

**a.** Obtain a relative-frequency histogram of these weight data.
**b.** Based on your histogram, do you think that, for the year in question, the birth weights of male babies who weighed under 5000 grams are approximately normally distributed? Explain your answer.

**21. Joint Fluids and Knee Surgery.** Proteins in the knee provide measures of lubrication and wear. In the article "Composition of Joint Fluid in Patients Undergoing Total Knee Replacement and Revision Arthroplasty" (*Biomaterials*, Vol. 25, No. 18, pp. 4433–4445), D. Mazzucco, et al. hypothesized that the protein make-up in the knee would change when patients undergo a total knee arthroplasty surgery. The mean concentration of hyaluronic acid in the knees of patients receiving total knee arthroplasty is 1.3 mg/ml; the standard deviation is 0.4 mg/ml. Assuming that hyaluronic acid concentration is normally distributed, find the percentage of patients receiving total knee arthroplasty who have a knee hyaluronic acid concentration
**a.** below 1.4 mg/ml.
**b.** between 1 and 2 mg/ml.
**c.** above 2.1 mg/ml.

**22. Verbal GRE Scores.** The Graduate Record Examination (GRE) is a standardized test that students usually take before entering graduate school. According to the Office of Educational Research and Improvement and the report *Digest of Education Statistics*, the scores on the verbal portion of the 2000 GRE are (approximately) normally distributed with mean 465 points and standard deviation 116 points.
**a.** Obtain and interpret the quartiles for these scores.
**b.** Find and interpret the 99th percentile for these scores.

**23. Verbal GRE Scores.** Refer to Problem 22 and fill in the following blanks.
**a.** 68.26% of students who took the verbal portion of the 2000 GRE scored between _____ and _____.
**b.** 95.44% of students who took the verbal portion of the 2000 GRE scored between _____ and _____.
**c.** 99.74% of students who took the verbal portion of the 2000 GRE scored between _____ and _____.

**24. Gas Prices.** According to the *AAA Daily Fuel Gauge Report*, the national average price for regular unleaded gasoline on December 6, 2005 was $2.127. That same day, a random sample of 12 gas stations across the country yielded the following prices for regular unleaded gasoline.

| | | | |
|---|---|---|---|
| 2.03 | 2.17 | 2.29 | 1.96 |
| 2.04 | 2.14 | 2.06 | 2.13 |
| 2.07 | 1.97 | 2.17 | 2.09 |

**a.** Use Table III to construct a normal probability plot for the gas-price data.
**b.** Use part (a) to identify any outliers.
**c.** Use part (a) to assess normality.
**d.** If you have access to technology, use it to obtain a normal probability plot for the gas-price data.

**25. Mortgage Industry Employees.** In an issue of *National Mortgage News*, a special report was published on publicly traded mortgage industry companies. A sample of 25 mortgage industry companies had the following numbers of employees.

| | | | | |
|---|---|---|---|---|
| 260 | 20,800 | 1,801 | 2,073 | 3,596 |
| 3,223 | 2,128 | 1,796 | 17,540 | 15 |
| 29,272 | 6,929 | 2,468 | 7,000 | 6,600 |
| 2,458 | 3,216 | 209 | 726 | 9,200 |
| 650 | 4,800 | 19,400 | 24,886 | 3,082 |

**a.** Obtain a normal probability plot of the data.
**b.** Use part (a) to identify any outliers.
**c.** Use part (a) to assess the normality of the variable under consideration.

# Focusing on Data Analysis  UWEC Undergraduates

Recall from Chapter 1 (see page 34) that the Focus database and Focus sample contain information on the undergraduate students at the University of Wisconsin - Eau Claire (UWEC). Now would be a good time for you to review the discussion about these data sets.

Begin by opening the Focus sample (FocusSample) in the statistical software package of your choice.

**a.** Obtain a normal probability plot of the sample data for each of the following variables: high school percentile, cumulative GPA, age, total earned credits, ACT English score, ACT math score, and ACT composite score.

**b.** Based on your results from part (a), which of the variables considered there appear to be approximately normally distributed?

**c.** Based on your results from part (a), which of the variables considered there appear to be far from normally distributed?

If your statistical software package will accommodate the entire Focus database (Focus), open that worksheet.

**d.** Obtain a histogram for each of the following variables: high school percentile, cumulative GPA, age, total earned credits, ACT English score, ACT math score, and ACT composite score.

**e.** In view of the histograms that you obtained in part (d), comment on your answers in parts (b) and (c).

# Case Study Discussion  Foot Length and Shoe Size for Women

On page 267, we discussed foot length and shoe size for women. Now that you have studied the normal distribution, you can solve several problems involving foot length and shoe size:

**a.** We mentioned on page 267 that the lengths of women's feet are normally distributed with a mean of 9.58 inches and a standard deviation of 0.51 inch. Based on that information, sketch the distribution of women's foot length and compare your result to the graph presented on page 267.

**b.** What percentage of women have foot lengths between 9 and 10 inches?

**c.** What percentage of women have foot lengths that exceed 11 inches?

**d.** Shoe manufacturers suggest that if a foot length is between two sizes, wear the larger size. Referring to the table on page 267, determine the percentage of women who wear size 8 shoes; size $11\frac{1}{2}$ shoes.

**e.** If an owner of a chain of shoe stores intends to purchase 10,000 pairs of women's shoes, roughly how many should he purchase of size 8? of size $11\frac{1}{2}$? Explain your reasoning.

# Biography    CARL FRIEDRICH GAUSS: Child Prodigy

**Carl Friedrich Gauss** was born on April 30, 1777, in Brunswick, Germany, the only son in a poor, semiliterate peasant family; he taught himself to calculate before he could talk. At the age of 3, he pointed out an error in his father's calculations of wages. In addition to his arithmetic experimentation, he taught himself to read. At the age of 8, Gauss instantly solved the summing of all numbers from 1 to 100. His father was persuaded to allow him to stay in school and to study after school instead of working to help support the family.

Impressed by Gauss's brilliance, the Duke of Brunswick supported him monetarily from the ages of 14 to 30. This patronage permitted Gauss to pursue his studies exclusively. He conceived most of his mathematical discoveries by the time he was 17. Gauss was granted a doctorate in absentia from the university at Helmstedt; his doctoral thesis developed the concept of complex numbers and proved the fun-

damental theorem of algebra, which had previously been only partially established. Shortly thereafter, Gauss published his theory of numbers, which is considered one of the most brilliant achievements in mathematics.

Gauss made important discoveries in mathematics, physics, astronomy, and statistics. Two of his major contributions to statistics were the development of the least-squares method and fundamental work with the normal distribution, often called the *Gaussian distribution* in his honor.

In 1807, Gauss accepted the directorship of the observatory at the University of Göttingen which ended his dependence on the Duke of Brunswick. He remained there the rest of his life. In 1833, Gauss and a colleague, Wilhelm Weber, invented a working electric telegraph, 5 years before Samuel Morse. Gauss died in Göttingen in 1855.

# StatCrunch in MyStatLab
## Analyzing Data Online

StatCrunch online statistical software offers an easy-to-use interface customized for this book. The StatCrunch feature for each chapter illustrates the use of the software to perform a statistical analysis discussed in the chapter. Exercises are provided to further apply StatCrunch to other statistical analyses examined in the chapter. Go to the WeissStats CD or to the Weiss Web site at www.aw-bc.com/weiss to access StatCrunch instructions and data sets. To access StatCrunch statistical software, go to the student content area of your Weiss MyStatLab course.

# Internet Projects
## Exploring Data Online

The Internet project for each chapter provides simulations, demonstrations, or activities that enhance the topics covered in the chapter. The project materials come from universities, individuals, governments, and companies from all over the world. To access the Internet projects on the Web, go to www.aw-bc.com/weiss. From this Web page, you can reach the Internet Projects Page, which we suggest that you bookmark for easy access in the future.

# 7

# The Sampling Distribution of the Sample Mean

## Chapter Objectives

In the preceding chapters, you have studied sampling, descriptive statistics, probability, and the normal distribution. Now you will learn how these seemingly diverse topics can be integrated to lay the groundwork for inferential statistics.

In Section 7.1, we introduce the concepts of *sampling error* and *sampling distribution* and explain the essential role these concepts play in the design of inferential studies. The *sampling distribution* of a statistic is the distribution of the statistic, that is, the distribution of all possible observations of the statistic for samples of a given size from a population. In this chapter, we concentrate on the sampling distribution of the sample mean.

In Sections 7.2 and 7.3, we provide the required background for applying the sampling distribution of the sample mean. Specifically, in Section 7.2, we present formulas for the mean and standard deviation of the sample mean. Then, in Section 7.3, we indicate that, under certain general conditions, the sampling distribution of the sample mean is a normal distribution, or at least approximately so.

We apply this momentous fact in Chapters 8 and 9 to develop two important statistical-inference procedures: using the mean, $\bar{x}$, of a sample from a population to estimate and to draw conclusions about the mean, $\mu$, of the entire population.

# The Chesapeake and Ohio Freight Study

Can relatively small samples actually provide results that are nearly as accurate as those obtained from a census? Statisticians have proven that such is the case, but a real study with sample and census results can be enlightening.

When a freight shipment travels over several railroads, the revenue from the freight charge is appropriately divided among those railroads. A *waybill*, which accompanies each freight shipment, provides information on the goods, route, and total charges. From the waybill, the amount due each railroad can be calculated.

Calculating these allocations for a large number of shipments is time consuming and costly. If the division of total revenue to the railroads could be done accurately on the basis of a sample—as statisticians contend—considerable savings could be realized in accounting and clerical costs.

To convince themselves of the validity of the sampling approach, officials of the Chesapeake and Ohio Railroad Company (C&O) undertook a study of freight shipments that had traveled over its Pere Marquette district and another railroad during a 6-month period. The total number of waybills for that period (22,984) and the total freight revenue were known.

The study used statistical theory to determine the smallest number of waybills needed to estimate, with a prescribed accuracy, the total freight revenue due C&O. In all, 2072 of the 22,984 waybills, roughly 9%, were sampled. For each waybill in the sample, the amount of freight revenue due C&O was calculated and, from those amounts, the total revenue due C&O was estimated to be $64,568.

How close was the estimate of $64,568, based on a sample of only 2072 waybills, to the total revenue actually due C&O for the 22,984 waybills? Take a guess! We'll discuss the answer at the end of this chapter.

## 7.1 Sampling Error; the Need for Sampling Distributions

We have already seen that using a sample to acquire information about a population is often preferable to conducting a census. Generally, sampling is less costly and can be done more quickly than a census; it is often the only practical way to gather information.

However, because a sample provides data for only a portion of an entire population, we cannot expect the sample to yield perfectly accurate information about the population. Thus we should anticipate that a certain amount of error—called *sampling error*—will result simply because we are sampling.

---

**Definition 7.1** | **Sampling Error**

**Sampling error** is the error resulting from using a sample to estimate a population characteristic.

---

**Example 7.1** | **Sampling Error and the Need for Sampling Distributions**

*Income Tax* The U.S. Internal Revenue Service (IRS) publishes annual figures on individual income tax returns in *Statistics of Income, Individual Income Tax Returns*. For the year 2003, the IRS reported that the mean tax of individual income tax returns was $8412. In actuality, the IRS reported the mean tax of a sample of 175,485 individual income tax returns from a total of more than 130 million such returns.

a. Identify the population under consideration.

b. Identify the variable under consideration.

c. Is the mean tax reported by the IRS a sample mean or the population mean?

d. Should we expect the mean tax, $\bar{x}$, of the 175,485 returns sampled by the IRS to be exactly the same as the mean tax, $\mu$, of all individual income tax returns for 2003?

e. How can we answer questions about sampling error? For instance, is the sample mean tax, $\bar{x}$, reported by the IRS likely to be within $100 of the population mean tax, $\mu$?

**Solution**

a. The population consists of all individual income tax returns for the year 2003.

b. The variable is "tax" (amount of income tax).

c. The mean tax reported is a sample mean, namely, the mean tax, $\bar{x}$, of the 175,485 returns sampled. It is not the population mean tax, $\mu$, of all individual income tax returns for 2003.

d. We certainly cannot expect the mean tax, $\bar{x}$, of the 175,485 returns sampled by the IRS to be exactly the same as the mean tax, $\mu$, of all individual income tax returns for 2003—some sampling error is to be anticipated.

e. To answer questions about sampling error, we need to know the distribution of all possible sample mean tax amounts (i.e., all possible $\bar{x}$-values) that could be obtained by sampling 175,485 individual income tax returns. That distribution is called the *sampling distribution of the sample mean*.

•  •  •

The distribution of a statistic (i.e., of all possible observations of the statistic for samples of a given size) is called the **sampling distribution** of the statistic. In this chapter, we concentrate on the *sampling distribution of the sample mean*, that is, of the statistic $\bar{x}$.

| Definition 7.2 | **Sampling Distribution of the Sample Mean** |
|---|---|

For a variable $x$ and a given sample size, the distribution of the variable $\bar{x}$ is called the **sampling distribution of the sample mean.**

**What Does It Mean?**

The sampling distribution of the sample mean is the distribution of all possible sample means for samples of a given size.

In statistics, the following terms and phrases are synonymous.

* Sampling distribution of the sample mean
* Distribution of the variable $\bar{x}$
* Distribution of all possible sample means of a given sample size

We, therefore, use these three terms interchangeably.

Introducing the sampling distribution of the sample mean with an example that is both realistic and concrete is difficult because even for moderately large populations the number of possible samples is enormous, thus prohibiting an actual listing of the possibilities.[†] Consequently, we use an unrealistically small population to introduce this concept.

| Example 7.2 | **Sampling Distribution of the Sample Mean** |
|---|---|

*Heights of Starting Players*  Suppose that the population of interest consists of the five starting players on a men's basketball team whom we will call A, B, C, D, and E. Further suppose that the variable of interest is height, in inches. Table 7.1 lists the players and their heights.

**TABLE 7.1**
Heights, in inches,
of the five starting players

| Player | A | B | C | D | E |
|---|---|---|---|---|---|
| Height | 76 | 78 | 79 | 81 | 86 |

a.  Obtain the sampling distribution of the sample mean for samples of size 2.

b.  Make some observations about sampling error when the mean height of a random sample of two players is used to estimate the population mean height.[‡]

c.  Find the probability that, for a random sample of size 2, the sampling error made in estimating the population mean by the sample mean will be 1 inch or less; that is, determine the probability that $\bar{x}$ will be within 1 inch of $\mu$.

**Solution**  For future reference we first compute the population mean height:

$$\mu = \frac{\Sigma x_i}{N} = \frac{76 + 78 + 79 + 81 + 86}{5} = 80 \text{ inches.}$$

---

[†]For example, the number of possible samples of size 50 from a population of size 10,000 is approximately equal to $3 \times 10^{135}$, a 3 followed by 135 zeros.

[‡]As we mentioned in Section 1.2, the statistical-inference techniques considered in this book are intended for use only with simple random sampling. Therefore, unless otherwise specified, when we say *random sample*, we mean *simple random sample*. Furthermore, we assume that sampling is without replacement unless explicitly stated otherwise.

**TABLE 7.2**

Possible samples and sample means for samples of size 2

| Sample | Heights | $\bar{x}$ |
|--------|---------|-----------|
| A, B | 76, 78 | 77.0 |
| A, C | 76, 79 | 77.5 |
| A, D | 76, 81 | 78.5 |
| A, E | 76, 86 | 81.0 |
| B, C | 78, 79 | 78.5 |
| B, D | 78, 81 | 79.5 |
| B, E | 78, 86 | 82.0 |
| C, D | 79, 81 | 80.0 |
| C, E | 79, 86 | 82.5 |
| D, E | 81, 86 | 83.5 |

**a.** The population is so small that we can list the possible samples of size 2. The first column of Table 7.2 gives the 10 possible samples, the second column the corresponding heights (values of the variable "height"), and the third column the sample means. Figure 7.1 is a dotplot for the distribution of the sample means (the sampling distribution of the sample mean for samples of size 2).

**b.** From Table 7.2 or Fig. 7.1, we see that the mean height of the two players selected isn't likely to equal the population mean of 80 inches. In fact, only 1 of the 10 samples has a mean of 80 inches, the eighth sample in Table 7.2. The chances are, therefore, only $\frac{1}{10}$, or 10%, that $\bar{x}$ will equal $\mu$; some sampling error is likely.

**c.** Figure 7.1 shows that 3 of the 10 samples have means within 1 inch of the population mean of 80 inches. So the probability is $\frac{3}{10}$, or 0.3, that the sampling error made in estimating $\mu$ by $\bar{x}$ will be 1 inch or less.

**FIGURE 7.1**

Dotplot for the sampling distribution of the sample mean for samples of size 2 ($n = 2$)

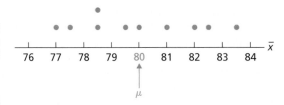

**You try it!**

Exercise 7.3 on page 312

**Interpretation** There is a 30% chance that the mean height of the two players selected will be within 1 inch of the population mean.

• • •

In the previous example, we determined the sampling distribution of the sample mean for samples of size 2. If we consider samples of another size—say, of size 4—we obtain a different sampling distribution of the sample mean, as demonstrated in the next example.

**Example 7.3** | **Sampling Distribution of the Sample Mean**

*Heights of Starting Players* Refer to Table 7.1, the heights of the five starting players on a men's basketball team.

**a.** Obtain the sampling distribution of the sample mean for samples of size 4.

**b.** Make some observations about sampling error when the mean height of a random sample of four players is used to estimate the population mean height.

**c.** Find the probability that, for a random sample of size 4, the sampling error made in estimating the population mean by the sample mean will be 1 inch or less; that is, determine the probability that $\bar{x}$ will be within 1 inch of $\mu$.

**TABLE 7.3**

Possible samples and sample means for samples of size 4

| Sample | Heights | $\bar{x}$ |
|--------|---------|-----------|
| A, B, C, D | 76, 78, 79, 81 | 78.50 |
| A, B, C, E | 76, 78, 79, 86 | 79.75 |
| A, B, D, E | 76, 78, 81, 86 | 80.25 |
| A, C, D, E | 76, 79, 81, 86 | 80.50 |
| B, C, D, E | 78, 79, 81, 86 | 81.00 |

**Solution**

**a.** There are five possible samples of size 4. The first column of Table 7.3 gives the possible samples, the second column the corresponding heights (values of the variable "height"), and the third column the sample means. Figure 7.2 is a dotplot for the distribution of the sample means.

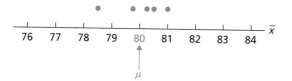

**FIGURE 7.2**

Dotplot for the sampling distribution
of the sample mean for samples
of size 4 ($n = 4$)

**b.** From Table 7.3 or Fig. 7.2, we see that none of the samples of size 4 has a mean equal to the population mean of 80 inches. Thus, some sampling error is certain.

**c.** Figure 7.2 shows that four of the five samples have means within 1 inch of the population mean of 80 inches. So the probability is $\frac{4}{5}$, or 0.8, that the sampling error made in estimating $\mu$ by $\bar{x}$ will be 1 inch or less.

Exercise 7.5
on page 313

**Interpretation** There is an 80% chance that the mean height of the four players selected will be within 1 inch of the population mean.

• • •

## Sample Size and Sampling Error

We continue our look at the sampling distributions of the sample mean for the heights of the five starting players on a basketball team. In Figs. 7.1 and 7.2, we drew dotplots for the sampling distributions of the sample mean for samples of sizes 2 and 4, respectively. Those two dotplots and dotplots for samples of sizes 1, 3, and 5 are displayed in Fig. 7.3.

Figure 7.3 vividly illustrates that the possible sample means cluster more closely around the population mean as the sample size increases. This result

**FIGURE 7.3**

Dotplots for the sampling distributions
of the sample mean for the heights
of the five starting players for samples
of sizes 1, 2, 3, 4, and 5

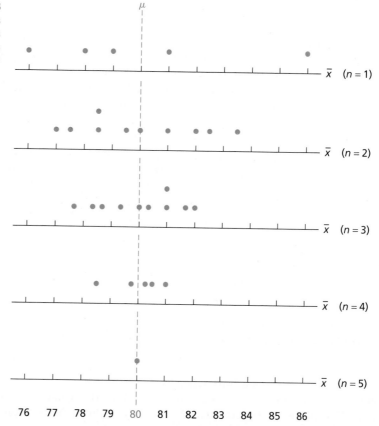

suggests that sampling error tends to be smaller for large samples than for small samples.

For example, Fig. 7.3 reveals that, for samples of size 1, two of five (40%) of the possible sample means lie within 1 inch of $\mu$. Likewise, for samples of sizes 2, 3, 4, and 5, respectively, three of ten (30%), five of ten (50%), four of five (80%), and one of one (100%) of the possible sample means lie within 1 inch of $\mu$. The first four columns of Table 7.4 summarize these results. The last two columns of that table provide other sampling-error results, easily obtained from Fig. 7.3.

**TABLE 7.4**

Sample size and sampling error illustrations for the heights of the basketball players ("No." is an abbreviation of "Number")

| Sample size $n$ | No. possible samples | No. within 1" of $\mu$ | % within 1" of $\mu$ | No. within 0.5" of $\mu$ | % within 0.5" of $\mu$ |
|---|---|---|---|---|---|
| 1 | 5 | 2 | 40% | 0 | 0% |
| 2 | 10 | 3 | 30% | 2 | 20% |
| 3 | 10 | 5 | 50% | 2 | 20% |
| 4 | 5 | 4 | 80% | 3 | 60% |
| 5 | 1 | 1 | 100% | 1 | 100% |

More generally, we can make the following qualitative statement.

**Key Fact 7.1**

## Sample Size and Sampling Error

The larger the sample size, the smaller the sampling error tends to be in estimating a population mean, $\mu$, by a sample mean, $\bar{x}$.

**What Does It Mean?**

The possible sample means cluster more closely around the population mean as the sample size increases.

## What We Do in Practice

We used the heights of a population of five basketball players to illustrate and explain the importance of the sampling distribution of the sample mean. For that small population with known population data, we easily determined the sampling distribution of the sample mean for any particular sample size by listing all possible sample means.

In practice, however, the populations with which we work are large and the population data are unknown, so proceeding as we did in the basketball-player example isn't possible. What do we do, then, in the usual case of a large and unknown population? Fortunately, we can use mathematical relationships to approximate the sampling distribution of the sample mean. We discuss those relationships in Sections 7.2 and 7.3.

# Exercises 7.1

## Understanding the Concepts and Skills

**7.1** Why is sampling often preferable to conducting a census for the purpose of obtaining information about a population?

**7.2** Why should you generally expect some error when estimating a parameter (e.g., a population mean) by a statistic (e.g., a sample mean)? What is this kind of error called?

*Exercises 7.3–7.15 are intended solely to provide concrete illustrations of the sampling distribution of the sample mean. For that reason, the populations considered are unrealistically small. In each exercise, assume that sampling is without replacement.*

**7.3 NBA Champs.** The winner of the 2002–2003 National Basketball Association (NBA) championship was the San

Antonio Spurs. The following table provides the usual starting players and their positions and heights.

| Player | Position | Height (in.) |
|---|---|---|
| Bruce Bowen (B) | Forward | 79 |
| Tim Duncan (D) | Forward | 84 |
| David Robinson (R) | Center | 85 |
| Manu Ginobli (G) | Guard | 78 |
| Tony Parker (P) | Guard | 74 |

**a.** Find the population mean height of the five players.
**b.** For samples of size 2, construct a table similar to Table 7.2 on page 310. Use the letter in parentheses after each player's name to represent each player.
**c.** Draw a dotplot for the sampling distribution of the sample mean for samples of size 2 like the one shown in Fig. 7.1 on page 310.
**d.** For a random sample of size 2, what is the chance that the sample mean will equal the population mean?
**e.** For a random sample of size 2, obtain the probability that the sampling error made in estimating the population mean by the sample mean will be 1 inch or less; that is, determine the probability that $\bar{x}$ will be within 1 inch of $\mu$. Interpret your result in terms of percentages.

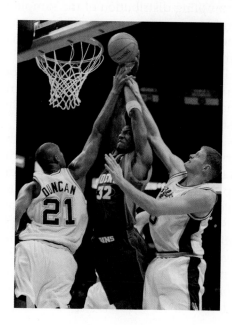

**7.4 NBA Champs.** Repeat parts (b)–(e) of Exercise 7.3 for samples of size 1.

**7.5 NBA Champs.** Repeat parts (b)–(e) of Exercise 7.3 for samples of size 3.

**7.6 NBA Champs.** Repeat parts (b)–(e) of Exercise 7.3 for samples of size 4.

**7.7 NBA Champs.** Repeat parts (b)–(e) of Exercise 7.3 for samples of size 5.

**7.8 NBA Champs.** This exercise requires that you have done Exercises 7.3–7.7.
**a.** Draw a graph similar to that shown in Fig. 7.3 on page 311 for sample sizes of 1, 2, 3, 4, and 5.
**b.** What does your graph in part (a) illustrate about the impact of increasing sample size on sampling error?
**c.** Construct a table similar to Table 7.4 on page 312 for some values of your choice.

**7.9 World's Richest.** Each year, *Forbes* magazine publishes a list of the world's richest people. In 2003, the six richest people, their countries, and their wealth (to the nearest billion dollars) are as shown in the following table. Consider these six people a population of interest.

| Person | Country | Wealth ($ billions) |
|---|---|---|
| William H. Gates III (G) | United States | 41 |
| Warren E. Buffett (B) | United States | 31 |
| Karl & Theo Albrecht (A) | Germany | 26 |
| Paul G. Allen (P) | United States | 20 |
| Prince Alwaleed Bin Talal Alsaud (T) | Saudi Arabia | 18 |
| Lawrence J. Ellison (E) | United States | 17 |

**a.** Calculate the mean wealth, $\mu$, of the six people.
**b.** For samples of size 2, construct a table similar to Table 7.2 on page 310. (There are 15 possible samples of size 2.)
**c.** Draw a dotplot for the sampling distribution of the sample mean for samples of size 2.
**d.** For a random sample of size 2, what is the chance that the sample mean will equal the population mean?
**e.** For a random sample of size 2, determine the probability that the mean wealth of the two people obtained will be within 2 (i.e., $2 billion) of the population mean. Interpret your result in terms of percentages.

**7.10 World's Richest.** Repeat parts (b)–(e) of Exercise 7.9 for samples of size 1.

**7.11 World's Richest.** Repeat parts (b)–(e) of Exercise 7.9 for samples of size 3. (There are 20 possible samples.)

**7.12 World's Richest.** Repeat parts (b)–(e) of Exercise 7.9 for samples of size 4. (There are 15 possible samples.)

**7.13 World's Richest.** Repeat parts (b)–(e) of Exercise 7.9 for samples of size 5. (There are six possible samples.)

**7.14 World's Richest.** Repeat parts (b)–(e) of Exercise 7.9 for samples of size 6. What is the relationship between the only possible sample here and the population?

**7.15 World's Richest.** Explain what the dotplots in part (c) of Exercises 7.9–7.14 illustrate about the impact of increasing sample size on sampling error.

## Extending the Concepts and Skills

**7.16** Suppose that a sample is to be taken without replacement from a finite population of size $N$. If the sample size is the same as the population size,
a. how many possible samples are there?
b. what are the possible sample means?
c. what is the relationship between the only possible sample and the population?

**7.17** Suppose that a random sample of size 1 is to be taken from a finite population of size $N$.
a. How many possible samples are there?
b. Identify the relationship between the possible sample means and the possible observations of the variable under consideration.
c. What is the difference between taking a random sample of size 1 from a population and selecting a member at random from the population?

## 7.2 The Mean and Standard Deviation of the Sample Mean

In Section 7.1, we discussed the sampling distribution of the sample mean—the distribution of all possible sample means for any specified sample size or, equivalently, the distribution of the variable $\bar{x}$. We use that distribution to make inferences about the population mean based on a sample mean.

As we said earlier, we generally do not know the sampling distribution of the sample mean exactly. Fortunately, however, we can often approximate that sampling distribution by a normal distribution; that is, under certain conditions, the variable $\bar{x}$ is approximately normally distributed.

Recall that a variable is normally distributed if its distribution has the shape of a normal curve and that a normal distribution is determined by the mean and standard deviation. Hence a first step in learning how to approximate the sampling distribution of the sample mean by a normal distribution is to obtain the mean and standard deviation of the sample mean, that is, of the variable $\bar{x}$. We describe how to do that in this section.

To begin, let's review the notation used for the mean and standard deviation of a variable. Recall that the mean of a variable is denoted $\mu$, subscripted if necessary with the letter representing the variable. So the mean of $x$ is written as $\mu_x$, the mean of $y$ as $\mu_y$, and so on. In particular, then, the mean of $\bar{x}$ is written as $\mu_{\bar{x}}$; similarly, the standard deviation of $\bar{x}$ is written as $\sigma_{\bar{x}}$.

### The Mean of the Sample Mean

There is a simple relationship between the mean of the variable $\bar{x}$ and the mean of the variable under consideration: They are equal, or $\mu_{\bar{x}} = \mu$. In other words, for any particular sample size, the mean of all possible sample means equals the population mean. This equality holds regardless of the size of the sample. In Example 7.4, we illustrate the relationship $\mu_{\bar{x}} = \mu$ by returning to the heights of the basketball players considered in Section 7.1.

| Example 7.4 | Mean of the Sample Mean |

*Heights of Starting Players*  The heights, in inches, of the five starting players on a men's basketball team are repeated in Table 7.5. Here the population is the five players and the variable is height.

**TABLE 7.5**
Heights of the five starting players

| Player | A | B | C | D | E |
|--------|---|---|---|---|---|
| Height | 76 | 78 | 79 | 81 | 86 |

a.  Determine the population mean, $\mu$.

b.  Obtain the mean, $\mu_{\bar{x}}$, of the variable $\bar{x}$ for samples of size 2. Verify that the relation $\mu_{\bar{x}} = \mu$ holds.

c.  Repeat part (b) for samples of size 4.

**Solution**

a.  To determine the population mean (the mean of the variable "height"), we apply Definition 3.11 on page 136 to the heights in Table 7.5:

$$\mu = \frac{\Sigma x_i}{N} = \frac{76 + 78 + 79 + 81 + 86}{5} = 80 \text{ inches.}$$

Thus the mean height of the five players is 80 inches.

b.  To obtain the mean of the variable $\bar{x}$ for samples of size 2, we again apply Definition 3.11, but this time to $\bar{x}$. Referring to the third column of Table 7.2 on page 310, we get

$$\mu_{\bar{x}} = \frac{77.0 + 77.5 + \cdots + 83.5}{10} = 80 \text{ inches.}$$

By part (a), $\mu = 80$ inches. So, for samples of size 2, $\mu_{\bar{x}} = \mu$.

**Interpretation**  For samples of size 2, the mean of all possible sample means equals the population mean.

c.  Proceeding as in part (b), but this time referring to the third column of Table 7.3 on page 310, we obtain the mean of the variable $\bar{x}$ for samples of size 4:

$$\mu_{\bar{x}} = \frac{78.50 + 79.75 + 80.25 + 80.50 + 81.00}{5} = 80 \text{ inches,}$$

which again is the same as $\mu$.

**Interpretation**  For samples of size 4, the mean of all possible sample means equals the population mean.

• • •

**You try it!**

Exercise 7.25
on page 319

For emphasis, we restate the relationship $\mu_{\bar{x}} = \mu$ in Formula 7.1.

| Formula 7.1 | **Mean of the Sample Mean** |

For samples of size $n$, the mean of the variable $\bar{x}$ equals the mean of the variable under consideration. In symbols,

$$\mu_{\bar{x}} = \mu.$$

**What Does It Mean?**

For each sample size, the mean of all possible sample means equals the population mean.

### The Standard Deviation of the Sample Mean

Next, we investigate the standard deviation of the variable $\bar{x}$ to discover any relationship it has to the standard deviation of the variable under consideration. We begin by returning to the basketball players.

**Example 7.5** | **Standard Deviation of the Sample Mean**

*Heights of Starting Players* Refer back to Table 7.5.

a. Determine the population standard deviation, $\sigma$.

b. Obtain the standard deviation, $\sigma_{\bar{x}}$, of the variable $\bar{x}$ for samples of size 2. Indicate any apparent relationship between $\sigma_{\bar{x}}$ and $\sigma$.

c. Repeat part (b) for samples of sizes 1, 3, 4, and 5.

d. Summarize and discuss the results obtained in parts (a)–(c).

**Solution**

a. To determine the population standard deviation (the standard deviation of the variable "height"), we apply Definition 3.12 on page 138 to the heights in Table 7.5. Recalling that $\mu = 80$ inches, we have

$$\sigma = \sqrt{\frac{\Sigma(x_i - \mu)^2}{N}}$$

$$= \sqrt{\frac{(76-80)^2 + (78-80)^2 + (79-80)^2 + (81-80)^2 + (86-80)^2}{5}}$$

$$= \sqrt{\frac{16+4+1+1+36}{5}} = \sqrt{11.6} = 3.41 \text{ inches.}$$

Thus the standard deviation of the heights of the five players is 3.41 inches.

b. To obtain the standard deviation of the variable $\bar{x}$ for samples of size 2, we again apply Definition 3.12, but this time to $\bar{x}$. Referring to the third column of Table 7.2 on page 310 and recalling that $\mu_{\bar{x}} = \mu = 80$ inches, we have

$$\sigma_{\bar{x}} = \sqrt{\frac{(77.0-80)^2 + (77.5-80)^2 + \cdots + (83.5-80)^2}{10}}$$

$$= \sqrt{\frac{9.00 + 6.25 + \cdots + 12.25}{10}} = \sqrt{4.35} = 2.09 \text{ inches,}$$

to two decimal places. Note that this result is not the same as the population standard deviation, which is $\sigma = 3.41$ inches. Also note that $\sigma_{\bar{x}}$ is smaller than $\sigma$.

c. Using the same procedure as in part (b), we compute $\sigma_{\bar{x}}$ for samples of sizes 1, 3, 4, and 5 and summarize the results in Table 7.6.

d. Table 7.6 suggests that the standard deviation of $\bar{x}$ gets smaller as the sample size gets larger. We could have predicted this result from the dotplots shown in Fig. 7.3 on page 311 and the fact that the standard deviation of a variable measures the variation of its possible values.

**TABLE 7.6**
The standard deviation of $\bar{x}$
for sample sizes 1, 2, 3, 4, and 5

| Sample size $n$ | Standard deviation of $\bar{x}$ $\sigma_{\bar{x}}$ |
|---|---|
| 1 | 3.41 |
| 2 | 2.09 |
| 3 | 1.39 |
| 4 | 0.85 |
| 5 | 0.00 |

• • •

Example 7.5 provides evidence that the standard deviation of $\bar{x}$ gets smaller as the sample size gets larger; that is, the variation of all possible sample means decreases as the sample size increases. The question now is whether there is a formula that relates the standard deviation of $\bar{x}$ to the sample size and standard deviation of the population. The answer is yes! In fact, two different formulas express the precise relationship.

When sampling is done without replacement from a finite population, as in Example 7.5, the appropriate formula is

$$\sigma_{\bar{x}} = \sqrt{\frac{N-n}{N-1}} \cdot \frac{\sigma}{\sqrt{n}} \qquad (7.1)$$

where, as usual, $n$ denotes the sample size and $N$ the population size. When sampling is done with replacement from a finite population or when it is done from an infinite population, the appropriate formula is

$$\sigma_{\bar{x}} = \frac{\sigma}{\sqrt{n}}. \qquad (7.2)$$

When the sample size is small relative to the population size, there is little difference between sampling with and without replacement.[†] So, in such cases, the two formulas for $\sigma_{\bar{x}}$ yield almost the same numbers. In most practical applications, the sample size is small relative to the population size, so in this book, we use the second formula only (with the understanding that the equality may be approximate).

---

**Formula 7.2**

## Standard Deviation of the Sample Mean

**What Does It Mean?**

For each sample size, the standard deviation of all possible sample means equals the population standard deviation divided by the square root of the sample size.

For samples of size $n$, the standard deviation of the variable $\bar{x}$ equals the standard deviation of the variable under consideration divided by the square root of the sample size. In symbols,

$$\sigma_{\bar{x}} = \frac{\sigma}{\sqrt{n}}.$$

**Note:** In the formula for the standard deviation of $\bar{x}$, the sample size, $n$, appears in the denominator. This explains mathematically why the standard deviation of $\bar{x}$ decreases as the sample size increases.

### Applying the Formulas

We have shown that simple formulas relate the mean and standard deviation of $\bar{x}$ to the mean and standard deviation of the population, namely, $\mu_{\bar{x}} = \mu$ and $\sigma_{\bar{x}} = \sigma/\sqrt{n}$ (at least approximately). We apply those formulas next.

---

**Example 7.6** | ## Mean and Standard Deviation of the Sample Mean

*Living Space of Homes* As reported by the U.S. Census Bureau in *Current Housing Reports*, the mean living space for single-family detached homes is 1742 square feet. The standard deviation is 568 square feet.

---

[†]As a rule of thumb, we say that the sample size is small relative to the population size if the size of the sample does not exceed 5% of the size of the population ($n \leq 0.05N$).

**a.** For samples of 25 single-family detached homes, determine the mean and standard deviation of the variable $\bar{x}$.

**b.** Repeat part (a) for a sample of size 500.

**Solution** Here the variable is living space, and the population consists of all single-family detached homes in the United States. From the given information, we know that $\mu = 1742$ sq. ft. and $\sigma = 568$ sq. ft.

**a.** We use Formula 7.1 (page 315) and Formula 7.2 to get

$$\mu_{\bar{x}} = \mu = 1742 \quad \text{and} \quad \sigma_{\bar{x}} = \frac{\sigma}{\sqrt{n}} = \frac{568}{\sqrt{25}} = 113.6.$$

**b.** We again use Formula 7.1 and Formula 7.2 to get

$$\mu_{\bar{x}} = \mu = 1742 \quad \text{and} \quad \sigma_{\bar{x}} = \frac{\sigma}{\sqrt{n}} = \frac{568}{\sqrt{500}} = 25.4.$$

You try it!

Exercise 7.31
on page 319

**Interpretation** For samples of 25 single-family detached homes, the mean and standard deviation of all possible sample mean living spaces are 1742 sq. ft. and 113.6 sq. ft., respectively. For samples of 500, these numbers are 1742 sq. ft. and 25.4 sq. ft., respectively.

• • •

### Sample Size and Sampling Error (Revisited)

Key Fact 7.1 states that the possible sample means cluster more closely around the population mean as the sample size increases, and therefore the larger the sample size, the smaller the sampling error tends to be in estimating a population mean by a sample mean. Here is why that key fact is true.

- The larger the sample size, the smaller is the standard deviation of $\bar{x}$.
- The smaller the standard deviation of $\bar{x}$, the more closely the possible values of $\bar{x}$ (the possible sample means) cluster around the mean of $\bar{x}$.
- The mean of $\bar{x}$ equals the population mean.

Because the standard deviation of $\bar{x}$ determines the amount of sampling error to be expected when a population mean is estimated by a sample mean, it is often referred to as the **standard error of the sample mean.** In general, the standard deviation of a statistic used to estimate a parameter is called the **standard error (SE)** of the statistic.

## Exercises 7.2

### Understanding the Concepts and Skills

**7.18** Although, in general, you cannot know the sampling distribution of the sample mean exactly, by what distribution can you often approximate it?

**7.19** Why is obtaining the mean and standard deviation of $\bar{x}$ a first step in approximating the sampling distribution of the sample mean by a normal distribution?

**7.20** Does the sample size have an effect on the mean of all possible sample means? Explain your answer.

**7.21** Does the sample size have an effect on the standard deviation of all possible sample means? Explain your answer.

**7.22** Explain why increasing the sample size tends to result in a smaller sampling error when a sample mean is used to estimate a population mean.

**7.23** What is another name for the standard deviation of the variable $\bar{x}$? What is the reason for that name?

**7.24** In this section, we stated that, when the sample size is small relative to the population size, there is little difference between sampling with and without replacement. Explain in your own words why that statement is true.

*Exercises 7.25–7.29 require that you have done Exercises 7.3–7.7.*

**7.25 NBA Champs.** The winner of the 2002–2003 National Basketball Association (NBA) championship was the San Antonio Spurs. The following table provides the usual starting players and their positions and heights.

| Player | Position | Height (in.) |
|--------|----------|--------------|
| Bruce Bowen (B) | Forward | 79 |
| Tim Duncan (D) | Forward | 84 |
| David Robinson (R) | Center | 85 |
| Manu Ginobli (G) | Guard | 78 |
| Tony Parker (P) | Guard | 74 |

a. Determine the population mean height, $\mu$, of the five players.
b. Consider samples of size 2 without replacement. Use your answer to Exercise 7.3(b) on page 313 and Definition 3.11 on page 136 to find the mean of the variable $\bar{x}$.
c. Find $\mu_{\bar{x}}$, using only the result of part (a).

**7.26 NBA Champs.** Repeat parts (b) and (c) of Exercise 7.25 for samples of size 1. For part (b), use your answer to Exercise 7.4(b).

**7.27 NBA Champs.** Repeat parts (b) and (c) of Exercise 7.25 for samples of size 3. For part (b), use your answer to Exercise 7.5(b).

**7.28 NBA Champs.** Repeat parts (b) and (c) of Exercise 7.25 for samples of size 4. For part (b), use your answer to Exercise 7.6(b).

**7.29 NBA Champs.** Repeat parts (b) and (c) of Exercise 7.25 for samples of size 5. For part (b), use your answer to Exercise 7.7(b).

**7.30 Working at Home.** According to the U.S. Bureau of Labor Statistics publication *News*, self-employed persons with home-based businesses work a mean of 23 hours per week at home with a standard deviation of 10 hours.
a. Identify the population and variable.
b. For samples of size 100, find the mean and standard deviation of all possible sample mean hours worked per week at home.
c. Repeat part (b) for samples of size 1000.

**7.31 Baby Weight.** The paper "Are Babies Normal?" by T. Clemons and M. Pagano (*The American Statistician*, Vol. 53, No. 4, pp. 298–302) focused on birth weights of babies. According to the article, the mean birth weight is 3369 grams (7 pounds, 6.5 ounces) with a standard deviation of 581 grams.

a. Identify the population and variable.
b. For samples of size 200, find the mean and standard deviation of all possible sample mean weights.
c. Repeat part (b) for samples of size 400.

**7.32 Menopause in Mexico.** In the journal article "Age at Menopause in Puebla, Mexico" (*Human Biology*, Vol. 75, No. 2, pp. 205–206), authors L. Sievert and S. Hautaniemi compared the age of menopause for different populations. Menopause, the last menstrual period, is a universal phenomenon among females. According to the article, the mean age of menopause, surgical or natural, in Puebla, Mexico is 44.8 years with a standard deviation of 5.87 years. Let $\bar{x}$ denote the mean age of menopause for a sample of females in Puebla, Mexico.
a. For samples of size 40, find the mean and standard deviation of $\bar{x}$. Interpret your results in words.
b. Repeat part (a) with $n = 120$.

**7.33 Mobile Homes.** According to the U.S. Census Bureau publication *Manufactured Housing Statistics*, the mean price of new mobile homes is $61,300. The standard deviation of the prices is $7200. Let $\bar{x}$ denote the mean price of a sample of new mobile homes.
a. For samples of size 50, find the mean and standard deviation of $\bar{x}$. Interpret your results in words.
b. Repeat part (a) with $n = 100$.

**7.34 The Self Employed.** S. Parker et al. analyzed the labor supply of self-employed individuals in the article "Wage Uncertainty and the Labour Supply of Self-Employed Workers" (*The Economic Journal*, Vol. 118, No. 502, pp. C190–C207). According to the article, the mean age of a self-employed individual is 46.6 years with a standard deviation of 10.8 years.
a. Identify the population and variable.
b. For samples of size 100, what is the mean and standard deviation of $\bar{x}$? Interpret your results in words.
c. Repeat part (b) with $n = 175$.

**7.35 Earthquakes.** According to *The Earth: Structure, Composition and Evolution* (The Open University, S237), for earthquakes with a magnitude of 7.5 or greater on the Richter scale, the time between successive earthquakes has a mean of 437 days and a standard deviation of 399 days. Suppose that you observe a sample of four times between successive earthquakes that have a magnitude of 7.5 or greater on the Richter scale.

a. On average, what would you expect to be the mean of the four times?

b. How much variation would you expect from your answer in part (a)? (*Hint:* Use the three-standard-deviations rule.)

**7.36** You have seen that the larger the sample size, the smaller the sampling error tends to be in estimating a population mean by a sample mean. This fact is reflected mathematically by the formula for the standard deviation of the sample mean: $\sigma_{\bar{x}} = \sigma/\sqrt{n}$. For a fixed sample size, explain what this formula implies about the relationship between the population standard deviation and sampling error.

## Working With Large Data Sets

**7.37 Provisional AIDS Cases.** The U.S. Department of Health and Human Services publishes information on AIDS in *Morbidity and Mortality Weekly Report*. During one year, the number of provisional cases of AIDS for each of the 50 states are as presented on the WeissStats CD. Use the technology of your choice to solve the following problems.

a. Obtain the standard deviation of the variable "number of provisional AIDS cases" for the population of 50 states.

b. Consider simple random samples without replacement from the population of 50 states. Strictly speaking, which is the correct formula for obtaining the standard deviation of the sample mean—Equation (7.1) or Equation (7.2)? Explain your answer.

c. Referring to part (b), obtain $\sigma_{\bar{x}}$ for simple random samples of size 30 by using both formulas. Why does Equation (7.2) provide such a poor estimate of the true value given by Equation (7.1)?

d. Referring to part (b), obtain $\sigma_{\bar{x}}$ for simple random samples of size 2 by using both formulas. Why does Equation (7.2) provide a somewhat reasonable estimate of the true value given by Equation (7.1)?

e. For simple random samples without replacement of sizes 1–50, construct a table to compare the true values of $\sigma_{\bar{x}}$—obtained by using Equation (7.1)—with the values of $\sigma_{\bar{x}}$ obtained by using Equation (7.2). Discuss your table in detail.

**7.38 SAT Scores.** Each year, thousands of high school students bound for college take the Scholastic Assessment Test (SAT). This test measures the verbal and mathematical abilities of prospective college students. Student scores are reported on a scale that ranges from a low of 200 to a high of 800. Summary results for the scores are published by the College Entrance Examination Board in *College Bound Seniors*. The SAT math scores for one high school graduating class are as provided on the WeissStats CD. Use the technology of your choice to solve the following problems.

a. Obtain the standard deviation of the variable "SAT math score" for this population of students.

b. For simple random samples without replacement of sizes 1–487, construct a table to compare the true values of $\sigma_{\bar{x}}$—obtained by using Equation (7.1)—with the values of $\sigma_{\bar{x}}$ obtained by using Equation (7.2). Explain why the results found by using Equation (7.2) are sometimes reasonably accurate and sometimes not.

## Extending the Concepts and Skills

**7.39 Unbiased and Biased Estimators.** A statistic is said to be an *unbiased estimator* of a parameter if the mean of all its possible values equals the parameter; otherwise, it is said to be a *biased estimator*. An unbiased estimator yields, on average, the correct value of the parameter, whereas a biased estimator does not.

a. Is the sample mean an unbiased estimator of the population mean? Explain your answer.

b. Is the sample median an unbiased estimator of the population median? (*Hint:* Refer to Example 7.2 on page 309. Consider samples of size 2.)

*For Exercises 7.40–7.42, refer to Equations (7.1) and (7.2) on page 317.*

**7.40** Suppose that a simple random sample is taken without replacement from a finite population of size $N$.

a. Show mathematically that Equations (7.1) and (7.2) are identical for samples of size 1.

b. Explain in words why part (a) is true.

c. Without doing any computations, determine $\sigma_{\bar{x}}$ for samples of size $N$ without replacement. Explain your reasoning.

d. Use Equation (7.1) to verify your answer in part (c).

**7.41 Heights of Starting Players.** In Example 7.5, we used the definition of the standard deviation of a variable (Definition 3.12 on page 138) to obtain the standard deviation of the heights of the five starting players on a men's basketball team and, also, the standard deviation of $\bar{x}$ for samples of sizes 1, 2, 3, 4, and 5. The results are summarized in Table 7.6 on page 316. Because the sampling is without replacement from a finite population, Equation (7.1) can also be used to obtain $\sigma_{\bar{x}}$.

a. Apply Equation (7.1) to compute $\sigma_{\bar{x}}$ for samples of sizes 1, 2, 3, 4, and 5. Compare your answers with those presented in Table 7.6.

b. Use the simpler formula, Equation (7.2), to compute $\sigma_{\bar{x}}$ for samples of sizes 1, 2, 3, 4, and 5. Compare your answers with those in Table 7.6. Why does Equation (7.2) generally yield such poor approximations to the true values?

c. What percentages of the population size are samples of sizes 1, 2, 3, 4, and 5?

**7.42 Finite Population Correction Factor.** Consider sample random samples of size $n$ without replacement from a population of size $N$.

**a.** Show that if $n \leq 0.05N$, then

$$0.97 \leq \sqrt{\frac{N-n}{N-1}} \leq 1.$$

**b.** Use part (a) to explain why there is little difference in the values provided by Equations (7.1) and (7.2) when the sample size is small relative to the population size—that is, when the size of the sample does not exceed 5% of the size of the population.

**c.** Explain why the finite population correction factor can be ignored and the simpler formula, Equation (7.2), can be used when the sample size is small relative to the population size.

**d.** The term $\sqrt{(N-n)/(N-1)}$ is known as the **finite population correction factor.** Can you explain why?

**7.43 Class Project Simulation.** This exercise can be done individually or, better yet, as a class project.

**a.** Use a random-number table or random-number generator to obtain a sample (with replacement) of four digits between 0 and 9. Do so a total of 50 times and compute the mean of each sample.

**b.** Theoretically, what are the mean and standard deviation of all possible sample means for samples of size 4?

**c.** Roughly what would you expect the mean and standard deviation of the 50 sample means you obtained in part (a) to be? Explain your answers.

**d.** Determine the mean and standard deviation of the 50 sample means you obtained in part (a).

**e.** Compare your answers in parts (c) and (d). Why are they different?

**7.44 Gestation Periods of Humans.** For humans, gestation periods are normally distributed with a mean of 266 days and a standard deviation of 16 days. Suppose that you observe the gestation periods for a sample of nine humans.

**a.** Theoretically, what are the mean and standard deviation of all possible sample means?

**b.** Use the technology of your choice to simulate 2000 samples of nine human gestation periods each.

**c.** Determine the mean of each of the 2000 samples you obtained in part (b).

**d.** Roughly what would you expect the mean and standard deviation of the 2000 sample means you obtained in part (c) to be? Explain your answers.

**e.** Determine the mean and standard deviation of the 2000 sample means you obtained in part (c).

**f.** Compare your answers in parts (d) and (e). Why are they different?

**7.45 Emergency Room Traffic.** Desert Samaritan Hospital in Mesa, Arizona, keeps records of emergency room traffic. Those records reveal that the times between arriving patients have a special type of reverse J-shaped distribution called an *exponential distribution.* They also indicate that the mean time between arriving patients is 8.7 minutes, as is the standard deviation. Suppose that you observe a sample of 10 interarrival times.

**a.** Theoretically, what are the mean and standard deviation of all possible sample means?

**b.** Use the technology of your choice to simulate 1000 samples of 10 interarrival times each.

**c.** Determine the mean of each of the 1000 samples you obtained in part (b).

**d.** Roughly what would you expect the mean and standard deviation of the 1000 sample means you obtained in part (c) to be? Explain your answers.

**e.** Determine the mean and standard deviation of the 1000 sample means you obtained in part (c).

**f.** Compare your answers in parts (d) and (e). Why are they different?

## 7.3 The Sampling Distribution of the Sample Mean

In Section 7.2, we took the first step in describing the sampling distribution of the sample mean, that is, the distribution of the variable $\bar{x}$. There, we showed that the mean and standard deviation of $\bar{x}$ can be expressed in terms of the sample size and the population mean and standard deviation: $\mu_{\bar{x}} = \mu$ and $\sigma_{\bar{x}} = \sigma/\sqrt{n}$.

In this section, we take the final step in describing the sampling distribution of the sample mean. In doing so, we distinguish between the case in which the variable under consideration is normally distributed and the case in which it may not be so.

## Sampling Distribution of the Sample Mean for Normally Distributed Variables

Although it is by no means obvious, if the variable under consideration is normally distributed, so is the variable $\bar{x}$. The proof of this fact requires advanced mathematics, but we can make it plausible by simulation, as shown next.

---

**Example 7.7**

**OUTPUT 7.1**

Histogram of the sample means for 1000 samples of four IQs with superimposed normal curve

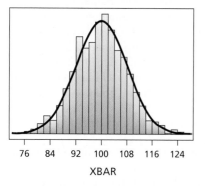

76   84   92   100   108   116   124

XBAR

### Sampling Distribution of the Sample Mean for a Normally Distributed Variable

*Intelligence Quotients* Intelligence quotients (IQs) measured on the Stanford Revision of the Binet–Simon Intelligence Scale are normally distributed with mean 100 and standard deviation 16. For a sample size of 4, use simulation to make plausible the fact that $\bar{x}$ is normally distributed.

**Solution** First, we apply Formula 7.1 (page 315) and Formula 7.2 (page 317) to conclude that $\mu_{\bar{x}} = \mu = 100$ and $\sigma_{\bar{x}} = \sigma/\sqrt{n} = 16/\sqrt{4} = 8$; that is, the variable $\bar{x}$ has mean 100 and standard deviation 8.

We simulated 1000 samples of four IQs each, determined the sample mean of each of the 1000 samples, and obtained a histogram (Output 7.1) of the 1000 sample means. We also superimposed on the histogram the normal distribution with mean 100 and standard deviation 8. The histogram is shaped roughly like a normal curve (with parameters 100 and 8).

**Interpretation** The histogram in Output 7.1 suggests that $\bar{x}$ is normally distributed, that is, that the possible sample mean IQs for samples of four people have a normal distribution.

• • •

---

**Key Fact 7.2**

**What Does It Mean?**

For a normally distributed variable, the possible sample means for samples of a given size are also normally distributed.

### Sampling Distribution of the Sample Mean for a Normally Distributed Variable

Suppose that a variable $x$ of a population is normally distributed with mean $\mu$ and standard deviation $\sigma$. Then, for samples of size $n$, the variable $\bar{x}$ is also normally distributed and has mean $\mu$ and standard deviation $\sigma/\sqrt{n}$.

We illustrate Key Fact 7.2 in the next example.

---

**Example 7.8**

### Sampling Distribution of the Sample Mean for a Normally Distributed Variable

*Intelligence Quotients* Consider again the variable IQ, which is normally distributed with mean 100 and standard deviation 16. Obtain the sampling distribution of the sample mean for samples of size

**a.** 4.                                        **b.** 16.

**Solution** The normal distribution for IQs is shown in Fig. 7.4(a). Because IQs are normally distributed, Key Fact 7.2 implies that, for any particular sample size $n$, the variable $\bar{x}$ is also normally distributed and has mean 100 and standard deviation $16/\sqrt{n}$.

**FIGURE 7.4**
(a) Normal distribution for IQs;
(b) sampling distribution of the sample
mean for *n* = 4; (c) sampling distribution
of the sample mean for *n* = 16

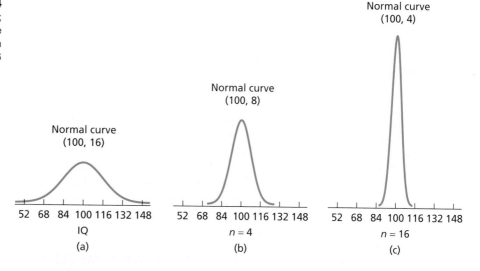

**FIGURE 7.4**
(a) Normal distribution for IQs;
(b) sampling distribution of the sample
mean for *n* = 4; (c) sampling distribution
of the sample mean for *n* = 16

a. For samples of size 4, we have $16/\sqrt{n} = 16/\sqrt{4} = 8$, and therefore the sampling distribution of the sample mean is a normal distribution with mean 100 and standard deviation 8. Figure 7.4(b) shows this normal distribution.

**Interpretation** The possible sample mean IQs for samples of four people have a normal distribution with mean 100 and standard deviation 8.

b. For samples of size 16, we have $16/\sqrt{n} = 16/\sqrt{16} = 4$, and therefore the sampling distribution of the sample mean is a normal distribution with mean 100 and standard deviation 4. Figure 7.4(c) shows this normal distribution.

**Interpretation** The possible sample mean IQs for samples of 16 people have a normal distribution with mean 100 and standard deviation 4.

**You try it!**

Exercise 7.53
on page 327

•  •  •

The normal curves in Figs. 7.4(b) and 7.4(c) are drawn to scale so that you can visualize two important things that you already know: both curves are centered at the population mean ($\mu_{\bar{x}} = \mu$), and the spread decreases as the sample size increases ($\sigma_{\bar{x}} = \sigma/\sqrt{n}$).

Figure 7.4 also illustrates something else that you already know: The possible sample means cluster more closely around the population mean as the sample size increases, and therefore the larger the sample size, the smaller the sampling error tends to be in estimating a population mean by a sample mean.

## Central Limit Theorem

According to Key Fact 7.2, if the variable $x$ is normally distributed, so is the variable $\bar{x}$. That key fact also holds approximately if $x$ is not normally distributed, provided only that the sample size is relatively large. This extraordinary fact, one of the most important theorems in statistics, is called the **central limit theorem**.

Key Fact 7.3

## The Central Limit Theorem (CLT)

For a relatively large sample size, the variable $\bar{x}$ is approximately normally distributed, regardless of the distribution of the variable under consideration. The approximation becomes better with increasing sample size.

**What Does It Mean?**

For a large sample size, the possible sample means are approximately normally distributed, regardless of the distribution of the variable under consideration.

Roughly speaking, the farther the variable under consideration is from being normally distributed, the larger the sample size must be for a normal distribution to provide an adequate approximation to the distribution of $\bar{x}$. Usually, however, a sample size of 30 or more ($n \geq 30$) is large enough.

The proof of the central limit theorem is difficult, but we can make it plausible by simulation, as shown next.

Example 7.9

## Checking the Plausibility of the CLT by Simulation

*Household Size* According to the U.S. Census Bureau publication *Current Population Reports*, a frequency distribution for the number of people per household in the United States is as displayed in Table 7.7. Frequencies are in millions of households.

**TABLE 7.7**
Frequency distribution
for U.S. household size

| Number of people | Frequency (millions) |
|:---:|:---:|
| 1 | 19.4 |
| 2 | 26.5 |
| 3 | 14.6 |
| 4 | 12.9 |
| 5 | 6.1 |
| 6 | 2.5 |
| 7 | 1.6 |

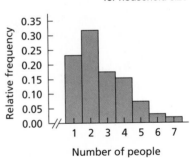

**FIGURE 7.5**
Relative-frequency histogram
for household size

Here, the variable is household size, and the population is all U.S. households. From Table 7.7, we find that the mean household size is $\mu = 2.685$ persons and the standard deviation is $\sigma = 1.47$ persons.

Figure 7.5 is a relative-frequency histogram for household size, obtained from Table 7.7. Note that household size is far from being normally distributed; it is right skewed. Nonetheless, according to the central limit theorem, the sampling distribution of the sample mean can be approximated by a normal distribution when the sample size is relatively large. Use simulation to make that fact plausible for a sample size of 30.

**Solution**   First, we apply Formula 7.1 (page 315) and Formula 7.2 (page 317) to conclude that, for samples of size 30,

$$\mu_{\bar{x}} = \mu = 2.685 \quad \text{and} \quad \sigma_{\bar{x}} = \sigma/\sqrt{n} = 1.47/\sqrt{30} = 0.27.$$

Thus the variable $\bar{x}$ has a mean of 2.685 and a standard deviation of 0.27.

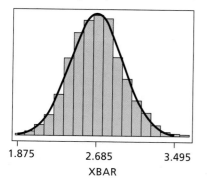

1.875          2.685          3.495

XBAR

We simulated 1000 samples of 30 households each, determined the sample mean of each of the 1000 samples, and obtained a histogram (Output 7.2) of the 1000 sample means. We also superimposed on the histogram the normal distribution with mean 2.685 and standard deviation 0.27. The histogram is shaped roughly like a normal curve (with parameters 2.685 and 0.27).

**Interpretation** The histogram in Output 7.2 suggests that $\bar{x}$ is approximately normally distributed, as guaranteed by the central limit theorem. Thus, for samples of 30 households, the possible sample mean household sizes have approximately a normal distribution.

• • •

## The Sampling Distribution of the Sample Mean

We now summarize the facts that we have learned about the sampling distribution of the sample mean.

**Key Fact 7.4**

### Sampling Distribution of the Sample Mean

Suppose that a variable $x$ of a population has mean $\mu$ and standard deviation $\sigma$. Then, for samples of size $n$,

- the mean of $\bar{x}$ equals the population mean, or $\mu_{\bar{x}} = \mu$;
- the standard deviation of $\bar{x}$ equals the population standard deviation divided by the square root of the sample size, or $\sigma_{\bar{x}} = \sigma/\sqrt{n}$;
- if $x$ is normally distributed, so is $\bar{x}$, regardless of sample size; and
- if the sample size is large, $\bar{x}$ is approximately normally distributed, regardless of the distribution of $x$.

**What Does It Mean?**

If either the variable under consideration is normally distributed or the sample size is large, then the possible sample means have, at least approximately, a normal distribution with mean $\mu$ and standard deviation $\sigma/\sqrt{n}$.

From Key Fact 7.4, we know that, if the variable under consideration is normally distributed, so is the variable $\bar{x}$, regardless of sample size, as illustrated by Fig. 7.6(a) at the top of the next page.

Additionally, we know that, if the sample size is large, the variable $\bar{x}$ is approximately normally distributed, regardless of the distribution of the variable under consideration. Figures 7.6(b) and 7.6(c) illustrate this fact for two nonnormal variables, one having a reverse-J-shaped distribution and the other having a uniform distribution.

In each of these latter two cases, for samples of size 2, the variable $\bar{x}$ is far from being normally distributed; for samples of size 10, it is already somewhat normally distributed; and for samples of size 30, it is very close to being normally distributed.

**Example 7.10** | ## Sampling Distribution of the Sample Mean

*Birth Weight* The U.S. National Center for Health Statistics publishes information about birth weights in *Vital Statistics of the United States*. According to that document, birth weights of male babies have a standard deviation of 1.33 lb. Determine the percentage of all samples of 400 male babies that have mean

**FIGURE 7.6**

Sampling distributions for (a) normal, (b) reverse-J-shaped, and (c) uniform variables

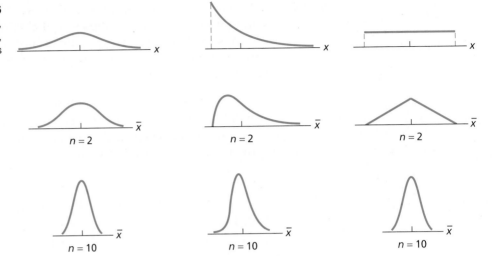

| (a) | (b) | (c) |

birth weights within 0.125 lb (2 oz) of the population mean birth weight of all male babies. Interpret your answer in terms of sampling error.

**Solution**  Let $\mu$ denote the population mean birth weight of all male babies. From Key Fact 7.4, for samples of size 400, the sample mean birth weight, $\bar{x}$, is approximately normally distributed with

$$\mu_{\bar{x}} = \mu \quad \text{and} \quad \sigma_{\bar{x}} = \frac{\sigma}{\sqrt{n}} = \frac{1.33}{\sqrt{400}} = 0.0665.$$

Thus, the percentage of all samples of 400 male babies that have mean birth weights within 0.125 lb of the population mean birth weight of all male babies is (approximately) equal to the area under the normal curve with parameters $\mu$ and 0.0665 that lies between $\mu - 0.125$ and $\mu + 0.125$. (See Fig. 7.7.) The corresponding z-scores are

$$z = \frac{(\mu - 0.125) - \mu}{0.0665} = \frac{-0.125}{0.0665} = -1.88$$

and

$$z = \frac{(\mu + 0.125) - \mu}{0.0665} = \frac{0.125}{0.0665} = 1.88.$$

Referring now to Table II, we find that the area under the standard normal curve between $-1.88$ and $1.88$ equals 0.9398. Consequently, 93.98% of all samples of

**FIGURE 7.7**

Percentage of all samples of 400 male babies that have mean birth weights within 0.125 lb of the population mean birth weight

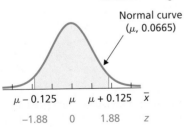

400 male babies have mean birth weights within 0.125 lb of the population mean birth weight of all male babies. You can already see the power of sampling.

**Interpretation** There is about a 94% chance that the sampling error made in estimating the mean birth weight of all male babies by that of a sample of 400 male babies will be at most 0.125 lb.

Exercise 7.57
on page 328

• • •

# Exercises 7.3

## Understanding the Concepts and Skills

**7.46** Identify the two different cases considered in discussing the sampling distribution of the sample mean. Why do we consider those two different cases separately?

**7.47** A variable of a population has a mean of $\mu = 100$ and a standard deviation of $\sigma = 28$.
**a.** Identify the sampling distribution of the sample mean for samples of size 49.
**b.** In answering part (a), what assumptions did you make about the distribution of the variable?
**c.** Can you answer part (a) if the sample size is 16 instead of 49? Why or why not?

**7.48** A variable of a population has a mean of $\mu = 35$ and a standard deviation of $\sigma = 42$.
**a.** If the variable is normally distributed, identify the sampling distribution of the sample mean for samples of size 9.
**b.** Can you answer part (a) if the distribution of the variable under consideration is unknown? Explain your answer.
**c.** Can you answer part (a) if the distribution of the variable under consideration is unknown but the sample size is 36 instead of 9? Why or why not?

**7.49** A variable of a population is normally distributed with mean $\mu$ and standard deviation $\sigma$.
**a.** Identify the distribution of $\bar{x}$.
**b.** Does your answer to part (a) depend on the sample size? Explain your answer.
**c.** Identify the mean and the standard deviation of $\bar{x}$.
**d.** Does your answer to part (c) depend on the assumption that the variable under consideration is normally distributed? Why or why not?

**7.50** A variable of a population has mean $\mu$ and standard deviation $\sigma$. For a large sample size $n$, answer the following questions.
**a.** Identify the distribution of $\bar{x}$.
**b.** Does your answer to part (a) depend on $n$ being large? Explain your answer.
**c.** Identify the mean and the standard deviation of $\bar{x}$.

**d.** Does your answer to part (c) depend on the sample size being large? Why or why not?

**7.51** Refer to Fig. 7.6 on page 326.
**a.** Why are the four graphs in Fig. 7.6(a) all centered at the same place?
**b.** Why does the spread of the graphs diminish with increasing sample size? How does this result affect the sampling error when you estimate a population mean, $\mu$, by a sample mean, $\bar{x}$?
**c.** Why are the graphs in Fig. 7.6(a) bell shaped?
**d.** Why do the graphs in Figs. 7.6(b) and (c) become bell shaped as the sample size increases?

**7.52** According to the central limit theorem, for a relatively large sample size, the variable $\bar{x}$ is approximately normally distributed.
**a.** What rule of thumb is used for deciding whether the sample size is relatively large?
**b.** Roughly speaking, what property of the distribution of the variable under consideration determines how large the sample size must be for a normal distribution to provide an adequate approximation to the distribution of $\bar{x}$?

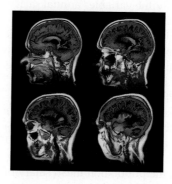

**7.53 Brain Weights.** In 1905, R. Pearl published the article "Biometrical Studies on Man. I. Variation and Correlation in Brain Weight" (*Biometrika*, Vol. 4, pp. 13–104). According to the study, brain weights of Swedish men are normally distributed with a mean of 1.40 kg and a standard deviation of 0.11 kg.

a. Determine the sampling distribution of the sample mean for samples of size 3. Interpret your answer in terms of the distribution of all possible sample mean brain weights for samples of three Swedish men.

b. Repeat part (a) for samples of size 12.

c. Construct graphs similar to those shown in Fig. 7.4 on page 323.

d. Determine the percentage of all samples of three Swedish men that have mean brain weights within 0.1 kg of the population mean brain weight of 1.40 kg. Interpret your answer in terms of sampling error.

e. Repeat part (d) for samples of size 12.

**7.54 New York City 10 km Run.** As reported by *Runner's World* magazine, the times of the finishers in the New York City 10 km run are normally distributed with a mean of 61 minutes and a standard deviation of 9 minutes. Do the following for the variable "finishing time" of finishers in the New York City 10 km run.

a. Find the sampling distribution of the sample mean for samples of size 4.

b. Repeat part (a) for samples of size 9.

c. Construct graphs similar to those shown in Fig. 7.4 on page 323.

d. Obtain the percentage of all samples of four finishers that have mean finishing times within 5 minutes of the population mean finishing time of 61 minutes. Interpret your answer in terms of sampling error.

e. Repeat part (d) for samples of size 9.

**7.55 Teacher Salaries.** Data on salaries in the public school system are published annually in *National Survey of Salaries and Wages in Public Schools* by the Education Research Service. The mean annual salary of (public) classroom teachers is $45,871. Assume a standard deviation of $9200. Do the following for the variable "annual salary" of classroom teachers.

a. Determine the sampling distribution of the sample mean for samples of size 64. Interpret your answer in terms of the distribution of all possible sample mean salaries for samples of 64 classroom teachers.

b. Repeat part (a) for samples of size 256.

c. Do you need to assume that classroom teacher salaries are normally distributed to answer parts (a) and (b)? Explain your answer.

d. What is the probability that the sampling error made in estimating the population mean salary of all classroom teachers by the mean salary of a sample of 64 classroom teachers will be at most $1000?

e. Repeat part (d) for samples of size 256.

**7.56 Loan Amounts.** Ciochetti et al. studied mortgage loans in the article "A Proportional Hazards Model of Commercial Mortgage Default with Originator Bias" (*Journal of Real Estate and Economics*, Vol. 27, No. 1, pp. 5–23). According to the article, the loan amounts of loans originated by a large insurance-company lender have a mean of $6.74 million with a standard deviation of $15.37 million. The variable "loan amount" is known to have a right-skewed distribution.

a. Using units of millions of dollars, determine the sampling distribution of the sample mean for samples of size 200. Interpret your result.

b. Repeat part (a) for samples of size 600.

c. Why can you still answer parts (a) and (b) when the distribution of loan amounts is not normal, but rather right skewed?

d. What is the probability that the sampling error made in estimating the population mean loan amount by the mean loan amount of a simple random sample of 200 loans will be at most $1 million?

e. Repeat part (d) for samples of size 600.

**7.57 Nurses and Hospital Stays.** In the article "A Multifactorial Intervention Program Reduces the Duration of Delirium, Length of Hospitalization, and Mortality in Delirious Patients" (*Journal of the American Geriatrics Society*, Vol. 53, No. 4, pp. 622–628), M. Lundstrom et al. investigate whether education programs for nurses improve the outcomes for their older patients. The standard deviation of the lengths of hospital stay on the intervention ward is 8.3 days.

a. For the variable "length of hospital stay," determine the sampling distribution of the sample mean for samples of 80 patients on the intervention ward.

b. The distribution of the length of hospital stay is right skewed. Does this invalidate your result in part (a)? Explain your answer.

c. Obtain the probability that the sampling error made in estimating the population mean length of stay on the intervention ward by the mean length of stay of a sample of 80 patients will be at most 2 days.

**7.58 Women at Work.** In the article "Job Mobility and Wage Growth" (*Monthly Labor Review*, Vol. 128, No. 2, pp. 33–39), A. Light examines data on employment and answers questions regarding why workers separate from their employers. According to the article, the standard deviation of the lengths of time that women with one job are employed during the first 8 years of their career is 92 weeks. Length of time employed during the first 8 years of career is a left-skewed variable. For that variable, do the following.

a. Determine the sampling distribution of the sample mean for simple random samples of 50 women with one job. Explain your reasoning.

b. Obtain the probability that the sampling error made in estimating the mean length of time employed by all women with one job by that of a random sample of 50 such women will be at most 20 weeks.

**7.59 Air Conditioning Service Contracts.** An air conditioning contractor is preparing to offer service contracts on the brand of compressor used in all of the units her company installs. Before she can work out the details, she must estimate how long those compressors last, on average. The contractor anticipated this need and has kept detailed records on the lifetimes of a random sample of 250 compressors. She plans to use the sample mean lifetime, $\bar{x}$, of those 250 compressors as her estimate for the population mean lifetime, $\mu$, of all such compressors. If the lifetimes of this brand of compressor have a standard deviation of 40 months, what is the probability that the contractor's estimate will be within 5 months of the true mean?

## Extending the Concepts and Skills

*Use the 68.26-95.44-99.74 rule (page 286) to answer the questions posed in parts (a)–(c) of Exercises 7.60 and 7.61.*

**7.60** A variable of a population is normally distributed with mean $\mu$ and standard deviation $\sigma$. For samples of size $n$, fill in the blanks. Justify your answers.
a. 68.26% of all possible samples have means that lie within _____ of the population mean, $\mu$.
b. 95.44% of all possible samples have means that lie within _____ of the population mean, $\mu$.
c. 99.74% of all possible samples have means that lie within _____ of the population mean, $\mu$.
d. $100(1 - \alpha)\%$ of all possible samples have means that lie within _____ of the population mean, $\mu$. (*Hint:* Draw a graph for the distribution of $\bar{x}$ and determine the $z$-scores dividing the area under the normal curve into a middle $1 - \alpha$ area and two outside areas of $\alpha/2$.)

**7.61** A variable of a population has mean $\mu$ and standard deviation $\sigma$. For a large sample size $n$, fill in the blanks. Justify your answers.
a. Approximately _____% of all possible samples have means within $\sigma/\sqrt{n}$ of the population mean, $\mu$.
b. Approximately _____% of all possible samples have means within $2\sigma/\sqrt{n}$ of the population mean, $\mu$.
c. Approximately _____% of all possible samples have means within $3\sigma/\sqrt{n}$ of the population mean, $\mu$.
d. Approximately _____% of all possible samples have means within $z_{\alpha/2}$ of the population mean, $\mu$.

**7.62 Testing for Content Accuracy.** A brand of water-softener salt comes in packages marked "net weight 40 lb." The company that packages the salt claims that the bags contain an average of 40 lb of salt and that the standard deviation of the weights is 1.5 lb. Assume that the weights are normally distributed.
a. Obtain the probability that the weight of one randomly selected bag of water-softener salt will be 39 lb or less, if the company's claim is true.

b. Determine the probability that the mean weight of 10 randomly selected bags of water-softener salt will be 39 lb or less, if the company's claim is true.
c. If you bought one bag of water-softener salt and it weighed 39 lb, would you consider this evidence that the company's claim is incorrect? Explain your answer.
d. If you bought 10 bags of water-softener salt and their mean weight was 39 lb, would you consider this evidence that the company's claim is incorrect? Explain your answer.

**7.63 Household Size.** In Example 7.9 on page 324, we conducted a simulation to check the plausibility of the central limit theorem. The variable under consideration there is household size and the population consists of all U.S. households. A frequency distribution for household size of U.S. households is presented in Table 7.7.
a. Suppose that you simulate 1000 samples of four households each, determine the sample mean of each of the 1000 samples, and obtain a histogram of the 1000 sample means. Would you expect the histogram to be bell shaped? Explain your answer.
b. Carry out the tasks in part (a) and note the shape of the histogram.
c. Repeat parts (a) and (b) for samples of size 10.
d. Repeat parts (a) and (b) for samples of size 100.

**7.64 Gestation Periods of Humans.** For humans, gestation periods are normally distributed with a mean of 266 days and a standard deviation of 16 days. Suppose that you observe the gestation periods for a sample of nine humans.
a. Use the technology of your choice to simulate 2000 samples of nine human gestation periods each.
b. Find the sample mean of each of the 2000 samples.
c. Obtain the mean, the standard deviation, and a histogram of the 2000 sample means.
d. Theoretically, what are the mean, standard deviation, and distribution of all possible sample means for samples of size 9?
e. Compare your results from parts (c) and (d).

**7.65 Emergency Room Traffic.** A variable is said to have an *exponential distribution* or to be *exponentially distributed* if its distribution has the shape of an exponential curve, that is, a curve of the form $y = e^{-x/\mu}/\mu$ for $x > 0$, where $\mu$ is the mean of the variable. The standard deviation of such a variable also equals $\mu$. At the emergency room at Desert Samaritan Hospital in Mesa, Arizona, the time from the arrival of one patient to the next, called an interarrival time, has an exponential distribution with a mean of 8.7 minutes.
a. Sketch the exponential curve for the distribution of the variable "interarrival time." Note that this variable is far from being normally distributed. What shape does its distribution have?

**b.** Use the technology of your choice to simulate 1000 samples of four interarrival times each.

**c.** Find the sample mean of each of the 1000 samples.

**d.** Determine the mean and standard deviation of the 1000 sample means.

**e.** Theoretically, what are the mean and the standard deviation of all possible sample means for samples of size 4? Compare your answers to those you obtained in part (d).

**f.** Obtain a histogram of the 1000 sample means. Is the histogram bell shaped? Would you necessarily expect it to be?

**g.** Repeat parts (b)–(f) for a sample size of 40.

## Chapter in Review

### You Should be Able to

1. use and understand the formulas in this chapter.

2. define sampling error and explain the need for sampling distributions.

3. find the mean and standard deviation of the variable $\bar{x}$, given the mean and standard deviation of the population and the sample size.

4. state and apply the central limit theorem.

5. determine the sampling distribution of the sample mean when the variable under consideration is normally distributed.

6. determine the sampling distribution of the sample mean when the sample size is relatively large.

### Key Terms

central limit theorem, *324*
sampling distribution, *309*

sampling distribution of the sample
   mean, *309*
sampling error, *308*

standard error (SE), *318*
standard error of the sample
   mean, *318*

## Review Problems

### Understanding the Concepts and Skills

**1.** Define sampling error.

**2.** What is the sampling distribution of a statistic? Why is it important?

**3.** Provide two synonyms for "the distribution of all possible sample means for samples of a given size."

**4.** Relative to the population mean, what happens to the possible sample means for samples of the same size as the sample size increases? Explain the relevance of this property in estimating a population mean by a sample mean.

**5. Income Tax and the IRS.** In 2003, the Internal Revenue Service (IRS) sampled 175,485 tax returns to obtain estimates of various parameters. Data were published in *Statistics of Income, Individual Income Tax Returns*. According to that document, the mean income tax per return for the returns sampled was $8412.

**a.** Explain the meaning of sampling error in this context.

**b.** If, in reality, the population mean income tax per return in 2003 was $8500, how much sampling error was made in estimating that parameter by the sample mean of $8412?

**c.** If the IRS had sampled 250,000 returns instead of 175,485, would the sampling error necessarily have been smaller? Explain your answer.

**d.** In future surveys, how can the IRS increase the likelihood of small sampling error?

**6. Officer Salaries.** The following table gives the monthly salaries (in $1000s) of the six officers of a company.

| Officer | A | B | C | D | E | F |
|---------|---|---|---|---|---|---|
| Salary  | 8 | 12 | 16 | 20 | 24 | 28 |

**a.** Calculate the population mean monthly salary, $\mu$.

There are 15 possible samples of size 4 from the population of six officers. They are listed in the first column of the following table.

| Sample | Salaries | $\bar{x}$ |
|--------|----------|-----------|
| A, B, C, D | 8, 12, 16, 20 | 14 |
| A, B, C, E | 8, 12, 16, 24 | 15 |
| A, B, C, F | 8, 12, 16, 28 | 16 |
| A, B, D, E | 8, 12, 20, 24 | 16 |
| A, B, D, F | 8, 12, 20, 28 | 17 |
| A, B, E, F | 8, 12, 24, 28 | 18 |
| A, C, D, E | | |
| A, C, D, F | | |
| A, C, E, F | | |
| A, D, E, F | | |
| B, C, D, E | | |
| B, C, D, F | | |
| B, C, E, F | | |
| B, D, E, F | | |
| C, D, E, F | | |

**b.** Complete the second and third columns of the table.
**c.** Complete the dotplot for the sampling distribution of the sample mean for samples of size 4. Locate the population mean on the graph.

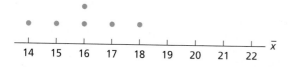

**d.** Obtain the probability that the mean salary of a random sample of four officers will be within 1 (i.e., $1000) of the population mean.
**e.** Use the answer you obtained in part (b) and Definition 3.11 on page 136 to find the mean of the variable $\bar{x}$. Interpret your answer.
**f.** Can you obtain the mean of the variable $\bar{x}$ without doing the calculation in part (e)? Explain your answer.

**7. New Car Passion.** Comerica Bank publishes information on new car prices in *Comerica Auto Affordability Index*.

In the year 2005, Americans spent an average of $27,958 for a new car. Assume a standard deviation of $10,200.
**a.** Identify the population and variable under consideration.
**b.** For samples of 50 new car sales in 2005, determine the mean and standard deviation of all possible sample mean prices.
**c.** Repeat part (b) for samples of size 100.
**d.** For samples of size 1000, answer the following question without doing any computations: Will the standard deviation of all possible sample mean prices be larger, smaller, or the same as that in part (c)? Explain your answer.

**8. Hours Actually Worked.** In the article "How Hours of Work Affect Occupational Earnings" (*Monthly Labor Review*, Vol. 121), D. Hecker discusses the number of hours actually worked as opposed to the number of hours paid for. The study examines both full-time men and full-time women in 87 different occupations. According to the article, the mean number of hours (actually) worked by female marketing and advertising managers is $\mu = 45$ hours. Assuming a standard deviation of $\sigma = 7$ hours, decide whether each of the following statements is true or false or whether the information is insufficient to decide. Give a reason for each of your answers.
**a.** For a random sample of 196 female marketing and advertising managers, chances are roughly 95.44% that the sample mean number of hours worked will be between 31 hours and 59 hours.
**b.** 95.44% of all possible observations of the number of hours worked by female marketing and advertising managers lie between 31 hours and 59 hours.
**c.** For a random sample of 196 female marketing and advertising managers, chances are roughly 95.44% that the sample mean number of hours worked will be between 44 hours and 46 hours.

**9. Hours Actually Worked.** Repeat Problem 8, assuming that the number of hours worked by female marketing and advertising managers is normally distributed.

**10. Antarctic Krill.** In the Southern Ocean food web, the krill species *Euphausia superba* is the most important prey species for many marine predators, from seabirds to the largest whales. Body lengths of the species are normally distributed with a mean of 40 mm and a standard deviation of 12 mm. [SOURCE: K. Reid et al., "Krill Population Dynamics at South Georgia 1991–1997 Based on Data From Predators and Nets," *Marine Ecology Progress Series*, Vol. 177, pp. 103–114]
**a.** Sketch the normal curve for the krill lengths.
**b.** Find the sampling distribution of the sample mean for samples of size 4. Draw a graph of the normal curve associated with $\bar{x}$.
**c.** Repeat part (b) for samples of size 9.

**11. Antarctic Krill.** Refer to Problem 10.
a. Determine the percentage of all samples of four krill that have mean lengths within 9 mm of the population mean length of 40 mm.
b. Obtain the probability that the mean length of four randomly selected krill will be within 9 mm of the population mean length of 40 mm.
c. Interpret the probability you obtained in part (b) in terms of sampling error.
d. Repeat parts (a)–(c) for samples of size 9.

**12.** The following graph shows the curve for a normally distributed variable. Superimposed are the curves for the sampling distributions of the sample mean for two different sample sizes.

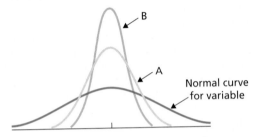

a. Explain why all three curves are centered at the same place.
b. Which curve corresponds to the larger sample size? Explain your answer.
c. Why is the spread of each curve different?
d. Which of the two sampling-distribution curves corresponds to the sample size that will tend to produce less sampling error? Explain your answer.
e. Why are the two sampling-distribution curves normal curves?

**13. Blood Glucose Level.** In the article "Drinking Glucose Improves Listening Span in Students Who Miss Breakfast" (*Educational Research*, Vol. 43, No. 2, pp. 201–207), authors N. Morris and P. Sarll explore the relationship between students who skip breakfast and their performance on a number of cognitive tasks. According to their findings, blood glucose levels in the morning, after a 9-hour fast, have a mean of 4.60 mmol/l with a standard deviation of 0.16 mmol/l. (*Note:* mmol/l is an abbreviation of millimoles/liter, which is the world standard unit for measuring glucose in blood).
a. Determine the sampling distribution of the sample mean for samples of size 60.
b. Repeat part (a) for samples of size 120.
c. Must you assume that the blood glucose levels are normally distributed to answer parts (a) and (b)? Explain your answer.

**14. Life Insurance in Force.** The American Council of Life Insurers provides information about life insurance in force per covered family in the *Life Insurers Fact Book*. Assume that the standard deviation of life insurance in force is $50,900.

a. Determine the probability that the sampling error made in estimating the population mean life insurance in force by that of a sample of 500 covered families will be $2000 or less.
b. Must you assume that life-insurance amounts are normally distributed in order to answer part (a)? What if the sample size is 20 instead of 500?
c. Repeat part (a) for a sample size of 5000.

**15. Paint Durability.** A paint manufacturer in Pittsburgh claims that his paint will last an average of 5 years. Assuming that paint life is normally distributed and has a standard deviation of 0.5 year, answer the following questions:
a. Suppose that you paint one house with the paint and that the paint lasts 4.5 years. Would you consider that evidence against the manufacturer's claim? (*Hint:* Assuming that the manufacturer's claim is correct, determine the probability that the paint life for a randomly selected house painted with the paint is 4.5 years or less.)
b. Suppose that you paint 10 houses with the paint and that the paint lasts an average of 4.5 years for the 10 houses. Would you consider that evidence against the manufacturer's claim?
c. Repeat part (b) if the paint lasts an average of 4.9 years for the 10 houses painted.

## Extending the Concepts and Skills

**16. Quantitative GRE Scores.** The Graduate Record Examination (GRE) is a standardized test that students usually take before entering graduate school. According to the Office of Educational Research and Improvement and the report *Digest of Education Statistics*, the scores on the quantitative portion of the 2000 GRE are (approximately) normally distributed with mean 578 points and standard deviation 147 points.
a. Use the technology of your choice to simulate 1000 samples of four GRE scores each.
b. Find the sample mean of each of the 1000 samples obtained in part (a).
c. Obtain the mean, the standard deviation, and a histogram of the 1000 sample means.
d. Theoretically, what are the mean, standard deviation, and distribution of all possible sample means for samples of size 4?
e. Compare your answers from parts (c) and (d).

**17. Random Numbers.** A variable is said to be *uniformly distributed* or to have a *uniform distribution* with parameters $a$ and $b$ if its distribution has the shape of the horizontal line segment $y = 1/(b - a)$, for $a < x < b$. The mean and standard deviation of such a variable are $(a + b)/2$ and $(b - a)/\sqrt{12}$, respectively. The basic random-number generator on a computer or calculator, which returns a number between 0 and 1, simulates a variable having a uniform distribution with parameters 0 and 1.
a. Sketch the distribution of a uniformly distributed variable with parameters 0 and 1. Observe from your sketch

that such a variable is far from being normally distributed.

**b.** Use the technology of your choice to simulate 2000 samples of two random numbers between 0 and 1.

**c.** Find the sample mean of each of the 2000 samples obtained in part (b).

**d.** Determine the mean and standard deviation of the 2000 sample means.

**e.** Theoretically, what are the mean and the standard deviation of all possible sample means for samples of size 2? Compare your answers to those you obtained in part (d).

**f.** Obtain a histogram of the 2000 sample means. Is the histogram bell shaped? Would you expect it to be?

**g.** Repeat parts (b)–(f) for a sample size of 35.

---

# Focusing on Data Analysis    UWEC Undergraduates

Recall from Chapter 1 (see page 34) that the Focus database and Focus sample contain information on the undergraduate students at the University of Wisconsin - Eau Claire (UWEC). Now would be a good time for you to review the discussion about these data sets.

Suppose that you want to conduct extensive interviews with a simple random sample of 25 UWEC undergraduate students. Use the technology of your choice to obtain such a sample and the corresponding data for the thirteen variables in the Focus database (Focus).

**Note:** If your statistical software package will not accommodate the entire Focus database, use the Focus sample (FocusSample) instead. Of course, in that case, your simple random sample of 25 UWEC undergraduate students will come from the 200 UWEC undergraduate students in the Focus sample rather than from all UWEC undergraduate students in the Focus database.

---

# Case Study Discussion    The Chesapeake and Ohio Freight Study

At the beginning of this chapter, we discussed a freight study commissioned by the Chesapeake and Ohio Railroad Company (C&O). A sample of 2072 waybills from a population of 22,984 waybills was used to estimate the total revenue due C&O. The estimate arrived at was $64,568.

Because all 22,984 waybills were available, a census could be taken to determine exactly the total revenue due C&O and thereby reveal the accuracy of the estimate obtained by sampling. The exact amount due C&O was found to be $64,651.

**a.** What percentage of the waybills constituted the sample?

**b.** What percentage error was made by using the sample to estimate the total revenue due C&O?

**c.** At the time of the study, the cost of a census was approximately $5000, whereas the cost for the sample estimate was only $1000. Knowing this information and your answers to parts (a) and (b), do you think that sampling was preferable to a census? Explain your answer.

**d.** In the study, the $83 error was against C&O. Could the error have been in C&O's favor?

---

# Biography    PIERRE-SIMON LAPLACE: The Newton of France

**Pierre-Simon Laplace** was born on March 23, 1749, at Beaumount-en-Auge, Normandy, France, the son of a peasant farmer. His early schooling was at the military academy at Beaumount, where he developed his mathematical abilities. At the age of 18, he went to Paris. Within 2 years he was recommended for a professorship at the École

Militaire by the French mathematician and philosopher Jean d'Alembert. (It is said that Laplace examined and passed Napoleon Bonaparte there in 1785.) In 1773 Laplace was granted membership in the Academy of Sciences.

Laplace held various positions in public life: He was president of the Bureau des Longitudes, professor at the

École Normale, Minister of the Interior under Napoleon for six weeks (at which time he was replaced by Napoleon's brother), and Chancellor of the Senate; he was also made a marquis.

Laplace's professional interests were also varied. He published several volumes on celestial mechanics (which the Scottish geologist and mathematician John Playfair said were "the highest point to which man has yet ascended in the scale of intellectual attainment"), a book entitled *Théorie analytique des probabilités* (Analytic Theory of Probability), and other works on physics and mathematics. Laplace's primary contribution to the field of probability and statistics was the remarkable and all-important central limit theorem, which appeared in an 1809 publication and was read to the Academy of Sciences on April 9, 1810.

Astronomy was Laplace's major work; approximately half of his publications were concerned with the solar system and its gravitational interactions. These interactions were so complex that even Sir Isaac Newton had concluded "divine intervention was periodically required to preserve the system in equilibrium." Laplace, however, proved that planets' average angular velocities are invariable and periodic, and thus made the most important advance in physical astronomy since Newton.

When Laplace died in Paris on March 5, 1827, he was eulogized by the famous French mathematician and physicist Simeon Poisson as "the Newton of France."

## StatCrunch in MyStatLab
### Analyzing Data Online

StatCrunch online statistical software offers an easy-to-use interface customized for this book. The StatCrunch feature for each chapter illustrates the use of the software to perform a statistical analysis discussed in the chapter. Exercises are provided to further apply StatCrunch to other statistical analyses examined in the chapter. Go to the WeissStats CD or to the Weiss Web site at www.aw-bc.com/weiss to access StatCrunch instructions and data sets. To access StatCrunch statistical software, go to the student content area of your Weiss MyStatLab course.

## Internet Projects
### Exploring Data Online

The Internet project for each chapter provides simulations, demonstrations, or activities that enhance the topics covered in the chapter. The project materials come from universities, individuals, governments, and companies from all over the world. To access the Internet projects on the Web, go to www.aw-bc.com/weiss. From this Web page, you can reach the Internet Projects Page, which we suggest that you bookmark for easy access in the future.

# PART IV

## Inferential Statistics

# 8

# Confidence Intervals for One Population Mean

**Chapter Objectives**

In this chapter, you begin your study of inferential statistics by examining methods for estimating the mean of a population. As you might suspect, the statistic used to estimate the population mean, $\mu$, is the sample mean, $\bar{x}$. Because of sampling error, you cannot expect $\bar{x}$ to equal $\mu$ exactly. Thus, providing information about the accuracy of the estimate is important, which leads to a discussion of confidence intervals, the main topic of this chapter.

In Section 8.1, we provide the intuitive foundation for confidence intervals. Then, in Section 8.2, we present confidence intervals for one population mean when the population standard deviation, $\sigma$, is known. Although, in practice, $\sigma$ is usually unknown, we first consider, for pedagogical reasons, the case where $\sigma$ is known.

In Section 8.3, we investigate the relationship between sample size and the precision with which a sample mean estimates the population mean. This investigation leads us to a discussion of the *margin of error*.

In Section 8.4, we discuss confidence intervals for one population when the population standard deviation is unknown. As a prerequisite to that topic, we introduce and describe one of the most important distributions in inferential statistics—*Student's t*.

# The Chips Ahoy! 1,000 Chips Challenge

Nabisco, the maker of Chips Ahoy! cookies, challenged students across the nation to confirm the cookie maker's claim that there are [at least] 1000 chocolate chips in every 18-ounce bag of Chips Ahoy! cookies. According to the folks at Nabisco, a chocolate chip is defined as "…any distinct piece of chocolate that is baked into or on top of the cookie dough regardless of whether or not it is 100% whole." Students competed for $25,000 in scholarships and other prizes for participating in the Challenge.

As reported by Brad Warner and Jim Rutledge in the paper "Checking the Chips Ahoy! Guarantee" (*Chance*, Vol. 12(1), pp. 10–14), one such group that participated in the Challenge was an introductory statistics class at the United States Air Force Academy. With chocolate chips on their minds, cadets and faculty accepted the Challenge. Friends and families of the cadets sent 275 bags of Chips Ahoy! cookies from all over the country. From the 275 bags, 42 were randomly selected for the study, while the other bags were used to keep cadet morale high during counting.

For each of the 42 bags selected for the study, the cadets dissolved the cookies in water to separate the chips, and then counted the chips. The following table gives the number of chips per bag for these 42 bags.

After studying confidence intervals in this chapter, you will be asked to analyze these data for the purpose of estimating the mean number of chips per bag for all bags of Chips Ahoy! cookies.

| | | | | | | |
|------|------|------|------|------|------|------|
| 1200 | 1219 | 1103 | 1213 | 1258 | 1325 | 1295 |
| 1247 | 1098 | 1185 | 1087 | 1377 | 1363 | 1121 |
| 1279 | 1269 | 1199 | 1244 | 1294 | 1356 | 1137 |
| 1545 | 1135 | 1143 | 1215 | 1402 | 1419 | 1166 |
| 1132 | 1514 | 1270 | 1345 | 1214 | 1154 | 1307 |
| 1293 | 1546 | 1228 | 1239 | 1440 | 1219 | 1191 |

## 8.1 Estimating a Population Mean

A common problem in statistics is to obtain information about the mean, $\mu$, of a population. For example, we might want to know

- the mean age of people in the civilian labor force,
- the mean cost of a wedding,
- the mean gas mileage of a new-model car, or
- the mean starting salary of liberal-arts graduates.

If the population is small, we can ordinarily determine $\mu$ exactly by first taking a census and then computing $\mu$ from the population data. But if the population is large, as it often is in practice, taking a census is generally impractical, extremely expensive, or impossible. Nonetheless, we can usually obtain sufficiently accurate information about $\mu$ by taking a sample from the population.

### Point Estimate

One way to obtain information about a population mean $\mu$ without taking a census is to estimate it by a sample mean $\bar{x}$, as illustrated in the next example.

| Example 8.1 | Point Estimate of a Population Mean |

*Prices of New Mobile Homes* The U.S. Census Bureau publishes annual price figures for new mobile homes in *Manufactured Housing Statistics*. The figures are obtained from sampling, not from a census. A simple random sample of 36 new mobile homes yielded the prices, in thousands of dollars, shown in Table 8.1. Use the data to estimate the population mean price, $\mu$, of all new mobile homes.

**TABLE 8.1**
Prices ($1000s) of 36 randomly selected new mobile homes

| 63.8 | 64.4 | 55.2 | 52.9 | 59.9 | 58.2 | 51.6 | 68.9 | 58.6 |
|------|------|------|------|------|------|------|------|------|
| 63.1 | 69.4 | 59.7 | 53.7 | 62.7 | 57.7 | 51.5 | 45.3 | 68.9 |
| 45.9 | 52.5 | 67.2 | 55.1 | 60.3 | 60.0 | 51.9 | 47.3 | 49.7 |
| 52.0 | 72.7 | 72.8 | 56.6 | 70.5 | 53.9 | 66.4 | 59.8 | 73.9 |

**Solution** We estimate the population mean price, $\mu$, of all new mobile homes by the sample mean price, $\bar{x}$, of the 36 new mobile homes sampled. From Table 8.1,

$$\bar{x} = \frac{\Sigma x_i}{n} = \frac{2134.0}{36} = 59.28.$$

**Interpretation** Based on the sample data, we estimate the mean price, $\mu$, of all new mobile homes to be approximately $59.28 thousand, that is, $59,280.

An estimate of this kind is called a **point estimate** for $\mu$ because it consists of a single number, or point.

**You try it!**

Exercise 8.3 on page 342

• • •

As indicated in the following definition, the term point estimate applies to the use of a statistic to estimate any parameter, not just a population mean.

**Definition 8.1**

**Point Estimate**

A **point estimate** of a parameter is the value of a statistic used to estimate the parameter.

In the previous example, the parameter is the mean price $\mu$ of all new mobile homes, which is unknown. The point estimate of that parameter is the mean price $\bar{x}$ of the 36 mobile homes sampled, which is \$59,280.

### Confidence-Interval Estimate

As you learned in Chapter 7, a sample mean is usually not equal to the population mean; generally there is sampling error. Therefore, we should accompany any point estimate of $\mu$ with information that indicates the accuracy of that estimate. This information is called a *confidence-interval estimate* for $\mu$, which we introduce in the next example.

**Example 8.2**

### Introducing Confidence Intervals

*Prices of New Mobile Homes* Consider again the problem of estimating the (population) mean price $\mu$ of all new mobile homes by using the sample data in Table 8.1. Let's assume that the population standard deviation of all such prices is \$7.2 thousand, that is, \$7200.[†]

a. Identify the distribution of the variable $\bar{x}$, that is, the sampling distribution of the sample mean for samples of size 36.

b. Use part (a) to show that 95.44% of all samples of 36 new mobile homes have the property that the interval from $\bar{x} - 2.4$ to $\bar{x} + 2.4$ contains $\mu$.

c. Use part (b) and the sample data in Table 8.1 to find a 95.44% *confidence interval* for $\mu$, that is, an interval of numbers that we can be 95.44% confident contains $\mu$.

### Solution

**FIGURE 8.1**

Normal probability plot of the price data in Table 8.1

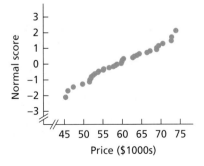

Price ($1000s)

a. Figure 8.1 is a normal probability plot of the price data in Table 8.1. The plot shows we can reasonably presume that prices of new mobile homes are normally distributed. Because $n = 36$, $\sigma = 7.2$, and prices of new mobile homes are normally distributed, Key Fact 7.4 on page 325 implies that

- $\mu_{\bar{x}} = \mu$ (which we don't know),
- $\sigma_{\bar{x}} = \sigma/\sqrt{n} = 7.2/\sqrt{36} = 1.2$, and
- $\bar{x}$ is normally distributed.

In other words, for samples of size 36, the variable $\bar{x}$ is normally distributed with mean $\mu$ and standard deviation 1.2.

b. The "95.44" part of the 68.26-95.44-99.74 rule states that, for a normally distributed variable, 95.44% of all possible observations lie within two standard deviations to either side of the mean. Applying this rule to the variable $\bar{x}$ and referring to part (a), we see that 95.44% of all samples of 36 new mobile homes have mean prices within $2 \cdot 1.2 = 2.4$ of $\mu$.

---

[†]We might know the population standard deviation from previous research or from a preliminary study of prices. We examine the more usual case, where $\sigma$ is unknown, in Section 8.4.

Equivalently, 95.44% of all samples of 36 new mobile homes have the property that the interval from $\bar{x} - 2.4$ to $\bar{x} + 2.4$ contains $\mu$.

c. Because we are taking a simple random sample, each possible sample of size 36 is equally likely to be the one obtained. From part (b), 95.44% of all such samples have the property that the interval from $\bar{x} - 2.4$ to $\bar{x} + 2.4$ contains $\mu$. Hence, chances are 95.44% that the sample we obtain has that property. Consequently, we can be 95.44% confident that the sample of 36 new mobile homes whose prices are shown in Table 8.1 has the property that the interval from $\bar{x} - 2.4$ to $\bar{x} + 2.4$ contains $\mu$. For that sample, $\bar{x} = 59.28$, so

$$\bar{x} - 2.4 = 59.28 - 2.4 = 56.88 \quad \text{and} \quad \bar{x} + 2.4 = 59.28 + 2.4 = 61.68.$$

Thus our 95.44% confidence interval is from 56.88 to 61.68.

**Interpretation**  We can be 95.44% confident that the mean price, $\mu$, of all new mobile homes is somewhere between \$56,880 and \$61,680.

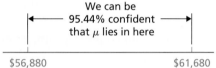

**Note:**  Although this or any other 95.44% confidence interval may or may not contain $\mu$, we can be 95.44% confident that it does.

**You try it!**

Exercise 8.5
on page 342

• • •

With the previous example in mind, we now define confidence-interval estimate and related terms. As indicated, the terms apply to estimating any parameter, not just a population mean.

---

**Definition 8.2**

**What Does It Mean?**

A confidence-interval estimate for a parameter provides a range of numbers along with a percentage confidence that the parameter lies in that range.

**Confidence-Interval Estimate**

**Confidence interval (CI):** An interval of numbers obtained from a point estimate of a parameter.

**Confidence level:** The confidence we have that the parameter lies in the confidence interval (i.e., that the confidence interval contains the parameter).

**Confidence-interval estimate:** The confidence level and confidence interval.

---

A confidence interval for a population mean depends on the sample mean, $\bar{x}$, which in turn depends on the sample selected. For example, suppose that the prices of the 36 new mobile homes sampled were as shown in Table 8.2 instead of as in Table 8.1. Then we would have $\bar{x} = 61.83$ so that

$$\bar{x} - 2.4 = 61.83 - 2.4 = 59.43 \quad \text{and} \quad \bar{x} + 2.4 = 61.83 + 2.4 = 64.23.$$

In this case, the 95.44% confidence interval for $\mu$ would be from 59.43 to 64.23. We could be 95.44% confident that the mean price, $\mu$, of all new mobile homes is somewhere between \$59,430 and \$64,230.

**TABLE 8.2**
Prices (\$1000s) of another sample of 36 randomly selected new mobile homes

| | | | | | | | | |
|---|---|---|---|---|---|---|---|---|
| 69.0 | 68.1 | 57.2 | 49.0 | 71.5 | 59.8 | 52.0 | 71.7 | 61.7 |
| 49.2 | 62.6 | 61.3 | 64.9 | 54.4 | 65.1 | 61.8 | 60.1 | 56.6 |
| 62.5 | 60.7 | 58.5 | 57.3 | 58.1 | 64.0 | 75.2 | 65.2 | 64.0 |
| 56.2 | 68.1 | 50.9 | 62.1 | 60.1 | 68.0 | 64.8 | 60.3 | 73.9 |

### Interpreting Confidence Intervals

The next example stresses the importance of interpreting a confidence interval correctly. It also illustrates that the population mean, $\mu$, may or may not lie in the confidence interval obtained.

**Example 8.3** | **Interpreting Confidence Intervals**

*Prices of New Mobile Homes*  Consider again the prices of new mobile homes. As demonstrated in part (b) of Example 8.2, 95.44% of all samples of 36 new mobile homes have the property that the interval from $\bar{x} - 2.4$ to $\bar{x} + 2.4$ contains $\mu$. In other words, if 36 new mobile homes are selected at random and their mean price, $\bar{x}$, is computed, the interval from

$$\bar{x} - 2.4 \quad \text{to} \quad \bar{x} + 2.4 \tag{8.1}$$

will be a 95.44% confidence interval for the mean price of all new mobile homes.

To illustrate that the mean price, $\mu$, of all new mobile homes may or may not lie in the 95.44% confidence interval obtained, we used a computer to simulate 20 samples of 36 new mobile home prices each. For the simulation, we assumed that $\mu = 61$ (i.e., \$61 thousand) and $\sigma = 7.2$ (i.e., \$7.2 thousand). In reality, we don't know $\mu$; we are assuming a value for $\mu$ to illustrate a point.

For each of the 20 samples of 36 new mobile home prices, we did three things: computed the sample mean price, $\bar{x}$; used Equation (8.1) to obtain the 95.44% confidence interval; and noted whether the population mean, $\mu = 61$, actually lies in the confidence interval.

Figure 8.2 summarizes our results. For each sample, we have drawn a graph on the right-hand side of Fig. 8.2. The dot represents the sample mean, $\bar{x}$,

**FIGURE 8.2**

Twenty confidence intervals for the mean price of all new mobile homes, each based on a sample of 36 new mobile homes

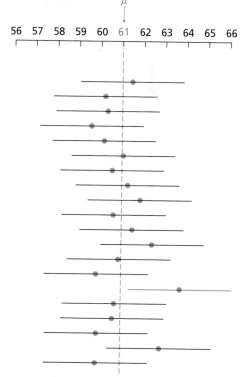

| Sample | $\bar{x}$ | 95.44% CI | $\mu$ in CI? |
|--------|-------|-----------|---------|
| 1 | 61.45 | 59.06 to 63.85 | yes |
| 2 | 60.21 | 57.81 to 62.61 | yes |
| 3 | 60.33 | 57.93 to 62.73 | yes |
| 4 | 59.59 | 57.19 to 61.99 | yes |
| 5 | 60.17 | 57.77 to 62.57 | yes |
| 6 | 61.07 | 58.67 to 63.47 | yes |
| 7 | 60.56 | 58.16 to 62.96 | yes |
| 8 | 61.28 | 58.88 to 63.68 | yes |
| 9 | 61.87 | 59.48 to 64.27 | yes |
| 10 | 60.61 | 58.22 to 63.01 | yes |
| 11 | 61.51 | 59.11 to 63.91 | yes |
| 12 | 62.45 | 60.05 to 64.85 | yes |
| 13 | 60.88 | 58.48 to 63.28 | yes |
| 14 | 59.85 | 57.45 to 62.25 | yes |
| 15 | 63.73 | 61.33 to 66.13 | no |
| 16 | 60.70 | 58.30 to 63.10 | yes |
| 17 | 60.60 | 58.20 to 63.00 | yes |
| 18 | 59.88 | 57.48 to 62.28 | yes |
| 19 | 62.82 | 60.42 to 65.22 | yes |
| 20 | 59.84 | 57.45 to 62.24 | yes |

in thousands of dollars, and the horizontal line represents the corresponding 95.44% confidence interval. Note that the population mean, $\mu$, lies in the confidence interval only when the horizontal line crosses the dashed line.

Figure 8.2 reveals that $\mu$ lies in the 95.44% confidence interval in 19 of the 20 samples, that is, in 95% of the samples. If, instead of 20 samples, we simulated 1000, we would probably find that the percentage of those 1000 samples for which $\mu$ lies in the 95.44% confidence interval would be even closer to 95.44%. Hence we can be 95.44% confident that any computed 95.44% confidence interval will contain $\mu$.

•  •  •

# Exercises 8.1

## Understanding the Concepts and Skills

**8.1** The value of a statistic used to estimate a parameter is called a _____ of the parameter.

**8.2** What is a confidence-interval estimate of a parameter? Why is such an estimate superior to a point estimate?

**8.3 Wedding Costs.** According to *Bride's Magazine*, getting married these days can be expensive when the costs of the reception, engagement ring, bridal gown, pictures—just to name a few—are included. A simple random sample of 20 recent U.S. weddings yielded the following data on wedding costs, in dollars.

| | | | | |
|---|---|---|---|---|
| 19,496 | 23,789 | 18,312 | 14,554 | 18,460 |
| 27,806 | 21,203 | 29,288 | 34,081 | 27,896 |
| 30,098 | 13,360 | 33,178 | 42,646 | 24,053 |
| 32,269 | 40,406 | 35,050 | 21,083 | 19,510 |

a. Use the data to obtain a point estimate for the population mean wedding cost, $\mu$, of all recent U.S. weddings. (*Note:* The sum of the data is $526,538.)
b. Is your point estimate in part (a) likely to equal $\mu$ exactly? Explain your answer.

**8.4 Cottonmouth Litter Size.** A study by Blem and Blem, published in the *Journal of Herpetology* (Vol. 29, pp. 391–398),

examined the reproductive characteristics of the eastern cottonmouth, a once widely distributed snake whose numbers have decreased recently due to encroachment by humans. A simple random sample of 44 female cottonmouths yielded the following data on number of young per litter.

| | | | | | | | | |
|---|---|---|---|---|---|---|---|---|
| 5 | 12 | 7 | 7 | 6 | 8 | 12 | 9 | 7 |
| 4 | 9 | 6 | 12 | 7 | 5 | 6 | 10 | 3 |
| 10 | 8 | 8 | 12 | 5 | 6 | 10 | 11 | 3 |
| 8 | 4 | 5 | 7 | 6 | 11 | 7 | 6 | 8 |
| 8 | 14 | 8 | 7 | 11 | 7 | 5 | 4 | |

a. Use the data to obtain a point estimate for the mean number of young per litter, $\mu$, of all female eastern cottonmouths. (*Note:* $\Sigma x_i = 334$.)
b. Is your point estimate in part (a) likely to equal $\mu$ exactly? Explain your answer.

*For Exercises 8.5–8.10, you may want to review Example 8.2, which begins on page 339.*

**8.5 Wedding Costs.** Refer to Exercise 8.3. Assume that recent wedding costs in the United States are normally distributed with a standard deviation of $8100.
a. Determine a 95.44% confidence interval for the mean cost, $\mu$, of all recent U.S. weddings.
b. Interpret your result in part (a).
c. Does the mean cost of all recent U.S. weddings lie in the confidence interval you obtained in part (a)? Explain your answer.

**8.6 Cottonmouth Litter Size.** Refer to Exercise 8.4. Assume that $\sigma = 2.4$.
a. Obtain an approximate 95.44% confidence interval for the mean number of young per litter of all female eastern cottonmouths.
b. Interpret your result in part (a).
c. Why is the 95.44% confidence interval that you obtained in part (a) not necessarily exact?

**8.7 Fuel Tank Capacity.** *Consumer Reports* provides information on new automobile models—including price, mileage ratings, engine size, body size, and indicators of features. A simple random sample of 35 new models yielded the following data on fuel tank capacity, in gallons.

| | | | | | | |
|---|---|---|---|---|---|---|
| 17.2 | 23.1 | 17.5 | 15.7 | 19.8 | 16.9 | 15.3 |
| 18.5 | 18.5 | 25.5 | 18.0 | 17.5 | 14.5 | 20.0 |
| 17.0 | 20.0 | 24.0 | 26.0 | 18.1 | 21.0 | 19.3 |
| 20.0 | 20.0 | 12.5 | 13.2 | 15.9 | 14.5 | 22.2 |
| 21.1 | 14.4 | 25.0 | 26.4 | 16.9 | 16.4 | 23.0 |

a. Find a point estimate for the mean fuel tank capacity of all new automobile models. Interpret your answer in words. (*Note:* $\Sigma x_i = 664.9$ gallons.)
b. Determine a 95.44% confidence interval for the mean fuel tank capacity of all new automobile models. Assume $\sigma = 3.50$ gallons.
c. How would you decide whether fuel tank capacities for new automobile models are approximately normally distributed?
d. Must fuel tank capacities for new automobile models be exactly normally distributed for the confidence interval that you obtained in part (b) to be approximately correct? Explain your answer.

**8.8 Home Improvements.** The *American Express Retail Index* provides information on budget amounts for home improvements. The following table displays the budgets, in dollars, of 45 randomly sampled home improvement jobs in the United States.

| | | | | | | | | |
|---|---|---|---|---|---|---|---|---|
| 3179 | 1032 | 1822 | 4093 | 2285 | 1478 | 955 | 2773 | 514 |
| 3915 | 4800 | 3843 | 5265 | 2467 | 2353 | 4200 | 3146 | 551 |
| 2659 | 4660 | 3570 | 1598 | 2605 | 3643 | 2816 | 3125 | 3104 |
| 4503 | 2911 | 3605 | 2948 | 1421 | 1910 | 5145 | 4557 | 2026 |
| 2750 | 2069 | 3056 | 2550 | 631 | 4550 | 5069 | 2124 | 1573 |

a. Determine a point estimate for the population mean budget, $\mu$, for such home improvement jobs. Interpret your answer in words. (*Note:* The sum of the data is $129,849.)
b. Obtain a 95.44% confidence interval for the population mean budget, $\mu$, for such home improvement jobs and interpret your result in words. Assume that the population standard deviation of budgets for home improvement jobs is $1350.
c. How would you decide whether budgets for such home improvement jobs are approximately normally distributed?
d. Must the budgets for such home improvement jobs be exactly normally distributed for the confidence interval that you obtained in part (b) to be approximately correct? Explain your answer.

**8.9 Giant Tarantulas.** A tarantula has two body parts. The anterior part of the body is covered above by a shell, or carapace. In the paper, "Reproductive Biology of Uruguayan Theraphosids" (*The Journal of Arachnology*, Vol. 30, No. 3, pp. 571–587), F. Costa and F. Perez–Miles discussed a large species of tarantula, whose common name is the Brazilian giant tawny red. A simple random sample of 15 of these adult male tarantulas provided the following data on carapace length, in millimeters (mm).

| | | | | |
|---|---|---|---|---|
| 15.7 | 18.3 | 19.7 | 17.6 | 19.0 |
| 19.2 | 19.8 | 18.1 | 18.0 | 20.9 |
| 16.4 | 16.8 | 18.9 | 18.5 | 19.5 |

a. Obtain a normal probability plot of the data.
b. Based on your result from part (a), is it reasonable to presume that carapace length of adult male Brazilian giant tawny red tarantulas is normally distributed? Explain your answer.
c. Find and interpret a 95.44% confidence interval for the mean carapace length of all adult male Brazilian giant tawny red tarantulas. The population standard deviation is 1.76 mm.
d. In Exercise 6.67, we noted that the mean carapace length of all adult male Brazilian giant tawny red tarantulas is 18.14 mm. Does your confidence interval in part (c) contain the population mean? Would it necessarily have to? Explain your answers.

**8.10 Serum Cholesterol Levels.** Information on serum total cholesterol level is published by the Centers for Disease Control and Prevention in *National Health and Nutrition Examination Survey*. A simple random sample of 12 U.S. females, 20 years old or older, provided the following data on serum total cholesterol level, in milligrams per deciliter (mg/dL).

| | | | | | |
|---|---|---|---|---|---|
| 260 | 289 | 190 | 214 | 110 | 241 |
| 169 | 173 | 191 | 178 | 129 | 185 |

a. Obtain a normal probability plot of the data.
b. Based on your result from part (a), is it reasonable to presume that serum total cholesterol level of U.S. females, 20 years old or older, is normally distributed? Explain your answer.
c. Find and interpret a 95.44% confidence interval for the mean serum total cholesterol level of U.S. females, 20 years old or older. The population standard deviation is 44.7 mg/dL.
d. In Exercise 6.68, we noted that the mean serum total cholesterol level of U.S. females, 20 years old or older, is 206 mg/dL. Does your confidence interval in part (c) contain the population mean? Would it necessarily have to? Explain your answers.

## Extending the Concepts and Skills

**8.11 New Mobile Homes.** Refer to Examples 8.1 and 8.2. Use the data in Table 8.1 on page 338 to obtain a 99.74% confidence interval for the mean price of all new mobile homes. (*Hint:* Proceed as in Example 8.2 but use the "99.74" part of the 68.26-95.44-99.74 rule instead of the "95.44" part.)

**8.12 New Mobile Homes.** Refer to Examples 8.1 and 8.2. Use the data in Table 8.1 on page 338 to obtain a 68.26% confidence interval for the mean price of all new mobile homes. (*Hint:* Proceed as in Example 8.2 but use the "68.26" part of the 68.26-95.44-99.74 rule instead of the "95.44" part.)

# 8.2 Confidence Intervals for One Population Mean When $\sigma$ Is Known

In Section 8.1, we showed how to find a 95.44% confidence interval for a population mean, that is, a confidence interval at a confidence level of 95.44%. In this section, we generalize the arguments used there to obtain a confidence interval for a population mean at any prescribed confidence level.

To begin, we introduce some general notation used with confidence intervals. Frequently, we want to write the confidence level in the form $1 - \alpha$, where $\alpha$ is a number between 0 and 1; that is, if the confidence level is expressed as a decimal, $\alpha$ is the number that must be subtracted from 1 to get the confidence level. To find $\alpha$, we simply subtract the confidence level from 1. If the confidence level is 95.44%, then $\alpha = 1 - 0.9544 = 0.0456$; if the confidence level is 90%, then $\alpha = 1 - 0.90 = 0.10$; and so on.

Next, recall from Section 6.2 that the symbol $z_\alpha$ denotes the z-score that has area $\alpha$ to its right under the standard normal curve. So, for example, $z_{0.05}$ denotes the z-score that has area 0.05 to its right and $z_{\alpha/2}$ denotes the z-score that has area $\alpha/2$ to its right.

### Obtaining Confidence Intervals for a Population Mean When $\sigma$ Is Known

We now develop a step-by-step procedure to obtain a confidence interval for a population mean when the population standard deviation is known. In doing so, we assume that the variable under consideration is normally distributed. Because of the central limit theorem, however, the procedure will also work to obtain an approximately correct confidence interval when the sample size is large, regardless of the distribution of the variable.

The basis of our confidence-interval procedure is stated in Key Fact 7.4: If $x$ is a normally distributed variable with mean $\mu$ and standard deviation $\sigma$, then, for samples of size $n$, the variable $\bar{x}$ is also normally distributed and has mean $\mu$ and standard deviation $\sigma/\sqrt{n}$. As in Section 8.1, we can use that fact and the "95.44" part of the 68.26-95.44-99.74 rule to conclude that 95.44% of all samples of size $n$ have means within $2 \cdot \sigma/\sqrt{n}$ of $\mu$, as depicted in Fig. 8.3(a).

More generally, we can say that $100(1 - \alpha)\%$ of all samples of size $n$ have means within $z_{\alpha/2} \cdot \sigma/\sqrt{n}$ of $\mu$, as depicted in Fig. 8.3(b). Equivalently, we can say that $100(1 - \alpha)\%$ of all samples of size $n$ have the property that the interval from

$$\bar{x} - z_{\alpha/2} \cdot \frac{\sigma}{\sqrt{n}} \quad \text{to} \quad \bar{x} + z_{\alpha/2} \cdot \frac{\sigma}{\sqrt{n}}$$

**FIGURE 8.3**
(a) 95.44% of all samples have means within 2 standard deviations of $\mu$;
(b) $100(1-\alpha)\%$ of all samples have means within $z_{\alpha/2}$ standard deviations of $\mu$

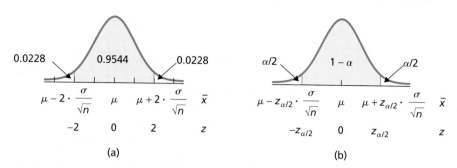

(a)                                         (b)

contains $\mu$. Consequently, we have Procedure 8.1, called the **one-mean z-interval procedure,** or, when no confusion can arise, simply the **z-interval procedure.**[†]

| Procedure 8.1 | **One-Mean z-Interval Procedure** |
|---|---|

*Purpose* To find a confidence interval for a population mean, $\mu$

*Assumptions*

1. Simple random sample
2. Normal population or large sample
3. $\sigma$ known

**STEP 1** For a confidence level of $1-\alpha$, use Table II to find $z_{\alpha/2}$.

**STEP 2** The confidence interval for $\mu$ is from

$$\bar{x} - z_{\alpha/2} \cdot \frac{\sigma}{\sqrt{n}} \quad \text{to} \quad \bar{x} + z_{\alpha/2} \cdot \frac{\sigma}{\sqrt{n}}$$

where $z_{\alpha/2}$ is found in Step 1, $n$ is the sample size, and $\bar{x}$ is computed from the sample data.

**STEP 3** Interpret the confidence interval.

The confidence interval is exact for normal populations and is approximately correct for large samples from nonnormal populations.

**Note:** By saying that the confidence interval is *exact,* we mean that the true confidence level equals $1-\alpha$; by saying that the confidence interval is *approximately correct,* we mean that the true confidence level only approximately equals $1-\alpha$.

Before applying Procedure 8.1, we need to make several comments about it and the assumptions for its use.

• We use the term **normal population** as an abbreviation for "the variable under consideration is normally distributed."

---

[†]The one-mean z-interval procedure is also known as the **one-sample z-interval procedure** and the **one-variable z-interval procedure.** We prefer "one-mean" because it makes clear the parameter being estimated.

- The z-interval procedure works reasonably well even when the variable is not normally distributed and the sample size is small or moderate, provided the variable is not too far from being normally distributed. Thus we say that the z-interval procedure is **robust** to moderate violations of the normality assumption.[†]
- Watch for outliers because their presence calls into question the normality assumption. Moreover, even for large samples, outliers can sometimes unduly affect a z-interval because the sample mean is not resistant to outliers.

Key Fact 8.1 lists some general guidelines for use of the z-interval procedure.

---

**Key Fact 8.1**

### When to Use the One-Mean z-Interval Procedure[‡]

- For small samples—say, of size less than 15—the z-interval procedure should be used only when the variable under consideration is normally distributed or very close to being so.
- For samples of moderate size—say, between 15 and 30—the z-interval procedure can be used unless the data contain outliers or the variable under consideration is far from being normally distributed.
- For large samples—say, of size 30 or more—the z-interval procedure can be used essentially without restriction. However, if outliers are present and their removal is not justified, you should compare the confidence intervals obtained with and without the outliers to see what effect the outliers have. If the effect is substantial, use a different procedure or take another sample.
- If outliers are present but their removal is justified and results in a data set for which the z-interval procedure is appropriate (as previously stated), the procedure can be used.

---

Key Fact 8.1 makes it clear that you should conduct preliminary data analyses before applying the z-interval procedure. More generally, the following fundamental principle of data analysis is relevant to all inferential procedures.

---

**Key Fact 8.2**

### A Fundamental Principle of Data Analysis

**What Does It Mean?**

Always look at the sample data (by constructing a histogram, normal probability plot, boxplot, etc.) prior to performing a statistical-inference procedure to help check whether the procedure is appropriate.

Before performing a statistical-inference procedure, examine the sample data. If any of the conditions required for using the procedure appear to be violated, do not apply the procedure. Instead use a different, more appropriate procedure, or, if you are unsure of one, consult a statistician.

---

Even for small samples, where graphical displays must be interpreted carefully, it is far better to examine the data than not to. Remember, though, to proceed cautiously when conducting graphical analyses of small samples, especially very small samples—say, of size 10 or less.

---

[†]A statistical procedure that works reasonably well even when one of its assumptions is violated (or moderately violated) is called a **robust procedure** relative to that assumption.

[‡]Statisticians also consider skewness. Roughly speaking, the more skewed the distribution of the variable under consideration, the larger is the sample size required for the validity of the z-interval procedure. See, for instance, the paper "How Large Does n Have to be for Z and t Intervals?" by D. Boos and J. Hughes-Oliver (*The American Statistician*, Vol. 54, No. 2, pp. 121–128).

**Example 8.4** | **The One-Mean $z$-Interval Procedure**

| | | | | |
|---|---|---|---|---|
| 22 | 58 | 40 | 42 | 43 |
| 32 | 34 | 45 | 38 | 19 |
| 33 | 16 | 49 | 29 | 30 |
| 43 | 37 | 19 | 21 | 62 |
| 60 | 41 | 28 | 35 | 37 |
| 51 | 37 | 65 | 57 | 26 |
| 27 | 31 | 33 | 24 | 34 |
| 28 | 39 | 43 | 26 | 38 |
| 42 | 40 | 31 | 34 | 38 |
| 35 | 29 | 33 | 32 | 33 |

*The Civilian Labor Force* The U.S. Bureau of Labor Statistics collects informa-tion on the ages of people in the civilian labor force and publishes the results in *Employment and Earnings*. Fifty people in the civilian labor force are randomly selected; their ages are displayed in Table 8.3. Find a 95% confidence interval for the mean age, $\mu$, of all people in the civilian labor force. Assume that the population standard deviation of the ages is 12.1 years.

**Solution** We constructed (but did not show) a normal probability plot, a his-togram, a stem-and-leaf diagram, and a boxplot for these age data. The boxplot indicated potential outliers, but in view of the other three graphs, we concluded that the data contain no outliers. Because the sample size is 50, which is large, and the population standard deviation is known, we can use Procedure 8.1 to find the required confidence interval.

**STEP 1  For a confidence level of $1 - \alpha$, use Table II to find $z_{\alpha/2}$.**

We want a 95% confidence interval, so $\alpha = 1 - 0.95 = 0.05$. From Table II, $z_{\alpha/2} = z_{0.05/2} = z_{0.025} = 1.96$.

**STEP 2  The confidence interval for $\mu$ is from**

$$\bar{x} - z_{\alpha/2} \cdot \frac{\sigma}{\sqrt{n}} \quad \text{to} \quad \bar{x} + z_{\alpha/2} \cdot \frac{\sigma}{\sqrt{n}}.$$

We know $\sigma = 12.1$, $n = 50$, and, from Step 1, $z_{\alpha/2} = 1.96$. To compute $\bar{x}$ for the data in Table 8.3, we apply the usual formula:

$$\bar{x} = \frac{\Sigma x_i}{n} = \frac{1819}{50} = 36.4,$$

to one decimal place. Consequently, a 95% confidence interval for $\mu$ is from

$$36.4 - 1.96 \cdot \frac{12.1}{\sqrt{50}} \quad \text{to} \quad 36.4 + 1.96 \cdot \frac{12.1}{\sqrt{50}}.$$

or 33.0 to 39.8.

**You try it!**

Exercise 8.31 on page 350

**STEP 3  Interpret the confidence interval.**

**Interpretation** We can be 95% confident that the mean age, $\mu$, of all people in the civilian labor force is somewhere between 33.0 years and 39.8 years.

• • •

## Confidence and Precision

The confidence level of a confidence interval for a population mean, $\mu$, signifies the confidence we have that $\mu$ actually lies in that interval. The length of the confidence interval indicates the precision of the estimate, or how well we have "pinned down" $\mu$. Long confidence intervals indicate poor precision; short confidence intervals indicate good precision.

How does the confidence level affect the length of the confidence inter-val? To answer this question, let's return to Example 8.4, where we found a 95% confidence interval for the mean age, $\mu$, of all people in the civilian labor force. The confidence level there is 0.95, and the confidence interval is from 33.0

to 39.8 years. If we change the confidence level from 0.95 to, say, 0.90, then $z_{\alpha/2}$ changes from $z_{0.05/2} = z_{0.025} = 1.96$ to $z_{0.10/2} = z_{0.05} = 1.645$. The resulting confidence interval, using the same sample data (Table 8.3), is from

$$36.4 - 1.645 \cdot \frac{12.1}{\sqrt{50}} \quad \text{to} \quad 36.4 + 1.645 \cdot \frac{12.1}{\sqrt{50}}$$

or from 33.6 to 39.2 years. Figure 8.4 shows both the 90% and 95% confidence intervals.

**FIGURE 8.4**
90% and 95% confidence intervals for $\mu$, using the data in Table 8.3

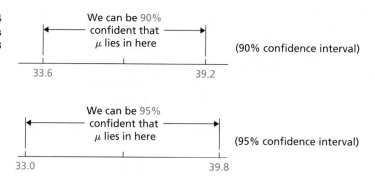

Thus, decreasing the confidence level decreases the length of the confidence interval, and vice versa. So, if we can settle for less confidence that $\mu$ lies in our confidence interval, we get a shorter interval. However, if we want more confidence that $\mu$ lies in our confidence interval, we must settle for a greater interval.

**Key Fact 8.3**

## Confidence and Precision

For a fixed sample size, decreasing the confidence level improves the precision, and vice versa.

# The Technology Center

Most statistical technologies have programs that automatically perform the one-mean $z$-interval procedure. In this subsection, we present output and step-by-step instructions for such programs.

**Example 8.5**

## Using Technology to Obtain a One-Mean $z$-Interval

*The Civilian Labor Force*   Table 8.3 on page 347 displays the ages of 50 randomly selected people in the civilian labor force. Use Minitab, Excel, or the TI-83/84 Plus to determine a 95% confidence interval for the mean age, $\mu$, of all people in the civilian labor force. Assume that the population standard deviation of the ages is 12.1 years.

**Solution**   We applied the one-mean $z$-interval programs to the data, resulting in Output 8.1. Steps for generating that output are presented in Instructions 8.1.

**OUTPUT 8.1**    One-mean z-interval for the sample of ages

**MINITAB**

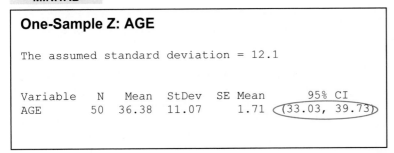

```
One-Sample Z: AGE

The assumed standard deviation = 12.1


Variable    N    Mean    StDev    SE Mean      95% CI
AGE        50    36.38   11.07     1.71    (33.03, 39.73)
```

**EXCEL**

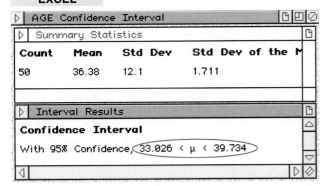

```
AGE Confidence Interval

Summary Statistics

Count    Mean    Std Dev    Std Dev of the M
50       36.38    12.1       1.711


Interval Results

Confidence Interval

With 95% Confidence, 33.026 < μ < 39.734
```

**TI-83/84 PLUS**

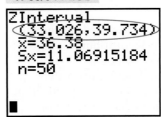

```
ZInterval
(33.026,39.734)
x̄=36.38
Sx=11.06915184
n=50
```

As shown in Output 8.1, the required 95% confidence interval is from 33.03 to 39.73. We can be 95% confident that the mean age of all people in the civilian labor force is somewhere between 33.0 years and 39.7 years. Compare this confidence interval to the one obtained in Example 8.4. Can you explain the slight discrepancy?

• • •

**INSTRUCTIONS 8.1**    Steps for generating Output 8.1

| MINITAB | EXCEL | TI-83/84 PLUS |
|---|---|---|
| 1  Store the data from Table 8.3 in a column named AGE | 1  Store the data from Table 8.3 in a range named AGE | 1  Store the data from Table 8.3 in a list named AGE |
| 2  Choose **Stat ➤ Basic Statistics ➤ 1-Sample Z...** | 2  Choose **DDXL ➤ Confidence Intervals** | 2  Press **STAT**, arrow over to **TESTS**, and press **7** |
| 3  Select the **Samples in columns** option button | 3  Select **1 Var z Interval** from the **Function type** drop-down box | 3  Highlight **Data** and press **ENTER** |
| 4  Click in the **Samples in columns** text box and specify AGE | 4  Specify AGE in the **Quantitative Variable** text box | 4  Press the down-arrow key, type 12.1 for $\sigma$, and press **ENTER** |
| 5  Click in the **Standard deviation** text box and type 12.1 | 5  Click **OK** | 5  Press **2nd ➤ LIST** |
| 6  Click the **Options...** button | 6  Click the **95%** button | 6  Arrow down to AGE and press **ENTER** three times |
| 7  Type 95 in the **Confidence level** text box | 7  Click in the **Type in the population standard deviation** text box and type 12.1 | 7  Type .95 for **C-Level** and press **ENTER** twice |
| 8  Click the arrow button at the right of the **Alternative** drop-down list box and select **not equal** | 8  Click the **Compute Interval** button | |
| 9  Click **OK** twice | | |

■ ■ ■

# Exercises 8.2

## Understanding the Concepts and Skills

**8.13** Find the confidence level and $\alpha$ for
a. a 90% confidence interval.
b. a 99% confidence interval.

**8.14** Find the confidence level and $\alpha$ for
a. an 85% confidence interval.
b. a 95% confidence interval.

**8.15** What is meant by saying that a $1 - \alpha$ confidence interval is
a. exact?                          b. approximately correct?

**8.16** In developing Procedure 8.1, we assumed that the variable under consideration is normally distributed.
a. Explain why we needed that assumption.
b. Explain why the procedure yields an approximately correct confidence interval for large samples, regardless of the distribution of the variable under consideration.

**8.17** For what is *normal population* an abbreviation?

**8.18** Refer to Procedure 8.1.
a. Explain in detail the assumptions required for using the z-interval procedure.
b. How important is the normality assumption? Explain your answer.

**8.19** What is meant by saying that a statistical procedure is robust?

**8.20** In each part, assume that the population standard deviation is known. Decide whether use of the z-interval procedure to obtain a confidence interval for the population mean is reasonable. Explain your answers.
a. The variable under consideration is very close to being normally distributed, and the sample size is 10.
b. The variable under consideration is very close to being normally distributed, and the sample size is 75.
c. The sample data contain outliers, and the sample size is 20.

**8.21** In each part, assume that the population standard deviation is known. Decide whether use of the z-interval procedure to obtain a confidence interval for the population mean is reasonable. Explain your answers.
a. The sample data contain no outliers, the variable under consideration is roughly normally distributed, and the sample size is 20.
b. The distribution of the variable under consideration is highly skewed, and the sample size is 20.
c. The sample data contain no outliers, the sample size is 250, and the variable under consideration is far from being normally distributed.

**8.22** Suppose that you have obtained data by taking a random sample from a population. Before performing a statistical inference, what should you do?

**8.23** Suppose that you have obtained data by taking a random sample from a population and that you intend to find a confidence interval for the population mean, $\mu$. Which confidence level, 95% or 99%, will result in the confidence interval's giving a more precise estimate of $\mu$?

**8.24** If a good typist can input 70 words per minute but a 99% confidence interval for the mean number of words input per minute by recent applicants lies entirely below 70, what can you conclude about the typing skills of recent applicants?

*In each of Exercises 8.25–8.30, we have provided a sample mean, sample size, population standard deviation, and confidence level. In each case, use the one-mean z-interval procedure to find a confidence interval for the mean of the population from which the sample was drawn.*

**8.25** $\bar{x} = 20$, $n = 36$, $\sigma = 3$, confidence level $= 95\%$

**8.26** $\bar{x} = 25$, $n = 36$, $\sigma = 3$, confidence level $= 95\%$

**8.27** $\bar{x} = 30$, $n = 25$, $\sigma = 4$, confidence level $= 90\%$

**8.28** $\bar{x} = 35$, $n = 25$, $\sigma = 4$, confidence level $= 90\%$

**8.29** $\bar{x} = 50$, $n = 16$, $\sigma = 5$, confidence level $= 99\%$

**8.30** $\bar{x} = 55$, $n = 16$, $\sigma = 5$, confidence level $= 99\%$

*Preliminary data analyses indicate that you can reasonably apply the z-interval procedure (Procedure 8.1 on page 345) in Exercises 8.31–8.36.*

**8.31 Venture-Capital Investments.** Data on investments in the high-tech industry by venture capitalists are compiled by VentureOne Corporation and published in *America's Network Telecom Investor Supplement*. A random sample of 18 venture-capital investments in the fiber optics business sector yielded the following data, in millions of dollars.

| | | | | | |
|---|---|---|---|---|---|
| 5.60 | 6.27 | 5.96 | 10.51 | 2.04 | 5.48 |
| 5.74 | 5.58 | 4.13 | 8.63 | 5.95 | 6.67 |
| 4.21 | 7.71 | 9.21 | 4.98 | 8.64 | 6.66 |

a. Determine a 95% confidence interval for the mean amount, $\mu$, of all venture-capital investments in the fiber optics business sector. Assume that the population standard deviation is $2.04 million. (*Note:* The sum of the data is $113.97 million.)
b. Interpret your answer from part (a).

**8.32 Poverty and Calcium.** Calcium is the most abundant mineral in the human body and has several important functions. Most body calcium is stored in the bones and teeth, where it functions to support their structure. Recommendations for calcium are provided in *Dietary Reference Intakes*, developed by the Institute of Medicine of the National Academy of Sciences. The recommended adequate intake (RAI) of calcium for adults (ages 19–50) is 1000 milligrams (mg) per day. A simple random sample of 18 adults with incomes below the poverty level gave the following daily calcium intakes.

| | | | | | |
|---|---|---|---|---|---|
| 886 | 633 | 943 | 847 | 934 | 841 |
| 1193 | 820 | 774 | 834 | 1050 | 1058 |
| 1192 | 975 | 1313 | 872 | 1079 | 809 |

a. Determine a 95% confidence interval for the mean calcium intake, $\mu$, of all adults with incomes below the poverty level. Assume that the population standard deviation is 188 mg. (*Note:* The sum of the data is 17,053 mg.)
b. Interpret your answer from part (a).

**8.33 Toxic Mushrooms?** Cadmium, a heavy metal, is toxic to animals. Mushrooms, however, are able to absorb and accumulate cadmium at high concentrations. The Czech and Slovak governments have set a safety limit for cadmium in dry vegetables at 0.5 part per million (ppm). M. Melgar et al. measured the cadmium levels in a random sample of the edible mushroom *Boletus pinicola* and published the results in the *Journal of Environmental Science and Health* (Vol. B33(4), pp. 439–455). Here are the data obtained by the researchers.

| | | | | | |
|---|---|---|---|---|---|
| 0.24 | 0.59 | 0.62 | 0.16 | 0.77 | 1.33 |
| 0.92 | 0.19 | 0.33 | 0.25 | 0.59 | 0.32 |

Find and interpret a 99% confidence interval for the mean cadmium level of all *Boletus pinicola* mushrooms. Assume a population standard deviation of cadmium levels in *Boletus pinicola* mushrooms of 0.37 ppm. (*Note:* The sum of the data is 6.31 ppm.)

**8.34 Smelling Out the Enemy.** Snakes deposit chemical trails as they travel through their habitats. These trails are often detected and recognized by lizards, which are potential prey. The ability to recognize their predators via tongue flicks can often mean life or death for lizards. Scientists from the University of Antwerp were interested in quantifying the responses of juveniles of the common lizard (*Lacerta vivipara*) to natural predator cues to determine whether the behavior is learned or congenital. Seventeen juvenile common lizards were exposed to the chemical cues of the viper snake. Their responses, in number of tongue flicks per 20 minutes, are presented in the following table.

[SOURCE: Van Damme et al., "Responses of Naïve Lizards to Predator Chemical Cues," *Journal of Herpetology*, Vol. 29(1), pp. 38–43.]

| | | | | | |
|---|---|---|---|---|---|
| 425 | 510 | 629 | 236 | 654 | 200 |
| 276 | 501 | 811 | 332 | 424 | 674 |
| 676 | 694 | 710 | 662 | 633 | |

Find and interpret a 90% confidence interval for the mean number of tongue flicks per 20 minutes for all juvenile common lizards. Assume a population standard deviation of 190.0.

**8.35 Political Prisoners.** Ehlers, Maercker, and Boos studied various characteristics of political prisoners from the former East Germany and presented their findings in the paper "Posttraumatic Stress Disorder (PTSD) Following Political Imprisonment: The Role of Mental Defeat, Alienation, and Perceived Permanent Change" (*Journal of Abnormal Psychology*, Vol. 109, pp. 45–55). According to the article, the mean duration of imprisonment for 32 patients with chronic PTSD was 33.4 months. Assuming that $\sigma = 42$ months, determine a 95% confidence interval for the mean duration of imprisonment, $\mu$, of all East German political prisoners with chronic PTSD. Interpret your answer in words.

**8.36 Keep on Rolling.** The Rolling Stones, a rock group formed in the 1960s, have toured extensively in support of new albums. *Pollstar* has collected data on the earnings from the Stones's North American tours. For 30 randomly selected Rolling Stones concerts, the mean gross earnings is $2.27 million. Assuming a population standard deviation gross earnings of $0.5 million, obtain a 99% confidence interval for the mean gross earnings of all Rolling Stones concerts. Interpret your answer in words.

**8.37 Venture-Capital Investments.** Refer to Exercise 8.31.
a. Find a 99% confidence interval for $\mu$.
b. Why is the confidence interval you found in part (a) longer than the one in Exercise 8.31?

**c.** Draw a graph similar to that shown in Fig. 8.4 on page 348 to display both confidence intervals.

**d.** Which confidence interval yields a more precise estimate of $\mu$? Explain your answer.

**8.38 Poverty and Calcium.** Refer to Exercise 8.32.

**a.** Find a 90% confidence interval for $\mu$.

**b.** Why is the confidence interval you found in part (a) shorter than the one in Exercise 8.32?

**c.** Draw a graph similar to that shown in Fig. 8.4 on page 348 to display both confidence intervals.

**d.** Which confidence interval yields a more precise estimate of $\mu$? Explain your answer.

**8.39 Doing Time.** The U.S. Bureau of Justice Statistics provides information on prison sentences in the document *National Corrections Reporting Program*. A random sample of 20 maximum sentences for murder yielded the data, in months, presented on the WeissStats CD. Use the technology of your choice to do the following.

**a.** Find a 95% confidence interval for the mean maximum sentence of all murders. Assume a population standard deviation of 30 months.

**b.** Obtain a normal probability plot, boxplot, histogram, and stem-and-leaf diagram of the data.

**c.** Remove the outliers (if any) from the data and then repeat part (a).

**d.** Comment on the advisability of using the z-interval procedure here.

**8.40 Ages of Diabetics.** According to the document *All About Diabetes*, found on the Web site of the American Diabetes Association, "...diabetes is a disease in which the body does not produce or properly use insulin, a hormone that is needed to convert sugar, starches, and other food into energy needed for daily life." A random sample of 15 diabetics yielded the data on ages, in years, presented on the WeissStats CD. Use the technology of your choice to do the following.

**a.** Find a 95% confidence interval for the mean age, $\mu$, of all people with diabetes. Assume that $\sigma = 21.2$ years.

**b.** Obtain a normal probability plot, boxplot, histogram, and stem-and-leaf diagram of the data.

**c.** Remove the outliers (if any) from the data and then repeat part (a).

**d.** Comment on the advisability of using the z-interval procedure here.

## Working With Large Data Sets

**8.41 Body Temperature.** A study by researchers at the University of Maryland addressed the question of whether the mean body temperature of humans is 98.6°F. The results of the study by P. Mackowiak et al. appeared in the article "A Critical Appraisal of 98.6°F, the Upper Limit of the Normal Body Temperature, and Other Legacies of Carl Reinhold August Wunderlich" (*Journal of the American Medical Association*, Vol. 268, pp. 1578–1580). Among other data, the researchers obtained the body temperatures of 93 healthy humans, as provided on the WeissStats CD. Use the technology of your choice to do the following.

**a.** Obtain a normal probability plot, boxplot, histogram, and stem-and-leaf diagram of the data.

**b.** Based on your results from part (a), can you reasonably apply the z-interval procedure to the data? Explain your reasoning.

**c.** Find and interpret a 99% confidence interval for the mean body temperature of all healthy humans. Assume that $\sigma = 0.63°$F. Does the result surprise you? Why?

**8.42 Malnutrition and Poverty.** R. Reifen et al. studied various nutritional measures of Ethiopian school children and published their findings in the paper "Ethiopian-Born and Native Israeli School Children Have Different Growth Patterns" (*Nutrition*, Vol. 19, pp. 427–431). The study, conducted in Azezo, North West Ethiopia, found that malnutrition is prevalent in primary and secondary school children because of economic poverty. The weights, in kilograms (kg), of 60 randomly selected male Ethiopian-born school children, ages 12–15, are presented on the WeissStats CD. Use the technology of your choice to do the following.

**a.** Obtain a normal probability plot, boxplot, histogram, and stem-and-leaf diagram of the data.

**b.** Based on your results from part (a), can you reasonably apply the z-interval procedure to the data? Explain your reasoning.

**c.** Find and interpret a 95% confidence interval for the mean weight of all male Ethiopian-born school children, ages 12–15 years. Assume that the population standard deviation is 4.5 kg.

**8.43 Clocking the Cheetah.** The cheetah (*Acinonyx jubatus*) is the fastest land mammal on earth and is highly specialized to run down prey. According to the Cheetah Conservation of Southern Africa *Trade Environment Database*, the cheetah often exceeds speeds of 60 mph and has been clocked at speeds of more than 70 mph. The WeissStats CD contains the top speeds, in miles per hour, for a sample of 35 cheetahs. Use the technology of your choice to do the following.

**a.** Find a 95% confidence interval for the mean top speed, $\mu$, of all cheetahs. Assume that the population standard deviation of top speeds is 3.2 mph.

**b.** Obtain a normal probability plot, boxplot, histogram, and stem-and-leaf diagram of the data.

**c.** Remove the outliers (if any) from the data and then repeat part (a).

**d.** Comment on the advisability of using the z-interval procedure here.

## Extending the Concepts and Skills

**8.44 Family Size.** The U.S. Census Bureau compiles data on family size and presents its findings in *Current Population Reports*. Suppose that 500 U.S. families are randomly selected to estimate the mean size, $\mu$, of all U.S. families. Further suppose that the results are as shown in the following frequency distribution.

| Size | 2 | 3 | 4 | 5 | 6 | 7 | 8 | 9 |
|------|---|---|---|---|---|---|---|---|
| Frequency | 198 | 118 | 101 | 59 | 12 | 3 | 8 | 1 |

a. If the population standard deviation of family sizes is 1.3, determine a 95% confidence interval for the mean size, $\mu$, of all U.S. families. (*Hint:* To find the sample mean, use the grouped-data formula on page 119.)
b. Interpret your answer from part (a).

**8.45** Key Fact 8.3 states that, for a fixed sample size, decreasing the confidence level improves the precision of the confidence-interval estimate of $\mu$ and vice versa.
a. Suppose that you want to increase the precision without reducing the level of confidence. What can you do?
b. Suppose that you want to increase the level of confidence without reducing the precision. What can you do?

**8.46 Class Project: Gestation Periods of Humans.** This exercise can be done individually or, better yet, as a class project. Gestation periods of humans are normally distributed with a mean of 266 days and a standard deviation of 16 days.
a. Simulate 100 samples of nine human gestation periods each.
b. For each sample in part (a), obtain a 95% confidence interval for the population mean gestation period.
c. For the 100 confidence intervals that you obtained in part (b), roughly how many would you expect to contain the population mean gestation period of 266 days?
d. For the 100 confidence intervals that you obtained in part (b), determine the number that contain the population mean gestation period of 266 days.
e. Compare your answers from parts (c) and (d) and comment on any observed difference.

*Another type of confidence interval is called a **one-sided confidence interval**. A one-sided confidence interval provides either a lower confidence bound or an upper confidence bound for the parameter in question. You are asked to examine one-sided confidence intervals in Exercises 8.47–8.49.*

**8.47 One-Sided One-Mean z-Intervals.** Presuming that the assumptions for a one-mean z-interval are satisfied, we have the following formulas for $(1-\alpha)$-level confidence bounds for a population mean $\mu$:

- Lower confidence bound: $\bar{x} - z_\alpha \cdot \sigma/\sqrt{n}$
- Upper confidence bound: $\bar{x} + z_\alpha \cdot \sigma/\sqrt{n}$

Interpret the preceding formulas for lower and upper confidence bounds in words.

**8.48 Poverty and Calcium.** Refer to Exercise 8.32.
a. Determine and interpret a 95% upper confidence bound for the mean calcium intake of all people with incomes below the poverty level.
b. Compare your one-sided confidence interval in part (a) to the (two-sided) confidence interval found in Exercise 8.32(a).

**8.49 Toxic Mushrooms?** Refer to Exercise 8.33.
a. Determine and interpret a 99% lower confidence bound for the mean cadmium level of all *Boletus pinicola* mushrooms.
b. Compare your one-sided confidence interval in part (a) to the (two-sided) confidence interval found in Exercise 8.33.

## 8.3 Margin of Error

Recall Key Fact 7.1, which states that the larger the sample size, the smaller the sampling error tends to be in estimating a population mean by a sample mean. Now that we have studied confidence intervals, we can determine exactly how sample size affects the accuracy of an estimate. We begin by introducing the concept of the *margin of error*.

**Example 8.6** | **Introducing Margin of Error**

*The Civilian Labor Force* In Example 8.4, we applied the one-mean z-interval procedure to the ages of a sample of 50 people in the civilian labor force to obtain a 95% confidence interval for the mean age, $\mu$, of all people in the civilian labor force.

a.  Discuss the precision with which $\bar{x}$ estimates $\mu$.

b.  What quantity determines this precision?

c.  As we saw in Section 8.2, we can decrease the length of the confidence interval and thereby improve the precision of the estimate by decreasing the confidence level from 95% to some lower level. But suppose that we want to retain the same level of confidence and still improve the precision. How can we do so?

d.  Explain why our answer to part (c) makes sense.

**Solution**  Recalling first that $z_{\alpha/2} = z_{0.05/2} = z_{0.025} = 1.96$, $n = 50$, $\sigma = 12.1$, and $\bar{x} = 36.4$, we found that a 95% confidence interval for $\mu$ is from

$$\bar{x} - z_{\alpha/2} \cdot \frac{\sigma}{\sqrt{n}} \quad \text{to} \quad \bar{x} + z_{\alpha/2} \cdot \frac{\sigma}{\sqrt{n}}$$

or

$$36.4 - 1.96 \cdot \frac{12.1}{\sqrt{50}} \quad \text{to} \quad 36.4 + 1.96 \cdot \frac{12.1}{\sqrt{50}}$$

or

$$36.4 - 3.4 \quad \text{to} \quad 36.4 + 3.4$$

or

$$33.0 \quad \text{to} \quad 39.8.$$

We can be 95% confident that the mean age, $\mu$, of all people in the civilian labor force is somewhere between 33.0 years and 39.8 years.

a.  The confidence interval has a wide range for the possible values of $\mu$. In other words, the precision of the estimate is poor.

b.  Let's look closely at the confidence interval, which we display in Fig. 8.5.

**FIGURE 8.5**
95% confidence interval for the mean age, $\mu$, of all people in the civilian labor force

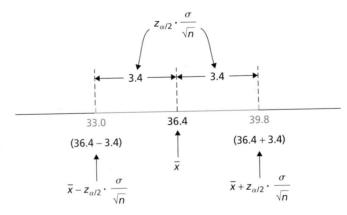

This figure shows that the estimate's precision is determined by the quantity

$$E = z_{\alpha/2} \cdot \frac{\sigma}{\sqrt{n}}$$

which is half the length of the confidence interval, or 3.4 in this case. The quantity $E$ is called the *margin of error*, also known as the *maximum error of the estimate*. We use this terminology because we are 95% confident that our error in estimating $\mu$ by $\bar{x}$ is at most 3.4 years. In newspapers and magazines, this phrase appears in sentences such as "The poll has a margin of error of 3.4 years," or "Theoretically, in 95 out of 100 such polls the margin of error will be 3.4 years."

c.  To improve the precision of the estimate, we need to decrease the margin of error, $E$. Because the sample size, $n$, occurs in the denominator of the formula for $E$, we can decrease $E$ by increasing the sample size.

d.  The answer to part (c) makes sense because we expect more precise information from larger samples.

• • •

---

**Definition 8.3**

**What Does It Mean?**

The margin of error is equal to half the length of the confidence interval, as depicted in Fig. 8.6.

**Margin of Error for the Estimate of $\mu$**

The **margin of error** for the estimate of $\mu$ is

$$E = z_{\alpha/2} \cdot \frac{\sigma}{\sqrt{n}}.$$

Figure 8.6 illustrates the margin of error.

**FIGURE 8.6**

Margin of error, $E = z_{\alpha/2} \cdot \dfrac{\sigma}{\sqrt{n}}$

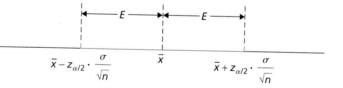

---

**Key Fact 8.4**

**Margin of Error, Precision, and Sample Size**

The length of a confidence interval for a population mean, $\mu$, and therefore the precision with which $\bar{x}$ estimates $\mu$, is determined by the margin of error, $E$. For a fixed confidence level, increasing the sample size improves the precision, and vice versa.

## Determining the Required Sample Size

If the margin of error and confidence level are given, then we must determine the sample size needed to meet those specifications. To find the formula for the required sample size, we solve the margin of error formula, $E = z_{\alpha/2} \cdot \sigma/\sqrt{n}$, for $n$.

**Formula 8.1**

**Sample Size for Estimating $\mu$**

The sample size required for a $(1 - \alpha)$-level confidence interval for $\mu$ with a specified margin of error, $E$, is given by the formula

$$n = \left( \frac{z_{\alpha/2} \cdot \sigma}{E} \right)^2$$

rounded up to the nearest whole number.

**Example 8.7** | **Sample Size for Estimating $\mu$**

*The Civilian Labor Force* Consider again the problem of estimating the mean age, $\mu$, of all people in the civilian labor force.

a. Determine the sample size needed in order to be 95% confident that $\mu$ is within 0.5 year of the estimate, $\bar{x}$. Recall that $\sigma = 12.1$ years.

b. Find a 95% confidence interval for $\mu$ if a sample of the size determined in part (a) has a mean age of 38.8 years.

**Solution**

a. To find the sample size, we use Formula 8.1. We know that $\sigma = 12.1$ and $E = 0.5$. The confidence level is 0.95, which means that $\alpha = 0.05$ and $z_{\alpha/2} = z_{0.025} = 1.96$. Thus

$$n = \left( \frac{z_{\alpha/2} \cdot \sigma}{E} \right)^2 = \left( \frac{1.96 \cdot 12.1}{0.5} \right)^2 = 2249.79,$$

which, rounded up to the nearest whole number, is 2250.

**Interpretation** If 2250 people in the civilian labor force are randomly selected, we can be 95% confident that the mean age of all people in the civilian labor force is within 0.5 year of the mean age of the people in the sample.

b. Applying Procedure 8.1 with $\alpha = 0.05$, $\sigma = 12.1$, $\bar{x} = 38.8$, and $n = 2250$, we get the confidence interval

$$38.8 - 1.96 \cdot \frac{12.1}{\sqrt{2250}} \quad \text{to} \quad 38.8 + 1.96 \cdot \frac{12.1}{\sqrt{2250}}$$

or 38.3 to 39.3.

**You try it!**

Exercise 8.65
on page 358

**Interpretation** We can be 95% confident that the mean age, $\mu$, of all people in the civilian labor force is somewhere between 38.3 years and 39.3 years.

• • •

**Note:** The sample size of 2250 was determined in part (a) of Example 8.7 to guarantee a margin of error of 0.5 year for a 95% confidence interval. According to Fig. 8.6 on page 355, we could have obtained the interval needed in part (b) simply by computing

$$\bar{x} \pm E = 38.8 \pm 0.5.$$

Doing so would give the same confidence interval, 38.3 to 39.3, but with much less work. The simpler method might have yielded a somewhat wider confidence interval because the sample size is rounded up. Hence, this simpler method gives, at worst, a slightly conservative estimate, so is acceptable in practice.

The formula for finding the required sample size, Formula 8.1, involves the population standard deviation, $\sigma$, which is usually unknown. In such cases, we can take a preliminary large sample, say, of size 30 or more, and use the sample standard deviation, $s$, in place of $\sigma$ in Formula 8.1.

## Exercises 8.3

### Understanding the Concepts and Skills

**8.50** Discuss the relationship between the margin of error and the standard error of the mean.

**8.51** Explain why the margin of error determines the precision with which a sample mean estimates a population mean.

**8.52** In each part, explain the effect on the margin of error and hence the effect on the precision of estimating a population mean by a sample mean.
a. Increasing the confidence level while keeping the same sample size.
b. Increasing the sample size while keeping the same confidence level.

**8.53** A confidence interval for a population mean has a margin of error of 3.4.
a. Determine the length of the confidence interval.
b. If the sample mean is 52.8, obtain the confidence interval.

**8.54** A confidence interval for a population mean has a margin of error of 0.047.
a. Determine the length of the confidence interval.
b. If the sample mean is 0.205, determine the confidence interval.

**8.55** A confidence interval for a population mean has length 20.
a. Determine the margin of error.
b. If the sample mean is 60, obtain the confidence interval.

**8.56** A confidence interval for a population mean has length 162.6.
a. Determine the margin of error.
b. If the sample mean is 643.1, determine the confidence interval.

**8.57** Answer true or false to each statement concerning a confidence interval for a population mean. Give reasons for your answers.
a. The length of a confidence interval can be determined if you know only the margin of error.
b. The margin of error can be determined if you know only the length of the confidence interval.
c. The confidence interval can be obtained if you know only the margin of error.
d. The confidence interval can be obtained if you know only the margin of error and the sample mean.

**8.58** Answer true or false to each statement concerning a confidence interval for a population mean. Give reasons for your answers.
a. The margin of error can be determined if you know only the confidence level.
b. The confidence level can be determined if you know only the margin of error.
c. The margin of error can be determined if you know only the confidence level, population standard deviation, and sample size.
d. The confidence level can be determined if you know only the margin of error, population standard deviation, and sample size.

**8.59** Formula 8.1 provides a method for computing the sample size required to obtain a confidence interval with a specified confidence level and margin of error. The number resulting from the formula should be rounded up to the nearest whole number.
a. Why do you want a whole number?
b. Why do you round up instead of down?

**8.60 Body Fat.** J. McWhorter et al. of the College of Health Sciences at the University of Nevada, Las Vegas, studied physical therapy students during their graduate-school years. The researchers were interested in the fact that, although graduate physical-therapy students are taught the principles of fitness, some have difficulty finding the time to implement those principles. In the study, published as "An Evaluation of Physical Fitness Parameters for Graduate Students" (*Journal of American College Health*, Vol. 51, No. 1, pp. 32–37), a sample of 27 female graduate physical-therapy students had a mean of 22.46 percent body fat.
a. Assuming that percent body fat of female graduate physical-therapy students is normally distributed with standard deviation 4.10 percent body fat, determine a 95% confidence interval for the mean percent body fat of all female graduate physical-therapy students.

**b.** Obtain the margin of error, $E$, for the confidence interval you found in part (a).

**c.** Explain the meaning of $E$ in this context in terms of the accuracy of the estimate.

**d.** Determine the sample size required to have a margin of error of 1.55 percent body fat with a 99% confidence level.

**8.61 Pulmonary Hypertension.** In the paper "Persistent Pulmonary Hypertension of the Neonate and Asymmetric Growth Restriction" (*Obstetrics & Gynecology*, Vol. 91, No. 3, pp. 336–341), M. Williams et al. reported on a study of characteristics of neonates. Infants treated for pulmonary hypertension, called the PH group, were compared with those not so treated, called the control group. One of the characteristics measured was head circumference. The mean head circumference of the 10 infants in the PH group was 34.2 centimeters (cm).

**a.** Assuming that head circumferences for infants treated for pulmonary hypertension are normally distributed with standard deviation 2.1 cm, determine a 90% confidence interval for the mean head circumference of all such infants.

**b.** Obtain the margin of error, $E$, for the confidence interval you found in part (a).

**c.** Explain the meaning of $E$ in this context in terms of the accuracy of the estimate.

**d.** Determine the sample size required to have a margin of error of 0.5 cm with a 95% confidence level.

**8.62 Fuel Expenditures.** In estimating the mean monthly fuel expenditure, $\mu$, per household vehicle, the U.S. Energy Information Administration takes a sample of size 6841. Assuming that $\sigma = \$20.65$, determine the margin of error in estimating $\mu$ at the 95% level of confidence.

**8.63 Venture-Capital Investments.** In Exercise 8.31, you found a 95% confidence interval for the mean amount of all venture-capital investments in the fiber optics business sector to be from $5.389 million to $7.274 million. Obtain the margin of error by

**a.** taking half the length of the confidence interval.

**b.** using the formula in Definition 8.3 on page 355. (Recall that $n = 18$ and $\sigma = \$2.04$ million.)

**8.64 Smelling Out the Enemy.** In Exercise 8.34, you found a 90% confidence interval for the mean number of tongue flicks per 20 minutes for all juvenile common lizards to be from 456.4 to 608.0. Obtain the margin of error by

**a.** taking half the length of the confidence interval.

**b.** using the formula in Definition 8.3 on page 355. (Recall that $n = 17$ and $\sigma = 190.0$.)

**8.65 Political Prisoners.** In Exercise 8.35, you found a 95% confidence interval of 18.8 months to 48.0 months for the mean duration of imprisonment, $\mu$, of all East German political prisoners with chronic PTSD.

**a.** Determine the margin of error, $E$.

**b.** Explain the meaning of $E$ in this context in terms of the accuracy of the estimate.

**c.** Find the sample size required to have a margin of error of 12 months and a 99% confidence level. (Recall that $\sigma = 42$ months.)

**d.** Find a 99% confidence interval for the mean duration of imprisonment, $\mu$, if a sample of the size determined in part (c) has a mean of 36.2 months.

**8.66 Keep on Rolling.** In Exercise 8.36, you found a 99% confidence interval of $2.03 million to $2.51 million for the mean gross earnings of all Rolling Stones concerts.

**a.** Determine the margin of error, $E$.

**b.** Explain the meaning of $E$ in this context in terms of the accuracy of the estimate.

**c.** Find the sample size required to have a margin of error of $0.1 million and a 95% confidence level. (Recall that $\sigma = \$0.5$ million.)

**d.** Obtain a 95% confidence interval for the mean gross earnings if a sample of the size determined in part (c) has a mean of $2.35 million.

## Extending the Concepts and Skills

**8.67 Millionaires.** Professor Thomas Stanley of Georgia State University has surveyed millionaires since 1973. Among other information, Professor Stanley obtains estimates for the mean age, $\mu$, of all U.S. millionaires. Suppose that one year's study involved a simple random sample of 36 U.S. millionaires whose mean age was 58.53 years with a sample standard deviation of 13.36 years.

**a.** If, for next year's study, a confidence interval for $\mu$ is to have a margin of error of 2 years and a confidence level of 95%, determine the required sample size.

**b.** Why did you use the sample standard deviation, $s = 13.36$, in place of $\sigma$ in your solution to part (a)? Why is it permissible to do so?

**8.68 Corporate Farms.** The U.S. Census Bureau estimates the mean value of the land and buildings per corporate farm. Those estimates are published in the *Census of Agriculture*. Suppose that an estimate, $\bar{x}$, is obtained and that the margin of error is $1000. Does this result imply that the true mean, $\mu$, is within $1000 of the estimate? Explain your answer.

**8.69** Suppose that a simple random sample is taken from a normal population having a standard deviation of 10 for the purpose of obtaining a 95% confidence interval for the mean of the population.

**a.** If the sample size is 4, obtain the margin of error.

**b.** Repeat part (a) for a sample size of 16.

**c.** Can you guess the margin of error for a sample size of 64? Explain your reasoning.

**8.70** For a fixed confidence level, show that (approximately) quadrupling the sample size is necessary to halve the margin of error. (*Hint:* Use Formula 8.1.)

## 8.4 Confidence Intervals for One Population Mean When $\sigma$ Is Unknown

In Section 8.2, you learned how to determine a confidence interval for a population mean, $\mu$, when the population standard deviation, $\sigma$, is known. The basis of the procedure is in Key Fact 7.4: If $x$ is a normally distributed variable with mean $\mu$ and standard deviation $\sigma$, then, for samples of size $n$, the variable $\bar{x}$ is also normally distributed and has mean $\mu$ and standard deviation $\sigma/\sqrt{n}$. Equivalently, the **standardized version of $\bar{x}$**

$$z = \frac{\bar{x} - \mu}{\sigma/\sqrt{n}} \qquad (8.2)$$

has the standard normal distribution.

But what if, as is usual in practice, the population standard deviation is unknown? Then we cannot base our confidence-interval procedure on the standardized version of $\bar{x}$. The best we can do is estimate the population standard deviation, $\sigma$, by the sample standard deviation, $s$; in other words, we replace $\sigma$ by $s$ in Equation (8.2) and base our confidence-interval procedure on the resulting variable

$$t = \frac{\bar{x} - \mu}{s/\sqrt{n}} \qquad (8.3)$$

called the **studentized version of $\bar{x}$.**

Unlike the standardized version, the studentized version of $\bar{x}$ does not have a normal distribution. To get an idea of how their distributions differ, we used statistical software to simulate each variable for samples of size 4, assuming that $\mu = 15$ and $\sigma = 0.8$. (Any sample size, population mean, and population standard deviation will do.)

1. We simulated 5000 samples of size 4 each.
2. For each of the 5000 samples, we obtained the sample mean and sample standard deviation.
3. For each of the 5000 samples, we determined the observed values of the standardized and studentized versions of $\bar{x}$.
4. We obtained histograms of the 5000 observed values of the standardized version of $\bar{x}$ and the 5000 observed values of the studentized version of $\bar{x}$, as shown in Output 8.2 at the top of the next page.

The two histograms suggest that the distributions of both the standardized version of $\bar{x}$—the variable $z$ in Equation (8.2)—and the studentized version of $\bar{x}$—the variable $t$ in Equation (8.3)—are bell shaped and symmetric about 0. But there is an important difference in the distributions: the studentized version has more spread than the standardized version. This difference is not surprising

**OUTPUT 8.2**

Histograms of z (standardized version of x̄) and t (studentized version of x̄) for 5000 samples of size 4

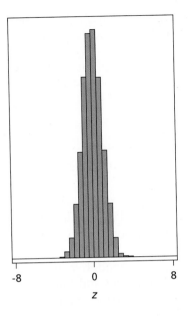
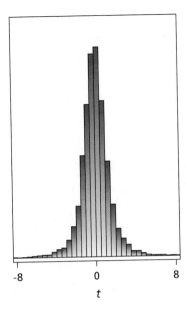

because the variation in the possible values of the standardized version is due solely to the variation of sample means, whereas that of the studentized version is due to the variation of both sample means and sample standard deviations.

As you know, the standardized version of $\bar{x}$ has the standard normal distribution. In 1908, William Gosset determined the distribution of the studentized version of $\bar{x}$, a distribution now called **Student's *t*-distribution** or, simply, the ***t*-distribution.** (The biography on page 377 has more on Gosset and the Student's *t*-distribution.)

## *t*-Distributions and *t*-Curves

There is a different *t*-distribution for each sample size. We identify a particular *t*-distribution by its number of **degrees of freedom (df).** For the studentized version of $\bar{x}$, the number of degrees of freedom is 1 less than the sample size, which we indicate symbolically by **df = *n* − 1.**

---

**Key Fact 8.5**

### Studentized Version of the Sample Mean

Suppose that a variable $x$ of a population is normally distributed with mean $\mu$. Then, for samples of size $n$, the variable

$$t = \frac{\bar{x} - \mu}{s/\sqrt{n}}$$

has the *t*-distribution with $n - 1$ degrees of freedom.

**What Does It Mean?**

For a normally distributed variable, the studentized version of the sample mean has the *t*-distribution with degrees of freedom 1 less than the sample size.

A variable with a *t*-distribution has an associated curve, called a ***t*-curve.** In this book, you need to understand the basic properties of a *t*-curve, but not its equation.

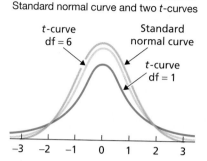

**FIGURE 8.7**
Standard normal curve and two $t$-curves

Although there is a different $t$-curve for each number of degrees of freedom, all $t$-curves are similar and resemble the standard normal curve, as illustrated in Fig. 8.7. That figure also illustrates the basic properties of $t$-curves, listed in Key Fact 8.6. Note that Properties 1–3 of $t$-curves are identical to those of the standard normal curve, as given in Key Fact 6.3 on page 277.

As mentioned earlier, and illustrated in Fig. 8.7, $t$-curves have more spread than the standard normal curve. This property follows from the fact that, for a $t$-curve with $\nu$ (pronounced "new") degrees of freedom, where $\nu > 2$, the standard deviation is $\sqrt{\nu/(\nu - 2)}$. This quantity always exceeds 1, which is the standard deviation of the standard normal curve.

**Key Fact 8.6**

## Basic Properties of $t$-Curves

**Property 1:** The total area under a $t$-curve equals 1.

**Property 2:** A $t$-curve extends indefinitely in both directions, approaching, but never touching, the horizontal axis as it does so.

**Property 3:** A $t$-curve is symmetric about 0.

**Property 4:** As the number of degrees of freedom becomes larger, $t$-curves look increasingly like the standard normal curve.

## Using the $t$-Table

Percentages (and probabilities) for a variable having a $t$-distribution equal areas under the variable's associated $t$-curve. For our purposes, one of which is obtaining confidence intervals for a population mean, we don't need a complete $t$-table for each $t$-curve; only certain areas will be important. Table IV, which appears in Appendix A and in abridged form inside the front cover, is sufficient for our purposes.

The two outside columns of Table IV, labeled df, display the number of degrees of freedom. As expected, the symbol $t_\alpha$ denotes the $t$-value having area $\alpha$ to its right under a $t$-curve. Thus the column headed $t_{0.10}$, for example, contains $t$-values having area 0.10 to their right.

**Example 8.8** | **Finding the $t$-Value Having a Specified Area to Its Right**

For a $t$-curve with 13 degrees of freedom, determine $t_{0.05}$; that is, find the $t$-value having area 0.05 to its right, as shown in Fig. 8.8(a).

**FIGURE 8.8**
Finding the $t$-value having area 0.05 to its right

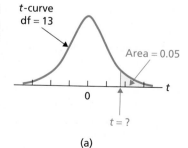

(a)

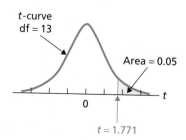

(b)

**Solution** To find the *t*-value in question, we use Table IV, a portion of which is given in Table 8.4.

**TABLE 8.4**
Values of $t_\alpha$

| df | $t_{0.10}$ | $t_{0.05}$ | $t_{0.025}$ | $t_{0.01}$ | $t_{0.005}$ | df |
|----|------------|------------|-------------|------------|-------------|----|
| .  | .          | .          | .           | .          | .           | .  |
| .  | .          | .          | .           | .          | .           | .  |
| .  | .          | .          | .           | .          | .           | .  |
| 12 | 1.356      | 1.782      | 2.179       | 2.681      | 3.055       | 12 |
| 13 | 1.350      | 1.771      | 2.160       | 2.650      | 3.012       | 13 |
| 14 | 1.345      | 1.761      | 2.145       | 2.624      | 2.977       | 14 |
| 15 | 1.341      | 1.753      | 2.131       | 2.602      | 2.947       | 15 |
| .  | .          | .          | .           | .          | .           | .  |
| .  | .          | .          | .           | .          | .           | .  |
| .  | .          | .          | .           | .          | .           | .  |

**You try it!**

Exercise 8.81
on page 368

The number of degrees of freedom is 13, so we first go down the outside columns, labeled df, to "13." Then, going across that row to the column labeled $t_{0.05}$, we reach 1.771. This number is the *t*-value having area 0.05 to its right, as shown in Fig. 8.8(b). In other words, for a *t*-curve with df = 13, $t_{0.05}$ = 1.771.

• • •

Note that Table IV in Appendix A contains degrees of freedom from 1 to 75, but then has only selected degrees of freedom. If the number of degrees of freedom you seek is not in Table IV, you could find a more detailed *t*-table, use technology, or use linear interpolation and Table IV. A less exact option is to use the degrees of freedom in Table IV closest to the one required.

As we noted earlier, *t*-curves look increasingly like the standard normal curve as the number of degrees of freedom gets larger. For degrees of freedom greater than 2000, a *t*-curve and the standard normal curve are virtually indistinguishable. Consequently, we stopped the *t*-table at df = 2000 and supplied the corresponding values of $z_\alpha$ beneath. These values can be used not only for the standard normal distribution, but also for any *t*-distribution having degrees of freedom greater than 2000.[†]

### Obtaining Confidence Intervals for a Population Mean When $\sigma$ Is Unknown

Having discussed *t*-distributions and *t*-curves, we can now develop a procedure for obtaining a confidence interval for a population mean when the population standard deviation is unknown. We proceed in essentially the same way as we did when the population standard deviation is known, except now we invoke a *t*-distribution instead of the standard normal distribution.

Hence we use $t_{\alpha/2}$ instead of $z_{\alpha/2}$ in the formula for the confidence interval. As a result, we have Procedure 8.2, which we call the **one-mean *t*-interval procedure** or, when no confusion can arise, simply the ***t*-interval procedure**.[‡]

---

[†]The values of $z_\alpha$ given at the bottom of Table IV are accurate to three decimal places and, because of that, some differ slightly from what you get by applying the method you learned for using Table II.

[‡]The one-mean *t*-interval procedure is also known as the **one-sample *t*-interval procedure** and the **one-variable *t*-interval procedure**. We prefer "one-mean" because it makes clear the parameter being estimated.

| Procedure 8.2 | **One-Mean *t*-Interval Procedure** |
|---|---|

*Purpose* To find a confidence interval for a population mean, $\mu$

*Assumptions*

1. Simple random sample
2. Normal population or large sample
3. $\sigma$ unknown

**STEP 1** For a confidence level of $1 - \alpha$, use Table IV to find $t_{\alpha/2}$ with df $= n - 1$, where $n$ is the sample size.

**STEP 2** The confidence interval for $\mu$ is from

$$\bar{x} - t_{\alpha/2} \cdot \frac{s}{\sqrt{n}} \quad \text{to} \quad \bar{x} + t_{\alpha/2} \cdot \frac{s}{\sqrt{n}}$$

where $t_{\alpha/2}$ is found in Step 1 and $\bar{x}$ and $s$ are computed from the sample data.

**STEP 3** Interpret the confidence interval.

The confidence interval is exact for normal populations and is approximately correct for large samples from nonnormal populations.

Properties and guidelines for use of the *t*-interval procedure are the same as those for the *z*-interval procedure, as given in Key Fact 8.1 on page 346. In particular, the *t*-interval procedure is robust to moderate violations of the normality assumption but, even for large samples, can sometimes be unduly affected by outliers because the sample mean and sample standard deviation are not resistant to outliers.

| Example 8.9 | **The One-Mean *t*-Interval Procedure** |
|---|---|

*Pickpocket Offenses* The U.S. Federal Bureau of Investigation (FBI) compiles data on robbery and property crimes and publishes the information in *Population-at-Risk Rates and Selected Crime Indicators*. A simple random sample of pickpocket offenses yielded the losses, in dollars, shown in Table 8.5. Use the data to find a 95% confidence interval for the mean loss, $\mu$, of all pickpocket offenses.

**TABLE 8.5**
Losses ($) for a sample of 25 pickpocket offenses

| | | | | |
|---|---|---|---|---|
| 447 | 207 | 627 | 430 | 883 |
| 313 | 844 | 253 | 397 | 214 |
| 217 | 768 | 1064 | 26 | 587 |
| 833 | 277 | 805 | 653 | 549 |
| 649 | 554 | 570 | 223 | 443 |

**Solution** Because the sample size, $n = 25$, is moderate, we first need to consider questions of normality and outliers. (See the second bulleted item in Key Fact 8.1 on page 346.) To do that, we constructed the normal probability plot in Fig. 8.9, shown at the top of the following page. The plot reveals no outliers and falls roughly in a straight line. So, we can apply Procedure 8.2 to find the confidence interval.

FIGURE 8.9

Normal probability plot
of the loss data in Table 8.5

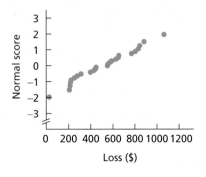

**STEP 1** For a confidence level of $1 - \alpha$, use Table IV to find $t_{\alpha/2}$ with $df = n - 1$, where $n$ is the sample size.

We want a 95% confidence interval, so $\alpha = 1 - 0.95 = 0.05$. For $n = 25$, we have $df = 25 - 1 = 24$. From Table IV, $t_{\alpha/2} = t_{0.05/2} = t_{0.025} = 2.064$.

**STEP 2** The confidence interval for $\mu$ is from

$$\bar{x} - t_{\alpha/2} \cdot \frac{s}{\sqrt{n}} \quad \text{to} \quad \bar{x} + t_{\alpha/2} \cdot \frac{s}{\sqrt{n}}.$$

From Step 1, $t_{\alpha/2} = 2.064$. Applying the usual formulas for $\bar{x}$ and $s$ to the data in Table 8.5 gives $\bar{x} = 513.32$ and $s = 262.23$. So a 95% confidence interval for $\mu$ is from

$$513.32 - 2.064 \cdot \frac{262.23}{\sqrt{25}} \quad \text{to} \quad 513.32 + 2.064 \cdot \frac{262.23}{\sqrt{25}}$$

or 405.07 to 621.57.

**You try it!**

Exercise 8.91
on page 368

**STEP 3** Interpret the confidence interval.

**Interpretation** We can be 95% confident that the mean loss of all pickpocket offenses is somewhere between $405.07 and $621.57.

• • •

**Example 8.10** | **The One-Mean *t*-Interval Procedure**

TABLE 8.6

Sample of year's chicken
consumption (lb)

| 47 | 39 | 62 | 49 | 50 | 70 |
|----|----|----|----|----|----|
| 59 | 53 | 55 | 0  | 65 | 63 |
| 53 | 51 | 50 | 72 | 45 |    |

*Chicken Consumption* The U.S. Department of Agriculture publishes data on chicken consumption in *Food Consumption, Prices, and Expenditures*. Table 8.6 shows a year's chicken consumption, in pounds, for 17 randomly selected people. Find a 90% confidence interval for the year's mean chicken consumption, $\mu$.

**Solution** A normal probability plot of the data, shown in Fig. 8.10(a), reveals an outlier (0 lb). Because the sample size is only moderate, applying Procedure 8.2 here is inappropriate.

FIGURE 8.10

Normal probability plots for chicken
consumption: (a) original data
and (b) data with outlier removed

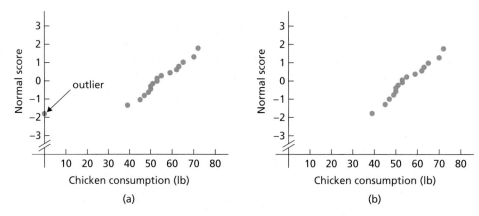

(a)

(b)

The outlier of 0 lb might be a recording error or it might reflect a person in the sample who does not eat chicken (e.g., a vegetarian). If we remove the outlier from the data, the normal probability plot for the abridged data shows no outliers and is roughly linear, as seen in Fig. 8.10(b).

Thus, if we are willing to take as our population only people who eat chicken, we can use Procedure 8.2 to obtain a confidence interval. Doing so yields a 90% confidence interval of 51.2 to 59.2.

**Interpretation**   We can be 90% confident that the year's mean chicken consumption, among people who eat chicken, is somewhere between 51.2 lb and 59.2 lb.

• • •

By restricting our population of interest to only those people who eat chicken, we were justified in removing the outlier of 0 lb. Generally, an outlier should not be removed without careful consideration. *Simply removing an outlier because it is an outlier is unacceptable statistical practice.*

In Example 8.10, if we had been careless in our analysis by blindly finding a confidence interval without first examining the data, our result would have been invalid and misleading.

### What If the Assumptions Are Not Satisfied?

Suppose you want to obtain a confidence interval for a population mean based on a small sample but preliminary data analyses indicate either the presence of outliers or that the variable under consideration is far from normally distributed. As neither the $z$-interval procedure nor the $t$-interval procedure is appropriate, what can you do?

Under certain conditions, you can use a *nonparametric method*.[†] For example, if the variable under consideration has a symmetric distribution, you can use a nonparametric method called the *Wilcoxon confidence-interval procedure* to find a confidence interval for the population mean.

Most nonparametric methods do not require even approximate normality, are resistant to outliers and other extreme values, and can be applied regardless of sample size. However, parametric methods, such as the $z$-interval and $t$-interval procedures, tend to give more accurate results than nonparametric methods when the normality assumption and other requirements for their use are met.

Although we do not cover nonparametric methods in this book, many basic statistics books do discuss them. See, for example, *Introductory Statistics, 8/e*, by Neil A. Weiss (Boston: Addison-Wesley, 2008).

| **Example 8.11** | **Choosing a Confidence-Interval Procedure** |

*Adjusted Gross Incomes*   The U.S. Internal Revenue Service (IRS) publishes data on federal individual income tax returns in *Statistics of Income, Individual Income Tax Returns*. A sample of 12 returns from a recent year revealed the adjusted gross incomes, in thousands of dollars, shown in Table 8.7. Which procedure should be used to obtain a confidence interval for the mean adjusted gross income, $\mu$, of all the year's individual income tax returns?

> **What Does It Mean?**
>
> Performing preliminary data analyses to check assumptions before applying inferential procedures is essential.

**TABLE 8.7**
Adjusted gross incomes ($1000)

| | | | |
|---|---|---|---|
| 9.7 | 93.1 | 33.0 | 21.2 |
| 81.4 | 51.1 | 43.5 | 10.6 |
| 12.8 | 7.8 | 18.1 | 12.7 |

---

[†]Recall that descriptive measures for a population, such as $\mu$ and $\sigma$, are called parameters. Technically, inferential methods concerned with parameters are called **parametric methods;** those that are not are called **nonparametric methods.** However, common practice is to refer to most methods that can be applied without assuming normality (regardless of sample size) as nonparametric. Thus the term *nonparametric method* as used in contemporary statistics is somewhat of a misnomer.

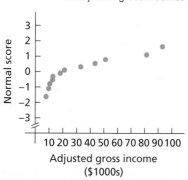

**FIGURE 8.11**
Normal probability plot for the sample
of adjusted gross incomes

**Solution** Because the sample size is small ($n = 12$), we must first consider questions of normality and outliers. A normal probability plot of the sample data, shown in Fig. 8.11, suggests that adjusted gross incomes are far from being normally distributed. Consequently, neither the $z$-interval procedure nor the $t$-interval procedure should be used; instead, some nonparametric confidence interval procedure should be applied.

• • •

**Note:** The normal probability plot in Fig. 8.11 further suggests that adjusted gross incomes do not have a symmetric distribution; so, using the Wilcoxon confidence-interval procedure also seems inappropriate. In cases like this, where no common procedure appears appropriate, consult a statistician.

# The Technology Center

Most statistical technologies have programs that automatically perform the one-mean $t$-interval procedure. In this subsection, we present output and step-by-step instructions for such programs.

**Example 8.12** **Using Technology to Obtain a One-Mean $t$-Interval**

*Pickpocket Offenses* The losses, in dollars, of 25 randomly selected pickpocket offenses are displayed in Table 8.5 on page 363. Use Minitab, Excel, or the TI-83/84 Plus to find a 95% confidence interval for the mean loss, $\mu$, of all pickpocket offenses.

**Solution** We applied the one-mean $t$-interval programs to the data, resulting in Output 8.3. Steps for generating that output are presented in Instructions 8.2.

**OUTPUT 8.3** One-mean $t$-interval for the sample of losses

**MINITAB**

**One-Sample T: LOSS**

| Variable | N | Mean | StDev | SE Mean | 95% CI |
|----------|----|-------|-------|---------|--------|
| LOSS | 25 | 513.3 | 262.2 | 52.4 | (405.1, 621.6) |

**EXCEL**

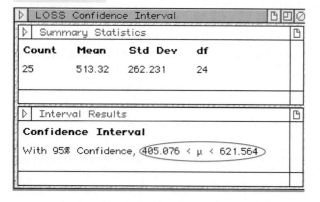

LOSS Confidence Interval

Summary Statistics

| Count | Mean | Std Dev | df |
|-------|--------|---------|----|
| 25 | 513.32 | 262.231 | 24 |

Interval Results

**Confidence Interval**

With 95% Confidence, 405.076 < $\mu$ < 621.564

**TI-83/84 PLUS**

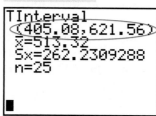

TInterval
(405.08,621.56)
x̄=513.32
Sx=262.2309288
n=25

As shown in Output 8.3, the required 95% confidence interval is from 405.1 to 621.6. We can be 95% confident that the mean loss of all pickpocket offenses is somewhere between $405.1 and $621.6.

• • •

**INSTRUCTIONS 8.2**   Steps for generating Output 8.3

| MINITAB | EXCEL | TI-83/84 PLUS |
|---|---|---|
| 1 Store the data from Table 8.5 in a column named LOSS<br>2 Choose **Stat ➤ Basic Statistics ➤ 1-Sample t...**<br>3 Select the **Samples in columns** option button<br>4 Click in the **Samples in columns** text box and specify LOSS<br>5 Click the **Options...** button<br>6 Type <u>95</u> in the **Confidence level** text box<br>7 Click the arrow button at the right of the **Alternative** drop-down list box and select **not equal**<br>8 Click **OK** twice | 1 Store the data from Table 8.5 in a range named LOSS<br>2 Choose **DDXL ➤ Confidence Intervals**<br>3 Select **1 Var t Interval** from the **Function type** drop-down box<br>4 Specify LOSS in the **Quantitative Variable** text box<br>5 Click **OK**<br>6 Click the **95%** button<br>7 Click the **Compute Interval** button | 1 Store the data from Table 8.5 in a list named LOSS<br>2 Press **STAT**, arrow over to **TESTS**, and press **8**<br>3 Highlight **Data** and press **ENTER**<br>4 Press the down-arrow key<br>5 Press **2nd ➤ LIST**<br>6 Arrow down to LOSS and press **ENTER** three times<br>7 Type <u>.95</u> for **C-Level** and press **ENTER** twice |

■ ■ ■

# Exercises 8.4

## Understanding the Concepts and Skills

**8.71** Explain the difference in the formulas for the standardized and studentized versions of $\bar{x}$.

**8.72** Why do you need to consider the studentized version of $\bar{x}$ to develop a confidence-interval procedure for a population mean when the population standard deviation is unknown?

**8.73** A variable has a mean of 100 and a standard deviation of 16. Four observations of this variable have a mean of 108 and a sample standard deviation of 12. Determine the observed value of the
**a.** standardized version of $\bar{x}$.
**b.** studentized version of $\bar{x}$.

**8.74** A variable of a population has a normal distribution. Suppose that you want to find a confidence interval for the population mean.
**a.** If you know the population standard deviation, which procedure would you use?
**b.** If you do not know the population standard deviation, which procedure would you use?

**8.75 Green Sea Urchins.** From the paper "Effects of Chronic Nitrate Exposure on Gonad Growth in Green

Sea Urchin *Strongylocentrotus Droebachiensis*" (*Aquaculture*, Vol. 242, No. 1–4, pp. 357–363) by S. Siikavuopio et al., the weights, $x$, of adult green sea urchins are normally distributed with mean 52.0 g and standard deviation 17.2 g. For samples of 12 such weights, identify the distribution of each of the following variables.

**a.** $\dfrac{\bar{x} - 52.0}{17.2/\sqrt{12}}$        **b.** $\dfrac{\bar{x} - 52.0}{s/\sqrt{12}}$

**8.76 Batting Averages.** An issue of *Scientific American* reveals that batting averages, $x$, of major-league baseball players are normally distributed and have a mean of 0.270 and a standard deviation of 0.031. For samples of 20 batting averages, identify the distribution of each variable.

**a.** $\dfrac{\bar{x} - 0.270}{0.031/\sqrt{20}}$        **b.** $\dfrac{\bar{x} - 0.270}{s/\sqrt{20}}$

**8.77** Explain why there is more variation in the possible values of the studentized version of $\bar{x}$ than in the possible values of the standardized version of $\bar{x}$.

**8.78** Two *t*-curves have degrees of freedom 12 and 20, respectively. Which one more closely resembles the standard normal curve? Explain your answer.

**8.79** For a $t$-curve with df $= 6$, use Table IV to find each $t$-value.

a. $t_{0.10}$      b. $t_{0.025}$      c. $t_{0.01}$

**8.80** For a $t$-curve with df $= 17$, use Table IV to find each $t$-value.

a. $t_{0.05}$      b. $t_{0.025}$      c. $t_{0.005}$

**8.81** For a $t$-curve with df $= 21$, find each $t$-value and illustrate your results graphically.
a. The $t$-value having area 0.10 to its right
b. $t_{0.01}$
c. The $t$-value having area 0.025 to its left (*Hint:* A $t$-curve is symmetric about 0.)
d. The two $t$-values that divide the area under the curve into a middle 0.90 area and two outside areas of 0.05

**8.82** For a $t$-curve with df $= 8$, find each $t$-value and illustrate your results graphically.
a. The $t$-value having area 0.05 to its right
b. $t_{0.10}$
c. The $t$-value having area 0.01 to its left (*Hint:* A $t$-curve is symmetric about 0.)
d. The two $t$-values that divide the area under the curve into a middle 0.95 area and two outside 0.025 areas

**8.83** A simple random sample of size 100 is taken from a population with unknown standard deviation. A normal probability plot of the data displays significant curvature but no outliers. Can you reasonably apply the $t$-interval procedure? Explain your answer.

**8.84** A simple random sample of size 17 is taken from a population with unknown standard deviation. A normal probability plot of the data reveals an outlier but is otherwise roughly linear. Can you reasonably apply the $t$-interval procedure? Explain your answer.

*In each of Exercises 8.85–8.90, we have provided a sample mean, sample size, sample standard deviation, and confidence level. In each case, use the one-mean t-interval procedure to find a confidence interval for the mean of the population from which the sample was drawn.*

**8.85** $\bar{x} = 20$, $n = 36$, $s = 3$, confidence level $= 95\%$

**8.86** $\bar{x} = 25$, $n = 36$, $s = 3$, confidence level $= 95\%$

**8.87** $\bar{x} = 30$, $n = 25$, $s = 4$, confidence level $= 90\%$

**8.88** $\bar{x} = 35$, $n = 25$, $s = 4$, confidence level $= 90\%$

**8.89** $\bar{x} = 50$, $n = 16$, $s = 5$, confidence level $= 99\%$

**8.90** $\bar{x} = 55$, $n = 16$, $s = 5$, confidence level $= 99\%$

*Preliminary data analyses indicate that you can reasonably apply the t-interval procedure (Procedure 8.2 on page 363) in Exercises 8.91–8.96.*

**8.91 Northeast Commutes.** According to Scarborough Research, more than 85% of working adults commute by car. Of all U.S. cities, Washington, D.C. and New York City have the longest commute times. A sample of 30 commuters in the Washington, D.C. area yielded the following commute times, in minutes.

| 24 | 28 | 31 | 29 | 54 | 28 |
|----|----|----|----|----|----|
| 27 | 38 | 24 | 14 | 46 | 38 |
| 31 | 16 | 21 | 11 | 21 | 15 |
| 30 | 29 | 17 | 23 | 27 | 18 |
| 29 | 44 | 19 | 35 | 34 | 38 |

a. Find a 90% confidence interval for the mean commute time of all commuters in Washington, D.C. (*Note:* $\bar{x} = 27.97$ minutes and $s = 10.04$ minutes.)
b. Interpret your answer from part (a).

**8.92 TV Viewing.** According to *Communications Industry Forecast*, published by Veronis Suhler Stevenson of New York, NY, the average person watched 4.47 hours of television per day in 2000. A random sample of 40 people gave the following number of hours of television watched per day for last year.

| 2.3 | 9.2 | 3.3 | 8.0 | 7.0 | 5.3 | 2.1 | 9.0 |
|-----|-----|-----|-----|-----|-----|-----|-----|
| 2.7 | 4.7 | 2.6 | 4.2 | 3.4 | 6.4 | 4.7 | 2.2 |
| 2.9 | 5.2 | 3.5 | 5.8 | 2.9 | 3.3 | 3.7 | 7.8 |
| 5.3 | 5.4 | 5.5 | 1.6 | 5.4 | 7.5 | 2.3 | 5.2 |
| 3.3 | 2.6 | 8.9 | 4.5 | 0.0 | 3.8 | 2.6 | 8.5 |

a. Find a 90% confidence interval for the amount of television watched per day last year by the average person. (*Note:* $\bar{x} = 4.615$ hr and $s = 2.277$ hr.)
b. Interpret your answer from part (a).

**8.93 Sleep.** In 1908, W. S. Gosset published the article "The Probable Error of a Mean" (*Biometrika*, Vol. 6, pp. 1–25). In this pioneering paper, written under the pseudonym "Student," Gosset introduced what later became known as Student's $t$-distribution. Gosset used the following data set, which gives the additional sleep in hours obtained by a sample of 10 patients using laevohysocyamine hydrobromide.

| 1.9 | 0.8 | 1.1 | 0.1 | −0.1 |
|-----|-----|-----|-----|------|
| 4.4 | 5.5 | 1.6 | 4.6 | 3.4 |

a. Obtain and interpret a 95% confidence interval for the additional sleep that would be obtained on average for all people using laevohysocyamine hydrobromide. (*Note:* $\bar{x} = 2.33$ hr; $s = 2.002$ hr.)
b. Was the drug effective in increasing sleep? Explain your answer.

**8.94 Family Fun?** Taking the family to an amusement park has become increasingly costly according to the industry publication *Amusement Business*, which provides figures on the cost for a family of four to spend the day at one of America's amusement parks. A random sample of 25 families of four that attended amusement parks yielded the following costs, rounded to the nearest dollar.

| | | | | |
|---|---|---|---|---|
| 156 | 212 | 218 | 189 | 172 |
| 221 | 175 | 208 | 152 | 184 |
| 209 | 195 | 207 | 179 | 181 |
| 202 | 166 | 213 | 221 | 237 |
| 130 | 217 | 161 | 208 | 220 |

Obtain and interpret a 95% confidence interval for the mean cost of a family of four to spend the day at an American amusement park. (*Note:* $\bar{x} = \$193.32$; $s = \$26.73$.)

**8.95 Lipid-Lowering Therapy.** In the paper "A Randomized Trial of Intensive Lipid-Lowering Therapy in Calcific Aortic Stenosis" (*New England Journal of Medicine*, Vol. 352, No. 23, pp. 2389–2397), S. Cowell et al. reported the results of a double-blind, placebo controlled trial designed to determine whether intensive lipid-lowering therapy would halt the progression of calcific aortic stenosis or induce its regression. The experiment group, which consisted of 77 patients with calcific aortic stenosis, received 80 mg of atorvastatin daily. The change in their aortic-jet velocity over the period of study (one of the measures used in evaluating the results) had a mean increase of 0.199 meters per second per year with a standard deviation of 0.210 meters per second per year.

a. Obtain and interpret a 95% confidence interval for the mean change in aortic-jet velocity of all such patients who receive 80 mg of atorvastatin daily.

b. Can you conclude that, on average, there is an increase in aortic-jet velocity for such patients? Explain your reasoning.

**8.96 Adrenomedullin and Pregnancy Loss.** Adrenomedullin, a hormone found in the adrenal gland, participates in blood-pressure and heart-rate control. The level of adrenomedullin is raised in a variety of diseases, and medical complications, including recurrent pregnancy loss, can result. In an article by M. Nakatsuka et al. titled "Increased Plasma Adrenomedullin in Women With Recurrent Pregnancy Loss" (*Obstetrics & Gynecology*, Vol. 102, No. 2, pp. 319–324), the plasma levels of adrenomedullin for 38 women with recurrent pregnancy loss had a mean of 5.6 pmol/l and a sample standard deviation of 1.9 pmol/l, where pmol/l is an abbreviation of picomoles per liter.

a. Find a 90% confidence interval for the mean plasma level of adrenomedullin for all women with recurrent pregnancy loss.

b. Interpret your answer from part (a).

*In each of Exercises 8.97–8.100, decide whether applying the t-interval procedure to obtain a confidence interval for the population mean in question appears reasonable. Explain your answers.*

**8.97 Oxygen Distribution.** In the article "Distribution of Oxygen in Surface Sediments from Central Sagami Bay, Japan: In Situ Measurements by Microelectrodes and Planar Optodes" (*Deep Sea Research Part I: Oceanographic Research Papers*, Vol. 52, Issue 10, pp. 1974–1987), R. Glud et al. explore the distributions of oxygen in surface sediments from central Sagami Bay. The oxygen distribution gives important information on the general biogeochemistry of marine sediments. Measurements were performed at 16 sites. A sample of 22 depths yielded the following data, in millimoles per square meter per day (mmol m$^{-2}$ d$^{-1}$), on diffusive oxygen uptake (DOU).

| | | | | | | | |
|---|---|---|---|---|---|---|---|
| 1.8 | 2.0 | 1.8 | 2.3 | 3.8 | 3.4 | 2.7 | 1.1 |
| 3.3 | 1.2 | 3.6 | 1.9 | 7.6 | 2.0 | 1.5 | 2.0 |
| 1.1 | 0.7 | 1.0 | 1.8 | 1.8 | 6.7 | | |

**8.98 Positively Selected Genes.** R. Nielsen et al. compared 13,731 annotated genes from humans with their chimpanzee orthologs to identify genes that show evidence of positive selection. The researchers published their findings in "A Scan for Positively Selected Genes in the Genomes of Humans and Chimpanzees" (*PLOS Biology*, Vol. 3, Issue 6, pp. 976–985). A simple random sample of 14 tissue types yielded the following number of genes.

| | | | | | | |
|---|---|---|---|---|---|---|
| 66 | 47 | 43 | 101 | 201 | 83 | 93 |
| 82 | 120 | 64 | 244 | 51 | 70 | 14 |

**8.99 Big Bucks.** In the article "The $350,000 Club" (*The Business Journal*, Vol. 24, Issue 14, pp. 80–82), J. Trunelle et al. examined Arizona public-company executives with salaries and bonuses totaling over $350,000. The following data provide the salaries, to the nearest thousand dollars, of a random sample of 20 such executives.

| | | | | |
|---|---|---|---|---|
| 516 | 574 | 560 | 623 | 600 |
| 770 | 680 | 672 | 745 | 450 |
| 450 | 545 | 630 | 650 | 461 |
| 836 | 404 | 428 | 620 | 604 |

**8.100 Shoe and Apparel E-Tailers.** In the special report "Mousetrap: The Most-Visited Shoe and Apparel E-tailers" (*Footwear News*, Vol. 58, No. 3, p. 18), we found the following data on the average time, in minutes, spent per user per month from January to June of one year for a sample of 15 shoe and apparel retail Web sites.

| | | | | |
|---|---|---|---|---|
| 13.3 | 9.0 | 11.1 | 9.1 | 8.4 |
| 15.6 | 8.1 | 8.3 | 13.0 | 17.1 |
| 16.3 | 13.5 | 8.0 | 15.1 | 5.8 |

## Working With Large Data Sets

**8.101 The Coruro's Burrow.** The subterranean coruro (*Spalacopus cyanus*) is a social rodent that lives in large colonies in underground burrows that can reach lengths of up to 600 meters. Zoologists S. Begall and M. Gallardo studied the characteristics of the burrow systems of the subterranean coruro in central Chile and published their findings in the *Journal of Zoology, London*, (Vol. 251, pp. 53–60). A sample of 51 burrows had the depths, in centimeters (cm), presented on the WeissStats CD. Use the technology of your choice to do the following.
a. Obtain a normal probability plot, boxplot, histogram, and stem-and-leaf diagram of the data.
b. Based on your results from part (a), can you reasonably apply the *t*-interval procedure to the data? Explain your reasoning.
c. Find and interpret a 90% confidence interval for the mean depth of all subterranean coruro burrows.

**8.102 Forearm Length.** In 1903, K. Pearson and A. Lee published the paper "On the Laws of Inheritance in Man. I. Inheritance of Physical Characters" (*Biometrika*, Vol. 2, pp. 357–462). The article examined and presented data on forearm length, in inches, for a sample of 140 men, which we have provided on the WeissStats CD. Use the technology of your choice to do the following.
a. Obtain a normal probability plot, boxplot, and histogram of the data.
b. Is it reasonable to apply the *t*-interval procedure to the data? Explain your answer.
c. If you answered "yes" to part (b), find a 95% confidence interval for the mean forearm length of men. Interpret your result.

**8.103 Blood Cholesterol and Heart Disease.** Numerous studies have shown that high blood cholesterol leads to artery clogging and subsequent heart disease. One such study by Scott et al. was published in the paper "Plasma Lipids as Collateral Risk Factors in Coronary Artery Disease: A Study of 371 Males With Chest Pain" (*Journal of Chronic Diseases*, Vol. 31, pp. 337–345). The research compared the plasma cholesterol concentrations of independent random samples of patients with and without evidence of heart disease. Evidence of heart disease was based on the degree of narrowing in the arteries. The data on plasma cholesterol concentrations, in milligrams/deciliter (mg/dl), are provided on the WeissStats CD. Use the technology of your choice to do the following.
a. Obtain a normal probability plot, boxplot, and histogram of the data for patients without evidence of heart disease.
b. Is it reasonable to apply the *t*-interval procedure to those data? Explain your answer.
c. If you answered "yes" to part (b), determine a 95% confidence interval for the mean plasma cholesterol concentration of all males without evidence of heart disease. Interpret your result.
d. Repeat parts (a)–(c) for males with evidence of heart disease.

## Extending the Concepts and Skills

**8.104 Bicycle Commuting Times.** A city planner working on bikeways designs a questionnaire to obtain information about local bicycle commuters. One of the questions asks how long it takes the rider to pedal from home to his or her destination. A sample of local bicycle commuters yields the following times, in minutes.

| | | | | | |
|---|---|---|---|---|---|
| 22 | 19 | 24 | 31 | 29 | 29 |
| 21 | 15 | 27 | 23 | 37 | 31 |
| 30 | 26 | 16 | 26 | 12 | |
| 23 | 48 | 22 | 29 | 28 | |

a. Find a 90% confidence interval for the mean commuting time of all local bicycle commuters in the city. (*Note:* The sample mean and sample standard deviation of the data are 25.82 minutes and 7.71 minutes, respectively.)
b. Interpret your result in part (a).
c. Graphical analyses of the data indicate that the time of 48 minutes may be an outlier. Remove this potential outlier and repeat part (a). (*Note:* The sample mean and sample standard deviation of the abridged data are 24.76 and 6.05, respectively.)
d. Should you have used the procedure that you did in part (a)? Explain your answer.

**8.105** Table IV in Appendix A contains degrees of freedom from 1 to 75 consecutively but then contains only selected degrees of freedom.
a. Why couldn't we provide entries for all possible degrees of freedom?
b. Why did we construct the table so that consecutive entries appear for smaller degrees of freedom but that only selected entries occur for larger degrees of freedom?
c. If you had only Table IV, what value would you use for $t_{0.05}$ with df = 87? with df = 125? with df = 650? with df = 3000? Explain your answers.

**8.106** As we mentioned earlier in this section, we stopped the $t$-table at df $= 2000$ and supplied the corresponding values of $z_\alpha$ beneath. Explain why that makes sense.

**8.107** A variable of a population has mean $\mu$ and standard deviation $\sigma$. For a sample of size $n$, under what conditions are the observed values of the studentized and standardized versions of $\bar{x}$ equal? Explain your answer.

**8.108** Let $0 < \alpha < 1$. For a $t$-curve, determine
a. the $t$-value having area $\alpha$ to its right in terms of $t_\alpha$.
b. the $t$-value having area $\alpha$ to its left in terms of $t_\alpha$.
c. the two $t$-values that divide the area under the curve into a middle $1 - \alpha$ area and two outside $\alpha/2$ areas.
d. Draw graphs to illustrate your results in parts (a)–(c).

**8.109 Batting Averages.** An issue of *Scientific American* reveals that the batting averages of major-league baseball players are normally distributed with mean .270 and standard deviation .031.
a. Simulate 2000 samples of five batting averages each.
b. Determine the sample mean and sample standard deviation of each of the 2000 samples.
c. For each of the 2000 samples, determine the observed value of the standardized version of $\bar{x}$.
d. Obtain a histogram of the 2000 observations in part (c).
e. Theoretically, what is the distribution of the standardized version of $\bar{x}$?
f. Compare your results from parts (d) and (e).
g. For each of the 2000 samples, determine the observed value of the studentized version of $\bar{x}$.
h. Obtain a histogram of the 2000 observations in part (g).
i. Theoretically, what is the distribution of the studentized version of $\bar{x}$?

j. Compare your results from parts (h) and (i).
k. Compare your histograms from parts (d) and (h). How and why do they differ?

*Another type of confidence interval is called a **one-sided confidence interval**. A one-sided confidence interval provides either a lower confidence bound or an upper confidence bound for the parameter in question. You are asked to examine one-sided confidence intervals in Exercises 8.110–8.112.*

**8.110 One-Sided One-Mean $t$-Intervals.** Presuming that the assumptions for a one-mean $t$-interval are satisfied, we have the following formulas for $(1 - \alpha)$-level confidence bounds for a population mean $\mu$:

- Lower confidence bound: $\bar{x} - t_\alpha \cdot s/\sqrt{n}$
- Upper confidence bound: $\bar{x} + t_\alpha \cdot s/\sqrt{n}$

Interpret the preceding formulas for lower and upper confidence bounds in words.

**8.111 Northeast Commutes.** Refer to Exercise 8.91.
a. Determine and interpret a 90% upper confidence bound for the mean commute time of all commuters in Washington, DC.
b. Compare your one-sided confidence interval in part (a) to the (two-sided) confidence interval found in Exercise 8.91(a).

**8.112 TV Viewing.** Refer to Exercise 8.92.
a. Determine and interpret a 90% lower confidence bound for the amount of television watched per day last year by the average person.
b. Compare your one-sided confidence interval in part (a) to the (two-sided) confidence interval found in Exercise 8.92(a).

## Chapter in Review

### You Should be Able to

1. use and understand the formulas in this chapter.

2. obtain a point estimate for a population mean.

3. find and interpret a confidence interval for a population mean when the population standard deviation is known.

4. compute and interpret the margin of error for the estimate of $\mu$.

5. understand the relationship between sample size, standard deviation, confidence level, and margin of error for a confidence interval for $\mu$.

6. determine the sample size required for a specified confidence level and margin of error for the estimate of $\mu$.

7. understand the difference between the standardized and studentized versions of $\bar{x}$.

8. state the basic properties of $t$-curves.

9. use Table IV to find $t_{\alpha/2}$ for df $= n - 1$ and selected values of $\alpha$.

10. find and interpret a confidence interval for a population mean when the population standard deviation is unknown.

11. decide whether it is appropriate to use the $z$-interval procedure, $t$-interval procedure, or neither.

### Key Terms

confidence interval (CI), *340*
confidence-interval estimate, *340*
confidence level, *340*
degrees of freedom (df), *360*
margin of error ($E$), *355*
nonparametric methods, *365*
normal population, *345*

one-mean $t$-interval procedure, *363*
one-mean $z$-interval procedure, *345*
parametric methods, *365*
point estimate, *339*
robust procedures, *346*
standardized version of $\bar{x}$, *359*
studentized version of $\bar{x}$, *359*

Student's $t$-distribution, *360*
$t_\alpha$, *361*
$t$-curve, *360*
$t$-distribution, *360*
$t$-interval procedure, *362*
$z_\alpha$, *344*
$z$-interval procedure, *345*

## Review Problems

### Understanding the Concepts and Skills

1. Explain the difference between a point estimate of a parameter and a confidence-interval estimate of a parameter.

2. Answer true or false to the following statement and give a reason for your answer: If a 95% confidence interval for a population mean, $\mu$, is from 33.8 to 39.0, the mean of the population must lie somewhere between 33.8 and 39.0.

3. Must the variable under consideration be normally distributed for you to use the $z$-interval procedure or $t$-interval procedure? Explain your answer.

4. If you obtained one thousand 95% confidence intervals for a population mean, $\mu$, roughly how many of the intervals would actually contain $\mu$?

5. Suppose that you have obtained a sample with the intent of performing a particular statistical-inference procedure.

What should you do before applying the procedure to the sample data? Why?

6. Suppose that you intend to find a 95% confidence interval for a population mean by applying the one-mean $z$-interval procedure to a sample of size 100.
   a. What would happen to the precision of the estimate if you used a sample of size 50 instead but kept the same confidence level of 0.95?
   b. What would happen to the precision of the estimate if you changed the confidence level to 0.90 but kept the same sample size of 100?

7. A confidence interval for a population mean has a margin of error of 10.7.
   a. Obtain the length of the confidence interval.
   b. If the mean of the sample is 75.2, determine the confidence interval.

8. Suppose that you plan to apply the one-mean $z$-interval procedure to obtain a 90% confidence interval for a population mean, $\mu$. You know that $\sigma = 12$ and that you are going to use a sample of size 9.
   a. What will be your margin of error?
   b. What else do you need to know in order to obtain the confidence interval?

9. A variable of a population has a mean of 266 and a standard deviation of 16. Ten observations of this variable have a mean of 262.1 and a sample standard deviation of 20.4. Obtain the observed value of the
   a. standardized version of $\bar{x}$.
   b. studentized version of $\bar{x}$.

10. **Baby Weight.** The paper "Are Babies Normal?" by T. Clemons and M. Pagano (*The American Statistician*, Vol. 53, No. 4, pp. 298–302) focused on birth weights of babies. According to the article, for babies born within the "normal" gestational range of 37–43 weeks, birth weights are normally distributed with a mean of 3432 grams (7 pounds 9 ounces) and a standard deviation of 482 grams (1 pound 1 ounce). For samples of 15 such birth weights, identify the distribution of each variable.

   a. $\dfrac{\bar{x} - 3432}{482/\sqrt{15}}$
   b. $\dfrac{\bar{x} - 3432}{s/\sqrt{15}}$

11. The following figure shows the standard normal curve and two $t$-curves. Which of the two $t$-curves has the larger degrees of freedom? Explain your answer.

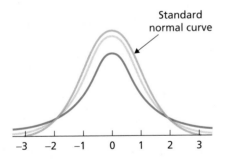

Standard normal curve

12. In each part of this problem, we have provided a scenario for a confidence interval. Decide whether the appropriate method for obtaining the confidence interval is the $z$-interval procedure, the $t$-interval procedure, or neither.
   a. A random sample of size 17 is taken from a population. A normal probability plot of the sample data is found to be very close to linear (straight line). The population standard deviation is unknown.
   b. A random sample of size 50 is taken from a population. A normal probability plot of the sample data is found to be roughly linear. The population standard deviation is known.
   c. A random sample of size 25 is taken from a population. A normal probability plot of the sample data shows three outliers but is otherwise roughly linear. Checking reveals that the outliers are due to recording errors. The population standard deviation is known.
   d. A random sample of size 20 is taken from a population. A normal probability plot of the sample data shows three outliers but is otherwise roughly linear. Removal of the outliers is questionable. The population standard deviation is unknown.
   e. A random sample of size 128 is taken from a population. A normal probability plot of the sample data shows no outliers but has significant curvature. The population standard deviation is known.
   f. A random sample of size 13 is taken from a population. A normal probability plot of the sample data shows no outliers but has significant curvature. The population standard deviation is unknown.

13. **Millionaires.** Dr. Thomas Stanley of Georgia State University has surveyed millionaires since 1973. Among other information, Stanley obtains estimates for the mean age, $\mu$, of all U.S. millionaires. Suppose that 36 randomly selected U.S. millionaires are the following ages.

| | | | | | | | | |
|---|---|---|---|---|---|---|---|---|
| 31 | 45 | 79 | 64 | 48 | 38 | 39 | 68 | 52 |
| 59 | 68 | 79 | 42 | 79 | 53 | 74 | 66 | 66 |
| 71 | 61 | 52 | 47 | 39 | 54 | 67 | 55 | 71 |
| 77 | 64 | 60 | 75 | 42 | 69 | 48 | 57 | 48 |

Determine a 95% confidence interval for the mean age, $\mu$, of all U.S. millionaires. Assume that the standard deviation of ages of all U.S. millionaires is 13.0 years. (*Note:* The mean of the data is 58.53 years.)

14. **Millionaires.** From Problem 13, we know that "a 95% confidence interval for the mean age of all U.S. millionaires is from 54.3 years to 62.8 years." Decide which of the following sentences provide a correct interpretation of the statement in quotes. Justify your answers.
   a. Ninety-five percent of all U.S. millionaires are between the ages of 54.3 years and 62.8 years.
   b. There is a 95% chance that the mean age of all U.S. millionaires is between 54.3 years and 62.8 years.
   c. We can be 95% confident that the mean age of all U.S. millionaires is between 54.3 years and 62.8 years.
   d. The probability is 0.95 that the mean age of all U.S. millionaires is between 54.3 years and 62.8 years.

15. **Sea Shell Morphology.** In a 1903 paper, Abigail Camp Dimon discussed the effect of environment on the shape and form of two sea snail species, *Nassa obsoleta* and *Nassa trivittata*. One of the variables that Dimon considered was length of shell. She found the mean shell length of 461 randomly selected specimens of *N. trivittata* to be 11.9 mm. [SOURCE: "Quantitative Study of the Effect of Environment Upon the Forms of *Nassa obsoleta* and *Nassa trivittata* from Cold Spring Harbor, Long Island," *Biometrika*, Vol. 2, pp. 24–43.]

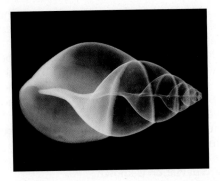

a. Assuming that $\sigma = 2.5$ mm, obtain a 90% confidence interval for the mean length, $\mu$, of all *N. trivittata*.
b. Interpret your answer from part (a).
c. What properties should a normal probability plot of the data have for it to be permissible to apply the procedure that you used in part (a)?

**16. Sea Shell Morphology.** Refer to Problem 15.
a. Find the margin of error, $E$.
b. Explain the meaning of $E$ as far as the accuracy of the estimate is concerned.
c. Determine the sample size required to have a margin of error of 0.1 mm and a 90% confidence level.
d. Find a 90% confidence interval for $\mu$ if a sample of the size determined in part (c) yields a mean of 12.0 mm.

**17.** For a $t$-curve with df $= 18$, obtain the $t$-value and illustrate your results graphically.
a. The $t$-value having area 0.025 to its right
b. $t_{0.05}$
c. The $t$-value having area 0.10 to its left
d. The two $t$-values that divide the area under the curve into a middle 0.99 area and two outside 0.005 areas

**18. Children of Diabetic Mothers.** A paper by Cho et al. in the May 2000 issue of *The Journal of Pediatrics* (Vol. 136(5), pp. 587–592) presented the results of research on various characteristics in children of diabetic mothers. Past studies have shown that maternal diabetes results in obesity, blood pressure, and glucose tolerance complications in the offspring. Following are the arterial blood pressures, in millimeters of mercury (mm Hg), for a random sample of 16 children of diabetic mothers.

| | | | | | | | |
|---|---|---|---|---|---|---|---|
| 81.6 | 84.1 | 87.6 | 82.8 | 82.0 | 88.9 | 86.7 | 96.4 |
| 84.6 | 101.9 | 90.8 | 94.0 | 69.4 | 78.9 | 75.2 | 91.0 |

a. Apply the $t$-interval procedure to these data to find a 95% confidence interval for the mean arterial blood pressure of all children of diabetic mothers. Interpret your result. (*Note:* $\bar{x} = 85.99$ mm Hg and $s = 8.08$ mm Hg.)
b. Obtain a normal probability plot, a boxplot, a histogram, and a stem-and-leaf diagram of the data.
c. Based on your graphs from part (b), is it reasonable to apply the $t$-interval procedure as you did in part (a)? Explain your answer.

**19. Diamond Pricing.** In a Singapore edition of *Business Times*, diamond pricing was explored. The price of a diamond is based on the diamond's weight, color, and clarity. A simple random sample of 18 one-half carat diamonds had the following prices, in dollars.

| | | | | | | | | |
|---|---|---|---|---|---|---|---|---|
| 1676 | 1442 | 1995 | 1718 | 1826 | 2071 | 1947 | 1983 | 2146 |
| 1995 | 1876 | 2032 | 1988 | 2071 | 2234 | 2108 | 1941 | 2316 |

a. Apply the $t$-interval procedure to these data to find a 90% confidence interval for the mean price of all one-half carat diamonds. Interpret your result. (*Note:* $\bar{x} = \$1964.7$ and $s = \$206.5$.)
b. Obtain a normal probability plot, a boxplot, a histogram, and a stem-and-leaf diagram of the data.
c. Based on your graphs from part (b), is it reasonable to apply the $t$-interval procedure as you did in part (a)? Explain your answer.

## Working With Large Data Sets

**20. Delaying Adulthood.** The convict surgeonfish is a common tropical reef fish that has been found to delay metamorphosis into adult by extending its larval phase. This delay often leads to enhanced survivorship in the species by increasing the chances of finding suitable habitat. In the paper, "Delayed Metamorphosis of a Tropical Reef Fish (*Acanthurus triostegus*): A Field Experiment" (*Marine Ecology Progress Series*, Vol. 176, pp. 25–38), Mark I. McCormick published data that he obtained on the larval duration, in days, of 90 convict surgeonfish. The data are contained on the WeissStats CD.
a. Import the data into the technology of your choice.
b. Use the technology of your choice to obtain a normal probability plot, boxplot, and histogram of the data.
c. Is it reasonable to apply the $t$-interval procedure to the data? Explain your answer.
d. If you answered "yes" to part (c), obtain a 99% confidence interval for the mean larval duration of convict surgeonfish. Interpret your result.

**21. Fuel Economy.** The U.S. Department of Energy collects fuel-economy information on new motor vehicles and publishes its findings in *Fuel Economy Guide*. The data included are the result of vehicle testing done at the Environmental Protection Agency's National Vehicle and Fuel Emissions Laboratory in Ann Arbor, Michigan, and by vehicle manufacturers themselves with oversight by the Environmental Protection Agency. On the WeissStats CD, we have provided the highway mileages, in miles per gallon (MPG), for 2006 cars. Use the technology of your choice to do the following.
a. Obtain a random sample of 35 of the mileages.
b. Use your data from part (b) and the $t$-interval procedure to find a 95% confidence interval for the mean highway gas mileage of all 2006 cars.
c. Does the mean highway gas mileage of all 2006 cars lie in the confidence interval that you found in part (c)? Would it necessarily have to? Explain your answers.

**22. Old Faithful Geyser.** In the article "Old Faithful at Yellowstone, a Bimodal Distribution," D. Howell examined various aspects of the Old Faithful Geyser at Yellowstone National Park. Despite its name, there is considerable variation in both the length of the eruptions and in the time interval between eruptions. The times between eruptions, in minutes, for 500 recent observations are provided on the WeissStats CD.

a. Identify the population and variable under consideration.

b. Use the technology of your choice to determine and interpret a 99% confidence interval for the mean time between eruptions.

c. Discuss the relevance of your confidence interval for future eruptions, say, 5 years from now.

**23. Booted Eagles.** The rare booted eagle of western Europe was the focus of a study by S. Suarez et al. to identify optimal nesting habitat for this raptor. According to their paper "Nesting Habitat Selection by Booted Eagles (*Hieraaetus pennatus*) and Implications for Management" (*Journal of*

*Applied Ecology*, Vol. 37, pp. 215–223), the distances of such nests to the nearest marshland are normally distributed with mean 4.66 km and standard deviation 0.75 km.

a. Simulate 3000 samples of four distances each.

b. Determine the sample mean and sample standard deviation of each of the 3000 samples.

c. For each of the 3000 samples, determine the observed value of the standardized version of $\bar{x}$.

d. Obtain a histogram of the 3000 observations in part (c).

e. Theoretically, what is the distribution of the standardized version of $\bar{x}$?

f. Compare your results from parts (d) and (e).

g. For each of the 3000 samples, determine the observed value of the studentized version of $\bar{x}$.

h. Obtain a histogram of the 3000 observations in part (g).

i. Theoretically, what is the distribution of the studentized version of $\bar{x}$?

j. Compare your results from parts (h) and (i).

k. Compare your histograms from parts (d) and (h). How and why do they differ?

## Focusing on Data Analysis    UWEC Undergraduates

Recall from Chapter 1 (see page 34) that the Focus database and Focus sample contain information on the undergraduate students at the University of Wisconsin - Eau Claire (UWEC). Now would be a good time for you to review the discussion about these data sets.

a. Open the Focus sample (FocusSample) in the statistical software package of your choice and then obtain and interpret a 95% confidence interval for the mean high school percentile of all UWEC undergraduate students. Interpret your result.

b. In practice, the (population) mean of the variable under consideration is unknown. However, in this case, we actually do have the population data, namely, in the

Focus database (Focus). If your statistical software package will accommodate the entire Focus database, open that worksheet and then obtain the mean high school percentile of all UWEC undergraduate students. (*Answer:* 74.0)

c. Does your confidence interval in part (a) contain the population mean found in part (b)? Would it necessarily have to? Explain your answers.

d. Repeat parts (a)–(c) for the variables cumulative GPA, age, total earned credits, ACT English score, ACT math score, and ACT composite score. (*Note:* The means of these variables are 3.055, 20.7, 70.2, 23.0, 23.5, and 23.6, respectively.)

## Case Study Discussion    The Chips Ahoy! 1,000 Chips Challenge

At the beginning of this chapter, on page 337, we presented data on the number of chocolate chips per bag for 42 bags of Chips Ahoy! cookies. These data were obtained by the students in an introductory statistics class at the United States Air Force Academy in response to the Chips Ahoy! 1,000 Chips Challenge sponsored by Nabisco, the makers of Chips Ahoy! cookies. Use the data collected by the students to answer the questions and conduct the analyses required in each part.

a. Obtain and interpret a point estimate for the mean number of chocolate chips per bag for all bags of Chips Ahoy! cookies. (*Note:* The sum of the data is 52,986.)

b. Construct and interpret a normal probability plot, boxplot, and histogram of the data.

c. Use the graphs in part (b) to identify outliers, if any.

d. Is it reasonable to use the one-mean $t$-interval procedure to obtain a confidence interval for the mean number of chocolate chips per bag for all bags of Chips Ahoy! cookies? Explain your answer.

e. Determine a 95% confidence interval for the mean number of chips per bag for all bags of Chips Ahoy! cookies, and interpret your result in words. (*Note:* $\bar{x} = 1261.6$; $s = 117.6$.)

## Biography  WILLIAM GOSSET: The "Student" in Student's *t*-Distribution

**William Sealy Gosset** was born in Canterbury, England, on June 13, 1876, the eldest son of Colonel Frederic Gosset and Agnes Sealy. He studied mathematics and chemistry at Winchester College and New College, Oxford, receiving a first-class degree in natural sciences in 1899.

After graduation Gosset began work with Arthur Guinness and Sons, a brewery in Dublin, Ireland. He saw the need for accurate statistical analyses of various brewing processes ranging from barley production to yeast fermentation, and pressed the firm to solicit mathematical advice. In 1906, the brewery sent him to work under Karl Pearson (see Biography in Chapter 12) at University College in London.

During the next few years, Gosset developed what has come to be known as Student's *t*-distribution. This distribution has proved to be fundamental in statistical analyses involving normal distributions. In particular, Student's *t*-distribution is used in performing inferences for a population mean when the population being sampled is (approximately) normally distributed and the population standard deviation is unknown. Although the statistical theory for large samples had been completed in the early 1800s, no small-sample theory was available before Gosset's work.

Because Guinness's brewery prohibited its employees from publishing any of their research, Gosset published his contributions to statistical theory under the pseudonym "Student"—consequently the name "Student" in Student's *t*-distribution.

Gosset remained with Guinness his entire working life. In 1935, he moved to London to take charge of a new brewery. But his tenure there was short lived; he died in Beaconsfield, England, on October 16, 1937.

## StatCrunch in MyStatLab
**Analyzing Data Online**

StatCrunch online statistical software offers an easy-to-use interface customized for this book. The StatCrunch feature for each chapter illustrates the use of the software to perform a statistical analysis discussed in the chapter. Exercises are provided to further apply StatCrunch to other statistical analyses examined in the chapter. Go to the WeissStats CD or to the Weiss Web site at www.aw-bc.com/weiss to access StatCrunch instructions and data sets. To access StatCrunch statistical software, go to the student content area of your Weiss MyStatLab course.

## Internet Projects
**Exploring Data Online**

The Internet project for each chapter provides simulations, demonstrations, or activities that enhance the topics covered in the chapter. The project materials come from universities, individuals, governments, and companies from all over the world. To access the Internet projects on the Web, go to www.aw-bc.com/weiss. From this Web page, you can reach the Internet Projects Page, which we suggest that you bookmark for easy access in the future.

# 9

# Hypothesis Tests for One Population Mean

## Chapter Objectives

In Chapter 8, we examined methods for obtaining confidence intervals for one population mean. We know that a confidence interval for a population mean, $\mu$, is based on a sample mean, $\bar{x}$. Now we show how that statistic can be used to make decisions about hypothesized values of a population mean.

For example, suppose that we want to decide whether the mean prison sentence, $\mu$, of all people imprisoned last year for drug offenses exceeds the 2000 mean of 75.5 months. To make that decision, we can take a random sample of people imprisoned last year for drug offenses, compute their sample mean sentence, $\bar{x}$, and then apply a statistical-inference technique called a *hypothesis test.*

In Sections 9.1 and 9.2, we present generic information about the nature, notation, and terminology of hypothesis testing. Then, in the remainder of the chapter, we consider two different hypothesis-testing procedures for one population mean—the *one-mean z-test* (Sections 9.3 and 9.4) and the *one-mean t-test* (Section 9.5). These two procedures are the hypothesis-test analogues of the one-mean *z*-interval and one-mean *t*-interval confidence-interval procedures, respectively, discussed in Chapter 8.

We also examine two different approaches to hypothesis testing, the *critical-value approach*—which we introduce and apply in Sections 9.2, 9.3, and 9.5—and the *P-value approach*—which we introduce and apply in Sections 9.4 and 9.5.

# Sex and Sense of Direction

Many of you have been there, a classic scene: mom yelling at dad to turn left while dad decides to do just the opposite. Well, who made the right call? And, more generally, who has a better sense of direction, women or men?

Dr. Jeanne Sholl et al. considered these and related questions in the paper "The Relation of Sex and Sense of Direction to Spatial Orientation in an Unfamiliar Environment" (*Journal of Environmental Psychology*, Vol. 20, pp. 17–28).

In their study, the spatial orientation skills of 30 male students and 30 female students from Boston College were challenged in Houghton Garden Park, a wooded park near the BC campus in Newton, Massachusetts. Before driving to the park, the participants were asked to rate their own sense of direction as either good or poor.

In the park, students were instructed to point to predesignated landmarks and also to the direction of south. Pointing was carried out by students moving a pointer attached to a 360° protractor; the angle of the pointing response was then recorded to the nearest degree. For the female students who had rated their sense of direction to be good, the following table displays the pointing errors (in degrees) when they attempted to point south.

| | | | | |
|----|-----|-----|-----|----|
| 14 | 122 | 128 | 109 | 12 |
| 91 | 8 | 78 | 31 | 36 |
| 27 | 68 | 20 | 69 | 18 |

Based on these data, can you conclude that, in general, women who consider themselves to have a good sense of direction really do better, on average, than they would by randomly guessing at the direction of south? To answer that question, you need to conduct a hypothesis test, which you will do after you study hypothesis testing in this chapter.

## 9.1 The Nature of Hypothesis Testing

We often use inferential statistics to make decisions or judgments about the value of a parameter, such as a population mean. For example, we might need to decide whether the mean weight, $\mu$, of all bags of pretzels packaged by a particular company differs from the advertised weight of 454 grams (g); or, we might want to determine whether the mean age, $\mu$, of all cars in use has increased from the 1995 mean of 8.5 years.

One of the most commonly used methods for making such decisions or judgments is to perform a *hypothesis test*. A **hypothesis** is a statement that something is true. For example, the statement "the mean weight of all bags of pretzels packaged differs from the advertised weight of 454 g" is a hypothesis.

Typically, a hypothesis test involves two hypotheses: the *null hypothesis* and the *alternative hypothesis* (or *research hypothesis*), which we define as follows.

**Definition 9.1**

### Null and Alternative Hypotheses; Hypothesis Test

**Null hypothesis:** A hypothesis to be tested. We use the symbol $H_0$ to represent the null hypothesis.

**Alternative hypothesis:** A hypothesis to be considered as an alternative to the null hypothesis. We use the symbol $H_a$ to represent the alternative hypothesis.

**Hypothesis test:** The problem in a hypothesis test is to decide whether the null hypothesis should be rejected in favor of the alternative hypothesis.

> **What Does It Mean?**
>
> Originally, the word *null* in *null hypothesis* stood for "no difference" or "the difference is null." Over the years, however, *null hypothesis* has come to mean simply a hypothesis to be tested.

For instance, in the pretzel-packaging example, the null hypothesis might be "the mean weight of all bags of pretzels packaged equals the advertised weight of 454 g," and the alternative hypothesis might be "the mean weight of all bags of pretzels packaged differs from the advertised weight of 454 g."

### Choosing the Hypotheses

The first step in setting up a hypothesis test is to decide on the null hypothesis and the alternative hypothesis. The following are some guidelines for choosing these two hypotheses. Although the guidelines refer specifically to hypothesis tests for one population mean, $\mu$, they apply to any hypothesis test concerning one parameter.

**Null Hypothesis**

In this book, the null hypothesis for a hypothesis test concerning a population mean, $\mu$, always specifies a single value for that parameter. Hence we can express the null hypothesis as

$$H_0: \mu = \mu_0$$

where $\mu_0$ is some number.

**Alternative Hypothesis**

The choice of the alternative hypothesis depends on and should reflect the purpose of the hypothesis test. Three choices are possible for the alternative hypothesis.

- If the primary concern is deciding whether a population mean, $\mu$, is *different from* a specified value $\mu_0$, we express the alternative hypothesis as

$$H_a\text{: } \mu \neq \mu_0.$$

A hypothesis test whose alternative hypothesis has this form is called a **two-tailed test.**

- If the primary concern is deciding whether a population mean, $\mu$, is *less than* a specified value $\mu_0$, we express the alternative hypothesis as

$$H_a\text{: } \mu < \mu_0.$$

A hypothesis test whose alternative hypothesis has this form is called a **left-tailed test.**

- If the primary concern is deciding whether a population mean, $\mu$, is *greater than* a specified value $\mu_0$, we express the alternative hypothesis as

$$H_a\text{: } \mu > \mu_0.$$

A hypothesis test whose alternative hypothesis has this form is called a **right-tailed test.**

A hypothesis test is called a **one-tailed test** if it is either left tailed or right tailed. In Section 9.2, we explain the term *tailed.* For now, let's consider Examples 9.1–9.3, which illustrate the preceding discussion.

**Example 9.1** | **Choosing the Null and Alternative Hypotheses**

*Quality Assurance* A snack-food company produces a 454 g bag of pretzels. Although the actual net weights deviate slightly from 454 g and vary from one bag to another, the company insists that the mean net weight of the bags be 454 g.

As part of its program, the quality assurance department periodically performs a hypothesis test to decide whether the packaging machine is working properly, that is, to decide whether the mean net weight of all bags packaged is 454 g.

a. Determine the null hypothesis for the hypothesis test.

b. Determine the alternative hypothesis for the hypothesis test.

c. Classify the hypothesis test as two tailed, left tailed, or right tailed.

**Solution** Let $\mu$ denote the mean net weight of all bags packaged.

a. The null hypothesis is that the packaging machine is working properly, that is, that the mean net weight, $\mu$, of all bags packaged *equals* 454 g. In symbols, $H_0\text{: } \mu = 454$ g.

b. The alternative hypothesis is that the packaging machine is not working properly, that is, that the mean net weight, $\mu$, of all bags packaged is *different from* 454 g. In symbols, $H_a\text{: } \mu \neq 454$ g.

c. This hypothesis test is two tailed because a does-not-equal sign ($\neq$) appears in the alternative hypothesis.

• • •

Example 9.2 | **Choosing the Null and Alternative Hypotheses**

*Prices of History Books* The R. R. Bowker Company of New York collects information on the retail prices of books and publishes the data in *The Bowker Annual Library and Book Trade Almanac*. In 2000, the mean retail price of history books was $51.46. Suppose that we want to perform a hypothesis test to decide whether this year's mean retail price of history books has increased from the 2000 mean.

a.  Determine the null hypothesis for the hypothesis test.

b.  Determine the alternative hypothesis for the hypothesis test.

c.  Classify the hypothesis test as two tailed, left tailed, or right tailed.

**Solution**   Let $\mu$ denote this year's mean retail price of history books.

a.  The null hypothesis is that this year's mean retail price of history books *equals* the 2000 mean of $51.46; that is, $H_0: \mu = \$51.46$.

b.  The alternative hypothesis is that this year's mean retail price of history books is *greater than* $51.46; that is, $H_a: \mu > \$51.46$.

c.  This hypothesis test is right tailed because a greater-than sign ($>$) appears in the alternative hypothesis.

• • •

Example 9.3 | **Choosing the Null and Alternative Hypotheses**

*Poverty and Calcium* Calcium is the most abundant mineral in the human body and has several important functions. Most body calcium is stored in the bones and teeth, where it functions to support their structure. Recommendations for calcium are provided in *Dietary Reference Intakes*, developed by the Institute of Medicine of the National Academy of Sciences. The recommended adequate intake (RAI) of calcium for adults (ages 19–50) is 1000 milligrams (mg) per day.

Suppose that we want to perform a hypothesis test to decide whether the average adult with an income below the poverty level gets less than the RAI of 1000 mg.

a.  Determine the null hypothesis for the hypothesis test.

b.  Determine the alternative hypothesis for the hypothesis test.

c.  Classify the hypothesis test as two tailed, left tailed, or right tailed.

**Solution**   Let $\mu$ denote the mean calcium intake (per day) of all adults with incomes below the poverty level.

a.  The null hypothesis is that the mean calcium intake of all adults with incomes below the poverty level *equals* the RAI of 1000 mg per day; that is, $H_0: \mu = 1000$ mg.

b.  The alternative hypothesis is that the mean calcium intake of all adults with incomes below the poverty level is *less than* the RAI of 1000 mg per day; that is, $H_a: \mu < 1000$ mg.

c. This hypothesis test is left tailed because a less-than sign ($<$) appears in the alternative hypothesis.

• • •

Exercise 9.5
on page 385

## The Logic of Hypothesis Testing

After we have chosen the null and alternative hypotheses, we must decide whether to reject the null hypothesis in favor of the alternative hypothesis. The procedure for deciding is roughly as follows.

> ### Basic Logic of Hypothesis Testing
>
> Take a random sample from the population. If the sample data are consistent with the null hypothesis, do not reject the null hypothesis; if the sample data are inconsistent with the null hypothesis (in the direction of the alternative hypothesis), reject the null hypothesis and conclude that the alternative hypothesis is true.

In practice, of course, we must have a precise criterion for deciding whether to reject the null hypothesis. Example 9.4 illustrates such a criterion for a two-tailed hypothesis test about a population mean. The example also introduces the logic and some of the terminology of hypothesis testing. Later in this chapter we give general procedures for performing hypothesis tests.

**Example 9.4** | **The Logic of Hypothesis Testing**

*Quality Assurance* A company that produces snack foods uses a machine to package 454 g bags of pretzels. We assume that the net weights are normally distributed and that the population standard deviation of all such weights is 7.8 g.[†] A simple random sample of 25 bags of pretzels has the net weights, in grams, displayed in Table 9.1. Do the data provide sufficient evidence to conclude that the packaging machine is not working properly? We use the following steps to answer the question.

**TABLE 9.1**
Weights, in grams, of 25 randomly selected bags of pretzels

| | | | | |
|---|---|---|---|---|
| 465 | 456 | 438 | 454 | 447 |
| 449 | 442 | 449 | 446 | 447 |
| 468 | 433 | 454 | 463 | 450 |
| 446 | 447 | 456 | 452 | 444 |
| 447 | 456 | 456 | 435 | 450 |

a. State the null and alternative hypotheses for the hypothesis test.

b. Discuss the logic of this hypothesis test.

c. Identify the distribution of the variable $\bar{x}$, that is, the sampling distribution of the sample mean for samples of size 25.

d. Obtain a precise criterion for deciding whether to reject the null hypothesis in favor of the alternative hypothesis.

e. Apply the criterion in part (d) to the sample data and state the conclusion.

**Solution**  Let $\mu$ denote the mean net weight of all bags packaged.

a. The null and alternative hypotheses, as stated in Example 9.1, are

$$H_0: \mu = 454 \text{ g (the packaging machine is working properly)}$$
$$H_a: \mu \neq 454 \text{ g (the packaging machine is not working properly)}.$$

---

[†]We might know the population standard deviation from previous research. In Section 9.5, we consider the more usual case of an unknown $\sigma$.

**b.** Basically, the logic of this hypothesis test is as follows: If the null hypothesis is true, then the mean weight, $\bar{x}$, of the sample of 25 bags of pretzels should approximately equal 454 g. We say "approximately equal" because we cannot expect a sample mean to equal exactly the population mean; some sampling error is anticipated. However, if the sample mean weight differs "too much" from 454 g, we would be inclined to reject the null hypothesis and conclude that the alternative hypothesis is true. As we show in part (d), we can use our knowledge of the sampling distribution of the sample mean to decide how much difference is "too much."

**c.** Because $n = 25$, $\sigma = 7.8$, and the weights are normally distributed, Key Fact 7.4 on page 325 implies that

- $\mu_{\bar{x}} = \mu$ (which we don't know),
- $\sigma_{\bar{x}} = \sigma/\sqrt{n} = 7.8/\sqrt{25} = 1.56$, and
- $\bar{x}$ is normally distributed.

In other words, for samples of size 25, the variable $\bar{x}$ is normally distributed with mean $\mu$ and standard deviation 1.56 g.

**d.** The "95.44" part of the 68.26-95.44-99.74 rule states that, for a normally distributed variable, 95.44% of all possible observations lie within two standard deviations to either side of the mean. Applying this part of the rule to the variable $\bar{x}$ and referring to part (c), we see that 95.44% of all samples of 25 bags of pretzels have mean weights within $2 \cdot 1.56 = 3.12$ g of $\mu$. Equivalently, only 4.56% of all samples of 25 bags of pretzels have mean weights that are not within 3.12 g of $\mu$, as illustrated in Fig. 9.1.

Thus, if the mean weight, $\bar{x}$, of the 25 bags of pretzels sampled is not within two standard deviations (3.12 g) of 454 g, we have evidence against the null hypothesis. Why? Because observing such a sample mean would occur by chance only 4.56% of the time if the null hypothesis is true.

In summary, then, we have obtained the following precise criterion for deciding whether to reject the null hypothesis. This criterion is portrayed graphically in Fig. 9.2(a).

If the mean weight, $\bar{x}$, of the 25 bags of pretzels sampled is more than two standard deviations (3.12 g) from 454 g, reject the null hypothesis and conclude that the alternative hypothesis is true. Otherwise, do not reject the null hypothesis.

**FIGURE 9.1**
95.44% of all samples of 25 bags of pretzels have mean weights within two standard deviations (3.12 g) of $\mu$

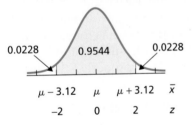

**FIGURE 9.2**
(a) Criterion for deciding whether to reject the null hypothesis; (b) normal curve associated with $\bar{x}$ if the null hypothesis is true, superimposed on the decision criterion

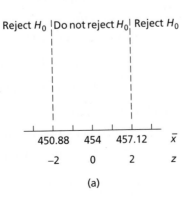

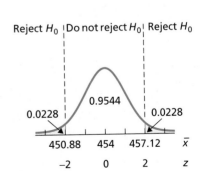

If the null hypothesis is true, the normal curve associated with $\bar{x}$ is the one with parameters 454 and 1.56; that normal curve is superimposed on Fig. 9.2(a) in Fig. 9.2(b).

**e.** The mean weight, $\bar{x}$, of the sample of 25 bags of pretzels whose weights are given in Table 9.1 is 450 g. Therefore,

$$z = \frac{\bar{x} - 454}{1.56} = \frac{450 - 454}{1.56} = -2.56.$$

That is, the sample mean of 450 g is 2.56 standard deviations below the null-hypothesis population mean of 454 g, as shown in Fig. 9.3.

Because the mean weight of the 25 bags of pretzels sampled is more than two standard deviations from 454 g, we reject the null hypothesis ($\mu = 454\,\text{g}$) and conclude that the alternative hypothesis ($\mu \neq 454\,\text{g}$) is true.

**Interpretation** The data provide sufficient evidence to conclude that the packaging machine is not working properly.

• • •

Example 9.4 contains all the elements of a hypothesis test, including the necessary theory. But don't worry too much about the details at this point. What you should understand now is (1) how to choose the null and alternative hypotheses for a hypothesis test, (2) how to classify a hypothesis test as two tailed, left tailed, or right tailed, and (3) the logic behind performing a hypothesis test.

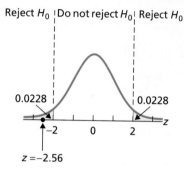

**FIGURE 9.3**
Graph showing the number of standard deviations that the sample mean of 450 g is from the null-hypothesis population mean of 454 g

## Exercises 9.1

### Understanding the Concepts and Skills

**9.1** Explain the meaning of the term *hypothesis* as used in inferential statistics.

**9.2** What role does the decision criterion play in a hypothesis test?

**9.3** Suppose that you want to perform a hypothesis test for a population mean $\mu$.
**a.** Express the null hypothesis both in words and in symbolic form.
**b.** Express each of the three possible alternative hypotheses in words and in symbolic form.

**9.4** Suppose that you are considering a hypothesis test for a population mean, $\mu$. In each part, express the alternative hypothesis symbolically and identify the hypothesis test as two tailed, left tailed, or right tailed.
**a.** You want to decide whether the population mean is different from a specified value $\mu_0$.
**b.** You want to decide whether the population mean is less than a specified value $\mu_0$.
**c.** You want to decide whether the population mean is greater than a specified value $\mu_0$.

*In Exercises 9.5–9.13, hypothesis tests are proposed. For each hypothesis test,*
*a. determine the null hypothesis.*
*b. determine the alternative hypothesis.*
*c. classify the hypothesis test as two tailed, left tailed, or right tailed.*

**9.5 Toxic Mushrooms?** Cadmium, a heavy metal, is toxic to animals. Mushrooms, however, are able to absorb and accumulate cadmium at high concentrations. The Czech and Slovak governments have set a safety limit for cadmium in dry vegetables at 0.5 part per million (ppm). M. Melgar et al. measured the cadmium levels in a random sample of the edible mushroom *Boletus pinicola* and published the results in the *Journal of Environmental Science and Health* (Vol. B33(4), pp. 439–455). A hypothesis test is to be performed to decide whether the mean cadmium level in *Boletus pinicola* mushrooms is greater than the government's recommended limit.

**9.6 Agriculture Books.** The R. R. Bowker Company of New York collects information on the retail prices of books and publishes the data in *The Bowker Annual Library and Book Trade Almanac*. In 2000, the mean retail price of agriculture books was $66.52. A hypothesis test is to be performed to decide whether this year's mean retail price of agriculture books has changed from the 2000 mean.

**9.7 Iron Deficiency?** Iron is essential to most life forms and to normal human physiology. It is an integral part of many proteins and enzymes that maintain good health. Recommendations for iron are provided in *Dietary Reference Intakes*, developed by the Institute of Medicine of the National Academy of Sciences. The recommended dietary allowance (RDA) of iron for adult females under the age of 51 is 18 milligrams (mg) per day. A hypothesis test is to be performed to decide whether adult females under the age of 51 are, on average, getting less than the RDA of 18 mg of iron.

**9.8 Early-Onset Dementia.** Dementia is the loss of the intellectual and social abilities severe enough to interfere with judgment, behavior, and daily functioning. Alzheimer's disease is the most common type of dementia. In the article "Living with Early Onset Dementia: Exploring the Experience and Developing Evidence-Based Guidelines for Practice" (*Alzheimer's Care Quarterly*, Vol. 5, Issue 2, pp. 111–122), P. Harris and J. Keady explored the experience and struggles of people diagnosed with dementia and their families. A hypothesis test is to be performed to decide whether the mean age at diagnosis of all people with early-onset dementia is less than 55 years old.

**9.9 Serving Time.** According to the Bureau of Crime Statistics and Research of Australia, as reported on *Lawlink*, the mean length of imprisonment for motor-vehicle theft offenders in Australia is 16.7 months. You want to perform a hypothesis test to decide whether the mean length of imprisonment for motor-vehicle theft offenders in Sydney differs from the national mean in Australia.

**9.10 Worker Fatigue.** A study by M. Chen et al. titled "Heat Stress Evaluation and Worker Fatigue in a Steel Plant" (*American Industrial Hygiene Association*, Vol. 64, pp. 352–359) assessed fatigue in steel-plant workers due to heat stress. Among other things, the researchers monitored the heart rates of a random sample of 29 casting workers. A hypothesis test is to be conducted to decide whether the mean post-work heart rate of casting workers exceeds the normal resting heart rate of 72 beats per minute (bpm).

**9.11 Body Temperature.** A study by researchers at the University of Maryland addressed the question of whether the mean body temperature of humans is 98.6°F. The results of the study by P. Mackowiak et al. appeared in the article "A Critical Appraisal of 98.6°F, the Upper Limit of the Normal Body Temperature, and Other Legacies of Carl Reinhold August Wunderlich" (*Journal of the American Medical Association*, Vol. 268, pp. 1578–1580). Among other data, the researchers obtained the body temperatures of 93 healthy humans. Suppose that you want to use those data to decide whether the mean body temperature of healthy humans differs from 98.6°F.

**9.12 Teacher Salaries.** The Educational Resource Service publishes information about wages and salaries in the public schools system in *National Survey of Salaries and Wages in Public Schools*. The mean annual salary of (public) classroom teachers is $45.9 thousand. A hypothesis test is to be performed to decide whether the mean annual salary of classroom teachers in Hawaii is less than the national mean.

**9.13 Cell Phones.** The number of cell phone users has increased dramatically since 1987. According to the *Semiannual Wireless Survey*, published by the Cellular Telecommunications & Internet Association, the mean local monthly bill for cell phone users in the United States was $47.37 in 2001. A hypothesis test is to be performed to determine whether last year's mean local monthly bill for cell phone users has increased from the 2001 mean of $47.37.

## Extending the Concepts and Skills

**9.14 Energy Use.** The U.S. Energy Information Administration compiles data on energy consumption and publishes its findings in *Residential Energy Consumption Survey: Household Energy Consumption and Expenditures*. The mean annual energy consumed per U.S. household is 92.2 million British thermal units (BTU). Twenty randomly selected households in the West have the following annual energy consumptions, in millions of BTU.

| | | | | |
|---|---|---|---|---|
| 104 | 84 | 72 | 95 | 69 |
| 80 | 78 | 74 | 76 | 81 |
| 82 | 61 | 94 | 65 | 100 |
| 70 | 65 | 83 | 76 | 84 |

Do the data provide sufficient evidence to conclude that the mean annual energy consumed by western households differs from that of all U.S. households? Assume that annual energy consumptions of western households are normally distributed and that the population standard deviation is 15 million BTU. Use the following steps to answer the question. You may want to refer to Example 9.4, which begins on page 383.

**a.** State the null and alternative hypotheses.
**b.** Discuss the logic of conducting the hypothesis test.
**c.** Identify the distribution of the variable $\bar{x}$, that is, the sampling distribution of the mean for samples of size 20.
**d.** Obtain a precise criterion for deciding whether to reject the null hypothesis in favor of the alternative hypothesis.
**e.** Apply the criterion in part (d) to the sample data and state your conclusion.

**9.15 Radios.** The Radio Advertising Bureau of New York reports in *Radio Facts* that, in 2000, the mean number of radios per U.S. household was 5.6. A random sample of 45 U.S. households taken this year yields the following data on number of radios owned.

| 4 | 10 | 4 | 7 | 4 | 4 | 5 | 10 | 6 |
|---|----|---|---|---|---|---|----|---|
| 8 | 6 | 9 | 7 | 5 | 4 | 5 | 6 | 9 |
| 7 | 5 | 3 | 4 | 9 | 5 | 4 | 4 | 7 |
| 8 | 4 | 9 | 8 | 5 | 9 | 1 | 3 | 2 |
| 8 | 6 | 4 | 4 | 4 | 10 | 7 | 9 | 3 |

Do the data provide sufficient evidence to conclude that this year's mean number of radios per U.S. household has changed from the 2000 mean of 5.6? Assume that the population standard deviation of this year's number of radios per U.S. household is 1.9. Use the following steps to answer the question. You may want to refer to Example 9.4, which begins on page 383.
**a.** State the null and alternative hypotheses.
**b.** Discuss the logic of conducting the hypothesis test.
**c.** Identify the distribution of the variable $\bar{x}$, that is, the sampling distribution of the mean for samples of size 45.
**d.** Obtain a precise criterion for deciding whether to reject the null hypothesis in favor of the alternative hypothesis.

**e.** Apply the criterion in part (d) to the sample data and state your conclusion.

**9.16 Quality Assurance.** Refer to Example 9.4, which begins on page 383. Suppose that, in the solution to part (d), we use the "68.26" part of the 68.26-95.44-99.74 rule.
**a.** Determine the resulting decision criterion and portray it graphically, using a graph similar to the one shown in Fig. 9.2(a) on page 384.
**b.** Construct a graph similar to the one shown in Fig. 9.2(b) that illustrates the implications of the decision criterion in part (a) if in fact the null hypothesis is true.
**c.** Apply the criterion from part (a) to the sample data in Table 9.1 on page 383 and state your conclusion.

**9.17 Quality Assurance.** Refer to Example 9.4, which begins on page 383. Suppose that, in the solution to part (d), we use the "99.74" part of the 68.26-95.44-99.74 rule.
**a.** Determine the resulting decision criterion and portray it graphically, using a graph similar to the one shown in Fig. 9.2(a) on page 384.
**b.** Construct a graph similar to the one shown in Fig. 9.2(b) that illustrates the implications of the decision criterion in part (a) if in fact the null hypothesis is true.
**c.** Apply the criterion from part (a) to the sample data in Table 9.1 on page 383 and state your conclusion.

**9.18 Quality Assurance.** Refer to Example 9.4, which begins on page 383. In that example, we rejected the null hypothesis that the mean net weight of all bags packaged is 454 g in favor of the alternative hypothesis that the mean net weight of all bags packaged differs from 454 g. If in fact the null hypothesis is true, what is the chance of incorrectly rejecting it using a sample of 25 bags of pretzels? (*Hint:* Refer to Fig. 9.2(b).)

# 9.2 Terms, Errors, and Hypotheses

To explain fully the nature of hypothesis testing, we need some additional terms and concepts. In this section, we define several more terms used in hypothesis testing, discuss the two types of errors that can occur in a hypothesis test, and interpret the possible conclusions for a hypothesis test.

### Some Additional Terminology

Recall from Example 9.4 on page 383 that the null and alternative hypotheses for the pretzel-packaging hypothesis test are

$$H_0: \mu = 454 \text{ g (the packaging machine is working properly)}$$
$$H_a: \mu \neq 454 \text{ g (the packaging machine is not working properly)},$$

where $\mu$ is the mean net weight of all bags of pretzels packaged.

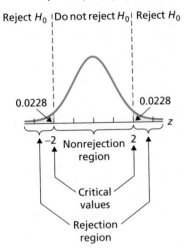

**FIGURE 9.5**

Rejection region, nonrejection region,
and critical values for the
pretzel-packaging illustration

As a basis for deciding whether to reject the null hypothesis, we used the variable

$$z = \frac{\bar{x} - \mu_0}{\sigma/\sqrt{n}} = \frac{\bar{x} - 454}{1.56}$$

which tells us how many standard deviations the sample mean is from the null hypothesis population mean of 454 g. That variable is called the **test statistic.** Figure 9.4 repeats the graph (from the previous section) that shows how to use the test statistic to decide whether to reject the null hypothesis.

The set of values for the test statistic that leads us to reject the null hypothesis is called the **rejection region.** In this case, the rejection region consists of all $z$-scores that lie either to the left of $-2$ or to the right of 2—that part of the horizontal axis under the shaded areas in Fig. 9.4.

The set of values for the test statistic that leads us not to reject the null hypothesis is called the **nonrejection region,** or **acceptance region.** Here, the nonrejection region consists of all $z$-scores that lie between $-2$ and 2—that part of the horizontal axis under the unshaded area in Fig. 9.4.

The values of the test statistic that separate the rejection and nonrejection regions are called the **critical values.** In this case, the critical values are $z = \pm 2$, as shown in Fig. 9.4. We summarize the preceding discussion in Fig. 9.5.

With the above discussion in mind, we present, in Definition 9.2, formal definitions of some important terms used in hypothesis testing. Before doing so, however, we note the following:

- The rejection region pictured in Fig. 9.5 is typical of that for a two-tailed hypothesis test. Shortly, we will discuss the form of the rejection regions for a left-tailed test and a right-tailed test.
- The terminology introduced so far in this section (and most of that which will be presented later) applies to any hypothesis test, not just to hypothesis tests for a population mean.

---

**Definition 9.2**

### Test Statistic, Rejection Region, Nonrejection Region, and Critical Values

**Test statistic:** The statistic used as a basis for deciding whether the null hypothesis should be rejected.

**Rejection region:** The set of values for the test statistic that leads to rejection of the null hypothesis.

**Nonrejection region:** The set of values for the test statistic that leads to nonrejection of the null hypothesis.[†]

**Critical values:** The values of the test statistic that separate the rejection and nonrejection regions. A critical value is considered part of the rejection region.

**What Does It Mean?**

If the value of the test statistic falls in the rejection region, reject the null hypothesis; otherwise, do not reject the null hypothesis.

For a two-tailed test, as in the pretzel-packaging illustration, the null hypothesis is rejected when the test statistic is either too small or too large. Thus the rejection region for such a test consists of two parts: one on the left and one on the right, as shown in Fig. 9.5 and Fig. 9.6(a).

---

[†]The reason that we prefer the term *nonrejection* to the term *acceptance* is explained in detail on page 392.

**FIGURE 9.6**

Graphical display of rejection regions for two-tailed, left-tailed, and right-tailed tests

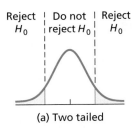

Reject $H_0$ | Do not reject $H_0$ | Reject $H_0$

(a) Two tailed

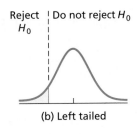

Reject $H_0$ | Do not reject $H_0$

(b) Left tailed

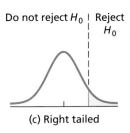

Do not reject $H_0$ | Reject $H_0$

(c) Right tailed

For a left-tailed test, as in Example 9.3 on page 382 (the calcium-intake illustration), the null hypothesis is rejected only when the test statistic is too small. Thus the rejection region for such a test consists of only one part, which is on the left, as shown in Fig. 9.6(b).

For a right-tailed test, as in Example 9.2 on page 382 (the history-book illustration), the null hypothesis is rejected only when the test statistic is too large. Thus the rejection region for such a test consists of only one part, which is on the right, as shown in Fig. 9.6(c).

Table 9.2 and Fig. 9.6 summarize our discussion. Figure 9.6 shows why the term *tailed* is used: The rejection region is in both tails for a two-tailed test, in the left tail for a left-tailed test, and in the right tail for a right-tailed test.

**TABLE 9.2**

Rejection regions for two-tailed, left-tailed, and right-tailed tests

|  | **Two-tailed test** | **Left-tailed test** | **Right-tailed test** |
|---|---|---|---|
| **Sign in $H_a$** | $\neq$ | $<$ | $>$ |
| **Rejection region** | Both sides | Left side | Right side |

## Type I and Type II Errors

Any decision we make based on a hypothesis test may be incorrect because we have used partial information, obtained from a sample, to draw conclusions about the entire population. There are two types of incorrect decisions—*Type I error* and *Type II error*, as indicated in Table 9.3 and Definition 9.3.

**TABLE 9.3**

Correct and incorrect decisions for a hypothesis test

|  |  | $H_0$ is: | |
|---|---|---|---|
|  |  | True | False |
| **Decision:** | Do not reject $H_0$ | Correct decision | Type II error |
|  | Reject $H_0$ | Type I error | Correct decision |

**Definition 9.3**

## Type I and Type II Errors

**Type I error:** Rejecting the null hypothesis when it is in fact true.

**Type II error:** Not rejecting the null hypothesis when it is in fact false.

| Example 9.5 | **Type I and Type II Errors** |
|---|---|

*Quality Assurance* Consider once again the pretzel-packaging hypothesis test. The null and alternative hypotheses are

$$H_0: \mu = 454 \text{ g (the packaging machine is working properly)}$$
$$H_a: \mu \neq 454 \text{ g (the packaging machine is not working properly),}$$

where $\mu$ is the mean net weight of all bags of pretzels packaged. Explain what each of the following would mean.

a. Type I error      b. Type II error      c. Correct decision

Recall from Example 9.4 that the results of sampling 25 bags of pretzels led to rejection of the null hypothesis, $\mu = 454$ g, that is, to the conclusion that $\mu \neq 454$ g. Classify that conclusion by error type or as a correct decision if

d. the mean net weight, $\mu$, is in fact 454 g.

e. the mean net weight, $\mu$, is in fact not 454 g.

**Solution**

a. A Type I error occurs when a true null hypothesis is rejected. In this case, a Type I error would occur if in fact $\mu = 454$ g but the results of the sampling lead to the conclusion that $\mu \neq 454$ g.

   **Interpretation** A Type I error occurs if we conclude that the packaging machine is not working properly when in fact it is working properly.

b. A Type II error occurs when a false null hypothesis is not rejected. In this case, a Type II error would occur if in fact $\mu \neq 454$ g but the results of the sampling fail to lead to that conclusion.

   **Interpretation** A Type II error occurs if we fail to conclude that the packaging machine is not working properly when in fact it is not working properly.

c. A correct decision can occur in either of two ways.

   - A true null hypothesis is not rejected. That would happen if in fact $\mu = 454$ g and the results of the sampling do not lead to the rejection of that fact.

   - A false null hypothesis is rejected. That would happen if in fact $\mu \neq 454$ g and the results of the sampling lead to that conclusion.

   **Interpretation** A correct decision occurs if either we fail to conclude that the packaging machine is not working properly when in fact it is working properly, or we conclude that the packaging machine is not working properly when in fact it is not working properly.

   *You try it!*

   Exercise 9.29 on page 393

d. If in fact $\mu = 454$ g, the null hypothesis is true. Consequently, by rejecting the null hypothesis, $\mu = 454$ g, a Type I error has been made—a true null hypothesis has been rejected.

e. If in fact $\mu \neq 454$ g, the null hypothesis is false. Consequently, by rejecting the null hypothesis, $\mu = 454$ g, a correct decision has been made—a false null hypothesis has been rejected.

• • •

## Probabilities of Type I and Type II Errors

Part of evaluating the effectiveness of a hypothesis test involves an analysis of the chances of making an incorrect decision. A Type I error occurs if the test statistic falls in the rejection region when in fact the null hypothesis is true. The probability of that happening, the **Type I error probability,** commonly called the **significance level** of the hypothesis test, is denoted $\alpha$ (the lowercase Greek letter alpha).

Figure 9.4 on page 388 shows the rejection and nonrejection regions for the pretzel-packaging hypothesis test in Example 9.4. It also shows the normal curve for the test statistic

$$z = \frac{\bar{x} - 454}{1.56}$$

under the assumption that the null hypothesis, $\mu = 454$ g, is true. Figure 9.4 reveals that, if the null hypothesis is true, the probability is $0.0228 + 0.0228$, or 0.0456, that the test statistic, $z$, will fall in the rejection region. Thus, for this hypothesis test, the significance level is 0.0456; in symbols, $\alpha = 0.0456$.

**Interpretation**   There is only a 4.56% chance of concluding that the packaging machine is not working properly when in fact it is working properly.

---

**Definition 9.4**

### Significance Level

The probability of making a Type I error, that is, of rejecting a true null hypothesis, is called the **significance level, $\alpha$,** of a hypothesis test.

---

A Type II error occurs if the test statistic falls in the nonrejection region when in fact the null hypothesis is false. The probability of that happening, the **Type II error probability,** is denoted $\beta$ (the lowercase Greek letter beta). It depends on the true value of $\mu$.

A concept closely related to that of Type II error probability is *power*. The **power** of a hypothesis test is the probability of not making a Type II error, that is, the probability of rejecting a false null hypothesis. From the complementation rule (Formula 5.2 on page 223),

$$\text{Power} = 1 - P(\text{Type II error}) = 1 - \beta.$$

Calculation of Type II error probabilities and power is examined briefly in Exercise 9.42. For a detailed discussion of the calculation of Type II error probabilities and power, refer to the text *Introductory Statistics, 8/e,* by Neil A. Weiss (Boston: Addison-Wesley, 2008).

Ideally, both Type I and Type II errors should have small probabilities. Then the chance of making an incorrect decision would be small, regardless of whether the null hypothesis is true or false. As we demonstrate in Section 9.3, we can design a hypothesis test to have any specified significance level. So, for instance, if not rejecting a true null hypothesis is important, we should specify a small value for $\alpha$. However, in making our choice for $\alpha$, we must keep Key Fact 9.1 in mind.

> **What Does It Mean?**
>
> The power of a hypothesis test is between 0 and 1 and measures the ability of the hypothesis test to detect a false null hypothesis. If the power is near 0, the hypothesis test is not very good at detecting a false null hypothesis; if the power is near 1, the hypothesis test is extremely good at detecting a false null hypothesis.

---

**Key Fact 9.1**

### Relation Between Type I and Type II Error Probabilities

For a fixed sample size, the smaller we specify the significance level, $\alpha$, the larger will be the probability, $\beta$, of not rejecting a false null hypothesis.

Consequently, we must always assess the risks involved in committing both types of errors and use that assessment as a method for balancing the Type I and Type II error probabilities.

## Possible Conclusions for a Hypothesis Test

The significance level, $\alpha$, is the probability of making a Type I error, that is, of rejecting a true null hypothesis. Therefore, if the hypothesis test is conducted at a small significance level (e.g., $\alpha = 0.05$), the chance of rejecting a true null hypothesis will be small. In this text, we generally specify a small significance level. Thus, if we do reject the null hypothesis, we can be reasonably confident that the null hypothesis is false and therefore that the alternative hypothesis is true.

However, we usually do not know the probability, $\beta$, of making a Type II error, that is, of not rejecting a false null hypothesis. Consequently, if we do not reject the null hypothesis, we simply reserve judgment about which hypothesis is true. In other words, if we do not reject the null hypothesis, we conclude only that the data did not provide sufficient evidence to support the alternative hypothesis; we do not conclude that the data provided sufficient evidence to support the null hypothesis.

| | |
|---|---|
| **Key Fact 9.2** | **Possible Conclusions for a Hypothesis Test**<br><br>Suppose that a hypothesis test is conducted at a small significance level.<br><br>• If the null hypothesis is rejected, we conclude that the alternative hypothesis is true.<br>• If the null hypothesis is not rejected, we conclude that the data do not provide sufficient evidence to support the alternative hypothesis. |

When the null hypothesis is rejected in a hypothesis test performed at the significance level $\alpha$, we frequently express that fact with the phrase "the test results are **statistically significant** at the $\alpha$ level." Similarly, when the null hypothesis is not rejected in a hypothesis test performed at the significance level $\alpha$, we often express that fact with the phrase "the test results are **not statistically significant** at the $\alpha$ level."

## Exercises 9.2

### Understanding the Concepts and Skills

**9.19** Decide whether each statement is true or false. Explain your answers.
a. If it is important not to reject a true null hypothesis, the hypothesis test should be performed at a small significance level.
b. For a fixed sample size, decreasing the significance level of a hypothesis test results in an increase in the probability of making a Type II error.

**9.20** Explain in your own words the meaning of each of the following terms.
a. test statistic
b. rejection region
c. nonrejection region
d. critical values
e. significance level

**9.21** Identify the two types of incorrect decisions in a hypothesis test. For each incorrect decision, what symbol is used to represent the probability of making that type of error?

**9.22** Suppose that a hypothesis test is performed at a small significance level. State the appropriate conclusion in each case by referring to Key Fact 9.2.
a. The null hypothesis is rejected.
b. The null hypothesis is not rejected.

*Exercises 9.23–9.28 contain graphs portraying the decision criterion for a hypothesis test for a population mean, μ. The null hypothesis for each test is $H_0: \mu = \mu_0$; the test statistic is*

$$z = \frac{\bar{x} - \mu_0}{\sigma/\sqrt{n}}.$$

*The curve in each graph is the normal curve for the test statistic under the assumption that the null hypothesis is true. For each exercise, determine the*

a. *rejection region.*          b. *nonrejection region.*
c. *critical value(s).*          d. *significance level.*
e. *Construct a graph similar to that in Fig. 9.5 on page 388 that depicts your results from parts (a)–(d).*
f. *Identify the hypothesis test as two tailed, left tailed, or right tailed.*

**9.23**

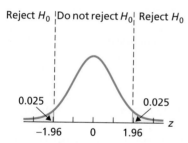

**9.24**

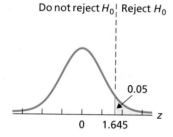

**9.25**

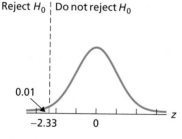

**9.26**

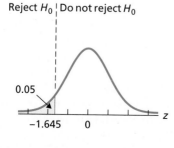

**9.27**

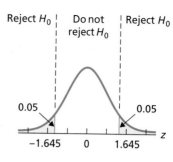

**9.28**

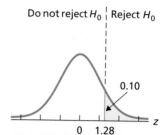

**9.29 Toxic Mushrooms?** Cadmium, a heavy metal, is toxic to animals. Mushrooms, however, are able to absorb and accumulate cadmium at high concentrations. The Czech and Slovak governments have set a safety limit for cadmium in dry vegetables at 0.5 part per million (ppm). M. Melgar et al. measured the cadmium levels in a random sample of the edible mushroom *Boletus pinicola* and published the results in the *Journal of Environmental Science and Health* (Vol. B33(4), pp. 439–455). A hypothesis test is to be performed to decide whether the mean cadmium level in *Boletus pinicola* mushrooms is greater than the government's recommended limit. The null and alternative hypotheses are

$$H_0: \mu = 0.5 \text{ ppm}$$
$$H_a: \mu > 0.5 \text{ ppm},$$

where μ is the mean cadmium level in *Boletus pinicola* mushrooms. Explain what each of the following would mean.
**a.** Type I error      **b.** Type II error      **c.** Correct decision

Now suppose that the results of carrying out the hypothesis test lead to nonrejection of the null hypothesis. Classify that conclusion by error type or as a correct decision if in fact the mean cadmium level in *Boletus pinicola* mushrooms
**d.** equals the safety limit of 0.5 ppm.
**e.** exceeds the safety limit of 0.5 ppm.

**9.30 Agriculture Books.** The R. R. Bowker Company of New York collects information on the retail prices of books and publishes the data in *The Bowker Annual Library and Book Trade Almanac*. In 2000, the mean retail price of agriculture books was $66.52. A hypothesis test is to be performed to decide whether this year's mean retail price of agriculture books has changed from the 2000 mean. The null and alternative hypotheses are

$$H_0: \mu = \$66.52$$
$$H_a: \mu \neq \$66.52,$$

where $\mu$ is this year's mean retail price of agriculture books. Explain what each of the following would mean.
**a.** Type I error    **b.** Type II error    **c.** Correct decision

Now suppose that the results of carrying out the hypothesis test lead to rejection of the null hypothesis. Classify that conclusion by error type or as a correct decision if in fact this year's mean retail price of agriculture books
**d.** equals the 2000 mean of $66.52.
**e.** differs from the 2000 mean of $66.52.

**9.31 Iron Deficiency?** Iron is essential to most life forms and to normal human physiology. It is an integral part of many proteins and enzymes that maintain good health. Recommendations for iron are provided in *Dietary Reference Intakes*, developed by the Institute of Medicine of the National Academy of Sciences. The recommended dietary allowance (RDA) of iron for adult females under the age of 51 is 18 milligrams (mg) per day. A hypothesis test is to be performed to decide whether adult females under the age of 51 are, on average, getting less than the RDA of 18 mg of iron. The null and alternative hypotheses are

$$H_0: \mu = 18 \text{ mg}$$
$$H_a: \mu < 18 \text{ mg},$$

where $\mu$ is the mean iron intake (per day) of all adult females under the age of 51. Explain what each of the following would mean.
**a.** Type I error    **b.** Type II error    **c.** Correct decision

Now suppose that the results of carrying out the hypothesis test lead to rejection of the null hypothesis. Classify that conclusion by error type or as a correct decision if in fact the mean iron intake of all adult females under the age of 51
**d.** equals the RDA of 18 mg per day.
**e.** is less than the RDA of 18 mg per day.

**9.32 Early-Onset Dementia.** Dementia is the loss of the intellectual and social abilities severe enough to interfere with judgment, behavior, and daily functioning. Alzheimer's disease is the most common type of dementia. In the article "Living with Early Onset Dementia: Exploring the Experience and Developing Evidence-Based Guidelines for Practice" (*Alzheimer's Care Quarterly*, Vol. 5, Issue 2, pp. 111–122), P. Harris and J. Keady explored the experience and struggles of people diagnosed with dementia and their families. A hypothesis test is to be performed to decide whether the mean age at diagnosis of all people with early-onset dementia is less than 55 years old. The null and alternative hypotheses are

$$H_0: \mu = 55 \text{ years old}$$
$$H_a: \mu < 55 \text{ years old},$$

where $\mu$ is the mean age at diagnosis of all people with early-onset dementia. Explain what each of the following would mean.
**a.** Type I error    **b.** Type II error    **c.** Correct decision

Now suppose that the results of carrying out the hypothesis test lead to nonrejection of the null hypothesis. Classify that conclusion by error type or as a correct decision if in fact the mean age at diagnosis of all people with early-onset dementia
**d.** is 55 years old.    **e.** is less than 55 years old.

**9.33 Serving Time.** According to the Bureau of Crime Statistics and Research of Australia, as reported on *Lawlink*, the mean length of imprisonment for motor-vehicle theft offenders in Australia is 16.7 months. You want to perform a hypothesis test to decide whether the mean length of imprisonment for motor-vehicle theft offenders in Sydney differs from the national mean in Australia. The null and alternative hypotheses are

$$H_0: \mu = 16.7 \text{ months}$$
$$H_a: \mu \neq 16.7 \text{ months},$$

where $\mu$ is the mean length of imprisonment for motor-vehicle theft offenders in Sydney, Australia. Explain what each of the following would mean.
**a.** Type I error    **b.** Type II error    **c.** Correct decision

Now suppose that the results of carrying out the hypothesis test lead to nonrejection of the null hypothesis. Classify that conclusion by error type or as a correct decision if in fact the mean length of imprisonment for motor-vehicle theft offenders in Sydney
**d.** equals the national mean of 16.7 months.
**e.** differs from the national mean of 16.7 months.

**9.34 Worker Fatigue.** A study by M. Chen et al. titled "Heat Stress Evaluation and Worker Fatigue in a Steel Plant" (*American Industrial Hygiene Association*, Vol. 64, pp. 352–359) assessed fatigue in steel-plant workers due to heat stress. Among other things, the researchers monitored the heart rates of a random sample of 29 casting workers. A hypothesis test is to be conducted to decide whether the mean post-work heart rate of casting workers exceeds the normal resting heart rate of 72 beats per minute (bpm). The null and alternative hypotheses are

$$H_0: \mu = 72 \text{ bpm}$$
$$H_a: \mu > 72 \text{ bpm},$$

where $\mu$ is the mean post-work heart rate of casting workers. Explain what each of the following would mean.
**a.** Type I error    **b.** Type II error    **c.** Correct decision

Now suppose that the results of the sampling lead to rejection of the null hypothesis. Classify that conclusion by error type or as a correct decision if in fact the mean post-work heart rate of casting workers
**d.** equals the normal resting heart rate of 72 bpm.
**e.** exceeds the normal resting heart rate of 72 bpm.

**9.35 Body Temperature.** A study by researchers at the University of Maryland addressed the question of whether the mean body temperature of humans is 98.6°F. The results of the study by P. Mackowiak et al. appeared in the article "A Critical Appraisal of 98.6°F, the Upper Limit of the Normal Body Temperature, and Other Legacies of Carl Reinhold August Wunderlich" (*Journal of the American Medical Association*, Vol. 268, pp. 1578–1580). Among other data, the researchers obtained the body temperatures of 93 healthy humans. Suppose that you want to use those data to decide whether the mean body temperature of healthy humans differs from 98.6°F. The null and alternative hypotheses are

$$H_0: \mu = 98.6°F$$
$$H_a: \mu \neq 98.6°F,$$

where $\mu$ is the mean body temperature of all healthy humans. Explain what each of the following would mean.
**a.** Type I error     **b.** Type II error     **c.** Correct decision

Now suppose that the sample of temperatures leads to rejection of the null hypothesis. Classify that conclusion by error type or as a correct decision if in fact the mean body temperature of all healthy humans
**d.** is 98.6°F.                    **e.** is not 98.6°F.

**9.36 Teacher Salaries.** The Educational Resource Service publishes information about wages and salaries in the public schools system in *National Survey of Salaries and Wages in Public Schools*. The mean annual salary of (public) classroom teachers is $45.9 thousand. A hypothesis test is to be performed to decide whether the mean annual salary of classroom teachers in Hawaii is less than the national mean. The null and alternative hypotheses are

$$H_0: \mu = 45.9 \text{ thousand}$$
$$H_a: \mu < 45.9 \text{ thousand},$$

where $\mu$ is the mean annual salary of classroom teachers in Hawaii. Explain what each of the following would mean.
**a.** Type I error     **b.** Type II error     **c.** Correct decision

Now suppose that the results of the sampling lead to nonrejection of the null hypothesis. Classify that conclusion by error type or as a correct decision if in fact the mean salary of classroom teachers in Hawaii
**d.** equals the national mean of $45.9 thousand.
**e.** is less than the national mean of $45.9 thousand.

**9.37 Cell Phones.** The number of cell phone users has increased dramatically since 1987. According to the *Semiannual Wireless Survey*, published by the Cellular Telecommunications & Internet Association, the mean local monthly bill for cell phone users in the United States was $47.37 in 2001. A hypothesis test is to be performed to determine whether last year's mean local monthly bill for cell phone users has increased from the 2001 mean of $47.37. The null and alternative hypotheses are

$$H_0: \mu = \$47.37$$
$$H_a: \mu > \$47.37,$$

where $\mu$ is last year's mean local monthly bill for cell phone users. Explain what each outcome would mean.
**a.** Type I error     **b.** Type II error     **c.** Correct decision

Now suppose that the results of performing the hypothesis test lead to nonrejection of the null hypothesis. Classify that conclusion by error type or as a correct decision if in fact last year's mean local monthly bill for cell phone users
**d.** equals the 2001 mean of $47.37.
**e.** exceeds the 2001 mean of $47.37.

**9.38** Suppose that you choose the significance level of a hypothesis test to be 0.
**a.** What is the probability of a Type I error?
**b.** What is the probability of a Type II error?

**9.39** Identify an exercise in this section for which it is important to have
**a.** a small $\alpha$ probability.     **b.** a small $\beta$ probability.
**c.** both $\alpha$ and $\beta$ probabilities small.

**9.40 Approving Nuclear Reactors.** Suppose that you are performing a statistical test to decide whether a nuclear reactor should be approved for use. Further suppose that failing to reject the null hypothesis corresponds to approval. What property would you want the Type II error probability, $\beta$, to have?

**9.41 Guilty or Innocent?** In the U.S. court system, a defendant is assumed innocent until proven guilty. Suppose that you regard a court trial as a hypothesis test with null and alternative hypotheses

$$H_0: \text{Defendant is innocent}$$
$$H_a: \text{Defendant is guilty.}$$

**a.** Explain the meaning of a Type I error.
**b.** Explain the meaning of a Type II error.
**c.** If you were the defendant, would you want $\alpha$ to be large or small? Explain your answer.
**d.** If you were the prosecuting attorney, would you want $\beta$ to be large or small? Explain your answer.
**e.** What are the consequences to the court system if you make $\alpha = 0$? $\beta = 0$?

## Extending the Concepts and Skills

**9.42 Type II Error Probabilities and Power.** For the pretzel-packaging hypothesis test (Example 9.4, page 383), the null and alternative hypotheses are

$H_0$: $\mu = 454$ g (machine is working properly)
$H_a$: $\mu \neq 454$ g (machine is not working properly),

where $\mu$ is the mean net weight of all bags of pretzels packaged. Recall that the net weights are normally distributed with a standard deviation of 7.8 g. Figure 9.2(a) on page 384 portrays the decision criterion for a hypothesis test at the 4.56% significance level ($\alpha = 0.0456$) using a sample size of 25.

**a.** Identify the probability of a Type I error.
**b.** Assuming that the mean net weight being packaged is in fact 447 g, identify the distribution of the variable $\bar{x}$, that is, the sampling distribution of the sample mean for samples of size 25.
**c.** Use part (b) to determine the probability, $\beta$, of a Type II error if in fact the mean net weight being packaged is 447 g. (*Hint:* Referring to Fig. 9.2(a), note that $\beta$ equals the percentage of all samples of 25 bags of pretzels whose mean weights are between 450.88 g and 457.12 g.)
**d.** Repeat parts (b) and (c) if in fact the mean net weight being packaged is 448 g, 449 g, 450 g, 451 g, 452 g, 453 g, 455 g, 456 g, 457 g, 458 g, 459 g, 460 g, and 461 g.

**e.** Use your answers from parts (b)–(d) to draw and interpret a graph of $\beta$ versus the true value of $\mu$. This type of graph is called an *operating characteristic (OC) curve*.
**f.** Use your answers from parts (b)–(d) to draw and interpret a graph of power versus the true value of $\mu$. This type of graph is called a *power curve*.

**9.43 Class Project: Quality Assurance.** This exercise can be done individually or, better yet, as a class project. For the pretzel-packaging hypothesis test in Example 9.4, the null and alternative hypotheses are

$H_0$: $\mu = 454$ g (machine is working properly)
$H_a$: $\mu \neq 454$ g (machine is not working properly),

where $\mu$ is the mean net weight of all bags of pretzels packaged. Recall that the net weights are normally distributed with a standard deviation of 7.8 g. Figure 9.4 on page 388 portrays the decision criterion for a test at the 4.56% significance level ($\alpha = 0.0456$). For a sample size of 25, the test statistic is

$$z = \frac{\bar{x} - \mu_0}{\sigma/\sqrt{n}} = \frac{\bar{x} - 454}{1.56}.$$

**a.** Assuming that $H_0$: $\mu = 454$ g is true, simulate 100 samples of 25 net weights each.
**b.** Determine the mean of each sample in part (a).
**c.** Use part (b) to determine the value of the test statistic, $z$, for each sample in part (a).
**d.** For the 100 samples obtained in part (a), roughly how many would you expect to lead to rejection of the null hypothesis? Explain your answer.
**e.** For the 100 samples obtained in part (a), determine the number that lead to rejection of the null hypothesis. (*Hint:* Refer to part (c) and Fig. 9.4.)
**f.** Compare your answers from parts (d) and (e) and comment on any observed difference.

## 9.3 Hypothesis Tests for One Population Mean When $\sigma$ Is Known

In this section, we describe how to conduct a hypothesis test for a population mean at any prescribed significance level. You have already learned most of what you need to know in order to do that. What remains is to discuss how to obtain the critical value(s) when the significance level is specified in advance.

Recall that the significance level of a hypothesis test is the probability of rejecting a true null hypothesis or, equivalently, the probability that the test statistic will fall in the rejection region when the null hypothesis is true. With this fact in mind, we state Key Fact 9.3, which applies to any hypothesis test.

**Key Fact 9.3** | **Obtaining Critical Values**

Suppose that a hypothesis test is to be performed at the significance level, $\alpha$. Then the critical value(s) must be chosen so that, if the null hypothesis is true, the probability is $\alpha$ that the test statistic will fall in the rejection region.

## Hypothesis Tests for a Population Mean When $\sigma$ Is Known

We now develop a simple step-by-step procedure for performing a hypothesis test for a population mean when the population standard deviation is known. In doing so, we assume that the variable under consideration is normally distributed. But keep in mind that, because of the central limit theorem, the procedure will work reasonably well when the sample size is large, regardless of the distribution of the variable.

As you have seen, the null hypothesis for a hypothesis test concerning one population mean, $\mu$, has the form $H_0$: $\mu = \mu_0$, where $\mu_0$ is some number. Recall, also, that the test statistic for the hypothesis test is

$$z = \frac{\bar{x} - \mu_0}{\sigma/\sqrt{n}}$$

which tells you how many standard deviations the observed sample mean, $\bar{x}$, is from $\mu_0$ (the value specified for the population mean in the null hypothesis).

The basis of the hypothesis testing procedure is in Key Fact 7.4: If $x$ is a normally distributed variable with mean $\mu$ and standard deviation $\sigma$, then, for samples of size $n$, the variable $\bar{x}$ is also normally distributed and has mean $\mu$ and standard deviation $\sigma/\sqrt{n}$. This fact and Key Fact 6.2 (page 271) applied to $\bar{x}$ imply that, if the null hypothesis is true, the test statistic $z$ has the standard normal distribution.

Consequently, in light of Key Fact 9.3, for a specified significance level, $\alpha$, we need to choose the critical value(s) so that the area under the standard normal curve that lies above the rejection region equals $\alpha$.

**Example 9.6** | **Obtaining the Critical Values**

Determine the critical value(s) for a hypothesis test at the 5% significance level ($\alpha = 0.05$) if the test is

**a.** two tailed.      **b.** left tailed.      **c.** right tailed.

**Solution** Because $\alpha = 0.05$, we need to choose the critical value(s) so that the area under the standard normal curve that lies above the rejection region equals 0.05.

**a.** For a two-tailed test, the rejection region is on both the left and right. So the critical values are the two $z$-scores that divide the area under the standard normal curve into a middle 0.95 area and two outside areas of 0.025. In other words, the critical values are $\pm z_{0.025}$. From Table II in Appendix A, $\pm z_{0.025} = \pm 1.96$, as shown in Fig. 9.7(a).

**FIGURE 9.7**

Critical value(s) for a hypothesis test at the 5% significance level if the test is (a) two tailed, (b) left tailed, or (c) right tailed

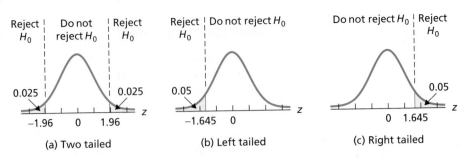

(a) Two tailed     (b) Left tailed     (c) Right tailed

**You try it!**

Exercise 9.49 on page 405

b.  For a left-tailed test, the rejection region is on the left. So the critical value is the $z$-score having area 0.05 to its left under the standard normal curve, which is $-z_{0.05}$. From Table II, $-z_{0.05} = -1.645$, as shown in Fig. 9.7(b).

c.  For a right-tailed test, the rejection region is on the right. So the critical value is the $z$-score having area 0.05 to its right under the standard normal curve, which is $z_{0.05}$. From Table II, $z_{0.05} = 1.645$, as shown in Fig. 9.7(c).

• • •

By reasoning as we did in the previous example, we can obtain the critical value(s) for any specified significance level, $\alpha$. As shown in Fig. 9.8, for a two-tailed test, the critical values are $\pm z_{\alpha/2}$; for a left-tailed test, the critical value is $-z_{\alpha}$; and for a right-tailed test, the critical value is $z_{\alpha}$.

**FIGURE 9.8**

Critical value(s) for a hypothesis test at the significance level $\alpha$ if the test is (a) two tailed, (b) left tailed, or (c) right tailed

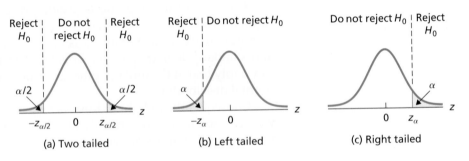

(a) Two tailed     (b) Left tailed     (c) Right tailed

The most commonly used significance levels are 0.10, 0.05, and 0.01. If we consider both one-tailed and two-tailed tests, these three significance levels give rise to five "tail areas." Using the standard-normal table, Table II, we obtained the value of $z_{\alpha}$ corresponding to each of those five tail areas, as shown in Table 9.4.

**TABLE 9.4**

Some important values of $z_{\alpha}$

| $z_{0.10}$ | $z_{0.05}$ | $z_{0.025}$ | $z_{0.01}$ | $z_{0.005}$ |
|------------|------------|-------------|------------|-------------|
| 1.28 | 1.645 | 1.96 | 2.33 | 2.575 |

Alternatively, we can find these five values of $z_{\alpha}$ at the bottom of the $t$-table, Table IV, where they are displayed to three decimal places. Can you explain the slight discrepancy between the values given for $z_{0.005}$ in the two tables?

### The One-Mean z-Test

Procedure 9.1 (next page) gives a step-by-step method for performing a hypothesis test for a population mean when the population standard deviation is known. The procedure summarizes what we have done in this and the previous two sections. We call Procedure 9.1 the **one-mean z-test** or, when no confusion can arise, simply the **z-test**.[†]

---

[†]The one-mean z-test is also known as the **one-sample z-test** and the **one-variable z-test**. We prefer "one-mean" because it makes clear the parameter being tested.

**Procedure 9.1**       **One-Mean z-Test (Critical-Value Approach)**

*Purpose* To perform a hypothesis test for a population mean, $\mu$

*Assumptions*

1. Simple random sample
2. Normal population or large sample
3. $\sigma$ known

**STEP 1** The null hypothesis is $H_0$: $\mu = \mu_0$, and the alternative hypothesis is

$$H_a: \mu \neq \mu_0 \qquad H_a: \mu < \mu_0 \qquad H_a: \mu > \mu_0$$
$$\text{(Two tailed)} \quad \text{or} \quad \text{(Left tailed)} \quad \text{or} \quad \text{(Right tailed)}$$

**STEP 2** Decide on the significance level, $\alpha$.

**STEP 3** Compute the value of the test statistic

$$z = \frac{\bar{x} - \mu_0}{\sigma/\sqrt{n}}.$$

**STEP 4** The critical value(s) are

$$\pm z_{\alpha/2} \qquad -z_\alpha \qquad z_\alpha$$
$$\text{(Two tailed)} \quad \text{or} \quad \text{(Left tailed)} \quad \text{or} \quad \text{(Right tailed)}$$

Use Table II to find the critical value(s).

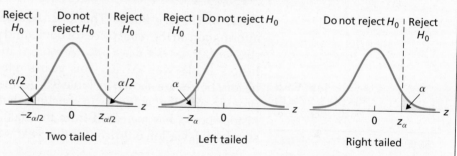

**STEP 5** If the value of the test statistic falls in the rejection region, reject $H_0$; otherwise, do not reject $H_0$.

**STEP 6** Interpret the results of the hypothesis test.

The hypothesis test is exact for normal populations and is approximately correct for large samples from nonnormal populations.

Properties and guidelines for use of the z-test are similar to those for the z-interval procedure. In particular, the z-test is robust to moderate violations of the normality assumption but, even for large samples, can sometimes be unduly affected by outliers because the sample mean is not resistant to outliers. Key Fact 9.4 lists some general guidelines for use of the z-test.

**Note:** By saying that the hypothesis test is *exact,* we mean that the true significance level equals $\alpha$; by saying that it is *approximately correct,* we mean that the true significance level only approximately equals $\alpha$.

**Key Fact 9.4**

## When to Use the One-Mean z-Test[†]

- For small samples—say, of size less than 15—the z-test should be used only when the variable under consideration is normally distributed or very close to being so.
- For samples of moderate size—say, between 15 and 30—the z-test can be used unless the data contain outliers or the variable under consideration is far from being normally distributed.
- For large samples—say, of size 30 or more—the z-test can be used essentially without restriction. However, if outliers are present and their removal is not justified, you should perform the hypothesis test once with the outliers and once without them to see what effect the outliers have. If the conclusion is affected, use a different procedure or take another sample.
- If outliers are present but their removal is justified and results in a data set for which the z-test is appropriate (as previously stated), the procedure can be used.

### Applying the One-Mean z-Test

Examples 9.7–9.9 illustrate use of the z-test, Procedure 9.1. We reexamine these examples in Section 9.4 when we discuss *P*-values.

**Example 9.7**

## The One-Mean z-Test

*Prices of History Books* The R. R. Bowker Company of New York collects information on the retail prices of books and publishes its findings in *The Bowker Annual Library and Book Trade Almanac*. In 2000, the mean retail price of all history books was $51.46. This year's retail prices for 40 randomly selected history books are shown in Table 9.5. At the 1% significance level, do the data provide sufficient evidence to conclude that this year's mean retail price of all history books has increased from the 2000 mean of $51.46? Assume that the standard deviation of prices for this year's history books is $7.61.

**TABLE 9.5**
This year's prices ($) for 40 history books

| | | | |
|---|---|---|---|
| 56.00 | 46.25 | 47.34 | 53.99 |
| 53.71 | 47.88 | 54.82 | 55.73 |
| 51.00 | 61.70 | 47.03 | 62.68 |
| 47.80 | 50.89 | 52.36 | 50.95 |
| 51.28 | 50.94 | 60.70 | 72.38 |
| 47.70 | 56.16 | 52.33 | 51.70 |
| 53.80 | 50.90 | 63.74 | 52.87 |
| 41.08 | 64.93 | 57.44 | 54.09 |
| 75.37 | 56.48 | 69.04 | 42.71 |
| 53.76 | 72.17 | 61.26 | 42.65 |

**Solution** We constructed (but did not show) a normal probability plot, a histogram, a stem-and-leaf diagram, and a boxplot for these price data. The boxplot indicated potential outliers, but in view of the other three graphs, we concluded that the data contain no outliers. Because the sample size is 40, which is large, and the population standard deviation is known, we can use Procedure 9.1 to conduct the required hypothesis test.

**STEP 1 State the null and alternative hypotheses.**

Let $\mu$ denote this year's mean retail price of all history books. We obtained the null and alternative hypotheses in Example 9.2 as

$$H_0: \mu = \$51.46 \text{ (mean price has not increased)}$$
$$H_a: \mu > \$51.46 \text{ (mean price has increased)}.$$

---

[†]We can refine these guidelines further by considering the impact of skewness. Roughly speaking, the more skewed the distribution of the variable under consideration, the larger is the sample size required to use the z-test.

Note that the hypothesis test is right tailed because a greater-than sign (>) appears in the alternative hypothesis.

**STEP 2　Decide on the significance level, $\alpha$.**

We are to perform the test at the 1% significance level, or $\alpha = 0.01$.

**STEP 3　Compute the value of the test statistic**

$$z = \frac{\bar{x} - \mu_0}{\sigma/\sqrt{n}}.$$

We have $\mu_0 = 51.46$, $\sigma = 7.61$, and $n = 40$. The mean of the sample data in Table 9.5 is $\bar{x} = 54.890$. Thus the value of the test statistic is

$$z = \frac{54.890 - 51.46}{7.61/\sqrt{40}} = 2.85.$$

This value of $z$ is marked with a dot in Fig. 9.9.

**STEP 4　The critical value for a right-tailed test is $z_\alpha$. Use Table II to find the critical value.**

Because $\alpha = 0.01$, the critical value is $z_{0.01}$. From Table II (or Table 9.4 on page 398), $z_{0.01} = 2.33$, as shown in Fig. 9.9.

**STEP 5　If the value of the test statistic falls in the rejection region, reject $H_0$; otherwise, do not reject $H_0$.**

The value of the test statistic, found in Step 3, is $z = 2.85$. Figure 9.9 reveals that this value falls in the rejection region, so we reject $H_0$. The test results are statistically significant at the 1% level.

**STEP 6　Interpret the results of the hypothesis test.**

**Interpretation**　At the 1% significance level, the data provide sufficient evidence to conclude that this year's mean retail price of all history books has increased from the 2000 mean of $51.46.

• • •

**FIGURE 9.9**
Criterion for deciding whether to reject the null hypothesis

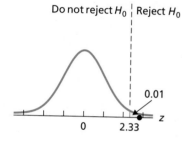

---

**Example 9.8** | **The One-Mean $z$-Test**

***Poverty and Calcium***　Calcium is the most abundant mineral in the human body and has several important functions. Most body calcium is stored in the bones and teeth, where it functions to support their structure. Recommendations for calcium are provided in *Dietary Reference Intakes*, developed by the Institute of Medicine of the National Academy of Sciences. The recommended adequate intake (RAI) of calcium for adults (ages 19–50) is 1000 milligrams (mg) per day.

A simple random sample of 18 adults with incomes below the poverty level gives the daily calcium intakes shown in Table 9.6. At the 5% significance level, do the data provide sufficient evidence to conclude that the mean calcium intake of all adults with incomes below the poverty level is less than the RAI of 1000 mg? Assume that $\sigma = 188$ mg.

**TABLE 9.6**
Daily calcium intakes (mg) for 18 adults with incomes below the poverty level

| | | | | | |
|---|---|---|---|---|---|
| 886 | 633 | 943 | 847 | 934 | 841 |
| 1193 | 820 | 774 | 834 | 1050 | 1058 |
| 1192 | 975 | 1313 | 872 | 1079 | 809 |

**Solution**  Because the sample size, $n = 18$, is moderate, we first need to consider questions of normality and outliers. (See the second bulleted item in Key Fact 9.4 on page 400.) Hence we constructed a normal probability plot for the data, shown in Fig. 9.10. The plot reveals no outliers and falls roughly in a straight line. Consequently, we can apply Procedure 9.1 to perform the required hypothesis test.

### STEP 1  State the null and alternative hypotheses.

Let $\mu$ denote the mean calcium intake (per day) of all adults with incomes below the poverty level. The null and alternative hypotheses are

$$H_0: \mu = 1000 \text{ mg (mean calcium intake is not less than the RAI)}$$
$$H_a: \mu < 1000 \text{ mg (mean calcium intake is less than the RAI)}.$$

Note that the hypothesis test is left tailed because a less-than sign ($<$) appears in the alternative hypothesis.

### STEP 2  Decide on the significance level, $\alpha$.

We are to perform the test at the 5% significance level, or $\alpha = 0.05$.

### STEP 3  Compute the value of the test statistic

$$z = \frac{\bar{x} - \mu_0}{\sigma/\sqrt{n}}.$$

We have $\mu_0 = 1000$, $\sigma = 188$, and $n = 18$. From the data in Table 9.6, we find that $\bar{x} = 947.4$. Thus the value of the test statistic is

$$z = \frac{947.4 - 1000}{188/\sqrt{18}} = -1.19.$$

This value of $z$ is marked with a dot in Fig. 9.11.

### STEP 4  The critical value for a left-tailed test is $-z_\alpha$. Use Table II to find the critical value.

As $\alpha = 0.05$, the critical value is $-z_{0.05}$. From Table II (or Table 9.4 or Table IV), $z_{0.05} = 1.645$. Hence the critical value is $-z_{0.05} = -1.645$, as shown in Fig. 9.11.

### STEP 5  If the value of the test statistic falls in the rejection region, reject $H_0$; otherwise, do not reject $H_0$.

The value of the test statistic, found in Step 3, is $z = -1.19$. Figure 9.11 reveals that this value does not fall in the rejection region, and so we do not reject $H_0$. The test results are not statistically significant at the 5% level.

### STEP 6  Interpret the results of the hypothesis test.

**Interpretation**  At the 5% significance level, the data do not provide sufficient evidence to conclude that the mean calcium intake of all adults with incomes below the poverty level is less than the RAI of 1000 mg per day.

• • •

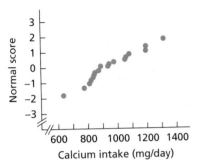

**FIGURE 9.10**

Normal probability plot of the calcium-intake data in Table 9.6

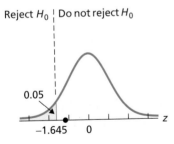

**FIGURE 9.11**

Criterion for deciding whether to reject the null hypothesis

**Example 9.9** | **The One-Mean $z$-Test**

*Clocking the Cheetah* The cheetah (*Acinonyx jubatus*) is the fastest land mammal on earth. According to the Cheetah Conservation of Southern Africa *Trade Environment Database*, the cheetah often exceeds speeds of 60 miles per hour (mph) and has been clocked at speeds of more than 70 mph.

One common estimate of mean top speed for cheetahs is 60 mph. Table 9.7 gives the top speeds, in miles per hour, for a sample of 35 cheetahs. At the 5% significance level, do the data provide sufficient evidence to conclude that the mean top speed of all cheetahs differs from 60 mph? Assume that the population standard deviation of top speeds is 3.2 mph.

**TABLE 9.7**

Top speeds, in miles per hour, for a sample of 35 cheetahs

| 57.3 | 57.5 | 59.0 | 56.5 | 61.3 |
|------|------|------|------|------|
| 57.6 | 59.2 | 65.0 | 60.1 | 59.7 |
| 62.6 | 52.6 | 60.7 | 62.3 | 65.2 |
| 54.8 | 55.4 | 55.5 | 57.8 | 58.7 |
| 57.8 | 60.9 | 75.3 | 60.6 | 58.1 |
| 55.9 | 61.6 | 59.6 | 59.8 | 63.4 |
| 54.7 | 60.2 | 52.4 | 58.3 | 66.0 |

**Solution** A normal probability plot of the data in Table 9.7, shown in Fig. 9.12, suggests that the top speed of 75.3 mph (third entry in the fifth row) is an outlier. A stem-and-leaf diagram, a boxplot, and a histogram further confirm that 75.3 is an outlier. Thus, as suggested in the third bulleted item in Key Fact 9.4 (page 400), we apply Procedure 9.1 first to the full data set in Table 9.7 and then to that data set with the outlier removed.

**STEP 1 State the null and alternative hypotheses.**

The null and alternative hypotheses are

$H_0$: $\mu = 60$ mph (mean top speed of cheetahs is 60 mph)
$H_a$: $\mu \neq 60$ mph (mean top speed of cheetahs is not 60 mph),

where $\mu$ denotes the mean top speed of all cheetahs. Note that the hypothesis test is two tailed because a does-not-equal sign ($\neq$) appears in the alternative hypothesis.

**FIGURE 9.12**

Normal probability plot of the top speeds in Table 9.7

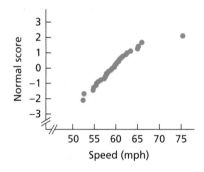

**STEP 2 Decide on the significance level, $\alpha$.**

We are to perform the hypothesis test at the 5% significance level, or $\alpha = 0.05$.

**STEP 3 Compute the value of the test statistic**

$$z = \frac{\bar{x} - \mu_0}{\sigma/\sqrt{n}}.$$

We have $\mu_0 = 60$, $\sigma = 3.2$, and $n = 35$. From the data in Table 9.7, we find that $\bar{x} = 59.526$. Thus the value of the test statistic is

$$z = \frac{59.526 - 60}{3.2/\sqrt{35}} = -0.88.$$

This value of $z$ is marked with a solid dot in Fig. 9.13.

**FIGURE 9.13**

Criterion for deciding whether to reject the null hypothesis

Reject $H_0$ | Do not reject $H_0$ | Reject $H_0$

0.025          0.025

−1.96     0     1.96     $z$

**STEP 4 The critical values for a two-tailed test are $\pm z_{\alpha/2}$. Use Table II to find the critical values.**

Because $\alpha = 0.05$, we find from Table II (or Table 9.4 or Table IV) the critical values of $\pm z_{0.05/2} = \pm z_{0.025} = \pm 1.96$, as shown in Fig. 9.13.

**STEP 5** **If the value of the test statistic falls in the rejection region, reject $H_0$; otherwise, do not reject $H_0$.**

From Step 3, the value of the test statistic is $z = -0.88$. This value does not fall in the rejection region shown in Fig. 9.13. Hence, we do not reject $H_0$. The test results are not statistically significant at the 5% level.

**STEP 6** **Interpret the results of the hypothesis test.**

**Interpretation** At the 5% significance level, the (unabridged) data do not provide sufficient evidence to conclude that the mean top speed of all cheetahs differs from 60 mph.

We have now completed the hypothesis test, using all 35 top speeds in Table 9.7. However, recall that the top speed of 75.3 mph is an outlier. Although in this case, we don't know whether removing this outlier is justified (a common situation), we can still remove it from the sample data and assess the effect on the hypothesis test.

With the outlier removed, the value of the test statistic is $z = -1.71$, which we have marked with a hollow dot in Fig. 9.13. This value also lies in the nonrejection region, although it is much closer to the rejection region than the value of the test statistic for the unabridged data, $z = -0.88$.

**Interpretation** In this case, removing the outlier therefore does not affect the conclusion of the hypothesis test. We can probably accept that the mean top speed of all cheetahs is roughly 60 mph.

Exercise 9.59
on page 405

• • •

## Statistical Significance Versus Practical Significance

Recall that the results of a hypothesis test are *statistically significant* if the null hypothesis is rejected at the chosen level of $\alpha$. Statistical significance means that the data provide sufficient evidence to conclude that the truth is different from the stated null hypothesis. However, it does not necessarily mean that the difference is important in any practical sense.

For example, the manufacturer of a new car, the Orion, claims that a typical car gets 26 miles per gallon. We think that the gas mileage is less. To test our suspicion, we perform the hypothesis test

$$H_0: \mu = 26 \text{ mpg (manufacturer's claim)}$$
$$H_a: \mu < 26 \text{ mpg (our suspicion)},$$

where $\mu$ is the mean gas mileage of all Orions.

We take a random sample of 1000 Orions and find that their mean gas mileage is 25.9 mpg. Assuming $\sigma = 1.4$ mpg, the value of the test statistic for a $z$-test is $z = -2.26$. This result is statistically significant at the 5% level. Thus, at the 5% significance level, we reject the manufacturer's claim.

Because the sample size, 1000, is so large, the sample mean, $\bar{x} = 25.9$ mpg, is probably nearly the same as the population mean. As a result, we rejected the manufacturer's claim because $\mu$ is about 25.9 mpg instead of 26 mpg. From a practical point of view, however, the difference between 25.9 mpg and 26 mpg is not important.

**What Does It Mean?**

Statistical significance does not necessarily imply practical significance!

### The Relation Between Hypothesis Tests and Confidence Intervals

Hypothesis tests and confidence intervals are closely related. Consider, for example, a two-tailed hypothesis test for a population mean at the significance level $\alpha$. In this case, the null hypothesis will be rejected if and only if the value $\mu_0$ given for the mean in the null hypothesis lies outside the $(1 - \alpha)$-level confidence interval for $\mu$. You can examine the relation between hypothesis tests and confidence intervals in greater detail in Exercises 9.70–9.72.

## Exercises 9.3

### Understanding the Concepts and Skills

*In Exercises 9.44–9.49, suppose that a hypothesis test is to be performed for a population mean, $\mu$, with a null hypothesis of $H_0: \mu = \mu_0$. Further suppose that the test statistic to be used is*

$$z = \frac{\bar{x} - \mu_0}{\sigma/\sqrt{n}}.$$

*For each exercise, determine the required critical value(s) and draw a graph that illustrates your answers.*

**9.44** A two-tailed test with $\alpha = 0.10$.

**9.45** A right-tailed test with $\alpha = 0.05$.

**9.46** A left-tailed test with $\alpha = 0.01$.

**9.47** A left-tailed test with $\alpha = 0.05$.

**9.48** A right-tailed test with $\alpha = 0.01$.

**9.49** A two-tailed test with $\alpha = 0.05$.

**9.50** Explain why considering outliers is important when you are conducting a one-mean $z$-test.

**9.51** Each part of this exercise provides a scenario for a hypothesis test for a population mean. Decide whether the $z$-test is an appropriate method for conducting the hypothesis test. Assume that the population standard deviation is known in each case.
a. Preliminary data analyses reveal that the sample data contain no outliers but that the distribution of the variable under consideration is probably highly skewed. The sample size is 24.
b. Preliminary data analyses reveal that the sample data contain no outliers but that the distribution of the variable under consideration is probably mildly skewed. The sample size is 70.

**9.52** Each part of this exercise provides a scenario for a hypothesis test for a population mean. Decide whether the $z$-test is an appropriate method for conducting the hypothesis test. Assume that the population standard deviation is known in each case.
a. A normal probability plot of the sample data shows no outliers and is quite linear. The sample size is 12.
b. Preliminary data analyses reveal that the sample data contain an outlier but is otherwise roughly linear. It is determined that the outlier is a legitimate observation and should not be removed. The sample size is 17.

*In each of Exercises 9.53–9.58, we have provided a sample mean, sample size, and population standard deviation. In each case, use the one-mean $z$-test to perform the required hypothesis test about the mean, $\mu$, of the population from which the sample was drawn.*

**9.53** $\bar{x} = 20, n = 32, \sigma = 4, H_0: \mu = 22, H_a: \mu < 22, \alpha = 0.05$

**9.54** $\bar{x} = 21, n = 32, \sigma = 4, H_0: \mu = 22, H_a: \mu < 22, \alpha = 0.05$

**9.55** $\bar{x} = 24, n = 15, \sigma = 4, H_0: \mu = 22, H_a: \mu > 22, \alpha = 0.05$

**9.56** $\bar{x} = 23, n = 15, \sigma = 4, H_0: \mu = 22, H_a: \mu > 22, \alpha = 0.05$

**9.57** $\bar{x} = 23, n = 24, \sigma = 4, H_0: \mu = 22, H_a: \mu \neq 22, \alpha = 0.05$

**9.58** $\bar{x} = 20, n = 24, \sigma = 4, H_0: \mu = 22, H_a: \mu \neq 22, \alpha = 0.05$

*Preliminary data analyses indicate that applying the $z$-test (Procedure 9.1 on page 399) in Exercises 9.59–9.64 is reasonable.*

**9.59 Toxic Mushrooms?** Cadmium, a heavy metal, is toxic to animals. Mushrooms, however, are able to absorb and accumulate cadmium at high concentrations. The Czech and Slovak governments have set a safety limit for cadmium in dry vegetables at 0.5 part per million (ppm). M. Melgar et al. measured the cadmium levels in a random sample of the edible mushroom *Boletus pinicola* and published the results in the *Journal of Environmental Science and Health* (Vol. B33(4), pp. 439–455). Here are the data.

| | | | | | |
|------|------|------|------|------|------|
| 0.24 | 0.59 | 0.62 | 0.16 | 0.77 | 1.33 |
| 0.92 | 0.19 | 0.33 | 0.25 | 0.59 | 0.32 |

At the 5% significance level, do the data provide sufficient evidence to conclude that the mean cadmium level in *Boletus pinicola* mushrooms is greater than the government's recommended limit of 0.5 ppm? Assume that the population standard deviation of cadmium levels in *Boletus pinicola* mushrooms is 0.37 ppm. (*Note:* The sum of the data is 6.31 ppm.)

**9.60 Agriculture Books.** The R. R. Bowker Company of New York collects information on the retail prices of books and publishes the data in *The Bowker Annual Library and Book Trade Almanac*. In 2000, the mean retail price of agriculture books was $66.52. This year's retail prices for 28 randomly selected agriculture books are shown in the following table.

| | | | | | | |
|---|---|---|---|---|---|---|
| 68.45 | 76.61 | 66.01 | 55.02 | 55.77 | 71.78 | 75.31 |
| 60.99 | 46.58 | 59.38 | 69.33 | 47.05 | 67.12 | 56.26 |
| 59.36 | 79.94 | 57.05 | 75.09 | 68.27 | 50.54 | 62.57 |
| 58.86 | 67.99 | 66.95 | 55.56 | 75.67 | 59.52 | 75.59 |

At the 10% significance level, do the data provide sufficient evidence to conclude that this year's mean retail price of agriculture books has changed from the 2000 mean? Assume that the standard deviation of prices for this year's agriculture books is $8.45. (*Note:* The sum of the data is $1788.62.)

**9.61 Iron Deficiency?** Iron is essential to most life forms and to normal human physiology. It is an integral part of many proteins and enzymes that maintain good health. Recommendations for iron are provided in *Dietary Reference Intakes*, developed by the Institute of Medicine of the National Academy of Sciences. The recommended dietary allowance (RDA) of iron for adult females under the age of 51 is 18 milligrams (mg) per day. The following iron intakes, in milligrams, were obtained during a 24-hour period for 45 randomly selected adult females under the age of 51.

| | | | | | | | | |
|---|---|---|---|---|---|---|---|---|
| 15.0 | 18.1 | 14.4 | 14.6 | 10.9 | 18.1 | 18.2 | 18.3 | 15.0 |
| 16.0 | 12.6 | 16.6 | 20.7 | 19.8 | 11.6 | 12.8 | 15.6 | 11.0 |
| 15.3 | 9.4 | 19.5 | 18.3 | 14.5 | 16.6 | 11.5 | 16.4 | 12.5 |
| 14.6 | 11.9 | 12.5 | 18.6 | 13.1 | 12.1 | 10.7 | 17.3 | 12.4 |
| 17.0 | 6.3 | 16.8 | 12.5 | 16.3 | 14.7 | 12.7 | 16.3 | 11.5 |

At the 1% significance level, do the data suggest that adult females under the age of 51 are, on average, getting less than the RDA of 18 mg of iron? Assume that the population standard deviation is 4.2 mg. (*Note:* $\bar{x} = 14.68$ mg.)

**9.62 Early-Onset Dementia.** Dementia is the loss of the intellectual and social abilities severe enough to interfere with judgment, behavior, and daily functioning. Alzheimer's disease is the most common type of dementia. In the article "Living with Early Onset Dementia: Exploring the Experience and Developing Evidence-Based Guidelines for Practice" (*Alzheimer's Care Quarterly*, Vol. 5, Issue 2, pp. 111–122),

P. Harris and J. Keady explored the experience and struggles of people diagnosed with dementia and their families. A simple random sample of 21 people with early-onset dementia gave the following data on age at diagnosis.

| | | | | | | |
|---|---|---|---|---|---|---|
| 60 | 58 | 52 | 58 | 59 | 58 | 51 |
| 61 | 54 | 59 | 55 | 53 | 44 | 46 |
| 47 | 42 | 56 | 57 | 49 | 41 | 43 |

At the 1% significance level, do the data provide sufficient evidence to conclude that the mean age at diagnosis of all people with early-onset dementia is less than 55 years old? Assume that the population standard deviation is 6.8 years. (*Note:* $\bar{x} = 52.5$ years.)

**9.63 Serving Time.** According to the Bureau of Crime Statistics and Research of Australia, as reported on *Lawlink*, the mean length of imprisonment for motor-vehicle theft offenders in Australia is 16.7 months. One hundred randomly selected motor-vehicle theft offenders in Sydney, Australia, had a mean length of imprisonment of 17.8 months. At the 5% significance level, do the data provide sufficient evidence to conclude that the mean length of imprisonment for motor-vehicle theft offenders in Sydney differs from the national mean in Australia? Assume that the population standard deviation of the lengths of imprisonment for motor-vehicle theft offenders in Sydney is 6.0 months.

**9.64 Worker Fatigue.** A study by M. Chen et al. titled "Heat Stress Evaluation and Worker Fatigue in a Steel Plant" (*American Industrial Hygiene Association*, Vol. 64, pp. 352–359) assessed fatigue in steel-plant workers due to heat stress. A random sample of 29 casting workers had a mean post-work heart rate of 78.3 beats per minute (bpm). At the 5% significance level, do the data provide sufficient evidence to conclude that the mean post-work heart rate for casting workers exceeds the normal resting heart rate of 72 bpm? Assume that the population standard deviation of post-work heart rates for casting workers is 11.2 bpm.

**9.65 Job Gains and Losses.** In the article "Business Employment Dynamics: New Data on Gross Job Gains and Losses" (*Monthly Labor Review*, Vol. 127, Issue 4, pp. 29–42), J. Spletzer et al. examined gross job gains and losses as a percentage of the average of previous and current employment figures. A simple random sample of 20 quarters provided the net percentage gains (losses are negative gains) for jobs as presented on the WeissStats CD. Use the technology of your choice to do the following.
a. Decide whether, on average, the net percentage gain for jobs exceeds 0.2. Assume a population standard deviation of 0.42. Apply the one-mean z-test with a 5% significance level.
b. Obtain a normal probability plot, boxplot, histogram, and stem-and-leaf diagram of the data.

c. Remove the outliers (if any) from the data and then repeat part (a).

d. Comment on the advisability of using the $z$-test here.

**9.66 Hotels and Motels.** The daily charges, in dollars, for a sample of 15 hotels and motels operating in South Carolina are provided on the WeissStats CD. The data were found in the report *South Carolina Statistical Abstract*, sponsored by the South Carolina Budget and Control Board.

a. Use the one-mean $z$-test to decide, at the 5% significance level, whether the data provide sufficient evidence to conclude that the mean daily charge for hotels and motels operating in South Carolina is less than $75. Assume a population standard deviation of $22.40.

b. Obtain a normal probability plot, boxplot, histogram, and stem-and-leaf diagram of the data.

c. Remove the outliers (if any) from the data and then repeat part (a).

d. Comment on the advisability of using the $z$-test here.

## Working With Large Data Sets

**9.67 Body Temperature.** A study by researchers at the University of Maryland addressed the question of whether the mean body temperature of humans is 98.6°F. The results of the study by P. Mackowiak et al. appeared in the article "A Critical Appraisal of 98.6°F, the Upper Limit of the Normal Body Temperature, and Other Legacies of Carl Reinhold August Wunderlich" (*Journal of the American Medical Association*, Vol. 268, pp. 1578–1580). Among other data, the researchers obtained the body temperatures of 93 healthy humans, which we have provided on the WeissStats CD. Use the technology of your choice to do the following.

a. Obtain a normal probability plot, boxplot, histogram, and stem-and-leaf diagram of the data.

b. Based on your results from part (a), can you reasonably apply the one-mean $z$-test to the data? Explain your reasoning.

c. At the 1% significance level, do the data provide sufficient evidence to conclude that the mean body temperature of healthy humans differs from 98.6°F? Assume that $\sigma = 0.63°F$. *Note:* You may need to consult Table II in Appendix A to obtain the critical values.

**9.68 Teacher Salaries.** The Educational Resource Service publishes information about wages and salaries in the public schools system in *National Survey of Salaries and Wages in Public Schools*. The mean annual salary of (public) classroom teachers is $45.9 thousand. A random sample of 90 classroom teachers in Hawaii yielded the annual salaries, in thousands of dollars, presented on the WeissStats CD. Use the technology of your choice to do the following.

a. Obtain a normal probability plot, boxplot, histogram, and stem-and-leaf diagram of the data.

b. Based on your results from part (a), can you reasonably apply the one-mean $z$-test to the data? Explain your reasoning.

c. At the 5% significance level, do the data provide sufficient evidence to conclude that the mean annual salary of classroom teachers in Hawaii is less than the national mean? Assume that the standard deviation of annual salaries for all classroom teachers in Hawaii is $9.2 thousand. *Note:* You may need to consult Table II in Appendix A to obtain the critical value.

**9.69 Cell Phones.** The number of cell phone users has increased dramatically since 1987. According to the *Semi-annual Wireless Survey*, published by the Cellular Telecommunications & Internet Association, the mean local monthly bill for cell phone users in the United States was $47.37 in 2001. Last year's local monthly bills, in dollars, for a random sample of 75 cell phone users are given on the WeissStats CD. Use the technology of your choice to do the following.

a. Obtain a normal probability plot, boxplot, histogram, and stem-and-leaf diagram of the data.

b. At the 5% significance level, do the data provide sufficient evidence to conclude that last year's mean local monthly bill for cell phone users has increased from the 2001 mean of $47.37? Assume that the standard deviation of last year's local monthly bills for cell phone users is $25. *Note:* You may need to consult Table II in Appendix A to obtain the critical value.

c. Remove the two outliers from the data and repeat parts (a) and (b).

d. State your conclusions regarding the hypothesis test.

## Extending the Concepts and Skills

**9.70 Two-Tailed Hypothesis Tests and CIs.** As we mentioned on page 405, the following relationship holds between hypothesis tests and confidence intervals for one-mean $z$-procedures: For a two-tailed hypothesis test at the significance level $\alpha$, the null hypothesis $H_0$: $\mu = \mu_0$ will be rejected in favor of the alternative hypothesis $H_a$: $\mu \neq \mu_0$ if and only if $\mu_0$ lies outside the $(1 - \alpha)$-level confidence interval for $\mu$. In each case, illustrate the preceding relationship by obtaining the appropriate one-mean $z$-interval (Procedure 8.1 on page 345) and comparing the result to the conclusion of the hypothesis test in the specified exercise.

a. Exercise 9.60

b. Exercise 9.63

**9.71 Left-Tailed Hypothesis Tests and CIs.** In Exercise 8.47 on page 353, we introduced one-sided one-mean $z$-intervals. The following relationship holds between hypothesis tests and confidence intervals for one-mean $z$-procedures: For a left-tailed hypothesis test at the significance level $\alpha$, the null hypothesis $H_0$: $\mu = \mu_0$ will be rejected in favor of the alternative hypothesis $H_a$: $\mu < \mu_0$ if and only if $\mu_0$ is greater than the $(1 - \alpha)$-level upper confidence bound for $\mu$. In each case, illustrate the preceding relationship by obtaining the appropriate upper confidence bound and comparing the result to the conclusion of the hypothesis test in the specified exercise.

a. Exercise 9.61      b. Exercise 9.62

**9.72 Right-Tailed Hypothesis Tests and CIs.** In Exercise 8.47 on page 353, we introduced one-sided one-mean $z$-intervals. The following relationship holds between hypothesis tests and confidence intervals for one-mean $z$-procedures: For a right-tailed hypothesis test at the signifi-

cance level $\alpha$, the null hypothesis $H_0$: $\mu = \mu_0$ will be rejected in favor of the alternative hypothesis $H_a$: $\mu > \mu_0$ if and only if $\mu_0$ is less than the $(1 - \alpha)$-level lower confidence bound for $\mu$. In each case, illustrate the preceding relationship by obtaining the appropriate lower confidence bound and comparing the result to the conclusion of the hypothesis test in the specified exercise.

a. Exercise 9.59      b. Exercise 9.64

## 9.4   *P*-Values

On page 399, we presented Procedure 9.1, a step-by-step method for performing a hypothesis test for a population mean when the population standard deviation is known. Step 4 of that procedure requires us to obtain critical values. Because of that requirement, Procedure 9.1 is said to use the **critical-value approach** to hypothesis testing.

In this section, we discuss another approach to hypothesis testing, called the ***P*-value approach.** Roughly speaking, the $P$-value indicates how likely observation of the value obtained for the test statistic would be if the null hypothesis is true. In particular, a small $P$-value (close to 0) indicates that observation of the value obtained for the test statistic would be unlikely if the null hypothesis is true.

We can define the ***P*-value** (also called the **observed significance level** or the **probability value**) as the percentage of samples that would yield a value of the test statistic as extreme as or more extreme than that observed if the null hypothesis is true. But, more commonly, the $P$-value is defined in the language of probability as follows.

**Definition 9.5**

### *P*-Value

To obtain the ***P*-value** of a hypothesis test, we assume that the null hypothesis is true and compute the probability of observing a value of the test statistic as extreme as or more extreme than that observed. By *extreme* we mean "far from what we would expect to observe if the null hypothesis is true." We use the letter $P$ to denote the $P$-value.

**What Does It Mean?**

Small $P$-values provide evidence against the null hypothesis; larger $P$-values do not. The smaller (closer to 0) the $P$-value, the stronger the evidence is against the null hypothesis.

In this section, we concentrate on $P$-values and the $P$-value approach for the one-mean $z$-test. But much of what we say here applies to $P$-values and the $P$-value approach for any hypothesis test.

### Obtaining *P*-Values for a One-Mean *z*-Test

Recall that the test statistic for a one-mean $z$-test with null hypothesis given by $H_0$: $\mu = \mu_0$ is

$$z = \frac{\bar{x} - \mu_0}{\sigma/\sqrt{n}}.$$

If the null hypothesis is true, this test statistic has the standard normal distribution, and its probabilities equal areas under the standard normal curve.

If we let $z_0$ denote the observed value of the test statistic $z$, we obtain the $P$-value as follows:

- *Two-tailed test:* The *P*-value is the probability of observing a value of the test statistic *z* that is at least as large in magnitude as the value actually observed, which is the area under the standard normal curve that lies outside the interval from $-|z_0|$ to $|z_0|$, as illustrated in Fig. 9.14(a).
- *Left-tailed test:* The *P*-value is the probability of observing a value of the test statistic *z* that is as small as or smaller than the value actually observed, which is the area under the standard normal curve that lies to the left of $z_0$, as illustrated in Fig. 9.14(b).
- *Right-tailed test:* The *P*-value is the probability of observing a value of the test statistic *z* that is as large as or larger than the value actually observed, which is the area under the standard normal curve that lies to the right of $z_0$, as illustrated in Fig. 9.14(c).

**FIGURE 9.14**

*P*-value for a *z*-test if the test is (a) two tailed, (b) left tailed, or (c) right tailed

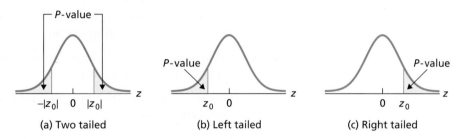

(a) Two tailed    (b) Left tailed    (c) Right tailed

**Example 9.10** | *P*-Values

*Prices of History Books*  Consider again the history book hypothesis test of Example 9.7, where we wanted to decide whether this year's mean (retail) price of all history books has increased from the 2000 mean of $51.46. The null and alternative hypotheses are

$$H_0: \mu = \$51.46 \text{ (mean price has not increased)}$$
$$H_a: \mu > \$51.46 \text{ (mean price has increased)},$$

where $\mu$ is this year's mean retail price of all history books. Note that the hypothesis test is right tailed.

Table 9.5 on page 400 displays this year's prices for 40 randomly selected history books; their mean price is $\bar{x} = \$54.890$. Using that sample mean and $\sigma = \$7.61$, we found the value of the test statistic to be 2.85. Obtain and interpret the *P*-value of the hypothesis test.

**FIGURE 9.15**

*P*-value for the history book hypothesis test

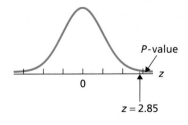

**Solution**  Because the hypothesis test is a right-tailed *z*-test, the *P*-value is the probability of observing a value of *z* of 2.85 or greater if the null hypothesis is true. That probability equals the area under the standard normal curve to the right of 2.85, the shaded area shown in Fig. 9.15. From Table II, we find that area to be $1 - 0.9978 = 0.0022$.

Therefore the *P*-value of this hypothesis test is 0.0022. If the null hypothesis is true, we would observe a value of the test statistic *z* of 2.85 or greater only about 2 times in 1000. In other words, if the null hypothesis is true, a random sample of 40 history books would have a mean price of $54.89 or greater only about 0.2% of the time.

**Interpretation**  The sample data provide very strong evidence against the null hypothesis.

You
try it!

Exercise 9.89
on page 417

• • •

## Example 9.11 | P-Values

*Clocking the Cheetah* In Example 9.9, we conducted a hypothesis test to decide whether the mean top speed of all cheetahs differs from 60 mph. The null and alternative hypotheses are

$$H_0: \mu = 60 \text{ mph (mean top speed of cheetahs is 60 mph)}$$
$$H_a: \mu \neq 60 \text{ mph (mean top speed of cheetahs is not 60 mph)},$$

where $\mu$ denotes the mean top speed of all cheetahs. Note that the hypothesis test is two tailed.

Table 9.7 on page 403 shows the top speeds of a random sample of 35 cheetahs. Recall that the top speed of 75.3 mph is an outlier.

**a.** Obtain and interpret the $P$-value of the hypothesis test, using the unabridged data (i.e., including the outlier).

**b.** Obtain and interpret the $P$-value of the hypothesis test, using the abridged data (i.e., with the outlier removed).

**c.** Comment on the effect that removing the outlier has on the evidence against the null hypothesis.

### Solution

**a.** In Example 9.9, we found the value of the test statistic for the unabridged data to be $-0.88$. Because the test is a two-tailed $z$-test, the $P$-value is the probability of observing a value of $z$ of 0.88 or greater in magnitude if the null hypothesis is true. That probability, as shown in Fig. 9.16(a), equals 0.3788.

Thus, if the null hypothesis is true, we would observe a value of the test statistic $z$ of 0.88 or greater in magnitude more than 37 times in 100. In other words, if the null hypothesis is true, a random sample of 35 cheetahs would have a mean top speed at least as far from 60 mph as that of our unabridged sample more than 37% of the time.

**Interpretation** The unabridged data do not provide evidence against the null hypothesis.

**b.** In Example 9.9, we found the value of the test statistic for the abridged data to be $-1.71$. In this case, the $P$-value, shown in Fig. 9.16(b), equals 0.0872.

Thus, if the null hypothesis is true, we would observe a value of the test statistic $z$ of 1.71 or greater in magnitude less than 9 times in 100. In other words, if the null hypothesis is true, a random sample of 34 cheetahs would have a mean top speed at least as far from 60 mph as that of our abridged sample less than 9% of the time.

**Interpretation** The abridged data provide moderate evidence against the null hypothesis.

**c.** Parts (a) and (b) indicate that the strength of the evidence against the null hypothesis depends on whether the outlier is retained or removed. If it is retained, there is virtually no evidence against the null hypothesis; if it is removed, there is moderate evidence against the null hypothesis.

**FIGURE 9.16**

*P*-value for the cheetah speed hypothesis test (a) including the outlier, and (b) with the outlier removed

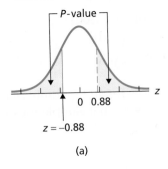

(a)

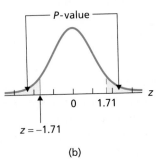

(b)

• • •

### The *P*-Value Approach to Hypothesis Testing

The *P*-value can be interpreted as the *observed significance level* of a hypothesis test. To illustrate, suppose that the value of the test statistic for a right-tailed *z*-test turns out to be 1.88. Then the *P*-value of the hypothesis test is 0.03 (actually 0.0301), as depicted by the shaded area in Fig. 9.17.

**FIGURE 9.17**
*P*-value as the observed significance level

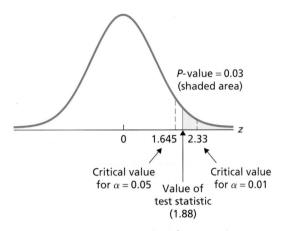

Figure 9.17 reveals that the null hypothesis would be rejected for a test at the 0.05 significance level but would not be rejected for a test at the 0.01 significance level. In fact, Fig. 9.17 makes it clear that the *P*-value is precisely the smallest significance level at which the null hypothesis would be rejected.

---

**Key Fact 9.5**     ***P*-Value as the Observed Significance Level**

The *P*-value of a hypothesis test equals the smallest significance level at which the null hypothesis can be rejected, that is, the smallest significance level for which the observed sample data results in rejection of $H_0$.

---

Key Fact 9.5 leads to the following criterion for deciding whether to reject the null hypothesis in favor of the alternative hypothesis.

---

**Key Fact 9.6**     **Decision Criterion for a Hypothesis Test Using the *P*-Value**

If the *P*-value is less than or equal to the specified significance level, reject the null hypothesis; otherwise, do not reject the null hypothesis.

---

Key Fact 9.6 provides a foundation for the *P*-value approach to hypothesis testing. For the one-mean *z*-test, then, we have Procedure 9.2 (next page).

In Example 9.8, we used the critical-value approach to perform a one-mean *z*-test to decide whether the mean calcium intake of all adults with incomes below the poverty level is less than the recommended adequate intake (RAI) of 1000 mg per day. We now perform that same hypothesis test by using the *P*-value approach.

| Procedure 9.2 | One-Mean z-Test (P-Value Approach) |
|---|---|

*Purpose* To perform a hypothesis test for a population mean, $\mu$

*Assumptions*

1. Simple random sample
2. Normal population or large sample
3. $\sigma$ known

**STEP 1** The null hypothesis is $H_0$: $\mu = \mu_0$, and the alternative hypothesis is

$$H_a: \mu \neq \mu_0 \qquad H_a: \mu < \mu_0 \qquad H_a: \mu > \mu_0$$
$$\text{(Two tailed)} \quad \text{or} \quad \text{(Left tailed)} \quad \text{or} \quad \text{(Right tailed)}$$

**STEP 2** Decide on the significance level, $\alpha$.

**STEP 3** Compute the value of the test statistic

$$z = \frac{\bar{x} - \mu_0}{\sigma/\sqrt{n}}$$

and denote that value $z_0$.

**STEP 4** Use Table II to obtain the P-value.

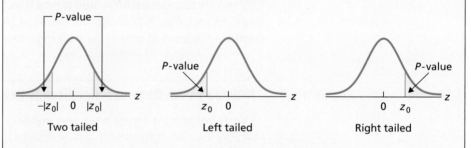

**STEP 5** If $P \leq \alpha$, reject $H_0$; otherwise, do not reject $H_0$.

**STEP 6** Interpret the results of the hypothesis test.

The hypothesis test is exact for normal populations and is approximately correct for large samples from nonnormal populations.

---

**Example 9.12**    **The One-Mean z-Test (P-Value Approach)**

**TABLE 9.8**

Daily calcium intakes (mg) for 18 adults with incomes below the poverty level

| | | | | | |
|---|---|---|---|---|---|
| 886 | 633 | 943 | 847 | 934 | 841 |
| 1193 | 820 | 774 | 834 | 1050 | 1058 |
| 1192 | 975 | 1313 | 872 | 1079 | 809 |

*Poverty and Calcium* A simple random sample of 18 adults with incomes below the poverty level gives the daily calcium intakes shown in Table 9.8. At the 5% significance level, do the data provide sufficient evidence to conclude that the mean calcium intake of all adults with incomes below the poverty level is less than the RAI of 1000 mg per day? Assume that daily calcium intakes of adults with incomes below the poverty level have a standard deviation of 188 milligrams (i.e., $\sigma = 188$ mg).

**Solution**   Recall that a normal probability plot of the data in Table 9.8 (Fig. 9.10 on page 402) reveals no outliers and falls roughly in a straight line. So, for this moderate sample size, we can apply Procedure 9.2 to perform the hypothesis test.

**STEP 1  State the null and alternative hypotheses.**

Let $\mu$ denote the mean calcium intake (per day) of all adults with incomes below the poverty level. The null and alternative hypotheses are

$$H_0: \mu = 1000 \text{ mg (mean calcium intake is not less than the RAI)}$$
$$H_a: \mu < 1000 \text{ mg (mean calcium intake is less than the RAI)}.$$

Note that the hypothesis test is left tailed.

**STEP 2  Decide on the significance level, $\alpha$.**

We are to perform the test at the 5% significance level, or $\alpha = 0.05$.

**STEP 3  Compute the value of the test statistic**

$$z = \frac{\bar{x} - \mu_0}{\sigma/\sqrt{n}}.$$

We found in Example 9.8 (page 401) that $z = -1.19$. This value is shown in Fig. 9.18.

**STEP 4  Use Table II to obtain the $P$-value.**

The test is left tailed, so the $P$-value is the probability of observing a value of $z$ of $-1.19$ or less if the null hypothesis is true. That probability equals the shaded area in Fig. 9.18, which by Table II is 0.1170. Hence $P = 0.1170$.

**STEP 5  If $P \leq \alpha$, reject $H_0$; otherwise, do not reject $H_0$.**

Because the $P$-value, 0.1170, exceeds the specified significance level of 0.05, we do not reject $H_0$. Consequently, the test results are not statistically significant at the 5% level.

**STEP 6  Interpret the results of the hypothesis test.**

**Interpretation**   At the 5% significance level, the data do not provide sufficient evidence to conclude that the mean calcium intake of all adults with incomes below the poverty level is less than the RAI of 1000 mg per day.

• • •

**FIGURE 9.18**
Value of the test statistic and the *P*-value for the calcium intake hypothesis test

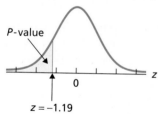

*P*-value

$z = -1.19$

*You try it!*

Exercise 9.95 on page 417

## Comparison of the Critical-Value and *P*-Value Approaches

We have now discussed both the critical-value and *P*-value approaches to hypothesis testing, but only in terms of the one-mean *z*-test. For future reference, the general steps involved in each approach appear in Table 9.9.

**TABLE 9.9** Comparison of critical-value and *P*-value approaches

| CRITICAL-VALUE APPROACH | or | *P*-VALUE APPROACH |
|---|---|---|

| **CRITICAL-VALUE APPROACH** | **P-VALUE APPROACH** |
|---|---|
| **STEP 1** State the null and alternative hypotheses. | **STEP 1** State the null and alternative hypotheses. |
| **STEP 2** Decide on the significance level, $\alpha$. | **STEP 2** Decide on the significance level, $\alpha$. |
| **STEP 3** Compute the value of the test statistic. | **STEP 3** Compute the value of the test statistic. |
| **STEP 4** Determine the critical value(s). | **STEP 4** Determine the *P*-value. |
| **STEP 5** If the value of the test statistic falls in the rejection region, reject $H_0$; otherwise, do not reject $H_0$. | **STEP 5** If $P \leq \alpha$, reject $H_0$; otherwise, do not reject $H_0$. |
| **STEP 6** Interpret the result of the hypothesis test. | **STEP 6** Interpret the result of the hypothesis test. |

## Using the *P*-Value to Assess the Evidence Against the Null Hypothesis

Key Fact 9.5 asserts that the *P*-value is the smallest significance level at which the null hypothesis can be rejected. Consequently, knowing the *P*-value allows us to assess significance at any level we desire. For instance, if the *P*-value of a hypothesis test is 0.03, the null hypothesis can be rejected at any significance level larger than 0.03 and it cannot be rejected at any significance level smaller than 0.03.

Knowing the *P*-value also allows us to evaluate the strength of the evidence against the null hypothesis: the smaller the *P*-value, the stronger will be the evidence against the null hypothesis. Table 9.10 presents guidelines for interpreting the *P*-value of a hypothesis test.

In Example 9.10 (page 409) and Example 9.11 (page 410), we used the *P*-value to evaluate the strength of the evidence against the null hypothesis without reference to critical values or significance levels. This practice is common among researchers.

**TABLE 9.10**
Guidelines for using the *P*-value to assess the evidence against the null hypothesis

| *P*-value | Evidence against $H_0$ |
|---|---|
| $P > 0.10$ | Weak or none |
| $0.05 < P \leq 0.10$ | Moderate |
| $0.01 < P \leq 0.05$ | Strong |
| $P \leq 0.01$ | Very strong |

**Hypothesis Tests Without Significance Levels:** Many researchers do not explicitly refer to significance levels or critical values. Instead, they simply obtain the *P*-value and use it (or let the reader use it) to assess the strength of the evidence against the null hypothesis.

# The Technology Center

Most statistical technologies have programs that automatically perform a one-mean *z*-test. In this subsection, we present output and step-by-step instructions for such programs.

## Example 9.13 Using Technology to Conduct a One-Mean *z*-Test

*Poverty and Calcium*  Table 9.8 on page 412 shows the daily calcium intakes for a simple random sample of 18 adults with incomes below the poverty level.

Use Minitab, Excel, or the TI-83/84 Plus to decide, at the 5% significance level, whether the data provide sufficient evidence to conclude that the mean calcium intake of all adults with incomes below the poverty level is less than the RAI of 1000 mg per day. Assume that $\sigma = 188$ mg.

**Solution** Let $\mu$ denote the mean calcium intake (per day) of all adults with incomes below the poverty level. We want to perform the hypothesis test

$$H_0: \mu = 1000 \text{ mg (mean calcium intake is not less than the RAI)}$$
$$H_a: \mu < 1000 \text{ mg (mean calcium intake is less than the RAI)}$$

at the 5% significance level ($\alpha = 0.05$). Note that the hypothesis test is left tailed.

We applied the one-mean *z*-test programs to the data, resulting in Output 9.1. Steps for generating that output are presented in Instructions 9.1 on the following page.

**OUTPUT 9.1**
One-mean *z*-test for the sample of calcium intakes

**MINITAB**

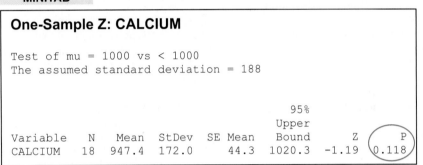

```
One-Sample Z: CALCIUM

Test of mu = 1000 vs < 1000
The assumed standard deviation = 188

                                      95%
                                    Upper
Variable   N    Mean   StDev  SE Mean  Bound     Z      P
CALCIUM   18   947.4   172.0    44.3  1020.3  -1.19  0.118
```

**EXCEL**

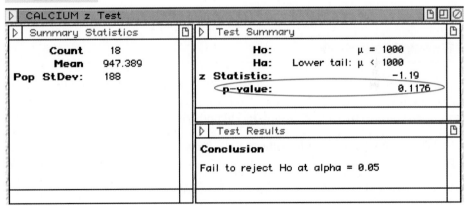

```
CALCIUM z Test

Summary Statistics          Test Summary

    Count    18                  Ho:          μ = 1000
    Mean    947.389              Ha:   Lower tail: μ < 1000
Pop StDev:   188          z Statistic:              -1.19
                            p-value:               0.1176


                            Test Results

                          Conclusion

                          Fail to reject Ho at alpha = 0.05
```

**TI-83/84 PLUS**

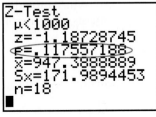

```
Z-Test
 μ<1000
 z=-1.18728745
 p=.117557188
 x̄=947.3888889
 Sx=171.9894453
 n=18
```

Using **Calculate**

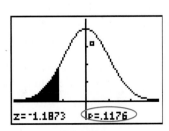

```
z=-1.1873    p=.1176
```

Using **Draw**

As shown in Output 9.1, the $P$-value for the hypothesis test is 0.118. Because the $P$-value exceeds the specified significance level of 0.05, we do not reject $H_0$. At the 5% significance level, the data do not provide sufficient evidence to conclude that the mean calcium intake of all adults with incomes below the poverty level is less than the RAI of 1000 mg per day.

• • •

**INSTRUCTIONS 9.1** Steps for generating Output 9.1

| MINITAB | EXCEL | TI-83/84 PLUS |
|---|---|---|
| 1 Store the data from Table 9.8 in a column named CALCIUM | 1 Store the data from Table 9.8 in a range named CALCIUM | 1 Store the data from Table 9.8 in a list named CALCI |
| 2 Choose **Stat ➤ Basic Statistics ➤ 1-Sample Z...** | 2 Choose **DDXL ➤ Hypothesis Tests** | 2 Press **STAT**, arrow over to **TESTS**, and press **1** |
| 3 Select the **Samples in columns** option button | 3 Select **1 Var z Test** from the **Function type** drop-down box | 3 Highlight **Data** and press **ENTER** |
| 4 Click in the **Samples in columns** text box and specify CALCIUM | 4 Specify CALCIUM in the **Quantitative Variable** text box | 4 Press the down-arrow key, type 1000 for $\mu_0$, and press **ENTER** |
| 5 Click in the **Standard deviation** text box and type 188 | 5 Click **OK** | 5 Type 188 for $\sigma$ and press **ENTER** |
| 6 Check the **Perform hypothesis test** check box | 6 Click the **Set $\mu$0 and sd** button | 6 Press **2nd ➤ LIST** |
| 7 Click in the **Hypothesized mean** text box and type 1000 | 7 Click in the **Hypothesized $\mu$0** text box and type 1000 | 7 Arrow down to CALCI and press **ENTER** three times |
| 8 Click the **Options...** button | 8 Click in the **Population std dev** text box and type 188 | 8 Highlight $< \mu_0$ and press **ENTER** |
| 9 Click the arrow button at the right of the **Alternative** drop-down list box and select **less than** | 9 Click **OK** | 9 Press the down-arrow key, highlight **Calculate** or **Draw**, and press **ENTER** |
| 10 Click **OK** twice | 10 Click the **0.05** button | |
| | 11 Click the $\mu < \mu$0 button | |
| | 12 Click the **Compute** button | |

▮ ▮ ▮

# Exercises 9.4

## Understanding the Concepts and Skills

**9.73** State two reasons why including the $P$-value is prudent when you are reporting the results of a hypothesis test.

**9.74** What is the $P$-value of a hypothesis test? When does it provide evidence against the null hypothesis?

**9.75** We presented two different approaches to hypothesis testing. Identify and compare these two approaches.

**9.76** Explain how the $P$-value is obtained for a one-mean $z$-test in case the hypothesis test is
**a.** left tailed.    **b.** right tailed.    **c.** two tailed.

**9.77** True or false: The $P$-value is the smallest significance level for which the observed sample data result in rejection of the null hypothesis.

**9.78** The $P$-value for a hypothesis test is 0.06. For each of the following significance levels, decide whether the null hypothesis should be rejected.
**a.** $\alpha = 0.05$    **b.** $\alpha = 0.10$    **c.** $\alpha = 0.06$

**9.79** The $P$-value for a hypothesis test is 0.083. For each of the following significance levels, decide whether the null hypothesis should be rejected.
**a.** $\alpha = 0.05$    **b.** $\alpha = 0.10$    **c.** $\alpha = 0.06$

**9.80** Which provides stronger evidence against the null hypothesis, a $P$-value of 0.02 or a $P$-value of 0.03? Explain your answer.

**9.81** In each part, we have given the $P$-value for a hypothesis test. For each case, refer to Table 9.10 on page 414 to determine the strength of the evidence against the null hypothesis.

**a.** $P = 0.06$      **b.** $P = 0.35$
**c.** $P = 0.027$      **d.** $P = 0.004$

**9.82** In each part, we have given the *P*-value for a hypothesis test. For each case, refer to Table 9.10 on page 414 to determine the strength of the evidence against the null hypothesis.

**a.** $P = 0.184$      **b.** $P = 0.086$
**c.** $P = 0.001$      **d.** $P = 0.012$

*In each of Exercises 9.83–9.88, we have provided a sample mean, sample size, and population standard deviation. In each case, do the following.*
*a. Use the P-value approach to perform a one-mean z-test about the mean, μ, of the population from which the sample was drawn.*
*b. Refer to Table 9.10 on page 414 to determine the strength of the evidence against the null hypothesis.*

**9.83** $\bar{x} = 20$, $n = 32$, $\sigma = 4$, $H_0$: $\mu = 22$, $H_a$: $\mu < 22$, $\alpha = 0.05$

**9.84** $\bar{x} = 21$, $n = 32$, $\sigma = 4$, $H_0$: $\mu = 22$, $H_a$: $\mu < 22$, $\alpha = 0.05$

**9.85** $\bar{x} = 24$, $n = 15$, $\sigma = 4$, $H_0$: $\mu = 22$, $H_a$: $\mu > 22$, $\alpha = 0.05$

**9.86** $\bar{x} = 23$, $n = 15$, $\sigma = 4$, $H_0$: $\mu = 22$, $H_a$: $\mu > 22$, $\alpha = 0.05$

**9.87** $\bar{x} = 23$, $n = 24$, $\sigma = 4$, $H_0$: $\mu = 22$, $H_a$: $\mu \neq 22$, $\alpha = 0.05$

**9.88** $\bar{x} = 20$, $n = 24$, $\sigma = 4$, $H_0$: $\mu = 22$, $H_a$: $\mu \neq 22$, $\alpha = 0.05$

*In Exercises 9.89–9.94, we have given the value obtained for the test statistic*

$$z = \frac{\bar{x} - \mu_0}{\sigma/\sqrt{n}}$$

*in a one-mean z-test. We have also specified whether the test is two tailed, left tailed, or right tailed. Determine the P-value in each case.*

**9.89** Right-tailed test:
**a.** $z = 2.03$      **b.** $z = -0.31$

**9.90** Left-tailed test:
**a.** $z = -1.84$      **b.** $z = 1.25$

**9.91** Left-tailed test:
**a.** $z = -0.74$      **b.** $z = 1.16$

**9.92** Two-tailed test:
**a.** $z = 3.08$      **b.** $z = -2.42$

**9.93** Two-tailed test:
**a.** $z = -1.66$      **b.** $z = 0.52$

**9.94** Right-tailed test:
**a.** $z = 1.24$      **b.** $z = -0.69$

*In Exercises 9.59–9.64 of Section 9.3, you were asked to use Procedure 9.1, which employs the critical-value approach to hypothesis testing, to perform a one-mean z-test. Now, in Exercises 9.95–9.100, you are asked to use Procedure 9.2 on page 412, which employs the P-value approach to hypothesis testing, to perform those same hypothesis tests. In addition, use Table 9.10 on page 414 to assess the strength of the evidence against the null hypotheses.*

**9.95 Toxic Mushrooms?** Cadmium, a heavy metal, is toxic to animals. Mushrooms, however, are able to absorb and accumulate cadmium at high concentrations. The Czech and Slovak governments have set a safety limit for cadmium in dry vegetables at 0.5 part per million (ppm). M. Melgar et al. measured the cadmium levels in a random sample of the edible mushroom *Boletus pinicola* and published the results in the *Journal of Environmental Science and Health* (Vol. B33(4), pp. 439–455). Here are the data.

| | | | | | |
|---|---|---|---|---|---|
| 0.24 | 0.59 | 0.62 | 0.16 | 0.77 | 1.33 |
| 0.92 | 0.19 | 0.33 | 0.25 | 0.59 | 0.32 |

At the 5% significance level, do the data provide sufficient evidence to conclude that the mean cadmium level in *Boletus pinicola* mushrooms is greater than the government's recommended limit of 0.5 ppm? Assume that the population standard deviation of cadmium levels in *Boletus pinicola* mushrooms is 0.37 ppm. (*Note:* The sum of the data is 6.31 ppm.)

**9.96 Agriculture Books.** The R. R. Bowker Company of New York collects information on the retail prices of books and publishes the data in *The Bowker Annual Library and Book Trade Almanac*. In 2000, the mean retail price of agriculture books was $66.52. This year's retail prices for 28 randomly selected agriculture books are shown in the following table.

| | | | | | | |
|---|---|---|---|---|---|---|
| 68.45 | 76.61 | 66.01 | 55.02 | 55.77 | 71.78 | 75.31 |
| 60.99 | 46.58 | 59.38 | 69.33 | 47.05 | 67.12 | 56.26 |
| 59.36 | 79.94 | 57.05 | 75.09 | 68.27 | 50.54 | 62.57 |
| 58.86 | 67.99 | 66.95 | 55.56 | 75.67 | 59.52 | 75.59 |

At the 10% significance level, do the data provide sufficient evidence to conclude that this year's mean retail price of agriculture books has changed from the 2000 mean? Assume that the standard deviation of prices for this year's agriculture books is $8.45. (*Note:* The sum of the data is $1788.62.)

**9.97 Iron Deficiency?** Iron is essential to most life forms and to normal human physiology. It is an integral part of many proteins and enzymes that maintain good health. Recommendations for iron are provided in *Dietary Reference Intakes*, developed by the Institute of Medicine of the National Academy of Sciences. The recommended dietary allowance (RDA) of iron for adult females under the age of 51 is 18 milligrams (mg) per day. The following iron intakes, in milligrams, were obtained during a 24-hour period for 45 randomly selected adult females under the age of 51.

| | | | | | | | | |
|---|---|---|---|---|---|---|---|---|
| 15.0 | 18.1 | 14.4 | 14.6 | 10.9 | 18.1 | 18.2 | 18.3 | 15.0 |
| 16.0 | 12.6 | 16.6 | 20.7 | 19.8 | 11.6 | 12.8 | 15.6 | 11.0 |
| 15.3 | 9.4 | 19.5 | 18.3 | 14.5 | 16.6 | 11.5 | 16.4 | 12.5 |
| 14.6 | 11.9 | 12.5 | 18.6 | 13.1 | 12.1 | 10.7 | 17.3 | 12.4 |
| 17.0 | 6.3 | 16.8 | 12.5 | 16.3 | 14.7 | 12.7 | 16.3 | 11.5 |

At the 1% significance level, do the data suggest that adult females under the age of 51 are, on average, getting less than the RDA of 18 mg of iron? Assume that the population standard deviation is 4.2 mg. (*Note:* $\bar{x} = 14.68$ mg.)

**9.98 Early-Onset Dementia.** Dementia is the loss of the intellectual and social abilities severe enough to interfere with judgment, behavior, and daily functioning. Alzheimer's disease is the most common type of dementia. In the article "Living with Early Onset Dementia: Exploring the Experience and Developing Evidence-Based Guidelines for Practice" (*Alzheimer's Care Quarterly*, Vol. 5, Issue 2, pp. 111–122), P. Harris and J. Keady explored the experience and struggles of people diagnosed with dementia and their families. A simple random sample of 21 people with early-onset dementia gave the following data on age at diagnosis.

| | | | | | | |
|---|---|---|---|---|---|---|
| 60 | 58 | 52 | 58 | 59 | 58 | 51 |
| 61 | 54 | 59 | 55 | 53 | 44 | 46 |
| 47 | 42 | 56 | 57 | 49 | 41 | 43 |

At the 1% significance level, do the data provide sufficient evidence to conclude that the mean age at diagnosis of all people with early-onset dementia is less than 55 years old? Assume that the population standard deviation is 6.8 years. (*Note:* $\bar{x} = 52.5$ years.)

**9.99 Serving Time.** According to the Bureau of Crime Statistics and Research of Australia, as reported on *Lawlink*, the mean length of imprisonment for motor-vehicle theft offenders in Australia is 16.7 months. One hundred randomly

selected motor-vehicle theft offenders in Sydney, Australia, had a mean length of imprisonment of 17.8 months. At the 5% significance level, do the data provide sufficient evidence to conclude that the mean length of imprisonment for motor-vehicle theft offenders in Sydney differs from the national mean in Australia? Assume that the population standard deviation of the lengths of imprisonment for motor-vehicle theft offenders in Sydney is 6.0 months.

**9.100 Worker Fatigue.** A study by M. Chen et al. titled "Heat Stress Evaluation and Worker Fatigue in a Steel Plant" (*American Industrial Hygiene Association*, Vol. 64, pp. 352–359) assessed fatigue in steel-plant workers due to heat stress. A random sample of 29 casting workers had a mean post-work heart rate of 78.3 beats per minute (bpm). At the 5% significance level, do the data provide sufficient evidence to conclude that the mean post-work heart rate for casting workers exceeds the normal resting heart rate of 72 bpm? Assume that the population standard deviation of post-work heart rates for casting workers is 11.2 bpm.

**9.101 Job Gains and Losses.** In the article "Business Employment Dynamics: New Data on Gross Job Gains and Losses" (*Monthly Labor Review*, Vol. 127, Issue 4, pp. 29–42), J. Spletzer et al. examined gross job gains and losses as a percentage of the average of previous and current employment figures. A simple random sample of 20 quarters provided the net percentage gains (losses are negative gains) for jobs as presented on the WeissStats CD. Use the technology of your choice to do the following.
a. Decide whether, on average, the net percentage gain for jobs exceeds 0.2. Assume a population standard deviation of 0.42. Apply the one-mean z-test with a 5% significance level. Use the P-value approach and assess the evidence against the null hypothesis.
b. Obtain a normal probability plot, boxplot, histogram, and stem-and-leaf diagram of the data.
c. Remove the outliers (if any) from the data and then repeat part (a).
d. Comment on the advisability of using the z-test here.

**9.102 Hotels and Motels.** The daily charges, in dollars, for a sample of 15 hotels and motels operating in South Carolina are provided on the WeissStats CD. The data were found in the report *South Carolina Statistical Abstract*, sponsored by the South Carolina Budget and Control Board.
a. Use the one-mean z-test to decide, at the 5% significance level, whether the data provide sufficient evidence to conclude that the mean daily charge for hotels and motels operating in South Carolina is less than $75. Assume a population standard deviation of $22.40. Use the P-value approach and assess the evidence against the null hypothesis.

b. Obtain a normal probability plot, boxplot, histogram, and stem-and-leaf diagram of the data.

c. Remove the outliers (if any) from the data and then repeat part (a).

d. Comment on the advisability of using the *z*-test here.

## Working With Large Data Sets

**9.103 Body Temperature.** A study by researchers at the University of Maryland addressed the question of whether the mean body temperature of humans is 98.6°F. The results of the study by P. Mackowiak et al. appeared in the article "A Critical Appraisal of 98.6°F, the Upper Limit of the Normal Body Temperature, and Other Legacies of Carl Reinhold August Wunderlich" (*Journal of the American Medical Association*, Vol. 268, pp. 1578–1580). Among other data, the researchers obtained the body temperatures of 93 healthy humans, which we have provided on the WeissStats CD. Use the technology of your choice to do the following.

a. Obtain a normal probability plot, boxplot, histogram, and stem-and-leaf diagram of the data.

b. Based on your results from part (a), can you reasonably apply the one-mean *z*-test to the data? Explain your reasoning.

c. At the 1% significance level, do the data provide sufficient evidence to conclude that the mean body temperature of healthy humans differs from 98.6°F? Assume that $\sigma = 0.63°F$. Use the *P*-value approach and assess the evidence against the null hypothesis.

**9.104 Teacher Salaries.** The Educational Resource Service publishes information about wages and salaries in the public schools system in *National Survey of Salaries and Wages in Public Schools*. The mean annual salary of (public) classroom teachers is $45.9 thousand. A random sample of 90 classroom teachers in Hawaii yielded the annual salaries, in thousands of dollars, presented on the WeissStats CD. Use the technology of your choice to do the following.

a. Obtain a normal probability plot, boxplot, histogram, and stem-and-leaf diagram of the data.

b. Based on your results from part (a), can you reasonably apply the one-mean *z*-test to the data? Explain your reasoning.

c. At the 5% significance level, do the data provide sufficient evidence to conclude that the mean annual salary of classroom teachers in Hawaii is less than the national mean? Assume that the standard deviation of annual salaries for all classroom teachers in Hawaii is $9.2 thousand. Use the *P*-value approach and assess the evidence against the null hypothesis.

**9.105 Cell Phones.** The number of cell phone users has increased dramatically since 1987. According to the *Semi-annual Wireless Survey*, published by the Cellular Telecommunications & Internet Association, the mean local monthly bill for cell phone users in the United States was $47.37 in 2001. Last year's local monthly bills, in dollars, for a random sample of 75 cell phone users are given on the WeissStats CD. Use the technology of your choice to do the following.

a. Obtain a normal probability plot, boxplot, histogram, and stem-and-leaf diagram of the data.

b. At the 5% significance level, do the data provide sufficient evidence to conclude that last year's mean local monthly bill for cell phone users has increased from the 2001 mean of $47.37? Assume that the standard deviation of last year's local monthly bills for cell phone nusers is $25. Use the *P*-value approach and assess the evidence against the null hypothesis.

c. Remove the two outliers from the data and repeat parts (a) and (b).

d. State your conclusions regarding the hypothesis test.

## Extending the Concepts and Skills

**9.106** Consider a one-mean *z*-test. Denote $z_0$ as the observed value of the test statistic *z*. If the test is right tailed, then the *P*-value can be expressed as $P(z \geq z_0)$. Determine the corresponding expression for the *P*-value if the test is

a. left tailed.          b. two tailed.

**9.107** The symbol $\Phi(z)$ is often used to denote the area under the standard normal curve that lies to the left of a specified value of *z*. Consider a one-mean *z*-test. Denote $z_0$ as the observed value of the test statistic *z*. Express the *P*-value of the hypothesis test in terms of $\Phi$ if the test is

a. left tailed.     b. right tailed.     c. two tailed.

**9.108 Obtaining the *P*-value.** Let *x* denote the test statistic for a hypothesis test and $x_0$ its observed value. Then the *P*-value of the hypothesis test equals

a. $P(x \geq x_0)$ for a right-tailed test,

b. $P(x \leq x_0)$ for a left-tailed test,

c. $2 \cdot \min\{P(x \leq x_0), P(x \geq x_0)\}$ for a two-tailed test,

where the probabilities are computed under the assumption that the null hypothesis is true. Suppose that you are considering a one-mean *z*-test. Verify that the probability expressions in parts (a)–(c) are equivalent to those obtained in Exercise 9.106.

**9.109** Discuss the relative advantages and disadvantages of using the *P*-value approach to hypothesis testing instead of the critical-value approach.

# 9.5 Hypothesis Tests for One Population Mean When $\sigma$ Is Unknown

In Section 9.3, you learned how to perform a hypothesis test for one population mean when the population standard deviation, $\sigma$, is known. However, as we have mentioned, the population standard deviation is usually not known.

To develop a hypothesis-testing procedure for a population mean when $\sigma$ is unknown, we begin by recalling Key Fact 8.5: If a variable $x$ of a population is normally distributed with mean $\mu$, then, for samples of size $n$, the studentized version of $\bar{x}$

$$t = \frac{\bar{x} - \mu}{s/\sqrt{n}}$$

has the $t$-distribution with $n - 1$ degrees of freedom.

Because of Key Fact 8.5, we can perform a hypothesis test for a population mean when the population standard deviation is unknown by proceeding in essentially the same way as when it is known. The only difference is that we invoke a $t$-distribution instead of the standard normal distribution. Specifically, for a test with null hypothesis $H_0$: $\mu = \mu_0$, we employ the variable

$$t = \frac{\bar{x} - \mu_0}{s/\sqrt{n}}$$

as our test statistic and use the $t$-table, Table IV, to obtain the critical value(s) or $P$-value. We call this hypothesis-testing procedure the **one-mean $t$-test** or, when no confusion can arise, simply the **$t$-test.**[†]

### P-Values for a t-Test

Before presenting a step-by-step procedure for conducting a (one-mean) $t$-test, we need to discuss $P$-values for such a test. $P$-values for a $t$-test are obtained in a manner similar to that for a $z$-test.

As we know, if the null hypothesis is true, the test statistic for a $t$-test has the $t$-distribution with $n - 1$ degrees of freedom; so its probabilities equal areas under the $t$-curve with df $= n - 1$. Thus, if we let $t_0$ be the observed value of the test statistic $t$, we determine the $P$-value as follows.

- *Two-tailed test:* The $P$-value is the probability of observing a value of the test statistic $t$ that is at least as large in magnitude as the value actually observed, which is the area under the $t$-curve that lies outside the interval from $-|t_0|$ to $|t_0|$, as shown in Fig. 9.19(a).
- *Left-tailed test:* The $P$-value is the probability of observing a value of the test statistic $t$ that is as small as or smaller than the value actually observed, which is the area under the $t$-curve that lies to the left of $t_0$, as shown in Fig. 9.19(b).
- *Right-tailed test:* The $P$-value is the probability of observing a value of the test statistic $t$ that is as large as or larger than the value actually observed, which is the area under the $t$-curve that lies to the right of $t_0$, as shown in Fig. 9.19(c).

---

[†]The one-mean $t$-test is also known as the **one-sample $t$-test** and the **one-variable $t$-test.** We prefer "one-mean" because it makes clear the parameter being tested.

**FIGURE 9.19**
P-value for a t-test if the test is
(a) two tailed, (b) left tailed,
or (c) right tailed

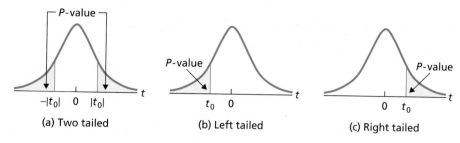

(a) Two tailed            (b) Left tailed            (c) Right tailed

## Estimating the P-Value of a t-Test

To obtain the exact P-value of a t-test, we need statistical software or a statistical calculator. However, we can use t-tables, such as Table IV, to estimate the P-value of a t-test, and an estimate of the P-value is usually sufficient for deciding whether to reject the null hypothesis.

For instance, consider a right-tailed t-test with $n = 15$, $\alpha = 0.05$, and a value of the test statistic of $t = 3.458$. For df $= 15 - 1 = 14$, the t-value 3.458 is larger than any t-value in Table IV, the largest one being $t_{0.005} = 2.977$ (which means that the area under the t-curve that lies to the right of 2.977 equals 0.005). This fact, in turn, implies that the area to the right of 3.458 is less than 0.005; in other words, $P < 0.005$. Because the P-value is less than the designated significance level of 0.05, we reject $H_0$.

Example 9.14 provides two more illustrations of how Table IV can be used to estimate the P-value of a t-test.

**Example 9.14** | **Using Table IV to Estimate the P-Value of a t-Test**

Use Table IV to estimate the P-value of each one-mean t-test.

a. Left-tailed test, $n = 12$, and $t = -1.938$

b. Two-tailed test, $n = 25$, and $t = -0.895$

**Solution**

a. Because the test is left tailed, the P-value is the area under the t-curve with df $= 12 - 1 = 11$ that lies to the left of $-1.938$, as shown in Fig. 9.20(a).

**FIGURE 9.20**
Estimating the P-value of a left-tailed
t-test with a sample size of 12
and test statistic $t = -1.938$

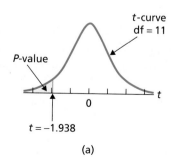

(a)

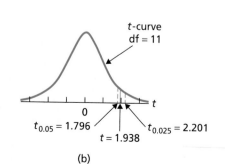

(b)

A $t$-curve is symmetric about 0, so the area to the left of $-1.938$ equals the area to the right of 1.938, which we can estimate by using Table IV. In the df $= 11$ row of Table IV, the two $t$-values that straddle 1.938 are $t_{0.05} = 1.796$ and $t_{0.025} = 2.201$. Therefore the area under the $t$-curve that lies to the right of 1.938 is between 0.025 and 0.05, as shown in Fig. 9.20(b).

Consequently, the area under the $t$-curve that lies to the left of $-1.938$ is also between 0.025 and 0.05, so $0.025 < P < 0.05$. Hence we can reject $H_0$ at any significance level of 0.05 or larger, and we cannot reject $H_0$ at any significance level of 0.025 or smaller. For significance levels between 0.025 and 0.05, Table IV is not sufficiently detailed to help us to decide whether to reject $H_0$.[†]

**b.** Because the test is two tailed, the $P$-value is the area under the $t$-curve with df $= 25 - 1 = 24$ that lies either to the left of $-0.895$ or to the right of 0.895, as shown in Fig. 9.21(a).

**FIGURE 9.21**
Estimating the $P$-value of a two-tailed $t$-test with a sample size of 25 and test statistic $t = -0.895$

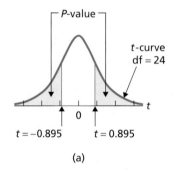

(a)

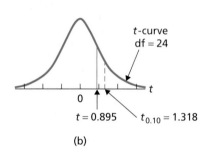

(b)

Because a $t$-curve is symmetric about 0, the areas to the left of $-0.895$ and to the right of 0.895 are equal. In the df $= 24$ row of Table IV, 0.895 is smaller than any other $t$-value, the smallest being $t_{0.10} = 1.318$. The area under the $t$-curve that lies to the right of 0.895, therefore, is greater than 0.10, as shown in Fig. 9.21(b).

Consequently, the area under the $t$-curve that lies either to the left of $-0.895$ or to the right of 0.895 is greater than 0.20, so $P > 0.20$. Hence we cannot reject $H_0$ at any significance level of 0.20 or smaller. For significance levels larger than 0.20, Table IV is not sufficiently detailed to help us to decide whether to reject $H_0$.

**You try it!**

Exercise 9.111 on page 428

• • •

## The One-Mean $t$-Test

We now present Procedure 9.3, a step-by-step method for performing a one-mean $t$-test. As you can see, Procedure 9.3 includes both the critical-value approach for a one-mean $t$-test and the $P$-value approach for a one-mean $t$-test.

---

[†]This latter case is an example of a $P$-value estimate that is not good enough. In such cases, use statistical software or a statistical calculator to find the exact $P$-value.

## Procedure 9.3  One-Mean *t*-Test

*Purpose* To perform a hypothesis test for a population mean, $\mu$

*Assumptions*

1. Simple random sample
2. Normal population or large sample
3. $\sigma$ unknown

**STEP 1**  The null hypothesis is $H_0$: $\mu = \mu_0$, and the alternative hypothesis is

$$H_a: \mu \neq \mu_0 \quad \text{or} \quad H_a: \mu < \mu_0 \quad \text{or} \quad H_a: \mu > \mu_0$$
$$\text{(Two tailed)} \qquad \text{(Left tailed)} \qquad \text{(Right tailed)}$$

**STEP 2**  Decide on the significance level, $\alpha$.

**STEP 3**  Compute the value of the test statistic

$$t = \frac{\bar{x} - \mu_0}{s/\sqrt{n}}$$

and denote that value $t_0$.

| CRITICAL-VALUE APPROACH | or | P-VALUE APPROACH |
|---|---|---|

**STEP 4**  The critical value(s) are

$$\pm t_{\alpha/2} \quad \text{or} \quad -t_\alpha \quad \text{or} \quad t_\alpha$$
$$\text{(Two tailed)} \qquad \text{(Left tailed)} \qquad \text{(Right tailed)}$$

with df $= n - 1$.  Use Table IV to find the critical value(s).

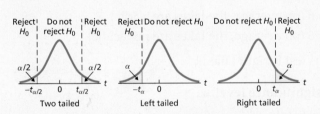

**STEP 4**  The *t*-statistic has df $= n - 1$.  Use Table IV to estimate the *P*-value, or obtain it exactly by using technology.

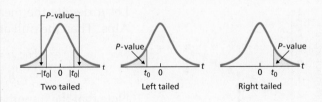

**STEP 5**  If $P \leq \alpha$, reject $H_0$; otherwise, do not reject $H_0$.

**STEP 5**  If the value of the test statistic falls in the rejection region, reject $H_0$; otherwise, do not reject $H_0$.

**STEP 6**  Interpret the results of the hypothesis test.

The hypothesis test is exact for normal populations and is approximately correct for large samples from nonnormal populations.

Properties and guidelines for use of the *t*-test are the same as those for the *z*-test, as given in Key Fact 9.4 on page 400. In particular, the *t*-test is robust to moderate violations of the normality assumption but, even for large samples, can sometimes be unduly affected by outliers because the sample mean and sample standard deviation are not resistant to outliers.

## Example 9.15 | The One-Mean *t*-Test

*Acid Rain and Lake Acidity* Acid rain from the burning of fossil fuels has caused many of the lakes around the world to become acidic. The biology in these lakes often collapses because of the rapid and unfavorable changes in water chemistry. A lake is classified as nonacidic if it has a pH greater than 6.

Aldo Marchetto and Andrea Lami measured the pH of high mountain lakes in the Southern Alps and reported their findings in the paper "Reconstruction of pH by Chrysophycean Scales in Some Lakes of the Southern Alps" (*Hydrobiologia*, Vol. 274, pp. 83–90). Table 9.11 shows the pH levels obtained by the researchers for 15 lakes. At the 5% significance level, do the data provide sufficient evidence to conclude that, on average, high mountain lakes in the Southern Alps are nonacidic?

**TABLE 9.11**
pH levels for 15 lakes

| | | | | |
|---|---|---|---|---|
| 7.2 | 7.3 | 6.1 | 6.9 | 6.6 |
| 7.3 | 6.3 | 5.5 | 6.3 | 6.5 |
| 5.7 | 6.9 | 6.7 | 7.9 | 5.8 |

**Solution** Figure 9.22, a normal probability plot of the data in Table 9.11, reveals no outliers and is quite linear. Consequently, we can apply Procedure 9.3 to conduct the required hypothesis test.

**FIGURE 9.22**
Normal probability plot of pH levels in Table 9.11

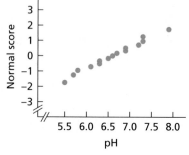

## STEP 1  State the null and alternative hypotheses.

Let $\mu$ denote the mean pH level of all high mountain lakes in the Southern Alps. Then the null and alternative hypotheses are

$$H_0: \mu = 6 \text{ (on average, the lakes are acidic)}$$
$$H_a: \mu > 6 \text{ (on average, the lakes are nonacidic)}.$$

Note that the hypothesis test is right tailed.

## STEP 2  Decide on the significance level, $\alpha$.

We are to perform the test at the 5% significance level, so $\alpha = 0.05$.

## STEP 3  Compute the value of the test statistic

$$t = \frac{\bar{x} - \mu_0}{s/\sqrt{n}}.$$

We have $\mu_0 = 6$ and $n = 15$ and calculate the mean and standard deviation of the sample data in Table 9.11 as 6.6 and 0.672, respectively. Hence the value of the test statistic is

$$t = \frac{6.6 - 6}{0.672/\sqrt{15}} = 3.458.$$

| CRITICAL-VALUE APPROACH | or | *P*-VALUE APPROACH |

**STEP 4** **The critical value for a right-tailed test is $t_\alpha$ with df $= n - 1$. Use Table IV to find the critical value.**

We have $n = 15$ and $\alpha = 0.05$. Table IV shows that for df $= 15 - 1 = 14$, $t_{0.05} = 1.761$, as shown in Fig. 9.23A.

**FIGURE 9.23A**

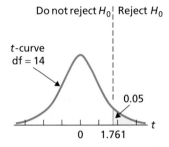

**STEP 5** **If the value of the test statistic falls in the rejection region, reject $H_0$; otherwise, do not reject $H_0$.**

The value of the test statistic, found in Step 3, is $t = 3.458$. Figure 9.23A reveals that it falls in the rejection region. Consequently, we reject $H_0$. The test results are statistically significant at the 5% level.

**STEP 4** **The $t$-statistic has df $= n - 1$. Use Table IV to estimate the $P$-value, or obtain it exactly by using technology.**

From Step 3, the value of the test statistic is $t = 3.458$. The test is right tailed, so the $P$-value is the probability of observing a value of $t$ of 3.458 or greater if the null hypothesis is true. That probability equals the shaded area in Fig. 9.23B.

**FIGURE 9.23B**

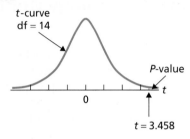

We have $n = 15$, and so df $= 15 - 1 = 14$. From Fig. 9.23B and Table IV, $P < 0.005$. (Using technology, we obtain $P = 0.00192$.)

**STEP 5** **If $P \le \alpha$, reject $H_0$; otherwise, do not reject $H_0$.**

From Step 4, $P < 0.005$. Because the $P$-value is less than the specified significance level of 0.05, we reject $H_0$. The test results are statistically significant at the 5% level and (see Table 9.10 on page 414) provide very strong evidence against the null hypothesis.

*You try it!*

Exercise 9.123
on page 428

**STEP 6** **Interpret the results of the hypothesis test.**

**Interpretation** At the 5% significance level, the data provide sufficient evidence to conclude that, on average, high mountain lakes in the Southern Alps are nonacidic.

• • •

## What If the Assumptions Are Not Satisfied?

Suppose you want to perform a hypothesis test for a population mean based on a small sample but preliminary data analyses indicate either the presence of outliers or that the variable under consideration is far from normally distributed. As neither the *z*-test nor the *t*-test is appropriate, what can you do?

Under certain conditions, you can use a nonparametric method. For example, if the variable under consideration has a symmetric distribution, you can use a nonparametric method called the *Wilcoxon signed-rank test* to perform a hypothesis test for the population mean.

As we said earlier, most nonparametric methods do not require even approximate normality, are resistant to outliers and other extreme values, and can be applied regardless of sample size. However, parametric methods, such as the $z$-test and $t$-test, tend to give more accurate results than nonparametric methods when the normality assumption and other requirements for their use are met.

We do not cover nonparametric methods in this book. But many basic statistics books do discuss them. See, for example, *Introductory Statistics, 8/e*, by Neil A. Weiss (Boston: Addison-Wesley, 2008).

# The Technology Center

Most statistical technologies have programs that automatically perform a one-mean $t$-test. In this subsection, we present output and step-by-step instructions for such programs.

**Example 9.16** ## Using Technology to Conduct a One-Mean *t*-Test

*Acid Rain and Lake Acidity*    Table 9.11 on page 424 gives the pH levels of a sample of 15 lakes in the Southern Alps. Use Minitab, Excel, or the TI-83/84 Plus to decide, at the 5% significance level, whether the data provide sufficient evidence to conclude that, on average, high mountain lakes in the Southern Alps are nonacidic.

**Solution**    Let $\mu$ denote the mean pH level of all high mountain lakes in the Southern Alps. We want to perform the hypothesis test

$$H_0: \mu = 6 \text{ (on average, the lakes are acidic)}$$
$$H_a: \mu > 6 \text{ (on average, the lakes are nonacidic)}$$

at the 5% significance level. Note that the hypothesis test is right tailed.

We applied the one-mean $t$-test programs to the data, resulting in Output 9.2. Steps for generating that output are presented in Instructions 9.2.

As shown in Output 9.2, the $P$-value for the hypothesis test is 0.002. The $P$-value is less than the specified significance level of 0.05, so we reject $H_0$. At the 5% significance level, the data provide sufficient evidence to conclude that, on average, high mountain lakes in the Southern Alps are nonacidic.

• • •

**OUTPUT 9.2**

One-mean *t*-test for the sample
of pH levels

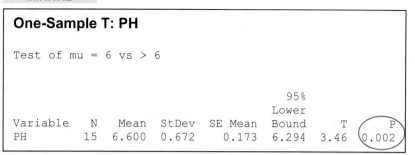

**MINITAB**

## One-Sample T: PH

```
Test of mu = 6 vs > 6

                                        95%
                                      Lower
Variable    N    Mean   StDev  SE Mean  Bound     T      P
PH         15   6.600   0.672    0.173  6.294   3.46   0.002
```

**EXCEL**

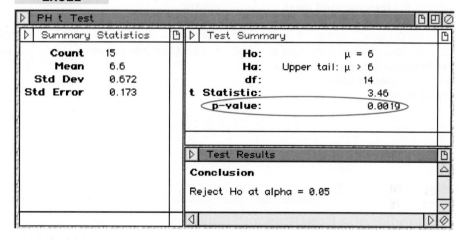

```
PH t Test

 Summary Statistics                    Test Summary

    Count    15                   Ho:              μ = 6
    Mean    6.6                   Ha:    Upper tail: μ > 6
  Std Dev   0.672                 df:               14
  Std Error 0.173            t Statistic:          3.46
                                  p-value:        0.0019


                              Test Results

                          Conclusion

                          Reject Ho at alpha = 0.05
```

**TI-83/84 PLUS**

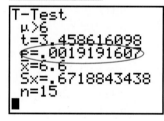

```
T-Test
 μ>6
 t=3.458616098
 p=.0019191607
 x̄=6.6
 Sx=.6718843438
 n=15
```

Using **Calculate**

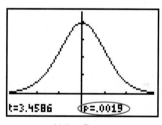

```
t=3.4586    p=.0019
```

Using **Draw**

**INSTRUCTIONS 9.2** Steps for generating Output 9.2

| MINITAB | EXCEL | TI-83/84 PLUS |
|---|---|---|
| 1 Store the data from Table 9.11 in a column named PH | 1 Store the data from Table 9.11 in a range named PH | 1 Store the data from Table 9.11 in a list named PH |
| 2 Choose **Stat ➤ Basic Statistics ➤ 1-Sample t...** | 2 Choose **DDXL ➤ Hypothesis Tests** | 2 Press **STAT**, arrow over to **TESTS**, and press **2** |
| 3 Select the **Samples in columns** option button | 3 Select **1 Var t Test** from the **Function type** drop-down box | 3 Highlight **Data** and press **ENTER** |
| 4 Click in the **Samples in columns** text box and specify PH | 4 Specify PH in the **Quantitative Variable** text box | 4 Press the down-arrow key, type 6 for $\mu_0$, and press **ENTER** |
| 5 Check the **Perform hypothesis test** check box | 5 Click **OK** | 5 Press **2nd ➤ LIST** |
| 6 Click in the **Hypothesized mean** text box and type 6 | 6 Click the **Set $\mu$0** button and type 6 | 6 Arrow down to PH and press **ENTER** three times |
| 7 Click the **Options...** button | 7 Click **OK** | 7 Highlight $> \mu_0$ and press **ENTER** |
| 8 Click the arrow button at the right of the **Alternative** drop-down list box and select **greater than** | 8 Click the **0.05** button | 8 Press the down-arrow key, highlight **Calculate** or **Draw**, and press **ENTER** |
| 9 Click **OK** twice | 9 Click the $\mu > \mu$0 button | |
| | 10 Click the **Compute** button | |

▮ ▮ ▮

# Exercises 9.5

## Understanding the Concepts and Skills

**9.110** What is the difference in assumptions between the one-mean $t$-test and the one-mean $z$-test?

*In Exercises 9.111–9.116, do the following for each of the specified one-mean t-tests.*
*a. Use Table IV in Appendix A to estimate the P-value.*
*b. Based on your estimate in part (a), state at which significance levels the null hypothesis can be rejected, at which significance levels it cannot be rejected, and at which significance levels it is not possible to decide.*

**9.111** Right-tailed test, $n = 20$, and $t = 2.235$

**9.112** Right-tailed test, $n = 11$, and $t = 1.246$

**9.113** Left-tailed test, $n = 10$, and $t = -3.381$

**9.114** Left-tailed test, $n = 30$, and $t = -1.572$

**9.115** Two-tailed test, $n = 17$, and $t = -2.733$

**9.116** Two-tailed test, $n = 8$, and $t = 3.725$

*In each of Exercises 9.117–9.122, we have provided a sample mean, sample standard deviation, and sample size. In each case, do the following.*
*a. Use the one-mean t-test to perform the required hypothesis test about the mean, $\mu$, of the population from which the sample was drawn.*

*b. Find (or estimate) the P-value and refer to Table 9.10 on page 414 to determine the strength of the evidence against the null hypothesis.*

**9.117** $\bar{x} = 20$, $s = 4$, $n = 32$, $H_0$: $\mu = 22$, $H_a$: $\mu < 22$, $\alpha = 0.05$

**9.118** $\bar{x} = 21$, $s = 4$, $n = 32$, $H_0$: $\mu = 22$, $H_a$: $\mu < 22$, $\alpha = 0.05$

**9.119** $\bar{x} = 24$, $s = 4$, $n = 15$, $H_0$: $\mu = 22$, $H_a$: $\mu > 22$, $\alpha = 0.05$

**9.120** $\bar{x} = 23$, $s = 4$, $n = 15$, $H_0$: $\mu = 22$, $H_a$: $\mu > 22$, $\alpha = 0.05$

**9.121** $\bar{x} = 23$, $s = 4$, $n = 24$, $H_0$: $\mu = 22$, $H_a$: $\mu \neq 22$, $\alpha = 0.05$

**9.122** $\bar{x} = 20$, $s = 4$, $n = 24$, $H_0$: $\mu = 22$, $H_a$: $\mu \neq 22$, $\alpha = 0.05$

*Preliminary data analyses indicate that you can reasonably use a t-test to conduct each of the hypothesis tests required in Exercises 9.123–9.128. Perform each t-test, using either the critical-value approach or the P-value approach.*

**9.123 TV Viewing.** According to *Communications Industry Forecast & Report*, published by Veronis Suhler Stevenson of New York, NY, the average person watched 4.66 hours of television per day in 2002. A random sample of 20 people gave the following number of hours of television watched per day for last year.

| 3.5 | 7.4 | 10.0 | 5.2 | 3.9 |
| 1.8 | 2.0 | 4.0 | 3.6 | 7.6 |
| 2.3 | 4.6 | 7.9 | 5.6 | 4.2 |
| 2.6 | 6.8 | 7.1 | 2.6 | 4.0 |

At the 10% significance level, do the data provide sufficient evidence to conclude that the amount of television watched per day last year by the average person differed from that in 2002? (*Note:* $\bar{x} = 4.835$ hours and $s = 2.291$ hours.)

**9.124 Golf Robots.** Serious golfers and golf equipment companies sometimes make use of golf equipment testing labs to obtain precise information about particular club heads, club shafts, and golf balls. One golfer requested information about the Jazz Fat Cat 5-iron from Golf Laboratories, Inc. The company tested the club by using a robot to hit a Titleist NXT Tour ball six times with a head velocity of 85 miles per hour. The golfer wanted a club that, on average, would hit the ball more than 180 yards at that club speed. The total yards each ball traveled was as follows.

| 180 | 187 | 181 | 182 | 185 | 181 |

**a.** At the 5% significance level, do the data provide sufficient evidence to conclude that the club does what the golfer wants? (*Note:* The sample mean and sample standard deviation of the data are 182.7 yards and 2.7 yards, respectively.)
**b.** Repeat part (a) for a test at the 1% significance level.

**9.125 Brewery Effluent and Crops.** Because many industrial wastes contain nutrients that enhance crop growth, efforts are being made for environmental purposes to use such wastes on agricultural soils. Two researchers, M. Ajmal and A. Khan, reported their findings on experiments with brewery wastes used for agricultural purposes in the article "Effects of Brewery Effluent on Agricultural Soil and Crop Plants" (*Environmental Pollution (Series A)*, 33, pp. 341–351). The researchers studied the physico-chemical properties of effluent from Mohan Meakin Breweries Ltd. (MMBL), Ghazibad, UP, India, and "…its effects on the physico-chemical characteristics of agricultural soil, seed germination pattern, and the growth of two common crop plants." They assessed the impact of using different concentrations of the effluent: 25%, 50%, 75%, and 100%. The following data, based on the results of the study, provide the percentages of limestone in the soil obtained by using 100% effluent.

| 2.41 | 2.31 | 2.54 | 2.28 | 2.72 |
| 2.60 | 2.51 | 2.51 | 2.42 | 2.70 |

Do the data provide sufficient evidence to conclude, at the 1% level of significance, that the mean available limestone in soil treated with 100% MMBL effluent exceeds 2.30%, the percentage ordinarily found? (*Note:* $\bar{x} = 2.5$ and $s = 0.149$.)

**9.126 Apparel and Services.** According to the document *Consumer Expenditures*, a publication of the U.S. Bureau of Labor Statistics, the average consumer unit spent $1749 on apparel and services in 2002. That same year, 25 consumer units in the Northeast had the following annual expenditures, in dollars, on apparel and services.

| 1292 | 1470 | 2033 | 1695 | 1286 |
| 2236 | 2246 | 2205 | 1624 | 1747 |
| 2701 | 2042 | 2179 | 1873 | 2457 |
| 1857 | 1778 | 2280 | 1535 | 2025 |
| 2003 | 1764 | 2126 | 2215 | 1725 |

At the 5% significance level, do the data provide sufficient evidence to conclude that the 2002 mean annual expenditure on apparel and services for consumer units in the Northeast differed from the national mean of $1749? (*Note:* The sample mean and sample standard deviation of the data are $1935.76 and $350.90, respectively.)

**9.127 Ankle Brachial Index.** The Ankle Brachial Index (ABI) compares the blood pressure of a patient's arm to the blood pressure of the patient's leg. The ABI can be an indicator of different diseases, including arterial diseases. A healthy (or normal) ABI is 0.9 or greater. In a study by M. McDermott et al. titled "Sex Differences in Peripheral Arterial Disease: Leg Symptoms and Physical Functioning" (*Journal of the American Geriatrics Society*, Vol. 51, No. 2, pp. 222–228), the researchers obtained the ABI of 187 women with peripheral arterial disease. The results were a mean ABI of 0.64 with a standard deviation of 0.15. At the 5% significance level, do the data provide sufficient evidence to conclude that, on average, women with peripheral arterial disease have an unhealthy ABI?

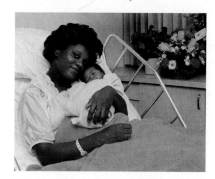

**9.128 Active Management of Labor.** Active management of labor (AML) is a group of interventions designed to help reduce the length of labor and the rate of cesarean deliveries. Physicians from the Department of Obstetrics and Gynecology at the University of New Mexico Health Sciences Center were interested in determining whether AML would also translate into a reduced cost for delivery. The results of their study can be found in Rogers et al., "Active Management of Labor: A Cost Analysis of a Randomized Controlled Trial" (*Western Journal of Medicine*, Vol. 172, pp. 240–243). According to the article, 200 AML deliveries had a mean cost of $2480 with a standard deviation of $766. At the time of the study, the average cost of having a baby in a U.S. hospital was $2528. At the 5% significance level, do the data provide sufficient evidence to conclude that, on average, AML reduces the cost of having a baby in a U.S. hospital?

*In each of Exercises 9.129–9.132, decide whether applying the t-test to perform a hypothesis test for the population mean in question appears reasonable. Explain your answers.*

**9.129 Cardiovascular Hospitalizations.** The Florida State Center for Health Statistics in the Agency for Health Care Administration reported in *Women and Cardiovascular Disease Hospitalizations* that, for cardiovascular hospitalizations, the mean age of women is 71.9 years. At one hospital, a random sample of 20 of its female cardiovascular patients had the following ages, in years.

| | | | | |
|---|---|---|---|---|
| 75.9 | 83.7 | 87.3 | 74.5 | 82.5 |
| 78.2 | 76.1 | 52.8 | 56.4 | 53.8 |
| 88.2 | 78.9 | 81.7 | 54.4 | 52.7 |
| 58.9 | 97.6 | 65.8 | 86.4 | 72.4 |

**9.130 Medieval Cremation Burials.** In the article "Material Culture as Memory: Combs and Cremations in Early Medieval Britain" (*Early Medieval Europe*, Vol. 12, Issue 2, pp. 89–128), H. Williams discussed the frequency of cremation burials found in 17 archaeological sites in eastern England. Here are the data.

| | | | | | | | | |
|---|---|---|---|---|---|---|---|---|
| 83 | 64 | 46 | 48 | 523 | 35 | 34 | 265 | 2484 |
| 46 | 385 | 21 | 86 | 429 | 51 | 258 | 119 | |

**9.131 Capital Spending.** An issue of *Brokerage Report* discussed the capital spending of telecommunications companies in the United States and Canada. The capital spending, in thousands of dollars, for each of 27 telecommunications companies is shown in the following table.

| | | | | | | |
|---|---|---|---|---|---|---|
| 9,310 | 2,515 | 3,027 | 1,300 | 1,800 | 70 | 3,634 |
| 656 | 664 | 5,947 | 649 | 682 | 1,433 | 389 |
| 17,341 | 5,299 | 195 | 8,543 | 4,200 | 7,886 | 11,189 |
| 1,006 | 1,403 | 1,982 | 21 | 125 | 2,205 | |

**9.132 Dating Artifacts.** In the paper "Reassessment of TL Age Estimates of Burnt Flint from the Paleolithic Site of Tabun Cave, Israel" (*Journal of Human Evolution*, Vol. 45, Issue 5, pp. 401–409), N. Mercier and H. Valladas discussed the re-dating of artifacts and human remains found at Tabun Cave by using new methodological improvements. A random sample of 18 excavated pieces yielded the following new thermoluminescence (TL) ages.

| | | | | | |
|---|---|---|---|---|---|
| 195 | 243 | 215 | 282 | 361 | 222 |
| 237 | 266 | 244 | 251 | 282 | 290 |
| 276 | 248 | 357 | 301 | 224 | 191 |

## Working With Large Data Sets

**9.133 Stressed-Out Bus Drivers.** Previous studies have shown that urban bus drivers have an extremely stressful job, and a large proportion of drivers retire prematurely with disabilities due to occupational stress. These stresses come from a combination of physical and social sources such as traffic congestion, incessant time pressure, and unruly passengers. In the paper, "Hassles on the Job: A Study of a Job Intervention With Urban Bus Drivers" (*Journal of Organizational Behavior*, Vol. 20, pp. 199–208), G. Evans et al. examined the effects of an intervention program to improve the conditions of urban bus drivers. Among other variables, the researchers monitored diastolic blood pressure of bus drivers in downtown Stockholm, Sweden. The data, in millimeters of mercury (mm Hg), on the WeissStats CD are based on the blood pressures obtained prior to intervention for the 41 bus drivers in the study. Use the technology of your choice to do the following.
a. Obtain a normal probability plot, boxplot, histogram, and stem-and-leaf diagram of the data.
b. Based on your results from part (a), can you reasonably apply the one-mean *t*-test to the data? Explain your reasoning.
c. At the 10% significance level, do the data provide sufficient evidence to conclude that the mean diastolic blood pressure of bus drivers in Stockholm exceeds the normal diastolic blood pressure of 80 mm Hg?

**9.134 How Far People Drive.** In 2000, the average car in the United States was driven 11.9 thousand miles, as reported by the U.S. Federal Highway Administration in *Highway Statistics*. On the WeissStats CD, we have provided last year's distance driven, in thousands of miles, by each of 500 randomly selected cars. Use the technology of your choice to do the following.
a. Obtain a normal probability plot and histogram of the data.
b. Based on your results from part (a), can you reasonably apply the one-mean *t*-test to the data? Explain your reasoning.
c. At the 5% significance level, do the data provide sufficient evidence to conclude that the mean distance driven last year differs from that in 2000?

**9.135 Fair Market Rent.** According to the document *Out of Reach*, published by the National Low Income Housing Coalition, the fair market rent (FMR) for a two-bedroom unit in Maine is $692. A sample of 100 randomly selected two-bedroom units in Maine yielded the data on monthly rents, in dollars, given on the WeissStats CD. Use the technology of your choice to do the following.

a. At the 5% significance level, do the data provide sufficient evidence to conclude that the mean monthly rent for two-bedroom units in Maine is greater than the FMR of $692? Apply the one-mean $t$-test.

b. Remove the outlier from the data and repeat the hypothesis test in part (a).

c. Comment on the effect that removing the outlier has on the hypothesis test.

d. State your conclusion regarding the hypothesis test and explain your answer.

## Extending the Concepts and Skills

**9.136** Suppose that you want to perform a hypothesis test for a population mean based on a small sample but that preliminary data analyses indicate either the presence of outliers or that the variable under consideration is far from normally distributed.

a. Is either the $z$-test or $t$-test appropriate?

b. If not, what type of procedure might be appropriate?

**9.137** Suppose that you want to perform a hypothesis test for a population mean. Assume that the variable under consideration is normally distributed and that the population standard deviation is unknown.

a. Is it permissible to use the $t$-test to perform the hypothesis test? Explain your answer.

b. Is it permissible to use the Wilcoxon signed-rank test to perform the hypothesis test? Explain your answer.

c. Which procedure is better to use, the $t$-test or the Wilcoxon signed-rank test? Explain your answer.

**9.138** Suppose that you want to perform a hypothesis test for a population mean. Assume that the variable under consideration has a symmetric nonnormal distribution and that the population standard deviation is unknown. Further assume that the sample size is large and that no outliers are present in the sample data.

a. Is it permissible to use the $t$-test to perform the hypothesis test? Explain your answer.

b. Is it permissible to use the Wilcoxon signed-rank test to perform the hypothesis test? Explain your answer.

c. Which procedure is better to use, the $t$-test or the Wilcoxon signed-rank test? Explain your answer.

**9.139 Two-Tailed Hypothesis Tests and CIs.** The following relationship holds between hypothesis tests and confidence intervals for one-mean $t$-procedures: For a two-tailed hypothesis test at the significance level $\alpha$, the null hypothesis $H_0$: $\mu = \mu_0$ will be rejected in favor of the alternative hypothesis $H_a$: $\mu \neq \mu_0$ if and only if $\mu_0$ lies outside the $(1 - \alpha)$-level confidence interval for $\mu$. In each case, illustrate the preceding relationship by obtaining the appropriate one-mean $t$-interval (Procedure 8.2 on page 363) and comparing the result to the conclusion of the hypothesis test in the specified exercise.

a. Exercise 9.123          b. Exercise 9.126

**9.140 Left-Tailed Hypothesis Tests and CIs.** In Exercise 8.110 on page 371, we introduced one-sided one-mean $t$-intervals. The following relationship holds between hypothesis tests and confidence intervals for one-mean $t$-procedures: For a left-tailed hypothesis test at the significance level $\alpha$, the null hypothesis $H_0$: $\mu = \mu_0$ will be rejected in favor of the alternative hypothesis $H_a$: $\mu < \mu_0$ if and only if $\mu_0$ is greater than the $(1 - \alpha)$-level upper confidence bound for $\mu$. In each case, illustrate the preceding relationship by obtaining the appropriate upper confidence bound and comparing the result to the conclusion of the hypothesis test in the specified exercise.

a. Exercise 9.127          b. Exercise 9.128

**9.141 Right-Tailed Hypothesis Tests and CIs.** In Exercise 8.110 on page 371, we introduced one-sided one-mean $t$-intervals. The following relationship holds between hypothesis tests and confidence intervals for one-mean $t$-procedures: For a right-tailed hypothesis test at the significance level $\alpha$, the null hypothesis $H_0$: $\mu = \mu_0$ will be rejected in favor of the alternative hypothesis $H_a$: $\mu > \mu_0$ if and only if $\mu_0$ is less than the $(1 - \alpha)$-level lower confidence bound for $\mu$. In each case, illustrate the preceding relationship by obtaining the appropriate lower confidence bound and comparing the result to the conclusion of the hypothesis test in the specified exercise.

a. Exercise 9.124 (both parts)

b. Exercise 9.125

## Chapter in Review

### You Should be Able to

1. use and understand the formulas in this chapter.

2. define the terms associated with hypothesis testing.

3. choose the null and alternative hypotheses for a hypothesis test.

4. explain the logic behind hypothesis testing.

5. identify the test statistic, rejection region, nonrejection region, and critical value(s) for a hypothesis test.

6. define and apply the concepts of Type I and Type II errors.

7. state and interpret the possible conclusions for a hypothesis test.

8. obtain the critical value(s) for a specified significance level.

9. perform a hypothesis test for a population mean when the population standard deviation is known.

10. obtain the $P$-value of a hypothesis test.

11. state and apply the steps for performing a hypothesis test, using the critical-value approach to hypothesis testing.

12. state and apply the steps for performing a hypothesis test, using the $P$-value approach to hypothesis testing.

13. perform a hypothesis test for a population mean when the population standard deviation is unknown.

### Key Terms

acceptance region, *388*
alternative hypothesis, *380*
critical-value approach to hypothesis
    testing, *408*
critical values, *388*
hypothesis, *380*
hypothesis test, *380*
left-tailed test, *381*
nonrejection region, *388*
not statistically significant, *392*
null hypothesis, *380*

observed significance level, *408*
one-mean *t*-test, *420, 423*
one-mean *z*-test, *399, 412*
one-tailed test, *381*
power, *391*
$P$-value (*P*), *408*
$P$-value approach to hypothesis
    testing, *408*
probability value, *408*
rejection region, *388*
research hypothesis, *380*

right-tailed test, *381*
significance level ($\alpha$), *391*
statistically significant, *392*
*t*-test, *420*
test statistic, *388*
two-tailed test, *381*
Type I error, *389*
Type I error probability ($\alpha$), *391*
Type II error, *389*
Type II error probability ($\beta$), *391*
*z*-test, *398*

## Review Problems

### Understanding the Concepts and Skills

1. Explain the meaning of each term.
a. Null hypothesis
b. Alternative hypothesis
c. Test statistic
d. Rejection region
e. Nonrejection region
f. Critical value(s)

2. The following statement appeared on a box of Tide laundry detergent: "Individual packages of Tide may weigh slightly more or less than the marked weight due to normal variations incurred with high speed packaging machines, but each day's production of Tide will average slightly above the marked weight."
a. Explain in statistical terms what the statement means.
b. Describe in words a hypothesis test for checking the statement.
c. Suppose that the marked weight is 76 ounces. State in words the null and alternative hypotheses for the hypothesis test. Then express those hypotheses in statistical terminology.

3. Regarding a hypothesis test:
a. What is the procedure, generally, for deciding whether the null hypothesis should be rejected?
b. How can the procedure identified in part (a) be made objective and precise?

4. There are three possible alternative hypotheses in a hypothesis test for a population mean. Identify them and explain when each is used.

5. Two types of incorrect decisions can be made in a hypothesis test: a Type I error and a Type II error.
a. Explain the meaning of each type of error.

**b.** Identify the letter used to represent the probability of each type of error.
**c.** If the null hypothesis is in fact true, only one type of error is possible. Which type is that? Explain your answer.
**d.** If you fail to reject the null hypothesis, only one type of error is possible. Which type is that? Explain your answer.

**6.** Suppose that you want to conduct a right-tailed hypothesis test at the 5% significance level. How must the critical value be chosen?

**7.** In each part, we have identified a hypothesis testing procedure for a population mean. State the assumptions required and the test statistic used in each case.
**a.** One-mean *t*-test          **b.** One-mean *z*-test

**8.** What is meant when we say that a hypothesis test is
**a.** exact?          **b.** approximately correct?

**9.** Discuss the difference between statistical significance and practical significance.

**10.** For a fixed sample size, what happens to the probability of a Type II error if the significance level is decreased from 0.05 to 0.01?

**11.** Regarding the *P*-value of a hypothesis test:
**a.** What is the *P*-value of a hypothesis test?
**b.** Answer true or false: A *P*-value of 0.02 provides more evidence against the null hypothesis than a *P*-value of 0.03. Explain your answer.
**c.** Answer true or false: A *P*-value of 0.74 provides essentially no evidence against the null hypothesis. Explain your answer.
**d.** Explain why the *P*-value of a hypothesis test is also referred to as the observed significance level.

**12.** Discuss the differences between the critical-value and *P*-value approaches to hypothesis testing.

**13. Cheese Consumption.** The U.S. Department of Agriculture reports in *Food Consumption, Prices, and Expenditures* that the average American consumed 30.0 lb of cheese in 2001. Cheese consumption has increased steadily since 1960 when the average American ate only 8.3 lb of cheese annually. Suppose that you want to decide whether last year's mean cheese consumption is greater than the 2001 mean.
**a.** Identify the null hypothesis.
**b.** Identify the alternative hypothesis.
**c.** Classify the hypothesis test as two tailed, left tailed, or right tailed.

**14.** The following graph portrays the decision criterion for a hypothesis test about a population mean, $\mu$. The null hypothesis for the test is $H_0$: $\mu = \mu_0$, and the test statistic is

$$z = \frac{\bar{x} - \mu_0}{\sigma/\sqrt{n}}.$$

The curve shown in the graph reveals the implications of the decision criterion if in fact the null hypothesis is true.

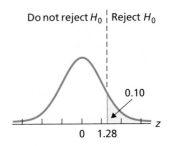

Determine the
**a.** rejection region.          **b.** nonrejection region.
**c.** critical value(s).          **d.** significance level.
**e.** Draw a graph that depicts the answers you obtained in parts (a)–(d).
**f.** Classify the hypothesis test as two tailed, left tailed, or right tailed.

**15. Cheese Consumption.** The null and alternative hypotheses for the hypothesis test in Problem 13 are

$H_0$: $\mu = 30.0$ lb (mean has not increased)
$H_a$: $\mu > 30.0$ lb (mean has increased),

where $\mu$ is last year's mean cheese consumption for all Americans. Explain what each of the following would mean.
**a.** Type I error          **b.** Type II error          **c.** Correct decision

Now suppose that the results of carrying out the hypothesis test lead to nonrejection of the null hypothesis. Classify that decision by error type or as a correct decision if in fact last year's mean cheese consumption
**d.** has not increased from the 2001 mean of 30.0 lb.
**e.** has increased from the 2001 mean of 30.0 lb.

**16. Cheese Consumption.** Refer to Problem 13. The following table provides last year's cheese consumption, in pounds, for 35 randomly selected Americans.

| 42 | 25 | 29 | 34 | 38 | 36 | 30 |
|----|----|----|----|----|----|----|
| 29 | 28 | 32 | 24 | 43 | 22 | 38 |
| 32 | 28 | 41 | 20 | 35 | 24 | 29 |
| 40 | 29 | 22 | 33 | 23 | 27 | 32 |
| 33 | 33 | 32 | 18 | 40 | 32 | 25 |

**a.** At the 10% significance level, do the data provide sufficient evidence to conclude that last year's mean cheese consumption for all Americans has increased over the 2001 mean? Assume that $\sigma = 6.9$ lb. For your hypothesis test, use a *z*-test and the critical-value approach. (*Note:* The sum of the data is 1078 lb.)
**b.** Given the conclusion in part (a), if an error has been made, what type must it be? Explain your answer.

**17. Cheese Consumption.** Refer to Problem 16.
**a.** Repeat the hypothesis test, using the *P*-value approach to hypothesis testing.
**b.** Use Table 9.10 on page 414 to assess the strength of the evidence against the null hypothesis.

**18. Purse Snatching.** The U.S. Federal Bureau of Investigation (FBI) compiles information on robbery and property crimes, by type and selected characteristic, and publishes its findings in *Population-at-Risk Rates and Selected Crime Indicators*. According to that document, the mean value lost to purse snatching was $332 in 2002. For last year, 12 randomly selected purse-snatching offenses yielded the following values lost, to the nearest dollar.

| | | | | | |
|---|---|---|---|---|---|
| 207 | 422 | 272 | 362 | 165 | 269 |
| 237 | 226 | 205 | 348 | 266 | 430 |

Use a *t*-test with either the critical-value approach or the *P*-value approach to decide, at the 5% significance level, whether last year's mean value lost to purse snatching has decreased from the 2002 mean. The mean and standard deviation of the data are $284.1 and $86.9, respectively.

**19. Betting the Spreads.** College basketball, and particularly the NCAA basketball tournament, is a popular venue for gambling, from novices in office betting pools to the high roller. To encourage uniform betting across teams, Las Vegas oddsmakers assign a point spread to each game. The *point spread* is the oddsmakers' prediction for the number of points by which the favored team will win. If you bet on the favorite, you win the bet provided the favorite wins by more than the point spread; otherwise, you lose the bet. Is the point spread a good measure of the relative ability of the two teams? H. Stern and B. Mock addressed this question in the paper "College Basketball Upsets: Will a 16-Seed Ever Beat a 1-Seed?" (*Chance*, Vol. 11(1), pp. 27–31). They obtained the difference between the actual margin of victory and the point spread, called the *point-spread error*, for 2109 college basketball games. The mean point-spread error was found to be −0.2 point with a standard deviation of 10.9 points. For a particular game, a point-spread error

of 0 indicates that the point spread was a perfect estimate of the two teams' relative abilities.

a. If, on average, the oddsmakers are estimating correctly, what is the (population) mean point-spread error?
b. Use the data to decide, at the 5% significance level, whether the (population) mean point-spread error differs from 0.
c. Interpret your answer in part (b).

*Problems 20–29 each include a normal probability plot and either a frequency histogram or a stem-and-leaf diagram for a set of sample data. The intent is to use the sample data to perform a hypothesis test for the mean of the population from which the data were obtained. In each case, consult the graphs provided to decide whether to use the z-test, the t-test, or neither. Explain your answer.*

**20.** The normal probability plot and histogram of the data are depicted in Fig. 9.24; $\sigma$ is known.

**21.** The normal probability plot and stem-and-leaf diagram of the data are depicted in Fig. 9.25; $\sigma$ is unknown.

**22.** The normal probability plot and stem-and-leaf diagram of the data are shown in Fig. 9.26; $\sigma$ is known.

**23.** The normal probability plot and histogram of the data are shown in Fig. 9.27; $\sigma$ is known.

**24.** The normal probability plot and histogram of the data are shown in Fig. 9.28; $\sigma$ is unknown.

**25.** The normal probability plot and stem-and-leaf diagram of the data are shown in Fig. 9.29; $\sigma$ is unknown.

**26.** The normal probability plot and stem-and-leaf diagram of the data are shown in Fig. 9.30; $\sigma$ is unknown.

**27.** The normal probability plot and stem-and-leaf diagram of the data are shown in Fig. 9.31; $\sigma$ is unknown. (*Note:* The decimal parts of the observations were removed before the stem-and-leaf diagram was constructed.)

**28.** The normal probability plot and stem-and-leaf diagram of the data are shown in Fig. 9.32; $\sigma$ is known.

**29.** The normal probability plot and stem-and-leaf diagram of the data are shown in Fig. 9.33; $\sigma$ is known.

**FIGURE 9.24**

Normal probability plot and histogram for Problem 20

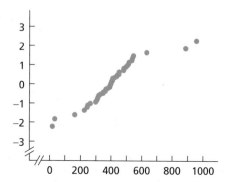
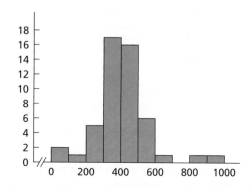

**FIGURE 9.25**

Normal probability plot
and stem-and-leaf diagram
for Problem 21

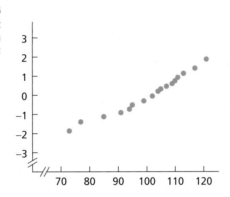

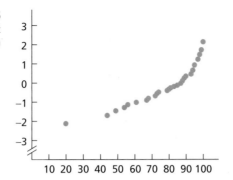

**FIGURE 9.26**

Normal probability plot
and stem-and-leaf diagram
for Problem 22

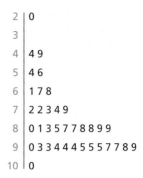

**FIGURE 9.27**

Normal probability plot and histogram
for Problem 23

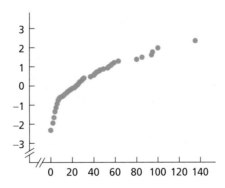

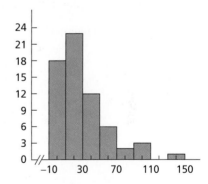

**FIGURE 9.28**

Normal probability plot and histogram
for Problem 24

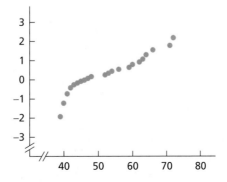

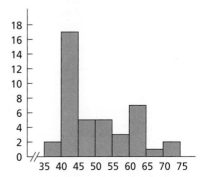

**FIGURE 9.29**
Normal probability plot
and stem-and-leaf diagram
for Problem 25

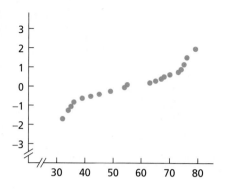

```
3 | 2 2 4
3 | 5 6 6 9
4 | 2
4 | 5 9 9
5 | 4 4
5 | 5
6 | 3
6 | 5 7 8
7 | 0 3 4
7 | 5 5 6 9
```

**FIGURE 9.30**
Normal probability plot
and stem-and-leaf diagram
for Problem 26

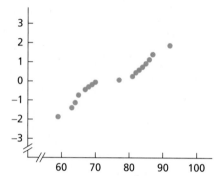

```
5 | 9
6 | 3 4
6 | 5 5 5 7 8 9
7 | 0
7 | 7
8 | 1 1 2 3 4
8 | 5 6 7
9 | 2
```

**FIGURE 9.31**
Normal probability plot
and stem-and-leaf diagram
for Problem 27

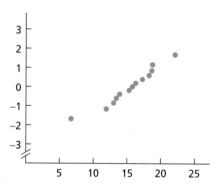

```
0 | 6
0 |
1 | 1
1 | 3 3 3
1 | 5 5
1 | 6 7
1 | 8 8 8
2 |
2 | 2
```

**FIGURE 9.32**
Normal probability plot
and stem-and-leaf diagram
for Problem 28

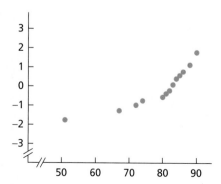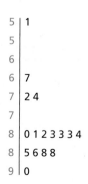

```
5 | 1
5 |
6 |
6 | 7
7 | 2 4
7 |
8 | 0 1 2 3 3 3 4
8 | 5 6 8 8
9 | 0
```

**FIGURE 9.33**
Normal probability plot
and stem-and-leaf diagram
for Problem 29

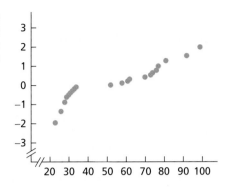

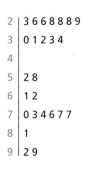

| 2 | 3 6 6 8 8 8 9 |
|---|---|
| 3 | 0 1 2 3 4 |
| 4 | |
| 5 | 2 8 |
| 6 | 1 2 |
| 7 | 0 3 4 6 7 7 |
| 8 | 1 |
| 9 | 2 9 |

## Working With Large Data Sets

**30. Beef Consumption.** According to *Food Consumption, Prices, and Expenditures*, published by the U.S. Department of Agriculture, the mean consumption of beef per person in 2002 was 64.5 lb (boneless, trimmed weight). A sample of 40 people taken this year yielded the data, in pounds, on last year's beef consumption given on the WeissStats CD. Use the technology of your choice to do the following.

a. Obtain a normal probability plot, a boxplot, a histogram, and a stem-and-leaf diagram of the data on beef consumptions.

b. Decide, at the 5% significance level, whether last year's mean beef consumption is less than the 2002 mean of 64.5 lb. Apply the one-mean $t$-test.

c. The sample data contain four potential outliers: 0, 0, 8, and 20. Remove those four observations, repeat the hypothesis test in part (b), and compare your result with that obtained in part (b).

d. Assuming that the four potential outliers are not recording errors, comment on the advisability of removing them from the sample data before performing the hypothesis test.

e. What action would you take regarding this hypothesis test?

**31. Body Mass Index.** Body Mass Index (BMI) is a measure of body fat based on height and weight. According to the document *Dietary Guidelines for Americans*, published by the U.S. Department of Agriculture and the U.S. Department of Health and Human Services, for adults, a BMI of greater than 25 indicates an above healthy weight. The BMIs of 75 randomly selected U.S. adults provided the data on the WeissStats CD. Use the technology of your choice to do the following.

a. Obtain a normal probability plot, a boxplot, and a histogram of the data.

b. Based on your graphs from part (a), is it reasonable to apply the one-mean $z$-test to the data? Explain your answer.

c. At the 5% significance level, do the data provide sufficient evidence to conclude that the average U.S. adult has an above healthy weight? Apply the one-mean $z$-test, assuming a standard deviation of 5.0 for the BMIs of all U.S. adults.

**32. Beer Drinking.** According to *The Beer Institute*, the mean annual consumption of beer per person in the United States is 22.0 gallons (roughly 235 twelve-ounce bottles). A random sample of 300 Washington D.C. residents yielded the annual beer consumptions provided on the WeissStats CD. Use the technology of your choice to do the following.

a. Obtain a histogram of the data.

b. Does your histogram in part (a) indicate any outliers?

c. At the 1% significance level, do the data provide sufficient evidence to conclude that the mean annual consumption of beer per person for the nation's capital differs from the national mean? (*Note:* See the third bulleted item in Key Fact 9.4 on page 400.)

# Focusing on Data Analysis   UWEC Undergraduates

Recall from Chapter 1 (see page 34) that the Focus database and Focus sample contain information on the undergraduate students at the University of Wisconsin - Eau Claire (UWEC). Now would be a good time for you to review the discussion about these data sets.

According to the document *High School Profile Report*, published by ACT, Inc., the national means for ACT composite, English, and math scores are 20.8, 20.3, and 20.6, respectively. You will use these national means in the following problems.

**a.** Apply the one-mean *t*-test to the ACT composite score data in the Focus sample (FocusSample) to decide, at the 5% significance level, whether the mean ACT composite score of UWEC undergraduates exceeds the national mean of 20.8 points. Interpret your result.

**b.** In practice, the (population) mean of the variable under consideration is unknown. However, in this case, we actually do have the population data, namely, in the Focus database (Focus). If your statistical software package will accommodate the entire Focus database, open that worksheet and then obtain the mean ACT composite score of all UWEC undergraduate students. (*Answer:* 23.6)

**c.** Was the decision concerning the hypothesis test in part (a) correct? Would it necessarily have to be? Explain your answers.

**d.** Repeat parts (a)–(c) for ACT English scores. (*Note:* The mean ACT English score of all UWEC undergraduate students is 23.0.)

**e.** Repeat parts (a)–(c) for ACT math scores. (*Note:* The mean ACT math score of all UWEC undergraduate students is 23.5.)

# Case Study Discussion   Sex and Sense of Direction

At the beginning of this chapter, we discussed research by J. Sholl et al. on the relationship between gender and sense of direction. Recall that, in their study, the spatial orientation skills of 30 male and 30 female students were challenged in a wooded park near the Boston College campus in Massachusetts. The participants were asked to rate their own sense of direction as either good or poor.

In the park, students were instructed to point to predesignated landmarks and also to the direction of south. For the female students who had rated their sense of direction to be good, the table on page 379 provides the pointing errors (in degrees) when they attempted to point south.

**a.** If, on average, women who consider themselves to have a good sense of direction do no better than they would by just randomly guessing at the direction of south, what would their mean pointing error be?

**b.** At the 1% significance level, do the data provide sufficient evidence to conclude that women who consider themselves to have a good sense of direction really do better, on average, than they would by just randomly guessing at the direction of south? Use a one-mean *t*-test.

**c.** Obtain a normal probability plot, boxplot, and stem-and-leaf diagram of the data. Based on these plots, is use of the *t*-test reasonable? Explain your answer.

**d.** Use the technology of your choice to perform the data analyses in parts (b) and (c).

**Biography**    JERZY NEYMAN: A Principal Founder of Modern Statistical Theory

**Jerzy Neyman** was born on April 16, 1894, in Russia. His father, Czeslaw, was a member of the Polish nobility, a lawyer, a judge, and an amateur archaeologist. Because Russian authorities prohibited the family from living in Poland, Jerzy Neyman grew up in various cities in Russia. He entered the university in Kharkov in 1912. At Kharkov he was at first interested in physics, but, because of his clumsiness in the laboratory, he decided to pursue mathematics.

After World War I, when Russia was at war with Poland over borders, Neyman was jailed as an enemy alien. In 1921, as a result of a prisoner exchange, he went to Poland for the first time. In 1924, he received his doctorate from the University of Warsaw. Between 1924 and 1934, Neyman worked with Karl Pearson (see Biography in Chapter 12) and his son Egon Pearson and held a position at the University of Kraków. In 1934, Neyman took a position in Karl Pearson's statistical laboratory at University College in London. He stayed in England, where he worked with Egon Pearson until 1938, at which time he accepted an offer to join the faculty at the University of California at Berkeley.

When the United States entered World War II, Neyman set aside development of a statistics program and did war work. After the war ended, Neyman organized a symposium to celebrate its end and "the return to theoretical research." That symposium, held in August 1945, and succeeding ones, held every 5 years until 1970, were instrumental in establishing Berkeley as a preeminent statistical center.

Neyman was a principal founder of the theory of modern statistics. His work on hypothesis testing, confidence intervals, and survey sampling transformed both the theory and the practice of statistics. His achievements were acknowledged by the granting of many honors and awards, including election to the United States National Academy of Sciences, the Guy Medal in Gold of the Royal Statistical Society, and the United States National Medal of Science.

Neyman remained active until his death of heart failure on August 5, 1981, at the age of 87, in Oakland, California.

 ## StatCrunch in MyStatLab
**Analyzing Data Online**

StatCrunch online statistical software offers an easy-to-use interface customized for this book. The StatCrunch feature for each chapter illustrates the use of the software to perform a statistical analysis discussed in the chapter. Exercises are provided to further apply StatCrunch to other statistical analyses examined in the chapter. Go to the WeissStats CD or to the Weiss Web site at www.aw-bc.com/weiss to access StatCrunch instructions and data sets. To access StatCrunch statistical software, go to the student content area of your Weiss MyStatLab course.

 ## Internet Projects
**Exploring Data Online**

The Internet project for each chapter provides simulations, demonstrations, or activities that enhance the topics covered in the chapter. The project materials come from universities, individuals, governments, and companies from all over the world. To access the Internet projects on the Web, go to www.aw-bc.com/weiss. From this Web page, you can reach the Internet Projects Page, which we suggest that you bookmark for easy access in the future.

# 10

# Inferences for Two Population Means

## Chapter Objectives

In Chapters 8 and 9, you learned how to obtain confidence intervals and perform hypothesis tests for one population mean. Frequently, however, inferential statistics is used to compare the means of two or more populations.

For example, we might want to perform a hypothesis test to decide whether the mean age of buyers of new domestic cars is greater than the mean age of buyers of new imported cars; or, we might want to find a confidence interval for the difference between the two mean ages.

Broadly speaking, in this chapter we examine two types of inferential procedures for comparing the means of two populations. The first type applies when the samples from the two populations are *independent,* meaning that the sample selected from one of the populations has no effect or bearing on the sample selected from the other population.

The second type of inferential procedure for comparing the means of two populations applies when the samples from the two populations are *paired.* A paired sample may be appropriate when there is a natural pairing of the members of the two populations such as husband and wife.

# Breast Milk and IQ

Considerable controversy exists over whether long-term neurodevelopment is affected by nutritional factors in early life. Five researchers summarized their findings on that question for preterm babies in the publication "Breast Milk and Subsequent Intelligence Quotient in Children Born Preterm" (*The Lancet*, Vol. 339, pp. 261–264). Their study was a continuation of work begun in January, 1982.

Previously, these five researchers had showed that a mother's decision to provide breast milk for preterm infants is associated with higher developmental scores for the children at age 18 months. In the article mentioned, they analyzed IQ data on the same children at age $7\frac{1}{2}$–8 years. IQ was measured for 300 children, using an abbreviated form of the Weschler Intelligence Scale for Children (revised Anglicized version: WISC-R UK).

The mothers of the children in the study had chosen whether to provide their infants with breast milk within 72 hours of delivery; 90 did not and 210 did. Of those 210 who chose to provide their infants with breast milk, 193 succeeded and 17 did not.

The children whose mothers declined to provide breast milk were designated by the researchers as Group I; those whose mothers had chosen but were unable to provide breast milk were designated as Group IIa; and those whose mothers had chosen and were able to provide breast milk were designated as Group IIb. The following table displays statistics for all three groups.

After studying the inferential methods discussed in this chapter, you will be able to conduct statistical analyses to examine how breast feeding affects subsequent IQ for children age $7\frac{1}{2}$–8 years who were born preterm.

| Group | Sample size | Mean IQ | St. Dev. |
|-------|-------------|---------|----------|
| I     | 90          | 92.8    | 15.2     |
| IIa   | 17          | 94.8    | 19.0     |
| IIb   | 193         | 103.7   | 15.3     |

# 10.1 The Sampling Distribution of the Difference Between Two Sample Means for Independent Samples

In this section, we lay the groundwork for making statistical inferences to compare the means of two populations. The methods that we first consider require not only that the samples selected from the two populations be simple random samples but also that they be **independent samples.** That is, the sample selected from one of the populations has no effect or bearing on the sample selected from the other population.

With **independent simple random samples,** each possible pair of samples (one from one population and one from the other) is equally likely to be the pair of samples selected. Example 10.1 provides an unrealistically simple illustration of independent samples, but it will help you understand the concept.

**Example 10.1** | **Introducing Independent Random Samples**

*Males and Females* Let's consider two small populations, one consisting of three men and the other of four women, as shown in the following figure.

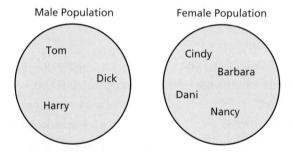

Suppose that we take a sample of size 2 from the male population and a sample of size 3 from the female population.

**a.** List the possible pairs of independent samples.

**b.** If the samples are selected at random, determine the chance of obtaining any particular pair of independent samples.

**Solution** For convenience, we use the first letter of each name as an abbreviation for the actual name.

**a.** In Table 10.1, the possible samples of size 2 from the male population are listed on the left; the possible samples of size 3 from the female population are listed on the right. To obtain the possible pairs of independent samples, we list each possible male sample of size 2 with each possible female sample of size 3, as shown in Table 10.2. There are 12 possible pairs of independent samples of two men and three women.

**b.** For independent simple random samples, each of the 12 possible pairs of samples shown in Table 10.2 is equally likely to be the pair selected. Therefore the chance of obtaining any particular pair of independent samples is $\frac{1}{12}$.

• • •

**TABLE 10.1**
Possible samples of size 2 from the male population and possible samples of size 3 from the female population

| Male sample of size 2 | Female sample of size 3 |
|:---:|:---:|
| T, D | C, B, D |
| T, H | C, B, N |
| D, H | C, D, N |
|  | B, D, N |

**TABLE 10.2**
Possible pairs of independent samples of two men and three women

| Male sample of size 2 | Female sample of size 3 |
|:---:|:---:|
| T, D | C, B, D |
| T, D | C, B, N |
| T, D | C, D, N |
| T, D | B, D, N |
| T, H | C, B, D |
| T, H | C, B, N |
| T, H | C, D, N |
| T, H | B, D, N |
| D, H | C, B, D |
| D, H | C, B, N |
| D, H | C, D, N |
| D, H | B, D, N |

The previous example provides a concrete illustration of independent samples and emphasizes that, for independent simple random samples of any given sizes, each possible pair of independent samples is equally likely to be the one selected. In practice, we neither obtain the number of possible pairs of independent samples nor explicitly compute the chance of selecting a particular pair of independent samples. But these concepts underlie the methods we do use.

**Note:** Recall that, when we say *random sample,* we mean *simple random sample* unless specifically stated otherwise. Likewise, when we say *independent random samples,* we mean *independent simple random samples,* unless specifically stated otherwise.

## Comparing Two Population Means, Using Independent Samples

We can now examine the process for comparing the means of two populations based on independent samples.

**Example 10.2** | **Comparing Two Population Means, Using Independent Samples**

*Faculty Salaries* The American Association of University Professors (AAUP) conducts salary studies of college professors and publishes its findings in *AAUP Annual Report on the Economic Status of the Profession.* Suppose that we want to decide whether the mean salaries of college faculty in public and private institutions are different.

a.  Pose the problem as a hypothesis test.

b.  Explain the basic idea for carrying out the hypothesis test.

c.  Suppose that 30 faculty members from public institutions and 35 faculty members from private institutions are randomly and independently selected and that their salaries are as shown in Table 10.3, in thousands of dollars rounded to the nearest hundred. Discuss the use of these data to make a decision concerning the hypothesis test.

| Sample 1 (public institutions) | | | | | | Sample 2 (private institutions) | | | | | | |
|------|------|-------|------|-------|------|-------|-------|-------|------|------|-------|------|
| 41.7 | 97.5 | 107.9 | 32.1 | 114.9 | 71.1 | 75.7 | 64.3 | 97.2 | 72.3 | 45.0 | 87.6 | 43.3 |
| 64.3 | 48.9 | 42.5 | 61.7 | 31.9 | 63.5 | 61.5 | 79.0 | 77.7 | 73.3 | 72.8 | 117.7 | 87.2 |
| 65.7 | 84.3 | 91.7 | 86.9 | 49.7 | 89.3 | 136.5 | 120.8 | 63.4 | 86.6 | 94.7 | 119.9 | 29.8 |
| 36.7 | 23.3 | 41.3 | 47.7 | 58.7 | 48.7 | 104.0 | 49.0 | 53.0 | 48.3 | 93.8 | 63.0 | 70.4 |
| 67.7 | 95.7 | 52.1 | 71.9 | 81.5 | 78.5 | 75.6 | 33.5 | 105.0 | 95.1 | 54.4 | 88.0 | 41.4 |

## Solution

a. We first note that we have one variable (salary) and two populations (faculty in public institutions and faculty in private institutions). Let the two populations in question be designated Populations 1 and 2, respectively:

Population 1: Faculty in public institutions

Population 2: Faculty in private institutions

Next, we denote the means of the variable "salary" for the two populations $\mu_1$ and $\mu_2$, respectively:

$\mu_1$ = mean salary of faculty in public institutions;

$\mu_2$ = mean salary of faculty in private institutions.

Then, we can state the hypothesis test we want to perform as

$H_0$: $\mu_1 = \mu_2$ (mean salaries are the same)

$H_a$: $\mu_1 \neq \mu_2$ (mean salaries are different).

b. Roughly speaking, we can carry out the hypothesis test as follows.

1. Independently and randomly take a sample of faculty members from public institutions (Population 1) and a sample of faculty members from private institutions (Population 2).
2. Compute the mean salary, $\bar{x}_1$, of the sample from public institutions and the mean salary, $\bar{x}_2$, of the sample from private institutions.
3. Reject the null hypothesis if the sample means, $\bar{x}_1$ and $\bar{x}_2$, differ by too much; otherwise, do not reject the null hypothesis.

This process is depicted in Fig. 10.1.

c. The means of the two samples in Table 10.3 are

$$\bar{x}_1 = \frac{\Sigma x_i}{n_1} = \frac{1949.4}{30} = 64.98 \quad \text{and} \quad \bar{x}_2 = \frac{\Sigma x_i}{n_2} = \frac{2680.8}{35} = 76.59.$$

The question now is, can the difference of 11.61 ($11,610) between these two sample means reasonably be attributed to sampling error or is the difference large enough to indicate that the two populations have different means? To answer that question, we need to know the distribution of the difference between two sample means—the *sampling distribution of the difference between two sample means*. We examine that sampling distribution in this section and complete the hypothesis test in the next section.

• • •

**FIGURE 10.1**

Process for comparing two population means, using independent samples

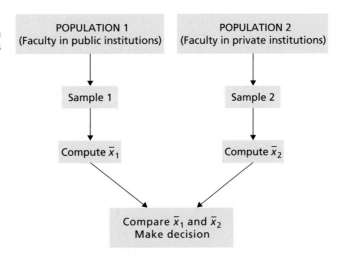

The Sampling Distribution of the Difference Between Two Sample Means for Independent Samples

First, we need to discuss the notation used for parameters and statistics when we are analyzing two populations. Let's call the two populations Population 1 and Population 2. Then, as indicated in the previous example, we use a subscript 1 when referring to parameters or statistics for Population 1 and a subscript 2 when referring to them for Population 2. See Table 10.4.

**TABLE 10.4**

Notation for parameters and statistics when considering two populations

|  | Population 1 | Population 2 |
|---|---|---|
| Population mean | $\mu_1$ | $\mu_2$ |
| Population standard deviation | $\sigma_1$ | $\sigma_2$ |
| Sample mean | $\bar{x}_1$ | $\bar{x}_2$ |
| Sample standard deviation | $s_1$ | $s_2$ |
| Sample size | $n_1$ | $n_2$ |

Armed with this notation, we describe in Key Fact 10.1 the **sampling distribution of the difference between two sample means.** Understanding Key Fact 10.1 is aided by recalling Key Fact 7.2: Suppose that a variable $x$ of a population is normally distributed and has mean $\mu$ and standard deviation $\sigma$. Then, for samples of size $n$,

- $\mu_{\bar{x}} = \mu$,
- $\sigma_{\bar{x}} = \sigma/\sqrt{n}$, and
- $\bar{x}$ is normally distributed.

---

**Key Fact 10.1**    **The Sampling Distribution of the Difference Between Two Sample Means for Independent Samples**

Suppose that $x$ is a normally distributed variable on each of two populations. Then, for independent samples of sizes $n_1$ and $n_2$ from the two populations,

- $\mu_{\bar{x}_1 - \bar{x}_2} = \mu_1 - \mu_2$,
- $\sigma_{\bar{x}_1 - \bar{x}_2} = \sqrt{(\sigma_1^2/n_1) + (\sigma_2^2/n_2)}$, and
- $\bar{x}_1 - \bar{x}_2$ is normally distributed.

In words, the first bulleted item says that the mean of all possible differences between the two sample means equals the difference between the two population means. The second bulleted item indicates that the standard deviation of all possible differences between the two sample means equals the square root of the sum of the population variances each divided by the corresponding sample size.

The formulas for the mean and standard deviation of $\bar{x}_1 - \bar{x}_2$ given in the first and second bulleted items, respectively, hold regardless of the distributions of the variable on the two populations. The assumption that the variable is normally distributed on each of the two populations is needed only to conclude that $\bar{x}_1 - \bar{x}_2$ is normally distributed (third bulleted item) and, because of the central limit theorem, that too holds approximately for large samples, regardless of distribution type.

Under the conditions of Key Fact 10.1, the standardized version of $\bar{x}_1 - \bar{x}_2$

$$z = \frac{(\bar{x}_1 - \bar{x}_2) - (\mu_1 - \mu_2)}{\sqrt{(\sigma_1^2/n_1) + (\sigma_2^2/n_2)}}$$

has the standard normal distribution. Using this fact, we can develop hypothesis-testing and confidence-interval procedures for comparing two population means when the population standard deviations are known.[†] However, because population standard deviations are usually unknown, we won't discuss those procedures. Instead, in Sections 10.2 and 10.3, we concentrate on the more usual situation where the population standard deviations are unknown.

## Exercises 10.1

### Understanding the Concepts and Skills

**10.1** Give an example of interest to you for comparing two population means. Identify the variable under consideration and the two populations.

**10.2** Define the phrase *independent samples.*

**10.3** Consider the quantities $\mu_1, \sigma_1, \bar{x}_1, s_1, \mu_2, \sigma_2, \bar{x}_2$, and $s_2$.
a. Which quantities represent parameters and which represent statistics?
b. Which quantities are fixed numbers and which are variables?

**10.4** Discuss the basic strategy for performing a hypothesis test to compare the means of two populations, based on independent samples.

**10.5** Why do you need to know the sampling distribution of the difference between two sample means in order to perform a hypothesis test to compare two population means?

**10.6** Identify the assumption for using the two-means $z$-test and the two-means $z$-interval procedure that renders those procedures generally impractical.

**10.7** Suppose that, in Example 10.2 on page 443, you want to decide whether the mean salary of faculty in public institutions is less than the mean salary of faculty in private institutions. State the null and alternative hypotheses for that hypothesis test.

**10.8** Suppose that, in Example 10.2 on page 443, you want to decide whether the mean salary of faculty in public institutions is greater than the mean salary of faculty in private institutions. State the null and alternative hypotheses for that hypothesis test.

*In Exercises 10.9–10.14, hypothesis tests are proposed. For each hypothesis test, do the following.*
*a. Identify the variable.*
*b. Identify the two populations.*
*c. Determine the null and alternative hypotheses.*
*d. Classify the hypothesis test as two tailed, left tailed, or right tailed.*

---

[†]We call these procedures the **two-means $z$-test** and the **two-means $z$-interval procedure,** respectively. The two-means $z$-test is also known as the **two-sample $z$-test** and the **two-variable $z$-test.** Likewise, the two-means $z$-interval procedure is also known as the **two-sample $z$-interval procedure** and the **two-variable $z$-interval procedure.**

**10.9 Offspring of Diabetic Mothers.** Samples of adolescent offspring of diabetic mothers (ODM) and nondiabetic mothers (ONM) were taken by Cho et al. and evaluated for potential differences in vital measurements, including blood pressure and glucose tolerance. The study was published in *The Journal of Pediatrics* (Vol. 136(5), pp. 587–592). A hypothesis test is to be performed to decide whether the mean systolic blood pressure of ODM adolescents exceeds that of ONM adolescents.

**10.10 Spending at the Mall.** An issue of *USA TODAY* discussed the amounts spent by teens and adults at shopping malls. Suppose that we want to perform a hypothesis test to decide whether the mean amount spent by teens is less than the mean amount spent by adults.

**10.11 Driving Distances.** Data on household vehicle miles of travel (VMT) are compiled annually by the Federal Highway Administration and are published in *National Household Travel Survey, Summary of Travel Trends*. A hypothesis test is to be performed to decide whether a difference exists in last year's mean VMT for households in the Midwest and South.

**10.12 Age of Car Buyers.** In the introduction to this chapter, we mentioned comparing the mean age of buyers of new domestic cars to the mean age of buyers of new imported cars. Suppose that we want to perform a hypothesis test to decide whether the mean age of buyers of new domestic cars is greater than the mean age of buyers of new imported cars.

**10.13 Neurosurgery Operative Times.** An ASU professor, R. Jacobowitz, Ph.D., along with G. Vishteh, M.D., and other neurosurgeons, obtained data on operative times, in minutes, for both a dynamic system (Z-plate) and a static system (ALPS plate). They wanted to perform a hypothesis test to decide whether the mean operative time is less with the dynamic system than with the static system.

**10.14 Wing Length.** D. Cristol et al. published results of their studies of two subspecies of dark-eyed juncos in *Animal Behaviour* (Vol. 66, pp. 317–328). One of the subspecies migrates each year and the other does not migrate. A hypothesis test is to be performed to decide whether the mean wing lengths for the two subspecies (migratory and nonmigratory) are different.

**10.15** A variable of two populations has a mean of 40 and a standard deviation of 12 for one of the populations and a mean of 40 and a standard deviation of 6 for the other population.
a. For independent samples of sizes 9 and 4, respectively, find the mean and standard deviation of $\bar{x}_1 - \bar{x}_2$.
b. Must the variable under consideration be normally distributed on each of the two populations for you to answer part (a)? Explain your answer.

c. Can you conclude that the variable $\bar{x}_1 - \bar{x}_2$ is normally distributed? Explain your answer.

**10.16** A variable of two populations has a mean of 7.9 and a standard deviation of 5.4 for one of the populations and a mean of 7.1 and a standard deviation of 4.6 for the other population.
a. For independent samples of sizes 3 and 6, respectively, find the mean and standard deviation of $\bar{x}_1 - \bar{x}_2$.
b. Must the variable under consideration be normally distributed on each of the two populations for you to answer part (a)? Explain your answer.
c. Can you conclude that the variable $\bar{x}_1 - \bar{x}_2$ is normally distributed? Explain your answer.

**10.17** A variable of two populations has a mean of 40 and a standard deviation of 12 for one of the populations and a mean of 40 and a standard deviation of 6 for the other population. Moreover, the variable is normally distributed on each of the two populations.
a. For independent samples of sizes 9 and 4, respectively, determine the mean and standard deviation of $\bar{x}_1 - \bar{x}_2$.
b. Can you conclude that the variable $\bar{x}_1 - \bar{x}_2$ is normally distributed? Explain your answer.
c. Determine the percentage of all pairs of independent samples of sizes 9 and 4, respectively, from the two populations with the property that the difference $\bar{x}_1 - \bar{x}_2$ between the sample means is between −10 and 10.

**10.18** A variable of two populations has a mean of 7.9 and a standard deviation of 5.4 for one of the populations and a mean of 7.1 and a standard deviation of 4.6 for the other population. Moreover, the variable is normally distributed on each of the two populations.
a. For independent samples of sizes 3 and 6, respectively, determine the mean and standard deviation of $\bar{x}_1 - \bar{x}_2$.
b. Can you conclude that the variable $\bar{x}_1 - \bar{x}_2$ is normally distributed? Explain your answer.
c. Determine the percentage of all pairs of independent samples of sizes 4 and 16, respectively, from the two populations with the property that the difference $\bar{x}_1 - \bar{x}_2$ between the sample means is between −3 and 4.

## Extending the Concepts and Skills

**10.19 Simulation.** To obtain the sampling distribution of the difference between two sample means for independent samples, as stated in Key Fact 10.1 on page 445, we need to know that, for independent observations, the difference of two normally distributed variables is also a normally distributed variable. In this exercise, you are to perform a computer simulation to make that fact plausible.
a. Simulate 2000 observations from a normally distributed variable with a mean of 100 and a standard deviation of 16.
b. Repeat part (a) for a normally distributed variable with a mean of 120 and a standard deviation of 12.

c. Determine the difference between each pair of observations in parts (a) and (b).
d. Obtain a histogram of the 2000 differences found in part (c). Why is the histogram bell shaped?

**10.20 Simulation.** In this exercise, you are to perform a computer simulation to illustrate the sampling distribution of the difference between two sample means for independent samples, Key Fact 10.1 on page 445.
a. Simulate 1000 samples of size 12 from a normally distributed variable with a mean of 640 and a standard deviation of 70. Obtain the sample mean of each of the 1000 samples.

b. Simulate 1000 samples of size 15 from a normally distributed variable with a mean of 715 and a standard deviation of 150. Obtain the sample mean of each of the 1000 samples.
c. Obtain the difference, $\bar{x}_1 - \bar{x}_2$, for each of the 1000 pairs of sample means obtained in parts (a) and (b).
d. Obtain the mean, the standard deviation, and a histogram of the 1000 differences found in part (c).
e. Theoretically, what are the mean, standard deviation, and distribution of all possible differences, $\bar{x}_1 - \bar{x}_2$?
f. Compare your answers from parts (d) and (e).

## 10.2 Inferences for Two Population Means, Using Independent Samples: Standard Deviations Assumed Equal

In Section 10.1, we laid the groundwork for developing inferential methods to compare the means of two populations based on independent samples. In this section, we develop such methods when the two populations have equal standard deviations; in Section 10.3, we develop such methods without that requirement.

### Hypothesis Tests for the Means of Two Populations With Equal Standard Deviations, Using Independent Samples

We now develop a procedure for performing a hypothesis test based on independent samples to compare the means of two populations with equal, but unknown, standard deviations. We must first find a test statistic for this test. In doing so, we assume that the variable under consideration is normally distributed on each population.

Let's use $\sigma$ to denote the common standard deviation of the two populations. We know from Key Fact 10.1 on page 445 that, for independent samples, the standardized version of $\bar{x}_1 - \bar{x}_2$

$$z = \frac{(\bar{x}_1 - \bar{x}_2) - (\mu_1 - \mu_2)}{\sqrt{(\sigma_1^2/n_1) + (\sigma_2^2/n_2)}}$$

has the standard normal distribution. Replacing $\sigma_1$ and $\sigma_2$ with their common value $\sigma$ and using some algebra, we obtain the variable

$$z = \frac{(\bar{x}_1 - \bar{x}_2) - (\mu_1 - \mu_2)}{\sigma\sqrt{(1/n_1) + (1/n_2)}}. \tag{10.1}$$

However, we cannot use this variable as a basis for the required test statistic because $\sigma$ is unknown.

Consequently, we need to use sample information to estimate $\sigma$, the unknown population standard deviation. We do so by first estimating the unknown population variance, $\sigma^2$. The best way to do that is to regard the sample variances, $s_1^2$ and $s_2^2$, as two estimates of $\sigma^2$ and then **pool** those estimates by weighting them according to sample size (actually by degrees of freedom). Thus our estimate of $\sigma^2$ is

$$s_p^2 = \frac{(n_1 - 1)s_1^2 + (n_2 - 1)s_2^2}{n_1 + n_2 - 2}$$

and hence that of $\sigma$ is

$$s_p = \sqrt{\frac{(n_1 - 1)s_1^2 + (n_2 - 1)s_2^2}{n_1 + n_2 - 2}}.$$

The subscript "p" stands for "pooled," and the quantity $s_p$ is called the **pooled sample standard deviation.**

Replacing $\sigma$ in Equation (10.1) with its estimate, $s_p$, we get the variable

$$\frac{(\bar{x}_1 - \bar{x}_2) - (\mu_1 - \mu_2)}{s_p\sqrt{(1/n_1) + (1/n_2)}}$$

which we can use as the required test statistic. Although the variable in Equation (10.1) has the standard normal distribution, this one has a $t$-distribution, with which you are already familiar.

---

**Key Fact 10.2**  **Distribution of the Pooled $t$-Statistic**

Suppose that $x$ is a normally distributed variable on each of two populations and that the population standard deviations are equal. Then, for independent samples of sizes $n_1$ and $n_2$ from the two populations, the variable

$$t = \frac{(\bar{x}_1 - \bar{x}_2) - (\mu_1 - \mu_2)}{s_p\sqrt{(1/n_1) + (1/n_2)}}$$

has the $t$-distribution with df $= n_1 + n_2 - 2$.

---

In light of Key Fact 10.2, for a hypothesis test that has null hypothesis $H_0$: $\mu_1 = \mu_2$ (population means are equal), we can use the variable

$$t = \frac{\bar{x}_1 - \bar{x}_2}{s_p\sqrt{(1/n_1) + (1/n_2)}}$$

as the test statistic and obtain the critical value(s) or $P$-value from the $t$-table, Table IV in Appendix A. We call this hypothesis-testing procedure the **pooled $t$-test.**[†] Procedure 10.1 (next page) provides a step-by-step method for performing a pooled $t$-test by using either the critical-value approach or the $P$-value approach.

---

[†]The pooled $t$-test is also known as the **two-sample $t$-test with equal variances assumed,** the **pooled two-variable $t$-test,** and the **pooled independent samples $t$-test.**

**Procedure 10.1    Pooled *t*-Test**

*Purpose*  To perform a hypothesis test to compare two population means, $\mu_1$ and $\mu_2$

*Assumptions*

1. Simple random samples
2. Independent samples
3. Normal populations or large samples
4. Equal population standard deviations

**STEP 1  The null hypothesis is $H_0$: $\mu_1 = \mu_2$, and the alternative hypothesis is**

$$H_a: \mu_1 \neq \mu_2 \quad \text{or} \quad H_a: \mu_1 < \mu_2 \quad \text{or} \quad H_a: \mu_1 > \mu_2$$
$$\text{(Two tailed)} \qquad \text{(Left tailed)} \qquad \text{(Right tailed)}$$

**STEP 2  Decide on the significance level, $\alpha$.**

**STEP 3  Compute the value of the test statistic**

$$t = \frac{\bar{x}_1 - \bar{x}_2}{s_p\sqrt{(1/n_1) + (1/n_2)}}$$

**where**

$$s_p = \sqrt{\frac{(n_1 - 1)s_1^2 + (n_2 - 1)s_2^2}{n_1 + n_2 - 2}}.$$

**Denote the value of the test statistic $t_0$.**

---

CRITICAL-VALUE APPROACH          or          P-VALUE APPROACH

**STEP 4  The critical value(s) are**

$$\pm t_{\alpha/2} \quad \text{or} \quad -t_\alpha \quad \text{or} \quad t_\alpha$$
$$\text{(Two tailed)} \qquad \text{(Left tailed)} \qquad \text{(Right tailed)}$$

**with df $= n_1 + n_2 - 2$. Use Table IV to find the critical value(s).**

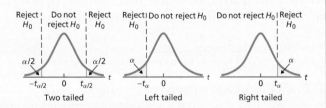

**STEP 4  The $t$-statistic has df $= n_1 + n_2 - 2$. Use Table IV to estimate the $P$-value, or obtain it exactly by using technology.**

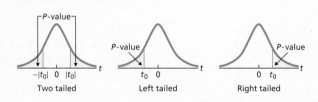

**STEP 5  If $P \leq \alpha$, reject $H_0$; otherwise, do not reject $H_0$.**

**STEP 5  If the value of the test statistic falls in the rejection region, reject $H_0$; otherwise, do not reject $H_0$.**

**STEP 6  Interpret the results of the hypothesis test.**

The hypothesis test is exact for normal populations and is approximately correct for large samples from nonnormal populations.

Regarding Assumptions 1 and 2, we note that the pooled $t$-test can also be used as a method for comparing two means with a designed experiment. Additionally, the pooled $t$-test is robust to moderate violations of Assumption 3 (normal populations) but, even for large samples, can sometimes be unduly affected by outliers because the sample mean and sample standard deviation are not resistant to outliers. The pooled $t$-test is also robust to moderate violations of Assumption 4 (equal population standard deviations) provided the sample sizes are roughly equal. We will say more about the robustness of the pooled $t$-test at the end of Section 10.3.

How can the conditions of normality and equal population standard deviations (Assumptions 3 and 4, respectively) be checked? As before, normality can be checked by using normal probability plots.

Checking equal population standard deviations can be difficult, especially when the sample sizes are small. As a rough rule of thumb, you can consider the condition of equal population standard deviations met if the ratio of the larger to the smaller sample standard deviation is less than 2. Comparing stem-and-leaf diagrams, histograms, or boxplots of the two samples is also helpful; be sure to use the same scales for each pair of graphs.[†]

**Example 10.3** | **The Pooled $t$-Test**

*Faculty Salaries* Let's return to the salary problem of Example 10.2, in which we want to perform a hypothesis test to decide whether the mean salaries of faculty in public institutions and private institutions are different.

Independent simple random samples of 30 faculty members in public institutions and 35 faculty members in private institutions yielded the data in Table 10.5. At the 5% significance level, do the data provide sufficient evidence to conclude that mean salaries for faculty in public and private institutions differ?

**TABLE 10.5**
Annual salaries ($1000s) for 30 faculty members in public institutions and 35 faculty members in private institutions

| Sample 1 (public institutions) | | | | | | Sample 2 (private institutions) | | | | | | |
|---|---|---|---|---|---|---|---|---|---|---|---|---|
| 41.7 | 97.5 | 107.9 | 32.1 | 114.9 | 71.1 | 75.7 | 64.3 | 97.2 | 72.3 | 45.0 | 87.6 | 43.3 |
| 64.3 | 48.9 | 42.5 | 61.7 | 31.9 | 63.5 | 61.5 | 79.0 | 77.7 | 73.3 | 72.8 | 117.7 | 87.2 |
| 65.7 | 84.3 | 91.7 | 86.9 | 49.7 | 89.3 | 136.5 | 120.8 | 63.4 | 86.6 | 94.7 | 119.9 | 29.8 |
| 36.7 | 23.3 | 41.3 | 47.7 | 58.7 | 48.7 | 104.0 | 49.0 | 53.0 | 48.3 | 93.8 | 63.0 | 70.4 |
| 67.7 | 95.7 | 52.1 | 71.9 | 81.5 | 78.5 | 75.6 | 33.5 | 105.0 | 95.1 | 54.4 | 88.0 | 41.4 |

**Solution** First, we find the required summary statistics for the two samples, as shown in Table 10.6. Next, we check the four conditions required for using the pooled $t$-test, as listed in Procedure 10.1.

**TABLE 10.6**
Summary statistics for the samples in Table 10.5

| Public institutions | Private institutions |
|---|---|
| $\bar{x}_1 = 64.98$ | $\bar{x}_2 = 76.59$ |
| $s_1 = 23.95$ | $s_2 = 26.21$ |
| $n_1 = 30$ | $n_2 = 35$ |

- The samples are given as simple random samples, hence Assumption 1 is satisfied.
- The samples are given as independent samples, hence Assumption 2 is satisfied.

---

[†]The assumption of equal population standard deviations is sometimes checked by performing a formal hypothesis test, called the two-standard-deviations $F$-test. We don't recommend that strategy because, although the pooled $t$-test is robust to moderate violations of normality, the two-standard-deviations $F$-test is extremely nonrobust to such violations. As the noted statistician George E. P. Box remarked: "To make a preliminary test on variances [standard deviations] is rather like putting to sea in a rowing boat to find out whether conditions are sufficiently calm for an ocean liner to leave port!"

- The sample sizes are 30 and 35, both of which are large; furthermore, Figs. 10.2 and 10.3 suggest no outliers for either sample. So, we can consider Assumption 3 satisfied.
- According to Table 10.6, the sample standard deviations are 23.95 and 26.21. These statistics are certainly close enough for us to consider Assumption 4 satisfied, as we also see from the boxplots in Fig. 10.3.

**FIGURE 10.2**

Normal probability plots of the sample data for faculty in (a) public institutions and (b) private institutions

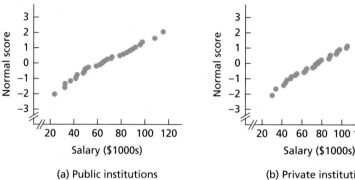

(a) Public institutions

(b) Private institutions

**FIGURE 10.3**

Boxplots of the salary data for faculty in public institutions and private institutions

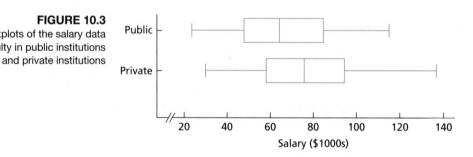

The preceding items suggest that the pooled $t$-test can be used to carry out the hypothesis test. We apply Procedure 10.1.

**STEP 1  State the null and alternative hypotheses.**

The null and alternative hypotheses are

$$H_0: \mu_1 = \mu_2 \text{ (mean salaries are the same)}$$
$$H_a: \mu_1 \neq \mu_2 \text{ (mean salaries are different)},$$

where $\mu_1$ and $\mu_2$ are the mean salaries of all faculty in public and private institutions, respectively. Note that the hypothesis test is two tailed.

**STEP 2  Decide on the significance level, $\alpha$.**

The test is to be performed at the 5% significance level, or $\alpha = 0.05$.

**STEP 3  Compute the value of the test statistic**

$$t = \frac{\bar{x}_1 - \bar{x}_2}{s_p\sqrt{(1/n_1) + (1/n_2)}}$$

**where**

$$s_p = \sqrt{\frac{(n_1 - 1)s_1^2 + (n_2 - 1)s_2^2}{n_1 + n_2 - 2}}.$$

To find the pooled sample standard deviation, $s_\text{p}$, we refer to Table 10.6:

$$s_\text{p} = \sqrt{\frac{(30-1) \cdot (23.95)^2 + (35-1) \cdot (26.21)^2}{30 + 35 - 2}} = 25.19.$$

Referring again to Table 10.6, we calculate the value of the test statistic:

$$t = \frac{\bar{x}_1 - \bar{x}_2}{s_\text{p}\sqrt{(1/n_1) + (1/n_2)}} = \frac{64.98 - 76.59}{25.19\sqrt{(1/30) + (1/35)}} = -1.852.$$

| CRITICAL-VALUE APPROACH | or | P-VALUE APPROACH |
|---|---|---|

**STEP 4** The critical values for a two-tailed test are $\pm t_{\alpha/2}$ with df $= n_1 + n_2 - 2$. Use Table IV to find the critical values.

From Table 10.6, $n_1 = 30$ and $n_2 = 35$, so df $= 30 + 35 - 2 = 63$. Also, from Step 2, we have $\alpha = 0.05$. In Table IV with df $= 63$, we find that the critical values are $\pm t_{\alpha/2} = \pm t_{0.05/2} = \pm t_{0.025} = \pm 1.998$, as shown in Fig. 10.4A.

**FIGURE 10.4A**

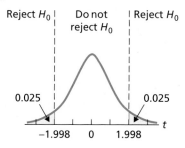

**STEP 5** If the value of the test statistic falls in the rejection region, reject $H_0$; otherwise, do not reject $H_0$.

From Step 3, the value of the test statistic is $t = -1.852$, which does not fall in the rejection region (see Fig. 10.4A). Thus we do not reject $H_0$. The test results are not statistically significant at the 5% level.

**STEP 4** The $t$-statistic has df $= n_1 + n_2 - 2$. Use Table IV to estimate the $P$-value or obtain it exactly by using technology.

From Step 3, the value of the test statistic is $t = -1.852$. The test is two tailed, so the $P$-value is the probability of observing a value of $t$ of 1.852 or greater in magnitude if the null hypothesis is true. That probability equals the shaded area in Fig. 10.4B.

**FIGURE 10.4B**

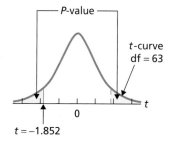

From Table 10.6, $n_1 = 30$ and $n_2 = 35$, so df $= 30 + 35 - 2 = 63$. Referring to Fig. 10.4B and to Table IV with df $= 63$, we find that $0.05 < P < 0.10$. (Using technology, we obtain $P = 0.0686$.)

**STEP 5** If $P \le \alpha$, reject $H_0$; otherwise, do not reject $H_0$.

From Step 4, $0.05 < P < 0.10$. Because the $P$-value exceeds the specified significance level of 0.05, we do not reject $H_0$. The test results are not statistically significant at the 5% level but (see Table 9.10 on page 414) the data provide moderate evidence against the null hypothesis.

**STEP 6** Interpret the results of the hypothesis test.

**Interpretation** At the 5% significance level, the data do not provide sufficient evidence to conclude that a difference exists between the mean salaries of faculty in public and private institutions.

*You try it!*

Exercise 10.33 on page 459

• • •

## Confidence Intervals for the Difference Between the Means of Two Populations With Equal Standard Deviations

We can also use Key Fact 10.2 on page 449 to derive a confidence-interval procedure, Procedure 10.2, for the difference between two population means, which we call the **pooled $t$-interval procedure.**[†]

---

| Procedure 10.2 | Pooled $t$-Interval Procedure |
|---|---|

*Purpose* To find a confidence interval for the difference between two population means, $\mu_1$ and $\mu_2$

*Assumptions*

1. Simple random samples
2. Independent samples
3. Normal populations or large samples
4. Equal population standard deviations

**STEP 1 For a confidence level of $1 - \alpha$, use Table IV to find $t_{\alpha/2}$ with df $= n_1 + n_2 - 2$.**

**STEP 2 The endpoints of the confidence interval for $\mu_1 - \mu_2$ are**

$$(\bar{x}_1 - \bar{x}_2) \pm t_{\alpha/2} \cdot s_p \sqrt{(1/n_1) + (1/n_2)}.$$

**STEP 3 Interpret the confidence interval.**

The confidence interval is exact for normal populations and is approximately correct for large samples from nonnormal populations.

---

| Example 10.4 | The Pooled $t$-Interval Procedure |
|---|---|

*Faculty Salaries* Obtain a 95% confidence interval for the difference, $\mu_1 - \mu_2$, between the mean salaries of faculty in public and private institutions.

**Solution** We apply Procedure 10.2.

**STEP 1 For a confidence level of $1 - \alpha$, use Table IV to find $t_{\alpha/2}$ with df $= n_1 + n_2 - 2$.**

For a 95% confidence interval, $\alpha = 0.05$. From Table 10.6, $n_1 = 30$ and $n_2 = 35$, so df $= n_1 + n_2 - 2 = 30 + 35 - 2 = 63$. In Table IV, we find that with df $= 63$, $t_{\alpha/2} = t_{0.05/2} = t_{0.025} = 1.998$.

---

[†]The pooled $t$-interval procedure is also known as the **two-sample $t$-interval procedure with equal variances assumed,** the **pooled two-variable $t$-interval procedure,** and the **pooled independent samples $t$-interval procedure.**

**STEP 2** The endpoints of the confidence interval for $\mu_1 - \mu_2$ are

$$(\bar{x}_1 - \bar{x}_2) \pm t_{\alpha/2} \cdot s_p\sqrt{(1/n_1) + (1/n_2)}.$$

From Step 1, $t_{\alpha/2} = 1.998$. Also, $n_1 = 30$, $n_2 = 35$, and, from Example 10.3, we know that $\bar{x}_1 = 64.98$, $\bar{x}_2 = 76.59$, and $s_p = 25.19$. Hence the endpoints of the confidence interval for $\mu_1 - \mu_2$ are

$$(64.98 - 76.59) \pm 1.998 \cdot 25.19\sqrt{(1/30) + (1/35)},$$

or $-11.61 \pm 12.52$. Thus the 95% confidence interval is from $-24.13$ to $0.91$.

**STEP 3** Interpret the confidence interval.

Exercise 10.39
on page 460

**Interpretation** We can be 95% confident that the difference between the mean salaries of faculty in public institutions and private institutions is somewhere between $-\$24,130$ and $\$910$.

• • •

## The Relation Between Hypothesis Tests and Confidence Intervals

Hypothesis tests and confidence intervals are closely related. Consider, for example, a two-tailed hypothesis test for comparing two population means at the significance level $\alpha$. In this case, the null hypothesis will be rejected if and only if the $(1 - \alpha)$-level confidence interval for $\mu_1 - \mu_2$ does not contain 0. You are asked to examine the relation between hypothesis tests and confidence intervals in greater detail in Exercises 10.51–10.53.

## What If the Assumptions Are Not Satisfied?

The pooled $t$-procedures (pooled $t$-test and pooled $t$-interval procedure) provide methods for comparing the means of two populations. As you know, the assumptions for using those procedures are (1) simple random samples, (2) independent samples, (3) normal populations or large samples, and (4) equal population standard deviations. If one or more of these conditions are not satisfied, then the pooled $t$-procedures should not be used.

If the samples are not independent but instead are paired (i.e., Assumption 2 is violated), then procedures designed for paired samples should be used. We discuss such procedures in Section 10.4.

If the populations are not normal and the samples are not large (i.e., Assumption 3 is violated), then a nonparametric method should be used. For example, if the samples are independent and the two distributions (one for each population) of the variable under consideration have the same shape, then you can use a nonparametric method called the *Mann–Whitney test* to perform a hypothesis test and a nonparametric method called the *Mann–Whitney confidence-interval procedure* to obtain a confidence interval.

If the population standard deviations are not equal (i.e., Assumption 4 is violated) but Assumptions 1–3 are satisfied, then you can use the nonpooled $t$-procedures. We present those procedures in the next section.

# The Technology Center

Most statistical technologies have programs that automatically perform pooled $t$-procedures. In this subsection, we present output and step-by-step instructions for such programs.

**Example 10.5** **Using Technology to Conduct Pooled *t*-Procedures**

*Faculty Salaries* Table 10.5 on page 451 shows the annual salaries, in thousands of dollars, for independent samples of 30 faculty members in public institutions and 35 faculty members in private institutions. Use Minitab, Excel, or the TI-83/84 Plus to perform the hypothesis test in Example 10.3 and obtain the confidence interval required in Example 10.4.

**Solution** Let $\mu_1$ and $\mu_2$ denote the mean salaries of all faculty in public and private institutions, respectively. The task in Example 10.3 is to perform the hypothesis test

$$H_0: \ \mu_1 = \mu_2 \ \text{(mean salaries are the same)}$$
$$H_a: \ \mu_1 \neq \mu_2 \ \text{(mean salaries are different)}$$

at the 5% significance level; the task in Example 10.4 is to obtain a 95% confidence interval for $\mu_1 - \mu_2$.

We applied the pooled $t$-procedures programs to the data, resulting in Output 10.1. Steps for generating that output are presented in Instructions 10.1 on page 458.

As shown in Output 10.1, the $P$-value for the hypothesis test is 0.069. Because the $P$-value exceeds the specified significance level of 0.05, we do not reject $H_0$. Output 10.1 also shows that a 95% confidence interval for the difference between the means is from $-24.14$ to $0.91$.

• • •

*Note to Minitab users:* Although Minitab simultaneously performs a hypothesis test and obtains a confidence interval, the type of confidence interval Minitab finds depends on the type of hypothesis test. Specifically, Minitab computes a two-sided confidence interval for a two-tailed test and a one-sided confidence interval for a one-tailed test. To perform a one-tailed hypothesis test and obtain a two-sided confidence interval, apply Minitab's pooled $t$-procedure twice: once for the one-tailed hypothesis test and once for the confidence interval specifying a two-tailed hypothesis test.

**OUTPUT 10.1**   Pooled $t$-procedures for the salary data

**MINITAB**

## Two-Sample T-Test and CI: PUBLIC, PRIVATE

```
Two-sample T for PUBLIC vs PRIVATE

              N   Mean   StDev   SE Mean
PUBLIC       30   65.0   24.0       4.4
PRIVATE      35   76.6   26.2       4.4


Difference = mu (PUBLIC) - mu (PRIVATE)
Estimate for difference:  -11.61
95% CI for difference: (-24.14, 0.91)
T-Test of difference = 0 (vs not =): T-Value = -1.85  P-Value = 0.069  DF = 63
Both use Pooled StDev = 25.1948
```

**EXCEL**

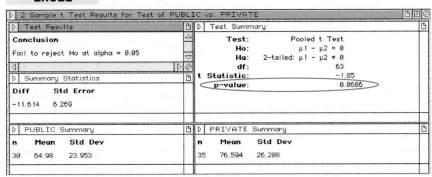

Using **2 Var t Test**

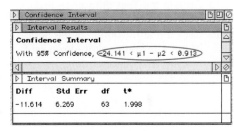

Using **2 Var t Interval**

**TI-83/84 PLUS**

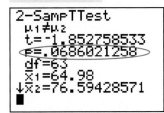

```
2-SampTTest
 μ₁≠μ₂
 t=-1.852758533
 p=.0686021258
 df=63
 x̄₁=64.98
↓x̄₂=76.59428571
■
```

```
2-SampTInt
 (-24.14,.9126)
 df=63
 x̄₁=64.98
 x̄₂=76.59428571
 Sx₁=23.9528042
↓Sx₂=26.2077553
■
```

```
2-SampTTest
 μ₁≠μ₂
↑Sx₁=23.9528042
 Sx₂=26.2077553
 SxP=25.1948429
 n₁=30
 n₂=35
■
```

```
2-SampTInt
 (-24.14,.9126)
↑Sx₁=23.9528042
 Sx₂=26.2077553
 SxP=25.1948429
 n₁=30
 n₂=35
■
```

Using **2-SampTTest**          Using **2-SampTInt**

**INSTRUCTIONS 10.1**  Steps for generating Output 10.1

| MINITAB | EXCEL | TI-83/84 PLUS |
|---|---|---|
| 1 Store the two samples of salary data from Table 10.5 in columns named PUBLIC and PRIVATE | Store the two samples of salary data from Table 10.5 in ranges named PUBLIC and PRIVATE. | Store the two samples of salary data from Table 10.5 in lists named PUBL and PRIV. |
| 2 Choose **Stat ➤ Basic Statistics ➤ 2-Sample t...** | **FOR THE HYPOTHESIS TEST:** | **FOR THE HYPOTHESIS TEST:** |
| 3 Select the **Samples in different columns** option button | 1 Choose **DDXL ➤ Hypothesis Tests** | 1 Press **STAT**, arrow over to **TESTS**, and press **4** |
| 4 Click in the **First** text box and specify PUBLIC | 2 Select **2 Var t Test** from the **Function type** drop-down box | 2 Highlight **Data** and press **ENTER** |
| 5 Click in the **Second** text box and specify PRIVATE | 3 Specify PUBLIC in the **1st Quantitative Variable** text box | 3 Press the down-arrow key |
| 6 Check the **Assume equal variances** check box | 4 Specify PRIVATE in the **2nd Quantitative Variable** text box | 4 Press **2nd ➤ LIST**, arrow down to PUBL, and press **ENTER** twice |
| 7 Click the **Options...** button | 5 Click **OK** | 5 Press **2nd ➤ LIST**, arrow down to PRIV, and press **ENTER** four times |
| 8 Click in the **Confidence level** text box and type 95 | 6 Click the **Pooled** button | 6 Highlight $\neq \mu 2$ and press **ENTER** |
| 9 Click in the **Test difference** text box and type 0 | 7 Click the **Set difference** button, type 0, and click **OK** | 7 Press the down-arrow key, highlight **Yes**, and press **ENTER** |
| 10 Click the arrow button at the right of the **Alternative** drop-down list box and select **not equal** | 8 Click the **0.05** button | 8 Press the down-arrow key, highlight **Calculate**, and press **ENTER** |
| 11 Click **OK** twice | 9 Click the $\mu 1 - \mu 2 \neq$ **diff** button | |
| | 10 Click the **Compute** button | **FOR THE CI:** |
| | | 1 Press **STAT**, arrow over to **TESTS**, and press **0** |
| | **FOR THE CI:** | 2 Highlight **Data** and press **ENTER** |
| | 1 Exit to Excel | 3 Press the down-arrow key |
| | 2 Choose **DDXL ➤ Confidence Intervals** | 4 Press **2nd ➤ LIST**, arrow down to PUBL, and press **ENTER** twice |
| | 3 Select **2 Var t Interval** from the **Function type** drop-down box | 5 Press **2nd ➤ LIST**, arrow down to PRIV, and press **ENTER** four times |
| | 4 Specify PUBLIC in the **1st Quantitative Variable** text box | 6 Type .95 for **C-Level** and press **ENTER** |
| | 5 Specify PRIVATE in the **2nd Quantitative Variable** text box | 7 Highlight **Yes**, and press **ENTER** |
| | 6 Click **OK** | 8 Press the down-arrow key and press **ENTER** |
| | 7 Click the **Pooled** button | |
| | 8 Click the **95%** button | |
| | 9 Click the **Compute Interval** button | |

■ ■ ■

# Exercises 10.2

## Understanding the Concepts and Skills

**10.21** Regarding the four conditions required for using the pooled *t*-procedures:
a. what are they?
b. how important is each condition?

**10.22** Explain why $s_p$ is called the pooled sample standard deviation.

*In each of Exercises 10.23–10.26, we have provided summary statistics for independent simple random samples from two pop-*ulations. *Preliminary data analyses indicate that the variable under consideration is normally distributed on each population. Decide, in each case, whether use of the pooled t-test and pooled t-interval procedure is reasonable. Explain your answer.*

**10.23** $\bar{x}_1 = 468.3$, $s_1 = 38.2$, $n_1 = 6$, $\bar{x}_2 = 394.6$, $s_2 = 84.7$, $n_2 = 14$

**10.24** $\bar{x}_1 = 115.1$, $s_1 = 79.4$, $n_1 = 51$, $\bar{x}_2 = 24.3$, $s_2 = 10.5$, $n_2 = 19$

**10.25** $\bar{x}_1 = 118$, $s_1 = 12.04$, $n_1 = 99$, $\bar{x}_2 = 110$, $s_2 = 11.25$, $n_2 = 80$

**10.26** $\bar{x}_1 = 39.04$, $s_1 = 18.82$, $n_1 = 51$, $\bar{x}_2 = 49.92$, $s_2 = 18.97$, $n_2 = 53$

*In each of Exercises 10.27–10.32, we have provided summary statistics for independent simple random samples from two populations. In each case, use the pooled t-test and the pooled t-interval procedure to conduct the required hypothesis test and obtain the specified confidence interval.*

**10.27** $\bar{x}_1 = 10$, $s_1 = 2.1$, $n_1 = 15$, $\bar{x}_2 = 12$, $s_2 = 2.3$, $n_2 = 15$
a. Two-tailed test, $\alpha = 0.05$
b. 95% confidence interval

**10.28** $\bar{x}_1 = 10$, $s_1 = 4$, $n_1 = 15$, $\bar{x}_2 = 12$, $s_2 = 5$, $n_2 = 15$
a. Two-tailed test, $\alpha = 0.05$
b. 95% confidence interval

**10.29** $\bar{x}_1 = 20$, $s_1 = 4$, $n_1 = 10$, $\bar{x}_2 = 18$, $s_2 = 5$, $n_2 = 15$
a. Right-tailed test, $\alpha = 0.05$
b. 90% confidence interval

**10.30** $\bar{x}_1 = 20$, $s_1 = 4$, $n_1 = 10$, $\bar{x}_2 = 23$, $s_2 = 5$, $n_2 = 15$
a. Left-tailed test, $\alpha = 0.05$
b. 90% confidence interval

**10.31** $\bar{x}_1 = 20$, $s_1 = 4$, $n_1 = 20$, $\bar{x}_2 = 24$, $s_2 = 5$, $n_2 = 15$
a. Left-tailed test, $\alpha = 0.05$
b. 90% confidence interval

**10.32** $\bar{x}_1 = 20$, $s_1 = 4$, $n_1 = 30$, $\bar{x}_2 = 18$, $s_2 = 5$, $n_2 = 40$
a. Right-tailed test, $\alpha = 0.05$
b. 90% confidence interval

*Preliminary data analyses indicate that you can reasonably consider the assumptions for using pooled t-procedures satisfied in Exercises 10.33–10.38. For each exercise, perform the required hypothesis test by using either the critical-value approach or the P-value approach.*

**10.33 Doing Time.** The U.S. Bureau of Prisons publishes data in *Prison Statistics* on the times served by prisoners released from federal institutions for the first time. Independent random samples of released prisoners in the fraud and firearms offense categories yielded the following information on time served, in months.

| Fraud | | Firearms | |
|---|---|---|---|
| 3.6 | 17.9 | 25.5 | 23.8 |
| 5.3 | 5.9 | 10.4 | 17.9 |
| 10.7 | 7.0 | 18.4 | 21.9 |
| 8.5 | 13.9 | 19.6 | 13.3 |
| 11.8 | 16.6 | 20.9 | 16.1 |

At the 5% significance level, do the data provide sufficient evidence to conclude that the mean time served for fraud

is less than that for firearms offenses? (*Note:* $\bar{x}_1 = 10.12$, $s_1 = 4.90$, $\bar{x}_2 = 18.78$, and $s_2 = 4.64$.)

**10.34 Sex and Direction.** In the paper "The Relation of Sex and Sense of Direction to Spatial Orientation in an Unfamiliar Environment" (*Journal of Environmental Psychology*, Vol. 20, pp. 17–28), Sholl et al. published the results of examining the sense of direction of 30 male and 30 female students. After being taken to an unfamiliar wooded park, the students were given some spatial orientation tests, including pointing to south, which tested their absolute frame of reference. The students pointed by moving a pointer attached to a 360° protractor. Following are the absolute pointing errors, in degrees, of the participants.

| Male | | | | | Female | | | | |
|---|---|---|---|---|---|---|---|---|---|
| 13 | 130 | 39 | 33 | 10 | 14 | 8 | 20 | 3 | 138 |
| 13 | 68 | 18 | 3 | 11 | 122 | 78 | 69 | 111 | 3 |
| 38 | 23 | 60 | 5 | 9 | 128 | 31 | 18 | 35 | 111 |
| 59 | 5 | 86 | 22 | 70 | 109 | 36 | 27 | 32 | 35 |
| 58 | 3 | 167 | 15 | 30 | 12 | 27 | 8 | 3 | 80 |
| 8 | 20 | 67 | 26 | 19 | 91 | 68 | 66 | 176 | 15 |

At the 1% significance level, do the data provide sufficient evidence to conclude that, on average, males have a better sense of direction and, in particular, a better frame of reference than females? (*Note:* $\bar{x}_1 = 37.6$, $s_1 = 38.5$, $\bar{x}_2 = 55.8$, and $s_2 = 48.3$.)

**10.35 Fortified Juice and PTH.** V. Tangpricha et al. did a study to determine whether fortifying orange juice with Vitamin D would result in changes in the blood levels of five biochemical variables. One of those variables was the concentration of parathyroid hormone (PTH), measured in picograms/milliliter (pg/mL). The researchers published their results in the paper "Fortification of Orange Juice with Vitamin D: A Novel Approach for Enhancing Vitamin D Nutritional Health" (*American Journal of Clinical Nutrition*, Vol. 77, pp. 1478–1483). A double-blind experiment was used in which 14 subjects drank 240 ml per day of orange juice fortified with 1000 IU of Vitamin D and 12 subjects drank 240 ml per day of unfortified orange juice. Concentration levels were recorded at the beginning of the experiment and again at the end of 12 weeks. The following data, based on the results of the study, provide the decrease (negative values indicate increase) in PTH levels for those drinking the fortified juice and for those drinking the unfortified juice.

| Fortified | | | | Unfortified | | |
|---|---|---|---|---|---|---|
| −7.7 | 11.2 | 65.8 | −45.6 | 65.1 | 0.0 | 40.0 |
| −4.8 | 26.4 | 55.9 | −15.5 | −48.8 | 15.0 | 8.8 |
| 34.4 | −5.0 | −2.2 | | 13.5 | −6.1 | 29.4 |
| −20.1 | −40.2 | 73.5 | | −20.5 | −48.4 | −28.7 |

At the 5% significance level, do the data provide sufficient evidence to conclude that drinking fortified orange juice reduces PTH level more than drinking unfortified orange juice? (*Note:* The mean and standard deviation for the data on fortified juice are 9.0 pg/mL and 37.4 pg/mL, respectively, and for the data on unfortified juice, they are 1.6 pg/mL and 34.6 pg/mL, respectively.)

**10.36 Driving Distances.** Data on household vehicle miles of travel (VMT) are compiled annually by the Federal Highway Administration and are published in *National Household Travel Survey, Summary of Travel Trends*. Independent random samples of 15 midwestern households and 14 southern households provided the following data on last year's VMT, in thousands of miles.

| Midwest | | | South | | |
|------|------|------|------|------|------|
| 16.2 | 12.9 | 17.3 | 22.2 | 19.2 | 9.3 |
| 14.6 | 18.6 | 10.8 | 24.6 | 20.2 | 15.8 |
| 11.2 | 16.6 | 16.6 | 18.0 | 12.2 | 20.1 |
| 24.4 | 20.3 | 20.9 | 16.0 | 17.5 | 18.2 |
| 9.6 | 15.1 | 18.3 | 22.8 | 11.5 | |

At the 5% significance level, does there appear to be a difference in last year's mean VMT for midwestern and southern households? (*Note:* $\bar{x}_1 = 16.23$, $s_1 = 4.06$, $\bar{x}_2 = 17.69$, and $s_2 = 4.42$.)

**10.37 Floral Diversity.** In the article "Floral Diversity in Relation to Playa Wetland Area and Watershed Disturbance" (*Conservation Biology*, Vol. 16, Issue 4, pp. 964–974), L. Smith and D. Haukos examined the relationship of species richness and diversity to playa area and watershed disturbance. Independent random samples of 126 playa with cropland and 98 playa with grassland in Southern Great Plains yielded the following summary statistics for the number of native species.

| Cropland | Wetland |
|------|------|
| $\bar{x}_1 = 14.06$ | $\bar{x}_2 = 15.36$ |
| $s_1 = 4.83$ | $s_2 = 4.95$ |
| $n_1 = 126$ | $n_2 = 98$ |

At the 5% significance level do the data provide sufficient evidence to conclude that a difference exists in the mean number of native species in the two regions?

**10.38 Dexamethasone and IQ.** In the paper "Outcomes at School Age After Postnatal Dexamethasone Therapy for Lung Disease of Prematurity" (*New England Journal of Medicine*, Vol. 350, No. 13, pp. 1304–1313), T. Yeh et al. studied the outcomes at school age in children who had participated in a double-blind, placebo-controlled trial of early postnatal dexamethasone therapy for the prevention of chronic lung disease of prematurity. One result reported in the study was that the control group of 74 children had a mean IQ score of 84.4 with standard deviation of 12.6, whereas the dexamethasone group of 72 children had a mean IQ score of 78.2 with a standard deviation of 15.0. Do the data provide sufficient evidence to conclude that early postnatal dexamethasone therapy has, on average, an adverse effect on IQ? Perform the required hypothesis test at the 1% level of significance.

*In Exercises 10.39–10.44, apply Procedure 10.2 on page 454 to obtain the required confidence interval. Interpret your result in each case.*

**10.39 Doing Time.** Refer to Exercise 10.33 and obtain a 90% confidence interval for the difference between the mean times served by prisoners in the fraud and firearms offense categories.

**10.40 Sex and Direction.** Refer to Exercise 10.34 and obtain a 98% confidence interval for the difference between the mean absolute pointing errors for males and females.

**10.41 Fortified Juice and PTH.** Refer to Exercise 10.35 and find a 90% confidence interval for the difference between the mean reductions in PTH levels for fortified and unfortified orange juice.

**10.42 Driving Distances.** Refer to Exercise 10.36 and determine a 95% confidence interval for the difference between last year's mean VMTs by midwestern and southern households.

**10.43 Floral Diversity.** Refer to Exercise 10.37 and determine a 95% confidence interval for the difference between the mean number of native species in the two regions.

**10.44 Dexamethasone and IQ.** Refer to Exercise 10.38 and find a 98% confidence interval for the difference between the mean IQs of school-age children without and with the dexamethasone therapy.

## Working With Large Data Sets

**10.45 Vegetarians and Omnivores.** Philosophical and health issues are prompting an increasing number of Taiwanese to switch to a vegetarian lifestyle. A study by Lu et al., published in the *Journal of Nutrition* (Vol. 130, pp. 1591–1596), compared the daily intake of nutrients by vegetarians and omnivores living in Taiwan. Among the nutrients considered was protein. Too little protein stunts growth and interferes with all bodily functions; too much protein puts a strain on the kidneys, can cause diarrhea and dehydration, and can leach calcium from bones and teeth. Independent random samples of 51 female vegetarians and 53 female omnivores yielded the data, in grams, on daily protein intake presented on the WeissStats CD. Use the technology of your choice to do the following.

a. Obtain normal probability plots, boxplots, and the standard deviations for the two samples.

b. Do the data provide sufficient evidence to conclude that the mean daily protein intakes of female vegetarians and female omnivores differ? Perform the required hypothesis test at the 1% significance level.

c. Find a 99% confidence interval for the difference between the mean daily protein intakes of female vegetarians and female omnivores.

d. Are your procedures in parts (b) and (c) justified? Explain your answer.

**10.46 Offspring of Diabetic Mothers.** Previous research indicates that children borne by diabetic mothers may suffer from obesity, high blood pressure, and glucose intolerance. Independent random samples of adolescent offspring of diabetic mothers (ODM) and nondiabetic mothers (ONM) were taken by Cho et al. and evaluated for potential differences in vital measurements, including blood pressure and glucose tolerance. The study was published in *The Journal of Pediatrics* (Vol. 136(5), pp. 587–592). The data for the systolic blood pressures, in mm Hg, of the 99 ODM participants and the 80 ONM participants can be found on the WeissStats CD.

a. Obtain normal probability plots, boxplots, and the standard deviations for the two samples.

b. At the 5% significance level, do the data provide sufficient evidence to conclude that the mean systolic blood pressure of ODM children exceeds that of ONM children?

c. Determine a 95% confidence interval for the difference between the mean systolic blood pressures of ODM and ONM children.

d. Are your procedures in parts (b) and (c) justified? Explain your answer.

**10.47 A Better Golf Tee?** An independent golf equipment testing facility compared the difference in the performance of golf balls hit off a regular 2-3/4" wooden tee to those hit off a 3" Stinger Competition golf tee. A Callaway Great Big Bertha driver with 10 degrees of loft was used for the test and a robot swung the club head at approximately 95 miles per hour. Data on total distance traveled (in yards) with

each type of tee, based on the test results, are provided on the WeissStats CD.

a. Obtain normal probability plots, boxplots, and the standard deviations for the two samples.

b. At the 1% significance level, do the data provide sufficient evidence to conclude that, on average, the Stinger tee improves total distance traveled?

c. Find a 99% confidence interval for the difference between the mean total distance traveled with the regular and Stinger tees.

d. Are your procedures in parts (b) and (c) justified? Why or why not?

## Extending the Concepts and Skills

**10.48** In this section, we introduced the pooled $t$-test, which provides a method for comparing two population means. In deriving the pooled $t$-test, we stated that the variable

$$z = \frac{(\bar{x}_1 - \bar{x}_2) - (\mu_1 - \mu_2)}{\sigma\sqrt{(1/n_1) + (1/n_2)}}$$

cannot be used as a basis for the required test statistic because $\sigma$ is unknown. Why can't that variable be used as a basis for the required test statistic?

**10.49** The formula for the pooled variance, $s_p^2$, is given on page 449. Show that, if the sample sizes, $n_1$ and $n_2$, are equal, then $s_p^2$ is the mean of $s_1^2$ and $s_2^2$.

**10.50 Simulation.** In this exercise, you are to perform a computer simulation to illustrate the distribution of the pooled $t$-statistic, given in Key Fact 10.2 on page 449.

a. Simulate 1000 random samples of size 4 from a normally distributed variable with a mean of 100 and a standard deviation of 16. Then obtain the sample mean and sample standard deviation of each of the 1000 samples.

b. Simulate 1000 random samples of size 3 from a normally distributed variable with a mean of 110 and a standard deviation of 16. Then obtain the sample mean and sample standard deviation of each of the 1000 samples.

c. Determine the value of the pooled $t$-statistic for each of the 1000 pairs of samples obtained in parts (a) and (b).

d. Obtain a histogram of the 1000 values found in part (c).

e. Theoretically, what is the distribution of all possible values of the pooled $t$-statistic?

f. Compare your results from parts (d) and (e).

**10.51 Two-Tailed Hypothesis Tests and CIs.** As we mentioned on page 455, the following relationship holds between hypothesis tests and confidence intervals for pooled $t$-procedures: For a two-tailed hypothesis test at the significance level $\alpha$, the null hypothesis $H_0: \mu_1 = \mu_2$ will be rejected in favor of the alternative hypothesis $H_a: \mu_1 \neq \mu_2$ if and only if the $(1 - \alpha)$-level confidence interval for $\mu_1 - \mu_2$ does not contain 0. In each case, illustrate the preceding

relationship by comparing the results of the hypothesis test and confidence interval in the specified exercises.
a. Exercises 10.36 and 10.42
b. Exercises 10.37 and 10.43

**10.52 Left-Tailed Hypothesis Tests and CIs.** Presuming that the assumptions for a pooled $t$-interval are satisfied, the formula for a $(1 - \alpha)$-level upper confidence bound for the difference, $\mu_1 - \mu_2$, between two population means is

$$(\bar{x}_1 - \bar{x}_2) + t_\alpha \cdot s_p\sqrt{(1/n_1) + (1/n_2)}.$$

For a left-tailed hypothesis test at the significance level $\alpha$, the null hypothesis $H_0: \mu_1 = \mu_2$ will be rejected in favor of the alternative hypothesis $H_a: \mu_1 < \mu_2$ if and only if the $(1 - \alpha)$-level upper confidence bound for $\mu_1 - \mu_2$ is negative. In each case, illustrate the preceding relationship by obtaining the appropriate upper confidence bound and comparing the result to the conclusion of the hypothesis test in the specified exercise.

a. Exercise 10.33          b. Exercise 10.34

**10.53 Right-Tailed Hypothesis Tests and CIs.** Presuming that the assumptions for a pooled $t$-interval are satisfied, the formula for a $(1 - \alpha)$-level lower confidence bound for the difference, $\mu_1 - \mu_2$, between two population means is

$$(\bar{x}_1 - \bar{x}_2) - t_\alpha \cdot s_p\sqrt{(1/n_1) + (1/n_2)}.$$

For a right-tailed hypothesis test at the significance level $\alpha$, the null hypothesis $H_0: \mu_1 = \mu_2$ will be rejected in favor of the alternative hypothesis $H_a: \mu_1 > \mu_2$ if and only if the $(1 - \alpha)$-level lower confidence bound for $\mu_1 - \mu_2$ is positive.

In each case, illustrate the preceding relationship by obtaining the appropriate lower confidence bound and comparing the result to the conclusion of the hypothesis test in the specified exercise.
a. Exercise 10.35          b. Exercise 10.38

**10.54** Suppose that you want to perform a hypothesis test to compare the means of two populations, using independent simple random samples. Assume that the two distributions (one for each population) of the variable under consideration are normally distributed and have equal standard deviations. Answer the following questions and explain your answers.
a. Is it permissible to use the pooled $t$-test to perform the hypothesis test?
b. Is it permissible to use the Mann–Whitney test to perform the hypothesis test?
c. Which procedure is preferable, the pooled $t$-test or the Mann–Whitney test?

**10.55** Suppose that you want to perform a hypothesis test to compare the means of two populations, using independent simple random samples. Assume that the two distributions of the variable under consideration have the same shape, but are not normal, and both sample sizes are large. Answer the following questions and explain your answers.
a. Is it permissible to use the pooled $t$-test to perform the hypothesis test?
b. Is it permissible to use the Mann–Whitney test to perform the hypothesis test?
c. Which procedure is preferable, the pooled $t$-test or the Mann–Whitney test?

## 10.3 Inferences for Two Population Means, Using Independent Samples: Standard Deviations Not Assumed Equal

In Section 10.2, we examined methods based on independent samples for performing inferences to compare the means of two populations. The methods discussed, called pooled $t$-procedures, require that the standard deviations of the two populations be equal.

In this section, we develop inferential procedures based on independent samples to compare the means of two populations that do not require the population standard deviations to be equal, even though they may be. As before, we assume that the population standard deviations are unknown, because that is usually the case in practice.

For our derivation, we also assume that the variable under consideration is normally distributed on each population. However, like the pooled $t$-procedures, the resulting inferential procedures are approximately correct for large samples, regardless of distribution type.

## Hypothesis Tests for the Means of Two Populations, Using Independent Samples

We begin by finding a test statistic. We know from Key Fact 10.1 on page 445 that, for independent samples, the standardized version of $\bar{x}_1 - \bar{x}_2$

$$z = \frac{(\bar{x}_1 - \bar{x}_2) - (\mu_1 - \mu_2)}{\sqrt{(\sigma_1^2/n_1) + (\sigma_2^2/n_2)}}$$

has the standard normal distribution. We are assuming that the population standard deviations, $\sigma_1$ and $\sigma_2$, are unknown, so we cannot use this variable as a basis for the required test statistic. We therefore replace $\sigma_1$ and $\sigma_2$ with their sample estimates, $s_1$ and $s_2$, and obtain the variable

$$\frac{(\bar{x}_1 - \bar{x}_2) - (\mu_1 - \mu_2)}{\sqrt{(s_1^2/n_1) + (s_2^2/n_2)}}$$

which we can use as a basis for the required test statistic. This variable does not have the standard normal distribution, but it does have roughly a $t$-distribution.

---

**Key Fact 10.3**

### Distribution of the Nonpooled $t$-Statistic

Suppose that $x$ is a normally distributed variable on each of two populations. Then, for independent samples of sizes $n_1$ and $n_2$ from the two populations, the variable

$$t = \frac{(\bar{x}_1 - \bar{x}_2) - (\mu_1 - \mu_2)}{\sqrt{(s_1^2/n_1) + (s_2^2/n_2)}}$$

has approximately a $t$-distribution. The degrees of freedom used is obtained from the sample data. It is denoted $\Delta$ and given by

$$\Delta = \frac{\left[(s_1^2/n_1) + (s_2^2/n_2)\right]^2}{\dfrac{\left(s_1^2/n_1\right)^2}{n_1 - 1} + \dfrac{\left(s_2^2/n_2\right)^2}{n_2 - 1}},$$

rounded down to the nearest integer.

---

In light of Key Fact 10.3, for a hypothesis test that has null hypothesis $H_0$: $\mu_1 = \mu_2$, we can use the variable

$$t = \frac{\bar{x}_1 - \bar{x}_2}{\sqrt{(s_1^2/n_1) + (s_2^2/n_2)}}$$

as the test statistic and obtain the critical value(s) or $P$-value from the $t$-table, Table IV. We call this hypothesis-testing procedure the **nonpooled $t$-test**.[†] Procedure 10.3 (next page) provides a step-by-step method for performing a nonpooled $t$-test by using either the critical-value approach or the $P$-value approach.

Regarding Assumptions 1 and 2, we note that the nonpooled $t$-test can also be used as a method for comparing two means with a designed experiment. Additionally, the pooled $t$-test is robust to moderate violations of Assumption 3 (normal populations) but, even for large samples, can sometimes be unduly

---

[†]The nonpooled $t$-test is also known as the **two-sample $t$-test** (with equal variances not assumed), the (nonpooled) **two-variable $t$-test,** and the (nonpooled) **independent samples $t$-test.**

affected by outliers because the sample mean and sample standard deviation are not resistant to outliers.

## Procedure 10.3 Nonpooled *t*-Test

*Purpose* To perform a hypothesis test to compare two population means, $\mu_1$ and $\mu_2$

*Assumptions*

1. Simple random samples
2. Independent samples
3. Normal populations or large samples

**STEP 1** **The null hypothesis is $H_0$: $\mu_1 = \mu_2$, and the alternative hypothesis is**

$$H_a: \mu_1 \neq \mu_2 \quad \text{or} \quad H_a: \mu_1 < \mu_2 \quad \text{or} \quad H_a: \mu_1 > \mu_2$$
$$\text{(Two tailed)} \qquad \text{(Left tailed)} \qquad \text{(Right tailed)}$$

**STEP 2** **Decide on the significance level, $\alpha$.**

**STEP 3** **Compute the value of the test statistic**

$$t = \frac{\bar{x}_1 - \bar{x}_2}{\sqrt{(s_1^2/n_1) + (s_2^2/n_2)}}.$$

**Denote the value of the test statistic $t_0$.**

---

| CRITICAL-VALUE APPROACH | or | P-VALUE APPROACH |

**STEP 4** **The critical value(s) are**

$$\pm t_{\alpha/2} \quad \text{or} \quad -t_\alpha \quad \text{or} \quad t_\alpha$$
$$\text{(Two tailed)} \qquad \text{(Left tailed)} \qquad \text{(Right tailed)}$$

**with df = $\Delta$, where**

$$\Delta = \frac{\left[(s_1^2/n_1) + (s_2^2/n_2)\right]^2}{\dfrac{(s_1^2/n_1)^2}{n_1 - 1} + \dfrac{(s_2^2/n_2)^2}{n_2 - 1}}$$

**rounded down to the nearest integer. Use Table IV to find the critical value(s).**

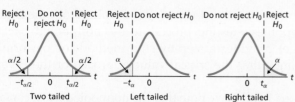

**STEP 5** **If the value of the test statistic falls in the rejection region, reject $H_0$; otherwise, do not reject $H_0$.**

**STEP 4** **The *t*-statistic has df = $\Delta$, where**

$$\Delta = \frac{\left[(s_1^2/n_1) + (s_2^2/n_2)\right]^2}{\dfrac{(s_1^2/n_1)^2}{n_1 - 1} + \dfrac{(s_2^2/n_2)^2}{n_2 - 1}}$$

**rounded down to the nearest integer. Use Table IV to estimate the *P*-value, or obtain it exactly by using technology.**

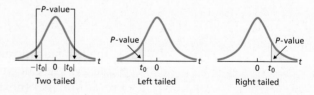

**STEP 5** **If $P \leq \alpha$, reject $H_0$; otherwise, do not reject $H_0$.**

---

**STEP 6** **Interpret the results of the hypothesis test.**

## Example 10.6 | The Nonpooled *t*-Test

*Neurosurgery Operative Times* Several neurosurgeons wanted to determine whether a dynamic system (Z-plate) reduced the operative time relative to a static system (ALPS plate). R. Jacobowitz, Ph.D., an ASU professor, along with G. Vishteh, M.D., and other neurosurgeons, obtained the data displayed in Table 10.7 on operative times, in minutes, for the two systems. At the 1% significance level, do the data provide sufficient evidence to conclude that the mean operative time is less with the dynamic system than with the static system?

**TABLE 10.7**
Operative times, in minutes, for dynamic and static systems

| Dynamic | | | | | | | Static | | |
|---|---|---|---|---|---|---|---|---|---|
| 370 | 360 | 510 | 445 | 295 | 315 | 490 | 430 | 445 | 455 |
| 345 | 450 | 505 | 335 | 280 | 325 | 500 | 455 | 490 | 535 |

**Solution** First, we find the required summary statistics for the two samples, as shown in Table 10.8. Because the two sample standard deviations are considerably different, as seen in Table 10.8 or Fig. 10.6, the pooled *t*-test is inappropriate here.

**TABLE 10.8**
Summary statistics for the samples in Table 10.7

| Dynamic | Static |
|---|---|
| $\bar{x}_1 = 394.6$ | $\bar{x}_2 = 468.3$ |
| $s_1 = 84.7$ | $s_2 = 38.2$ |
| $n_1 = 14$ | $n_2 = 6$ |

Next, we check the three conditions required for using the nonpooled *t*-test. These data were obtained from a randomized comparative experiment, a type of designed experiment. Therefore, we can consider Assumptions 1 and 2 satisfied.

To check Assumption 3, we refer to the normal probability plots and boxplots in Figs. 10.5 and 10.6, respectively. These graphs reveal no outliers and, keeping in mind that the nonpooled *t*-test is robust to moderate violations of normality, show that we can consider Assumption 3 satisfied.

**FIGURE 10.5**
Normal probability plots of the sample data for the (a) dynamic system and (b) static system

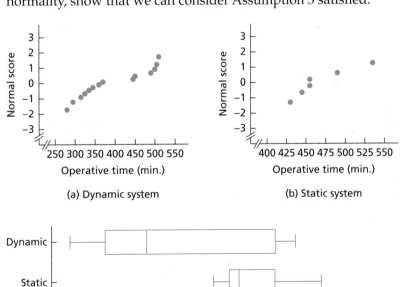

(a) Dynamic system

(b) Static system

**FIGURE 10.6**
Boxplots of the operative times for the dynamic and static systems

The preceding two paragraphs suggest that the nonpooled *t*-test can be used to carry out the hypothesis test. We apply Procedure 10.3.

**STEP 1** **State the null and alternative hypotheses.**

Let $\mu_1$ and $\mu_2$ denote the mean operative times for the dynamic and static systems, respectively. Then the null and alternative hypotheses are

$H_0$: $\mu_1 = \mu_2$ (mean dynamic time is not less than mean static time)
$H_a$: $\mu_1 < \mu_2$ (mean dynamic time is less than mean static time).

Note that the hypothesis test is left tailed.

**STEP 2** **Decide on the significance level, $\alpha$.**

The test is to be performed at the 1% significance level, or $\alpha = 0.01$.

**STEP 3** **Compute the value of the test statistic**

$$t = \frac{\bar{x}_1 - \bar{x}_2}{\sqrt{(s_1^2/n_1) + (s_2^2/n_2)}}.$$

Referring to Table 10.8, we get

$$t = \frac{394.6 - 468.3}{\sqrt{(84.7^2/14) + (38.2^2/6)}} = -2.681.$$

| CRITICAL-VALUE APPROACH | or | P-VALUE APPROACH |
|---|---|---|

**STEP 4** **The critical value for a left-tailed test is $-t_\alpha$ with df = $\Delta$. Use Table IV to find the critical value.**

From Step 2, $\alpha = 0.01$. Also, from Table 10.8, we see that

$$df = \Delta = \frac{\left[(84.7^2/14) + (38.2^2/6)\right]^2}{\dfrac{\left(84.7^2/14\right)^2}{14 - 1} + \dfrac{\left(38.2^2/6\right)^2}{6 - 1}}$$

which equals 17 when rounded down. From Table IV with df = 17, we find that the critical value is $-t_\alpha = -t_{0.01} = -2.567$, as shown in Fig. 10.7A.

**FIGURE 10.7A**

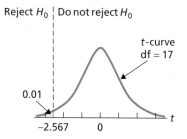

Reject $H_0$ | Do not reject $H_0$

$t$-curve
df = 17

0.01

−2.567    0    $t$

**STEP 4** **The $t$-statistic has df = $\Delta$. Use Table IV to estimate the P-value or obtain it exactly by using technology.**

From Step 3, the value of the test statistic is $t = -2.681$. The test is left tailed, so the P-value is the probability of observing a value of $t$ of $-2.681$ or less if the null hypothesis is true. That probability equals the shaded area shown in Fig. 10.7B.

**FIGURE 10.7B**

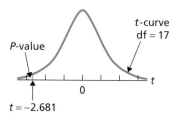

P-value

$t$-curve
df = 17

0

$t = -2.681$

From Table 10.8, we find that

$$df = \Delta = \frac{\left[(84.7^2/14) + (38.2^2/6)\right]^2}{\dfrac{\left(84.7^2/14\right)^2}{14 - 1} + \dfrac{\left(38.2^2/6\right)^2}{6 - 1}}$$

which equals 17 when rounded down. Referring to Fig. 10.7B and Table IV with df = 17, we find that $0.005 < P < 0.01$. (Using technology, $P = 0.008$.)

| CRITICAL-VALUE APPROACH | or | P-VALUE APPROACH |
|---|---|---|

**STEP 5** **If the value of the test statistic falls in the rejection region, reject $H_0$; otherwise, do not reject $H_0$.**

From Step 3, the value of the test statistic is $t = -2.681$, which, as we see from Fig. 10.7A, falls in the rejection region. Thus we reject $H_0$. The test results are statistically significant at the 1% level.

**STEP 5** **If $P \leq \alpha$, reject $H_0$; otherwise, do not reject $H_0$.**

From Step 4, $0.005 < P < 0.01$. Because the $P$-value is less than the specified significance level of 0.01, we reject $H_0$. The test results are statistically significant at the 1% level and (see Table 9.10 on page 414) provide very strong evidence against the null hypothesis.

Exercise 10.65
on page 473

**STEP 6** **Interpret the results of the hypothesis test.**

**Interpretation**   At the 1% significance level, the data provide sufficient evidence to conclude that the mean operative time is less with the dynamic system than with the static system.

• • •

## Confidence Intervals for the Difference Between the Means of Two Populations, Using Independent Samples

Key Fact 10.3 on page 463 can also be used to derive a confidence-interval procedure for the difference between two means. We call this procedure the **nonpooled $t$-interval procedure.**[†]

| Procedure 10.4 | **Nonpooled $t$-Interval Procedure** |
|---|---|

*Purpose*  To find a confidence interval for the difference between two population means, $\mu_1$ and $\mu_2$

*Assumptions*

1. Simple random samples
2. Independent samples
3. Normal populations or large samples

**STEP 1**  **For a confidence level of $1 - \alpha$, use Table IV to find $t_{\alpha/2}$ with df $= \Delta$, where**

$$\Delta = \frac{\left[(s_1^2/n_1) + (s_2^2/n_2)\right]^2}{\dfrac{(s_1^2/n_1)^2}{n_1 - 1} + \dfrac{(s_2^2/n_2)^2}{n_2 - 1}}$$

**rounded down to the nearest integer.**

**STEP 2**  **The endpoints of the confidence interval for $\mu_1 - \mu_2$ are**

$$(\bar{x}_1 - \bar{x}_2) \pm t_{\alpha/2} \cdot \sqrt{(s_1^2/n_1) + (s_2^2/n_2)}.$$

**STEP 3**  **Interpret the confidence interval.**

---

[†]The nonpooled $t$-interval procedure is also known as the **two-sample $t$-interval procedure** (with equal variances not assumed), the (nonpooled) **two-variable $t$-interval procedure,** and the (nonpooled) **independent samples $t$-interval procedure.**

## Example 10.7 | The Nonpooled *t*-Interval Procedure

*Neurosurgery Operative Times* Use the sample data in Table 10.7 on page 465 to obtain a 99% confidence interval for the difference, $\mu_1 - \mu_2$, between the mean operative times of the dynamic and static systems.

**Solution** We apply Procedure 10.4.

**STEP 1 For a confidence level of $1 - \alpha$, use Table IV to find $t_{\alpha/2}$ with df = $\Delta$.**

For a 99% confidence interval, $\alpha = 0.01$. From Example 10.6, df = 17. In Table IV, with df = 17, $t_{\alpha/2} = t_{0.01/2} = t_{0.005} = 2.898$.

**STEP 2 The endpoints of the confidence interval for $\mu_1 - \mu_2$ are**

$$(\bar{x}_1 - \bar{x}_2) \pm t_{\alpha/2} \cdot \sqrt{(s_1^2/n_1) + (s_2^2/n_2)}.$$

From Step 1, $t_{\alpha/2} = 2.898$. Referring to Table 10.8 on page 465, we conclude that the endpoints of the confidence interval for $\mu_1 - \mu_2$ are

$$(394.6 - 468.3) \pm 2.898 \cdot \sqrt{(84.7^2/14) + (38.2^2/6)}$$

or −153.4 to 6.0.

You
try it!

Exercise 10.71
on page 474

**STEP 3 Interpret the confidence interval.**

**Interpretation** We can be 99% confident that the difference between the mean operative times of the dynamic and static systems is somewhere between −153.4 minutes and 6.0 minutes.

• • •

### Pooled Versus Nonpooled *t*-Procedures

Suppose that we want to perform a hypothesis test based on independent simple random samples to compare the means of two populations. Further suppose that either the variable under consideration is normally distributed on each of the two populations or the sample sizes are large. Then two tests are candidates for the job: the pooled *t*-test or the nonpooled *t*-test.

In theory, the pooled *t*-test requires that the population standard deviations be equal, but what if they are not? The answer depends on several factors. If the population standard deviations are not too unequal and the sample sizes are nearly the same, using the pooled *t*-test will not cause serious difficulties. If the population standard deviations are quite different, however, using the pooled *t*-test can result in a significantly larger Type I error probability than the specified one.

In contrast, the nonpooled *t*-test applies whether or not the population standard deviations are equal. Then why use the pooled *t*-test at all? The reason is that, if the population standard deviations are equal or nearly so, then, on average, the pooled *t*-test is slightly more powerful; that is, the probability of making a Type II error is somewhat smaller. Similar remarks apply to the pooled *t*-interval and nonpooled *t*-interval procedures.

**Key Fact 10.4**

## Choosing Between a Pooled and Nonpooled *t*-Procedure

Suppose you want to use independent simple random samples to compare the means of two populations. To decide between a pooled *t*-procedure and a nonpooled *t*-procedure, follow these guidelines: If you are reasonably sure that the populations have nearly equal standard deviations, use a pooled *t*-procedure; otherwise, use a nonpooled *t*-procedure.

### What If the Assumptions Are Not Satisfied?

The nonpooled *t*-procedures (nonpooled *t*-test and nonpooled *t*-interval procedure) provide methods for comparing the means of two populations. As you know, the assumptions for using those procedures are (1) simple random samples, (2) independent samples, and (3) normal populations or large samples. If one or more of these conditions are not satisfied, then the nonpooled *t*-procedures should not be used.

If the samples are not independent but instead are paired (i.e., Assumption 2 is violated), then procedures designed for paired samples should be used. We discuss such procedures in the next section.

If the populations are not normal and the samples are not large (i.e., Assumption 3 is violated), then a nonparametric method should be used. For example, if the samples are independent and the two distributions (one for each population) of the variable under consideration have the same shape, then you can use a nonparametric method called the *Mann–Whitney test* to perform a hypothesis test and a nonparametric method called the *Mann–Whitney confidence-interval procedure* to obtain a confidence interval.

# The Technology Center

Most statistical technologies have programs that automatically perform nonpooled *t*-procedures. In this subsection, we present output and step-by-step instructions for such programs.

**Example 10.8**

## Using Technology to Conduct Nonpooled *t*-Procedures

*Neurosurgery Operative Times* Table 10.7 on page 465 displays samples of neurosurgery operative times, in minutes, for dynamic and static systems. Use Minitab, Excel, or the TI-83/84 Plus to perform the hypothesis test in Example 10.6 and obtain the confidence interval required in Example 10.7.

**Solution**  Let $\mu_1$ and $\mu_2$ denote, respectively, the mean operative times of the dynamic and static systems. The task in Example 10.6 is to perform the hypothesis test

$H_0$: $\mu_1 = \mu_2$ (mean dynamic time is not less than mean static time)

$H_a$: $\mu_1 < \mu_2$ (mean dynamic time is less than mean static time)

at the 1% significance level; the task in Example 10.7 is to obtain a 99% confidence interval for $\mu_1 - \mu_2$.

We applied the nonpooled $t$-procedures programs to the data, resulting in Output 10.2. Steps for generating that output are presented in Instructions 10.2 on page 472.

As shown in Output 10.2, the $P$-value for the hypothesis test is 0.008. Because the $P$-value is less than the specified significance level of 0.01, we reject $H_0$. Output 10.2 also shows that a 99% confidence interval for the difference between the means is from $-153$ to 6.

**OUTPUT 10.2** Nonpooled $t$-procedures for the operative-time data

**MINITAB**

**Two-Sample T-Test and CI: DYNAMIC, STATIC**   [FOR THE HYPOTHESIS TEST]

```
Two-sample T for DYNAMIC vs STATIC

                         SE
          N   Mean  StDev  Mean
DYNAMIC  14  394.6   84.7    23
STATIC    6  468.3   38.2    16


Difference = mu (DYNAMIC) - mu (STATIC)
Estimate for difference:  -73.7
99% upper bound for difference:  -3.1
T-Test of difference = 0 (vs <): T-Value = -2.68  P-Value = 0.008  DF = 17
```

**Two-Sample T-Test and CI: DYNAMIC, STATIC**   [FOR THE CONFIDENCE INTERVAL]

```
Two-sample T for DYNAMIC vs STATIC

                         SE
          N   Mean  StDev  Mean
DYNAMIC  14  394.6   84.7    23
STATIC    6  468.3   38.2    16


Difference = mu (DYNAMIC) - mu (STATIC)
Estimate for difference:  -73.7
99% CI for difference:  (-153.4, 6.0)
T-Test of difference = 0 (vs not =): T-Value = -2.68  P-Value = 0.016  DF = 17
```

**EXCEL**

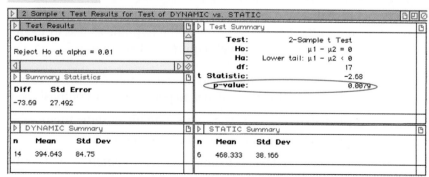

Using **2 Var t Test**

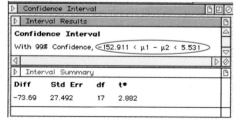

Using **2 Var t Interval**

**OUTPUT 10.2 (cont.)**
Nonpooled *t*-procedures
for the operative-time data

**TI-83/84 PLUS**

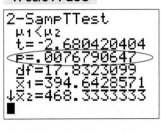

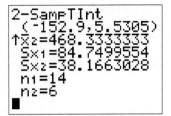

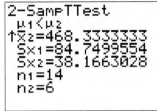

Using **2-SampTTest**                    Using **2-SampTInt**

**Note:** For nonpooled *t*-procedures, discrepancies may occur among results provided by statistical technologies because some round the number of degrees of freedom and others do not.

• • •

*Note to Minitab users:* As we noted on page 456, Minitab computes a two-sided confidence interval for a two-tailed test and a one-sided confidence interval for a one-tailed test. To perform a one-tailed hypothesis test and obtain a two-sided confidence interval, apply Minitab's nonpooled *t*-procedure twice: once for the one-tailed hypothesis test and once for the confidence interval specifying a two-tailed hypothesis test.

**INSTRUCTIONS 10.2**    Steps for generating Output 10.2

| MINITAB | EXCEL | TI-83/84 PLUS |
|---|---|---|
| Store the two samples of operative-time data from Table 10.7 in columns named DYNAMIC and STATIC. | Store the two samples of operative-time data from Table 10.7 in ranges named DYNAMIC and STATIC. | Store the two samples of operative-time data from Table 10.7 in lists named DYNA and STAT. |

**MINITAB**

FOR THE HYPOTHESIS TEST:
1 Choose **Stat ➤ Basic Statistics ➤ 2-Sample t...**
2 Select the **Samples in different columns** option button
3 Click in the **First** text box and specify DYNAMIC
4 Click in the **Second** text box and specify STATIC
5 Uncheck the **Assume equal variances** check box
6 Click the **Options...** button
7 Click in the **Confidence level** text box and type 99
8 Click in the **Test difference** text box and type 0
9 Click the arrow button at the right of the **Alternative** drop-down list box and select **less than**
10 Click **OK** twice

FOR THE CI:
1 Choose **Edit ➤ Edit Last Dialog**
2 Click the **Options...** button
3 Click the arrow button at the right of the **Alternative** drop-down list box and select **not equal**
4 Click **OK** twice

**EXCEL**

FOR THE HYPOTHESIS TEST:
1 Choose **DDXL ➤ Hypothesis Tests**
2 Select **2 Var t Test** from the **Function type** drop-down box
3 Specify DYNAMIC in the **1st Quantitative Variable** text box
4 Specify STATIC in the **2nd Quantitative Variable** text box
5 Click **OK**
6 Click the **2-sample** button
7 Click the **Set difference** button, type 0, and click **OK**
8 Click the **0.01** button
9 Click the $\mu 1 - \mu 2 <$ **diff** button
10 Click the **Compute** button

FOR THE CI:
1 Exit to Excel
2 Choose **DDXL ➤ Confidence Intervals**
3 Select **2 Var t Interval** from the **Function type** drop-down box
4 Specify DYNAMIC in the **1st Quantitative Variable** text box
5 Specify STATIC in the **2nd Quantitative Variable** text box
6 Click **OK**
7 Click the **2-sample** button
8 Click the **99%** button
9 Click the **Compute Interval** button

**TI-83/84 PLUS**

FOR THE HYPOTHESIS TEST:
1 Press **STAT**, arrow over to **TESTS**, and press **4**
2 Highlight **Data** and press **ENTER**
3 Press the down-arrow key
4 Press **2nd ➤ LIST**, arrow down to DYNA, and press **ENTER** twice
5 Press **2nd ➤ LIST**, arrow down to STAT, and press **ENTER** four times
6 Highlight $< \mu 2$ and press **ENTER**
7 Press the down-arrow key, highlight **No**, and press **ENTER**
8 Press the down-arrow key, highlight **Calculate**, and press **ENTER**

FOR THE CI:
1 Press **STAT**, arrow over to **TESTS**, and press **0**
2 Highlight **Data** and press **ENTER**
3 Press the down-arrow key
4 Press **2nd ➤ LIST**, arrow down to DYNA, and press **ENTER** twice
5 Press **2nd ➤ LIST**, arrow down to STAT, and press **ENTER** four times
6 Type .99 for **C-Level** and press **ENTER**
7 Highlight **No** and press **ENTER**
8 Press the down-arrow key and press **ENTER**

---

# Exercises 10.3

## Understanding the Concepts and Skills

**10.56** What is the difference in assumptions between the pooled and nonpooled *t*-procedures?

**10.57** Suppose that you know that a variable is normally distributed on each of two populations. Further suppose that you want to perform a hypothesis test based on independent simple random samples to compare the two population means. In each case, decide whether you would use the pooled or nonpooled *t*-test and give a reason for your answer.
**a.** You know that the population standard deviations are equal.

**b.** You know that the population standard deviations are not equal.
**c.** The sample standard deviations are 23.6 and 25.2, and each sample size is 25.
**d.** The sample standard deviations are 23.6 and 59.2.

**10.58** Discuss the relative advantages and disadvantages of using pooled and nonpooled *t*-procedures.

*In each of Exercises **10.59–10.64**, we have provided summary statistics for independent simple random samples from two populations. In each case, use the nonpooled t-test and the nonpooled t-interval procedure to conduct the required hypothesis test and obtain the specified confidence interval.*

**10.59** $\bar{x}_1 = 10, s_1 = 2, n_1 = 15, \bar{x}_2 = 12, s_2 = 5, n_2 = 15$
**a.** Two-tailed test, $\alpha = 0.05$    **b.** 95% confidence interval

**10.60** $\bar{x}_1 = 15, s_1 = 2, n_1 = 15, \bar{x}_2 = 12, s_2 = 5, n_2 = 15$
**a.** Two-tailed test, $\alpha = 0.05$    **b.** 95% confidence interval

**10.61** $\bar{x}_1 = 20, s_1 = 4, n_1 = 10, \bar{x}_2 = 18, s_2 = 5, n_2 = 15$
**a.** Right-tailed test, $\alpha = 0.05$    **b.** 90% confidence interval

**10.62** $\bar{x}_1 = 20, s_1 = 4, n_1 = 10, \bar{x}_2 = 23, s_2 = 5, n_2 = 15$
**a.** Left-tailed test, $\alpha = 0.05$    **b.** 90% confidence interval

**10.63** $\bar{x}_1 = 20, s_1 = 6, n_1 = 20, \bar{x}_2 = 24, s_2 = 2, n_2 = 15$
**a.** Left-tailed test, $\alpha = 0.05$    **b.** 90% confidence interval

**10.64** $\bar{x}_1 = 20, s_1 = 2, n_1 = 30, \bar{x}_2 = 18, s_2 = 5, n_2 = 40$
**a.** Right-tailed test, $\alpha = 0.05$    **b.** 90% confidence interval

*Preliminary data analyses indicate that you can reasonably use nonpooled t-procedures in Exercises 10.65–10.70. For each exercise, apply a nonpooled t-test to perform the required hypothesis test, using either the critical-value approach or the P-value approach.*

**10.65 Political Prisoners.** According to the American Psychiatric Association, posttraumatic stress disorder (PTSD) is a common psychological consequence of traumatic events that involve a threat to life or physical integrity. During the Cold War, some 200,000 people in East Germany were imprisoned for political reasons. Many were subjected to physical and psychological torture during their imprisonment, resulting in PTSD. Ehlers, Maercker, and Boos studied various characteristics of political prisoners from the former East Germany and presented their findings in the paper "Posttraumatic Stress Disorder (PTSD) Following Political Imprisonment: The Role of Mental Defeat, Alienation, and Perceived Permanent Change" (*Journal of Abnormal Psychology*, Vol. 109, pp. 45–55). The researchers randomly and independently selected 32 former prisoners diagnosed with chronic PTSD and 20 former prisoners that were diagnosed with PTSD after release from prison but had since recovered (remitted). The ages, in years, at arrest yielded the following summary statistics.

| Chronic | Remitted |
|---|---|
| $\bar{x}_1 = 25.8$ | $\bar{x}_2 = 22.1$ |
| $s_1 = 9.2$ | $s_2 = 5.7$ |
| $n_1 = 32$ | $n_2 = 20$ |

At the 10% significance level, is there sufficient evidence to conclude that a difference exists in the mean age at arrest of East German prisoners with chronic PTSD and remitted PTSD?

**10.66 Nitrogen and Seagrass.** The seagrass *Thalassia testudinum* is an integral part of the Texas coastal ecosystem.

Essential to the growth of *T. testudinum* is ammonium. Researchers K. Lee and K. Dunton of the Marine Science Institute of the University of Texas at Austin noticed that the seagrass beds in Corpus Christi Bay (CCB) were taller and thicker than those in Lower Laguna Madre (LLM). They compared the sediment ammonium concentrations in the two locations and published their findings in *Marine Ecology Progress Series* (Vol. 196, pp. 39–48). Following are the summary statistics on sediment ammonium concentrations, in micromoles, obtained by the researchers.

| CCB | LLM |
|---|---|
| $\bar{x}_1 = 115.1$ | $\bar{x}_2 = 24.3$ |
| $s_1 = 79.4$ | $s_2 = 10.5$ |
| $n_1 = 51$ | $n_2 = 19$ |

At the 1% significance level, is there sufficient evidence to conclude that the mean sediment ammonium concentration in CCB exceeds that in LLM?

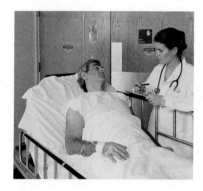

**10.67 Acute Postoperative Days.** Refer to Example 10.6 on page 465. The researchers also obtained the following data on the number of acute postoperative days in the hospital using the dynamic and static systems.

| Dynamic | | | | | | | Static | | |
|---|---|---|---|---|---|---|---|---|---|
| 7 | 5 | 8 | 8 | 6 | 7 | 7 | 6 | 18 | 9 |
| 9 | 10 | 7 | 7 | 7 | 7 | 8 | 7 | 14 | 9 |

At the 5% significance level, do the data provide sufficient evidence to conclude that the mean number of acute postoperative days in the hospital is smaller with the dynamic system than with the static system? (*Note:* $\bar{x}_1 = 7.36, s_1 = 1.22, \bar{x}_2 = 10.50,$ and $s_2 = 4.59$.)

**10.68 Stressed-Out Bus Drivers.** Frustrated passengers, congested streets, time schedules, and air and noise pollution are just some of the physical and social pressures that lead many urban bus drivers to retire prematurely with disabilities such as coronary heart disease and stomach disorders. An intervention program designed by the Stockholm Transit District was implemented to improve the work conditions of the city's bus drivers. Improvements were evaluated by Evans et al. who collected physiological and psychological data for bus drivers who drove on the improved routes (intervention) and for drivers who were assigned the normal routes (control). Their findings were published in the article "Hassles on the Job: a Study of a Job Intervention With Urban Bus Drivers" (*Journal of Organizational Behavior* Vol. 20, pp. 199–208). Following are data, based on the results of the study, for the heart rates, in beats per minute, of the intervention and control drivers.

| Intervention | | Control | | | | | | |
|---|---|---|---|---|---|---|---|---|
| 68 | 66 | 74 | 52 | 67 | 63 | 77 | 57 | 80 |
| 74 | 58 | 77 | 53 | 76 | 54 | 73 | 54 | |
| 69 | 63 | 60 | 77 | 63 | 60 | 68 | 64 | |
| 68 | 73 | 66 | 71 | 66 | 55 | 71 | 84 | |
| 64 | 76 | 63 | 73 | 59 | 68 | 64 | 82 | |

a. At the 5% significance level, do the data provide sufficient evidence to conclude that the intervention program reduces mean heart rate of urban bus drivers in Stockholm? (*Note:* $\bar{x}_1 = 67.90$, $s_1 = 5.49$, $\bar{x}_2 = 66.81$, and $s_2 = 9.04$.)
b. Can you provide an explanation for the somewhat surprising results of the study?
c. Is the study a designed experiment or an observational study? Explain your answer.

**10.69 Schizophrenia and Dopamine.** Previous research has suggested that changes in the activity of dopamine, a neurotransmitter in the brain, may be a causative factor for schizophrenia. In the paper "Schizophrenia: Dopamine b-Hydroxylase Activity and Treatment Response" (*Science*, Vol. 216, pp. 1423–1425), Sternberg et al. published the results of their study in which they examined 25 schizophrenic patients who had been classified as either psychotic or not psychotic by hospital staff. The activity of dopamine was measured in each patient by using the enzyme dopamine b-hydroxylase to assess differences in dopamine activity between the two groups. The following are the data, in nanomoles per milliliter-hour per milligram (nmol/ml-h/mg).

| Psychotic | | Not psychotic | | |
|---|---|---|---|---|
| 0.0150 | 0.0222 | 0.0104 | 0.0230 | 0.0145 |
| 0.0204 | 0.0275 | 0.0200 | 0.0116 | 0.0180 |
| 0.0306 | 0.0270 | 0.0210 | 0.0252 | 0.0154 |
| 0.0320 | 0.0226 | 0.0105 | 0.0130 | 0.0170 |
| 0.0208 | 0.0245 | 0.0112 | 0.0200 | 0.0156 |

At the 1% significance level, do the data suggest that dopamine activity is higher, on average, in psychotic patients? (*Note:* $\bar{x}_1 = 0.02426$, $s_1 = 0.00514$, $\bar{x}_2 = 0.01643$, and $s_2 = 0.00470$.)

**10.70 Wing Length.** D. Cristol et al. published results of their studies of two subspecies of dark-eyed juncos in the article "Migratory Dark-Eyed Juncos, *Junco Hyemalis*, Have Better Spatial Memory and Denser Hippocampal Neurons than Nonmigratory Conspecifics" (*Animal Behaviour*, Vol. 66, pp. 317–328). One of the subspecies migrates each year and the other does not migrate. Several physical characteristics of 14 birds of each subspecies were measured, one of which was wing length. The following data, based on results obtained by the researchers, provide the wing lengths, in millimeters (mm), for the samples of two subspecies.

| Migratory | | | Nonmigratory | | |
|---|---|---|---|---|---|
| 84.5 | 81.0 | 82.6 | 82.1 | 82.4 | 83.9 |
| 82.8 | 84.5 | 81.2 | 87.1 | 84.6 | 85.1 |
| 80.5 | 82.1 | 82.3 | 86.3 | 86.6 | 83.9 |
| 80.1 | 83.4 | 81.7 | 84.2 | 84.3 | 86.2 |
| 83.0 | 79.7 | | 87.8 | 84.1 | |

a. At the 1% significance level, do the data provide sufficient evidence to conclude that the mean wing lengths for the two subspecies are different? (*Note:* The mean and standard deviation for the migratory-bird data are 82.1 mm and 1.501 mm, respectively, and that for the nonmigratory-bird data are 84.9 mm and 1.698 mm, respectively.)
b. Would it be reasonable to use a pooled *t*-test here? Explain your answer.
c. If your answer to part (b) was *yes*, then perform a pooled *t*-test to answer the question in part (a) and compare your results to that found in part (a) by using a nonpooled *t*-test.

*In Exercises 10.71–10.76, apply Procedure 10.4 on page 467 to obtain the required confidence interval. Interpret your result in each case.*

**10.71 Political Prisoners.** Refer to Exercise 10.65 and obtain a 90% confidence interval for the difference, $\mu_1 - \mu_2$, between the mean ages at arrest of East German prisoners with chronic PTSD and remitted PTSD.

**10.72 Nitrogen and Seagrass.** Refer to Exercise 10.66 and determine a 98% confidence interval for the difference, $\mu_1 - \mu_2$, between the mean sediment ammonium concentrations in CCB and LLM.

**10.73 Acute Postoperative Days.** Refer to Exercise 10.67 and find a 90% confidence interval for the difference between the mean numbers of acute postoperative days in the hospital with the dynamic and static systems.

**10.74 Stressed-Out Bus Drivers.** Refer to Exercise 10.68 and find a 90% confidence interval for the difference between the mean heart rates of urban bus drivers in Stockholm in the two environments.

**10.75 Schizophrenia and Dopamine.** Refer to Exercise 10.69 and determine a 98% confidence interval for the difference between the mean dopamine activities of psychotic and non-psychotic patients.

**10.76 Wing Length.** Refer to Exercise 10.70 and find a 99% confidence interval for the difference between the mean wing lengths of the two subspecies.

**10.77 Sleep Apnea.** In the article "Sleep Apnea in Adults With Traumatic Brain Injury: A Preliminary Investigation" (*Archives of Physical Medicine and Rehabilitation*, Vol. 82, Issue 3, pp. 316–321), J. Webster et al. investigated sleep-related breathing disorders in adults with traumatic brain injuries (TBI). The respiratory disturbance index (RDI), which is the number of apneic and hypopneic episodes per hour of sleep, was used as a measure of severity of sleep apnea. An RDI of 5 or more indicates sleep-related breathing disturbances. The RDIs for the females and males in the study are as follows.

| Female | | | | Male | | | | | | |
|---|---|---|---|---|---|---|---|---|---|---|
| 0.1 | 0.5 | 0.3 | 2.3 | 2.6 | 19.3 | 1.4 | 1.0 | 0.0 | 39.2 | 4.1 |
| 2.0 | 1.4 | 0.0 | | 0.0 | 2.1 | 1.1 | 5.6 | 5.0 | 7.0 | 2.3 |
| | | | | 4.3 | 7.5 | 16.5 | 7.8 | 3.3 | 8.9 | 7.3 |

Use the technology of your choice to answer the following questions. Explain your answers.
a. If you had to choose between the use of pooled *t*-procedures and nonpooled *t*-procedures here, which would you choose?
b. Is it reasonable to use the type of procedure that you selected in part (a)?

**10.78 Mandate Perceptions.** L. Grossback et al. examined mandate perceptions and their causes in the paper "Comparing Competing Theories on the Causes of Mandate Perceptions" (*American Journal of Political Science*, Vol. 49, Issue 2, pp. 406–419). Following are data on the percentage of members in each chamber of Congress who reacted to mandates in various years.

| House | | | | Senate | | | | |
|---|---|---|---|---|---|---|---|---|
| 30.3 | 41.1 | 15.6 | 10.1 | 21 | 38 | 40 | 39 | 27 |
| 23.9 | 15.2 | 11.7 | | 27 | 17 | 25 | 25 | |

Use the technology of your choice to answer the following questions. Explain your answers.
a. If you had to choose between the use of pooled *t*-procedures and nonpooled *t*-procedures here, which would you choose?
b. Is it reasonable to use the type of procedure that you selected in part (a)?

**10.79 Acute Postoperative Days.** In Exercise 10.67, you conducted a nonpooled *t*-test to decide whether the mean number of acute postoperative days spent in the hospital is smaller with the dynamic system than with the static system.
a. Using a pooled *t*-test, repeat that hypothesis test.
b. Compare your results from the pooled and nonpooled *t*-tests.
c. Which test do you think is more appropriate, the pooled or nonpooled *t*-test? Explain your answer.

**10.80 Neurosurgery Operative Times.** In Example 10.6 on page 465, we conducted a nonpooled *t*-test to decide whether the mean operative time is less with the dynamic system than with the static system.
a. Using a pooled *t*-test, repeat that hypothesis test.
b. Compare your results from the pooled and nonpooled *t*-tests.
c. Which test do you think is more appropriate, the pooled or nonpooled *t*-test? Explain your answer.

## Working With Large Data Sets

**10.81 Treating Psychotic Illness.** L. Petersen et al. evaluated the effects of integrated treatment for patients with a first episode of psychotic illness in the paper "A Randomised Multicentre Trial of Integrated Versus Standard Treatment for Patients With a First Episode of Psychotic Illness" (*British Medical Journal*, Vol. 331, (7517):602). Part of the study included a questionnaire that was designed to measure client satisfaction for both the integrated treatment and a standard treatment. The data on the WeissStats CD are based on the results of the client questionnaire. Use the technology of your choice to do the following.
a. Obtain normal probability plots, boxplots, and the standard deviations for the two samples.
b. Based on your results from part (a), which would you be inclined to use to compare the population means: a pooled or nonpooled *t*-procedure? Explain your answer.
c. Do the data provide sufficient evidence to conclude that, on average, clients preferred the integrated treatment? Perform the required hypothesis test at the 1% significance level by using both the pooled *t*-test and the nonpooled *t*-test. Compare your results.
d. Find a 98% confidence interval for the difference between mean client satisfaction scores for the two treatments. Obtain the required confidence interval by using both the pooled *t*-interval procedure and the nonpooled *t*-interval procedure. Compare your results.

**10.82 A Better Golf Tee?** An independent golf equipment testing facility compared the difference in the performance of golf balls hit off a regular 2-3/4" wooden tee to those hit off a 3" Stinger Competition golf tee. A Callaway Great Big Bertha driver with 10 degrees of loft was used for the test and a robot swung the club head at approximately 95 miles per hour. Data on ball velocity (in miles per hour) with each type of tee, based on the test results, are provided on the WeissStats CD. Use the technology of your choice to do the following.

a. Obtain normal probability plots, boxplots, and the standard deviations for the two samples.

b. Based on your results from part (a), which would you be inclined to use to compare the population means: a pooled or nonpooled $t$-procedure? Explain your answer.

c. At the 5% significance level, do the data provide sufficient evidence to conclude that, on average, ball velocity is less with the regular tee than with the Stinger tee? Perform the required hypothesis test by using both the pooled $t$-test and the nonpooled $t$-test, and compare results.

d. Find a 90% confidence interval for the difference between the mean ball velocities with the regular and Stinger tees. Obtain the required confidence interval by using both the pooled $t$-interval procedure and the nonpooled $t$-interval procedure. Compare your results.

**10.83 The Etruscans.** Anthropologists are still trying to unravel the mystery of the origins of the Etruscan empire, a highly advanced Italic civilization formed around the eighth century B.C. in central Italy. Were they native to the Italian peninsula or, as many aspects of their civilization suggest, did they migrate from the East by land or sea? The maximum head breadth, in millimeters, of 70 modern Italian male skulls and 84 preserved Etruscan male skulls were analyzed to help researchers decide whether the Etruscans were native to Italy. The resulting data can be found on the WeissStats CD. [SOURCE: N. Barnicot and D. Brothwell. "The Evaluation of Metrical Data in the Comparison of Ancient and Modern Bones." In *Medical Biology and Etruscan Origins*, G. Wolstenholme and C. O'Connor, eds., Little, Brown & Co., 1959.]

a. Obtain normal probability plots, boxplots, and the standard deviations for the two samples.

b. Based on your results from part (a), which would you be inclined to use to compare the population means: a pooled or nonpooled $t$-procedure? Explain your answer.

c. Do the data provide sufficient evidence to conclude that a difference exists between the mean maximum head breadths of modern Italian males and Etruscan males?

Perform the required hypothesis test at the 5% significance level by using both the pooled $t$-test and the nonpooled $t$-test. Compare your results.

d. Find a 95% confidence interval for the difference between the mean maximum head breadths of modern Italian males and Etruscan males. Obtain the required confidence interval by using both the pooled $t$-interval procedure and the nonpooled $t$-interval procedure. Compare your results.

## Extending the Concepts and Skills

**10.84** Each pair of graphs in Fig. 10.8 shows the distributions of a variable on two populations. Suppose that, in each case, you want to perform a small-sample hypothesis test based on independent simple random samples to compare the means of the two populations. In each case, decide which, if any, of the following tests is preferable: the pooled $t$-test, the nonpooled $t$-test, or the Mann–Whitney test. Explain your answers.

**10.85** Suppose that the sample sizes, $n_1$ and $n_2$, are equal for independent simple random samples from two populations.

a. Show that the values of the pooled and nonpooled $t$-statistics will be identical. (*Hint:* Refer to Exercise 10.49 on page 461.)

b. Explain why part (a) does not imply that the two $t$-tests are equivalent (i.e., will necessarily lead to the same conclusion) when the sample sizes are equal.

**10.86 Tukey's Quick Test.** In this exercise, we examine an alternative method, conceived by the late Professor John Tukey, for performing a two-tailed hypothesis test for two population means based on independent random samples. To apply this procedure, one of the samples must contain the largest observation (high group) and the other sample must contain the smallest observation (low group). Here are the steps for performing Tukey's quick test.

*Step 1* Count the number of observations in the high group that are greater than or equal to the largest observation in the low group. Count ties as $1/2$.

**FIGURE 10.8**
Figure for Exercise 10.84

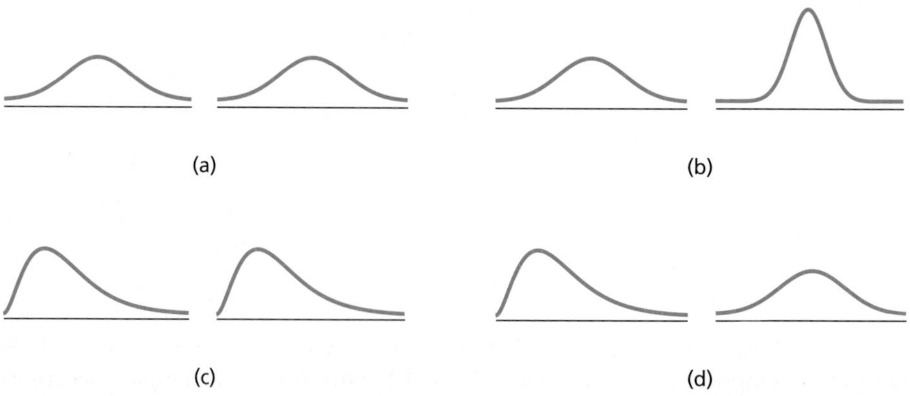

(a)

(b)

(c)

(d)

*Step 2* Count the number of observations in the low group that are less than or equal to the smallest observation in the high group. Count ties as 1/2.

*Step 3* Add the two counts obtained in Steps 1 and 2, and denote the sum $c$.

*Step 4* Reject the null hypothesis at the 5% significance level if and only if $c \geq 7$; reject it at the 1% significance level if and only if $c \geq 10$; and reject it at the 0.1% significance level if and only if $c \geq 13$.

**a.** Can Tukey's quick test be applied to Exercise 10.36 on page 460? Explain your answer.

**b.** If your answer to part (a) was *yes*, apply Tukey's quick test and compare your result to that found in Exercise 10.36, where a *t*-test was used.

**c.** Can Tukey's quick test be applied to Exercise 10.70? Explain your answer.

**d.** If your answer to part (c) was *yes*, apply Tukey's quick test and compare your result to that found in Exercise 10.70, where a *t*-test was used.

For more details about Tukey's quick test, see John Tukey's paper "A Quick, Compact, Two-Sample Test to Duckworth's Specifications" (*Technometrics*, Vol. 1, No. 1, pp. 31–48).

## 10.4 Inferences for Two Population Means, Using Paired Samples

So far, we have compared the means of two populations by using independent samples. In this section, we will compare such means by using a *paired sample*. A paired sample may be appropriate when the members of the two populations have a natural pairing.

Each pair in a **paired sample** consists of a member of one population and that member's corresponding member in the other population. With a **simple random paired sample,** each possible paired sample is equally likely to be the one selected. Example 10.9 provides an unrealistically simple illustration of paired samples, but it will help you understand the concept.

**Example 10.9** | **Introducing Random Paired Samples**

*Husbands and Wives* Let's consider two small populations, one consisting of five married women and the other of their five husbands, as shown in the following figure. The arrows in the figure indicate the married couples, which constitute the pairs for these two populations.

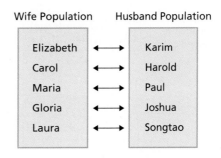

Suppose that we take a paired sample of size 3 (i.e., a sample of three pairs) from these two populations.

**a.** List the possible paired samples.

**b.** If a paired sample is selected at random (simple random paired sample), find the chance of obtaining any particular paired sample.

**TABLE 10.9**

Possible paired samples of size 3
from the wife and husband populations

| Paired sample |
|---|
| (E, K), (C, H), (M, P) |
| (E, K), (C, H), (G, J) |
| (E, K), (C, H), (L, S) |
| (E, K), (M, P), (G, J) |
| (E, K), (M, P), (L, S) |
| (E, K), (G, J), (L, S) |
| (C, H), (M, P), (G, J) |
| (C, H), (M, P), (L, S) |
| (C, H), (G, J), (L, S) |
| (M, P), (G, J), (L, S) |

**Solution** We designated a wife-husband pair by using the first letter of each name. For example, (E, K) represents the couple Elizabeth and Karim.

a. There are 10 possible paired samples of size 3, as displayed in Table 10.9.

b. For a simple random paired sample of size 3, each of the 10 possible paired samples listed in Table 10.9 is equally likely to be the one selected. Therefore the chance of obtaining any particular paired sample of size 3 is $\frac{1}{10}$.

• • •

The previous example provides a concrete illustration of paired samples and emphasizes that, for simple random paired samples of any given size, each possible paired sample is equally likely to be the one selected. In practice, we neither obtain the number of possible paired samples nor explicitly compute the chance of selecting a particular paired sample. However, these concepts underlie the methods we do use.

## Comparing Two Population Means, Using a Paired Sample

We are now ready to examine a process for comparing the means of two populations by using a paired sample.

**Example 10.10** | **Comparing Two Means, Using a Paired Sample**

*Gasoline Additive* Suppose that we want to decide whether a newly developed gasoline additive increases gas mileage.

a. Formulate the problem statistically by posing it as a hypothesis test.

b. Explain the basic idea for carrying out the hypothesis test.

c. Suppose that 10 cars are selected at random and that the cars sampled are driven both with and without the additive, yielding a paired sample of size 10. Further suppose that the resulting gas mileages, in miles per gallon (mpg), are as displayed in the second and third columns of Table 10.10. Discuss the use of these data to make a decision concerning the hypothesis test.

**TABLE 10.10**

Gas mileages, with and without additive, for 10 randomly selected cars

| Car | Gas mileage with additive | Gas mileage w/o additive | Paired difference $d$ |
|---|---|---|---|
| 1 | 25.7 | 24.9 | 0.8 |
| 2 | 20.0 | 18.8 | 1.2 |
| 3 | 28.4 | 27.7 | 0.7 |
| 4 | 13.7 | 13.0 | 0.7 |
| 5 | 18.8 | 17.8 | 1.0 |
| 6 | 12.5 | 11.3 | 1.2 |
| 7 | 28.4 | 27.8 | 0.6 |
| 8 | 8.1 | 8.2 | −0.1 |
| 9 | 23.1 | 23.1 | 0.0 |
| 10 | 10.4 | 9.9 | 0.5 |
| | | | 6.6 |

**Solution**

a. To formulate the problem statistically, we first note that we have one variable—gas mileage—and two populations:

> Population 1: All cars when the additive is used
> Population 2: All cars when the additive is not used

Let $\mu_1$ and $\mu_2$ denote the means of the variable "gas mileage" for Population 1 and Population 2, respectively:

> $\mu_1$ = mean gas mileage of all cars when the additive is used;
> $\mu_2$ = mean gas mileage of all cars when the additive is not used.

We want to perform the hypothesis test

> $H_0$: $\mu_1 = \mu_2$ (mean gas mileage with additive is not greater)
> $H_a$: $\mu_1 > \mu_2$ (mean gas mileage with additive is greater).

b. Independent samples could be used to carry out the hypothesis test: Take independent simple random samples of, say, 10 cars each; have the sample from Population 1 driven with the additive and the sample from Population 2 driven without the additive; and then apply a pooled or nonpooled $t$-test to the gas-mileage data obtained.

   However, in this case, a paired sample is more appropriate. Here, a pair consists of a car driven with the additive and the same car driven without the additive. The variable we analyze is the difference between the gas mileages of a car driven with and without the additive.

   By using a paired sample, we can remove an extraneous source of variation: the variation in the gas mileages of cars. The sampling error thus made in estimating the difference between the population means will generally be smaller and, therefore, we are more likely to detect differences between the population means when such differences exist.

c. The last column of Table 10.10 contains the difference, $d$, between the gas mileages with and without the additive for each of the 10 cars sampled. We refer to each difference as a **paired difference** because it is the difference of a pair of observations. For example, the first car got 25.7 mpg with the additive and 24.9 mpg without, giving a paired difference of 0.8 mpg, an increase in gas mileage of 0.8 mpg with the additive.

   If the null hypothesis is true, the paired differences of the gas mileages for the cars sampled should average about zero; that is, the sample mean $\bar{d}$ of the paired differences should be roughly zero. If $\bar{d}$ is too much greater than zero, we would take this as evidence that the null hypothesis is false.

   From the last column of Table 10.10, we find that the sample mean of the paired differences is

$$\bar{d} = \frac{\Sigma d_i}{n} = \frac{6.6}{10} = 0.66 \text{ mpg,}$$

a mean increase of 0.66 mpg when the additive is used. The question now is, can this mean increase in gas mileage be reasonably attributed to sampling error or is it large enough to indicate that, on average, the additive improves gas mileage (i.e., $\mu_1 > \mu_2$)? To answer that question, we need to know the distribution of the variable $\bar{d}$, which we discuss next.

• • •

### The Paired *t*-Statistic

Suppose that $x$ is a variable on each of two populations whose members can be paired. For each pair, we let $d$ denote the difference between the values of the variable $x$ on the members of the pair. We call $d$ the **paired-difference variable.**

It can be shown that the mean of the paired differences equals the difference between the two population means. In symbols,

$$\mu_d = \mu_1 - \mu_2.$$

Furthermore, if $d$ is normally distributed, we can apply this equation and our knowledge of the studentized version of a sample mean (Key Fact 8.5 on page 360) to obtain Key Fact 10.5.

---

**Key Fact 10.5** | **Distribution of the Paired *t*-Statistic**

Suppose that $x$ is a variable on each of two populations whose members can be paired. Further suppose that the paired-difference variable $d$ is normally distributed. Then, for paired samples of size $n$, the variable

$$t = \frac{\bar{d} - (\mu_1 - \mu_2)}{s_d/\sqrt{n}}$$

has the *t*-distribution with df $= n - 1$.

---

**Note:** We use the phrase **normal differences** as an abbreviation of "the paired-difference variable is normally distributed."

### Hypothesis Tests for the Means of Two Populations, Using a Paired Sample

We now present a hypothesis-testing procedure based on a paired sample for comparing the means of two populations when the paired-difference variable is normally distributed. In light of Key Fact 10.5, for a hypothesis test with null hypothesis $H_0$: $\mu_1 = \mu_2$, we can use the variable

$$t = \frac{\bar{d}}{s_d/\sqrt{n}}$$

as the test statistic and obtain the critical value(s) or $P$-value from the *t*-table, Table IV.

We call this hypothesis-testing procedure the **paired *t*-test.** Note that the paired *t*-test is simply the one-mean *t*-test applied to the paired-difference variable with null hypothesis $H_0$: $\mu_d = 0$. Procedure 10.5 provides a step-by-step method for performing a paired *t*-test by using either the critical-value approach or the $P$-value approach.

## Procedure 10.5 Paired *t*-Test

*Purpose* To perform a hypothesis test to compare two population means, $\mu_1$ and $\mu_2$

*Assumptions*

1. Simple random paired sample
2. Normal differences or large sample

**STEP 1** The null hypothesis is $H_0$: $\mu_1 = \mu_2$, and the alternative hypothesis is

$$H_a: \mu_1 \neq \mu_2 \quad \text{or} \quad H_a: \mu_1 < \mu_2 \quad \text{or} \quad H_a: \mu_1 > \mu_2$$
$$\text{(Two tailed)} \qquad \text{(Left tailed)} \qquad \text{(Right tailed)}$$

**STEP 2** Decide on the significance level, $\alpha$.

**STEP 3** Calculate the paired differences of the sample pairs.

**STEP 4** Compute the value of the test statistic

$$t = \frac{\bar{d}}{s_d / \sqrt{n}}$$

and denote that value $t_0$.

| CRITICAL-VALUE APPROACH | or | P-VALUE APPROACH |
|---|---|---|

**STEP 5** The critical value(s) are

$$\pm t_{\alpha/2} \quad \text{or} \quad -t_\alpha \quad \text{or} \quad t_\alpha$$
$$\text{(Two tailed)} \qquad \text{(Left tailed)} \qquad \text{(Right tailed)}$$

with df = $n - 1$. Use Table IV to find the critical value(s).

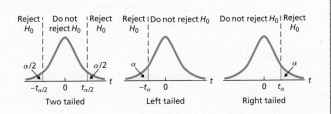

**STEP 6** If the value of the test statistic falls in the rejection region, reject $H_0$; otherwise, do not reject $H_0$.

**STEP 5** The *t*-statistic has df = $n - 1$. Use Table IV to estimate the *P*-value, or obtain it exactly by using technology.

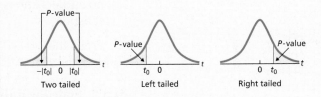

**STEP 6** If $P \leq \alpha$, reject $H_0$; otherwise, do not reject $H_0$.

**STEP 7** Interpret the results of the hypothesis test.

The hypothesis test is exact for normal differences and is approximately correct for large samples and nonnormal differences.

Properties and guidelines for use of the paired $t$-test are the same as those given for the one-mean $z$-test in Key Fact 9.4 on page 400 when applied to paired differences. In particular, the paired $t$-test is robust to moderate violations of the normality assumption but, even for large samples, can sometimes be unduly affected by outliers because the sample mean and sample standard deviation are not resistant to outliers. Here are two other important points:

- Do not apply the paired $t$-test to independent samples and, likewise, do not apply a pooled or nonpooled $t$-test to a paired sample.
- The normality assumption for a paired $t$-test refers to the distribution of the paired-difference variable, not to the two distributions of the variable under consideration.

**Example 10.11** | **The Paired $t$-Test**

*Gasoline Additive* We now return to the gas mileage problem posed in Example 10.10. The gas mileages of 10 randomly selected cars, both with and without a new gasoline additive, are displayed in the second and third columns of Table 10.10 on page 478. At the 5% significance level, do the data provide sufficient evidence to conclude that, on average, the gasoline additive improves gas mileage?

**Solution** First, we check the two conditions required for using the paired $t$-test, as listed in Procedure 10.5.

- Assumption 1 is satisfied because we have a simple random paired sample. Each pair consists of a car driven both with and without the additive.
- Because the sample size, $n = 10$, is small, we need to examine issues of normality and outliers. (See the first bulleted item in Key Fact 9.4 on page 400.) To do so, we constructed, in Fig. 10.9, a normal probability plot for the sample of paired differences in the last column of Table 10.10. This plot reveals no outliers and is roughly linear. Therefore, we can consider Assumption 2 satisfied.

**FIGURE 10.9**

Normal probability plot of the paired differences in Table 10.10

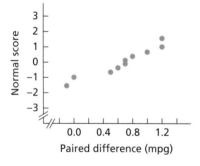

From the preceding items, we see that the paired $t$-test can be used to conduct the required hypothesis test. We apply Procedure 10.5.

**STEP 1** **State the null and alternative hypotheses.**

Let $\mu_1$ denote the mean gas mileage of all cars when the additive is used and $\mu_2$ denote the mean gas mileage of all cars when the additive is not used. Then the null and alternative hypotheses are

$$H_0: \mu_1 = \mu_2 \text{ (mean gas mileage with additive is not greater)}$$
$$H_a: \mu_1 > \mu_2 \text{ (mean gas mileage with additive is greater).}$$

Note that the hypothesis test is right tailed.

**STEP 2** **Decide on the significance level, $\alpha$.**

The test is to be performed at the 5% significance level, or $\alpha = 0.05$.

**STEP 3** **Calculate the paired differences of the sample pairs.**

We have already done so in Table 10.10 on page 478.

**STEP 4  Compute the value of the test statistic**

$$t = \frac{\bar{d}}{s_d/\sqrt{n}}.$$

We first need to determine the sample mean and sample standard deviation of the paired differences. We do so in the usual manner:

$$\bar{d} = \frac{\Sigma d_i}{n} = \frac{6.6}{10} = 0.66$$

and

$$s_d = \sqrt{\frac{\Sigma d_i^2 - (\Sigma d_i)^2/n}{n-1}} = \sqrt{\frac{6.12 - (6.6)^2/10}{10-1}} = 0.443.$$

Consequently, the value of the test statistic is

$$t = \frac{\bar{d}}{s_d/\sqrt{n}} = \frac{0.66}{0.443/\sqrt{10}} = 4.711.$$

| CRITICAL-VALUE APPROACH | or | P-VALUE APPROACH |
|---|---|---|

**STEP 5  The critical value for a right-tailed test is $t_\alpha$ with df $= n - 1$. Use Table IV to find the critical value.**

We have $n = 10$ and $\alpha = 0.05$. Table IV reveals that, for df $= 10 - 1 = 9$, we have $t_{0.05} = 1.833$, as shown in Fig. 10.10A.

**FIGURE 10.10A**

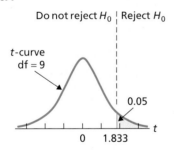

Do not reject $H_0$ | Reject $H_0$

$t$-curve
df $= 9$

0.05

0    1.833

**STEP 6  If the value of the test statistic falls in the rejection region, reject $H_0$; otherwise, do not reject $H_0$.**

From Step 4, the value of the test statistic is $t = 4.711$, which falls in the rejection region, as shown in Fig. 10.10A. Hence we reject $H_0$. The test results are statistically significant at the 5% level.

**STEP 5  The $t$-statistic has df $= n - 1$. Use Table IV to estimate the P-value or obtain it exactly by using technology.**

From Step 4, the value of the test statistic is $t = 4.711$. The test is right tailed, so the P-value is the probability of observing a value of $t$ of 4.711 or greater if the null hypothesis is true. That probability equals the shaded area shown in Fig. 10.10B.

**FIGURE 10.10B**

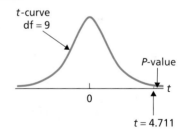

$t$-curve
df $= 9$

P-value

0

$t = 4.711$

Because $n = 10$, we have df $= 10 - 1 = 9$. Referring to Fig. 10.10B and to Table IV, we find that $P < 0.005$. (Using technology, we found that $P = 0.000549$.)

**STEP 6  If $P \leq \alpha$, reject $H_0$; otherwise, do not reject $H_0$.**

From Step 5, $P < 0.005$. Because the P-value is less than the specified significance level of 0.05, we reject $H_0$. The test results are statistically significant at the 5% level and (see Table 9.10 on page 414) provide very strong evidence against the null hypothesis.

Exercise 10.95
on page 489

**STEP 7 Interpret the results of the hypothesis test.**

**Interpretation** At the 5% significance level, the data provide sufficient evidence to conclude that, on average, the additive is effective in increasing gas mileage.

• • •

## Confidence Intervals for the Difference Between the Means of Two Populations, Using a Paired Sample

We can also use Key Fact 10.5 on page 480 to derive a confidence-interval procedure for the difference between two population means. We call that confidence-interval procedure the **paired *t*-interval procedure.**

---

**Procedure 10.6** | **Paired *t*-Interval Procedure**

*Purpose* To find a confidence interval for the difference between two population means, $\mu_1$ and $\mu_2$.

*Assumptions*

1. Simple random paired sample
2. Normal differences or large sample

**STEP 1 For a confidence level of $1 - \alpha$, use Table IV to find $t_{\alpha/2}$ with df $= n - 1$.**

**STEP 2 The endpoints of the confidence interval for $\mu_1 - \mu_2$ are**

$$\bar{d} \pm t_{\alpha/2} \cdot \frac{s_d}{\sqrt{n}}.$$

**STEP 3 Interpret the confidence interval.**

The confidence interval is exact for normal differences and is approximately correct for large samples and nonnormal differences.

---

**Example 10.12** | **The Paired *t*-Interval Procedure**

*Gasoline Additive* Use the sample data in Table 10.10 on page 478 to obtain a 90% confidence interval for the difference, $\mu_1 - \mu_2$, between the mean gas mileage of all cars when the additive is used and the mean gas mileage of all cars when the additive is not used.

**Solution** We apply Procedure 10.6.

**STEP 1 For a confidence level of $1 - \alpha$, use Table IV to find $t_{\alpha/2}$ with df $= n - 1$.**

For a 90% confidence interval, $\alpha = 0.10$. From Table IV, we determine that, for df $= n - 1 = 10 - 1 = 9$, we have $t_{\alpha/2} = t_{0.10/2} = t_{0.05} = 1.833$.

STEP 2  **The endpoints of the confidence interval for $\mu_1 - \mu_2$ are**

$$\bar{d} \pm t_{\alpha/2} \cdot \frac{s_d}{\sqrt{n}}.$$

We know that $t_{\alpha/2} = 1.833$, $n = 10$, and (from Example 10.11) $\bar{d} = 0.66$ and $s_d = 0.443$. So, the endpoints of the confidence interval for $\mu_1 - \mu_2$ are

$$0.66 \pm 1.833 \cdot \frac{0.443}{\sqrt{10}}$$

or 0.40 to 0.92.

STEP 3  **Interpret the confidence interval.**

Exercise 10.101
on page 491

**Interpretation**  We can be 90% confident that the difference between the mean gas mileages of all cars with and without the additive is somewhere between 0.40 mpg and 0.92 mpg.  In particular, we can be confident that, on average, the additive increases gas mileage by at least 0.40 mpg.

• • •

### What If the Assumptions Are Not Satisfied?

The paired $t$-procedures (paired $t$-test and paired $t$-interval procedure) provide methods for comparing the means of two populations.  As you know, the assumptions for using those procedures are (1) simple random paired sample and (2) normal differences or large sample.  If one or more of these conditions are not satisfied, then the paired $t$-procedures should not be used.

If the paired-difference variable is not normally distributed and the sample size is not large (i.e., Assumption 2 is violated), then a nonparametric method should be used.  For example, with a paired sample and a paired-difference variable that has a symmetric distribution, you can use a nonparametric method called the *paired Wilcoxon signed-rank test* to perform a hypothesis test and a nonparametric method called the *paired Wilcoxon confidence-interval procedure* to obtain a confidence interval.

## The Technology Center

Most statistical technologies have programs that automatically perform paired $t$-procedures.  In this subsection, we present output and step-by-step instructions for such programs.

**Example 10.13**  **Using Technology to Conduct Paired $t$-Procedures**

*Gasoline Additive*  Table 10.10 on page 478 gives the gas mileages of 10 randomly selected cars, both with and without a new gasoline additive.  Use Minitab, Excel, or the TI-83/84 Plus to perform the hypothesis test in Example 10.11 and obtain the confidence interval required in Example 10.12.

**Solution** Let $\mu_1$ denote the mean gas mileage of all cars when the additive is used and let $\mu_2$ denote the mean gas mileage of all cars when the additive is not used. The task in Example 10.11 is to perform the hypothesis test

$$H_0: \mu_1 = \mu_2 \text{ (mean gas mileage with additive is not greater)}$$
$$H_a: \mu_1 > \mu_2 \text{ (mean gas mileage with additive is greater),}$$

at the 5% significance level; the task in Example 10.12 is to obtain a 90% confidence interval for $\mu_1 - \mu_2$.

We applied the paired $t$-procedures programs to the data, resulting in Output 10.3 on pages 487 and 488. Steps for generating that output are presented in Instructions 10.3 on page 488.

As shown in Output 10.3, the $P$-value for the hypothesis test is about 0.001. Because the $P$-value is less than the specified significance level of 0.05, we reject $H_0$. Output 10.3 also shows that a 90% confidence interval for the difference between the means is from 0.403 to 0.917.

• • •

*Note to Minitab users:* As we noted on page 456, Minitab computes a two-sided confidence interval for a two-tailed test and a one-sided confidence interval for a one-tailed test. To perform a one-tailed hypothesis test and obtain a two-sided confidence interval, apply Minitab's paired $t$-procedure twice: once for the one-tailed hypothesis test and once for the confidence interval specifying a two-tailed hypothesis test.

**OUTPUT 10.3**   Paired *t*-procedures for the mileage data

## Paired T-Test and CI: WITH, WITHOUT      [FOR THE HYPOTHESIS TEST]

```
Paired T for WITH - WITHOUT

            N   Mean   StDev   SE Mean
WITH       10   18.91  7.47    2.36
WITHOUT    10   18.25  7.42    2.35
Difference 10   0.660  0.443   0.140


90% lower bound for mean difference: 0.466
T-Test of mean difference = 0 (vs > 0): T-Value = 4.71   P-Value = 0.001
```

## Paired T-Test and CI: WITH, WITHOUT      [FOR THE CONFIDENCE INTERVAL]

```
Paired T for WITH - WITHOUT

            N   Mean   StDev   SE Mean
WITH       10   18.91  7.47    2.36
WITHOUT    10   18.25  7.42    2.35
Difference 10   0.660  0.443   0.140


90% CI for mean difference: (0.403, 0.917)
T-Test of mean difference = 0 (vs not = 0): T-Value = 4.71   P-Value = 0.001
```

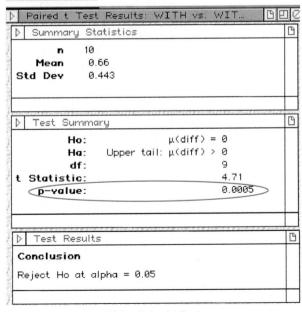

Using **Paired t Test**

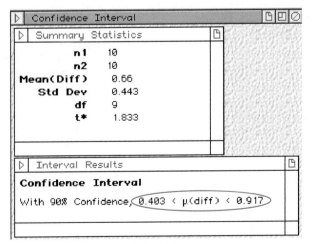

Using **Paired t Interval**

**OUTPUT 10.3 (cont.)**

Paired *t*-procedures for the mileage data

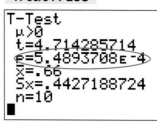

Using **T-Test**

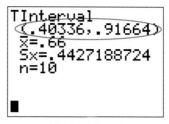

Using **TInterval**

**INSTRUCTIONS 10.3**   Steps for generating Output 10.3

| MINITAB | EXCEL | TI-83/84 PLUS |
|---|---|---|
| Store the gas-mileage data from the second and third columns of Table 10.10 in columns named WITH and WITHOUT. | Store the gas-mileage data from the second and third columns of Table 10.10 in ranges named WITH and WITHOUT. | Store the gas-mileage data from the second and third columns of Table 10.10 in lists named WITH and WOUT. |

**MINITAB**

Store the gas-mileage data from the second and third columns of Table 10.10 in columns named WITH and WITHOUT.

FOR THE HYPOTHESIS TEST:
1. Choose **Stat ➤ Basic Statistics ➤ Paired t...**
2. Select the **Samples in columns** option button
3. Click in the **First sample** text box and specify WITH
4. Click in the **Second sample** text box and specify WITHOUT
5. Click the **Options...** button
6. Click in the **Confidence level** text box and type 90
7. Click in the **Test mean** text box and type 0
8. Click the arrow button at the right of the **Alternative** drop-down list box and select **greater than**
9. Click **OK** twice

FOR THE CI:
1. Choose **Edit ➤ Edit Last Dialog**
2. Click the **Options...** button
3. Click the arrow button at the right of the **Alternative** drop-down list box and select **not equal**
4. Click **OK** twice

**EXCEL**

Store the gas-mileage data from the second and third columns of Table 10.10 in ranges named WITH and WITHOUT.

FOR THE HYPOTHESIS TEST:
1. Choose **DDXL ➤ Hypothesis Tests**
2. Select **Paired t Test** from the **Function type** drop-down box
3. Specify WITH in the **1st Quantitative Variable** text box
4. Specify WITHOUT in the **2nd Quantitative Variable** text box
5. Click **OK**
6. Click the **Set $\mu$(diff)0** button, type 0, and click **OK**
7. Click the **0.05** button
8. Click the **$\mu$(diff) > $\mu$(diff)0** button
9. Click the **Compute** button

FOR THE CI:
1. Exit to Excel
2. Choose **DDXL ➤ Confidence Intervals**
3. Select **Paired t Interval** from the **Function type** drop-down box
4. Specify WITH in the **1st Quantitative Variable** text box
5. Specify WITHOUT in the **2nd Quantitative Variable** text box
6. Click **OK**
7. Click the **90%** button
8. Click the **Compute Interval** button

**TI-83/84 PLUS**

Store the gas-mileage data from the second and third columns of Table 10.10 in lists named WITH and WOUT.

FOR THE PAIRED DIFFERENCES:
1. Press **2nd ➤ LIST**, arrow down to WITH, and press **ENTER**
2. Press −
3. Press **2nd ➤ LIST**, arrow down to WOUT, and press **ENTER**
4. Press **STO ▶**
5. Press **2nd ➤ A-LOCK**, type DIFF, and press **ENTER**

FOR THE HYPOTHESIS TEST:
1. Press **STAT**, arrow over to **TESTS**, and press **2**
2. Highlight **Data** and press **ENTER**
3. Press the down-arrow key, type 0 for $\mu_0$, and press **ENTER**
4. Press **2nd ➤ LIST**, arrow down to DIFF, and press **ENTER** three times
5. Highlight > $\mu_0$ and press **ENTER**
6. Press the down-arrow key, highlight **Calculate**, and press **ENTER**

FOR THE CI:
1. Press **STAT**, arrow over to **TESTS**, and press **8**
2. Highlight **Data** and press **ENTER**
3. Press the down-arrow key
4. Press **2nd ➤ LIST**, arrow down to DIFF, and press **ENTER** three times
5. Type .9 for **C-Level** and press **ENTER** twice

# Exercises 10.4

## Understanding the Concepts and Skills

**10.87** State one possible advantage of using paired samples instead of independent samples.

**10.88** What constitutes each pair in a paired sample?

**10.89** State the two conditions required for performing a paired *t*-procedure. How important are those conditions?

**10.90** Provide an example (different from the ones considered in this section) of a procedure based on a paired sample being more appropriate than one based on independent samples.

*In Exercises 10.91–10.94, hypothesis tests are proposed. For each hypothesis test, do the following.*
*a. Identify the variable.*
*b. Identify the two populations.*
*c. Identify the pairs.*
*d. Identify the paired-difference variable.*
*e. Determine the null and alternative hypotheses.*
*f. Classify the hypothesis test as two tailed, left tailed, or right tailed.*

**10.91 Ages of Married People.** The U.S. Census Bureau publishes information on the ages of married people in *Current Population Reports*. Suppose that you want to use a paired sample to decide whether the mean age of married men exceeds the mean age of married women.

**10.92 TV Viewing.** The A. C. Nielsen Company collects data on the TV viewing habits of Americans and publishes the information in *Nielsen Report on Television*. Suppose that you want to use a paired sample to decide whether the mean viewing time of married men is less than that of married women.

**10.93 Sports Stadiums and Home Values.** In the paper "Housing Values Near New Sporting Stadiums" (*Land Economics*, Vol. 81, Issue 3, pp. 379–395), C. Tu examined the effects of construction of new sports stadiums on home values. Suppose that you want to use a paired sample to decide whether construction of new sports stadiums affects the mean price of neighboring homes.

**10.94 Fiber Density.** In the article "Comparison of Fiber Counting by TV Screen and Eyepieces of Phase Contrast Microscopy" (*American Industrial Hygiene Association Journal*, Vol. 63, pp. 756–761), I. Moa et al. reported on determining fiber density by two different methods. The fiber density of 10 samples with varying fiber density was obtained by using both an eyepiece method and a TV-screen method. A hypothesis test is to be performed to decide whether, on average, the eyepiece method gives a greater fiber density reading than the TV-screen method.

*Preliminary data analyses indicate that use of a paired t-test is reasonable in Exercises 10.95–10.100. Perform each hypothesis test by using either the critical-value approach or the P-value approach.*

**10.95 Zea Mays.** Charles Darwin, author of *Origin of Species*, investigated the effect of cross-fertilization on the heights of plants. In one study he planted 15 pairs of Zea mays plants. Each pair consisted of one cross-fertilized plant and one self-fertilized plant grown in the same pot. The following table gives the height differences, in eighths of an inch, for the 15 pairs. Each difference is obtained by subtracting the height of the self-fertilized plant from that of the cross-fertilized plant.

| | | | | |
|---|---|---|---|---|
| 49 | −67 | 8 | 16 | 6 |
| 23 | 28 | 41 | 14 | 29 |
| 56 | 24 | 75 | 60 | −48 |

**a.** Identify the variable under consideration.
**b.** Identify the two populations.
**c.** Identify the paired-difference variable.
**d.** Are the numbers in the table paired differences? Why or why not?
**e.** At the 5% significance level, do the data provide sufficient evidence to conclude that the mean heights of cross-fertilized and self-fertilized Zea mays differ? (*Note:* $\bar{d} = 20.93$ and $s_d = 37.74$.)
**f.** Repeat part (e) at the 1% significance level.

**10.96 Sleep.** In 1908, W. S. Gosset published "The Probable Error of a Mean" (*Biometrika*, Vol. 6, pp. 1–25). In this pioneering paper, published under the pseudonym "Student," he introduced what later became known as Student's *t*-distribution. Gosset used the following data set, which gives the additional sleep in hours obtained by 10 patients who used laevohysocyamine hydrobromide.

| | | | | |
|---|---|---|---|---|
| 1.9 | 0.8 | 1.1 | 0.1 | −0.1 |
| 4.4 | 5.5 | 1.6 | 4.6 | 3.4 |

a. Identify the variable under consideration.
b. Identify the two populations.
c. Identify the paired-difference variable.
d. Are the numbers in the table paired differences? Why or why not?
e. At the 5% significance level, do the data provide sufficient evidence to conclude that laevohysocyamine hydrobromide is effective in increasing sleep?
(*Note:* $\bar{d} = 2.33$ and $s_d = 2.002$.)
f. Repeat part (e) at the 1% significance level.

**10.97 Anorexia Treatment.** Anorexia nervosa is a serious eating disorder, particularly among young women. The following data provide the weights, in pounds, of 17 anorexic young women before and after receiving a family therapy treatment for anorexia nervosa. [SOURCE: Hand et al. (ed.) *A Handbook of Small Data Sets*, London: Chapman & Hall, 1994. Raw data from B. Everitt (personal communication).]

| Before | After | Before | After | Before | After |
|--------|-------|--------|-------|--------|-------|
| 83.3 | 94.3 | 76.9 | 76.8 | 82.1 | 95.5 |
| 86.0 | 91.5 | 94.2 | 101.6 | 77.6 | 90.7 |
| 82.5 | 91.9 | 73.4 | 94.9 | 83.5 | 92.5 |
| 86.7 | 100.3 | 80.5 | 75.2 | 89.9 | 93.8 |
| 79.6 | 76.7 | 81.6 | 77.8 | 86.0 | 91.7 |
| 87.3 | 98.0 | 83.8 | 95.2 | | |

Does family therapy appear to be effective in helping anorexic young women gain weight? Perform the appropriate hypothesis test at the 5% significance level.

**10.98 Measuring Treadwear.** Stichler, Richey, and Mandel compared two methods of measuring treadwear in their paper "Measurement of Treadwear of Commercial Tires" (*Rubber Age*, Vol. 73:2). Eleven tires were each measured for treadwear by two methods, one based on weight and the other on groove wear. The following are the data, in thousands of miles.

| Weight method | Groove method | Weight method | Groove method |
|---------------|---------------|---------------|---------------|
| 30.5 | 28.7 | 24.5 | 16.1 |
| 30.9 | 25.9 | 20.9 | 19.9 |
| 31.9 | 23.3 | 18.9 | 15.2 |
| 30.4 | 23.1 | 13.7 | 11.5 |
| 27.3 | 23.7 | 11.4 | 11.2 |
| 20.4 | 20.9 | | |

At the 5% significance level, do the data provide sufficient evidence to conclude that, on average, the two measurement methods give different results?

**10.99 Glaucoma and Corneal Thickness.** Glaucoma is a leading cause of blindness in the United States. N. Ehlers measured the corneal thickness of eight patients who had glaucoma in one eye but not in the other. The results of the study were published as the paper "On Corneal Thickness and Intraocular Pressure, II" (*Acta Opthalmologica*, Vol. 48, pp. 1107–1112). The following are the data on corneal thickness, in microns.

| Patient | Normal | Glaucoma |
|---------|--------|----------|
| 1 | 484 | 488 |
| 2 | 478 | 478 |
| 3 | 492 | 480 |
| 4 | 444 | 426 |
| 5 | 436 | 440 |
| 6 | 398 | 410 |
| 7 | 464 | 458 |
| 8 | 476 | 460 |

At the 10% significance level, do the data provide sufficient evidence to conclude that mean corneal thickness is greater in normal eyes than in eyes with glaucoma?

**10.100 Fortified Orange Juice.** V. Tangpricha et al. conducted a study to determine whether fortifying orange juice with Vitamin D would increase serum 25-hydroxyvitamin D [25(OH)D] concentration in the blood. The researchers reported their findings in the paper "Fortification of Orange Juice with Vitamin D: A Novel Approach for Enhancing Vitamin D Nutritional Health" (*American Journal of Clinical Nutrition*, Vol. 77, pp. 1478–1483). A double-blind experiment was used in which 14 subjects drank 240 ml per day of orange juice fortified with 1000 IU of Vitamin D and 12 subjects drank 240 ml per day of unfortified orange juice. Concentration levels were recorded at the beginning of the experiment and again at the end of 12 weeks. The following data, based on the results of the study, provide the before and after serum 25(OH)D concentrations in the blood, in nanomoles per liter (nmo/L), for the group that drank the fortified juice.

| Before | After | Before | After |
|--------|-------|--------|-------|
| 8.6 | 33.8 | 3.9 | 75.0 |
| 32.3 | 137.0 | 1.5 | 83.3 |
| 60.7 | 110.6 | 18.1 | 71.5 |
| 20.4 | 52.7 | 100.9 | 142.0 |
| 39.4 | 110.5 | 84.3 | 171.4 |
| 15.7 | 39.1 | 32.3 | 52.1 |
| 58.3 | 124.1 | 41.7 | 112.9 |

At the 1% significance level, do the data provide sufficient evidence to conclude that, on average, drinking fortified orange juice increases the serum 25(OH)D concentration in the blood? (*Note:* The mean and standard deviation of the paired differences are −56.99 nmo/L and 26.20 nmo/L, respectively.)

*In Exercises 10.101–10.106, apply Procedure 10.6 on page 484 to obtain the required confidence interval. Interpret your result in each case.*

**10.101 Zea Mays.** Refer to Exercise 10.95.
a. Determine a 95% confidence interval for the difference between the mean heights of cross-fertilized and self-fertilized Zea mays.
b. Repeat part (a) for a 99% confidence level.

**10.102 Sleep.** Refer to Exercise 10.96.
a. Determine a 90% confidence interval for the additional sleep that would be obtained, on average, by using laevo-hysocyamine hydrobromide.
b. Repeat part (a) for a 98% confidence level.

**10.103 Anorexia Treatment.** Refer to Exercise 10.97 and find a 90% confidence interval for the weight gain that would be obtained, on average, by using the family therapy treatment.

**10.104 Measuring Treadwear.** Refer to Exercise 10.98 and find a 95% confidence interval for the mean difference in measurement by the weight and groove methods.

**10.105 Glaucoma and Corneal Thickness.** Refer to Exercise 10.99 and obtain an 80% confidence interval for the difference between the mean corneal thickness of normal eyes and that of eyes with glaucoma.

**10.106 Fortified Orange Juice.** Refer to Exercise 10.100 and obtain a 98% confidence interval for the mean increase of the serum 25(OH)D concentration after 12 weeks of drinking fortified orange juice.

**10.107 Tobacco Mosaic Virus.** To assess the effects of two different strains of the tobacco mosaic virus, W. Youden and H. Beale randomly selected eight tobacco leaves. Half of each leaf was subjected to one of the strains of tobacco mosaic virus and the other half to the other strain. The researchers then counted the number of local lesions apparent on each half of each leaf. The results of their study were published as the paper "A Statistical Study of the Local Lesion Method for Estimating Tobacco Mosaic Virus" (*Contributions to Boyce Thompson Institute*, Vol. 6, p. 437). Here are the data.

| Leaf | 1 | 2 | 3 | 4 | 5 | 6 | 7 | 8 |
|---|---|---|---|---|---|---|---|---|
| Virus 1 | 31 | 20 | 18 | 17 | 9 | 8 | 10 | 7 |
| Virus 2 | 18 | 17 | 14 | 11 | 10 | 7 | 5 | 6 |

Suppose that you want to perform a hypothesis test to determine whether a difference exists between the mean numbers of local lesions resulting from the two viral strains. Conduct preliminary graphical analyses to decide whether applying the paired *t*-test is reasonable. Explain your decision.

**10.108 Improving Car Emissions?** The makers of the MAGNETIZER Engine Energizer System (EES) claim that it improves gas mileage and reduces emissions in automobiles by using magnetic free energy to increase the amount of oxygen in the fuel for greater combustion efficiency. Following are test results, performed under International and U.S. Government agency standards, on a random sample of 14 vehicles. The data give the carbon monoxide (CO) levels, in parts per million, of each vehicle tested, both before installation of EES and after installation. [SOURCE: *Global Source Marketing.*]

| Before | After | Before | After |
|---|---|---|---|
| 1.60 | 0.15 | 2.60 | 1.60 |
| 0.30 | 0.20 | 0.15 | 0.06 |
| 3.80 | 2.80 | 0.06 | 0.16 |
| 6.20 | 3.60 | 0.60 | 0.35 |
| 3.60 | 1.00 | 0.03 | 0.01 |
| 1.50 | 0.50 | 0.10 | 0.00 |
| 2.00 | 1.60 | 0.19 | 0.00 |

Suppose that you want to perform a hypothesis test to determine whether, on average, EES reduces CO emissions. Conduct preliminary graphical data analyses to decide whether applying the paired *t*-test is reasonable. Explain your decision.

**10.109 Antiviral Therapy.** In the article "Improved Outcome for Children With Disseminated Adenoviral Infection Following Allogeneic Stem Cell Transplantation" (*British Journal of Haematology*, Vol. 130, Issue 4, p. 595), B. Kampmann et al. examined children who received stem cell transplants and subsequently became infected with a variety of ailments. A new antiviral therapy was administered to 11 patients. Their ABS lymphs ($\times 10^9$/L) at onset and resolution were as follows.

| Onset | Resolution | Onset | Resolution |
|---|---|---|---|
| 0.08 | 0.59 | 0.31 | 0.38 |
| 0.02 | 0.37 | 0.23 | 0.39 |
| 0.03 | 0.07 | 0.09 | 0.02 |
| 0.64 | 0.81 | 0.10 | 0.38 |
| 0.03 | 0.76 | 0.04 | 0.60 |
| 0.15 | 0.44 | | |

a. Obtain normal probability plots and boxplots of the onset data, the resolution data, and the paired differences of those data.

b. Based on your results from part (a), is applying a one-mean $t$-procedure to the onset data reasonable?

c. Based on your results from part (a), is applying a one-mean $t$-procedure to the resolution data reasonable?

d. Based on your results from part (a), is applying a paired $t$-procedure to the data reasonable?

e. What do your answers from parts (b)–(d) imply about the conditions for using a paired $t$-procedure?

## Working With Large Data Sets

**10.110 Faculty Salaries.** The American Association of University Professors (AAUP) conducts salary studies of college professors and publishes its findings in *AAUP Annual Report on the Economic Status of the Profession*. In Example 10.3 on page 451, we performed a hypothesis test based on independent samples to decide whether mean salaries differ for faculty in public and private institutions. Now you are to perform that same hypothesis test based on a paired sample. Pairs were formed by matching faculty in public and private institutions by rank and specialty. A random sample of 30 pairs yielded the data, in thousands of dollars, presented on the WeissStats CD. Use the technology of your choice to do the following.

a. Decide, at the 5% significance level, whether the data provide sufficient evidence to conclude that mean salaries differ for faculty in public institutions and private institutions.

b. Compare your result in part (a) to the one obtained in Example 10.3.

c. Which test do you think is preferable: the pooled $t$-test of Example 10.3 or the paired $t$-test done in part (a)? Explain your answer.

d. Find and interpret a 95% confidence interval for the difference between the mean salaries of faculty in public and private institutions.

e. Compare your result in part (d) to the one obtained in Example 10.4 on page 454.

f. Obtain a normal probability plot and a boxplot of the paired differences.

g. Based on your graphs from part (f), do you think that applying paired $t$-procedures here is reasonable?

**10.111 Marriage Ages.** In the *Statistics Norway* on-line article "The Times They are a Changing," J. Kristiansen discussed the changes in age at the time of marriage in Norway. The ages, in years, at the time of marriage for 75 Norwegian couples are presented on the WeissStats CD. Use the technology of your choice to do the following.

a. Decide, at the 1% significance level, whether the data provide sufficient evidence to conclude that the mean age of Norwegian men at the time of marriage exceeds that of Norwegian women.

b. Find and interpret a 99% confidence interval for the difference between the mean ages at the time of marriage for Norwegian men and women.

c. Remove the two paired-difference outliers and repeat parts (a) and (b). Compare your results to those in parts (a) and (b).

**10.112 Storm Hydrology and Clear Cutting.** In the document "Peak Discharge from Unlogged and Logged Watersheds," J. Jones and G. Grant compiled (paired) data on peak discharge from storms in two watersheds, one unlogged and one logged (100% clear-cut). If there is an effect due to clear-cutting, one would expect that the runoff would be greater in the logged area than in the unlogged area. The runoffs, in cubic meters per second per square kilometer ($m^3/s/km^2$) are provided on the WeissStats CD. Use the technology of your choice to do the following.

a. Formulate the null and alternative hypotheses to reflect the expectation expressed above.

b. Perform the required hypothesis test at the 1% significance level.

c. Obtain and interpret a 99% confidence interval for the difference between mean runoffs in the logged and unlogged watersheds.

d. Construct a histogram of the sample data to identify the approximate shape of the paired-difference variable.

e. Based on your result from part (d), do you think that applying the paired $t$-procedures in parts (b) and (c) is reasonable? Explain your answer.

## Extending the Concepts and Skills

**10.113** Explain exactly how a paired $t$-test can be formulated as a one-mean $t$-test. (*Hint:* Work solely with the paired-difference variable.)

**10.114 Gasoline Additive.** This exercise shows what can happen when a hypothesis-testing procedure designed for use with independent samples is applied to perform a hypothesis test on a paired sample. In Example 10.11 on page 482, we applied the paired $t$-test to decide whether a gasoline additive is effective in increasing gas mileage.

a. Apply the nonpooled $t$-test to the sample data in the second and third columns of Table 10.10 on page 478 to perform the hypothesis test. Use $\alpha = 0.05$.

b. Why is performing the hypothesis test the way you did in part (a) inappropriate?

c. Compare your result in part (a) to the one obtained in Example 10.11.

**10.115** A hypothesis test is to be performed to compare the means of two populations, using a paired sample. The sample of 15 paired differences contains an outlier but otherwise is roughly bell shaped. Assuming that it is not legitimate to remove the outlier, which test is better to use—the paired *t*-test or the paired Wilcoxon signed-rank test? Explain your answer.

**10.116** Suppose that you want to perform a hypothesis test to compare the means of two populations, using a paired sample. For each part, decide whether you would use the paired *t*-test, the paired Wilcoxon signed-rank test, or neither of these tests, if preliminary data analyses of the sample of paired differences suggest that the distribution of the paired-difference variable is
**a.** approximately normal.
**b.** highly skewed; the sample size is 20.
**c.** symmetric bimodal.

**10.117** Suppose that you want to perform a hypothesis test to compare the means of two populations, using a paired sample. For each part, decide whether you would use the paired *t*-test, the paired Wilcoxon signed-rank test, or neither of these tests, if preliminary data analyses of the sample of paired differences suggest that the distribution of the paired-difference variable is
**a.** uniform.
**b.** neither symmetric nor normal; the sample size is 132.
**c.** moderately skewed but otherwise roughly bell shaped.

## Chapter in Review

### You Should be Able to

1. use and understand the formulas in this chapter.

2. perform inferences based on independent simple random samples to compare the means of two populations when the population standard deviations are unknown but are assumed to be equal.

3. perform inferences based on independent simple random samples to compare the means of two populations when the population standard deviations are unknown but are not assumed to be equal.

4. perform inferences based on a simple random paired sample to compare the means of two populations.

### Key Terms

independent samples, *442*
independent simple random
    samples, *442*
nonpooled *t*-interval procedure, *467*
nonpooled *t*-test, *463, 464*
normal differences, *480*
paired difference, *479*

paired-difference variable, *480*
paired sample, *477*
paired *t*-interval procedure, *484*
paired *t*-test, *480, 481*
pool, *449*
pooled sample standard
    deviation ($s_p$), *449*

pooled *t*-interval procedure, *454*
pooled *t*-test, *449, 450*
sampling distribution of the
    difference between two sample
    means, *445*
simple random paired sample, *477*

## Review Problems

### Understanding the Concepts and Skills

**1.** Discuss the basic strategy for comparing the means of two populations based on independent simple random samples.

**2.** Discuss the basic strategy for comparing the means of two populations based on a simple random paired sample.

**3.** Regarding the pooled and nonpooled *t*-procedures,
**a.** what is the difference in assumptions between the two procedures?
**b.** how important is the assumption of independent simple random samples for these procedures?

**c.** how important is the normality assumption for these procedures?
**d.** Suppose that the variable under consideration is normally distributed on each of the two populations and that you are going to use independent simple random samples to compare the population means. Fill in the blank and explain your answer: Unless you are quite sure that the _____ are equal, the nonpooled *t*-procedures should be used instead of the pooled *t*-procedures.

**4.** Explain one possible advantage of using a paired sample instead of independent samples.

**5. Grip and Leg Strength.** In the paper, "Sex Differences in Static Strength and Fatigability in Three Different Muscle Groups" (*Research Quarterly for Exercise and Sport*, Vol 61(3), pp. 238–242), J. Misner et al. published results of a study on grip and leg strength of males and females. The following data, in newtons, is based on their measurements of right-leg strength.

| Male | | | Female | | |
|---|---|---|---|---|---|
| 2632 | 1796 | 2256 | 1344 | 1351 | 1369 |
| 2235 | 2298 | 1917 | 2479 | 1573 | 1665 |
| 1105 | 1926 | 2644 | 1791 | 1866 | 1544 |
| 1569 | 3129 | 2167 | 2359 | 1694 | 2799 |
| 1977 | | | 1868 | 2098 | |

Preliminary data analyses indicate that you can reasonably presume leg strength is normally distributed for both males and females and that the standard deviations of leg strength are approximately equal.

**a.** At the 5% significance level, do the data provide sufficient evidence to conclude that mean right-leg strength of males exceeds that of females? (*Note:* $\bar{x}_1 = 2127$, $s_1 = 513$, $\bar{x}_2 = 1843$, and $s_2 = 446$.)

**b.** Estimate the *P*-value of the hypothesis test, and use that estimate and Table 9.10 on page 414 to assess the strength of the evidence against the null hypothesis.

**6. Grip and Leg Strength.** Refer to Problem 5. Determine a 90% confidence interval for the difference between the mean right-leg strengths of males and females. Interpret your result.

**7. Cottonmouth Litter Size.** A study published by Blem and Blem in the *Journal of Herpetology* (Vol. 29, pp. 391–398) examined the reproductive characteristics of the eastern cottonmouth. The data in the following table, based on the results of the researchers' study, give the number of young per litter for 24 female cottonmouths in Florida and 44 female cottonmouths in Virginia.

| Florida | | | Virginia | | | | | |
|---|---|---|---|---|---|---|---|---|
| 8 | 6 | 7 | 5 | 12 | 7 | 7 | 6 | 8 |
| 7 | 4 | 3 | 12 | 9 | 7 | 4 | 9 | 6 |
| 1 | 7 | 5 | 12 | 7 | 5 | 6 | 10 | 3 |
| 6 | 6 | 5 | 10 | 8 | 8 | 12 | 5 | 6 |
| 6 | 8 | 5 | 10 | 11 | 3 | 8 | 4 | 5 |
| 5 | 7 | 4 | 7 | 6 | 11 | 7 | 6 | 8 |
| 6 | 6 | 5 | 8 | 14 | 8 | 7 | 11 | 7 |
| 5 | 5 | 4 | 5 | 4 | | | | |

Preliminary data analyses indicate that you can reasonably presume that litter sizes of cottonmouths in both states are approximately normally distributed. At the 1% significance level, do the data provide sufficient evidence to conclude that, on average, the number of young per litter of cottonmouths in Florida is less than that in Virginia? Do not assume that the population standard deviations are equal. (*Note:* $\bar{x}_1 = 5.46$, $s_1 = 1.59$, $\bar{x}_2 = 7.59$, and $s_2 = 2.68$.)

**8. Cottonmouth Litter Size.** Refer to Problem 7. Find a 98% confidence interval for the difference between the mean litter sizes of cottonmouths in Florida and Virginia. Interpret your result.

**9. Ecosystem Response.** In the on-line paper "Changes in Lake Ice: Ecosystem Response to Global Change" (*Teaching Issues and Experiments in Ecology*, tiee.ecoed.net, Vol. 3), R. Bohanan et al. questioned whether there is evidence for global warming in long term data on changes in dates of ice cover in three Wisconsin Lakes. The following table gives data, for a sample of eight years, on the number of days that ice stayed on two lakes in Madison, WI—Lake Mendota and Lake Monona.

| Year | Mendota | Monona |
|---|---|---|
| 1 | 119 | 107 |
| 2 | 115 | 108 |
| 3 | 53 | 52 |
| 4 | 108 | 108 |
| 5 | 74 | 85 |
| 6 | 47 | 47 |
| 7 | 102 | 96 |
| 8 | 87 | 91 |

**a.** Obtain a normal probability plot and boxplot of the paired differences.

**b.** Based on your results from part (a), is performing a paired *t*-test on the data reasonable? Explain your answer.

**c.** At the 10% significance level, do the data provide sufficient evidence to conclude that a difference exists in the mean length of time that ice stays on these two lakes?

**10. Ecosystem Response.** Refer to Problem 9 and find a 90% confidence interval for the difference in the mean lengths of time that ice stays on the two lakes. Interpret your result.

*Each of Problems **11–16** provides a type of sampling (independent or paired), sample size(s), and a figure showing the results of preliminary data analyses on the sample(s). For independent samples, the graphs are for the two samples; for a paired sample, the graphs are for the paired differences. The intent is to employ the sample data to perform a hypothesis test to compare the means of the two populations from which the data were obtained. In each case, decide which, if any, of the procedures that you have studied should be applied.*

11. Paired; $n = 75$; Fig. 10.11

12. Independent; $n_1 = 25$, $n_2 = 20$; Fig. 10.12

13. Independent; $n_1 = 17$, $n_2 = 17$; Fig. 10.13

14. Independent; $n_1 = 40$, $n_2 = 45$; Fig. 10.14

15. Independent; $n_1 = 20$, $n_2 = 15$; Fig. 10.15

16. Paired; $n = 18$; Fig. 10.16

## Working With Large Data Sets

**17. Drink and Be Merry?** In the paper, "Drink and Be Merry? Gender, Life Satisfaction, and Alcohol Consumption Among College Students" (*Psychology of Addictive Behaviors*, Vol. 19, Issue 2, pp. 184–191), J. Murphy et al. examined the impact of alcohol use and alcohol-related problems on several domains of life satisfaction (LS) in a sample of 353 college students. All LS items were rated on a 7-point Likert scale that ranged from 1 (strongly disagree) to 7 (strongly agree). On the WeissStats CD you will find data for dating satisfaction, based on the results of the study. Use the technology of your choice to do the following.

**a.** Obtain normal probability plots, boxplots, and the standard deviations for the two samples.

**b.** Based on your results from part (a), which are preferable here, pooled or nonpooled *t*-procedures? Explain your reasoning.

**c.** At the 5% significance level, do the data provide sufficient evidence to conclude that a difference exists in mean dating satisfaction of male and female college students?

**d.** Determine a 95% confidence interval for the difference between mean dating satisfaction of male and female college students.

**e.** Are your procedures in parts (c) and (d) justified? Explain your answer.

**18. Insulin and BMD.** I. Ertuğrul et al. conducted a study to determine the association between insulin growth factor 1 (IGF-1) and bone mineral density (BMD) in men over 65 years of age. The researchers published their results in the paper "Relationship Between Insulin-Like Growth Factor 1 and Bone Mineral Density in Men Aged Over 65 Years"

**FIGURE 10.11** Results of preliminary data analyses in Problem 11

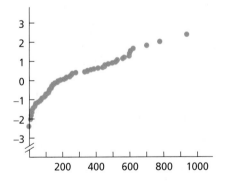 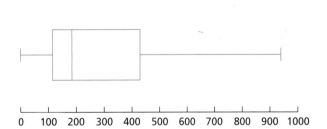

**FIGURE 10.12** Results of preliminary data analyses in Problem 12

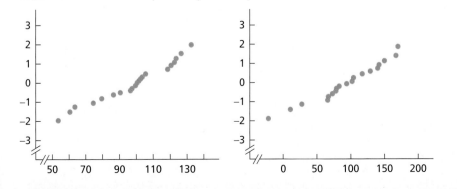 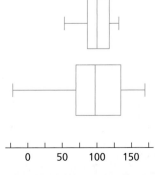

**FIGURE 10.13** Results of preliminary data analyses in Problem 13

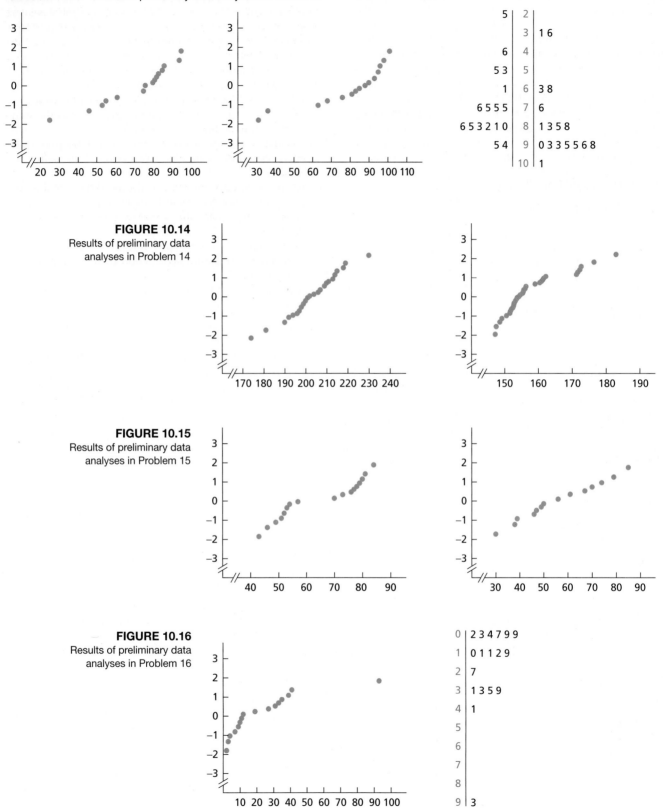

**FIGURE 10.14**
Results of preliminary data
analyses in Problem 14

**FIGURE 10.15**
Results of preliminary data
analyses in Problem 15

**FIGURE 10.16**
Results of preliminary data
analyses in Problem 16

(*Medical Principles and Practice*, Vol. 12, pp. 231–236). Forty-one men over 65 years old were enrolled in the study, as was a control group consisting of 20 younger men, ages 19–62. On the WeissStats CD, we have provided data on IGF-1 levels (in ng/ml), based on the results of the study. Use the technology of your choice to do the following.

a. Obtain normal probability plots, boxplots, and the standard deviations for the two samples.

b. Based on your results from part (a), which are preferable here, pooled or nonpooled *t*-procedures? Explain your reasoning.

c. At the 1% significance level, do the data provide sufficient evidence to conclude that, on average, men over 65 have a lower IGF-1 level than younger men?

d. Find and interpret a 99% confidence interval for the difference between the mean IGF-1 levels of men over 65 and younger men.

e. Are your procedures in parts (c) and (d) justified? Explain your answer.

**19. Weekly Earnings.** The U.S. Bureau of Labor Statistics publishes data on weekly earnings of full-time wage and salary workers in *Employment and Earnings*. Male and female workers were paired according to occupation and experience. Their weekly earnings, in dollars, are provided on the WeissStats CD. Use the technology of your choice to do the following.

a. Apply the paired *t*-test to decide, at the 5% significance level, whether the data provide sufficient evidence to conclude that, on average, the weekly earnings of male full-time wage and salary workers exceed that of women.

b. Find and interpret a 90% confidence interval for the difference between the mean weekly earnings of male and female full-time wage and salary workers. Use the paired *t*-interval procedure.

c. Obtain a normal probability plot, boxplot, and stem-and-leaf diagram of the paired differences.

d. Based on your results in part (c), are your procedures in parts (a) and (b) justified? Explain your answer.

# Focusing on Data Analysis    UWEC Undergraduates

Recall from Chapter 1 (see page 34) that the Focus database and Focus sample contain information on the undergraduate students at the University of Wisconsin - Eau Claire (UWEC). Now would be a good time for you to review the discussion about these data sets.

Open the Focus sample worksheet (FocusSample) in the technology of your choice and then do the following.

a.  Obtain normal probability plots, boxplots, and the sample standard deviations of the ACT composite scores for the sampled males and the sampled females.
b.  At the 5% significance level, do the data provide sufficient evidence to conclude that mean ACT composite scores differ for male and female UWEC undergraduates? Justify the use of the procedure you chose to carry out the hypothesis test.
c.  Determine and interpret a 95% confidence interval for the difference between the mean ACT composite scores of male and female UWEC undergraduates.

d.  Repeat parts (a)–(c) for cumulative GPA.
e.  Obtain a normal probability plot, boxplot, and histogram of the paired differences of the ACT English scores and ACT math scores for the sampled UWEC undergraduates.
f.  At the 5% significance level, do the data provide sufficient evidence to conclude that, for UWEC undergraduates, the mean ACT English score is less than the mean ACT math score? Justify the use of the procedure you chose to carry out the hypothesis test.
g.  Repeat part (f) at the 10% significance level.
h.  Find and interpret a 90% confidence interval for the difference between the mean ACT English score and the mean ACT math score of UWEC undergraduates.
i.  Repeat part (h), using an 80% confidence level.

# Case Study Discussion    Breast Milk and IQ

On page 441, we presented data obtained by five researchers studying the effect of breast feeding on subsequent IQ of preterm babies at age $7\frac{1}{2}$–8 years. Three categories were considered: children whose mothers declined to provide breast milk (Group I); those whose mothers had chosen but were unable to provide breast milk (Group IIa); and those whose mothers had chosen and were able to provide breast milk (Group IIb).

**1.**  Assume that IQs are normally distributed in all three categories. In each part, decide, at the 5% significance level, whether the data provide sufficient evidence to draw the specified conclusion for children age $7\frac{1}{2}$–8 years who are born preterm.
a.  A difference exists in mean IQ between those whose mothers decline to provide breast milk and those whose mothers choose but are unable to provide breast milk.

b.  The mean IQ of those whose mothers decline to provide breast milk is less than that of those whose mothers choose and are able to provide breast milk.
c.  The mean IQ of those whose mothers choose but are unable to provide breast milk is less than that of those whose mothers choose and are able to provide breast milk.

**2.**  Is this study observational, or is it a designed experiment? Explain your answer.

**3.**  Based on your answer to Problem 2, interpret your results in Problem 1.

## Biography   GERTRUDE COX: Spreading the Gospel According to St. Gertrude

**Gertrude Mary Cox** was born on January 13, 1900, in Dayton, Iowa, the daughter of John and Emmaline Cox. She graduated from Perry High School, Perry, Iowa, in 1918. Between 1918 and 1925, Cox prepared to become a deaconess in the Methodist Episcopal Church. However, in 1925, she decided to continue her education at Iowa State College in Ames, where she studied mathematics and statistics.

In 1929 and 1931, Cox received a B.S. and an M.S., respectively. Her work was directed by George W. Snedecor, and her degree was the first master's degree in statistics given by the department of mathematics at Iowa State.

From 1931 to 1933, Cox studied psychological statistics at the University of California at Berkeley. Snedecor meanwhile had established a new Statistical Laboratory at Iowa State, and in 1933 he asked her to be his assistant. This position launched her internationally influential career in statistics. Cox worked in the lab until becoming an Iowa State assistant professor in 1939.

In 1940, the committee in charge of filling a newly created position as head of the department of experimental statistics at North Carolina State College in Raleigh asked Snedecor for recommendations; he first named several male statisticians, then wrote, "...but if you would consider a woman for this position I would recommend Gertrude Cox of my staff." They did consider a woman and Cox accepted their offer.

In 1945, Cox organized and became director of the Institute of Statistics, which combined the teaching of statistics at the University of North Carolina and North Carolina State. Work conferences that Cox organized established the Institute as an international center for statistics. She also developed statistical programs throughout the South, referred to as "spreading the gospel according to St. Gertrude."

Cox's area of expertise was experimental design. She, with W. G. Cochran, wrote *Experimental Designs* (1950), recognized as the classic textbook on design and analysis of replicated experiments.

From 1960–1964, Cox was director of the Statistics Section of the Research Triangle Institute in Durham, North Carolina. She then retired, working only as a consultant. She died on October 17, 1978, in Durham.

## StatCrunch in MyStatLab
### Analyzing Data Online

StatCrunch online statistical software offers an easy-to-use interface customized for this book. The StatCrunch feature for each chapter illustrates the use of the software to perform a statistical analysis discussed in the chapter. Exercises are provided to further apply StatCrunch to other statistical analyses examined in the chapter. Go to the WeissStats CD or to the Weiss Web site at www.aw-bc.com/weiss to access StatCrunch instructions and data sets. To access StatCrunch statistical software, go to the student content area of your Weiss MyStatLab course.

## Internet Projects
### Exploring Data Online

The Internet project for each chapter provides simulations, demonstrations, or activities that enhance the topics covered in the chapter. The project materials come from universities, individuals, governments, and companies from all over the world. To access the Internet projects on the Web, go to www.aw-bc.com/weiss. From this Web page, you can reach the Internet Projects Page, which we suggest that you bookmark for easy access in the future.

# 11

# Inferences for Population Proportions

## Chapter Objectives

In Chapters 8–10, we discussed methods for finding confidence intervals and performing hypothesis tests for one or two population means. Now we describe how to conduct those inferences for one or two population proportions.

A *population proportion* is the proportion (percentage) of a population that has a specified attribute. For example, if the population under consideration consists of all Americans and the specified attribute is "retired," the population proportion is the proportion of all Americans who are retired.

In Section 11.1, we begin by introducing notation and terminology needed to perform proportion inferences; then we discuss confidence intervals for one population proportion. Next, in Section 11.2, we examine a method for conducting a hypothesis test for one population proportion.

In Section 11.3, we investigate how to perform a hypothesis test to compare two population proportions and how to construct a confidence interval for the difference between two population proportions.

An *Associated Press* article from July, 2005, discussed two independent polls about American and Japanese views on world wars, nuclear weapons, and related issues. The poll of 1000 adults in the United States was conducted from July 5–10 for the Associated Press by Ipsos, an international polling company; the poll of 1045 adults in Japan was conducted from July 1–3 for Kyodo by the Public Opinion Research Center.

According to the AP article, "...Americans are far more likely than the Japanese to expect another world war in their lifetime." More precisely, 6 in 10 of the Americans surveyed thought that such a war was likely, whereas only one-third of the Japanese surveyed thought so.

In an attempt to end World War II, President Truman made the decision to drop atomic bombs on Hiroshima on August 6, 1945, and on Nagasaki 3 days later. The bombs immediately killed thousands of people. Two-thirds of the Americans surveyed said that the use of atomic bombs on Japan in World War II was unavoidable; only 20% of the Japanese surveyed felt that way.

People in both countries now perceive the other country favorably. Four in five of the Americans surveyed had a positive view of Japan and two-thirds of the Japanese surveyed felt that way about the United States.

After you study this chapter, you will be asked to statistically analyze the results of the two polls reported on by the Associated Press.

# 11.1 Confidence Intervals for One Population Proportion

Statisticians often need to determine the proportion (percentage) of a population that has a specified attribute. Some examples are

- the percentage of U.S. adults who have health insurance,
- the percentage of cars in the United States that are imports,
- the percentage of U.S. adults who favor stricter clean air health standards, or
- the percentage of Canadian women in the labor force.

In the first case, the population consists of all U.S. adults and the specified attribute is "has health insurance." For the second case, the population consists of all cars in the United States and the specified attribute is "is an import." The population in the third case is all U.S. adults and the specified attribute is "favors stricter clean air health standards." And, in the fourth case, the population consists of all Canadian women and the specified attribute is "is in the labor force."

We know that it is often impractical or impossible to take a census of a large population. In practice, therefore, we use data from a sample to make inferences about the population proportion. We introduce proportion notation and terminology in the next example.

## Example 11.1 | Proportion Notation and Terminology

*Playing Hooky From Work* Many employers are concerned about the problem of employees who call in sick when they are not ill. The Hilton Hotels Corporation commissioned a survey to investigate this issue. One question asked the respondents whether they call in sick at least once a year when they simply need time to relax. For brevity, we use the phrase *play hooky* to refer to that practice.

The survey polled 1010 randomly selected U.S. employees. The proportion of the 1010 employees sampled who play hooky was used to estimate the proportion of all U.S. employees who play hooky. Discuss the statistical notation and terminology used in this and similar studies on proportions.

**Solution** We use $p$ to denote the proportion of all U.S. employees who play hooky; it represents the **population proportion** and is the parameter whose value is to be estimated. The proportion of the 1010 U.S. employees sampled who play hooky is designated $\hat{p}$ (read "p hat") and represents a **sample proportion;** it is the statistic used to estimate the unknown population proportion, $p$.

Although unknown, the population proportion, $p$, is a fixed number. In contrast, the sample proportion, $\hat{p}$, is a variable; its value varies from sample to sample. For instance, if 202 of the 1010 employees sampled play hooky, then

$$\hat{p} = \frac{202}{1010} = 0.2,$$

that is, 20.0% of the employees sampled play hooky. If 184 of the 1010 employees sampled play hooky, however, then

$$\hat{p} = \frac{184}{1010} = 0.182,$$

that is, 18.2% of the employees sampled play hooky.

**You try it!**

Exercise 11.5(a)–(b)
on page 510

These two calculations also reveal how to compute a sample proportion: Divide the number of employees sampled who play hooky, denoted $x$, by the total number of employees sampled, $n$. In symbols, $\hat{p} = x/n$. We generalize these new concepts below.

• • •

---

**Definition 11.1**

### Population Proportion and Sample Proportion

Consider a population in which each member either has or does not have a specified attribute. Then we use the following notation and terminology.

**Population proportion, $p$:** The proportion (percentage) of the entire population that has the specified attribute.

**Sample proportion, $\hat{p}$:** The proportion (percentage) of a sample from the population that has the specified attribute.

---

**Formula 11.1**

### Sample Proportion

A sample proportion, $\hat{p}$, is computed by using the formula

$$\hat{p} = \frac{x}{n}$$

where $x$ denotes the number of members in the sample that have the specified attribute and, as usual, $n$ denotes the sample size.

**What Does It Mean?**

A sample proportion is obtained by dividing the number of members sampled that have the specified attribute by the total number of members sampled.

**Note:** For convenience, we sometimes refer to $x$ (the number of members in the sample that have the specified attribute) as the **number of successes** and to $n - x$ (the number of members in the sample that do not have the specified attribute) as the **number of failures.** In this context, the words *success* and *failure* may not have their ordinary meanings.

**TABLE 11.1**
Correspondence between notations for means and proportions

|  | Parameter | Statistic |
|---|---|---|
| **Means** | $\mu$ | $\bar{x}$ |
| **Proportions** | $p$ | $\hat{p}$ |

Table 11.1 shows the correspondence between the notation for means and the notation for proportions. Recall that a sample mean, $\bar{x}$, can be used to make inferences about a population mean, $\mu$. Similarly, a sample proportion, $\hat{p}$, can be used to make inferences about a population proportion, $p$.

### The Sampling Distribution of the Sample Proportion

To make inferences about a population mean, $\mu$, we must know the sampling distribution of the sample mean, that is, the distribution of the variable $\bar{x}$. The same is true for proportions: To make inferences about a population proportion, $p$, we need to know the **sampling distribution of the sample proportion,** that is, the distribution of the variable $\hat{p}$.

Because a proportion can always be regarded as a mean, we can use our knowledge of the sampling distribution of the sample mean to derive the sampling distribution of the sample proportion. (See Exercise 11.46 for details.) In practice, the sample size usually is large, so we concentrate on that case.

**Key Fact 11.1**

### The Sampling Distribution of the Sample Proportion

For samples of size $n$,

- the mean of $\hat{p}$ equals the population proportion: $\mu_{\hat{p}} = p$;
- the standard deviation of $\hat{p}$ equals the square root of the product of the population proportion and one minus the population proportion divided by the sample size: $\sigma_{\hat{p}} = \sqrt{p(1-p)/n}$; and
- $\hat{p}$ is approximately normally distributed for large $n$.

The accuracy of the normal approximation depends on $n$ and $p$. If $p$ is close to 0.5, the approximation is quite accurate, even for moderate $n$. The farther $p$ is from 0.5, the larger $n$ must be for the approximation to be accurate. As a rule of thumb, we use the normal approximation when *np and n(1 − p) are both 5 or greater.*[†] In this chapter, when we say that $n$ is large, we mean that $np$ and $n(1 - p)$ are both 5 or greater.

**Example 11.2** | ### The Sampling Distribution of the Sample Proportion

**OUTPUT 11.1**

Histogram of $\hat{p}$ for 2000 samples of size 1010 with superimposed normal curve

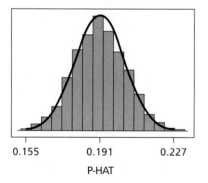

0.155        0.191        0.227

P-HAT

*Playing Hooky From Work* In Example 11.1, suppose that 19.1% of all U.S. employees play hooky; that is, that the population proportion is $p = 0.191$. Then, according to Key Fact 11.1, for samples of size 1010, the variable $\hat{p}$ is approximately normally distributed with mean $\mu_{\hat{p}} = p = 0.191$ and standard deviation

$$\sigma_{\hat{p}} = \sqrt{\frac{p(1-p)}{n}} = \sqrt{\frac{0.191(1 - 0.191)}{1010}} = 0.012.$$

Use simulation to make that fact plausible.

**Solution** We first simulated 2000 samples of 1010 U.S. employees each. Then, for each of those 2000 samples, we determined the sample proportion, $\hat{p}$, of those who play hooky. Output 11.1 shows a histogram of those 2000 values of $\hat{p}$, which is shaped like the superimposed normal curve with parameters 0.191 and 0.012.

• • •

### Large-Sample Confidence Intervals for a Population Proportion

Procedure 11.1 gives a step-by-step method for finding a confidence interval for a population proportion. We call this method the **one-proportion z-interval procedure.**[‡] It is based on Key Fact 11.1 and is derived in a way similar to the one-mean z-interval procedure (Procedure 8.1 on page 345).

---

[†]Another commonly used rule of thumb is that $np$ and $n(1 - p)$ are both 10 or greater; still another is that $np(1 - p)$ is 25 or greater. However, our rule of thumb, which is less conservative than either of those two, is consistent with the conditions required for performing a chi-square goodness-of-fit test (discussed in Chapter 12).

[‡]The one-proportion z-interval procedure is also known as the **one-sample z-interval procedure for a population proportion** and the **one-variable proportion interval procedure.**

| Procedure 11.1 | **One-Proportion z-Interval Procedure** |

*Purpose* To find a confidence interval for a population proportion, $p$

*Assumptions*

1. Simple random sample
2. The number of successes, $x$, and the number of failures, $n - x$, are both 5 or greater.

**STEP 1** For a confidence level of $1 - \alpha$, use Table II to find $z_{\alpha/2}$.

**STEP 2** The confidence interval for $p$ is from

$$\hat{p} - z_{\alpha/2} \cdot \sqrt{\hat{p}(1 - \hat{p})/n} \quad \text{to} \quad \hat{p} + z_{\alpha/2} \cdot \sqrt{\hat{p}(1 - \hat{p})/n}$$

where $z_{\alpha/2}$ is found in Step 1, $n$ is the sample size, and $\hat{p} = x/n$ is the sample proportion.

**STEP 3** Interpret the confidence interval.

**Note:** As stated in Assumption 2 of Procedure 11.1, a condition for using that procedure is that "the number of successes, $x$, and the number of failures, $n - x$, are both 5 or greater." We can restate this condition as "$n\hat{p}$ and $n(1 - \hat{p})$ are both 5 or greater," which, for an unknown $p$, corresponds to the rule of thumb for using the normal approximation given after Key Fact 11.1.

| Example 11.3 | **The One-Proportion z-Interval Procedure** |

*Playing Hooky From Work* A poll was taken of 1010 U.S. employees. The employees sampled were asked whether they "play hooky," that is, call in sick at least once a year when they simply need time to relax; 202 responded "yes." Use these data to find a 95% confidence interval for the proportion, $p$, of all U.S. employees who play hooky.

**Solution** The attribute in question is "plays hooky," the sample size is 1010, and the number of employees sampled who play hooky is 202. We have $n = 1010$. Also, $x = 202$ and $n - x = 1010 - 202 = 808$, both of which are 5 or greater. We can therefore apply Procedure 11.1 to obtain the required confidence interval.

**STEP 1** For a confidence level of $1 - \alpha$, use Table II to find $z_{\alpha/2}$.

We want a 95% confidence interval, which means that $\alpha = 0.05$. In Table II or at the bottom of Table IV, we find that $z_{\alpha/2} = z_{0.05/2} = z_{0.025} = 1.96$.

**STEP 2** The confidence interval for $p$ is from

$$\hat{p} - z_{\alpha/2} \cdot \sqrt{\hat{p}(1 - \hat{p})/n} \quad \text{to} \quad \hat{p} + z_{\alpha/2} \cdot \sqrt{\hat{p}(1 - \hat{p})/n}.$$

We have $n = 1010$ and, from Step 1, $z_{\alpha/2} = 1.96$. Also, because 202 of the 1010 employees sampled play hooky, $\hat{p} = x/n = 202/1010 = 0.2$. Consequently, a 95% confidence interval for $p$ is from

$$0.2 - 1.96 \cdot \sqrt{(0.2)(1 - 0.2)/1010} \quad \text{to} \quad 0.2 + 1.96 \cdot \sqrt{(0.2)(1 - 0.2)/1010},$$

or

$$0.2 - 0.025 \quad \text{to} \quad 0.2 + 0.025,$$

or 0.175 to 0.225.

Exercise 11.25
on page 512

**STEP 3** **Interpret the confidence interval.**

**Interpretation**   We can be 95% confident that the percentage of all U.S. employees who play hooky is somewhere between 17.5% and 22.5%.

• • •

### Margin of Error

In Section 8.3, we discussed the margin of error in estimating a population mean by a sample mean. In general, the **margin of error** of an estimator represents the precision with which it estimates the parameter in question. The confidence-interval formula in Step 2 of Procedure 11.1 indicates that the margin of error, $E$, in estimating a population proportion by a sample proportion is $z_{\alpha/2} \cdot \sqrt{\hat{p}(1 - \hat{p})/n}$.

---

**Definition 11.2**

**? What Does It Mean?**

The margin of error is equal to half the length of the confidence interval. It represents the precision with which a sample proportion estimates the population proportion at the specified confidence level.

### Margin of Error for the Estimate of $p$

The **margin of error** for the estimate of $p$ is

$$E = z_{\alpha/2} \cdot \sqrt{\hat{p}(1 - \hat{p})/n}.$$

---

In Example 11.3, the margin of error is

$$E = z_{\alpha/2} \cdot \sqrt{\hat{p}(1 - \hat{p})/n} = 1.96 \cdot \sqrt{(0.2)(1 - 0.2)/1010} = 0.025,$$

which can also be obtained by taking one-half the length of the confidence interval: $(0.225 - 0.175)/2 = 0.025$. Therefore we can be 95% confident that the error in estimating the proportion $p$ of all U.S. employees who play hooky by the proportion, 0.2, of those in the sample who play hooky is at most 0.025, that is, plus or minus 2.5 percentage points.

On the one hand, given a confidence interval, we can find the margin of error by taking half the length of the confidence interval. On the other hand, given the sample proportion and the margin of error, we can determine the confidence interval—its endpoints are $\hat{p} \pm E$.

Most newspaper and magazine polls provide the sample proportion and the margin of error associated with a 95% confidence interval. For example, a survey of U.S. women conducted by Gallup for the CNBC cable network stated, "36% of those polled believe their gender will hurt them; the margin of error for the poll is plus or minus 4 percentage points."

Translated into our terminology, $\hat{p} = 0.36$ and $E = 0.04$. Thus the confidence interval has endpoints $\hat{p} \pm E = 0.36 \pm 0.04$, or 0.32 to 0.40. As a result, we can be 95% confident that the percentage of all U.S. women who believe that their gender will hurt them is somewhere between 32% and 40%.

### Determining the Required Sample Size

If the margin of error and confidence level are given, then we must determine the sample size required to meet those specifications. Solving for $n$ in the formula for the margin of error, we get

$$n = \hat{p}(1 - \hat{p}) \left(\frac{z_{\alpha/2}}{E}\right)^2. \tag{11.1}$$

This formula cannot be used to obtain the required sample size because the sample proportion, $\hat{p}$, is not known prior to sampling.

There are two ways around this problem. To begin, we examine the graph of $\hat{p}(1 - \hat{p})$ versus $\hat{p}$ shown in Fig. 11.1, which reveals that the largest $\hat{p}(1 - \hat{p})$ can be is 0.25, or when $\hat{p} = 0.5$. The farther $\hat{p}$ is from 0.5, the smaller will be the value of $\hat{p}(1 - \hat{p})$.

Because the largest possible value of $\hat{p}(1 - \hat{p})$ is 0.25, the most conservative approach for determining sample size is to use that value in Equation (11.1). The sample size obtained then will generally be larger than necessary and the margin of error less than required. Nonetheless, this approach guarantees that the specifications will at least be met.

However, because sampling tends to be time consuming and expensive, we usually do not want to take a larger sample than necessary. If we can make an educated guess for the observed value of $\hat{p}$—say, from a previous study or theoretical considerations—we can use that guess to obtain a more realistic sample size.

In this same vein, if we have in mind a likely range for the observed value of $\hat{p}$, then, in light of Fig. 11.1, we should take as our educated guess for $\hat{p}$ the value in the range closest to 0.5. But, in either case, we should be aware that, if the observed value of $\hat{p}$ is closer to 0.5 than is our educated guess, the margin of error will be larger than desired.

**FIGURE 11.1**
Graph of $\hat{p}(1 - \hat{p})$ versus $\hat{p}$

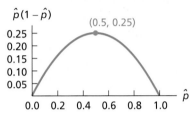

---

Formula 11.2 | **Sample Size for Estimating $p$**

A $(1 - \alpha)$-level confidence interval for a population proportion that has a margin of error of at most $E$ can be obtained by choosing

$$n = 0.25 \left(\frac{z_{\alpha/2}}{E}\right)^2$$

rounded up to the nearest whole number. If you can make an educated guess, $\hat{p}_g$ ($g$ for guess), for the observed value of $\hat{p}$, then you should instead choose

$$n = \hat{p}_g(1 - \hat{p}_g) \left(\frac{z_{\alpha/2}}{E}\right)^2$$

rounded up to the nearest whole number.

---

Example 11.4 | **Sample Size for Estimating $p$**

*Playing Hooky From Work* Consider again the problem of estimating the proportion of all U.S. employees who play hooky.

**a.** Obtain a sample size that will ensure a margin of error of at most 0.01 for a 95% confidence interval.

**b.** Find a 95% confidence interval for $p$ if, for a sample of the size determined in part (a), the proportion of those who play hooky is 0.194.

**c.** Determine the margin of error for the estimate in part (b) and compare it to the margin of error specified in part (a).

**d.** Repeat parts (a)–(c) if the proportion of those sampled who play hooky can reasonably be presumed to be between 0.1 and 0.3.

**e.** Compare the results obtained in parts (a)–(c) with those obtained in part (d).

### Solution

**a.** We apply the first equation in Formula 11.2. To do so, we must identify $z_{\alpha/2}$ and the margin of error, $E$. The confidence level is stipulated to be 0.95, so $z_{\alpha/2} = z_{0.05/2} = z_{0.025} = 1.96$, and the margin of error is specified at 0.01. Thus a sample size that will ensure a margin of error of at most 0.01 for a 95% confidence interval is

$$n = 0.25 \left( \frac{z_{\alpha/2}}{E} \right)^2 = 0.25 \left( \frac{1.96}{0.01} \right)^2 = 9604.$$

**Interpretation**   If we take a sample of 9604 U.S. employees, the margin of error for our estimate of the proportion of all U.S. employees who play hooky will be 0.01 or less—that is, plus or minus at most 1 percentage point.

**b.** We find, by applying Procedure 11.1 (page 505) with $\alpha = 0.05$, $n = 9604$, and $\hat{p} = 0.194$, that a 95% confidence interval for $p$ has endpoints

$$0.194 \pm 1.96 \cdot \sqrt{(0.194)(1 - 0.194)/9604},$$

or $0.194 \pm 0.008$, or 0.186 to 0.202.

**Interpretation**   Based on a sample of 9604 U.S. employees, we can be 95% confident that the percentage of all U.S. employees who play hooky is somewhere between 18.6% and 20.2%.

**c.** The margin of error for the estimate in part (b) is 0.008. Not surprisingly, this is less than the margin of error of 0.01 specified in part (a).

**d.** If we can reasonably presume that the proportion of those sampled who play hooky will be between 0.1 and 0.3, we use the second equation in Formula 11.2, with $\hat{p}_g = 0.3$ (the value in the range closest to 0.5), to determine the sample size:

$$n = \hat{p}_g(1 - \hat{p}_g) \left( \frac{z_{\alpha/2}}{E} \right)^2 = (0.3)(1 - 0.3) \left( \frac{1.96}{0.01} \right)^2 = 8068 \text{ (rounded up).}$$

Applying Procedure 11.1 with $\alpha = 0.05$, $n = 8068$, and $\hat{p} = 0.194$, we find that a 95% confidence interval for $p$ has endpoints

$$0.194 \pm 1.96 \cdot \sqrt{(0.194)(1 - 0.194)/8068},$$

or $0.194 \pm 0.009$, or 0.185 to 0.203.

**Interpretation**   Based on a sample of 8068 U.S. employees, we can be 95% confident that the percentage of all U.S. employees who play hooky is somewhere between 18.5% and 20.3%. The margin of error for the estimate is 0.009.

You
try it!

Exercise 11.33
on page 513

**e.** By using the educated guess for $\hat{p}$ in part (d), we reduced the required sample size by more than 1500 (from 9604 to 8068). Moreover, only 0.1% (0.001) of precision was lost—the margin of error rose from 0.008 to 0.009. The risk of using the guess 0.3 for $\hat{p}$ is that, if the observed value of $\hat{p}$ had turned out to be larger than 0.3 (but smaller than 0.7), the achieved margin of error would have exceeded the specified 0.01.

• • •

## The Technology Center

Most statistical technologies have programs that automatically perform the one-proportion $z$-interval procedure. In this subsection, we present output and step-by-step instructions for such programs.

**Example 11.5** **Using Technology to Obtain a One-Proportion *z*-Interval**

*Playing Hooky From Work*    Of 1010 randomly selected U.S. employees asked whether they play hooky from work, 202 said they do. Use Minitab, Excel, or the TI-83/84 Plus to find a 95% confidence interval for the proportion, $p$, of all U.S. employees who play hooky.

**Solution**    We applied the one-proportion $z$-interval programs to the data, resulting in Output 11.2. Steps for generating that output are presented in Instructions 11.1 at the top of the next page.

**OUTPUT 11.2**    One-proportion *z*-interval for the data on playing hooky from work

**MINITAB**

**Test and CI for One Proportion**

```
Sample    X      N    Sample p          95% CI
1        202   1010   0.200000   (0.175331, 0.224669)

Using the normal approximation.
```

**EXCEL**

| ▷ | x Confidence Interval | | | | | | 🗗🗖⊘ |

| ▷ | Summary Statistics | | 🗗 | ▷ | Interval Results | | 🗗 |
|---|---|---|---|---|---|---|---|
| n | 1010 | | | **Confidence Interval** | | | |
| p-hat | 0.2 | | | With 95% Confidence, 0.175 < p < 0.225 | | | |
| Std Err | 0.0126 | | | | | | |
| z* | 1.96 | | | | | | |

**TI-83/84 PLUS**

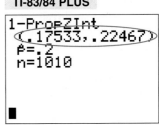

As shown in Output 11.2, the required 95% confidence interval is from 0.175 to 0.225. We can be 95% confident that the percentage of all U.S. employees who play hooky is somewhere between 17.5% and 22.5%.

• • •

**INSTRUCTIONS 11.1** Steps for generating Output 11.2

| MINITAB | EXCEL | TI-83/84 PLUS |
|---|---|---|
| 1 Choose **Stat ➤ Basic Statistics ➤ 1 Proportion...**<br>2 Select the **Summarized data** option button<br>3 Click in the **Number of events** text box and type 202<br>4 Click in the **Number of trials** text box and type 1010<br>5 Click the **Options...** button<br>6 Click in the **Confidence level** text box and type 95<br>7 Check the **Use test and interval based on normal distribution** check box<br>8 Click **OK** twice | 1 Store the sample size, 1010, and the number of successes, 202, in ranges named n and x, respectively<br>2 Choose **DDXL ➤ Confidence Intervals**<br>3 Select **Summ 1 Var Prop Interval** from the **Function type** drop-down list box<br>4 Specify x in the **Num Successes** text box<br>5 Specify n in the **Num Trials** text box<br>6 Click **OK**<br>7 Click the **95%** button<br>8 Click the **Compute Interval** button | 1 Press **STAT**, arrow over to **TESTS**, and press **ALPHA ➤ A**<br>2 Type 202 for **x** and press **ENTER**<br>3 Type 1010 for **n** and press **ENTER**<br>4 Type .95 for **C-Level** and press **ENTER** twice |

▮ ▮ ▮

# Exercises 11.1

## Understanding the Concepts and Skills

**11.1** In a newspaper or magazine of your choice, find a statistical study that contains an estimated population proportion.

**11.2** Why is statistical inference generally used to obtain information about a population proportion?

**11.3** Is a population proportion a parameter or a statistic? What about a sample proportion? Explain your answers.

**11.4** Answer the following questions about the basic notation and terminology for proportions.
a. What is a population proportion?
b. What symbol is used for a population proportion?
c. What is a sample proportion?
d. What symbol is used for a sample proportion?
e. For what is the phrase "number of successes" an abbreviation? What symbol is used for the number of successes?
f. For what is the phrase "number of failures" an abbreviation?
g. Explain the relationships among the sample proportion, the number of successes, and the sample size.

**11.5** This exercise involves the use of an unrealistically small population to provide a concrete illustration for the exact distribution of a sample proportion. A population consists of three men and two women. The first names of the men are Jose, Pete, and Carlo; the first names of the women are Gail and Frances. Suppose that the specified attribute is "female."
a. Determine the population proportion, $p$.

b. The first column of the following table provides the possible samples of size 2, where each person is represented by the first letter of his or her first name; the second column gives the number of successes—the number of females obtained—for each sample, and the third column shows the sample proportion. Complete the table.

| Sample | Number of females<br>$x$ | Sample proportion<br>$\hat{p}$ |
|---|---|---|
| J, G | 1 | 0.5 |
| J, P | 0 | 0.0 |
| J, C | 0 | 0.0 |
| J, F | 1 | 0.5 |
| G, P | | |
| G, C | | |
| G, F | | |
| P, C | | |
| P, F | | |
| C, F | | |

c. Construct a dotplot for the sampling distribution of the proportion for samples of size 2. Mark the position of the population proportion on the dotplot.
d. Use the third column of the table to obtain the mean of the variable $\hat{p}$.
e. Compare your answers from parts (a) and (d). Why are they the same?

**11.6** Repeat parts (b)–(e) of Exercise 11.5 for samples of size 1.

**11.7** Repeat parts (b)–(e) of Exercise 11.5 for samples of size 3. (There are 10 possible samples.)

**11.8** Repeat parts (b)–(e) of Exercise 11.5 for samples of size 4. (There are five possible samples.)

**11.9** Repeat parts (b)–(e) of Exercise 11.5 for samples of size 5.

**11.10** Prerequisite to this exercise are Exercises 11.5–11.9. What do your graphs in parts (c) of those exercises illustrate about the impact of increasing sample size on sampling error? Explain your answer.

**11.11 NBA Draft Picks.** Since 1966, 45% of the No. 1 draft picks in the National Basketball Association have been centers.
a. Identify the population.
b. Identify the specified attribute.
c. Is the proportion 0.45 (45%) a population proportion or a sample proportion? Explain your answer.

**11.12 Staying Single.** According to an article in *Time* magazine, women are staying single longer these days, by choice. In 1963, 83% of women in the United States between the ages of 25 and 54 were married, compared to 68% in 2000. For 2000,
a. identify the population.
b. identify the specified attribute.
c. Under what circumstances is the proportion 0.68 a population proportion? a sample proportion? Explain your answers.

**11.13 Random Drug Testing.** A *Harris Poll* asked Americans whether states should be allowed to conduct random drug tests on elected officials. Of 21,355 respondents, 79% said "yes."
a. Determine the margin of error for a 99% confidence interval.
b. Without doing any calculations, indicate whether the margin of error is larger or smaller for a 90% confidence interval. Explain your answer.

**11.14 Genetic Binge Eating.** According to an article in *Science News*, binge eating has been associated with a mutation of the gene for a brain protein called melanocortin 4

receptor (MC4R). In one study, F. Horber of the Hirslanden Clinic in Zurich and his colleagues genetically analyzed the blood of 469 obese people and found that 24 carried a mutated MC4R gene. Suppose that you want to estimate the proportion of all obese people who carry a mutated MC4R gene.
a. Determine the margin of error for a 90% confidence interval.
b. Without doing any calculations, indicate whether the margin of error is larger or smaller for a 95% confidence interval. Explain your answer.

**11.15** In each of parts (a)–(c), we have given a likely range for the observed value of a sample proportion $\hat{p}$. Based on the given range, identify the educated guess that should be used for the observed value of $\hat{p}$ to calculate the required sample size for a prescribed confidence level and margin of error.
a. 0.2 to 0.4    b. 0.2 or less    c. 0.4 or greater
d. In each of parts (a)–(c), which observed values of the sample proportion will yield a larger margin of error than the one specified if the educated guess is used for the sample size computation?

**11.16** In each of parts (a)–(c), we have given a likely range for the observed value of a sample proportion $\hat{p}$. Based on the given range, identify the educated guess that should be used for the observed value of $\hat{p}$ to calculate the required sample size for a prescribed confidence level and margin of error.
a. 0.4 to 0.7    b. 0.7 or greater    c. 0.7 or less
d. In each of parts (a)–(c), which observed values of the sample proportion will yield a larger margin of error than the one specified if the educated guess is used for the sample size computation?

*In each of Exercises 11.17–11.22, we have given the number of successes and the sample size for a simple random sample from a population. In each case, do the following.*
*a. Determine the sample proportion.*
*b. Decide whether using the one-proportion z-interval procedure is appropriate.*
*c. If appropriate, use the one-proportion z-interval procedure to find the confidence interval at the specified confidence level.*

**11.17** $x = 8$, $n = 40$, 95% level.

**11.18** $x = 10$, $n = 40$, 90% level.

**11.19** $x = 35$, $n = 50$, 99% level.

**11.20** $x = 40$, $n = 50$, 95% level.

**11.21** $x = 16$, $n = 20$, 90% level.

**11.22** $x = 3$, $n = 100$, 99% level.

*In Exercises 11.23–11.28, use Procedure 11.1 on page 505 to find the required confidence interval. Be sure to check the conditions for using that procedure.*

**11.23 Shopping Online.** A recent issue of *Time Style and Design* reported on a poll conducted by Schulman Ronca & Bucuvalas Public Affairs about the shopping habits of wealthy Americans. A total of 603 interviews were conducted among a national sample of adults with household incomes of at least $150,000. Of the adults interviewed, 410 said they had purchased clothing, accessories, or books online in the past year. Find a 95% confidence interval for the proportion of all U.S. adults with household incomes of at least $150,000 who purchased clothing, accessories, or books online in the past year.

**11.24 Life Support.** In 2005, the Terri Schiavo case focused national attention on the issue of withdrawal of life support from terminally ill patients or those in a vegetative state. A *Harris Poll* of 1010 U.S. adults was conducted by telephone on April 5–10, 2005. Of those surveyed, 140 had experienced the death of at least one family member or close friend within the last 10 years who died after the removal of life support. Find a 90% confidence interval for the proportion of all U.S. adults who had experienced the death of at least one family member or close friend within the last 10 years after life support had been withdrawn.

**11.25 Asthmatics and Sulfites.** Studies are performed to estimate the percentage of the nation's 10 million asthmatics who are allergic to sulfites. In one survey, 38 of 500 randomly selected U.S. asthmatics were found to be allergic to sulfites.
**a.** Find a 95% confidence interval for the proportion, $p$, of all U.S. asthmatics who are allergic to sulfites.
**b.** Interpret your result from part (a).

**11.26 Drinking Habits.** A *Reader's Digest/Gallup Survey* on the drinking habits of Americans estimated the percentage of adults across the country who drink beer, wine, or hard liquor, at least occasionally. Of the 1516 adults interviewed, 985 said that they drank.
**a.** Determine a 95% confidence interval for the proportion, $p$, of all Americans who drink beer, wine, or hard liquor, at least occasionally.
**b.** Interpret your result from part (a).

**11.27 Factory Farming Funk.** The U.S. Environmental Protection Agency recently reported that confined animal feeding operations (CAFOs) dump 2 trillion pounds of waste into the environment annually, contaminating the ground water in 17 states and polluting more than 35,000 miles of our nation's rivers. In a recent survey of 1000 registered voters by Snell, Perry and Associates, 80% favored the creation of standards to limit such pollution and, in general, viewed CAFOs unfavorably.
**a.** Find a 99% confidence interval for the percentage of all registered voters who favor the creation of standards on CAFO pollution and, in general, view CAFOs unfavorably.
**b.** Interpret your answer in part (a).

**11.28 The Nipah Virus.** From fall 1998 through mid 1999, Malaysia was the site of an encephalitis outbreak caused by the Nipah virus, a paramyxovirus that appears to spread from pigs to workers on pig farms. As reported by Goh et al. in the *New England Journal of Medicine* (Vol. 342(17), p. 1229), neurologists from the University of Malaysia found that, among 94 patients infected with the Nipah virus, 30 died from encephalitis.
**a.** Find a 90% confidence interval for the percentage of Malaysians infected with the Nipah virus who will die from encephalitis.
**b.** Interpret your answer in part (a).

**11.29 Literate Adults.** Suppose that you have been hired to estimate the percentage of adults in your state who are literate. You take a random sample of 100 adults and find that 96 are literate. You then obtain a 95% confidence interval of

$$0.96 \pm 1.96 \cdot \sqrt{(0.96)(0.04)/100},$$

or 0.922 to 0.998. From it you conclude that you can be 95% confident that the percentage of all adults in your state who are literate is somewhere between 92.2% and 99.8%. Is anything wrong with this reasoning?

**11.30 IMR in Singapore.** Recall that the infant mortality rate (IMR) is the number of infant deaths per 1000 live births. Suppose that you have been commissioned to estimate the IMR in Singapore. From a random sample of 1109 live births in Singapore, you find that 0.361% of them resulted in infant deaths. You next find a 90% confidence interval:

$$0.00361 \pm 1.645 \cdot \sqrt{(0.00361)(0.99639)/1109},$$

or 0.000647 to 0.00657. You then conclude that, "I can be 90% confident that the IMR in Singapore is somewhere between 0.647 and 6.57." How did you do?

**11.31 Warming to Russia.** An *ABCNEWS Poll* found that Americans now have relatively warm feelings toward Russia, a former adversary. The poll, conducted by telephone among a random sample of 1043 adults, found that 647 of those sampled consider the two countries friends. The margin of error for the poll was plus or minus 2.9 percentage points (for a 0.95 confidence level). Use this information to obtain a 95% confidence interval for the percentage of all Americans who consider the two countries friends.

**11.32 Online Tax Returns.** According to the U.S. Internal Revenue Service, among people entitled to tax refunds, those who file online receive their refunds twice as fast as paper filers. A study conducted by ICR of Media, Pennsylvania, found that 57% of those polled said that they are not worried about the privacy of their financial information when filing their tax returns online. The telephone survey of 1002 people had a margin of error of plus or minus 3 percentage points (for a 0.95 confidence level). Use this information to determine a 95% confidence interval for the percentage of all people who are not worried about the privacy of their financial information when filing their tax returns online.

**11.33 Asthmatics and Sulfites.** Refer to Exercise 11.25.
a. Determine the margin of error for the estimate of $p$.
b. Obtain a sample size that will ensure a margin of error of at most 0.01 for a 95% confidence interval without making a guess for the observed value of $\hat{p}$.
c. Find a 95% confidence interval for $p$ if, for a sample of the size determined in part (b), the proportion of asthmatics sampled who are allergic to sulfites is 0.071.
d. Determine the margin of error for the estimate in part (c) and compare it to the margin of error specified in part (b).
e. Repeat parts (b)–(d) if you can reasonably presume that the proportion of asthmatics sampled who are allergic to sulfites will be at most 0.10.
f. Compare the results you obtained in parts (b)–(d) with those obtained in part (e).

**11.34 Drinking Habits.** Refer to Exercise 11.26.
a. Find the margin of error for the estimate of $p$.
b. Obtain a sample size that will ensure a margin of error of at most 0.02 for a 95% confidence interval without making a guess for the observed value of $\hat{p}$.
c. Find a 95% confidence interval for $p$ if, for a sample of the size determined in part (b), 63% of those sampled drink alcoholic beverages.
d. Determine the margin of error for the estimate in part (c) and compare it to the margin of error specified in part (b).
e. Repeat parts (b)–(d) if you can reasonably presume that the percentage of adults sampled who drink alcoholic beverages will be at least 60%.
f. Compare the results you obtained in parts (b)–(d) with those obtained in part (e).

**11.35 Factory Farming Funk.** Refer to Exercise 11.27.
a. Determine the margin of error for the estimate of the percentage.
b. Obtain a sample size that will ensure a margin of error of at most 1.5 percentage points for a 99% confidence interval without making a guess for the observed value of $\hat{p}$.
c. Find a 99% confidence interval for $p$ if, for a sample of the size determined in part (b), 82.2% of the registered voters sampled favor the creation of standards on CAFO pollution and, in general, view CAFOs unfavorably.
d. Determine the margin of error for the estimate in part (c) and compare it to the margin of error specified in part (b).
e. Repeat parts (b)–(d) if you can reasonably presume that the percentage of registered voters sampled who favor the creation of standards on CAFO pollution and, in general, view CAFOs unfavorably will be between 75% and 85%.
f. Compare the results you obtained in parts (b)–(d) with those obtained in part (e).

**11.36 The Nipah Virus.** Refer to Exercise 11.28.
a. Find the margin of error for the estimate of the percentage.
b. Obtain a sample size that will ensure a margin of error of at most 5 percentage points for a 90% confidence interval without making a guess for the observed value of $\hat{p}$.
c. Find a 90% confidence interval for $p$ if, for a sample of the size determined in part (b), 28.8% of the sampled Malaysians infected with the Nipah virus die from encephalitis.
d. Determine the margin of error for the estimate in part (c) and compare it to the margin of error specified in part (b).
e. Repeat parts (b)–(d) if you can reasonably presume that the percentage of sampled Malaysians infected with the Nipah virus who will die from encephalitis will be between 25% and 40%.
f. Compare the results you obtained in parts (b)–(d) with those obtained in part (e).

**11.37 Product Response Rate.** A company manufactures goods that are sold exclusively by mail order. The director of market research needed to test market a new product. She planned to send brochures to a random sample of households and use the proportion of orders obtained as an estimate of the true proportion, known as the *product response rate*. The results of the market research were to be utilized as a primary source for advance production planning, so the director wanted the figures she presented to be as accurate as possible. Specifically, she wanted to be 95% confident that the estimate of the product response rate would be accurate to within 1%.
a. Without making any assumptions, determine the sample size required.
b. Historically, product response rates for products sold by this company have ranged from 0.5% to 4.9%. If the director had been willing to assume that the sample product response rate for this product would also fall in that range, find the required sample size.
c. Compare the results from parts (a) and (b).
d. Discuss the possible consequences if the assumption made in part (b) turns out to be incorrect.

**11.38 Indicted Governor.** On Thursday, June 13, 1996, then-Arizona Governor Fife Symington was indicted on 23 counts of fraud and extortion. Just hours after the federal prosecutors announced the indictment, several polls were conducted of Arizonans asking whether they thought Symington should resign. One poll, conducted by Research Resources, Inc., which appeared in the *Phoenix Gazette*, revealed that 58% of Arizonans felt that Symington should resign; it had a margin of error of plus or minus 4.9 percentage points. Another poll, conducted by Phoenix-based Behavior Research Center and appearing in the *Tempe Daily News*, reported that 54% of Arizonans felt that Symington should resign; it had a margin of error of plus or minus 4.4 percentage points. Can the conclusions of both polls be correct? Explain your answer.

*In each of Exercises 11.39–11.42, use the technology of your choice to find the required confidence interval.*

**11.39 President's Job Rating.** The headline read "President's Job Ratings Fall to Lowest Point of His Presidency." A *Harris Poll* taken April 5–10, 2005, of 1010 U.S. adults found that 444 of them approved of the way that President George W. Bush was doing his job. Find and interpret a 95% confidence interval for the proportion of all U.S. adults who, at the time, approved of President Bush.

**11.40 Major Hurricanes.** A major hurricane is a category 3, 4, or 5 hurricane on the Saffir/Simpson Hurricane Scale. From the document "The Deadliest, Costliest, and Most Intense United States Tropical Cyclones From 1851 to 2004" (*NOAA Technical Memorandum*, NWS TPC-4, Updated 2005) by E. Blake et al., we found that of the 273 hurricanes affecting the continental United States, 92 were major.
a. Based on these data, find and interpret a 90% confidence interval for the probability, $p$, that a hurricane affecting the continental United States will be a major hurricane.
b. Discuss the possible problems with this analysis.

**11.41 Bankrupt Automakers.** In a nationwide survey of U.S. adults by the Cincinnati-based research firm Directions Research Inc., only 276 of the 1063 respondents said they would purchase or lease a new car from a manufacturer that had declared bankruptcy. Determine and interpret a 90% confidence interval for the percentage of all U.S. adults who would purchase or lease a new car from a manufacturer that had declared bankruptcy.

**11.42 Mineral Waters.** In the article "Bottled Natural Mineral Waters in Romania" (*Environmental Geology Journal*, Vol. 46, Issue 5, pp. 670–674), A. Feru compared the mineral, ionic, and carbon dioxide content of mineral-water source locations in Romania. Of 31 randomly selected source locations, 22 had natural carbonated natural (NCN) mineral water. Determine a 95% confidence interval for the proportion of all mineral-water source locations in Romania that have NCN mineral water.

### Extending the Concepts and Skills

**11.43** What important theorem in statistics implies that, for a large sample size, the possible sample proportions of that size have approximately a normal distribution?

**11.44** In discussing the sample size required for obtaining a confidence interval with a prescribed confidence level and margin of error, we made the following statement: "If we have in mind a likely range for the observed value of $\hat{p}$, then, in light of Fig. 11.1, we should take as our educated guess for $\hat{p}$ the value in the range closest to 0.5." Explain why.

**11.45** In discussing the sample size required for obtaining a confidence interval with a prescribed confidence level and margin of error, we made the following statement: "…we should be aware that, if the observed value of $\hat{p}$ is closer to 0.5 than is our educated guess, the margin of error will be larger than desired." Explain why.

**11.46** Consider a population in which the proportion of members having a specified attribute is $p$. Let $y$ be the variable whose value is 1 if a member has the specified attribute and 0 if a member does not.
a. If the size of the population is $N$, how many members of the population have the specified attribute?
b. Use part (a) and Definition 3.11 on page 136 to show that $\mu_y = p$.
c. Use part (b) and the computing formula in Definition 3.12 on page 138 to show that $\sigma_y = \sqrt{p(1-p)}$.
d. Explain why $\bar{y} = \hat{p}$.
e. Use parts (b)–(d) and Key Fact 7.4 on page 325, to justify Key Fact 11.1.

## 11.2 Hypothesis Tests for One Population Proportion

In Section 11.1, we showed how to obtain confidence intervals for a population proportion. Now we show how to perform hypothesis tests for a population proportion. This procedure is actually a special case of the one-mean $z$-test.

From Key Fact 11.1 on page 504, we deduce that, for large $n$, the standardized version of $\hat{p}$,

$$z = \frac{\hat{p} - p}{\sqrt{p(1-p)/n}}$$

has approximately the standard normal distribution. Consequently, to perform a large-sample hypothesis test with null hypothesis $H_0: p = p_0$, we can use the variable

$$z = \frac{\hat{p} - p_0}{\sqrt{p_0(1-p_0)/n}}$$

as the test statistic and obtain the critical value(s) or $P$-value from the standard normal table, Table II.

We call this hypothesis-testing procedure the **one-proportion $z$-test.**[†] Procedure 11.2 provides a step-by-step method for performing a one-proportion $z$-test by using either the critical-value approach or the $P$-value approach.

---

### Procedure 11.2  One-Proportion $z$-Test

*Purpose*  To perform a hypothesis test for a population proportion, $p$

*Assumptions*

1. Simple random sample
2. Both $np_0$ and $n(1 - p_0)$ are 5 or greater

**STEP 1**  The null hypothesis is $H_0$: $p = p_0$, and the alternative hypothesis is

$$H_a: p \neq p_0 \quad \text{or} \quad H_a: p < p_0 \quad \text{or} \quad H_a: p > p_0$$
$$\text{(Two tailed)} \qquad \text{(Left tailed)} \qquad \text{(Right tailed)}$$

**STEP 2**  Decide on the significance level, $\alpha$.

**STEP 3**  Compute the value of the test statistic

$$z = \frac{\hat{p} - p_0}{\sqrt{p_0(1 - p_0)/n}}$$

and denote that value $z_0$.

---

| CRITICAL-VALUE APPROACH | or | P-VALUE APPROACH |
|---|---|---|

**STEP 4**  The critical value(s) are

$$\pm z_{\alpha/2} \quad \text{or} \quad -z_{\alpha} \quad \text{or} \quad z_{\alpha}$$
$$\text{(Two tailed)} \qquad \text{(Left tailed)} \qquad \text{(Right tailed)}$$

Use Table II to find the critical value(s).

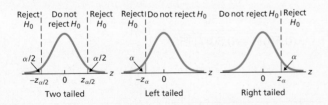

**STEP 5**  If the value of the test statistic falls in the rejection region, reject $H_0$; otherwise, do not reject $H_0$.

**STEP 4**  Use Table II to obtain the $P$-value.

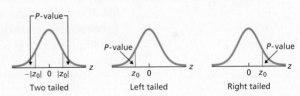

**STEP 5**  If $P \leq \alpha$, reject $H_0$; otherwise, do not reject $H_0$.

---

**STEP 6**  Interpret the results of the hypothesis test.

---

[†]The one-proportion $z$-test is also known as the **one-sample $z$-test for a population proportion** and the **one-variable proportion test.**

## Example 11.6 | The One-Proportion z-Test

*Gun Control* One of the more controversial issues in the United States is gun control; there are many avid proponents and opponents of banning handgun sales. A *Harris Poll* asked 1250 U.S. adults their views on banning handgun sales. Of those sampled, 650 favored a ban. At the 5% significance level, do the data provide sufficient evidence to conclude that a majority (more than 50%) of U.S. adults favor banning handgun sales?

**Solution**   Because $n = 1250$ and $p_0 = 0.50$ (50%), we have

$$np_0 = 1250 \cdot 0.50 = 625 \quad \text{and} \quad n(1 - p_0) = 1250 \cdot (1 - 0.50) = 625.$$

Because both $np_0$ and $n(1 - p_0)$ are 5 or greater, we can apply Procedure 11.2.

**STEP 1  State the null and alternative hypotheses.**

Let $p$ denote the proportion of all U.S. adults who favor banning handgun sales. Then the null and alternative hypotheses are

$$H_0: \ p = 0.50 \text{ (it is not true that a majority favor a ban)}$$
$$H_a: \ p > 0.50 \text{ (a majority favor a ban).}$$

Note that the hypothesis test is right tailed.

**STEP 2  Decide on the significance level, $\alpha$.**

We are to perform the hypothesis test at the 5% significance level; so, $\alpha = 0.05$.

**STEP 3  Compute the value of the test statistic**

$$z = \frac{\hat{p} - p_0}{\sqrt{p_0(1 - p_0)/n}}.$$

We have $n = 1250$ and $p_0 = 0.50$. The number of U.S. adults surveyed who favor banning handgun sales is 650. Therefore the proportion of those surveyed who favor a ban is $\hat{p} = x/n = 650/1250 = 0.520$ (52.0%). So, the value of the test statistic is

$$z = \frac{0.520 - 0.50}{\sqrt{(0.50)(1 - 0.50)/1250}} = 1.41.$$

| CRITICAL-VALUE APPROACH | or | P-VALUE APPROACH |
|---|---|---|

**STEP 4** The critical value for a right-tailed test is $z_\alpha$. Use Table II to find the critical value.

For $\alpha = 0.05$, the critical value is $z_{0.05} = 1.645$, as shown in Fig. 11.2A.

**STEP 4** Use Table II to obtain the P-value.

From Step 3, the value of the test statistic is $z = 1.41$. The test is right tailed, so the P-value is the probability of observing a value of $z$ of 1.41 or greater if the null hypothesis is true. That probability equals the shaded area in Fig. 11.2B, which by Table II is 0.0793.

| CRITICAL-VALUE APPROACH | or | P-VALUE APPROACH |
|---|---|---|

**FIGURE 11.2A**

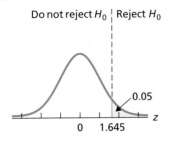

Do not reject $H_0$ | Reject $H_0$

0.05

0    1.645

z

**STEP 5** If the value of the test statistic falls in the rejection region, reject $H_0$; otherwise, do not reject $H_0$.

From Step 3, the value of the test statistic is $z = 1.41$, which, as Fig. 11.2A shows, does not fall in the rejection region. Thus we do not reject $H_0$. The test results are not statistically significant at the 5% level.

**FIGURE 11.2B**

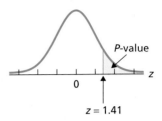

P-value

0

$z = 1.41$

z

**STEP 5** If $P \leq \alpha$, reject $H_0$; otherwise, do not reject $H_0$.

From Step 4, $P = 0.0793$. Because the P-value exceeds the specified significance level of 0.05, we do not reject $H_0$. The test results are not statistically significant at the 5% level but (see Table 9.10 on page 414) the data do provide moderate evidence against the null hypothesis.

**You try it!**

Exercise 11.55
on page 519

**STEP 6** Interpret the results of the hypothesis test.

**Interpretation** At the 5% significance level, the data do not provide sufficient evidence to conclude that a majority of U.S. adults favor banning handgun sales.

• • •

Example 11.6 illustrates how statistical results are sometimes misstated. The headline for the newspaper article that featured the survey read "Fear prompts 52% in U.S. to back pistol-sale ban, poll says." In fact, the poll says no such thing. It says only that 52% of those *sampled* back a pistol-sale ban. And, as we have demonstrated, at the 5% significance level the poll does not provide sufficient evidence to conclude that a majority of U.S. adults back a pistol-sale ban.

## The Technology Center

Most statistical technologies have programs that automatically perform the one-proportion z-test. In this subsection, we present output and step-by-step instructions for such programs.

**Example 11.7**    **Using Technology to Conduct a One-Proportion z-Test**

*Gun Control*    Of 1250 randomly selected U.S. adults asked whether they favor banning handgun sales, 650 said they do. Use Minitab, Excel, or the TI-83/84 Plus to decide, at the 5% significance level, whether the data provide sufficient evidence to conclude that a majority of U.S. adults favor banning handgun sales.

**Solution** Let $p$ denote the proportion of all U.S. adults who favor banning handgun sales. The task is to perform the hypothesis test

$$H_0\text{: } p = 0.50 \text{ (it is not true that a majority favor a ban)}$$
$$H_a\text{: } p > 0.50 \text{ (a majority favor a ban)}$$

at the 5% significance level. Note that the hypothesis test is right tailed.

We applied the one-proportion $z$-test programs to the data, resulting in Output 11.3. Steps for generating that output are presented in Instructions 11.2.

**OUTPUT 11.3** One-proportion $z$-test for the data on banning handguns

**MINITAB**

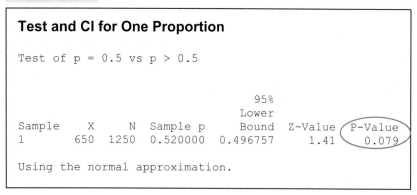

```
Test and CI for One Proportion

Test of p = 0.5 vs p > 0.5

                                    95%
                                  Lower
Sample    X     N  Sample p     Bound   Z-Value   P-Value
1       650  1250  0.520000  0.496757     1.41     0.079

Using the normal approximation.
```

**EXCEL**

```
x  Proportion Test
  Summary Statistics          Test Summary
      n    1250              p0:              0.5
  p-hat   0.52              Ho:         p = 0.5
 Std Dev  0.0141            Ha:  Upper tail: p > 0.5
                       z Statistic:          1.414
                          p-value:           0.0786

                          Test Results
                     Conclusion
                     Fail to reject Ho at alpha =  0.05
```

**TI-83/84 PLUS**

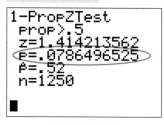

```
1-PropZTest
 prop>.5
 z=1.414213562
 P=.0786496525
 p=.52
 n=1250
```

Using **Calculate**

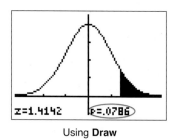

```
z=1.4142    P=.0786
```

Using **Draw**

As shown in Output 11.3, the $P$-value for the hypothesis test is 0.079. Because the $P$-value exceeds the specified significance level of 0.05, we do not reject $H_0$. At the 5% significance level, the data do not provide sufficient evidence to conclude that a majority of U.S. adults favor banning handgun sales.

• • •

**INSTRUCTIONS 11.2**   Steps for generating Output 11.3

| MINITAB | EXCEL | TI-83/84 PLUS |
|---|---|---|
| 1 Choose **Stat ➤ Basic Statistics ➤ 1 Proportion...** <br> 2 Select the **Summarized data** option button <br> 3 Click in the **Number of events** text box and type 650 <br> 4 Click in the **Number of trials** text box and type 1250 <br> 5 Check the **Perform hypothesis test** check box <br> 6 Click in the **Hypothesized proportion** text box and type 0.50 <br> 7 Click the **Options...** button <br> 8 Click the arrow button at the right of the **Alternative** drop-down list box and select **greater than** <br> 9 Check the **Use test and interval based on normal distribution** check box <br> 10 Click **OK** twice | 1 Store the sample size, 1250, and the number of successes, 650, in ranges named n and x, respectively <br> 2 Choose **DDXL ➤ Hypothesis Tests** <br> 3 Select **Summ 1 Var Prop Test** from the **Function type** drop-down list box <br> 4 Specify x in the **Num Successes** text box <br> 5 Specify n in the **Num Trials** text box <br> 6 Click **OK** <br> 7 Click the **Set p0** button <br> 8 Click in the **Hypothesized Population Proportion** text box and type 0.50 <br> 9 Click **OK** <br> 10 Click the **.05** button <br> 11 Click the **p > p0** button <br> 12 Click the **Compute** button | 1 Press **STAT**, arrow over to **TESTS**, and press **5** <br> 2 Type 0.50 for $p_0$ and press **ENTER** <br> 3 Type 650 for **x** and press **ENTER** <br> 4 Type 1250 for **n** and press **ENTER** <br> 5 Highlight > $p_0$ and press **ENTER** <br> 6 Press the down-arrow key, highlight **Calculate** or **Draw**, and press **ENTER** |

▮▮▮

# Exercises 11.2

## Understanding the Concepts and Skills

**11.47** Of what procedure is Procedure 11.2 a special case? Why do you think that is so?

**11.48** The paragraph immediately following Example 11.6 discusses how statistical results are sometimes misstated. Find an article in a newspaper, magazine, or on the Internet that misstates a statistical result in a similar way.

*In each of Exercises 11.49–11.54, we have given the number of successes and the sample size for a simple random sample from a population. In each case, do the following.*
*a. Determine the sample proportion.*
*b. Decide whether using the one-proportion z-test is appropriate.*
*c. If appropriate, use the one-proportion z-test to perform the specified hypothesis test.*

**11.49** $x = 8$, $n = 40$, $H_0$: $p = 0.3$, $H_a$: $p < 0.3$, $\alpha = 0.10$

**11.50** $x = 10$, $n = 40$, $H_0$: $p = 0.3$, $H_a$: $p < 0.3$, $\alpha = 0.05$

**11.51** $x = 35$, $n = 50$, $H_0$: $p = 0.6$, $H_a$: $p > 0.6$, $\alpha = 0.05$

**11.52** $x = 40$, $n = 50$, $H_0$: $p = 0.6$, $H_a$: $p > 0.6$, $\alpha = 0.01$

**11.53** $x = 16$, $n = 20$, $H_0$: $p = 0.7$, $H_a$: $p \neq 0.7$, $\alpha = 0.05$

**11.54** $x = 3$, $n = 100$, $H_0$: $p = 0.04$, $H_a$: $p \neq 0.04$, $\alpha = 0.10$

*In Exercises 11.55–11.60, use Procedure 11.2 on page 515 to find the required confidence interval. Be sure to check the conditions for using that procedure.*

**11.55 Generation Y Online.** People who were born between 1978 and 1983 are sometimes classified by demographers as belonging to Generation Y. According to a recent Forrester Research survey published in *American Demographics* (Vol. 22(1), p. 12), of 850 Generation Y Web users, 459 reported using the Internet to download music.
**a.** Determine the sample proportion.
**b.** At the 5% significance level, do the data provide sufficient evidence to conclude that a majority of Generation Y Web users use the Internet to download music?

**11.56 Christmas Presents.** The *Arizona Republic* conducted a telephone poll of 758 Arizona adults who celebrate Christmas. The question asked was, "In your family, do you open presents on Christmas Eve or Christmas Day?" Of those surveyed, 394 said they wait until Christmas Day.

a. Determine the sample proportion.
b. At the 5% significance level, do the data provide sufficient evidence to conclude that a majority of Arizona families who celebrate Christmas wait until Christmas Day to open their presents?

**11.57 Marijuana and Hashish.** The U.S. Substance Abuse and Mental Health Services Administration conducts surveys on drug use by type of drug and age group. Results are published in *National Household Survey on Drug Abuse*. According to that publication, 13.6% of 18–25-year-olds were current users of marijuana or hashish in 2000. A recent poll of 1283 randomly selected 18–25-year-olds revealed that 205 currently use marijuana or hashish. At the 10% significance level, do the data provide sufficient evidence to conclude that the percentage of 18–25-year-olds who currently use marijuana or hashish has changed from the 2000 percentage of 13.6%?

**11.58 Families in Poverty.** In 2000, 8.6% of all U.S. families had incomes below the poverty level, as reported by the U.S. Census Bureau in *Current Population Reports*. During that same year, of 400 randomly selected families whose householder had at least a Bachelor's degree, 9 had incomes below the poverty level. Do the data provide sufficient evidence to conclude that, in 2000, the percentage of families that earned incomes below the poverty level was lower among those whose householders had at least a Bachelor's degree than among all U.S. families? Use $\alpha = 0.01$.

**11.59 Labor-Union Support.** Labor Day was created by the U.S. labor movement over 100 years ago. It was subsequently adopted by most states as an official holiday. In a recent *Gallup Poll* 1003 randomly selected adults were asked whether they approve of labor unions; 65% said yes.
a. In 1936, about 72% of Americans approved of labor unions. At the 5% significance level, do the data provide sufficient evidence to conclude that the percentage of Americans who approve of labor unions now has decreased since 1936?
b. In 1963, roughly 67% of Americans approved of labor unions. At the 5% significance level, do the data provide sufficient evidence to conclude that the percentage of Americans who approve of labor unions now has decreased since 1963?

**11.60 An Edge in Roulette?** Of the 38 numbers on an American roulette wheel, 18 are red, 18 are black, and 2 are green. If the wheel is balanced, the probability of the ball landing on red is $\frac{18}{38} = 0.474$. A gambler has been studying a roulette wheel. If the wheel is out of balance, he can improve his odds of winning. The gambler observes 200 spins of the wheel and finds that the ball lands on red 93 times. At the 10% significance level, do the data provide sufficient evidence to conclude that the ball is not landing on red the correct percentage of the time for a balanced wheel?

*In each of Exercises **11.61–11.64**, use the technology of your choice to conduct the required hypothesis test.*

**11.61 Recovering From Katrina.** A *CNN/USA TODAY/ Gallup Poll*, conducted in September, 2005, had the headline "Most Americans Believe New Orleans Will Never Recover." Of 609 adults polled by telephone, 341 said they believe the hurricane devastated the city beyond repair. At the 1% significance level, do the data provide sufficient evidence to justify the headline? Explain your answer.

**11.62 Delayed Perinatal Stroke.** In the article "Prothrombotic Factors in Children With Stroke or Porencephaly" (*Pediatrics Journal*, Vol. 116, Issue 2, pp. 447–453), J. Lynch et al. compared differences and similarities in children with arterial ischemic stroke and porencephaly. Three classification categories were used: perinatal stroke, delayed perinatal stroke, and childhood stroke. Of 59 children, 25 were diagnosed with delayed perinatal stroke. At the 5% significance level, do the data provide sufficient evidence to conclude that delayed perinatal stroke does not comprise one-third of the cases among the three categories?

**11.63 Drowning Deaths.** In the article "Drowning Deaths of Zero to Five Year Old Children in Victorian Dams, 1989–2001" (*Australian Journal of Rural Health*, Vol. 13, Issue 5, pp. 300–308), L. Bugeja and R. Franklin examined drowning deaths of young children in Victorian dams to identify common contributing factors and develop strategies for future prevention. Of 11 young children who drowned in Victorian dams located on farms, 5 were girls. At the 5% significance level, do the data provide sufficient evidence to conclude that, of all young children drowning in Victorian dams located on farms, less than half are girls?

**11.64 U.S. Troops in Iraq.** In a *Zogby International Poll*, conducted in early 2006 in conjunction with Le Moyne College's Center for Peace and Global Studies, roughly 29% of the 944 military respondents, serving in Iraq in various branches of the armed forces, said the United States should leave Iraq immediately. Do the data provide sufficient evidence to conclude that, at the time, more than one-quarter of all U.S. troops in Iraq were in favor of leaving immediately? Use $\alpha = 0.01$.

## 11.3   Inferences for Two Population Proportions

In Sections 11.1 and 11.2, you studied inferences for one population proportion. Now we examine inferences for comparing two population proportions. In this case, we have two populations and one specified attribute; the problem is to compare the proportion of one population that has the specified attribute to the proportion of the other population that has the specified attribute. We begin by discussing hypothesis testing.

**Example 11.8** | ### Hypothesis Tests for Two Population Proportions

*Eating Out Vegetarian*  Zogby International surveyed 1181 U.S. adults to gauge the demand for vegetarian meals in restaurants. The study, commissioned by the Vegetarian Resource Group and published in the *Vegetarian Journal*, polled independent random samples of 747 men and 434 women. Of those sampled, 276 men and 195 women said that they sometimes order a dish without meat, fish, or fowl when they eat out.

Suppose we want to use the data to decide whether, in the United States, the percentage of men who sometimes order a dish without meat, fish, or fowl is smaller than the percentage of women who sometimes order a dish without meat, fish, or fowl.

**a.** Formulate the problem statistically by posing it as a hypothesis test.

**b.** Explain the basic idea for carrying out the hypothesis test.

**c.** Discuss the use of the data to make a decision concerning the hypothesis test.

**Solution**

**a.** The specified attribute is "sometimes orders a dish without meat, fish, or fowl," which we abbreviate throughout this section as "sometimes orders veg." The two populations are

$$\text{Population 1: All U.S. men}$$
$$\text{Population 2: All U.S. women.}$$

Let $p_1$ and $p_2$ denote the population proportions for the two populations:

$$p_1 = \text{proportion of all U.S. men who sometimes order veg}$$
$$p_2 = \text{proportion of all U.S. women who sometimes order veg.}$$

We want to perform the hypothesis test

$H_0$: $p_1 = p_2$ (percentage for men is not less than that for women)

$H_a$: $p_1 < p_2$ (percentage for men is less than that for women).

**b.** Roughly speaking, we can carry out the hypothesis test as follows:

**1.** Compute the proportion of the men sampled who sometimes order veg, $\hat{p}_1$, and compute the proportion of the women sampled who sometimes order veg, $\hat{p}_2$.

**2.** If $\hat{p}_1$ is too much smaller than $\hat{p}_2$, reject $H_0$; otherwise, do not reject $H_0$.

**c.** To use the data to make a decision concerning the hypothesis test, we apply the two steps just listed. The first step is easy. Because 276 of the 747 men sampled sometimes order veg and 195 of the 434 women sampled sometimes order veg,

$$\hat{p}_1 = \frac{x_1}{n_1} = \frac{276}{747} = 0.369 \ (36.9\%)$$

and

$$\hat{p}_2 = \frac{x_2}{n_2} = \frac{195}{434} = 0.449 \ (44.9\%).$$

For the second step, we must decide whether the sample proportion $\hat{p}_1 = 0.369$ is less than the sample proportion $\hat{p}_2 = 0.449$ by a sufficient amount to warrant rejecting the null hypothesis in favor of the alternative hypothesis. To make that decision, we need to know the distribution of the difference between two sample proportions.

• • •

## The Sampling Distribution of the Difference Between Two Sample Proportions for Large and Independent Samples

Let's begin by summarizing the required notation in Table 11.2.

**TABLE 11.2**
Notation for parameters and statistics when two population proportions are being considered

|  | Population 1 | Population 2 |
|---|---|---|
| Population proportion | $p_1$ | $p_2$ |
| Sample size | $n_1$ | $n_2$ |
| Number of successes | $x_1$ | $x_2$ |
| Sample proportion | $\hat{p}_1$ | $\hat{p}_2$ |

**You try it!**

Exercise 11.67
on page 532

Recall that the *number of successes* refers to the number of members sampled that have the specified attribute. Consequently, we compute the sample proportions by using the formulas

$$\hat{p}_1 = \frac{x_1}{n_1} \quad \text{and} \quad \hat{p}_2 = \frac{x_2}{n_2}.$$

Armed with the notation in Table 11.2, we now describe the **sampling distribution of the difference between two sample proportions.**

---

**Key Fact 11.2**

**What Does It Mean?**

For large independent samples, the possible differences between two sample proportions have approximately a normal distribution with mean $p_1 - p_2$ and standard deviation $\sqrt{p_1(1-p_1)/n_1 + p_2(1-p_2)/n_2}$.

### The Sampling Distribution of the Difference Between Two Sample Proportions for Independent Samples

For independent samples of sizes $n_1$ and $n_2$ from the two populations,

- $\mu_{\hat{p}_1 - \hat{p}_2} = p_1 - p_2$,
- $\sigma_{\hat{p}_1 - \hat{p}_2} = \sqrt{p_1(1-p_1)/n_1 + p_2(1-p_2)/n_2}$, and
- $\hat{p}_1 - \hat{p}_2$ is approximately normally distributed for large $n_1$ and $n_2$.

## Large-Sample Hypothesis Tests for Two Population Proportions, Using Independent Samples

Now we can develop a hypothesis-testing procedure for comparing two population proportions. Our immediate goal is to identify a variable that we can use as the test statistic. From Key Fact 11.2, we know that, for large, independent samples, the standardized variable

$$z = \frac{(\hat{p}_1 - \hat{p}_2) - (p_1 - p_2)}{\sqrt{p_1(1 - p_1)/n_1 + p_2(1 - p_2)/n_2}} \tag{11.2}$$

has approximately the standard normal distribution.

The null hypothesis for a hypothesis test to compare two population proportions is

$$H_0: p_1 = p_2 \text{ (population proportions are equal).}$$

If the null hypothesis is true, then $p_1 - p_2 = 0$ and, consequently, the variable in Equation (11.2) becomes

$$z = \frac{\hat{p}_1 - \hat{p}_2}{\sqrt{p(1 - p)/n_1 + p(1 - p)/n_2}} \tag{11.3}$$

where $p$ denotes the common value of $p_1$ and $p_2$. Factoring $p(1 - p)$ out of the denominator of Equation (11.3) yields the variable

$$z = \frac{\hat{p}_1 - \hat{p}_2}{\sqrt{p(1 - p)}\sqrt{(1/n_1) + (1/n_2)}}. \tag{11.4}$$

However, because $p$ is unknown, we cannot use this variable as the test statistic.

Consequently, we must estimate $p$ by using sample information. The best estimate of $p$ is obtained by pooling the data to get the proportion of successes in both samples combined; that is, we estimate $p$ by

$$\hat{p}_p = \frac{x_1 + x_2}{n_1 + n_2}.$$

We call $\hat{p}_p$ the **pooled sample proportion.**

Replacing $p$ in Equation (11.4) with its estimate $\hat{p}_p$ yields the variable

$$\frac{\hat{p}_1 - \hat{p}_2}{\sqrt{\hat{p}_p(1 - \hat{p}_p)}\sqrt{(1/n_1) + (1/n_2)}}$$

which can be used as the test statistic and, like the variable in Equation (11.4), has approximately the standard normal distribution for large samples if the null hypothesis is true. Therefore we have Procedure 11.3 (next page), which we call the **two-proportions z-test.**[†] Note that Procedure 11.3 and its confidence-interval counterpart (Procedure 11.4 on page 527) also apply to designed experiments with two treatments.

---

[†]The two-proportions z-test is also known as the **two-sample z-test for two population proportions** and the **two-variable proportions test.**

**Procedure 11.3   Two-Proportions z-Test**

*Purpose* To perform a hypothesis test to compare two population proportions, $p_1$ and $p_2$

*Assumptions*

1. Simple random samples
2. Independent samples
3. $x_1$, $n_1 - x_1$, $x_2$, and $n_2 - x_2$ are all 5 or greater

**STEP 1** The null hypothesis is $H_0$: $p_1 = p_2$, and the alternative hypothesis is

$$H_a: p_1 \neq p_2 \quad\text{or}\quad H_a: p_1 < p_2 \quad\text{or}\quad H_a: p_1 > p_2$$
$$\text{(Two tailed)} \qquad\qquad \text{(Left tailed)} \qquad\qquad \text{(Right tailed)}$$

**STEP 2** Decide on the significance level, $\alpha$.

**STEP 3** Compute the value of the test statistic

$$z = \frac{\hat{p}_1 - \hat{p}_2}{\sqrt{\hat{p}_p(1 - \hat{p}_p)}\sqrt{(1/n_1) + (1/n_2)}}$$

where $\hat{p}_p = (x_1 + x_2)/(n_1 + n_2)$. Denote the value of the test statistic $z_0$.

---

CRITICAL-VALUE APPROACH   *or*   P-VALUE APPROACH

**STEP 4** The critical value(s) are

$$\pm z_{\alpha/2} \quad\text{or}\quad -z_\alpha \quad\text{or}\quad z_\alpha$$
$$\text{(Two tailed)} \qquad \text{(Left tailed)} \qquad \text{(Right tailed)}$$

Use Table II to find the critical value(s).

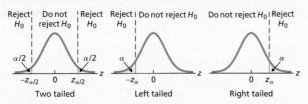

Two tailed      Left tailed      Right tailed

**STEP 5** If the value of the test statistic falls in the rejection region, reject $H_0$; otherwise, do not reject $H_0$.

**STEP 4** Use Table II to obtain the *P*-value.

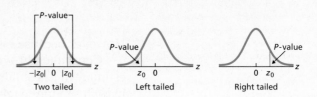

Two tailed      Left tailed      Right tailed

**STEP 5** If $P \leq \alpha$, reject $H_0$; otherwise, do not reject $H_0$.

**STEP 6** Interpret the results of the hypothesis test.

---

**Example 11.9** | **The Two-Proportions z-Test**

*Eating Out Vegetarian* Let's solve the problem posed in Example 11.8: Do the data from the Zogby International poll provide sufficient evidence to conclude that the percentage of U.S. men who sometimes order veg is smaller than the percentage of U.S. women who sometimes order veg? Use a 5% level of significance.

**Solution**  We apply Procedure 11.3, noting first that the assumptions for its use are satisfied.

**STEP 1  State the null and alternative hypotheses.**

Let $p_1$ and $p_2$ denote the proportions of all U.S. men and all U.S. women who sometimes order veg, respectively. The null and alternative hypotheses are

$$H_0\text{: } p_1 = p_2 \text{ (percentage for men is not less than that for women)}$$
$$H_a\text{: } p_1 < p_2 \text{ (percentage for men is less than that for women).}$$

Note that the hypothesis test is left tailed.

**STEP 2  Decide on the significance level, $\alpha$.**

The test is to be performed at the 5% significance level, or $\alpha = 0.05$.

**STEP 3  Compute the value of the test statistic**

$$z = \frac{\hat{p}_1 - \hat{p}_2}{\sqrt{\hat{p}_p(1 - \hat{p}_p)}\sqrt{(1/n_1) + (1/n_2)}}$$

**where $\hat{p}_p = (x_1 + x_2)/(n_1 + n_2)$.**

We first obtain $\hat{p}_1$, $\hat{p}_2$, and $\hat{p}_p$. Because 276 of the 747 men sampled and 195 of the 434 women sampled sometimes order veg, $x_1 = 276$, $n_1 = 747$, $x_2 = 195$, and $n_2 = 434$. Therefore,

$$\hat{p}_1 = \frac{x_1}{n_1} = \frac{276}{747} = 0.369, \qquad \hat{p}_2 = \frac{x_2}{n_2} = \frac{195}{434} = 0.449,$$

and

$$\hat{p}_p = \frac{x_1 + x_2}{n_1 + n_2} = \frac{276 + 195}{747 + 434} = \frac{471}{1181} = 0.399.$$

Consequently, the value of the test statistic is

$$z = \frac{\hat{p}_1 - \hat{p}_2}{\sqrt{\hat{p}_p(1 - \hat{p}_p)}\sqrt{(1/n_1) + (1/n_2)}}$$

$$= \frac{0.369 - 0.449}{\sqrt{(0.399)(1 - 0.399)}\sqrt{(1/747) + (1/434)}} = -2.71.$$

| CRITICAL-VALUE APPROACH | or | P-VALUE APPROACH |
|---|---|---|

**STEP 4  The critical value for a left-tailed test is $-z_\alpha$. Use Table II to find the critical value.**

For $\alpha = 0.05$, we find from Table II that the critical value is $-z_{0.05} = -1.645$, as shown in Fig. 11.3A.

**STEP 4  Use Table II to obtain the $P$-value.**

From Step 3, the value of the test statistic is $z = -2.71$. The test is left tailed, so the $P$-value is the probability of observing a value of $z$ of $-2.71$ or less if the null hypothesis is true. That probability equals the shaded area in Fig. 11.3B, which by Table II is 0.0034.

| CRITICAL-VALUE APPROACH | or | P-VALUE APPROACH |

**FIGURE 11.3A**

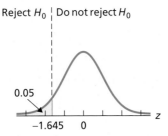

Reject $H_0$ ¦ Do not reject $H_0$

0.05

−1.645   0   z

**STEP 5** **If the value of the test statistic falls in the rejection region, reject $H_0$; otherwise, do not reject $H_0$.**

From Step 3, the value of the test statistic is $z = -2.71$, which, as Fig. 11.3A shows, falls in the rejection region. Thus we reject $H_0$. The test results are statistically significant at the 5% level.

**FIGURE 11.3B**

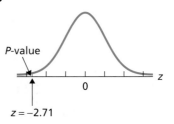

P-value

0   z

$z = -2.71$

**STEP 5** **If $P \leq \alpha$, reject $H_0$; otherwise, do not reject $H_0$.**

From Step 4, $P = 0.0034$. Because the P-value is less than the specified significance level of 0.05, we reject $H_0$. The test results are statistically significant at the 5% level and (see Table 9.10 on page 414) provide very strong evidence against the null hypothesis.

**You try it!**

Exercise 11.77
on page 532

**STEP 6** **Interpret the results of the hypothesis test.**

**Interpretation** At the 5% significance level, the data provide sufficient evidence to conclude that, in the United States, the percentage of men who sometimes order veg is smaller than the percentage of women who sometimes order veg.

• • •

## Large-Sample Confidence Intervals for the Difference Between Two Population Proportions

We can also use Key Fact 11.2 on page 522 to derive a confidence-interval procedure for the difference between two population proportions, called the **two-proportions $z$-interval procedure** (Procedure 11.4 on the next page).[†]

**Example 11.10** | **The Two-Proportions $z$-Interval Procedure**

*Eating Out Vegetarian* Refer to Example 11.9 and find a 90% confidence interval for the difference, $p_1 - p_2$, between the proportions of U.S. men and U.S. women who sometimes order veg.

**Solution** We apply Procedure 11.4, noting first that the conditions for its use are met.

**STEP 1** **For a confidence level of $1 - \alpha$, use Table II to find $z_{\alpha/2}$.**

For a 90% confidence interval, we have $\alpha = 0.10$. From Table II, we determine that $z_{\alpha/2} = z_{0.10/2} = z_{0.05} = 1.645$.

---

[†]The two-proportions z-interval procedure is also known as the **two-sample z-interval procedure for two population proportions** and the **two-variable proportions interval procedure**.

| Procedure 11.4 | **Two-Proportions $z$-Interval Procedure** |
|---|---|

*Purpose* To find a confidence interval for the difference between two population proportions, $p_1$ and $p_2$

*Assumptions*

1. Simple random samples
2. Independent samples
3. $x_1$, $n_1 - x_1$, $x_2$, and $n_2 - x_2$ are all 5 or greater

**STEP 1** For a confidence level of $1 - \alpha$, use Table II to find $z_{\alpha/2}$.

**STEP 2** The endpoints of the confidence interval for $p_1 - p_2$ are

$$(\hat{p}_1 - \hat{p}_2) \pm z_{\alpha/2} \cdot \sqrt{\hat{p}_1(1 - \hat{p}_1)/n_1 + \hat{p}_2(1 - \hat{p}_2)/n_2}.$$

**STEP 3** Interpret the confidence interval.

**STEP 2** The endpoints of the confidence interval for $p_1 - p_2$ are

$$(\hat{p}_1 - \hat{p}_2) \pm z_{\alpha/2} \cdot \sqrt{\hat{p}_1(1 - \hat{p}_1)/n_1 + \hat{p}_2(1 - \hat{p}_2)/n_2}.$$

From Step 1, $z_{\alpha/2} = 1.645$. As we found in Example 11.9, $\hat{p}_1 = 0.369$, $n_1 = 747$, $\hat{p}_2 = 0.449$, and $n_2 = 434$. Therefore the endpoints of the 90% confidence interval for $p_1 - p_2$ are

$$(0.369 - 0.449) \pm 1.645 \cdot \sqrt{(0.369)(1 - 0.369)/747 + (0.449)(1 - 0.449)/434},$$

or $-0.080 \pm 0.049$, or $-0.129$ to $-0.031$.

**STEP 3** Interpret the confidence interval.

**Interpretation** We can be 90% confident that, in the United States, the difference between the proportions of men and women who sometimes order veg is somewhere between $-0.129$ and $-0.031$. In other words, we can be 90% confident that the percentage of U.S. women who sometimes order veg exceeds the percentage of U.S. men who sometimes order veg by somewhere between 3.1 and 11.9 percentage points.

**You try it!**

Exercise 11.83
on page 533

• • •

## Margin of Error and Sample Size

We can obtain the **margin of error** in estimating the difference between two population proportions by referring to Step 2 of Procedure 11.4. Specifically, we have the following formula.

**Formula 11.3**

### Margin of Error for the Estimate of $p_1 - p_2$

The margin of error for the estimate of $p_1 - p_2$ is

$$E = z_{\alpha/2} \cdot \sqrt{\hat{p}_1(1 - \hat{p}_1)/n_1 + \hat{p}_2(1 - \hat{p}_2)/n_2}.$$

**What Does It Mean?**

The margin of error equals half the length of the confidence interval. It represents the precision with which the difference between the sample proportions estimates the difference between the population proportions at the specified confidence level.

From the formula for the margin of error, we can determine the sample sizes required to obtain a confidence interval with a specified confidence level and margin of error.

**Formula 11.4**

### Sample Size for Estimating $p_1 - p_2$

A $(1 - \alpha)$-level confidence interval for the difference between two population proportions that has a margin of error of at most $E$ can be obtained by choosing

$$n_1 = n_2 = 0.5 \left(\frac{z_{\alpha/2}}{E}\right)^2$$

rounded up to the nearest whole number. If you can make educated guesses, $\hat{p}_{1g}$ and $\hat{p}_{2g}$, for the observed values of $\hat{p}_1$ and $\hat{p}_2$, you should instead choose

$$n_1 = n_2 = \left(\hat{p}_{1g}(1 - \hat{p}_{1g}) + \hat{p}_{2g}(1 - \hat{p}_{2g})\right) \left(\frac{z_{\alpha/2}}{E}\right)^2$$

rounded up to the nearest whole number.

The first formula in Formula 11.4 provides sample sizes that ensure obtaining a $(1 - \alpha)$-level confidence interval with a margin of error of at most $E$, but it may yield sample sizes that are unnecessarily large. The second formula in Formula 11.4 yields smaller sample sizes, but it should not be used unless the guesses for the sample proportions are considered reasonably accurate.

If you know likely ranges for the observed values of the two sample proportions, use the values in the ranges closest to 0.5 as the educated guesses. For further discussion of these ideas and for applications of Formulas 11.3 and 11.4, see Exercise 11.91.

# The Technology Center

Most statistical technologies have programs that automatically perform two-proportions z-procedures. In this subsection, we present output and step-by-step instructions for such programs.

**Example 11.11**

### Using Technology to Conduct Two-Proportions z-Procedures

*Eating Out Vegetarian*  Independent random samples of 747 U.S. men and 434 U.S. women were taken. Of those sampled, 276 men and 195 women said that they sometimes order veg. Use Minitab, Excel, or the TI-83/84 Plus to perform the hypothesis test in Example 11.9 and obtain the confidence interval in Example 11.10.

**Solution** Let $p_1$ and $p_2$ denote the proportions of all U.S. men and all U.S. women who sometimes order veg, respectively. The task in Example 11.9 is to perform the hypothesis test

$$H_0: p_1 = p_2 \text{ (percentage for men is not less than that for women)}$$
$$H_a: p_1 < p_2 \text{ (percentage for men is less than that for women)}$$

at the 5% significance level; the task in Example 11.10 is to obtain a 90% confidence interval for $p_1 - p_2$.

We applied the two-proportions z-procedures programs to the data, resulting in Output 11.4, shown on this and the following page. Steps for generating that output are presented in Instructions 11.3 on page 531.

**OUTPUT 11.4**  Two-proportions z-test and z-interval for the study on ordering vegetarian

**MINITAB**

**Test and CI for Two Proportions**     [FOR THE HYPOTHESIS TEST]

```
Sample   X    N   Sample p
1       276  747  0.369478
2       195  434  0.449309


Difference = p (1) - p (2)
Estimate for difference:  -0.0798308
90% upper bound for difference:  -0.0417711
Test for difference = 0 (vs < 0):  Z = -2.70  P-Value = 0.003
```

**Test and CI for Two Proportions**     [FOR THE CONFIDENCE INTERVAL]

```
Sample   X    N   Sample p
1       276  747  0.369478
2       195  434  0.449309


Difference = p (1) - p (2)
Estimate for difference:  -0.0798308
90% CI for difference:  (-0.128680, -0.0309817)
Test for difference = 0 (vs not = 0):  Z = -2.70  P-Value = 0.007
```

**EXCEL**

```
Proportion Test for the Difference Between x_1 and n_1

Summary Statistics              Test Summary
        n1    747                 p:              0
    p-hat1    0.369              Ho:     p1 - p2 = 0
        n2    434                Ha: Lower tail: p1 - p2 < 0
    p-hat2    0.449       z Statistic:           -2.7
Difference    -0.0798           p-value:          0.0035
  pooled n    1181
pooled p-hat  0.399
   Std Err    0.0296         Test Results

                            Conclusion

                            Reject Ho at alpha =  0.05
```

Using **Summ 2 Var Prop Test**

**OUTPUT 11.4 (cont.)**

Two-proportions *z*-test and *z*-interval for the study on ordering vegetarian

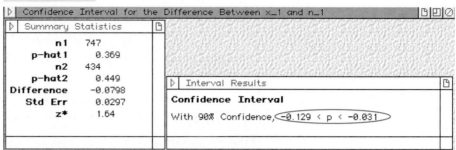

Using **Summ 2 Var Prop Interval**

**TI-83/84 PLUS**

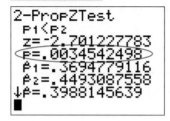

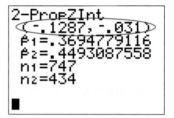

Using **2-PropZInt**

Using **2-PropZTest**

As shown in Output 11.4, the *P*-value for the hypothesis test is 0.003. Because the *P*-value is less than the specified significance level of 0.05, we reject $H_0$. Output 11.4 also shows that a 90% confidence interval for the difference between the population proportions is from $-0.129$ to $-0.031$.

• • •

**INSTRUCTIONS 11.3**   Steps for generating Output 11.4

| MINITAB | EXCEL | TI-83/84 PLUS |
|---|---|---|

**MINITAB**

FOR THE HYPOTHESIS TEST:
1 Choose **Stat ➤ Basic Statistics ➤ 2 Proportions...**
2 Select the **Summarized data** option button
3 Click in the **Trials** text box for **First** and type 747
4 Click in the **Events** text box for **First** and type 276
5 Click in the **Trials** text box for **Second** and type 434
6 Click in the **Events** text box for **Second** and type 195
7 Click the **Options...** button
8 Click in the **Confidence level** text box and type 90
9 Click in the **Test difference** text box and type 0
10 Click the arrow button at the right of the **Alternative** drop-down list box and select **less than**
11 Check the **Use pooled estimate of p for test** check box
12 Click **OK** twice

FOR THE CI:
1 Choose **Edit ➤ Edit Last Dialog**
2 Click the **Options...** button
3 Click the arrow button at the right of the **Alternative** drop-down list box and select **not equal**
4 Click **OK** twice

**EXCEL**

Store the sample sizes, 747 and 434, in ranges named n_1 and n_2, respectively, and store the numbers of successes, 276 and 195, in ranges named x_1 and x_2, respectively.

FOR THE HYPOTHESIS TEST:
1 Choose **DDXL ➤ Hypothesis tests**
2 Select **Summ 2 Var Prop Test** from the **Function type** drop-down list box
3 Specify x_1 in the **Num Successes 1** text box, n_1 in the **Num Trials 1** text box, x_2 in the **Num Successes 2** text box, and n_2 in the **Num Trials 2** text box
4 Click **OK**
5 Click the **Set p** button
6 Click in the **Specify p** text box and type 0
7 Click **OK**
8 Click the **.05** button
9 Click the **p1 − p2 < p** button
10 Click the **Compute** button

FOR THE CI:
1 Choose **DDXL ➤ Confidence Intervals**
2 Select **Summ 2 Var Prop Interval** from the **Function type** drop-down list box
3 Specify x_1 in the **Num Successes 1** text box, n_1 in the **Num Trials 1** text box, x_2 in the **Num Successes 2** text box, and n_2 in the **Num Trials 2** text box
4 Click **OK**
5 Click the **90%** button
6 Click the **Compute Interval** button

**TI-83/84 PLUS**

FOR THE HYPOTHESIS TEST:
1 Press **STAT**, arrow over to **TESTS**, and press **6**
2 Type 276 for **x1** and press **ENTER**
3 Type 747 for **n1** and press **ENTER**
4 Type 195 for **x2** and press **ENTER**
5 Type 434 for **n2** and press **ENTER**
6 Highlight **<p2** and press **ENTER**
7 Press the down-arrow key, highlight **Calculate**, and press **ENTER**

FOR THE CI:
1 Press **STAT**, arrow over to **TESTS**, and press **ALPHA ➤ B**
2 Type 276 for **x1** and press **ENTER**
3 Type 747 for **n1** and press **ENTER**
4 Type 195 for **x2** and press **ENTER**
5 Type 434 for **n2** and press **ENTER**
6 Type .90 for **C-Level** and press **ENTER** twice

▌ ▌ ▌

# Exercises 11.3

## Understanding the Concepts and Skills

**11.65** Explain the basic idea for performing a hypothesis test, based on independent samples, to compare two population proportions.

**11.66 Kids Attending Church.** A Roper Starch Worldwide for A.B.C. Global Kids Study conducted surveys in various countries to estimate the percentage of children who attend church at least once a week. Two of the countries in the survey were the United States and Germany. Considering these two countries only,
a. identify the specified attribute.
b. identify the two populations.
c. What are the two population proportions under consideration?

**11.67 Sunscreen Use.** Industry Research polled teenagers on sunscreen use. The survey revealed that 46% of teenage girls and 30% of teenage boys regularly use sunscreen before going out in the sun.
a. Identify the specified attribute.
b. Identify the two populations.
c. Are the proportions 0.46 (46%) and 0.30 (30%) sample proportions or population proportions? Explain your answer.

**11.68** Consider a hypothesis test for two population proportions with the null hypothesis $H_0$: $p_1 = p_2$. What parameter is being estimated by the
a. sample proportion $\hat{p}_1$?
b. sample proportion $\hat{p}_2$?
c. pooled sample proportion $\hat{p}_p$?

**11.69** Of the quantities $p_1$, $p_2$, $x_1$, $x_2$, $\hat{p}_1$, $\hat{p}_2$, and $\hat{p}_p$,
a. which represent parameters and which represent statistics?
b. which are fixed numbers and which are variables?

*In each of Exercises 11.70–11.75, we have provided the numbers of successes and the sample sizes for independent simple random samples from two populations. In each case, do the following.*
*a. Determine the sample proportions.*
*b. Decide whether using the two-proportions z-procedures is appropriate. If so, also do parts (c) and (d).*
*c. Use the two-proportions z-test to conduct the required hypothesis test.*
*d. Use the two-proportions z-interval procedure to find the specified confidence interval.*

**11.70** $x_1 = 10$, $n_1 = 20$, $x_2 = 18$, $n_2 = 30$; left-tailed test, $\alpha = 0.10$; 80% confidence interval

**11.71** $x_1 = 18$, $n_1 = 40$, $x_2 = 30$, $n_2 = 40$; left-tailed test, $\alpha = 0.10$; 80% confidence interval

**11.72** $x_1 = 14$, $n_1 = 20$, $x_2 = 8$, $n_2 = 20$; right-tailed test, $\alpha = 0.05$; 90% confidence interval

**11.73** $x_1 = 15$, $n_1 = 20$, $x_2 = 18$, $n_2 = 30$; right-tailed test, $\alpha = 0.05$; 90% confidence interval

**11.74** $x_1 = 18$, $n_1 = 30$, $x_2 = 10$, $n_2 = 20$; two-tailed test, $\alpha = 0.05$; 95% confidence interval

**11.75** $x_1 = 30$, $n_1 = 80$, $x_2 = 15$, $n_2 = 20$; two-tailed test, $\alpha = 0.05$; 95% confidence interval

*For Exercises 11.76–11.81, use either the critical-value approach or the P-value approach to perform the required hypothesis test.*

**11.76 Vasectomies and Prostate Cancer.** Approximately 450,000 vasectomies are performed each year in the United States. In this surgical procedure for contraception, the tube carrying sperm from the testicles is cut and tied. Several studies have been conducted to analyze the relationship between vasectomies and prostate cancer. The results of one such study by E. Giovannucci et al. appeared in the paper "A Retrospective Cohort Study of Vasectomy and Prostate Cancer in U.S. Men" (*Journal of the American Medical Association*, Vol. 269(7), pp. 878–882). Of 21,300 men who had not had a vasectomy, 69 were found to have prostate cancer; of 22,000 men who had had a vasectomy, 113 were found to have prostate cancer.
a. At the 1% significance level, do the data provide sufficient evidence to conclude that men who have had a vasectomy are at greater risk of having prostate cancer?
b. Is this study a designed experiment or an observational study? Explain your answer.
c. In view of your answers to parts (a) and (b), could you reasonably conclude that having a vasectomy causes an increased risk of prostate cancer? Explain your answer.

**11.77 Folic Acid and Birth Defects.** For several years, evidence had been mounting that folic acid reduces major birth defects. A. Czeizel and I. Dudas of the National Institute of Hygiene in Budapest directed a study that provided the strongest evidence to date. Their results were published in the paper "Prevention of the First Occurrence of Neural-Tube Defects by Periconceptional Vitamin Supplementation" (*New England Journal of Medicine*, Vol. 327(26), p. 1832). For the study, the doctors enrolled women prior to conception, and divided them randomly into two groups. One group, consisting of 2701 women, took daily multivitamins containing 0.8 mg of folic acid; the other group, consisting of 2052 women, received only trace elements. Major birth defects occurred in 35 cases when the women took folic acid and in 47 cases when the women did not.

a. At the 1% significance level, do the data provide sufficient evidence to conclude that women who take folic acid are at lesser risk of having children with major birth defects?

b. Is this study a designed experiment or an observational study? Explain your answer.

c. In view of your answers to parts (a) and (b), could you reasonably conclude that taking folic acid causes a reduction in major birth defects? Explain your answer.

**11.78 Racial Crossover.** In the paper "The Racial Crossover in Comorbidity, Disability, and Mortality," (*Demography*, Vol. 37(3), pp. 267–283), Nan E. Johnson investigated the health of independent random samples of white and African-American elderly (aged 70 or older). Of the 4989 white elderly surveyed, 529 had at least one stroke, whereas 103 of the 906 African-American elderly surveyed reported at least one stroke. At the 5% significance level, do the data suggest that there is a difference in stroke incidence between white and African-American elderly?

**11.79 Buckling Up.** Response Insurance collects data on seat-belt use among U.S. drivers. Of 1000 drivers 25–34 years old, 27% said that they buckle up, whereas 330 of 1100 drivers 45–64 years old said that they did. At the 10% significance level, do the data suggest that there is a difference in seat-belt use between drivers 25–34 years old and those 45–64 years old? [SOURCE: *USA TODAY Online*.]

**11.80 Ballistic Fingerprinting.** Guns make unique markings on bullets they fire and their shell casings. These markings are called *ballistic fingerprints*. An *ABCNEWS Poll* examined the opinions of Americans on the enactment of a law "…that would require every gun sold in the United States to be test-fired first, so law enforcement would have its fingerprint in case it were ever used in a crime." The following problem is based on the results of that poll. Independent simple random samples were taken of 537 women and 495 men. When asked whether they support a ballistic fingerprinting law, 446 of the women and 307 of the men said "yes." At the 1% significance level, do the data provide sufficient evidence to conclude that women tend to favor ballistic fingerprinting more than men?

**11.81 Body Mass Index.** Body Mass Index (BMI) is a measure of body fat based on height and weight. According to the document *Dietary Guidelines for Americans*, published by the U.S. Department of Agriculture and the U.S. Department of Health and Human Services, for adults, a BMI of greater than 25 indicates an above healthy weight (i.e., overweight or obese). Of 750 randomly selected adults whose highest degree is a bachelors, 386 have an above healthy weight; and of 500 randomly selected adults with a graduate degree, 237 have an above healthy weight.

a. What assumptions are required for using the two-proportions z-test here?

b. Apply the two-proportions z-test to determine, at the 5% significance level, whether the percentage of adults who have an above healthy weight is greater for those whose highest degree is a bachelors than for those with a graduate degree.

c. Repeat part (b) at the 10% significance level.

*In Exercises 11.82–11.87, apply Procedure 11.4 on page 527 to find the required confidence interval.*

**11.82 Vasectomies and Prostate Cancer.** Refer to Exercise 11.76 and determine and interpret a 98% confidence interval for the difference between the prostate cancer rates of men who have had a vasectomy and those who have not.

**11.83 Folic Acid and Birth Defects.** Refer to Exercise 11.77 and determine and interpret a 98% confidence interval for the difference between the rates of major birth defects for babies born to women who have taken folic acid and those born to women who have not.

**11.84 Racial Crossover.** Refer to Exercise 11.78 and find and interpret a 95% confidence interval for the difference between the stroke incidences of white and African-American elderly.

**11.85 Buckling Up.** Refer to Exercise 11.79 and find and interpret a 90% confidence interval for the difference between the proportions of seat-belt users for drivers in the age groups 25–34 years and 45–64 years.

**11.86 Ballistic Fingerprinting.** Refer to Exercise 11.80 and find and interpret a 98% confidence interval for the difference between the percentages of women and men who favor ballistic fingerprinting.

**11.87 Body Mass Index.** Refer to Exercise 11.81.

a. Determine and interpret a 90% confidence interval for the difference between the percentages of adults in the two degree categories who have an above healthy weight.

b. Repeat part (a) for an 80% confidence interval.

*In each of Exercises **11.88–11.90**, use the technology of your choice to conduct the required analyses.*

**11.88 Hormone Therapy and Dementia.** An issue of *Science News* (Vol. 163, No. 22, pp. 341–342) reported that the Women's Health Initiative cast doubts on the benefit of hormone-replacement therapy. Researchers randomly divided 4532 healthy women over the age of 65 into two groups. One group, consisting of 2229 women, received hormone-replacement therapy; the other group, consisting of 2303 women, received placebo. Over 5 years, 40 of the women receiving the hormone-replacement therapy were diagnosed with dementia, compared with 21 of those getting placebo.

**a.** At the 5% significance level, do the data provide sufficient evidence to conclude that healthy women over 65 years old who take hormone-replacement therapy are at greater risk for dementia than those who do not?

**b.** Referring to part (a), assess the strength of the evidence against the null hypothesis and hence in favor of the alternative hypothesis that healthy women over 65 years old who take hormone-replacement therapy are at greater risk for dementia than those who do not.

**c.** Determine and interpret a 95% confidence interval for the difference in dementia risk rates for healthy women over 65 years old who take hormone-replacement therapy and those who do not.

**11.89 Women in the Labor Force.** The Organization for Economic Cooperation and Development, Paris, France, summarizes data on labor-force participation rates in *OECD in Figures*. Independent simple random samples were taken of 300 U.S. women and 250 Canadian women. Of the U.S. women, 219 were found to be in the labor force; and of the Canadian women, 174 were found to be in the labor force.

**a.** At the 5% significance level, do the data suggest that there is a difference between the labor-force participation rates of U.S. and Canadian women?

**b.** Find and interpret a 95% confidence interval for the difference between the labor-force participation rates of U.S. and Canadian women.

**11.90 Neutropenia.** Neutropenia is an abnormally low number of neutrophils (a type of white blood cell) in the blood. Chemotherapy often reduces the number of neutrophils to a level that makes patients susceptible to fever and infections. G. Bucaneve et al. published a study of such cancer patients in the paper "Levofloxacin to Prevent

Bacterial Infection in Patients with Cancer and Neutropenia" (*New England Journal of Medicine*, Vol. 353, No. 10, pp. 977–987). For the study, 375 patients were randomly assigned to receive a daily dose of levofloxacin and 363 were given placebo. In the group receiving levofloxacin, fever was present in 243 patients for the duration of neutropenia, whereas fever was experienced by 308 patients in the placebo group.

**a.** At the 1% significance level, do the data provide sufficient evidence to conclude that levofloxacin is effective in reducing the occurrence of fever in such patients?

**b.** Find a 99% confidence level for the difference in the proportions of such cancer patients who would experience fever for the duration of neutropenia.

## Extending the Concepts and Skills

**11.91 Eating Out Vegetarian.** In this exercise, apply Formulas 11.3 and 11.4 on page 528 to the study on ordering vegetarian considered in Examples 11.8–11.10.

**a.** Obtain the margin of error for the estimate of the difference between the proportions of men and women who sometimes order veg by taking half the length of the confidence interval found in Example 11.10 on page 526. Interpret your answer in words.

**b.** Obtain the margin of error for the estimate of the difference between the proportions of men and women who sometimes order veg by applying Formula 11.3.

**c.** Without making a guess for the observed values of the sample proportions, find the common sample size that will ensure a margin of error of at most 0.01 for a 90% confidence interval.

**d.** Find a 90% confidence interval for $p_1 - p_2$ if, for samples of the size determined in part (c), 38.3% of the men and 43.7% of the women sometimes order veg.

**e.** Determine the margin of error for the estimate in part (d) and compare it to the required margin of error specified in part (c).

**f.** Repeat parts (c)–(e) if you can reasonably presume that at most 41% of the men sampled and at most 49% of the women sampled will be people who sometimes order veg.

**g.** Compare the results obtained in parts (c)–(e) to those obtained in part (f).

**11.92** Identify the formula in this section that shows that the difference between the two sample proportions is an unbiased estimator of the difference between the two population proportions.

## Chapter in Review

### You Should be Able to

1. use and understand the formulas in this chapter.

2. find a large-sample confidence interval for a population proportion.

3. compute the margin of error for the estimate of a population proportion.

4. understand the relationship between the sample size, confidence level, and margin of error for a confidence interval for a population proportion.

5. determine the sample size required for a specified confidence level and margin of error for the estimate of a population proportion.

6. perform a large-sample hypothesis test for a population proportion.

7. perform large-sample inferences (hypothesis tests and confidence intervals) to compare two population proportions.

8. understand the relationship between the sample sizes, confidence level, and margin of error for a confidence interval for the difference between two population proportions.

9. determine the sample sizes required for a specified confidence level and margin of error for the estimate of the difference between two population proportions.

### Key Terms

margin of error, *506, 528*
number of failures, *503*
number of successes, *503*
one-proportion *z*-interval
    procedure, *505*
one-proportion *z*-test, *515*

pooled sample proportion ($\hat{p}_p$), *523*
population proportion (*p*), *503*
sample proportion ($\hat{p}$), *503*
sampling distribution of the
    difference between two sample
    proportions, *522*

sampling distribution of the sample
    proportion, *504*
two-proportions *z*-interval
    procedure, *526, 527*
two-proportions *z*-test, *523, 524*

## Review Problems

### Understanding the Concepts and Skills

**1. Medical Marijuana?** A *Harris Poll* was conducted to estimate the proportion of Americans who feel that marijuana should be legalized for medicinal use in patients with cancer and other painful and terminal diseases. Identify the
a. specified attribute.          b. population.
c. population proportion.
d. According to the poll, 80% of the 83,957 respondents said that marijuana should be legalized for medicinal use. Is the proportion 0.80 (80%) a sample proportion or a population proportion? Explain your answer.

**2.** Why is a sample proportion generally used to estimate a population proportion instead of obtaining the population proportion directly?

**3.** Explain what each phrase means in the context of inferences for a population proportion.
a. Number of successes
b. Number of failures

**4.** Fill in the blanks.
a. The mean of all possible sample proportions is equal to the _____.

b. For large samples, the possible sample proportions have approximately a _____ distribution.
c. A rule of thumb for using a normal distribution to approximate the distribution of all possible sample proportions is that both _____ and _____ are _____ or greater.

**5.** What does the margin of error for the estimate of a population proportion tell you?

**6. Holiday Blues.** A poll was conducted by *Opinion Research Corporation* to estimate the proportions of men and women who get the "holiday blues." Identify the
a. specified attribute.          b. two populations.
c. two population proportions.
d. two sample proportions.
e. According to the poll, 34% of men and 44% of women get the "holiday blues." Are the proportions 0.34 and 0.44 sample proportions or population proportions? Explain your answer.

**7.** Suppose that you are using independent samples to compare two population proportions. Fill in the blanks.

**a.** The mean of all possible differences between the two sample proportions equals the _____.

**b.** For large samples, the possible differences between the two sample proportions have approximately a _____ distribution.

**8. Smallpox Vaccine.** In December of 2002, *ABCNEWS.com* published the results of a poll that asked U.S. adults whether they would get a smallpox shot if it were available. Sampling, data collection, and tabulation were done by TNS Intersearch of Horsham, PA. When the risk of the vaccine was described in detail, 4 in 10 of those surveyed said they would take the smallpox shot. According to the article, "the results have a three-point margin of error" (for a 0.95 confidence level). Use the information provided to obtain a 95% confidence interval for the percentage of all U.S. adults who would take a smallpox shot, knowing the risk of the vaccine.

**9.** Suppose that you want to find a 95% confidence interval based on independent samples for the difference between two population proportions and that you want a margin of error of at most 0.01.

**a.** Without making an educated guess for the observed sample proportions, find the required common sample size.

**b.** Suppose that, from past experience, you are quite sure that the two sample proportions will be 0.75 or greater. What common sample size should you use?

**10. Getting a Job.** The National Association of Colleges and Employers sponsors the *Graduating Student and Alumni Survey*. Part of the survey gauges student optimism in landing a job after graduation. According to one year's survey results, published in *American Demographics*, among the 1218 respondents, 733 said that they expected difficulty finding a job. Use these data to obtain and interpret a 95% confidence interval for the proportion of students who expect difficulty finding a job.

**11. Getting a Job.** Refer to Problem 10.

**a.** Find the margin of error for the estimate of $p$.

**b.** Obtain a sample size that will ensure a margin of error of at most 0.02 for a 95% confidence interval without making a guess for the observed value of $\hat{p}$.

**c.** Find a 95% confidence interval for $p$ if, for a sample of the size determined in part (b), 58.7% of those surveyed say that they expect difficulty finding a job.

**d.** Determine the margin of error for the estimate in part (c) and compare it to the required margin of error specified in part (b).

**e.** Repeat parts (b)–(d) if you can reasonably presume that the percentage of those surveyed who say that they expect difficulty finding a job will be at least 56%.

**f.** Compare the results obtained in parts (b)–(d) with those obtained in part (e).

**12. Justice in the Courts?** In an issue of *Parade Magazine*, the editors reported on a national survey on law and order. One question asked of the 2512 U.S. adults who took part was whether they believed that juries "almost always" convict the guilty and free the innocent. Only 578 said that they did. Do the data provide sufficient evidence to conclude that less than one in four Americans believe that juries "almost always" convict the guilty and free the innocent?

**a.** Use either the critical-value approach or the *P*-value approach to perform the appropriate hypothesis test at the 5% significance level.

**b.** Assess the strength of the evidence against the null hypothesis by referring to Table 9.10 on page 414.

**13. Height and Breast Cancer.** An article published in an issue of the *Annals of Epidemiology* discussed the relationship between height and breast cancer. The study by the National Cancer Institute, which took 5 years and involved more than 1500 women with breast cancer and 2000 women without breast cancer, revealed a trend between height and breast cancer: "...taller women have a 50 to 80 percent greater risk of getting breast cancer than women who are closer to 5 feet tall." But Christine Swanson, a nutritionist who was involved with the study, added that "...height may be associated with the culprit, ...but no one really knows" the exact relationship between height and the risk of breast cancer.

**a.** Classify this study as either an observational study or a designed experiment. Explain your answer.

**b.** Interpret the statement made by Christine Swanson in light of your answer to part (a).

**14. Views on the Economy.** State and local governments often poll their constituents about their views on the economy. In two polls, taken approximately 1 year apart, O'Neil Associates asked 600 Maricopa County, Arizona, residents whether they thought the state's economy would improve over the next 2 years. In the first poll, 48% said "yes"; in the second poll, 60% said "yes." Do the data provide sufficient evidence to conclude that the percentage of Maricopa County residents who thought the state's economy would improve over the next 2 years was less during the time of the first poll than during the time of the second?

**a.** Use either the critical-value approach or the *P*-value approach to perform the appropriate hypothesis test at the 1% significance level.

**b.** Assess the strength of the evidence against the null hypothesis by referring to Table 9.10 on page 414.

**15. Views on the Economy.** Refer to Problem 14.

**a.** Determine a 98% confidence interval for the difference, $p_1 - p_2$, between the proportions of Maricopa County residents who thought that the state's economy would improve over the next 2 years during the time of the first poll and during the time of the second poll.

**b.** Interpret your answer from part (a).

**16. Views on the Economy.** Refer to Problems 14 and 15.

**a.** Take half the length of the confidence interval found in Problem 15(a) to obtain the margin of error for the estimate of the difference between the two population proportions. Interpret your result in words.

**b.** Solve part (a) by applying Formula 11.3 on page 528.

**c.** Obtain the common sample size that will ensure a margin of error of at most 0.03 for a 98% confidence interval without making a guess for the observed values of the sample proportions.

**d.** Find a 98% confidence interval for $p_1 - p_2$ if, for samples of the size determined in part (c), the sample proportions are 0.475 and 0.603, respectively.

**e.** Determine the margin of error for the estimate in part (d) and compare it to the required margin of error specified in part (c).

**17. Bulletproof Vests.** In the article "Common Police Vest Fails Bulletproof Test," E. Lichtblau reported on a 2005 Justice Department study of 103 bulletproof vests containing a fiber known as Zylon. In ballistics tests, only 4 of these vests produced acceptable safety results. Applying the one-proportion z-interval procedure, we found that a 95% confidence interval for the proportion of all such vests that would produce acceptable safety results is from 0.002 to 0.076. Using an exact method (which we have not presented), we found the 95% confidence interval to be from 0.011 to 0.096. Explain the large discrepancy between the two methods.

*In each of Problems 18–21, use the technology of your choice to conduct the required analyses.*

**18. March Madness.** The NCAA Men's Division I Basketball Championship is held each spring and features 65 college basketball teams. This 20-day tournament is colloquially known as "March Madness." A *Harris Poll*, conducted in March, 2006, asked 2435 randomly selected U.S. adults whether they would participate in an office pool for March Madness; 317 said they would. Use these data to find and interpret a 95% confidence interval for the percentage of U.S. adults who would participate in an office pool for March Madness.

**19. Abstinence and AIDS.** In a January, 2006 *Harris Poll* of 1961 randomly selected U.S. adults, 1137 said that they do not believe that abstinence programs are effective in reducing or preventing AIDS. At the 5% significance level, do the data provide sufficient evidence to conclude that a majority of all U.S. adults feel that way?

**20. Bug Buster.** N. Hill et al. conducted a clinical study to compare the standard treatment for head lice infestation with the Bug Buster kit, which involves using a fine-toothed comb on thoroughly wet hair four times at 4-day intervals. The researchers published their findings in the *British Medical Journal* (Vol. 331, pp. 384–387). For the study, 56 patients were randomly assigned to use the Bug Buster kit and 70 were assigned to use the standard treatment. Thirty-two patients in the Bug Buster kit group were cured, whereas nine of those in the standard treatment group were cured.

**a.** At the 5% significance level, do these data provide sufficient evidence to conclude that a difference exists in the cure rates of the two types of treatment.

**b.** Determine a 95% confidence interval for the difference in cure rates for the two types of treatment.

**21. Finasteride and Prostate Cancer.** The results of a major study to examine the effect of finasteride in reducing the risk of prostate cancer appeared on June 24, 2003, on the Web site of the *New England Journal of Medicine*. The study, known as the Prostate Cancer Prevention Trial (PCPT), was sponsored by the U.S. Public Health Service and the National Cancer Institute. In the PCPT trial, 18,832 men, 55 years old or older with normal physical exams and prostate-specific antigen (PSA) levels of 3.0 nanograms per milliliter or lower, were randomly assigned to receive 5 mg of finasteride daily or a placebo. At 7 years, of the 9060 men included in the final analysis, 4368 had taken finasteride and 4692 had received a placebo. For those who took finasteride, 803 cases of prostate cancer were diagnosed, compared with 1147 cases for those who took a placebo. Use the technology of your choice to decide, at the 1% significance level, whether finasteride reduces the risk of prostate cancer. (*Note:* As reported in the August 2003 issue of the *Public Citizen's Health Research Group Newsletter*, most of the aforementioned detected cancers were "low-grade cancers of little clinical significance." Moreover, the risk of high-grade cancers was determined to be elevated for those taking finasteride.)

## Focusing on Data Analysis  UWEC Undergraduates

Recall from Chapter 1 (see page 34) that the Focus database and Focus sample contain information on the undergraduate students at the University of Wisconsin - Eau Claire (UWEC). Now would be a good time for you to review the discussion about these data sets.

Open the Focus sample worksheet (FocusSample) in the technology of your choice and then do the following.

a. At the 5% significance level, do the data provide sufficient evidence to conclude that more than half of UWEC undergraduates are females?

b. Repeat part (a) using a 1% significance level.

c. Determine and interpret a 95% confidence interval for the percentage of UWEC undergraduates who are females.

d. At the 5% significance level, do the data provide sufficient evidence to conclude that a difference exists in the percentages of females among resident and nonresident UWEC undergraduates?

e. Repeat part (d) using a 10% significance level.

f. Determine and interpret a 95% confidence interval for the difference between the percentages of females among resident and nonresident UWEC undergraduates.

## Case Study Discussion  World War III?

At the beginning of this chapter, we discussed two independent polls about American and Japanese views on world wars, nuclear weapons, and related issues. Independent samples were taken of 1000 adults in the United States and 1045 adults in Japan. Use the information on page 501 to solve the following problems. (*Note:* The number in a sample that have a specified attribute equals the sample proportion times the sample size, rounded to the nearest integer.)

a. At the 5% significance level, do the data provide sufficient evidence to conclude that a majority of all U.S. adults feel that another world war is likely in their lifetime?

b. Find and interpret a 95% confidence interval for the proportion of all adults in Japan who feel that another world war is likely in their lifetime.

c. Determine and interpret a 90% confidence interval for the difference between the proportions of all people in the two countries who feel that the use of atomic bombs on Japan in World War II was unavoidable.

d. At the 1% significance level, do the data provide sufficient evidence to conclude that the percentage of all adults in Japan who have a positive view of the United States is smaller than the percentage of all adults in the United States who have a positive view of Japan?

## Biography  ABRAHAM DE MOIVRE: Paving the Way for Proportion Inferences

**Abraham de Moivre** was born in Vitry-le-Francois, France, on May 26, 1667, the son of a country surgeon. He was educated in the Catholic school in his village and at the Protestant Academy at Sedan. In 1684, he went to Paris to study under Jacques Ozanam.

In late 1685, de Moivre, a French Huguenot (Protestant), was imprisoned in Paris because of his religion. (In October, 1685, Louis XIV revoked an edict that had allowed Protestantism in addition to the Catholicism favored by the French Court.) The duration of his incarceration is unclear, but de Moivre was probably jailed 1 to 3 years. In any case, upon his release he fled to London where he began tutoring students in mathematics.

In London, de Moivre mastered Sir Isaac Newton's *Principia* and became a close friend of Newton's and of Edmond Halley's, an English astronomer (in whose honor, incidentally, Halley's Comet is named). In Newton's later years, he would refuse to take new students, saying, "Go to Mr. de Moivre; he knows these things better than I do."

De Moivre's contributions to probability theory, mathematics, and statistics range from the definition of statistical independence to analytical trigonometric formulas to his major discovery: the normal approximation to the binomial distribution—of monumental importance in its own right, precursor to the central limit theorem, and fundamental to proportion inferences. The definition of statistical independence appeared in *The Doctrine of Chances*, published in 1718 and dedicated to Newton; the normal approximation to the binomial distribution was contained in a Latin pamphlet published in 1733. Many of his other papers were published in *Philosophical Transactions of the Royal Society*.

De Moivre also did research on the analysis of mortality statistics and the theory of annuities. In 1725, the first edition of his *Annuities on Lives*, in which he derived annuity formulas and addressed other annuity problems, was published.

De Moivre was elected to the Royal Society in 1697, to the Berlin Academy of Sciences in 1735, and to the Paris Academy in 1754. But, despite his obvious talents as a mathematician and his many champions, he was never able to obtain a position in any of England's universities. Instead, he had to rely on his meager earnings as a tutor in mathematics and a consultant on gambling and insurance, supplemented by the sales of his books. De Moivre died in London on November 27, 1754.

## StatCrunch in MyStatLab
### Analyzing Data Online

StatCrunch online statistical software offers an easy-to-use interface customized for this book. The StatCrunch feature for each chapter illustrates the use of the software to perform a statistical analysis discussed in the chapter. Exercises are provided to further apply StatCrunch to other statistical analyses examined in the chapter. Go to the WeissStats CD or to the Weiss Web site at www.aw-bc.com/weiss to access StatCrunch instructions and data sets. To access StatCrunch statistical software, go to the student content area of your Weiss MyStatLab course.

## Internet Projects
### Exploring Data Online

The Internet project for each chapter provides simulations, demonstrations, or activities that enhance the topics covered in the chapter. The project materials come from universities, individuals, governments, and companies from all over the world. To access the Internet projects on the Web, go to www.aw-bc.com/weiss. From this Web page, you can reach the Internet Projects Page, which we suggest that you bookmark for easy access in the future.

# 12 Chi-Square Procedures

## Chapter Objectives

The statistical-inference techniques presented so far have dealt exclusively with hypothesis tests and confidence intervals for population parameters, such as population means and population proportions. In this chapter, we consider two widely used inferential procedures that are not concerned with population parameters. These two procedures are often called **chi-square procedures** because they rely on a distribution called the *chi-square distribution.*

Before we study chi-square procedures, we discuss the chi-square distribution in Section 12.1. In Section 12.2, we present the chi-square goodness-of-fit test, a hypothesis test that can be used to make inferences about the distribution of a variable. For instance, we could apply that test to a sample of university students to decide whether the political preference distribution of all university students differs from that of the population as a whole.

In Section 12.3, as a preliminary to the study of a second chi-square procedure, we discuss contingency tables—frequency distributions for bivariate data—and related topics. Then, in Section 12.4, we present the chi-square independence test, a hypothesis test used to decide whether an association exists between two characteristics of a population. For instance, we could apply that test to a sample of U.S. adults to decide whether an association exists between annual income and educational level for all U.S. adults.

# Road Rage

The report *Controlling Road Rage: A Literature Review and Pilot Study* was prepared for the AAA Foundation for Traffic Safety by Daniel B. Rathbone, Ph.D., and Jorg C. Huckabee, MSCE. The authors discuss the results of a literature review and pilot study on how to prevent aggressive driving and road rage.

*Road rage* is defined as "…an incident in which an angry or impatient motorist or passenger intentionally injures or kills another motorist, passenger, or pedestrian, or attempts or threatens to injure or kill another motorist, passenger, or pedestrian."

One of the goals of the study was to determine when road rage occurs most often. The following table gives the days on which 69 road-rage incidents occurred. In this sample, more incidents occurred on Friday than any other day. This information, of course, is a descriptive statistic and, by itself, does not imply that Friday is the mode for all road-rage incidents.

We can, however, use the data to make inferences about all road-rage incidents. For instance, do the data provide sufficient evidence to conclude that some days are more likely than others for the occurrence of road-rage incidents? At the end of this chapter, you will answer this question by applying the chi-square goodness-of-fit test.

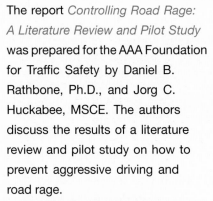

| F  | Sa | W  | M  | Tu | F  | Th | M  |
|----|----|----|----|----|----|----|----|
| Tu | F  | Tu | F  | Su | W  | Th | F  |
| Th | W  | Th | Sa | W  | W  | F  | F  |
| Tu | Su | Tu | Th | W  | Sa | Tu | Th |
| F  | W  | F  | F  | Su | F  | Th | Tu |
| F  | Tu | Tu | Tu | Sa | W  | W  | Sa |
| F  | Sa | Th | W  | F  | Th | F  | M  |
| F  | M  | F  | Su | W  | Th | M  | Tu |
| Sa | Th | F  | Su | W  |    |    |    |

# 12.1 The Chi-Square Distribution

**FIGURE 12.1**
$\chi^2$-curves for df = 5, 10, and 19

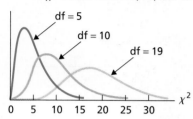

**FIGURE 12.1**
$\chi^2$-curves for df = 5, 10, and 19

The statistical-inference procedures discussed in this chapter rely on a distribution called the *chi-square distribution*. Chi (pronounced "kī") is a Greek letter whose lowercase form is $\chi$.

A variable has a **chi-square distribution** if its distribution has the shape of a special type of right-skewed curve, called a **chi-square ($\chi^2$) curve.** Actually, there are infinitely many chi-square distributions, and we identify the chi-square distribution (and $\chi^2$-curve) in question by its number of degrees of freedom, just as we did for *t*-distributions. Figure 12.1 shows three $\chi^2$-curves and illustrates some basic properties of $\chi^2$-curves.

**Key Fact 12.1**

### Basic Properties of $\chi^2$-Curves

**Property 1:** The total area under a $\chi^2$-curve equals 1.

**Property 2:** A $\chi^2$-curve starts at 0 on the horizontal axis and extends indefinitely to the right, approaching, but never touching, the horizontal axis.

**Property 3:** A $\chi^2$-curve is right skewed.

**Property 4:** As the number of degrees of freedom becomes larger, $\chi^2$-curves look increasingly like normal curves.

## Using the $\chi^2$-Table

Percentages (and probabilities) for a variable that has a chi-square distribution are equal to areas under its associated $\chi^2$-curve. To perform a chi-square test, we need to know how to find the $\chi^2$-value that has a specified area to its right. Table V in Appendix A provides $\chi^2$-values that correspond to several areas for various degrees of freedom.

The $\chi^2$-table (Table V) is similar to the *t*-table (Table IV). The two outside columns of Table V, labeled df, display the number of degrees of freedom. As expected, the symbol $\chi^2_\alpha$ denotes the $\chi^2$-value that has area $\alpha$ to its right under a $\chi^2$-curve. Thus the column headed $\chi^2_{0.05}$, for example, contains $\chi^2$-values that have area 0.05 to their right.

**Example 12.1** | **Finding the $\chi^2$-Value Having a Specified Area to Its Right**

For a $\chi^2$-curve with 12 degrees of freedom, find $\chi^2_{0.025}$; that is, find the $\chi^2$-value that has area 0.025 to its right, as shown in Fig. 12.2(a).

**FIGURE 12.2**
Finding the $\chi^2$-value that has area 0.025 to its right

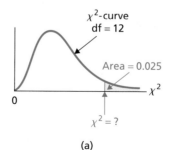

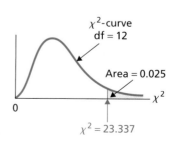

(a)

(b)

Exercise 12.5
on page 543

**Solution** To find this $\chi^2$-value, we use Table V. The number of degrees of freedom is 12, so we first go down the outside columns, labeled df, to "12." Then, going across that row to the column labeled $\chi^2_{0.025}$, we reach 23.337. This number is the $\chi^2$-value having area 0.025 to its right, as shown in Fig. 12.2(b). In other words, for a $\chi^2$-curve with df = 12, $\chi^2_{0.025} = 23.337$.

• • •

## Exercises 12.1

### Understanding the Concepts and Skills

**12.1** What is meant by saying that a variable has a chi-square distribution?

**12.2** How do you identify different chi-square distributions?

**12.3** Consider two $\chi^2$-curves with degrees of freedom 12 and 20, respectively. Which one more closely resembles a normal curve? Explain your answer.

**12.4** The $t$-table has entries for areas of 0.10, 0.05, 0.025, 0.01, and 0.005. In contrast, the $\chi^2$-table has entries for those areas and for 0.995, 0.99, 0.975, 0.95, and 0.90. Explain why the $t$-values corresponding to these additional areas can be obtained from the existing $t$-table, but must be provided explicitly in the $\chi^2$-table.

*In Exercises 12.5–12.8, use Table V to determine the required $\chi^2$-values. Illustrate your work graphically.*

**12.5** For a $\chi^2$-curve with 19 degrees of freedom, find the $\chi^2$-value that has area
a. 0.025 to its right.          b. 0.95 to its right.

**12.6** For a $\chi^2$-curve with 22 degrees of freedom, find the $\chi^2$-value that has area
a. 0.01 to its right.          b. 0.995 to its right.

**12.7** For a $\chi^2$-curve with df = 10, determine
a. $\chi^2_{0.05}$.          b. $\chi^2_{0.975}$.

**12.8** For a $\chi^2$-curve with df = 4, determine
a. $\chi^2_{0.005}$.          b. $\chi^2_{0.99}$.

### Extending the Concepts and Skills

**12.9** Explain how you would use Table V to find the $\chi^2$-value that has area 0.05 to its left. Obtain this $\chi^2$-value for a $\chi^2$-curve with df = 26.

**12.10** Explain how you would use Table V to find the two $\chi^2$-values that divide the area under a $\chi^2$-curve into a middle 0.95 area and two outside 0.025 areas. Find these two $\chi^2$-values for a $\chi^2$-curve with df = 14.

## 12.2 Chi-Square Goodness-of-Fit Test

Our first chi-square procedure is called the **chi-square goodness-of-fit test.** We can use this procedure to perform a hypothesis test about the distribution of a qualitative (categorical) variable or a discrete quantitative variable that has only finitely many possible values.[†] We introduce and explain the reasoning behind the chi-square goodness-of-fit test next.

**Example 12.2** | **Introduces the Chi-Square Goodness-of-Fit Test**

*Violent Crimes* The U.S. Federal Bureau of Investigation (FBI) compiles data on crimes and crime rates and publishes the information in *Crime in the United States*. A violent crime is classified by the FBI as murder, forcible rape, robbery,

---

[†]Actually, the chi-square goodness-of-fit test can be applied to any variable whose possible values have been grouped into a finite number of categories.

**TABLE 12.1**

Distribution of violent crimes
in the United States, 2000

| Type of violent crime | Relative frequency |
|---|---|
| Murder | 0.011 |
| Forcible rape | 0.063 |
| Robbery | 0.286 |
| Agg. assault | 0.640 |
| | 1.000 |

**TABLE 12.2**

Sample results for 500 randomly
selected violent-crime
reports from last year

| Type of violent crime | Frequency |
|---|---|
| Murder | 3 |
| Forcible rape | 37 |
| Robbery | 154 |
| Agg. assault | 306 |
| | 500 |

or aggravated assault. Table 12.1 gives a relative-frequency distribution for (reported) violent crimes in 2000. For instance, in 2000, 28.6% of violent crimes were robberies.

A random sample of 500 violent-crime reports from last year yielded the frequency distribution shown in Table 12.2. Suppose that we want to use the data in Tables 12.1 and 12.2 to decide whether last year's distribution of violent crimes has changed from the 2000 distribution.

**a.** Formulate the problem statistically by posing it as a hypothesis test.

**b.** Explain the basic idea for carrying out the hypothesis test.

**c.** Discuss the details for making a decision concerning the hypothesis test.

**Solution**

**a.** The population is last year's (reported) violent crimes. The variable is "type of violent crime," and its possible values are murder, forcible rape, robbery, and aggravated assault. We want to perform the hypothesis test

$H_0$: Last year's violent-crime distribution is the same as the 2000 distribution.

$H_a$: Last year's violent-crime distribution is different from the 2000 distribution.

**b.** The idea behind the chi-square goodness-of-fit test is to compare the **observed frequencies** in the second column of Table 12.2 to the frequencies that would be expected—the **expected frequencies**—if last year's violent-crime distribution is the same as the 2000 distribution. If the observed and expected frequencies match fairly well, (i.e., each observed frequency is roughly equal to its corresponding expected frequency), we do not reject the null hypothesis; otherwise, we reject the null hypothesis.

**c.** To formulate a precise procedure for carrying out the hypothesis test, we need to answer two questions:

1. What frequencies should we expect from a random sample of 500 violent-crime reports from last year if last year's violent-crime distribution is the same as the 2000 distribution?
2. How do we decide whether the observed and expected frequencies match fairly well?

The first question is easy to answer, which we illustrate with robberies. If last year's violent-crime distribution is the same as the 2000 distribution, then, according to Table 12.1, 28.6% of last year's violent crimes would have been robberies. Therefore, in a random sample of 500 violent-crime reports from last year, we would expect about 28.6% of the 500, or 143, to be robberies.

In general, we compute each expected frequency, denoted $E$, by using the formula $E = np$, where $n$ is the sample size and $p$ is the appropriate relative frequency from the second column of Table 12.1. Calculations of the expected frequencies for all four types of violent crime are shown in Table 12.3.

The third column of Table 12.3 answers the first question. It gives the frequencies that we would expect if last year's violent-crime distribution is the same as the 2000 distribution.

**TABLE 12.3**
Expected frequencies if last year's violent-crime distribution is the same as the 2000 distribution

| Type of violent crime | Relative frequency $p$ | Expected frequency $np = E$ |
|---|---|---|
| Murder | 0.011 | $500 \cdot 0.011 = \phantom{0}5.5$ |
| Forcible rape | 0.063 | $500 \cdot 0.063 = \phantom{0}31.5$ |
| Robbery | 0.286 | $500 \cdot 0.286 = 143.0$ |
| Agg. assault | 0.640 | $500 \cdot 0.640 = 320.0$ |

The second question—whether the observed and expected frequencies match fairly well—is harder to answer. We need to calculate a number that measures the goodness of fit.

In Table 12.4, the second column repeats the observed frequencies from the second column of Table 12.2. The third column of Table 12.4 repeats the expected frequencies from the third column of Table 12.3.

**TABLE 12.4**
Calculating the goodness of fit

| Type of violent crime $x$ | Observed frequency $O$ | Expected frequency $E$ | Difference $O - E$ | Square of difference $(O - E)^2$ | Chi-square subtotal $(O - E)^2/E$ |
|---|---|---|---|---|---|
| Murder | 3 | 5.5 | −2.5 | 6.25 | 1.136 |
| Forcible rape | 37 | 31.5 | 5.5 | 30.25 | 0.960 |
| Robbery | 154 | 143.0 | 11.0 | 121.00 | 0.846 |
| Agg. assault | 306 | 320.0 | −14.0 | 196.00 | 0.613 |
|  | 500 | 500.0 | 0 |  | 3.555 |

To measure the goodness of fit of the observed and expected frequencies, we look at the differences, $O - E$, shown in the fourth column of Table 12.4. Summing these differences to obtain a measure of goodness of fit isn't very useful because the sum is 0. Instead, we square each difference (shown in the fifth column) and then divide by the corresponding expected frequency. Doing so gives the values $(O - E)^2/E$, called **chi-square subtotals,** shown in the sixth column. The sum of the chi-square subtotals,

$$\Sigma(O - E)^2/E = 3.555,$$

is the statistic used to measure the goodness of fit of the observed and expected frequencies.[†]

If the null hypothesis is true, the observed and expected frequencies should be roughly equal, resulting in a small value of the test statistic, $\Sigma(O - E)^2/E$. In other words, large values of $\Sigma(O - E)^2/E$ provide evidence against the null hypothesis.

As we have seen, $\Sigma(O - E)^2/E = 3.555$. Can this value be reasonably attributed to sampling error, or is it large enough to suggest that the null hypothesis is false? To answer this question, we need to know the distribution of the test statistic $\Sigma(O - E)^2/E$.

• • •

---

[†]Using subscripts alone or both subscripts and indices, we would write $\Sigma(O - E)^2/E$ as

$$\Sigma(O_i - E_i)^2/E_i \quad \text{or} \quad \sum_{i=1}^{k}(O_i - E_i)^2/E_i$$

where $k$ denotes the number of possible values for the variable, in this case, four ($k = 4$). However, because no confusion can arise, we use the simpler notation without subscripts or indices.

**Key Fact 12.2**

## Distribution of the $\chi^2$-Statistic for a Goodness-of-Fit Test

For a chi-square goodness-of-fit test, the test statistic

$$\chi^2 = \Sigma(O - E)^2/E$$

has approximately a chi-square distribution if the null hypothesis is true. The number of degrees of freedom is 1 less than the number of possible values for the variable under consideration.

**What Does It Mean?**

To obtain a chi-square subtotal, square the difference between an observed and expected frequency and divide the result by the expected frequency. Adding the chi-square subtotals gives the $\chi^2$-statistic, which has approximately a chi-square distribution.

### Procedure for the Chi-Square Goodness-of-Fit Test

In light of Key Fact 12.2, we now present, in Procedure 12.1 on the next page, a step-by-step method for conducting a chi-square goodness-of-fit test. Because the null hypothesis is rejected only when the test statistic is too large, a chi-square goodness-of-fit test is always right tailed.

**Example 12.3**

### The Chi-Square Goodness-of-Fit Test

*Violent Crimes* We can now complete the hypothesis test introduced in Example 12.2. Table 12.5 repeats the relative-frequency distribution for violent crimes in the United States in 2000.

A random sample of 500 violent-crime reports from last year yielded the frequency distribution shown in Table 12.6. At the 5% significance level, do the data provide sufficient evidence to conclude that last year's violent-crime distribution is different from the 2000 distribution?

**TABLE 12.5**

Distribution of violent crimes in the United States, 2000

| Type of violent crime | Relative frequency |
|---|---|
| Murder | 0.011 |
| Forcible rape | 0.063 |
| Robbery | 0.286 |
| Agg. assault | 0.640 |

**Solution** We apply Procedure 12.1.

**STEP 1 State the null and alternative hypotheses.**

The null and alternative hypotheses are

$H_0$: Last year's violent-crime distribution is the same as the 2000 distribution.

$H_a$: Last year's violent-crime distribution is different from the 2000 distribution.

**STEP 2 Calculate the expected frequency for each possible value of the variable by using the formula $E = np$, where $n$ is the sample size and $p$ is the relative frequency given for the value in the null hypothesis.**

We have $n = 500$ and the values for $p$ are in the second column of Table 12.5. We already calculated the expected frequencies in Table 12.3 on page 545.

**TABLE 12.6**

Sample results for 500 randomly selected violent-crime reports from last year

| Type of violent crime | Observed frequency |
|---|---|
| Murder | 3 |
| Forcible rape | 37 |
| Robbery | 154 |
| Agg. assault | 306 |

**STEP 3 Determine whether the expected frequencies satisfy Assumptions 1 and 2.**

1. Are all expected frequencies 1 or greater? Yes, as shown in Table 12.3.
2. Are at most 20% of the expected frequencies less than 5? Yes, in fact, none of the expected frequencies are less than 5, as shown in Table 12.3.

**STEP 4 Decide on the significance level, $\alpha$.**

We are to perform the test at the 5% significance level, so $\alpha = 0.05$.

**Procedure 12.1    Chi-Square Goodness-of-Fit Test**

*Purpose* To perform a hypothesis test for the distribution of a variable

*Assumptions*[†]
1. All expected frequencies are 1 or greater
2. At most 20% of the expected frequencies are less than 5
3. Simple random sample

**STEP 1**  The null and alternative hypotheses are

$H_0$: The variable has the specified distribution.

$H_a$: The variable does not have the specified distribution.

**STEP 2**  Calculate the expected frequency for each possible value of the variable by using the formula $E = np$, where $n$ is the sample size and $p$ is the relative frequency (or probability) given for the value in the null hypothesis.

**STEP 3**  Determine whether the expected frequencies satisfy Assumptions 1 and 2. If they do not, this procedure should not be used.

**STEP 4**  Decide on the significance level, $\alpha$.

**STEP 5**  Compute the value of the test statistic

$$\chi^2 = \Sigma(O - E)^2/E,$$

where $O$ and $E$ represent observed and expected frequencies, respectively. Denote the value of the test statistic $\chi_0^2$.

| CRITICAL-VALUE APPROACH | or | P-VALUE APPROACH |
|---|---|---|

**STEP 6**  The critical value is $\chi_\alpha^2$ with df $= k - 1$, where $k$ is the number of possible values for the variable. Use Table V to find the critical value.

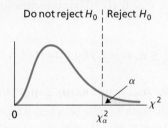

Do not reject $H_0$ ┊ Reject $H_0$

**STEP 7**  If the value of the test statistic falls in the rejection region, reject $H_0$; otherwise, do not reject $H_0$.

**STEP 6**  The $\chi^2$-statistic has df $= k - 1$, where $k$ is the number of possible values for the variable. Use Table V to estimate the $P$-value, or obtain it exactly by using technology.

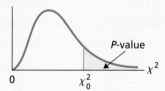

P-value

**STEP 7**  If $P \leq \alpha$, reject $H_0$; otherwise, do not reject $H_0$.

**STEP 8**  Interpret the results of the hypothesis test.

---

[†]Regarding Assumptions 1 and 2, in many texts the rule given is that all expected frequencies be 5 or greater. Research by the noted statistician W. G. Cochran shows that the "rule of 5" is too restrictive.

**STEP 5** Compute the value of the test statistic

$$\chi^2 = \Sigma(O - E)^2/E,$$

where $O$ and $E$ represent observed and expected frequencies, respectively.

We already calculated the value of the test statistic in Table 12.4 on page 545:

$$\chi^2 = \Sigma(O - E)^2/E = 3.555.$$

| CRITICAL-VALUE APPROACH | or | P-VALUE APPROACH |
|---|---|---|

**STEP 6** The critical value is $\chi_\alpha^2$ with df $= k - 1$, where $k$ is the number of possible values for the variable. Use Table V to find the critical value.

From Step 4, $\alpha = 0.05$. The variable is "type of violent crime." There are four types of violent crime, so $k = 4$. In Table V, we find that, for df $= k - 1 = 4 - 1 = 3$, $\chi_{0.05}^2 = 7.815$, as shown in Fig. 12.3A.

**FIGURE 12.3A**

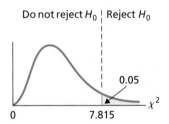

**STEP 7** If the value of the test statistic falls in the rejection region, reject $H_0$; otherwise, do not reject $H_0$.

From Step 5, the value of the test statistic is $\chi^2 = 3.555$. Because it does not fall in the rejection region, as shown in Fig. 12.3A, we do not reject $H_0$. The test results are not statistically significant at the 5% level.

**STEP 6** The $\chi^2$-statistic has df $= k - 1$, where $k$ is the number of possible values for the variable. Use Table V to estimate the $P$-value, or obtain it exactly by using technology.

From Step 5, the value of the test statistic is $\chi^2 = 3.555$. The test is right tailed, so the $P$-value is the probability of observing a value of $\chi^2$ of 3.555 or greater if the null hypothesis is true. That probability equals the shaded area in Fig. 12.3B.

**FIGURE 12.3B**

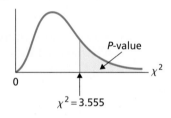

The variable is "type of violent crime." Because there are four types of violent crime, $k = 4$. Referring to Fig. 12.3B and to Table V with df $= k - 1 = 4 - 1 = 3$, we find that $P > 0.10$. (Using technology, we obtain $P = 0.314$.)

**STEP 7** If $P \leq \alpha$, reject $H_0$; otherwise, do not reject $H_0$.

From Step 6, $P > 0.10$. Because the $P$-value exceeds the specified significance level of 0.05, we do not reject $H_0$. The test results are not statistically significant at the 5% level and (see Table 9.10 on page 414) provide essentially no evidence against the null hypothesis.

**STEP 8** Interpret the results of the hypothesis test.

**Interpretation** At the 5% significance level, the data do not provide sufficient evidence to conclude that last year's violent-crime distribution differs from the 2000 distribution.

*You try it!*

Exercise 12.27
on page 552

• • •

## The Technology Center

Some statistical technologies have programs that automatically perform a chi-square goodness-of-fit test, but others do not. In this subsection, we present output and step-by-step instructions for such programs.

*Note to TI-83/84 Plus users:* At the time of this writing, the TI-83/84 Plus does not have a built-in program for a chi-square goodness-of-fit test. However, a TI program (macro) for this procedure, called chigft, is supplied in the TI Programs folder on the WeissStats CD. To download that program to your calculator, right-click the chigft file icon and then select **Send To TI Device….**

**Example 12.4**    **Using Technology to Perform a Goodness-of-Fit Test**

*Violent Crimes*    Table 12.5 on page 546 shows a relative-frequency distribution for violent crimes in the United States in 2000, and Table 12.6 on page 546 gives a frequency distribution for a random sample of 500 violent-crime reports from last year. Use Minitab, Excel, or the TI-83/84 Plus to decide, at the 5% significance level, whether the data provide sufficient evidence to conclude that last year's violent-crime distribution is different from the 2000 distribution.

**Solution**    We want to perform the hypothesis test

$H_0$: Last year's violent-crime distribution is the same as the 2000 distribution

$H_a$: Last year's violent-crime distribution is different from the 2000 distribution

at the 5% significance level.

We applied the chi-square goodness-of-fit programs to the data, resulting in Output 12.1 on the following page. Steps for generating that output are presented in Instructions 12.1 on page 551.

As shown in Output 12.1, the *P*-value for the hypothesis test is 0.314. Because the *P*-value exceeds the specified significance level of 0.05, we do not reject $H_0$. At the 5% significance level, the data do not provide sufficient evidence to conclude that last year's violent-crime distribution differs from the 2000 distribution.

• • •

**OUTPUT 12.1** Goodness-of-fit for violent-crime data

**MINITAB**

## Chi-Square Goodness-of-Fit Test for Observed Counts in Variable: O

Using category names in CRIME

| Category | Observed | Test Proportion | Expected | Contribution to Chi-Sq |
|---|---|---|---|---|
| Agg. assault | 306 | 0.640 | 320.0 | 0.61250 |
| Forcible rape | 37 | 0.063 | 31.5 | 0.96032 |
| Murder | 3 | 0.011 | 5.5 | 1.13636 |
| Robbery | 154 | 0.286 | 143.0 | 0.84615 |

| N | DF | Chi-Sq | P-Value |
|---|---|---|---|
| 500 | 3 | 3.55533 | 0.314 |

**EXCEL**

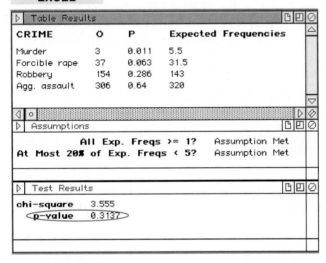

```
Table Results
CRIME        O      P       Expected Frequencies

Murder       3    0.011    5.5
Forcible rape 37   0.063    31.5
Robbery      154  0.286    143
Agg. assault 306  0.64     320

Assumptions

        All Exp. Freqs >= 1?    Assumption Met
At Most 20% of Exp. Freqs < 5?  Assumption Met

Test Results

chi-square   3.555
p-value      0.3137
```

**TI-83/84 PLUS**

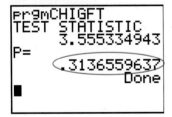

```
prgmCHIGFT
TEST STATISTIC
        3.555334943
P=
        .3136559637
            Done
```

**INSTRUCTIONS 12.1**    Steps for generating Output 12.1

| MINITAB | EXCEL | TI-83/84 PLUS |
|---|---|---|
| 1 Store the violent-crime types, relative frequencies, and observed frequencies in columns named CRIME, P, and O, respectively<br>2 Choose **Stat ▶ Tables ▶ Chi-Square Goodness-of-Fit Test...**<br>3 Select the **Observed counts** option button<br>4 Specify O in the **Observed counts** text box<br>5 Specify CRIME in the **Category names** text box<br>6 Select the **Specific proportions** option button from the **Test** list<br>7 Specify P in the **Specific proportions** text box<br>8 Click **OK** | 1 Store the violent-crime types, relative frequencies, and observed frequencies in ranges named CRIME, P, and O, respectively<br>2 Choose **DDXL ▶ Tables**<br>3 Select **Goodness of Fit** from the **Function type** drop-down list box<br>4 Specify CRIME in the **Category Names** text box<br>5 Specify O in the **Observed Counts** text box<br>6 Specify P in the **Test Distribution** text box<br>7 Click **OK** | 1 Store the observed frequencies and relative frequencies in Lists 1 and 2, respectively<br>2 Press **PRGM**<br>3 Arrow down to CHIGFT and press **ENTER** twice |

▮▮▮

# Exercises 12.2

## Understanding the Concepts and Skills

**12.11** Why is the phrase "goodness of fit" used to describe the type of hypothesis test considered in this section?

**12.12** Are the observed frequencies variables? What about the expected frequencies? Explain your answers.

*In each of Exercises 12.13–12.18, we have given the relative frequencies for the null hypothesis of a chi-square goodness-of-fit test and the sample size. In each case, decide whether Assumptions 1 and 2 for using that test are satisfied.*

**12.13** Sample size: $n = 100$.
Relative frequencies: 0.65, 0.30, 0.05.

**12.14** Sample size: $n = 50$.
Relative frequencies: 0.65, 0.30, 0.05.

**12.15** Sample size: $n = 50$.
Relative frequencies: 0.20, 0.20, 0.25, 0.30, 0.05.

**12.16** Sample size: $n = 50$.
Relative frequencies: 0.22, 0.21, 0.25, 0.30, 0.02.

**12.17** Sample size: $n = 50$.
Relative frequencies: 0.22, 0.22, 0.25, 0.30, 0.01.

**12.18** Sample size: $n = 100$.
Relative frequencies: 0.44, 0.25, 0.30, 0.01.

**12.19 Primary Heating Fuel.** According to *Current Housing Reports*, published by the U.S. Census Bureau, the primary heating fuel for all occupied housing units is distributed as follows.

| Primary heating fuel | Percentage |
|---|---|
| Utility gas | 51.5 |
| Fuel oil, kerosene | 9.8 |
| Electricity | 30.7 |
| Bottled, tank, or LPG | 5.7 |
| Wood and other fuel | 1.9 |
| None | 0.4 |

Suppose that you want to determine whether the distribution of primary heating fuel for occupied housing units built after 2000 differs from that of all occupied housing units. To decide, you take a random sample of housing units built after 2000 and obtain a frequency distribution of their primary heating fuel.

a. Identify the population and variable under consideration here.

b. For each of the following sample sizes, determine whether conducting a chi-square goodness-of-fit test is appropriate and explain your answers: 200; 250; 300.

c. Strictly speaking, what is the smallest sample size for which conducting a chi-square goodness-of-fit test is appropriate?

*In each of Exercises 12.20–12.25, we have provided a distribution and the observed frequencies of the values of a variable from a simple random sample of a population. In each case, use the chi-square goodness-of-fit test to decide, at the specified significance level, whether the distribution of the variable differs from the given distribution.*

**12.20** Distribution: 0.2, 0.4, 0.3, 0.1;
Observed frequencies: 39, 78, 64, 19;
Significance level = 0.05

**12.21** Distribution: 0.2, 0.4, 0.3, 0.1;
Observed frequencies: 85, 215, 130, 70;
Significance level = 0.05

**12.22** Distribution: 0.2, 0.1, 0.1, 0.3, 0.3;
Observed frequencies: 29, 13, 5, 25, 28;
Significance level = 0.10

**12.23** Distribution: 0.2, 0.1, 0.1, 0.3, 0.3;
Observed frequencies: 9, 7, 1, 12, 21;
Significance level = 0.10

**12.24** Distribution: 0.5, 0.3, 0.2;
Observed frequencies: 45, 39, 16;
Significance level = 0.01

**12.25** Distribution: 0.5, 0.3, 0.2;
Observed frequencies: 147, 115, 88;
Significance level = 0.01

*In each of Exercises 12.26–12.31, apply the chi-square goodness-of-fit test, using either the critical-value approach or the P-value approach, to perform the required hypothesis test.*

**12.26 Population by Region.** According to the U.S. Census Bureau publication *Demographic Profiles*, a relative-frequency distribution of the U.S. resident population by region in 2000 was as follows.

| Region | Northeast | Midwest | South | West |
|---|---|---|---|---|
| Rel. freq. | 0.190 | 0.229 | 0.356 | 0.225 |

A simple random sample of this year's U.S. residents gave the following frequency distribution.

| Region | Northeast | Midwest | South | West |
|---|---|---|---|---|
| Frequency | 50 | 43 | 111 | 46 |

At the 5% significance level, do the data provide sufficient evidence to conclude that this year's resident population distribution by region has changed from the 2000 distribution?

**12.27 Freshmen Politics.** The Higher Education Research Institute of the University of California, Los Angeles, publishes information on characteristics of incoming college freshmen in *The American Freshman*. In 2000, 27.7% of incoming freshmen characterized their political views as liberal, 51.9% as moderate, and 20.4% as conservative. For this year, a random sample of 500 incoming college freshmen yielded the following frequency distribution for political views.

| Political view | Frequency |
|---|---|
| Liberal | 160 |
| Moderate | 246 |
| Conservative | 94 |

**a.** Identify the population and variable under consideration here.
**b.** At the 5% significance level, do the data provide sufficient evidence to conclude that this year's distribution of political views for incoming college freshmen has changed from the 2000 distribution?
**c.** Repeat part (b), using a significance level of 10%.

**12.28 Car Sales.** The American Automobile Manufacturers Association compiles data on U.S. car sales by type of car. Following is the 1990 distribution, as reported in the *World Almanac*.

| Type of car | Small | Midsize | Large | Luxury |
|---|---|---|---|---|
| Percentage | 32.8 | 44.8 | 9.4 | 13.0 |

A random sample of last year's U.S. car sales yielded the following data.

| Type of car | Small | Midsize | Large | Luxury |
|---|---|---|---|---|
| Frequency | 133 | 249 | 47 | 71 |

**a.** Identify the population and variable under consideration here.
**b.** At the 5% significance level, do the data provide sufficient evidence to conclude that last year's type-of-car distribution for U.S. car sales differs from the 1990 distribution?

**12.29 M&M Colors.** Observing that the proportion of blue M&Ms in his bowl of candy appeared to be less than that of the other colors, Ronald D. Fricker, Jr., decided to compare the color distribution in randomly chosen bags of M&Ms to the theoretical distribution reported by M&M/MARS consumer affairs. Fricker published his findings in the article "The Mysterious Case of the Blue M&Ms"

(*Chance*, Vol. 9(4), pp. 19–22). The following is the theoretical distribution.

| Color | Percentage |
|---|---|
| Brown | 30 |
| Yellow | 20 |
| Red | 20 |
| Orange | 10 |
| Green | 10 |
| Blue | 10 |

For his study, Fricker bought three bags of M&Ms from local stores and counted the number of each color. The average number of each color in the three bags was distributed as follows.

| Color | Frequency |
|---|---|
| Brown | 152 |
| Yellow | 114 |
| Red | 106 |
| Orange | 51 |
| Green | 43 |
| Blue | 43 |

Do the data provide sufficient evidence to conclude that the color distribution of M&Ms differs from that reported by M&M/MARS consumer affairs? Use $\alpha = 0.05$.

**12.30 An Edge in Roulette?** An American roulette wheel contains 18 red numbers, 18 black numbers, and 2 green numbers. The following table shows the frequency with which the ball landed on each color in 200 trials.

| Number | Red | Black | Green |
|---|---|---|---|
| Frequency | 88 | 102 | 10 |

At the 5% significance level, do the data suggest that the wheel is out of balance?

**12.31 Loaded Die?** A gambler thinks a die may be loaded, that is, that the six numbers are not equally likely. To test his suspicion, he rolled the die 150 times and obtained the data shown in the following table.

| Number | 1 | 2 | 3 | 4 | 5 | 6 |
|---|---|---|---|---|---|---|
| Frequency | 23 | 26 | 23 | 21 | 31 | 26 |

Do the data provide sufficient evidence to conclude that the die is loaded? Perform the hypothesis test at the 0.05 level of significance.

*In each of Exercises* **12.32–12.35**, *use the technology of your choice to conduct the required chi-square goodness-of-fit test.*

**12.32 Japanese Exports.** The Japan Automobile Manufacturer's Association provides data on exported vehicles in *Japan's Motor Vehicle Statistics, Total Exports by Year*. In 2000, cars, trucks, and buses constituted 85.2%, 13.9%, and 0.9% of vehicle exports, respectively. This year, a simple random sample of 750 vehicle exports yielded 628 cars, 111 trucks, and 11 buses. At the 10% significance level, do the data provide sufficient evidence to conclude that this year's distribution for exported vehicles differs from the 2000 distribution?

**12.33 World Series.** For most of its history, the World Series of baseball has been a best-of-seven October event. If two teams are evenly matched, the probabilities of the series lasting 4, 5, 6, or 7 games are as given in the second column of the following table. From the *Information Please Almanac*, we found that, through the 2004 World Series, the actual numbers of times that the series lasted 4, 5, 6, or 7 games are as shown in the third column of the table.

| Games | Probability | Actual |
|---|---|---|
| 4 | 0.1250 | 17 |
| 5 | 0.2500 | 23 |
| 6 | 0.3125 | 22 |
| 7 | 0.3125 | 35 |

**a.** At the 5% significance level, do the data provide sufficient evidence to conclude that World Series' teams are not evenly matched?

**b.** Discuss the appropriateness of using the chi-square goodness-of-fit test here.

**12.34 Credit Card Marketing.** According to market research by Brittain Associates, published in an issue of *American Demographics*, the income distribution of adult Internet users closely mirrors that of credit card applicants. That is exactly what many major credit card issuers want to hear because they hope to replace direct mail marketing with more efficient Web-based marketing. Following is an income distribution for credit card applicants.

| Income ($1000) | < 30 | 30 < 50 | 50 < 70 | 70+ |
|---|---|---|---|---|
| Percentage | 28 | 33 | 21 | 18 |

A random sample of 109 adult Internet users yielded the following income distribution.

| Income ($1000) | < 30 | 30 < 50 | 50 < 70 | 70+ |
|---|---|---|---|---|
| Frequency | 25 | 29 | 26 | 29 |

a. Decide, at the 5% significance level, whether the data do not support the claim by Brittain Associates.
b. Repeat part (a) at the 10% significance level.

**12.35 Migrating Women.** In the article "Waves of Rural Brides: Female Marriage Migration in China" (*Annals of the Association of American Geographers*, Vol. 88(2), pp. 227–251), C. Fan and Y. Huang report on the reasons that women in China migrate within the country to new places of residence. The percentages for reasons given by 15–29-year-old women for migrating within the same province are presented in the second column of the following table. For a random sample of 500 women in the same age group who migrated to a different province, the number giving each of the reasons is recorded in the third column of the table.

| Reason | Intraprovincial migrants (%) | Interprovincial migrants |
|---|---|---|
| Job transfer | 4.8 | 20 |
| Job assignment | 7.2 | 23 |
| Industry/business | 17.8 | 108 |
| Study/training | 16.9 | 47 |
| Help from friends/ relatives | 6.2 | 43 |
| Joining family | 6.8 | 45 |
| Marriage | 36.8 | 205 |
| Other | 3.5 | 9 |

Decide, at the 1% significance level, whether the data provide sufficient evidence to conclude that the distribution of reasons for migration between provinces is different from that for migration within provinces.

## Extending the Concepts and Skills

**12.36** Table 12.4 on page 545 showed the calculated sums of the observed frequencies, the expected frequencies, and their differences. Strictly speaking, those sums are not needed. However, they serve as a check for computational errors.

a. In general, what common value should the sum of the observed frequencies and the sum of the expected frequencies equal? Explain your answer.
b. Fill in the blank. The sum of the differences between each observed and expected frequency should equal _____.
c. Suppose that you are conducting a chi-square goodness-of-fit test. If the sum of the expected frequencies does not equal the sample size, what do you conclude?
d. Suppose that you are conducting a chi-square goodness-of-fit test. If the sum of the expected frequencies equals the sample size, can you conclude that you made no error in calculating the expected frequencies? Explain your answer.

**12.37** The chi-square goodness-of-fit test provides a method for performing a hypothesis test about the distribution of a variable that has $k$ possible values. If the number of possible values is 2, that is, $k = 2$, the chi-square goodness-of-fit test is equivalent to a procedure that you studied earlier.

a. Which procedure is that? Explain your answer.
b. Suppose that you want to perform a hypothesis test to decide whether the proportion of a population that has a specified attribute is different from $p_0$. Discuss the method for performing such a test if you use (1) the one-proportion z-test (page 515) or (2) the chi-square goodness-of-fit test.

## 12.3 Contingency Tables; Association

The next chi-square procedure we present is the *chi-square independence test*, which is used to decide, based on sample data, whether two variables of a population are statistically related. Before we can do that, however, we need to discuss two prerequisite concepts: *contingency tables* and *association*.

### Contingency Tables

In Section 2.2, you learned how to group data from one variable into a frequency distribution. Data from one variable of a population are called **univariate data.**

Now, we show how to simultaneously group data from two variables into a frequency distribution. Data from two variables of a population are called **bivariate data,** and a frequency distribution for bivariate data is called a **contingency table** or **two-way table.**

**Example 12.5** | **Introducing Contingency Tables**

*Political Party and Class Level*  In Example 2.8 on page 50, we considered data on political party affiliation for the students in Professor Weiss's introductory statistics course. These are univariate data from the single variable "political party affiliation."

Now, we simultaneously consider data on political party affiliation and on class level for the students in Professor Weiss's introductory statistics course, as shown in Table 12.7. These are bivariate data from the two variables "political party affiliation" and "class level." Group these bivariate data into a contingency table.

**TABLE 12.7**
Political party affiliation and class level for students in introductory statistics

| Student | Political party | Class level | Student | Political party | Class level |
|---|---|---|---|---|---|
| 1 | Democratic | Freshman | 21 | Democratic | Junior |
| 2 | Other | Junior | 22 | Democratic | Senior |
| 3 | Democratic | Senior | 23 | Republican | Freshman |
| 4 | Other | Sophomore | 24 | Democratic | Sophomore |
| 5 | Democratic | Sophomore | 25 | Democratic | Senior |
| 6 | Republican | Sophomore | 26 | Republican | Sophomore |
| 7 | Republican | Junior | 27 | Republican | Junior |
| 8 | Other | Freshman | 28 | Other | Junior |
| 9 | Other | Sophomore | 29 | Other | Junior |
| 10 | Republican | Sophomore | 30 | Democratic | Sophomore |
| 11 | Republican | Sophomore | 31 | Republican | Sophomore |
| 12 | Republican | Junior | 32 | Democratic | Junior |
| 13 | Republican | Sophomore | 33 | Republican | Junior |
| 14 | Democratic | Junior | 34 | Other | Senior |
| 15 | Republican | Sophomore | 35 | Other | Sophomore |
| 16 | Republican | Senior | 36 | Republican | Freshman |
| 17 | Democratic | Sophomore | 37 | Republican | Freshman |
| 18 | Democratic | Junior | 38 | Republican | Freshman |
| 19 | Other | Senior | 39 | Democratic | Junior |
| 20 | Republican | Sophomore | 40 | Republican | Senior |

**Solution**  A contingency table must accommodate each possible pair of values for the two variables. The contingency table for these two variables has the form shown in Table 12.8. The small boxes inside the rectangle formed by the heavy lines are called **cells,** which hold the frequencies.

**TABLE 12.8**
Preliminary contingency table for political party affiliation and class level

| Party | | Freshman | Sophomore | Junior | Senior | Total |
|---|---|---|---|---|---|---|
| | Democratic | I | IIII | ШΙ | III | |
| | Republican | IIII | ШΙ III | IIII | II | |
| | Other | I | III | III | II | |
| | Total | | | | | |

*(Column group header: Class level)*

To complete the contingency table, we first go through the data in Table 12.7 and place a tally mark in the appropriate cell of Table 12.8 for each student. For instance, the first student is both a Democrat and a freshman, so this calls for a tally mark in the upper left cell of Table 12.8. The results of the tallying procedure are shown in Table 12.8. Replacing the tallies in Table 12.8 by the frequencies (counts of the tallies), we obtain the required contingency table, as shown in Table 12.9.

**TABLE 12.9**
Contingency table for political party affiliation and class level

| | | Class level | | | | |
|---|---|:---:|:---:|:---:|:---:|:---:|
| | | Freshman | Sophomore | Junior | Senior | **Total** |
| | Democratic | 1 | 4 | 5 | 3 | 13 |
| Party | Republican | 4 | 8 | 4 | 2 | 18 |
| | Other | 1 | 3 | 3 | 2 | 9 |
| | **Total** | 6 | 15 | 12 | 7 | 40 |

The upper left cell of Table 12.9 shows that one student in the course is both a Democrat and a freshman. The cell diagonally below and to the right of that cell shows that eight students in the course are both Republicans and sophomores.

According to the first row total, 13 $(1 + 4 + 5 + 3)$ of the students are Democrats. Similarly, the third column total shows that 12 of the students are juniors. The lower right corner gives the total number of students in the course, 40. You can find that total by summing the row totals, the column totals, or the frequencies in the 12 cells.

**You try it!**

Exercise 12.45(a) on page 562

• • •

Grouping bivariate data into a contingency table by hand, as we did in Example 12.5, is a useful teaching tool. In practice, however, computers are almost always used to accomplish such tasks.

## Association Between Variables

Next, we need to discuss the concept of *association* between two variables. We do so for variables that are either categorical or quantitative with only finitely many possible values. Roughly speaking, two variables of a population are associated if knowing the value of one of the variables imparts information about the value of the other variable.

**Example 12.6** | **Introduces Association Between Variables**

*Political Party and Class Level* In Example 12.5, we presented data on political party affiliation and class level for the students in Professor Weiss's introductory statistics course. Consider those students a population of interest.

a. Find the distribution of political party affiliation within each class level.

b. Use the result of part (a) to decide whether the variables "political party affiliation" and "class level" are associated.

c. What would it mean if the variables "political party affiliation" and "class level" were not associated?

**d.** Explain how a *segmented bar graph* represents whether the variables "political party affiliation" and "class level" are associated.

**e.** Discuss another method for deciding whether the variables "political party affiliation" and "class level" are associated.

## Solution

**a.** To obtain the distribution of political party affiliation within each class level, divide each entry in a column of the contingency table in Table 12.9 by its column total. Table 12.10 shows the results.

**TABLE 12.10**

Conditional distributions of political party affiliation by class level

| | | Class level | | | | |
|---|---|---|---|---|---|---|
| | | Freshman | Sophomore | Junior | Senior | **Total** |
| Party | Democratic | 0.167 | 0.267 | 0.417 | 0.429 | 0.325 |
| | Republican | 0.667 | 0.533 | 0.333 | 0.286 | 0.450 |
| | Other | 0.167 | 0.200 | 0.250 | 0.286 | 0.225 |
| | **Total** | 1.000 | 1.000 | 1.000 | 1.000 | 1.000 |

The first column of Table 12.10 gives the distribution of political party affiliation for freshman: 16.7% are Democrats, 66.7% are Republicans, and 16.7% are Other. This distribution is called the **conditional distribution** of the variable "political party affiliation" corresponding to the value "freshman" of the variable "class level"; or, more simply, the conditional distribution of political party affiliation for freshmen.

Similarly, the second, third, and fourth columns give the conditional distributions of political party affiliation for sophomores, juniors, and seniors, respectively. The "Total" column provides the (unconditional) distribution of political party affiliation for the entire population which, in this context, is called the **marginal distribution** of the variable "political party affiliation." This distribution is the same as the one we found in Example 2.8 (Table 2.11 on page 50).

**b.** Table 12.10 reveals that the variables "political party affiliation" and "class level" are associated because knowing the value of the variable "class level" imparts information about the value of the variable "political party affiliation." For instance, as shown in Table 12.10, if we do not know the class level of a student in the course, there is a 32.5% chance that the student is a Democrat. But, if we know that the student is a junior, there is a 41.7% chance that the student is a Democrat.

**c.** If the variables "political party affiliation" and "class level" were not associated, the four conditional distributions of political party affiliation would be the same as each other and as the marginal distribution of political party affiliation; in other words, all five columns of Table 12.10 would be identical.

**d.** A **segmented bar graph** lets us visualize the concept of association. The first four bars of the segmented bar graph in Fig. 12.4 show the conditional distributions of political party affiliation for freshmen, sophomores, juniors, and seniors, respectively, and the fifth bar gives the marginal distribution of political party affiliation. This segmented bar graph is derived from Table 12.10.

**FIGURE 12.4**

Segmented bar graph for the conditional distributions and marginal distribution of political party affiliation

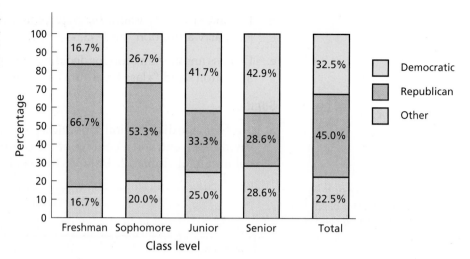

If political party affiliation and class level were not associated, the four bars displaying the conditional distributions of political party affiliation would be the same as each other and as the bar displaying the marginal distribution of political party affiliation; in other words, all five bars in Fig. 12.4 would be identical. That political party affiliation and class level are in fact associated is illustrated by the nonidentical bars.

**You try it!**

Exercise 12.45(b)–(d) on page 562

e. Alternatively, we could decide whether the two variables are associated by obtaining the conditional distribution of class level within each political party affiliation. The conclusion regarding association (or nonassociation) will be the same, regardless of which variable's conditional distributions we examined.

• • •

---

**Definition 12.1**

**What Does It Mean?**

Roughly speaking, two variables of a population are associated if knowing the value of one variable imparts information about the value of the other variable.

### Association Between Variables

We say that two variables of a population are **associated** (or that an **association** exists between the two variables) if the conditional distributions of one variable given the other are not identical.

---

**Note:** Two associated variables are also called **statistically dependent variables.** Similarly, two nonassociated variables are often called **statistically independent variables.**

# The Technology Center

Some statistical technologies have programs that automatically group bivariate data into a contingency table and also obtain conditional and marginal distributions. In this subsection, we present output and step-by-step instructions for such programs. (*Note to TI-83/84 Plus users:* At the time of this writing, the TI-83/84 Plus does not have a built-in program for conducting these analyses.)

**Example 12.7    Using Technology to Group Bivariate Data**

*Political Party and Class Level*    Table 12.7 on page 555 gives the political party affiliations and class levels for the students in Professor Weiss's introductory statistics course. Use Minitab or Excel to group these data into a contingency table.

**Solution**    We applied the bivariate grouping programs to the data, resulting in Output 12.2. Steps for generating that output are presented in Instructions 12.2.

**OUTPUT 12.2**
Contingency table for political-party and class-level data

**MINITAB**

**Tabulated statistics: PARTY, CLASS**

```
Rows: PARTY    Columns: CLASS

              Freshman   Junior   Senior   Sophomore   All

Democratic        1        5        3          4        13
Other             1        3        2          3         9
Republican        4        4        2          8        18
All               6       12        7         15        40

Cell Contents:        Count
```

**EXCEL**

| ▷ Contingency Table: Counts | | | | | |
|---|---|---|---|---|---|
| Rows are levels of **PARTY** | | | | | |
| Columns are levels of **CLASS** | | | | | |
| No Selector | | | | | |
| | Freshman | Junior | Senior | Sophomore | total |
| Democratic | 1 | 5 | 3 | 4 | 13 |
| Other | 1 | 3 | 2 | 3 | 9 |
| Republican | 4 | 4 | 2 | 8 | 18 |
| total | 6 | 12 | 7 | 15 | 40 |

table contents:
Count

**INSTRUCTIONS 12.2**
Steps for generating Output 12.2

**MINITAB**

1 Store the political-party and class-level data from Table 12.7 in columns named PARTY and CLASS, respectively
2 Choose **Stat ➤ Tables ➤ Cross Tabulation and Chi-Square...**
3 Specify PARTY in the **For rows** text box
4 Specify CLASS in the **For columns** text box
5 In the **Display** list, check only the **Counts** check box
6 Click **OK**

**EXCEL**

1 Store the political-party and class-level data from Table 12.7 in ranges named PARTY and CLASS, respectively
2 Choose **DDXL ➤ Tables**
3 Select **Contingency Table** from the **Function type** drop-down list box
4 Specify PARTY in the **1st Categorical Variable** text box
5 Specify CLASS in the **2nd Categorical Variable** text box
6 Click **OK**

Compare Output 12.2 to Table 12.9 on page 556. (*Note to Minitab users:* By using **Column ➤ Value Order...**, available from the **Editor** menu when in the Worksheet window, you can order the rows and columns of the Minitab output to match that in Table 12.9.)

• • •

**Example 12.8**

## Using Technology to Get Conditional and Marginal Distributions

*Political Party and Class Level*   Table 12.7 on page 555 gives the political party affiliations and class levels for the students in Professor Weiss's introductory statistics course. Use Minitab or Excel to determine the conditional distribution of party within each class level and the marginal distribution of party.

**Solution**   We applied the appropriate programs to the data, resulting in Output 12.3. Steps for generating that output are presented in Instructions 12.3.

**OUTPUT 12.3**
Conditional distribution of party within each class level and marginal distribution of party

**MINITAB**

```
Tabulated statistics: PARTY, CLASS

Rows: PARTY    Columns: CLASS

                Freshman   Junior   Senior   Sophomore      All

Democratic        16.67    41.67    42.86       26.67    32.50
Other             16.67    25.00    28.57       20.00    22.50
Republican        66.67    33.33    28.57       53.33    45.00
All              100.00   100.00   100.00      100.00   100.00

Cell Contents:       % of Column
```

**EXCEL**

```
▷ Contingency Table: Column8                              ▢▣⊘
Rows are levels of       PARTY                             △
Columns are levels of    CLASS
No Selector

               Freshman    Junior    Senior    Sophomore    total

Democratic       16.7       41.7      42.9        26.7       32.5

Other            16.7       25        28.6        20         22.5

Republican       66.7       33.3      28.6        53.3       45

total            100        100       100         100        100

table contents:
Percent of Column Total                                   ▽
◁                                                   ▷ ◇
```

Compare Output 12.3 to Table 12.10 on page 557. (*Note to Minitab users:* By using **Column ➤ Value Order…**, available from the **Editor** menu when in the Worksheet window, you can order the rows and columns of the Minitab output to match that in Table 12.10.)

• • •

**INSTRUCTIONS 12.3**
Steps for generating Output 12.3

| MINITAB | EXCEL |
|---|---|
| 1 Store the political-party and class-level data from Table 12.7 in columns named PARTY and CLASS, respectively | 1 Store the political-party and class-level data from Table 12.7 in ranges named PARTY and CLASS, respectively |
| 2 Choose **Stat ➤ Tables ➤ Cross Tabulation and Chi-Square…** | 2 Choose **DDXL ➤ Tables** |
| 3 Specify PARTY in the **For rows** text box | 3 Select **Contingency Table** from the **Function type** drop-down list box |
| 4 Specify CLASS in the **For columns** text box | 4 Specify PARTY in the **1st Categorical Variable** text box |
| 5 In the **Display** list, check only the **Column percents** check box | 5 Specify CLASS in the **2nd Categorical Variable** text box |
| 6 Click **OK** | 6 Click **OK** |
| | 7 Click **Column Percents** |

▮ ▮ ▮

## Exercises 12.3

### Understanding the Concepts and Skills

**12.38** Identify the type of table that is used to group bivariate data.

**12.39** What are the small boxes inside the heavy lines of a contingency table called?

**12.40** Suppose that bivariate data are to be grouped into a contingency table. Determine the number of cells that the contingency table will have if the number of possible values for the two variables are
**a.** two and three. **b.** four and three. **c.** $m$ and $n$.

**12.41** Identify three ways in which the total number of observations of bivariate data can be obtained from the frequencies in a contingency table.

**12.42 Presidential Election.** According to *Dave Leip's Atlas of U.S. Presidential Elections*, in the 2004 presidential election, 50.7% of those voting voted for the Republican candidate (George W. Bush), whereas 61.1% of those voting who lived in Texas did so. For that presidential election, does an association exist between the variables "party of presidential candidate voted for" and "state of residence" for those who voted? Explain your answer.

**12.43 Physicians in Residency.** According to the article "Women Physicians Statistics" (*Journal of the American Medical Association*, Vol. 290, No. 9, pp. 1234–1236), in 2002, 8.2% of male physicians in residency specialized in family practice and 12.2% of female physicians in residency specialized in family practice. Does an association exist between the variables "gender" and "specialty" for physicians in residency in 2002? Explain your answer.

*Table 12.11 at the top of the next page provides data on gender, class level, and college for the students in one section of the course Introduction to Computer Science during one semester at Arizona State University. In the table, we have used the abbreviations BUS for Business, ENG for Engineering and Applied Sciences, and LIB for Liberal Arts and Sciences.*

**TABLE 12.11**

Gender, class level, and college for students in Introduction to Computer Science

| Gender | Class | College | Gender | Class | College |
|--------|-------|---------|--------|-------|---------|
| M | Junior | ENG | F | Soph | BUS |
| M | Soph | ENG | F | Junior | ENG |
| F | Senior | BUS | M | Junior | LIB |
| F | Junior | BUS | F | Junior | BUS |
| M | Junior | ENG | M | Soph | BUS |
| F | Junior | LIB | M | Junior | BUS |
| M | Senior | LIB | M | Soph | ENG |
| M | Soph | ENG | M | Junior | ENG |
| M | Junior | ENG | M | Junior | ENG |
| M | Soph | ENG | M | Soph | LIB |
| F | Soph | BUS | F | Senior | ENG |
| F | Junior | BUS | F | Senior | BUS |
| M | Junior | ENG | | | |

*In Exercises **12.44–12.46**, use the data in Table 12.11.*

**12.44 Gender and Class Level.** Refer to Table 12.11. Consider the variables "gender" and "class level."
a. Group the bivariate data for these two variables into a contingency table.
b. Determine the conditional distribution of gender within each class level and the marginal distribution of gender.
c. Determine the conditional distribution of class level within each gender and the marginal distribution of class level.
d. Does an association exist between the variables "gender" and "class level" for this population? Explain your answer.

**12.45 Gender and College.** Refer to Table 12.11. Consider the variables "gender" and "college."
a. Group the bivariate data for these two variables into a contingency table.
b. Determine the conditional distribution of gender within each college and the marginal distribution of gender.
c. Determine the conditional distribution of college within each gender and the marginal distribution of college.
d. Does an association exist between the variables "gender" and "college" for this population? Explain your answer.

**12.46 Class level and College.** Refer to Table 12.11. Consider the variables "class level" and "college."
a. Group the bivariate data for these two variables into a contingency table.
b. Determine the conditional distribution of class level within each college and the marginal distribution of class level.
c. Determine the conditional distribution of college within each class level and the marginal distribution of college.

d. Does an association exist between the variables "class level" and "college" for this population? Explain your answer.

*Table 12.12 provides hypothetical data on political party affiliation and class level for the students in a night-school course.*

**TABLE 12.12**

Political party affiliation and class level for the students in a night-school course (hypothetical data)

| Party | Class | Party | Class | Party | Class |
|-------|-------|-------|-------|-------|-------|
| Rep | Jun | Rep | Soph | Rep | Jun |
| Dem | Soph | Other | Jun | Rep | Soph |
| Dem | Jun | Dem | Soph | Rep | Soph |
| Other | Jun | Rep | Soph | Rep | Fresh |
| Dem | Jun | Dem | Sen | Rep | Soph |
| Dem | Fresh | Rep | Jun | Rep | Jun |
| Dem | Soph | Dem | Jun | Rep | Sen |
| Dem | Sen | Dem | Jun | Rep | Jun |
| Other | Sen | Rep | Sen | Dem | Soph |
| Dem | Fresh | Rep | Fresh | Rep | Jun |
| Rep | Jun | Rep | Jun | Other | Jun |
| Rep | Jun | Dem | Jun | Dem | Jun |
| Dem | Sen | Rep | Sen | Other | Soph |
| Rep | Jun | Rep | Sen | Rep | Sen |
| Dem | Sen | Rep | Sen | Other | Soph |
| Rep | Jun | Dem | Soph | Rep | Soph |
| Rep | Soph | Other | Fresh | Other | Soph |
| Rep | Fresh | Rep | Soph | Other | Sen |
| Rep | Jun | Other | Jun | Rep | Soph |
| Dem | Soph | Dem | Jun | Dem | Jun |

*In Exercises **12.47** and **12.48**, use the data in Table 12.12.*

**12.47 Party and Class.** Refer to Table 12.12.
a. Group the bivariate data for the two variables into a contingency table.
b. Determine the conditional distribution of political party affiliation within each class level.
c. Are the variables "political party affiliation" and "class level" for this population of night-school students associated? Explain your answer.
d. Without doing any further calculation, determine the marginal distribution of political party affiliation.
e. Without doing further calculation, respond true or false to the following statement and explain your answer: "The conditional distributions of class level within political party affiliations are identical to each other and to the marginal distribution of class level."

**12.48 Party and Class.** Refer to Table 12.12.
a. If you have not done Exercise 12.47, group the bivariate data for the two variables into a contingency table.

**b.** Determine the conditional distribution of class level within each political party affiliation.

**c.** Are the variables "political party affiliation" and "class level" for this population of night-school students associated? Explain your answer.

**d.** Without doing any further calculation, determine the marginal distribution of class level.

**e.** Without doing further calculation, respond true or false to the following statement and explain your answer: "The conditional distributions of political party affiliation within class levels are identical to each other and to the marginal distribution of political party affiliation."

**12.49 AIDS Cases.** According to the Centers for Disease Control (*HIV/AIDS Surveillance Report*, Vol. 16), the number of AIDS cases in the United States in 2004, by gender and race/ethnicity, is as shown in the following contingency table.

| | | Gender | | |
| --- | --- | --- | --- | --- |
| | | Male | Female | Total |
| Race/ethnicity | White | 10,118 | | 11,978 |
| | Black | 13,398 | 7,395 | 20,793 |
| | Hispanic | 6,041 | 1,643 | |
| | Other | 520 | 156 | 676 |
| | Total | | 11,054 | |

**a.** How many cells does this contingency table have?
**b.** Fill in the missing entries.
**c.** What was the total number of AIDS cases in the United States in 2004?
**d.** How many AIDS cases were Hispanics?
**e.** How many AIDS cases were males?
**f.** How many AIDS cases were white females?

**12.50 Vehicles in Use.** As reported by the Motor Vehicle Manufacturers Association of the United States in *Motor Vehicle Facts and Figures*, the number of cars and trucks in use by age are as shown in the following contingency table. Frequencies are in millions.

| | | Type | | |
| --- | --- | --- | --- | --- |
| | | Car | Truck | Total |
| Age (yr) | Under 6 | 46.2 | 27.8 | 74.0 |
| | 6–8 | 26.9 | | 40.0 |
| | 9–11 | 23.3 | 10.7 | |
| | 12 & over | 26.8 | 18.6 | 45.4 |
| | Total | 123.2 | | |

**a.** How many cells does this contingency table have?
**b.** Fill in the missing entries.
**c.** What is the total number of cars and trucks in use?
**d.** How many vehicles are trucks?
**e.** How many vehicles are between 6 and 8 years old?
**f.** How many vehicles are trucks that are between 9 and 11 years old?

**12.51 Education of Prisoners.** In the article "Education and Correctional Populations" (*Bureau of Justice Statistics Special Report*, NCJ 195670), C. Harlow examined the educational attainment of prisoners by type of prison facility. The following contingency table was adapted from Table 1 of the article. Frequencies are in thousands, rounded to the nearest hundred.

| | | Prison facility | | | |
| --- | --- | --- | --- | --- | --- |
| | | State | Federal | Local | Total |
| Educational attainment | 8th grade or less | 149.9 | 10.6 | 66.0 | 226.5 |
| | Some high school | 269.1 | 12.9 | 168.2 | 450.2 |
| | GED | 300.8 | 20.1 | 71.0 | 391.9 |
| | High school diploma | 216.4 | 24.0 | 130.4 | 370.8 |
| | Postsecondary | 95.0 | 14.0 | 51.9 | 160.9 |
| | College grad or more | 25.3 | 7.2 | 16.1 | 48.6 |
| | Total | 1056.5 | 88.8 | 503.6 | 1648.9 |

How many prisoners
**a.** are in state facilities?
**b.** have at least a college education?
**c.** are in federal facilities and have at most an 8th grade education?
**d.** are in federal facilities or have at most an 8th grade education?
**e.** in local facilities have a postsecondary educational attainment?
**f.** with a postsecondary educational attainment are in local facilities?
**g.** are not in federal facilities?

**12.52 U.S. Hospitals.** The American Hospital Association publishes information about U.S. hospitals and nursing homes in *Hospital Statistics*. The following contingency table provides a cross-classification of U.S. hospitals and nursing homes by type of facility and number of beds.

| | Number of beds | | | |
|---|---|---|---|---|
| Facility | 24 or fewer | 25–74 | 75 or more | Total |
| General | 260 | 1586 | 3557 | 5403 |
| Psychiatric | 24 | 242 | 471 | 737 |
| Chronic | 1 | 3 | 22 | 26 |
| Tuberculosis | 0 | 2 | 2 | 4 |
| Other | 25 | 177 | 208 | 410 |
| Total | 310 | 2010 | 4260 | 6580 |

In the following questions, the term *hospital* refers to either a hospital or nursing home.

a. How many hospitals have at least 75 beds?
b. How many hospitals are psychiatric facilities?
c. How many hospitals are psychiatric facilities with at least 75 beds?
d. How many hospitals either are psychiatric facilities or have at least 75 beds?
e. How many general facilities have between 25 and 74 beds?
f. How many hospitals with between 25 and 74 beds are chronic facilities?
g. How many hospitals have more than 24 beds?

**12.53 Farms.** The U.S. Department of Agriculture, National Agricultural Statistics Service, publishes information about U.S. farms in *Census of Agriculture*. A joint frequency distribution for number of farms, by acreage and tenure of operator, is provided in the following contingency table. Frequencies are in thousands.

| | Tenure of operator | | | |
|---|---|---|---|---|
| Acreage | Full owner | Part owner | Tenant | Total |
| Under 50 | | 64 | 41 | |
| 50 < 180 | 487 | 131 | 41 | 659 |
| 180 < 500 | 203 | | | 389 |
| 500 < 1000 | 54 | 91 | 17 | 162 |
| 1000 & over | 46 | 112 | 18 | 176 |
| Total | 1429 | 551 | | |

a. Fill in the six missing entries.
b. How many cells does this contingency table have?
c. How many farms have under 50 acres?
d. How many farms are tenant operated?
e. How many farms are operated by part owners and have between 500 and 1000 acres?

f. How many farms are not full-owner operated?
g. How many tenant-operated farms have 180 acres or more?

**12.54 Housing Units.** The U.S. Census Bureau publishes information about housing units in *American Housing Survey for the United States*. The following table cross-classifies occupied housing units by number of persons and tenure of occupier. The frequencies are in thousands.

| | Tenure | |
|---|---|---|
| Persons | Owner | Renter |
| 1 | 15,283 | 12,656 |
| 2 | 25,255 | 9,132 |
| 3 | 11,881 | 5,168 |
| 4 | 11,669 | 3,822 |
| 5 | 5,017 | 1,771 |
| 6 | 1,628 | 718 |
| 7+ | 975 | 459 |

a. How many occupied housing units are occupied by exactly three persons?
b. How many occupied housing units are owner occupied?
c. How many occupied housing units are rented and have seven or more persons in them?
d. How many occupied housing units are occupied by more than one person?
e. How many occupied housing units are either owner occupied or have only one person in them?

**12.55 AIDS Cases.** Refer to Exercise 12.49. Here, we abbreviate "race/ethnicity" by "race." For AIDS cases in the United States in 2004, answer the following questions:
a. Find and interpret the conditional distribution of gender by race.
b. Find and interpret the marginal distribution of gender.
c. Are the variables "gender" and "race" associated? Explain your answer.
d. What percentage of AIDS cases were females?
e. What percentage of AIDS cases among whites were females?
f. Without doing further calculations, respond true or false to the following statement and explain your answer: "The conditional distributions of race by gender are not identical."
g. Find and interpret the marginal distribution of race and the conditional distributions of race by gender.

**12.56 Vehicles in Use.** Refer to Exercise 12.50. Here, the term "vehicle" refers to either a U.S. car or truck currently in use.

**a.** Determine the conditional distribution of age group for each type of vehicle.

**b.** Determine the marginal distribution of age group for vehicles.

**c.** Are the variables "type" and "age group" for vehicles associated? Explain your answer.

**d.** Find the percentage of vehicles under 6 years old.

**e.** Find the percentage of cars under 6 years old.

**f.** Without doing any further calculations, respond true or false to the following statement and explain your answer: "The conditional distributions of type of vehicle within age groups are not identical."

**g.** Determine and interpret the marginal distribution of type of vehicle and the conditional distributions of type of vehicle within age groups.

**12.57 Education of Prisoners.** Refer to Exercise 12.51.

**a.** Find the conditional distribution of educational attainment within each type of prison facility.

**b.** Does an association exist between educational attainment and type of prison facility for prisoners? Explain your answer.

**c.** Determine the marginal distribution of educational attainment for prisoners.

**d.** Construct a segmented bar graph for the conditional distributions of educational attainment and marginal distribution of educational attainment that you obtained in parts (a) and (c), respectively. Interpret the graph in light of your answer to part (b).

**e.** Without doing any further calculations, respond true or false to the following statement and explain your answer: "The conditional distributions of facility type within educational attainment categories are identical."

**f.** Determine the marginal distribution of facility type and the conditional distributions of facility type within educational attainment categories.

**g.** Find the percentage of prisoners who are in federal facilities.

**h.** Find the percentage of prisoners with at most an 8th grade education who are in federal facilities.

**i.** Find the percentage of prisoners in federal facilities who have at most an 8th grade education.

**12.58 U.S. Hospitals.** Refer to Exercise 12.52.

**a.** Determine the conditional distribution of number of beds within each facility type.

**b.** Does an association exist between facility type and number of beds for U.S. hospitals? Explain your answer.

**c.** Determine the marginal distribution of number of beds for U.S. hospitals.

**d.** Construct a segmented bar graph for the conditional distributions and marginal distribution of number of beds. Interpret the graph in light of your answer to part (b).

**e.** Without doing any further calculations, respond true or false to the following statement and explain your answer: "The conditional distributions of facility type within number-of-beds categories are identical."

**f.** Obtain the marginal distribution of facility type and the conditional distributions of facility type within number-of-beds categories.

**g.** What percentage of hospitals are general facilities?

**h.** What percentage of hospitals that have at least 75 beds are general facilities?

**i.** What percentage of general facilities have at least 75 beds?

## Working With Large Data Sets

*In each of Exercises 12.59–12.61, use the technology of your choice to solve the specified problems.*

**12.59 Governors.** The National Governors Association publishes information on U.S. governors in *Governors' Political Affiliations & Terms of Office*. Based on that document, we obtained the data on region of residence and political party given on the WeissStats CD.

**a.** Group the bivariate data for these two variables into a contingency table.

**b.** Determine the conditional distribution of region within each party and the marginal distribution of region.

**c.** Determine the conditional distribution of party within each region and the marginal distribution of party.

**d.** Are the variables "region" and "party" for U.S. governors associated? Explain your answer.

**12.60 Motorcycle Accidents.** The *Scottish Executive*, Analytical Services Division Transport Statistics, compiles information on motorcycle accidents in Scotland. During one year, data on the number of motorcycle accidents, by day of the week and type of road (built up or non built up), are as presented on the WeissStats CD.

**a.** Group the bivariate data for these two variables into a contingency table.

**b.** Determine the conditional distribution of day of the week within each type-of-road category and the marginal distribution of day of the week.

**c.** Determine the conditional distribution of type of road within each day of the week and the marginal distribution of type of road.

**d.** Does an association exist between the variables "day of the week" and "type of road" for these motorcycle accidents? Explain your answer.

**12.61 Senators.** The U.S. Congress, Joint Committee on Printing, provides information on the composition of Congress in *Congressional Directory*. On the WeissStats CD, we have presented data on party and class for the senators in the 109th Congress.

**a.** Group the bivariate data for these two variables into a contingency table.

b. Determine the conditional distribution of party within each class and the marginal distribution of party.

c. Determine the conditional distribution of class within each party and the marginal distribution of class.

d. Are the variables "party" and "class" for U.S. senators in the 109th Congress associated? Explain your answer.

## Extending the Concepts and Skills

**12.62** In this exercise, you are to consider two variables, $x$ and $y$, defined on a hypothetical population. Following are the conditional distributions of the variable $y$ corresponding to each value of the variable $x$.

| | | $x$ | | | |
|---|---|---|---|---|---|
| | | $A$ | $B$ | $C$ | Total |
| | 0 | 0.316 | 0.316 | 0.316 | |
| | 1 | 0.422 | 0.422 | 0.422 | |
| $y$ | 2 | 0.211 | 0.211 | 0.211 | |
| | 3 | 0.047 | 0.047 | 0.047 | |
| | 4 | 0.004 | 0.004 | 0.004 | |
| | Total | 1.000 | 1.000 | 1.000 | |

a. Are the variables $x$ and $y$ associated? Explain your answer.

b. Determine the marginal distribution of $y$.

c. Can you determine the marginal distribution of $x$? Explain your answer.

**12.63 Age and Gender.** The U.S. Census Bureau publishes census data on the resident population of the United States in *Current Population Reports*. According to that document, 7.2% of male residents are in the age group 20–24 years.

a. If no association exists between age group and gender, what percentage of the resident population would be in the age group 20–24 years? Explain your answer.

b. If no association exists between age group and gender, what percentage of female residents would be in the age group 20–24 years? Explain your answer.

c. There are about 145.0 million female residents of the United States. If no association exists between age group and gender, how many female residents would there be in the age group 20–24 years?

d. In fact there are some 9.6 million female residents in the age group 20–24 years. Given this number and your answer to part (c), what do you conclude?

## 12.4 Chi-Square Independence Test

In Section 12.3, you learned how to determine whether an association exists between two variables of a population if you have the bivariate data for the entire population. However, because, in most cases, data for an entire population are not available, you must usually apply inferential methods to decide whether an association exists between two variables.

One of the most commonly used procedures for making such decisions is the **chi-square independence test.** In the next example, we introduce and explain the reasoning behind the chi-square independence test.

**Example 12.9** | **Introducing the Chi-Square Independence Test**

*Marital Status and Drinking* A national survey was conducted to obtain information on the alcohol consumption patterns of U.S. adults by marital status. A random sample of 1772 residents, 18 years old and older, yielded the data displayed in Table 12.13.[†]

Suppose we want to use the data in Table 12.13 to decide whether marital status and alcohol consumption are associated.

a. Formulate the problem statistically by posing it as a hypothesis test.

b. Explain the basic idea for carrying out the hypothesis test.

[†]Adapted from research by W. Clark and L. Midanik. In: National Institute on Alcohol Abuse and Alcoholism. *Alcohol Consumption and Related Problems: Alcohol and Health Monograph 1.* DHHS Pub. No. (ADM) 82–1190.

**TABLE 12.13**

Contingency table of marital status and alcohol consumption for 1772 randomly selected U.S. adults

| Marital status | Drinks per month | | | |
| --- | --- | --- | --- | --- |
| | Abstain | 1–60 | Over 60 | **Total** |
| Single | 67 | 213 | 74 | 354 |
| Married | 411 | 633 | 129 | 1173 |
| Widowed | 85 | 51 | 7 | 143 |
| Divorced | 27 | 60 | 15 | 102 |
| **Total** | 590 | 957 | 225 | 1772 |

**c.** Develop a formula for computing the expected frequencies.

**d.** Construct a table that provides both the observed frequencies in Table 12.13 and the expected frequencies.

**e.** Discuss the details for making a decision concerning the hypothesis test.

### Solution

**a.** For a chi-square independence test, the null hypothesis is that the two variables are not associated; the alternative hypothesis is that the two variables are associated. Thus, we want to perform the hypothesis test

$H_0$: Marital status and alcohol consumption are not associated.

$H_a$: Marital status and alcohol consumption are associated.

**b.** The idea behind the chi-square independence test is to compare the observed frequencies in Table 12.13 with the frequencies we would expect if the null hypothesis of nonassociation is true. The test statistic for making the comparison has the same form as the one used for the goodness-of-fit test: $\chi^2 = \Sigma (O - E)^2 / E$, where $O$ represents observed frequency and $E$ represents expected frequency.

**c.** To develop a formula for computing the expected frequencies, consider, for instance, the cell of Table 12.13 corresponding to "Married *and* Abstain," the cell in the second row and first column. We note that the population proportion of all adults who abstain can be estimated by the sample proportion of the 1772 adults sampled who abstain, that is, by

Number sampled who abstain ↘

$$\frac{590}{1772} = 0.333 \quad \text{or} \quad 33.3\%.$$

Total number sampled ↗

If no association exists between marital status and alcohol consumption (i.e., if $H_0$ is true), then the proportion of married adults who abstain is the same as the proportion of all adults who abstain. Therefore, of the 1173 married adults sampled, we would expect about

$$\frac{590}{1772} \cdot 1173 = 390.6$$

to abstain from alcohol.

Let's rewrite the left side of this expected-frequency computation in a slightly different way. By using algebra and referring to Table 12.13, we obtain

$$\text{Expected frequency} = \frac{590}{1772} \cdot 1173$$

$$= \frac{1173 \cdot 590}{1772}$$

$$= \frac{(\text{Row total}) \cdot (\text{Column total})}{\text{Sample size}}.$$

If we let $R$ denote "Row total" and $C$ denote "Column total," we can write this equation as

$$E = \frac{R \cdot C}{n}, \tag{12.1}$$

where, as usual, $E$ denotes expected frequency and $n$ denotes sample size.

**d.** Using Equation (12.1), we can calculate the expected frequencies for all the cells in Table 12.13. For the cell in the upper right corner of the table, we get

$$E = \frac{R \cdot C}{n} = \frac{354 \cdot 225}{1772} = 44.9.$$

In Table 12.14, we have modified Table 12.13 by including each expected frequency beneath the corresponding observed frequency. Table 12.14 shows, for instance, that of the adults sampled, 74 were observed to be single and consume more than 60 drinks per month, whereas if marital status and alcohol consumption are not associated, the expected frequency is 44.9.

**TABLE 12.14**

Observed and expected frequencies for marital status and alcohol consumption (expected frequencies printed below observed frequencies)

| Marital status | | Drinks per month | | | |
| --- | --- | --- | --- | --- | --- |
| | | Abstain | 1–60 | Over 60 | **Total** |
| | Single | 67 / 117.9 | 213 / 191.2 | 74 / 44.9 | 354 |
| | Married | 411 / 390.6 | 633 / 633.5 | 129 / 148.9 | 1173 |
| | Widowed | 85 / 47.6 | 51 / 77.2 | 7 / 18.2 | 143 |
| | Divorced | 27 / 34.0 | 60 / 55.1 | 15 / 13.0 | 102 |
| | **Total** | 590 | 957 | 225 | 1772 |

**e.** If the null hypothesis of nonassociation is true, the observed and expected frequencies should be approximately equal, which would result in a relatively small value of the test statistic, $\chi^2 = \Sigma(O - E)^2/E$. Consequently, if $\chi^2$ is too large, we reject the null hypothesis and conclude that an association exists between marital status and alcohol consumption. From Table 12.14, we find that

$$\chi^2 = \Sigma(O - E)^2/E$$
$$= (67 - 117.9)^2/117.9 + (213 - 191.2)^2/191.2 + (74 - 44.9)^2/44.9$$
$$+ (411 - 390.6)^2/390.6 + (633 - 633.5)^2/633.5 + (129 - 148.9)^2/148.9$$
$$+ (85 - 47.6)^2/47.6 + (51 - 77.2)^2/77.2 + (7 - 18.2)^2/18.2$$
$$+ (27 - 34.0)^2/34.0 + (60 - 55.1)^2/55.1 + (15 - 13.0)^2/13.0$$
$$= 21.952 + 2.489 + 18.776 + 1.070 + 0.000 + 2.670$$
$$+ 29.358 + 8.908 + 6.856 + 1.427 + 0.438 + 0.324$$
$$= 94.269.^\dagger$$

Can this value be reasonably attributed to sampling error, or is it large enough to indicate that marital status and alcohol consumption are associated? Before we can answer that question, we must know the distribution of the $\chi^2$-statistic.

• • •

---

**Key Fact 12.3**

### Distribution of the $\chi^2$-Statistic for a Chi-Square Independence Test

For a chi-square independence test, the test statistic

$$\chi^2 = \Sigma(O - E)^2/E$$

has approximately a chi-square distribution if the null hypothesis of non-association is true. The number of degrees of freedom is $(r - 1)(c - 1)$, where $r$ and $c$ are the number of possible values for the two variables under consideration.

**What Does It Mean?**

To obtain a chi-square subtotal, square the difference between an observed and expected frequency and divide the result by the expected frequency. Adding the chi-square subtotals gives the $\chi^2$-statistic, which has approximately a chi-square distribution.

### Procedure for the Chi-Square Independence Test

In light of Key Fact 12.3, we present, in Procedure 12.2 on the following page, a step-by-step method for conducting a chi-square independence test by using either the critical-value approach or the *P*-value approach. Because the null hypothesis is rejected only when the test statistic is too large, a chi-square independence test is always right tailed.

---

†Although we have displayed the expected frequencies to one decimal place and the chi-square subtotals to three decimal places, the calculations were made at full calculator accuracy.

## Procedure 12.2  Chi-Square Independence Test

*Purpose* To perform a hypothesis test to decide whether two variables are associated

*Assumptions*

1. All expected frequencies are 1 or greater
2. At most 20% of the expected frequencies are less than 5
3. Simple random sample

**STEP 1** The null and alternative hypotheses are

$H_0$: The two variables are not associated.

$H_a$: The two variables are associated.

**STEP 2** Calculate the expected frequencies by using the formula $E = RC/n$, where $R$ = row total, $C$ = column total, and $n$ = sample size. Place each expected frequency below its corresponding observed frequency in the contingency table.

**STEP 3** Determine whether the expected frequencies satisfy Assumptions 1 and 2. If they do not, this procedure should not be used.

**STEP 4** Decide on the significance level, $\alpha$.

**STEP 5** Compute the value of the test statistic

$$\chi^2 = \Sigma (O - E)^2/E,$$

where $O$ and $E$ represent observed and expected frequencies, respectively. Denote the value of the test statistic $\chi_0^2$.

---

| CRITICAL-VALUE APPROACH | or | P-VALUE APPROACH |

**STEP 6** The critical value is $\chi_\alpha^2$ with $\text{df} = (r - 1) \times (c - 1)$, where $r$ and $c$ are the number of possible values for the two variables. Use Table V to find the critical value.

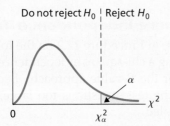

**STEP 7** If the value of the test statistic falls in the rejection region, reject $H_0$; otherwise, do not reject $H_0$.

**STEP 6** The $\chi^2$-statistic has $\text{df} = (r - 1)(c - 1)$, where $r$ and $c$ are the number of possible values for the two variables. Use Table V to estimate the $P$-value, or obtain it exactly by using technology.

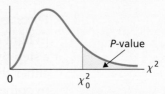

**STEP 7** If $P \leq \alpha$, reject $H_0$; otherwise, do not reject $H_0$.

**STEP 8** Interpret the results of the hypothesis test.

## Example 12.10 | The Chi-Square Independence Test

*Marital Status and Drinking* A random sample of 1772 U.S. adults yielded the data on marital status and alcohol consumption displayed in Table 12.13 on page 567. At the 5% significance level, do the data provide sufficient evidence to conclude that an association exists between marital status and alcohol consumption?

**Solution**   We apply Procedure 12.2.

**STEP 1  State the null and alternative hypotheses.**

The null and alternative hypotheses are

> $H_0$: Marital status and alcohol consumption are not associated.
>
> $H_a$: Marital status and alcohol consumption are associated.

**STEP 2  Calculate the expected frequencies by using the formula $E = RC/n$, where $R$ = row total, $C$ = column total, and $n$ = sample size. Place each expected frequency below its corresponding observed frequency in the contingency table.**

We did so earlier and displayed the results in Table 12.14 on page 568.

**STEP 3  Determine whether the expected frequencies satisfy Assumptions 1 and 2.**

1. Are all expected frequencies 1 or greater? Yes, as we can verify by referring to Table 12.14.
2. Are at most 20% of the expected frequencies less than 5? Yes, in fact, none of the expected frequencies are less than 5, as we can verify by referring to Table 12.14.

**STEP 4  Decide on the significance level, $\alpha$.**

The test is to be performed at the 5% significance level, so $\alpha = 0.05$.

**STEP 5  Compute the value of the test statistic**

$$\chi^2 = \Sigma (O - E)^2 / E,$$

**where $O$ and $E$ represent observed and expected frequencies, respectively.**

The observed and expected frequencies are displayed in Table 12.14. Using them, we compute the value of the test statistic:

$$\chi^2 = (67 - 117.9)^2/117.9 + (213 - 191.2)^2/191.2 + \cdots$$
$$+ (15 - 13.0)^2/13.0$$
$$= 21.952 + 2.489 + \cdots + 0.324$$
$$= 94.269.$$

| CRITICAL-VALUE APPROACH | or | P-VALUE APPROACH |
| --- | --- | --- |

**STEP 6** The critical value is $\chi_\alpha^2$ with df $= (r-1) \times (c-1)$, where $r$ and $c$ are the number of possible values for the two variables. Use Table V to find the critical value.

The number of marital status categories is four, and the number of drinks-per-month categories is three. Hence $r = 4, c = 3$, and

$$\text{df} = (r-1)(c-1) = 3 \cdot 2 = 6.$$

For $\alpha = 0.05$, Table V reveals that the critical value is $\chi_{0.05}^2 = 12.592$, as shown in Fig. 12.5A.

**FIGURE 12.5A**

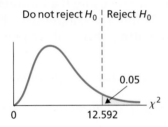

Do not reject $H_0$ | Reject $H_0$

0.05

0          12.592        $\chi^2$

**STEP 7** If the value of the test statistic falls in the rejection region, reject $H_0$; otherwise, do not reject $H_0$.

From Step 5, we see that the value of the test statistic is $\chi^2 = 94.269$, which falls in the rejection region, as shown in Fig. 12.5A. Thus we reject $H_0$. The test results are statistically significant at the 5% level.

**STEP 6** The $\chi^2$-statistic has df $= (r-1)(c-1)$, where $r$ and $c$ are the number of possible values for the two variables. Use Table V to estimate the $P$-value, or obtain it exactly by using technology.

From Step 5, we see that the value of the test statistic is $\chi^2 = 94.269$. Because the test is right tailed, the $P$-value is the probability of observing a value of $\chi^2$ of 94.269 or greater if the null hypothesis is true. That probability equals the shaded area shown in Fig. 12.5B.

**FIGURE 12.5B**

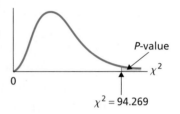

$P$-value

$\chi^2$

0

$\chi^2 = 94.269$

The number of marital status categories is four and the number of drinks-per-month categories is three. Hence $r = 4, c = 3$, and

$$\text{df} = (r-1)(c-1) = 3 \cdot 2 = 6.$$

From Fig. 12.5B and Table V with df $= 6$, we find that $P < 0.005$. (Using technology, we determined that $P = 0.000$ to three decimal places.)

**STEP 7** If $P \le \alpha$, reject $H_0$; otherwise, do not reject $H_0$.

From Step 6, $P < 0.005$. Because the $P$-value is less than the specified significance level of 0.05, we reject $H_0$. The test results are statistically significant at the 5% level and (see Table 9.10 on page 414) provide very strong evidence against the null hypothesis.

You try it!

Exercise 12.71 on page 575

**STEP 8** Interpret the results of the hypothesis test.

**Interpretation** At the 5% significance level, the data provide sufficient evidence to conclude that marital status and alcohol consumption are associated.

• • •

## Concerning the Assumptions

In Procedure 12.2, we made two assumptions about expected frequencies:

1. All expected frequencies are 1 or greater.
2. At most 20% of the expected frequencies are less than 5.

What can we do if one or both of these assumptions are violated? Three approaches are possible. We can combine rows or columns to increase the expected frequencies in those cells in which they are too small; we can eliminate certain rows or columns in which the small expected frequencies occur; or we can increase the sample size.

### Association and Causation

**What Does It Mean?**

Association does not imply causation!

Two variables may be associated without being causally related. In Example 12.10, we concluded that the variables marital status and alcohol consumption are associated. This result means that knowing the marital status of a person imparts information about the alcohol consumption of that person, and vice versa. But it does not necessarily mean, for instance, that being single causes a person to drink more.

Although we must keep in mind that association does not imply causation, we must also note that, if two variables are not associated, there is no point in looking for a causal relationship. In other words, association is a necessary but not sufficient condition for causation.

## The Technology Center

Most statistical technologies have programs that automatically perform a chi-square independence test. In this subsection, we present output and step-by-step instructions for such programs.

**Example 12.11**    **Using Technology to Perform an Independence Test**

*Marital Status and Drinking*    A random sample of 1772 U.S. adults yielded the data on marital status and alcohol consumption shown in Table 12.13 on page 567. Use Minitab, Excel, or the TI-83/84 Plus to decide, at the 5% significance level, whether the data provide sufficient evidence to conclude that an association exists between marital status and alcohol consumption.

**Solution**    We want to perform the hypothesis test

$H_0$: Marital status and alcohol consumption are not associated.

$H_a$: Marital status and alcohol consumption are associated.

at the 5% significance level.

We applied the chi-square independence test programs to the data, resulting in Output 12.4 on the following page. Steps for generating that output are presented in Instructions 12.4 on page 575.

As shown in Output 12.4, the $P$-value for the hypothesis test is 0.000 to three decimal places. Because the $P$-value is less than the specified significance level of 0.05, we reject $H_0$. At the 5% significance level, the data provide sufficient evidence to conclude that marital status and alcohol consumption are associated.

• • •

**OUTPUT 12.4** Chi-square independence test for the data on marital status and alcohol consumption

---

**MINITAB**

### Chi-Square Test: Abstain, 1-60, Over 60

Expected counts are printed below observed counts
Chi-Square contributions are printed below expected counts

|   | Abstain | 1-60 | Over 60 | Total |
|---|---|---|---|---|
| 1 | 67 | 213 | 74 | 354 |
|   | 117.87 | 191.18 | 44.95 | |
|   | 21.952 | 2.489 | 18.776 | |
| 2 | 411 | 633 | 129 | 1173 |
|   | 390.56 | 633.50 | 148.94 | |
|   | 1.070 | 0.000 | 2.670 | |
| 3 | 85 | 51 | 7 | 143 |
|   | 47.61 | 77.23 | 18.16 | |
|   | 29.358 | 8.908 | 6.856 | |
| 4 | 27 | 60 | 15 | 102 |
|   | 33.96 | 55.09 | 12.95 | |
|   | 1.427 | 0.438 | 0.324 | |
| Total | 590 | 957 | 225 | 1772 |

Chi-Sq = 94.269, DF = 6, P-Value = 0.000

---

**EXCEL**

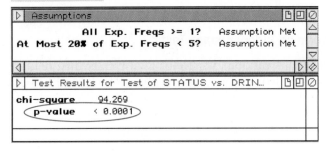

**TI-83/84 PLUS**

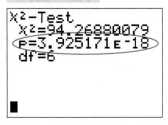

Using **Calculate**

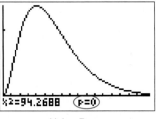

Using **Draw**

**INSTRUCTIONS 12.4**    Steps for generating Output 12.4

| MINITAB | EXCEL | TI-83/84 PLUS |
|---|---|---|
| 1 Store the cell data from Table 12.13 in columns named Abstain, 1-60, and Over 60.<br>2 Choose **Stat ➤ Tables ➤ Chi-Square Test (Table in Worksheet)...**<br>3 Specify Abstain, '1-60', and 'Over 60' in the **Columns containing the table** text box<br>4 Click **OK** | 1 Store the 12 possible combinations of marital status and drinks per month in ranges named STATUS and DRINKS, respectively, with the corresponding counts in a range named COUNT<br>2 Choose **DDXL ➤ Tables**<br>3 Select **Indep. Test for Summ Data** from the **Function type** drop-down list box<br>4 Specify STATUS in the **Variable One Names** text box<br>5 Specify DRINKS in the **Variable Two Names** text box<br>6 Specify COUNTS in the **Counts** text box<br>7 Click **OK** | 1 Press **2nd ➤ MATRIX**, arrow over to **EDIT**, and press **1**<br>2 Type 4 and press **ENTER**<br>3 Type 3 and press **ENTER**<br>4 Enter the cell data from Table 12.13, pressing **ENTER** after each entry<br>5 Press **STAT**, arrow over to **TESTS**, and press **ALPHA ➤ C**<br>6 Press **2nd ➤ MATRIX**, press **1**, and press **ENTER**<br>7 Press **2nd ➤ MATRIX**, press **2**, and press **ENTER**<br>8 Highlight **Calculate** or **Draw**, and press **ENTER** |

■ ■ ■

# Exercises 12.4

## Understanding the Concepts and Skills

**12.64** To decide whether two variables of a population are associated, we usually need to resort to inferential methods such as the chi-square independence test. Why?

**12.65** Step 1 of Procedure 12.2 gives generic statements for the null and alternative hypotheses of a chi-square independence test. Use the terms *statistically dependent* and *statistically independent,* introduced on page 558, to restate those hypotheses.

**12.66** In Example 12.9, we made the following statement: If no association exists between marital status and alcohol consumption, the proportion of married adults who abstain is the same as the proportion of all adults who abstain. Explain why that statement is true.

**12.67** A chi-square independence test is to be conducted to decide whether an association exists between two variables of a population. One variable has six possible values and the other variable has four. What is the degrees of freedom for the $\chi^2$-statistic?

**12.68 Education and Salary.** Studies have shown that a positive association exists between educational level and annual salary; in other words, people with more education tend to make more money.
**a.** Does this finding mean that more education *causes* a person to make more money? Explain your answer.

**b.** Do you think there is a causal relationship between educational level and annual salary? Explain your answer.

**12.69** We stated earlier that, if two variables are not associated, there is no point in looking for a causal relationship. Why is that so?

**12.70** Identify three techniques that can be tried as a remedy when one or more of the expected-frequency assumptions for a chi-square independence test are violated.

*In Exercises **12.71–12.78**, use either the critical-value approach or the P-value approach to perform a chi-square independence test, provided the conditions for using the test are met.*

**12.71 Siskel and Ebert.** In the late Gene Siskel and Roger Ebert's TV show *Sneak Preview,* the two Chicago movie critics reviewed the week's new movie releases and then rated them thumbs up (positive), mixed, or thumbs down (negative). These two critics often saw the merits of a movie differently. But, in general, were the ratings given by Siskel and Ebert associated? The answer to this question was the focus of the paper "Evaluating Agreement and Disagreement Among Movie Reviewers" by Alan Agresti and Larry Winner that appeared in *Chance* (Vol. 10(2), pp. 10–14). Following is a contingency table that summarizes the ratings by Siskel and Ebert for 160 movies.

| | Ebert's rating | | | |
|---|---|---|---|---|
| Siskel's rating | Thumbs down | Mixed | Thumbs up | Total |
| Thumbs down | 24 | 8 | 13 | 45 |
| Mixed | 8 | 13 | 11 | 32 |
| Thumbs up | 10 | 9 | 64 | 83 |
| Total | 42 | 30 | 88 | 160 |

At the 1% significance level, do the data provide sufficient evidence to conclude that an association exists between the ratings of Siskel and Ebert?

**12.72 Diabetes in Native Americans.** Preventable chronic diseases are increasing rapidly in Native American populations, particularly diabetes. Gilliland et al. examined the diabetes issue in the paper, "Preventative Health Care among Rural American Indians in New Mexico" (*Preventative Medicine*, Vol. 28, pp. 194–202). Following is a contingency table showing cross-classification of educational attainment and diabetic state for a sample of 1273 Native Americans.

| | Diabetic state | | |
|---|---|---|---|
| Education | Diabetes | No diabetes | Total |
| Less than HS | 33 | 218 | 251 |
| HS grad | 25 | 389 | 414 |
| Some college | 20 | 393 | 413 |
| College grad | 17 | 178 | 195 |
| Total | 95 | 1178 | 1273 |

At the 1% significance level, do the data provide sufficient evidence to conclude that an association exists between educational level and diabetic state for Native Americans?

**12.73 Learning at Home.** M. Stuart et al. studied various aspects of grade-school children and their mothers and reported their findings in the article "Learning to Read at Home and at School" (*British Journal of Educational Psychology*, 68(1), pp. 3–14). The researchers gave a questionnaire

to parents of 66 children in kindergarten through second grade. Two social-class groups, middle and working, were identified based on the mother's occupation.

**a.** One of the questions dealt with the children's knowledge of nursery rhymes. The following data were obtained.

| | | Nursery-rhyme knowledge | | |
|---|---|---|---|---|
| | | A few | Some | Lots |
| Social class | Middle | 4 | 13 | 15 |
| | Working | 5 | 11 | 18 |

Are Assumptions 1 and 2 satisfied for a chi-square independence test? If so, conduct the test and interpret your results. Use $\alpha = 0.01$.

**b.** Another question dealt with whether the parents played "I Spy" games with their children. The following data were obtained.

| | | Frequency of games | | |
|---|---|---|---|---|
| | | Never | Sometimes | Often |
| Social class | Middle | 2 | 8 | 22 |
| | Working | 11 | 10 | 13 |

Are Assumptions 1 and 2 satisfied for a chi-square independence test? If so, conduct the test and interpret your results. Use $\alpha = 0.01$.

**12.74 Thoughts of Suicide.** A study reported by D. Goldberg in *The Detection of Psychiatric Illness by Questionnaire* (Oxford University, London, p. 126, 1972) examined the relationship between mental-health classification and thoughts of suicide. The mental health of each person in a sample of 295 was classified as normal, mild psychiatric illness, or severe psychiatric illness. Also, each person was asked, "Have you recently found that the idea of taking your own life kept coming into your mind?" Following are the results.

| | | Mental health | | | |
|---|---|---|---|---|---|
| | | Normal | Mild illness | Severe illness | Total |
| Response | Definitely not | 90 | 43 | 34 | 167 |
| | Don't think so | 5 | 18 | 8 | 31 |
| | Crossed my mind | 3 | 21 | 21 | 45 |
| | Definitely yes | 1 | 15 | 36 | 52 |
| | Total | 99 | 97 | 99 | 295 |

Is there evidence that an association exists between response to the suicide question and mental-health classification? Use $\alpha = 0.05$.

**12.75 Lawyers.** The American Bar Foundation publishes information on the characteristics of lawyers in *The Lawyer Statistical Report*. The following contingency table cross-classifies 307 randomly selected U.S. lawyers by status in practice and the size of the city in which they practice.

| Status in practice | Size of city | | | |
|---|---|---|---|---|
| | Less than 250,000 | 250,000–499,999 | 500,000 or more | Total |
| Government | 12 | 4 | 14 | 30 |
| Judicial | 8 | 1 | 2 | 11 |
| Private practice | 122 | 31 | 69 | 222 |
| Salaried | 19 | 7 | 18 | 44 |
| Total | 161 | 43 | 103 | 307 |

At the 5% significance level, do the data provide sufficient evidence to conclude that size of city and status in practice are statistically dependent for U.S. lawyers?

**12.76 Lack of Support.** In the article, "Advanced Practice Clinicians' Interest in Providing Medical Abortion: Results of a California Survey" (*Perspectives on Sexual and Reproductive Health*, Vol. 2, pp. 1–5), A. Hwang et al. examined the views on abortion of practicing clinicians in California. The following contingency table provides a cross-classification of a sample of clinicians in California who received lack of support after performing a medical abortion.

| Type | Source | |
|---|---|---|
| | Colleagues | Friends |
| Physician assistant | 3 | 6 |
| Nurse-midwife | 6 | 3 |

At the 5% significance level, do the data provide sufficient evidence to conclude that type of clinician and source of lack of support are statistically dependent for practicing clinicians in California?

**12.77 BMD and Depression.** In the paper "Depression and Bone Mineral Density: Is There a Relationship in Elderly Asian Men?" (*Osteoporosis International*, Vol. 16, pp. 610–615), S. Wong et al. published results of their study on bone mineral density (BMD) and depression for

1999 Hong Kong men aged 65 to 92 years. Here are the cross-classified data.

| BMD | Depression | | |
|---|---|---|---|
| | Depressed | Not depressed | Total |
| Osteoporitic | 3 | 35 | 38 |
| Low BMD | 69 | 533 | 602 |
| Normal | 97 | 1262 | 1359 |
| Total | 169 | 1830 | 1999 |

At the 1% significance level, do the data provide sufficient evidence to conclude that BMD and depression are statistically dependent for elderly Asian men?

**12.78 Ballot Preference.** In Issue 338 of the *Amstat News*, then president of the American Statistical Association, Fritz Scheuren, reported the results of a survey on how members would prefer to receive ballots in annual elections.

a. Following are the results of the survey, cross-classified by gender. At the 5% significance level, do the data provide sufficient evidence to conclude that gender and preference are associated?

| Preference | Gender | | |
|---|---|---|---|
| | Male | Female | Total |
| Mail | 58 | 26 | 84 |
| Email | 151 | 86 | 237 |
| Both | 72 | 40 | 112 |
| N/A | 76 | 50 | 126 |
| Total | 357 | 202 | 559 |

b. Following are the results of the survey, cross-classified by age. At the 5% significance level, do the data provide sufficient evidence to conclude that age and preference are associated?

| Preference | Age | | |
|---|---|---|---|
| | Under 40 | 40 or over | Total |
| Mail | 22 | 56 | 78 |
| Email | 95 | 135 | 230 |
| Both | 38 | 72 | 110 |
| N/A | 42 | 80 | 122 |
| Total | 197 | 343 | 540 |

**c.** Following are the results of the survey, cross-classified by degree. At the 5% significance level, do the data provide sufficient evidence to conclude that degree and preference are associated?

| | | PhD | MA | Other | Total |
|---|---|---|---|---|---|
| | | **Degree** | | | |
| **Preference** | Mail | 65 | 18 | 3 | 56 |
| | Email | 166 | 71 | 2 | 239 |
| | Both | 84 | 23 | 5 | 112 |
| | N/A | 73 | 55 | 1 | 129 |
| | **Total** | 388 | 167 | 11 | 566 |

**12.79 Job Satisfaction.** A *CNN/USA TODAY* poll conducted by Gallup asked a sample of employed Americans the following question: "Which do you enjoy more, the hours when you are on your job, or the hours when you are not on your job?" The responses to this question were cross-tabulated against several characteristics, among which were gender, age, type of community, educational attainment, income, and type of employer. The data are provided on the WeissStats CD. Use the technology of your choice to decide, at the 5% significance level, whether an association exists between each of the following pairs of variables.
**a.** gender and response (to the question)
**b.** age and response
**c.** type of community and response
**d.** educational attainment and response
**e.** income and response
**f.** type of employer and response

## Extending the Concepts and Skills

**12.80 Lawyers.** In Exercise 12.75, you couldn't perform the chi-square independence test because the assumptions regarding expected frequencies were not met. As mentioned in the text, three approaches are available for remedying the situation: (1) combine rows or columns; (2) eliminate rows or columns; or (3) increase the sample size.
**a.** Combine the first two rows of the contingency table in Exercise 12.75 to form a new contingency table.
**b.** Use the table obtained in part (a) to perform the hypothesis test required in Exercise 12.75, if possible.
**c.** Eliminate the second row of the contingency table in Exercise 12.75 to form a new contingency table.
**d.** Use the table obtained in part (c) to perform the hypothesis test required in Exercise 12.75, if possible.

**12.81 Ballot Preference.** In part (c) of Exercise 12.78, you couldn't perform the chi-square independence test because the assumptions regarding expected frequencies were not met. Combine the MA and Other categories and then attempt to perform the hypothesis test again.

**Chi-Square Homogeneity Test.** The **chi-square homogeneity test** is used to decide whether a difference exists among the distributions of a variable of two or more populations. In particular, it can be used to decide whether a difference exists among two or more population proportions. The assumptions for use of the chi-square homogeneity test are (1) simple random samples, (2) independent samples, and (3) the two expected-frequency assumptions for the chi-square independence test. The test statistic is exactly the same as the one used for the chi-square independence test. Note that the chi-square homogeneity test is also valid for designed experiments.

**12.82** Show that, for two populations, the assumptions for the chi-square homogeneity test are the same as those for the two-proportions z-test, Procedure 11.3 on page 524. *Note:* As demonstrated in the paper "Equivalence of Different Statistical Tests for Common Problems" (*The AMATYC Review*, Vol. 4, No. 2, pp. 5–13) by M. Hassett and N. Weiss, for two populations, the chi-square homogeneity test and the two-tailed two-proportions z-test are equivalent.

*In Exercises 12.83 and 12.84, perform a chi-square homogeneity test and interpret your results.*

**12.83 Region and Race.** The U.S. Census Bureau compiles data on the U.S. population by region and race, and publishes its findings in *Current Population Reports*. Independent simple random samples of residents in the four U.S. regions gave the following data on race.

| | | White | Black | Other | Total |
|---|---|---|---|---|---|
| | | **Race** | | | |
| **Region** | Northeast | 93 | 14 | 6 | 113 |
| | Midwest | 118 | 14 | 4 | 136 |
| | South | 167 | 42 | 7 | 216 |
| | West | 113 | 7 | 15 | 135 |
| | **Total** | 491 | 77 | 32 | 600 |

At the 1% significance level, do the data provide sufficient evidence to conclude that a difference exists in race distributions among the four U.S. regions?

**12.84 Scoliosis.** Scoliosis is a condition involving curvature of the spine. In a study by A. Nachemson and L. Peterson, reported in *The Journal of Bone and Joint Surgery* (Vol. 77, Issue 6, pp. 815–822), 286 girls aged 10 to 15 years, were followed to determine the effect of observation only (129 patients), an underarm plastic brace (111 patients), and nighttime surface electrical stimulation (46 patients). A treatment was deemed to have failed if the curvature of the spine increased by 6° on two successive examinations. The table at the right summarizes the results obtained by the researchers.

At the 5% significance level, do the data provide sufficient evidence to conclude that a difference in failure rate exists among the three types of treatments?

|  | **Result** | | |
|---|---|---|---|
|  | Not failure | Failure | **Total** |
| Brace | 94 | 17 | 111 |
| Stimulation | 24 | 22 | 46 |
| Observation | 71 | 58 | 129 |
| **Total** | 189 | 97 | 286 |

*Treatment*

## Chapter in Review

### You Should be Able to

1. use and understand the formulas in this chapter.

2. identify the basic properties of $\chi^2$-curves.

3. use the chi-square table, Table V.

4. explain the reasoning behind the chi-square goodness-of-fit test.

5. perform a chi-square goodness-of-fit test.

6. group bivariate data into a contingency table.

7. find and graph marginal and conditional distributions.

8. decide whether an association exists between two variables of a population, given bivariate data for the entire population.

9. explain the reasoning behind the chi-square independence test.

10. perform a chi-square independence test to decide whether an association exists between two variables of a population, given bivariate data for a sample of the population.

### Key Terms

associated variables, *558*
association, *558*
bivariate data, *554*
cells, *555*
$\chi^2_\alpha$, *542*
chi-square ($\chi^2$) curve, *542*
chi-square distribution, *542*
chi-square goodness-of-fit test, *543, 547*

chi-square independence test, *566, 570*
chi-square procedures, *540*
chi-square subtotals, *545*
conditional distribution, *557*
contingency table, *554*
expected frequencies, *544*
marginal distribution, *557*
observed frequencies, *544*

segmented bar graph, *557*
statistically dependent variables, *558*
statistically independent
    variables, *558*
two-way table, *554*
univariate data, *554*

## Review Problems

### Understanding the Concepts and Skills

**1.** How do you distinguish among the infinitely many different chi-square distributions and their corresponding $\chi^2$-curves?

**2.** Regarding a $\chi^2$-curve,
**a.** at what point on the horizontal axis does the curve begin?
**b.** what shape does it have?

**c.** As the number of degrees of freedom increases, a $\chi^2$-curve begins to look like another type of curve. What type of curve is that?

**3.** Recall that the number of degrees of freedom for the $t$-distribution used in a one-mean $t$-test depends on the sample size.

**a.** Is that true for the chi-square distribution used in a chi-square goodness-of-fit test? Explain your answer.
**b.** Is that true for the chi-square distribution used in a chi-square independence test? Explain your answer.

**4.** Explain why a chi-square goodness-of-fit test or a chi-square independence test is always right tailed.

**5.** If the observed and expected frequencies for a chi-square goodness-of-fit test or a chi-square independence test matched perfectly, what would be the value of the test statistic?

**6.** Regarding the expected-frequency assumptions for a chi-square goodness-of-fit test or a chi-square independence test,
**a.** state them.          **b.** how important are they?

**7. Race and Region.** T. G. Exter's book *Regional Markets, Vol. 2/Households* (Ithaca, NY: New Strategist Publications, Inc.) provides data on U.S. households by region of the country. One table in the book cross-classifies households by race (of the householder) and region of residence. The table shows that 7.8% of all U.S. households are Hispanic.
**a.** If race and region of residence are not associated, what percentage of Midwest households would be Hispanic?
**b.** There are 24.7 million Midwest households. If race and region of residence are not associated, how many Midwest households would be Hispanic?
**c.** In fact, there are 645 thousand Midwest Hispanic households. Given this information and your answer to part (b), what can you conclude?

**8.** Suppose that you have bivariate data for an entire population.
**a.** How would you decide whether an association exists between the two variables under consideration?
**b.** Assuming that you make no calculation mistakes, could your conclusion be in error? Explain your answer.

**9.** Suppose that you have bivariate data for a sample of a population.
**a.** How would you decide whether an association exists between the two variables under consideration?
**b.** Assuming that you make no calculation mistakes, could your conclusion be in error? Explain your answer.

**10.** Consider a $\chi^2$-curve with 17 degrees of freedom. Use Table V to determine
**a.** $\chi^2_{0.99}$.          **b.** $\chi^2_{0.01}$.
**c.** the $\chi^2$-value that has area 0.05 to its right.
**d.** the $\chi^2$-value that has area 0.05 to its left.
**e.** the two $\chi^2$-values that divide the area under the curve into a middle 0.95 area and two outside 0.025 areas.

**11. Educational Attainment.** The U.S. Census Bureau compiles census data on educational attainment of Americans. From the document *Current Population Reports*, we obtained the 2000 distribution of educational attainment for U.S. adults 25 years old and older. Here is that distribution.

| Highest level | Percentage |
|---|---|
| Not HS graduate | 15.8 |
| HS graduate | 33.2 |
| Some college | 17.6 |
| Associate's degree | 7.8 |
| Bachelor's degree | 17.0 |
| Advanced degree | 8.6 |

A random sample of 500 U.S. adults (25 years old and older) taken this year gave the following frequency distribution.

| Highest level | Frequency |
|---|---|
| Not HS graduate | 84 |
| HS graduate | 160 |
| Some college | 88 |
| Associate's degree | 32 |
| Bachelor's degree | 87 |
| Advanced degree | 49 |

**a.** Decide, at the 5% significance level, whether this year's distribution of educational attainment differs from the 2000 distribution.
**b.** Estimate the $P$-value of the test and use that estimate to assess the evidence against the null hypothesis.

**12. Presidents.** From the *Information Please Almanac*, we compiled the following table on U.S. region of birth and political party of the first 43 U.S. presidents. The table uses these abbreviations: F = Federalist, DR = Democratic-Republican, D = Democratic, W = Whig, R = Republican, U = Union; NE = Northeast, MW = Midwest, SO = South, WE = West.

| Region | Party | Region | Party | Region | Party |
|---|---|---|---|---|---|
| SO | F | SO | R | NE | R |
| NE | F | SO | U | MW | R |
| SO | DR | MW | R | NE | D |
| SO | DR | MW | R | MW | D |
| SO | DR | MW | R | SO | R |
| NE | DR | NE | R | NE | D |
| SO | D | NE | D | SO | D |
| NE | D | MW | R | WE | R |
| SO | W | NE | D | MW | R |
| SO | W | MW | R | SO | D |
| SO | D | NE | R | MW | R |
| SO | W | MW | R | NE | R |
| NE | W | SO | D | SO | D |
| NE | D | MW | R | NE | R |
| NE | D | | | | |

**a.** What is the population under consideration?
**b.** What are the two variables under consideration?
**c.** Group the bivariate data for the variables "birth region" and "party" into a contingency table.

**13. Presidents.** Refer to Problem 12.
a. Find the conditional distributions of birth region by party and the marginal distribution of birth region.
b. Find the conditional distributions of party by birth region and the marginal distribution of party.
c. Does an association exist between the variables "birth region" and "party" for the U.S. presidents? Explain your answer.
d. What percentage of presidents are Republicans?
e. If no association existed between birth region and party, what percentage of presidents born in the South would be Republicans?
f. In reality, what percentage of presidents born in the South are Republicans?
g. What percentage of presidents were born in the South?
h. If no association existed between birth region and party, what percentage of Republican presidents would have been born in the South?
i. In reality, what percentage of Republican presidents were born in the South?

**14. Hospitals.** From data in *Hospital Statistics*, published by the American Hospital Association, we obtained the following contingency table for U.S. hospitals and nursing homes, by type of facility and type of control. We used the abbreviations Gov for Government, Prop for Proprietary, and NP for nonprofit.

| | | Control | | | |
|---|---|---|---|---|---|
| | | Gov | Prop | NP | Total |
| **Facility** | General | 1697 | 660 | 3046 | 5403 |
| | Psychiatric | 266 | 358 | 113 | 737 |
| | Chronic | 21 | 1 | 4 | 26 |
| | Tuberculosis | 3 | 0 | 1 | 4 |
| | Other | 59 | 148 | 203 | 410 |
| | **Total** | 2046 | 1167 | 3367 | 6580 |

In the following questions, the term *hospital* refers to either a hospital or nursing home:

a. How many hospitals are government controlled?
b. How many hospitals are psychiatric facilities?
c. How many hospitals are government controlled psychiatric facilities?
d. How many general facilities are nonprofit?
e. How many hospitals are not under proprietary control?
f. How many hospitals are either general facilities or under proprietary control?

**15. Hospitals.** Refer to Problem 14.
a. Obtain the conditional distribution of control type within each facility type.
b. Does an association exist between facility type and control type for U.S. hospitals? Explain your answer.
c. Determine the marginal distribution of control type for U.S. hospitals.
d. Construct a segmented bar graph for the conditional distributions and marginal distribution of control type. Interpret the graph in light of your answer to part (b).
e. Without doing any further calculations, respond true or false to the following statement and explain your answer: "The conditional distributions of facility type within control types are identical."
f. Determine the marginal distribution of facility type and the conditional distributions of facility type within control types.
g. What percentage of hospitals are under proprietary control?
h. What percentage of psychiatric hospitals are under proprietary control?
i. What percentage of hospitals under proprietary control are psychiatric hospitals?

**16. Hodgkin's Disease.** Hodgkin's disease is a malignant, progressive, sometimes fatal disease of unknown cause, and characterized by enlargement of the lymph nodes, spleen, and liver. The following contingency table summarizes data collected during a study by Hancock et al. (*Journal of Clinical Oncology*, Vol. 5(4), pp. 283–297) of 538 patients with Hodgkin's disease. The table cross-classifies the histological types of patients and their responses to treatment three months prior to the study.

| | | Response | | | |
|---|---|---|---|---|---|
| | | Positive | Partial | None | Total |
| **Histological type** | Lymphocyte depletion | 18 | 10 | 44 | 72 |
| | Lymphocyte predominance | 74 | 18 | 12 | 104 |
| | Mixed cellularity | 154 | 54 | 58 | 266 |
| | Nodular sclerosis | 68 | 16 | 12 | 96 |
| | **Total** | 314 | 98 | 126 | 538 |

At the 1% significance level, do the data provide sufficient evidence to conclude that histological type and treatment response are statistically dependent?

## Working With Large Data Sets

**17. Presidents.** From the *Information Please Almanac*, we compiled information on U.S. region of birth and age group at the time of inauguration for the first 43 U.S. presidents. The data are presented on the WeissStats CD. Use the technology of your choice to do the following.
a. Group the bivariate data for the variables "birth region" and "inauguration age group" into a contingency table.
b. Find the conditional distributions of birth region by inauguration age group and the marginal distribution of birth region.
c. Find the conditional distributions of inauguration age group by birth region and the marginal distribution of inauguration age group.

d. Does an association exist between the variables "birth region" and "inauguration age group" for the U.S. presidents? Explain your answer.

**18. Withholding Treatment.** Several years ago, a *Gallup Poll* asked 1528 adults the following question: "The New Jersey Supreme Court recently ruled that all life-sustaining medical treatment may be withheld or withdrawn from terminally ill patients, provided that is what the patients want or would want if they were able to express their wishes. Would you like to see such a ruling in the state in which you live, or not?" The data on the WeissStats CD gives the responses by opinion and educational level. Use the technology of your choice to decide, at the 1% significance level, whether the data provide sufficient evidence to conclude that opinion on this issue and educational level are associated.

## Focusing on Data Analysis    UWEC Undergraduates

Recall from Chapter 1 (see page 34) that the Focus database and Focus sample contain information on the undergraduate students at the University of Wisconsin - Eau Claire (UWEC). Now would be a good time for you to review the discussion about these data sets.

Open the Focus sample worksheet (FocusSample) in the technology of your choice. In each part, apply the chi-square independence test to decide, at the 5% significance level, whether the data provide sufficient evidence to conclude that an association exists between the indicated vari-

ables for the population of all UWEC undergraduates. Be sure to check whether the assumptions for performing each test are satisfied. Interpret your results.

a. sex and classification
b. sex and residency
c. sex and college
d. classification and residency
e. classification and college
f. college and residency

## Case Study Discussion    Road Rage

At the beginning of this chapter, we discussed the report *Controlling Road Rage: A Literature Review and Pilot Study*, prepared for the AAA Foundation for Traffic Safety by Daniel B. Rathbone, Ph.D., and Jorg C. Huckabee, MSCE. The authors examined the results of a literature review and pilot study on how to prevent aggressive driving and road rage.

One aspect of the study was to investigate road rage as a function of the day of the week. The table on page 541 indicates the day of the week on which road rage occurred for a random sample of 69 road-rage incidents. Use those data to decide, at the 5% significance level, whether road-rage incidents are more likely to occur on some days than on others.

## Biography    KARL PEARSON: The Founding Developer of Chi-Square Tests

**Karl Pearson** was born on March 27, 1857, in London, the second son of William Pearson, a prominent lawyer, and his wife, Fanny Smith. Karl Pearson's early education took place at home. At the age of 9, he was sent to University College School in London, where he remained for the next 7 years. Because of ill health, Pearson was then privately tutored for a year. He received a scholarship at King's College, Cambridge, in 1875. There he earned a B.A. (with honors) in mathematics in 1879 and an M.A. in law in 1882. He then studied physics and metaphysics in Heidelberg, Germany.

In addition to his expertise in mathematics, law, physics, and metaphysics, Pearson was competent in literature and knowledgeable about German history, folklore, and philosophy. He was also considered somewhat of a political radical because of his interest in the ideas of Karl Marx and the rights of women.

In 1884, Pearson was appointed Goldsmid professor of applied mathematics and mechanics at University College; from 1891–1894, he was also a lecturer in geometry at Gresham College, London. In 1911, he gave up the Goldsmid chair to become the first Galton professor of Eugenics at University College. Pearson was elected to the Royal Society—a prestigious association of scientists—in 1896 and was awarded the society's Darwin Medal in 1898.

Pearson really began his pioneering work in statistics in 1893, mainly through an association with Walter Weldon (a zoology professor at University College), Francis Edgeworth (a professor of logic at University College), and Sir Francis Galton (see the Chapter 15 Biography). An analysis of published data on roulette wheels at Monte Carlo led to Pearson's discovery of the chi-square goodness-of-fit test. He also coined the term *standard deviations*, introduced his amazingly diverse skew curves, and developed the most widely used measure of correlation, the correlation coefficient.

Pearson, Weldon, and Galton cofounded the statistical journal *Biometrika,* of which Pearson was editor (1901–1936) and a major contributor. Pearson retired from University College in 1933. He died in London on April 27, 1936.

## StatCrunch in MyStatLab
### Analyzing Data Online

StatCrunch online statistical software offers an easy-to-use interface customized for this book. The StatCrunch feature for each chapter illustrates the use of the software to perform a statistical analysis discussed in the chapter. Exercises are provided to further apply StatCrunch to other statistical analyses examined in the chapter. Go to the WeissStats CD or to the Weiss Web site at www.aw-bc.com/weiss to access StatCrunch instructions and data sets. To access StatCrunch statistical software, go to the student content area of your Weiss MyStatLab course.

## Internet Projects
### Exploring Data Online

The Internet project for each chapter provides simulations, demonstrations, or activities that enhance the topics covered in the chapter. The project materials come from universities, individuals, governments, and companies from all over the world. To access the Internet projects on the Web, go to www.aw-bc.com/weiss. From this Web page, you can reach the Internet Projects Page, which we suggest that you bookmark for easy access in the future.

# 13

# Analysis of Variance (ANOVA)

## Chapter Objectives

In Chapter 10, you studied inferential methods for comparing the means of two populations. Now you will study **analysis of variance,** or **ANOVA,** which provides methods for comparing the means of more than two populations. For instance, you could use ANOVA to compare the mean energy consumption by households among the four U.S. regions. Just as there are several different procedures for comparing two population means, there are several different ANOVA procedures.

In Section 13.1, to prepare for the study of ANOVA, we consider the *F*-distribution. Next, in Section 13.2, we introduce one-way analysis of variance and examine the logic behind it. Then we discuss the one-way ANOVA procedure itself in Section 13.3.

# Heavy Drinking Among College Students

Professor Kate Carey of Syracuse University surveyed 78 college students, all of whom were regular drinkers of alcohol. Her purpose was twofold—to identify interpersonal and intrapersonal situations associated with excessive drinking among college students and to detect situations that differentiate heavy drinkers from light and moderate drinkers. She published her findings in the paper "Situational Determinants of Heavy Drinking Among College Students"

(*Journal of Counseling Psychology*, Vol. 40, pp. 217–220).

To assess the frequency of excessive drinking in interpersonal and intrapersonal situations, Carey utilized the short form of the Inventory of Drinking Situations (IDS). The following table gives the sample size, sample mean, and sample standard deviation of IDS scores for each drinking category and situational context.

At the end of this chapter, you will analyze these data to decide, for each IDS category, whether a difference exists in mean IDS scores among the three drinker categories.

| IDS subscale | Light drinkers ($n_1 = 16$) | | Moderate drinkers ($n_2 = 47$) | | Heavy drinkers ($n_3 = 15$) | |
|---|---|---|---|---|---|---|
| | $\bar{x}_1$ | $s_1$ | $\bar{x}_2$ | $s_2$ | $\bar{x}_3$ | $s_3$ |
| *Interpersonal situations* | | | | | | |
| Conflict with others | 1.23 | 0.27 | 1.53 | 0.49 | 1.79 | 0.49 |
| Social pressure to drink | 2.64 | 0.80 | 2.91 | 0.55 | 3.51 | 0.51 |
| Pleasant times with others | 2.21 | 0.67 | 2.53 | 0.51 | 3.03 | 0.38 |
| *Intrapersonal situations* | | | | | | |
| Unpleasant emotions | 1.22 | 0.35 | 1.61 | 0.69 | 1.68 | 0.46 |
| Physical discomfort | 1.03 | 0.08 | 1.19 | 0.29 | 1.40 | 0.32 |
| Pleasant emotions | 2.09 | 0.73 | 2.61 | 0.58 | 3.03 | 0.30 |
| Testing personal control | 1.52 | 0.74 | 1.56 | 0.56 | 1.53 | 0.48 |
| Urges and temptations | 1.80 | 0.56 | 1.96 | 0.51 | 2.33 | 0.58 |

# 13.1 The *F*-Distribution

Analysis-of-variance procedures rely on a distribution called the *F-distribution*, named in honor of Sir Ronald Fisher. See the Biography at the end of this chapter for more information about Fisher.

A variable is said to have an **F-distribution** if its distribution has the shape of a special type of right-skewed curve, called an **F-curve.** There are infinitely many *F*-distributions, and we identify an *F*-distribution (and *F*-curve) by its number of degrees of freedom, just as we did for *t*-distributions and chi-square distributions.

But an *F*-distribution has two numbers of degrees of freedom instead of one. Figure 13.1 depicts two different *F*-curves; one has df = (10, 2), and the other has df = (9, 50).

The first number of degrees of freedom for an *F*-curve is called the **degrees of freedom for the numerator,** and the second is called the **degrees of freedom for the denominator.** (The reason for this terminology will become clear in Section 13.2.) Thus, for the *F*-curve in Fig. 13.1 with df = (10, 2), we have

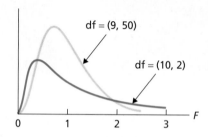

**FIGURE 13.1**
Two different *F*-curves

$$df = (10, 2)$$

Degrees of freedom for the numerator     Degrees of freedom for the denominator

**Key Fact 13.1** | **Basic Properties of *F*-Curves**

**Property 1:** The total area under an *F*-curve equals 1.

**Property 2:** An *F*-curve starts at 0 on the horizontal axis and extends indefinitely to the right, approaching, but never touching, the horizontal axis as it does so.

**Property 3:** An *F*-curve is right skewed.

## Using the *F*-Table

Percentages (and probabilities) for a variable having an *F*-distribution equal areas under its associated *F*-curve. To perform an ANOVA test, we need to know how to find the *F*-value having a specified area to its right. The symbol $F_\alpha$ denotes the *F*-value having area $\alpha$ to its right.

Table VI in Appendix A provides *F*-values corresponding to several areas for various degrees of freedom. The degrees of freedom for the denominator (dfd) are displayed in the outside columns of the table, the values of $\alpha$ in the next columns, and the degrees of freedom for the numerator (dfn) along the top.

**Example 13.1** | **Finding the *F*-Value Having a Specified Area to Its Right**

For an *F*-curve with df = (4, 12), find $F_{0.05}$; that is, find the *F*-value having area 0.05 to its right, as shown in Fig. 13.2(a).

**Solution** We use Table VI to find the *F*-value. In this case, $\alpha = 0.05$, the degrees of freedom for the numerator is 4, and the degrees of freedom for the denominator is 12.

**FIGURE 13.2**
Finding the F-value having
area 0.05 to its right

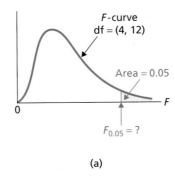

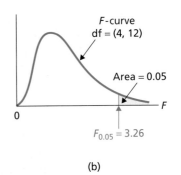

(a)                                                          (b)

**You
try it!**

Exercise 13.7
on page 587

We first go down the dfd column to "12." Next, we concentrate on the row for $\alpha$ labeled 0.05. Then, going across that row to the column labeled "4," we reach 3.26. This number is the F-value having area 0.05 to its right, as shown in Fig. 13.2(b). In other words, for an F-curve with df $= (4, 12)$, $F_{0.05} = 3.26$.

• • •

## Exercises 13.1

### Understanding the Concepts and Skills

**13.1** How do we identify an F-distribution and its corresponding F-curve?

**13.2** How many degrees of freedom does an F-curve have? What are those degrees of freedom called?

**13.3** What symbol is used to denote the F-value having area 0.05 to its right? 0.025 to its right? $\alpha$ to its right?

**13.4** Using the $F_\alpha$-notation, identify the F-value having area 0.975 to its left.

**13.5** An F-curve has df $= (12, 7)$. What is the number of degrees of freedom for the
**a.** numerator?                **b.** denominator?

**13.6** An F-curve has df $= (8, 19)$. What is the number of degrees of freedom for the
**a.** denominator?              **b.** numerator?

*In Exercises 13.7–13.10, use Table VI in Appendix A to find the required F-values. Illustrate your work with graphs similar to that shown in Fig. 13.2.*

**13.7** An F-curve has df $= (24, 30)$. In each case, find the F-value having the specified area to its right.
**a.** 0.05          **b.** 0.01          **c.** 0.025

**13.8** An F-curve has df $= (12, 5)$. In each case, find the F-value having the specified area to its right.
**a.** 0.01          **b.** 0.05          **c.** 0.005

**13.9** For an F-curve with df $= (20, 21)$, find
**a.** $F_{0.01}$.          **b.** $F_{0.05}$.          **c.** $F_{0.10}$.

**13.10** For an F-curve with df $= (6, 10)$, find
**a.** $F_{0.05}$.          **b.** $F_{0.01}$.          **c.** $F_{0.025}$.

### Extending the Concepts and Skills

**13.11** Refer to Table VI in Appendix A. Because of space restrictions, the numbers of degrees of freedom are not consecutive. For instance, the degrees of freedom for the numerator skips from 24 to 30. If you had only Table VI and you needed to find $F_{0.05}$ for df $= (25, 20)$, how would you do it?

## 13.2 One-Way ANOVA: The Logic

In Chapter 10, you learned how to compare two population means, that is, the means of a single variable for two different populations. You studied various methods for making such comparisons, one being the pooled $t$-procedure.

Analysis of variance (ANOVA) provides methods for comparing several population means, that is, the means of a single variable for several populations. In this section and Section 13.3, we present the simplest kind of ANOVA,

**one-way analysis of variance.** This type of ANOVA is called *one-way* analysis of variance because it compares the means of a variable for populations that result from a classification by *one* other variable, called the **factor.** The possible values of the factor are referred to as the **levels** of the factor.

For example, suppose that you want to compare the mean energy consumption by households among the four regions of the United States. The variable under consideration is "energy consumption" and there are four populations: households in the Northeast, Midwest, South, and West. The four populations result from classifying households in the United States by the factor "region," whose levels are Northeast, Midwest, South, and West.

One-way analysis of variance is the generalization to more than two populations of the pooled *t*-procedure. As in the pooled *t*-procedure, we make the following assumptions.

---

**Key Fact 13.2**

### Assumptions (Conditions) for One-Way ANOVA

1. **Simple random samples:** The samples taken from the populations under consideration are simple random samples.
2. **Independent samples:** The samples taken from the populations under consideration are independent of one another.
3. **Normal populations:** For each population, the variable under consideration is normally distributed.
4. **Equal standard deviations:** The standard deviations of the variable under consideration are the same for all the populations.

---

Regarding Assumptions 1 and 2, we note that one-way ANOVA can also be used as a method for comparing several means with a designed experiment. Additionally, like the pooled *t*-procedure, one-way ANOVA is robust to moderate violations of Assumption 3 (normal populations) and is also robust to moderate violations of Assumption 4 (equal standard deviations) provided the sample sizes are roughly equal.

How can the conditions of normal populations and equal standard deviations be checked? Normal probability plots of the sample data are effective in detecting gross violations of normality. Checking equal population standard deviations, however, can be difficult, especially when the sample sizes are small; as a rule of thumb, you can consider that condition met if *the ratio of the largest to the smallest sample standard deviation is less than 2.* We call that rule of thumb the **rule of 2.**

Another way to assess the normality and equal standard deviations assumptions is to perform a **residual analysis.** In ANOVA, the **residual** of an observation is the difference between the observation and the mean of the sample containing it. If the normality and equal standard deviations assumptions are met, a normal probability plot of (all) the residuals should be roughly linear. Moreover, a plot of the residuals against the sample means should fall roughly in a horizontal band centered and symmetric about the horizontal axis.

### The Logic Behind One-Way ANOVA

The reason for the word *variance* in *analysis of variance* is that the procedure for comparing the means analyzes the variation in the sample data. To examine how this procedure works, let's suppose that independent random samples are taken from two populations—say, Populations 1 and 2—with means $\mu_1$ and $\mu_2$. Further, let's suppose that the means of the two samples are $\bar{x}_1 = 20$ and $\bar{x}_2 = 25$. Can we reasonably conclude from these statistics that $\mu_1 \neq \mu_2$, that is, that the population means are different? To answer this question, we must consider the variation within the samples.

Suppose, for instance, that the sample data are as displayed in Table 13.1 and depicted in Fig. 13.3.

**TABLE 13.1**
Sample data from Populations 1 and 2

| Sample from Population 1 | 21 | 37 | 11 | 20 | 8 | 23 |
|---|---|---|---|---|---|---|
| Sample from Population 2 | 24 | 31 | 29 | 40 | 9 | 17 |

**FIGURE 13.3**
Dotplots for sample data in Table 13.1

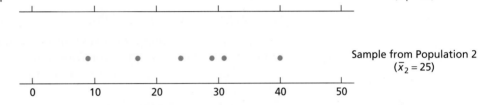

For these two samples, $\bar{x}_1 = 20$ and $\bar{x}_2 = 25$. But here we cannot infer that $\mu_1 \neq \mu_2$ because it is not clear whether the difference between the sample means is due to a difference between the population means or to the variation within the populations.

However, suppose that the sample data are as displayed in Table 13.2 and depicted in Fig. 13.4.

**What Does It Mean?**

Intuitively speaking, because the variation between the sample means is not large relative to the variation within the samples, we cannot conclude that $\mu_1 \neq \mu_2$.

**TABLE 13.2**
Sample data from Populations 1 and 2

| Sample from Population 1 | 21 | 21 | 20 | 18 | 20 | 20 |
|---|---|---|---|---|---|---|
| Sample from Population 2 | 25 | 28 | 25 | 24 | 24 | 24 |

**FIGURE 13.4**
Dotplots for sample data in Table 13.2

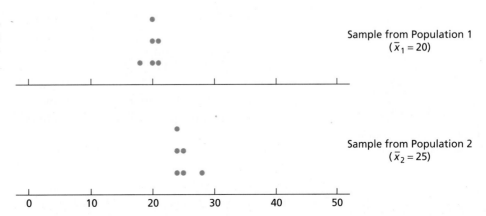

**What Does It Mean?**

Intuitively speaking, because the variation between the sample means is large relative to the variation within the samples, we can conclude that $\mu_1 \neq \mu_2$.

Again, for these two samples, $\bar{x}_1 = 20$ and $\bar{x}_2 = 25$. But this time, we *can* infer that $\mu_1 \neq \mu_2$ because it seems clear that the difference between the sample means is due to a difference between the population means, not to the variation within the populations.

The preceding two illustrations reveal the basic idea for performing a one-way analysis of variance to compare the means of several populations:

1.  Take independent simple random samples from the populations.
2.  Compute the sample means.
3.  If the variation among the sample means is large relative to the variation within the samples, conclude that the means of the populations are not all equal.

To make this process precise, we need quantitative measures of the variation among the sample means and the variation within the samples. We also need an objective method for deciding whether the variation among the sample means is large relative to the variation within the samples.

## Mean Squares and *F*-Statistic in One-Way ANOVA

As before, when dealing with several populations, we use subscripts on parameters and statistics. Thus, for Population $j$, we use $\mu_j, \bar{x}_j, s_j$, and $n_j$ to denote the population mean, sample mean, sample standard deviation, and sample size, respectively.

We first consider the measure of variation among the sample means. In hypothesis tests for two population means, we measure the variation between the two sample means by calculating their difference, $\bar{x}_1 - \bar{x}_2$. When more than two populations are involved, we cannot measure the variation among the sample means simply by taking a difference. However, we can measure that variation by computing the standard deviation or variance of the sample means or by computing any descriptive statistic that measures variation.

In one-way ANOVA, we measure the variation among the sample means by a weighted average of their squared deviations about the mean, $\bar{x}$, of all the sample data. That measure of variation is called the **treatment mean square, MSTR,** and is defined as

**What Does It Mean?**

*MSTR* measures the variation among the sample means.

$$MSTR = \frac{SSTR}{k-1}$$

where $k$ denotes the number of populations being sampled and

$$SSTR = n_1(\bar{x}_1 - \bar{x})^2 + n_2(\bar{x}_2 - \bar{x})^2 + \cdots + n_k(\bar{x}_k - \bar{x})^2.$$

The quantity **SSTR** is called the **treatment sum of squares.**

We note that *MSTR* is similar to the sample variance of the sample means. In fact, if all the sample sizes are identical, then *MSTR* equals that common sample size times the sample variance of the sample means.

Next we consider the measure of variation within the samples. This measure is the pooled estimate of the common population variance, $\sigma^2$. It is called the **error mean square, MSE,** and is defined as

**What Does It Mean?**

*MSE* measures the variation within the samples.

$$MSE = \frac{SSE}{n-k}$$

where $n$ denotes the total number of observations and

$$SSE = (n_1 - 1)s_1^2 + (n_2 - 1)s_2^2 + \cdots + (n_k - 1)s_k^2.$$

The quantity $SSE$ is called the **error sum of squares.**[†][‡]

Finally, we consider how to compare the variation among the sample means, $MSTR$, to the variation within the samples, $MSE$. To do so, we use the statistic $F = MSTR/MSE$, which we refer to as the **F-statistic.** Large values of $F$ indicate that the variation among the sample means is large relative to the variation within the samples and hence that the null hypothesis of equal population means should be rejected.

<table>
<tr><td>

**What Does It Mean?**

The F-statistic compares the variation among the sample means to the variation within the samples.

</td></tr>
</table>

**Definition 13.1**

## Mean Squares and F-Statistic in One-Way ANOVA

**Treatment mean square, MSTR:** The variation among the sample means: $MSTR = SSTR/(k - 1)$, where $SSTR$ is the treatment sum of squares and $k$ is the number of populations under consideration.

**Error mean square, MSE:** The variation within the samples: $MSE = SSE/(n - k)$, where $SSE$ is the error sum of squares and $n$ is the total number of observations.

**F-statistic, F:** The ratio of the variation among the sample means to the variation within the samples: $F = MSTR/MSE$.

**Example 13.2** | ## Introducing One-Way ANOVA

*Energy Consumption* The U.S. Energy Information Administration gathers data on residential energy consumption and expenditures and publishes its findings in *Residential Energy Consumption Survey: Consumption and Expenditures*. Suppose that we want to decide whether a difference exists in mean annual energy consumption by households among the four U.S. regions.

Let $\mu_1$, $\mu_2$, $\mu_3$, and $\mu_4$ denote last year's mean energy consumptions by households in the Northeast, Midwest, South, and West, respectively. Then the hypotheses to be tested are

$H_0$: $\mu_1 = \mu_2 = \mu_3 = \mu_4$ (mean energy consumptions are all equal)

$H_a$: Not all the means are equal.

The basic strategy for carrying out this hypothesis test follows the three steps mentioned on page 590 and is illustrated in Fig. 13.5 on the next page.

**Step 1.** Independently and randomly take samples of households in the four U.S. regions.

**Step 2.** Compute last year's mean energy consumptions, $\bar{x}_1$, $\bar{x}_2$, $\bar{x}_3$, and $\bar{x}_4$, of the four samples.

---

[†]The terms **treatment** and **error** arose from the fact that many ANOVA techniques were first developed to analyze agricultural experiments. In any case, the treatments refer to the different populations and the errors pertain to the variation within the populations.

[‡]For two populations (i.e., $k = 2$), $MSE$ is the pooled variance, $s_p^2$, defined in Section 10.2 on page 449.

**Step 3.** Reject the null hypothesis if the variation among the sample means is large relative to the variation within the samples; otherwise, do not reject the null hypothesis.

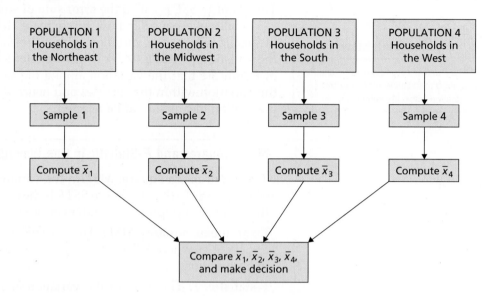

**FIGURE 13.5**
Process for comparing four population means

In Steps 1 and 2, we obtain the sample data and compute the sample means. Suppose that the results of those steps are as shown in Table 13.3, where the data are displayed to the nearest 10 million BTU.

**TABLE 13.3**
Samples and their means of last year's energy consumptions for households in the four U.S. regions

| Northeast | Midwest | South | West |
|-----------|---------|-------|------|
| 15 | 17 | 11 | 10 |
| 10 | 12 | 7 | 12 |
| 13 | 18 | 9 | 8 |
| 14 | 13 | 13 | 7 |
| 13 | 15 | | 9 |
| | 12 | | |
| 13.0 | 14.5 | 10.0 | 9.2    ← *Means* |

In Step 3, we compare the variation among the four sample means (see bottom of Table 13.3) to the variation within the samples. To accomplish that, we need to compute the treatment mean square ($MSTR$), the error mean square ($MSE$), and the $F$-statistic.

First, we determine $MSTR$. We have $k = 4$, $n_1 = 5$, $n_2 = 6$, $n_3 = 4$, $n_4 = 5$, $\bar{x}_1 = 13.0$, $\bar{x}_2 = 14.5$, $\bar{x}_3 = 10.0$, and $\bar{x}_4 = 9.2$. To find the overall mean, $\bar{x}$, we need to divide the sum of all the observations in Table 13.3 by the total number of observations:

$$\bar{x} = \frac{\Sigma x_i}{n} = \frac{15 + 10 + 13 + \cdots + 7 + 9}{20} = \frac{238}{20} = 11.9.$$

Thus,

$$SSTR = n_1(\bar{x}_1 - \bar{x})^2 + n_2(\bar{x}_2 - \bar{x})^2 + n_3(\bar{x}_3 - \bar{x})^2 + n_4(\bar{x}_4 - \bar{x})^2$$
$$= 5(13.0 - 11.9)^2 + 6(14.5 - 11.9)^2 + 4(10.0 - 11.9)^2 + 5(9.2 - 11.9)^2$$
$$= 97.5.$$

So,

$$MSTR = \frac{SSTR}{k-1} = \frac{97.5}{4-1} = 32.5.$$

Next, we determine $MSE$. We have $k = 4$, $n_1 = 5$, $n_2 = 6$, $n_3 = 4$, $n_4 = 5$, and $n = 20$. Computing the variance of each sample gives $s_1^2 = 3.5$, $s_2^2 = 6.7$, $s_3^2 = 6.\overline{6}$, and $s_4^2 = 3.7$. Consequently,

$$SSE = (n_1 - 1)s_1^2 + (n_2 - 1)s_2^2 + (n_3 - 1)s_3^2 + (n_4 - 1)s_4^2$$
$$= (5-1) \cdot 3.5 + (6-1) \cdot 6.7 + (4-1) \cdot 6.\overline{6} + (5-1) \cdot 3.7 = 82.3.$$

So,

$$MSE = \frac{SSE}{n-k} = \frac{82.3}{20-4} = 5.144.$$

Finally, we determine $F$. As $MSTR = 32.5$ and $MSE = 5.144$, the value of the $F$-statistic is

$$F = \frac{MSTR}{MSE} = \frac{32.5}{5.144} = 6.32.$$

Is this value of $F$ large enough to conclude that the null hypothesis of equal population means is false? To answer that question, we need to know the distribution of the $F$-statistic, which we discuss in Section 13.3.

Exercise 13.25
on page 594

• • •

## Exercises 13.2

### Understanding the Concepts and Skills

**13.12** State the four assumptions required for one-way ANOVA. How crucial are these assumptions?

**13.13** One-way ANOVA is a procedure for comparing the means of several populations. It is the generalization of what procedure for comparing the means of two populations?

**13.14** If we define $s = \sqrt{MSE}$, of which parameter is $s$ an estimate?

**13.15** Explain the reason for the word *variance* in the phrase *analysis of variance*.

**13.16** The null and alternative hypotheses for a one-way ANOVA test are

$$H_0: \mu_1 = \mu_2 = \cdots = \mu_k$$
$$H_a: \text{Not all means are equal.}$$

Suppose that, in reality, the null hypothesis is false. Does that mean that no two of the populations have the same mean? If not, what does it mean?

**13.17** In one-way ANOVA, identify the statistic used
**a.** as a measure of variation among the sample means.
**b.** as a measure of variation within the samples.

**c.** to compare the variation among the sample means to the variation within the samples.

**13.18** Explain the logic behind one-way ANOVA.

**13.19** What does the term *one-way* signify in the phrase *one-way ANOVA*?

**13.20** Figure 13.6 (next page) shows side-by-side boxplots of independent samples from three normally distributed populations having equal standard deviations. Based on these boxplots, would you be inclined to reject the null hypothesis of equal population means? Explain your answer.

**13.21** Figure 13.7 (next page) shows side-by-side boxplots of independent samples from three normally distributed populations having equal standard deviations. Based on these boxplots, would you be inclined to reject the null hypothesis of equal population means? Explain your answer.

**13.22** Discuss two methods for checking the assumptions of normal populations and equal standard deviations for a one-way ANOVA.

**13.23** In one-way ANOVA, what is the residual of an observation?

**FIGURE 13.6**
Side-by-side boxplots for Exercise 13.20

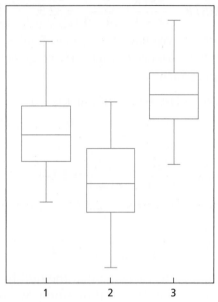

**FIGURE 13.7**
Side-by-side boxplots for Exercise 13.21

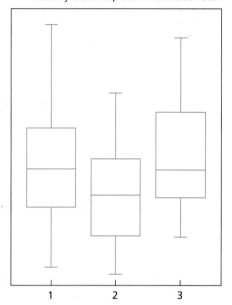

*In Exercises **13.24–13.27**, we have provided data from independent simple random samples from several populations. In each case, determine the following.*
*a. SSTR     b. MSTR     c. SSE     d. MSE     e. F*

**13.24**

| Sample 1 | Sample 2 | Sample 3 |
|----------|----------|----------|
| 1 | 10 | 4 |
| 9 | 4 | 16 |
|  | 8 | 10 |
|  | 6 |  |
|  | 2 |  |

**13.25**

| Sample 1 | Sample 2 | Sample 3 |
|----------|----------|----------|
| 8 | 2 | 4 |
| 4 | 1 | 3 |
| 6 | 3 | 6 |
|  |  | 3 |

**13.26**

| Sample 1 | Sample 2 | Sample 3 | Sample 4 |
|----------|----------|----------|----------|
| 6 | 9 | 4 | 8 |
| 3 | 5 | 4 | 4 |
| 3 | 7 | 2 | 6 |
|  | 8 | 2 |  |
|  | 6 | 3 |  |

**13.27**

| Sample 1 | Sample 2 | Sample 3 | Sample 4 | Sample 5 |
|----------|----------|----------|----------|----------|
| 7 | 5 | 6 | 3 | 7 |
| 4 | 9 | 7 | 7 | 9 |
| 5 | 4 | 5 | 7 | 11 |
| 4 |  | 4 | 4 |  |
|  |  |  | 8 |  |
|  |  |  | 4 |  |

## Extending the Concepts and Skills

**13.28** Show that, for two populations, $MSE = s_p^2$, where $s_p^2$ is the pooled variance defined in Section 10.2 on page 449. Conclude that $\sqrt{MSE}$ is the pooled sample standard deviation, $s_p$.

**13.29** Suppose that the variable under consideration is normally distributed on each of two populations and that the population standard deviations are equal. Further suppose that you want to perform a hypothesis test to decide whether the populations have different means, that is, whether $\mu_1 \neq \mu_2$. If independent simple random samples are used, identify two hypothesis-testing procedures that you can use to carry out the hypothesis test.

## 13.3 One-Way ANOVA: The Procedure

In this section, we present a step-by-step procedure for performing a one-way ANOVA to compare the means of several populations. To begin, we need to identify the distribution of the variable $F = MSTR/MSE$, introduced in Section 13.2.

**Key Fact 13.3**

### Distribution of the *F*-Statistic for One-Way ANOVA

Suppose that the variable under consideration is normally distributed on each of $k$ populations and that the population standard deviations are equal. Then, for independent samples from the $k$ populations, the variable

$$F = \frac{MSTR}{MSE}$$

has the $F$-distribution with $df = (k - 1, n - k)$ if the null hypothesis of equal population means is true. Here, $n$ denotes the total number of observations.

Although we have now covered all the elements required to formulate a procedure for performing a one-way ANOVA, we still need to consider two additional concepts.

### One-Way ANOVA Identity

First, we define another sum of squares—one that provides a measure of total variation among all the sample data. It is called the **total sum of squares, *SST,*** and is defined by

$$SST = \Sigma(x_i - \bar{x})^2,$$

**What Does It Mean?**

*SST* measures the total variation among all the sample data.

where the sum extends over all $n$ observations. If we divide $SST$ by $n - 1$, we get the sample variance of all the observations.

For the energy consumption data in Table 13.3 on page 592, $\bar{x} = 11.9$, and therefore

$$SST = \Sigma(x_i - \bar{x})^2 = (15 - 11.9)^2 + (10 - 11.9)^2 + \cdots + (9 - 11.9)^2$$
$$= 9.61 + 3.61 + \cdots + 8.41 = 179.8.$$

In Section 13.2, we found that, for the energy consumption data, $SSTR = 97.5$ and $SSE = 82.3$. Because $179.8 = 97.5 + 82.3$, we have $SST = SSTR + SSE$. This equation is always true and is called the **one-way ANOVA identity.**

**Key Fact 13.4**

### One-Way ANOVA Identity

The total sum of squares equals the treatment sum of squares plus the error sum of squares: $SST = SSTR + SSE$.

**What Does It Mean?**

The total variation among all the sample data can be partitioned into two components, one representing variation among the sample means and the other representing variation within the samples.

**Note:** The one-way ANOVA identity shows that the total variation among all the observations can be partitioned into two components. The partitioning of the total variation among all the observations into two or more components is fundamental not only in one-way ANOVA but in all types of ANOVA.

We provide a graphical representation of the one-way ANOVA identity in Fig. 13.8.

**FIGURE 13.8**
Partitioning of the total sum of squares into the treatment sum of squares and the error sum of squares

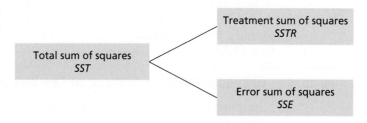

## One-Way ANOVA Tables

To organize and summarize the quantities required for performing a one-way analysis of variance, we use a **one-way ANOVA table.** The general format of such a table is as shown in Table 13.4.

**TABLE 13.4**
ANOVA table format for a one-way analysis of variance

| Source | df | SS | MS = SS/df | F-statistic |
|---|---|---|---|---|
| Treatment | $k - 1$ | $SSTR$ | $MSTR = \dfrac{SSTR}{k - 1}$ | $F = \dfrac{MSTR}{MSE}$ |
| Error | $n - k$ | $SSE$ | $MSE = \dfrac{SSE}{n - k}$ | |
| Total | $n - 1$ | $SST$ | | |

For the energy consumption data in Table 13.3, we have already computed all quantities appearing in the one-way ANOVA table. See Table 13.5.

**TABLE 13.5**
One-way ANOVA table for the energy consumption data

| Source | df | SS | MS = SS/df | F-statistic |
|---|---|---|---|---|
| Treatment | 3 | 97.5 | 32.500 | 6.32 |
| Error | 16 | 82.3 | 5.144 | |
| Total | 19 | 179.8 | | |

## Performing a One-Way ANOVA

To perform a one-way ANOVA, we need to determine the three sums of squares, $SST$, $SSTR$, and $SSE$. We can do so by using the defining formulas introduced earlier. Generally, however, when calculating by hand from the raw data, computing formulas are more accurate and easier to use. Both sets of formulas are presented next.

**Formula 13.1**    **Sums of Squares in One-Way ANOVA**

For a one-way ANOVA of $k$ population means, the defining and computing formulas for the three sums of squares are as follows.

| Sum of squares | Defining formula | Computing formula |
|---|---|---|
| Total, SST | $\Sigma(x_i - \bar{x})^2$ | $\Sigma x_i^2 - (\Sigma x_i)^2/n$ |
| Treatment, SSTR | $\Sigma n_j(\bar{x}_j - \bar{x})^2$ | $\Sigma(T_j^2/n_j) - (\Sigma x_i)^2/n$ |
| Error, SSE | $\Sigma(n_j - 1)s_j^2$ | $SST - SSTR$ |

In this table, we used the notation

$$n = \text{total number of observations}$$
$$\bar{x} = \text{mean of all } n \text{ observations;}$$

and, for $j = 1, 2, ..., k,$

$$n_j = \text{size of sample from Population } j$$
$$\bar{x}_j = \text{mean of sample from Population } j$$
$$s_j^2 = \text{variance of sample from Population } j$$
$$T_j = \text{sum of sample data from Population } j.$$

Note that summations involving subscript $i$s are over all $n$ observations; those involving subscript $j$s are over the $k$ populations.

Keep the following facts in mind when you use Formula 13.1.

- Only two of the three sums of squares need ever be calculated; the remaining one can always be found by using the one-way ANOVA identity.
- When using the computing formulas, the most efficient formula for calculating the sum of all $n$ observations is $\Sigma x_i = \Sigma T_j$.

Procedure 13.1 on the next page gives a step-by-step method for conducting a **one-way ANOVA test** by using either the critical-value approach or the $P$-value approach. Because the null hypothesis is rejected only when the test statistic, $F$, is too large, a one-way ANOVA test is always right tailed.

**Example 13.3**    **The One-Way ANOVA Test**

*Energy Consumption* Recall that independent simple random samples of households in the four U.S. regions yielded the data on last year's energy consumptions shown in Table 13.6 on the top of page 599. At the 5% significance level, do the data provide sufficient evidence to conclude that a difference exists in last year's mean energy consumption by households among the four U.S. regions?

**Solution** First, we check the four conditions required for performing a one-way ANOVA test, as listed in Procedure 13.1.

## Procedure 13.1  **One-Way ANOVA Test**

*Purpose* To perform a hypothesis test to compare $k$ population means, $\mu_1, \mu_2, \ldots, \mu_k$

*Assumptions*

1. Simple random samples
2. Independent samples
3. Normal populations
4. Equal population standard deviations

**STEP 1** The null and alternative hypotheses are

$$H_0: \mu_1 = \mu_2 = \cdots = \mu_k$$
$$H_a: \text{Not all the means are equal.}$$

**STEP 2** Decide on the significance level, $\alpha$.

**STEP 3** Obtain the three sums of squares, $SST$, $SSTR$, and $SSE$.

**STEP 4** Construct a one-way ANOVA table to obtain the value of the $F$-statistic. Denote that value $F_0$.

| Source | df | SS | MS = SS/df | F-statistic |
|--------|------|------|-----------------------|--------------------|
| Treatment | $k-1$ | $SSTR$ | $MSTR = \dfrac{SSTR}{k-1}$ | $F = \dfrac{MSTR}{MSE}$ |
| Error | $n-k$ | $SSE$ | $MSE = \dfrac{SSE}{n-k}$ | |
| Total | $n-1$ | $SST$ | | |

| CRITICAL-VALUE APPROACH | or | *P*-VALUE APPROACH |
|---|---|---|

**STEP 5** The critical value is $F_\alpha$ with df $= (k-1, n-k)$. Use Table VI to find the critical value.

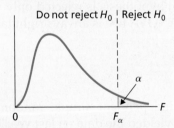

**STEP 6** If the value of the $F$-statistic falls in the rejection region, reject $H_0$; otherwise, do not reject $H_0$.

**STEP 5** The $F$-statistic has df $= (k-1, n-k)$. Use Table VI to estimate the $P$-value or obtain it exactly by using technology.

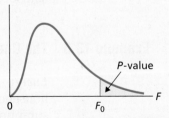

**STEP 6** If $P \leq \alpha$, reject $H_0$; otherwise, do not reject $H_0$.

**STEP 7** Interpret the results of the hypothesis test.

**TABLE 13.6**
Last year's energy consumptions for samples of households in the four U.S. regions

| Northeast | Midwest | South | West |
|:---:|:---:|:---:|:---:|
| 15 | 17 | 11 | 10 |
| 10 | 12 | 7 | 12 |
| 13 | 18 | 9 | 8 |
| 14 | 13 | 13 | 7 |
| 13 | 15 |  | 9 |
|  | 12 |  |  |

- The samples are given as simple random samples; therefore, Assumption 1 is satisfied.
- The samples are given as independent samples; therefore, Assumption 2 is satisfied.
- Normal probability plots of the four samples, presented in Fig. 13.9, show no outliers and are roughly linear, indicating no gross violations of the normality assumption; thus we can consider Assumption 3 satisfied.
- The sample standard deviations of the four samples are 1.87, 2.59, 2.58, and 1.92, respectively. The ratio of the largest to the smallest standard deviation is $2.59/1.87 = 1.39$, which is less than 2. Thus, by the rule of 2, we can consider Assumption 4 satisfied.

**FIGURE 13.9**
Normal probability plots of the energy-consumption data: (a) Northeast, (b) Midwest, (c) South, (d) West

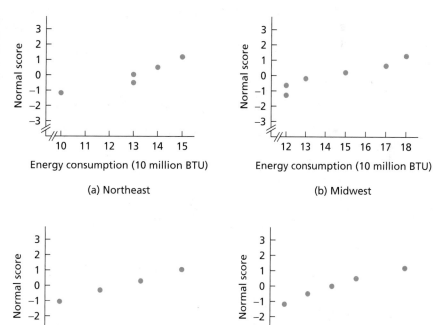

As it is reasonable to presume that the four assumptions for performing a one-way ANOVA test are satisfied, we now apply Procedure 13.1 to carry out the required hypothesis test.

### STEP 1  State the null and alternative hypotheses.

Let $\mu_1, \mu_2, \mu_3,$ and $\mu_4$ denote last year's mean energy consumptions for households in the Northeast, Midwest, South, and West, respectively. Then the null and alternative hypotheses are

$H_0$: $\mu_1 = \mu_2 = \mu_3 = \mu_4$ (mean energy consumptions are equal)
$H_a$: Not all the means are equal.

### STEP 2  Decide on the significance level, $\alpha$.

We are to perform the test at the 5% significance level; so, $\alpha = 0.05$.

### STEP 3  Obtain the three sums of squares, SST, SSTR, and SSE.

Although we determined these sums earlier by using the defining formulas, we determine them again to illustrate use of the computing formulas. Referring to Formula 13.1 on page 597 and Table 13.6, we find that

$$k = 4$$

$$n_1 = 5 \qquad n_2 = 6 \qquad n_3 = 4 \qquad n_4 = 5$$
$$T_1 = 65 \qquad T_2 = 87 \qquad T_3 = 40 \qquad T_4 = 46$$

and

$$n = \Sigma n_j = 5 + 6 + 4 + 5 = 20$$
$$\Sigma x_i = \Sigma T_j = 65 + 87 + 40 + 46 = 238.$$

Summing the squares of all the data in Table 13.6 yields

$$\Sigma x_i^2 = (15)^2 + (10)^2 + (13)^2 + \cdots + (7)^2 + (9)^2 = 3012.$$

Consequently,

$$SST = \Sigma x_i^2 - (\Sigma x_i)^2/n = 3012 - (238)^2/20 = 3012 - 2832.2 = 179.8,$$

$$SSTR = \Sigma(T_j^2/n_j) - (\Sigma x_i)^2/n$$
$$= (65)^2/5 + (87)^2/6 + (40)^2/4 + (46)^2/5 - (238)^2/20$$
$$= 2929.7 - 2832.2 = 97.5,$$

and

$$SSE = SST - SSTR = 179.8 - 97.5 = 82.3.$$

### STEP 4  Construct a one-way ANOVA table to obtain the value of the F-statistic.

Table 13.5 on page 596 is the one-way ANOVA table for the energy consumption data. It reveals that $F = 6.32$.

| CRITICAL-VALUE APPROACH | or | P-VALUE APPROACH |
|---|---|---|

**STEP 5** The critical value is $F_\alpha$ with df $= (k - 1, n - k)$. Use Table VI to find the critical value.

From Step 2, $\alpha = 0.05$. Also, Table 13.6 shows that four populations are under consideration, or $k = 4$, and that the number of observations total 20, or $n = 20$. Hence, df $= (k - 1, n - k) = (4 - 1, 20 - 4) = (3, 16)$. From Table VI, the critical value is $F_{0.05} = 3.24$, as shown in Fig. 13.10A.

**FIGURE 13.10A**

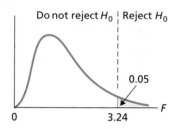

**STEP 6** If the value of the $F$-statistic falls in the rejection region, reject $H_0$; otherwise, do not reject $H_0$.

From Step 4, the value of the $F$-statistic is $F = 6.32$, which, as Fig. 13.10A shows, falls in the rejection region. Thus we reject $H_0$. The test results are statistically significant at the 5% level.

**STEP 5** The $F$-statistic has df $= (k - 1, n - k)$. Use Table VI to estimate the $P$-value or obtain it exactly by using technology.

From Step 4, the value of the $F$-statistic is $F = 6.32$. Because the test is right tailed, the $P$-value is the probability of observing a value of $F$ of 6.32 or greater if the null hypothesis is true. That probability equals the shaded area in Fig. 13.10B.

**FIGURE 13.10B**

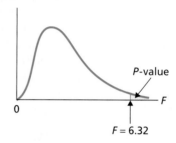

From Table 13.6, four populations are under consideration, or $k = 4$, and the number of observations total 20, or $n = 20$. Thus, we have df $= (k - 1, n - k) = (4 - 1, 20 - 4) = (3, 16)$. Referring to Fig. 13.10B and to Table VI with df $= (3, 16)$, we find $P < 0.005$. (Using technology, we get $P = 0.00495$.)

**STEP 6** If $P \le \alpha$, reject $H_0$; otherwise, do not reject $H_0$.

From Step 5, $P < 0.005$. Because the $P$-value is less than the specified significance level of 0.05, we reject $H_0$. The test results are statistically significant at the 5% level and (see Table 9.10 on page 414) provide very strong evidence against the null hypothesis.

**STEP 7** Interpret the results of the hypothesis test.

**You try it!**

Exercise 13.45 on page 606

**Interpretation** At the 5% significance level, the data provide sufficient evidence to conclude that a difference exists in last year's mean energy consumption by households among the four U.S. regions. Evidently, at least two of the regions have different mean energy consumptions.

• • •

## Multiple Comparisons

The results of the one-way ANOVA in Example 13.3 show that (at the 5% significance level) a difference exists in last year's mean energy consumption by households among the four U.S. regions. But the analysis does not tell us which means are different, which mean is largest, or, more generally, the relationship among the four population means. Such questions are answered by using techniques called *multiple comparisons*.

Although we do not cover multiple comparisons in this book, many basic statistics texts discuss these techniques. See, for example, *Introductory Statistics, 8/e,* by Neil A. Weiss (Boston: Addison-Wesley, 2008).

### What If the Assumptions Are Not Satisfied?

One-way ANOVA provides methods for comparing the means of several populations. As you know, the assumptions for using that procedure are (1) simple random samples, (2) independent samples, (3) normal populations, and (4) equal population standard deviations. If one or more of these conditions are not satisfied, then the one-way ANOVA test should not be used. The test that should be used depends on which conditions are violated and which hold.

For example, suppose that you have independent simple random samples and that the distributions (one for each population) of the variable under consideration have the same shape. Then you can use a nonparametric method called the *Kruskal–Wallis test*, regardless of normality.[†]

### Other Types of ANOVA

We can consider one-way ANOVA to be a method for comparing the means of populations classified according to one factor. Put another way, it is a method for analyzing the effect of one factor on the mean of the variable under consideration, called the **response variable.**

For instance, in Example 13.3, we compared last year's mean energy consumption by households among the four U.S. regions (Northeast, Midwest, South, and West). Here, the factor is "region" and the response variable is "energy consumption." One-way ANOVA permits us to analyze the effect of region on mean energy consumption.

Other ANOVA procedures provide methods for comparing the means of populations classified according to two or more factors. Put another way, these are methods for simultaneously analyzing the effect of two or more factors on the mean of a response variable.

For example, suppose that you want to consider the effect of "region" and "home type" (the two factors) on energy consumption (the response variable). Two-way ANOVA permits you to determine simultaneously whether region affects mean energy consumption, whether home type affects mean energy consumption, and whether region and home type interact in their effect on mean energy consumption (e.g., whether the effect of home type on mean energy consumption depends on region).

Two-way ANOVA and other ANOVA procedures, such as randomized block ANOVA, are treated in detail in the chapter *Design of Experiments and Analysis of Variance*, on the WeissStats CD accompanying this book.

---

[†]Although we will not cover nonparametric methods in this book, you can find a discussion of the Kruskal–Wallis test and other nonparametric procedures in the book *Introductory Statistics, 8/e,* by Neil A. Weiss (Boston: Addison-Wesley, 2008).

# The Technology Center

Most statistical technologies have programs that automatically perform a one-way analysis of variance. In this subsection, we present output and step-by-step instructions for such programs.

**Example 13.4** **Using Technology to Conduct a One-Way ANOVA Test**

*Energy Consumption* Table 13.6 on page 599 shows last year's energy consumptions for independent simple random samples of households in the four U.S. regions. Use Minitab, Excel, or the TI-83/84 Plus to decide, at the 5% significance level, whether the data provide sufficient evidence to conclude that a difference exists in last year's mean energy consumption by households among the four U.S. regions.

**Solution** Let $\mu_1, \mu_2, \mu_3$, and $\mu_4$ denote last year's mean energy consumptions for households in the Northeast, Midwest, South, and West, respectively. We want to perform the hypothesis test

$H_0$: $\mu_1 = \mu_2 = \mu_3 = \mu_4$ (mean energy consumptions are equal)
$H_a$: Not all the means are equal

at the 5% significance level.

We applied the one-way ANOVA programs to the data, resulting in Output 13.1, shown at the top of the following page. Steps for generating that output are presented in Instructions 13.1, also on the next page.

As shown in Output 13.1, the $P$-value for the hypothesis test is 0.005. Because the $P$-value is less than the specified significance level of 0.05, we reject $H_0$. At the 5% significance level, the data provide sufficient evidence to conclude that last year's mean energy consumptions for households in the four U.S. regions are not all the same.

• • •

**OUTPUT 13.1** One-way ANOVA test for the energy consumption data

**MINITAB**

## One-way ANOVA: ENERGY versus REGION

```
Source  DF      SS     MS    F       P
REGION   3   97.50  32.50  6.32   0.005
Error   16   82.30   5.14
Total   19  179.80

S = 2.268   R-Sq = 54.23%   R-Sq(adj) = 45.64%
```

**EXCEL**

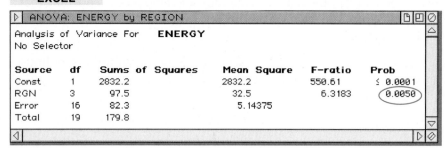

```
ANOVA: ENERGY by REGION

Analysis of Variance For    ENERGY
No Selector

Source   df   Sums of Squares   Mean Square   F-ratio   Prob
Const     1   2832.2            2832.2        550.61    ≤ 0.0001
RGN       3     97.5              32.5          6.3183     0.0050
Error    16     82.3               5.14375
Total    19    179.8
```

**TI-83/84 PLUS**

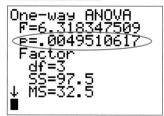

```
One-way ANOVA
 F=6.318347509
 P=.0049510617
 Factor
  df=3
  SS=97.5
↓ MS=32.5
■
```

```
One-way ANOVA
↑ MS=32.5
 Error
  df=16
  SS=82.3
  MS=5.14375
 Sxp=2.26798369
```

**INSTRUCTIONS 13.1** Steps for generating Output 13.1

**MINITAB**

1. Store all 20 energy consumptions from Table 13.6 in a column named ENERGY
2. Store the regions corresponding to the energy consumptions in a column named REGION
3. Choose **Stat ➤ ANOVA ➤ One-Way...**
4. Specify ENERGY in the **Response** text box
5. Specify REGION in the **Factor** text box
6. Click **OK**

**EXCEL**

1. Store all 20 energy consumptions from Table 13.6 in a range named ENERGY
2. Store the regions corresponding to the energy consumptions in a range named REGION
3. Choose **DDXL ➤ ANOVA**
4. Select **1 Way ANOVA** from the **Function type** drop-down box
5. Specify ENERGY in the **Response Variable** text box
6. Specify REGION in the **Factor Variable** text box
7. Click **OK**

**TI-83/84 PLUS**

1. Store the four samples from Table 13.6 in lists named NE, MW, SO, and WE
2. Press **STAT**, arrow over to **TESTS**, and press **ALPHA ➤ F**
3. Press **2nd ➤ LIST**, arrow down to NE and press **ENTER**
4. Press **, ➤ 2nd ➤ LIST**, arrow down to MW, and press **ENTER**
5. Press **, ➤ 2nd ➤ LIST**, arrow down to SO, and press **ENTER**
6. Press **, ➤ 2nd ➤ LIST**, arrow down to WE, and press **ENTER**
7. Press **)** and then **ENTER**

# Exercises 13.3

## Understanding the Concepts and Skills

**13.30** Suppose that a one-way ANOVA is being performed to compare the means of three populations and that the sample sizes are 10, 12, and 15. Determine the degrees of freedom for the $F$-statistic.

**13.31** We stated earlier that a one-way ANOVA test is always right tailed because the null hypothesis is rejected only when the test statistic, $F$, is too large. Why is the null hypothesis rejected only when $F$ is too large?

**13.32** Following are the notations for the three sums of squares. State the name of each sum of squares and the source of variation each sum of squares represents.
**a.** $SSE$          **b.** $SSTR$          **c.** $SST$

**13.33** State the one-way ANOVA identity and interpret its meaning with regard to partitioning the total variation in the data.

**13.34** True or false: If you know any two of the three sums of squares, $SST$, $SSTR$, and $SSE$, you can determine the remaining one. Explain your answer.

**13.35** In each part, specify what type of analysis you might use.
**a.** To study the effect of one factor on the mean of a response variable
**b.** To study the effect of two factors on the mean of a response variable

*In Exercises 13.36–13.39, fill in the missing entries in the partially completed one-way ANOVA tables.*

**13.36**

| Source | df | SS | MS = SS/df | F-statistic |
|---|---|---|---|---|
| Treatment | 2 | | 21.652 | |
| Error | | 84.400 | | |
| Total | 14 | | | |

**13.37**

| Source | df | SS | MS = SS/df | F-statistic |
|---|---|---|---|---|
| Treatment | | 2.124 | 0.708 | 0.75 |
| Error | 20 | | | |
| Total | | | | |

**13.38**

| Source | df | SS | MS = SS/df | F-statistic |
|---|---|---|---|---|
| Treatment | 4 | | | |
| Error | 20 | | 6.76 | |
| Total | | 173.04 | | |

**13.39**

| Source | df | SS | MS = SS/df | F-statistic |
|---|---|---|---|---|
| Treatment | | | 1.4 | |
| Error | 12 | | 0.9 | |
| Total | 14 | | | |

*In Exercises 13.40–13.43, we provide data from independent simple random samples from several populations. In each case, do the following.*
*a. Compute SST, SSTR, and SSE by using the computing formulas given in Formula 13.1 on page 597.*
*b. Compare your results in part (a) for SSTR and SSE with those in Exercises 13.24–13.27, where you employed the defining formulas.*
*c. Construct a one-way ANOVA table.*
*d. Decide, at the 5% significance level, whether the data provide sufficient evidence to conclude that the means of the populations from which the samples were drawn are not all the same.*

**13.40**

| Sample 1 | Sample 2 | Sample 3 |
|---|---|---|
| 1 | 10 | 4 |
| 9 | 4 | 16 |
| | 8 | 10 |
| | 6 | |
| | 2 | |

**13.41**

| Sample 1 | Sample 2 | Sample 3 |
|---|---|---|
| 8 | 2 | 4 |
| 4 | 1 | 3 |
| 6 | 3 | 6 |
| | | 3 |

**13.42**

| Sample 1 | Sample 2 | Sample 3 | Sample 4 |
|---|---|---|---|
| 6 | 9 | 4 | 8 |
| 3 | 5 | 4 | 4 |
| 3 | 7 | 2 | 6 |
| | 8 | 2 | |
| | 6 | 3 | |

**13.43**

| Sample 1 | Sample 2 | Sample 3 | Sample 4 | Sample 5 |
|---|---|---|---|---|
| 7 | 5 | 6 | 3 | 7 |
| 4 | 9 | 7 | 7 | 9 |
| 5 | 4 | 5 | 7 | 11 |
| 4 | | 4 | 4 | |
| | | 8 | 4 | |

*In Exercises 13.44–13.49, apply Procedure 13.1 on page 598 to perform a one-way ANOVA test by using either the critical-value approach or the P-value approach.*

**13.44 Movie Guide.** Movie fans use the annual *Leonard Martin Movie Guide* for facts, cast members, and reviews of over 21,000 films. The movies are rated from 4 stars (4*), indicating a very good movie, to 1 star (1*), which Leonard Martin refers to as a BOMB. The following table gives the running times, in minutes, of a random sample of films listed in the 2006 guide.

| 1* or 1.5* | 2* or 2.5* | 3* or 3.5* | 4* |
|---|---|---|---|
| 75 | 97 | 101 | 101 |
| 95 | 70 | 89 | 135 |
| 84 | 105 | 97 | 93 |
| 86 | 119 | 103 | 117 |
| 58 | 87 | 86 | 126 |
| 85 | 95 | 100 | 119 |

At the 1% significance level, do the data provide sufficient evidence to conclude that a difference exists in mean running times among films in the four rating groups? (*Note:* $T_1 = 483, T_2 = 573, T_3 = 576, T_4 = 691$, and $\Sigma x_i^2 = 232{,}117$).

**13.45 Copepod Cuisine.** Copepods are tiny crustaceans that are an essential link in the estuarine food web. Marine scientists G. Weiss, G. McManus, and H. Harvey at the Chesapeake Biological Laboratory in Maryland designed an experiment to determine whether dietary lipid (fat) content is important in the population growth of a Chesapeake Bay copepod. Their findings were published as the paper "Development and Lipid Composition of the Harpacticoid Copepod Nitocra Spinipes Reared on Different Diets" (*Marine Ecology Progress Series*, Vol. 132, pp. 57–61). Independent random samples of copepods were placed in containers containing lipid-rich diatoms, bacteria, or leafy macroalgae. There were 12 containers total with four replicates per diet. Five gravid (egg-bearing) females were placed in each container. After 14 days, the number of copepods in each container were as follows.

| Diatoms | Bacteria | Macroalgae |
|---|---|---|
| 426 | 303 | 277 |
| 467 | 301 | 324 |
| 438 | 293 | 302 |
| 497 | 328 | 272 |

At the 5% significance level, do the data provide sufficient evidence to conclude that a difference exists in mean number of copepods among the three different diets? (*Note:* $T_1 = 1828, T_2 = 1225, T_3 = 1175$, and $\Sigma x_i^2 = 1{,}561{,}154$.)

**13.46** In Section 13.2, we considered two hypothetical examples to explain the logic behind one-way ANOVA. Now, you are to further examine those examples.

**a.** Refer to Table 13.1 on page 589. Perform a one-way ANOVA on the data and compare your conclusion to that stated in the corresponding "What Does it Mean?" box. Use $\alpha = 0.05$.

**b.** Repeat part (a) for the data in Table 13.2 on page 589.

**13.47 Staph Infections.** In the article "Using EDE, ANOVA and Regression to Optimize Some Microbiology Data" (*Journal of Statistics Education*, Vol. 12, No. 2, online), N. Binnie analyzes bacteria-culture data collected by Gavin Cooper at the Auckland University of Technology. Five strains of cultured *Staphylococcus aureus*—bacteria that cause staph infections—were observed for 24 hours at 27 degrees Celsius. The following table reports bacteria counts, in millions, for different cases from each of the five strains.

| Strain A | Strain B | Strain C | Strain D | Strain E |
|---|---|---|---|---|
| 9 | 3 | 10 | 14 | 33 |
| 27 | 32 | 47 | 18 | 43 |
| 22 | 37 | 50 | 17 | 28 |
| 30 | 45 | 52 | 29 | 59 |
| 16 | 12 | 26 | 20 | 31 |

At the 5% significance level, do the data provide sufficient evidence to conclude that a difference exists in mean bacteria counts among the five strains of *Staphylococcus aureus*? (*Note:* $T_1 = 104, T_2 = 129, T_3 = 185, T_4 = 98, T_5 = 194$, and $\Sigma x_i^2 = 25{,}424$.)

**13.48 Permeation Sampling.** Permeation sampling is a method of sampling air in buildings for pollutants. It can be used over a long period of time and is not affected by humidity, air currents, or temperature. In the paper "Calibration of Permeation Passive Samplers With Silicone Membranes Based on Physicochemical Properties of the Analytes" (*Analytical Chemistry*, Vol. 75, No. 13, pp. 3182–3192), B. Zabiegata, T. Gorecki, and J. Namiesnik obtained calibration constants experimentally for samples of compounds in each of four compound groups. The following data summarize their results.

| Esters | Alcohols | Aliphatic hydrocarbons | Aromatic hydrocarbons |
|---|---|---|---|
| 0.185 | 0.185 | 0.230 | 0.166 |
| 0.155 | 0.160 | 0.184 | 0.144 |
| 0.131 | 0.142 | 0.160 | 0.117 |
| 0.103 | 0.122 | 0.132 | 0.072 |
| 0.064 | 0.117 | 0.100 | |
| | 0.115 | 0.064 | |
| | 0.110 | | |
| | 0.095 | | |
| | 0.085 | | |
| | 0.075 | | |

At the 5% significance level, do the data provide sufficient evidence to conclude that a difference exists in mean calibration constant among the four compound groups? (*Note:* $T_1 = 0.638$, $T_2 = 1.206$, $T_3 = 0.870$, $T_4 = 0.499$, and $\Sigma x_i^2 = 0.456919$.)

**13.49 Wheat Resistance.** J. Engle et al. explore wheat's resistance to a disease that causes leaf-spotting, shriveling, and reduced yield in the article "Reaction of Commercial Soft Red Winter Wheat Cultivars to *Stagonospora nodorum* in the Greenhouse and Field" (*Plant Disease, An International Journal of Applied Plant Pathology*, Vol. 90, No. 5, pp. 576–582). Fields of soft red winter wheat were subjected to the pathogen *Stagonospora nodorum* during the summers of 2000–2002. Afterwards, the fields were rated on a scale of 0 (resistant to the disease) to 10 (susceptible to the disease). The following table gives the rankings of the fields tested during the three summers.

| 2000 | | 2001 | | 2002 | |
|------|------|------|------|------|------|
| 7.0 | 4.0 | 7.0 | 7.3 | 5.0 | 4.7 |
| 5.0 | 3.0 | 7.3 | 7.0 | 4.7 | 5.0 |
| 8.0 | 8.0 | 6.0 | 7.0 | 6.0 | 3.7 |
| 3.0 | 6.0 | 5.0 | 7.0 | 5.0 | 4.3 |
| 2.0 | 2.0 | 8.0 | 6.3 | 3.3 | 4.3 |
| 7.0 | | 3.3 | 6.7 | | |
| | | 6.3 | | | |

At the 5% significance level, do the data provide sufficient evidence to conclude that a difference exists in mean resistance to *Stagonospora nodorum* among the three years of wheat harvests? (*Note:* $T_1 = 55$, $T_2 = 84.2$, $T_3 = 46$, and $\Sigma x_i^2 = 1,108.48$.)

*In Exercises 13.50–13.55, use the technology of your choice to do the following.*

*a. Conduct a one-way ANOVA test on the data.*

*b. Interpret your results from part (a).*

*c. Decide whether presuming that the assumptions of normal populations and equal population standard deviations are met is reasonable.*

**13.50 Empty Stomachs.** In the publication "How Often Do Fishes 'Run on Empty'?" (*Ecology*, Vol. 83, No 8, pp. 2145–2151), D. Arrington et al. examined almost 37,000 fish of 254 species from the waters of Africa, South and Central America, and North America to determine the percentage of fish with empty stomachs. The fish were classified as piscivores (fish-eating), invertivores (invertibrate-eating), omnivores (anything-eating) and algivores/detritivores (algae and other organic matter). For those fish in African waters, the data on the WeissStats CD give the proportions of each species of fish with empty stomachs. At the 1% significance level, do the data provide sufficient evidence to conclude that a difference exists in the mean percentages of fish with empty stomachs among the four different types of feeders?

**13.51 Monthly Rents.** The U.S. Census Bureau collects data on monthly rents of newly completed apartments and publishes the results in *Current Housing Reports*. Independent random samples of newly completed apartments in the four U.S. regions yielded the data on monthly rents, in dollars, given on the WeissStats CD. At the 5% significance level, do the data provide sufficient evidence to conclude that a difference exists in mean monthly rents among newly completed apartments in the four U.S. regions?

**13.52 Ground Water.** The U.S. Geological Survey, in cooperation with the Florida Department of Environmental Protection, investigated the effects of waste disposal practices on ground water quality at five poultry farms in north-central Florida. At one site, they drilled four monitoring wells, numbered 1, 2, 3, and 4. Over a period of 9 months, water samples were collected from the last three wells and analyzed for a variety of chemicals, including potassium, chlorides, nitrates, and phosphorus. The concentrations, in milligrams per liter, are provided on the WeissStats CD. For each of the four chemicals, decide whether the data provide sufficient evidence to conclude that a difference exists in mean concentration among the three wells. Use $\alpha = 0.01$. [SOURCE: USGS Water Resources Investigations Report 95-4064, *Effects of Waste-Disposal Practices on Ground-Water Quality at Five Poultry (Broiler) Farms in North-Central Florida*, H. Hatzell, U.S. Geological Survey.]

**13.53 Rock Sparrows.** Rock sparrows breeding in northern Italy are the subject of a long-term ecology and conservation study due to their wide variety of breeding patterns. Both males and females have a yellow patch on their breasts that is thought to play a significant role in their sexual behavior. A. Pilastro et al. conducted an experiment in which they increased or reduced the size of a female's breast patch by dying feathers at the edge of a patch and then observed several characteristics of the behavior of the male. Their results were published in the paper "Male Rock Sparrows Adjust Their Breeding Strategy According to Female Ornamentation: Parental or Mating Investment?" (*Animal Behaviour*, Vol. 66, Issue 2, pp. 265–271). Eight mating pairs were observed in each of three groups: a reduced patch size group, a control group, and an enlarged patch size group. The data on the WeissStats CD, based on the results reported by the researchers, give the number of minutes per hour that males sang in the vicinity of the nest after the patch size manipulation was done on the females. At the 1% significance level, do the data provide sufficient evidence to conclude that a difference exists in the mean singing rates among male rock sparrows exposed to the three types of breast treatments?

**13.54 Artificial Teeth: Wear.** In a study by J. Zeng et al. (*Journal of Prosthetic Dentistry*, Vol. 94, Issue 5, pp. 453–457), three materials for making artificial teeth—Endura, Duradent, and Duracross—were tested for wear. Using a machine that simulated grinding by two right first molars at 60 strokes per minute for a total of 50,000 strokes, the researchers measured the volume of material worn away, in cubic millimeters. Six pairs of teeth were tested for each material. The data on the WeissStats CD are based on the results obtained by the researchers. At the 5% significance level, do the data provide sufficient evidence to conclude that there is a difference in mean wear among the three materials?

**13.55 Artificial Teeth: Hardness.** In a study by J. Zeng et al. (*Journal of Prosthetic Dentistry*, Vol. 94, Issue 5, pp. 453–457), three materials for making artificial teeth—Endura, Duradent, and Duracross—were tested for hardness. The Vickers microhardness (VHN) of the occlusal surfaces was measured with a load of 50 grams and a loading time of 30 seconds. Six pairs of teeth were tested for each material. The data on the WeissStats CD are based on the results obtained by the researchers. At the 5% significance level, do the data provide sufficient evidence to conclude that there is a difference in mean hardness among the three materials?

## Working With Large Data Sets

*In Exercises 13.56–13.64, use the technology of your choice to do the following.*
*a. Obtain individual normal probability plots and the standard deviations of the samples.*
*b. Perform a residual analysis.*
*c. Use your results from parts (a) and (b) to decide whether conducting a one-way ANOVA test on the data is reasonable. If so, also do parts (d) and (e).*
*d. Use a one-way ANOVA test to decide, at the 5% significance level, whether the data provide sufficient evidence to conclude that a difference exists among the means of the populations from which the samples were taken.*
*e. Interpret your results from part (d).*

**13.56 Daily TV Viewing Time.** Nielsen Media Research collects information on daily TV viewing time, in hours, and publishes its findings in *Time Spent Viewing*. The WeissStats CD provides data on daily viewing times of independent simple random samples of men, women, teens, and children.

**13.57 Earnings by Degree.** The U.S. Census Bureau publishes information on annual earnings of persons in the document *Current Population Reports*. Independent simple random samples of people with various amounts of college education gave the data on annual earnings, in thousands of dollars, by highest degree obtained, presented on the WeissStats CD. We have used "Some" to represent "Some college, but no college degree."

**13.58 Popular Diets.** In the article "Comparison of the Atkins, Ornish, Weight Watchers, and Zone Diets for Weight Loss and Heart Disease Risk Reduction" (*Journal of the American Medical Association*, Vol. 293, No. 1, pp. 43–53), M. Dansinger et al. conducted a randomized trial to assess the effectiveness of four popular diets for weight loss. Overweight adults with average body mass index of 35 and ages 22–72 years participated in the randomized trial for 1 year. The weight losses, in kilograms, based on the results of the experiment, are given on the WeissStats CD. Negative losses are gains. WW = Weight Watchers.

**13.59 Cuckoo Care.** Many species of cuckoos are brood parasites. The females lay their eggs in the nests of smaller bird species that then raise the young cuckoos at the expense of their own young. The question might be asked, "Do the cuckoos lay the same size eggs regardless of the size of the bird whose nest they use?" Data on the lengths, in millimeters, of cuckoo eggs found in the nests of six bird species—Meadow Pipit, Tree Pipit, Hedge Sparrow, Robin, Pied Wagtail, and Wren—are provided on the WeissStats CD. These data were collected by the late O. Latter in 1902 and used by L. Tippett in his text *The Methods of Statistics* (New York: Wiley, 1952, p. 176).

**13.60 Doing Time.** The U.S. Bureau of Prisons publishes data in *Statistical Report* on the times served by prisoners released from federal institutions for the first time. Independent simple random samples of released prisoners for five different offense categories yielded the data on time served, in months, shown on the WeissStats CD.

**13.61 Book Prices.** The R. R. Bowker Company collects data on book prices and publishes its finding in *The Bowker Annual Library and Book Trade Almanac*. Independent simple random samples of hardcover books in law, science, medicine, and technology gave the data, in dollars, on the WeissStats CD.

**13.62 Magazine Ads.** Advertising researchers Shuptrine and McVicker wanted to determine whether there were significant differences in the readability of magazine advertisements. Thirty magazines were classified based on their educational level—high, mid, or low—and then three magazines were randomly selected from each level. From each magazine, six advertisements were randomly chosen and examined for readability. In this particular case, readability was characterized by the numbers of words, sentences, and words of three syllables or more in each ad. The researchers published their findings in the *Journal of Advertising Research* (Vol. 21, p. 47). The number of words of three syllables or more in each ad are provided on the WeissStats CD.

**13.63 Sickle Cell Disease.** A study published by E. Anionwu et al. in the *British Medical Journal* (Vol. 282, pp. 283–286) measured the steady state hemoglobin levels of patients with three different types of sickle cell disease: HB SS, HB ST, and HB SC. The data are presented on the WeissStats CD.

**13.64 Prolonging Life.** Vitamin C (ascorbate) boosts the human immune system and is effective in preventing a variety of illnesses. In a study by E. Cameron and L. Pauling, published in the *Proceedings of the National Academy of Science USA* (Vol. 75, pp. 4538–4542), patients in advanced stages of cancer were given a Vitamin C supplement. Patients were grouped according to the organ affected by cancer: stomach, bronchus, colon, ovary, or breast. The study yielded the survival times, in days, given on the WeissStats CD.

## Extending the Concepts and Skills

**13.65 Political Prisoners.** Journal articles and other sources frequently provide only summary statistics of data. This exercise gives you practice in working with such data in the context of ANOVA. According to the American Psychiatric Association, posttraumatic stress disorder (PTSD) is a common psychological consequence of traumatic events that involve threat to life or physical integrity. During the Cold War, some 200,000 people in East Germany were imprisoned for political reasons. Many of these prisoners were subjected to physical and psychological torture during their imprisonment, resulting in PTSD. Ehlers, Maercker, and Boos studied various characteristics of political prisoners from the former East Germany and presented their findings in the paper "Posttraumatic Stress Disorder (PTSD) Following Political Imprisonment: The Role of Mental Defeat, Alienation, and Perceived Permanent Change" (*Journal of Abnormal Psychology*, Vol. 109, pp. 45–55). The researchers randomly and independently selected 32 former prisoners diagnosed with chronic PTSD, 20 former prisoners diagnosed with PTSD after release from prison but subsequently recovered (remitted), and 29 diagnosed with no signs of PTSD. The ages, in years, of these people at time of arrest were as follows.

| PTSD | $n_j$ | $\bar{x}_j$ | $s_j$ |
|------|-------|-------------|-------|
| Chronic | 32 | 25.8 | 9.2 |
| Remitted | 20 | 22.1 | 5.7 |
| None | 29 | 26.6 | 9.6 |

At the 1% significance level, do the data provide sufficient evidence to conclude that a difference exists in mean age at time of arrest among the three types of former prisoners? (*Note:* Use the formula

$$\bar{x} = \frac{n_1\bar{x}_1 + n_2\bar{x}_2 + n_3\bar{x}_3}{n_1 + n_2 + n_3}$$

to determine the mean of all the observations.)

**Confidence Intervals in One-Way ANOVA.** Assume that the conditions for one-way ANOVA are satisfied and let $s = \sqrt{MSE}$. Then we have the following confidence-interval formulas.

- A $(1 - \alpha)$-level confidence interval for any particular population mean, say, $\mu_i$, has endpoints

$$\bar{x}_i \pm t_{\alpha/2} \cdot \frac{s}{\sqrt{n_i}}.$$

- A $(1 - \alpha)$-level confidence interval for the difference between any two particular population means, say, $\mu_i$ and $\mu_j$, has endpoints

$$(\bar{x}_i - \bar{x}_j) \pm t_{\alpha/2} \cdot s\sqrt{(1/n_i) + (1/n_j)}.$$

In both formulas, df $= n - k$, where, as usual, $k$ denotes the number of populations and $n$ denotes the total number of observations. Apply these formulas in Exercise 13.66.

**13.66 Monthly Rents.** Refer to Exercise 13.51. The data on monthly rents, in dollars, for independent random samples of newly completed apartments in the four U.S. regions are presented in the following table.

| Northeast | Midwest | South | West |
|-----------|---------|-------|------|
| 1005 | 870 | 891 | 1025 |
| 898 | 748 | 630 | 1012 |
| 948 | 699 | 861 | 1090 |
| 1181 | 814 | 1036 | 926 |
| 1244 | 721 | | 1269 |
| | 606 | | |

a. Find and interpret a 95% confidence interval for the mean monthly rent of newly completed apartments in the Midwest.
b. Find and interpret a 95% confidence interval for the difference between the mean monthly rents of newly completed apartments in the Northeast and South.
c. What assumptions are you making in solving parts (a) and (b)?

**13.67 Monthly Rents.** Refer to Exercise 13.66. Suppose that you have obtained a 95% confidence interval for each of the two differences, $\mu_1 - \mu_2$ and $\mu_1 - \mu_3$. Can you be 95% confident of both results simultaneously, that is, that both differences are contained in their corresponding confidence intervals? Explain your answer.

## Chapter in Review

### You Should be Able to

1. use and understand the formulas in this chapter.

2. use the $F$-table, Table VI in Appendix A.

3. explain the essential ideas behind a one-way analysis of variance.

4. state and check the assumptions required for a one-way ANOVA.

5. obtain the sums of squares for a one-way ANOVA by using the defining formulas.

6. obtain the sums of squares for a one-way ANOVA by using the computing formulas.

7. compute the mean squares and the $F$-statistic for a one-way ANOVA.

8. construct a one-way ANOVA table.

9. perform a one-way ANOVA test.

### Key Terms

analysis of variance (ANOVA), *584*
degrees of freedom for the denominator, *586*
degrees of freedom for the numerator, *586*
error, *591*
error mean square (MSE), *591*
error sum of squares (SSE), *591*
$F_\alpha$, *586*

$F$-curve, *586*
$F$-distribution, *586*
$F$-statistic, *591*
factor, *588*
levels, *588*
one-way analysis of variance, *588*
one-way ANOVA identity, *595*
one-way ANOVA table, *596*
one-way ANOVA test, *597, 598*

residual, *588*
residual analysis, *588*
response variable, *602*
rule of 2, *588*
total sum of squares (SST), *595*
treatment, *591*
treatment mean square (MSTR), *591*
treatment sum of squares (SSTR), *590*

## Review Problems

### Understanding the Concepts and Skills

1. For what is one-way ANOVA used?

2. State the four assumptions for one-way ANOVA and explain how those assumptions can be checked.

3. On what distribution does one-way ANOVA rely?

4. Suppose that you want to compare the means of three populations by using one-way ANOVA. If the sample sizes are 5, 6, and 6, determine the degrees of freedom for the appropriate $F$-curve.

5. In one-way ANOVA, identify a statistic that measures
a. the variation among the sample means.
b. the variation within the samples.

6. In one-way ANOVA,
a. list and interpret the three sums of squares.
b. state the one-way ANOVA identity and interpret its meaning with regard to partitioning the total variation among all the data.

7. For a one-way ANOVA,
a. identify one purpose of one-way ANOVA tables.
b. construct a generic one-way ANOVA table.

8. Consider an $F$-curve with df $= (2, 14)$.
a. Identify the degrees of freedom for the numerator.
b. Identify the degrees of freedom for the denominator.
c. Determine $F_{0.05}$.
d. Find the $F$-value having area 0.01 to its right.
e. Find the $F$-value having area 0.05 to its right.

9. Consider the following hypothetical samples.

| A | B | C |
|---|---|---|
| 1 | 0 | 3 |
| 3 | 6 | 12 |
| 5 | 2 | 6 |
|   | 5 | 3 |
|   | 2 |   |

a. Obtain the sample mean and sample standard deviation of each of the three samples.
b. Obtain $SST$, $SSTR$, and $SSE$ by using the defining formulas and verify that the one-way ANOVA identity holds.
c. Obtain $SST$, $SSTR$, and $SSE$ by using the computing formulas.
d. Construct the one-way ANOVA table.

10. **Losses to Robbery.** The U.S. Federal Bureau of Investigation conducts surveys to obtain information on the value of losses from various types of robberies. Results of the surveys are published in *Population-at-Risk Rates and Selected Crime Indicators*. Independent simple random samples of reports for three types of robberies—highway, gas station, and convenience store—gave the following data, in dollars, on value of losses.

| Highway | Gas station | Convenience store |
|---------|-------------|-------------------|
| 797 | 825 | 665 |
| 841 | 722 | 742 |
| 684 | 701 | 701 |
| 933 | 640 | 527 |
| 869 | 480 | 423 |
|     | 807 | 435 |

a. What does $MSTR$ measure?
b. What does $MSE$ measure?
c. Suppose that you want to perform a one-way ANOVA to compare the mean losses among the three types of robberies. What conditions are necessary? How crucial are those conditions?

11. **Losses to Robbery.** Refer to Problem 10.
a. Obtain individual normal probability plots and the standard deviations of the samples.
b. Perform a residual analysis.
c. Decide whether presuming that the assumptions of normal populations and equal standard deviations are met is reasonable.

12. **Losses to Robbery.** Refer to Problem 10. At the 5% significance level, do the data provide sufficient evidence to conclude that a difference in mean losses exists among the three types of robberies? Use one-way ANOVA to perform the required hypothesis test. (*Note:* $T_1 = 4124$, $T_2 = 4175$, $T_3 = 3493$, and $\Sigma x^2 = 8{,}550{,}628$.)

## Working With Large Data Sets

*In Problems 13–15, use the technology of your choice to do the following.*
  *a. Obtain individual normal probability plots and the standard deviations of the samples.*
  *b. Perform a residual analysis.*
  *c. Use your results from parts (a) and (b) to decide whether conducting a one-way ANOVA test on the data is reasonable. If so, also do parts (d)–(f).*
  *d. Use a one-way ANOVA test to decide, at the 5% significance level, whether the data provide sufficient evidence to conclude that a difference exists among the means of the populations from which the samples were taken.*
  *e. Interpret your results from part (d).*

13. **Weight Loss and BMI.** In the paper "Voluntary Weight Reduction in Older Men Increases Hip Bone Loss: The Osteoporotic Fractures in Men Study" (*Journal of Clinical Endocrinology & Metabolism*, Vol. 90, Issue 4, pp. 1998–2004), K. Ensrud et al. reported on the effect of voluntary weight reduction on hip bone loss in older men. In the study, 1342 older men participated in two physical examinations, an average of 1.8 years apart. After the second exam, they were categorized into three groups according to their change in weight between exams: weight loss of more than 5%, weight gain of more than 5%, and stable weight (between 5% loss and 5% gain). For purposes of the hip bone density study, other characteristics were compared, one such being body mass index (BMI). On the WeissStats CD, we have provided the BMI data for the three groups, based on the results obtained by the researchers.

14. **Weight Loss and Leg Power.** Another characteristic compared in the hip bone density study, discussed in Problem 13, was Maximum Nottingham leg power, in watts. On the WeissStats CD, we have provided the leg-power data for the three groups, based on the results obtained by the researchers.

15. **Income by Age.** The U.S. Census Bureau collects information on incomes of employed persons and publishes the results in *Historical Income Tables*. Independent simple random samples of 100 employed persons in each of four age groups gave the data on annual income, in thousands of dollars, presented on the WeissStats CD.

# Focusing on Data Analysis   UWEC Undergraduates

Recall from Chapter 1 (see page 34) that the Focus database and Focus sample contain information on the undergraduate students at the University of Wisconsin - Eau Claire (UWEC). Now would be a good time for you to review the discussion about these data sets.

Open the Focus sample worksheet (FocusSample) in the technology of your choice and do the following.

**a.** At the 10% significance level, do the data provide sufficient evidence to conclude that a difference exists among mean cumulative GPA for freshmen, sophomores, juniors, and seniors at UWEC? Use the one-way ANOVA procedure.

**b.** Obtain individual normal probability plots and the sample standard deviations of the GPAs of the sampled students in each class level. Based on your results, decide whether conducting a one-way ANOVA test on the data is reasonable.

**c.** Perform a residual analysis of the GPAs by class level. Based on your results, decide whether conducting a one-way ANOVA test on the data is reasonable.

**d.** Repeat parts (a)–(c) for mean cumulative GPA by college.

# Case Study Discussion   Heavy Drinking Among College Students

As you learned at the beginning of this chapter, Professor Kate Carey of Syracuse University surveyed 78 college students from an introductory psychology class, all of whom were regular drinkers of alcohol. Her purpose was to identify interpersonal and intrapersonal situations associated with excessive drinking among college students and to detect those situations that differentiate heavy drinkers from light and moderate drinkers.

To quantify drinking patterns, Carey used the time-line follow-back procedure (TLFB). The TLFB is a structured interview that provides reliable estimates of daily drinking. In this case, each student filled in each day of a blank calendar covering the previous month with the number of standard drink equivalents (SDEs) consumed on that day. One SDE is defined to be 1 fluid ounce of hard liquor, 12 fluid ounces of beer, or 4 fluid ounces of wine. Based on the results of the TLFB, the students were divided into three categories according to average quantity of alcohol consumed per drinking day: light drinkers ($\leq$ 3 SDEs), moderate drinkers (4–6 SDEs), and heavy drinkers (> 6 SDEs).

To assess the frequency of excessive drinking in interpersonal and intrapersonal situations, Carey utilized the short form of the Inventory of Drinking Situations (IDS). This form consists of 42 items, each of which is rated on a 4-point scale ranging from (1) never drink heavily in that type of situation to (4) almost always drink heavily in that type of situation. The 42 items are divided into eight subscales, three interpersonal and five intrapersonal, as displayed in the first column of the table on page 585. A subscale score represents the average rating for the items constituting the subscale.

**a.** For each of the eight IDS subscales, perform a (separate) one-way ANOVA to decide whether a difference exists in mean IDS scores among the three drinker categories. Use $\alpha = 0.05$. (*Note:* For each ANOVA, use the formula

$$\bar{x} = \frac{n_1\bar{x}_1 + n_2\bar{x}_2 + n_3\bar{x}_3}{n_1 + n_2 + n_3}$$

to determine the mean of all the observations.)

**b.** Based on the data in the table on page 585, should any of the eight ANOVAs perhaps not have been carried out? Explain your answer. (*Hint:* Rule of 2.)

## Biography   SIR RONALD FISHER: Mr. ANOVA

**Ronald Fisher** was born on February 17, 1890, in London, England; he was a surviving twin in a family of eight children; his father was a prominent auctioneer. Fisher graduated from Cambridge in 1912 with degrees in mathematics and physics.

From 1912 to 1919, Fisher worked at an investment house, did farm chores in Canada, and taught high school. In 1919, he took a position as a statistician at Rothamsted Experimental Station in Harpenden, West Hertford, England. His charge was to sort and reassess a 66-year accumulation of data on manurial field trials and weather records.

Fisher's work at Rothamsted during the next 15 years earned him the reputation as the leading statistician of his day and as a top-ranking geneticist. It was there, in 1925, that he published *Statistics for Research Workers,* a book that remained in print for 50 years. Fisher made important contributions to analysis of variance (ANOVA), exact tests of

significance for small samples, and maximum-likelihood solutions. He developed experimental designs to address issues in biological research, such as small samples, variable materials, and fluctuating environments.

Fisher has been described as "slight, bearded, eloquent, reactionary, and quirkish; genial to his disciples and hostile to his dissenters." He was also a prolific writer—over a span of 50 years, he wrote an average of one paper every 2 months!

In 1933, Fisher became Galton professor of Eugenics at University College in London and, in 1943, Balfour professor of genetics at Cambridge. In 1952, he was knighted. Fisher "retired" in 1959, moved to Australia, and spent the last 3 years of his life working at the Division of Mathematical Statistics of the Commonwealth Scientific and Industrial Research Organization. He died in 1962 in Adelaide, Australia.

## StatCrunch in MyStatLab
### Analyzing Data Online

StatCrunch online statistical software offers an easy-to-use interface customized for this book. The StatCrunch feature for each chapter illustrates the use of the software to perform a statistical analysis discussed in the chapter. Exercises are provided to further apply StatCrunch to other statistical analyses examined in the chapter. Go to the WeissStats CD or to the Weiss Web site at www.aw-bc.com/weiss to access StatCrunch instructions and data sets. To access StatCrunch statistical software, go to the student content area of your Weiss MyStatLab course.

## Internet Projects
### Exploring Data Online

The Internet project for each chapter provides simulations, demonstrations, or activities that enhance the topics covered in the chapter. The project materials come from universities, individuals, governments, and companies from all over the world. To access the Internet projects on the Web, go to www.aw-bc.com/weiss. From this Web page, you can reach the Internet Projects Page, which we suggest that you bookmark for easy access in the future.

# 14

# Inferential Methods in Regression and Correlation

## Chapter Outline

## Chapter Objectives

In Chapter 4, you studied descriptive methods in regression and correlation. You discovered how to determine the regression equation for a set of data points and how to use that equation to make predictions. You also learned how to compute and interpret the coefficient of determination and the linear correlation coefficient for a set of data points.

In this chapter, you will study inferential methods in regression and correlation. In Section 14.1, we examine the conditions required for performing such inferences and the methods for checking whether those conditions are satisfied. In presenting the first inferential method, in Section 14.2, we show how to decide whether a regression equation is useful for making predictions.

In Section 14.3, we investigate two additional inferential methods: one for estimating the mean of the response variable corresponding to a particular value of the predictor variable, the other for predicting the value of the response variable for a particular value of the predictor variable. We also discuss, in Section 14.4, the use of the linear correlation coefficient of a set of data points to decide whether the two variables under consideration are linearly correlated and, if so, the nature of the linear correlation.

# Shoe Size and Height

As mentioned in the Chapter 4 case study, most of us have heard that tall people generally have larger feet than short people. To examine the relationship between shoe size and height, Professor Dennis Young obtained data on those two variables from a sample of students at Arizona State University. We presented the data obtained by Professor Young in the Chapter 4 case study and repeat them here in the table below. Height is measured in inches.

At the end of Chapter 4, on page 199, you were asked to conduct regression and correlation analyses on these shoe-size and height data. The analyses done there were descriptive. At the end of this chapter, you will be asked to return to the data to make regression and correlation inferences.

| Shoe size | Height | Sex | Shoe size | Height | Sex |
|-----------|--------|-----|-----------|--------|-----|
| 6.5 | 66.0 | F | 13.0 | 77.0 | M |
| 9.0 | 68.0 | F | 11.5 | 72.0 | M |
| 8.5 | 64.5 | F | 8.5 | 59.0 | F |
| 8.5 | 65.0 | F | 5.0 | 62.0 | F |
| 10.5 | 70.0 | M | 10.0 | 72.0 | M |
| 7.0 | 64.0 | F | 6.5 | 66.0 | F |
| 9.5 | 70.0 | F | 7.5 | 64.0 | F |
| 9.0 | 71.0 | F | 8.5 | 67.0 | M |
| 13.0 | 72.0 | M | 10.5 | 73.0 | M |
| 7.5 | 64.0 | F | 8.5 | 69.0 | F |
| 10.5 | 74.5 | M | 10.5 | 72.0 | M |
| 8.5 | 67.0 | F | 11.0 | 70.0 | M |
| 12.0 | 71.0 | M | 9.0 | 69.0 | M |
| 10.5 | 71.0 | M | 13.0 | 70.0 | M |

## 14.1 The Regression Model; Analysis of Residuals

Before we can perform statistical inferences in regression and correlation, we must know whether the variables under consideration satisfy certain conditions. In this section, we discuss those conditions and examine methods for deciding whether they hold.

### The Regression Model

**TABLE 14.1**
Age and price data for a sample of 11 Orions

| Age (yr) $x$ | Price ($100) $y$ |
|:---:|:---:|
| 5 | 85 |
| 4 | 103 |
| 6 | 70 |
| 5 | 82 |
| 5 | 89 |
| 5 | 98 |
| 6 | 66 |
| 6 | 95 |
| 2 | 169 |
| 7 | 70 |
| 7 | 48 |

Let's return to the Orion illustration used throughout Chapter 4. In Table 14.1, we reproduce the data on age and price for a sample of 11 Orions.

With age as the predictor variable and price as the response variable, the regression equation for these data is $\hat{y} = 195.47 - 20.26x$, as we found in Chapter 4 on page 165. Recall that the regression equation can be used to predict the price of an Orion from its age. However, we cannot expect such predictions to be completely accurate because prices vary even for Orions of the same age.

For instance, the sample data in Table 14.1 include four 5-year-old Orions. Their prices are $8500, $8200, $8900, and $9800. We expect this variation in price for 5-year-old Orions because such cars generally have different mileages, interior conditions, paint quality, and so forth.

We use the population of all 5-year-old Orions to introduce some important regression terminology. The distribution of their prices is called the **conditional distribution** of the response variable "price" corresponding to the value 5 of the predictor variable "age." Likewise, their mean price is called the **conditional mean** of the response variable "price" corresponding to the value 5 of the predictor variable "age." Similar terminology applies to the standard deviation and other parameters.

Of course, there is a population of Orions for each age. The distribution, mean, and standard deviation of prices for that population are called the *conditional distribution, conditional mean*, and *conditional standard deviation*, respectively, of the response variable "price" corresponding to the value of the predictor variable "age."

The terminology of conditional distributions, means, and standard deviations is used in general for any predictor variable and response variable. Using that terminology, we now state the conditions required for applying inferential methods in regression analysis.

**Key Fact 14.1**

### Assumptions (Conditions) for Regression Inferences

1. **Population regression line:** There are constants $\beta_0$ and $\beta_1$ such that, for each value $x$ of the predictor variable, the conditional mean of the response variable is $\beta_0 + \beta_1 x$.

2. **Equal standard deviations:** The conditional standard deviations of the response variable are the same for all values of the predictor variable. We denote this common standard deviation $\sigma$.

3. **Normal populations:** For each value of the predictor variable, the conditional distribution of the response variable is a normal distribution.

4. **Independent observations:** The observations of the response variable are independent of one another.

**What Does It Mean?**

Assumptions 1–3 require that there are constants $\beta_0$, $\beta_1$, and $\sigma$ so that, for each value $x$ of the predictor variable, the conditional distribution of the response variable, $y$, is a normal distribution with mean $\beta_0 + \beta_1 x$ and standard deviation $\sigma$. These assumptions are often referred to as the **regression model**.

**Note:** We refer to the line $y = \beta_0 + \beta_1 x$—on which the conditional means of the response variable lie—as the **population regression line** and to its equation as the **population regression equation.**[†]

The inferential procedures in regression are robust to moderate violations of Assumptions 1–3 for regression inferences. In other words, the inferential procedures work reasonably well provided the variables under consideration don't violate any of those assumptions too badly.

## Example 14.1 | Assumptions for Regression Inferences

*Age and Price of Orions* For Orions, with age as the predictor variable and price as the response variable, what would it mean for the regression-inference Assumptions 1–3 to be satisfied? Display those assumptions graphically.

**Solution** Satisfying regression-inference Assumptions 1–3 requires that there are constants $\beta_0$, $\beta_1$, and $\sigma$ so that for each age, $x$, the prices of all Orions of that age are normally distributed with mean $\beta_0 + \beta_1 x$ and standard deviation $\sigma$. Thus the prices of all 2-year-old Orions must be normally distributed with mean $\beta_0 + \beta_1 \cdot 2$ and standard deviation $\sigma$, the prices of all 3-year-old Orions must be normally distributed with mean $\beta_0 + \beta_1 \cdot 3$ and standard deviation $\sigma$, and so on.

To display the assumptions for regression inferences graphically, let's first consider Assumption 1. This assumption requires that for each age, the mean price of all Orions of that age lies on the line $y = \beta_0 + \beta_1 x$, as shown in Fig. 14.1.

**FIGURE 14.1**
Population regression line

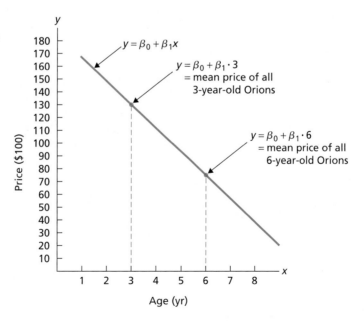

Assumptions 2 and 3 require that the price distributions for the various ages of Orions are all normally distributed with the same standard deviation, $\sigma$.

---

[†]The condition of equal standard deviations is called **homoscedasticity.** When that condition fails, we have what is called **heteroscedasticity.**

**FIGURE 14.2**
Price distributions for 2-, 5-,
and 7-year-old Orions
under Assumptions 2 and 3 (The
means shown for the three normal
distributions reflect Assumption 1)

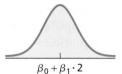

$\beta_0 + \beta_1 \cdot 2$

Prices of 2-year-old Orions

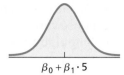

$\beta_0 + \beta_1 \cdot 5$

Prices of 5-year-old Orions

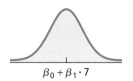

$\beta_0 + \beta_1 \cdot 7$

Prices of 7-year-old Orions

Figure 14.2 illustrates those two assumptions for the price distributions of 2-, 5-, and 7-year-old Orions. The shapes of the three normal curves in Fig. 14.2 are identical because normal distributions that have the same standard deviation have the same shape.

Assumptions 1–3 for regression inferences, as they pertain to the variables age and price of Orions, can be portrayed graphically by combining Figs. 14.1 and 14.2 into a three-dimensional graph, as shown in Fig. 14.3. Whether those assumptions actually hold remains to be seen.

Exercise 14.17
on page 626

• • •

**FIGURE 14.3** Graphical portrayal of Assumptions 1–3 for regression inferences pertaining to age and price of Orions

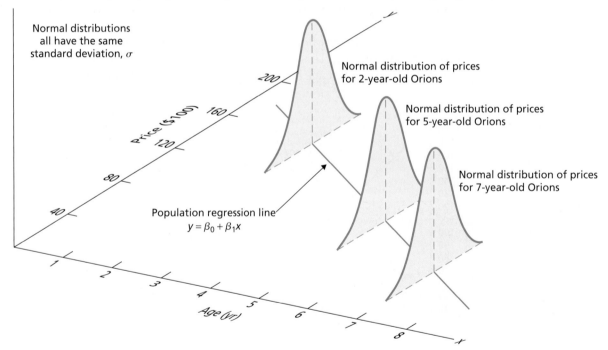

### Estimating the Regression Parameters

Suppose that we are considering two variables, $x$ and $y$, for which the assumptions for regression inferences are met. Then there are constants $\beta_0$, $\beta_1$, and $\sigma$ so that, for each value $x$ of the predictor variable, the conditional distribution of the response variable is a normal distribution with mean $\beta_0 + \beta_1 x$ and standard deviation $\sigma$.

Because the parameters $\beta_0$, $\beta_1$, and $\sigma$ are usually unknown, we must estimate them from sample data. We use the $y$-intercept and slope of a sample regression line as point estimates of the $y$-intercept and slope, respectively, of the population regression line; that is, we use $b_0$ and $b_1$ to estimate $\beta_0$ and $\beta_1$, respectively.

Equivalently, we use a sample regression line to estimate the unknown population regression line. Of course, a sample regression line ordinarily will not be the same as the population regression line, just as a sample mean generally will not equal the population mean. In Fig. 14.4, we illustrate this situation for the Orion example. Although the population regression line is unknown, we have drawn it to illustrate the difference between the population regression line and a sample regression line.

**FIGURE 14.4**
Population regression line and sample regression line for age and price of Orions

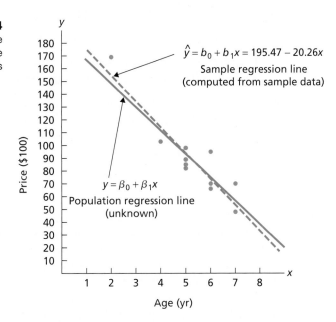

In Fig. 14.4, the sample regression line (the dashed line) is the best approximation that can be made to the population regression line (the solid line) by using the sample data in Table 14.1 on page 616. A different sample of Orions would almost certainly yield a different sample regression line.

The statistic used to obtain a point estimate for the common conditional standard deviation $\sigma$ is called the **standard error of the estimate.**

---

**Definition 14.1**

**Standard Error of the Estimate**

The **standard error of the estimate, $s_e$,** is defined by

$$s_e = \sqrt{\frac{SSE}{n-2}}$$

where $SSE$ is the error sum of squares.

**What Does It Mean?**

Roughly speaking, the standard error of the estimate indicates how much, on average, the predicted values of the response variable differ from the observed values of the response variable.

In the next example, we illustrate the computation and interpretation of the standard error of the estimate.

## Example 14.2 | Standard Error of the Estimate

*Age and Price of Orions*  Refer to the age and price data for a sample of 11 Orions, given in Table 14.1 on page 616.

**a.**  Compute and interpret the standard error of the estimate.

**b.**  Presuming that the variables age and price for Orions satisfy the assumptions for regression inferences, interpret the result from part (a).

### Solution

**a.**  On page 182, we found that $SSE = 1423.5$.  So the standard error of the estimate is

$$s_e = \sqrt{\frac{SSE}{n-2}} = \sqrt{\frac{1423.5}{11-2}} = 12.58.$$

**Interpretation**  Roughly speaking, the predicted price of an Orion in the sample differs, on average, from the observed price by $1258.

**b.**  Presuming that the variables age and price for Orions satisfy the assumptions for regression inferences, the standard error of the estimate, $s_e = 12.58$, or $1258, provides an estimate for the common population standard deviation, $\sigma$, of prices for all Orions of any particular age.

**You try it!**

Exercise 14.23(a)–(b)
on page 627

• • •

### Analysis of Residuals

Next we discuss how to use sample data to decide whether we can reasonably presume that the assumptions for regression inferences are met.  We concentrate on Assumptions 1–3; checking Assumption 4 is more involved and is best left for a second course in statistics.

The method for checking Assumptions 1–3 relies on an analysis of the errors made by using the regression equation to predict the observed values of the response variable, that is, on the differences between the observed and predicted values of the response variable.  Each such difference is called a **residual,** generically denoted **e.**  Thus,

$$\text{Residual} = e_i = y_i - \hat{y}_i.$$

Figure 14.5 shows the residual of a single data point.

**FIGURE 14.5**
Residual of a data point

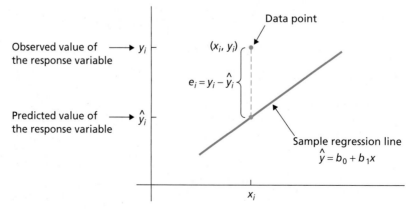

We can express the standard error of the estimate in terms of the residuals:

$$s_e = \sqrt{\frac{SSE}{n-2}} = \sqrt{\frac{\Sigma(y_i - \hat{y}_i)^2}{n-2}} = \sqrt{\frac{\Sigma e_i^2}{n-2}}.$$

We can show that the sum of the residuals is always 0, which, in turn, implies that $\bar{e} = 0$. Consequently, the standard error of the estimate is essentially the same as the standard deviation of the residuals.[†] Thus the standard error of the estimate is sometimes called the **residual standard deviation.**

We can analyze the residuals to decide whether Assumptions 1–3 for regression inferences are met because those assumptions can be translated into conditions on the residuals. To show how, let's consider a sample of data points obtained from two variables that satisfy the assumptions for regression inferences.

In light of Assumption 1, the data points should be scattered about the (sample) regression line, which means that the residuals should be scattered about the x-axis. In light of Assumption 2, the variation of the observed values of the response variable should remain approximately constant from one value of the predictor variable to the next, which means the residuals should fall roughly in a horizontal band. In light of Assumption 3, for each value of the predictor variable, the distribution of the corresponding observed values of the response variable should be approximately bell shaped, which implies that the horizontal band should be centered and symmetric about the x-axis.

Furthermore, considering all four regression assumptions simultaneously, we can regard the residuals as independent observations of a variable having a normal distribution with mean 0 and standard deviation $\sigma$. Thus a normal probability plot of the residuals should be roughly linear.

---

**Key Fact 14.2**    **Residual Analysis for the Regression Model**

If the assumptions for regression inferences are met, the following two conditions should hold:

- A plot of the residuals against the values of the predictor variable should fall roughly in a horizontal band centered and symmetric about the x-axis.

- A normal probability plot of the residuals should be roughly linear.

Failure of either of these two conditions casts doubt on the validity of one or more of the assumptions for regression inferences for the variables under consideration.

---

A plot of the residuals against the values of the predictor variable, called a **residual plot,** provides approximately the same information as does a scatterplot of the data points. However, a residual plot makes spotting patterns such as curvature and nonconstant standard deviation easier.

---

[†]The exact standard deviation of the residuals is obtained by dividing by $n-1$ instead of $n-2$.

To illustrate the use of residual plots for regression diagnostics, let's consider the three plots in Fig. 14.6.

- Fig. 14.6(a): In this plot, the residuals are scattered about the x-axis (residuals = 0) and fall roughly in a horizontal band, so Assumptions 1 and 2 appear to be met.
- Fig. 14.6(b): This plot suggests that the relation between the variables is curved, indicating that Assumption 1 may be violated.
- Fig. 14.6(c): This plot suggests that the conditional standard deviations increase as x increases, indicating that Assumption 2 may be violated.

**FIGURE 14.6**
Residual plots suggesting (a) no violation of linearity or constant standard deviation, (b) violation of linearity, and (c) violation of constant standard deviation

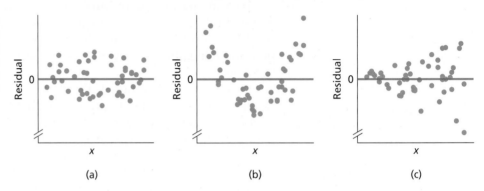

(a)        (b)        (c)

**Example 14.3** | **Analysis of Residuals**

*Age and Price of Orions* Perform a residual analysis to decide whether we can reasonably consider the assumptions for regression inferences to be met by the variables age and price of Orions.

**Solution** We apply the criteria presented in Key Fact 14.2. The ages and residuals for the Orion data are displayed in the first and fourth columns of Table 4.8 on page 182, respectively. We repeat that information in Table 14.2.

**TABLE 14.2** Age and residual data for Orions

| Age $x$ | 5 | 4 | 6 | 5 | 5 | 5 | 6 | 6 | 2 | 7 | 7 |
|---|---|---|---|---|---|---|---|---|---|---|---|
| Residual $e$ | −9.16 | −11.42 | −3.90 | −12.16 | −5.16 | 3.84 | −7.90 | 21.10 | 14.05 | 16.36 | −5.64 |

Figure 14.7(a) shows a plot of the residuals against age, and Fig. 14.7(b) shows a normal probability plot for the residuals.

Taking into account the small sample size, we can say that the residuals fall roughly in a horizontal band that is centered and symmetric about the x-axis. We can also say that the normal probability plot for the residuals is (very) roughly linear, although the departure from linearity is sufficient for some concern.[†]

**You try it!**

Exercise 14.23(c)–(d)
on page 627

**Interpretation** There are no obvious violations of the assumptions for regression inferences for the variables age and price of 2- to 7-year-old Orions.

• • •

---

[†]Recall, though, that the inferential procedures in regression analysis are robust to moderate violations of Assumptions 1–3 for regression inferences.

**FIGURE 14.7**
(a) Residual plot; (b) normal
probability plot for residuals

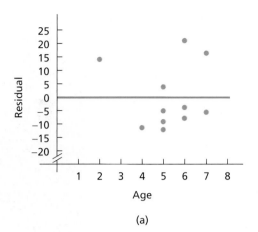

(a)

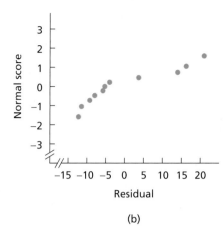

(b)

## The Technology Center

Most statistical technologies provide the standard error of the estimate as part of their regression analysis output. For instance, consider the Minitab and Excel regression analysis in Output 4.2 on page 172 for the age and price data of 11 Orions. The items circled in green give the standard error of the estimate, so $s_e = 12.58$. (*Note to TI-83/84 Plus users:* At the time of this writing, the TI-83/84 Plus does not display the standard error of the estimate. However, it can be found after running the regression procedure. See the *TI-83/84 Plus Manual* for details.)

We can also use statistical technology to obtain a residual plot and a normal probability plot of the residuals.

**Example 14.4** | **Using Technology to Obtain Plots of Residuals**

*Age and Price of Orions*   Use Minitab, Excel, or the TI-83/84 Plus to obtain a residual plot and a normal probability plot of the residuals for the age and price data of Orions given in Table 14.1 on page 616.

**Solution** | We applied the plots-of-residuals programs to the data, resulting in Output 14.1 on the following page. Steps for generating that output are presented in Instructions 14.1 on page 625.

Note the following:

- Minitab's default normal probability plot uses percents instead of normal scores on the vertical axis.
- Excel plots the residuals against the predicted values of the response variable rather than against the observed values of the predictor variable.

These and similar modifications, however, do not affect the use of the plots as diagnostic tools to help assess the appropriateness of regression inferences.

• • •

**OUTPUT 14.1** Residual plots and normal probability plots of the residuals for the age and price data of 11 Orions

**MINITAB**

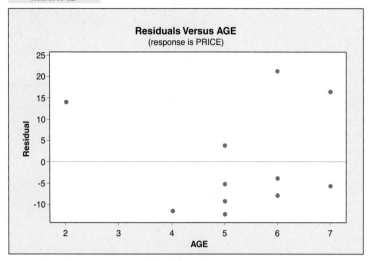

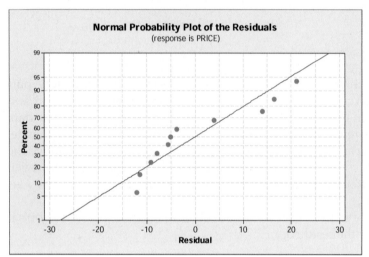

**EXCEL**

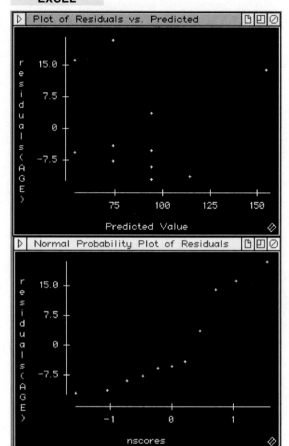

**TI-83/84 PLUS**

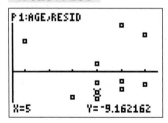

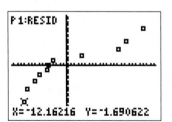

**INSTRUCTIONS 14.1**    Steps for generating Output 14.1

| MINITAB | EXCEL | TI-83/84 PLUS |
|---|---|---|

**MINITAB**

1 Store the age and price data from Table 14.1 in columns named AGE and PRICE, respectively
2 Choose **Stat ➤ Regression ➤ Regression...**
3 Specify PRICE in the **Response** text box
4 Specify AGE in the **Predictors** text box
5 Click the **Graphs...** button
6 Select the **Regular** option button from the **Residuals for Plots** list
7 Select the **Individual plots** option button from the **Residual Plots** list
8 Select the **Normal plot of residuals** check box from the **Individual plots** list
9 Click in the **Residuals versus the variables** text box and specify AGE
10 Click **OK** twice

**EXCEL**

1 Store the age and price data from Table 14.1 in ranges named AGE and PRICE, respectively
2 Choose **DDXL ➤ Regression**
3 Select **Simple regression** from the **Function type** drop-down list box
4 Specify PRICE in the **Response Variable** text box
5 Specify AGE in the **Explanatory Variable** text box
6 Click **OK**
7 Click the **Check the Residuals** button

**TI-83/84 PLUS**

1 Store the age and price data from Table 14.1 in lists named AGE and PRICE, respectively
2 Clear the **Y=** screen or turn off any equations located there
3 Press **STAT**, arrow over to **CALC**, and press **8**
4 Press **2nd ➤ LIST**, arrow down to AGE, and press **ENTER**
5 Press **, ➤ 2nd ➤ LIST**, arrow down to PRICE, and press **ENTER** twice
6 Press **2nd ➤ STAT PLOT** and then press **ENTER** twice
7 Arrow to the first graph icon and press **ENTER**
8 Press the down-arrow key
9 Press **2nd ➤ LIST**, arrow down to AGE, and press **ENTER** twice
10 Press **2nd ➤ LIST**, arrow down to RESID, and press **ENTER** twice
11 Press **ZOOM** and then **9** (and then **TRACE**, if desired)
12 Press **2nd ➤ STAT PLOT** and then press **ENTER** twice
13 Arrow to the sixth graph icon and press **ENTER**
14 Press the down-arrow key
15 Press **2nd ➤ LIST**, arrow down to RESID, and press **ENTER** twice
16 Press **ZOOM** and then **9** (and then **TRACE**, if desired)

■ ■ ■

# Exercises 14.1

## Understanding the Concepts and Skills

**14.1** Suppose that $x$ and $y$ are predictor and response variables, respectively, of a population. Consider the population that consists of all members of the original population that have a specified value of the predictor variable. The distribution, mean, and standard deviation of the response variable for this population are called the _____, _____, and _____, respectively, corresponding to the specified value of the predictor variable.

**14.2** State the four conditions required for making regression inferences.

*In Exercises **14.3–14.6**, assume that the variables under consideration satisfy the assumptions for regression inferences.*

**14.3** Fill in the blanks.
**a.** The line $y = \beta_0 + \beta_1 x$ is called the _____.
**b.** The common conditional standard deviation of the response variable is denoted _____.
**c.** For $x = 6$, the conditional distribution of the response variable is a _____ distribution having mean _____ and standard deviation _____.

**14.4** What statistic is used to estimate
**a.** the $y$-intercept of the population regression line?
**b.** the slope of the population regression line?
**c.** the common conditional standard deviation, $\sigma$, of the response variable?

**14.5** Based on a sample of data points, what is the best estimate of the population regression line?

**14.6** Regarding the standard error of the estimate,
a. give two interpretations of it.
b. identify another name used for it and explain the rationale for that name.
c. which one of the three sums of squares figures in its computation?

**14.7** The difference between an observed value and a predicted value of the response variable is called a _____.

**14.8** Identify two graphs used in a residual analysis to check the Assumptions 1–3 for regression inferences and explain the reasoning behind their use.

**14.9** Which graph used in a residual analysis provides roughly the same information as a scatterplot? What advantages does it have over a scatterplot?

*In Exercises 14.10–14.15, we repeat the data and provide the sample regression equations for Exercises 4.44–4.49. Here do the following.*
*a. Determine the standard error of the estimate.*
*b. Construct a residual plot.*
*c. Construct a normal probability plot of the residuals.*

**14.10**

| x | 2 | 4 | 3 |
|---|---|---|---|
| y | 3 | 5 | 7 |

$\hat{y} = 2 + x$

**14.11**

| x | 3 | 1 | 2 |
|---|---|---|---|
| y | −4 | 0 | −5 |

$\hat{y} = 1 - 2x$

**14.12**

| x | 0 | 4 | 3 | 1 | 2 |
|---|---|---|---|---|---|
| y | 1 | 9 | 8 | 4 | 3 |

$\hat{y} = 1 + 2x$

**14.13**

| x | 3 | 4 | 1 | 2 |
|---|---|---|---|---|
| y | 4 | 5 | 0 | −1 |

$\hat{y} = -3 + 2x$

**14.14**

| x | 1 | 1 | 5 | 5 |
|---|---|---|---|---|
| y | 1 | 3 | 2 | 4 |

$\hat{y} = 1.75 + 0.25x$

**14.15**

| x | 0 | 2 | 2 | 5 | 6 |
|---|---|---|---|---|---|
| y | 4 | 2 | 0 | −2 | 1 |

$\hat{y} = 2.875 - 0.625x$

*In Exercises 14.16–14.21, we repeat the information from Exercises 4.50–4.55. For each exercise here, discuss what satisfying Assumptions 1–3 for regression inferences by the variables under consideration would mean.*

**14.16 Tax Efficiency.** *Tax efficiency* is a measure—ranging from 0 to 100—of how much tax due to capital gains stock or mutual funds investors pay on their investments each year; the higher the tax efficiency, the lower is the tax. The paper "At the Mercy of the Manager" (*Financial Planning*, Vol. 30(5), pp. 54–56) by Craig Israelsen examined the relationship between investments in mutual fund portfolios and their associated tax efficiencies. The following table shows percentage of investments in energy securities (x) and tax efficiency (y) for 10 mutual fund portfolios.

| x | 3.1 | 3.2 | 3.7 | 4.3 | 4.0 | 5.5 | 6.7 | 7.4 | 7.4 | 10.6 |
|---|---|---|---|---|---|---|---|---|---|---|
| y | 98.1 | 94.7 | 92.0 | 89.8 | 87.5 | 85.0 | 82.0 | 77.8 | 72.1 | 53.5 |

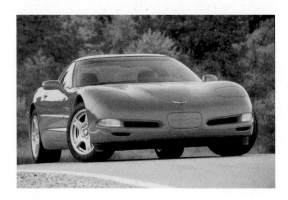

**14.17 Corvette Prices.** The *Kelley Blue Book* provides information on wholesale and retail prices of cars. Following are age and price data for 10 randomly selected Corvettes between 1 and 6 years old. Here, x denotes age, in years, and y denotes price, in hundreds of dollars.

| x | 6 | 6 | 6 | 2 | 2 | 5 | 4 | 5 | 1 | 4 |
|---|---|---|---|---|---|---|---|---|---|---|
| y | 270 | 260 | 275 | 405 | 364 | 295 | 335 | 308 | 405 | 305 |

**14.18 Custom Homes.** Hanna Properties specializes in custom-home resales in the Equestrian Estates, an exclusive subdivision in Phoenix, Arizona. A random sample of nine custom homes currently listed for sale provided the following information on size and price. Here, x denotes size, in hundreds of square feet, rounded to the nearest hundred, and y denotes price, in thousands of dollars, rounded to the nearest thousand.

| x | 26 | 27 | 33 | 29 | 29 | 34 | 30 | 40 | 22 |
|---|---|---|---|---|---|---|---|---|---|
| y | 540 | 555 | 575 | 577 | 606 | 661 | 738 | 804 | 496 |

**14.19 Plant Emissions.** Plants emit gases that trigger the ripening of fruit, attract pollinators, and cue other physiological responses. N. G. Agelopolous, K. Chamberlain, and J. A. Pickett examined factors that affect the emission of volatile compounds by the potato plant *Solanum tuberosom* and published their findings in the *Journal of Chemical*

*Ecology* (Vol. 26(2), pp. 497–511). The volatile compounds analyzed were hydrocarbons used by other plants and animals. Following are data on plant weight (*x*), in grams, and quantity of volatile compounds emitted (*y*), in hundreds of nanograms, for 11 potato plants.

| *x* | 57 | 85 | 57 | 65 | 52 | 67 | 62 | 80 | 77 | 53 | 68 |
|---|---|---|---|---|---|---|---|---|---|---|---|
| *y* | 8.0 | 22.0 | 10.5 | 22.5 | 12.0 | 11.5 | 7.5 | 13.0 | 16.5 | 21.0 | 12.0 |

**14.20 Crown-Rump Length.** In the article "The Human Vomeronasal Organ. Part II: Prenatal Development" (*Journal of Anatomy*, Vol. 197, Issue 3, pp. 421–436), T. Smith and K. Bhatnagar examined the controversial issue of the human vomeronasal organ, regarding its structure, function, and identity. The following table shows the age of fetuses (*x*), in weeks, and length of crown-rump (*y*), in millimeters.

| *x* | 10 | 10 | 13 | 13 | 18 | 19 | 19 | 23 | 25 | 28 |
|---|---|---|---|---|---|---|---|---|---|---|
| *y* | 66 | 66 | 108 | 106 | 161 | 166 | 177 | 228 | 235 | 280 |

**14.21 Study Time and Score.** An instructor at Arizona State University asked a random sample of eight students to record their study times in a beginning calculus course. She then made a table for total hours studied (*x*) over 2 weeks and test score (*y*) at the end of the 2 weeks. Here are the results.

| *x* | 10 | 15 | 12 | 20 | 8 | 16 | 14 | 22 |
|---|---|---|---|---|---|---|---|---|
| *y* | 92 | 81 | 84 | 74 | 85 | 80 | 84 | 80 |

*In Exercises **14.22–14.27**, do the following.*
*a. Compute the standard error of the estimate and interpret your answer.*
*b. Interpret your result from part (a) if the assumptions for regression inferences hold.*
*c. Obtain a residual plot and a normal probability plot of the residuals.*
*d. Decide whether you can reasonably consider Assumptions 1–3 for regression inferences to be met by the variables under consideration. (The answer here is subjective, especially in view of the extremely small sample sizes.)*

**14.22 Tax Efficiency.** Use the data on percentage of investments in energy securities and tax efficiency from Exercise 14.16.

**14.23 Corvette Prices.** Use the age and price data for Corvettes from Exercise 14.17.

**14.24 Custom Homes.** Use the size and price data for custom homes from Exercise 14.18.

**14.25 Plant Emissions.** Use the data on plant weight and quantity of volatile emissions from Exercise 14.19.

**14.26 Crown-Rump Length.** Use the data on age of fetuses and length of crown-rump from Exercise 14.20.

**14.27 Study Time and Score.** Use the data on total hours studied over 2 weeks and test score at the end of the 2 weeks from Exercise 14.21.

**14.28** Figure 14.8 shows three residual plots and a normal probability plot of residuals. For each part, decide whether the graph suggests violation of one or more of the assumptions for regression inferences. Explain your answers.

**FIGURE 14.8**
Plots for Exercise 14.28

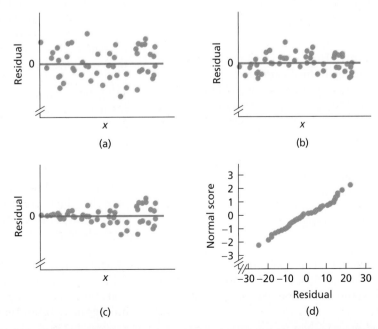

(a)

(b)

(c)

(d)

**FIGURE 14.9**
Plots for Exercise 14.29

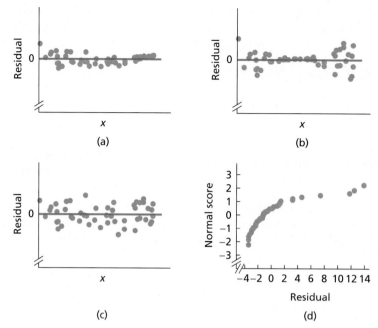

(a)

(b)

(c)

(d)

**14.29** Figure 14.9 shows three residual plots and a normal probability plot of residuals. For each part, decide whether the graph suggests violation of one or more of the assumptions for regression inferences. Explain your answers.

## Working With Large Data Sets

*In Exercises **14.30–14.39**, use the technology of your choice to do the following.*
*a. Obtain and interpret the standard error of the estimate.*
*b. Obtain a residual plot and a normal probability plot of the residuals.*
*c. Decide whether you can reasonably consider Assumptions 1–3 for regression inferences met by the two variables under consideration.*

**14.30 Birdies and Score.** How important are birdies (one under par) in determining the final score of a woman golfer? From the *U.S. Women's Open* Web site, we obtained data on number of birdies during a tournament and final score for 63 women golfers. The data are presented on the WeissStats CD.

**14.31 U.S. Presidents.** The *Information Please Almanac* provides data on the ages at inauguration and of death for the presidents of the United States. We have given those data on the WeissStats CD for those presidents who are not still living at the time of this writing.

**14.32 Health Care.** From the *Statistical Abstract of the United States*, we obtained data on percentage of gross domestic product (GDP) spent on health care and life expectancy, in years, for selected countries. Those data are provided on the WeissStats CD.

**14.33 Acreage and Value.** The document *Arizona Residential Property Valuation System*, published by the Arizona Department of Revenue, describes how county assessors use computerized systems to value single family residential properties for property tax purposes. On the WeissStats CD are data on lot size (in acres) and assessed value (in thousands of dollars) for a sample of homes in a particular area.

**14.34 Home Size and Value.** On the WeissStats CD are data on home size (in square feet) and assessed value (in thousands of dollars) for the same homes as in Exercise 14.33.

**14.35 High and Low Temperature.** The U.S. National Oceanic and Atmospheric Administration publishes temperature information of cities around the world in *Climates of the World*. A random sample of 50 cities gave the data on average high and low temperatures in January shown on the WeissStats CD.

**14.36 PCBs and Pelicans.** Polychlorinated biphenyls (PCBs), industrial pollutants, are a great danger to natural ecosystems. In a study by R. W. Risebrough titled "Effects of Environmental Pollutants Upon Animals Other Than Man" (*Proceedings of the 6th Berkeley Symposium on Mathematics and Statistics, VI*, University of California Press, pp. 443–463), 60 Anacapa pelican eggs were collected and measured for their shell thickness, in millimeters (mm), and concentration of PCBs, in parts per million (ppm). The data are presented on the WeissStats CD.

**14.37 Gas Guzzlers.** The magazine *Consumer Reports* publishes information on automobile gas mileage and variables that affect gas mileage. In the April 1999 issue, data on gas mileage (in mpg) and engine displacement (in liters, L) were published for 121 vehicles. Those data are stored on the WeissStats CD.

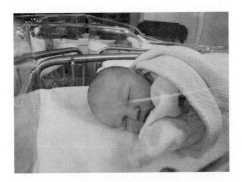

**14.38 Estriol Level and Birth Weight.** J. Greene and J. Touchstone conducted a study on the relationship between the estriol levels of pregnant women and the birth weights of their children. Their findings, "Urinary Tract Estriol: An Index of Placental Function," were published in the *American Journal of Obstetrics and Gynecology* (Vol. 85(1), pp. 1–9). The data points are provided on the WeissStats CD, where estriol levels are in mg/24 hr and birth weights are in hectograms (hg).

**14.39 Shortleaf Pines.** The ability to estimate the volume of a tree based on a simple measurement, such as the diameter of the tree, is important to the lumber industry, ecologists, and conservationists. Data on volume, in cubic feet, and diameter at breast height, in inches, for 70 shortleaf pines was reported in C. Bruce and F. X. Schumacher's *Forest Mensuration* (New York: McGraw-Hill, 1935) and analyzed by A. C. Akinson in the article "Transforming Both Sides of a Tree" (*The American Statistician*, Vol. 48, pp. 307–312). The data are provided on the WeissStats CD.

# 14.2 Inferences for the Slope of the Population Regression Line

In this and the next section, we examine several inferential procedures used in regression analysis. Strictly speaking, these inferential techniques require that the assumptions given in Key Fact 14.1 on page 616 be satisfied. However, as we noted earlier, these techniques are robust to moderate violations of those assumptions.

The first inferential methods we present concern the slope, $\beta_1$, of the population regression line. To begin, we consider hypothesis testing.

### Hypothesis Tests for the Slope of the Population Regression Line

Suppose that the variables $x$ and $y$ satisfy the assumptions for regression inferences. Then, for each value $x$ of the predictor variable, the conditional distribution of the response variable is a normal distribution with mean $\beta_0 + \beta_1 x$ and standard deviation $\sigma$.

Of particular interest is whether the slope, $\beta_1$, of the population regression line equals 0. If $\beta_1 = 0$, then, for each value $x$ of the predictor variable, the conditional distribution of the response variable is a normal distribution having mean $\beta_0$ ($= \beta_0 + 0 \cdot x$) and standard deviation $\sigma$. Because $x$ does not appear in either of those two parameters, it is useless as a predictor of $y$.[†]

Hence, we can decide whether $x$ is useful as a (linear) predictor of $y$—that is, whether the regression equation has utility—by performing the hypothesis test

$$H_0: \beta_1 = 0 \ (x \text{ is not useful for predicting } y)$$
$$H_a: \beta_1 \neq 0 \ (x \text{ is useful for predicting } y).$$

[†]Although $x$ alone may not be useful for predicting $y$, it may be useful in conjunction with another variable or variables. Thus, in this section, when we say that $x$ is not useful for predicting $y$, we really mean that the regression equation with $x$ as the only predictor variable is not useful for predicting $y$. Conversely, although $x$ alone may be useful for predicting $y$, it may not be useful in conjunction with another variable or variables. Thus, in this section, when we say that $x$ is useful for predicting $y$, we really mean that the regression equation with $x$ as the only predictor variable is useful for predicting $y$.

**TABLE 14.3**
Age and price data
for a sample of 11 Orions

| Age (yr) $y$ | Price ($100) $x$ |
|---|---|
| 5 | 85 |
| 4 | 103 |
| 6 | 70 |
| 5 | 82 |
| 5 | 89 |
| 5 | 98 |
| 6 | 66 |
| 6 | 95 |
| 2 | 169 |
| 7 | 70 |
| 7 | 48 |

We base hypothesis tests for $\beta_1$ (the slope of the population regression line) on the statistic $b_1$ (the slope of a sample regression line). To explain how this method works, let's return to the Orion illustration. The data on age and price for a sample of 11 Orions are repeated in Table 14.3.

With age as the predictor variable and price as the response variable, the regression equation for these data is $\hat{y} = 195.47 - 20.26x$, as we found in Chapter 4. In particular, the slope, $b_1$, of the sample regression line is $-20.26$.

We now consider all possible samples of 11 Orions whose ages are the same as those given in the first column of Table 14.3. For such samples, the slope, $b_1$, of the sample regression line varies from one sample to another and is therefore a variable. Its distribution is called the **sampling distribution of the slope of the regression line.** From the assumptions for regression inferences, we can show that this distribution is a normal distribution whose mean is the slope, $\beta_1$, of the population regression line. More generally, we have Key Fact 14.3.

---

**Key Fact 14.3**

**What Does It Mean?**

For fixed values of the predictor variable, the slopes of all possible sample regression lines have a normal distribution with mean $\beta_1$ and standard deviation $\sigma/\sqrt{S_{xx}}$.

### The Sampling Distribution of the Slope of the Regression Line

Suppose that the variables $x$ and $y$ satisfy the four assumptions for regression inferences. Then, for samples of size $n$, each with the same values $x_1, x_2, \ldots, x_n$, for the predictor variable, the following properties hold for the slope, $b_1$, of the sample regression line:

- The mean of $b_1$ equals the slope of the population regression line; that is, we have $\mu_{b_1} = \beta_1$.
- The standard deviation of $b_1$ is $\sigma_{b_1} = \sigma/\sqrt{S_{xx}}$.
- The variable $b_1$ is normally distributed.

As a consequence of Key Fact 14.3, the standardized variable

$$z = \frac{b_1 - \beta_1}{\sigma/\sqrt{S_{xx}}}$$

has the standard normal distribution. But this variable cannot be used as a basis for the required test statistic because the common conditional standard deviation, $\sigma$, is unknown. We therefore replace $\sigma$ with its sample estimate $s_e$, the standard error of the estimate. As you might suspect, the resulting variable has a $t$-distribution.

---

**Key Fact 14.4**

### $t$-Distribution for Inferences for $\beta_1$

Suppose that the variables $x$ and $y$ satisfy the four assumptions for regression inferences. Then, for samples of size $n$, each with the same values $x_1, x_2, \ldots, x_n$ for the predictor variable, the variable

$$t = \frac{b_1 - \beta_1}{s_e/\sqrt{S_{xx}}}$$

has the $t$-distribution with df $= n - 2$.

In light of Key Fact 14.4, for a hypothesis test with the null hypothesis $H_0$: $\beta_1 = 0$, we can use the variable

$$t = \frac{b_1}{s_e / \sqrt{S_{xx}}}$$

as the test statistic and obtain the critical values or $P$-value from the $t$-table, Table IV in Appendix A. We call this hypothesis-testing procedure the **regression $t$-test**. Procedure 14.1 provides a step-by-step method for performing a regression $t$-test by using either the critical-value approach or the $P$-value approach. *Note:* By "the four assumptions for regression inferences," we mean the four conditions stated in Key Fact 14.1 on page 616.

---

### Procedure 14.1   Regression *t*-Test

*Purpose* To perform a hypothesis test to decide whether a predictor variable is useful for making predictions

*Assumptions* The four assumptions for regression inferences

**STEP 1  The null and alternative hypotheses are**

$H_0$: $\beta_1 = 0$ (predictor variable is not useful for making predictions)

$H_a$: $\beta_1 \neq 0$ (predictor variable is useful for making predictions).

**STEP 2  Decide on the significance level, $\alpha$.**

**STEP 3  Compute the value of the test statistic**

$$t = \frac{b_1}{s_e / \sqrt{S_{xx}}}$$

**and denote that value $t_0$.**

| CRITICAL-VALUE APPROACH | or | P-VALUE APPROACH |

**STEP 4  The critical values are $\pm t_{\alpha/2}$ with df $= n - 2$. Use Table IV to find the critical values.**

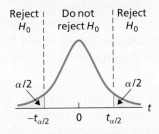

**STEP 5  If the value of the test statistic falls in the rejection region, reject $H_0$; otherwise, do not reject $H_0$.**

**STEP 4  The $t$-statistic has df $= n - 2$. Use Table IV to estimate the $P$-value, or obtain it exactly by using technology.**

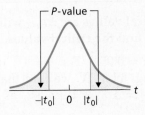

**STEP 5  If $P \leq \alpha$, reject $H_0$; otherwise, do not reject $H_0$.**

**STEP 6  Interpret the results of the hypothesis test.**

## Example 14.5 | The Regression $t$-Test

*Age and Price of Orions* The data on age and price for a sample of 11 Orions are displayed in Table 14.3 on page 630. At the 5% significance level, do the data provide sufficient evidence to conclude that age is useful as a (linear) predictor of price for Orions?

**Solution**  As we discovered in Example 14.3, we can reasonably consider the assumptions for regression inferences to be satisfied by the variables age and price for Orions, at least for Orions between 2 and 7 years old. So we apply Procedure 14.1 to carry out the required hypothesis test.

**STEP 1  State the null and alternative hypotheses.**

Let $\beta_1$ denote the slope of the population regression line that relates price to age for Orions. Then the null and alternative hypotheses are

$$H_0: \beta_1 = 0 \text{ (age is not useful for predicting price)}$$
$$H_a: \beta_1 \neq 0 \text{ (age is useful for predicting price)}.$$

**STEP 2  Decide on the significance level, $\alpha$.**

We are to perform the hypothesis test at the 5% significance level, or $\alpha = 0.05$.

**STEP 3  Compute the value of the test statistic**

$$t = \frac{b_1}{s_e/\sqrt{S_{xx}}}.$$

In Example 4.4 on page 165, we found that $b_1 = -20.26$, $\Sigma x_i^2 = 326$, and $\Sigma x_i = 58$. Also, in Example 14.2 on page 620, we determined that $s_e = 12.58$. Therefore, because $n = 11$, the value of the test statistic is

$$t = \frac{b_1}{s_e/\sqrt{S_{xx}}} = \frac{b_1}{s_e/\sqrt{\Sigma x_i^2 - (\Sigma x_i)^2/n}} = \frac{-20.26}{12.58/\sqrt{326 - (58)^2/11}} = -7.235.$$

| CRITICAL-VALUE APPROACH | or | P-VALUE APPROACH |
|---|---|---|

**STEP 4  The critical values are $\pm t_{\alpha/2}$ with df $= n - 2$. Use Table IV to find the critical values.**

From Step 2, $\alpha = 0.05$. For $n = 11$, df $= n - 2 = 11 - 2 = 9$. Using Table IV, we find that the critical values are $\pm t_{\alpha/2} = \pm t_{0.025} = \pm 2.262$, as depicted in Fig. 14.10A.

**STEP 4  The $t$-statistic has df $= n - 2$. Use Table IV to estimate the $P$-value or obtain it exactly by using technology.**

From Step 3, the value of the test statistic is $t = -7.235$. Because the test is two tailed, the $P$-value is the probability of observing a value of $t$ of 7.235 or greater in magnitude if the null hypothesis is true. That probability equals the shaded area shown in Fig. 14.10B.

| CRITICAL-VALUE APPROACH | or | P-VALUE APPROACH |

**FIGURE 14.10A**

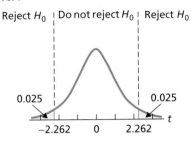

Reject $H_0$ | Do not reject $H_0$ | Reject $H_0$

0.025          0.025

−2.262    0    2.262    $t$

**FIGURE 14.10B**

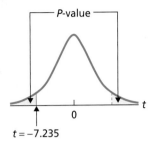

$P$-value

$t = -7.235$     0     $t$

**STEP 5 If the value of the test statistic falls in the rejection region, reject $H_0$; otherwise, do not reject $H_0$.**

The value of the test statistic, found in Step 3, is $t = -7.235$. Because this value falls in the rejection region, we reject $H_0$. The test results are statistically significant at the 5% level.

For $n = 11$, df $= 11 - 2 = 9$. Referring to Fig. 14.10B and to Table IV with df $= 9$, we find that $P < 0.01$. (Using technology, we obtain $P = 0.0000488$.)

**STEP 5 If $P \leq \alpha$, reject $H_0$; otherwise, do not reject $H_0$.**

From Step 4, $P < 0.01$. Because the $P$-value is less than the specified significance level of 0.05, we reject $H_0$. The test results are statistically significant at the 5% level and (see Table 9.10 on page 414) provide very strong evidence against the null hypothesis.

You try it!

Exercise 14.51 on page 636

**STEP 6 Interpret the results of the hypothesis test.**

**Interpretation** At the 5% significance level, the data provide sufficient evidence to conclude that the slope of the population regression line is not 0 and hence that age is useful as a (linear) predictor of price for Orions.

• • •

## Other Procedures for Testing Utility of the Regression

We use Procedure 14.1 on page 631, which is based on the statistic $b_1$, to perform a hypothesis test to decide whether the slope of the population regression line is not 0 or, equivalently, whether the regression equation is useful for making predictions.

In Section 4.3, we introduced the coefficient of determination, $r^2$, as a descriptive measure of the utility of the regression equation for making predictions. We should therefore also be able to use the statistic $r^2$ as a basis for performing a hypothesis test to decide whether the regression equation is useful for making predictions—and indeed we can. However, we do not cover the hypothesis test based on $r^2$ because it is equivalent to the hypothesis test based on $b_1$.

We can also use the linear correlation coefficient, $r$, introduced in Section 4.4, as a basis for performing a hypothesis test to decide whether the regression equation is useful for making predictions. That test too is equivalent to the hypothesis test based on $b_1$ but, because it has other uses, we discuss it in Section 14.4.

## Confidence Intervals for the Slope of the Population Regression Line

Recall that the slope of a line represents the change in the dependent variable, $y$, resulting from an increase in the independent variable, $x$, by 1 unit. Also recall that the population regression line, whose slope is $\beta_1$, gives the conditional means of the response variable. Therefore $\beta_1$ represents the change in the conditional mean of the response variable for each increase in the value of the predictor variable by 1 unit.

For instance, consider the variables age and price of Orions. In this case, $\beta_1$ is the amount that the mean price decreases for every increase in age by 1 year. In other words, $\beta_1$ is the mean yearly depreciation of Orions.

Consequently, obtaining an estimate for the slope of the population regression line is worthwhile. We know that a point estimate for $\beta_1$ is provided by $b_1$. To determine a confidence-interval estimate for $\beta_1$, we apply Key Fact 14.4 on page 630 to obtain Procedure 14.2, called the **regression $t$-interval procedure.**

---

**Procedure 14.2**    **Regression $t$-Interval Procedure**

*Purpose* To find a confidence interval for the slope, $\beta_1$, of the population regression line

*Assumptions* The four assumptions for regression inferences

**STEP 1 For a confidence level of $1 - \alpha$, use Table IV to find $t_{\alpha/2}$ with df $= n - 2$.**

**STEP 2 The endpoints of the confidence interval for $\beta_1$ are**

$$b_1 \pm t_{\alpha/2} \cdot \frac{s_e}{\sqrt{S_{xx}}}.$$

**STEP 3 Interpret the confidence interval.**

---

**Example 14.6** | **The Regression $t$-Interval Procedure**

*Age and Price of Orions* Use the data in Table 14.3 on page 630 to determine a 95% confidence interval for the slope of the population regression line that relates price to age for Orions.

**Solution** We apply Procedure 14.2.

**STEP 1 For a confidence level of $1 - \alpha$, use Table IV to find $t_{\alpha/2}$ with df $= n - 2$.**

For a 95% confidence interval, $\alpha = 0.05$. Because $n = 11$, df $= 11 - 2 = 9$. From Table IV, $t_{\alpha/2} = t_{0.05/2} = t_{0.025} = 2.262$.

**STEP 2** The endpoints of the confidence interval for $\beta_1$ are

$$b_1 \pm t_{\alpha/2} \cdot \frac{s_e}{\sqrt{S_{xx}}}.$$

From Example 4.4, $b_1 = -20.26$, $\Sigma x_i^2 = 326$, and $\Sigma x_i = 58$. Also, from Example 14.2, $s_e = 12.58$. Hence the endpoints of the confidence interval for $\beta_1$ are

$$-20.26 \pm 2.262 \cdot \frac{12.58}{\sqrt{326 - (58)^2/11}}$$

or $-20.26 \pm 6.33$, or $-26.59$ to $-13.93$.

**STEP 3** Interpret the confidence interval.

You try it!

Exercise 14.57 on page 636

**Interpretation**   We can be 95% confident that the slope of the population regression line is somewhere between $-26.59$ and $-13.93$. In other words, we can be 95% confident that the yearly decrease in mean price for Orions is somewhere between $1393 and $2659.

• • •

## The Technology Center

Most statistical technologies provide the information needed to perform a regression $t$-test as part of their regression analysis output. For instance, consider the Minitab and Excel regression analysis in Output 4.2 on page 172 for the age and price data of 11 Orions. The items circled in orange give the $t$-statistic and the $P$-value for the regression $t$-test.

To perform a regression $t$-test with the TI-83/84 Plus, we use the **LinRegTTest** program. See the *TI-83/84 Plus Manual* for details.

▪ ▪ ▪

# Exercises 14.2

## Understanding the Concepts and Skills

**14.40** Explain why the predictor variable is useless as a predictor of the response variable if the slope of the population regression line is 0.

**14.41** For two variables satisfying Assumptions 1–3 for regression inferences, the population regression equation is $y = 20 - 3.5x$. For samples of size 10 and given values of the predictor variable, the distribution of slopes of all possible sample regression lines is a _____ distribution with mean _____.

**14.42** Consider the standardized variable

$$z = \frac{b_1 - \beta_1}{\sigma/\sqrt{S_{xx}}}.$$

a. Identify its distribution.
b. Why can't it be used as the test statistic for a hypothesis test concerning $\beta_1$?
c. What statistic is used? What is the distribution of that statistic?

**14.43** In this section, we used the statistic $b_1$ as a basis for conducting a hypothesis test to decide whether a regression equation is useful for prediction. Identify two other statistics that can be used as a basis for such a test.

*In Exercises **14.44–14.49**, we repeat the information from Exercises 14.10–14.15. Do the following.*
a. *Decide, at the 10% significance level, whether the data provide sufficient evidence to conclude that x is useful for predicting y.*
b. *Find a 90% confidence interval for the slope of the population regression line.*

**14.44**

| x | 2 | 4 | 3 |
|---|---|---|---|
| y | 3 | 5 | 7 |

$\hat{y} = 2 + x$

**14.45**

| x | 3 | 1 | 2 |
|---|---|---|---|
| y | −4 | 0 | −5 |

$\hat{y} = 1 - 2x$

**14.46**

| x | 0 | 4 | 3 | 1 | 2 |
|---|---|---|---|---|---|
| y | 1 | 9 | 8 | 4 | 3 |

$\hat{y} = 1 + 2x$

**14.47**

| x | 3 | 4 | 1 | 2 |
|---|---|---|---|---|
| y | 4 | 5 | 0 | −1 |

$\hat{y} = -3 + 2x$

**14.48**

| x | 1 | 1 | 5 | 5 |
|---|---|---|---|---|
| y | 1 | 3 | 2 | 4 |

$\hat{y} = 1.75 + 0.25x$

**14.49**

| x | 0 | 2 | 2 | 5 | 6 |
|---|---|---|---|---|---|
| y | 4 | 2 | 0 | −2 | 1 |

$\hat{y} = 2.875 - 0.625x$

*In Exercises **14.50–14.55**, we repeat the information from Exercises 14.16–14.21. Presuming that the assumptions for regression inferences are met, decide at the specified significance level whether the data provide sufficient evidence to conclude that the predictor variable is useful for predicting the response variable.*

**14.50 Tax Efficiency.** Following are the data on percentage of investments in energy securities and tax efficiency from Exercise 14.16. Use $\alpha = 0.05$.

| x | 3.1 | 3.2 | 3.7 | 4.3 | 4.0 | 5.5 | 6.7 | 7.4 | 7.4 | 10.6 |
|---|-----|-----|-----|-----|-----|-----|-----|-----|-----|------|
| y | 98.1 | 94.7 | 92.0 | 89.8 | 87.5 | 85.0 | 82.0 | 77.8 | 72.1 | 53.5 |

**14.51 Corvette Prices.** Following are the age and price data for Corvettes from Exercise 14.17. Use $\alpha = 0.10$.

| x | 6 | 6 | 6 | 2 | 2 | 5 | 4 | 5 | 1 | 4 |
|---|---|---|---|---|---|---|---|---|---|---|
| y | 270 | 260 | 275 | 405 | 364 | 295 | 335 | 308 | 405 | 305 |

**14.52 Custom Homes.** Following are the size and price data for custom homes from Exercise 14.18. Use $\alpha = 0.01$.

| x | 26 | 27 | 33 | 29 | 29 | 34 | 30 | 40 | 22 |
|---|----|----|----|----|----|----|----|----|----|
| y | 540 | 555 | 575 | 577 | 606 | 661 | 738 | 804 | 496 |

**14.53 Plant Emissions.** Following are the data on plant weight and quantity of volatile emissions from Exercise 14.19. Use $\alpha = 0.05$.

| x | 57 | 85 | 57 | 65 | 52 | 67 | 62 | 80 | 77 | 53 | 68 |
|---|----|----|----|----|----|----|----|----|----|----|----|
| y | 8.0 | 22.0 | 10.5 | 22.5 | 12.0 | 11.5 | 7.5 | 13.0 | 16.5 | 21.0 | 12.0 |

**14.54 Crown-Rump Length.** Following are the data on age of fetuses and length of crown-rump from Exercise 14.20. Use $\alpha = 0.10$.

| x | 10 | 10 | 13 | 13 | 18 | 19 | 19 | 23 | 25 | 28 |
|---|----|----|----|----|----|----|----|----|----|----|
| y | 66 | 66 | 108 | 106 | 161 | 166 | 177 | 228 | 235 | 280 |

**14.55 Study Time and Score.** Following are the data on total hours studied over 2 weeks and test score at the end of the 2 weeks from Exercise 14.21. Use $\alpha = 0.01$.

| x | 10 | 15 | 12 | 20 | 8 | 16 | 14 | 22 |
|---|----|----|----|----|---|----|----|----|
| y | 92 | 81 | 84 | 74 | 85 | 80 | 84 | 80 |

*In each of Exercises **14.56–14.61**, apply Procedure 14.2 on page 634 to find and interpret a confidence interval, at the specified confidence level, for the slope of the population regression line that relates the response variable to the predictor variable.*

**14.56 Tax Efficiency.** Refer to Exercise 14.50; 95%.

**14.57 Corvette Prices.** Refer to Exercise 14.51; 90%.

**14.58 Custom Homes.** Refer to Exercise 14.52; 99%.

**14.59 Plant Emissions.** Refer to Exercise 14.53; 95%.

**14.60 Crown-Rump Length.** Refer to Exercise 14.54; 90%.

**14.61 Study Time and Score.** Refer to Exercise 14.55; 99%.

## Working With Large Data Sets

*In Exercises **14.62–14.72**, use the technology of your choice to do the following.*
a. *Decide whether you can reasonably apply the regression t-test. If so, then also do parts (b) and (c).*
b. *Decide, at the 5% significance level, whether the data provide sufficient evidence to conclude that the predictor variable is useful for predicting the response variable.*
c. *Use Table 9.10 on page 414 to determine the strength of the evidence against the null hypothesis and hence in favor of the utility of the regression equation for making predictions.*

**14.62 Birdies and Score.** The data from Exercise 14.30 for number of birdies during a tournament and final score for 63 women golfers are on the WeissStats CD.

**14.63 U.S. Presidents.** The data from Exercise 14.31 for the ages at inauguration and of death for the presidents of the United States are on the WeissStats CD.

**14.64 Health Care.** The data from Exercise 14.32 for percentage of gross domestic product (GDP) spent on health care and life expectancy, in years, for selected countries are on the WeissStats CD.

**14.65 Acreage and Value.** The data from Exercise 14.33 for lot size (in acres) and assessed value (in thousands of dollars) for a sample of homes in a particular area are on the WeissStats CD.

**14.66 Home Size and Value.** The data from Exercise 14.34 for home size (in square feet) and assessed value (in thousands of dollars) for the same homes as in Exercise 14.65 are on the WeissStats CD.

**14.67 High and Low Temperature.** The data from Exercise 14.35 for average high and low temperatures in January for a random sample of 50 cities are on the WeissStats CD.

**14.68 PCBs and Pelicans.** Use the data points given on the WeissStats CD for shell thickness and concentration of PCBs for 60 Anacapa pelican eggs referred to in Exercise 14.36.

**14.69 Gas Guzzlers.** Use the data on the WeissStats CD for gas mileage and engine displacement for 121 vehicles referred to in Exercise 14.37.

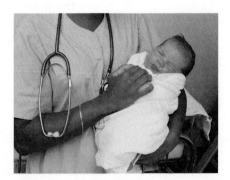

**14.70 Estriol Level and Birth Weight.** Use the data on the WeissStats CD for estriol levels of pregnant women and birth weights of their children referred to in Exercise 14.38.

**14.71 Shortleaf Pines.** The data from Exercise 14.39 for volume, in cubic feet, and diameter at breast height, in inches, for 70 shortleaf pines are on the WeissStats CD.

**14.72 Body Fat.** In the paper "Total Body Composition by Dual-Photon ($^{153}$Gd) Absorptiometry" (*American Journal of Clinical Nutrition*, Vol. 40, pp. 834–839), R. Mazess et al. studied methods for quantifying body composition. Eighteen randomly selected adults were measured for percentage of body fat, using dual-photon absorptiometry. Each adult's age and percentage of body fat are shown on the WeissStats CD.

## 14.3 Estimation and Prediction

In this section, we examine how a sample regression equation can be used to make two important inferences:

- Estimate the conditional mean of the response variable corresponding to a particular value of the predictor variable.
- Predict the value of the response variable for a particular value of the predictor variable.

We again use the Orion data to illustrate the pertinent ideas. In doing so, we presume that the assumptions for regression inferences (Key Fact 14.1 on page 616) are satisfied by the variables age and price for Orions. Example 14.3 on page 622 shows that to presume so is not unreasonable.

## Example 14.7 | Estimating Conditional Means in Regression

**TABLE 14.4**
Age and price data
for a sample of 11 Orions

| Age (yr) $x$ | Price ($100) $y$ |
|:---:|:---:|
| 5 | 85 |
| 4 | 103 |
| 6 | 70 |
| 5 | 82 |
| 5 | 89 |
| 5 | 98 |
| 6 | 66 |
| 6 | 95 |
| 2 | 169 |
| 7 | 70 |
| 7 | 48 |

*Age and Price of Orions* Use the data on age and price for a sample of 11 Orions, repeated in Table 14.4, to estimate the mean price of all 3-year-old Orions.

**Solution** By Assumption 1 for regression inferences, the population regression line gives the mean prices for the various ages of Orions. In particular, the mean price of all 3-year-old Orions is $\beta_0 + \beta_1 \cdot 3$. Because $\beta_0$ and $\beta_1$ are unknown, we estimate the mean price of all 3-year-old Orions ($\beta_0 + \beta_1 \cdot 3$) by the corresponding value on the sample regression line, namely, $b_0 + b_1 \cdot 3$.

Recalling that the sample regression equation for the age and price data in Table 14.4 is $\hat{y} = 195.47 - 20.26x$, we estimate that the mean price of all 3-year-old Orions is

$$\hat{y} = 195.47 - 20.26 \cdot 3 = 134.69,$$

or $13,469. Note that the estimate for the mean price of all 3-year-old Orions is the same as the predicted price for a 3-year-old Orion. Both are obtained by substituting $x = 3$ into the sample regression equation.

• • •

**You try it!**

Exercise 14.81(a)
on page 645

### Confidence Intervals for Conditional Means in Regression

The estimate of $13,469 for the mean price of all 3-year-old Orions, found in the previous example, is a point estimate. Providing a confidence-interval estimate for the mean price of all 3-year-old Orions would be more informative.

To that end, consider all possible samples of 11 Orions whose ages are the same as those given in the first column of Table 14.4. For such samples, the predicted price of a 3-year-old Orion varies from one sample to another and is therefore a variable. Using the assumptions for regression inferences, we can show that its distribution is a normal distribution whose mean equals the mean price of all 3-year-old Orions. More generally, we have Key Fact 14.5.

### Key Fact 14.5 | Distribution of the Predicted Value of a Response Variable

Suppose that the variables $x$ and $y$ satisfy the four assumptions for regression inferences. Let $x_p$ denote a particular value of the predictor variable and let $\hat{y}_p$ be the corresponding value predicted for the response variable by the sample regression equation; that is, $\hat{y}_p = b_0 + b_1 x_p$. Then, for samples of size $n$, each with the same values, $x_1, x_2, \ldots, x_n$, for the predictor variable, the following properties hold for $\hat{y}_p$.

- The mean of $\hat{y}_p$ equals the conditional mean of the response variable corresponding to the value $x_p$ of the predictor variable: $\mu_{\hat{y}_p} = \beta_0 + \beta_1 x_p$.
- The standard deviation of $\hat{y}_p$ is

$$\sigma_{\hat{y}_p} = \sigma \sqrt{\frac{1}{n} + \frac{(x_p - \Sigma x_i/n)^2}{S_{xx}}}.$$

- The variable $\hat{y}_p$ is normally distributed.

In particular, for fixed values of the predictor variable, the possible predicted values of the response variable corresponding to $x_p$ have a normal distribution with mean $\beta_0 + \beta_1 x_p$.

In light of Key Fact 14.5, if we standardize the variable $\hat{y}_p$, the resulting variable has the standard normal distribution. However, because the standardized variable contains the unknown parameter $\sigma$, it cannot be used as a basis for a confidence-interval formula. Therefore we replace $\sigma$ by its estimate $s_e$, the standard error of the estimate. The resulting variable has a $t$-distribution.

---

**Key Fact 14.6**

### $t$-Distribution for Confidence Intervals for Conditional Means in Regression

Suppose that the variables $x$ and $y$ satisfy the four assumptions for regression inferences. Then, for samples of size $n$, each with the same values, $x_1, x_2, \ldots, x_n$, for the predictor variable, the variable

$$t = \frac{\hat{y}_p - (\beta_0 + \beta_1 x_p)}{s_e \sqrt{\dfrac{1}{n} + \dfrac{(x_p - \Sigma x_i/n)^2}{S_{xx}}}}$$

has the $t$-distribution with df $= n - 2$.

---

Recalling that $\beta_0 + \beta_1 x_p$ is the conditional mean of the response variable corresponding to the value $x_p$ of the predictor variable, we can apply Key Fact 14.6 to derive a confidence-interval procedure for means in regression. We call that procedure the **conditional mean $t$-interval procedure.**

---

**Procedure 14.3**

### Conditional Mean $t$-Interval Procedure

*Purpose* To find a confidence interval for the conditional mean of the response variable corresponding to a particular value of the predictor variable, $x_p$

*Assumptions* The four assumptions for regression inferences

**STEP 1 For a confidence level of $1 - \alpha$, use Table IV to find $t_{\alpha/2}$ with df $= n - 2$.**

**STEP 2 Compute the point estimate, $\hat{y}_p = b_0 + b_1 x_p$.**

**STEP 3 The endpoints of the confidence interval for the conditional mean of the response variable are**

$$\hat{y}_p \pm t_{\alpha/2} \cdot s_e \sqrt{\frac{1}{n} + \frac{(x_p - \Sigma x_i/n)^2}{S_{xx}}}.$$

**STEP 4 Interpret the confidence interval.**

---

**Example 14.8** | ### The Conditional Mean $t$-Interval Procedure

*Age and Price of Orions* Use the sample data in Table 14.4 on page 638 to obtain a 95% confidence interval for the mean price of all 3-year-old Orions.

**Solution** We apply Procedure 14.3.

**STEP 1 For a confidence level of $1 - \alpha$, use Table IV to find $t_{\alpha/2}$ with df $= n - 2$.**

We want a 95% confidence interval, or $\alpha = 0.05$. Because $n = 11$, we have df $= 9$. From Table IV, $t_{\alpha/2} = t_{0.05/2} = t_{0.025} = 2.262$.

**STEP 2 Compute the point estimate, $\hat{y}_p = b_0 + b_1 x_p$.**

Here, $x_p = 3$ (3-year-old Orions). From Example 14.7, the point estimate for the mean price of all 3-year-old Orions is

$$\hat{y}_p = 195.47 - 20.26 \cdot 3 = 134.69.$$

**STEP 3 The endpoints of the confidence interval for the conditional mean of the response variable are**

$$\hat{y}_p \pm t_{\alpha/2} \cdot s_e \sqrt{\frac{1}{n} + \frac{(x_p - \Sigma x_i/n)^2}{S_{xx}}}.$$

In Example 4.4, we found that $\Sigma x_i = 58$ and $\Sigma x_i^2 = 326$; in Example 14.2, we determined that $s_e = 12.58$. Also, from Step 1, $t_{\alpha/2} = 2.262$ and, from Step 2, $\hat{y}_p = 134.69$. Consequently, the endpoints of the confidence interval for the conditional mean are

$$134.69 \pm 2.262 \cdot 12.58 \sqrt{\frac{1}{11} + \frac{(3 - 58/11)^2}{326 - (58)^2/11}}$$

or $134.69 \pm 16.76$, or $117.93$ to $151.45$.

You try it!

Exercise 14.81(b) on page 645

**STEP 4 Interpret the confidence interval.**

**Interpretation** We can be 95% confident that the mean price of all 3-year-old Orions is somewhere between $11,793 and $15,145.

• • •

## Prediction Intervals

A primary use of a sample regression equation is to make predictions. As we have seen, for the Orion data in Table 14.4 on page 638, the sample regression equation is $\hat{y} = 195.47 - 20.26x$. Substituting $x = 3$ into that equation, we get the predicted price for a 3-year-old Orion of 134.69, or $13,469. Because the prices of such cars vary, finding a **prediction interval** for the price of a 3-year-old Orion makes more sense than giving a single predicted value.[†]

To that end, we first recall that, from the assumptions for regression inferences, the price of a 3-year-old Orion has a normal distribution with mean $\beta_0 + \beta_1 \cdot 3$ and standard deviation $\sigma$. Because $\beta_0$ and $\beta_1$ are unknown, we estimate the mean price by its point estimate $b_0 + b_1 \cdot 3$, which is also the predicted price of a 3-year-old Orion.

---

[†]Prediction intervals are similar to confidence intervals. The term *confidence* is usually reserved for interval estimates of parameters, such as the mean price of all 3-year-old Orions. The term *prediction* is used for interval estimates of variables, such as the price of a 3-year-old Orion.

Thus, to find a prediction interval, we need the distribution of the difference between the price of a 3-year-old Orion and the predicted price of a 3-year-old Orion. Using the assumptions for regression inferences, we can show that this distribution is normal. More generally, we have Key Fact 14.7.

---

**Key Fact 14.7**

**Distribution of the Difference Between the Observed and Predicted Values of the Response Variable**

Suppose that the variables $x$ and $y$ satisfy the four assumptions for regression inferences. Let $x_p$ denote a particular value of the predictor variable and let $\hat{y}_p$ be the corresponding value predicted for the response variable by the sample regression equation. Furthermore, let $y_p$ be an independently observed value of the response variable corresponding to the value $x_p$ of the predictor variable. Then, for samples of size $n$, each with the same values, $x_1, x_2, \ldots, x_n$, for the predictor variable, the following properties hold for $y_p - \hat{y}_p$, the difference between the observed and predicted values.

- The mean of $y_p - \hat{y}_p$ equals zero: $\mu_{y_p - \hat{y}_p} = 0$.
- The standard deviation of $y_p - \hat{y}_p$ is

$$\sigma_{y_p - \hat{y}_p} = \sigma\sqrt{1 + \frac{1}{n} + \frac{(x_p - \Sigma x_i/n)^2}{S_{xx}}}.$$

- The variable $y_p - \hat{y}_p$ is normally distributed.

In particular, for fixed values of the predictor variable, the possible differences between the observed and predicted values of the response variable corresponding to $x_p$ have a normal distribution with a mean of 0.

---

In light of Key Fact 14.7, if we standardize the variable $y_p - \hat{y}_p$, the resulting variable has the standard normal distribution. However, because the standardized variable contains the unknown parameter $\sigma$, it cannot be used as a basis for a prediction-interval formula. So we replace $\sigma$ by its estimate $s_e$, the standard error of the estimate. The resulting variable has a $t$-distribution.

---

**Key Fact 14.8**

***t*-Distribution for Prediction Intervals in Regression**

Suppose that the variables $x$ and $y$ satisfy the four assumptions for regression inferences. Then, for samples of size $n$, each with the same values, $x_1, x_2, \ldots, x_n$, for the predictor variable, the variable

$$t = \frac{y_p - \hat{y}_p}{s_e\sqrt{1 + \frac{1}{n} + \frac{(x_p - \Sigma x_i/n)^2}{S_{xx}}}}$$

has the $t$-distribution with df $= n - 2$.

---

Using Key Fact 14.8, we can derive a prediction-interval procedure, called the **predicted value $t$-interval procedure.**

| Procedure 14.4 | **Predicted Value $t$-Interval Procedure** |
|---|---|

*Purpose* To find a prediction interval for the value of the response variable corresponding to a particular value of the predictor variable, $x_p$

*Assumptions* The four assumptions for regression inferences

**STEP 1** For a prediction level of $1 - \alpha$, use Table IV to find $t_{\alpha/2}$ with df $= n - 2$.

**STEP 2** Compute the predicted value, $\hat{y}_p = b_0 + b_1 x_p$.

**STEP 3** The endpoints of the prediction interval for the value of the response variable are

$$\hat{y}_p \pm t_{\alpha/2} \cdot s_e \sqrt{1 + \frac{1}{n} + \frac{(x_p - \Sigma x_i/n)^2}{S_{xx}}}.$$

**STEP 4** Interpret the prediction interval.

---

| Example 14.9 | **The Predicted Value $t$-Interval Procedure** |
|---|---|

*Age and Price of Orions* Using the sample data in Table 14.4 on page 638, find a 95% prediction interval for the price of a 3-year-old Orion.

**Solution** We apply Procedure 14.4.

**STEP 1** For a prediction level of $1 - \alpha$, use Table IV to find $t_{\alpha/2}$ with df $= n - 2$.

We want a 95% prediction interval, or $\alpha = 0.05$. Also, because $n = 11$, we have df $= 9$. From Table IV, $t_{\alpha/2} = t_{0.05/2} = t_{0.025} = 2.262$.

**STEP 2** Compute the predicted value, $\hat{y}_p = b_0 + b_1 x_p$.

As previously shown, the sample regression equation for the data in Table 14.4 is $\hat{y} = 195.47 - 20.26x$. Therefore, the predicted price for a 3-year-old Orion is

$$\hat{y}_p = 195.47 - 20.26 \cdot 3 = 134.69.$$

**STEP 3** The endpoints of the prediction interval for the value of the response variable are

$$\hat{y}_p \pm t_{\alpha/2} \cdot s_e \sqrt{1 + \frac{1}{n} + \frac{(x_p - \Sigma x_i/n)^2}{S_{xx}}}.$$

From Example 4.4, $\Sigma x_i = 58$ and $\Sigma x_i^2 = 326$; from Example 14.2, we know that $s_e = 12.58$. Also, $n = 11$, $t_{\alpha/2} = 2.262$, $x_p = 3$, and $\hat{y}_p = 134.69$. Consequently, the endpoints of the prediction interval are

$$134.69 \pm 2.262 \cdot 12.58 \sqrt{1 + \frac{1}{11} + \frac{(3 - 58/11)^2}{326 - (58)^2/11}}$$

or $134.69 \pm 33.02$, or 101.67 to 167.71.

**You try it!**

Exercise 14.81(d)
on page 645

**STEP 4** **Interpret the prediction interval.**

**Interpretation** We can be 95% certain that the price of a 3-year-old Orion will be somewhere between $10,167 and $16,771.

•  •  •

We just demonstrated that a 95% prediction interval for the observed price of a 3-year-old Orion is from $10,167 to $16,771. In Example 14.8, we found that a 95% confidence interval for the mean price of all 3-year-old Orions is from $11,793 to $15,145. We show both intervals in Fig. 14.11.

**FIGURE 14.11**
Prediction and confidence
intervals for 3-year-old Orions

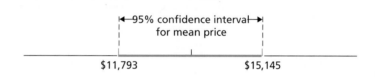

**What Does It Mean?**

More error is involved in predicting the price of a single 3-year-old Orion than in estimating the mean price of all 3-year-old Orions.

Note that the prediction interval is wider than the confidence interval, a result to be expected, for the following reason: The error in the estimate of the mean price of all 3-year-old Orions is due only to the fact that the population regression line is being estimated by a sample regression line, whereas the error in the prediction of the price of one particular 3-year-old Orion is due to the error in estimating the mean price plus the variation in prices of 3-year-old Orions.

# The Technology Center

Some statistical technologies have programs that automatically perform conditional mean and predicted value *t*-interval procedures. In this subsection, we present output and step-by-step instructions for such programs. (*Note to TI-83/84 Plus users:* At the time of this writing, the TI-83/84 Plus does not have built-in programs for conducting conditional mean and predicted value *t*-interval procedures.)

**Example 14.10** **Using Technology to Obtain Conditional Mean and Predicted Value *t*-Intervals**

*Age and Price of Orions*    Table 14.4 on page 638 gives the age and price data for a sample of 11 Orions. Use Minitab or Excel to determine a 95% confidence interval for the mean price of all 3-year-old Orions and a 95% prediction interval for the price of a 3-year-old Orion.

**Solution**    We applied the conditional mean and predicted value *t*-interval programs to the data. Output 14.2 on the next page shows only the portion of the regression output relevant to the confidence and prediction intervals. Steps for generating that output are presented in Instructions 14.2, also on the next page.

**OUTPUT 14.2** Confidence and prediction intervals for 3-year-old Orions

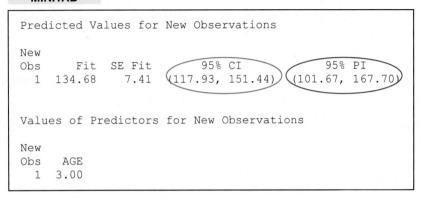

MINITAB

```
Predicted Values for New Observations

New
Obs      Fit   SE Fit        95% CI              95% PI
  1   134.68     7.41   (117.93, 151.44)   (101.67, 167.70)


Values of Predictors for New Observations

New
Obs    AGE
  1   3.00
```

EXCEL

| PRICE | AGE | Predicted Value | Lower Cond. Mean Limit | Upper Cond. Mean Limit | Lower Prediction Limit | Upper Prediction Limit |
|-------|-----|-----------------|------------------------|------------------------|------------------------|------------------------|
| 85  | 5 | 94.162162  | 85.411957  | 102.91237 | 64.396763 | 123.92756 |
| 103 | 4 | 114.42342  | 102.65278  | 126.19407 | 83.634447 | 145.2124  |
| 70  | 6 | 73.900901  | 64.164572  | 83.63723  | 43.830832 | 103.97097 |
| 82  | 5 | 94.162162  | 85.411957  | 102.91237 | 64.396763 | 123.92756 |
| 89  | 5 | 94.162162  | 85.411957  | 102.91237 | 64.396763 | 123.92756 |
| 98  | 5 | 94.162162  | 85.411957  | 102.91237 | 64.396763 | 123.92756 |
| 66  | 6 | 73.900901  | 64.164572  | 83.63723  | 43.830832 | 103.97097 |
| 95  | 6 | 73.900901  | 64.164572  | 83.63723  | 43.830832 | 103.97097 |
| 169 | 2 | 154.94595  | 132.51497  | 177.37692 | 118.71666 | 191.17524 |
| 70  | 7 | 53.63964   | 39.738625  | 67.540655 | 21.974973 | 85.304307 |
| 48  | 7 | 53.63964   | 39.738625  | 67.540655 | 21.974973 | 85.304307 |
| .   | 3 | 134.68468  | 117.92932  | 151.44005 | 101.56719 | 167.70218 |

In Output 14.2, the items circled in red and blue give the required 95% confidence and prediction intervals, respectively, in hundreds of dollars.

• • •

**INSTRUCTIONS 14.2**
Steps for generating Output 14.2

MINITAB

1 Store the age and price data from Table 14.4 in columns named AGE and PRICE, respectively
2 Choose **Stat ➤ Regression ➤ Regression...**
3 Specify PRICE in the **Response** text box
4 Specify AGE in the **Predictors** text box
5 Click the **Results...** button
6 Select the **Regression equation, table of coefficients, s, R-squared, and basic analysis of variance** option button
7 Click **OK**
8 Click the **Options...** button
9 Type 3 in the **Prediction intervals for new observations** text box
10 Type 95 in the **Confidence level** text box
11 Click **OK** twice

EXCEL

1 Append a row to the age and price data in Table 14.4 with a 3 for age and a period for price; store the extended data in ranges named AGE and PRICE, respectively
2 Choose **DDXL ➤ Regression**
3 Select **Simple regression** from the **Function type** drop-down list box
4 Specify PRICE in the **Response Variable** text box
5 Specify AGE in the **Explanatory Variable** text box
6 Click **OK**
7 Click the **95% Confidence and Prediction Intervals** button

# Exercises 14.3

## Understanding the Concepts and Skills

**14.73** Without doing any calculations, fill in the blank and explain your answer. Based on the sample data in Table 14.4, the predicted price for a 4-year-old Orion is $11,443. A point estimate for the mean price of all 4-year-old Orions, based on the same sample data, is _____.

*In Exercises 14.74–14.79, we repeat the data from Exercises 14.10–14.15 and specify a value of the predictor variable. For each exercise, do the following.*

*a. Determine a point estimate for the conditional mean of the response variable corresponding to the specified value of the predictor variable.*

*b. Find a 95% confidence interval for the conditional mean of the response variable corresponding to the specified value of the predictor variable.*

*c. Determine the predicted value of the response variable corresponding to the specified value of the predictor variable.*

*d. Find a 95% prediction interval for the value of the response variable corresponding to the specified value of the predictor variable.*

**14.74**

| x | 2 | 4 | 3 |
|---|---|---|---|
| y | 3 | 5 | 7 |

$x = 3$

**14.75**

| x | 3 | 1 | 2 |
|---|---|---|---|
| y | −4 | 0 | −5 |

$x = 2$

**14.76**

| x | 0 | 4 | 3 | 1 | 2 |
|---|---|---|---|---|---|
| y | 1 | 9 | 8 | 4 | 3 |

$x = 1$

**14.77**

| x | 3 | 4 | 1 | 2 |
|---|---|---|---|---|
| y | 4 | 5 | 0 | −1 |

$x = 4$

**14.78**

| x | 1 | 1 | 5 | 5 |
|---|---|---|---|---|
| y | 1 | 3 | 2 | 4 |

$x = 2$

**14.79**

| x | 0 | 2 | 2 | 5 | 6 |
|---|---|---|---|---|---|
| y | 4 | 2 | 0 | −2 | 1 |

$x = 3$

*In Exercises 14.80–14.85, we repeat the information from Exercises 14.16–14.21. Presuming that the assumptions for regression inferences are met, determine the required confidence and prediction intervals.*

**14.80 Tax Efficiency.** Following are the data on percentage of investments in energy securities and tax efficiency from Exercise 14.16.

| x | 3.1 | 3.2 | 3.7 | 4.3 | 4.0 | 5.5 | 6.7 | 7.4 | 7.4 | 10.6 |
|---|-----|-----|-----|-----|-----|-----|-----|-----|-----|------|
| y | 98.1 | 94.7 | 92.0 | 89.8 | 87.5 | 85.0 | 82.0 | 77.8 | 72.1 | 53.5 |

a. Obtain a point estimate for the mean tax efficiency of all mutual fund portfolios with 6% of their investments in energy securities.

b. Determine a 95% confidence interval for the mean tax efficiency of all mutual fund portfolios with 6% of their investments in energy securities.

c. Find the predicted tax efficiency of a mutual fund portfolio with 6% of its investments in energy securities.

d. Determine a 95% prediction interval for the tax efficiency of a mutual fund portfolio with 6% of its investments in energy securities.

e. Draw graphs similar to those in Fig. 14.11 on page 643, showing both the 95% confidence interval from part (b) and the 95% prediction interval from part (d).

f. Why is the prediction interval wider than the confidence interval?

**14.81 Corvette Prices.** Following are the age and price data for Corvettes from Exercise 14.17.

| x | 6 | 6 | 6 | 2 | 2 | 5 | 4 | 5 | 1 | 4 |
|---|---|---|---|---|---|---|---|---|---|---|
| y | 270 | 260 | 275 | 405 | 364 | 295 | 335 | 308 | 405 | 305 |

a. Obtain a point estimate for the mean price of all 4-year-old Corvettes.

b. Determine a 90% confidence interval for the mean price of all 4-year-old Corvettes.

c. Find the predicted price of a 4-year-old Corvette.

d. Determine a 90% prediction interval for the price of a 4-year-old Corvette.

e. Draw graphs similar to those in Fig. 14.11 on page 643, showing both the 90% confidence interval from part (b) and the 90% prediction interval from part (d).

f. Why is the prediction interval wider than the confidence interval?

**14.82 Custom Homes.** Following are the size and price data for custom homes from Exercise 14.18.

| x | 26 | 27 | 33 | 29 | 29 | 34 | 30 | 40 | 22 |
|---|----|----|----|----|----|----|----|----|----|
| y | 540 | 555 | 575 | 577 | 606 | 661 | 738 | 804 | 496 |

a. Determine a point estimate for the mean price of all 2800-sq-ft Equestrian Estate homes.

b. Find a 99% confidence interval for the mean price of all 2800-sq-ft Equestrian Estate homes.

c. Find the predicted price of a 2800-sq-ft Equestrian Estate home.

d. Determine a 99% prediction interval for the price of a 2800-sq-ft Equestrian Estate home.

**14.83 Plant Emissions.** Following are the data on plant weight and quantity of volatile emissions from Exercise 14.19.

| x | 57 | 85 | 57 | 65 | 52 | 67 | 62 | 80 | 77 | 53 | 68 |
|---|----|----|----|----|----|----|----|----|----|----|----|
| y | 8.0 | 22.0 | 10.5 | 22.5 | 12.0 | 11.5 | 7.5 | 13.0 | 16.5 | 21.0 | 12.0 |

a. Obtain a point estimate for the mean quantity of volatile emissions of all (*Solanum tuberosom*) plants that weigh 60 g.

b. Find a 95% confidence interval for the mean quantity of volatile emissions of all plants that weigh 60 g.

c. Find the predicted quantity of volatile emissions for a plant that weighs 60 g.

d. Determine a 95% prediction interval for the quantity of volatile emissions for a plant that weighs 60 g.

**14.84 Crown-Rump Length.** Following are the data on age of fetuses and length of crown-rump from Exercise 14.20.

| x | 10 | 10 | 13 | 13 | 18 | 19 | 19 | 23 | 25 | 28 |
|---|----|----|----|----|----|----|----|----|----|----|
| y | 66 | 66 | 108 | 106 | 161 | 166 | 177 | 228 | 235 | 280 |

a. Determine a point estimate for the mean crown-rump length of all 19-week-old fetuses.

b. Find a 90% confidence interval for the mean crown-rump length of all 19-week-old fetuses.

c. Find the predicted crown-rump length of a 19-week-old fetus.

d. Determine a 90% prediction interval for the crown-rump length of a 19-week-old fetus.

**14.85 Study Time and Score.** Following are the data on total hours studied over 2 weeks and test score at the end of the 2 weeks from Exercise 14.21.

| x | 10 | 15 | 12 | 20 | 8 | 16 | 14 | 22 |
|---|----|----|----|----|----|----|----|----|
| y | 92 | 81 | 84 | 74 | 85 | 80 | 84 | 80 |

a. Determine a point estimate for the mean test score of all beginning calculus students who study for 15 hours.

b. Find a 99% confidence interval for the mean test score of all beginning calculus students who study for 15 hours.

c. Find the predicted test score of a beginning calculus student who studies for 15 hours.

d. Determine a 99% prediction interval for the test score of a beginning calculus student who studies for 15 hours.

## Working With Large Data Sets

*In Exercises 14.86–14.96, use the technology of your choice to do the following.*

*a. Decide whether you can reasonably apply the conditional mean and predicted value t-interval procedures to the data. If so, then also do parts (b)–(f).*

*b. Determine and interpret a point estimate for the conditional mean of the response variable corresponding to the specified value of the predictor variable.*

*c. Find and interpret a 95% confidence interval for the conditional mean of the response variable corresponding to the specified value of the predictor variable.*

*d. Determine and interpret the predicted value of the response variable corresponding to the specified value of the predictor variable.*

*e. Find and interpret a 95% prediction interval for the value of the response variable corresponding to the specified value of the predictor variable.*

*f. Compare and discuss the differences between the confidence interval that you obtained in part (c) and the prediction interval that you obtained in part (e).*

**14.86 Birdies and Score.** The data from Exercise 14.30 for number of birdies during a tournament and final score of 63 women golfers are on the WeissStats CD. Specified value of the predictor variable: 12 birdies.

**14.87 U.S. Presidents.** The data from Exercise 14.31 for the ages at inauguration and of death of the presidents of the United States are on the WeissStats CD. Specified value of the predictor variable: 53 years.

**14.88 Health Care.** The data from Exercise 14.32 for percentage of gross domestic product (GDP) spent on health care and life expectancy, in years, of selected countries are on the WeissStats CD. Specified value of the predictor variable: 8.6%.

**14.89 Acreage and Value.** The data from Exercise 14.33 for lot size (in acres) and assessed value (in thousands of dollars) of a sample of homes in a particular area are on the WeissStats CD. Specified value of the predictor variable: 2.5 acres.

**14.90 Home Size and Value.** The data from Exercise 14.34 for home size (in square feet) and assessed value (in thousands of dollars) for the same homes as in Exercise 14.89 are on the WeissStats CD. Specified value of the predictor variable: 3000 sq ft.

**14.91 High and Low Temperature.** The data from Exercise 14.35 for average high and low temperatures in January of a random sample of 50 cities are on the WeissStats CD. Specified value of the predictor variable: 55°F.

**14.92 PCBs and Pelicans.** The data from Exercise 14.36 for shell thickness and concentration of PCBs of 60 Anacapa pelican eggs are on the WeissStats CD. Specified value of the predictor variable: 220 ppm.

**14.93 Gas Guzzlers.** The data from Exercise 14.37 for gas mileage and engine displacement of 121 vehicles are on the WeissStats CD. Specified value of the predictor variable: 3.0 L.

**14.94 Estriol Level and Birth Weight.** The data from Exercise 14.38 for estriol levels of pregnant women and birth weights of their children are on the WeissStats CD. Specified value of the predictor variable: 18 mg/24 hr.

**14.95 Shortleaf Pines.** The data from Exercise 14.39 for volume, in cubic feet, and diameter at breast height, in inches, of 70 shortleaf pines are on the WeissStats CD. Specified value of the predictor variable: 11 inches.

**14.96 Body Fat.** The data from Exercise 14.72 for age and body fat of 18 randomly selected adults are on the WeissStats CD. Specified value of the predictor variable: 30 years.

## Extending the Concepts and Skills

**Margin of Error in Regression.** In Exercises 14.97 and 14.98, you will examine the magnitude of the margin of error of confidence intervals and prediction intervals in regression as a function of how far the specified value of the predictor variable is from the mean of the observed values of the predictor variable.

**14.97 Age and Price of Orions.** Refer to the data on age and price of a sample of 11 Orions given in Table 14.4 on page 638.
a. For each age between 2 and 7 years, obtain a 95% confidence interval for the mean price of all Orions of that age. Plot the confidence intervals against age and discuss your results.
b. Determine the margin of error for each confidence interval that you obtained in part (a). Plot the margins of error against age and discuss your results.
c. Repeat parts (a) and (b) for prediction intervals.

**14.98** Refer to the confidence interval and prediction interval formulas in Procedures 14.3 and 14.4, respectively.
a. Explain why, for a fixed confidence level, the margin of error for the estimate of the conditional mean of the response variable increases as the value of the predictor variable moves farther from the mean of the observed values of the predictor variable.
b. Explain why, for a fixed prediction level, the margin of error for the estimate of the predicted value of the response variable increases as the value of the predictor variable moves farther from the mean of the observed values of the predictor variable.

## 14.4 Inferences in Correlation

Frequently, we want to decide whether two variables are linearly correlated, that is, whether there is a linear relationship between the two variables. In the context of regression, we can make that decision by performing a hypothesis test for the slope of the population regression line, as discussed in Section 14.2.

Alternatively, we can perform a hypothesis test for the **population linear correlation coefficient, $\rho$** (rho). This parameter measures the linear correlation of all possible pairs of observations of two variables in the same way that a sample linear correlation coefficient, $r$, measures the linear correlation of a sample of pairs. Thus, $\rho$ actually describes the strength of the linear relationship between two variables; $r$ is only an estimate of $\rho$ obtained from sample data.

The population linear correlation coefficient of two variables, $x$ and $y$, always lies between $-1$ and $1$. Values of $\rho$ near $-1$ or $1$ indicate a strong linear relationship between the variables, whereas values of $\rho$ near $0$ indicate a weak linear relationship between the variables. Note the following:

- If $\rho = 0$, the variables are **linearly uncorrelated,** meaning that there is no linear relationship between the variables.
- If $\rho > 0$, the variables are **positively linearly correlated,** meaning that $y$ tends to increase linearly as $x$ increases (and vice versa), with the tendency being greater the closer $\rho$ is to $1$.
- If $\rho < 0$, the variables are **negatively linearly correlated,** meaning that $y$ tends to decrease linearly as $x$ increases (and vice versa), with the tendency being greater the closer $\rho$ is to $-1$.

- If $\rho \neq 0$, the variables are **linearly correlated.** Linearly correlated variables are either positively linearly correlated or negatively linearly correlated.

As we mentioned, a sample linear correlation coefficient, $r$, is an estimate of the population linear correlation coefficient, $\rho$. Consequently, we can use $r$ as a basis for performing a hypothesis test for $\rho$. To do so, we require the following fact.

---

**Key Fact 14.9** | **_t_-Distribution for a Correlation Test**

Suppose that the variables $x$ and $y$ satisfy the four assumptions for regression inferences and that $\rho = 0$. Then, for samples of size $n$, the variable

$$t = \frac{r}{\sqrt{\dfrac{1 - r^2}{n - 2}}}$$

has the $t$-distribution with df $= n - 2$.

---

In light of Key Fact 14.9, for a hypothesis test with the null hypothesis $H_0$: $\rho = 0$, we can use the variable

$$t = \frac{r}{\sqrt{\dfrac{1 - r^2}{n - 2}}}$$

as the test statistic and obtain the critical values or $P$-value from the $t$-table, Table IV. We call this hypothesis-testing procedure the **correlation _t_-test.** Procedure 14.5 provides a step-by-step method for performing a correlation $t$-test by using either the critical-value approach or the $P$-value approach.

**Example 14.11** | **The Correlation _t_-Test**

**TABLE 14.5**
Age and price data
for a sample of 11 Orions

| Age (yr) $x$ | Price ($100) $y$ |
|:---:|:---:|
| 5 | 85 |
| 4 | 103 |
| 6 | 70 |
| 5 | 82 |
| 5 | 89 |
| 5 | 98 |
| 6 | 66 |
| 6 | 95 |
| 2 | 169 |
| 7 | 70 |
| 7 | 48 |

*Age and Price of Orions* The data on age and price for a sample of 11 Orions are repeated in Table 14.5. At the 5% significance level, do the data provide sufficient evidence to conclude that age and price of Orions are negatively linearly correlated?

**Solution** As we discovered in Example 14.3 on page 622, considering that the assumptions for regression inferences are met by the variables age and price for Orions is not unreasonable, at least for Orions between 2 and 7 years old. Consequently, we apply Procedure 14.5 to carry out the required hypothesis test.

**STEP 1 State the null and alternative hypotheses.**

Let $\rho$ denote the population linear correlation coefficient for the variables age and price of Orions. Then the null and alternative hypotheses are

$$H_0: \rho = 0 \text{ (age and price are linearly uncorrelated)}$$
$$H_a: \rho < 0 \text{ (age and price are negatively linearly correlated)}.$$

Note that the hypothesis test is left tailed.

**STEP 2 Decide on the significance level, $\alpha$.**

We are to use $\alpha = 0.05$.

## Procedure 14.5 Correlation *t*-Test

*Purpose* To perform a hypothesis test for a population linear correlation coefficient, $\rho$

*Assumptions* The four assumptions for regression inferences

**STEP 1** The null hypothesis is $H_0$: $\rho = 0$, and the alternative hypothesis is

$$
\begin{array}{ccc}
H_a\text{: } \rho \neq 0 & & H_a\text{: } \rho < 0 & & H_a\text{: } \rho > 0 \\
\text{(Two tailed)} & \text{or} & \text{(Left tailed)} & \text{or} & \text{(Right tailed)}
\end{array}
$$

**STEP 2** Decide on the significance level, $\alpha$.

**STEP 3** Compute the value of the test statistic

$$
t = \frac{r}{\sqrt{\dfrac{1 - r^2}{n - 2}}}
$$

and denote that value $t_0$.

---

| CRITICAL-VALUE APPROACH | or | P-VALUE APPROACH |

**STEP 4** The critical value(s) are

$$
\begin{array}{ccc}
\pm t_{\alpha/2} & & -t_\alpha & & t_\alpha \\
\text{(Two tailed)} & \text{or} & \text{(Left tailed)} & \text{or} & \text{(Right tailed)}
\end{array}
$$

with df $= n - 2$. Use Table IV to find the critical value(s).

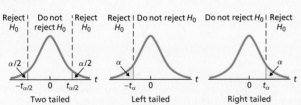

**STEP 5** If the value of the test statistic falls in the rejection region, reject $H_0$; otherwise, do not reject $H_0$.

**STEP 4** The *t*-statistic has df $= n - 2$. Use Table IV to estimate the *P*-value, or obtain it exactly by using technology.

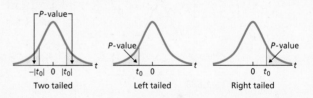

**STEP 5** If $P \leq \alpha$, reject $H_0$; otherwise, do not reject $H_0$.

**STEP 6** Interpret the results of the hypothesis test.

**STEP 3** Compute the value of the test statistic

$$t = \frac{r}{\sqrt{\dfrac{1-r^2}{n-2}}}.$$

In Example 4.10 on page 189, we found that $r = -0.924$, so the value of the test statistic is

$$t = \frac{-0.924}{\sqrt{\dfrac{1-(-0.924)^2}{11-2}}} = -7.249.$$

| CRITICAL-VALUE APPROACH | or | P-VALUE APPROACH |

**STEP 4** The critical value for a left-tailed test is $-t_\alpha$ with df $= n - 2$. Use Table IV to find the critical value.

For $n = 11$, df $= 9$. Also, $\alpha = 0.05$. From Table IV, for df $= 9$, $t_{0.05} = 1.833$. Consequently, the critical value is $-t_{0.05} = -1.833$, as shown in Fig. 14.12A.

**FIGURE 14.12A**

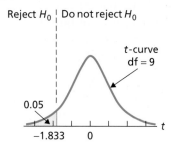

**STEP 5** If the value of the test statistic falls in the rejection region, reject $H_0$; otherwise, do not reject $H_0$.

The value of the test statistic, found in Step 3, is $t = -7.249$. Figure 14.12A shows that this value falls in the rejection region, so we reject $H_0$. The test results are statistically significant at the 5% level.

**STEP 4** The $t$-statistic has df $= n - 2$. Use Table IV to estimate the $P$-value or obtain it exactly by using technology.

From Step 3, the value of the test statistic is $t = -7.249$. Because the test is left tailed, the $P$-value is the probability of observing a value of $t$ of $-7.249$ or less if the null hypothesis is true. That probability equals the shaded area shown in Fig. 14.12B.

**FIGURE 14.12B**

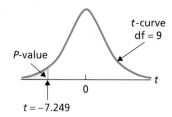

For $n = 11$, df $= 9$. Referring to Fig. 14.12B and Table IV, we find that $P < 0.005$. (Using technology, we obtain $P = 0.0000244$.)

**STEP 5** If $P \le \alpha$, reject $H_0$; otherwise, do not reject $H_0$.

From Step 4, $P < 0.005$. Because the $P$-value is less than the specified significance level of 0.05, we reject $H_0$. The test results are statistically significant at the 5% level and (see Table 9.10 on page 414) provide very strong evidence against the null hypothesis.

Exercise 14.109
on page 653

**STEP 6** Interpret the results of the hypothesis test.

**Interpretation** At the 5% significance level, the data provide sufficient evidence to conclude that age and price of Orions are negatively linearly correlated. Prices for 2- to 7-year-old Orions tend to decrease linearly with increasing age.

● ● ●

# The Technology Center

Most statistical technologies have programs that automatically perform a correlation $t$-test. In this subsection, we present output and step-by-step instructions for such programs.

*Note to Minitab users:* At the time of this writing, Minitab does only a two-tailed correlation $t$-test. However, we can get a one-tailed $P$-value from the provided two-tailed $P$-value by using the result of Exercise 9.108 on page 419. This result implies, for instance, that if the sign of the sample linear correlation coefficient is in the same direction as the alternative hypothesis, then the one-tailed $P$-value equals one-half of the two-tailed $P$-value.

**Example 14.12** **Using Technology to Conduct a Correlation $t$-Test**

*Age and Price of Orions* Table 14.5 on page 648 gives the age and price data for a sample of 11 Orions. Use Minitab, Excel, or the TI-83/84 Plus to decide, at the 5% significance level, whether the data provide sufficient evidence to conclude that age and price of Orions are negatively linearly correlated.

**Solution** Let $\rho$ denote the population linear correlation coefficient for the variables age and price of Orions. We want to perform the hypothesis test

$H_0$: $\rho = 0$ (age and price are linearly uncorrelated)

$H_a$: $\rho < 0$ (age and price are negatively linearly correlated)

at the 5% significance level. Note that the hypothesis test is left tailed.

We applied the correlation $t$-test programs to the data, resulting in Output 14.3. Steps for generating that output are presented in Instructions 14.3 on the following page.

**OUTPUT 14.3**  Correlation $t$-test for the Orion data

| MINITAB | EXCEL |
|---|---|

**Correlations: AGE, PRICE**

Pearson correlation of AGE and PRICE = -0.924
P-Value = 0.000

Correlation t-test Results

|  | Ha | Left Tailed |
|---|---|---|
| t-statistic | | -7.24 |
| p-value | | < .0001 |

**TI-83/84 PLUS**

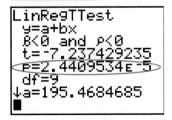

```
LinRegTTest
 y=a+bx
 β<0 and ρ<0
 t=-7.237429235
 P=2.4409534E-5
 df=9
↓a=195.4684685
```

```
LinRegTTest
 y=a+bx
 β<0 and ρ<0
↑b=-20.26126126
 s=12.57657227
 r²=.8533733464
 r=-.9237820881
```

As shown in Output 14.3, the $P$-value is less than the specified significance level of 0.05, so we reject $H_0$. At the 5% significance level, the data provide sufficient evidence to conclude that age and price of Orions are negatively linearly correlated.

• • •

**INSTRUCTIONS 14.3**  Steps for generating Output 14.3

| MINITAB | EXCEL | TI-83/84 PLUS |
|---|---|---|
| 1 Store the age and price data from Table 14.5 in columns named AGE and PRICE, respectively | 1 Store the age and price data from Table 14.5 in ranges named AGE and PRICE, respectively | 1 Store the age and price data from Table 14.5 in lists named AGE and PRICE, respectively |
| 2 Choose **Stat ➤ Basic Statistics ➤ Correlation...** | 2 Choose **DDXL ➤ Regression** | 2 Press **STAT**, arrow over to **TESTS**, and press **ALPHA ➤ E** |
| 3 Specify AGE and PRICE in the **Variables** text box | 3 Select **Correlation** from the **Function type** drop-down list box | 3 Press **2nd ➤ LIST**, arrow down to AGE, and press **ENTER** twice |
| 4 Cleck the **Display p-values** text box | 4 Specify AGE in the **x-Axis Quantitative Variable** text box | 4 Press **2nd ➤ LIST**, arrow down to PRICE, and press **ENTER** three times |
| 5 Click **OK** | 5 Specify PRICE in the **y-Axis Quantitative Variable** text box | 5 Highlight **<0** and press **ENTER** |
| | 6 Click **OK** | 6 Highlight **Calculate** and press **ENTER** |
| | 7 Click the **Perform a Left Tailed Test** button | |

▮ ▮ ▮

# Exercises 14.4

## Understanding the Concepts and Skills

**14.99** Identify the statistic used to estimate the population linear correlation coefficient.

**14.100** Suppose that, for a sample of pairs of observations from two variables, the linear correlation coefficient, $r$, is positive. Does this result necessarily imply that the variables are positively linearly correlated? Explain.

**14.101** Fill in the blanks.
a. If $\rho = 0$, the two variables under consideration are linearly _____.
b. If two variables are positively linearly correlated, one of the variables tends to increase as the other _____.
c. If two variables are _____ linearly correlated, one of the variables tends to decrease as the other increases.

*In Exercises **14.102–14.107**, we repeat the data from Exercises 14.10–14.15 and specify an alternative hypothesis for a correlation t-test. For each exercise, decide, at the 10% significance level, whether the data provide sufficient evidence to reject the null hypothesis in favor of the alternative hypothesis.*

**14.102**

| $x$ | 2 | 4 | 3 |
|---|---|---|---|
| $y$ | 3 | 5 | 7 |

$H_a: \rho > 0$

**14.103**

| $x$ | 3 | 1 | 2 |
|---|---|---|---|
| $y$ | $-4$ | 0 | $-5$ |

$H_a: \rho < 0$

**14.104**

| $x$ | 0 | 4 | 3 | 1 | 2 |
|---|---|---|---|---|---|
| $y$ | 1 | 9 | 8 | 4 | 3 |

$H_a: \rho \neq 0$

**14.105**

| $x$ | 3 | 4 | 1 | 2 |
|---|---|---|---|---|
| $y$ | 4 | 5 | 0 | $-1$ |

$H_a: \rho > 0$

**14.106**

| $x$ | 1 | 1 | 5 | 5 |
|---|---|---|---|---|
| $y$ | 1 | 3 | 2 | 4 |

$H_a: \rho < 0$

**14.107**

| x | 0 | 2 | 2 | 5 | 6 |
|---|---|---|---|---|---|
| y | 4 | 2 | 0 | −2 | 1 |

$H_a: \rho \neq 0$

*In Exercises 14.108–14.113, we repeat the information from Exercises 14.16–14.21. Presuming that the assumptions for regression inferences are met, perform the required correlation t-tests, using either the critical-value approach or the P-value approach.*

**14.108 Tax Efficiency.** Following are the data on percentage of investments in energy securities and tax efficiency from Exercise 14.16.

| x | 3.1 | 3.2 | 3.7 | 4.3 | 4.0 | 5.5 | 6.7 | 7.4 | 7.4 | 10.6 |
|---|---|---|---|---|---|---|---|---|---|---|
| y | 98.1 | 94.7 | 92.0 | 89.8 | 87.5 | 85.0 | 82.0 | 77.8 | 72.1 | 53.5 |

At the 2.5% significance level, do the data provide sufficient evidence to conclude that percentage of investments in energy securities and tax efficiency are negatively linearly correlated for mutual fund portfolios?

**14.109 Corvette Prices.** Following are the age and price data for Corvettes from Exercise 14.17.

| x | 6 | 6 | 6 | 2 | 2 | 5 | 4 | 5 | 1 | 4 |
|---|---|---|---|---|---|---|---|---|---|---|
| y | 270 | 260 | 275 | 405 | 364 | 295 | 335 | 308 | 405 | 305 |

At the 5% level of significance, do the data provide sufficient evidence to conclude that age and price of Corvettes are negatively linearly correlated?

**14.110 Custom Homes.** Following are the size and price data for custom homes from Exercise 14.18.

| x | 26 | 27 | 33 | 29 | 29 | 34 | 30 | 40 | 22 |
|---|---|---|---|---|---|---|---|---|---|
| y | 540 | 555 | 575 | 577 | 606 | 661 | 738 | 804 | 496 |

At the 0.5% significance level, do the data provide sufficient evidence to conclude that, for custom homes in the Equestrian Estates, size and price are positively linearly correlated?

**14.111 Plant Emissions.** Following are the data on plant weight and quantity of volatile emissions from Exercise 14.19.

| x | 57 | 85 | 57 | 65 | 52 | 67 | 62 | 80 | 77 | 53 | 68 |
|---|---|---|---|---|---|---|---|---|---|---|---|
| y | 8.0 | 22.0 | 10.5 | 22.5 | 12.0 | 11.5 | 7.5 | 13.0 | 16.5 | 21.0 | 12.0 |

Do the data suggest that, for the potato plant *Solanum tuberosom*, weight and quantity of volatile emissions are linearly correlated? Use $\alpha = 0.05$.

**14.112 Crown-Rump Length.** Following are the data on age of fetuses and length of crown-rump from Exercise 14.20.

| x | 10 | 10 | 13 | 13 | 18 | 19 | 19 | 23 | 25 | 28 |
|---|---|---|---|---|---|---|---|---|---|---|
| y | 66 | 66 | 108 | 106 | 161 | 166 | 177 | 228 | 235 | 280 |

At the 10% significance level, do the data provide sufficient evidence to conclude that age and crown-rump length are linearly correlated?

**14.113 Study Time and Score.** Following are the data on total hours studied over 2 weeks and test score at the end of the 2 weeks from Exercise 14.21.

| x | 10 | 15 | 12 | 20 | 8 | 16 | 14 | 22 |
|---|---|---|---|---|---|---|---|---|
| y | 92 | 81 | 84 | 74 | 85 | 80 | 84 | 80 |

a. At the 1% significance level, do the data provide sufficient evidence to conclude that a negative linear correlation exists between study time and test score for beginning calculus students?

b. Repeat part (a) using a 5% significance level.

**14.114 Height and Score.** A random sample of 10 students was taken from an introductory statistics class. The following data were obtained, where x denotes height, in inches, and y denotes score on the final exam.

| x | 71 | 68 | 71 | 65 | 66 | 68 | 68 | 64 | 62 | 65 |
|---|---|---|---|---|---|---|---|---|---|---|
| y | 87 | 96 | 66 | 71 | 71 | 55 | 83 | 67 | 86 | 60 |

At the 5% significance level, do the data provide sufficient evidence to conclude that, for students in introductory statistics courses, height and final exam score are linearly correlated?

**14.115** Is $\rho$ a parameter or a statistic? What about $r$? Explain your answers.

## Working With Large Data Sets

*In each of Exercises 14.116–14.126, use the technology of your choice to decide whether you can reasonably apply the correlation t-test. If so, perform and interpret the required correlation t-test(s) at the 5% significance level.*

**14.116 Birdies and Score.** The data from Exercise 14.30 for number of birdies during a tournament and final score of 63 women golfers are on the WeissStats CD. Do the data provide sufficient evidence to conclude that, for women golfers, number of birdies and score are negatively linearly correlated?

**14.117 U.S. Presidents.** The data from Exercise 14.31 for the ages at inauguration and of death of the presidents of the United States are on the WeissStats CD. Do the data provide sufficient evidence to conclude that, for U.S. presidents, age at inauguration and age at death are positively linearly correlated?

**14.118 Health Care.** The data from Exercise 14.32 for percentage of gross domestic product (GDP) spent on health care and life expectancy, in years, of selected countries are on the WeissStats CD.
a. Do the data provide sufficient evidence to conclude that, for countries, percentage of GDP spent on health care and life expectancy are positively linearly correlated?
b. Remove the potential outliers (observation numbers 1 and 12) and then repeat part (a).

**14.119 Acreage and Value.** The data from Exercise 14.33 for lot size (in acres) and assessed value (in thousands of dollars) of a sample of homes in a particular area are on the WeissStats CD. Do the data provide sufficient evidence to conclude that, for homes in this particular area, lot size and assessed value are positively linearly correlated?

**14.120 Home Size and Value.** The data from Exercise 14.34 for home size (in square feet) and assessed value (in thousands of dollars) for the same homes as in Exercise 14.119 are on the WeissStats CD. Do the data provide sufficient evidence to conclude that, for homes in this particular area, lot size and assessed value are positively linearly correlated?

**14.121 High and Low Temperature.** The data from Exercise 14.35 for average high and low temperatures in January of a random sample of 50 cities are on the WeissStats CD. Do the data provide sufficient evidence to conclude that, for cities, average high and low temperatures in January are linearly correlated?

**14.122 PCBs and Pelicans.** The data from Exercise 14.36 for shell thickness and concentration of PCBs of 60 Anacapa pelican eggs are on the WeissStats CD. Do the data provide sufficient evidence to conclude that concentration of PCBs and shell thickness are linearly correlated for Anacapa pelican eggs?

**14.123 Gas Guzzlers.** The data from Exercise 14.37 for gas mileage and engine displacement of 121 vehicles are on the WeissStats CD. Do the data provide sufficient evidence to conclude that engine displacement and gas mileage are negatively linearly correlated?

**14.124 Estriol Level and Birth Weight.** The data from Exercise 14.38 for estriol levels of pregnant women and birth weights of their children are on the WeissStats CD. Do the data provide sufficient evidence to conclude that estriol level and birth weight are positively linearly correlated?

**14.125 Shortleaf Pines.** The data from Exercise 14.39 for volume, in cubic feet, and diameter at breast height, in inches, of 70 shortleaf pines are on the WeissStats CD. Do the data provide sufficient evidence to conclude that diameter at breast height and volume are positively linearly correlated for shortleaf pines?

**14.126 Body Fat.** The data from Exercise 14.72 for age and body fat of 18 randomly selected adults are on the WeissStats CD.
a. Do the data provide sufficient evidence to conclude that, for adults, age and percentage of body fat are positively linearly correlated?
b. Remove the potential outlier and repeat part (a).
c. Compare your results with and without the removal of the potential outlier and state your conclusions.

## Chapter in Review

### You Should be Able to

1. use and understand the formulas in this chapter.

2. state the assumptions for regression inferences.

3. understand the difference between the population regression line and a sample regression line.

4. estimate the regression parameters $\beta_0$, $\beta_1$, and $\sigma$.

5. determine the standard error of the estimate.

6. perform a residual analysis to check the assumptions for regression inferences.

7. perform a hypothesis test to decide whether the slope, $\beta_1$, of the population regression line is not 0 and hence whether $x$ is useful for predicting $y$.

8. obtain a confidence interval for $\beta_1$.

9. determine a point estimate and a confidence interval for the conditional mean of the response variable corresponding to a particular value of the predictor variable.

10. determine a predicted value and a prediction interval for the response variable corresponding to a particular value of the predictor variable.

11. understand the difference between the population correlation coefficient and a sample correlation coefficient.

12. perform a hypothesis test for a population linear correlation coefficient.

## Key Terms

conditional distribution, *616*
conditional mean, *616*
conditional mean *t*-interval
    procedure, *639*
correlation *t*-test, *648, 649*
linearly correlated variables, *648*
linearly uncorrelated variables, *647*
negatively linearly correlated
    variables, *647*

population linear correlation
    coefficient ($\rho$), *647*
population regression equation, *617*
population regression line, *617*
positively linearly correlated
    variables, *647*
predicted value *t*-interval
    procedure, *642*
prediction interval, *640*

regression model, *616*
regression *t*-interval procedure, *634*
regression *t*-test, *631*
residual (*e*), *620*
residual plot, *621*
residual standard deviation, *621*
sampling distribution of the slope of
    the regression line, *630*
standard error of the estimate ($s_e$), *619*

## Review Problems

### Understanding the Concepts and Skills

**1.** Suppose that $x$ and $y$ are two variables of a population with $x$ a predictor variable and $y$ a response variable.
**a.** The distribution of all possible values of the response variable $y$ corresponding to a particular value of the predictor variable $x$ is called a _____ distribution of the response variable.
**b.** State the four assumptions for regression inferences.

**2.** Suppose that $x$ and $y$ are two variables of a population and that the assumptions for regression inferences are met with $x$ as the predictor variable and $y$ as the response variable.
**a.** What statistic is used to estimate the slope of the population regression line?
**b.** What statistic is used to estimate the $y$-intercept of the population regression line?
**c.** What statistic is used to estimate the common conditional standard deviation of the response variable corresponding to fixed values of the predictor variable?

**3.** What two plots did we use in this chapter to decide whether we can reasonably presume that the assumptions for regression inferences are met by two variables of a population? What properties should those plots have?

**4.** Regarding analysis of residuals, decide in each case which assumption for regression inferences may be violated.
**a.** A residual plot—that is, a plot of the residuals against the observed values of the predictor variable—shows curvature.
**b.** A residual plot becomes wider with increasing values of the predictor variable.
**c.** A normal probability plot of the residuals shows extreme curvature.
**d.** A normal probability plot of the residuals shows outliers but is otherwise roughly linear.

**5.** Suppose that you perform a hypothesis test for the slope of the population regression line with the null hypothesis $H_0: \beta_1 = 0$ and the alternative hypothesis $H_a: \beta_1 \neq 0$. If you

reject the null hypothesis, what can you say about the utility of the regression equation for making predictions?

**6.** Identify three statistics that can be used as a basis for testing the utility of a regression.

**7.** For a particular value of a predictor variable, is there a difference between the predicted value of the response variable and the point estimate for the conditional mean of the response variable? Explain your answer.

**8.** Generally speaking, what is the difference between a confidence interval and a prediction interval?

**9.** Fill in the blank: $\bar{x}$ is to $\mu$ as $r$ is to _____.

**10.** Identify the relationship between two variables and the terminology used to describe that relationship if
**a.** $\rho > 0$.      **b.** $\rho = 0$.      **c.** $\rho < 0$.

**11. Graduation Rates.** Graduation rate—the percentage of entering freshmen, attending full time and graduating within 5 years—and what influences it have become a concern in U.S. colleges and universities. *U.S. News and World Report*'s "College Guide" provides data on graduation rates for colleges and universities as a function of the percentage of freshmen in the top 10% of their high school class, total spending per student, and student-to-faculty ratio. A random sample of 10 universities gave the following data on student-to-faculty ratio (S/F ratio) and graduation rate (grad rate).

| S/F ratio $x$ | Grad rate $y$ | S/F ratio $x$ | Grad rate $y$ |
|---|---|---|---|
| 16 | 45 | 17 | 46 |
| 20 | 55 | 17 | 50 |
| 17 | 70 | 17 | 66 |
| 19 | 50 | 10 | 26 |
| 22 | 47 | 18 | 60 |

Discuss what satisfying the assumptions for regression inferences would mean with student-to-faculty ratio as the predictor variable and graduation rate as the response variable.

**12. Graduation Rates.** Refer to Problem 11.
**a.** Determine the regression equation for the data.
**b.** Compute and interpret the standard error of the estimate.
**c.** Presuming that the assumptions for regression inferences are met, interpret your answer to part (b).

**13. Graduation Rates.** Refer to Problems 11 and 12. Perform a residual analysis to decide whether considering the assumptions for regression inferences to be met by the variables student-to-faculty ratio and graduation rate is reasonable.

*For the remainder of these review problems, presume that the variables student-to-faculty ratio and graduation rate satisfy the assumptions for regression inferences.*

**14. Graduation Rates.** Refer to Problems 11 and 12.
**a.** At the 5% significance level, do the data provide sufficient evidence to conclude that student-to-faculty ratio is useful as a predictor of graduation rate?
**b.** Determine a 95% confidence interval for the slope, $\beta_1$, of the population regression line that relates graduation rate to student-to-faculty ratio. Interpret your answer.

**15. Graduation Rates.** Refer to Problems 11 and 12.
**a.** Find a point estimate for the mean graduation rate of all universities that have a student-to-faculty ratio of 17.
**b.** Determine a 95% confidence interval for the mean graduation rate of all universities that have a student-to-faculty ratio of 17.
**c.** Find the predicted graduation rate for a university that has a student-to-faculty ratio of 17.
**d.** Find a 95% prediction interval for the graduation rate of a university that has a student-to-faculty ratio of 17.
**e.** Explain why the prediction interval in part (d) is wider than the confidence interval in part (b).

**16. Graduation Rates.** Refer to Problem 11. At the 2.5% significance level, do the data provide sufficient evidence to conclude that the variables student-to-faculty ratio and graduation rate are positively linearly correlated?

## Working With Large Data Sets

*In Problems 17–20, use the technology of your choice to do the following.*
*a. Determine the sample regression equation.*
*b. Find and interpret the standard error of the estimate.*
*c. Decide, at the 5% significance level, whether the data provide sufficient evidence to conclude that the predictor variable is useful for predicting the response variable.*
*d. Determine and interpret a point estimate for the conditional mean of the response variable corresponding to the specified value of the predictor variable.*
*e. Find and interpret a 95% confidence interval for the conditional mean of the response variable corresponding to the specified value of the predictor variable.*
*f. Determine and interpret the predicted value of the response variable corresponding to the specified value of the predictor variable.*
*g. Find and interpret a 95% prediction interval for the value of the response variable corresponding to the specified value of the predictor variable.*
*h. Compare and discuss the differences between the confidence interval that you obtained in part (e) and the prediction interval that you obtained in part (g).*
*i. Perform and interpret the required correlation t-test at the 5% significance level.*
*j. Perform a residual analysis to decide whether making the preceding inferences is reasonable. Explain your answer.*

**17. IMR and Life Expectancy.** From the *Statistical Abstract of the United States*, we obtained data on infant mortality rate (IMR) and life expectancy, in years, for countries with 12 million or more population in 2003. The data are presented on the WeissStats CD.

- For the estimations and predictions, use an IMR of 30.
- For the correlation test, decide whether IMR and life expectancy are negatively linearly correlated.

**18. High Temperature and Precipitation.** The U.S. National Oceanic and Atmospheric Administration publishes temperature and precipitation information for cities around the world in *Climates of the World*. Data on average high temperature (in degrees Fahrenheit) in July and average precipitation (in inches) in July for 48 cities are on the WeissStats CD.

- For the estimations and predictions, use an average July temperature of 83°F.
- For the correlation test, decide whether average high temperature in July and average precipitation in July are linearly correlated.

**19. Fat Consumption and Prostate Cancer.** Researchers have asked whether there is a relationship between nutrition and cancer, and many studies have shown that there is. In fact, one of the conclusions of a study by B. Reddy et al., "Nutrition and Its Relationship to Cancer" (*Advances in Cancer Research*, Vol. 32, pp. 237–345), was that "…none of the risk factors for cancer is probably more significant than diet and nutrition." One dietary factor that has been studied for its relationship with prostate cancer is fat consumption. On the WeissStats CD, you will find data on per capita fat consumption (in grams per day) and prostate cancer death rate (per 100,000 males) for nations of the world. The data were obtained from a graph—adapted from information in the article mentioned—in John Robbins' classic book *Diet for a New America* (Walpole, NH: Stillpoint, 1987, p. 271).

- For the estimations and predictions, use a per capita fat consumption of 92 grams per day.
- For the correlation test, decide whether per capita fat consumption and prostate cancer death rate are positively linearly correlated.

**20. Payroll and Success.** Steve Galbraith, Morgan Stanley's Chief Investment Officer for U.S. Equity Research, examined several financial aspects of baseball in the article "Finding the Financial Equivalent of a Walk" (*Perspectives*, October 2003, pp. 1–3). The article includes a table containing the 2003 payrolls (in millions of dollars) and winning percentages (through August 3) of major league baseball teams. We have reproduced the data on the WeissStats CD.

- For the estimations and predictions, use a payroll of $50 million.
- For the correlation test, decide whether payroll and winning percentage are positively linearly correlated.

# Focusing on Data Analysis UWEC Undergraduates

Recall from Chapter 1 (see page 34) that the Focus database and Focus sample contain information on the undergraduate students at the University of Wisconsin - Eau Claire (UWEC). Now would be a good time for you to review the discussion about these data sets.

Open the Focus sample worksheet (FocusSample) in the technology of your choice and do the following.

a. Perform a residual analysis to decide whether considering the assumptions for regression inferences met by the variables high school percentile and cumulative GPA appears reasonable.
b. With high school percentile as the predictor variable and cumulative GPA as the response variable, determine and interpret the standard error of the estimate.
c. At the 5% significance level, do the data provide sufficient evidence to conclude that high school percentile is useful for predicting cumulative GPA of UWEC undergraduates?
d. Determine a point estimate for the mean cumulative GPA of all UWEC undergraduates who had high school percentiles of 74.
e. Find a 95% confidence interval for the mean cumulative GPA of all UWEC undergraduates who had high school percentiles of 74.
f. Determine the predicted cumulative GPA of a UWEC undergraduate who had a high school percentile of 74.
g. Find a 95% prediction interval for the cumulative GPA of a UWEC undergraduate who had a high school percentile of 74.
h. At the 5% significance level, do the data provide sufficient evidence to conclude that high school percentile and cumulative GPA are positively linearly correlated?

## Case Study Discussion Shoe Size and Height

At the beginning of this chapter, we repeated data from Chapter 4 on shoe size and height for a sample of students at Arizona State University. In Chapter 4, you used those data to perform some descriptive regression and correlation analyses. Now you are to employ those same data to carry out several inferential procedures in regression and correlation. We recommend that you use statistical software or a graphing calculator to solve the following problems, but they can also be done by hand:

a. Separate the data in the table on page 615 into two tables, one for males and the other for females. Parts (b)–(j) are for the male data.
b. Determine the sample regression equation with shoe size as the predictor variable for height.
c. Perform a residual analysis to decide whether considering Assumptions 1–3 for regression inferences to be satisfied by the variables shoe size and height appears reasonable.
d. Find and interpret the standard error of the estimate.
e. Determine the $P$-value for a test of whether shoe size is useful for predicting height. Then refer to Table 9.10 on page 414 to assess the evidence in favor of utility.

f. Find a point estimate for the mean height of all males who wear a size $10\frac{1}{2}$ shoe.
g. Obtain a 95% confidence interval for the mean height of all males who wear a size $10\frac{1}{2}$ shoe. Interpret your answer.
h. Determine the predicted height of a male who wears a size $10\frac{1}{2}$ shoe.
i. Find a 95% prediction interval for the height of a male who wears a size $10\frac{1}{2}$ shoe. Interpret your answer.
j. At the 5% significance level, do the data provide sufficient evidence to conclude that shoe size and height are positively linearly correlated?
k. Repeat parts (b)–(j) for the unabridged data on shoe size and height for females. Do the estimation and prediction problems for a size 8 shoe.
l. Repeat part (k) for the data on shoe size and height for females with the outlier removed. Compare your results with those obtained in part (k).

## Biography  SIR FRANCIS GALTON: Discoverer of Regression and Correlation

**Francis Galton** was born on February 16, 1822, into a wealthy Quaker family of bankers and gunsmiths on his father's side and as a cousin of Charles Darwin's on his mother's side. Although his IQ was estimated to be about 200, his formal education was unfinished.

He began training in medicine in Birmingham and London, but quit when, in his words, "A passion for travel seized me as if I had been a migratory bird." After a tour through Germany and southeastern Europe, he went to Trinity College in Cambridge to study mathematics. He left Cambridge in his third year, broken from overwork. He recovered quickly and resumed his medical studies in London. However, his father died before he had finished medical school and left to him, at 22, "a sufficient fortune to make me independent of the medical profession."

Galton held no professional or academic positions; nearly all his experiments were conducted at his home or performed by friends. He was curious about almost ev-

erything, and carried out research in fields that included meteorology, biology, psychology, statistics, and genetics.

The origination of the concepts of regression and correlation, developed by Galton as tools for measuring the influence of heredity, are summed up in his work *Natural Inheritance*. He discovered regression during experiments with sweet-pea seeds to determine the law of inheritance of size. He made his other great discovery, correlation, while applying his techniques to the problem of measuring the degree of association between the sizes of two different body organs of an individual.

In his later years, Galton was associated with Karl Pearson, who became his champion and an extender of his ideas. Pearson was the first holder of the chair of eugenics at University College in London, which Galton had endowed in his will. Galton was knighted in 1909. He died in Haslemere, Surrey, England, in 1911.

## StatCrunch in MyStatLab
### Analyzing Data Online

StatCrunch online statistical software offers an easy-to-use interface customized for this book. The StatCrunch feature for each chapter illustrates the use of the software to perform a statistical analysis discussed in the chapter. Exercises are provided to further apply StatCrunch to other statistical analyses examined in the chapter. Go to the WeissStats CD or to the Weiss Web site at www.aw-bc.com/weiss to access StatCrunch instructions and data sets. To access StatCrunch statistical software, go to the student content area of your Weiss MyStatLab course.

## Internet Projects
### Exploring Data Online

The Internet project for each chapter provides simulations, demonstrations, or activities that enhance the topics covered in the chapter. The project materials come from universities, individuals, governments, and companies from all over the world. To access the Internet projects on the Web, go to www.aw-bc.com/weiss. From this Web page, you can reach the Internet Projects Page, which we suggest that you bookmark for easy access in the future.

# Appendixes

# A

# Statistical Tables

**TABLE I**
Random numbers

| Line number | 00–09 | | 10–19 | | 20–29 | | 30–39 | | 40–49 | |
|---|---|---|---|---|---|---|---|---|---|---|
| 00 | 15544 | 80712 | 97742 | 21500 | 97081 | 42451 | 50623 | 56071 | 28882 | 28739 |
| 01 | 01011 | 21285 | 04729 | 39986 | 73150 | 31548 | 30168 | 76189 | 56996 | 19210 |
| 02 | 47435 | 53308 | 40718 | 29050 | 74858 | 64517 | 93573 | 51058 | 68501 | 42723 |
| 03 | 91312 | 75137 | 86274 | 59834 | 69844 | 19853 | 06917 | 17413 | 44474 | 86530 |
| 04 | 12775 | 08768 | 80791 | 16298 | 22934 | 09630 | 98862 | 39746 | 64623 | 32768 |
| 05 | 31466 | 43761 | 94872 | 92230 | 52367 | 13205 | 38634 | 55882 | 77518 | 36252 |
| 06 | 09300 | 43847 | 40881 | 51243 | 97810 | 18903 | 53914 | 31688 | 06220 | 40422 |
| 07 | 73582 | 13810 | 57784 | 72454 | 68997 | 72229 | 30340 | 08844 | 53924 | 89630 |
| 08 | 11092 | 81392 | 58189 | 22697 | 41063 | 09451 | 09789 | 00637 | 06450 | 85990 |
| 09 | 93322 | 98567 | 00116 | 35605 | 66790 | 52965 | 62877 | 21740 | 56476 | 49296 |
| 10 | 80134 | 12484 | 67089 | 08674 | 70753 | 90959 | 45842 | 59844 | 45214 | 36505 |
| 11 | 97888 | 31797 | 95037 | 84400 | 76041 | 96668 | 75920 | 68482 | 56855 | 97417 |
| 12 | 92612 | 27082 | 59459 | 69380 | 98654 | 20407 | 88151 | 56263 | 27126 | 63797 |
| 13 | 72744 | 45586 | 43279 | 44218 | 83638 | 05422 | 00995 | 70217 | 78925 | 39097 |
| 14 | 96256 | 70653 | 45285 | 26293 | 78305 | 80252 | 03625 | 40159 | 68760 | 84716 |
| 15 | 07851 | 47452 | 66742 | 83331 | 54701 | 06573 | 98169 | 37499 | 67756 | 68301 |
| 16 | 25594 | 41552 | 96475 | 56151 | 02089 | 33748 | 65289 | 89956 | 89559 | 33687 |
| 17 | 65358 | 15155 | 59374 | 80940 | 03411 | 94656 | 69440 | 47156 | 77115 | 99463 |
| 18 | 09402 | 31008 | 53424 | 21928 | 02198 | 61201 | 02457 | 87214 | 59750 | 51330 |
| 19 | 97424 | 90765 | 01634 | 37328 | 41243 | 33564 | 17884 | 94747 | 93650 | 77668 |

**TABLE II**
Areas under the standard normal curve

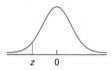

| 0.09 | 0.08 | 0.07 | 0.06 | 0.05 | 0.04 | 0.03 | 0.02 | 0.01 | 0.00 | z |
|---|---|---|---|---|---|---|---|---|---|---|
| | | | | | | | | | 0.0000[†] | −3.9 |
| 0.0001 | 0.0001 | 0.0001 | 0.0001 | 0.0001 | 0.0001 | 0.0001 | 0.0001 | 0.0001 | 0.0001 | −3.8 |
| 0.0001 | 0.0001 | 0.0001 | 0.0001 | 0.0001 | 0.0001 | 0.0001 | 0.0001 | 0.0001 | 0.0001 | −3.7 |
| 0.0001 | 0.0001 | 0.0001 | 0.0001 | 0.0001 | 0.0001 | 0.0001 | 0.0001 | 0.0002 | 0.0002 | −3.6 |
| 0.0002 | 0.0002 | 0.0002 | 0.0002 | 0.0002 | 0.0002 | 0.0002 | 0.0002 | 0.0002 | 0.0002 | −3.5 |
| 0.0002 | 0.0003 | 0.0003 | 0.0003 | 0.0003 | 0.0003 | 0.0003 | 0.0003 | 0.0003 | 0.0003 | −3.4 |
| 0.0003 | 0.0004 | 0.0004 | 0.0004 | 0.0004 | 0.0004 | 0.0004 | 0.0005 | 0.0005 | 0.0005 | −3.3 |
| 0.0005 | 0.0005 | 0.0005 | 0.0006 | 0.0006 | 0.0006 | 0.0006 | 0.0006 | 0.0007 | 0.0007 | −3.2 |
| 0.0007 | 0.0007 | 0.0008 | 0.0008 | 0.0008 | 0.0008 | 0.0009 | 0.0009 | 0.0009 | 0.0010 | −3.1 |
| 0.0010 | 0.0010 | 0.0011 | 0.0011 | 0.0011 | 0.0012 | 0.0012 | 0.0013 | 0.0013 | 0.0013 | −3.0 |
| 0.0014 | 0.0014 | 0.0015 | 0.0015 | 0.0016 | 0.0016 | 0.0017 | 0.0018 | 0.0018 | 0.0019 | −2.9 |
| 0.0019 | 0.0020 | 0.0021 | 0.0021 | 0.0022 | 0.0023 | 0.0023 | 0.0024 | 0.0025 | 0.0026 | −2.8 |
| 0.0026 | 0.0027 | 0.0028 | 0.0029 | 0.0030 | 0.0031 | 0.0032 | 0.0033 | 0.0034 | 0.0035 | −2.7 |
| 0.0036 | 0.0037 | 0.0038 | 0.0039 | 0.0040 | 0.0041 | 0.0043 | 0.0044 | 0.0045 | 0.0047 | −2.6 |
| 0.0048 | 0.0049 | 0.0051 | 0.0052 | 0.0054 | 0.0055 | 0.0057 | 0.0059 | 0.0060 | 0.0062 | −2.5 |
| 0.0064 | 0.0066 | 0.0068 | 0.0069 | 0.0071 | 0.0073 | 0.0075 | 0.0078 | 0.0080 | 0.0082 | −2.4 |
| 0.0084 | 0.0087 | 0.0089 | 0.0091 | 0.0094 | 0.0096 | 0.0099 | 0.0102 | 0.0104 | 0.0107 | −2.3 |
| 0.0110 | 0.0113 | 0.0116 | 0.0119 | 0.0122 | 0.0125 | 0.0129 | 0.0132 | 0.0136 | 0.0139 | −2.2 |
| 0.0143 | 0.0146 | 0.0150 | 0.0154 | 0.0158 | 0.0162 | 0.0166 | 0.0170 | 0.0174 | 0.0179 | −2.1 |
| 0.0183 | 0.0188 | 0.0192 | 0.0197 | 0.0202 | 0.0207 | 0.0212 | 0.0217 | 0.0222 | 0.0228 | −2.0 |
| 0.0233 | 0.0239 | 0.0244 | 0.0250 | 0.0256 | 0.0262 | 0.0268 | 0.0274 | 0.0281 | 0.0287 | −1.9 |
| 0.0294 | 0.0301 | 0.0307 | 0.0314 | 0.0322 | 0.0329 | 0.0336 | 0.0344 | 0.0351 | 0.0359 | −1.8 |
| 0.0367 | 0.0375 | 0.0384 | 0.0392 | 0.0401 | 0.0409 | 0.0418 | 0.0427 | 0.0436 | 0.0446 | −1.7 |
| 0.0455 | 0.0465 | 0.0475 | 0.0485 | 0.0495 | 0.0505 | 0.0516 | 0.0526 | 0.0537 | 0.0548 | −1.6 |
| 0.0559 | 0.0571 | 0.0582 | 0.0594 | 0.0606 | 0.0618 | 0.0630 | 0.0643 | 0.0655 | 0.0668 | −1.5 |
| 0.0681 | 0.0694 | 0.0708 | 0.0721 | 0.0735 | 0.0749 | 0.0764 | 0.0778 | 0.0793 | 0.0808 | −1.4 |
| 0.0823 | 0.0838 | 0.0853 | 0.0869 | 0.0885 | 0.0901 | 0.0918 | 0.0934 | 0.0951 | 0.0968 | −1.3 |
| 0.0985 | 0.1003 | 0.1020 | 0.1038 | 0.1056 | 0.1075 | 0.1093 | 0.1112 | 0.1131 | 0.1151 | −1.2 |
| 0.1170 | 0.1190 | 0.1210 | 0.1230 | 0.1251 | 0.1271 | 0.1292 | 0.1314 | 0.1335 | 0.1357 | −1.1 |
| 0.1379 | 0.1401 | 0.1423 | 0.1446 | 0.1469 | 0.1492 | 0.1515 | 0.1539 | 0.1562 | 0.1587 | −1.0 |
| 0.1611 | 0.1635 | 0.1660 | 0.1685 | 0.1711 | 0.1736 | 0.1762 | 0.1788 | 0.1814 | 0.1841 | −0.9 |
| 0.1867 | 0.1894 | 0.1922 | 0.1949 | 0.1977 | 0.2005 | 0.2033 | 0.2061 | 0.2090 | 0.2119 | −0.8 |
| 0.2148 | 0.2177 | 0.2206 | 0.2236 | 0.2266 | 0.2296 | 0.2327 | 0.2358 | 0.2389 | 0.2420 | −0.7 |
| 0.2451 | 0.2483 | 0.2514 | 0.2546 | 0.2578 | 0.2611 | 0.2643 | 0.2676 | 0.2709 | 0.2743 | −0.6 |
| 0.2776 | 0.2810 | 0.2843 | 0.2877 | 0.2912 | 0.2946 | 0.2981 | 0.3015 | 0.3050 | 0.3085 | −0.5 |
| 0.3121 | 0.3156 | 0.3192 | 0.3228 | 0.3264 | 0.3300 | 0.3336 | 0.3372 | 0.3409 | 0.3446 | −0.4 |
| 0.3483 | 0.3520 | 0.3557 | 0.3594 | 0.3632 | 0.3669 | 0.3707 | 0.3745 | 0.3783 | 0.3821 | −0.3 |
| 0.3859 | 0.3897 | 0.3936 | 0.3974 | 0.4013 | 0.4052 | 0.4090 | 0.4129 | 0.4168 | 0.4207 | −0.2 |
| 0.4247 | 0.4286 | 0.4325 | 0.4364 | 0.4404 | 0.4443 | 0.4483 | 0.4522 | 0.4562 | 0.4602 | −0.1 |
| 0.4641 | 0.4681 | 0.4721 | 0.4761 | 0.4801 | 0.4840 | 0.4880 | 0.4920 | 0.4960 | 0.5000 | −0.0 |

[†] For $z \leq -3.90$, the areas are 0.0000 to four decimal places.

**TABLE II** **(cont.)**
Areas under the
standard normal curve

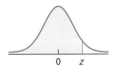

| z | 0.00 | 0.01 | 0.02 | 0.03 | 0.04 | 0.05 | 0.06 | 0.07 | 0.08 | 0.09 |
|---|---|---|---|---|---|---|---|---|---|---|
| | | | | | Second decimal place in z | | | | | |
| 0.0 | 0.5000 | 0.5040 | 0.5080 | 0.5120 | 0.5160 | 0.5199 | 0.5239 | 0.5279 | 0.5319 | 0.5359 |
| 0.1 | 0.5398 | 0.5438 | 0.5478 | 0.5517 | 0.5557 | 0.5596 | 0.5636 | 0.5675 | 0.5714 | 0.5753 |
| 0.2 | 0.5793 | 0.5832 | 0.5871 | 0.5910 | 0.5948 | 0.5987 | 0.6026 | 0.6064 | 0.6103 | 0.6141 |
| 0.3 | 0.6179 | 0.6217 | 0.6255 | 0.6293 | 0.6331 | 0.6368 | 0.6406 | 0.6443 | 0.6480 | 0.6517 |
| 0.4 | 0.6554 | 0.6591 | 0.6628 | 0.6664 | 0.6700 | 0.6736 | 0.6772 | 0.6808 | 0.6844 | 0.6879 |
| 0.5 | 0.6915 | 0.6950 | 0.6985 | 0.7019 | 0.7054 | 0.7088 | 0.7123 | 0.7157 | 0.7190 | 0.7224 |
| 0.6 | 0.7257 | 0.7291 | 0.7324 | 0.7357 | 0.7389 | 0.7422 | 0.7454 | 0.7486 | 0.7517 | 0.7549 |
| 0.7 | 0.7580 | 0.7611 | 0.7642 | 0.7673 | 0.7704 | 0.7734 | 0.7764 | 0.7794 | 0.7823 | 0.7852 |
| 0.8 | 0.7881 | 0.7910 | 0.7939 | 0.7967 | 0.7995 | 0.8023 | 0.8051 | 0.8078 | 0.8106 | 0.8133 |
| 0.9 | 0.8159 | 0.8186 | 0.8212 | 0.8238 | 0.8264 | 0.8289 | 0.8315 | 0.8340 | 0.8365 | 0.8389 |
| 1.0 | 0.8413 | 0.8438 | 0.8461 | 0.8485 | 0.8508 | 0.8531 | 0.8554 | 0.8577 | 0.8599 | 0.8621 |
| 1.1 | 0.8643 | 0.8665 | 0.8686 | 0.8708 | 0.8729 | 0.8749 | 0.8770 | 0.8790 | 0.8810 | 0.8830 |
| 1.2 | 0.8849 | 0.8869 | 0.8888 | 0.8907 | 0.8925 | 0.8944 | 0.8962 | 0.8980 | 0.8997 | 0.9015 |
| 1.3 | 0.9032 | 0.9049 | 0.9066 | 0.9082 | 0.9099 | 0.9115 | 0.9131 | 0.9147 | 0.9162 | 0.9177 |
| 1.4 | 0.9192 | 0.9207 | 0.9222 | 0.9236 | 0.9251 | 0.9265 | 0.9279 | 0.9292 | 0.9306 | 0.9319 |
| 1.5 | 0.9332 | 0.9345 | 0.9357 | 0.9370 | 0.9382 | 0.9394 | 0.9406 | 0.9418 | 0.9429 | 0.9441 |
| 1.6 | 0.9452 | 0.9463 | 0.9474 | 0.9484 | 0.9495 | 0.9505 | 0.9515 | 0.9525 | 0.9535 | 0.9545 |
| 1.7 | 0.9554 | 0.9564 | 0.9573 | 0.9582 | 0.9591 | 0.9599 | 0.9608 | 0.9616 | 0.9625 | 0.9633 |
| 1.8 | 0.9641 | 0.9649 | 0.9656 | 0.9664 | 0.9671 | 0.9678 | 0.9686 | 0.9693 | 0.9699 | 0.9706 |
| 1.9 | 0.9713 | 0.9719 | 0.9726 | 0.9732 | 0.9738 | 0.9744 | 0.9750 | 0.9756 | 0.9761 | 0.9767 |
| 2.0 | 0.9772 | 0.9778 | 0.9783 | 0.9788 | 0.9793 | 0.9798 | 0.9803 | 0.9808 | 0.9812 | 0.9817 |
| 2.1 | 0.9821 | 0.9826 | 0.9830 | 0.9834 | 0.9838 | 0.9842 | 0.9846 | 0.9850 | 0.9854 | 0.9857 |
| 2.2 | 0.9861 | 0.9864 | 0.9868 | 0.9871 | 0.9875 | 0.9878 | 0.9881 | 0.9884 | 0.9887 | 0.9890 |
| 2.3 | 0.9893 | 0.9896 | 0.9898 | 0.9901 | 0.9904 | 0.9906 | 0.9909 | 0.9911 | 0.9913 | 0.9916 |
| 2.4 | 0.9918 | 0.9920 | 0.9922 | 0.9925 | 0.9927 | 0.9929 | 0.9931 | 0.9932 | 0.9934 | 0.9936 |
| 2.5 | 0.9938 | 0.9940 | 0.9941 | 0.9943 | 0.9945 | 0.9946 | 0.9948 | 0.9949 | 0.9951 | 0.9952 |
| 2.6 | 0.9953 | 0.9955 | 0.9956 | 0.9957 | 0.9959 | 0.9960 | 0.9961 | 0.9962 | 0.9963 | 0.9964 |
| 2.7 | 0.9965 | 0.9966 | 0.9967 | 0.9968 | 0.9969 | 0.9970 | 0.9971 | 0.9972 | 0.9973 | 0.9974 |
| 2.8 | 0.9974 | 0.9975 | 0.9976 | 0.9977 | 0.9977 | 0.9978 | 0.9979 | 0.9979 | 0.9980 | 0.9981 |
| 2.9 | 0.9981 | 0.9982 | 0.9982 | 0.9983 | 0.9984 | 0.9984 | 0.9985 | 0.9985 | 0.9986 | 0.9986 |
| 3.0 | 0.9987 | 0.9987 | 0.9987 | 0.9988 | 0.9988 | 0.9989 | 0.9989 | 0.9989 | 0.9990 | 0.9990 |
| 3.1 | 0.9990 | 0.9991 | 0.9991 | 0.9991 | 0.9992 | 0.9992 | 0.9992 | 0.9992 | 0.9993 | 0.9993 |
| 3.2 | 0.9993 | 0.9993 | 0.9994 | 0.9994 | 0.9994 | 0.9994 | 0.9994 | 0.9995 | 0.9995 | 0.9995 |
| 3.3 | 0.9995 | 0.9995 | 0.9995 | 0.9996 | 0.9996 | 0.9996 | 0.9996 | 0.9996 | 0.9996 | 0.9997 |
| 3.4 | 0.9997 | 0.9997 | 0.9997 | 0.9997 | 0.9997 | 0.9997 | 0.9997 | 0.9997 | 0.9997 | 0.9998 |
| 3.5 | 0.9998 | 0.9998 | 0.9998 | 0.9998 | 0.9998 | 0.9998 | 0.9998 | 0.9998 | 0.9998 | 0.9998 |
| 3.6 | 0.9998 | 0.9998 | 0.9999 | 0.9999 | 0.9999 | 0.9999 | 0.9999 | 0.9999 | 0.9999 | 0.9999 |
| 3.7 | 0.9999 | 0.9999 | 0.9999 | 0.9999 | 0.9999 | 0.9999 | 0.9999 | 0.9999 | 0.9999 | 0.9999 |
| 3.8 | 0.9999 | 0.9999 | 0.9999 | 0.9999 | 0.9999 | 0.9999 | 0.9999 | 0.9999 | 0.9999 | 0.9999 |
| 3.9 | 1.0000[†] | | | | | | | | | |

[†] For $z \geq 3.90$, the areas are 1.0000 to four decimal places.

**TABLE III**

Normal scores

| Ordered position | n | | | | | | | | |
|---|---|---|---|---|---|---|---|---|---|
| | 5 | 6 | 7 | 8 | 9 | 10 | 11 | 12 | 13 |
| 1 | −1.18 | −1.28 | −1.36 | −1.43 | −1.50 | −1.55 | −1.59 | −1.64 | −1.68 |
| 2 | −0.50 | −0.64 | −0.76 | −0.85 | −0.93 | −1.00 | −1.06 | −1.11 | −1.16 |
| 3 | 0.00 | −0.20 | −0.35 | −0.47 | −0.57 | −0.65 | −0.73 | −0.79 | −0.85 |
| 4 | 0.50 | 0.20 | 0.00 | −0.15 | −0.27 | −0.37 | −0.46 | −0.53 | −0.60 |
| 5 | 1.18 | 0.64 | 0.35 | 0.15 | 0.00 | −0.12 | −0.22 | −0.31 | −0.39 |
| 6 | | 1.28 | 0.76 | 0.47 | 0.27 | 0.12 | 0.00 | −0.10 | −0.19 |
| 7 | | | 1.36 | 0.85 | 0.57 | 0.37 | 0.22 | 0.10 | 0.00 |
| 8 | | | | 1.43 | 0.93 | 0.65 | 0.46 | 0.31 | 0.19 |
| 9 | | | | | 1.50 | 1.00 | 0.73 | 0.53 | 0.39 |
| 10 | | | | | | 1.55 | 1.06 | 0.79 | 0.60 |
| 11 | | | | | | | 1.59 | 1.11 | 0.85 |
| 12 | | | | | | | | 1.64 | 1.16 |
| 13 | | | | | | | | | 1.68 |

**TABLE III   (cont.)**

Normal scores

| Ordered position | n | | | | | | | | |
|---|---|---|---|---|---|---|---|---|---|
| | 14 | 15 | 16 | 17 | 18 | 19 | 20 | 21 | 22 |
| 1 | −1.71 | −1.74 | −1.77 | −1.80 | −1.82 | −1.85 | −1.87 | −1.89 | −1.91 |
| 2 | −1.20 | −1.24 | −1.28 | −1.32 | −1.35 | −1.38 | −1.40 | −1.43 | −1.45 |
| 3 | −0.90 | −0.94 | −0.99 | −1.03 | −1.06 | −1.10 | −1.13 | −1.16 | −1.18 |
| 4 | −0.66 | −0.71 | −0.76 | −0.80 | −0.84 | −0.88 | −0.92 | −0.95 | −0.98 |
| 5 | −0.45 | −0.51 | −0.57 | −0.62 | −0.66 | −0.70 | −0.74 | −0.78 | −0.81 |
| 6 | −0.27 | −0.33 | −0.39 | −0.45 | −0.50 | −0.54 | −0.59 | −0.63 | −0.66 |
| 7 | −0.09 | −0.16 | −0.23 | −0.29 | −0.35 | −0.40 | −0.45 | −0.49 | −0.53 |
| 8 | 0.09 | 0.00 | −0.08 | −0.15 | −0.21 | −0.26 | −0.31 | −0.36 | −0.40 |
| 9 | 0.27 | 0.16 | 0.08 | 0.00 | −0.07 | −0.13 | −0.19 | −0.24 | −0.28 |
| 10 | 0.45 | 0.33 | 0.23 | 0.15 | 0.07 | 0.00 | −0.06 | −0.12 | −0.17 |
| 11 | 0.66 | 0.51 | 0.39 | 0.29 | 0.21 | 0.13 | 0.06 | 0.00 | −0.06 |
| 12 | 0.90 | 0.71 | 0.57 | 0.45 | 0.35 | 0.26 | 0.19 | 0.12 | 0.06 |
| 13 | 1.20 | 0.94 | 0.76 | 0.62 | 0.50 | 0.40 | 0.31 | 0.24 | 0.17 |
| 14 | 1.71 | 1.24 | 0.99 | 0.80 | 0.66 | 0.54 | 0.45 | 0.36 | 0.28 |
| 15 | | 1.74 | 1.28 | 1.03 | 0.84 | 0.70 | 0.59 | 0.49 | 0.40 |
| 16 | | | 1.77 | 1.32 | 1.06 | 0.88 | 0.74 | 0.63 | 0.53 |
| 17 | | | | 1.80 | 1.35 | 1.10 | 0.92 | 0.78 | 0.66 |
| 18 | | | | | 1.82 | 1.38 | 1.13 | 0.95 | 0.81 |
| 19 | | | | | | 1.85 | 1.40 | 1.16 | 0.98 |
| 20 | | | | | | | 1.87 | 1.43 | 1.18 |
| 21 | | | | | | | | 1.89 | 1.45 |
| 22 | | | | | | | | | 1.91 |

**TABLE III**  (cont.)

Normal scores

| Ordered position | | n | | | | | | |
|---|---|---|---|---|---|---|---|---|
| | *23* | *24* | *25* | *26* | *27* | *28* | *29* | *30* |
| *1* | −1.93 | −1.95 | −1.97 | −1.98 | −2.00 | −2.01 | −2.03 | −2.04 |
| *2* | −1.48 | −1.50 | −1.52 | −1.54 | −1.56 | −1.58 | −1.59 | −1.61 |
| *3* | −1.21 | −1.24 | −1.26 | −1.28 | −1.30 | −1.32 | −1.34 | −1.36 |
| *4* | −1.01 | −1.04 | −1.06 | −1.09 | −1.11 | −1.13 | −1.15 | −1.17 |
| *5* | −0.84 | −0.87 | −0.90 | −0.93 | −0.95 | −0.98 | −1.00 | −1.02 |
| *6* | −0.70 | −0.73 | −0.76 | −0.79 | −0.82 | −0.84 | −0.87 | −0.89 |
| *7* | −0.57 | −0.60 | −0.63 | −0.66 | −0.69 | −0.72 | −0.75 | −0.77 |
| *8* | −0.44 | −0.48 | −0.52 | −0.55 | −0.58 | −0.61 | −0.64 | −0.67 |
| *9* | −0.33 | −0.37 | −0.41 | −0.44 | −0.48 | −0.51 | −0.54 | −0.57 |
| *10* | −0.22 | −0.26 | −0.30 | −0.34 | −0.38 | −0.41 | −0.44 | −0.47 |
| *11* | −0.11 | −0.15 | −0.20 | −0.24 | −0.28 | −0.31 | −0.35 | −0.38 |
| *12* | 0.00 | −0.05 | −0.10 | −0.14 | −0.18 | −0.22 | −0.26 | −0.29 |
| *13* | 0.11 | 0.05 | 0.00 | −0.05 | −0.09 | −0.13 | −0.17 | −0.21 |
| *14* | 0.22 | 0.15 | 0.10 | 0.05 | 0.00 | −0.04 | −0.09 | −0.12 |
| *15* | 0.33 | 0.26 | 0.20 | 0.14 | 0.09 | 0.04 | 0.00 | −0.04 |
| *16* | 0.44 | 0.37 | 0.30 | 0.24 | 0.18 | 0.13 | 0.09 | 0.04 |
| *17* | 0.57 | 0.48 | 0.41 | 0.34 | 0.28 | 0.22 | 0.17 | 0.12 |
| *18* | 0.70 | 0.60 | 0.52 | 0.44 | 0.38 | 0.31 | 0.26 | 0.21 |
| *19* | 0.84 | 0.73 | 0.63 | 0.55 | 0.48 | 0.41 | 0.35 | 0.29 |
| *20* | 1.01 | 0.87 | 0.76 | 0.66 | 0.58 | 0.51 | 0.44 | 0.38 |
| *21* | 1.21 | 1.04 | 0.90 | 0.79 | 0.69 | 0.61 | 0.54 | 0.47 |
| *22* | 1.48 | 1.24 | 1.06 | 0.93 | 0.82 | 0.72 | 0.64 | 0.57 |
| *23* | 1.93 | 1.50 | 1.26 | 1.09 | 0.95 | 0.84 | 0.75 | 0.67 |
| *24* | | 1.95 | 1.52 | 1.28 | 1.11 | 0.98 | 0.87 | 0.77 |
| *25* | | | 1.97 | 1.54 | 1.30 | 1.13 | 1.00 | 0.89 |
| *26* | | | | 1.98 | 1.56 | 1.32 | 1.15 | 1.02 |
| *27* | | | | | 2.00 | 1.58 | 1.34 | 1.17 |
| *28* | | | | | | 2.01 | 1.59 | 1.36 |
| *29* | | | | | | | 2.03 | 1.61 |
| *30* | | | | | | | | 2.04 |

**TABLE IV**
Values of $t_\alpha$

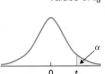

| df | $t_{0.10}$ | $t_{0.05}$ | $t_{0.025}$ | $t_{0.01}$ | $t_{0.005}$ | df |
|----|------|------|------|------|------|----|
| 1 | 3.078 | 6.314 | 12.706 | 31.821 | 63.657 | 1 |
| 2 | 1.886 | 2.920 | 4.303 | 6.965 | 9.925 | 2 |
| 3 | 1.638 | 2.353 | 3.182 | 4.541 | 5.841 | 3 |
| 4 | 1.533 | 2.132 | 2.776 | 3.747 | 4.604 | 4 |
| 5 | 1.476 | 2.015 | 2.571 | 3.365 | 4.032 | 5 |
| 6 | 1.440 | 1.943 | 2.447 | 3.143 | 3.707 | 6 |
| 7 | 1.415 | 1.895 | 2.365 | 2.998 | 3.499 | 7 |
| 8 | 1.397 | 1.860 | 2.306 | 2.896 | 3.355 | 8 |
| 9 | 1.383 | 1.833 | 2.262 | 2.821 | 3.250 | 9 |
| 10 | 1.372 | 1.812 | 2.228 | 2.764 | 3.169 | 10 |
| 11 | 1.363 | 1.796 | 2.201 | 2.718 | 3.106 | 11 |
| 12 | 1.356 | 1.782 | 2.179 | 2.681 | 3.055 | 12 |
| 13 | 1.350 | 1.771 | 2.160 | 2.650 | 3.012 | 13 |
| 14 | 1.345 | 1.761 | 2.145 | 2.624 | 2.977 | 14 |
| 15 | 1.341 | 1.753 | 2.131 | 2.602 | 2.947 | 15 |
| 16 | 1.337 | 1.746 | 2.120 | 2.583 | 2.921 | 16 |
| 17 | 1.333 | 1.740 | 2.110 | 2.567 | 2.898 | 17 |
| 18 | 1.330 | 1.734 | 2.101 | 2.552 | 2.878 | 18 |
| 19 | 1.328 | 1.729 | 2.093 | 2.539 | 2.861 | 19 |
| 20 | 1.325 | 1.725 | 2.086 | 2.528 | 2.845 | 20 |
| 21 | 1.323 | 1.721 | 2.080 | 2.518 | 2.831 | 21 |
| 22 | 1.321 | 1.717 | 2.074 | 2.508 | 2.819 | 22 |
| 23 | 1.319 | 1.714 | 2.069 | 2.500 | 2.807 | 23 |
| 24 | 1.318 | 1.711 | 2.064 | 2.492 | 2.797 | 24 |
| 25 | 1.316 | 1.708 | 2.060 | 2.485 | 2.787 | 25 |
| 26 | 1.315 | 1.706 | 2.056 | 2.479 | 2.779 | 26 |
| 27 | 1.314 | 1.703 | 2.052 | 2.473 | 2.771 | 27 |
| 28 | 1.313 | 1.701 | 2.048 | 2.467 | 2.763 | 28 |
| 29 | 1.311 | 1.699 | 2.045 | 2.462 | 2.756 | 29 |
| 30 | 1.310 | 1.697 | 2.042 | 2.457 | 2.750 | 30 |
| 31 | 1.309 | 1.696 | 2.040 | 2.453 | 2.744 | 31 |
| 32 | 1.309 | 1.694 | 2.037 | 2.449 | 2.738 | 32 |
| 33 | 1.308 | 1.692 | 2.035 | 2.445 | 2.733 | 33 |
| 34 | 1.307 | 1.691 | 2.032 | 2.441 | 2.728 | 34 |
| 35 | 1.306 | 1.690 | 2.030 | 2.438 | 2.724 | 35 |
| 36 | 1.306 | 1.688 | 2.028 | 2.434 | 2.719 | 36 |
| 37 | 1.305 | 1.687 | 2.026 | 2.431 | 2.715 | 37 |
| 38 | 1.304 | 1.686 | 2.024 | 2.429 | 2.712 | 38 |
| 39 | 1.304 | 1.685 | 2.023 | 2.426 | 2.708 | 39 |
| 40 | 1.303 | 1.684 | 2.021 | 2.423 | 2.704 | 40 |
| 41 | 1.303 | 1.683 | 2.020 | 2.421 | 2.701 | 41 |
| 42 | 1.302 | 1.682 | 2.018 | 2.418 | 2.698 | 42 |
| 43 | 1.302 | 1.681 | 2.017 | 2.416 | 2.695 | 43 |
| 44 | 1.301 | 1.680 | 2.015 | 2.414 | 2.692 | 44 |
| 45 | 1.301 | 1.679 | 2.014 | 2.412 | 2.690 | 45 |
| 46 | 1.300 | 1.679 | 2.013 | 2.410 | 2.687 | 46 |
| 47 | 1.300 | 1.678 | 2.012 | 2.408 | 2.685 | 47 |
| 48 | 1.299 | 1.677 | 2.011 | 2.407 | 2.682 | 48 |
| 49 | 1.299 | 1.677 | 2.010 | 2.405 | 2.680 | 49 |

**TABLE IV   (cont.)**
Values of $t_\alpha$

| df | $t_{0.10}$ | $t_{0.05}$ | $t_{0.025}$ | $t_{0.01}$ | $t_{0.005}$ | df |
|----|-----------|-----------|------------|-----------|------------|------|
| 50 | 1.299 | 1.676 | 2.009 | 2.403 | 2.678 | 50 |
| 51 | 1.298 | 1.675 | 2.008 | 2.402 | 2.676 | 51 |
| 52 | 1.298 | 1.675 | 2.007 | 2.400 | 2.674 | 52 |
| 53 | 1.298 | 1.674 | 2.006 | 2.399 | 2.672 | 53 |
| 54 | 1.297 | 1.674 | 2.005 | 2.397 | 2.670 | 54 |
| 55 | 1.297 | 1.673 | 2.004 | 2.396 | 2.668 | 55 |
| 56 | 1.297 | 1.673 | 2.003 | 2.395 | 2.667 | 56 |
| 57 | 1.297 | 1.672 | 2.002 | 2.394 | 2.665 | 57 |
| 58 | 1.296 | 1.672 | 2.002 | 2.392 | 2.663 | 58 |
| 59 | 1.296 | 1.671 | 2.001 | 2.391 | 2.662 | 59 |
| 60 | 1.296 | 1.671 | 2.000 | 2.390 | 2.660 | 60 |
| 61 | 1.296 | 1.670 | 2.000 | 2.389 | 2.659 | 61 |
| 62 | 1.295 | 1.670 | 1.999 | 2.388 | 2.657 | 62 |
| 63 | 1.295 | 1.669 | 1.998 | 2.387 | 2.656 | 63 |
| 64 | 1.295 | 1.669 | 1.998 | 2.386 | 2.655 | 64 |
| 65 | 1.295 | 1.669 | 1.997 | 2.385 | 2.654 | 65 |
| 66 | 1.295 | 1.668 | 1.997 | 2.384 | 2.652 | 66 |
| 67 | 1.294 | 1.668 | 1.996 | 2.383 | 2.651 | 67 |
| 68 | 1.294 | 1.668 | 1.995 | 2.382 | 2.650 | 68 |
| 69 | 1.294 | 1.667 | 1.995 | 2.382 | 2.649 | 69 |
| 70 | 1.294 | 1.667 | 1.994 | 2.381 | 2.648 | 70 |
| 71 | 1.294 | 1.667 | 1.994 | 2.380 | 2.647 | 71 |
| 72 | 1.293 | 1.666 | 1.993 | 2.379 | 2.646 | 72 |
| 73 | 1.293 | 1.666 | 1.993 | 2.379 | 2.645 | 73 |
| 74 | 1.293 | 1.666 | 1.993 | 2.378 | 2.644 | 74 |
| 75 | 1.293 | 1.665 | 1.992 | 2.377 | 2.643 | 75 |
| 80 | 1.292 | 1.664 | 1.990 | 2.374 | 2.639 | 80 |
| 85 | 1.292 | 1.663 | 1.988 | 2.371 | 2.635 | 85 |
| 90 | 1.291 | 1.662 | 1.987 | 2.368 | 2.632 | 90 |
| 95 | 1.291 | 1.661 | 1.985 | 2.366 | 2.629 | 95 |
| 100 | 1.290 | 1.660 | 1.984 | 2.364 | 2.626 | 100 |
| 200 | 1.286 | 1.653 | 1.972 | 2.345 | 2.601 | 200 |
| 300 | 1.284 | 1.650 | 1.968 | 2.339 | 2.592 | 300 |
| 400 | 1.284 | 1.649 | 1.966 | 2.336 | 2.588 | 400 |
| 500 | 1.283 | 1.648 | 1.965 | 2.334 | 2.586 | 500 |
| 600 | 1.283 | 1.647 | 1.964 | 2.333 | 2.584 | 600 |
| 700 | 1.283 | 1.647 | 1.963 | 2.332 | 2.583 | 700 |
| 800 | 1.283 | 1.647 | 1.963 | 2.331 | 2.582 | 800 |
| 900 | 1.282 | 1.647 | 1.963 | 2.330 | 2.581 | 900 |
| 1000 | 1.282 | 1.646 | 1.962 | 2.330 | 2.581 | 1000 |
| 2000 | 1.282 | 1.646 | 1.961 | 2.328 | 2.578 | 2000 |

| $z_{0.10}$ | $z_{0.05}$ | $z_{0.025}$ | $z_{0.01}$ | $z_{0.005}$ |
|-----------|-----------|------------|-----------|------------|
| 1.282 | 1.645 | 1.960 | 2.326 | 2.576 |

**TABLE V**
Values of $\chi_\alpha^2$

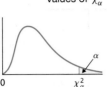

| df | $\chi_{0.995}^2$ | $\chi_{0.99}^2$ | $\chi_{0.975}^2$ | $\chi_{0.95}^2$ | $\chi_{0.90}^2$ |
|----|------------------|-----------------|------------------|-----------------|-----------------|
| 1 | 0.000 | 0.000 | 0.001 | 0.004 | 0.016 |
| 2 | 0.010 | 0.020 | 0.051 | 0.103 | 0.211 |
| 3 | 0.072 | 0.115 | 0.216 | 0.352 | 0.584 |
| 4 | 0.207 | 0.297 | 0.484 | 0.711 | 1.064 |
| 5 | 0.412 | 0.554 | 0.831 | 1.145 | 1.610 |
| 6 | 0.676 | 0.872 | 1.237 | 1.635 | 2.204 |
| 7 | 0.989 | 1.239 | 1.690 | 2.167 | 2.833 |
| 8 | 1.344 | 1.646 | 2.180 | 2.733 | 3.490 |
| 9 | 1.735 | 2.088 | 2.700 | 3.325 | 4.168 |
| 10 | 2.156 | 2.558 | 3.247 | 3.940 | 4.865 |
| 11 | 2.603 | 3.053 | 3.816 | 4.575 | 5.578 |
| 12 | 3.074 | 3.571 | 4.404 | 5.226 | 6.304 |
| 13 | 3.565 | 4.107 | 5.009 | 5.892 | 7.042 |
| 14 | 4.075 | 4.660 | 5.629 | 6.571 | 7.790 |
| 15 | 4.601 | 5.229 | 6.262 | 7.261 | 8.547 |
| 16 | 5.142 | 5.812 | 6.908 | 7.962 | 9.312 |
| 17 | 5.697 | 6.408 | 7.564 | 8.672 | 10.085 |
| 18 | 6.265 | 7.015 | 8.231 | 9.390 | 10.865 |
| 19 | 6.844 | 7.633 | 8.907 | 10.117 | 11.651 |
| 20 | 7.434 | 8.260 | 9.591 | 10.851 | 12.443 |
| 21 | 8.034 | 8.897 | 10.283 | 11.591 | 13.240 |
| 22 | 8.643 | 9.542 | 10.982 | 12.338 | 14.041 |
| 23 | 9.260 | 10.196 | 11.689 | 13.091 | 14.848 |
| 24 | 9.886 | 10.856 | 12.401 | 13.848 | 15.659 |
| 25 | 10.520 | 11.524 | 13.120 | 14.611 | 16.473 |
| 26 | 11.160 | 12.198 | 13.844 | 15.379 | 17.292 |
| 27 | 11.808 | 12.879 | 14.573 | 16.151 | 18.114 |
| 28 | 12.461 | 13.565 | 15.308 | 16.928 | 18.939 |
| 29 | 13.121 | 14.256 | 16.047 | 17.708 | 19.768 |
| 30 | 13.787 | 14.953 | 16.791 | 18.493 | 20.599 |
| 40 | 20.707 | 22.164 | 24.433 | 26.509 | 29.051 |
| 50 | 27.991 | 29.707 | 32.357 | 34.764 | 37.689 |
| 60 | 35.534 | 37.485 | 40.482 | 43.188 | 46.459 |
| 70 | 43.275 | 45.442 | 48.758 | 51.739 | 55.329 |
| 80 | 51.172 | 53.540 | 57.153 | 60.391 | 64.278 |
| 90 | 59.196 | 61.754 | 65.647 | 69.126 | 73.291 |
| 100 | 67.328 | 70.065 | 74.222 | 77.930 | 82.358 |

**TABLE V   (cont.)**
Values of $\chi_\alpha^2$

| $\chi_{0.10}^2$ | $\chi_{0.05}^2$ | $\chi_{0.025}^2$ | $\chi_{0.01}^2$ | $\chi_{0.005}^2$ | df |
|---|---|---|---|---|---|
| 2.706 | 3.841 | 5.024 | 6.635 | 7.879 | 1 |
| 4.605 | 5.991 | 7.378 | 9.210 | 10.597 | 2 |
| 6.251 | 7.815 | 9.348 | 11.345 | 12.838 | 3 |
| 7.779 | 9.488 | 11.143 | 13.277 | 14.860 | 4 |
| 9.236 | 11.070 | 12.833 | 15.086 | 16.750 | 5 |
| 10.645 | 12.592 | 14.449 | 16.812 | 18.548 | 6 |
| 12.017 | 14.067 | 16.013 | 18.475 | 20.278 | 7 |
| 13.362 | 15.507 | 17.535 | 20.090 | 21.955 | 8 |
| 14.684 | 16.919 | 19.023 | 21.666 | 23.589 | 9 |
| 15.987 | 18.307 | 20.483 | 23.209 | 25.188 | 10 |
| 17.275 | 19.675 | 21.920 | 24.725 | 26.757 | 11 |
| 18.549 | 21.026 | 23.337 | 26.217 | 28.300 | 12 |
| 19.812 | 22.362 | 24.736 | 27.688 | 29.819 | 13 |
| 21.064 | 23.685 | 26.119 | 29.141 | 31.319 | 14 |
| 22.307 | 24.996 | 27.488 | 30.578 | 32.801 | 15 |
| 23.542 | 26.296 | 28.845 | 32.000 | 34.267 | 16 |
| 24.769 | 27.587 | 30.191 | 33.409 | 35.718 | 17 |
| 25.989 | 28.869 | 31.526 | 34.805 | 37.156 | 18 |
| 27.204 | 30.143 | 32.852 | 36.191 | 38.582 | 19 |
| 28.412 | 31.410 | 34.170 | 37.566 | 39.997 | 20 |
| 29.615 | 32.671 | 35.479 | 38.932 | 41.401 | 21 |
| 30.813 | 33.924 | 36.781 | 40.290 | 42.796 | 22 |
| 32.007 | 35.172 | 38.076 | 41.638 | 44.181 | 23 |
| 33.196 | 36.415 | 39.364 | 42.980 | 45.559 | 24 |
| 34.382 | 37.653 | 40.647 | 44.314 | 46.928 | 25 |
| 35.563 | 38.885 | 41.923 | 45.642 | 48.290 | 26 |
| 36.741 | 40.113 | 43.195 | 46.963 | 49.645 | 27 |
| 37.916 | 41.337 | 44.461 | 48.278 | 50.994 | 28 |
| 39.087 | 42.557 | 45.722 | 49.588 | 52.336 | 29 |
| 40.256 | 43.773 | 46.979 | 50.892 | 53.672 | 30 |
| 51.805 | 55.759 | 59.342 | 63.691 | 66.767 | 40 |
| 63.167 | 67.505 | 71.420 | 76.154 | 79.490 | 50 |
| 74.397 | 79.082 | 83.298 | 88.381 | 91.955 | 60 |
| 85.527 | 90.531 | 95.023 | 100.424 | 104.213 | 70 |
| 96.578 | 101.879 | 106.628 | 112.328 | 116.320 | 80 |
| 107.565 | 113.145 | 118.135 | 124.115 | 128.296 | 90 |
| 118.499 | 124.343 | 129.563 | 135.811 | 140.177 | 100 |

**TABLE VI**
Values of $F_\alpha$

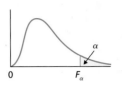

| dfd | $\alpha$ | 1 | 2 | 3 | 4 | 5 | 6 | 7 | 8 | 9 |
|-----|------|---|---|---|---|---|---|---|---|---|
| | | | | | | dfn | | | | |
| 1 | 0.10 | 39.86 | 49.50 | 53.59 | 55.83 | 57.24 | 58.20 | 58.91 | 59.44 | 59.86 |
| | 0.05 | 161.45 | 199.50 | 215.71 | 224.58 | 230.16 | 233.99 | 236.77 | 238.88 | 240.54 |
| | 0.025 | 647.79 | 799.50 | 864.16 | 899.58 | 921.85 | 937.11 | 948.22 | 956.66 | 963.28 |
| | 0.01 | 4052.2 | 4999.5 | 5403.4 | 5624.6 | 5763.6 | 5859.0 | 5928.4 | 5981.1 | 6022.5 |
| | 0.005 | 16211 | 20000 | 21615 | 22500 | 23056 | 23437 | 23715 | 23925 | 24091 |
| 2 | 0.10 | 8.53 | 9.00 | 9.16 | 9.24 | 9.29 | 9.33 | 9.35 | 9.37 | 9.38 |
| | 0.05 | 18.51 | 19.00 | 19.16 | 19.25 | 19.30 | 19.33 | 19.35 | 19.37 | 19.38 |
| | 0.025 | 38.51 | 39.00 | 39.17 | 39.25 | 39.30 | 39.33 | 39.36 | 39.37 | 39.39 |
| | 0.01 | 98.50 | 99.00 | 99.17 | 99.25 | 99.30 | 99.33 | 99.36 | 99.37 | 99.39 |
| | 0.005 | 198.50 | 199.00 | 199.17 | 199.25 | 199.30 | 199.33 | 199.36 | 199.37 | 199.39 |
| 3 | 0.10 | 5.54 | 5.46 | 5.39 | 5.34 | 5.31 | 5.28 | 5.27 | 5.25 | 5.24 |
| | 0.05 | 10.13 | 9.55 | 9.28 | 9.12 | 9.01 | 8.94 | 8.89 | 8.85 | 8.81 |
| | 0.025 | 17.44 | 16.04 | 15.44 | 15.10 | 14.88 | 14.73 | 14.62 | 14.54 | 14.47 |
| | 0.01 | 34.12 | 30.82 | 29.46 | 28.71 | 28.24 | 27.91 | 27.67 | 27.49 | 27.35 |
| | 0.005 | 55.55 | 49.80 | 47.47 | 46.19 | 45.39 | 44.84 | 44.43 | 44.13 | 43.88 |
| 4 | 0.10 | 4.54 | 4.32 | 4.19 | 4.11 | 4.05 | 4.01 | 3.98 | 3.95 | 3.94 |
| | 0.05 | 7.71 | 6.94 | 6.59 | 6.39 | 6.26 | 6.16 | 6.09 | 6.04 | 6.00 |
| | 0.025 | 12.22 | 10.65 | 9.98 | 9.60 | 9.36 | 9.20 | 9.07 | 8.98 | 8.90 |
| | 0.01 | 21.20 | 18.00 | 16.69 | 15.98 | 15.52 | 15.21 | 14.98 | 14.80 | 14.66 |
| | 0.005 | 31.33 | 26.28 | 24.26 | 23.15 | 22.46 | 21.97 | 21.62 | 21.35 | 21.14 |
| 5 | 0.10 | 4.06 | 3.78 | 3.62 | 3.52 | 3.45 | 3.40 | 3.37 | 3.34 | 3.32 |
| | 0.05 | 6.61 | 5.79 | 5.41 | 5.19 | 5.05 | 4.95 | 4.88 | 4.82 | 4.77 |
| | 0.025 | 10.01 | 8.43 | 7.76 | 7.39 | 7.15 | 6.98 | 6.85 | 6.76 | 6.68 |
| | 0.01 | 16.26 | 13.27 | 12.06 | 11.39 | 10.97 | 10.67 | 10.46 | 10.29 | 10.16 |
| | 0.005 | 22.78 | 18.31 | 16.53 | 15.56 | 14.94 | 14.51 | 14.20 | 13.96 | 13.77 |
| 6 | 0.10 | 3.78 | 3.46 | 3.29 | 3.18 | 3.11 | 3.05 | 3.01 | 2.98 | 2.96 |
| | 0.05 | 5.99 | 5.14 | 4.76 | 4.53 | 4.39 | 4.28 | 4.21 | 4.15 | 4.10 |
| | 0.025 | 8.81 | 7.26 | 6.60 | 6.23 | 5.99 | 5.82 | 5.70 | 5.60 | 5.52 |
| | 0.01 | 13.75 | 10.92 | 9.78 | 9.15 | 8.75 | 8.47 | 8.26 | 8.10 | 7.98 |
| | 0.005 | 18.63 | 14.54 | 12.92 | 12.03 | 11.46 | 11.07 | 10.79 | 10.57 | 10.39 |
| 7 | 0.10 | 3.59 | 3.26 | 3.07 | 2.96 | 2.88 | 2.83 | 2.78 | 2.75 | 2.72 |
| | 0.05 | 5.59 | 4.74 | 4.35 | 4.12 | 3.97 | 3.87 | 3.79 | 3.73 | 3.68 |
| | 0.025 | 8.07 | 6.54 | 5.89 | 5.52 | 5.29 | 5.12 | 4.99 | 4.90 | 4.82 |
| | 0.01 | 12.25 | 9.55 | 8.45 | 7.85 | 7.46 | 7.19 | 6.99 | 6.84 | 6.72 |
| | 0.005 | 16.24 | 12.40 | 10.88 | 10.05 | 9.52 | 9.16 | 8.89 | 8.68 | 8.51 |
| 8 | 0.10 | 3.46 | 3.11 | 2.92 | 2.81 | 2.73 | 2.67 | 2.62 | 2.59 | 2.56 |
| | 0.05 | 5.32 | 4.46 | 4.07 | 3.84 | 3.69 | 3.58 | 3.50 | 3.44 | 3.39 |
| | 0.025 | 7.57 | 6.06 | 5.42 | 5.05 | 4.82 | 4.65 | 4.53 | 4.43 | 4.36 |
| | 0.01 | 11.26 | 8.65 | 7.59 | 7.01 | 6.63 | 6.37 | 6.18 | 6.03 | 5.91 |
| | 0.005 | 14.69 | 11.04 | 9.60 | 8.81 | 8.30 | 7.95 | 7.69 | 7.50 | 7.34 |

**TABLE VI   (cont.)**
Values of $F_\alpha$

| | | | | dfn | | | | | | | |
|---|---|---|---|---|---|---|---|---|---|---|---|
| *10* | *12* | *15* | *20* | *24* | *30* | *40* | *60* | *120* | *α* | *dfd* |
| 60.19 | 60.71 | 61.22 | 61.74 | 62.00 | 62.26 | 62.53 | 62.79 | 63.06 | *0.10* | |
| 241.88 | 243.91 | 245.95 | 248.01 | 249.05 | 250.10 | 251.14 | 252.20 | 253.25 | *0.05* | |
| 968.63 | 976.71 | 984.87 | 993.10 | 997.25 | 1001.41 | 1005.60 | 1009.80 | 1014.02 | *0.025* | *1* |
| 6055.8 | 6106.3 | 6157.3 | 6208.7 | 6234.6 | 6260.6 | 6286.7 | 631.9 | 6339.4 | *0.01* | |
| 24224 | 24426 | 24630 | 24836 | 24940 | 25044 | 25148 | 25253 | 25359 | *0.005* | |
| 9.39 | 9.41 | 9.42 | 9.44 | 9.45 | 9.46 | 9.47 | 9.47 | 9.48 | *0.10* | |
| 19.40 | 19.41 | 19.43 | 19.45 | 19.45 | 19.46 | 19.47 | 19.48 | 19.49 | *0.05* | |
| 39.40 | 39.41 | 39.43 | 39.45 | 39.46 | 39.46 | 39.47 | 39.48 | 39.49 | *0.025* | *2* |
| 99.40 | 99.42 | 99.43 | 99.45 | 99.46 | 99.47 | 99.47 | 99.48 | 99.49 | *0.01* | |
| 199.40 | 199.42 | 199.43 | 199.45 | 199.46 | 199.47 | 199.47 | 199.48 | 199.49 | *0.005* | |
| 5.23 | 5.22 | 5.20 | 5.18 | 5.18 | 5.17 | 5.16 | 5.15 | 5.14 | *0.10* | |
| 8.79 | 8.74 | 8.70 | 8.66 | 8.64 | 8.62 | 8.59 | 8.57 | 8.55 | *0.05* | |
| 14.42 | 14.34 | 14.25 | 14.17 | 14.12 | 14.08 | 14.04 | 13.99 | 13.95 | *0.025* | *3* |
| 27.23 | 27.05 | 26.87 | 26.69 | 26.60 | 26.50 | 26.41 | 26.32 | 26.22 | *0.01* | |
| 43.69 | 43.39 | 43.08 | 42.78 | 42.62 | 42.47 | 42.31 | 42.15 | 41.99 | *0.005* | |
| 3.92 | 3.90 | 3.87 | 3.84 | 3.83 | 3.82 | 3.80 | 3.79 | 3.78 | *0.10* | |
| 5.96 | 5.91 | 5.86 | 5.80 | 5.77 | 5.75 | 5.72 | 5.69 | 5.66 | *0.05* | |
| 8.84 | 8.75 | 8.66 | 8.56 | 8.51 | 8.46 | 8.41 | 8.36 | 8.31 | *0.025* | *4* |
| 14.55 | 14.37 | 14.20 | 14.02 | 13.93 | 13.84 | 13.75 | 13.65 | 13.56 | *0.01* | |
| 20.97 | 20.70 | 20.44 | 20.17 | 20.03 | 19.89 | 19.75 | 19.61 | 19.47 | *0.005* | |
| 3.30 | 3.27 | 3.24 | 3.21 | 3.19 | 3.17 | 3.16 | 3.14 | 3.12 | *0.10* | |
| 4.74 | 4.68 | 4.62 | 4.56 | 4.53 | 4.50 | 4.46 | 4.43 | 4.40 | *0.05* | |
| 6.62 | 6.52 | 6.43 | 6.33 | 6.28 | 6.23 | 6.18 | 6.12 | 6.07 | *0.025* | *5* |
| 10.05 | 9.89 | 9.72 | 9.55 | 9.47 | 9.38 | 9.29 | 9.20 | 9.11 | *0.01* | |
| 13.62 | 13.38 | 13.15 | 12.90 | 12.78 | 12.66 | 12.53 | 12.40 | 12.27 | *0.005* | |
| 2.94 | 2.90 | 2.87 | 2.84 | 2.82 | 2.80 | 2.78 | 2.76 | 2.74 | *0.10* | |
| 4.06 | 4.00 | 3.94 | 3.87 | 3.84 | 3.81 | 3.77 | 3.74 | 3.70 | *0.05* | |
| 5.46 | 5.37 | 5.27 | 5.17 | 5.12 | 5.07 | 5.01 | 4.96 | 4.90 | *0.025* | *6* |
| 7.87 | 7.72 | 7.56 | 7.40 | 7.31 | 7.23 | 7.14 | 7.06 | 6.97 | *0.01* | |
| 10.25 | 10.03 | 9.81 | 9.59 | 9.47 | 9.36 | 9.24 | 9.12 | 9.00 | *0.005* | |
| 2.70 | 2.67 | 2.63 | 2.59 | 2.58 | 2.56 | 2.54 | 2.51 | 2.49 | *0.10* | |
| 3.64 | 3.57 | 3.51 | 3.44 | 3.41 | 3.38 | 3.34 | 3.30 | 3.27 | *0.05* | |
| 4.76 | 4.67 | 4.57 | 4.47 | 4.41 | 4.36 | 4.31 | 4.25 | 4.20 | *0.025* | *7* |
| 6.62 | 6.47 | 6.31 | 6.16 | 6.07 | 5.99 | 5.91 | 5.82 | 5.74 | *0.01* | |
| 8.38 | 8.18 | 7.97 | 7.75 | 7.64 | 7.53 | 7.42 | 7.31 | 7.19 | *0.005* | |
| 2.54 | 2.50 | 2.46 | 2.42 | 2.40 | 2.38 | 2.36 | 2.34 | 2.32 | *0.10* | |
| 3.35 | 3.28 | 3.22 | 3.15 | 3.12 | 3.08 | 3.04 | 3.01 | 2.97 | *0.05* | |
| 4.30 | 4.20 | 4.10 | 4.00 | 3.95 | 3.89 | 3.84 | 3.78 | 3.73 | *0.025* | *8* |
| 5.81 | 5.67 | 5.52 | 5.36 | 5.28 | 5.20 | 5.12 | 5.03 | 4.95 | *0.01* | |
| 7.21 | 7.01 | 6.81 | 6.61 | 6.50 | 6.40 | 6.29 | 6.18 | 6.06 | *0.005* | |

**TABLE VI   (cont.)**
Values of $F_\alpha$

| dfd | $\alpha$ | dfn | | | | | | | | |
|---|---|---|---|---|---|---|---|---|---|---|
| | | 1 | 2 | 3 | 4 | 5 | 6 | 7 | 8 | 9 |
| | 0.10 | 3.36 | 3.01 | 2.81 | 2.69 | 2.61 | 2.55 | 2.51 | 2.47 | 2.44 |
| | 0.05 | 5.12 | 4.26 | 3.86 | 3.63 | 3.48 | 3.37 | 3.29 | 3.23 | 3.18 |
| 9 | 0.025 | 7.21 | 5.71 | 5.08 | 4.72 | 4.48 | 4.32 | 4.20 | 4.10 | 4.03 |
| | 0.01 | 10.56 | 8.02 | 6.99 | 6.42 | 6.06 | 5.80 | 5.61 | 5.47 | 5.35 |
| | 0.005 | 13.61 | 10.11 | 8.72 | 7.96 | 7.47 | 7.13 | 6.88 | 6.69 | 6.54 |
| | 0.10 | 3.29 | 2.92 | 2.73 | 2.61 | 2.52 | 2.46 | 2.41 | 2.38 | 2.35 |
| | 0.05 | 4.96 | 4.10 | 3.71 | 3.48 | 3.33 | 3.22 | 3.14 | 3.07 | 3.02 |
| 10 | 0.025 | 6.94 | 5.46 | 4.83 | 4.47 | 4.24 | 4.07 | 3.95 | 3.85 | 3.78 |
| | 0.01 | 10.04 | 7.56 | 6.55 | 5.99 | 5.64 | 5.39 | 5.20 | 5.06 | 4.94 |
| | 0.005 | 12.83 | 9.43 | 8.08 | 7.34 | 6.87 | 6.54 | 6.30 | 6.12 | 5.97 |
| | 0.10 | 3.23 | 2.86 | 2.66 | 2.54 | 2.45 | 2.39 | 2.34 | 2.30 | 2.27 |
| | 0.05 | 4.84 | 3.98 | 3.59 | 3.36 | 3.20 | 3.09 | 3.01 | 2.95 | 2.90 |
| 11 | 0.025 | 6.72 | 5.26 | 4.63 | 4.28 | 4.04 | 3.88 | 3.76 | 3.66 | 3.59 |
| | 0.01 | 9.65 | 7.21 | 6.22 | 5.67 | 5.32 | 5.07 | 4.89 | 4.74 | 4.63 |
| | 0.005 | 12.23 | 8.91 | 7.60 | 6.88 | 6.42 | 6.10 | 5.86 | 5.68 | 5.54 |
| | 0.10 | 3.18 | 2.81 | 2.61 | 2.48 | 2.39 | 2.33 | 2.28 | 2.24 | 2.21 |
| | 0.05 | 4.75 | 3.89 | 3.49 | 3.26 | 3.11 | 3.00 | 2.91 | 2.85 | 2.80 |
| 12 | 0.025 | 6.55 | 5.10 | 4.47 | 4.12 | 3.89 | 3.73 | 3.61 | 3.51 | 3.44 |
| | 0.01 | 9.33 | 6.93 | 5.95 | 5.41 | 5.06 | 4.82 | 4.64 | 4.50 | 4.39 |
| | 0.005 | 11.75 | 8.51 | 7.23 | 6.52 | 6.07 | 5.76 | 5.52 | 5.35 | 5.20 |
| | 0.10 | 3.14 | 2.76 | 2.56 | 2.43 | 2.35 | 2.28 | 2.23 | 2.20 | 2.16 |
| | 0.05 | 4.67 | 3.81 | 3.41 | 3.18 | 3.03 | 2.92 | 2.83 | 2.77 | 2.71 |
| 13 | 0.025 | 6.41 | 4.97 | 4.35 | 4.00 | 3.77 | 3.60 | 3.48 | 3.39 | 3.31 |
| | 0.01 | 9.07 | 6.70 | 5.74 | 5.21 | 4.86 | 4.62 | 4.44 | 4.30 | 4.19 |
| | 0.005 | 11.37 | 8.19 | 6.93 | 6.23 | 5.79 | 5.48 | 5.25 | 5.08 | 4.94 |
| | 0.10 | 3.10 | 2.73 | 2.52 | 2.39 | 2.31 | 2.24 | 2.19 | 2.15 | 2.12 |
| | 0.05 | 4.60 | 3.74 | 3.34 | 3.11 | 2.96 | 2.85 | 2.76 | 2.70 | 2.65 |
| 14 | 0.025 | 6.30 | 4.86 | 4.24 | 3.89 | 3.66 | 3.50 | 3.38 | 3.29 | 3.21 |
| | 0.01 | 8.86 | 6.51 | 5.56 | 5.04 | 4.69 | 4.46 | 4.28 | 4.14 | 4.03 |
| | 0.005 | 11.06 | 7.92 | 6.68 | 6.00 | 5.56 | 5.26 | 5.03 | 4.86 | 4.72 |
| | 0.10 | 3.07 | 2.70 | 2.49 | 2.36 | 2.27 | 2.21 | 2.16 | 2.12 | 2.09 |
| | 0.05 | 4.54 | 3.68 | 3.29 | 3.06 | 2.90 | 2.79 | 2.71 | 2.64 | 2.59 |
| 15 | 0.025 | 6.20 | 4.77 | 4.15 | 3.80 | 3.58 | 3.41 | 3.29 | 3.20 | 3.12 |
| | 0.01 | 8.68 | 6.36 | 5.42 | 4.89 | 4.56 | 4.32 | 4.14 | 4.00 | 3.89 |
| | 0.005 | 10.80 | 7.70 | 6.48 | 5.80 | 5.37 | 5.07 | 4.85 | 4.67 | 4.54 |
| | 0.10 | 3.05 | 2.67 | 2.46 | 2.33 | 2.24 | 2.18 | 2.13 | 2.09 | 2.06 |
| | 0.05 | 4.49 | 3.63 | 3.24 | 3.01 | 2.85 | 2.74 | 2.66 | 2.59 | 2.54 |
| 16 | 0.025 | 6.12 | 4.69 | 4.08 | 3.73 | 3.50 | 3.34 | 3.22 | 3.12 | 3.05 |
| | 0.01 | 8.53 | 6.23 | 5.29 | 4.77 | 4.44 | 4.20 | 4.03 | 3.89 | 3.78 |
| | 0.005 | 10.58 | 7.51 | 6.30 | 5.64 | 5.21 | 4.91 | 4.69 | 4.52 | 4.38 |

**TABLE VI   (cont.)**
Values of $F_\alpha$

| 10 | 12 | 15 | 20 | 24 | 30 | 40 | 60 | 120 | $\alpha$ | dfd |
|----|----|----|----|----|----|----|----|-----|----------|-----|
| 2.42 | 2.38 | 2.34 | 2.30 | 2.28 | 2.25 | 2.23 | 2.21 | 2.18 | 0.10 | |
| 3.14 | 3.07 | 3.01 | 2.94 | 2.90 | 2.86 | 2.83 | 2.79 | 2.75 | 0.05 | |
| 3.96 | 3.87 | 3.77 | 3.67 | 3.61 | 3.56 | 3.51 | 3.45 | 3.39 | 0.025 | 9 |
| 5.26 | 5.11 | 4.96 | 4.81 | 4.73 | 4.65 | 4.57 | 4.48 | 4.40 | 0.01 | |
| 6.42 | 6.23 | 6.03 | 5.83 | 5.73 | 5.62 | 5.52 | 5.41 | 5.30 | 0.005 | |
| 2.32 | 2.28 | 2.24 | 2.20 | 2.18 | 2.16 | 2.13 | 2.11 | 2.08 | 0.10 | |
| 2.98 | 2.91 | 2.85 | 2.77 | 2.74 | 2.70 | 2.66 | 2.62 | 2.58 | 0.05 | |
| 3.72 | 3.62 | 3.52 | 3.42 | 3.37 | 3.31 | 3.26 | 3.20 | 3.14 | 0.025 | 10 |
| 4.85 | 4.71 | 4.56 | 4.41 | 4.33 | 4.25 | 4.17 | 4.08 | 4.00 | 0.01 | |
| 5.85 | 5.66 | 5.47 | 5.27 | 5.17 | 5.07 | 4.97 | 4.86 | 4.75 | 0.005 | |
| 2.25 | 2.21 | 2.17 | 2.12 | 2.10 | 2.08 | 2.05 | 2.03 | 2.00 | 0.10 | |
| 2.85 | 2.79 | 2.72 | 2.65 | 2.61 | 2.57 | 2.53 | 2.49 | 2.45 | 0.05 | |
| 3.53 | 3.43 | 3.33 | 3.23 | 3.17 | 3.12 | 3.06 | 3.00 | 2.94 | 0.025 | 11 |
| 4.54 | 4.40 | 4.25 | 4.10 | 4.02 | 3.94 | 3.86 | 3.78 | 3.69 | 0.01 | |
| 5.42 | 5.24 | 5.05 | 4.86 | 4.76 | 4.65 | 4.55 | 4.45 | 4.34 | 0.005 | |
| 2.19 | 2.15 | 2.10 | 2.06 | 2.04 | 2.01 | 1.99 | 1.96 | 1.93 | 0.10 | |
| 2.75 | 2.69 | 2.62 | 2.54 | 2.51 | 2.47 | 2.43 | 2.38 | 2.34 | 0.05 | |
| 3.37 | 3.28 | 3.18 | 3.07 | 3.02 | 2.96 | 2.91 | 2.85 | 2.79 | 0.025 | 12 |
| 4.30 | 4.16 | 4.01 | 3.86 | 3.78 | 3.70 | 3.62 | 3.54 | 3.45 | 0.01 | |
| 5.09 | 4.91 | 4.72 | 4.53 | 4.43 | 4.33 | 4.23 | 4.12 | 4.01 | 0.005 | |
| 2.14 | 2.10 | 2.05 | 2.01 | 1.98 | 1.96 | 1.93 | 1.90 | 1.88 | 0.10 | |
| 2.67 | 2.60 | 2.53 | 2.46 | 2.42 | 2.38 | 2.34 | 2.30 | 2.25 | 0.05 | |
| 3.25 | 3.15 | 3.05 | 2.95 | 2.89 | 2.84 | 2.78 | 2.72 | 2.66 | 0.025 | 13 |
| 4.10 | 3.96 | 3.82 | 3.66 | 3.59 | 3.51 | 3.43 | 3.34 | 3.25 | 0.01 | |
| 4.82 | 4.64 | 4.46 | 4.27 | 4.17 | 4.07 | 3.97 | 3.87 | 3.76 | 0.005 | |
| 2.10 | 2.05 | 2.01 | 1.96 | 1.94 | 1.91 | 1.89 | 1.86 | 1.83 | 0.10 | |
| 2.60 | 2.53 | 2.46 | 2.39 | 2.35 | 2.31 | 2.27 | 2.22 | 2.18 | 0.05 | |
| 3.15 | 3.05 | 2.95 | 2.84 | 2.79 | 2.73 | 2.67 | 2.61 | 2.55 | 0.025 | 14 |
| 3.94 | 3.80 | 3.66 | 3.51 | 3.43 | 3.35 | 3.27 | 3.18 | 3.09 | 0.01 | |
| 4.60 | 4.43 | 4.25 | 4.06 | 3.96 | 3.86 | 3.76 | 3.66 | 3.55 | 0.005 | |
| 2.06 | 2.02 | 1.97 | 1.92 | 1.90 | 1.87 | 1.85 | 1.82 | 1.79 | 0.10 | |
| 2.54 | 2.48 | 2.40 | 2.33 | 2.29 | 2.25 | 2.20 | 2.16 | 2.11 | 0.05 | |
| 3.06 | 2.96 | 2.86 | 2.76 | 2.70 | 2.64 | 2.59 | 2.52 | 2.46 | 0.025 | 15 |
| 3.80 | 3.67 | 3.52 | 3.37 | 3.29 | 3.21 | 3.13 | 3.05 | 2.96 | 0.01 | |
| 4.42 | 4.25 | 4.07 | 3.88 | 3.79 | 3.69 | 3.58 | 3.48 | 3.37 | 0.005 | |
| 2.03 | 1.99 | 1.94 | 1.89 | 1.87 | 1.84 | 1.81 | 1.78 | 1.75 | 0.10 | |
| 2.49 | 2.42 | 2.35 | 2.28 | 2.24 | 2.19 | 2.15 | 2.11 | 2.06 | 0.05 | |
| 2.99 | 2.89 | 2.79 | 2.68 | 2.63 | 2.57 | 2.51 | 2.45 | 2.38 | 0.025 | 16 |
| 3.69 | 3.55 | 3.41 | 3.26 | 3.18 | 3.10 | 3.02 | 2.93 | 2.84 | 0.01 | |
| 4.27 | 4.10 | 3.92 | 3.73 | 3.64 | 3.54 | 3.44 | 3.33 | 3.22 | 0.005 | |

dfn

**TABLE VI   (cont.)**
Values of $F_\alpha$

| dfd | $\alpha$ | 1 | 2 | 3 | 4 | 5 | 6 | 7 | 8 | 9 |
|---|---|---|---|---|---|---|---|---|---|---|
| | | | | | | dfn | | | | |
| | 0.10 | 3.03 | 2.64 | 2.44 | 2.31 | 2.22 | 2.15 | 2.10 | 2.06 | 2.03 |
| | 0.05 | 4.45 | 3.59 | 3.20 | 2.96 | 2.81 | 2.70 | 2.61 | 2.55 | 2.49 |
| 17 | 0.025 | 6.04 | 4.62 | 4.01 | 3.66 | 3.44 | 3.28 | 3.16 | 3.06 | 2.98 |
| | 0.01 | 8.40 | 6.11 | 5.18 | 4.67 | 4.34 | 4.10 | 3.93 | 3.79 | 3.68 |
| | 0.005 | 10.38 | 7.35 | 6.16 | 5.50 | 5.07 | 4.78 | 4.56 | 4.39 | 4.25 |
| | 0.10 | 3.01 | 2.62 | 2.42 | 2.29 | 2.20 | 2.13 | 2.08 | 2.04 | 2.00 |
| | 0.05 | 4.41 | 3.55 | 3.16 | 2.93 | 2.77 | 2.66 | 2.58 | 2.51 | 2.46 |
| 18 | 0.025 | 5.98 | 4.56 | 3.95 | 3.61 | 3.38 | 3.22 | 3.10 | 3.01 | 2.93 |
| | 0.01 | 8.29 | 6.01 | 5.09 | 4.58 | 4.25 | 4.01 | 3.84 | 3.71 | 3.60 |
| | 0.005 | 10.22 | 7.21 | 6.03 | 5.37 | 4.96 | 4.66 | 4.44 | 4.28 | 4.14 |
| | 0.10 | 2.99 | 2.61 | 2.40 | 2.27 | 2.18 | 2.11 | 2.06 | 2.02 | 1.98 |
| | 0.05 | 4.38 | 3.52 | 3.13 | 2.90 | 2.74 | 2.63 | 2.54 | 2.48 | 2.42 |
| 19 | 0.025 | 5.92 | 4.51 | 3.90 | 3.56 | 3.33 | 3.17 | 3.05 | 2.96 | 2.88 |
| | 0.01 | 8.18 | 5.93 | 5.01 | 4.50 | 4.17 | 3.94 | 3.77 | 3.63 | 3.52 |
| | 0.005 | 10.07 | 7.09 | 5.92 | 5.27 | 4.85 | 4.56 | 4.34 | 4.18 | 4.04 |
| | 0.10 | 2.97 | 2.59 | 2.38 | 2.25 | 2.16 | 2.09 | 2.04 | 2.00 | 1.96 |
| | 0.05 | 4.35 | 3.49 | 3.10 | 2.87 | 2.71 | 2.60 | 2.51 | 2.45 | 2.39 |
| 20 | 0.025 | 5.87 | 4.46 | 3.86 | 3.51 | 3.29 | 3.13 | 3.01 | 2.91 | 2.84 |
| | 0.01 | 8.10 | 5.85 | 4.94 | 4.43 | 4.10 | 3.87 | 3.70 | 3.56 | 3.46 |
| | 0.005 | 9.94 | 6.99 | 5.82 | 5.17 | 4.76 | 4.47 | 4.26 | 4.09 | 3.96 |
| | 0.10 | 2.96 | 2.57 | 2.36 | 2.23 | 2.14 | 2.08 | 2.02 | 1.98 | 1.95 |
| | 0.05 | 4.32 | 3.47 | 3.07 | 2.84 | 2.68 | 2.57 | 2.49 | 2.42 | 2.37 |
| 21 | 0.025 | 5.83 | 4.42 | 3.82 | 3.48 | 3.25 | 3.09 | 2.97 | 2.87 | 2.80 |
| | 0.01 | 8.02 | 5.78 | 4.87 | 4.37 | 4.04 | 3.81 | 3.64 | 3.51 | 3.40 |
| | 0.005 | 9.83 | 6.89 | 5.73 | 5.09 | 4.68 | 4.39 | 4.18 | 4.01 | 3.88 |
| | 0.10 | 2.95 | 2.56 | 2.35 | 2.22 | 2.13 | 2.06 | 2.01 | 1.97 | 1.93 |
| | 0.05 | 4.30 | 3.44 | 3.05 | 2.82 | 2.66 | 2.55 | 2.46 | 2.40 | 2.34 |
| 22 | 0.025 | 5.79 | 4.38 | 3.78 | 3.44 | 3.22 | 3.05 | 2.93 | 2.84 | 2.76 |
| | 0.01 | 7.95 | 5.72 | 4.82 | 4.31 | 3.99 | 3.76 | 3.59 | 3.45 | 3.35 |
| | 0.005 | 9.73 | 6.81 | 5.65 | 5.02 | 4.61 | 4.32 | 4.11 | 3.94 | 3.81 |
| | 0.10 | 2.94 | 2.55 | 2.34 | 2.21 | 2.11 | 2.05 | 1.99 | 1.95 | 1.92 |
| | 0.05 | 4.28 | 3.42 | 3.03 | 2.80 | 2.64 | 2.53 | 2.44 | 2.37 | 2.32 |
| 23 | 0.025 | 5.75 | 4.35 | 3.75 | 3.41 | 3.18 | 3.02 | 2.90 | 2.81 | 2.73 |
| | 0.01 | 7.88 | 5.66 | 4.76 | 4.26 | 3.94 | 3.71 | 3.54 | 3.41 | 3.30 |
| | 0.005 | 9.63 | 6.73 | 5.58 | 4.95 | 4.54 | 4.26 | 4.05 | 3.88 | 3.75 |
| | 0.10 | 2.93 | 2.54 | 2.33 | 2.19 | 2.10 | 2.04 | 1.98 | 1.94 | 1.91 |
| | 0.05 | 4.26 | 3.40 | 3.01 | 2.78 | 2.62 | 2.51 | 2.42 | 2.36 | 2.30 |
| 24 | 0.025 | 5.72 | 4.32 | 3.72 | 3.38 | 3.15 | 2.99 | 2.87 | 2.78 | 2.70 |
| | 0.01 | 7.82 | 5.61 | 4.72 | 4.22 | 3.90 | 3.67 | 3.50 | 3.36 | 3.26 |
| | 0.005 | 9.55 | 6.66 | 5.52 | 4.89 | 4.49 | 4.20 | 3.99 | 3.83 | 3.69 |

**TABLE VI** (cont.)
Values of $F_\alpha$

| | | | | dfn | | | | | | | |
|---|---|---|---|---|---|---|---|---|---|---|---|
| *10* | *12* | *15* | *20* | *24* | *30* | *40* | *60* | *120* | *α* | *dfd* |
| 2.00 | 1.96 | 1.91 | 1.86 | 1.84 | 1.81 | 1.78 | 1.75 | 1.72 | *0.10* | |
| 2.45 | 2.38 | 2.31 | 2.23 | 2.19 | 2.15 | 2.10 | 2.06 | 2.01 | *0.05* | |
| 2.92 | 2.82 | 2.72 | 2.62 | 2.56 | 2.50 | 2.44 | 2.38 | 2.32 | *0.025* | *17* |
| 3.59 | 3.46 | 3.31 | 3.16 | 3.08 | 3.00 | 2.92 | 2.83 | 2.75 | *0.01* | |
| 4.14 | 3.97 | 3.79 | 3.61 | 3.51 | 3.41 | 3.31 | 3.21 | 3.10 | *0.005* | |
| 1.98 | 1.93 | 1.89 | 1.84 | 1.81 | 1.78 | 1.75 | 1.72 | 1.69 | *0.10* | |
| 2.41 | 2.34 | 2.27 | 2.19 | 2.15 | 2.11 | 2.06 | 2.02 | 1.97 | *0.05* | |
| 2.87 | 2.77 | 2.67 | 2.56 | 2.50 | 2.44 | 2.38 | 2.32 | 2.26 | *0.025* | *18* |
| 3.51 | 3.37 | 3.23 | 3.08 | 3.00 | 2.92 | 2.84 | 2.75 | 2.66 | *0.01* | |
| 4.03 | 3.86 | 3.68 | 3.50 | 3.40 | 3.30 | 3.20 | 3.10 | 2.99 | *0.005* | |
| 1.96 | 1.91 | 1.86 | 1.81 | 1.79 | 1.76 | 1.73 | 1.70 | 1.67 | *0.10* | |
| 2.38 | 2.31 | 2.23 | 2.16 | 2.11 | 2.07 | 2.03 | 1.98 | 1.93 | *0.05* | |
| 2.82 | 2.72 | 2.62 | 2.51 | 2.45 | 2.39 | 2.33 | 2.27 | 2.20 | *0.025* | *19* |
| 3.43 | 3.30 | 3.15 | 3.00 | 2.92 | 2.84 | 2.76 | 2.67 | 2.58 | *0.01* | |
| 3.93 | 3.76 | 3.59 | 3.40 | 3.31 | 3.21 | 3.11 | 3.00 | 2.89 | *0.005* | |
| 1.94 | 1.89 | 1.84 | 1.79 | 1.77 | 1.74 | 1.71 | 1.68 | 1.64 | *0.10* | |
| 2.35 | 2.28 | 2.20 | 2.12 | 2.08 | 2.04 | 1.99 | 1.95 | 1.90 | *0.05* | |
| 2.77 | 2.68 | 2.57 | 2.46 | 2.41 | 2.35 | 2.29 | 2.22 | 2.16 | *0.025* | *20* |
| 3.37 | 3.23 | 3.09 | 2.94 | 2.86 | 2.78 | 2.69 | 2.61 | 2.52 | *0.01* | |
| 3.85 | 3.68 | 3.50 | 3.32 | 3.22 | 3.12 | 3.02 | 2.92 | 2.81 | *0.005* | |
| 1.92 | 1.87 | 1.83 | 1.78 | 1.75 | 1.72 | 1.69 | 1.66 | 1.62 | *0.10* | |
| 2.32 | 2.25 | 2.18 | 2.10 | 2.05 | 2.01 | 1.96 | 1.92 | 1.87 | *0.05* | |
| 2.73 | 2.64 | 2.53 | 2.42 | 2.37 | 2.31 | 2.25 | 2.18 | 2.11 | *0.025* | *21* |
| 3.31 | 3.17 | 3.03 | 2.88 | 2.80 | 2.72 | 2.64 | 2.55 | 2.46 | *0.01* | |
| 3.77 | 3.60 | 3.43 | 3.24 | 3.15 | 3.05 | 2.95 | 2.84 | 2.73 | *0.005* | |
| 1.90 | 1.86 | 1.81 | 1.76 | 1.73 | 1.70 | 1.67 | 1.64 | 1.60 | *0.10* | |
| 2.30 | 2.23 | 2.15 | 2.07 | 2.03 | 1.98 | 1.94 | 1.89 | 1.84 | *0.05* | |
| 2.70 | 2.60 | 2.50 | 2.39 | 2.33 | 2.27 | 2.21 | 2.14 | 2.08 | *0.025* | *22* |
| 3.26 | 3.12 | 2.98 | 2.83 | 2.75 | 2.67 | 2.58 | 2.50 | 2.40 | *0.01* | |
| 3.70 | 3.54 | 3.36 | 3.18 | 3.08 | 2.98 | 2.88 | 2.77 | 2.66 | *0.005* | |
| 1.89 | 1.84 | 1.80 | 1.74 | 1.72 | 1.69 | 1.66 | 1.62 | 1.59 | *0.10* | |
| 2.27 | 2.20 | 2.13 | 2.05 | 2.01 | 1.96 | 1.91 | 1.86 | 1.81 | *0.05* | |
| 2.67 | 2.57 | 2.47 | 2.36 | 2.30 | 2.24 | 2.18 | 2.11 | 2.04 | *0.025* | *23* |
| 3.21 | 3.07 | 2.93 | 2.78 | 2.70 | 2.62 | 2.54 | 2.45 | 2.35 | *0.01* | |
| 3.64 | 3.47 | 3.30 | 3.12 | 3.02 | 2.92 | 2.82 | 2.71 | 2.60 | *0.005* | |
| 1.88 | 1.83 | 1.78 | 1.73 | 1.70 | 1.67 | 1.64 | 1.61 | 1.57 | *0.10* | |
| 2.25 | 2.18 | 2.11 | 2.03 | 1.98 | 1.94 | 1.89 | 1.84 | 1.79 | *0.05* | |
| 2.64 | 2.54 | 2.44 | 2.33 | 2.27 | 2.21 | 2.15 | 2.08 | 2.01 | *0.025* | *24* |
| 3.17 | 3.03 | 2.89 | 2.74 | 2.66 | 2.58 | 2.49 | 2.40 | 2.31 | *0.01* | |
| 3.59 | 3.42 | 3.25 | 3.06 | 2.97 | 2.87 | 2.77 | 2.66 | 2.55 | *0.005* | |

**TABLE VI   (cont.)**
Values of $F_\alpha$

| dfd | $\alpha$ | 1 | 2 | 3 | 4 | 5 | 6 | 7 | 8 | 9 |
|-----|------|------|------|------|------|------|------|------|------|------|
|     |      |      |      |      |      | dfn  |      |      |      |      |
|     | 0.10 | 2.92 | 2.53 | 2.32 | 2.18 | 2.09 | 2.02 | 1.97 | 1.93 | 1.89 |
|     | 0.05 | 4.24 | 3.39 | 2.99 | 2.76 | 2.60 | 2.49 | 2.40 | 2.34 | 2.28 |
| 25  | 0.025| 5.69 | 4.29 | 3.69 | 3.35 | 3.13 | 2.97 | 2.85 | 2.75 | 2.68 |
|     | 0.01 | 7.77 | 5.57 | 4.68 | 4.18 | 3.85 | 3.63 | 3.46 | 3.32 | 3.22 |
|     | 0.005| 9.48 | 6.60 | 5.46 | 4.84 | 4.43 | 4.15 | 3.94 | 3.78 | 3.64 |
|     | 0.10 | 2.91 | 2.52 | 2.31 | 2.17 | 2.08 | 2.01 | 1.96 | 1.92 | 1.88 |
|     | 0.05 | 4.23 | 3.37 | 2.98 | 2.74 | 2.59 | 2.47 | 2.39 | 2.32 | 2.27 |
| 26  | 0.025| 5.66 | 4.27 | 3.67 | 3.33 | 3.10 | 2.94 | 2.82 | 2.73 | 2.65 |
|     | 0.01 | 7.72 | 5.53 | 4.64 | 4.14 | 3.82 | 3.59 | 3.42 | 3.29 | 3.18 |
|     | 0.005| 9.41 | 6.54 | 5.41 | 4.79 | 4.38 | 4.10 | 3.89 | 3.73 | 3.60 |
|     | 0.10 | 2.90 | 2.51 | 2.30 | 2.17 | 2.07 | 2.00 | 1.95 | 1.91 | 1.87 |
|     | 0.05 | 4.21 | 3.35 | 2.96 | 2.73 | 2.57 | 2.46 | 2.37 | 2.31 | 2.25 |
| 27  | 0.025| 5.63 | 4.24 | 3.65 | 3.31 | 3.08 | 2.92 | 2.80 | 2.71 | 2.63 |
|     | 0.01 | 7.68 | 5.49 | 4.60 | 4.11 | 3.78 | 3.56 | 3.39 | 3.26 | 3.15 |
|     | 0.005| 9.34 | 6.49 | 5.36 | 4.74 | 4.34 | 4.06 | 3.85 | 3.69 | 3.56 |
|     | 0.10 | 2.89 | 2.50 | 2.29 | 2.16 | 2.06 | 2.00 | 1.94 | 1.90 | 1.87 |
|     | 0.05 | 4.20 | 3.34 | 2.95 | 2.71 | 2.56 | 2.45 | 2.36 | 2.29 | 2.24 |
| 28  | 0.025| 5.61 | 4.22 | 3.63 | 3.29 | 3.06 | 2.90 | 2.78 | 2.69 | 2.61 |
|     | 0.01 | 7.64 | 5.45 | 4.57 | 4.07 | 3.75 | 3.53 | 3.36 | 3.23 | 3.12 |
|     | 0.005| 9.28 | 6.44 | 5.32 | 4.70 | 4.30 | 4.02 | 3.81 | 3.65 | 3.52 |
|     | 0.10 | 2.89 | 2.50 | 2.28 | 2.15 | 2.06 | 1.99 | 1.93 | 1.89 | 1.86 |
|     | 0.05 | 4.18 | 3.33 | 2.93 | 2.70 | 2.55 | 2.43 | 2.35 | 2.28 | 2.22 |
| 29  | 0.025| 5.59 | 4.20 | 3.61 | 3.27 | 3.04 | 2.88 | 2.76 | 2.67 | 2.59 |
|     | 0.01 | 7.60 | 5.42 | 4.54 | 4.04 | 3.73 | 3.50 | 3.33 | 3.20 | 3.09 |
|     | 0.005| 9.23 | 6.40 | 5.28 | 4.66 | 4.26 | 3.98 | 3.77 | 3.61 | 3.48 |
|     | 0.10 | 2.88 | 2.49 | 2.28 | 2.14 | 2.05 | 1.98 | 1.93 | 1.88 | 1.85 |
|     | 0.05 | 4.17 | 3.32 | 2.92 | 2.69 | 2.53 | 2.42 | 2.33 | 2.27 | 2.21 |
| 30  | 0.025| 5.57 | 4.18 | 3.59 | 3.25 | 3.03 | 2.87 | 2.75 | 2.65 | 2.57 |
|     | 0.01 | 7.56 | 5.39 | 4.51 | 4.02 | 3.70 | 3.47 | 3.30 | 3.17 | 3.07 |
|     | 0.005| 9.18 | 6.35 | 5.24 | 4.62 | 4.23 | 3.95 | 3.74 | 3.58 | 3.45 |
|     | 0.10 | 2.79 | 2.39 | 2.18 | 2.04 | 1.95 | 1.87 | 1.82 | 1.77 | 1.74 |
|     | 0.05 | 4.00 | 3.15 | 2.76 | 2.53 | 2.37 | 2.25 | 2.17 | 2.10 | 2.04 |
| 60  | 0.025| 5.29 | 3.93 | 3.34 | 3.01 | 2.79 | 2.63 | 2.51 | 2.41 | 2.33 |
|     | 0.01 | 7.08 | 4.98 | 4.13 | 3.65 | 3.34 | 3.12 | 2.95 | 2.82 | 2.72 |
|     | 0.005| 8.49 | 5.79 | 4.73 | 4.14 | 3.76 | 3.49 | 3.29 | 3.13 | 3.01 |
|     | 0.10 | 2.75 | 2.35 | 2.13 | 1.99 | 1.90 | 1.82 | 1.77 | 1.72 | 1.68 |
|     | 0.05 | 3.92 | 3.07 | 2.68 | 2.45 | 2.29 | 2.18 | 2.09 | 2.02 | 1.96 |
| 120 | 0.025| 5.15 | 3.80 | 3.23 | 2.89 | 2.67 | 2.52 | 2.39 | 2.30 | 2.22 |
|     | 0.01 | 6.85 | 4.79 | 3.95 | 3.48 | 3.17 | 2.96 | 2.79 | 2.66 | 2.56 |
|     | 0.005| 8.18 | 5.54 | 4.50 | 3.92 | 3.55 | 3.28 | 3.09 | 2.93 | 2.81 |

**TABLE VI (cont.)**
Values of $F_\alpha$

| | | | | dfn | | | | | | |
|---|---|---|---|---|---|---|---|---|---|---|
| 10 | 12 | 15 | 20 | 24 | 30 | 40 | 60 | 120 | α | dfd |
| 1.87 | 1.82 | 1.77 | 1.72 | 1.69 | 1.66 | 1.63 | 1.59 | 1.56 | 0.10 | |
| 2.24 | 2.16 | 2.09 | 2.01 | 1.96 | 1.92 | 1.87 | 1.82 | 1.77 | 0.05 | |
| 2.61 | 2.51 | 2.41 | 2.30 | 2.24 | 2.18 | 2.12 | 2.05 | 1.98 | 0.025 | 25 |
| 3.13 | 2.99 | 2.85 | 2.70 | 2.62 | 2.54 | 2.45 | 2.36 | 2.27 | 0.01 | |
| 3.54 | 3.37 | 3.20 | 3.01 | 2.92 | 2.82 | 2.72 | 2.61 | 2.50 | 0.005 | |
| 1.86 | 1.81 | 1.76 | 1.71 | 1.68 | 1.65 | 1.61 | 1.58 | 1.54 | 0.10 | |
| 2.22 | 2.15 | 2.07 | 1.99 | 1.95 | 1.90 | 1.85 | 1.80 | 1.75 | 0.05 | |
| 2.59 | 2.49 | 2.39 | 2.28 | 2.22 | 2.16 | 2.09 | 2.03 | 1.95 | 0.025 | 26 |
| 3.09 | 2.96 | 2.81 | 2.66 | 2.58 | 2.50 | 2.42 | 2.33 | 2.23 | 0.01 | |
| 3.49 | 3.33 | 3.15 | 2.97 | 2.87 | 2.77 | 2.67 | 2.56 | 2.45 | 0.005 | |
| 1.85 | 1.80 | 1.75 | 1.70 | 1.67 | 1.64 | 1.60 | 1.57 | 1.53 | 0.10 | |
| 2.20 | 2.13 | 2.06 | 1.97 | 1.93 | 1.88 | 1.84 | 1.79 | 1.73 | 0.05 | |
| 2.57 | 2.47 | 2.36 | 2.25 | 2.19 | 2.13 | 2.07 | 2.00 | 1.93 | 0.025 | 27 |
| 3.06 | 2.93 | 2.78 | 2.63 | 2.55 | 2.47 | 2.38 | 2.29 | 2.20 | 0.01 | |
| 3.45 | 3.28 | 3.11 | 2.93 | 2.83 | 2.73 | 2.63 | 2.52 | 2.41 | 0.005 | |
| 1.84 | 1.79 | 1.74 | 1.69 | 1.66 | 1.63 | 1.59 | 1.56 | 1.52 | 0.10 | |
| 2.19 | 2.12 | 2.04 | 1.96 | 1.91 | 1.87 | 1.82 | 1.77 | 1.71 | 0.05 | |
| 2.55 | 2.45 | 2.34 | 2.23 | 2.17 | 2.11 | 2.05 | 1.98 | 1.91 | 0.025 | 28 |
| 3.03 | 2.90 | 2.75 | 2.60 | 2.52 | 2.44 | 2.35 | 2.26 | 2.17 | 0.01 | |
| 3.41 | 3.25 | 3.07 | 2.89 | 2.79 | 2.69 | 2.59 | 2.48 | 2.37 | 0.005 | |
| 1.83 | 1.78 | 1.73 | 1.68 | 1.65 | 1.62 | 1.58 | 1.55 | 1.51 | 0.10 | |
| 2.18 | 2.10 | 2.03 | 1.94 | 1.90 | 1.85 | 1.81 | 1.75 | 1.70 | 0.05 | |
| 2.53 | 2.43 | 2.32 | 2.21 | 2.15 | 2.09 | 2.03 | 1.96 | 1.89 | 0.025 | 29 |
| 3.00 | 2.87 | 2.73 | 2.57 | 2.49 | 2.41 | 2.33 | 2.23 | 2.14 | 0.01 | |
| 3.38 | 3.21 | 3.04 | 2.86 | 2.76 | 2.66 | 2.56 | 2.45 | 2.33 | 0.005 | |
| 1.82 | 1.77 | 1.72 | 1.67 | 1.64 | 1.61 | 1.57 | 1.54 | 1.50 | 0.10 | |
| 2.16 | 2.09 | 2.01 | 1.93 | 1.89 | 1.84 | 1.79 | 1.74 | 1.68 | 0.05 | |
| 2.51 | 2.41 | 2.31 | 2.20 | 2.14 | 2.07 | 2.01 | 1.94 | 1.87 | 0.025 | 30 |
| 2.98 | 2.84 | 2.70 | 2.55 | 2.47 | 2.39 | 2.30 | 2.21 | 2.11 | 0.01 | |
| 3.34 | 3.18 | 3.01 | 2.82 | 2.73 | 2.63 | 2.52 | 2.42 | 2.30 | 0.005 | |
| 1.71 | 1.66 | 1.60 | 1.54 | 1.51 | 1.48 | 1.44 | 1.40 | 1.35 | 0.10 | |
| 1.99 | 1.92 | 1.84 | 1.75 | 1.70 | 1.65 | 1.59 | 1.53 | 1.47 | 0.05 | |
| 2.27 | 2.17 | 2.06 | 1.94 | 1.88 | 1.82 | 1.74 | 1.67 | 1.58 | 0.025 | 60 |
| 2.63 | 2.50 | 2.35 | 2.20 | 2.12 | 2.03 | 1.94 | 1.84 | 1.73 | 0.01 | |
| 2.90 | 2.74 | 2.57 | 2.39 | 2.29 | 2.19 | 2.08 | 1.96 | 1.83 | 0.005 | |
| 1.65 | 1.60 | 1.55 | 1.48 | 1.45 | 1.41 | 1.37 | 1.32 | 1.26 | 0.10 | |
| 1.91 | 1.83 | 1.75 | 1.66 | 1.61 | 1.55 | 1.50 | 1.43 | 1.35 | 0.05 | |
| 2.16 | 2.05 | 1.94 | 1.82 | 1.76 | 1.69 | 1.61 | 1.53 | 1.43 | 0.025 | 120 |
| 2.47 | 2.34 | 2.19 | 2.03 | 1.95 | 1.86 | 1.76 | 1.66 | 1.53 | 0.01 | |
| 2.71 | 2.54 | 2.37 | 2.19 | 2.09 | 1.98 | 1.87 | 1.75 | 1.61 | 0.005 | |

**TABLE VII**

Binomial probabilities:

$$\binom{n}{x} p^x (1-p)^{n-x}$$

| n | x | p 0.1 | 0.2 | 0.25 | 0.3 | 0.4 | 0.5 | 0.6 | 0.7 | 0.75 | 0.8 | 0.9 |
|---|---|------|------|------|------|------|------|------|------|------|------|------|
| 1 | 0 | 0.900 | 0.800 | 0.750 | 0.700 | 0.600 | 0.500 | 0.400 | 0.300 | 0.250 | 0.200 | 0.100 |
|   | 1 | 0.100 | 0.200 | 0.250 | 0.300 | 0.400 | 0.500 | 0.600 | 0.700 | 0.750 | 0.800 | 0.900 |
| 2 | 0 | 0.810 | 0.640 | 0.563 | 0.490 | 0.360 | 0.250 | 0.160 | 0.090 | 0.063 | 0.040 | 0.010 |
|   | 1 | 0.180 | 0.320 | 0.375 | 0.420 | 0.480 | 0.500 | 0.480 | 0.420 | 0.375 | 0.320 | 0.180 |
|   | 2 | 0.010 | 0.040 | 0.063 | 0.090 | 0.160 | 0.250 | 0.360 | 0.490 | 0.563 | 0.640 | 0.810 |
| 3 | 0 | 0.729 | 0.512 | 0.422 | 0.343 | 0.216 | 0.125 | 0.064 | 0.027 | 0.016 | 0.008 | 0.001 |
|   | 1 | 0.243 | 0.384 | 0.422 | 0.441 | 0.432 | 0.375 | 0.288 | 0.189 | 0.141 | 0.096 | 0.027 |
|   | 2 | 0.027 | 0.096 | 0.141 | 0.189 | 0.288 | 0.375 | 0.432 | 0.441 | 0.422 | 0.384 | 0.243 |
|   | 3 | 0.001 | 0.008 | 0.016 | 0.027 | 0.064 | 0.125 | 0.216 | 0.343 | 0.422 | 0.512 | 0.729 |
| 4 | 0 | 0.656 | 0.410 | 0.316 | 0.240 | 0.130 | 0.063 | 0.026 | 0.008 | 0.004 | 0.002 | 0.000 |
|   | 1 | 0.292 | 0.410 | 0.422 | 0.412 | 0.346 | 0.250 | 0.154 | 0.076 | 0.047 | 0.026 | 0.004 |
|   | 2 | 0.049 | 0.154 | 0.211 | 0.265 | 0.346 | 0.375 | 0.346 | 0.265 | 0.211 | 0.154 | 0.049 |
|   | 3 | 0.004 | 0.026 | 0.047 | 0.076 | 0.154 | 0.250 | 0.346 | 0.412 | 0.422 | 0.410 | 0.292 |
|   | 4 | 0.000 | 0.002 | 0.004 | 0.008 | 0.026 | 0.063 | 0.130 | 0.240 | 0.316 | 0.410 | 0.656 |
| 5 | 0 | 0.590 | 0.328 | 0.237 | 0.168 | 0.078 | 0.031 | 0.010 | 0.002 | 0.001 | 0.000 | 0.000 |
|   | 1 | 0.328 | 0.410 | 0.396 | 0.360 | 0.259 | 0.156 | 0.077 | 0.028 | 0.015 | 0.006 | 0.000 |
|   | 2 | 0.073 | 0.205 | 0.264 | 0.309 | 0.346 | 0.312 | 0.230 | 0.132 | 0.088 | 0.051 | 0.008 |
|   | 3 | 0.008 | 0.051 | 0.088 | 0.132 | 0.230 | 0.312 | 0.346 | 0.309 | 0.264 | 0.205 | 0.073 |
|   | 4 | 0.000 | 0.006 | 0.015 | 0.028 | 0.077 | 0.156 | 0.259 | 0.360 | 0.396 | 0.410 | 0.328 |
|   | 5 | 0.000 | 0.000 | 0.001 | 0.002 | 0.010 | 0.031 | 0.078 | 0.168 | 0.237 | 0.328 | 0.590 |
| 6 | 0 | 0.531 | 0.262 | 0.178 | 0.118 | 0.047 | 0.016 | 0.004 | 0.001 | 0.000 | 0.000 | 0.000 |
|   | 1 | 0.354 | 0.393 | 0.356 | 0.303 | 0.187 | 0.094 | 0.037 | 0.010 | 0.004 | 0.002 | 0.000 |
|   | 2 | 0.098 | 0.246 | 0.297 | 0.324 | 0.311 | 0.234 | 0.138 | 0.060 | 0.033 | 0.015 | 0.001 |
|   | 3 | 0.015 | 0.082 | 0.132 | 0.185 | 0.276 | 0.313 | 0.276 | 0.185 | 0.132 | 0.082 | 0.015 |
|   | 4 | 0.001 | 0.015 | 0.033 | 0.060 | 0.138 | 0.234 | 0.311 | 0.324 | 0.297 | 0.246 | 0.098 |
|   | 5 | 0.000 | 0.002 | 0.004 | 0.010 | 0.037 | 0.094 | 0.187 | 0.303 | 0.356 | 0.393 | 0.354 |
|   | 6 | 0.000 | 0.000 | 0.000 | 0.001 | 0.004 | 0.016 | 0.047 | 0.118 | 0.178 | 0.262 | 0.531 |
| 7 | 0 | 0.478 | 0.210 | 0.133 | 0.082 | 0.028 | 0.008 | 0.002 | 0.000 | 0.000 | 0.000 | 0.000 |
|   | 1 | 0.372 | 0.367 | 0.311 | 0.247 | 0.131 | 0.055 | 0.017 | 0.004 | 0.001 | 0.000 | 0.000 |
|   | 2 | 0.124 | 0.275 | 0.311 | 0.318 | 0.261 | 0.164 | 0.077 | 0.025 | 0.012 | 0.004 | 0.000 |
|   | 3 | 0.023 | 0.115 | 0.173 | 0.227 | 0.290 | 0.273 | 0.194 | 0.097 | 0.058 | 0.029 | 0.003 |
|   | 4 | 0.003 | 0.029 | 0.058 | 0.097 | 0.194 | 0.273 | 0.290 | 0.227 | 0.173 | 0.115 | 0.023 |
|   | 5 | 0.000 | 0.004 | 0.012 | 0.025 | 0.077 | 0.164 | 0.261 | 0.318 | 0.311 | 0.275 | 0.124 |
|   | 6 | 0.000 | 0.000 | 0.001 | 0.004 | 0.017 | 0.055 | 0.131 | 0.247 | 0.311 | 0.367 | 0.372 |
|   | 7 | 0.000 | 0.000 | 0.000 | 0.000 | 0.002 | 0.008 | 0.028 | 0.082 | 0.133 | 0.210 | 0.478 |

**TABLE VII  (cont.)**
Binomial probabilities:

$$\binom{n}{x} p^x (1-p)^{n-x}$$

| n | x | 0.1 | 0.2 | 0.25 | 0.3 | 0.4 | 0.5 | 0.6 | 0.7 | 0.75 | 0.8 | 0.9 |
|---|---|-----|-----|------|-----|-----|-----|-----|-----|------|-----|-----|
| 8 | 0 | 0.430 | 0.168 | 0.100 | 0.058 | 0.017 | 0.004 | 0.001 | 0.000 | 0.000 | 0.000 | 0.000 |
|   | 1 | 0.383 | 0.336 | 0.267 | 0.198 | 0.090 | 0.031 | 0.008 | 0.001 | 0.000 | 0.000 | 0.000 |
|   | 2 | 0.149 | 0.294 | 0.311 | 0.296 | 0.209 | 0.109 | 0.041 | 0.010 | 0.004 | 0.001 | 0.000 |
|   | 3 | 0.033 | 0.147 | 0.208 | 0.254 | 0.279 | 0.219 | 0.124 | 0.047 | 0.023 | 0.009 | 0.000 |
|   | 4 | 0.005 | 0.046 | 0.087 | 0.136 | 0.232 | 0.273 | 0.232 | 0.136 | 0.087 | 0.046 | 0.005 |
|   | 5 | 0.000 | 0.009 | 0.023 | 0.047 | 0.124 | 0.219 | 0.279 | 0.254 | 0.208 | 0.147 | 0.033 |
|   | 6 | 0.000 | 0.001 | 0.004 | 0.010 | 0.041 | 0.109 | 0.209 | 0.296 | 0.311 | 0.294 | 0.149 |
|   | 7 | 0.000 | 0.000 | 0.000 | 0.001 | 0.008 | 0.031 | 0.090 | 0.198 | 0.267 | 0.336 | 0.383 |
|   | 8 | 0.000 | 0.000 | 0.000 | 0.000 | 0.001 | 0.004 | 0.017 | 0.058 | 0.100 | 0.168 | 0.430 |
| 9 | 0 | 0.387 | 0.134 | 0.075 | 0.040 | 0.010 | 0.002 | 0.000 | 0.000 | 0.000 | 0.000 | 0.000 |
|   | 1 | 0.387 | 0.302 | 0.225 | 0.156 | 0.060 | 0.018 | 0.004 | 0.000 | 0.000 | 0.000 | 0.000 |
|   | 2 | 0.172 | 0.302 | 0.300 | 0.267 | 0.161 | 0.070 | 0.021 | 0.004 | 0.001 | 0.000 | 0.000 |
|   | 3 | 0.045 | 0.176 | 0.234 | 0.267 | 0.251 | 0.164 | 0.074 | 0.021 | 0.009 | 0.003 | 0.000 |
|   | 4 | 0.007 | 0.066 | 0.117 | 0.172 | 0.251 | 0.246 | 0.167 | 0.074 | 0.039 | 0.017 | 0.001 |
|   | 5 | 0.001 | 0.017 | 0.039 | 0.074 | 0.167 | 0.246 | 0.251 | 0.172 | 0.117 | 0.066 | 0.007 |
|   | 6 | 0.000 | 0.003 | 0.009 | 0.021 | 0.074 | 0.164 | 0.251 | 0.267 | 0.234 | 0.176 | 0.045 |
|   | 7 | 0.000 | 0.000 | 0.001 | 0.004 | 0.021 | 0.070 | 0.161 | 0.267 | 0.300 | 0.302 | 0.172 |
|   | 8 | 0.000 | 0.000 | 0.000 | 0.000 | 0.004 | 0.018 | 0.060 | 0.156 | 0.225 | 0.302 | 0.387 |
|   | 9 | 0.000 | 0.000 | 0.000 | 0.000 | 0.000 | 0.002 | 0.010 | 0.040 | 0.075 | 0.134 | 0.387 |
| 10 | 0 | 0.349 | 0.107 | 0.056 | 0.028 | 0.006 | 0.001 | 0.000 | 0.000 | 0.000 | 0.000 | 0.000 |
|   | 1 | 0.387 | 0.268 | 0.188 | 0.121 | 0.040 | 0.010 | 0.002 | 0.000 | 0.000 | 0.000 | 0.000 |
|   | 2 | 0.194 | 0.302 | 0.282 | 0.233 | 0.121 | 0.044 | 0.011 | 0.001 | 0.000 | 0.000 | 0.000 |
|   | 3 | 0.057 | 0.201 | 0.250 | 0.267 | 0.215 | 0.117 | 0.042 | 0.009 | 0.003 | 0.001 | 0.000 |
|   | 4 | 0.011 | 0.088 | 0.146 | 0.200 | 0.251 | 0.205 | 0.111 | 0.037 | 0.016 | 0.006 | 0.000 |
|   | 5 | 0.001 | 0.026 | 0.058 | 0.103 | 0.201 | 0.246 | 0.201 | 0.103 | 0.058 | 0.026 | 0.001 |
|   | 6 | 0.000 | 0.006 | 0.016 | 0.037 | 0.111 | 0.205 | 0.251 | 0.200 | 0.146 | 0.088 | 0.011 |
|   | 7 | 0.000 | 0.001 | 0.003 | 0.009 | 0.042 | 0.117 | 0.215 | 0.267 | 0.250 | 0.201 | 0.057 |
|   | 8 | 0.000 | 0.000 | 0.000 | 0.001 | 0.011 | 0.044 | 0.121 | 0.233 | 0.282 | 0.302 | 0.194 |
|   | 9 | 0.000 | 0.000 | 0.000 | 0.000 | 0.002 | 0.010 | 0.040 | 0.121 | 0.188 | 0.268 | 0.387 |
|   | 10 | 0.000 | 0.000 | 0.000 | 0.000 | 0.000 | 0.001 | 0.006 | 0.028 | 0.056 | 0.107 | 0.349 |
| 11 | 0 | 0.314 | 0.086 | 0.042 | 0.020 | 0.004 | 0.000 | 0.000 | 0.000 | 0.000 | 0.000 | 0.000 |
|   | 1 | 0.384 | 0.236 | 0.155 | 0.093 | 0.027 | 0.005 | 0.001 | 0.000 | 0.000 | 0.000 | 0.000 |
|   | 2 | 0.213 | 0.295 | 0.258 | 0.200 | 0.089 | 0.027 | 0.005 | 0.001 | 0.000 | 0.000 | 0.000 |
|   | 3 | 0.071 | 0.221 | 0.258 | 0.257 | 0.177 | 0.081 | 0.023 | 0.004 | 0.001 | 0.000 | 0.000 |
|   | 4 | 0.016 | 0.111 | 0.172 | 0.220 | 0.236 | 0.161 | 0.070 | 0.017 | 0.006 | 0.002 | 0.000 |
|   | 5 | 0.002 | 0.039 | 0.080 | 0.132 | 0.221 | 0.226 | 0.147 | 0.057 | 0.027 | 0.010 | 0.000 |
|   | 6 | 0.000 | 0.010 | 0.027 | 0.057 | 0.147 | 0.226 | 0.221 | 0.132 | 0.080 | 0.039 | 0.002 |
|   | 7 | 0.000 | 0.002 | 0.006 | 0.017 | 0.070 | 0.161 | 0.236 | 0.220 | 0.172 | 0.111 | 0.016 |
|   | 8 | 0.000 | 0.000 | 0.001 | 0.004 | 0.023 | 0.081 | 0.177 | 0.257 | 0.258 | 0.221 | 0.071 |
|   | 9 | 0.000 | 0.000 | 0.000 | 0.001 | 0.005 | 0.027 | 0.089 | 0.200 | 0.258 | 0.295 | 0.213 |
|   | 10 | 0.000 | 0.000 | 0.000 | 0.000 | 0.001 | 0.005 | 0.027 | 0.093 | 0.155 | 0.236 | 0.384 |
|   | 11 | 0.000 | 0.000 | 0.000 | 0.000 | 0.000 | 0.000 | 0.004 | 0.020 | 0.042 | 0.086 | 0.314 |

**TABLE VII   (cont.)**
Binomial probabilities:

$$\binom{n}{x}p^x(1-p)^{n-x}$$

| n | x | 0.1 | 0.2 | 0.25 | 0.3 | 0.4 | 0.5 | 0.6 | 0.7 | 0.75 | 0.8 | 0.9 |
|---|---|-----|-----|------|-----|-----|-----|-----|-----|------|-----|-----|
| 12 | 0 | 0.282 | 0.069 | 0.032 | 0.014 | 0.002 | 0.000 | 0.000 | 0.000 | 0.000 | 0.000 | 0.000 |
| | 1 | 0.377 | 0.206 | 0.127 | 0.071 | 0.017 | 0.003 | 0.000 | 0.000 | 0.000 | 0.000 | 0.000 |
| | 2 | 0.230 | 0.283 | 0.232 | 0.168 | 0.064 | 0.016 | 0.002 | 0.000 | 0.000 | 0.000 | 0.000 |
| | 3 | 0.085 | 0.236 | 0.258 | 0.240 | 0.142 | 0.054 | 0.012 | 0.001 | 0.000 | 0.000 | 0.000 |
| | 4 | 0.021 | 0.133 | 0.194 | 0.231 | 0.213 | 0.121 | 0.042 | 0.008 | 0.002 | 0.001 | 0.000 |
| | 5 | 0.004 | 0.053 | 0.103 | 0.158 | 0.227 | 0.193 | 0.101 | 0.029 | 0.011 | 0.003 | 0.000 |
| | 6 | 0.000 | 0.016 | 0.040 | 0.079 | 0.177 | 0.226 | 0.177 | 0.079 | 0.040 | 0.016 | 0.000 |
| | 7 | 0.000 | 0.003 | 0.011 | 0.029 | 0.101 | 0.193 | 0.227 | 0.158 | 0.103 | 0.053 | 0.004 |
| | 8 | 0.000 | 0.001 | 0.002 | 0.008 | 0.042 | 0.121 | 0.213 | 0.231 | 0.194 | 0.133 | 0.021 |
| | 9 | 0.000 | 0.000 | 0.000 | 0.001 | 0.012 | 0.054 | 0.142 | 0.240 | 0.258 | 0.236 | 0.085 |
| | 10 | 0.000 | 0.000 | 0.000 | 0.000 | 0.002 | 0.016 | 0.064 | 0.168 | 0.232 | 0.283 | 0.230 |
| | 11 | 0.000 | 0.000 | 0.000 | 0.000 | 0.000 | 0.003 | 0.017 | 0.071 | 0.127 | 0.206 | 0.377 |
| | 12 | 0.000 | 0.000 | 0.000 | 0.000 | 0.000 | 0.000 | 0.002 | 0.014 | 0.032 | 0.069 | 0.282 |
| 13 | 0 | 0.254 | 0.055 | 0.024 | 0.010 | 0.001 | 0.000 | 0.000 | 0.000 | 0.000 | 0.000 | 0.000 |
| | 1 | 0.367 | 0.179 | 0.103 | 0.054 | 0.011 | 0.002 | 0.000 | 0.000 | 0.000 | 0.000 | 0.000 |
| | 2 | 0.245 | 0.268 | 0.206 | 0.139 | 0.045 | 0.010 | 0.001 | 0.000 | 0.000 | 0.000 | 0.000 |
| | 3 | 0.100 | 0.246 | 0.252 | 0.218 | 0.111 | 0.035 | 0.006 | 0.001 | 0.000 | 0.000 | 0.000 |
| | 4 | 0.028 | 0.154 | 0.210 | 0.234 | 0.184 | 0.087 | 0.024 | 0.003 | 0.001 | 0.000 | 0.000 |
| | 5 | 0.006 | 0.069 | 0.126 | 0.180 | 0.221 | 0.157 | 0.066 | 0.014 | 0.005 | 0.001 | 0.000 |
| | 6 | 0.001 | 0.023 | 0.056 | 0.103 | 0.197 | 0.209 | 0.131 | 0.044 | 0.019 | 0.006 | 0.000 |
| | 7 | 0.000 | 0.006 | 0.019 | 0.044 | 0.131 | 0.209 | 0.197 | 0.103 | 0.056 | 0.023 | 0.001 |
| | 8 | 0.000 | 0.001 | 0.005 | 0.014 | 0.066 | 0.157 | 0.221 | 0.180 | 0.126 | 0.069 | 0.006 |
| | 9 | 0.000 | 0.000 | 0.001 | 0.003 | 0.024 | 0.087 | 0.184 | 0.234 | 0.210 | 0.154 | 0.028 |
| | 10 | 0.000 | 0.000 | 0.000 | 0.001 | 0.006 | 0.035 | 0.111 | 0.218 | 0.252 | 0.246 | 0.100 |
| | 11 | 0.000 | 0.000 | 0.000 | 0.000 | 0.001 | 0.010 | 0.045 | 0.139 | 0.206 | 0.268 | 0.245 |
| | 12 | 0.000 | 0.000 | 0.000 | 0.000 | 0.000 | 0.002 | 0.011 | 0.054 | 0.103 | 0.179 | 0.367 |
| | 13 | 0.000 | 0.000 | 0.000 | 0.000 | 0.000 | 0.000 | 0.001 | 0.010 | 0.024 | 0.055 | 0.254 |
| 14 | 0 | 0.229 | 0.044 | 0.018 | 0.007 | 0.001 | 0.000 | 0.000 | 0.000 | 0.000 | 0.000 | 0.000 |
| | 1 | 0.356 | 0.154 | 0.083 | 0.041 | 0.007 | 0.001 | 0.000 | 0.000 | 0.000 | 0.000 | 0.000 |
| | 2 | 0.257 | 0.250 | 0.180 | 0.113 | 0.032 | 0.006 | 0.001 | 0.000 | 0.000 | 0.000 | 0.000 |
| | 3 | 0.114 | 0.250 | 0.240 | 0.194 | 0.085 | 0.022 | 0.003 | 0.000 | 0.000 | 0.000 | 0.000 |
| | 4 | 0.035 | 0.172 | 0.220 | 0.229 | 0.155 | 0.061 | 0.014 | 0.001 | 0.000 | 0.000 | 0.000 |
| | 5 | 0.008 | 0.086 | 0.147 | 0.196 | 0.207 | 0.122 | 0.041 | 0.007 | 0.002 | 0.000 | 0.000 |
| | 6 | 0.001 | 0.032 | 0.073 | 0.126 | 0.207 | 0.183 | 0.092 | 0.023 | 0.008 | 0.002 | 0.000 |
| | 7 | 0.000 | 0.009 | 0.028 | 0.062 | 0.157 | 0.209 | 0.157 | 0.062 | 0.028 | 0.009 | 0.000 |
| | 8 | 0.000 | 0.002 | 0.008 | 0.023 | 0.092 | 0.183 | 0.207 | 0.126 | 0.073 | 0.032 | 0.001 |
| | 9 | 0.000 | 0.000 | 0.002 | 0.007 | 0.041 | 0.122 | 0.207 | 0.196 | 0.147 | 0.086 | 0.008 |
| | 10 | 0.000 | 0.000 | 0.000 | 0.001 | 0.014 | 0.061 | 0.155 | 0.229 | 0.220 | 0.172 | 0.035 |
| | 11 | 0.000 | 0.000 | 0.000 | 0.000 | 0.003 | 0.022 | 0.085 | 0.194 | 0.240 | 0.250 | 0.114 |
| | 12 | 0.000 | 0.000 | 0.000 | 0.000 | 0.001 | 0.006 | 0.032 | 0.113 | 0.180 | 0.250 | 0.257 |
| | 13 | 0.000 | 0.000 | 0.000 | 0.000 | 0.000 | 0.001 | 0.007 | 0.041 | 0.083 | 0.154 | 0.356 |
| | 14 | 0.000 | 0.000 | 0.000 | 0.000 | 0.000 | 0.000 | 0.001 | 0.007 | 0.018 | 0.044 | 0.229 |

**TABLE VII   (cont.)**
Binomial probabilities:

$$\binom{n}{x} p^x (1-p)^{n-x}$$

| n | x | 0.1 | 0.2 | 0.25 | 0.3 | 0.4 | 0.5 | 0.6 | 0.7 | 0.75 | 0.8 | 0.9 |
|---|---|-----|-----|------|-----|-----|-----|-----|-----|------|-----|-----|
| 15 | 0 | 0.206 | 0.035 | 0.013 | 0.005 | 0.000 | 0.000 | 0.000 | 0.000 | 0.000 | 0.000 | 0.000 |
| | 1 | 0.343 | 0.132 | 0.067 | 0.031 | 0.005 | 0.000 | 0.000 | 0.000 | 0.000 | 0.000 | 0.000 |
| | 2 | 0.267 | 0.231 | 0.156 | 0.092 | 0.022 | 0.003 | 0.000 | 0.000 | 0.000 | 0.000 | 0.000 |
| | 3 | 0.129 | 0.250 | 0.225 | 0.170 | 0.063 | 0.014 | 0.002 | 0.000 | 0.000 | 0.000 | 0.000 |
| | 4 | 0.043 | 0.188 | 0.225 | 0.219 | 0.127 | 0.042 | 0.007 | 0.001 | 0.000 | 0.000 | 0.000 |
| | 5 | 0.010 | 0.103 | 0.165 | 0.206 | 0.186 | 0.092 | 0.024 | 0.003 | 0.001 | 0.000 | 0.000 |
| | 6 | 0.002 | 0.043 | 0.092 | 0.147 | 0.207 | 0.153 | 0.061 | 0.012 | 0.003 | 0.001 | 0.000 |
| | 7 | 0.000 | 0.014 | 0.039 | 0.081 | 0.177 | 0.196 | 0.118 | 0.035 | 0.013 | 0.003 | 0.000 |
| | 8 | 0.000 | 0.003 | 0.013 | 0.035 | 0.118 | 0.196 | 0.177 | 0.081 | 0.039 | 0.014 | 0.000 |
| | 9 | 0.000 | 0.001 | 0.003 | 0.012 | 0.061 | 0.153 | 0.207 | 0.147 | 0.092 | 0.043 | 0.002 |
| | 10 | 0.000 | 0.000 | 0.001 | 0.003 | 0.024 | 0.092 | 0.186 | 0.206 | 0.165 | 0.103 | 0.010 |
| | 11 | 0.000 | 0.000 | 0.000 | 0.001 | 0.007 | 0.042 | 0.127 | 0.219 | 0.225 | 0.188 | 0.043 |
| | 12 | 0.000 | 0.000 | 0.000 | 0.000 | 0.002 | 0.014 | 0.063 | 0.170 | 0.225 | 0.250 | 0.129 |
| | 13 | 0.000 | 0.000 | 0.000 | 0.000 | 0.000 | 0.003 | 0.022 | 0.092 | 0.156 | 0.231 | 0.267 |
| | 14 | 0.000 | 0.000 | 0.000 | 0.000 | 0.000 | 0.000 | 0.005 | 0.031 | 0.067 | 0.132 | 0.343 |
| | 15 | 0.000 | 0.000 | 0.000 | 0.000 | 0.000 | 0.000 | 0.000 | 0.005 | 0.013 | 0.035 | 0.206 |
| 20 | 0 | 0.122 | 0.012 | 0.003 | 0.001 | 0.000 | 0.000 | 0.000 | 0.000 | 0.000 | 0.000 | 0.000 |
| | 1 | 0.270 | 0.058 | 0.021 | 0.007 | 0.000 | 0.000 | 0.000 | 0.000 | 0.000 | 0.000 | 0.000 |
| | 2 | 0.285 | 0.137 | 0.067 | 0.028 | 0.003 | 0.000 | 0.000 | 0.000 | 0.000 | 0.000 | 0.000 |
| | 3 | 0.190 | 0.205 | 0.134 | 0.072 | 0.012 | 0.001 | 0.000 | 0.000 | 0.000 | 0.000 | 0.000 |
| | 4 | 0.090 | 0.218 | 0.190 | 0.130 | 0.035 | 0.005 | 0.000 | 0.000 | 0.000 | 0.000 | 0.000 |
| | 5 | 0.032 | 0.175 | 0.202 | 0.179 | 0.075 | 0.015 | 0.001 | 0.000 | 0.000 | 0.000 | 0.000 |
| | 6 | 0.009 | 0.109 | 0.169 | 0.192 | 0.124 | 0.037 | 0.005 | 0.000 | 0.000 | 0.000 | 0.000 |
| | 7 | 0.002 | 0.055 | 0.112 | 0.164 | 0.166 | 0.074 | 0.015 | 0.001 | 0.000 | 0.000 | 0.000 |
| | 8 | 0.000 | 0.022 | 0.061 | 0.114 | 0.180 | 0.120 | 0.035 | 0.004 | 0.001 | 0.000 | 0.000 |
| | 9 | 0.000 | 0.007 | 0.027 | 0.065 | 0.160 | 0.160 | 0.071 | 0.012 | 0.003 | 0.000 | 0.000 |
| | 10 | 0.000 | 0.002 | 0.010 | 0.031 | 0.117 | 0.176 | 0.117 | 0.031 | 0.010 | 0.002 | 0.000 |
| | 11 | 0.000 | 0.000 | 0.003 | 0.012 | 0.071 | 0.160 | 0.160 | 0.065 | 0.027 | 0.007 | 0.000 |
| | 12 | 0.000 | 0.000 | 0.001 | 0.004 | 0.035 | 0.120 | 0.180 | 0.114 | 0.061 | 0.022 | 0.000 |
| | 13 | 0.000 | 0.000 | 0.000 | 0.001 | 0.015 | 0.074 | 0.166 | 0.164 | 0.112 | 0.055 | 0.002 |
| | 14 | 0.000 | 0.000 | 0.000 | 0.000 | 0.005 | 0.037 | 0.124 | 0.192 | 0.169 | 0.109 | 0.009 |
| | 15 | 0.000 | 0.000 | 0.000 | 0.000 | 0.001 | 0.015 | 0.075 | 0.179 | 0.202 | 0.175 | 0.032 |
| | 16 | 0.000 | 0.000 | 0.000 | 0.000 | 0.000 | 0.005 | 0.035 | 0.130 | 0.190 | 0.218 | 0.090 |
| | 17 | 0.000 | 0.000 | 0.000 | 0.000 | 0.000 | 0.001 | 0.012 | 0.072 | 0.134 | 0.205 | 0.190 |
| | 18 | 0.000 | 0.000 | 0.000 | 0.000 | 0.000 | 0.000 | 0.003 | 0.028 | 0.067 | 0.137 | 0.285 |
| | 19 | 0.000 | 0.000 | 0.000 | 0.000 | 0.000 | 0.000 | 0.000 | 0.007 | 0.021 | 0.058 | 0.270 |
| | 20 | 0.000 | 0.000 | 0.000 | 0.000 | 0.000 | 0.000 | 0.000 | 0.001 | 0.003 | 0.012 | 0.122 |

# B

# Answers to Selected Exercises

# Chapter 1

## EXERCISES 1.1

**1.1** See Definition 1.2 on page 5.

**1.3** Descriptive statistics includes the construction of graphs, charts, and tables, and the calculation of various descriptive measures such as averages, measures of variation, and percentiles.

**1.5**
a. In an observational study, researchers simply observe characteristics and take measurements, as in a sample survey.
b. In a designed experiment, researchers impose treatments and controls and then observe characteristics and take measurements.

**1.7** Inferential

**1.9** Descriptive

**1.11** Descriptive

**1.13**
a. Inferential
b. The sample consists of those U.S. adults who were interviewed; the population consists of all U.S. adults.

**1.15**
a. Descriptive          b. Inferential

**1.17** Designed experiment

**1.19** Observational study

**1.21** Designed experiment

## EXERCISES 1.2

**1.27** Conducting a census may be time consuming, costly, impractical, or even impossible.

**1.29** Because the sample will be used to draw conclusions about the entire population.

**1.31** Dentists form a high-income group whose incomes are not representative of the incomes of Seattle residents in general.

**1.33**
a. In probability sampling, a random device—such as tossing a coin, consulting a table of random numbers, or employing a random-number generator—is used to decide which members of the population will constitute the sample instead of leaving such decisions to human judgment.
b. No. Because probability sampling uses a random device, it is possible to obtain a nonrepresentative sample.
c. Probability sampling eliminates unintentional selection bias and permits the researcher to control the chance of obtaining a nonrepresentative sample. Also, use of probability sampling guarantees that the techniques of inferential statistics can be applied.

**1.35** Simple random sampling

**1.37**
a.

| | | | | |
|---|---|---|---|---|
| G, L, S | G, L, A | G, L, T | G, S, A | G, S, T |
| G, A, T | L, S, A | L, S, T | L, A, T | S, A, T |

b. $\frac{1}{10}, \frac{1}{10}, \frac{1}{10}$

**1.39**
a.

| | | | | |
|---|---|---|---|---|
| E, M, P, L | E, M, P, A | E, M, P, B | E, M, L, A | E, M, L, B |
| E, M, A, B | E, P, L, A | E, P, L, B | E, P, A, B | E, L, A, B |
| M, P, L, A | M, P, L, B | M, P, A, B | M, L, A, B | P, L, A, B |

b. Write the initials of the six artists on separate pieces of paper, place the six slips of paper in a box, and then, while blindfolded, pick four of the slips of paper.
c. $\frac{1}{15}, \frac{1}{15}$

**1.41**
a.

| | | | | |
|---|---|---|---|---|
| C, W, H | C, W, V | C, W, A | C, H, V | C, H, A |
| C, V, A | W, H, V | W, H, A | W, V, A | H, V, A |

b. $\frac{1}{10}, \frac{1}{10}$

**1.43**
a. Answers will vary.          b. Answers will vary.

## EXERCISES 1.3

**1.49**
a. Answers will vary.
b. Systematic random sampling
c. Answers will vary.

**1.51**
a. Number the suites from 1 to 48, use a table of random numbers to randomly select 3 of the 48 suites, and take as the sample the 24 dormitory residents living in the 3 suites obtained.
b. Probably not, because friends often have similar opinions.
c. Proportional allocation dictates that the number of freshmen, sophomores, juniors, and seniors selected be 8, 7, 6, and 3, respectively. Thus a stratified sample of 24 dormitory residents can be obtained as follows: Number the freshman dormitory residents from 1 through 128 and use a table of random numbers to randomly select 8 of the 128 freshman dormitory residents; number the sophomore dormitory residents from 1 through 112 and use a table of random numbers to randomly select 7 of the 112 sophomore dormitory residents; and so on.

**1.53** Stratified sampling

## EXERCISES 1.4

**1.59**
a. The individuals or items on which the experiment is performed
b. Subject

**1.61**
a. Three
b. The pharmacologic therapy alone group

**c.** Two.   Pharmacologic therapy with a pacemaker; pharmacologic therapy with a pacemaker–defibrillator combination

**d.** 304 in the pharmacologic therapy alone group; 608 each in the pharmacologic therapy with a pacemaker group and the pharmacologic therapy with a pacemaker–defibrillator combination group

**1.63**

**a.** Batches of the product being sold (Some might say that the stores are the experimental units.)

**b.** Unit sales of the product

**c.** Display type and pricing scheme

**d.** Display type has three levels: normal display space interior to an aisle, normal display space at the end of an aisle, and enlarged display space.  Pricing scheme has three levels: regular price, reduced price, and cost.

**e.** Each treatment is a combination of a level of display type and a level of pricing scheme.

**1.65**

**a.** The female lions

**b.** Whether or not (yes or no) the female lions approached a male dummy

**c.** Mane length and mane color

**d.** Mane length has two levels: long and short.  Mane color has two levels: blonde and dark.

**e.** The four different possible combinations of the two mane lengths and the two mane colors

## REVIEW PROBLEMS FOR CHAPTER 1

1. Answers will vary.

2. It is almost always necessary to invoke techniques of descriptive statistics to organize and summarize the information obtained from a sample before carrying out an inferential analysis.

3. Descriptive

4. Descriptive

5. Inferential

6. **a.** The statement regarding 18% of youths abusing Vicodin is a descriptive statement, indicating the percentage of youths in the sample that had abused Vicodin.

   **b.** The statement regarding 4.3 million youths abusing Vicodin is an inferential statement, extending the sample percentage of abusers to the entire teen population.

7. **a.** In an observational study, researchers simply observe characteristics and take measurements, as in a sample survey.  In a designed experiment, researchers impose treatments and controls and then observe characteristics and take measurements.

   **b.** Observational studies can reveal only association, whereas designed experiments can help establish causation.

8. Observational study

9. Designed experiment

10. A literature search

11. **a.** A representative sample is a sample that reflects as closely as possible the relevant characteristics of the population under consideration.

    **b.** In probability sampling, a random device, such as tossing a coin or consulting a table of random numbers, is used to decide which members of the population will constitute the sample instead of leaving such decisions to human judgement.

    **c.** Simple random sampling is a sampling procedure for which each possible sample of a given size is equally likely to be the one obtained from the population.

12. No, because parents of students at Yale tend to have higher incomes than parents of college students in general.

13. Only (b)

14. **a.**

| HA, 3C, G4 | HA, 3C, ZV | HA, 3C, F9 |
| HA, G4, ZV | HA, G4, F9 | HA, ZV, F9 |
| 3C, G4, ZV | 3C, G4, F9 | 3C, ZV, F9 |
| G4, ZV, F9 | | |

**b.** $\frac{1}{10}$, $\frac{1}{10}$, $\frac{1}{10}$

**c.** Answers will vary.     **d.** Answers will vary.

15. **a.** Number the athletes from 1 to 100, use Table I to obtain 15 different numbers between 1 and 100, and take as the sample the 15 athletes who are numbered with the numbers obtained.

    **b.** 082, 008, 016, 001, 047, 094, 097, 074, 052, 076, 098, 003, 089, 041, 063

    **c.** Answers will vary.

16. See Section 1.3 and, in particular, (a) Procedure 1.1 on page 19, (b) Procedure 1.2 on page 21, and (c) Procedure 1.3 on page 22.

17. **a.** Answers will vary.

    **b.** Yes, unless for some reason there is a cyclical pattern in the listing of the athletes.

18. **a.** Proportional allocation dictates that 10 full professors, 16 associate professors, 12 assistant professors, and 2 instructors be selected.

    **b.** The procedure is as follows: Number the full professors from 1 to 205, and use Table I to randomly select 10 of the 205 full professors; number the associate professors from 1 to 328, and use Table I to randomly select 16 of the 328 associate professors; and so on.

19. The statement under the vote tally is a disclaimer as to the validity of the survey.  Because the results reflect only responses of Internet users, they cannot be regarded as representative of the public in general.  Moreover, because the sample was not chosen at random from Internet users—but rather was obtained only from volunteers— the results cannot even be considered representative of Internet users.

20. **a.** Designed experiment

    **b.** The treatment group consists of the 158 patients who took AVONEX. The control group consists of the 143 patients who were given placebo. The treatments are AVONEX and placebo.

**21.** See Key Fact 1.1 on page 25.

**22. a.** The tomato plants in the study (Some might say the plots of land are the experimental units.)
    **b.** Yield of tomato plants
    **c.** Tomato variety and planting density
    **d.** Different tomato varieties and different planting densities
    **e.** Each treatment is a combination of a level of tomato variety and a level of planting density.

**23. a.** The children on the panel
    **b.** Whether the bottle is opened or not (yes or no)
    **c.** Design type
    **d.** The three design types
    **e.** The three design types

**24.** Completely randomized design

**25. a.** Completely randomized design
    **b.** Randomized block design; the six different car models
    **c.** The randomized block design in part (b)

**26.** Answers will vary.

**27.** Answers will vary.

# Chapter 2

## EXERCISES 2.1

**2.1** Answers will vary.

**2.3** See Definition 2.2 on page 41.

**2.5** Qualitative variable

**2.7**
**a.** Quantitative, discrete; rank of city by highest temperature
**b.** Quantitative, continuous; highest temperature, in degrees Fahrenheit
**c.** Qualitative; state in which a U.S. city is located

**2.9**
**a.** Quantitative, discrete; rank of wired local market by percentage that uses broadband
**b.** Qualitative; name of wired local market
**c.** Quantitative, discrete; percentage of wired local market that uses broadband

**2.11** Rank is quantitative, discrete data; season to date is quantitative, discrete data; show title is qualitative data; network name is qualitative data; number of viewers is quantitative, discrete data.

**2.13** Brand of smartphone is qualitative data; size is quantitative, continuous data; battery life is quantitative, continuous data; voice recognition (yes or no) is qualitative data; weight is quantitative, continuous data.

## EXERCISES 2.2

**2.15** Grouping can help to make a large and complicated set of data more compact and easier to understand.

**2.17** See 1–3 on pages 45 and 46.

**2.19**
**a.** True                 **b.** False
**c.** Relative frequencies always lie between 0 and 1 and hence provide a standard for comparison.

**2.23** The midpoint of each class is the same as the class.

**2.25**

| Speed (mph) | Frequency | Relative frequency | Midpoint |
|---|---|---|---|
| 52 ◁ 54 | 2 | 0.057 | 53 |
| 54 ◁ 56 | 5 | 0.143 | 55 |
| 56 ◁ 58 | 6 | 0.171 | 57 |
| 58 ◁ 60 | 8 | 0.229 | 59 |
| 60 ◁ 62 | 7 | 0.200 | 61 |
| 62 ◁ 64 | 3 | 0.086 | 63 |
| 64 ◁ 66 | 2 | 0.057 | 65 |
| 66 ◁ 68 | 1 | 0.029 | 67 |
| 68 ◁ 70 | 0 | 0.000 | 69 |
| 70 ◁ 72 | 0 | 0.000 | 71 |
| 72 ◁ 74 | 0 | 0.000 | 73 |
| 74 ◁ 76 | 1 | 0.029 | 75 |
| | 35 | 1.001 | |

**2.27**

| Speed (mph) | Frequency | Relative frequency | Mark |
|---|---|---|---|
| 52–53.9 | 2 | 0.057 | 52.95 |
| 54–55.9 | 5 | 0.143 | 54.95 |
| 56–57.9 | 6 | 0.171 | 56.95 |
| 58–59.9 | 8 | 0.229 | 58.95 |
| 60–61.9 | 7 | 0.200 | 60.95 |
| 62–63.9 | 3 | 0.086 | 62.95 |
| 64–65.9 | 2 | 0.057 | 64.95 |
| 66–67.9 | 1 | 0.029 | 66.95 |
| 68–69.9 | 0 | 0.000 | 68.95 |
| 70–71.9 | 0 | 0.000 | 70.95 |
| 72–73.9 | 0 | 0.000 | 72.95 |
| 74–75.9 | 1 | 0.029 | 74.95 |
| | 35 | 1.001 | |

**2.29**

| Audience (millions) | Frequency | Relative frequency | Midpoint |
|---|---|---|---|
| 11 ◁ 14 | 5 | 0.25 | 12.5 |
| 14 ◁ 17 | 4 | 0.20 | 15.5 |
| 17 ◁ 20 | 6 | 0.30 | 18.5 |
| 20 ◁ 23 | 2 | 0.10 | 21.5 |
| 23 ◁ 26 | 2 | 0.10 | 24.5 |
| 26 ◁ 29 | 1 | 0.05 | 27.5 |
| | 20 | 1.00 | |

**2.31**

| Siblings | Frequency | Relative frequency |
|---|---|---|
| 0 | 8 | 0.200 |
| 1 | 17 | 0.425 |
| 2 | 11 | 0.275 |
| 3 | 3 | 0.075 |
| 4 | 1 | 0.025 |
|  | 40 | 1.000 |

**2.33**

| Champion | Frequency | Relative frequency |
|---|---|---|
| Arizona St. | 1 | 0.04 |
| Iowa | 15 | 0.60 |
| Iowa St. | 1 | 0.04 |
| Minnesota | 2 | 0.08 |
| Oklahoma St. | 6 | 0.24 |
|  | 25 | 1.00 |

## EXERCISES 2.3

**2.43** A frequency histogram displays the class frequencies on the vertical axis, whereas a relative-frequency histogram displays the class relative frequencies on the vertical axis.

**2.45** Answers will vary.

**2.47** Histogram. Stem-and-leaf diagrams are generally not useful with large data sets.

**2.49** To avoid confusing bar graphs and histograms

**2.51**
a. The heights of the bars in the histogram would be comparable to the heights of the columns of dots in the dotplot.
b. No. Each bar in the histogram would represent a range of possible values, whereas, in the dotplot, each column of dots represents one possible value.

**2.53**
a.

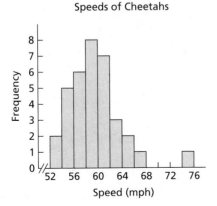
Speeds of Cheetahs

b.

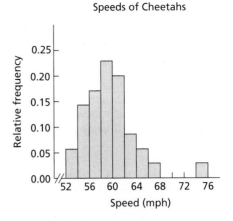
Speeds of Cheetahs

**2.55**
a.

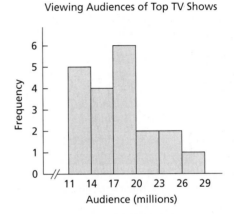
Viewing Audiences of Top TV Shows

b.

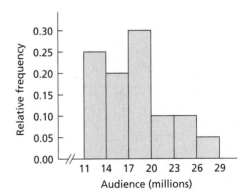
Viewing Audiences of Top TV Shows

**2.57**

**a.**

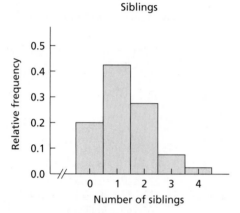

Siblings

**b.**

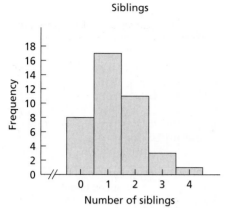

Siblings

**2.59**

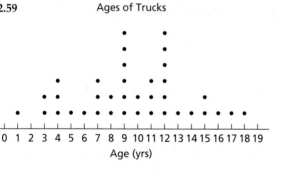

Ages of Trucks

**2.61**

**a.**

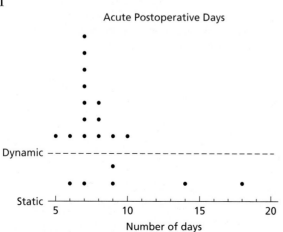

Acute Postoperative Days

**b.** For these data, the number of acute postoperative days is, on average, less with the dynamic system than with the static system. Also, more variation exists in the number of acute postoperative days with the static system than with the dynamic system.

**2.63**

| 1 | 2 3 8 |
|---|---|
| 2 | 1 6 7 8 8 9 9 |
| 3 | 3 4 4 5 9 |
| 4 | 0 4 |

**2.65**

**a.**

| 26 | 6 |
|---|---|
| 27 | 7 9 9 9 |
| 28 | 1 4 5 6 7 8 9 |
| 29 | 1 3 3 4 4 7 8 8 9 |
| 30 | 3 5 7 9 |
| 31 | 0 0 8 |
| 32 | 7 |
| 33 | |
| 34 | 0 |

**b.**

| 26 | 6 |
|---|---|
| 27 | |
| 27 | 7 9 9 9 |
| 28 | 1 4 |
| 28 | 5 6 7 8 9 |
| 29 | 1 3 3 4 4 |
| 29 | 7 8 8 9 |
| 30 | 3 |
| 30 | 5 7 9 |
| 31 | 0 0 |
| 31 | 8 |
| 32 | |
| 32 | 7 |
| 33 | |
| 33 | |
| 34 | 0 |

**c.** Answers will vary.

**2.67**

**a.**

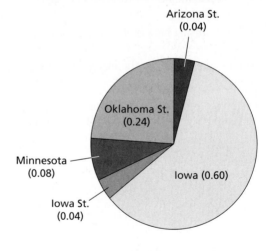

NCAA Wrestling Champs

**b.**

NCAA Wrestling Champs

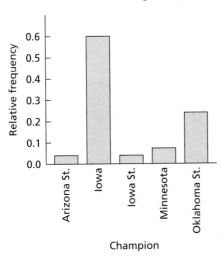

**2.69**

**a.**

M&M Colors

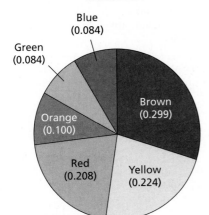

**b.**

M&M Colors

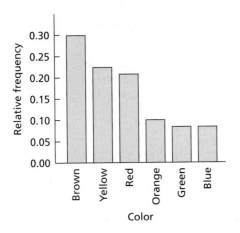

**2.71**
**a.** 20%          **b.** 25%          **c.** 7

## EXERCISES 2.4

**2.85**
**a.** The *distribution of a data set* is a table, graph, or formula that provides the values of the observations and how often they occur.
**b.** *Sample data* are the values of a variable for a sample of the population.
**c.** *Population data* are the values of a variable for an entire population.
**d.** *Census data* is another name for population data.
**e.** A *sample distribution* is the distribution of sample data.
**f.** The *population distribution* is the distribution of population data.
**g.** *Distribution of a variable* is another name for population distribution.

**2.87**  Roughly a bell shape

**2.89**  Answers will vary.

**2.91**
**a.** Right skewed (Bell shaped is also an acceptable answer.)
**b.** Right skewed (Symmetric is also an acceptable answer.)

**2.93**
**a.** Left skewed          **b.** Left skewed

**2.95**
**a.** Bell shaped          **b.** Symmetric

**2.97**
**a.** Left skewed          **b.** Left skewed

**2.99**
**a.** Right skewed          **b.** Right skewed

**2.101**
**a.** Right skewed; reverse J shaped
**b.** Right skewed; right skewed

## EXERCISES 2.5

**2.113**
**a.** Part of the vertical axis of the graph has been cut off, or truncated.
**b.** It may allow relevant information to be conveyed more easily.
**c.** Start the axis at 0 and put slashes in the axis to indicate that part of the axis is missing.

**2.115**
**c.** They give the misleading impression that the district average is much greater relative to the national average than it actually is.

**2.117**

**a.** It is a truncated graph.

**b.**

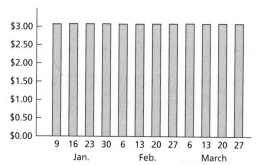

Money Supply
(weekly average of M2 in trillions)

**c.**

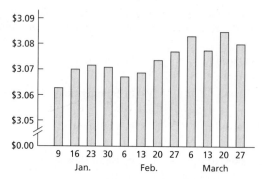

Money Supply
(weekly average of M2 in trillions)

**2.119**

**b.** That the price has dropped by roughly 70%.

**c.** 9.93%

**d.** Because it is a truncated graph.

**e.** Start the graph at 0 instead of 60, or use some method (such as slashes) to warn the reader that the vertical scale has been modified.

## REVIEW PROBLEMS FOR CHAPTER 2

1. **a.** A variable is a characteristic that varies from one person or thing to another.
   **b.** Quantitative variables and qualitative (or categorical) variables
   **c.** Discrete variables and continuous variables
   **d.** Values of a variable
   **e.** By the type of variable

2. It helps organize the data, making them much simpler to comprehend.

3. Qualitative data. It makes no sense to look for cutpoints or midpoints for nonnumerical data (or for numerical data obtained by coding nonnumerical data).

4. **a.** 6 and 14        **b.** 18
   **c.** 22 and 30       **d.** The third class

5. **a.** 10        **b.** 20        **c.** 25 and 35

6. When grouping discrete data in which there are only relatively few distinct observations

7. **a.** The bar for each class extends from the lower cutpoint to the upper cutpoint of the class.
   **b.** The bar for each class is centered over the midpoint of the class.

8. Pie charts and bar graphs

9. Histogram. Stem-and-leaf diagrams are generally not useful with large data sets.

10. See Fig. 2.9 on page 73.

11. Answers will vary.

12. **a.** Left skewed. The distribution of a random sample taken from a population approximates the population distribution. The larger the sample, the better the approximation tends to be.
    **b.** No. Sample distributions vary from sample to sample.
    **c.** Yes. Left skewed. The overall shapes of the two sample distributions should be similar to that of the population distribution and hence to each other.

13. **a.** Discrete quantitative
    **b.** Continuous quantitative
    **c.** Qualitative

14. **a.**

| Age at inauguration | Frequency | Relative frequency | Mark |
|---|---|---|---|
| 40–44 | 2 | 0.047 | 42 |
| 45–49 | 6 | 0.140 | 47 |
| 50–54 | 13 | 0.302 | 52 |
| 55–59 | 12 | 0.279 | 57 |
| 60–64 | 7 | 0.163 | 62 |
| 65–69 | 3 | 0.070 | 67 |
|  | 43 | 1.001 | |

**b.** 40 and 45        **c.** 5

**d.**

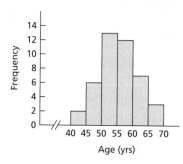

Ages at Inauguration
for First 43 U.S. Presidents

**e.** Bell shaped        **f.** Symmetric

**15.**

Ages at Inauguration
for First 43 U.S. Presidents

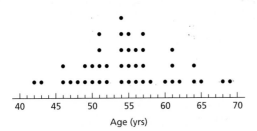

Age (yrs)

**e.**

Busy Bank Tellers

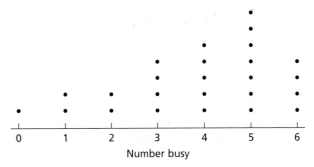

Number busy

**f.** They have identical shapes.

**16. a. 4** | 2 3 6 6 7 8 9 9
**5** | 0 0 1 1 1 1 2 2 4 4 4 4 4 5 5 5 5 5 6 6 6 7 7 7 7 8
**6** | 0 1 1 1 2 4 4 5 8 9

**b. 4** | 2 3
**4** | 6 6 7 8 9 9
**5** | 0 0 1 1 1 1 2 2 4 4 4 4 4
**5** | 5 5 5 5 6 6 6 7 7 7 7 8
**6** | 0 1 1 1 2 4 4
**6** | 5 8 9

**c.** The one in part (b)

**18. a.**

Oldest Players

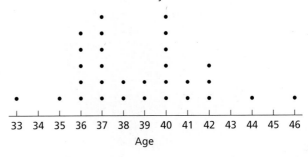

Age

**b.** Bimodal or multimodal
**c.** Symmetric

**17. a.**

| Number busy | Frequency | Relative frequency |
|---|---|---|
| 0 | 1 | 0.04 |
| 1 | 2 | 0.08 |
| 2 | 2 | 0.08 |
| 3 | 4 | 0.16 |
| 4 | 5 | 0.20 |
| 5 | 7 | 0.28 |
| 6 | 4 | 0.16 |
| | 25 | 1.00 |

**19. a.**

Buybacks

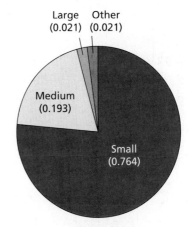

Large (0.021)  Other (0.021)

Medium (0.193)

Small (0.764)

**b.**

Busy Tellers

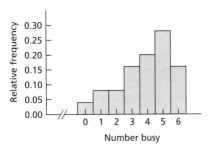

Number busy

**c.** Left skewed  **d.** Left skewed

**b.**

Homicides

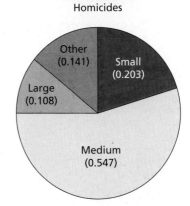

Other (0.141)  Small (0.203)

Large (0.108)

Medium (0.547)

**20. a.**

| Class level | Frequency | Relative frequency |
|---|---|---|
| Freshman | 6 | 0.150 |
| Sophomore | 15 | 0.375 |
| Junior | 12 | 0.300 |
| Senior | 7 | 0.175 |
| | 40 | 1.000 |

**b.**

Class Levels for Statistics Students

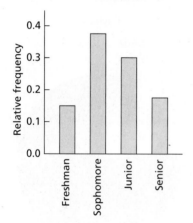

**c.**

Class Levels for Statistics Students

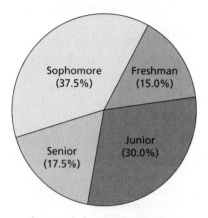

**21. a.**

| High | Freq. | Relative frequency | Midpoint |
|---|---|---|---|
| 0 ≮ 1000 | 8 | 0.222 | 500 |
| 1000 ≮ 2000 | 10 | 0.278 | 1500 |
| 2000 ≮ 3000 | 4 | 0.111 | 2500 |
| 3000 ≮ 4000 | 4 | 0.111 | 3500 |
| 4000 ≮ 5000 | 0 | 0.000 | 4500 |
| 5000 ≮ 6000 | 1 | 0.028 | 5500 |
| 6000 ≮ 7000 | 1 | 0.028 | 6500 |
| 7000 ≮ 8000 | 0 | 0.000 | 7500 |
| 8000 ≮ 9000 | 1 | 0.028 | 8500 |
| 9000 ≮ 10000 | 1 | 0.028 | 9500 |
| 10000 ≮ 11000 | 3 | 0.083 | 10500 |
| 11000 ≮ 12000 | 3 | 0.083 | 11500 |
| | 36 | 1.000 | |

**b.**

Dow Jones Highs, 1969 – 2004*

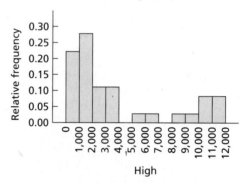

* Data from the *World Almanac*

**22.** Answers will vary, but here is one possibility:

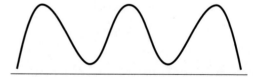

**23. a.** To warn the reader that part of it has been removed
   **b.** To enable the reader to see differences among the amounts of $CO_2$ that can be kept in different geological spaces without causing misinterpretation

**24. b.** Having followed the directions in part (a), you might conclude that the percentage of women in the labor force for 2000 is about 3.5 times that for 1960.
   **c.** Not covering up the vertical axis, you would find that the percentage of women in the labor force for 2000 is about 1.8 times that for 1960.
   **d.** The graph is potentially misleading because it is truncated. Note that the vertical axis begins at 30 rather than at 0.
   **e.** To make the graph less potentially misleading, start it at 0 instead of at 30.

# Chapter 3

## EXERCISES 3.1

**3.1** To indicate where the center or most typical value of a data set lies

**3.3** The mode

**3.5**
a. Mean = 5; median = 5.
b. Mean = 15; median = 5. The median is a better measure of center because it is not influenced by the one unusually large value, 99.
c. Resistance

**3.7** Median. Unlike the mean, the median is not affected strongly by the relatively few homes that have extremely large or small floor spaces.

**3.9**
a. 7.3 days          b. 6.0 days          c. 5, 6, 11 days

**3.11**
a. 78.4 tornadoes          b. 77.0 tornadoes
c. no mode

**3.13**
a. $29.70 billion     b. $22.50 billion     c. no mode

**3.15**
a. 14.1 mpg          b. 14.0 mpg          c. 14 mpg

**3.17**
a. 292.8, 83.0, 46 cremation burials
b. The median, because of its resistance to extreme observations

**3.19**
a. 88.4, 88.0
b. 70.3, 59.0
c. Friday for built up; Sunday for non built up
d. Friday is a work day, so it is likely that people involved in accidents are commuters using the built-up roads; note that Friday has the second lowest number of accidents on the non-built-up roads. Sunday is a day when riders may be more likely to be cruising around the countryside off the built-up roads; note that Sunday has the lowest number of accidents on the built-up roads.

**3.21** No; the population mean is a constant. Yes; the sample mean is a variable because it varies from sample to sample.

**3.23**
a. 4          b. 46          c. 11.5

**3.25**
a. 10          b. 23.3 hr          c. 2.33 hr

**3.27**
a. 14          b. 16,458 tons          c. 1175.6 tons

**3.29**
a. Iowa          b. Inappropriate

**3.31**
a. Brown          b. Inappropriate

## EXERCISES 3.2

**3.45** To indicate the amount of variation in a data set

**3.47** The mean

**3.49**
a. 2.7          b. 31.6          c. Resistance

**3.51**
a. 45 years          b. 19.9 years          c. 19.9 years

**3.53** range = 6 days; $s$ = 2.6 days

**3.55** range = 202 tornadoes; $s$ = 53.9 tornadoes

**3.57** range = $34 billion; $s$ = $15.08 billion

**3.59** range = 6 mpg; $s$ = 1.5 mpg

**3.61**
a. 586.3 cremation burials
b. No, because of its lack of resistance

**3.63**
a. Non built up
b. Built up: range = 34 accidents; $s$ = 12.8 accidents
   Non built up: range = 49 accidents; $s$ = 19.8 accidents

## EXERCISES 3.3

**3.85** The median and interquartile range are resistant measures, whereas the mean and standard deviation are not.

**3.87** No. It may, for example, be an indication of skewness.

**3.89**
a. A measure of variation
b. Roughly, the range of the middle 50% of the observations

**3.91** When both the minimum and maximum observations lie within the lower and upper limits

*Note:* If you use technology to obtain your results for Exercises 3.93–3.101, they may differ from those presented here because different technologies often use different rules for computing quartiles.

**3.93** Units are in games.
a. $Q_1 = 73.5$, $Q_2 = 79$, $Q_3 = 80$
b. 6.5
c. 45, 73.5, 79, 80, 82
d. 45 and 48
e.

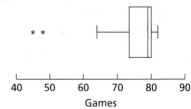

**3.95** Units are in days.
a. $Q_1 = 4$, $Q_2 = 7$, $Q_3 = 12$
b. 8
c. 1, 4, 7, 12, 55
d. 55

**e.**

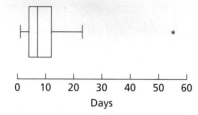

**3.97** Units are in kilograms per hectare per year.
**a.** $Q_1 = 88$, $Q_2 = 131.5$, $Q_3 = 154$
**b.** 66
**c.** 57, 88, 131.5, 154, 175
**d.** No potential outliers
**e.**

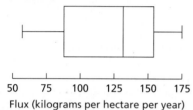

**3.99** Units are in thousands of dollars.
**a.** $Q_1 = 660$, $Q_2 = 1800$, $Q_3 = 4749.5$
**b.** 4089.5
**c.** 21, 660, 1800, 4749.5, 17,341
**d.** 11,189 and 17,341
**e.**

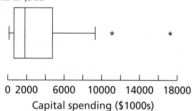

**3.101**
**a.** $Q_1 = 8$ cigs/day, $Q_2 = 9$ cigs/day, $Q_3 = 10$ cigs/day
**b.** The quartiles for this data set are not particularly useful because of its small range and the relatively large number of identical values. Note, for instance, that $Q_3$ and Max are equal.

**3.103** The weight losses for the two groups are, on average, roughly the same. However, there is less variation in the weight losses of Group 1 than of Group 2.

**3.105** On average, the hemoglobin levels for HB SC and HB ST are roughly the same, and both exceed that for HB SS. Also, the variation in hemoglobin levels appears to be greatest for HB ST and least for HB SC.

**3.107** It is symmetric (about the median).

## EXERCISES 3.4

**3.119** To describe the entire population

**3.121**
**a.** 0, 1
**b.** the number of standard deviations that the observation is from the mean, that is, how far the observation is from the mean in units of standard deviation
**c.** above (greater than); below (less than)

**3.123** Parameter. A parameter is a descriptive measure of a population.

**3.125**
**a.** The variable is age and the population consists of all U.S. residents.
**b.** Mean $= 32.5$ years; median $= 37.0$ years. Statistics. $\bar{x} = 32.5$ years; $M = 37.0$ years.
**c.** Parameters. $\mu = 35.8$ years; $\eta = 35.3$ years.

**3.127**
**a.** $\mu = 92.3$ mph          **b.** $\sigma = 41.6$ mph
**c.** $\eta = 75.0$ mph          **d.** Mode $= 50$ mph
**e.** IQR $= 80$ mph

**3.129**
**a.** 820.0 cases; 1552.6 cases
**b.** The standard deviation is smaller for Phoenix because there is less variation in the numbers of cases for Phoenix.
**c.** 85.5 cases; 251.3 cases
**d.** Yes.

**3.131**
**a.** $z = (x - 16.3)/17.9$          **b.** 0; 1
**c.** 2.70; −0.68
**d.** The time served of 64.7 months is 2.70 standard deviations above the mean time served of 16.3 months; the time served of 4.2 months is 0.68 standard deviations below the mean time served of 16.3 months.

**3.133**
**a.** $z = (x - 6.71)/0.67$
**b.** −2.25; 2.07. The thumb length of 5.2 mm is 2.25 standard deviations below the mean thumb length of 6.71 mm; the thumb length of 8.1 mm is 2.07 standard deviations above the mean thumb length of 6.71 mm.

**3.135**
**a.** −3.13
**b.** Yes. Assuming the advertised claim is correct, the three-standard-deviations rule implies that your car's mileage is lower than most other cars of that model.

## REVIEW PROBLEMS FOR CHAPTER 3

1. **a.** Numbers that are used to describe data sets are called descriptive measures.
   **b.** Descriptive measures that indicate where the center or most typical value of a data set lies are called measures of center.
   **c.** Descriptive measures that indicate the amount of variation, or spread, in a data set are called measures of variation.

2. Mean and median. The median is a resistant measure, whereas the mean is not. The mean takes into account the actual numerical value of all observations, whereas the median does not.

3. The mode

4. **a.** Standard deviation          **b.** Interquartile range

5. **a.** $\bar{x}$     **b.** $s$          **c.** $\mu$          **d.** $\sigma$

6. **a.** Not necessarily true          **b.** Necessarily true

7. three

**8. a.** Minimum, quartiles, and maximum; that is, Min, $Q_1, Q_2, Q_3$, Max
  **b.** $Q_2$ can be used to describe center. Max − Min, $Q_1$ − Min, Max − $Q_3$, $Q_2$ − $Q_1$, $Q_3$ − $Q_2$, and $Q_3$ − $Q_1$ are all measures of variation for different portions of the data.
  **c.** Boxplot

**9. a.** An outlier is an observation that falls well outside the overall pattern of the data.
  **b.** First, determine the lower and upper limits—the numbers 1.5 IQRs below the first quartile and 1.5 IQRs above the third quartile, respectively. Observations that lie outside the lower and upper limits—either below the lower limit or above the upper limit—are potential outliers.

**10. a.** Subtract from $x$ its mean and then divide by its standard deviation.
  **b.** The $z$-score of an observation gives the number of standard deviations that the observation is from the mean, that is, how far the observation is from the mean in units of standard deviation.
  **c.** The observation is 2.9 standard deviations above the mean. It is larger than most of the other observations.

**11. a.** 2.35 drinks; 2.0 drinks; 1, 2 drinks
  **b.** Answers will vary.

**12.** The median, because it is resistant to outliers and other extreme values.

**13.** The mode; neither the mean nor the median can be used as a measure of center for qualitative data.

**14.** 30.53 mm; 32.50 mm; 33 mm

**15. a.** $\bar{x} = 45.7$ kg  **b.** Range = 17 kg  **c.** $s = 5.0$ kg

**16. a.**
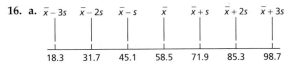

| $\bar{x} - 3s$ | $\bar{x} - 2s$ | $\bar{x} - s$ | $\bar{x}$ | $\bar{x} + s$ | $\bar{x} + 2s$ | $\bar{x} + 3s$ |
|---|---|---|---|---|---|---|
| 18.3 | 31.7 | 45.1 | 58.5 | 71.9 | 85.3 | 98.7 |

  **b.** 18.3 yr, 98.7 yr

**17. a.** $Q_1 = 48.0$ yr, $Q_2 = 59.5$ yr, $Q_3 = 68.5$ yr
  **b.** 20.5 yr; roughly speaking, the middle 50% of the ages has a range of 20.5 yr.
  **c.** 31, 48.0, 59.5, 68.5, 79 yr
  **d.** Lower limit: 17.25 yr. Upper limit: 99.25 yr.
  **e.** No potential outliers
  **f.**

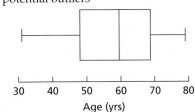

Age (yrs)

**18.** Units are in millimoles per square meter per day.
  **a.** 0.7, 1.50, 1.95, 3.30, 7.6
  **b.** 6.7 and 7.6

**c.**

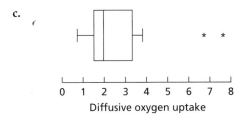

Diffusive oxygen uptake

**19.** During the years in question, more traffic fatalities occurred, on average, in Wisconsin than in New Mexico; in fact, the greatest number of annual fatalities in New Mexico was less than the least number of annual fatalities in Wisconsin. However, the variations in the numbers of annual traffic fatalities that occurred in the two states appear to be comparable.

**20. a.** 19.79 thousand students
  **b.** 3.69 thousand students
  **c.** $z = (x - 19.79)/3.69$   **d.** 0; 1
  **f.** 1.38; −1.24. The enrollment at Los Angeles is 1.38 standard deviations above the UC campuses' mean enrollment of 19.79 thousand students; the enrollment at Riverside is 1.24 standard deviations below that mean.

**21. a.** A sample mean
  **b.** $\bar{x}$
  **c.** A statistic

# Chapter 4

## EXERCISES 4.1

**4.1**
**a.** $y = b_0 + b_1 x$
**b.** $b_0$ and $b_1$ represent constants; $x$ and $y$ represent variables.
**c.** $x$ is the independent variable; $y$ is the dependent variable.

**4.3**
**a.** The number $b_0$ is the $y$-intercept. It is the $y$-value of the point of intersection of the line and the $y$-axis.
**b.** The number $b_1$ is the slope. It measures the steepness of the line; more precisely, $b_1$ indicates how much the $y$-value changes (increases or decreases) when the $x$-value increases by 1 unit.

**4.5**
**a.** $y = 68.22 + 0.25x$   **b.** $b_0 = 68.22, b_1 = 0.25$
**c.**

| $x$ | 50 | 100 | 250 |
|---|---|---|---|
| $y$ | 80.72 | 93.22 | 130.72 |

**d.**

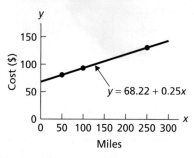

**e.** About \$105; exact cost is \$105.72

**4.7**
**a.** $b_0 = 32, b_1 = 1.8$         **b.** $-40, 32, 68, 212$
**c.**

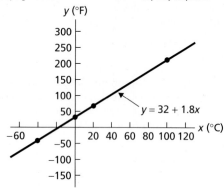

**d.** About 80° F; exact temperature is 82.4° F

**4.9**
**a.** $b_0 = 68.22, b_1 = 0.25$
**b.** The $y$-intercept, $b_0 = 68.22$, gives the $y$-value at which the line, $y = 68.22 + 0.25x$, intersects the $y$-axis. The slope, $b_1 = 0.25$, indicates that the $y$-value increases by 0.25 unit for every increase in $x$ of 1 unit.
**c.** The $y$-intercept, $b_0 = 68.22$, is the cost (in dollars) for driving the car 0 miles. The slope, $b_1 = 0.25$, represents the fact that the cost per mile is \$0.25; it is the amount the total cost increases for each additional mile driven.

**4.11**
**a.** $b_0 = 32, b_1 = 1.8$
**b.** The $y$-intercept, $b_0 = 32$, gives the $y$-value at which the line, $y = 32 + 1.8x$, intersects the $y$-axis. The slope, $b_1 = 1.8$, indicates that the $y$-value increases by 1.8 units for every increase in $x$ of 1 unit.
**c.** The $y$-intercept, $b_0 = 32$, is the Fahrenheit temperature corresponding to 0° C. The slope, $b_1 = 1.8$, represents the fact that the Fahrenheit temperature increases by 1.8° for every increase of the Celsius temperature of 1°.

**4.13**
**a.** $b_0 = 3, b_1 = 4$         **b.** Slopes upward

**c.**

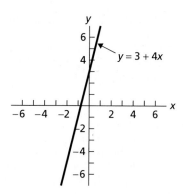

**4.15**
**a.** $b_0 = 6, b_1 = -7$         **b.** Slopes downward
**c.**

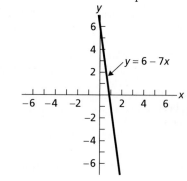

**4.17**
**a.** $b_0 = -2, b_1 = 0.5$         **b.** Slopes upward
**4.19**
**a.** $b_0 = 2, b_1 = 0$         **b.** Horizontal
**4.21**
**a.** $b_0 = 0, b_1 = 1.5$         **b.** Slopes upward
**4.23**
**a.** Slopes upward         **b.** $y = 5 + 2x$
**c.**

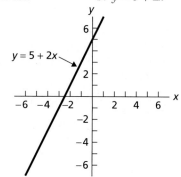

**4.25**
**a.** Slopes downward         **b.** $y = -2 - 3x$

**c.**

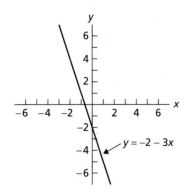

$y = -2 - 3x$

**4.27**
**a.** Slopes downward      **b.** $y = -0.5x$

**4.29**
**a.** Horizontal      **b.** $y = 3$

## EXERCISES 4.2

**4.35**
**a.** Least-squares criterion
**b.** The line that best fits a set of data points is the one having the smallest possible sum of squared errors.

**4.37**
**a.** Response variable
**b.** Predictor variable, or explanatory variable

**4.39**
**a.** Outlier
**b.** Influential observation

**4.41**
**a.**      Line A: $y = 3 - 0.6x$

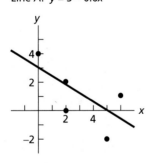

Line B: $y = 4 - x$

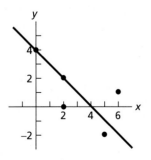

**b.** Line A: $y = 3 - 0.6x$

| $x$ | $y$ | $\hat{y}$ | $e$ | $e^2$ |
|---|---|---|---|---|
| 0 | 4 | 3.0 | 1.0 | 1.00 |
| 2 | 2 | 1.8 | 0.2 | 0.04 |
| 2 | 0 | 1.8 | −1.8 | 3.24 |
| 5 | −2 | 0.0 | −2.0 | 4.00 |
| 6 | 1 | −0.6 | 1.6 | 2.56 |
|  |  |  |  | 10.84 |

Line B: $y = 4 - x$

| $x$ | $y$ | $\hat{y}$ | $e$ | $e^2$ |
|---|---|---|---|---|
| 0 | 4 | 4 | 0 | 0 |
| 2 | 2 | 2 | 0 | 0 |
| 2 | 0 | 2 | −2 | 4 |
| 5 | −2 | −1 | −1 | 1 |
| 6 | 1 | −2 | 3 | 9 |
|  |  |  |  | 14 |

**c.** Line $A$

**4.43**
**a.** $\hat{y} = -1 + x$      **b.** $\hat{y} = 4.5 - 1.5x$

**4.45**
**a.** $\hat{y} = 1 - 2x$
**b.**

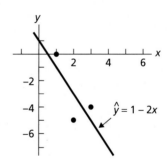

$\hat{y} = 1 - 2x$

**4.47**
**a.** $\hat{y} = -3 + 2x$
**b.**

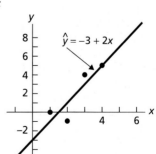

$\hat{y} = -3 + 2x$

**4.49**
**a.** $\hat{y} = 2.875 - 0.625x$

**b.**

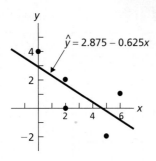

Note: Recall the second bulleted item on page A-27.

**4.51**

**a.** $\hat{y} = 436.6 - 27.9x$

**b.**

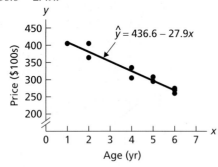

**c.** Price tends to decrease as age increases.

**d.** Corvettes depreciate an estimated $2790 per year, at least in the 1- to 6-year-old range.

**e.** The predictor variable is age (in years); the response variable is price (in hundreds of dollars).

**f.** None

**g.** $38,080; $35,289

**4.53**

**a.** $\hat{y} = 3.52 + 0.16x$

**b.**

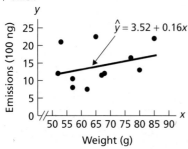

**c.** Quantity of volatile compounds emitted tends to increase as potato plant weight increases.

**d.** The quantity of volatile compounds emitted increases an estimated 16 nanograms for each increase in potato plant weight of 1 g.

**e.** The predictor variable is potato plant weight (in grams); the response variable is quantity of volatile compounds emitted (in hundreds of nanograms).

**f.** None

**g.** 1574 nanograms

**4.55**

**a.** $\hat{y} = 94.9 - 0.8x$

**b.**

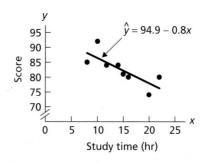

**c.** Test score in beginning calculus courses tends to decrease as study time increases.

**d.** Test score in beginning calculus courses decreases an estimated 0.8 points for each increase in study time of 1 hour.

**e.** The predictor variable is study time (in hours); the response variable is test score.

**f.** None

**g.** 82.2 points

**4.57** Only the second one

**4.59**

**a.** It is acceptable to use the regression equation to predict the price of a 4-year-old Corvette because that age lies within the range of ages in the sample data. It is not acceptable (and would be extrapolation) to use the regression equation to predict the price of a 10-year-old Corvette because that age lies outside the range of the ages in the sample data.

**b.** Ages between 1 and 6 years, inclusive

**4.61** Answers will vary. One possible explanation is that students with an aptitude for calculus will not need to study as long to master the material.

**4.63**

**a.**

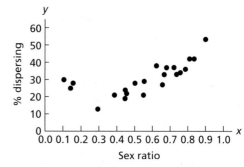

**b.** No, because the data points are scattered about a curve, not a line.

## EXERCISES 4.3

**4.79**

**a.** The coefficient of determination, $r^2$

**b.** The proportion of variation in the observed values of the response variable explained by the regression

**4.81**

**a.** $r^2 = 0.920$; 92.0% of the variation in the observed values of the response variable is explained by the regression. The fact that $r^2$ is near 1 indicates that the regression equation is extremely useful for making predictions.

**b.** 664.4

**4.83**
a. $SST = 14$, $SSR = 8$, $SSE = 6$
b. $14 = 8 + 6$   c. $r^2 = 0.571$
d. 57.1%   e. Moderately useful

**4.85**
a. $SST = 26$, $SSR = 20$, $SSE = 6$
b. $26 = 20 + 6$   c. $r^2 = 0.769$
d. 76.9%   e. Useful

**4.87**
a. $SST = 20$, $SSR = 9.375$, $SSE = 10.625$
b. $20 = 9.375 + 10.625$   c. $r^2 = 0.469$
d. 46.9%   e. Moderately useful

**4.89**
a. $SST = 25{,}681.6$, $SSR = 24{,}057.9$, $SSE = 1623.7$
b. 0.937
c. 93.7%; 93.7% of the variation in the price data is explained by age.
d. Extremely useful

**4.91**
a. $SST = 296.68$, $SSR = 32.52$, $SSE = 264.16$
b. 0.110
c. 11.0%; 11.0% of the variation in the quantity of volatile emissions is explained by potato plant weight.
d. Not very useful

**4.93**
a. $SST = 188.0$, $SSR = 112.9$, $SSE = 75.1$
b. 0.600
c. 60.0%; 60.0% of the variation in the score data is explained by study time.
d. Moderately useful

## EXERCISES 4.4

**4.109** Pearson product moment correlation coefficient

**4.111**
a. $\pm 1$   b. Not very useful

**4.113** False. Correlation does not imply causation.

**4.115** $r = -0.842$

**4.117** $r = -0.756$

**4.119** $r = 0.877$

**4.121** $r = -0.685$

**4.123**
a. $r = -0.968$
b. Suggests an extremely strong negative linear relationship between age and price of Corvettes.
c. Data points are clustered closely about the regression line.
d. $r^2 = 0.937$. This value of $r^2$ is the same as the one obtained in Exercise 4.89(b).

**4.125**
a. $r = 0.331$
b. Suggests a weak positive linear relationship between potato plant weight and quantity of volatile emissions.
c. Data points are scattered widely about the regression line.
d. $r^2 = 0.110$. This value of $r^2$ is the same as the one obtained in Exercise 4.91(b).

**4.127**
a. $r = -0.775$
b. Suggests a moderately strong negative linear relationship between study time and score for students in beginning calculus courses.
c. Data points are clustered moderately closely about the regression line.
d. $r^2 = 0.601$. From Exercise 4.93(b), $r^2 = 0.600$. The discrepancy is due to the error resulting from rounding $r$ to three decimal places before squaring.

**4.129**
a. $r = 0$
b. No. Only that there is no *linear* relationship between the variables
d. No. Because the data points are not scattered about a line
e. For each data point $(x, y)$, the relation $y = x^2$ holds.

**4.131**
a. Approximately 0   b. Negative
c. Positive

## REVIEW PROBLEMS FOR CHAPTER 4

1. a. $x$   b. $y$   c. $b_1$   d. $b_0$
2. a. $y = 4$   b. $x = 0$   c. $-3$
   d. $-3$ units   e. 6 units
3. a. True. The $y$-intercept indicates only where the line crosses the $y$-axis; that is, it is the $y$-value when $x = 0$.
   b. False. Its slope is 0.
   c. True. This is equivalent to saying: If a line has a positive slope, then $y$-values on the line increase as the $x$-values increase.
4. Scatterplot
5. Within the range of the observed values of the predictor variable, we can use the regression equation to make predictions for the response variable.
6. a. Predictor variable, or explanatory variable
   b. Response variable
7. a. Smallest   b. Regression   c. Extrapolation
8. a. An outlier is a data point that lies far from the regression line, relative to the other data points.
   b. An influential observation is a data point whose removal causes the regression equation (and line) to change considerably.
9. It is a descriptive measure of the utility of the regression equation for making predictions.
10. a. $SST$ is the total sum of squares. It measures the variation in the observed values of the response variable.
    b. $SSR$ is the regression sum of squares. It measures the variation in the observed values of the response variable explained by the regression.
    c. $SSE$ is the error sum of squares. It measures the variation in the observed values of the response variable not explained by the regression.
11. a. Linear   b. Increases
    c. Negative   d. 0
12. True

**13. a.** $y = 72 - 12x$      **b.** $b_0 = 72, b_1 = -12$
  **c.** The line slopes downward because $b_1 < 0$.
  **d.** \$4800; \$1200
  **e.**

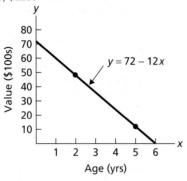

  **f.** About \$2500; exact value is \$2400.

**14. a.**

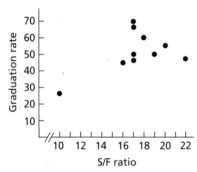

  **b.** It is reasonable to find a regression line for the data because the data points appear to be scattered about a line.
  **c.** $\hat{y} = 16.4 + 2.03x$

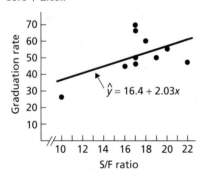

  **d.** Graduation rate tends to increase as student-to-faculty ratio increases.
  **e.** Graduation rate increases by an estimated 2.03 percentage points for each increase of 1 in the student-to-faculty ratio.
  **f.** 50.9%
  **g.** There are no outliers. The data point $(10, 26)$ is a potential influential observation.

**15. a.** $SST = 1384.50$; $SSR = 361.66$; $SSE = 1022.84$
  **b.** $r^2 = 0.261$      **c.** 26.1%
  **d.** Not very useful

**16. a.** $r = 0.511$
  **b.** Suggests a moderately weak positive linear relationship between student-to-faculty ratio and graduation rate.

  **c.** Data points are rather widely scattered about the regression line.
  **d.** $r^2 = (0.511)^2 = 0.261$

# Chapter 5

## EXERCISES 5.1

**5.1** An experiment is an action whose outcome cannot be predicted with certainty. An event is some specified result that may or may not occur when an experiment is performed.

**5.3** There is no difference.

**5.5** The probability of an event is the proportion of times it occurs in a large number of repetitions of the experiment.

**5.7** (b) and (e) because the probability of an event must always be between 0 and 1, inclusive.

**5.9**
**a.**

| G, L, S, A | G, L, S, T | G, L, A, T | G, S, A, T | L, S, A, T |
|---|---|---|---|---|

**b.** 0.4      **c.** 0.6      **d.** 0.8

**5.11**
**a.** 0.745      **b.** 0.029      **c.** 0.255

**5.13**
**a.** 0.193      **b.** 0.700      **c.** 0.016
**d.** 0      **e.** 1

**5.15**
**a.** 0.189    **b.** 0.176    **c.** 0.239    **d.** 0.761

**5.17**
**a.** 0.137      **b.** 0.371      **c.** 0.840

**5.19**
**a.** 0.139    **b.** 0.500    **c.** 0.222    **d.** 0.111

**5.21** The event in part (e) is certain; the event in part (d) is impossible.

## EXERCISES 5.2

**5.29** Venn diagrams

**5.31** Two or more events are mutually exclusive if at most one of them can occur when the experiment is performed. Thus two events are mutually exclusive if they do not have outcomes in common. Three events are mutually exclusive if no two of them have outcomes in common.

**5.33** $A =$ ⚁ ⚃ ⚅

     $B =$ ⚃ ⚄ ⚅

     $C =$ ⚀ ⚄

     $D =$ ⚁

**5.35** *A*: JM, WM, JS, WS, JH, WH, JW, WJ;
*B*: HM, HS, HJ, HW;
*C*: MW, SW, HW, JW;
*D*: MS, SM, HM, MH, SH, HS

**5.37**
**a.** (not *A*) = [die 1] [die 3] [die 5]

The event that the die comes up odd

**b.** (*A* & *B*) = [die 4] [die 6]

The event that the die comes up 4 or 6

**c.** (*B* or *C*) = [die 1] [die 2] [die 4] [die 5] [die 6]

The event that the die does not come up 3

**5.39**
**a.** (not *A*) = MS, SM, HM, MH, SH, HS, MJ, SJ, HJ, MW, SW, HW; the event that a female is appointed chairperson.
**b.** (*B* & *D*) = HM, HS; the event that Holly is appointed chairperson and either Maria or Susan is appointed secretary.
**c.** (*B* or *C*) = HM, HS, HJ, HW, MW, SW, JW; the event that either Holly is appointed chairperson or Will is appointed secretary (or both).

**5.41**
**a.** (not *C*) is the event that the state has a diabetes prevalence percentage of less than 5% or at least 10%; nine states satisfy that property.
**b.** (*A* & *B*) is the event that the state has a diabetes prevalence percentage of at least 8%, but less than 7%, which is impossible; no states satisfy that property.
**c.** (*C* or *D*) is the event that the state has a diabetes prevalence percentage of less than 10%; 49 states satisfy that property.
**d.** (*C* & *B*) is the event that the state has a diabetes prevalence percentage of at least 5% but less than 7%; 25 states satisfy that property.

**5.43** Note that Medicare and Medicaid are government agencies.
**a.** (*A* or *D*) is the event that either Medicare or the patient or a charity paid the bill; 15,495 of the bills were paid that way.
**b.** (not *C*) is the event that private insurance paid the bill; 26,825 of the bills were paid that way.
**c.** (*B* & (not *A*)) is the event that some government agency other than Medicare paid the bill; 9919 of the bills were paid that way.
**d.** (not (*C* or *D*)) is the event that private insurance paid the bill; 26,825 of the bills were paid that way.

**5.45**
**a.** (not *A*) is the event the unit has at least five rooms; 84,526 thousand units have that property.
**b.** (*A* & *B*) is the event the unit has two, three, or four rooms; 35,731 thousand units have that property.
**c.** (*C* or *D*) is the event the unit has at least five rooms; 84,526 thousand units have that property. (*Note:* From part (a), (not *A*) = (*C* or *D*).)

**5.47**
**a.** No        **b.** Yes        **c.** No
**d.** Yes, events *B*, *C*, and *D*. No.

**5.49** *A* and *C*; *A* and *D*; *C* and *D*; *A*, *C*, and *D*

**EXERCISES 5.3**

**5.55**
**a.** 0.77                    **b.** *S* = (*A* or *B* or *C*)
**c.** 0.12, 0.33, 0.32        **d.** 0.77

**5.57**
**a.** 0.267        **b.** 0.169        **c.** 0.088

**5.59**
**a.** 7.5%        **b.** 13.7%        **c.** 68.3%

**5.61**
**a.** 0.88                    **b.** 0.77

**5.63**
**a.** 0.93                    **b.** 0.93

**5.65**
**a.** 0.167, 0.056, 0.028, 0.056, 0.028, 0.139, 0.167
**b.** 0.223    **c.** 0.112    **d.** 0.278    **e.** 0.278

**5.67** 91.0%

**5.69**
**a.** No, because *P*(*A* or *B*) ≠ *P*(*A*) + *P*(*B*).
**b.** $\frac{1}{12}$ or about 0.083

**EXERCISES 5.4**

**5.73**
**a.** Probability                **b.** Probability

**5.75** {*X* = 3} is the event that the student has three siblings; *P*(*X* = 3) is the probability of the event that the student has three siblings.

**5.77** The probability distribution of the random variable

**5.79**
**a.** 2, 3, 4, 5, 6, 7, 8        **b.** {*X* = 7}
**c.** 0.021. 2.1% of the shuttle missions between April 1981 and July 2000 had a crew size of 4.
**d.**

| *x* | 2 | 3 | 4 | 5 | 6 | 7 | 8 |
|---|---|---|---|---|---|---|---|
| *P*(*X* = *x*) | 0.042 | 0.010 | 0.021 | 0.375 | 0.188 | 0.344 | 0.021 |

**e.**

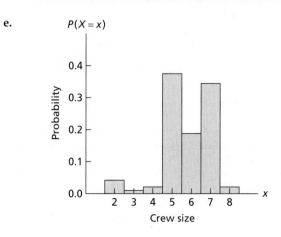

**5.81**
**a.** {*Y* ≥ 1}                **b.** {*Y* = 2}
**c.** {1 ≤ *Y* ≤ 3}            **d.** {*Y* = 1 or 3 or 5}

**e.** 0.991    **f.** 0.371    **g.** 0.914    **h.** 0.559

**5.83**
**a.** 2, 3, 4, 5, 6, 7, 8, 9, 10, 11, 12
**b.** $\{Y = 7\}$      **c.** $\frac{1}{6}$

**d.**

| $y$ | 2 | 3 | 4 | 5 | 6 | 7 | 8 | 9 | 10 | 11 | 12 |
|---|---|---|---|---|---|---|---|---|---|---|---|
| $P(Y = y)$ | $\frac{1}{36}$ | $\frac{1}{18}$ | $\frac{1}{12}$ | $\frac{1}{9}$ | $\frac{5}{36}$ | $\frac{1}{6}$ | $\frac{5}{36}$ | $\frac{1}{9}$ | $\frac{1}{12}$ | $\frac{1}{18}$ | $\frac{1}{36}$ |

**f.** $\frac{2}{9}$          **g.** $\frac{1}{9}$

**5.85**
**a.**

| $s$ | 0 | 5 | 10 |
|---|---|---|---|
| $P(S = s)$ | 0.651 | 0.262 | 0.087 |

**b.** 0.262; 0.349; 0.913; 0.087; 1; 0

## EXERCISES 5.5

**5.91** The mean of a variable of a finite population (population mean)

**5.93**
**a.** 5.8 crew members      **b.** 1.3 crew members

**5.95**
**a.** 1.9 color TVs      **b.** 1.0 color TVs

**5.97**
**a.** 7      **b.** 2.4

**5.99**
**a.** 2.18 points      **b.** 3.24 points

**5.101**
**b.** −0.052    **c.** 5.2¢      **d.** $5.20, $52

**5.103**
**a.** $\mu_W = 0.25$, $\sigma_W = 0.536$    **b.** 0.25
**c.** 62.5

## EXERCISES 5.6

**5.107** Answers will vary.

**5.109** 6; 5040; 40,320; 362,880

**5.111**
**a.** 10    **b.** 35    **c.** 120    **d.** 792

**5.113**
**a.** 4    **b.** 15    **c.** 56    **d.** 84

**5.115**
**a.** Each trial consists of observing whether a child with pinworm is cured by treatment with pyrantel pamoate and has two possible outcomes: cured or not cured. The trials are independent. The success probability is 0.9; that is, $p = 0.9$.

**b.**

| Outcome | Probability |
|---|---|
| sss | $(0.9)(0.9)(0.9) = 0.729$ |
| ssf | $(0.9)(0.9)(0.1) = 0.081$ |
| sfs | $(0.9)(0.1)(0.9) = 0.081$ |
| sff | $(0.9)(0.1)(0.1) = 0.009$ |
| fss | $(0.1)(0.9)(0.9) = 0.081$ |
| fsf | $(0.1)(0.9)(0.1) = 0.009$ |
| ffs | $(0.1)(0.1)(0.9) = 0.009$ |
| fff | $(0.1)(0.1)(0.1) = 0.001$ |

**d.** $ssf$, $sfs$, $fss$
**e.** 0.081. Because each probability is obtained by multiplying two success probabilities of 0.9 and one failure probability of 0.1.
**f.** 0.243
**g.**

| $x$ | 0 | 1 | 2 | 3 |
|---|---|---|---|---|
| $P(X = x)$ | 0.001 | 0.027 | 0.243 | 0.729 |

**5.117** The appropriate binomial probability formula is

$$P(X = x) = \binom{3}{x}(0.9)^x(0.1)^{3-x}.$$

Applying this formula for $x = 0$, 1, 2, and 3, gives the same result as in part (g) of Exercise 5.115.

**5.119**
**a.** $p = 0.5$          **b.** $p < 0.5$

**5.121**
**a.** 0.161    **b.** 0.332    **c.** 0.468    **d.** 0.821
**e.**

| $x$ | $P(X = x)$ |
|---|---|
| 0 | 0.004 |
| 1 | 0.040 |
| 2 | 0.161 |
| 3 | 0.328 |
| 4 | 0.332 |
| 5 | 0.135 |

**f.** Left skewed
**g.**

**h.** $\mu = 3.35$ times; $\sigma = 1.05$ times.

i.  $\mu = 3.35$ times; $\sigma = 1.05$ times.
j.  On average, the favorite will finish in the money 3.35 times for every 5 races.

**5.123**
a.  0.279; 0.685; 0.594          b.  0.720
c.  3.2 traffic fatalities; on average, 3.2 of every 8 traffic fatalities involve an intoxicated or alcohol-impaired driver or nonoccupant.
d.  1.4 traffic fatalities

**5.125**
a.  0.118; 0.946; 0.172
b.  0.979; 0.121          c.  0.0535
d.

| $x$ | $P(X = x)$ |
|---|---|
| 0 | 0.0207 |
| 1 | 0.1004 |
| 2 | 0.2162 |
| 3 | 0.2716 |
| 4 | 0.2194 |
| 5 | 0.1181 |
| 6 | 0.0424 |
| 7 | 0.0098 |
| 8 | 0.0013 |
| 9 | 0.0001 |

e.  Because the sampling is done without replacement from a finite population. Hypergeometric distribution.

**5.127**
a.

| $x$ | $P(X = x)$ |
|---|---|
| 0 | 0.4861 |
| 1 | 0.3842 |
| 2 | 0.1139 |
| 3 | 0.0150 |
| 4 | 0.0007 |

b.  0.66; on average, we would expect about 0.66 of four people under the age of 65 to have no health insurance.
c.  Yes, because if the uninsured rate today were the same as in 2002, there is only a 1.6% chance that three or more of the four people would not be covered.
d.  Probably not, because if the uninsured rate today were the same as in 2002, there is a 13.0% chance that two or more of the four people would not be covered.

## REVIEW PROBLEMS FOR CHAPTER 5

1.  It enables you to evaluate and control the likelihood that a statistical inference is correct. More generally, probability theory provides the mathematical basis for inferential statistics.

2.  a.  The experiment has a finite number of possible outcomes, all equally likely.
    b.  The probability of an event equals the ratio of the number of ways that the event can occur to the total number of possible outcomes.

3.  It is the proportion of times the event occurs in a large number of repetitions of the experiment.

4.  (b) and (c), because the probability of an event must always be between 0 and 1, inclusive.

5.  Venn diagrams

6.  Two or more events are said to be mutually exclusive if at most one of them can occur when the experiment is performed, that is, if no two of them have outcomes in common.

7.  a.  $P(E)$          b.  $P(E) = 0.436$

8.  a.  False          b.  True

9.  It is sometimes easier to compute the probability that an event does not occur than the probability that it does occur.

10.  a.  0.199          b.  0.394
    c.  0.199, 0.179, 0.141, 0.107, 0.080, 0.206, 0.088

11.  a.  (not $J$) is the event that the return shows an AGI of at least $100K. There are 11,415 thousand such returns.
    b.  ($H \& I$) is the event that the return shows an AGI of between $20K and $50K. There are 42,782 thousand such returns.
    c.  ($H$ or $K$) is the event that the return shows an AGI of at least $20K. There are 81,112 thousand such returns.
    d.  ($H \& K$) is the event that the return shows an AGI of between $50K and $100K. There are 26,915 thousand such returns.

12.  a.  Not mutually exclusive
    b.  Mutually exclusive
    c.  Mutually exclusive
    d.  Not mutually exclusive

13.  a.  0.534, 0.706, 0.912, 0.294
    b.  $H = (C$ or $D$ or $E$ or $F)$
       $I = (A$ or $B$ or $C$ or $D$ or $E)$
       $J = (A$ or $B$ or $C$ or $D$ or $E$ or $F)$
       $K = (F$ or $G)$
    c.  0.534, 0.706, 0.912, 0.294

14.  a.  0.088, 0.328, 0.622, 0.206
    b.  0.912          c.  0.622
    d.  They are the same.

15.  a.  random variable
    b.  can be listed

16.  The possible values and corresponding probabilities of the discrete random variable

17.  Probability histogram

18.  1

19.  a.  $P(X = 2) = 0.386$          b.  38.6%
    c.  19.3; 193

20.  3.6

21.  $X$, because it has a smaller standard deviation, therefore less variation.

**22.** Each trial has the same two possible outcomes; the trials are independent; the probability of a success remains the same from trial to trial.

**23.** The binomial distribution is the probability distribution for the number of successes in a finite sequence of Bernoulli trials.

**24.** 120

**25.** Substitute the binomial probability formula into the formulas for the mean and standard deviation of a discrete random variable and then simplify mathematically.

**26.** **a.** Binomial distribution
  **b.** Hypergeometric distribution
  **c.** When the sample size does not exceed 5% of the population size because, under this condition, there is little difference between sampling with and without replacement.

**27.** **a.** 1, 2, 3, 4  **b.** $\{X = 3\}$
  **c.** 0.253; 25.3% of undergraduates at ASU are juniors.
  **d.**

| $x$ | 1 | 2 | 3 | 4 |
|---|---|---|---|---|
| $P(X = x)$ | 0.220 | 0.215 | 0.253 | 0.312 |

  **e.**

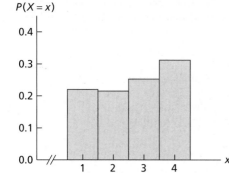

**28.** **a.** $\{Y = 4\}$  **b.** $\{Y \geq 4\}$  **c.** $\{2 \leq Y \leq 4\}$
  **d.** $\{Y \geq 1\}$  **e.** 0.174  **f.** 0.322
  **g.** 0.646  **h.** 0.948

**29.** **a.** 2.817 lines  **b.** 2.817 lines  **c.** 1.504 lines

**30.** 1, 6, 24, 5040

**31.** **a.** 56  **b.** 56  **c.** 1
  **d.** 45  **e.** 91,390  **f.** 1

**32.** **a.** $p = 0.493$
  **b.**

| Outcome | Probability |
|---|---|
| sss | $(0.493)(0.493)(0.493) = 0.120$ |
| ssf | $(0.493)(0.493)(0.507) = 0.123$ |
| sfs | $(0.493)(0.507)(0.493) = 0.123$ |
| sff | $(0.493)(0.507)(0.507) = 0.127$ |
| fss | $(0.507)(0.493)(0.493) = 0.123$ |
| fsf | $(0.507)(0.493)(0.507) = 0.127$ |
| ffs | $(0.507)(0.507)(0.493) = 0.127$ |
| fff | $(0.507)(0.507)(0.507) = 0.130$ |

  **d.** ssf, sfs, fss
  **e.** 0.123. Each probability is obtained by multiplying two success probabilities of 0.493 and one failure probability of 0.507.
  **f.** 0.369
  **g.**

| $y$ | 0 | 1 | 2 | 3 |
|---|---|---|---|---|
| $P(Y = y)$ | 0.130 | 0.381 | 0.369 | 0.120 |

  **h.** Binomial with parameters $n = 3$ and $p = 0.493$

**33.** **a.** 0.3456  **b.** 0.4752  **c.** 0.8704
  **d.**

| $x$ | $P(X = x)$ |
|---|---|
| 0 | 0.0256 |
| 1 | 0.1536 |
| 2 | 0.3456 |
| 3 | 0.3456 |
| 4 | 0.1296 |

  **e.** Left skewed
  **f.**

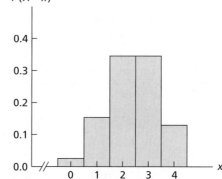

  **g.** The probability distribution is only approximately correct because the sampling is without replacement; a hypergeometric distribution.
  **h.** 2.4 households; on average, 2.4 of every 4 U.S. households live with one or more pets.
  **i.** 0.98 households

**34.** **a.** $p > 0.5$  **b.** $p = 0.5$

# Chapter 6

## EXERCISES 6.1

**6.1** Roughly bell shaped

**6.3** They are the same. A normal distribution is completely determined by the mean and standard deviation.

**6.5**
**a.** True. They have the same shape because their standard deviations are equal.
**b.** False. A normal distribution is centered at its mean, which is different for these two distributions.

**6.7** True. The shape of a normal distribution is completely determined by its standard deviation.

**6.9**

**a.**

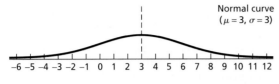

Normal curve
($\mu = 3, \sigma = 3$)

-6 -5 -4 -3 -2 -1 0 1 2 3 4 5 6 7 8 9 10 11 12

**b.**

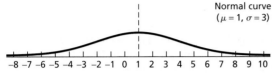

Normal curve
($\mu = 1, \sigma = 3$)

-8 -7 -6 -5 -4 -3 -2 -1 0 1 2 3 4 5 6 7 8 9 10

**c.**

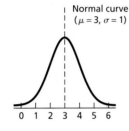

Normal curve
($\mu = 3, \sigma = 1$)

0 1 2 3 4 5 6

**6.11** They are equal. They are approximately equal.

**6.13**
**a.** 55.70%
**b.** 0.5570; This is only an estimate because the distribution of heights is only approximately normally distributed.

**6.15**
**a.**

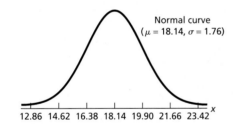

Normal curve
($\mu = 18.14, \sigma = 1.76$)

12.86  14.62  16.38  18.14  19.90  21.66  23.42  $x$

**b.** $z = (x - 18.14)/1.76$
**c.** Standard normal distribution

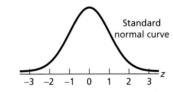

Standard normal curve

-3  -2  -1  0  1  2  3  $z$

**d.** $-1.22; -0.65$          **e.** right; 0.49

**6.17**
**a.**

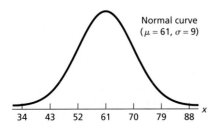

Normal curve
($\mu = 61, \sigma = 9$)

34  43  52  61  70  79  88  $x$

**b.** $z = (x - 61)/9$

**c.** Standard normal distribution; see the graph in the answer to Exercise 6.15(c).
**d.** $-1.22; 1$          **e.** left; 1.56

**6.19**
**a.**

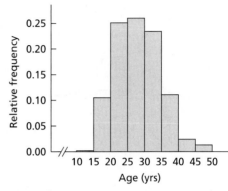

Age (yrs)

**b.** Yes, because the histogram has very roughly the shape of a normal curve.

## EXERCISES 6.2

**6.27** For a normally distributed variable, you can determine the percentage of all possible observations that lie within any specified range by first converting to $z$-scores and then obtaining the corresponding area under the standard normal curve.

**6.29** The total area under the standard normal curve equals 1, and the standard normal curve is symmetric about 0. So the area to the right of 0 is one-half of 1, or 0.5.

**6.31** 0.3336. The total area under the curve is 1, so the area to the right of 0.43 equals 1 minus the area to its left, which is $1 - 0.6664 = 0.3336$.

**6.33** 99.74%

**6.35**
**a.** Read the area directly from the table.
**b.** Subtract the table area from 1.
**c.** Subtract the smaller table area from the larger.

**6.37**
**a.** 0.9875          **b.** 0.0594          **c.** 0.5
**d.** 0.0000 (to four decimal places)

**6.39**
**a.** 0.8577          **b.** 0.2743          **c.** 0.5
**d.** 0.0000 (to four decimal places)

**6.41**
**a.** 0.9105     **b.** 0.0440     **c.** 0.2121     **d.** 0.1357

**6.43**
**a.** 0.0645                    **b.** 0.7975

**6.45**
**a.** 0.7994     **b.** 0.8990     **c.** 0.0500     **d.** 0.0198

**6.47**
**a.** 0.6826

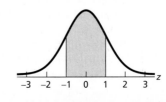

-3  -2  -1  0  1  2  3  $z$

**b.** 0.9544

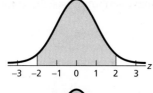

**c.** 0.9974

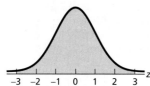

**6.49** −1.96

**6.51** 0.67

**6.53** −1.645

**6.55** 0.44

**6.57**

**a.** 1.88                    **b.** 2.575

**6.59** ±1.645

**6.61** The four missing entries are 1.645, 1.96, 2.33, and 2.575.

## EXERCISES 6.3

**6.65** The z-scores corresponding to the x-values that lie two standard deviations below and above the mean are −2 and 2, respectively.

*Note:* In the remainder of this chapter, your answers may vary from those given here depending on whether you use Table II or technology.

**6.67**

**a.** 14.66%                    **b.** 31.21%

**c.** 16.96 mm, 18.14 mm, 19.32 mm

**d.** 21.04 mm; 95% of adult male *G. mollicoma* have carapace lengths less than 21.04 mm and 5% have carapace lengths greater than 21.04 mm.

**6.69**

**a.** 73.01%                    **b.** 94.06%

**c.** 58.8 minutes; 40% of finishers in the New York City 10-km run have times less than 58.8 minutes and 60% have times greater than 58.8 minutes.

**d.** 68.6 minutes; 80% of finishers in the New York City 10-km run have times less than 68.6 minutes and 20% have times greater than 68.6 minutes.

**6.71**

**a.** 76.47%                    **b.** 0.03%

**6.73**

**a.** 68.26%          **b.** 95.44%          **c.** 99.74%

**6.75**

**a.** 1.29 kg; 1.51 kg                    **b.** 1.18 kg; 1.62 kg

**c.** 1.07 kg; 1.73 kg

**d.** See the graphs shown in Fig. A.1.

**FIGURE A.1**   Graphs for Exercise 6.75(d)

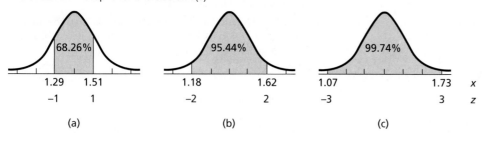

**6.77**

**a.** (i) 11.70%    (ii) 12.23%

**b.** (i) 39.83%    (ii) 39.14%

## EXERCISES 6.4

**6.85** Decisions about whether a variable is normally distributed often are important in subsequent analyses—from percentage or percentile calculations to statistical inferences.

**6.87**

**a.** A normal probability plot is a plot of the observed values of the variable versus the normal scores—the observations expected for a variable having the standard normal distribution. If the variable is normally distributed, the normal probability plot should be roughly linear and vice versa.

**b.** In a normal probability plot, outliers lie outside the overall pattern formed by the other points in the plot.

**6.89**

**a.**

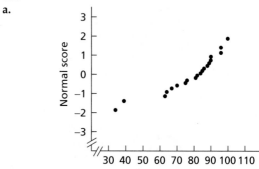

**b.** 34 and 39 are outliers.

**c.** Final-exam scores in this introductory statistics class do not appear to be normally distributed.

**6.91**

**a.**

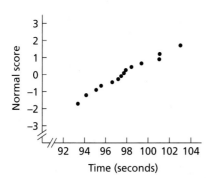

**b.** No outliers

**c.** It appears plausible that finishing times for the winners of 1-mile thoroughbred horse races are (approximately) normally distributed.

**6.93**

**a.**

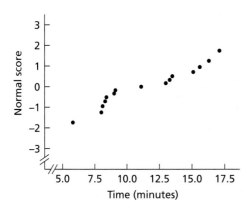

**b.** No outliers

**c.** It appears plausible that the average times spent per user per month from January to June of the year in question are (approximately) normally distributed.

**6.95**

**a.**

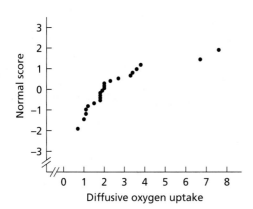

**b.** 6.7 and 7.6 are outliers

**c.** Diffusive oxygen uptakes in surface sediments from central Sagami Bay do not appear to be normally distributed.

## REVIEW PROBLEMS FOR CHAPTER 6

1. It appears again and again in both theory and practice.

2. **a.** A variable is said to be normally distributed if its distribution has the shape of a normal curve.
   **b.** If a variable of a population is normally distributed and is the only variable under consideration, common practice is to say that the population is a normally distributed population.
   **c.** The parameters for a normal curve are the corresponding mean and standard deviation of the variable.

3. **a.** False
   **b.** True. A normal distribution is completely determined by its mean and standard deviation.

4. They are the same when areas are expressed as percentages.

5. Standard normal distribution

6. **a.** True          **b.** True

7. **a.** The second curve
   **b.** The first and second curves
   **c.** The first and third curves
   **d.** The third curve          **e.** The fourth curve

8. Key Fact 6.2, which states that the standardized version of a normally distributed variable has the standard normal distribution

9. **a.** Read the area directly from the table.
   **b.** Subtract the table area from 1.
   **c.** Subtract the smaller table area from the larger.

10. **a.** Locate the table entry closest to the specified area and read the corresponding $z$-score.
    **b.** Locate the table entry closest to 1 minus the specified area and read the corresponding $z$-score.

11. The $z$-score having area $\alpha$ to its right under the standard normal curve

12. See Key Fact 6.4.

13. The observations expected for a sample of the same size from a variable that has the standard normal distribution

14. Linear

15. **a.**

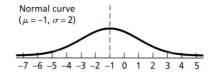

   **b.**

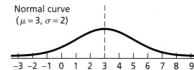

**c.**

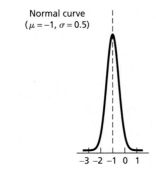

Normal curve
($\mu = -1$, $\sigma = 0.5$)

**16. a.**

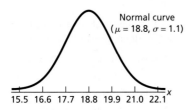

Normal curve
($\mu = 18.8$, $\sigma = 1.1$)

**b.** $z = (x - 18.8)/1.1$
**c.** Standard normal distribution
**d.** 0.8115            **e.** left; $-2.55$

**17. a.** 0.1469        **b.** 0.1469            **c.** 0.7062

**18. a.** 0.0013        **b.** 0.2709            **c.** 0.1305
**d.** 0.9803        **e.** 0.0668            **f.** 0.8426

**19. a.** $-0.52$                        **b.** 1.28
**c.** 1.96; 1.645; 2.33; 2.575        **d.** $\pm 2.575$

**20. a.**

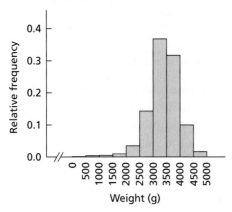

**b.** No, because the histogram is left skewed.

**21. a.** 59.87%        **b.** 73.33%            **c.** 2.28%

**22. a.** 387.3, 465.0, 542.7 points; 25% of the scores are less than 387.3 points, 25% are between 387.3 points and 465.0 points, 25% are between 465.0 points and 542.7 points, and 25% exceed 542.7 points.
**b.** 735.3 points; 99% of the scores are less than 735.3 points and 1% are greater than 735.3 points.

**23. a.** 349 points; 581 points
**b.** 233 points; 697 points
**c.** 117 points; 813 points

**24. a.**

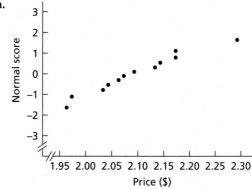

**b.** No outliers
**c.** It appears plausible that prices for unleaded regular gasoline on December 6, 2005 are (approximately) normally distributed.

**25. a.**

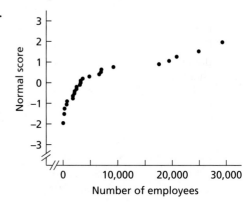

**b.** No outliers
**c.** The numbers of employees of publicly traded mortgage industry companies do not appear to be normally distributed.

# Chapter 7

## EXERCISES 7.1

**7.1** Generally, sampling is less costly and can be done more quickly than a census.

**7.3**
**a.** $\mu = 80$ inches

**b.**

| Sample | Heights (in.) | $\bar{x}$ |
|--------|--------------|-----------|
| B, D | 79, 84 | 81.5 |
| B, R | 79, 85 | 82.0 |
| B, G | 79, 78 | 78.5 |
| B, P | 79, 74 | 76.5 |
| D, R | 84, 85 | 84.5 |
| D, G | 84, 78 | 81.0 |
| D, P | 84, 74 | 79.0 |
| R, G | 85, 78 | 81.5 |
| R, P | 85, 74 | 79.5 |
| G, P | 78, 74 | 76.0 |

**c.**

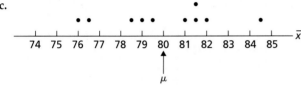

**d.** 0

**e.** 0.3. If a random sample of two players is taken, there is a 30% chance that the mean height of the two players selected will be within 1 inch of the population mean height.

**7.5**

**b.**

| Sample | Heights (in.) | $\bar{x}$ |
|--------|--------------|-----------|
| B, D, R | 79, 84, 85 | 82.7 |
| B, D, G | 79, 84, 78 | 80.3 |
| B, D, P | 79, 84, 74 | 79.0 |
| B, R, G | 79, 85, 78 | 80.7 |
| B, R, P | 79, 85, 74 | 79.3 |
| B, G, P | 79, 78, 74 | 77.0 |
| D, R, G | 84, 85, 78 | 82.3 |
| D, R, P | 84, 85, 74 | 81.0 |
| D, G, P | 84, 78, 74 | 78.7 |
| R, G, P | 85, 78, 74 | 79.0 |

**c.**

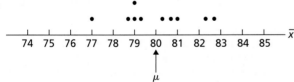

**d.** 0

**e.** 0.6. If a random sample of three players is taken, there is a 60% chance that the mean height of the three players selected will be within 1 inch of the population mean height.

**7.7**

**b.**

| Sample | Heights (in.) | $\bar{x}$ |
|--------|--------------|-----------|
| B, D, R, G, P | 79, 84, 85, 78, 74 | 80 |

**c.**

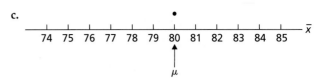

**d.** 1

**e.** 1. If a random sample of five players is taken, it is certain (a 100% chance) that the mean height of the five players selected will be within 1 inch of the population mean height.

**7.9**

**a.** $\mu = \$25.5$ billion

**b.**

| Sample | Wealth ($billions) | $\bar{x}$ |
|--------|-------------------|-----------|
| G, B | 41, 31 | 36.0 |
| G, A | 41, 26 | 33.5 |
| G, P | 41, 20 | 30.5 |
| G, T | 41, 18 | 29.5 |
| G, E | 41, 17 | 29.0 |
| B, A | 31, 26 | 28.5 |
| B, P | 31, 20 | 25.5 |
| B, T | 31, 18 | 24.5 |
| B, E | 31, 17 | 24.0 |
| A, P | 26, 20 | 23.0 |
| A, T | 26, 18 | 22.0 |
| A, E | 26, 17 | 21.5 |
| P, T | 20, 18 | 19.0 |
| P, E | 20, 17 | 18.5 |
| T, E | 18, 17 | 17.5 |

**d.** $\frac{1}{15}$

**e.** 0.2. If a random sample of two of the six richest people is taken, there is a 20% chance that the mean wealth of the two people selected will be within 2 (i.e., $2 billion) of the population mean wealth.

**7.11**

**b.**

| Sample | Wealth ($billions) | $\bar{x}$ |
|--------|--------------------|-----------|
| G, B, A | 41, 31, 26 | 32.7 |
| G, B, P | 41, 31, 20 | 30.7 |
| G, B, T | 41, 31, 18 | 30.0 |
| G, B, E | 41, 31, 17 | 29.7 |
| G, A, P | 41, 26, 20 | 29.0 |
| G, A, T | 41, 26, 18 | 28.3 |
| G, A, E | 41, 26, 17 | 28.0 |
| G, P, T | 41, 20, 18 | 26.3 |
| G, P, E | 41, 20, 17 | 26.0 |
| G, T, E | 41, 18, 17 | 25.3 |
| B, A, P | 31, 26, 20 | 25.7 |
| B, A, T | 31, 26, 18 | 25.0 |
| B, A, E | 31, 26, 17 | 24.7 |
| B, P, T | 31, 20, 18 | 23.0 |
| B, P, E | 31, 20, 17 | 22.7 |
| B, T, E | 31, 18, 17 | 22.0 |
| A, P, T | 26, 20, 18 | 21.3 |
| A, P, E | 26, 20, 17 | 21.0 |
| A, T, E | 26, 18, 17 | 20.3 |
| P, T, E | 20, 18, 17 | 18.3 |

**d.** 0

**e.** 0.3. If a random sample of three of the six richest people is taken, there is a 30% chance that the mean wealth of the three people selected will be within 2 (i.e., $2 billion) of the population mean wealth.

**7.13**

**b.**

| Sample | Wealth ($billions) | $\bar{x}$ |
|--------|--------------------|-----------|
| G, B, A, P, T | 41, 31, 26, 20, 18 | 27.2 |
| G, B, A, P, E | 41, 31, 26, 20, 17 | 27.0 |
| G, B, A, T, E | 41, 31, 26, 18, 17 | 26.6 |
| G, B, P, T, E | 41, 31, 20, 18, 17 | 25.4 |
| G, A, P, T, E | 41, 26, 20, 18, 17 | 24.4 |
| B, A, P, T, E | 31, 26, 20, 18, 17 | 22.4 |

**d.** 0

**e.** $\frac{5}{6}$. If a random sample of five of the six richest people is taken, there is an 83.3% chance that the mean wealth of the five people selected will be within 2 (i.e., $2 billion) of the population mean wealth.

**7.15** Sampling error tends to be smaller for large samples than for small samples.

## EXERCISES 7.2

**7.19** A normal distribution is determined by the mean and standard deviation. Hence a first step in learning how to approximate the sampling distribution of the mean by a normal distribution is to obtain the mean and standard deviation of the variable $\bar{x}$.

**7.21** Yes. The standard deviation of all possible sample means (i.e., of the variable $\bar{x}$) gets smaller as the sample size gets larger.

**7.23** Standard error (SE) of the mean. Because the standard deviation of $\bar{x}$ determines the amount of sampling error to be expected when a population mean is estimated by a sample mean.

**7.25**

**a.** $\mu = 80$ inches       **b.** $\mu_{\bar{x}} = 80$ inches

**c.** $\mu_{\bar{x}} = \mu = 80$ inches

**7.27**

**b.** $\mu_{\bar{x}} = 80$ inches       **c.** $\mu_{\bar{x}} = \mu = 80$ inches

**7.29**

**b.** $\mu_{\bar{x}} = 80$ inches       **c.** $\mu_{\bar{x}} = \mu = 80$ inches

**7.31**

**a.** The population consists of all babies born in 1991. The variable is birth weight.

**b.** 3369 g; 41.1 g       **c.** 3369 g; 29.1 g

**7.33**

**a.** $\mu_{\bar{x}} = \$61,300$, $\sigma_{\bar{x}} = \$1018.2$. For samples of 50 new mobile homes, the mean and standard deviation of all possible sample mean prices are $61,300 and $1018.2, respectively.

**b.** $\mu_{\bar{x}} = \$61,300$, $\sigma_{\bar{x}} = \$720.0$. For samples of 100 new mobile homes, the mean and standard deviation of all possible sample mean prices are $61,300 and $720.0, respectively.

**7.35**

**a.** 437 days       **b.** ±598.5 days

## EXERCISES 7.3

**7.47**

**a.** Approximately normally distributed with a mean of 100 and a standard deviation of 4

**b.** None

**c.** No. Because the distribution of the variable under consideration is not specified, a sample size of at least 30 is needed to apply Key Fact 7.4.

**7.49**

**a.** Normal with mean $\mu$ and standard deviation $\sigma/\sqrt{n}$

**b.** No. Because the variable under consideration is normally distributed.

**c.** $\mu$ and $\sigma/\sqrt{n}$

**d.** Essentially, no. For any variable, the mean of $\bar{x}$ equals the population mean, and the standard deviation of $\bar{x}$ equals (at least approximately) the population standard deviation divided by the square root of the sample size.

**7.51**

**a.** All four graphs are centered at the same place because $\mu_{\bar{x}} = \mu$ and normal distributions are centered at their means.

**b.** Because $\sigma_{\bar{x}} = \sigma/\sqrt{n}$, $\sigma_{\bar{x}}$ decreases as $n$ increases. This fact results in a diminishing of the spread because the spread of a distribution is determined by its standard deviation. As a consequence, the larger the sample size, the greater is the likelihood for small sampling error.

**c.** If the variable under consideration is normally distributed, so is the sampling distribution of the mean, regardless of sample size.

**d.** The central limit theorem indicates that, if the sample size is relatively large, the sampling distribution of the mean is approximately a normal distribution, regardless of the distribution of the variable under consideration.

**7.53**

**a.** A normal distribution with a mean of 1.40 and a standard deviation of 0.064. Thus, for samples of three Swedish men, the possible sample mean brain weights have a normal distribution with a mean of 1.40 kg and a standard deviation of 0.064 kg.

**b.** A normal distribution with a mean of 1.40 and a standard deviation of 0.032. Thus, for samples of 12 Swedish men, the possible sample mean brain weights have a normal distribution with a mean of 1.40 kg and a standard deviation of 0.032 kg.

**c.**

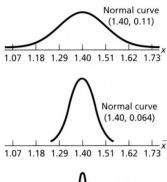

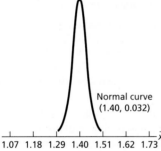

**d.** 88.12%. Chances are 88.12% that the sampling error made in estimating the mean brain weight of all Swedish men by that of a sample of three Swedish men will be at most 0.1 kg.

**e.** 99.82%. Chances are 99.82% that the sampling error made in estimating the mean brain weight of all Swedish men by that of a sample of 12 Swedish men will be at most 0.1 kg.

**7.55**

**a.** Approximately a normal distribution with a mean of 45,871 and a standard deviation of 1150. Thus, for samples of 64 classroom teachers in the public school system, the possible sample mean annual salaries are approximately normally distributed with a mean of $45,871 and a standard deviation of $1150.

**b.** Approximately a normal distribution with a mean of 45,871 and a standard deviation of 575. Thus, for samples of 256 classroom teachers in the public school system, the possible sample mean annual salaries are approximately normally distributed with a mean of $45,871 and a standard deviation of $575.

**c.** No. Because, in each case, the sample size exceeds 30.

**d.** 0.6156          **e.** 0.9182

**7.57** Let $\mu$ denote the mean length of hospital stay on the intervention ward.

**a.** Approximately a normal distribution with mean $\mu$ and standard deviation 0.93 days.

**b.** No, because the sample size is well in excess of 30.

**c.** 0.9684

**7.59** 0.9522

## REVIEW PROBLEMS FOR CHAPTER 7

1. Sampling error is the error resulting from using a sample to estimate a population characteristic.

2. The distribution of a statistic (i.e., of all possible observations of the statistic for samples of a given size) is called the sampling distribution of the statistic.

3. Sampling distribution of the sample mean; distribution of the variable $\bar{x}$

4. The possible sample means cluster closer around the population mean as the sample size increases. Thus, the larger the sample size, the smaller the sampling error tends to be in estimating a population mean, $\mu$, by a sample mean, $\bar{x}$.

5. **a.** The error resulting from using the mean income tax, $\bar{x}$, of the 175,485 tax returns selected as an estimate of the mean income tax, $\mu$, of all 2003 tax returns.
   **b.** $88
   **c.** No, not necessarily. However, increasing the sample size from 175,485 to 250,000 would increase the likelihood for small sampling error.
   **d.** Increase the sample size.

6. **a.** $\mu = \$18$ thousand
   **b.** The completed table is as follows.

| Sample | Salaries | $\bar{x}$ |
|--------|----------|-----------|
| A, B, C, D | 8, 12, 16, 20 | 14 |
| A, B, C, E | 8, 12, 16, 24 | 15 |
| A, B, C, F | 8, 12, 16, 28 | 16 |
| A, B, D, E | 8, 12, 20, 24 | 16 |
| A, B, D, F | 8, 12, 20, 28 | 17 |
| A, B, E, F | 8, 12, 24, 28 | 18 |
| A, C, D, E | 8, 16, 20, 24 | 17 |
| A, C, D, F | 8, 16, 20, 28 | 18 |
| A, C, E, F | 8, 16, 24, 28 | 19 |
| A, D, E, F | 8, 20, 24, 28 | 20 |
| B, C, D, E | 12, 16, 20, 24 | 18 |
| B, C, D, F | 12, 16, 20, 28 | 19 |
| B, C, E, F | 12, 16, 24, 28 | 20 |
| B, D, E, F | 12, 20, 24, 28 | 21 |
| C, D, E, F | 16, 20, 24, 28 | 22 |

**c.**

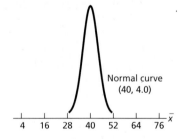

**d.** $\frac{7}{15}$

**e.** $18 thousand. For samples of four officers from the six, the mean of all possible sample mean monthly salaries equals $18 thousand.

**f.** Yes. Because $\mu_{\bar{x}} = \mu$ and, from part (a), $\mu = \$18$ thousand.

**7. a.** The population consists of all new cars sold in the United States in 2005. The variable is the amount spent on a new car.

**b.** $27,958; $1442.5

**c.** $27,958; $1020.0

**d.** Smaller, because $\sigma_{\bar{x}} = \sigma/\sqrt{n}$ and hence $\sigma_{\bar{x}}$ decreases with increasing sample size.

**8. a.** False                **b.** Not possible to tell
**c.** True

**9. a.** False        **b.** True            **c.** True

**10. a.** See the first graph that follows.

**b.** Normal distribution with a mean of 40 mm and a standard deviation of 6.0 mm, as shown in the second graph that follows.

**c.** Normal distribution with a mean of 40 mm and a standard deviation of 4.0 mm, as shown in the third graph that follows.

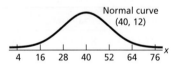

Normal curve (40, 12)

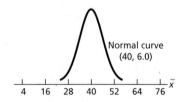

Normal curve (40, 6.0)

Normal curve (40, 4.0)

**11. a.** 86.64%                **b.** 0.8664

**c.** The probability that the sampling error will be at most 9 mm in estimating the population mean length of all krill by the mean length of a random sample of four krill is 0.8664.

**d.** 97.56%. 0.9756. The probability that the sampling error will be at most 9 mm in estimating the population mean length of all krill by the mean length of a random sample of nine krill is 0.9756.

**12. a.** For a normally distributed variable, the sampling distribution of the mean is a normal distribution, regardless of the sample size. Also, we know that $\mu_{\bar{x}} = \mu$. Consequently, because the normal curve for a normally distributed variable is centered at the mean, all three curves are centered at the same place.

**b.** Curve B. Because $\sigma_{\bar{x}} = \sigma/\sqrt{n}$, the larger the sample size, the smaller is the value of $\sigma_{\bar{x}}$ and hence the smaller is the spread of the normal curve for $\bar{x}$. Thus, Curve B, which has the smaller spread, corresponds to the larger sample size.

**c.** Because $\sigma_{\bar{x}} = \sigma/\sqrt{n}$ and the spread of a normal curve is determined by the standard deviation, different sample sizes result in normal curves with different spreads.

**d.** Curve B. The smaller the value of $\sigma_{\bar{x}}$, the smaller the sampling error tends to be.

**e.** Because the variable under consideration is normally distributed and, hence, so is the sampling distribution of the mean, regardless of sample size.

**13. a.** Approximately normally distributed with mean 4.60 and standard deviation 0.021.

**b.** Approximately normally distributed with mean 4.60 and standard deviation 0.015.

**c.** No, because, in each case, the sample size exceeds 30.

**14. a.** 0.6212

**b.** No. Because the sample size is large and therefore $\bar{x}$ is approximately normally distributed, regardless of the distribution of life insurance amounts. Yes.

**c.** 0.9946

**15. a.** No. If the manufacturer's claim is correct, the probability that the paint life for a randomly selected house painted with this paint will be 4.5 years or less is 0.1587; that is, such an event would occur roughly 16% of the time.

**b.** Yes. If the manufacturer's claim is correct, the probability that the mean paint life for 10 randomly selected houses painted with this paint will be 4.5 years or less is 0.0008; that is, such an event would occur less than 0.1% of the time.

**c.** No. If the manufacturer's claim is correct, the probability that the mean paint life for 10 randomly selected houses painted with this paint will be 4.9 years or less is 0.2643; that is, such an event would occur roughly 26% of the time.

# Chapter 8

## EXERCISES 8.1

**8.1** Point estimate

**8.3**
a. $26,326.9
b. No. It is unlikely that a sample mean, $\bar{x}$, will exactly equal the population mean, $\mu$; some sampling error is to be anticipated.

**8.5**
a. $22,704.5 to $29,949.3
b. We can be 95.44% confident that the mean cost, $\mu$, of all recent U.S. weddings is somewhere between $22,704.5 and $29,949.3.
c. It may or may not, but we can be 95.44% confident that it does.

**8.7**
a. 19.00 gallons. Based on the sample data, the mean fuel tank capacity of all 2003 automobile models is estimated to be 19.00 gallons.
b. 17.82 to 20.18. We can be 95.44% confident that the mean fuel tank capacity of all 2003 automobile models is somewhere between 17.82 gallons and 20.18 gallons.
c. Obtain a normal probability plot of the data.
d. No. Because the sample size is large.

**8.9**
a.

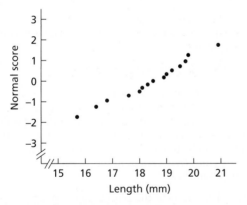

b. Yes, the plot is roughly linear and shows no outliers.
c. 17.52 mm to 19.34 mm. We can be 95.44% confident that the mean carapace length of all adult male Brazilian giant tawny red tarantulas is somewhere between 17.52 mm and 19.34 mm.
d. Yes. No.

## EXERCISES 8.2

**8.13**
a. Confidence level $= 0.90$; $\alpha = 0.10$
b. Confidence level $= 0.99$; $\alpha = 0.01$

**8.15**
a. Saying that the CI is exact means that the true confidence level is equal to $1 - \alpha$.
b. Saying that the CI is approximately correct means that the true confidence level is only approximately equal to $1 - \alpha$.

**8.17** The variable under consideration is normally distributed on the population of interest.

**8.19** A statistical procedure is said to be *robust* if it is insensitive to departures from the assumptions on which it is based.

**8.21** Key Fact 8.1 yields the following answers:
a. Reasonable          b. Not reasonable
c. Reasonable

**8.23** a 95% confidence level

**8.25** 19.0 to 21.0

**8.27** 28.7 to 31.3

**8.29** 46.8 to 53.2

**8.31**
a. $5.389 million to $7.274 million
b. We can be 95% confident that the mean amount of all venture-capital investments in the fiber optics business sector is somewhere between $5.389 million and $7.274 million.

**8.33**
a. 0.251 ppm to 0.801 ppm
b. We can be 99% confident that the mean cadmium level of all *Boletus pinicola* mushrooms is somewhere between 0.251 ppm and 0.801 ppm.

**8.35** 18.8 to 48.0 months. We can be 95% confident that the mean duration of imprisonment, $\mu$, of all East German political prisoners with chronic PTSD is somewhere between 18.8 and 48.0 months.

**8.37**
a. $5.093 million to $7.570 million
b. It is longer because the confidence level is greater.
c.

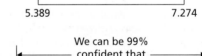

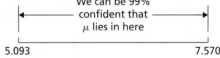

d. The 95% CI is a more precise estimate of $\mu$ because it is narrower than the 99% CI.

**8.39**
a. 276.8 months to 303.1 months
c. 272.0 months to 299.0 months
d. Although removal of the outlier does not appreciably affect the confidence interval, using the z-interval procedure here is not advisable because the sample size is moderate and the data contain an outlier.

## EXERCISES 8.3

**8.51** Because the margin of error equals half the length of a CI, it determines the precision with which a sample mean estimates a population mean.

**8.53**
a. 6.8                    b. 49.4 to 56.2

**8.55**
a. 10                           b. 50 to 70

**8.57**
a. True.  Because the margin of error is half the length of a CI, you can determine the length of a CI by doubling the margin of error.
b. True.  By taking half the length of a CI, you can determine the margin of error.
c. False. You need to know the sample mean as well.
d. True. Because the CI is from $\bar{x} - E$ to $\bar{x} + E$, you can obtain a CI by knowing only the margin of error, $E$, and the sample mean, $\bar{x}$.

**8.59**
a. The sample size (number of observations) cannot be fractional; it must be a whole number.
b. The number resulting from Formula 8.1 is the smallest value that will provide the required margin of error. If that value were rounded down, the sample size thus obtained would be insufficient to ensure the required margin of error.

**8.61**
a. 33.1 cm to 35.3 cm
b. 1.1 cm
c. We can be 90% confident that the error made in estimating $\mu$ by $\bar{x}$ is at most 1.1 cm.
d. 68

**8.63**
a. $0.94 million              b. $0.9424 million

**8.65**
a. 14.6 months
b. We can be 95% confident that the error made in estimating $\mu$ by $\bar{x}$ is at most 14.6 months.
c. 82 prisoners              d. 24.3 to 48.1 months

## EXERCISES 8.4

**8.71** The difference in the formulas lies in their denominators. The denominator of the standardized version of $\bar{x}$ uses the population standard deviation, $\sigma$, whereas the denominator of the studentized version of $\bar{x}$ uses the sample standard deviation, $s$.

**8.73**
a. $z = 1$                     b. $t = 1.333$

**8.75**
a. The standard normal distribution
b. $t$-distribution with df = 11

**8.77** The variation in the possible values of the standardized version is due solely to the variation of sample means, whereas that of the studentized version is due to the variation of both sample means and sample standard deviations.

**8.79**
a. 1.440          b. 2.447          c. 3.143

**8.81**
a. 1.323     b. 2.518     c. −2.080     d. ±1.721

**8.83** Yes. Because the sample size exceeds 30 and there are no outliers.

**8.85** 19.0 to 21.0

**8.87** 28.6 to 31.4

**8.89** 46.3 to 53.7

**8.91**
a. 24.9 minutes to 31.1 minutes
b. We can be 90% confident that the mean commute time of all commuters in Washington, D.C., is somewhere between 24.9 minutes and 31.1 minutes.

**8.93**
a. 0.90 hr to 3.76 hr.  We can be 95% confident that the additional sleep that would be obtained on average for all people using laevohysocyamine hydrobromide is somewhere between 0.90 hr and 3.76 hr.
b. It appears so because, based on the confidence interval, we can be 95% confident that the mean additional sleep is somewhere between 0.90 hr and 3.76 hr and that, in particular, the mean is positive.

**8.95**
a. 0.151 m/s to 0.247 m/s.  We can be 95% confident that the mean change in aortic-jet velocity of all such patients who receive 80 mg of atorvastatin daily is somewhere between 0.151 m/s and 0.247 m/s.
b. It appears so because, based on the confidence interval, we can be 95% confident that the mean change in aortic-jet velocity is somewhere between 0.151 m/s and 0.247 m/s and that, in particular, the mean is positive.

**8.97** No, not reasonable. The sample size is only moderate, the data contain outliers, and a normal probability plot indicates that the variable under consideration is far from normally distributed.

**8.99** Yes, it appears reasonable. The sample size is moderate and a normal probability plot of the data shows no outliers and is roughly linear.

## REVIEW PROBLEMS FOR CHAPTER 8

1. A point estimate of a parameter is the value of a statistic that is used to estimate the parameter; it consists of a single number, or point. A confidence-interval estimate of a parameter consists of an interval of numbers obtained from a point estimate of the parameter and a percentage that specifies how confident we are that the parameter lies in the interval.

2. False.  The mean of the population may or may not lie somewhere between 33.8 and 39.0, but we can be 95% confident that it does.

3. No. See the guidelines in Key Fact 8.1 on page 346.

4. Roughly 950 intervals would actually contain $\mu$.

5. Look at graphical displays of the data to ascertain whether the conditions required for using the procedure appear to be satisfied.

6. a. The precision of the estimate would decrease because the CI would be wider for a sample of size 50.
   b. The precision of the estimate would increase because the CI would be narrower for a 90% confidence level.

7. a. Because the length of a CI is twice the margin of error, the length of the CI is 21.4.
   b. 64.5 to 85.9

8. **a.** 6.58
   **b.** The sample mean, $\bar{x}$

9. **a.** $z = -0.77$          **b.** $t = -0.605$

10. **a.** Standard normal distribution
    **b.** $t$-distribution with 14 degrees of freedom

11. From Property 4 of Key Fact 8.6 (page 361), as the number of degrees of freedom becomes larger, $t$-curves look increasingly like the standard normal curve. So the curve that is closer to the standard normal curve has the larger degrees of freedom.

12. **a.** $t$-interval procedure        **b.** $z$-interval procedure
    **c.** $z$-interval procedure        **d.** Neither procedure
    **e.** $z$-interval procedure        **f.** Neither procedure

13. 54.3 yr to 62.8 yr

14. Part (c) provides the correct interpretation of the statement in quotes.

15. **a.** 11.7 mm to 12.1 mm
    **b.** We can be 90% confident that the mean length, $\mu$, of all *N. trivittata* is somewhere between 11.7 mm and 12.1 mm.
    **c.** A normal probability plot of the data should fall roughly in a straight line.

16. **a.** 0.2 mm
    **b.** We can be 90% confident that the error made in estimating $\mu$ by $\bar{x}$ is at most 0.2 mm.
    **c.** $n = 1692$          **d.** 11.9 mm to 12.1 mm

17. **a.** 2.101          **b.** 1.734
    **c.** $-1.330$        **d.** $\pm 2.878$

18. **a.** 81.69 mm Hg to 90.30 mm Hg. We can be 95% confident that the mean arterial blood pressure of all children of diabetic mothers is somewhere between 81.69 mm Hg and 90.30 mm Hg.
    **c.** Yes, the sample size is moderate, none of the graphs show any outliers, and the normal probability plot is linear.

19. **a.** $1880.1 to $2049.4. We can be 90% confident that the mean price of all one-half carat diamonds is somewhere between $1880.1 and $2049.4.
    **c.** This one is a tough call, but using the $t$-interval procedure is probably reasonable. The sample size is moderate and, although the boxplot shows a potential outlier, the other three plots suggest that the potential outlier may, in fact, not be an outlier. Furthermore, the normal probability plot is roughly linear.

# Chapter 9

## EXERCISES 9.1

**9.1** A hypothesis is a statement that something is true.

**9.3**
**a.** The population mean, $\mu$, equals some specified number, $\mu_0$; $H_0$: $\mu = \mu_0$.

**b.** Two tailed: The population mean, $\mu$, differs from $\mu_0$; $H_a$: $\mu \neq \mu_0$.
Left tailed: The population mean, $\mu$, is less than $\mu_0$; $H_a$: $\mu < \mu_0$.
Right tailed: The population mean, $\mu$, is greater than $\mu_0$; $H_a$: $\mu > \mu_0$.

**9.5** Let $\mu$ denote the mean cadmium level in *Boletus pinicola* mushrooms.
**a.** $H_0$: $\mu = 0.5$ ppm          **b.** $H_a$: $\mu > 0.5$ ppm
**c.** Right-tailed test

**9.7** Let $\mu$ denote the mean iron intake (per day) of all adult females under the age of 51.
**a.** $H_0$: $\mu = 18$ mg          **b.** $H_a$: $\mu < 18$ mg
**c.** Left-tailed test

**9.9** Let $\mu$ denote the mean length of imprisonment for motor-vehicle theft offenders in Sydney, Australia.
**a.** $H_0$: $\mu = 16.7$ months          **b.** $H_a$: $\mu \neq 16.7$ months
**c.** Two-tailed test

**9.11** Let $\mu$ denote the mean body temperature of all healthy humans.
**a.** $H_0$: $\mu = 98.6°$F          **b.** $H_a$: $\mu \neq 98.6°$F
**c.** Two-tailed test

**9.13** Let $\mu$ denote last year's mean local monthly bill for cell phone users.
**a.** $H_0$: $\mu = \$47.37$          **b.** $H_a$: $\mu > \$47.37$
**c.** Right-tailed test

## EXERCISES 9.2

**9.19**
**a.** True. Because the significance level, $\alpha$, is the probability of making a Type I error, it is unlikely that a true null hypothesis will be rejected if the hypothesis test is conducted at a small significance level.
**b.** True. By Key Fact 9.1 (page 391), for a fixed sample size, the smaller you specify the significance level, $\alpha$, the larger will be the probability, $\beta$, of not rejecting a false null hypothesis.

**9.21** The two types of incorrect decisions are a Type I error (rejection of a true null hypothesis) and a Type II error (nonrejection of a false null hypothesis). The probabilities of these two errors are denoted $\alpha$ and $\beta$, respectively.

**9.23**
**a.** $z \geq 1.645$        **b.** $z < 1.645$        **c.** $z = 1.645$
**d.** $\alpha = 0.05$
**e.**

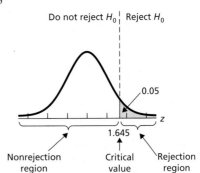

**f.** Right-tailed test

**9.25**
**a.** $z \le -2.33$       **b.** $z > -2.33$
**c.** $z = -2.33$       **d.** $\alpha = 0.01$
**e.**

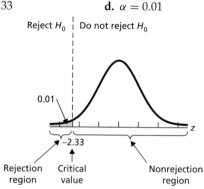

**f.** Left-tailed test

**9.27**
**a.** $z \le -1.645$ or $z \ge 1.645$    **b.** $-1.645 < z < 1.645$
**c.** $z = \pm 1.645$       **d.** $\alpha = 0.10$
**e.**

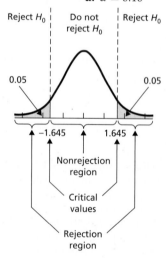

**f.** Two-tailed test

**9.29**
**a.** A Type I error would occur if in fact $\mu = 0.5$ ppm, but the results of the sampling lead to the conclusion that $\mu > 0.5$ ppm.
**b.** A Type II error would occur if in fact $\mu > 0.5$ ppm, but the results of the sampling fail to lead to that conclusion.
**c.** A correct decision would occur if in fact $\mu = 0.5$ ppm and the results of the sampling do not lead to the rejection of that fact; or if in fact $\mu > 0.5$ ppm and the results of the sampling lead to that conclusion.
**d.** Correct decision       **e.** Type II error

**9.31**
**a.** A Type I error would occur if in fact $\mu = 18$ mg, but the results of the sampling lead to the conclusion that $\mu < 18$ mg.
**b.** A Type II error would occur if in fact $\mu < 18$ mg, but the results of the sampling fail to lead to that conclusion.
**c.** A correct decision would occur if in fact $\mu = 18$ mg and the results of the sampling do not lead to the rejection of that fact; or if in fact $\mu < 18$ mg and the results of the sampling lead to that conclusion.

**d.** Type I error       **e.** Correct decision

**9.33**
**a.** A Type I error would occur if in fact $\mu = 16.7$ months, but the results of the sampling lead to the conclusion that $\mu \ne 16.7$ months.
**b.** A Type II error would occur if in fact $\mu \ne 16.7$ months, but the results of the sampling fail to lead to that conclusion.
**c.** A correct decision would occur if in fact $\mu = 16.7$ months and the results of the sampling do not lead to the rejection of that fact; or if in fact $\mu \ne 16.7$ months and the results of the sampling lead to that conclusion.
**d.** Correct decision       **e.** Type II error

**9.35**
**a.** A Type I error would occur if in fact $\mu = 98.6°$F, but the results of the sampling lead to the conclusion that $\mu \ne 98.6°$F.
**b.** A Type II error would occur if in fact $\mu \ne 98.6°$F, but the results of the sampling fail to lead to that conclusion.
**c.** A correct decision would occur if in fact $\mu = 98.6°$F and the results of the sampling do not lead to the rejection of that fact; or if in fact $\mu \ne 98.6°$F and the results of the sampling lead to that conclusion.
**d.** Type I error       **e.** Correct decision

**9.37**
**a.** A Type I error would occur if in fact $\mu = \$47.37$, but the results of the sampling lead to the conclusion that $\mu > \$47.37$.
**b.** A Type II error would occur if in fact $\mu > \$47.37$, but the results of the sampling fail to lead to that conclusion.
**c.** A correct decision would occur if in fact $\mu = \$47.37$ and the results of the sampling do not lead to the rejection of that fact; or if in fact $\mu > \$47.37$ and the results of the sampling lead to that conclusion.
**d.** Correct decision       **e.** Type II error

**9.39** Answers will vary.

**9.41**
**a.** Concluding that the defendant is guilty when in fact he or she is not.
**b.** Concluding that the defendant is not guilty when in fact he or she is.
**c.** Small (close to 0).       **d.** Small (close to 0).
**e.** An innocent person is never convicted; a guilty person is always convicted.

## EXERCISES 9.3

**9.45** Critical value: $z_{0.05} = 1.645$

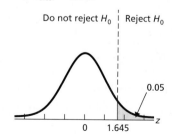

**9.47** Critical value: $-z_{0.05} = -1.645$

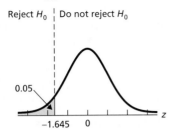

**9.49** Critical values: $\pm z_{0.025} = \pm 1.96$

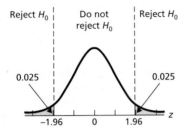

**9.51**
a. Inappropriate          b. Appropriate

**9.53** $z = -2.83$; critical value $= -1.645$; reject $H_0$

**9.55** $z = 1.94$; critical value $= 1.645$; reject $H_0$

**9.57** $z = 1.22$; critical values $= \pm 1.96$; do not reject $H_0$

**9.59** $H_0: \mu = 0.5$ ppm, $H_a: \mu > 0.5$ ppm; $\alpha = 0.05$; $z = 0.24$; critical value $= 1.645$; do not reject $H_0$; at the 5% significance level, the data do not provide sufficient evidence to conclude that the mean cadmium level in *Boletus pinicola* mushrooms is greater than the government's recommended limit of 0.5 ppm.

**9.61** $H_0: \mu = 18$ mg, $H_a: \mu < 18$ mg; $\alpha = 0.01$; $z = -5.30$; critical value $= -2.33$; reject $H_0$; at the 1% significance level, the data provide sufficient evidence to conclude that adult females under the age of 51 are, on average, getting less than the RDA of 18 mg of iron.

**9.63** $H_0: \mu = 16.7$ months, $H_a: \mu \neq 16.7$ months; $\alpha = 0.05$; $z = 1.83$; critical values $= \pm 1.96$; do not reject $H_0$; at the 5% significance level, the data do not provide sufficient evidence to conclude that the mean length of imprisonment for motor-vehicle theft offenders in Sydney differs from the national mean in Australia.

**9.65**
a. At the 5% significance level, the data do not provide sufficient evidence to conclude that, on average, the net percentage gain for jobs exceeds 0.2.
c. Removing the potential outlier ($-1.1$), we conclude, at the 5% significance level, that, on average, the net percentage gain for jobs exceeds 0.2.
d. The sample size is moderate, there is a potential outlier in the data, and the variable under consideration appears to be left skewed. Furthermore, removal of the potential outlier affects the conclusion of the hypothesis test. Using the z-test here is not advisable.

**EXERCISES 9.4**

**9.73** (1) It allows you to assess significance at any desired level. (2) It permits you to evaluate the strength of the evidence against the null hypothesis.

**9.75** The two different approaches to hypothesis testing are the critical-value approach and the *P*-value approach. For a comparison of the two approaches, see Table 9.9 on page 414.

**9.77** True

**9.79**
a. Do not reject the null hypothesis.
b. Reject the null hypothesis.
c. Do not reject the null hypothesis.

**9.81**
a. Moderate                    b. Weak or none
c. Strong                      d. Very strong

**9.83**
a. $z = -2.83$; *P*-value $= 0.002$; reject $H_0$
b. Very strong

**9.85**
a. $z = 1.94$; *P*-value $= 0.026$; reject $H_0$
b. Strong

**9.87**
a. $z = 1.22$; *P*-value $= 0.221$; do not reject $H_0$
b. Weak or none

**9.89**
a. 0.0212                      b. 0.6217

**9.91**
a. 0.2296                      b. 0.8770

**9.93**
a. 0.0970                      b. 0.6030

**9.95** $H_0: \mu = 0.5$ ppm, $H_a: \mu > 0.5$ ppm; $\alpha = 0.05$; $z = 0.24$; $P = 0.404$; do not reject $H_0$; at the 5% significance level, the data do not provide sufficient evidence to conclude that the mean cadmium level in *Boletus pinicola* mushrooms is greater than the government's recommended limit of 0.5 ppm. The evidence against the null hypothesis is weak or none.

**9.97** $H_0: \mu = 18$ mg, $H_a: \mu < 18$ mg; $\alpha = 0.01$; $z = -5.30$; $P = 0.0000$ (to four decimal places); reject $H_0$; at the 1% significance level, the data provide sufficient evidence to conclude that adult females under the age of 51 are, on average, getting less than the RDA of 18 mg of iron. The evidence against the null hypothesis is very strong.

**9.99** $H_0: \mu = 16.7$ months, $H_a: \mu \neq 16.7$ months; $\alpha = 0.05$; $z = 1.83$; $P = 0.067$; do not reject $H_0$; at the 5% significance level, the data do not provide sufficient evidence to conclude that the mean length of imprisonment for motor-vehicle theft offenders in Sydney differs from the national mean in Australia. The evidence against the null hypothesis is moderate.

**9.101**
a. $P = 0.156$. At the 5% significance level, the data do not provide sufficient evidence to conclude that, on average, the net percentage gain for jobs exceeds 0.2. The evidence against the null hypothesis is weak or none.

**c.** Removing the potential outlier $(-1.1)$, we get $P = 0.040$ and conclude, at the 5% significance level, that, on average, the net percentage gain for jobs exceeds 0.2. The evidence against the null hypothesis is strong.

**d.** The sample size is moderate, there is a potential outlier in the data, and the variable under consideration appears to be left skewed. Furthermore, removal of the potential outlier affects the conclusion of the hypothesis test. Using the $z$-test here is not advisable.

## EXERCISES 9.5

**9.111**
**a.** $0.01 < P < 0.025$
**b.** We can reject $H_0$ at any significance level of 0.025 or larger, and we cannot reject $H_0$ at any significance level of 0.01 or smaller. For significance levels between 0.01 and 0.025, Table IV is not sufficiently detailed to help us to decide whether to reject $H_0$.

**9.113**
**a.** $P < 0.005$
**b.** We can reject $H_0$ at any significance level of 0.005 or larger. For significance levels smaller than 0.005, Table IV is not sufficiently detailed to help us to decide whether to reject $H_0$.

**9.115**
**a.** $0.01 < P < 0.02$
**b.** We can reject $H_0$ at any significance level of 0.02 or larger, and we cannot reject $H_0$ at any significance level of 0.01 or smaller. For significance levels between 0.01 and 0.02, Table IV is not sufficiently detailed to help us to decide whether to reject $H_0$.

**9.117**
**a.** $t = -2.83$; critical value $= -1.696$; $P$-value $= 0.004$ ($P < 0.005$); reject $H_0$
**b.** Very strong

**9.119**
**a.** $t = 1.94$; critical value $= 1.761$; $P$-value $= 0.037$ ($0.025 < P < 0.05$); reject $H_0$
**b.** Strong

**9.121**
**a.** $t = 1.22$; critical values $= \pm 2.069$; $P$-value $= 0.233$ ($P > 0.20$); do not reject $H_0$
**b.** Weak or none

**9.123** $H_0$: $\mu = 4.66$ hr, $H_a$: $\mu \neq 4.66$ hr; $\alpha = 0.10$; $t = 0.34$; critical values $= \pm 1.729$; $P > 0.20$; do not reject $H_0$; at the 10% significance level, the data do not provide sufficient evidence to conclude that the amount of television watched per day last year by the average person differed from that in 2002.

**9.125** $H_0$: $\mu = 2.30\%$, $H_a$: $\mu > 2.30\%$; $\alpha = 0.01$; $t = 4.251$; critical value $= 2.821$; $P < 0.005$; reject $H_0$; at the 1% significance level, the data provide sufficient evidence to conclude that the mean available limestone in soil treated with 100% MMBL effluent exceeds 2.30%.

**9.127** $H_0$: $\mu = 0.9$, $H_a$: $\mu < 0.9$; $\alpha = 0.05$; $t = -23.703$; critical value $= -1.653$; $P < 0.005$; reject $H_0$; at the 5% significance level, the data provide sufficient evidence to

conclude that, on average, women with peripheral arterial disease have an unhealthy ABI.

**9.129** Yes, it appears reasonable. The sample size is moderate and a normal probability plot shows no outliers and is (very) roughly linear.

**9.131** No, not reasonable. The sample size is only moderate, and it appears that the variable under consideration is highly right skewed and hence far from normally distributed.

## REVIEW PROBLEMS FOR CHAPTER 9

**1. a.** The null hypothesis is a hypothesis to be tested.
**b.** The alternative hypothesis is a hypothesis to be considered as an alternate to the null hypothesis.
**c.** The test statistic is the statistic used as a basis for deciding whether the null hypothesis should be rejected.
**d.** The rejection region is the set of values for the test statistic that leads to rejection of the null hypothesis.
**e.** The nonrejection region is the set of values for the test statistic that leads to nonrejection of the null hypothesis.
**f.** The critical values are the values of the test statistic that separate the rejection and nonrejection regions. The critical values are considered part of the rejection region.

**2. a.** The weight of a package of Tide is a variable. A particular package may weigh slightly more or less than the marked weight. The mean weight of all packages produced on any specified day (the population mean weight for that day) exceeds the marked weight.
**b.** The null hypothesis would be that the population mean weight for a specified day equals the marked weight; the alternative hypothesis would be that the population mean weight for the specified day exceeds the marked weight.
**c.** The null hypothesis would be that the population mean weight for a specified day equals the marked weight of 76 oz; the alternative hypothesis would be that the population mean weight for the specified day exceeds the marked weight of 76 oz. In statistical terminology, the hypothesis test would be $H_0$: $\mu = 76$ oz and $H_a$: $\mu > 76$ oz, where $\mu$ is the mean weight of all packages produced on the specified day.

**3. a.** Obtain the data from a random sample of the population or from a designed experiment. If the data are consistent with the null hypothesis, do not reject the null hypothesis; if the data are inconsistent with the null hypothesis, reject the null hypothesis and conclude that the alternative hypothesis is true.
**b.** We establish a precise criterion for deciding whether to reject the null hypothesis prior to obtaining the data.

**4.** Two-tailed test, $H_a$: $\mu \neq \mu_0$. Used when the primary concern is deciding whether a population mean, $\mu$, is different from a specified value $\mu_0$.

Left-tailed test, $H_a$: $\mu < \mu_0$. Used when the primary concern is deciding whether a population mean, $\mu$, is less than a specified value $\mu_0$.

Right-tailed test, $H_a: \mu > \mu_0$. Used when the primary concern is deciding whether a population mean, $\mu$, is greater than a specified value $\mu_0$.

5.  **a.** A Type I error is the incorrect decision of rejecting a true null hypothesis. A Type II error is the incorrect decision of not rejecting a false null hypothesis.
    **b.** $\alpha$ and $\beta$, respectively
    **c.** A Type I error          **d.** A Type II error

6.  It must be chosen so that, if the null hypothesis is true, the probability equals 0.05 that the test statistic will fall in the rejection region, in this case, to the right of the critical value.

7.  **a.** Assumptions: simple random sample; normal population or large sample; $\sigma$ unknown. Test statistic: $t = (\bar{x} - \mu_0)/(s/\sqrt{n})$.
    **b.** Assumptions: simple random sample; normal population or large sample; $\sigma$ known. Test statistic: $z = (\bar{x} - \mu_0)/(\sigma/\sqrt{n})$.

8.  **a.** The true significance level equals $\alpha$.
    **b.** The true significance level only approximately equals $\alpha$.

9.  The results of a hypothesis test are statistically significant if the null hypothesis is rejected at the specified significance level. Statistical significance means that the data provide sufficient evidence to conclude that the truth is different from the stated null hypothesis. It does not necessarily mean that the difference is important in any practical sense.

10. It increases.

11. **a.** The $P$-value is the probability, calculated under the assumption that the null hypothesis is true, of observing a value of the test statistic as extreme as or more extreme than that observed. By *extreme* we mean "far from what we would expect to observe if the null hypothesis is true."
    **b.** True          **c.** True
    **d.** Because it is the smallest significance level for which the observed sample data result in rejection of the null hypothesis.

13. Let $\mu$ denote last year's mean cheese consumption by Americans.
    **a.** $H_0: \mu = 30.0$ lb          **b.** $H_a: \mu > 30.0$ lb
    **c.** Right tailed

14. **a.** $z \geq 1.28$          **b.** $z < 1.28$
    **c.** $z = 1.28$          **d.** $\alpha = 0.10$
    **e.**

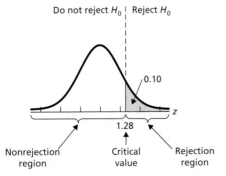

**f.** Right tailed

15. **a.** A Type I error would occur if in fact $\mu = 30.0$ lb, but the results of the sampling lead to the conclusion that $\mu > 30.0$ lb.
    **b.** A Type II error would occur if in fact $\mu > 30.0$ lb, but the results of the sampling fail to lead to that conclusion.
    **c.** A correct decision would occur if in fact $\mu = 30.0$ lb and the results of the sampling do not lead to the rejection of that fact; or if in fact $\mu > 30.0$ lb and the results of the sampling lead to that conclusion.
    **d.** Correct decision          **e.** Type II error

16. **a.** $H_0: \mu = 30.0$ lb, $H_a: \mu > 30.0$ lb; $\alpha = 0.10$; $z = 0.69$; critical value $= 1.28$; do not reject $H_0$; at the 10% significance level, the data do not provide sufficient evidence to conclude that last year's mean cheese consumption for all Americans has increased over the 2001 mean of 30.0 lb.
    **b.** A Type II error because, given that the null hypothesis was not rejected, the only error that could be made is the error of not rejecting a false null hypothesis.

17. **a.** $H_0: \mu = 30.0$ lb, $H_a: \mu > 30.0$ lb; $\alpha = 0.10$; $z = 0.69$; $P = 0.2451$; do not reject $H_0$; at the 10% significance level, the data do not provide sufficient evidence to conclude that last year's mean cheese consumption for all Americans has increased over the 2001 mean of 30.0 lb.
    **b.** The data provide at most weak evidence against the null hypothesis.

18. $H_0: \mu = \$332$, $H_a: \mu < \$332$; $\alpha = 0.05$; $t = -1.909$; critical value $= -1.796$; $0.025 < P < 0.05$; reject $H_0$; at the 5% significance level, the data provide sufficient evidence to conclude that last year's mean value lost to purse snatching has decreased from the 2002 mean of $332.

19. **a.** 0 points
    **b.** $H_0: \mu = 0$ points, $H_a: \mu \neq 0$ points; $\alpha = 0.05$; $t = -0.843$; critical values $= \pm 1.96$; $P > 0.20$; do not reject $H_0$.
    **c.** At the 5% significance level, the data do not provide sufficient evidence to conclude that the population mean point-spread error differs from 0. In fact, because $P > 0.20$, there is virtually no evidence against the null hypothesis that the population mean point-spread error equals 0.

*Note:* Although in some of Problems 20–29, a different type of procedure (such as a nonparametric method) might be preferable, the stated procedures can be considered acceptable.

20. $z$-test

21. $t$-test

22. $z$-test

23. $z$-test

24. $t$-test

25. Neither (although some statisticians would consider the $t$-test acceptable)

26. Neither (although some statisticians would consider the $t$-test acceptable)

27. Neither

28. Neither

29. Neither

# Chapter 10

## EXERCISES 10.1

**10.1** Answers will vary.

**10.3**

**a.** $\mu_1$, $\sigma_1$, $\mu_2$, and $\sigma_2$ are parameters; $\bar{x}_1$, $s_1$, $\bar{x}_2$, and $s_2$ are statistics.

**b.** $\mu_1$, $\sigma_1$, $\mu_2$, and $\sigma_2$ are fixed numbers; $\bar{x}_1$, $s_1$, $\bar{x}_2$, and $s_2$ are variables.

**10.5** So that you can determine whether the observed difference between the two sample means can be reasonably attributed to sampling error or whether that difference suggests that the null hypothesis of equal population means is false and the alternative hypothesis is true.

**10.7** Let $\mu_1$ and $\mu_2$ denote the mean salaries of faculty in public and private institutions, respectively. The null and alternative hypotheses are $H_0$: $\mu_1 = \mu_2$ and $H_a$: $\mu_1 < \mu_2$.

**10.9**

**a.** Systolic blood pressure

**b.** ODM adolescents and ONM adolescents

**c.** Let $\mu_1$ and $\mu_2$ denote the mean systolic blood pressures of ODM adolescents and ONM adolescents, respectively. The null and alternative hypotheses are $H_0$: $\mu_1 = \mu_2$ and $H_a$: $\mu_1 > \mu_2$.

**d.** Right tailed

**10.11**

**a.** Last year's vehicle miles of travel (VMT)

**b.** Households in the Midwest and households in the South

**c.** Let $\mu_1$ and $\mu_2$ denote last year's mean VMT for households in the Midwest and South, respectively. The null and alternative hypotheses are $H_0$: $\mu_1 = \mu_2$ and $H_a$: $\mu_1 \neq \mu_2$.

**d.** Two tailed

**10.13**

**a.** Operative time

**b.** Dynamic-system operations and static-system operations

**c.** Let $\mu_1$ and $\mu_2$ denote the mean operative times with the dynamic and static systems, respectively. The null and alternative hypotheses are $H_0$: $\mu_1 = \mu_2$ and $H_a$: $\mu_1 < \mu_2$.

**d.** Left tailed

**10.15**

**a.** 0 and 5     **b.** No.     **c.** No.

**10.17**

**a.** 0 and 5     **b.** Yes.     **c.** 95.44%

## EXERCISES 10.2

**10.21**

**a.** Simple random samples, independent samples, normal populations or large samples, and equal population standard deviations

**b.** Simple random samples and independent samples are essential assumptions. Moderate violations of the normality assumption are permissible even for small or moderate size samples. Moderate violations of the equal standard deviations requirement are not serious provided the two sample sizes are roughly equal.

*Note:* From the instructions for Exercises 10.23–10.26, the only assumption for pooled $t$-procedures we need to address is that of equal population standard deviations.

**10.23** No, not reasonable, because the sample standard deviations suggest that the two population standard deviations differ and the sample sizes are not roughly equal.

**10.25** Yes, because the sample standard deviations are close to being equal, suggesting that assuming the population standard deviations are equal is reasonable.

**10.27**

**a.** $t = -2.49$; critical values $= \pm 2.048$; $0.01 < P < 0.02$; reject $H_0$

**b.** $-3.65$ to $-0.35$

**10.29**

**a.** $t = 1.06$; critical value $= 1.714$; $P > 0.10$; do not reject $H_0$

**b.** $-1.24$ to $5.24$

**10.31**

**a.** $t = -2.63$; critical value $= -1.692$; $0.005 < P < 0.01$; reject $H_0$

**b.** $-6.57$ to $-1.43$

**10.33** $H_0$: $\mu_1 = \mu_2$, $H_a$: $\mu_1 < \mu_2$; $\alpha = 0.05$; $t = -4.058$; critical value $= -1.734$; $P < 0.005$; reject $H_0$; at the 5% significance level, the data provide sufficient evidence to conclude that the mean time served for fraud is less than that for firearms offenses.

**10.35** $H_0$: $\mu_1 = \mu_2$, $H_a$: $\mu_1 > \mu_2$; $\alpha = 0.05$; $t = 0.520$; critical value $= 1.711$; $P > 0.10$; do not reject $H_0$; at the 5% significance level, the data do not provide sufficient evidence to conclude that drinking fortified orange juice reduces PTH level more than drinking unfortified orange juice.

**10.37** $H_0$: $\mu_1 = \mu_2$, $H_a$: $\mu_1 \neq \mu_2$; $\alpha = 0.05$; $t = -1.98$; critical values $= \pm 1.971$; $0.02 < P < 0.05$; reject $H_0$; at the 5% significance level, the data provide sufficient evidence to conclude that a difference exists in the mean number of native species in the two regions.

**10.39** $-12.36$ to $-4.96$ months. We can be 90% confident that the difference between the mean times served by prisoners in the fraud and firearms offense categories is somewhere between $-12.36$ and $-4.96$ months.

**10.41** $-16.92$ to $31.72$ pg/mL. We can be 90% confident that the difference between the mean reductions in PTH levels for fortified and unfortified orange juice is somewhere between $-16.92$ and $31.72$ pg/mL.

**10.43** $-2.596$ to $-0.004$ native species. We can be 95% confident that the difference between the mean number of native species in the two regions is somewhere between $-2.596$ and $-0.004$.

## EXERCISES 10.3

**10.57**

**a.** Pooled $t$-test     **b.** Nonpooled $t$-test

**c.** Pooled $t$-test     **d.** Nonpooled $t$-test

**10.59**
a. $t = -1.44$; critical values $= \pm 2.101$; $0.10 < P < 0.20$; do not reject $H_0$
b. $-4.92$ to $0.92$

**10.61**
a. $t = 1.11$; critical value $= 1.717$; $P > 0.10$; do not reject $H_0$
b. $-1.10$ to $5.10$

**10.63**
a. $t = -2.78$; critical value $= -1.711$; $0.005 < P < 0.01$; reject $H_0$
b. $-6.46$ to $-1.54$

**10.65** $H_0$: $\mu_1 = \mu_2$, $H_a$: $\mu_1 \neq \mu_2$; $\alpha = 0.10$; $t = 1.791$; critical values $= \pm 1.677$; $0.05 < P < 0.10$; reject $H_0$; at the 10% significance level, the data provide sufficient evidence to conclude that a difference exists in the mean age at arrest of East German prisoners with chronic PTSD and remitted PTSD.

**10.67** $H_0$: $\mu_1 = \mu_2$, $H_a$: $\mu_1 < \mu_2$; $\alpha = 0.05$; $t = -1.651$; critical value $= -2.015$; $0.05 < P < 0.10$; do not reject $H_0$; at the 5% significance level, the data do not provide sufficient evidence to conclude that the mean number of acute postoperative days in the hospital is smaller with the dynamic system than with the static system.

**10.69** $H_0$: $\mu_1 = \mu_2$, $H_a$: $\mu_1 > \mu_2$; $\alpha = 0.01$; $t = 3.863$; critical value $= 2.552$; $P < 0.005$; reject $H_0$; at the 1% significance level, the data provide sufficient evidence to conclude that dopamine activity is higher, on average, in psychotic patients.

**10.71** 0.2 to 7.2 years. We can be 90% confident that the difference between the mean ages at arrest of East German prisoners with chronic PTSD and remitted PTSD is somewhere between 0.2 and 7.2 yr.

**10.73** $-6.97$ to $0.69$ days. We can be 90% confident that the difference between the mean number of acute postoperative days in the hospital with the dynamic and static systems is somewhere between $-6.97$ and $0.69$ days.

**10.75** 0.00266 to 0.01301 nmol/ml-h/mg. We can be 98% confident that the difference between the mean dopamine activities of psychotic and nonpsychotic patients is somewhere between 0.00266 and 0.01301 nmol/ml-h/mg.

**10.77**
a. Nonpooled $t$-procedures because the sample standard deviations indicate that the population standard deviations are far from equal and the sample sizes are quite different.
b. No, because a normal probability plot for the males' data is far from linear and indicates the presence of outliers.

**10.79**
a. $H_0$: $\mu_1 = \mu_2$, $H_a$: $\mu_1 < \mu_2$; $\alpha = 0.05$; $t = -2.45$; critical value $= -1.734$; $0.01 < P < 0.025$; reject $H_0$; at the 5% significance level, the data provide sufficient evidence to conclude that the mean number of acute postoperative days in the hospital is smaller with the dynamic system than with the static system.
b. The null hypothesis is not rejected using the nonpooled $t$-test, whereas it is rejected using the pooled $t$-test.
c. The nonpooled $t$-test, because the sample standard deviations strongly suggest that the population standard deviations are not equal.

## EXERCISES 10.4

**10.87** By using a paired sample, extraneous sources of variation can be removed. The sampling error thus made in estimating the difference between the population means will generally be smaller. As a result, detecting differences between the population means is more likely when such differences exist.

**10.89** Simple random paired sample, and normal differences or large sample. The simple-random-paired-sample assumption is essential. Moderate violations of the normal-differences assumption are permissible even for small or moderate size samples.

**10.91**
a. Age
b. Married men and married women
c. Married couples
d. The difference between the ages of a married couple
e. Let $\mu_1$ and $\mu_2$ denote the mean ages of married men and married women, respectively. The null and alternative hypotheses are $H_0$: $\mu_1 = \mu_2$ and $H_a$: $\mu_1 > \mu_2$.
f. Right tailed

**10.93**
a. Home price
b. Homes neighboring and homes not neighboring newly constructed sports stadiums
c. A pair of comparable homes, one neighboring and the other not neighboring a newly constructed sports stadium
d. The difference between the prices of a pair of comparable homes, one neighboring and the other not neighboring a newly constructed sports stadium
e. Let $\mu_1$ and $\mu_2$ denote the mean prices of homes neighboring and not neighboring newly constructed sports stadiums, respectively. The null and alternative hypotheses are $H_0$: $\mu_1 = \mu_2$ and $H_a$: $\mu_1 \neq \mu_2$.
f. Two tailed

**10.95**
a. Height (of Zea mays)
b. Cross-fertilized Zea mays and self-fertilized Zea mays
c. The difference between the heights of a cross-fertilized Zea may and a self-fertilized Zea may grown in the same pot
d. Yes. Because each number is the difference between the heights of a cross-fertilized Zea may and a self-fertilized Zea may grown in the same pot
e. $H_0$: $\mu_1 = \mu_2$, $H_a$: $\mu_1 \neq \mu_2$; $\alpha = 0.05$; $t = 2.148$; critical values $= \pm 2.145$; $0.02 < P < 0.05$; reject $H_0$; at the 5% significance level, the data provide sufficient evidence to conclude that the mean heights of cross-fertilized and self-fertilized Zea mays differ.
f. $H_0$: $\mu_1 = \mu_2$, $H_a$: $\mu_1 \neq \mu_2$; $\alpha = 0.01$; $t = 2.148$; critical values $= \pm 2.977$; $0.02 < P < 0.05$; do not reject $H_0$; at the 1% significance level, the data do not provide sufficient evidence to conclude that the mean heights of cross-fertilized and self-fertilized Zea mays differ.

**10.97** $H_0$: $\mu_1 = \mu_2$, $H_a$: $\mu_1 < \mu_2$; $\alpha = 0.05$; $t = -4.185$; critical value $= -1.746$; $P < 0.005$; reject $H_0$; at the 5% significance level, the data provide sufficient evidence to conclude that family therapy is effective in helping anorexic young women gain weight.

**10.99** $H_0$: $\mu_1 = \mu_2$, $H_a$: $\mu_1 > \mu_2$; $\alpha = 0.10$; $t = 1.053$; critical value $= 1.415$; $P > 0.10$; do not reject $H_0$; at the 10% significance level, the data do not provide sufficient evidence to conclude that mean corneal thickness is greater in normal eyes than in eyes with glaucoma.

**10.101**
**a.** 0.03 to 41.84 eighths of an inch
**b.** −8.08 to 49.94 eighths of an inch

**10.103** −10.30 to −4.23 lb. We can be 90% confident that the weight gain that would be obtained, on average, by using the family therapy treatment is somewhere between 4.23 and 10.30 lb.

**10.105** −1.4 to 9.4 microns. We can be 80% confident that the difference between the mean corneal thickness of normal eyes and that of eyes with glaucoma is somewhere between −1.4 and 9.4 microns.

**10.107** Evidently, the first paired difference (13) is an outlier. Therefore, in view of the small sample size, applying the paired $t$-test is not reasonable.

**10.109**
**b.** The normal probability plot of the onset data indicates extreme deviation from normality. Therefore, in view of the small sample size, applying a one-mean $t$-procedure is not reasonable.
**c.** The normal probability plot of the resolution data is only roughly linear and the boxplot suggests a potential outlier. Therefore, in view of the small sample size, applying a one-mean $t$-procedure is probably not reasonable.
**d.** Neither the normal probability plot nor the boxplot of the paired differences suggest the presence of outliers and, furthermore, the normal probability plot of the paired differences is quite linear. Therefore, applying a paired $t$-procedure is reasonable.
**e.** Whether applying a paired $t$-procedure is reasonable depends on the properties of the paired-difference variable and not on those of the individual variables that constitute the paired-difference variable.

## REVIEW PROBLEMS FOR CHAPTER 10

1. Independently and randomly take samples from the two populations; compute the two sample means; compare the two sample means; and make the decision.

2. Randomly take a paired sample from the two populations; calculate the paired differences of the sample pairs; compute the mean of the sample of paired differences; compare that sample mean to 0; and make the decision.

3. **a.** The pooled $t$-procedures require equal population standard deviations, whereas the nonpooled $t$-procedures do not.
   **b.** It is essential that the assumption of independence be satisfied.
   **c.** For very small sample sizes, the normality assumption is essential for both $t$-procedures. However, for larger samples, the normality assumption is less important.
   **d.** Population standard deviations

4. By using a paired sample, extraneous sources of variation can be removed. As a consequence, the sampling error made in estimating the difference between the population means will generally be smaller. This fact, in turn, makes it more likely that differences between the population means will be detected when such differences exist.

5. **a.** $H_0$: $\mu_1 = \mu_2$, $H_a$: $\mu_1 > \mu_2$; $\alpha = 0.05$; $t = 1.538$; critical value $= 1.708$; $0.05 < P < 0.10$; do not reject $H_0$; at the 5% significance level, the data do not provide sufficient evidence to conclude that the mean right-leg strength of males exceeds that of females.
   **b.** $0.05 < P < 0.10$; the evidence against the null hypothesis is moderate.

6. −31.3 to 599.3 newtons (N). We can be 90% confident that the difference between the mean right-leg strengths of males and females is somewhere between −31.3 and 599.3 N.

7. $H_0$: $\mu_1 = \mu_2$, $H_a$: $\mu_1 < \mu_2$; $\alpha = 0.01$; $t = -4.118$; critical value $= -2.385$; $P < 0.005$; reject $H_0$; at the 1% significance level, the data provide sufficient evidence to conclude that, on average, the number of young per litter of cottonmouths in Florida is less than that in Virginia.

8. −3.4 to −0.9 young per litter. We can be 98% confident that the difference between the mean litter sizes of cottonmouths in Florida and Virginia is somewhere between −3.4 and −0.9. With 98% confidence, we can say that, on average, cottonmouths in Virginia have somewhere between 0.9 and 3.4 more young per litter than those in Florida.

9. **b.** Yes, the normal probability plot is quite linear and neither that plot nor the boxplot reveal any outliers.
   **c.** $H_0$: $\mu_1 = \mu_2$, $H_a$: $\mu_1 \neq \mu_2$; $\alpha = 0.10$; $t = 0.55$; critical values $= \pm 1.895$; $P > 0.20$; do not reject $H_0$; at the 10% significance level, the data do not provide sufficient evidence to conclude that a difference exists in the mean length of time that ice stays on these two lakes.

10. −3.4 to 6.1 days. We can be 90% confident that the difference in the mean lengths of time that ice stays on the two lakes is somewhere between −3.4 and 6.1 days.

11. paired $t$-test

12. nonpooled $t$-test

13. none that you have studied

14. nonpooled $t$-test

15. none that you have studied

16. none that you have studied

# Chapter 11

## EXERCISES 11.1

**11.1** Answers will vary.

**11.3** A population proportion is a parameter because it is a descriptive measure for a population. A sample proportion is a statistic because it is a descriptive measure for a sample.

**11.5**

**a.** $p = 0.4$

**b.**

| Sample | No. of females $x$ | Sample proportion $\hat{p}$ |
|--------|--------------------|-----------------------------|
| J, G   | 1                  | 0.5                         |
| J, P   | 0                  | 0.0                         |
| J, C   | 0                  | 0.0                         |
| J, F   | 1                  | 0.5                         |
| G, P   | 1                  | 0.5                         |
| G, C   | 1                  | 0.5                         |
| G, F   | 2                  | 1.0                         |
| P, C   | 0                  | 0.0                         |
| P, F   | 1                  | 0.5                         |
| C, F   | 1                  | 0.5                         |

**c.**

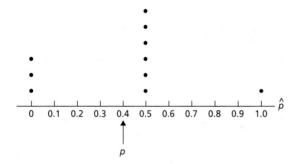

**d.** 0.4

**e.** They are the same because the mean of the variable $\hat{p}$ equals the population proportion; in symbols, $\mu_{\hat{p}} = p$.

**11.7**

**b.**

| Sample | No. of females $x$ | Sample proportion $\hat{p}$ |
|--------|--------------------|-----------------------------|
| J, P, C | 0                 | 0.00                        |
| J, P, G | 1                 | 0.33                        |
| J, P, F | 1                 | 0.33                        |
| J, C, G | 1                 | 0.33                        |
| J, C, F | 1                 | 0.33                        |
| J, G, F | 2                 | 0.67                        |
| P, C, G | 1                 | 0.33                        |
| P, C, F | 1                 | 0.33                        |
| P, G, F | 2                 | 0.67                        |
| C, G, F | 2                 | 0.67                        |

**c.**

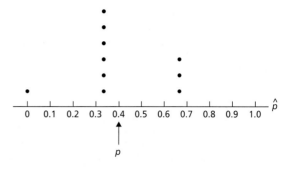

**d.** 0.4

**e.** They are the same because the mean of the variable $\hat{p}$ equals the population proportion; in symbols, $\mu_{\hat{p}} = p$.

**11.9**

**b.**

| Sample | No. of females $x$ | Sample proportion $\hat{p}$ |
|--------|--------------------|-----------------------------|
| J, P, C, G, F | 2           | 0.4                         |

**c.**

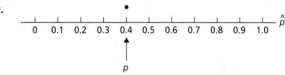

**d.** 0.4

**e.** They are the same because the mean of the variable $\hat{p}$ equals the population proportion; in symbols, $\mu_{\hat{p}} = p$.

**11.11**

**a.** The No. 1 draft picks in the NBA since 1966

**b.** Being a center

**c.** Population proportion. It is the proportion of the population of No. 1 draft picks in the NBA since 1996 who are centers.

**11.13**

**a.** 0.00718          **b.** Smaller

**11.15**

**a.** 0.4          **b.** 0.2          **c.** 0.5

**d.** (a) $0.4 < \hat{p} < 0.6$  (b) $0.2 < \hat{p} < 0.8$  (c) None

**11.17**

**a.** $\hat{p} = 0.2$          **b.** Appropriate

**c.** 0.076 to 0.324

**11.19**

**a.** $\hat{p} = 0.7$          **b.** Appropriate

**c.** 0.533 to 0.867

**11.21**

**a.** $\hat{p} = 0.8$          **b.** Not appropriate

**11.23** 0.643 to 0.717. We can be 95% confident that the proportion of all U.S. adults with household incomes of at least $150,000 who purchased clothing, accessories, or books online in the past year is somewhere between 0.643 and 0.717.

**11.25**

**a.** 0.0528 to 0.0992

**b.** We can be 95% confident that the proportion of all U.S. asthmatics who are allergic to sulfites is somewhere between 0.0528 and 0.0992.

**11.27**

**a.** 76.7% to 83.3%

**b.** We can be 99% confident that the percentage of all registered voters who favor the creation of standards on CAFO pollution and, in general, view CAFOs unfavorably is somewhere between 76.7% and 83.3%.

**11.29** Yes! Procedure 11.1 was applied without checking one of the assumptions for its use; namely, that the number of successes, $x$, and the number of failures, $n - x$, are both 5 or greater. Because the number of failures here is only 4, Procedure 11.1 should not have been used.

**11.31** 59.1% to 64.9%

**11.33**

**a.** 0.0232      **b.** 9604      **c.** 0.0659 to 0.0761

**d.** 0.0051, which is less than 0.01

**e.** 3458; 0.0624 to 0.0796; 0.0086, which is less than 0.01

**f.** By using the guess for $\hat{p}$ in part (e), the required sample size is reduced by 6146. Moreover, only 0.35% of precision is lost—the margin of error rises from 0.0051 to 0.0086.

**11.35**

**a.** 3.3% (i.e., 3.3 percentage points)      **b.** 7368

**c.** 0.811 to 0.833

**d.** 0.011, which is less than 0.015 (1.5%)

**e.** 5526; 0.809 to 0.835; 0.013, which is less than 0.015 (1.5%)

**f.** By using the guess for $\hat{p}$ in part (e), the required sample size is reduced by 1842. Moreover, only 0.2% of precision is lost—the margin of error rises from 0.011 to 0.013.

**11.37**

**a.** 9604      **b.** 1791

**c.** By using the guess for $\hat{p}$ in part (b), the required sample size is reduced by 7813.

**d.** If the observed value of $\hat{p}$ turns out to be larger than 0.049 (but smaller than 0.951), the achieved margin of error will exceed the specified 0.01.

**11.39** 0.409 to 0.470. We can be 95% confident that the proportion of all U.S. adults who, at the time, approved of President Bush was somewhere between 0.409 and 0.470.

**11.41** 23.8% to 28.2%. We can be 90% confident that the percentage of all U.S. adults who would purchase or lease a new car from a manufacturer that had declared bankruptcy is somewhere between 23.8% and 28.2%.

## EXERCISES 11.2

**11.47** The one-mean $z$-test. Because proportions can be regarded as means. Indeed, define the variable $y$ to equal 1 or 0 according to whether a member of the population has or does not have the specified attribute. Then $p = \mu_y$ and $\hat{p} = \bar{y}$.

**11.49**

**a.** $\hat{p} = 0.2$      **b.** Appropriate

**c.** $z = -1.38$; critical value $= -1.28$; $P$-value $= 0.084$; reject $H_0$

**11.51**

**a.** $\hat{p} = 0.7$      **b.** Appropriate

**c.** $z = 1.44$; critical value $= 1.645$; $P$-value $= 0.074$; do not reject $H_0$

**11.53**

**a.** $\hat{p} = 0.8$      **b.** Appropriate

**c.** $z = 0.98$; critical values $= \pm 1.96$; $P$-value $= 0.329$; do not reject $H_0$

**11.55**

**a.** 0.54

**b.** $H_0: p = 0.5$, $H_a: p > 0.5$; $\alpha = 0.05$; $z = 2.33$; critical value $= 1.645$; $P = 0.0099$; reject $H_0$; at the 5% significance level, the data provide sufficient evidence to conclude that a majority of Generation Y Web users use the Internet to download music.

**11.57** $H_0: p = 0.136$, $H_a: p \neq 0.136$; $\alpha = 0.10$; $z = 2.49$; critical values $= \pm 1.645$; $P = 0.0128$; reject $H_0$; at the 10% significance level, the data provide sufficient evidence to conclude that the percentage of 18–25-year-olds who currently use marijuana or hashish has changed from the 2000 percentage of 13.6%.

**11.59**

**a.** $H_0: p = 0.72$, $H_a: p < 0.72$; $\alpha = 0.05$; $z = -4.93$; critical value $= -1.645$; $P = 0.0000$ (to four decimal places); reject $H_0$; at the 5% significance level, the data provide sufficient evidence to conclude that the percentage of Americans who approve of labor unions now has decreased since 1936.

**b.** $H_0: p = 0.67$, $H_a: p < 0.67$; $\alpha = 0.05$; $z = -1.34$; critical value $= -1.645$; $P = 0.0901$; do not reject $H_0$; at the 5% significance level, the data do not provide sufficient evidence to conclude that the percentage of Americans who approve of labor unions now has decreased since 1963.

**11.61** $H_0: p = 0.5$, $H_a: p > 0.5$; $\alpha = 0.01$; $z = 2.96$; critical value $= 2.33$; $P = 0.002$; reject $H_0$; at the 1% significance level, the data provide sufficient evidence to conclude that most Americans believe New Orleans will never recover.

**11.63** $H_0: p = 0.5$, $H_a: p < 0.5$; $\alpha = 0.05$; $z = -0.30$; critical value $= -1.645$; $P = 0.382$; do not reject $H_0$; at the 5% significance level, the data do not provide sufficient evidence to conclude that, of all young children drowning in Victorian dams located on farms, less than half are girls.

## EXERCISES 11.3

**11.65** For a two-tailed test, the basic strategy is as follows: (1) independently and randomly take samples from the two populations under consideration; (2) compute the sample proportions, $\hat{p}_1$ and $\hat{p}_2$; and (3) reject the null hypothesis if the sample proportions differ by too much—otherwise, do not reject the null hypothesis. The process is the same for a one-tailed test except that, for a left-tailed test, the null hypothesis is rejected only when $\hat{p}_1$ is too much smaller than $\hat{p}_2$, and, for a right-tailed test, the null hypothesis is rejected only when $\hat{p}_1$ is too much larger than $\hat{p}_2$.

**11.67**

**a.** Uses sunscreen before going out in the sun

**b.** Teenage girls and teenage boys

**c.** Sample proportions. Industry Research acquired those proportions by polling samples of the populations of all teenage girls and all teenage boys.

**11.69**

**a.** $p_1$ and $p_2$ are parameters, and the other quantities are statistics.

**b.** $p_1$ and $p_2$ are fixed numbers, and the other quantities are variables.

**11.71**

**a.** $\hat{p}_1 = 0.45$, $\hat{p}_2 = 0.75$      **b.** Appropriate

**c.** $z = -2.74$; critical value $= -1.28$; $P$-value $= 0.003$; reject $H_0$
**d.** $-0.434$ to $-0.166$

**11.73**
**a.** $\hat{p}_1 = 0.75$, $\hat{p}_2 = 0.60$      **b.** Appropriate
**c.** $z = 1.10$; critical value $= 1.645$; $P$-value $= 0.136$; do not reject $H_0$
**d.** $-0.067$ to $0.367$

**11.75**
**a.** $\hat{p}_1 = 0.375$, $\hat{p}_2 = 0.750$      **b.** Appropriate
**c.** $z = -3.02$; critical values $= \pm 1.96$; $P$-value $= 0.003$; reject $H_0$
**d.** $-0.592$ to $-0.158$

**11.77**
**a.** $H_0: p_1 = p_2$, $H_a: p_1 < p_2$; $\alpha = 0.01$; $z = -2.61$; critical value $= -2.33$; $P = 0.0045$; reject $H_0$; at the 1% significance level, the data provide sufficient evidence to conclude that women who take folic acid are at lesser risk of having children with major birth defects.
**b.** Designed experiment
**c.** Yes. Because for a designed experiment, it is reasonable to interpret statistical significance as a causal relationship.

**11.79** $H_0: p_1 = p_2$, $H_a: p_1 \neq p_2$; $\alpha = 0.10$; $z = -1.52$; critical values $= \pm 1.645$; $P = 0.1286$; do not reject $H_0$; at the 10% significance level, the data do not provide sufficient evidence to conclude that there is a difference in seat-belt use between drivers who are 25–34 years old and drivers who are 45–64 years old.

**11.81**
**a.** The samples must be independent simple random samples; of those sampled whose highest degree is a bachelors, at least five must be overweight and at least five must not be overweight; and of those sampled with a graduate degree, at least five must be overweight and at least five must not be overweight.
**b.** $H_0: p_1 = p_2$, $H_a: p_1 > p_2$; $\alpha = 0.05$; $z = 1.41$; critical value $= 1.645$; $P = 0.0793$; do not reject $H_0$; at the 5% significance level, the data do not provide sufficient evidence to conclude that the percentage who are overweight is greater for those whose highest degree is a bachelors than for those with a graduate degree.
**c.** $H_0: p_1 = p_2$, $H_a: p_1 > p_2$; $\alpha = 0.10$; $z = 1.41$; critical value $= 1.28$; $P = 0.0793$; reject $H_0$; at the 10% significance level, the data provide sufficient evidence to conclude that the percentage who are overweight is greater for those whose highest degree is a bachelors than for those with a graduate degree.

**11.83**
**a.** $-0.0191$ to $-0.000746$, or about $-0.019$ to $-0.001$
**b.** Roughly, we can be 98% confident that the rate of major birth defects for babies born to women who have not taken folic acid is somewhere between 1 per 1000 and 19 per 1000 higher than for babies born to women who have taken folic acid.

**11.85** $-0.0624$ to $0.00240$. We can be 90% confident that the difference between the proportions of seat-belt users for drivers in the age groups 25–34 years and 45–64 years is somewhere between $-0.0624$ and $0.00240$.

**11.87**
**a.** $-0.68\%$ to $8.81\%$      **b.** $0.37\%$ to $7.76\%$

**11.89**
**a.** $H_0: p_1 = p_2$, $H_a: p_1 \neq p_2$; $\alpha = 0.05$; $z = 0.88$; critical values $= \pm 1.96$; $P = 0.379$; do not reject $H_0$; at the 5% significance level, the data do not provide sufficient evidence to conclude that there is a difference between the labor-force participation rates of U.S. and Canadian women.
**b.** $-0.042$ to $0.110$. We can be 95% confident that the difference between the labor-force participation rates of U.S. and Canadian women is somewhere between $-4.2\%$ and $11.0\%$.

**REVIEW PROBLEMS FOR CHAPTER 11**

**1. a.** Feeling that marijuana should be legalized for medicinal use in patients with cancer and other painful and terminal diseases
**b.** Americans
**c.** Proportion of all Americans who feel that marijuana should be legalized for medicinal use in patients with cancer and other painful and terminal diseases
**d.** Sample proportion. It is the proportion of Americans sampled who feel that marijuana should be legalized for medicinal use in patients with cancer and other painful and terminal diseases.

**2.** Generally, obtaining a sample proportion can be done more quickly and is less costly than obtaining the population proportion. Sampling is often the only practical way to proceed.

**3. a.** The number of members in the sample that have the specified attribute
**b.** The number of members in the sample that do not have the specified attribute

**4. a.** Population proportion
**b.** Normal
**c.** $np$, $n(1-p)$, 5

**5.** The precision with which a sample proportion, $\hat{p}$, estimates the population proportion, $p$, at the specified confidence level

**6. a.** Getting the "holiday blues"
**b.** All men, all women
**c.** The proportion of all men who get the "holiday blues" and the proportion of all women who get the "holiday blues"
**d.** The proportion of all sampled men who get the "holiday blues" and the proportion of all sampled women who get the "holiday blues"
**e.** Sample proportions. The poll used samples of men and women to obtain the proportions.

**7. a.** Difference between the population proportions
**b.** Normal

**8.** 37.0% to 43.0%

**9. a.** 19,208      **b.** 14,406

**10.** 0.574 to 0.629. We can be 95% confident that the percentage of students who expect difficulty finding a job is somewhere between 57.4% and 62.9%.

**11. a.** 0.028      **b.** 2401      **c.** 0.567 to 0.607
**d.** 0.020, which is the same as that specified in part (b)

e. 2367; 0.567 to 0.607; 0.020

f. By using the guess for $\hat{p}$ in part (e), the required sample size is reduced by 34 with (virtually) no sacrifice in precision.

12. a. $H_0$: $p = 0.25$, $H_a$: $p < 0.25$; $\alpha = 0.05$; $z = -2.30$; critical value $= -1.645$; $P = 0.0107$; reject $H_0$; at the 5% significance level, the data provide sufficient evidence to conclude that less than one in four Americans believe that juries "almost always" convict the guilty and free the innocent.

b. The data provide strong evidence against the null hypothesis and hence in favor of the alternative hypothesis that less than one in four Americans believe that juries "almost always" convict the guilty and free the innocent.

13. a. Observational study

b. Being observational, the study established only an association between height and breast cancer; no causal relationship can be inferred, although there may be one.

14. a. $H_0$: $p_1 = p_2$, $H_a$: $p_1 < p_2$; $\alpha = 0.01$; $z = -4.17$; critical value $= -2.33$; $P = 0.0000$ (to four decimal places); reject $H_0$; at the 1% significance level, the data provide sufficient evidence to conclude that the percentage of Maricopa County residents who thought Arizona's economy would improve over the next 2 years was less during the time of the first poll than during the time of the second poll.

b. The data provide very strong evidence against the null hypothesis and hence in favor of the alternative hypothesis that the percentage of Maricopa County residents who thought Arizona's economy would improve over the next 2 years was less during the time of the first poll than during the time of the second poll.

15. a. $-0.186$ to $-0.054$

b. We can be 98% confident that the difference between the percentages of Maricopa County residents who thought Arizona's economy would improve over the next 2 years during the time of the first poll and during the time of the second poll is somewhere between $-18.6\%$ and $-5.4\%$.

16. a. 0.066; we can be 98% confident that the error in estimating the difference between the two population proportions, $p_1 - p_2$, by the difference between the two sample proportions, $-0.12$, is at most 0.066.

b. 0.066    c. 3006    d. $-0.158$ to $-0.098$

e. 0.03, which is the same as that specified in part (c)

17. The number of "successes" is less than 5, so the one-proportion $z$-interval procedure should not be used here and is unreliable.

18. 11.7% to 14.4%. We can be 95% confident that the percentage of U.S. adults who would participate in an office pool for March Madness is somewhere between 11.7% and 14.4%.

19. $H_0$: $p = 0.5$, $H_a$: $p > 0.5$; $\alpha = 0.05$; $z = 7.07$; critical value $= 1.645$; $P = 0.0000$ (to four decimal places); reject $H_0$; at the 5% significance level, the data provide sufficient evidence to conclude that a majority of U.S. adults do not believe that abstinence programs are effective in reducing or preventing AIDS.

20. a. $H_0$: $p_1 = p_2$, $H_a$: $p_1 \neq p_2$; $\alpha = 0.05$; $z = 5.27$; critical values $= \pm 1.96$; $P = 0.0000$ (to four decimal places); reject $H_0$; at the 5% significance level, the data provide sufficient evidence to conclude that a difference exists in the cure rates of the two types of treatment.

b. 0.291 to 0.594. We can be 95% confident that use of the Bug Buster kit will increase the proportion of those cured by somewhere between 0.291 and 0.594.

21. $H_0$: $p_1 = p_2$, $H_a$: $p_1 < p_2$; $\alpha = 0.01$; $z = -7.02$; critical value $= -2.33$; $P = 0.0000$ (to four decimal places); reject $H_0$; at the 1% significance level, the data provide sufficient evidence to conclude that finasteride reduces the risk of prostate cancer.

# Chapter 12

## EXERCISES 12.1

**12.1** A variable is said to have a chi-square distribution if its distribution has the shape of a special type of right-skewed curve, called a chi-square curve.

**12.3** The $\chi^2$-curve with 20 degrees of freedom more closely resembles a normal curve. As the number of degrees of freedom becomes larger, $\chi^2$-curves look increasingly like normal curves.

**12.5**

a. 32.852    b. 10.117

**12.7**

a. 18.307    b. 3.247

## EXERCISES 12.2

**12.11** Because the hypothesis test is carried out by determining how well the observed frequencies fit the expected frequencies

**12.13** Both assumptions are satisfied.

**12.15** Both assumptions are satisfied. Note that 20% of the expected frequencies are less than 5.

**12.17** Assumption 2 is satisfied because only 20% of the expected frequencies are less than 5, but Assumption 1 fails because there is an expected frequency of 0.5 (which is less than 1).

**12.19**

a. The population consists of occupied housing units built after 2000; the variable is primary heating fuel.

b. In the following table, the first column gives the sample size, the second column shows the number of expected frequencies less than 1 and parenthetically whether Assumption 1 is satisfied, the third column shows the percentage of expected frequencies less than 5 and parenthetically whether Assumption 2 is satisfied, and the fourth column states whether both assumptions for a chi-square goodness-of-fit test are satisfied.

| Sample size | Number less than 1 | Percentage less than 5 | Both satisfied? |
|---|---|---|---|
| 200 | 1 (no) | 33.3 (no) | no |
| 250 | 0 (yes) | 33.3 (no) | no |
| 300 | 0 (yes) | 16.7 (yes) | yes |

**c.** 264

*Note:* In each of Exercises 12.21–12.25, the null hypothesis is that the variable has the distribution given in the problem statement and the alternative hypothesis is that the variable does not have that distribution.

**12.21** $\chi^2 = 14.042$; critical value $= 7.815$; $P < 0.005$; reject $H_0$

**12.23** $\chi^2 = 7.1$; critical value $= 7.779$; $P > 0.10$; do not reject $H_0$

**12.25** $\chi^2 = 10.061$; critical value $= 9.210$; $0.005 < P < 0.01$; reject $H_0$

**12.27**
**a.** The population consists of all this year's incoming college freshmen in the United States; the variable is political view.
**b.** $H_0$: This year's distribution of political views for incoming college freshmen is the same as the 2000 distribution. $H_a$: This year's distribution of political views for incoming college freshmen has changed from the 2000 distribution. $\alpha = 0.05$; $\chi^2 = 4.667$; critical value $= 5.991$; $0.05 < P < 0.10$; do not reject $H_0$; at the 5% significance level, the data do not provide sufficient evidence to conclude that this year's distribution of political views for incoming college freshmen has changed from the 2000 distribution.
**c.** $H_0$: This year's distribution of political views for incoming college freshmen is the same as the 2000 distribution. $H_a$: This year's distribution of political views for incoming college freshmen has changed from the 2000 distribution. $\alpha = 0.10$; $\chi^2 = 4.667$; critical value $= 4.605$; $0.05 < P < 0.10$; reject $H_0$; at the 10% significance level, the data provide sufficient evidence to conclude that this year's distribution of political views for incoming college freshmen has changed from the 2000 distribution.

**12.29** $H_0$: The color distribution of M&Ms is that reported by M&M/MARS consumer affairs. $H_a$: The color distribution of M&Ms differs from that reported by M&M/MARS consumer affairs. $\alpha = 0.05$; $\chi^2 = 4.091$; critical value $= 11.070$; $P > 0.10$; do not reject $H_0$; at the 5% significance level, the data do not provide sufficient evidence to conclude that the color distribution of M&Ms differs from that reported by M&M/MARS consumer affairs.

**12.31** $H_0$: The die is not loaded. $H_a$: The die is loaded. $\alpha = 0.05$; $\chi^2 = 2.48$; critical value $= 11.070$; $P > 0.10$; do not reject $H_0$; at the 5% significance level, the data do not provide sufficient evidence to conclude that the die is loaded.

**12.33**
**a.** $H_0$: World Series' teams are evenly matched. $H_a$: World Series' teams are not evenly matched. $\alpha = 0.05$; $\chi^2 = 5.029$; $P = 0.170$; do not reject $H_0$; at the 5% significance level, the data do not provide sufficient evidence to conclude that World Series' teams are not evenly matched.
**b.** The data are not from a simple random sample, so using the chi-square goodness-of-fit test here is inappropriate.

**12.35** $H_0$: The distribution of reasons for migration between provinces is the same as that for migration within provinces. $H_a$: The distribution of reasons for migration between provinces is different from that for migration within provinces. $\alpha = 0.01$; $\chi^2 = 40.789$; $P = 0.000$ (to three decimal places); reject $H_0$; at the 1% significance level, the data provide sufficient evidence to conclude that the distribution of reasons for migration between provinces is different from that for migration within provinces.

**EXERCISES 12.3**

**12.39** Cells

**12.41** Summing the row totals, summing the column totals, or summing the frequencies in the cells

**12.43** Yes. If no association existed between "gender" and "specialty," the percentage of male physicians in residency who specialized in family practice would be identical to the percentage of female physicians in residency who specialized in family practice. As that is not the case, an association exists between the two variables.

**12.45**
**a.**

| Gender | College | | | |
|---|---|---|---|---|
| | Bus. | Engr. | Lib. Arts | Total |
| Male | 2 | 10 | 3 | 15 |
| Female | 7 | 2 | 1 | 10 |
| Total | 9 | 12 | 4 | 25 |

**b.**

| Gender | College | | | |
|---|---|---|---|---|
| | Bus. | Engr. | Lib. Arts | Total |
| Male | 0.222 | 0.833 | 0.750 | 0.600 |
| Female | 0.778 | 0.167 | 0.250 | 0.400 |
| Total | 1.000 | 1.000 | 1.000 | 1.000 |

**c.**

| Gender | College | | | |
|---|---|---|---|---|
| | Bus. | Engr. | Lib. Arts | Total |
| Male | 0.133 | 0.667 | 0.200 | 1.000 |
| Female | 0.700 | 0.200 | 0.100 | 1.000 |
| Total | 0.360 | 0.480 | 0.160 | 1.000 |

**d.** Yes. The tables in parts (b) and (c) show that the conditional distributions of one variable given the other are not identical.

**12.47**

**a.**

| | | Class | | | |
|---|---|---|---|---|---|
| | | Fresh. | Soph. | Junior | Senior | **Total** |
| **Party** | Republican | 3 | 9 | 12 | 6 | 30 |
| | Democrat | 2 | 6 | 8 | 4 | 20 |
| | Other | 1 | 3 | 4 | 2 | 10 |
| | **Total** | 6 | 18 | 24 | 12 | 60 |

**b.**

| | | Class | | | |
|---|---|---|---|---|---|
| | | Fresh. | Soph. | Junior | Senior |
| **Party** | Republican | 0.500 | 0.500 | 0.500 | 0.500 |
| | Democrat | 0.333 | 0.333 | 0.333 | 0.333 |
| | Other | 0.167 | 0.167 | 0.167 | 0.167 |
| | **Total** | 1.000 | 1.000 | 1.000 | 1.000 |

**c.** No. The table in part (b) shows that the conditional distributions of political party affiliation within class levels are identical.

**d.** Republican 0.500, Democrat 0.333, Other 0.167, Total 1.000

**e.** True. From part (c), political party affiliation and class level are not associated. Therefore the conditional distributions of class level within political party affiliations are identical to each other and to the marginal distribution of class level.

**12.49**

**a.** 8

**b.** The missing entries, from top to bottom and left to right, are: 1860, 7684, 30,077, and 41,131.

**c.** 41,131    **d.** 7684    **e.** 30,077    **f.** 1860

**12.51**

**a.** 1056.5 thousand    **b.** 48.6 thousand
**c.** 10.6 thousand    **d.** 304.7 thousand
**e.** 51.9 thousand    **f.** 51.9 thousand
**g.** 1560.1 thousand

**12.53**

**a.** The missing entries, from top to bottom and left to right, are: 639, 744, 153, 33, 150, and 2130.

**b.** 15

**c.** 744 thousand    **d.** 150 thousand    **e.** 91 thousand
**f.** 701 thousand    **g.** 68 thousand

**12.55**

**a.**

| | | Gender | | |
|---|---|---|---|---|
| | | Male | Female | Total |
| **Race/Ethnicity** | White | 0.845 | 0.155 | 1.000 |
| | Black | 0.644 | 0.356 | 1.000 |
| | Hispanic | 0.786 | 0.214 | 1.000 |
| | Other | 0.769 | 0.231 | 1.000 |

**b.** Male 0.731; Female 0.269; Total 1.000

**c.** Yes. Because the conditional distributions of gender within races are not identical

**d.** 26.9%    **e.** 15.5%

**f.** True. Because by part (c), an association exists between the variables "gender" and "race."

**g.**

| | | Gender | | |
|---|---|---|---|---|
| | | Male | Female | Total |
| **Race/Ethnicity** | White | 0.336 | 0.168 | 0.291 |
| | Black | 0.445 | 0.669 | 0.506 |
| | Hispanic | 0.201 | 0.149 | 0.187 |
| | Other | 0.017 | 0.014 | 0.016 |
| | **Total** | 1.000 | 1.000 | 1.000 |

**12.57**

**a.**

| | | Prison facility | | |
|---|---|---|---|---|
| | | State | Federal | Local |
| **Educational attainment** | 8th grade or less | 0.142 | 0.119 | 0.131 |
| | Some high school | 0.255 | 0.145 | 0.334 |
| | GED | 0.285 | 0.226 | 0.141 |
| | High school diploma | 0.205 | 0.270 | 0.259 |
| | Postsecondary | 0.090 | 0.158 | 0.103 |
| | College grad or more | 0.024 | 0.081 | 0.032 |
| | **Total** | 1.000 | 1.000 | 1.000 |

**b.** Yes. Because the conditional distributions of educational attainment within type of prison facility categories are not identical.

**c.** 8th grade or less: 0.137; Some high school: 0.273; GED: 0.238; High school diploma: 0.225; Postsecondary: 0.098; College grad or more: 0.029

**d.**

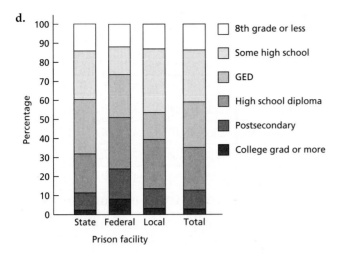

That the bars are not identical reflects the fact that there is an association between educational attainment and type of prison facility.

**e.** False. Because by part (b), there is an association between educational attainment and type of prison facility.

**f.**

| | Prison facility | | | |
|---|---|---|---|---|
| **Educational attainment** | | State | Federal | Local | Total |
| 8th grade or less | 0.662 | 0.047 | 0.291 | 1.000 |
| Some high school | 0.598 | 0.029 | 0.374 | 1.000 |
| GED | 0.768 | 0.051 | 0.181 | 1.000 |
| High school diploma | 0.584 | 0.065 | 0.352 | 1.000 |
| Postsecondary | 0.590 | 0.087 | 0.323 | 1.000 |
| College grad or more | 0.521 | 0.148 | 0.331 | 1.000 |
| **Total** | 0.641 | 0.054 | 0.305 | 1.000 |

**g.**  5.4%        **h.**  4.7%        **i.**  11.9%

### EXERCISES 12.4

**12.65** $H_0$: The two variables under consideration are statistically independent. $H_a$: The two variables under consideration are statistically dependent.

**12.67** 15

**12.69** If a causal relationship exists between two variables, they are necessarily associated. In other words, if no association exists between two variables, they could not possibly be causally related.

**12.71** $H_0$: An association does not exist between the ratings of Siskel and Ebert. $H_a$: An association exists between the ratings of Siskel and Ebert. Assumptions 1 and 2 are satisfied because all expected frequencies are 5 or greater. $\alpha = 0.01$; $\chi^2 = 45.357$; critical value = 13.277; $P < 0.005$; reject $H_0$; at the 1% significance level, the data provide sufficient evidence

to conclude that an association exists between the ratings of Siskel and Ebert.

**12.73**
**a.** No. Assumption 2 fails because 33.3% of the expected frequencies are less than 5.
**b.** Yes. $H_0$: Social class and frequency of games are not associated. $H_a$: Social class and frequency of games are associated. Assumptions 1 and 2 are satisfied because all expected frequencies are 5 or greater. $\alpha = 0.01$; $\chi^2 = 8.715$; critical value = 9.210; $0.01 < P < 0.025$; do not reject $H_0$; at the 1% significance level, the data do not provide sufficient evidence to conclude that an association exists between social class and frequency of games.

**12.75** $H_0$: Size of city and status in practice are statistically independent for U.S. lawyers. $H_a$: Size of city and status in practice are statistically dependent for U.S. lawyers. Assumption 1 is satisfied, but Assumption 2 is not because 25% (3 of 12) of the expected frequencies are less than 5. Consequently, the chi-square independence test should not be applied here.

**12.77** $H_0$: BMD and depression are statistically independent for elderly Asian men. $H_a$: BMD and depression are statistically dependent for elderly Asian men. Assumption 1 is satisfied; Assumption 2 is also satisfied because 16.7% (1 of 6) of the expected frequencies are less than 5. $\alpha = 0.01$; $\chi^2 = 10.095$; critical value = 9.210; $0.005 < P < 0.01$; reject $H_0$; at the 1% significance level, the data provide sufficient evidence to conclude that BMD and depression are statistically dependent for elderly Asian men.

**12.79** In each part: the null hypothesis is that no association exists between the two specified variables; the alternative hypothesis is that an association exists between the two specified variables; and $\alpha = 0.05$.
**a.** $\chi^2 = 1.350$; $P = 0.509$; do not reject $H_0$
**b.** $\chi^2 = 41.430$; $P = 0.000$ (to three decimal places); reject $H_0$
**c.** $\chi^2 = 6.952$; $P = 0.138$; do not reject $H_0$
**d.** $\chi^2 = 13.651$; $P = 0.034$; reject $H_0$
**e.** $\chi^2 = 16.634$; $P = 0.011$; reject $H_0$
**f.** $\chi^2 = 49.665$; $P = 0.000$ (to three decimal places); reject $H_0$

### REVIEW PROBLEMS FOR CHAPTER 12

1. By their degrees of freedom

2. **a.** 0        **b.** Right skewed      **c.** Normal curve

3. **a.** No. The degrees of freedom for the chi-square goodness-of-fit test depends on the number of possible values for the variable under consideration, not on the sample size.
**b.** No. The degrees of freedom for the chi-square independence test depends on the number of possible values for the two variables under consideration, not on the sample size.

4. For both tests, the null hypothesis is rejected only when the observed and expected frequencies match up poorly, which corresponds to large values of the chi-square test statistic. Thus both tests are always right tailed.

5. 0

**6. a.** (1) All expected frequencies are 1 or greater. (2) At most 20% of the expected frequencies are less than 5.
   **b.** They are very important. If the assumptions are not met, the results could be invalid.

**7. a.** 7.8%        **b.** Roughly 1.9 million
   **c.** Race and region of residence are associated.

**8. a.** Obtain the conditional distribution of one of the variables for each possible value of the other variable. If all these conditional distributions are identical, no association exists between the two variables; otherwise, an association exists between the two variables.
   **b.** No. Because the data are for an entire population, no inference is being made from a sample to the population. The conclusion is a fact.

**9. a.** Perform a chi-square independence test.
   **b.** Yes. As in any inference, it is always possible that the conclusion is in error.

**10. a.** 6.408     **b.** 33.409     **c.** 27.587
   **d.** 8.672     **e.** 7.564, 30.191

**11. a.** $H_0$: This year's distribution of educational attainment is the same as the 2000 distribution. $H_a$: This year's distribution of educational attainment differs from the 2000 distribution. Assumptions 1 and 2 are satisfied because all expected frequencies are 5 or greater. $\alpha = 0.05$; $\chi^2 = 2.674$; critical value $= 11.070$; $P > 0.10$; do not reject $H_0$; at the 5% significance level, the data do not provide sufficient evidence to conclude that this year's distribution of educational attainment differs from the 2000 distribution.
   **b.** $P > 0.10$. The evidence against the null hypothesis is weak or none.

**12. a.** The first 43 presidents of the United States
   **b.** Region of birth and political party
   **c.**

| | | Region | | | | |
|---|---|---|---|---|---|---|
| | | NE | MW | SO | WE | Total |
| **Party** | Federalist | 1 | 0 | 1 | 0 | 2 |
| | DR | 1 | 0 | 3 | 0 | 4 |
| | Democratic | 7 | 1 | 6 | 0 | 14 |
| | Whig | 1 | 0 | 3 | 0 | 4 |
| | Republican | 5 | 10 | 2 | 1 | 18 |
| | Union | 0 | 0 | 1 | 0 | 1 |
| | Total | 15 | 11 | 16 | 1 | 43 |

**13. a.**

| | | Region | | | | |
|---|---|---|---|---|---|---|
| | | NE | MW | SO | WE | Total |
| **Party** | Federalist | 0.500 | 0.000 | 0.500 | 0.000 | 1.000 |
| | DR | 0.250 | 0.000 | 0.750 | 0.000 | 1.000 |
| | Democratic | 0.500 | 0.071 | 0.429 | 0.000 | 1.000 |
| | Whig | 0.250 | 0.000 | 0.750 | 0.000 | 1.000 |
| | Republican | 0.278 | 0.556 | 0.111 | 0.056 | 1.000 |
| | Union | 0.000 | 0.000 | 1.000 | 0.000 | 1.000 |
| | Total | 0.349 | 0.256 | 0.372 | 0.023 | 1.000 |

**b.**

| | | Region | | | | |
|---|---|---|---|---|---|---|
| | | NE | MW | SO | WE | Total |
| **Party** | Federalist | 0.067 | 0.000 | 0.063 | 0.000 | 0.047 |
| | DR | 0.067 | 0.000 | 0.188 | 0.000 | 0.093 |
| | Democratic | 0.467 | 0.091 | 0.375 | 0.000 | 0.326 |
| | Whig | 0.067 | 0.000 | 0.188 | 0.000 | 0.093 |
| | Republican | 0.333 | 0.909 | 0.125 | 1.000 | 0.419 |
| | Union | 0.000 | 0.000 | 0.063 | 0.000 | 0.023 |
| | Total | 1.000 | 1.000 | 1.000 | 1.000 | 1.000 |

**c.** Yes, because the conditional distributions in either part (a) or part (b) are not identical.
   **d.** 41.9%    **e.** 41.9%    **f.** 12.5%
   **g.** 37.2%    **h.** 37.2%    **i.** 11.1%

**14. a.** 2046    **b.** 737    **c.** 266
   **d.** 3046    **e.** 5413    **f.** 5910

**15. a.**

| | | Control | | | |
|---|---|---|---|---|---|
| | | Gov | Prop | NP | Total |
| **Facility** | General | 0.314 | 0.122 | 0.564 | 1.000 |
| | Psychiatric | 0.361 | 0.486 | 0.153 | 1.000 |
| | Chronic | 0.808 | 0.038 | 0.154 | 1.000 |
| | Tuberculosis | 0.750 | 0.000 | 0.250 | 1.000 |
| | Other | 0.144 | 0.361 | 0.495 | 1.000 |

**b.** Yes. Because the conditional distributions of control type within facility types are not identical
**c.** Gov 0.311, Prop 0.177, NP 0.512, Total 1.000

**d.**

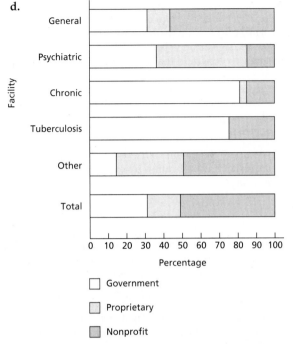

That the bars are not identical reflects the fact that an association exists between facility type and control type.

**e.** False. By part (b), an association exists between facility type and control type.

**f.**

| | | Control | | |
|---|---|---|---|---|
| | | Gov | Prop | NP | Total |
| Facility | General | 0.829 | 0.566 | 0.905 | 0.821 |
| | Psychiatric | 0.130 | 0.307 | 0.034 | 0.112 |
| | Chronic | 0.010 | 0.001 | 0.001 | 0.004 |
| | Tuberculosis | 0.001 | 0.000 | 0.000 | 0.001 |
| | Other | 0.029 | 0.127 | 0.060 | 0.062 |
| | Total | 1.000 | 1.000 | 1.000 | 1.000 |

**g.** 17.7%      **h.** 48.6%      **i.** 30.7%

**16.** $H_0$: Histological type and treatment response are statistically independent. $H_a$: Histological type and treatment response are statistically dependent. Assumptions 1 and 2 are satisfied because all expected frequencies are 5 or greater. $\alpha = 0.01$; $\chi^2 = 75.890$; critical value = 16.812; $P < 0.005$; reject $H_0$; at the 1% significance level, the data provide sufficient evidence to conclude that histological type and treatment response are statistically dependent.

# Chapter 13

## EXERCISES 13.1

**13.1** By stating its two numbers of degrees of freedom

**13.3** $F_{0.05}, F_{0.025}, F_\alpha$

**13.5**
**a.** 12                              **b.** 7

**13.7**
**a.** 1.89        **b.** 2.47        **c.** 2.14

**13.9**
**a.** 2.88        **b.** 2.10        **c.** 1.78

## EXERCISES 13.2

**13.13** The pooled $t$-procedure of Section 10.2

**13.15** The procedure for comparing the means analyzes the variation in the sample data.

**13.17**
**a.** The treatment mean square, $MSTR$
**b.** The error mean square, $MSE$
**c.** The $F$-statistic, $F = MSTR/MSE$

**13.19** It signifies that the ANOVA compares the means of a variable for populations that result from a classification by *one* other variable (called the *factor*).

**13.21** No. Because the variation among the sample means is not large relative to the variation within the samples

**13.23** The difference between the observation and the mean of the sample containing it

**13.25**
**a.** 24        **b.** 12        **c.** 16
**d.** 2.29        **e.** 5.25

**13.27**
**a.** 36        **b.** 9        **c.** 52
**d.** 3.47        **e.** 2.60

## EXERCISES 13.3

**13.31** A small value of $F$ results when $SSTR$ is small relative to $SSE$, that is, when the variation among sample means is small relative to the variation within samples. This result describes what is expected when the null hypothesis is true; thus it doesn't constitute evidence against the null hypothesis. Only when the variation among sample means is large relative to the variation within samples (i.e., only when $F$ is large), is there evidence that the null hypothesis is false.

**13.33** $SST = SSTR + SSE$. The total variation among all the sample data can be partitioned into a component representing variation among the sample means and a component representing variation within the samples.

**13.35**
**a.** One-way ANOVA                **b.** Two-way ANOVA

**13.37** The missing entries are as follows: In the first row, it is 3; in the second row, they are 18.880 and 0.944; and in the third row, they are 23 and 21.004.

**13.39** The missing entries are as follows: In the first row, they are 2, 2.8, and 1.56; in the second row, it is 10.8.

**13.41**
**a.** 40, 24, 16
**b.** They are the same.

**c.**

| Source | df | SS | MS | F |
|---|---|---|---|---|
| Treatment | 2 | 24 | 12 | 5.25 |
| Error | 7 | 16 | 2.29 | |
| Total | 9 | 40 | | |

**d.** $H_0: \mu_1 = \mu_2 = \mu_3$, $H_a$: Not all the means are equal. $\alpha = 0.05$; $F = 5.25$; critical value $= 4.74$; $0.025 < P < 0.05$; reject $H_0$.

**13.43**
**a.** 88, 36, 52
**b.** They are the same.
**c.**

| Source | df | SS | MS | F |
|---|---|---|---|---|
| Treatment | 4 | 36 | 9 | 2.60 |
| Error | 15 | 52 | 3.47 | |
| Total | 19 | 88 | | |

**d.** $H_0: \mu_1 = \mu_2 = \mu_3 = \mu_4 = \mu_5$, $H_a$: Not all the means are equal. $\alpha = 0.05$; $F = 2.60$; critical value $= 3.06$; $0.05 < P < 0.10$; do not reject $H_0$.

**13.45** $H_0: \mu_1 = \mu_2 = \mu_3$, $H_a$: Not all the means are equal. $\alpha = 0.05$; $F = 54.58$; critical value $= 4.26$; $P < 0.005$; reject $H_0$; at the 5% significance level, the data provide sufficient evidence to conclude that a difference exists in mean number of copepods among the three different diets.

**13.47** $H_0: \mu_1 = \mu_2 = \mu_3 = \mu_4 = \mu_5$, $H_a$: Not all the means are equal. $\alpha = 0.05$; $F = 2.23$; critical value $= 2.87$; $P > 0.10$; do not reject $H_0$; at the 5% significance level, the data do not provide sufficient evidence to conclude that a difference exists in mean bacteria counts among the five strains of *Staphylococcus aureus*.

**13.49** $H_0: \mu_1 = \mu_2 = \mu_3$, $H_a$: Not all the means are equal. $\alpha = 0.05$; $F = 4.69$; critical value $\approx 3.32$; $0.01 < P < 0.025$; reject $H_0$; at the 5% significance level, the data provide sufficient evidence to conclude that a difference exists in mean resistance to *Stagonospora nodorum* among the three years of wheat harvests.

**13.51**
**a.** $H_0: \mu_1 = \mu_2 = \mu_3 = \mu_4$, $H_a$: Not all the means are equal. $\alpha = 0.05$; $F = 7.54$; $P = 0.002$; reject $H_0$.
**b.** At the 5% significance level, the data provide sufficient evidence to conclude that a difference exists in mean monthly rents among newly completed apartments in the four U.S. regions.
**c.** It appears reasonable to presume that the assumptions of normal populations and equal population standard deviations are both met.

**13.53**
**a.** $H_0: \mu_1 = \mu_2 = \mu_3$, $H_a$: Not all the means are equal. $\alpha = 0.01$; $F = 6.09$; $P = 0.008$; reject $H_0$.
**b.** At the 1% significance level, the data provide sufficient evidence to conclude that a difference exists in the mean singing rates among male rock sparrows exposed to the three types of breast treatments.
**c.** It appears reasonable to presume that the assumption of normal populations is met, but the assumption of equal population standard deviations appears to be violated.

**13.55**
**a.** $H_0: \mu_1 = \mu_2 = \mu_3$, $H_a$: Not all the means are equal. $\alpha = 0.05$; $F = 114.71$; $P = 0.000$ (to three decimal places); reject $H_0$.
**b.** At the 5% significance level, the data provide sufficient evidence to conclude that there is a difference in mean hardness among the three materials.
**c.** The assumptions of normal populations and equal population standard deviations both appear to be violated.

## REVIEW PROBLEMS FOR CHAPTER 13

**1.** To compare the means of a variable for populations that result from a classification by one other variable (called the *factor*)

**2.** *Simple random samples:* Check by carefully studying the way the sampling was done. *Independent samples:* Check by carefully studying the way the sampling was done. *Normal populations:* Check by constructing normal probability plots. *Equal standard deviations:* As a rule of thumb, this assumption is considered to be satisfied if the ratio of the largest sample standard deviation to the smallest sample standard deviation is less than 2.

Also, the normality and equal-standard-deviations assumptions can be assessed by performing a residual analysis.

**3.** The $F$-distribution

**4.** df $= (2, 14)$

**5. a.** *MSTR* (or *SSTR*)    **b.** *MSE* (or *SSE*)

**6. a.** The total sum of squares, *SST*, represents the total variation among all the sample data; the treatment sum of squares, *SSTR*, represents the variation among the sample means; and the error sum of squares, *SSE*, represents the variation within the samples.
**b.** $SST = SSTR + SSE$; the one-way ANOVA identity shows that the total variation among all the sample data can be partitioned into a component representing variation among the sample means and a component representing variation within the samples.

**7. a.** For organizing and summarizing the quantities required for performing a one-way analysis of variance
**b.**

| Source | df | SS | MS = SS/df | F |
|---|---|---|---|---|
| Treatment | $k-1$ | SSTR | $MSTR = \dfrac{SSTR}{k-1}$ | $F = \dfrac{MSTR}{MSE}$ |
| Error | $n-k$ | SSE | $MSE = \dfrac{SSE}{n-k}$ | |
| Total | $n-1$ | SST | | |

**8. a.** 2    **b.** 14    **c.** 3.74
   **d.** 6.51    **e.** 3.74

**9. a.** The sample means are 3, 3, and 6, respectively; the sample standard deviations are 2, 2.449, and 4.243, respectively.
**b.** $SST = 110$, $SSTR = 24$, $SSE = 86$; $110 = 24 + 86$
**c.** $SST = 110$, $SSTR = 24$, $SSE = 86$

d.

| Source | df | SS | MS = SS/df | F |
|--------|----|----|-----------|---|
| Treatment | 2 | 24 | 12 | 1.26 |
| Error | 9 | 86 | 9.556 | |
| Total | 11 | 110 | | |

10. **a.** The variation among the sample means
**b.** The variation within the samples
**c.** Simple random samples, independent samples, normal populations, and equal (population) standard deviations. One-way ANOVA is robust to moderate violations of the normality assumption. It is also reasonably robust to moderate violations of the equal-standard-deviations assumption if the sample sizes are roughly equal.

11. **a.** $s_1 = \$92.9$, $s_2 = \$126.1$, $s_3 = \$139.0$. Figure A.2 shows individual normal probability plots of the three samples.
**b.** See Fig. A.3.
**c.** Referring to the results of either part (a) or part (b), we conclude that presuming that the assumptions of normal populations and equal standard deviations are met is reasonable.

**FIGURE A.2**   Normal probability plots for Problem 11(a)

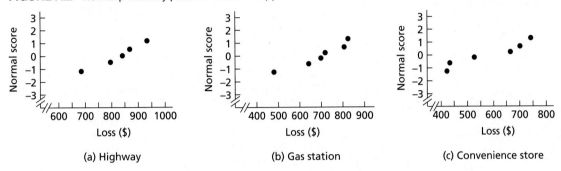

(a) Highway          (b) Gas station          (c) Convenience store

**FIGURE A.3**   (a) Residual plot and (b) normal probability plot of the residuals for Problem 11(b)

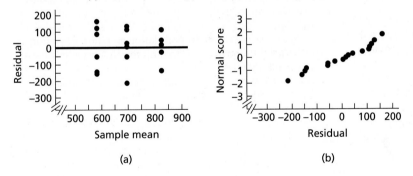

(a)          (b)

12. $H_0$: $\mu_1 = \mu_2 = \mu_3$, $H_a$: Not all the means are equal. $\alpha = 0.05$; $F = 5.34$; critical value $= 3.74$; $0.01 < P < 0.025$; reject $H_0$; at the 5% significance level, the data provide sufficient evidence to conclude that a difference in mean losses exists among the three types of robberies.

# Chapter 14

## EXERCISES 14.1

**14.1** Conditional distribution, conditional mean, conditional standard deviation

**14.3**
**a.** Population regression line
**b.** $\sigma$          **c.** Normal; $\beta_0 + 6\beta_1$; $\sigma$

**14.5** The sample regression line, $\hat{y} = b_0 + b_1 x$

**14.7** Residual

**14.9** A residual plot, that is, a plot of the residuals against the values of the predictor variable. A residual plot makes it easier to spot patterns such as curvature and nonconstant standard deviation than does a scatterplot.

**14.11**
**a.** 2.45
**b.**

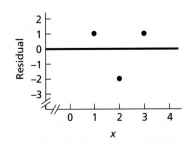

**c.**

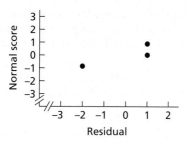

**14.13**
**a.** 1.73
**b.**

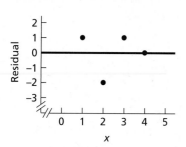

**c.**

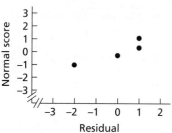

**14.15**
**a.** 1.88
**b.**

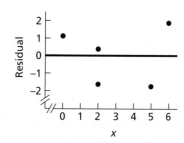

**c.**

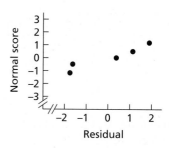

**14.17** There are constants, $\beta_0$, $\beta_1$, and $\sigma$, such that, for each age, $x$, the prices of all Corvettes of that age are normally distributed with mean $\beta_0 + \beta_1 x$ and standard deviation $\sigma$.

**14.19** There are constants, $\beta_0$, $\beta_1$, and $\sigma$, such that, for each weight, $x$, the quantities of volatile compounds emitted by all potato plants of that weight are normally distributed with mean $\beta_0 + \beta_1 x$ and standard deviation $\sigma$.

**14.21** There are constants, $\beta_0$, $\beta_1$, and $\sigma$, such that, for each total number of hours studied, $x$, the test scores of all students in beginning calculus courses who study that number of hours are normally distributed with mean $\beta_0 + \beta_1 x$ and standard deviation $\sigma$.

**14.23**

**a.** $s_e = 14.25$; very roughly speaking, on average, the predicted price of a Corvette in the sample differs from the observed price by about $1425.
**b.** Presuming that, for Corvettes, the variables age ($x$) and price ($y$) satisfy the assumptions for regression inferences, the standard error of the estimate, $s_e = 14.25$, provides an estimate for the common population standard deviation, $\sigma$, of prices (in hundreds of dollars) for all Corvettes of any particular age.
**c.** See Fig. A.4.          **d.** It appears reasonable.

**FIGURE A.4**
(a) Residual plot and (b) normal probability plot of residuals for Exercise 14.23(c)

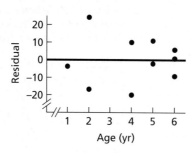

**FIGURE A.5**
(a) Residual plot and (b) normal probability plot of residuals for Exercise 14.25(c)

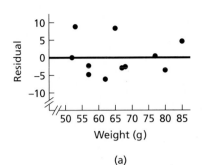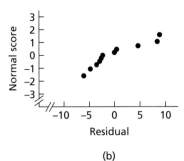

(a)            (b)

**FIGURE A.6**
(a) Residual plot and (b) normal probability plot of residuals for Exercise 14.27(c)

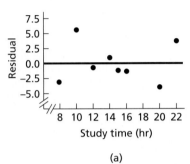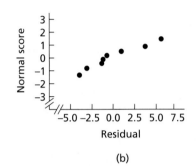

(a)            (b)

**14.25**

**a.** $s_e = 5.42$; very roughly speaking, on average, the predicted quantity of volatile compounds emitted by a potato plant in the sample differs from the observed quantity by about 542 nanograms.

**b.** Presuming that, for potato plants, the variables weight ($x$) and quantity of volatile compounds emitted ($y$) satisfy the assumptions for regression inferences, the standard error of the estimate, $s_e = 5.42$, provides an estimate for the common population standard deviation, $\sigma$, of quantities of volatile compounds emitted (in hundreds of nanograms) for all potato plants of any particular weight.

**c.** See Fig. A.5.

**d.** Although Fig. A.5(b) shows some curvature, it is probably not sufficiently curved to call into question the validity of the normality assumption (Assumption 3).

**14.27**

**a.** $s_e = 3.54$; very roughly speaking, on average, the predicted test score of a student in the sample differs from the observed score by about 3.54 points.

**b.** Presuming that, for students in beginning calculus courses, the variables study time ($x$) and test score ($y$) satisfy the assumptions for regression inferences, the standard error of the estimate, $s_e = 3.54$, provides an estimate for the common population standard deviation, $\sigma$, of test scores for all students who study for any particular amount of time.

**c.** See Fig. A.6.      **d.** It appears reasonable.

**14.29** Part (a) is a tough call, but the assumption of linearity (Assumption 1) may be violated, as may be the assumption of equal standard deviations (Assumption 2). In part (b), it appears that the assumption of equal standard deviations (Assumption 2) is violated. In part (d), it appears that the normality assumption (Assumption 3) is violated.

## EXERCISES 14.2

**14.41** normal, $-3.5$

**14.43** $r^2, r$

*Note:* In each of Exercises 14.45–14.49, the null hypothesis is that $x$ is not useful for predicting $y$ and the alternative hypothesis is that $x$ is useful for predicting $y$.

**14.45**

**a.** $t = -1.15$; critical values $= \pm 6.314$; $P > 0.20$; do not reject $H_0$

**b.** $-12.94$ to $8.94$

**14.47**

**a.** $t = 2.58$; critical values $= \pm 2.920$; $0.10 < P < 0.20$; do not reject $H_0$

**b.** $-0.26$ to $4.26$

**14.49**

**a.** $t = -1.63$; critical values $= \pm 2.353$; $P > 0.20$; do not reject $H_0$

**b.** $-1.53$ to $0.28$

**14.51** $H_0: \beta_1 = 0$, $H_a: \beta_1 \neq 0$; $\alpha = 0.10$; $t = -10.887$; critical values $= \pm 1.860$; $P < 0.01$; reject $H_0$; at the 10% significance level, the data provide sufficient evidence to conclude that age is useful as a predictor of price for Corvettes.

**14.53** $H_0: \beta_1 = 0$, $H_a: \beta_1 \neq 0$; $\alpha = 0.05$; $t = 1.053$; critical values $= \pm 2.262$; $P > 0.20$; do not reject $H_0$; at the 5% significance level, the data do not provide sufficient evidence to conclude that weight is useful as a predictor of quantity of volatile emissions for the potato plant *Solanum tuberosom*.

**14.55** $H_0: \beta_1 = 0$, $H_a: \beta_1 \neq 0$; $\alpha = 0.01$; $t = -3.00$; critical values $= \pm 3.707$; $0.02 < P < 0.05$; do not reject $H_0$; at the 1% significance level, the data do not provide sufficient evidence to conclude that study time is useful as a predictor of test score for students in beginning calculus courses.

**14.57** −32.7 to −23.1. We can be 90% confident that, for Corvettes, the decrease in mean price per 1-year increase in age (i.e., the mean annual depreciation) is somewhere between $2310 and $3270.

**14.59** −0.19 to 0.51. We can be 95% confident that, for the potato plant *Solanum tuberosom*, the change in the mean quantity of volatile emissions per 1-gram increase of weight is somewhere between −19 and 51 nanograms.

**14.61** −1.89 to 0.20. We can be 99% confident that, for students in beginning calculus courses, the change in mean test score per increase of 1 hour studied is somewhere between −1.89 and 0.20 points.

## EXERCISES 14.3

**14.73** $11,443. A point estimate for the mean price is the same as the predicted price.

**14.75**
a. −3
c. −3
b. −20.97 to 14.97
d. −38.94 to 32.94

**14.77**
a. 5
c. 5
b. −1.24 to 11.24
d. −4.72 to 14.72

**14.79**
a. 1
c. 1
b. −1.68 to 3.68
d. −5.56 to 7.56

**14.81**
a. 324.99 ($32,499)
b. 316.60 to 333.38. We can be 90% confident that the mean price of all 4-year-old Corvettes is somewhere between $31,660 and $33,338.
c. 324.99 ($32,499)
d. 297.20 to 352.78. We can be 90% certain that the price of a 4-year-old Corvette will be somewhere between $29,720 and $35,278.
e. See Fig. A.7.
f. The error in the estimate of the mean price of all 4-year-old Corvettes is due only to the fact that the population regression line is being estimated by a sample regression line. In contrast, the error in the prediction of the price of a 4-year-old Corvette is due to the error in estimating the mean price plus the variation in prices of 4-year-old Corvettes.

**FIGURE A.7**   90% confidence and prediction intervals for Exercise 14.81(e)

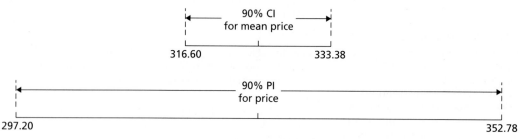

**14.83**
a. 13.29 (1329 ng)
b. 9.09 to 17.50. We can be 95% confident that the mean quantity of volatile emissions of all plants that weigh 60 g is somewhere between 909 and 1750 ng.
c. 13.29 (1329 ng)
d. 0.34 to 26.25. We can be 95% certain that the quantity of volatile emissions of a plant that weighs 60 g will be somewhere between 34 and 2625 ng.

**14.85**
a. 82.2 points
b. 77.5 to 86.8. We can be 99% confident that the mean test score of all beginning calculus students who study for 15 hours is somewhere between 77.5 and 86.8 points.
c. 82.2 points
d. 68.3 to 96.1. We can be 99% certain that the test score of a beginning calculus student who studies for 15 hours will be somewhere between 68.3 and 96.1 points.

## EXERCISES 14.4

**14.99** The (sample) linear correlation coefficient, $r$

**14.101**
a. Uncorrelated      b. Increases      c. Negatively

**14.103** $t = -1.15$; critical value $= -3.078$; $P > 0.10$; do not reject $H_0$

**14.105** $t = 2.58$; critical value $= 1.886$; $0.05 < P < 0.10$; reject $H_0$

**14.107** $t = -1.63$; critical values $= \pm2.353$; $P > 0.20$; do not reject $H_0$

**14.109** $H_0$: $\rho = 0$, $H_a$: $\rho < 0$; $\alpha = 0.05$; $t = -10.887$; critical value $= -1.860$; $P < 0.005$; reject $H_0$; at the 5% significance level, the data provide sufficient evidence to conclude that, for Corvettes, age and price are negatively linearly correlated.

**14.111** $H_0$: $\rho = 0$, $H_a$: $\rho \neq 0$; $\alpha = 0.05$; $t = 1.053$; critical values $= \pm2.262$; $P > 0.20$; do not reject $H_0$; at the 5% significance level, the data do not provide sufficient evidence to conclude that, for the potato plant *Solanum tuberosom*, weight and quantity of volatile emissions are linearly correlated.

**14.113**

**a.** $H_0: \rho = 0$, $H_a: \rho < 0$; $\alpha = 0.01$; $t = -3.00$; critical value $= -3.143$; $0.01 < P < 0.025$; do not reject $H_0$; at the 1% significance level, the data do not provide sufficient evidence to conclude that a negative linear correlation exists between study time and test score for beginning calculus students.

**b.** $H_0: \rho = 0$, $H_a: \rho < 0$; $\alpha = 0.05$; $t = -3.00$; critical value $= -1.943$; $0.01 < P < 0.025$; reject $H_0$; at the 5% significance level, the data provide sufficient evidence to conclude that a negative linear correlation exists between study time and test score for beginning calculus students.

**14.115** $\rho$ is a parameter; $r$ is a statistic

## REVIEW PROBLEMS FOR CHAPTER 14

1. **a.** conditional
   **b.** See Key Fact 14.1 on page 616.

2. **a.** $b_1$        **b.** $b_0$        **c.** $s_e$

3. A residual plot (i.e., a plot of the residuals against the observed values of the predictor variable) and a normal probability plot of the residuals. A plot of the residuals against the observed values of the predictor variable should fall roughly in a horizontal band, centered and symmetric about the $x$-axis. A normal probability plot of the residuals should be roughly linear.

4. **a.** Assumption 1        **b.** Assumption 2
   **c.** Assumption 3        **d.** Assumption 3

5. The regression equation is useful for making predictions.

6. $b_1, r, r^2$

7. No. Both equal the number obtained by substituting the specified value of the predictor variable into the sample regression equation.

8. The term *confidence* is usually reserved for interval estimates of parameters, whereas the term *prediction* is used for interval estimates of variables.

9. $\rho$

10. **a.** The variables are positively linearly correlated, meaning that $y$ tends to increase linearly as $x$ increases (and vice versa), with the tendency being greater the closer that $\rho$ is to 1.
    **b.** The variables are linearly uncorrelated, meaning that there is no linear relationship between the variables.
    **c.** The variables are negatively linearly correlated, meaning that $y$ tends to decrease linearly as $x$ increases (and vice versa), with the tendency being greater the closer that $\rho$ is to $-1$.

11. There are constants, $\beta_0$, $\beta_1$, and $\sigma$, such that, for each student-to-faculty ratio, $x$, the graduation rates for all universities with that student-to-faculty ratio are normally distributed with mean $\beta_0 + \beta_1 x$ and standard deviation $\sigma$.

12. **a.** $\hat{y} = 16.4 + 2.03x$
    **b.** $s_e = 11.31\%$; very roughly speaking, on average, the predicted graduation rate for a university in the sample differs from the observed graduation rate by about 11.31 percentage points.

**c.** Presuming that, for universities, the variables student-to-faculty ratio ($x$) and graduation rate ($y$) satisfy the assumptions for regression inferences, the standard error of the estimate, $s_e = 11.31\%$, provides an estimate for the common population standard deviation, $\sigma$, of graduation rates for all universities with any particular student-to-faculty ratio.

13.

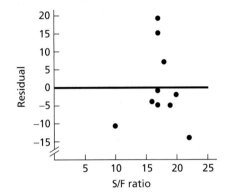

It appears reasonable.

14. **a.** $H_0: \beta_1 = 0$, $H_a: \beta_1 \neq 0$; $\alpha = 0.05$; $t = 1.682$; critical values $= \pm 2.306$; $0.10 < P < 0.20$; do not reject $H_0$; at the 5% significance level, the data do not provide sufficient evidence to conclude that, for universities, student-to-faculty ratio is useful as a predictor of graduation rate.
    **b.** $-0.75\%$ to $4.80\%$. We can be 95% confident that, for universities, the change in mean graduation rate per increase by 1 in the student-to-faculty ratio is somewhere between $-0.75$ and $4.80$ percentage points.

15. **a.** 50.9%
    **b.** 42.6% to 59.2%. We can be 95% confident that the mean graduation rate of all universities that have a student-to-faculty ratio of 17 is somewhere between 42.6% and 59.2%.
    **c.** 50.9%
    **d.** 23.5% to 78.3%. We can be 95% certain that the observed graduation rate of a university that has a student-to-faculty ratio of 17 will be somewhere between 23.5% and 78.3%.

**e.** The error in the estimate of the mean graduation rate of all universities that have a student-to-faculty ratio of 17 is due only to the fact that the population regression line is being estimated by a sample regression line, whereas the error in the prediction of the observed graduation rate of a university that has a student-to-faculty ratio of 17 is due to the error in estimating the mean graduation rate plus the variation in graduation rates of universities that have a student-to-faculty ratio of 17.

**16.** $H_0: \rho = 0$, $H_a: \rho > 0$; $\alpha = 0.025$; $t = 1.682$; critical value $= 2.306$; $0.05 < P < 0.10$; do not reject $H_0$; at the 2.5% significance level, the data do not provide sufficient evidence to conclude that, for universities, the variables student-to-faculty ratio and graduation rate are positively linearly correlated.

# Index

## Photo Credits

Pages 3 and 34, movie poster for "African Queen" © Everett Collection; page 5, photo of Harry Truman © Bettman/Corbis; page 11, photo of iron lung © Hulton-Deutsch Collection/Corbis; page 35, image of Florence Nightingale from www.spartacus.schoolnet.co.uk/REnightingale.htm; pages 39 and 90, photo of infants © Digital Vision; page 41, 109th Boston Marathon © Getty Editorial; page 53, photo of cheetahs © PhotoDisc; page 54, photo of great white shark, Corbis RF; page 55, photo of stockbrokers © PhotoDisc; page 68, photo of M & Ms courtesy Beth Anderson; pages 70 and 395, photo of boy with thermometer © PhotoDisc; page 91, image of Adolphe Quetelet courtesy St. Andrews University; pages 93 and 152, Triple Crown horse racing © Corbis; page 105, photo of driver courtesy AAA Foundation for Traffic Safety; page 117, photo of Hurricane Hugo damage © PhotoDisc; page 119, photo of Beatles © Getty Editorial; page 136, photo of U.S. Women's World Cup Soccer Team © Reuters/Corbis; page 145, photo of hurricane eye © PhotoDisc; page 149, photo of party © PhotoDisc; page 151, photo of Chicago White Sox © Getty Editorial; page 153, image of John Tukey courtesy St. Andrews University; pages 155 and 199, photo of people © Corbis RF; page 160, photo of thermometer © PhotoDisc; page 185, photo of Corvette © PhotoDisc Vol. 2; page 186, photo of pelicans © PhotoDisc; page 198, photo of fatty food © PhotoDisc; page 200, image of Adrien Legendre courtesy St. Andrews University; pages 203 and 264, Powerball winner 2006 © Corbis; page 212, photo of Funny Cide © Jason Szenes/Corbis; page 227, photo of oil spill courtesy of NOAA; page 232, photo of elementary school children © PhotoDisc; page 235, photo of space shuttle launch courtesy of NASA; page 237, photo of solar eclipse © PhotoDisc; page 242, photo of factory © PhotoDisc; page 249, photo of couple © PhotoDisc; page 261, photo of meeting © PhotoDisc; page 263, photo of craps table © Digital Vision; page 265, image of Andrei Kolmogorov courtesy of St. Andrews University; pages 267 and 304, photo of feet © Stockbyte Platinum/Getty RF; page 274, giant tarantula © Corbis RF; page 293, photo of Jingdong black gibbon © Rod Williams/Bruce Coleman, PictureQuest; page 305, image of Carl Friedrich Gauss courtesy St. Andrews University; pages 307 and 333, photo of freight train © Corbis RF; page 313, photo of San Antonio Spurs © Joe Mitchell/Stringer/Reuters New Media Inc./Corbis; page 313, photo of Bill Gates © Reuters NewMedia, Inc./Corbis; page 319, photo of earthquake damage © CARE; page 327, image of brain scans © PhotoDisc; page 330, photo of tax preparation courtesy of Beth Anderson; page 333, image of Pierre-Simon Laplace courtesy St. Andrews University; pages 337 and 376, photo of cookies courtesy of Beth Anderson; page 342, wedding photo courtesy Aaron and Carla Weiss; page 351, photo of the Rolling Stones © Lynn Goldsmith/Corbis; page 358, photo of common lizard © Corbis RF; page 369, photo of amusement park ride © PhotoDisc; page 374, photo of seashell © PhotoDisc; page 377, image of William Gosset courtesy of St. Andrews University; pages 379 and 438, photo of students in park © PhotoDisc Red; page 386, photo of cell-phone user © PhotoDisc; page 395, photo of trial, Corbis RF; page 429, photo of mother and newborn, Corbis RF; page 434, photo of an arrest © Jonathan Blair/Corbis; page 439, image of Jerzy Neyman courtesy St. Andrews University; pages 441 and 498, photo of reader © Digital Vision; page 461, photo of robotic golf ball driver © Stinger Tees; page 473, photo of patient © PhotoDisc Vol 18; page 489, photo of person sleeping © PhotoDisc; page 494, photo of snake, Corbis RF; page 498, image of Gertrude Cox courtesy of Research Triangle Institute; pages 501 and 538, photo of bomb explosion © PhotoDisc; page 511, photo of women, Corbis RF; page 520, photo of a family opening presents © PhotoDisc; page 533, photo of "Buckle Up" highway sign © PhotoDisc; page 536, photo of jurors, Corbis RF; page 538, image of Abraham de Moivre courtesy of St. Andrews University; pages 541 and 582, photo of angry driver © PhotoDisc Red; page 553, photo of M&M's courtesy of Beth Anderson; page 575, photo of Siskel and Ebert © AP Wideworld Photos; page 581, photo of hospital © Corbis RF; page 582, image of Karl Pearson © Brown Brothers; pages 585 and 612, photo of college students © PhotoDisc; page 607, photo of couple in new apartment, Corbis RF; page 609, photo of Linus Pauling with Vitamin C molecule © Roger Ressmeyer/Corbis; page 612, image of Sir Ronald Fisher courtesy St. Andrews University; pages 615 and 658, photo of people © Corbis RF; page 626, photo of Corvette © PhotoDisc Vol. 2; page 629, photo of newborn, Corbis RF; page 636, photo of potato plants © Patrick Johns/Corbis; page 637, photo of newborn © PhotoDisc; page 656, photo of graduate courtesy of Greg Weiss; page 658, image of Sir Francis Galton courtesy of St. Andrews University.

# Indexes for
# Case Studies & Biographical Sketches

**TABLE II**

Areas under the
standard normal curve

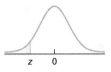

| | | | | Second decimal place in $z$ | | | | | | |
|---|---|---|---|---|---|---|---|---|---|---|
| 0.09 | 0.08 | 0.07 | 0.06 | 0.05 | 0.04 | 0.03 | 0.02 | 0.01 | 0.00 | $z$ |
| | | | | | | | | | 0.0000 | −3.9 |
| 0.0001 | 0.0001 | 0.0001 | 0.0001 | 0.0001 | 0.0001 | 0.0001 | 0.0001 | 0.0001 | 0.0001 | −3.8 |
| 0.0001 | 0.0001 | 0.0001 | 0.0001 | 0.0001 | 0.0001 | 0.0001 | 0.0001 | 0.0001 | 0.0001 | −3.7 |
| 0.0001 | 0.0001 | 0.0001 | 0.0001 | 0.0001 | 0.0001 | 0.0001 | 0.0001 | 0.0002 | 0.0002 | −3.6 |
| 0.0002 | 0.0002 | 0.0002 | 0.0002 | 0.0002 | 0.0002 | 0.0002 | 0.0002 | 0.0002 | 0.0002 | −3.5 |
| 0.0002 | 0.0003 | 0.0003 | 0.0003 | 0.0003 | 0.0003 | 0.0003 | 0.0003 | 0.0003 | 0.0003 | −3.4 |
| 0.0003 | 0.0004 | 0.0004 | 0.0004 | 0.0004 | 0.0004 | 0.0004 | 0.0005 | 0.0005 | 0.0005 | −3.3 |
| 0.0005 | 0.0005 | 0.0005 | 0.0006 | 0.0006 | 0.0006 | 0.0006 | 0.0006 | 0.0007 | 0.0007 | −3.2 |
| 0.0007 | 0.0007 | 0.0008 | 0.0008 | 0.0008 | 0.0008 | 0.0009 | 0.0009 | 0.0009 | 0.0010 | −3.1 |
| 0.0010 | 0.0010 | 0.0011 | 0.0011 | 0.0011 | 0.0012 | 0.0012 | 0.0013 | 0.0013 | 0.0013 | −3.0 |
| 0.0014 | 0.0014 | 0.0015 | 0.0015 | 0.0016 | 0.0016 | 0.0017 | 0.0018 | 0.0018 | 0.0019 | −2.9 |
| 0.0019 | 0.0020 | 0.0021 | 0.0021 | 0.0022 | 0.0023 | 0.0023 | 0.0024 | 0.0025 | 0.0026 | −2.8 |
| 0.0026 | 0.0027 | 0.0028 | 0.0029 | 0.0030 | 0.0031 | 0.0032 | 0.0033 | 0.0034 | 0.0035 | −2.7 |
| 0.0036 | 0.0037 | 0.0038 | 0.0039 | 0.0040 | 0.0041 | 0.0043 | 0.0044 | 0.0045 | 0.0047 | −2.6 |
| 0.0048 | 0.0049 | 0.0051 | 0.0052 | 0.0054 | 0.0055 | 0.0057 | 0.0059 | 0.0060 | 0.0062 | −2.5 |
| 0.0064 | 0.0066 | 0.0068 | 0.0069 | 0.0071 | 0.0073 | 0.0075 | 0.0078 | 0.0080 | 0.0082 | −2.4 |
| 0.0084 | 0.0087 | 0.0089 | 0.0091 | 0.0094 | 0.0096 | 0.0099 | 0.0102 | 0.0104 | 0.0107 | −2.3 |
| 0.0110 | 0.0113 | 0.0116 | 0.0119 | 0.0122 | 0.0125 | 0.0129 | 0.0132 | 0.0136 | 0.0139 | −2.2 |
| 0.0143 | 0.0146 | 0.0150 | 0.0154 | 0.0158 | 0.0162 | 0.0166 | 0.0170 | 0.0174 | 0.0179 | −2.1 |
| 0.0183 | 0.0188 | 0.0192 | 0.0197 | 0.0202 | 0.0207 | 0.0212 | 0.0217 | 0.0222 | 0.0228 | −2.0 |
| 0.0233 | 0.0239 | 0.0244 | 0.0250 | 0.0256 | 0.0262 | 0.0268 | 0.0274 | 0.0281 | 0.0287 | −1.9 |
| 0.0294 | 0.0301 | 0.0307 | 0.0314 | 0.0322 | 0.0329 | 0.0336 | 0.0344 | 0.0351 | 0.0359 | −1.8 |
| 0.0367 | 0.0375 | 0.0384 | 0.0392 | 0.0401 | 0.0409 | 0.0418 | 0.0427 | 0.0436 | 0.0446 | −1.7 |
| 0.0455 | 0.0465 | 0.0475 | 0.0485 | 0.0495 | 0.0505 | 0.0516 | 0.0526 | 0.0537 | 0.0548 | −1.6 |
| 0.0559 | 0.0571 | 0.0582 | 0.0594 | 0.0606 | 0.0618 | 0.0630 | 0.0643 | 0.0655 | 0.0668 | −1.5 |
| 0.0681 | 0.0694 | 0.0708 | 0.0721 | 0.0735 | 0.0749 | 0.0764 | 0.0778 | 0.0793 | 0.0808 | −1.4 |
| 0.0823 | 0.0838 | 0.0853 | 0.0869 | 0.0885 | 0.0901 | 0.0918 | 0.0934 | 0.0951 | 0.0968 | −1.3 |
| 0.0985 | 0.1003 | 0.1020 | 0.1038 | 0.1056 | 0.1075 | 0.1093 | 0.1112 | 0.1131 | 0.1151 | −1.2 |
| 0.1170 | 0.1190 | 0.1210 | 0.1230 | 0.1251 | 0.1271 | 0.1292 | 0.1314 | 0.1335 | 0.1357 | −1.1 |
| 0.1379 | 0.1401 | 0.1423 | 0.1446 | 0.1469 | 0.1492 | 0.1515 | 0.1539 | 0.1562 | 0.1587 | −1.0 |
| 0.1611 | 0.1635 | 0.1660 | 0.1685 | 0.1711 | 0.1736 | 0.1762 | 0.1788 | 0.1814 | 0.1841 | −0.9 |
| 0.1867 | 0.1894 | 0.1922 | 0.1949 | 0.1977 | 0.2005 | 0.2033 | 0.2061 | 0.2090 | 0.2119 | −0.8 |
| 0.2148 | 0.2177 | 0.2206 | 0.2236 | 0.2266 | 0.2296 | 0.2327 | 0.2358 | 0.2389 | 0.2420 | −0.7 |
| 0.2451 | 0.2483 | 0.2514 | 0.2546 | 0.2578 | 0.2611 | 0.2643 | 0.2676 | 0.2709 | 0.2743 | −0.6 |
| 0.2776 | 0.2810 | 0.2843 | 0.2877 | 0.2912 | 0.2946 | 0.2981 | 0.3015 | 0.3050 | 0.3085 | −0.5 |
| 0.3121 | 0.3156 | 0.3192 | 0.3228 | 0.3264 | 0.3300 | 0.3336 | 0.3372 | 0.3409 | 0.3446 | −0.4 |
| 0.3483 | 0.3520 | 0.3557 | 0.3594 | 0.3632 | 0.3669 | 0.3707 | 0.3745 | 0.3783 | 0.3821 | −0.3 |
| 0.3859 | 0.3897 | 0.3936 | 0.3974 | 0.4013 | 0.4052 | 0.4090 | 0.4129 | 0.4168 | 0.4207 | −0.2 |
| 0.4247 | 0.4286 | 0.4325 | 0.4364 | 0.4404 | 0.4443 | 0.4483 | 0.4522 | 0.4562 | 0.4602 | −0.1 |
| 0.4641 | 0.4681 | 0.4721 | 0.4761 | 0.4801 | 0.4840 | 0.4880 | 0.4920 | 0.4960 | 0.5000 | −0.0 |

For $z \leq -3.90$, the areas are 0.0000 to four decimal places.